Springer Tracts in Advanced Robotics 109

Editors

Prof. Bruno Siciliano
Dipartimento di Ingegneria Elettrica
e Tecnologie dell'Informazione
Università degli Studi di Napoli
Federico II
Via Claudio 21, 80125 Napoli
Italy
E-mail: siciliano@unina.it

Prof. Oussama Khatib
Artificial Intelligence Laboratory
Department of Computer Science
Stanford University
Stanford, CA 94305-9010
USA
E-mail: khatib@cs.stanford.edu

More information about this series at http://www.springer.com/series/5208

STAR (Springer Tracts in Advanced Robotics) has been promoted under the auspices of EURON (European Robotics Research Network)

M. Ani Hsieh · Oussama Khatib
Vijay Kumar
Editors

Experimental Robotics

The 14th International Symposium on Experimental Robotics

Springer

Editors
M. Ani Hsieh
Drexel University
Philadelphia, PA
USA

Oussama Khatib
Stanford University
Stanford, CA
USA

Vijay Kumar
School of Engineering and Applied Science
University of Pennsylvania
Philadelphia, PA
USA

ISSN 1610-7438 ISSN 1610-742X (electronic)
Springer Tracts in Advanced Robotics
ISBN 978-3-319-23777-0 ISBN 978-3-319-23778-7 (eBook)
DOI 10.1007/978-3-319-23778-7

Library of Congress Control Number: 2015950030

Springer Cham Heidelberg New York Dordrecht London

Printed on acid-free paper

Springer International Publishing AG Switzerland is part of Springer Science+Business Media (www.springer.com)

Foreword

Robotics is undergoing a major transformation in scope and dimension. From a largely dominant industrial focus, robotics is rapidly expanding into human environments and is vigorously engaged in its new challenges. Interacting with, assisting, serving, and exploring with humans, the emerging robots will increasingly touch people and their lives.

Beyond its impact on physical robots, the body of knowledge robotics has produced is revealing a much wider range of applications reaching across diverse research areas and scientific disciplines, such as biomechanics, haptics, neurosciences, virtual simulation, animation, surgery, and sensor networks among others. In return, the challenges of the new emerging areas are proving an abundant source of stimulation and insights into the field of robotics. It is indeed at the intersection of disciplines that the most striking advances happen.

The *Springer Tracts in Advanced Robotics* (*STAR*) is devoted to bringing to the research community the latest advances in the robotics field on the basis of their significance and quality. Through a wide and timely dissemination of critical research developments in robotics, our objective with this series is to promote more exchanges and collaborations among the researchers in the community and contribute to further advancements in this rapidly growing field.

As one of robotics pioneering symposia, the International Symposium on Experimental Robotics (ISER) has established over the past two decades some of the field's most fundamental and lasting contributions. Since the launching of STAR, ISER and several other thematic symposia in robotics have found an important platform for closer links and extended reach within the robotics community.

The fourteenth edition of *Experimental Robotics* edited by Ani Hsieh, Oussama Khatib, and Vijay Kumar offers in its 12-part volume a collection of a broad range of topics in the field and human-centered robotics. The contents of these contributions represent a cross-section of the current state of robotics research from one particular aspect: experimental work and how it reflects on the theoretical basis of future developments. Experimental validation of algorithms, concepts, or techniques is the common thread running through this large collection of widely diverse contributions, spanning from mechanisms to locomotion, from manipulation to human–robot interaction, from haptics to sensor networks, from perception and planning to mapping and localization.

From its warm social program to its excellent technical program, which included the novelty of interactive technical presentations, ISER culminates with this unique reference on the current developments and new directions of experimental robotics—a genuine tribute to its contributors and organizers!

Naples, Italy
July 2015

Bruno Siciliano
STAR Editor

Preface

The International Symposium on Experimental Robotics (ISER) is a series of biennial symposia which began in 1989, and is sponsored by the International Foundation of Robotics Research (IFRR). ISER emphasizes experimental work while providing the robotics community with a forum for presenting research driven by creative ideas, bold visions, new systems, and novel applications of robotics. The tradition in ISER is to foster scholarly work that either addresses validation of theoretical paradigms through careful experimentation or contributes to the creation of novel experimental platforms that in turn inspire new theoretical developments. The ISER symposia are conceived to bring together in a small group setting researchers from around the world who are at the forefront of experimental robotics research, to assess and share their views and ideas about the state of the art, and to discuss promising new avenues for future research exploration in experimental robotics. The ISER meetings are organized around oral and interactive technical presentations in a single-track format.

The Fourteenth Symposium was held during June 15–18, 2014 in Marrakech and Essaouira, Morocco. The symposium was chaired by M. Ani Hsieh (Drexel University, USA), Oussama Khatib (Stanford University, USA), and Vijay Kumar (University of Pennsylvania, USA). The local organizing committee was chaired by Philippe Bidaud (ONERA French Aerospace Lab/University Pierre et Marie Curie, France) and Said Zeghloul (University of Poitiers, France). The International Steering Committee for ISER is chaired by Oussama Khatib and includes Marcelo Ang (Singapore), Herman Bruyninckx (Belgium), Alicia Casals (Spain), Raja Chatila (France), Peter Corke (Australia), John Craig (USA), Jaydev Desai (USA), Paolo Dario (Italy), Greg Dudeck (Canada), Vincent Hayward (Canada, France), Gerd Hirzinger (Germany), Yoshihiko Nakamura (Japan), Paul Newman (UK), Daniela Rus (USA), Kenneth Salisbury (USA), Bruno Siciliano (Italy), Sanjiv Singh (USA), James Trevelyan (Australia), Tsuneo Yoshikawa (Japan), and Alex Zelinsky (Australia).

The program of the Fourteenth Symposium included 59 technical papers, selected from open submission through a review process organized by the

International Steering Committee. The symposium contributions report on a variety of new theoretical and experimental results, and point to new visions and trends in the field. The topics of the technical sessions covered a broad spectrum of experimental robotics research activities. This year 19 papers were presented in interactive format on electronic displays. The symposium sessions were Locomotion; Haptics; Manipulation; Perception; Human–robot Interaction; Mapping and Localization; Mechanisms; Perception and Planning; Sensor Networks; Many Robot Systems. The program also included a plenary talk delivered remotely by Michel L'Hour who is the Scientific and Technical Advisor for UNESCO's Department of Underwater Archaeological Research and Underwater General Curator of Heritage and Vincent Creuze from CNRS/University of Montpellier. Lastly, the Fourteenth Symposium also featured the Robotics Workshop: Trends and Challenges which was organized by Fatima Bouyahia (University of Cadi Ayyad, Morocco), Nabil Elmarzouqi (University of Cadi Ayyad, Morocco), Abdellah Ait Ouahman (University of Cadi Ayyad, Morocco), Med Amine Laribi (University of Poitiers, France), Sad Zeghloul, Philippe Bidaud, and Oussama Khatib. The workshop brought together robotics experts and Moroccan master and Ph.D. students and researchers in areas related to robotics for a full day of engaging talks and discussions.

This volume includes the complete collection of the contributions presented at the symposium, with authoritative introductions to each section by the chairs of the corresponding sessions. We are grateful to the authors and the participants who have all contributed to the success of this symposium by bringing an outstanding program, excellent technical presentations, and stimulating and insightful discussions. We would like also to express our thanks and gratitude to the local organizing team that created the perfect environment for fostering technical discussions and promoting intellectual debates in a relaxed setting.

Marrakech/Essaouira, Morocco
June 2014

M. Ani Hsieh
Oussama Khatib
Vijay Kumar

Contents

Part III Manipulation

Part IV Perception

Part V Human–Robot Interaction

Part VI Mapping and Localization

Part VII Mechanisms

Part VIII Perception and Planning

Part IX Sensor Networks

Part X Many-Robot Systems

Part XI Interactive Presentations

Part I
Locomotion

Katie Byl
University of California, Santa Barbara

Locomotion remains a fundamental challenge in developing robots that can do useful work in the real world, outside of office, factory, and laboratory environments. Even today, as self-driving cars and autonomous quadrotors seem teasingly close to transforming our daily lives, robots clearly fall far short of the incredible capabilities of animal locomotion. In particular, it is difficult to achieve a balanced combination of speed and agility, low energy use, and high reliability. The four papers in this session all focus on legged robot morphologies, toward providing unique mobility across rough and/or discontinuous terrain, and they each consider some combination of trade-offs between agility, energetics and stability. Two of the papers (Duperret et al. and Miller et al.) study highly dynamic systems with an aerial phase during locomotion, while the other two (Satzinger et al. and Grand et al.) focus on trajectory planning for redundant limbs to produce quasi-static motions to negotiate extreme terrain.

The first paper in this session, *Towards a Comparative Measure of Legged Agility* by J.M. Duperret, G.D. Kenneally, J.L. Pusey, and D.E. Koditschek, introduces a new metric for specific agility and uses it to quantify agility versus endurance for two different legged robots during leaping trials: the four-legged Canid robot, which has a flexible spine, and the six-legged XRL robot, which has a rigid body. Their experimental data support two hypotheses. First, the robot with the spine achieves greater agility, and second, both active and passive dynamics of the spine improve agility.

The second paper, *On Prismatic and Torsional Actuation for Running Legged Robots* by Bruce Miller, Jason Brown and Jonathan Clark, studies hybrid mechanisms for robots with spring-loaded inverted pendulum (SLIP) dynamics to exploit both prismatic and torsional actuation together to achieve speed, stability, and efficiency. Specifically, they investigate the coupled interplay of these actuation sources and discover near-optimal gait characteristics that simultaneously achieve each of these three performance goals in idealized models, and they also demonstrate similar characteristics in experiments with a hexapedal robot with a design similar to iSprawl.

Experimental Results for Dexterous Quadruped Locomotion Planning with RoboSimian by Brian Satzinger, Chelsea Lau, Marten Byl, and Katie Byl presents a practical solution for resolving kinematic redundancy for a dexterous, four-limbed robot. Their approach combines rapidly-exploring random tree (RRT) searches over the degrees of freedom of either one or two of the legs with heuristic solutions for inverse kinematics to constrain the (x,y,z) positions of the remaining end effectors to remain in place on the ground during locomotion. They explore the planning time required and quantify dexterity in terms of the additional feasible workspace reachable by the robot by allowing body motion during a swing leg trajectory, and they test the approach through experimental trials with RoboSimian, demonstrating both agile and highly reliable walking on terrain designed for the DARPA Robotics Challenge (DRC).

The final paper in this group, *Experimental Evaluation of Obstacle Clearance by a Hybrid Wheel-Legged Robot* by Christophe Grand, Pierre Jarrault, Faiz BenAmar, and Philippe Bidaud, presents a control approach that allows a redundantly actuated vehicle with four wheel-legs to cross a step that is taller than the wheel diameter. Their approach optimizes for the center of mass position and distribution of internal forces such that torque and friction constraints are met using a minimax optimization approach, toward maximizing robustness while simultaneously achieving highly agile mobility, and they demonstrate the approach in experimental trials on steep, step-like terrain.

Towards a Comparative Measure of Legged Agility

J.M. Duperret, G.D. Kenneally, J.L. Pusey and D.E. Koditschek

Abstract We introduce an agility measure enabling the comparison of two very different leaping-from-rest transitions by two comparably powered but morphologically different legged robots. We use the measure to show that a flexible spine outperforms a rigid back in the leaping-from-rest task. The agility measure also sheds light on the source of this benefit: core actuation through a sufficiently powerful parallel elastic actuated spine outperforms a similar power budget applied either only to preload the spine or only to actuate the spine during the leap, as well as a rigid backed configuration of the identical machine.

Keywords Legged locomotion · Experimental metric · Agile mobility

1 Introduction

The past decades' slow trickle of dynamical legged robots has grown to a stream of academic [1] and commercial [2] advances yielding an emerging set of design and control principles sufficient for steady-state locomotion [3–9]. In contrast, leaping, dodging, recovering and similar transitional mobility behaviors characteristic of animals' explosive agility—the intuitive motivation for legs—has received much less attention. Recent interest in such transitional legged behaviors [10–15] is impeded by the lack of a well-formulated theory alongside the absence of appropriate performance metrics.

J.M. Duperret (✉) · D.E. Koditschek
Department of Electrical and Systems Engineering, University of Pennsylvania, 200 South 33rd Street, Philadelphia, PA 19104, USA
e-mail: jdup@seas.upenn.edu

G.D. Kenneally
Department of Mechanical Engineering and Applied Mechanics, University of Pennsylvania, 220 South 33rd Street, Philadelphia, PA 19104, USA

J.L. Pusey
Army Research Laboratory, Aberdeen, MD, USA

M.A. Hsieh et al. (eds.), *Experimental Robotics*, Springer Tracts in Advanced Robotics 109, DOI 10.1007/978-3-319-23778-7_1

In this paper we propose a pair of measures for nimble legged transitions that help organize a suite of experiments designed to test hypotheses about the comparative benefits of specific morphological features. In Sect. 2 we introduce a candidate measure of *specific agility*, counterposed with a measure of *endurance* with the goal of quantifying the transitional performance of legged platforms across different scales, morphologies, power resources, and operating points. We use these measures in Sect. 3 to compare the empirical performance of two comparably powered but morphologically different robots, Canid [16] and XRL [17], in a leaping-from-rest transition, and to reach the judgement in Section Sect. 4 that Canid's parallel elastic-actuated spine confers greater leaping agility. We review the main experimental insights in Sect. 5 and comment on future work.

2 Technical Approach: Specific Agility and Endurance

Legged agility has not yet been formally defined in the robotics literature so for this paper we explore the implications of a well-cited definition within the sports science community holding that agility is "a rapid whole-body movement with change of velocity or direction in response to a stimulus" [18].

Notwithstanding the many informative and inspiring studies of legged animal performance, e.g. [19–25], we have not been able to find any formalization of this idea suitable for comparing robots of different morphologies and different sizes over different tasks. Perhaps the most common measure for acceleration and leaping used in the legged biology literature is specific power (watts per kilogram taken over a gait cycle of leg power output relative to leg muscle mass or body mass) [22, 26–28] but it is not scale invariant as we observe in Appendix 2. Specific work has been proposed as a measure for legged leaping with respect to muscle mass [22], and this seems closest to the body mass normalized measure we will introduce below. In contrast, characterizing directional aspects of agility performance seems trickier. Animal turning maneuvers have been studied in robotics [29] as well as biology [30] yielding a variety of useful associated performance measures such as turning radius at speed, leg effectiveness, linear maneuverability number [31], and usage of braking/acceleration forces [32]. But it is not clear to us how to generalize such measures for reasons we will discuss below as well.

Many intuitive measures for a legged platform involving, say, jumping height or the magnitude of linear acceleration, are equivalent to a change in kinetic and gravitational potential energy during the stance phase of locomotion. Thus, we focus our proposed measure on the change in what we term the extrinsic body energy, the sum of the mass center's kinetic and gravitational potential energy, relative to the natural unit over which a legged platform can adjust it, a single, isolated stance. We use the qualifier "extrinsic" to distinguish this notion from the body energy introduced in [16] that is sensitive to the state of a platform's internal mechanical springs. Catapult-like elastic energy storage used to augment muscle power in leaping from rest has been shown to occur in animals across widely different scales [21, 33] and, intuitively, we feel such use of initially stored spring energy should not count against the agility of a transition. We also avoid the notion of "stride" which connotes a regularity of stance and swing that may not prevail in sudden legged maneuvers characterized by combinatorial sequences of leg contacts [10]. Instead, we construe "stance" as the dimensionless event characterized by some number of legs in ground contact, punctuated either by a prior or subsequent aerial phase (or both).

Thus, for present purposes, we find it useful to introduce a working notion of *specific agility* during stance in terms of the mass-normalized change in extrinsic body energy:

$$\alpha := \frac{\Delta W}{m} / stance\ event, \tag{1}$$

where ΔW is the extrinsic body energy (the sum of the mass center's kinetic and gravitational potential energy) at the end of stance minus the extrinsic body energy at the start stance[1,2] and m is the mass of the agent. The SI units of α are $(\mathrm{m/s})^2$ and can be interpreted as mass-specific work in the equivalent units of (J/kg).

As we have tried to suggest in our brief survey of the extensive literature, and seems most carefully summarized in [34], it does not appear straightforward to find a single dimensionless group capable of capturing all relevant aspects of maneuverability and agility. We tolerate the lack of a dimensionless measure in our quantification of agility because mass-specific work seems to be the fundamental quantity of interest—at least for changes in velocity magnitude. For example, measuring work done on the body during stance is sensitive to accelerations along a velocity vector fixed in the inertial frame and takes into account the operating point, capturing the greater energetic cost of accelerating a given amount at higher relative to lower speeds (such energetic costs are consistent with biological observations of animal accelerating and braking [26]). However, it does not reward purely directional changes even though, intuitively, rapid turns ought to represent a similarly important concomitant of any comprehensive "agility" measure. Any attempt to reconcile nimble turning with energetic expressions of performance must address the fact that fixed rate circular motion entails no work since the direction of motion is orthogonal to the force.

The proposed measure (1) does appear to confer some scale invariance. In biology, this is predicted by arguments found in [35] and empirical observations of vertical jumping height known as Borelli's law [36]. This 'law' is demonstrated in animals across eight orders of magnitude mass variation which are shown to have vertical jumping heights (proportional to specific agility if air resistance is neglected) within a factor of three—ranging from around 20 to 60 cm or a specifc agility of around

[1]Steady state motions such as running or hopping that can be approximated with Hamiltonian systems will have negligible agility according to our metric in accordance with biological observations that these motions require significantly less muscle power output as compared to leaping accelerations [22, 28].

[2]Likely it will be useful in later work to consider a notion of integrated specific agility accumulated over a sequence of stance events, such as when evaluating the agility of an accelerating bound containing a brief aerial phase between front and rear leg-ground contacts.

2 to 6 m^2/s^2. Similar arguments about the scale invariance of this measure with electromagnetic actuators in a robotic leg are detailed in Appendix 2.

The operational utility of an agile motion will generally depend on the number of times n it can be performed in succession—which we term *endurance*. Given resource constraints present in executing a movement, we expect endurance to decrease with increasing specific agility. For example a robot that heats its motors to its thermal limits in a single leap cannot immediately perform the same leap on the next step; it must wait until its motors cool before completing the action again, giving it an n of 1. A robot capable of performing an agile motion an infinite number of times (unlikely with current technology given limited energy storage) would have an n of ∞. Although "stance event" was introduced as taking integer values, we find it convenient to recast the measure as taking (extended) real values. Specifically, we outline in Appendix 1 our appeal to a motor thermal model as a means of estimating how much time our actuators might be able to sustain the maneuver under consideration, and thereby back out an equivalent real estimate of the predicted number of viable stance events. Thus we will consider the ordered real pair (α, n) when evaluating agile motions in an experimental setting.

3 Experiments and Results

We use this framework to compare the performance of Canid [16] and XRL [17], in the open-loop leaping-from-rest task, a transitional behavior of near ubiquitous value, e.g. in gap crossing or rapid preparation [37] of high energy steady-state gait basins [38]. We use this comparison to examine the relevant benefit of distal versus core actuation as the quadrupedal Canid uses two motors to actuate its spine while the hexapedal XRL uses these two motors to actuate a pair of additional legs. This comparison seems particularly apt because of the close relationship between the two machines described in [17]: both robots have the same electronics, use similar motors and gearing, and are capable of comparable (respecting speed and specific resistance) steady-state locomotion as suggested in the accompanying video and partially documented in [16]. Disregarding the spine, the platforms differ primarily in their mass—Canid weighs 11.3 kg while XRL weights 7.3 kg—and leg actuation as Canid's four hip actuators drive their C-legs through a four-bar linkage while XRL directly actuates its six C-legs. It is worth noting the enabling role the specific agility measure plays in allowing this comparison that requires somehow normalizing for the very different actuation strategies used by these two nominally similar machines during forward leaping. Canid only uses 3 of its 6 motors (actuating its rear 2 legs and the top spine cables), while XRL uses 4 of its 6 motors (2 are not used since they contribute little to leaping [10]).

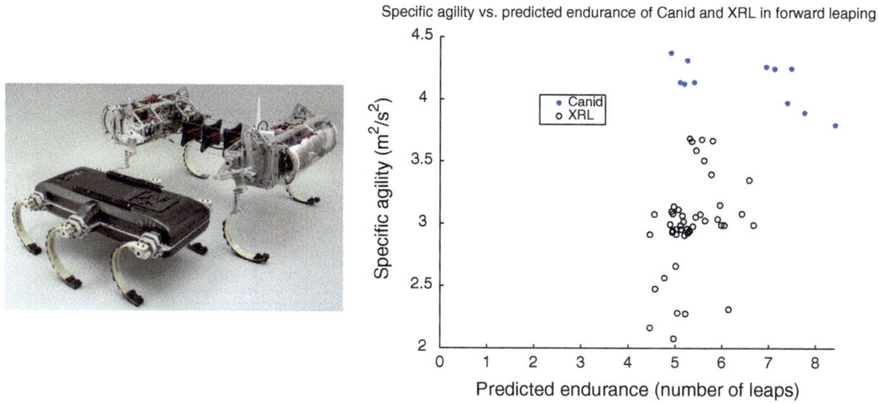

Fig. 1 Canid (*top*) and XRL (*bottom*) specific agility versus predicted endurance during forward leaping. The methods used for calculating these quantities are explained in Footnote 3

Canid leapt 11 times under a motion capture system,[3] including 5 times across an 85 cm gap which is close to the observed limit of its repeatable leaping ability from standstill (leaps over gaps up to 1 m across have been achieved however not in a repeatable fashion). XRL leaping data for this paper was taken from [39] during which parameters for quadrupedal forward leaping were systematically varied to search for various high extrinsic body energy forward leaps. The best XRL forward leap crossed a 50 cm gap, which is likely very close to the limit of its leaping ability from standstill.

The resulting specific agility and endurance for each Canid and XRL leap is shown in Fig. 1. Canid has a better maximum observed specific agility than XRL at a comparable endurance. Although it is likely that we could tune both machines to perform incrementally better, such adjustments would likely further advantage Canid, since these are the very first leaping experiments with Canid whereas XRL leaping has already benefitted from extensive past study and tuning [39]. These results indicate that at least one of the salient morphological differences between Canid and XRL confers upon Canid a significant agility advantage, particularly in light of its reduced (1 fewer) number of actuators used during liftoff.

Additional experiments summarized in Fig. 2 were conducted on Canid to quantify the relative agility benefit conferred by Canid's parallel elastic actuated spine (rather than its four-bar leg transmission) while leaping. Forward leaping data was collected on Canid using 5 different spine stiffnesses varying from rigid to negligible stiffness. Zero agility is recorded in the case where the robot was unable to achieve an aerial phase due to insufficient spine power. For each spine stiffness, Canid was run multiple times while systematically varying its spine motor current limit from 15 to 0 A in

[3] Vicon motion capture data is used to back out the kinetic and potential energy of the robots. Neglecting air resistance, the apex specific extrinsic body energy minus the starting specific extrinsic body energy gives a very close approximation to the specific agility (1) of the leap. The method used to calculate endurance is given in Appendix 1.

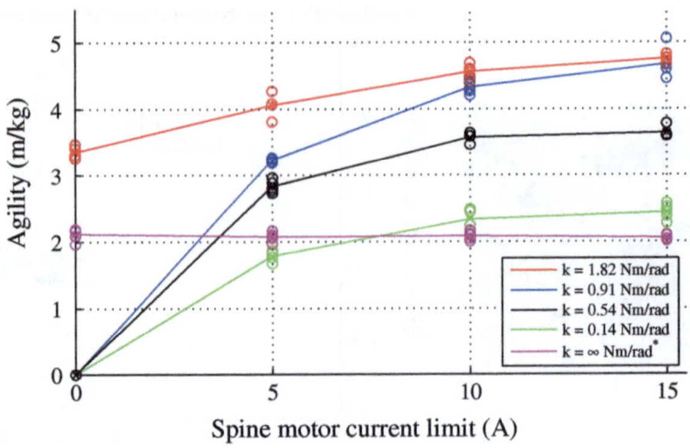

Fig. 2 Canid leaping agility with a variety of spine stiffnesses and spine motor current limits. A total of 80 runs are shown. Canid was allowed to preload its spine to the same angular displacement in every run (except for the rigid case) before setting the lower spine current limit and leaping. Zero agility is recorded in the case where the robot was unable to achieve an aerial phase due to insufficient spine power. *The $k = \infty$ case is approximated and was not empirically measured for fear of damaging the spine. Rigidity was achieved by locking the spine mechanism with minimal added mass

increments of 5 A—always from the same initial condition for every run (except for the rigid case) characterized by a spine preloaded to the same angular displacement prior to its release with lowered current limit reset at the onset of leaping. While successively more compliant spines afford the possibility of successively greater spine pre-loading by a given actuator, we chose to fix the preloading angle because: (1) this avoids the confounding effects of varied initial posture (and attendant variations in control strategy); and (2) the spine motors are capable of breaking the elastic fiberglass plate and it is not yet clear at which point plastic deformation begins. The spine stiffness is approximated by empirical data fit to a torsional Hooke's law spring as discussed in [16]. Current limits above 15 A are tenable in principle for our actuators, but do not result in substantially different results in any of these cases because the rapidly extending spine quickly brings them into the no-load regime, as discussed in Sect. 4.

4 Experimental Insights: Spine Agility Hypotheses

We now discuss in detail the manner in which these data support the overarching hypotheses regarding the agility benefits of the spine listed in Table 1.

Hypothesis 1: Replacing a Rigid Back with a Spine Can Increase Leaping Agility

The results in Fig. 2 indicate that—all else being equal—replacing a rigid back with a sufficiently powered spine mechanism (either through releasing initially stored elastic energy or through actuation) can provide a significant morphological

Table 1 Hypotheses regarding the performance of Canid and XRL and the proposed series of experiments to support or refute them

	Hypotheses	Supporting evidence
H1	Replacing a rigid back with a sufficiently powered spine mechanism can provide a significant morphological advantage for forward leaping agility	Sufficiently powered (either through actuation or releasing initially stored elastic energy) non-rigid spine leaping performance is significantly better than with a rigid spine
H2	Spine elastic energy release and spine actuation benefit leaping agility both individually as well as in combination	Motor energy output alone cannot account for the change in extrinsic body energy upon leaping with a non-rigid elastic spine, and increasing spine motor current limits monotonically increases specific agility. Additionally leaping with a full-powered spine and a "tuned" stiffness significantly outperforms leaping with a purely actuated or purely passive spine
H3	A four-bar transmission increases agility in forward leaping at the expense of reducing the number of other behaviors the robot can perform well	H3 would be supported if the four-bar diminishes agility of XRL leaping along particular directions relative to others

Hypotheses 1–2 were shown to be consistent with experimental data from this paper and Hypothesis 3 is the subject of further experiments presently underway

advantage for forward leaping agility. The average rigid spine specific agility over the runs was 2.1 m^2/s^2 which was bested by all actuated spines except for the severely underpowered negligible stiffness $k = 0.14$ Nm/rad spine with a 5 A current limit. In the case of unactuated spines, the $k = 1.82$ Nm/rad spine achieves an average specific agility of 3.4 m^2/s^2 and is thus endowed with enough initially stored elastic energy to outperform the rigid back by 62 %. The increase of available mechanical power through the spine's elastic energy release and/or actuation (discussed further in Hypothesis 2) is likely a primary source of this specific agility benefit.

We note that sufficiently-powered core actuation substantially increases rear leg loading in forward leaping. Rear leg stance duration during forward leaping on Canid was observed to be approximately constant over all runs (varying only by a few milliseconds), much to the surprise of the authors given the wide performance range of leaps shown in Fig. 2. Thus the larger agility achieved by core actuation must have generated higher rear leg forces during stance when compared to the rigid case. Specifically, the more than 2-fold increase in average specific agility between the rigid back (2.1 m^2/s^2) and the best spined runs (4.8 m^2/s^2) must have been accompanied by a more than 2-fold increase in average rear leg forces. Canid avoids torque-saturating the rear leg motors with spine forces by operating the rear legs near their kinematic singularity when the spine is doing work. Similar consideration of rear-leg kinematics may be required in general if a spine is added to a legged machine.

Hypothesis 2: Both Spine Elastic Energy Release and Spine Actuation Benefit Leaping Agility

Active and Passive Spine Elements in Isolation Figure 2 shows that the $k = 1.8$ Nm/rad spine with no actuation outperforms the fully actuated negligible stiffness $k = 0.14$ Nm/rad spine as well as the rigid spine (both of which initially store a negligible amount of initial elastic energy). The rear legs in the $k = 1.82$ Nm/rad spine case with no spine actuation output on average 65 J of work per leap of which less than 49 J get transferred into the extrinsic body energy due to the rear leg maximum gearbox efficiency of 75 %—an overestimate of the true transmission efficiency because we are not accounting the actual gearbox efficiency nor other sources of transmission friction as they are difficult to measure. However the change in extrinsic body energy of these runs averaged 54 J, leaving at least 5 J unaccounted for by the rear legs. Since the only other source of energy in the $k = 1.82$ Nm/rad spine case is the initially stored spine elastic energy, this indicates that initial elastic energy stored in spine bending contributes to forward leaping agility.

There is a monotonic average increase in agility with increased spine actuation power for the runs shown in Fig. 2. The decreasing efficacy of motor torque attested by the saturating contours of Fig. 2 reflects the no-load speed regime into which the actuators are quickly driven by the rapidly extending spine. Clearly the spine motors can be geared lower to achieve higher agility at the expense of decreasing endurance for this behavior—as will be taken into account in future Canid design iterations. Notice, as well, at the low end of spine stiffness, that Canid is unable to leap at all without spine actuation. Both observations indicate that the spine motors are directly contributing to forward leaping agility irrespective of spine elastic stiffness—except for of course in the rigid case.

The above results show an individual leaping benefit of spine elastic energy and spine actuation. This should come as no surprise since spine elastic energy release and actuation both augment the available mechanical power output of the machine.

Active and Passive Spine Elements in Parallel Combination The greatest observed forward leaping performance was achieved with the $k = 0.91$ Nm/rad and $k = 1.82$ Nm/rad spines using the highest spine actuator current limits, averaging a specific agility of $4.7\,\mathrm{m}^2/\mathrm{s}^2$. The best purely actuated spine and purely elastic spine experiments on the other hand were only able to achieve specific agilities of 2.6 and $3.5\,\mathrm{m}^2/\mathrm{s}^2$, respectively. The best "tuned" parallel elastic-actuated spines thus outperform the purely actuated spine by 81 % and the purely elastic spine by 34 %. This is likely because the nature of parallel elastic-actuations allows the release of the elastic energy stored in the spine to augment the spine motor power during the leap. These results support our hypothesis that using parallel elastic spine actuation outperforms both a purely actuated and purely elastic spine in forward leaping.

However if efficiency is deemed more important than raw agility then the performance using the unactuated $k = 1.82$ Nm/rad spine shown in Fig. 3 demonstrates that a sufficiently stiff spine, if properly pre-loaded in stance (either by motors or by a prior maneuver) may offer almost similar agility with considerably greater efficiency. We also note that although at lower actuator limits the $k = 1.82$ Nm/rad spine clearly

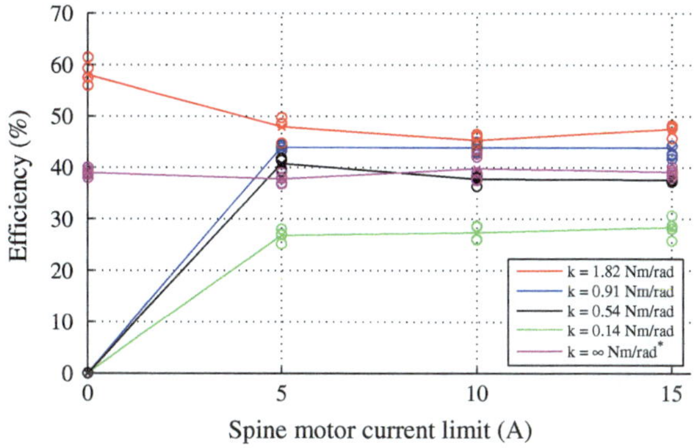

Fig. 3 Energetic efficiency of the experiments shown in Fig. 2. Energetic efficiency was calculated by dividing the total change in extrinsic body energy during the leap by the combined mechanical energy output of the motors (comprising of Canid's two rear leg motors and the spine motor actuating the top spine cables). This motor energy output is calculated at the output shaft before the gearbox and thus doesn't directly take into account gearbox or transmission inefficiencies

outperforms the $k = 0.91$ Nm/rad spine, this advantage diminishes as the actuator limits are increased. We are not sure if this reflects the beginning of the "crest" of the "sweet spot" specific agility ridge whose diminishing "other side" is evidenced in these preliminary experiments only by the most extreme $k = \infty$ Nm/rad case. This "sweet spot" is defined by spine stiffness that results in motor torque saturation at the initial spine angular displacement flexion, as this stiffness will store the maximal amount of initial elastic energy. Further experiments are now in progress with still stiffer (but not quite rigid) spines to better fill in the other side of the ridge.

5 Conclusions and Future Work

Specific agility, the mass-specific change in extrinsic body energy accomplished during a stance event, provides a comparative measure for quantifying performance of transitional behaviors such as jumping and accelerating across different platforms using different power budgets. Pairing this measure with endurance, the number of times a transition can be repeated given resource limitations, provides a clearer picture of a robot's useful operational agility.

Leaping experiments suggest that a significant benefit is conferred by adding a four-bar and a parallel elastic actuated spine [16] to the base XRL robot at no cost to endurance. Further investigation into characterizing the isolated benefit of the spine concluded that—all else being equal—replacing a rigid back with a sufficiently powered spine mechanism (either through releasing initially stored elastic energy or through actuation) can provide a significant morphological advantage for forward leaping agility. There was a measurable individual specific agility contribution from both spine actuation and releasing stored elastic energy in the spine. Furthermore, a parallel-elastic actuated spine confers a larger specific agility benefit to forward leaping than does a purely actuated or purely passive spine. In conclusion, a parallel elastic actuated spine morphology shows a significant agility advantage in forward leaping as compared to a rigid back. Experiments are now in progress to quantify the relative benefit conferred by the addition of a leg four-bar transmission in Canid. Future work will concentrate on comparing the relative benefit of core actuation using a spine to using the same motors instead for additional distal leg actuation.

We are still in the early stages of understanding how to characterize legged agility. Following the tradition of the more mature aircraft [40], aquatic [34], and wheeled [41] vehicle literatures (wherein variously dimensioned agility and maneuverability measures are introduced for different purposes and at different operating points), we explore the utility of a dimensional measure (m^2/s^2) that at the very least proves useful for comparing legged leaps from rest of different machines. Given its (rough) invariance across animal leaping maneuvers, this measure may also have relevance for probing biological energetics. Most immediately, we aim to apply insights provided by the empirical support or refutation of our stated hypotheses toward the design of more agile machines.

Our narrow focus on legged performance presently ignores the fascinating broader question of how to compare agility of such hybrid locomotory platforms against those employing a persistent stance (e.g. cars [41–43] or boats [34]) or aerial (e.g. jets [34, 40, 44–47]) phase—or even against legged platforms whose limbed manipulation of inertia or momentum in flight significantly enhances their terrestrial locomotory prowess [11]. We trust that further debate and study within the robotics research community along the lines we introduce here will help advance that important goal.

Acknowledgments This work is supported by the National Science Foundation Graduate Research Fellowship under Grant No. DGE-0822, by the Army Research Laboratory under Cooperative Agreement Number W911NF-10-2-0016, and by the Fonds Quebecois de la Recherche sur la Nature et les Technologies B1 168461. We would like to thank Shai Revzen and Robert Full for conversations regarding Borelli's law as well as Ben Kramer for experiments characterizing Canid's motor thermal properties.

Appendix 1: Endurance Calculations

The endurance of each leap is calculated as follows. The thermal temperature rise ΔT_i incurred by each motor $i \in I$ during the leap is calculated via the thermal model described in Fig. 5 of [48]. Let T_F denote the failure temperature of motor i and let T_{i0} denote the motor i's initial temperature before the leap. The number of times n_i that motor i can perform the leap is approximated by:

$$n_i = \frac{T_F - T_{i0}}{\Delta T_i}.$$

The endurance of the leap is then given by the lowest individual motor endurance, or:

$$
\begin{aligned}
n &= \inf_{i\in I}\,(n_i)\\
&= \inf_{i\in I}\Big(\frac{T_F - T_{i0}}{\Delta T_i}\Big),
\end{aligned}
$$

so as to extrapolate how many times the leap can be performed sequentially before thermal failure since thermal capacity represents the limiting resource for both Canid and XRL. This method allows us to sidestep the need to run repeated experiments pushing the thermal limits for each machine in order to calculate endurance which would risk motor damage.

Appendix 2: Energy and Power Density for Legged EM Actuators

Assuming that EM motors produce a magnetic field of uniform density, the motor creates force by having this field interact with permanent magnets. This interaction occurs over some area (the air gap) and so is proportional to l^2. Assuming that the motor does work by rotating through a fixed angle, the transformed displacement through a leg of arbitrary geometry will scale according to the characteristic length, l. The energy produced by the motor (the work done) is therefore proportional to l^3, so for constant density, specific energy is scale invariant.

Power density scaling is originally presented in [35], pp. 176–181, but will be reworked below with more detailed scaling analysis. Assuming energy density is mass-invariant in an actuator, the power density scaling will be considered for a hopping task. Neglecting air resistance the apex height will be constant, and so it follows that the liftoff velocity, v_f, will also be constant. Assuming the system starts crouched at rest, the leg must go through a fixed extension, l, and accelerate the body to v_f. Assuming constant acceleration, a, $v_f = at$ and $l = \frac{1}{2}at^2$ where t is the time the system is in contact with the ground. Substituting for a, $l = \frac{1}{2}v_f t$. Since v_f is constant, t scales according to l. Given constant energy density, power density then scales according to l^{-1}. This means that for specific energy to remain performance limiting, specific power must scale according to l^{-1}. This is in sharp contrast to [49] where specific power scales according to $l^{0.5}$ in support of maintaining dynamic similarity with respect to the pendulous motion of a swinging body characteristic of certain animal climbers [50].

References

1. Zhou, X., Bi, S.: A survey of bio-inspired compliant legged robot designs. Bioinspiration Biomimetics **7**(4), 041001 (2012)
2. "Boston dynamics": http://www.bostondynamics.com
3. Seok, S., Wang, A., Chuah, M.Y., Otten, D., Lang, J., Kim, S.: Design principles for highly efficient quadrupeds and implementation on the mit cheetah robot. In: Proceedings—IEEE International Conference on Robotics and Automation (2013) pp. 3307–3312
4. Ananthanarayanan, A., Azadi, M., Kim, S.: Towards a bio-inspired leg design for high-speed running. Bioinspiration Biomimetics **7**(4), 046005 (2012)
5. Sreenath, K., Park, H., Poulakakis, I., Grizzle, J.W.: A compliant hybrid zero dynamics controller for stable, efficient and fast bipedal walking on mabel. Int. J. Robot. Res. **30**(9), 1170–1193 (2011)
6. Grimes J.A., Hurst, J.W.: The design of atrias 1.0 a unique monopod, hopping robot. In: Adaptive Mobile Robotics—Proceedings of the 15th International Conference on Climbing and Walking Robots and the Support Technologies for Mobile Machines, CLAWAR 2012 (2012) pp. 548–554
7. Pasupuleti, M., Nadubettu Yadukumar, S., Ames, A.: Human-inspired underactuated bipedal robotic walking with amber on flat-ground, up-slope and uneven terrain. In: IEEE/RSJ International Conference on Intelligent Robots and Systems (IROS), Algarve, Portugal (2012)
8. Holmes, P., Full, R.J., Koditschek, D.E., Guckenheimer, J.: The dynamics of legged locomotion: models, analyses, and challenges. SIAM Rev. **48**(2), 207304 (2006)
9. Koditschek, D.E., Full, R.J., Buehler, M.: Mechanical aspects of legged locomotion control. Arthropod Struct. Dev. **33**(3), 251272 (2004)
10. Johnson, A.M., Koditschek, D.E.: Toward a vocabulary of legged leaping. In: Proceedings of the 2013 IEEE International Conference on Robotics and Automation, May 2013, pp. 2553–2560
11. Libby, T., Moore, T.Y., Chang-Siu, E., Li, D., Cohen, D.J., Jusufi, A., Full, R.J.: Tail-assisted pitch control in lizards, robots and dinosaurs. Nature **481**(7380), 181–184 (2012)
12. Degani, A., Feng, S., Brown, H., Lynch, K., Choset, H., Mason, M.: The parkourbot-a dynamic bowleg climbing robot. In: 2011 IEEE International Conference on IEEE, Robotics and Automation (ICRA) (2011) pp. 795–801
13. Urata J., Nakanishi, Y., Okada, K., Inaba, M.: Design of high torque and high speed leg module for high power humanoid. In: 2010 IEEE/RSJ International Conference on IEEE, Intelligent Robots and Systems (IROS) (2010), pp. 4497–4502
14. Dickson, J.D., Patel, J., Clark, J.E.: Towards maneuverability in plane with a dynamic climbing platform. In: 2013 IEEE International Conference on IEEE, Robotics and Automation (ICRA), (2013), pp. 1355–1361
15. Bowling, A.P.: Dynamic performance, mobility, and agility of multilegged robots. J. Dyn. Syst. Meas. Control, Trans. ASME **128**(4), 765–777 (2006)
16. Pusey, J.L., Duperret, J.M., Haynes, G.C., Knopf, R., Koditschek, D.E.: Free-standing leaping experiments with a power-autonomous elastic-spined quadruped. In: SPIE Defense, Security, and Sensing, International Society for Optics and Photonics, (2013), vol. 8741, pp. 87 410W–87 410W
17. Haynes, G.C., Pusey, J., Knopf, R., Johnson, A.M., Koditschek, D.E.: Laboratory on legs: an architecture for adjustable morphology with legged robots. In: Karlsen, R.E., Gage, D.W., Shoemaker, C.M., Gerhart, G.R. (eds.) Unmanned Systems Technology XIV, vol. 8387, p. 83870W. SPIE (2012)
18. Sheppard, J., Young, W.: Agility literature review: classifications, training and testing. J. Sports Sci. **24**(9), 919–932 (2006)
19. Jindrich, D.L., Qiao, M.: Maneuvers during legged locomotion. Chaos: An Interdisc. J. Nonlinear Sci. **19**(2), 026 105–026 105 (2009)
20. Bennet, H.C.: Clark, The energetics of the jump of the locust schistocerca gregaria. J. Exp. Biol. **63**(1), 53–83 (1975)

21. Bennet-Clark, H.C., Lucey, E.C.: The jump of the flea: a study of the energetics and a model of the mechanism. J. Exp. Biol. **47**(1), 59–67 (1967)
22. Roberts, T.J., Abbott, E.M., Azizi, E.: The weak link: do muscle properties determine locomotor performance in frogs? Philos. Trans. R. Soc. B: Biol. Sci. **366**(1570), 1488–1495 (2011)
23. Moore, D., Deuel, N., Drevemo, S., Van den Bogert, A.: Kinematic analysis of world championship three-day event horses jumping a cross-country drop fence. J. Equine Vet. Sci. **15**(12), 527–531 (1995)
24. Pfau, T., Garland de Rivaz, A., Brighton, S., Weller, R.: Kinetics of jump landing in agility dogs. Vet. J. **190**(2), 278–283 (2011)
25. Walker, J.A.: Does a rigid body limit maneuverability? J. Exp. Biol. **203**(22), 3391–3396 (2000)
26. Williams, S.B., Tan, H., Usherwood, J.R., Wilson, A.M.: Pitch then power: limitations to acceleration in quadrupeds. Biol. Lett. **5**(5), 610–613 (2009)
27. Roberts, T.J., Scales, J.A.: Mechanical power output during running accelerations in wild turkeys. J. Exp. Biol. **205**(10), 1485–1494 (2002)
28. McGowan, C.P., Baudinette, R.V., Usherwood, J.R., Biewener, A.A.: The mechanics of jumping versus steady hopping in yellow-footed rock wallabies. J. Exp. Biol. **208**(14), 2741–2751 (2005)
29. Hoover, A.M., Burden, S., Fu, X., Sastry, S.S., Fearing, R.S.: Bio-inspired design and dynamic maneuverability of a minimally actuated six-legged robot. In: 2010 3rd IEEE RAS and EMBS International Conference on Biomedical Robotics and Biomechatronics, BioRob 2010, (2010), pp. 869–876
30. Jindrich, D.L., Smith, N.C., Jespers, K., Wilson, A.M.: Mechanics of cutting maneuvers by ostriches (struthio camelus). J. Exp. Biol. **210**(8), 1378–1390 (2007)
31. Jindrich, D.L., Full, R.J.: Many-legged maneuverability: dynamics of turning in hexapods. J. Exp. Biol. **202**(12), 1603–1623 (1999)
32. Jindrich, D.L., Besier, T.F., Lloyd, D.G.: A hypothesis for the function of braking forces during running turns. J. Biomech. **39**(9), 1611–1620 (2006)
33. Astley, H.C., Roberts, T.J.: Evidence for a vertebrate catapult: elastic energy storage in the plantaris tendon during frog jumping. Biol. Lett. **8**(3), 386–389 (2012)
34. Webb, P.W.: Maneuverability-general issues. IEEE J. Oceanic Eng. **29**(3), 547–555 (2004)
35. Schmidt-Nielsen, K.: Scaling: Why is Animal Size so Important?. Cambridge University Press, Cambridge, (1984)
36. d'Arcy, W.T.: On Growth and Form. Cambridge Univ Press, Cambridge (1963)
37. Burridge, R.R., Rizzi, A.A., Koditschek, D.E.: Sequential composition of dynamically dexterous robot behaviors. Int. J. Robot. Res. **18**(6), 534555 (1999)
38. Ghigliazza, R.M., Altendorfer, R., Holmes, P., Koditschek, D.: A simply stabilized running model. SIAM J. Appl. Dyn. Syst. **2**(2), 187218 (2003)
39. Johnson, A.M., Koditschek, D.E.: Parametric Jumping Dataset on the Rhex Robot. University of Pennsylvania, Tech. Rep. (2012)
40. Paranjape, A.A., Ananthkrishnan, N.: Combat aircraft agility metrics-a review. J. Aerosp. Sci. Technol. **58**(2), 143–154 (2006)
41. Cho, W., Choi, J., Kim, C., Choi, S., Yi, K.: Unified chassis control for the improvement of agility, maneuverability, and lateral stability. IEEE Trans. Veh. Technol. **61**(3), 1008–1020 (2012)
42. Yi, J., Li, J., Lu, J., Liu, Z.: On the stability and agility of aggressive vehicle maneuvers: a pendulum-turn maneuver example. IEEE Trans. Control Syst. Technol. **20**(3), 663–676 (2012)
43. Nie, C., Van Dooren, S.C., Shah, J., Spenko, M.: "Increasing agility in unmanned ground vehicles using variable internal mass and inertial properties," in. Proc. SPIE **7332**, 733218 (2009)
44. Paranjape, A.A., Chung, S.-J., Selig, M.S.: Flight mechanics of a tailless articulated wing aircraft. Bioinspiration Biomimetics **6**(2), 026005 (2011)
45. Wise, K.A., Roy, D.J.B.: Agile missile dynamics and control. J. Guidance, Control, Dyn. **21**(3), 441–449 (1998)
46. P. W.G., F.M.: Agard advisory report 314 on operational agility. http://ftp.rta.nato.int/public//PubFullText/AGARD/AR/AGARD-AR-314/AGARDAR314.pdf (1994)

47. Bitten, R.: Qualitative and quantitative comparison of government and industry agility metrics. J. Aircr. **27**(3), 276–282 (1990)
48. Galloway, K.C., Haynes, G.C., Ilhan, B.D., Johnson, A.M., Knopf, R., Lynch, G., Plotnick, B., White, M., Koditschek, D.E.: X-rhex: A Highly Mobile Hexapedal Robot for Sensorimotor Tasks. University of Pennsylvania, Tech. Rep. (2010)
49. Lynch, G.A., Clark, J.E., Lin, P.-C., Koditschek, D.E.: A bioinspired dynamical vertical climbing robot. Int. J. Robot. Res. **31**(8), 974–996 (2012)
50. Goldman, D.I., Chen, T.S., Dudek, D.M., Full, R.J.: Dynamics of rapid vertical climbing in cockroaches reveals a template. J. Exp. Biol. **209**(15), 29903000 (2006)

On Prismatic and Torsional Actuation for Running Legged Robots

Bruce D. Miller, Jason M. Brown and Jonathan E. Clark

Abstract Among the challenges faced when developing dynamic, legged platforms are the manner and mechanisms utilized to modulate system energy. A great deal of success has been demonstrated by low degree of freedom platforms that rely on either pure torsion or thrusting to provide the requisite locomotive power. However, means of synergizing these approaches and the potential benefits thereof are not well understood. In this study, the effects of torsional and prismatic energy addition on running performance are investigated, both in isolation and as a hybrid approach. By allowing both mechanisms to be used in tandem, improvements to speed, stability and efficiency are noted. Additionally, these results suggest that rather than utilizing prismatic and torsional actuation to provide an even distribution of power, inhomogenous power generation may lead to futher performance benefits. This study not only examines the degree of actuator hybridization that leads to improved running, but also identifies the fundamental mechanisms by which these two approaches affect performance. These insights, in turn, provide physical intuition for the design of future legged platforms of more complex morphologies.

1 Introduction

Using legs, animals are able to dexterously negotiate a multitude of environments that often stymie conventional mobile robots. This has led to numerous studies aimed at understanding the fundamentals of legged locomotion (e.g. [6, 10, 13]) and developing robust and versatile technologies for application on robotic platforms (e.g. [23, 27]). Among the challenges facing these mechanical, legged systems is the manner by which they are actuated. Animals are able to leverage an over-actuated composition of controlled and passive degrees of freedom to produce effective locomotory behaviors utilizing a combination of highly-tuned joint compliance and coordinated

B.D. Miller · J.M. Brown · J.E. Clark (✉)
FAMU & FSU College of Engineering, 2003 Levy Avenue, Tallahassee, Fl, USA
e-mail: Jeclark@fsu.edu

B.D. Miller
e-mail: bdmiller@fsu.edu

M.A. Hsieh et al. (eds.), *Experimental Robotics*, Springer Tracts in Advanced Robotics 109, DOI 10.1007/978-3-319-23778-7_2

muscle activation [15]. However, the implementation of similarly complex artificial systems, both from a mechanical design and a controls standpoint, poses a significant technological hurdle.

An alternate solution is to model the animal behaviors resulting from complex neuromusculoskeletal interactions with simple, reduced-order dynamical templates [8], which can in turn be anchored by low degree of freedom platforms. This approach has resulted in remarkable success, as evidenced by robots such as RHex [22] and iSprawl [14], along with many others. These platforms in particular epitomize two of the most common approaches for actuating legged runners, which can be categorized as *torsionally* and *prismatically* actuated systems. Torsional actuation, as demonstrated by the RHex-family, is characterized by incorporating locomotive energy via torque applied at the hips while linear motion between the hip and ground contact is passively regulated as a function of the tuned limb compliance. Conversely, prismatically-actuated platforms, such as the Sprawl-family, modulate system energy via extension and retraction of the nominal limb length while relying on tuned linear and rotational passive compliant elements to redirect the energy into stable, forward locomotion.

Both of these techniques have been utilized with a high degree of success. However, the particular advantages afforded by each method of actuation are not well understood. Preliminary investigations comparing these approaches have only considered particular instantiations of strictly prismatic or strictly torsional actuation [16]. Conversely, studies that examined hybrid systems have primarily utilized multi-variate optimization [21] or have limited consideration to particular coordination schemes [11]. Additionally, many applications intrinsically require the use of limbs with multiple controlled degrees-of-freedom (e.g. dexterous manipulation or reconfiguration) and an improved understanding of energy incorporation may facilitate hybrid actuation schemes that leverage the already available actuators. In this study, we explore the underlying effects of simple and feed-forward prismatic, torsional and hybrid actuation on the gait characteristics of running systems and aim to ascertain the mechanisms by which they improve or hinder performance. Furthermore, we present a novel, dynamical legged platform able to implement and investigate this hybrid approach on a physical system.

2 Modeling and Simulation

To investigate the effects of torsional and prismatic actuation on running, we turn to reduced-order templates. Such models provide a lens for exploring the dynamics of running while reducing the confounding couplings that arise from the additional free parameters and degrees of freedom in more complex models. In this work, attention is focused on the spring loaded inverted pendulum (SLIP) model. This model captures the sagittal plane dynamics of running [4] and has been shown to produce similar whole-body dynamics to biological [5, 7, 25] and robotic runners [2, 3, 20].

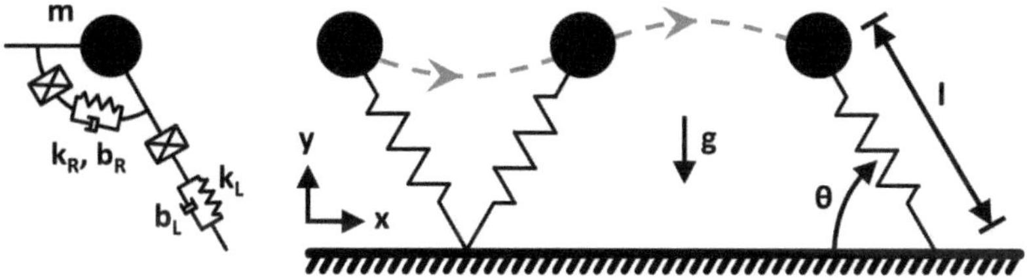

Fig. 1 Depiction of the SLIP model and the nominal trajectory that it follows over the course of a step. The parameters m, k_L, b_L, k_R, b_R and g correspond to the system mass, linear stiffness coefficient, linear damping coefficient, rotational stiffness coefficient, rotational damping coefficient and gravitational acceleration. The crossed boxes in series with the linear and rotational components represent the linear and rotational actuators driving the rest leg length l_{rest} and rest leg angle θ_{rest}, respectively. The variables x and y are the coordinates used for flight phase dynamics while l and θ are the coordinates used for stance phase dynamics

2.1 The Spring Loaded Inverted Pendulum Model

In its most basic form, the SLIP model is composed of two simple elements: a body, modeled as a point-mass, and a leg, modeled as a linear spring that is affixed to the body. A modified form of the model is utilized in this work that allows for both passive and active modulation of the system energy, as shown in Fig. 1. A damper element is added in parallel with the linear spring and a linear actuator is added in series to modulate energy along the leg axis. Additionally, a spring-damper element in series with a torsional actuator is added at the joint connecting the body and the leg. These extensions to the model provide both linear and rotational mechanisms to add or remove energy during locomotion.

The trajectory of the model progresses following the hybrid dynamics of stance and flight phases. Stance begins when the distal end of the unsprung leg contacting the running surface. This initiates a point-contact that acts as a pin joint about which the system rotates. As the body moves forward, the leg spring compresses and decompresses, redirecting the body and accelerating it to the point of lift-off. This event occurs when the ground reaction force at the leg-surface contact drops to zero. Following this occurrence, the system moves forward through the air following ballistic dynamics as the leg resets prior to the next touch-down event.

While the governing dynamics during flight can be simply described by

$$\begin{aligned} \ddot{x} &= 0, \\ \ddot{y} &= g, \end{aligned} \tag{1}$$

expressing the equations of motion that govern the stance behavior requires a more detailed examination of the SLIP model. Using the Euler Lagrange approach, we begin by describing the kinetic and potential energies of the system

$$
\begin{aligned}
T &= \frac{1}{2} m \dot{l}^2 + \frac{1}{2} m \left(l \dot{\theta}\right)^2, \\
V &= \frac{1}{2} k_L \left(l_{rest} - l\right)^2 + \frac{1}{2} k_R \left(\theta_{rest} - \theta\right)^2 + mgl \sin\theta,
\end{aligned} \tag{2}
$$

which can be used to calculate the Lagrangian

$$
\begin{aligned}
\mathscr{L} &= T - V, \\
&= \frac{1}{2} m \left(\dot{l}^2 + l^2 \dot{\theta}^2\right) - \frac{1}{2} \left(k_L \left(l_{rest} - l\right)^2 + k_R \left(\theta_{rest} - \theta\right)^2\right) - mgl \sin\theta .
\end{aligned} \tag{3}
$$

Additionally, the nonconservative terms (i.e. damping) can be expressed as

$$
\begin{aligned}
Q_{nc_L} &= b_L \left(\dot{l}_{rest} - \dot{l}\right), \\
Q_{nc_R} &= b_R \left(\dot{\theta}_{rest} - \dot{\theta}\right).
\end{aligned} \tag{4}
$$

Substituting (3) and (4) into the Euler-Lagrange equations, the equations of motion governing the stance dynamics can be derived as

$$
\begin{aligned}
\ddot{l} &= l \dot{\theta}^2 + \frac{k_L}{m} \left(l_{rest} - l\right) - \frac{b_L}{m} \left(\dot{l}_{rest} - \dot{l}\right) - g \sin\theta, \\
\ddot{\theta} &= -\frac{2 \dot{l} \dot{\theta}}{l} + \frac{k_R}{m l^2} \left(\theta_{rest} - \theta\right) - \frac{b_R}{m l^2} \left(\dot{\theta}_{rest} - \dot{\theta}\right) - \frac{g}{l} \cos\theta .
\end{aligned} \tag{5}
$$

2.2 Control

While the conservative, point-mass SLIP model (i.e. $\{b_L, k_R, b_R\} = 0$) can be self-stabilizing with appropriate parameter tuning and requires no actuation, the nonconservative analog necessitates a means of energy incorporation to offset the losses due to damping. Furthermore, both feedforward and feedback actuation have been shown to afford better stability than the model does on its own [1, 18, 19, 24].

In this study, we adopt feedforward controllers to drive both the prismatic and torsional actuators. This simplifies the control structure and reduces the computational requirements once instantiated on a physical system. An additional caveat for our controller is the assumption that the resulting platform is bipedal, rather than monopedal. This allows the results to be extended beyond single-legged, hopping systems and more readily apply to running platforms that have two sets of linked appendages (i.e. trotting pairs).[1]

[1]Without this assumption, all viable gaits require large aerial phases that are typically not desired as part of steady-state running.

The feedforward trajectories for the prismatic and torsional actuators were selected to sinusoidally adjust the rest leg length and leg angle. These trajectories can be described by

$$\begin{aligned} l_{rest_k} &= l_{nom} + l_d \sin\left(2\pi\left(\omega t + \phi_a + \phi_b + 0.5k\right)\right), \\ \theta_{rest_k} &= \theta_{nom} + \frac{\theta_d}{2}\left(1 - \cos\left(2\pi\left(\omega t + \phi_b + 0.5k\right)\right)\right), \end{aligned} \tag{6}$$

where l_{nom} is the nominal length of the prismatic actuator, θ_{nom} is the nominal touch-down angle of the torsional actuator, l_d is the stroke length prismatic actuator, θ_d is sweep angle of the torsional actuator, ω is the actuator frequency, ϕ_a is the phase offset between the actuators, ϕ_b is the phase offset of the torsional actuator from the touch-down event, and k is a counting parameter that switches between 0 and 1 with each successive step.

2.3 Simulation

Due to the intractability of an exact analytical solution to the SLIP model, a numerical simulation was utilized to explore and characterize the model behavior. The simulation was developed in MATLAB 2013b (Mathworks, Inc.) and employs the built-in Runga-Kutta numerical integrator *ode45* with absolute and relative tolerances of 10^{-10}. A Newton-Raphson fixed point search was implemented to find periodic orbits of the model with a tolerance of 10^{-8}. Additional scripts were written and utilized to capture the energetics and stability of the model for various parameter settings and initial conditions.[2]

3 Simulation Experiments

3.1 Simulation Setup

To examine the effects of prismatic and torsional actuation on running performance, a simulation study was conducted. Using the SLIP model and MATLAB implementation described in Sect. 2.3, a parameter sweep the energy incorporation parameters was performed. This allowed for the quantification of the effects of varying torsional and prismatic actuation on the locomotion speed, stability and efficiency.

[2]Energetics was quantified by prismatic and torsional actuator power. Stability was determined as the maximum eigenvalue of the Jacobian at the Poincaré section coinciding with the touch-down event. Additional details are found in Sect. 3.1.

Table 1 Parameter settings for SLIP parameter sweeps

Parameter	Model values
Mass (m)	0.3 kg
Nominal leg length (l_{nom})	0.03 m
Nominal leg angle (θ_{nom})	30°, 70°and110°
Linear stiffness (k_L)	1200 N m^{-1}
Rotational stiffness (k_R)	3 N m rad^{-1}
Linear damping (b_L)	3.95 N s m^{-1}
Rotational damping (b_R)	0.1 N m s rad^{-1}
Prismatic actuator stroke length (l_d)	0–0.03 m
Torsional actuator sweep angle (θ_d)	0°–125°
Actuator frequency (ω)	3.5 Hz
Actuator phase offset (ϕ_a)	0.75

Parameter values for the simulation were selected to correlate to the future instantiation of a physical platform (see Sect. 5). A mass of 0.3 kg and a nominal leg length of 0.03 m were chosen based on physical measurements. Leg stiffness, both linear and rotational were calculated to allow for resonant running when actuated at 5 Hz [26]. Linear and rotational damping were computed such that the damping ratio was 0.1.

Nominal values for control parameters needed to be determined as well. The actuator displacements l_d and θ_d were the primary parameters considered in the sweep. A range of 0–0.03 m for l_d and 0 °–125° for θ_d were chosen, as values less than zero are not physically meaningful and above these upper limits, stable gaits could not be found. Multiple values of the nominal leg angle θ_{nom} also had to be investigated, as using a steeper or shallower nominal leg angle was found to strongly skew results towards increased prismatic or rotational actuation, respectively. With this in mind, parameter sweeps were performed on three nominal leg angles, 110°, 70° and 30°. Additional control parameters include ω, ϕ_a and ϕ_b. ω was chosen to be 3.5 Hz and ϕ_a was set to 0.75. However, ϕ_b was found to be gait dependent and was determined as part of the Newton-Raphson search. A list of all physical and control parameters is tabulated in Table 1.

To quantify the running performance of the SLIP model under the various conditions of the parameter sweep, several performance indicators were computed. Mean fore-aft velocity $\bar{v}$ was calculated from the fore-aft distance traveled and the stride period. Though this measure is dimensional, and thus subject to the effects of scale, the nominal leg length remained consistent for all parameter settings so normalization is not required. The second measure of running performance was stability, which was quantified for small perturbations by the maximum eigenvalue λ_{max} of the Jacobian of the linearized return map [12]. As the maximum eigenvalue correlates to the rate at which the system returns to steady-state, small values for λ_{max} equate to better stability than large ones. The third measure, efficiency, was expressed by the specific resistance SR, describing the ratio of mean actuator power to the locomotive

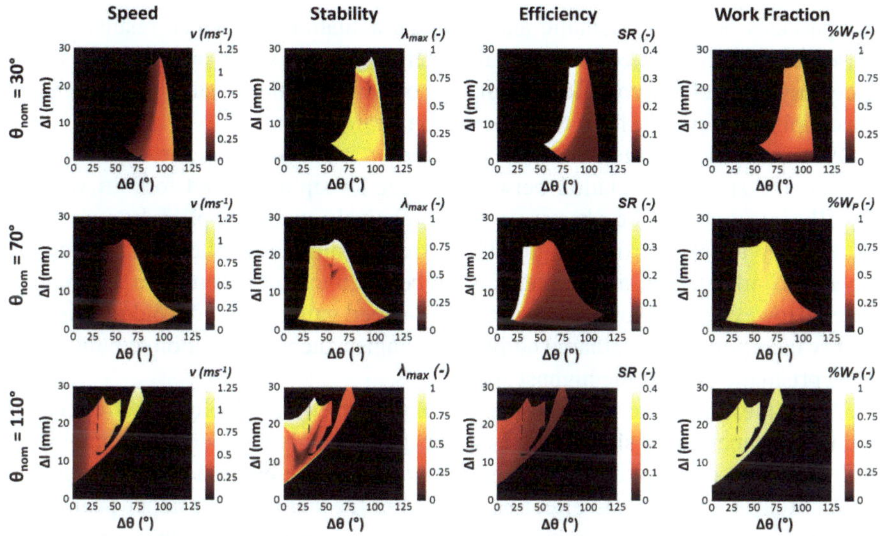

Fig. 2 Locomotive performance for the reduced-order running model as a function of the energy incorporation method. The plots (from *left* to *right*) show the speed, stability, efficiency and fractional work done by the linear actuator when running with varying degrees of prismatic and torsional energy input. *Black regions* around each plot show actuation settings for which stable gaits were not found

power required to move the system a given distance [9, 22]. As with λ_{max}, efficiency is maximized when SR is minimized. Additional measures, including the work done by the prismatic and torsional actuators and parasitic energy lost to the linear and rotational springs, were recorded as well.

3.2 Simulation Results

The calculated gait characteristics for each parameter set, speed, stability and efficiency, are shown in the left three columns of Fig. 2. Additionally, the work fraction, presented as the fractional work performed by the prismatic actuator, is included in the right column of Fig. 2. A review of these results can lead to the identification of several actuator-dependent effects.

As remarked on in the simulation setup, there is no nominal leg angle that can produce stable running gaits for both purely prismatic and purely torsional actuation. Furthermore, the range of fractional work performed by the actuators varies as a function of the nominal leg angle. This implies a significant coupling between the actuation approach and the parameters of the physical system and may merit the investigation of additional parameter couplings (e.g. linear/rotational stiffness and damping) in future studies. Additionally, it alludes to a challenge for generalized comparisons between prismatic and torsional actuation, for which results may be biased due to the choice of nominal system parameters. To overcome this, it is necessary to carefully consider the impacts of non-actuation system parameters and the couplings they may have to the actuation strategies.

When considering the effects of actuation on running speed, the most apparent trend is the strong correlation between the leg sweep angle and fore-aft velocity. Equally intriguing is the observation of a relatively weak coupling between fore-aft velocity and leg stroke length. These trends follow regardless of the nominal leg angle, indicative of a fundamental link between torsional actuation and running speed. However, it is worth noting that high-speed gaits are not devoid of prismatic energy incorporation, as at large nominal leg angles, the work fraction remains above 70 % prismatic even at the highest speed gaits.

While the leg stroke may only be weakly coupled to velocity, its effect on the stability is much more significant. Leg sweep is strongly coupled to the stability as well. It is also noteworthy that increasing or decreasing either actuation parameter does not have a monotonic effect on improving the stability, but rather peaks at an intermediate level. Additionally, the maximum eigenvalues are much smaller and the region for which the maximum eigenvalues remains small is much larger when utilizing a steep nominal leg angle.

Though speed and stability appear to correlate with the magnitudes of the leg sweep and leg stroke, efficiency demonstrates a stronger relationship to the ratio between the two. Aside from the extremely inefficient left edge seen at $\theta_{nom} = 30°$ and 70°,[3] the specific resistance closely mirrors the work fraction rather than either actuation parameter independently, with more efficient gaits noted with a smaller prismatic work fraction (or, equivalently, larger torsional work fraction).

To provide insight into the energetic effects, the negative work done by the actuators and parasitic behaviors of the springs were considered and the results are presented in Fig. 3. The energetic losses have been scaled by the steady-state system energy to provide a means of normalization between dissimilar gait types. For the actuators in the two left columns, it appears that more negative work is done by the torsional actuator than by the prismatic actuator at high speeds, while the negative work done by the prismatic actuation peaks at low speed, high stroke length gaits. Additionally, for the prismatic actuator, a region is present for which it generates a negligible amount of negative work. Furthermore, this region tends to include a section of the maximally stable set of gaits.

In the right two columns, the amount of elastic potential remaining in the springs at the point of lift-off can be observed. Parasitic losses to the linear spring are typically an order of magnitude or more smaller than the torsional. However, the amount of energy parasitically lost to both springs falls significantly at high-speed gaits and is indicative of a more efficient exchange of energy at these speeds. It is also notable

[3]For this region, the efficiency is actually worse than reported (up to and greater than 20) but has been limited to a maximum of 0.4 so the trends in the rest of the region is still discernible.

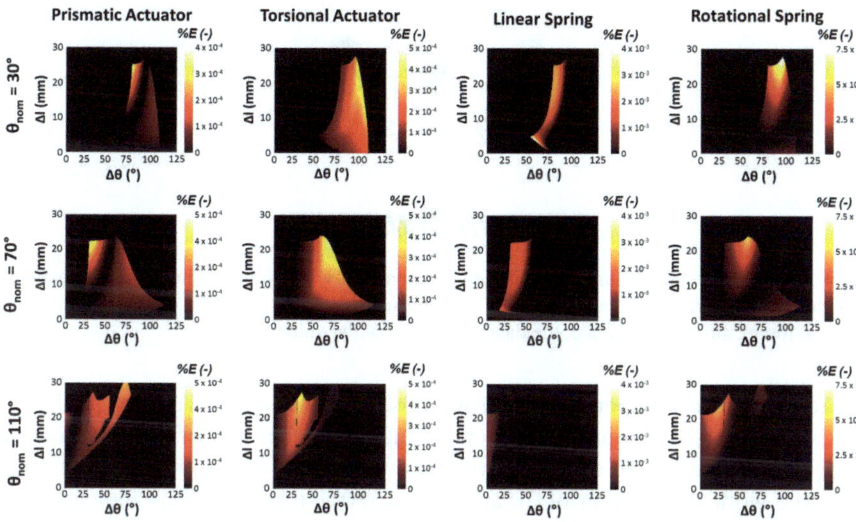

Fig. 3 Fractional sources of energetic losses during running. The two left columns show the amount of negative work done by the prismatic and torsional actuators. The two right columns depict the parasitic losses to the linear and torsional springs that arise due to lift-off occurring prior to the respective springs returning the elastic potential they have stored. All of the plots have been normalized to the steady-state system energy (i.e. kinetic and potential energy during the steady-state flight phase) and are expressed as the fraction of system energy

that the energetic losses due to the springs is considerably larger than for negative actuator work and that the drop in energetic losses coincides with the decrease in specific resistance.

4 Discussion

4.1 Prismatic, Torsional and Hybrid Actuation

In considering the independent and hybrid utilization of both prismatic and torsional actuation, it is evident that running performance is largely influenced by the total actuator input as well as the ratio of work provided by the prismatic and torsional actuators. It is clear that the entire spectrum, from purely prismatic to purely torsional actuation, is not available to produce viable gaits for any particular set of physical and control parameters. However, proper tuning of the ratio and magnitude of the actuator work can produce fast, stable and efficient behaviors at the nominal leg angles examined.

Perhaps the most striking result is that when utilizing a steep leg angle with a significant degree of hybridization ($\Delta L = 18$ mm, $\Delta\theta = 45°$ and $\theta_{nom} = 110°$), all three performance characteristics concurrently settle close to their 'optimal' values.

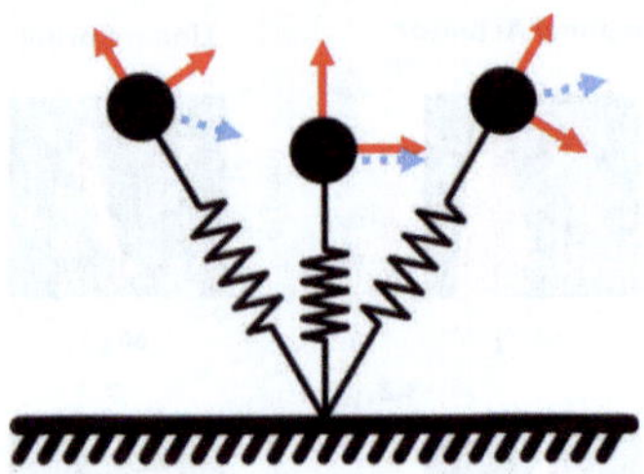

Fig. 4 Schematic depicting the hypothesized mechanism by which linear and prismatic actuator work contribute to running performance. The solid lines show the direction of the prismatic and torsional forces throughout stance; the dashes lines show the velocity heading

This result may be of import for several reasons. First, this suggests that few trade-offs may be necessary to optimize for any particular goal when properly tuned. Second, rather than settling at an evenly distributed work fraction, the 'optimal' point has a relative work distribution of 70 % prismatic/30 % torsional. This indicates that during actuator selection, inhomogeneous sets of prismatic and torsional actuators may allow for the best performance while maximizing specific power.

4.2 Mechanisms for Speed

As discussed in the simulation results, for a given nominal leg angle, running speed seems to correlate more strongly to leg sweep than to leg stroke. A likely mechanism for this behavior results from the orientation of the velocity vector with respect to the forces produced by the prismatic and torsional actuators, as depicted in Fig. 4. For prismatic actuation, extension of the leg during stance results in a force along the axis of the leg. For most nominal leg angles, this force is the negative direction immediately following touch-down and then progresses to accelerate the body forward after mid-stance. While still positive work, this method of energy input can results in slower speeds while increasing the apex height during flight. Conversely, the force produced by sweeping the leg is perpendicular to the leg axis and can accelerate the system throughout the stance phase. Thus, while increasing the leg stroke produces large decelerating forces in addition to accelerating ones, increasing leg sweep only increases accelerating forces (and potentially decreases decelerating ones, if present).

4.3 Mechanisms for Stability

In contrast to speed, stability appears to be influenced significantly by both leg sweep and leg stroke. A closer examination of the most stable parameter settings reveals that gaits with improved stability demonstrate a trajectory phasing for which reaching the maximal stroke and sweep angle coincides with the end of stance. This may be a key factor in the underlying mechanism contributing to the stability of these running gaits.

We hypothesize that this stabilizing mechanism utilizes a two-step approach. The mechanism stabilizes the gait rapidly due to the change in actuator positions at touch-down in the step following a perturbation. This effectively reduces the stroke length and sweep angle for the step following a perturbation that increases the system energy while increasing the stroke and sweep if energy is depressed.

For example, when a perturbation increases the energy during one step, the vertical velocity at lift-off will be increased. This will result in a lengthened flight phase, during which the prismatic and torsional actuators will advance past their nominal touch-down conditions. In the following step, both the stroke and sweep of the leg will progress as normal; however, they will reach their maximal lengths prior to the end of stance due to the lag introduced by the increased flight phase. Thus, at the end of stance they will begin to retract, removing energy and bringing the system back to its nominal energy level.

This mechanism favors gaits in which the maximal leg extension and sweep coincide at the same time as lift-off. However, it will still allow for stabilization when this does not occur, albeit at a slower rate. Since a sinusoidal trajectory is followed, the rate at which the leg extends or sweeps slows as the maximum extension is reached. Thus, even if energy is not actively removed by the actuators at the end of stance, less energy will be added during a high-energy stride (and more will be added if the energy is low).

4.4 Mechanisms for Efficiency

Rather than being linked directly to the work done by either (or both) actuators, efficiency appears to be influenced by the ratio between the two. The mechanism to describe this effect may be closely related to that for fore-aft velocity. As discussed for that mechanism, the work done by the torsional actuator is primarily funneled into the speed of travel while a significant portion of the work done by the prismatic actuator contributes to the apex height as well. Thus, increasing the torsional work results in an analogous increase in speed. This contributes to a decrease in the specific resistance (or at the very least, maintains it at a constant value). However, when increasing prismatic work, a significant portion of the work done translates to vertical motion rather than fore-aft speed, and it follows that an increase in specific resistance should result.

5 Physical Platform

Analysis of the simulation study demonstrates the efficacy of the hybrid utilization of both prismatic and torsional actuators for running. This motivates the development of legged platforms that can employ both actuation strategies to leverage these

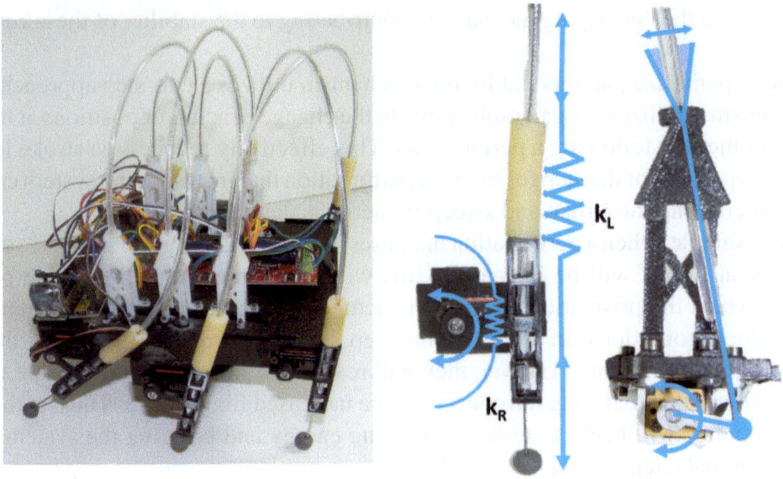

Fig. 5 Photograph of the hexapedal platform developed as a part of this study. This platform is designed to enable both prismatic and torsional actuation to improve performance and allow for modal transitions and limb reorientation in autonomous field operation

new-found insights. One particular application that dovetails with this finding are legged platforms designed for multi-modal running and climbing [17]. Purely prismatic actuation has shown to be effective at producing rapid and robust running and climbing behaviors. However, autonomous transitions between the disparate modalities are impossible without the presence of torsional, 'shape-changing' actuators to alter the limb morphology. Coupled with the simulation insights, we aim to develop a platform that can embed shape-changing actuators to allow for transitions, as well as employ them during steady-state behaviors for hybridly actuated gaits.

Several design criterion were considered in developing this physical platform, including: proximal location of the actuators to minimize leg inertia, a simple and robust transmission for prismatic and torsional actuation, readily tunable linear and rotational limb compliance, and an appropriate distribution of actuator power for both steady-state locomotion and transitional behaviors. The platform itself, shown in Fig. 5, draws substantial inspiration from iSprawl [14], as evidenced by the hexapedal morphology and cable-drive system. The cable drive, in particular, is beneficial as it allows the prismatic actuators to be situated on the central body rather than mounted on the individual legs. Thus, while the mass of each leg, including the actuators, is 33 g ($\sim$8 % of the total platform mass), over half of the weight is affixed to the central body, significantly reducing the leg inertia.

To drive each leg, two actuators are utilized: a DC motor (Pololu, 75:1 Micro Metal Gearmotor HP), which extends and retracts the cable drive via a modified four-bar linkage, and a servo motor (PowerHD, DSM44 Digital Servo), mounted at the hip joint, which sweeps the leg. The modified four-bar uses a crank to directly drive the cable, eliminating the need for a separate coupler by allowing the joint at which the drive cable enters the sheath to rotate, as can be seen in the right panel of Fig. 5. Compliant elements are also readily incorporated into this design. A soft section of silicon tubing at the joint between the sheath and the leg provides linear compliance that can be adjusted by modifying the thickness and length of this element. Rotational compliance is incorporated by a shaft linking the hip and the leg, which acts as a cantilever beam.

With the presence of 12 active DoFs, it is necessary to utilize a robust control architecture that can handle this quantity of actuators. Since the control law itself is clock-based and feed-forward, minimal coordination is needed between the individual limbs. This allows for off-the-shelf components to be utilized that are capable of driving the actuators at fixed rates. Three dual motor controllers (Orion Robotics, 2×5A RoboClaw) are used to drive the leg stroke actuators. The DC motors are equipped with quadrature encoders to provide position feedback for the velocity PID controllers. The servos are connected to a controller (Pololu, Mini Maestro 12-Channel) that is capable of prescribing set positions and sweep rates for each servo. These components are connected to a Wixel microcontroller (Pololu) that serves as the on-board processor as well as a communication link, via wireless RF, to the human operator.

A final aspect of the design process related to the power distribution of the actuators. While chosen partially for the functionality they provided, the DC and servo motors also drew from the observation of inhomogeneous power generation in the simulation study. While a wide range of hybrid strategies demonstrated the capacity for high performance running, those at the 110° nominal leg angle showed the highest levels of speed and stability. Furthermore, a fairly consistent actuator work distribution of approximately 75 %\25 % prismatic to torsional was found for stable running gaits. This supports the motor selection for this platform, for which the drive motors (driving the four-bar linkage, i.e. prismatic actuators) can supply approximately 9.6 w apiece while the servo motors (rotating the hips, i.e. torsional actuators) can provide 4.2 w. This distribution of 69 %\31 % prismatic to torsional is in line with the observed simulation behavior.

6 Conclusion

In this work, we have investigated running with prismatic and torsional, as well as hybrid actuation to examine the effects of each strategy on the overall running performance. We used simulations of reduced-order models to investigate the variation in actuator parameters. From this study, we can determine 'optimal' degrees of hybridization for a given system. Furthermore, we considered the trends that effect the performance criterion to hypothesize several mechanisms that may govern the observed performance characteristics, including speed, stability and efficiency. Additionally, we developed and presented the design of a novel, hexapedal robot with the capacity to actuate prismatically, torsionally and using hybrid modes.

Future studies will extend these findings in a number of ways. First, the platform will be used to more extensively investigate how various degrees of hybridization affect the performance of a physical platform. Additionally, we will investigate how other parameters interact, as a significant parameter dependence was observed with regards to the nominal leg angles and hypothesized dependences to the leg and hip compliance are expected as well. These studies will enable the novel platform, as well as future legged system, both simple in nature and with more complex morphologies, to better leverage their on-board actuators for high performance locomotion.

References

1. Altendorfer, R., Koditschek, D.E., Holmes, P.: Stability analysis of a clock-driven rigid-body SLIP model for RHex. Int. J. Robot. Res. **23**, 1001–1012 (2004)
2. Altendorfer, R., Saranli, U., Komsuo?lu, H., Koditschek, D., Brown Jr, H.B., Buehler, M., Moore, N., McMordie, D., Full, R.: Evidence for spring loaded inverted pendulum running in a hexapod robot. In: Rus, D., Singh, S. (eds.) Experimental Robotics VII. Lecture Notes in Control and Information Sciences, vol. 271, pp. 291–302. Springer, Berlin Heidelberg (2001)
3. Birkmeyer, P., Peterson, K., Fearing, R.S.: DASH: a dynamic 16g hexapedal robot. In: IEEE/RSJ International Conference on Intelligent Robots and Systems (IROS), St. Louis, MO (2009), pp. 2683–2689
4. Blickhan, R.: The spring-mass model for running and hopping. J. Biomech. **22**, 1217–1227 (1989)
5. Blickhan, R., Full, R.J.: Similarity in multilegged locomotion: bouncing like a monopode. J. Comp. Physiol. A: Sens. Neural Behav. Physiol. **173**, 509–517 (1993)
6. Dickinson, M.H., Farley, C.T., Full, R.J., Koehl, M.A.R., Kram, R., Lehman, S.: How animals move: an integrative view. Science **288**, 100–106 (2000)
7. Farley, C.T., Glasheen, J., McMahon, T.A.: Running springs: speed and animal size. J. Exp. Biol. **185**, 71–86 (1993)
8. Full, R.J., Koditschek, D.E.: Templates and anchors: neuromechanical hypotheses of legged locomotion on land. J. Exp. Biol. **202**, 3325–3332 (1999)
9. Gabrielli, G., von Kármán, T.: What price speed? specific power requirements for propulsion of vehicles. Mech. Eng. **72**, 775–781 (1950)
10. Geyer, H., Seyfarth, A., Blickhan, R.: Compliant leg behavior explains basic dynamics of walking and running. Proc. R. Soc. B: Biol. Sci. **273**, 2861–2867 (2006)
11. Görner, M., Albu-Schäffer, A.: A robust sagittal plan hexapedal running model with serial elastic actuation and simple periodic feedforward control. In: IEEE/RSJ International Conference on Intelligent Robots and Systems (IROS), Tokyo, Japan (2013), pp. 5586–5592
12. Guckenheimer, J., Holmes, P.: Nonlinear oscillations, dynamical systems, and bifurcations of vector fields. Springer (1983)
13. Holmes, P., Full, R.J., Koditschek, D., Guckenheimer, J.: The dynamics of legged locomotion: models, analyses, and challenges. SIAM Rev. **48**, 207–304 (2006)
14. Kim, S., Clark, J.E., Cutkosky, M.R.: iSprawl: design and tuning for high-speed autonomous open-loop running. Int. J. Robot. Res. **25**, 903–912 (2006)
15. Koditschek, D.E., Full, R.J., Buehler, M.: Mechanical aspects of legged locomotion control. Arthropod Struct. Dev. **33**, 251–272 (2004)
16. Larson, P., Seipel, J.: A spring-loaded inverted pendulum locomotion model with radial forcing. In: ASME International Design Engineering Tecnical Conferences & Computers and Information in Engineering Conference (IDETC/CIE). vol. 4, Chicago, IL (2012), pp. 877–883
17. Miller, B., Darnell, A., Clark, J.: Running in the horizontal plane with a multi-modal dynamical robot. In: IEEE International Conference on Robotics and Automation (ICRA) (2013)

18. Poulakakis, I., Grizzle, J.W.: The spring loaded inverted pendulum as the hybrid zero dynamics of an asymmetric hopper. IEEE Trans. Autom. Control **54**, 1779–1793 (2009)
19. Raibert, M.: Hopping in legged systems - modeling and simulation for the two-dimensional one-legged case. In: IEEE Transactions on Systems, Man and Cybernetics **SMC-14** (1984), pp. 451–463
20. Raibert, M.H., Chepponis, M., Brown Jr, H.B.: Running on four legs as though they were one. IEEE J. Robot. Autom. **2**, 70–82 (1986)
21. Remy, C.D., Buffinton, K., Siegwart, R.: Comparison of cost functions for electrically driven running robots. In: IEEE International Conference on Robotics and Automation (ICRA), Saint Paul, MN (2012), pp. 2343–2350
22. Saranli, U., Buehler, M., Koditschek, D.E.: RHex: a simple and highly mobile hexapod robot. Int. J. Robot. Res. **20**, 616–631 (2001)
23. Sayyad, A., Seth, B., Seshu, P.: Single-legged hopping robotics research - a review. Robotica **25**, 587–613 (2007)
24. Schmitt, J., Clark, J.: Modeling posture-dependent leg actuation in sagittal plane locomotion. Bioinspiration Biomimetics **4** (2009) 046005 (17)
25. Srinivasan, M., Holmes, P.: How well can spring-mass-like telescoping leg models fit multipedal sagittal-plane locomotion data? J. Theoretical Biol. **255**, 1–7 (2008)
26. Thompson, C.M., Raibert, M.H.: Passive dynamic running. In: Hawyward, V., Khatib, O. (eds.) Experimental Robotics I. Lecture Note in Control and Information Sciences, vol. 139, pp. 74–83. Springer, Berlin Heidelberg (1990)
27. Zhou, X., Bi, S.: A survey of bio-inspired compliant legged robot designs. Bioinspiration Biomimetics **7** (2012) 041001 (20)

Experimental Results for Dexterous Quadruped Locomotion Planning with RoboSimian

Brian W. Satzinger, Chelsea Lau, Marten Byl and Katie Byl

Abstract RoboSimian is a quadruped robot inspired by an ape-like morphology, with four symmetric limbs that provide a large dexterous workspace and high torque output capabilities. Advantages of using RoboSimian for rough terrain locomotion include (1) its large, stable base of support, and (2) existence of redundant kinematic solutions, toward avoiding collisions with complex terrain obstacles. However, these same advantages provide significant challenges in experimental implementation of walking gaits. Specifically: (1) a wide support base results in high variability of required body pose and foothold heights, in particular when compared with planning for humanoid robots, (2) the long limbs on RoboSimian have a strong proclivity for self-collision and terrain collision, requiring particular care in trajectory planning, and (3) having rear limbs outside the field of view requires adequate perception with respect to a world map. In our results, we present a tractable means of planning statically stable and collision-free gaits, which combines practical heuristics for kinematics with traditional randomized (RRT) search algorithms. In planning experiments, our method outperforms other tested methodologies. Finally, real-world testing indicates that perception limitations provide the greatest challenge in real-world implementation.

Keywords Kinodynamic planning · Dexterous locomotion · Quadruped

B.W. Satzinger (✉) · C. Lau · M. Byl · K. Byl
Robotics Laboratory, University of California at Santa Barbara (UCSB),
Santa Barbara, USA
e-mail: bsatzinger@gmail.com

C. Lau
e-mail: cslau12@gmail.com

M. Byl
e-mail: marten.byl@gmail.com

K. Byl
e-mail: katiebyl@gmail.com

M.A. Hsieh et al. (eds.), *Experimental Robotics*, Springer Tracts
in Advanced Robotics 109, DOI 10.1007/978-3-319-23778-7_3

1 Introduction and Problem Statement

RoboSimian (Fig. 1) is a human-scale robot designed and built by JPL to compete in the DARPA Robotics Challenge (DRC) [1]. In this paper, we address the problem of planning desired joint reference trajectories for this high-dimensional quadruped to walk on rough terrain. This is an example of kinodynamic planning [2, 3], simultaneously considering kinematic constraints as well as dynamics. For RoboSimian, the primary kinematic challenges involve selecting among redundant solutions and avoiding collisions of the robot with terrain obstacles and with itself, while the main dynamic constraints are joint velocity limits and static balance requirements. For locomotion more generally, consideration of joint accelerations and allowable center of pressure (aka ZMP) location are also key considerations, but in practice, a low joint velocity limit (1.5 rad/s) makes these constraints relatively trivial to achieve for RoboSimian's current design.

Comparing with past work in planning quadruped locomotion on rough terrain for LittleDog [4–7], two particular challenges for RoboSimian are that it has seven degrees of freedom (DOFs) per limb, rather than three, and that perception relies solely on on-board sensing, rather than the use of motion capture (Vicon) along with saved (point-cloud) terrain maps.

Each of RoboSimian's four identical limbs consists of a kinematic chain of six rotational DOFs to define the (6 DOF) position and orientation of a lower leg segment, shown in green in Fig. 2, relative to the body frame. A final (7th) rotational joint simply allows the most distal end, or foot, of the lower leg to twist relative to the leg, so that the L-shaped lower leg segment itself can yaw while the foot remains fixed with respect to the ground. Even with only six actuators to set the 6-DOF pose of the lower leg, there are frequently redundant solutions. Qualitatively, each solution involves making one of two geometric choices (akin to "which way to bend

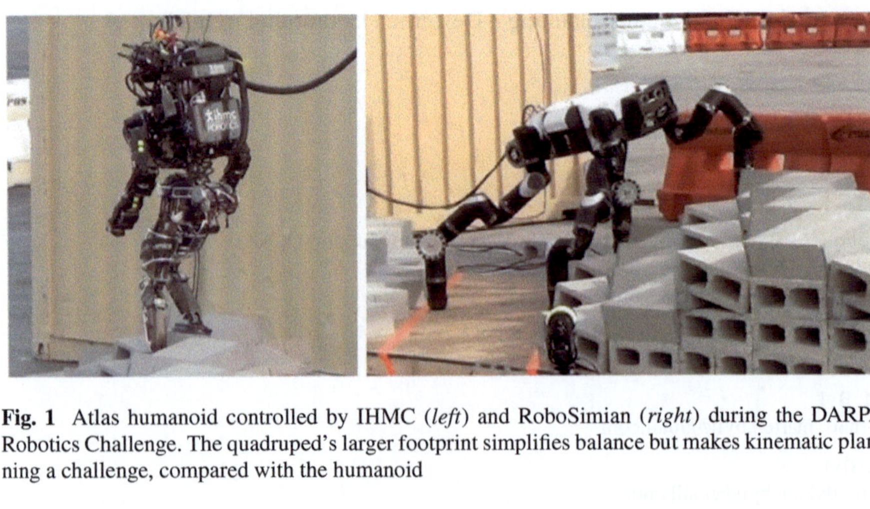

Fig. 1 Atlas humanoid controlled by IHMC (*left*) and RoboSimian (*right*) during the DARPA Robotics Challenge. The quadruped's larger footprint simplifies balance but makes kinematic planning a challenge, compared with the humanoid

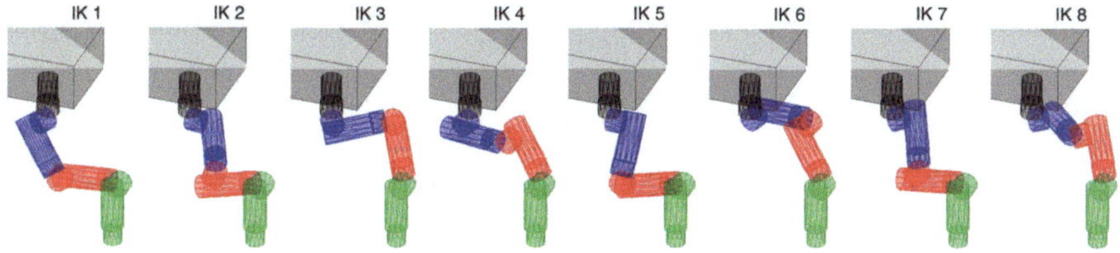

Fig. 2 Redundant inverse kinematic (IK) solutions for RoboSimian

an elbow") at each of three points along the chain: 2^3 results in a total of 8 IK families, as depicted in Fig. 2. The workspace and proclivity for self-collision of each family is different, and solutions for continuous trajectories in task space within a single family sometimes require discrete jumps in joint angles, so that kinematic planning is quite complex. In our problem formulation, we seek tractable methods to design trajectories for all 28 actuated joints, for slow walking with high-torque joints, given a set of candidate footholds on complex terrain.

2 Technical Approach

Our approach begins with a set of candidate footholds locations. We use a graph search to find a specific foothold plan, consisting of a series of steps that will be taken. Because our approach allows body motion during a step, we next search for body poses for the initial and final pose for each step, over a horizon of the next several steps. These poses are passed to our RRT-Connect implementation, which finds a path between the steps, respecting static stability, kinematic feasibility, and collision constraints. The motion plan is then executed on the robot (or a simulation). If more steps remain to be planned, control passes back to the RRT-Connect planner. Otherwise, control passes back to the pose finder. This process repeats until an error has occurred, or the goal has been reached. These phases are described in more detail below (Fig. 3).

2.1 Foothold Graph Search

Although A* search guarantees an optimal solution, it is suggested by [8] that sub-optimal solutions may be found by related best-first search algorithms in much less time. In particular, they propose a K-best-first search algorithm that expands the K best nodes at once (A* corresponding to the special case of $K = 1$). We have implemented a modified approach where K threads expand nodes asynchronously. The empirical performance of this algorithm relative to alternative graph search algorithms is not a focus of this paper, but some performance data will be presented

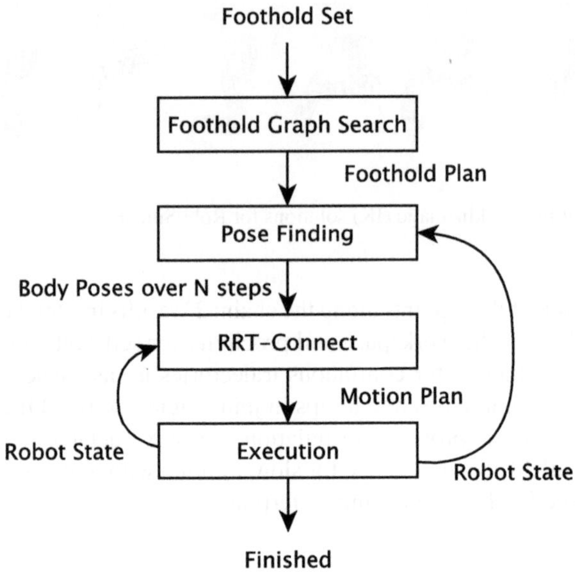

Fig. 3 Planning phases in our approach

in Sect. 3.2 in the context of the entire system. We expect other foothold planning methods to be applicable as well [5–7, 9].

We search for a feasible sequence of steps to bring the robot to a goal location, using a pre-determined set of feasible foothold locations (e.g., based on perception and classification of the terrain, a priori knowledge of terrain shape as in DRC simulation, or manual selection). In our formulation, a node identifies a particular stance among the set of possible footholds. Our cost heuristic is the linear distance between the centroids of the footholds of two nodes (or between a node and the goal location). We expect future work to modify this cost heuristic to reflect other planning preferences, such as the preferential use of certain footholds over others.

The search can enforce a gait order or allow free gaits with steps in any order. In general, free gaits increase the complexity of the search because each node has more potential child nodes. However, for the DRC terrain the free gait search is advantageous because it enables the negotiation of difficult parts of the terrain. An experiment with gait order fixed did not find a solution.

2.2 IK Tables

Once a foothold plan is determined, we must choose body poses at the beginning and end of each step. This process requires an inverse kinematics solution that addresses the difficulties inherent in high-DOF robot limbs. We used an IK table rather than an IK solver because, although an IK solver such as ikfast [10] can easily provide an arbitrary number of solutions to achieve a given 6-DOF pose with RoboSimian's 7-DOF limb, many such solutions result in joint reconfigurations once continuous limb motions are planned. Also, ikfast does not distinguish among "families" of kinematics when giving joint solutions; our grouping of solutions depends upon a customized but relatively slow IK solver, written in-house at UCSB. We precompute an IK table, in terms of only the relative 3-DOF position of a limb with respect to the body coordinate system. This exploits the fact that the lower leg need not be exactly normal to the ground during stance and greatly simplifies planning for body pitch and roll. Methods to produce and optimize IK table solutions for RoboSimian are described in more detail in [11].

2.3 Body Pose Search

Another advantage of pre-computing an IK table for the (x, y, z) coordinates of a limb is that we can also test potential body poses for feasibility very rapidly. Given a set of either 3 stance legs and 1 swing leg, we set a nominal body orientation (roll, pitch, and yaw) heuristically, to match the underlying foothold locations and heights. Then, we search numerically over a 6-DOF $7 \times 7 \times 9 \times 5 \times 5 \times 5$ (x, y, z, roll, pitch, yaw) grid of potential body poses centered on the heuristic pose. We search in an order that tests poses closer to the nominal pose first, and we terminate as soon as a single feasible pose is found. A feasible pose consists of one that is kinematically feasible for all four limbs, with static stability on a support triangle given by the 3 stance footholds.

In order to handle uncertainty in the terrain while planning, we also test that the swing foot will be able to reach above and below the planned foothold location while remaining kinematically feasible. This ground penetration distance can be set as a parameter. Choosing a larger value will allow greater uncertainty in the terrain, but also forces the planner to choose only conservative motions.

RoboSimian's four limbs together limbs account for roughly 60 % of its total mass. Because limb motions affect center of mass location significantly, testing a step for feasibility requires performing two body pose searches, one with the swing foot at the initial pose, and one with the swing foot at the final pose. This also allows us to plan steps with different body poses at the beginning and end. In this paper, we analyze on the effect of allowing body motion during a step on the volume of reachable footholds. Our results demonstrate that this capability does not have much advantage when planning a regular forward crawl gait on flat ground, but does significantly increase the reach on complex terrain with irregular steps or height changes.

We expect future work to address several shortcomings with this method of finding body poses. The approach is somewhat computationally expensive. In our results, we quantify the time spent searching for body poses in comparison to other planning processes. The search also does not guarantee that the solution is better than other possible solutions or that it is far away from infeasibility. The authors plan to implement either a more sophisticated search or a subsequent pose optimization step to address one or both of these issues.

2.4 Motion Planning

Our general trajectory planning framework is described in more detail in [12]. We use RRT-Connect [13] to solve for a feasible path to a previously established goal position. Kuffner [14] has demonstrated this method to plan locomotion for a humanoid with 6-DOF limbs, but, in practice, this required a search over an apparently much smaller configuration space (e.g., C^3) than in our case (C^{16}). Several works plan locomotion by first searching over a graph and then filling in allowable motions [15–17]. In particular, Bretl [17] developed a non-gaited motion planner for the LEMUR quadruped, which has 3 DOF per limb. Hauser [16] solved for non-gaited motions on a 36-DOF humanoid by focusing on clever (contact-before-motion) sampling, but a single step still required several minutes, and a plan for climbing a ladder took a few hours, computationally.

As described in [12], our approach uses RRT-Connect to solve for paths between the initial and final goal. We will briefly summarize this to provide context for our experimental results. We parameterize the configuration space in a way that allows us to reduce the number of dimensions substantially when compared to a naive approach while simultaneously addressing kinematic closure of the stance legs and allowing the use of all degrees of freedom on the swing limb in order to allow dexterous motion.

Figure 4 illustrates this approach. During a swing motion, the complete pose of the robot can be specified by the 7 joint angles of the dominant limb and the 7 joint angles of the swing limb. We can also allow rotation at the contact between the dominant limb and the ground by introducing 2 additional degrees of freedom to give roll and pitch at this contact; recall that the most distal actuated joint already allows for yaw near the ground contact. This gives a total of 16 degrees of freedom.

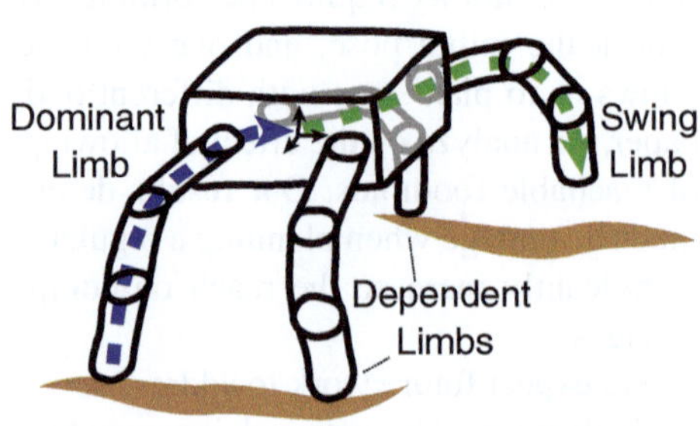

Fig. 4 Cartoon sketch from [12] illustrating the design of our RRT-Connect configuration space parameterization for RoboSimian

The remaining two dependent limbs are not directly represented in the parameterization. We determine their positions using IK table lookups. This requires knowing the location of the footholds relative to RoboSimian's body. The body position, in turn, is defined relative to the dominant foothold by the forward kinematics of the dominant limb. It is possible to generate poses where one or more dependent limbs do not have a kinematically feasible IK table solution. These poses are considered to be infeasible and are not used.

A similar approach can be used to allow a body shift with all four feet in contact with the ground along with different initial and final body poses. This still requires one dominant limb to determine the body pose and the remaining three limbs to be dependent limbs. This gives 9 degrees of freedom, including the roll/pitch contact with the dominant foothold.

All dimensions are given in radians, and angles are not wrapped at multiples of 2π. Although RoboSimian's actuators do not have hard joint limits and can continuously rotate, accumulating several rotations could damage cabling passing through the actuators. To avoid the accumulation of rotations during planning, we treat all joint angles (e.g., 0 and 2π) as distinct.

3 Results

In this section, we present results from various experiments that demonstrate the capabilities and benefits of our approach to trajectory planning for Robosimian.

3.1 *Feasible Step Volume*

In contrast with the planning approach used on RoboSimian during the 2013 DRC trials, our RRT configuration space design supports movement of the body during a step, from an initial pose to a final pose. We performed an analysis to consider the effects of this design decision on the ability for our system to plan steps on difficult terrain.

For the analysis, the robot's initial footholds are placed in a crawl-gait position corresponding to a steady state step length of 0.50 m. Then, we consider planning a step with one of the limbs from its initial position to a set of goal positions on a 3D grid. The search grid had 32 points in x and y, and 18 points in z. We compare the volumes reachable when the initial and final poses are forced to be the same, and when they are allowed (but not required) to be different. Therefore, the "same pose"volume S will always be a subset of the "different pose" volume D, and we can quantify the benefits of this approach by the size and shape of the relative complement of S in D.

The analysis found that 2541 of the destination foothold locations were likely reachable based on finding a feasible body pose for the end of the motion. Our

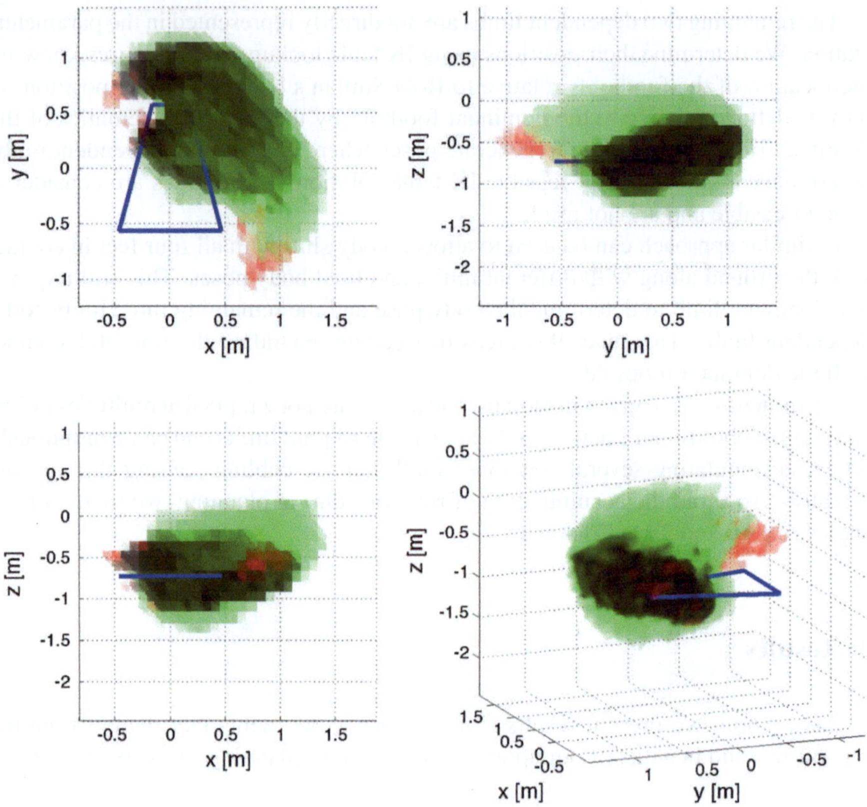

Fig. 5 Volumetric rendering of the reachable volume from several perspectives. *Black* shows the volume S, and *green* shows the relative complement of S in D (showing the region reachable only with body motion during a step). *Red* shows regions contained in S and D where the RRT solvers were not able to find a path, despite the existence of feasible initial and final poses. *Blue* shows the outline of the initial support region, with the initial swing foothold embedded in the reachable volume

RRT-Connect implementation was able to find paths to 2499 of them (98.3 %). The remaining 42 points where the RRT-Connect search failed to find a path are shown in red in Fig. 5. The 2499 successful paths represent 1.92 Km of total motion, with an average step length (linear distance from start to finish) of 76.8 cm, and a maximum step length of 169.7 cm.

In contrast, when the body position was required to be the same at the beginning and end of the step, only 1065 steps were possible, with an average step length of only 55.7 cm, and a maximum step length of only 130.6 cm.

As Fig. 5 shows, the reachable volume is substantially increased by allowing the body to move during a step. Notably, the maximum distance that it is possible to move the foot directly forward (positive x direction) does not vary much between S and D, so there is little benefit for a regular crawl gait on approximately flat terrain.

However, S is much larger in other directions, suggesting benefits for irregular gaits, especially on terrain with large changes in height, or when the robot is turning.

3.2 Simulated Traversal of DRC Terrain

We performed a simulation experiment to quantify the performance of our approach. A simulated DRC terrain was created by placing footholds on a $16 \times 16 \times 6$ inch grid, to match the spacing created by the cinderblocks used in the trial. An additional foothold was placed on the highest terrain level in order to allow the A* planner to find a feasible solution. The goal location was set 9 m from the starting location. In order to reduce the A* search space, only two rows of the terrain grid were populated with footholds, as this was sufficient to allow a solution (Fig. 6). We also did not consider the orientation of the footholds, which are on locally sloped surfaces for the second half of the terrain.

Although collision detection with the terrain is used during RRT planning, integration into body pose finding and A* planning is not complete as of writing. If the initial and final body poses chosen for a step happen to be in collision with the terrain, the RRT solvers will be unable to find a solution. Therefore, for this simulation, a terrain collision model was not used. However, checks for self collisions between different parts of the robot are always made.

The simulation was performed using RoboSimian's control software in a special offline mode, which provides the software interfaces, interpolation algorithms,

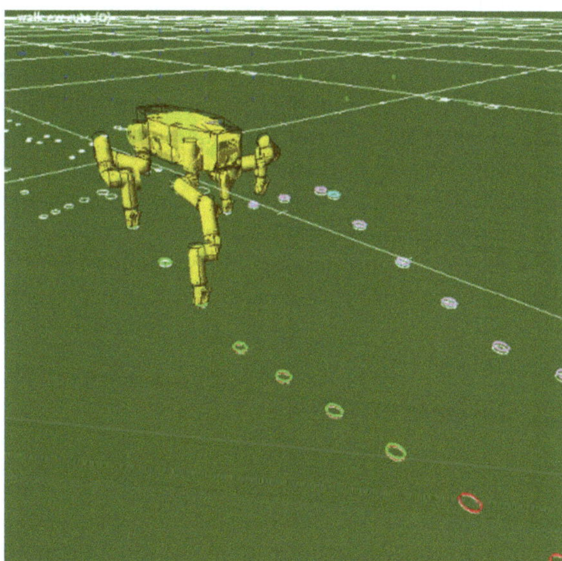

Fig. 6 RoboSimian straddles the second ridge on a simulated crossing of the DRC terrain

Table 1 Time spent in various subtasks (non-overlapping) while simulating DRC terrain crossing

Subtask	Time (s)	Time (%)
Foothold graph search	169.9	13.1
Pose finding	192.4	14.9
RRT-connect	212.8	16.5
Execution	647.2	50.0
Other	71.1	5.5
Total	1293.5	100.0

and error checking that are used with the robot hardware. Therefore, the planning/execution cycle occured in real time with representative communications overhead. All software involved in the simulation was run on a single laptop with an Intel i7-4900MQ 2.8 GHz CPU and 16 GB of memory. All planning was done autonomously, without operator input.

The A* search was executed once at the beginning for the entire terrain. This process was multithreaded using 7 threads to expand the search tree in parallel. Pose finding was performed every four steps, and chose 8 body poses (for the beginning and end of each step). The pose finding search used 8 threads.

RRT based pathfinding algorithms were used to find feasible and stable paths between the initial and final poses given by each step or body movement. Our RRT formulation gives a special role to one of the three (during a step) or four (during a body shift) stance limbs, which we call the 'dominant' limb. We call the RRT solver three (or four) times with each choice of dominant limb in turn. The solution with the shortest time (respecting the robot's actuator velocity and acceleration limits) is used. Because these calls are independent, in principle, they can be parallelized. However, for this experiment they were done serially.

For this offline experiment, we did not include uncertainty in the terrain height relative to the given foothold locations. This would have allowed us to use pipelining and, for example, call the RRT for the next step while the current step was still executing. However, when running on hardware we discover the true height of the terrain at the end of a step when contact is made with the ground and incorporate that knowledge into planned sequence of footholds. Therefore, we postpone calling the RRT solvers until just before the motion is to be executed so that motions are planned with the most up to date information possible (Table 1).

3.3 Obstacle Avoidance

In the prior simulation, we did not include a terrain collision model. In order to demonstrate the capability to avoid complex obstacle geometry, we generated a set of obstacles for Robosimian to avoid when taking one step forward. Boxes were arranged in the formation specified in Table 2 and shown in Fig. 7. A foothold plan

Table 2 Body poses, foothold locations, and box poses and sizes for the RRT demonstration in Fig. 7

	Position (m)			Rotation (°)			Size (m)		
	x	y	z	Roll	Ritch	Yaw	x	y	z
Body—entry	0.10	0.10	0.00	0	0	0	–	–	–
Body—exit	0.10	0.10	0.00	10	−5	10	–	–	–
Foothold—FR	0.51	0.51	0.65	–	–	–	–	–	–
Foothold—BR	−0.51	0.51	0.65	–	–	–	–	–	–
Foothold—BL (start)	−0.51	−0.51	0.55	–	–	–	–	–	–
Foothold—FL	−0.51	−0.51	0.65	–	–	–	–	–	–
Foothold—BL (end)	−0.01	−0.51	0.55	–	–	–	–	–	–
Box 1	−0.22	−0.50	0.56	0	0	0	0.04	0.61	0.19
Box 2	−0.29	−0.75	0.26	0	0	0	0.04	0.41	0.76
Box 3	−0.22	−0.64	0.31	−10	30	5	0.20	0.61	0.03
Box 4	−0.27	−0.30	0.56	0	50	0	0.49	0.20	0.03

The acronyms FR, BR, BL, and FL refer to the *front right limb*, *back right limb*, *back left limb*, and *front left limb*, respectively

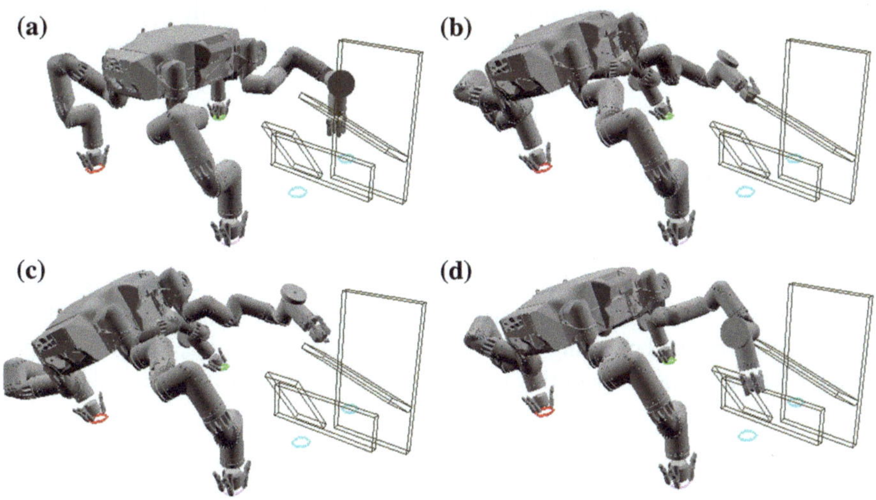

Fig. 7 Demonstration of an RRT generated swing trajectory with four *boxes* arranged as obstacles. RoboSimian's back *left* limb successfully avoids the *boxes* as it swings to its next foothold 0.50 m forward in the x-direction

was generated to move Robosimian's back left leg forward 0.50 m. Such a situation would likely be encountered on rough terrain, especially those involving fallen debris.

Because our pose finding algorithm does not yet account for terrain collisions, the initial and final body poses were chosen manually to keep Robosimian's center of pressure within the support polygon during swing. A rotation was added to the final body pose to allow for body adjustments required to step over the obstacles.

The footholds for the non-swing limbs were set in the crawl gait formation with end effectors approximately 1 m apart in both the x- and y-directions. The end effector of the swing leg was set 0.10 m above its nominal walk pose to account for contact behavior.

A swing trajectory to maneuver Robosimian's back left leg over the obstacles was successfully found using RRT-Connect. The generated trajectory included 241 waypoints with a playback time of 12.13 s. Such complicated maneuvers would not be possible using only IK table solutions since many of the poses required of Robosimian to swing its leg over the boxes are atypical. As demonstrated by this example, when traversing rough terrain, many situations require unique and complicated movements beyond those that are reasonable to store in an IK table. The combined RRT plus IK tables approach allows us to generate feasible solutions for unpredictable situations within an acceptable time frame. Computation time for the RRT to generate the swing trajectory depicted in Fig. 7 was 17.12 s. The RRT required 8227 nodes from the starting foothold and 7912 nodes from the goal foothold for a total of 16139 nodes.

Another important aspect of this approach to solving for swing trajectories is the flexibility of the body pose during swing. The movement of the body is integral to the RRT's ability to find a solution for the foothold as the contortions required of the swing limb without movement of the body would result in collisions between joints and the body.

3.4 DARPA Robotics Challenge

Our methods have been tested for short foothold plans in lab on RoboSimian in preparation for the DARPA Robotics Challenge (DRC). However, the most significant (and disappointing) result for us was that drift in perception of our world map made careful foothold planning a significant challenge. In practice, RoboSimian performed the locomotion task during the DRC using RRT-Connect for a set of heuristic

Fig. 8 RoboSimian pitches its body and stretches to a near-singular configuration to traverse terrain at the DRC

footholds planned blindly on terrain, using force feedback to detect ground contact. Results were still good enough to place 5th in the 2013 DRC trials and qualify for the final competition, scheduled for June 2015. Since then, we have successfully crossed the DRC terrain shown in Fig. 8 using our combined IK Table and RRT approach; however, practical implementation currently requires that an operator periodically corrects for drift in the estimation of body pose with respect to the world map. On this extreme terrain, combined computation and execution time allows RoboSimian to walk at about 1.2 ft/min, and walking speed increases to around 5 ft/min on milder portions of the DRC terrain. For reference, the fastest pre-planned walking gait we can currently obtain on RoboSimian goes about 15 ft/min on flat ground, which benchmarks limitations due to joint velocity limits, versus for computation and terrain roughness.

4 Experimental Insights and Future Work

Our key experimental insights are (1) that our approach blending IK tables and RRT-Connect provides a computationally practical kinodynamic planning method with a high rate of success, (2) that allowing for body motions during a swing leg motion significantly increases the set of reachable next footholds compared with sequentially moving either a leg or the body during a crawl gait, and (3) that adequate perception is a strong requirement for real-world implementation on RoboSimian. Specifically, although three pairs of forward-facing stereo cameras can view front limbs, rear limbs are often well over a meter behind the front limbs, requiring an accurate world map of previously viewed terrain. Future upgrades to the robot plan to incorporate LIDAR sensing to improve mapping and localization significantly.

Acknowledgments This work is supported by DARPA. The authors would also like to thank the entire RoboSimian team for their efforts in designing the robotic system hardware and software.

References

1. Hebert, P., Bajracharya, M., Ma, J., Hudson, N., Aydemir, A., Reid, J., Bergh, C., Borders, J., Frost, M., Hagman, M., Leichty, J., Backes, P., Kennedy, B., Karplus, P., Byl, K., Satzinger, B., Shandar, K., Burdick, J.: Mobile manipulation and mobility as manipulation—design and algorithms of RoboSimian. J. Field Robot. (JFR), Special Issue on the DRC (to appear) (2014)
2. Donald, B., Xavier, P., Canny, J., Reif, J.: Kinodynamic motion planning. J. ACM **40**(5), 1048–1066 (1993)
3. Donald, B.R., Xavier, P.: Provably good approximation algorithms for optimal kinodynamic planning for Cartesian robots and open-chain manipulators. Algorithmica **14**(6), 480–530 (1995)
4. Byl, K., Shkolnik, A., Prentice, S., Roy, N., Tedrake, R.: Reliable dynamic motions for a stiff quadruped. Proc. ISER **54**(2009), 319–328 (2008)
5. Byl, K.: Metastable legged-robot locomotion. Ph.D. dissertation, MIT, 2008

6. Kolter, J.Z., Ng, A.Y.: The stanford littledog: a learning and rapid replanning approach to quadruped locomotion. Int. J. Robot. Res. (IJRR) **30**(2), 150–174 (2011)
7. Zucker, M., Ratliff, N., Stolle, M., Chestnutt, J., Bagnell, J.A., Atkeson, C.G., Kuffner, J.: Optimization and learning for rough terrain legged locomotion. Int. J. Robot. Res. **30**(2), 175–191 (2011)
8. Felner, A., Kraus, S., Korf, R.: Kbfs: K-best-first search. Ann. Math. Artif. Intell. **39**(1–2), 19–39 (2003). http://dx.doi.org/10.1023/A%3A1024452529781
9. Vernaza, P., Likhachev, M., Bhattacharya, S., Chitta, S., Kushleyev, A., Lee, D.D.: Search-based planning for a legged robot over rough terrain. In: Proceedings of IEEE International Conference on Robotics and Automation (ICRA). IEEE, 2009, pp. 2380–2387
10. Diankov, R.: Automated construction of robotic manipulation programs. Ph.D. dissertation, Carnegie Mellon University, Robotics Institute, Aug 2010
11. Byl, K., Byl, M., Satzinger, B.: Algorithmic optimization of inverse kinematics tables for high degree-of-freedom limbs. In: Proceedings of ASME Dynamic Systems and Control Conference (DSCC), 2014 (to appear)
12. Satzinger, B., Byl, K.: More solutions means more problems: Resolving kinematic redundancy in robot locomotion on complex terrain. Submitted to IEEE/RSJ International Conference on Intelligent Robots and Systems (IROS), 2014
13. Kuffner, J., LaValle, S.: RRT-connect: an efficient approach to single-query path planning. In: Proceedings IEEE International Conference on Robotics and Automation (ICRA), vol. 2, pp. 995–1001 (2000)
14. Kuffner, J., Kagami, S., Nishiwaki, K., Inaba, M., Inoue, H.: Dynamically-stable motion planning for humanoid robots. Auton. Robots **12**(1), 105–118 (2002)
15. Bouyarmane, K., Kheddar, A.: Humanoid robot locomotion and manipulation step planning. Adv. Robot. Int. J. Robot. Soc. Jpn, Special Issue on the Cutting Edge of Robotics in Japan 2012, **26**(10), 1099–1126 (2012)
16. Hauser, K., Bretl, T., Latombe, J.C.: Non-gaited humanoid locomotion planning. In: Proceedings of International Conference on Humanoid Robots. IEEE, 2005, pp. 7–12
17. Bretl, T.W.: Multi-step motion planning: application to free-climbing robots. Ph.D. dissertation, Citeseer, 2005

Experimental Evaluation of Obstacle Clearance by a Hybrid Wheel-Legged Robot

Christophe Grand, Pierre Jarrault, Faiz Ben Amar and Philippe Bidaud

Abstract This paper deals with the problem of frontal obstacle crossing by a poly-articulated wheeled robot. We focus on the particular architecture of hybrid wheel-legged robots that are redundantly actuated systems. In this paper, experimental results that show the climbing capabilities of such system when crossing large obstacle are presented. We focus on the case of a step like obstacle whose height is superior to the diameter of the wheels. In this case, the adhesion properties have a large impact on the crossing capabilities. First, we introduce our control methodology which is based on the optimization of both the robot posture and the distribution of internal forces. The optimization criterion represents the maximum allowable force disturbance that the system can support before violating the frictional contact constraint. Then, our experimental prototype, the robot Hylos2 and the experimental setup are presented. As our approach is based on the control of the contact forces, experiments used to quantify the level of friction in the mechanical transmissions are first reported. Then a step-crossing trial on the real system is presented.

Keywords Mobile robot · Wheel-legged · High mobility

1 Introduction

In this paper, we address the motion control of poly-articulated mobile robots during obstacle clearance. We are focusing on a special class of mobile systems, often called hybrid wheel-legged robots, which are designed in order to increase both obstacle crossing and terrain adaptation capabilities.

The kinematic architecture of hybrid wheel-legged systems we consider in this work, is made of wheels mounted at the end of actuated legs. Many robots have already been developed based on this kinematic. Most of them are based on a 4

C. Grand (✉) · P. Jarrault · F. Ben Amar · P. Bidaud
ISIR (UMR-7222) UPMC/CNRS, Paris, France
e-mail: christophe.grand@upmc.fr
URL: http://www.isir.upmc.fr

M.A. Hsieh et al. (eds.), *Experimental Robotics*, Springer Tracts in Advanced Robotics 109, DOI 10.1007/978-3-319-23778-7_4

Fig. 1 Hylos2 prototype

wheel-legs arrangement such as the Hylos robot [4], the Workpartner [11] or the PAW robot [9]. Some concepts use only three wheels, like the family of robots Tri-Star dedicated to planetary exploration [1], or the robot ROAMeR [3] that uses three active limbs with a steerable wheel at the end to form a planar reconfigurable omnidirectional mobile robot. Some few other systems use more than four legs like both the Athlete, developed by NASA/JPL [10], and the Asterisk-H, from the Arai Laboratory [12] which are using 6 wheel-legs.

The control method developed in this paper is dedicated to the specific case of 4 wheel-legs robots and is validated on our experimental prototype Hylos2 (see Fig. 1). Its mechanical architecture is composed of 4 wheel-legs, each wheel-leg being a multi-dof serial chain ended by a driven and steerable wheel. The nominal mechanical parameters are given in Table 1. This robot have the ability to change the position of its center of mass (CoM) and to modify the distribution of its contact forces. Furthermore, it is redundantly actuated systems exhibiting internal forces that should be optimized. The proposed motion controller is based on a torque control at the joints level that addresses the combined optimization of the internal forces and the CoM position, in order to maximize the contact stability (increasing traction and avoiding tip-over).

Table 1 Nominal parameters of the Hylos2 robot

Mass	20 kg
Nominal velocity	0.6 m/s
Wheels diameter	150 mm
Ground clearance	100–300 mm
Nominal length	700 mm
Width	450 mm

2 Approach

2.1 System Modeling

We consider the general case of a system supported by n wheel-legs (Fig. 2). The frame $\mathcal{R}_p = (G, \boldsymbol{x}_p, \boldsymbol{y}_p, \boldsymbol{z}_p)$ is attached to the main body of the system allowing to describe its motion with respect to the ground reference frame $\mathcal{R}_0$, and G is the center of mass of the main body.

The ith leg is in a frictional contact with the ground at point P_i which coordinates $\boldsymbol{p}_i$ are expressed in the frame $\mathcal{R}_p$ (Fig. 2). The contact force of the ground on each leg is denoted $\boldsymbol{f}_i = \left[\, f_{x_i} \; f_{y_i} \; f_{z_i} \,\right]^t$ where f_{x_i}, f_{y_i} and f_{z_i} are the components of the force along the contact frame's axis $\mathcal{R}_{c_i} = (P_i, \boldsymbol{x}_i, \boldsymbol{y}_i, \boldsymbol{z}_i)$, such that $\boldsymbol{z}_i$ is the contact normal and $\boldsymbol{x}_i, \boldsymbol{y}_i$ are the tangential directions.

The equations describing the equilibrium of the system are given by:

$$\mathbf{G}\boldsymbol{f} = \boldsymbol{F} \tag{1}$$

where $\boldsymbol{f}^{\,t} = \left[\boldsymbol{f}_1{}^t \; ... \, \boldsymbol{f}_n{}^t\right]$ is a $[3n \times 1]$ vector containing all contact forces, expressed in each contact frame $\mathcal{R}_{c_i}$, and $\boldsymbol{F}$ is the set of external and inertial wrench, expressed in the local frame $\mathcal{R}_p$, that are applied to the platform at point G. $\mathbf{G}$ is a $[6 \times 3n]$ matrix giving the equivalent wrench to the contact forces at the center of mass in the frame $\mathcal{R}_p$ (see [5] for more details):

$$\mathbf{G} = \begin{bmatrix} \mathbf{R}_{c_1}^{P} & \cdots & \mathbf{R}_{c_n}^{P} \\ \widetilde{\mathbf{p}}_1\mathbf{R}_{c_1}^{P} & \cdots & \widetilde{\mathbf{p}}_n\mathbf{R}_{c_n}^{P} \end{bmatrix} \tag{2}$$

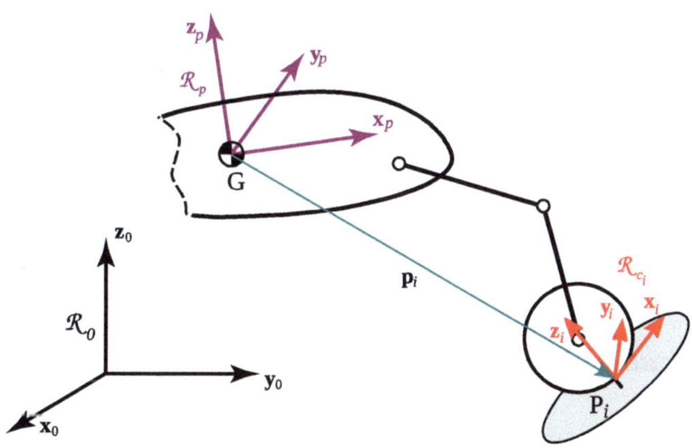

Fig. 2 Description of the ith leg parameters

where $\mathbf{R}_{c_i}^P$ is the rotation matrix of the ith contact frame $\mathcal{R}_{c_i}$ with respect to the platform $\mathcal{R}_p$, and $\widetilde{\mathbf{p}}_i$ is the skew-symmetric matrix of the cross product operator.

The contact forces must respect constraints related to actuators saturation and Coulomb friction law. The actuators limits are defined as follow:

$$\begin{cases} \mathbf{J}^T \boldsymbol{f} < \boldsymbol{\tau}_{\max} \\ -\mathbf{J}^T \boldsymbol{f} < \boldsymbol{\tau}_{\max} \end{cases} \tag{3}$$

where $\mathbf{J} = blockdiag(\mathbf{J}_i)$, $\mathbf{J}_i$ being the Jacobian matrix of the ith leg and $\boldsymbol{\tau}_{\max}$ is the vector of the actuators torque limits.

The contact constraints are defined by Coulomb's friction law as a function of the contact forces:

$$\begin{cases} \mu_i f_{z_i}^2 < f_{x_i}^2 + f_{y_i}^2 \\ f_{z_i} > 0 \end{cases} \tag{4}$$

where μ_i is static friction coefficient at the contact. These constraints can be advantageously represented using a conservative pyramidal form adapted form [7] and given by:

$$\mathbf{A}_i \boldsymbol{f}_i < 0 \tag{5}$$

where

$$\mathbf{A}_i = \begin{bmatrix} 0 & 1 & \frac{\mu_i}{\sqrt{2}} \\ 0 & -1 & \frac{\mu_i}{\sqrt{2}} \\ 1 & 0 & \frac{\mu_i}{\sqrt{2}} \\ -1 & 0 & \frac{\mu_i}{\sqrt{2}} \end{bmatrix}$$

Thus we propose in this paper a stability criterion based on the smallest perturbation allowed at the contact level. This leads to to maximize robustness with respect to modeling errors affecting the contact force control. Let us define vector $\boldsymbol{d}$ as:

$$\boldsymbol{d} = \mathbf{A}\boldsymbol{f} \tag{6}$$

where $\mathbf{A}$ is a matrix defined by $\mathbf{A} = blockdiag(\mathbf{A}_i)$.

2.2 Optimization Procedure

Considering the elements given in the previous section, the forces distribution problem is formulated as a "minimax" optimization procedure. Indeed, the objective is to maximize the smallest acceptable perturbation ($\phi = \min(\mathbf{A}\boldsymbol{f})$) subject to the constraints (1) and (3):

$$\max_{f \in \mathbb{R}^{3n}} \min(\mathbf{A}f) \quad s.t. \begin{cases} \mathbf{G}f = \boldsymbol{F} \\ \mathbf{J}^T f < \boldsymbol{\tau}_{\max} \\ -\mathbf{J}^T f < \boldsymbol{\tau}_{\max} \end{cases} \tag{7}$$

This problem can be transformed into its compact primal form [2]. The solution f of the optimization (7) should satisfy the force equilibrium equation constraint (1). This solution is divided in a particular solution $\boldsymbol{f}_p$ and the homogeneous solution $\boldsymbol{f}_h$:

$$f = \boldsymbol{f}_p + \boldsymbol{f}_h$$

The particular solution is chosen arbitrarily as the solution given by the weighted pseudo-inverse. The homogeneous solution, that correspond to the internal forces, is belong the null space of the matrix $\mathbf{G}$ expressing the force equilibrium equation. Let us denote $\{\boldsymbol{g}_i\}_{i\in[1,m]}$ the vectors defining a basis of the null space of $\mathbf{G}$, with $m = \dim(\mathbf{G}) - \mathrm{rank}(\mathbf{G})$. The particular and homogeneous solutions can then be written as:

$$\begin{aligned} \boldsymbol{f}_p &= \mathbf{G}^{+}_{\mathrm{norm}} \boldsymbol{F} \\ \boldsymbol{f}_h &= \mathbf{N}_g \boldsymbol{x}_h \end{aligned} \tag{8}$$

where $\mathbf{N}_g = [\boldsymbol{g}_1 \ldots \boldsymbol{g}_m] = \ker(\mathbf{G})$.

This new formulation allows us to rewrite the optimization problem (7) by implicitly including the force equilibrium constraint, and to reduce the dimension of the optimization search space ($m < 3n$):

$$\max_{\boldsymbol{x}_f \in \mathbb{R}^m} \min\left(\mathbf{A}\left[\boldsymbol{f}_p + \mathbf{N}_g \boldsymbol{x}_f\right]\right) \quad s.t. \begin{cases} (\mathbf{J}^T\mathbf{N}_g)\boldsymbol{x}_f < \boldsymbol{\tau}_{\max} - \mathbf{J}^T\boldsymbol{f}_p \\ -(\mathbf{J}^T\mathbf{N}_g)\boldsymbol{x}_f < \boldsymbol{\tau}_{\max} + \mathbf{J}^T\boldsymbol{f}_p \end{cases} \tag{9}$$

We consider in this paper the special case of a frontal obstacle crossing, as depicted in Fig. 3, where the motion along the lateral direction has a negligible effect on the robot stability. Thus, we only consider the reconfiguration of the posture in the sagittal plane (G, z_p, $\boldsymbol{x}_p$). For these reasons and for sake of simplicity, we consider the motion of left and right sides of the robot as symmetrical.

Thus, we define $\boldsymbol{p} = [X\ Z]^T$ as the vector containing the longitudinal and vertical coordinates of the CoM position relatively to the rear wheel-soil contact point. To this purpose, an intermediate reference frame $\mathcal{R}_r$ aligned with the inertial frame $\mathcal{R}_0$ but attached to this point is defined.

In this case, the null space of $\mathbf{G}$ is entirely defined by the contact geometry (the angle α in the schema). For a fixed distance between contact points (the wheelbase e_t),

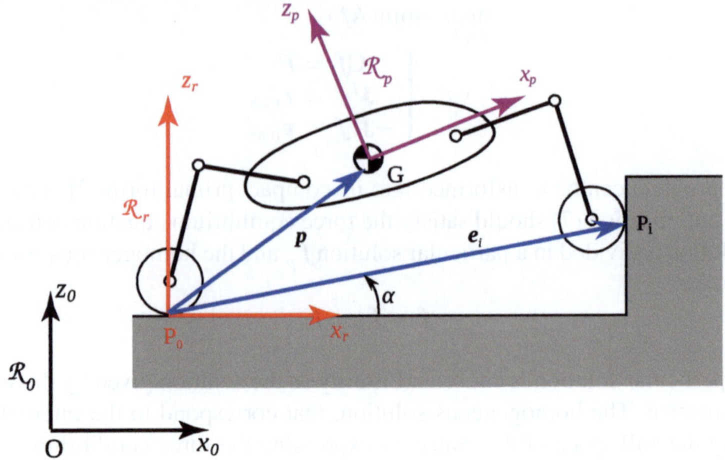

Fig. 3 Local frame and posture definition during frontal obstacle crossing

the null space of **G** is not modified by a change of X or/and Z. For a frontal crossing, *i.e.* the rotation between the ground and the contact's frame of the obstacle is around the lateral axis of the robot ($\boldsymbol{y}_p$), the particular solution of force equilibrium equation can be defined as a combination of the CoM position (X, Z) and the configuration of the contact angles (represented by the particular solution $\boldsymbol{f}_0$):

$$\boldsymbol{f}_p = \mathbf{G}^{+}_{\mathrm{norm}}\boldsymbol{F} = \boldsymbol{c}_x\ X + \boldsymbol{c}_z\ Z + \boldsymbol{f}_0 \tag{10}$$

leading to the following general solution of the force equilibrium constraint:

$$\boldsymbol{f} = \mathbf{N}\boldsymbol{x} + \boldsymbol{f}_0 \tag{11}$$

where

$$\begin{cases} \mathbf{N} = \begin{bmatrix} \mathbf{N}_g & \boldsymbol{c}_x & \boldsymbol{c}_z \end{bmatrix} \\ \boldsymbol{x}^T = \begin{bmatrix} \boldsymbol{x}_f^T & \boldsymbol{p}^T \end{bmatrix} \end{cases}$$

Thus, the CoM position $\boldsymbol{p}$ is added to the vector of optimization variables and the combined optimization of both the internal forces and the posture can be formulated from (9) as:

$$\begin{aligned} &\max_{x\in\mathbb{R}^{m+2}} \ \min\left(\overline{\mathbf{A}}\boldsymbol{x} + \overline{\boldsymbol{f}_0}\right) \\ &s.t. \begin{cases} \overline{\mathbf{J}}^T\boldsymbol{x} < \boldsymbol{\tau}_{\max} - \mathbf{J}^T\boldsymbol{f}_0 \\ -\overline{\mathbf{J}}^T\boldsymbol{x} < \boldsymbol{\tau}_{\max} + \mathbf{J}^T\boldsymbol{f}_0 \end{cases} \end{aligned} \tag{12}$$

where

$$\begin{cases} \overline{\mathbf{A}} &= \mathbf{A}\,\mathbf{N} \\ \overline{\mathbf{J}}^T &= \mathbf{N}^T\mathbf{J} \\ \overline{\boldsymbol{f}_0} &= \mathbf{A}\boldsymbol{f}_0 \end{cases}$$

This optimization algorithm computes at each time step the optimal posture (position of the CoM) and the optimal distribution of internal forces to be applied. The control of the robot posture requires the application of a position or velocity input at the joints level, whereas the control of internal forces needs a torque control. Thus, we have to face up with the problem of dual force/position control. In the present work, we propose to implement our controller by following the classical impedance control approach [6]. Indeed, the results obtained from the optimization do not necessarily involve that desired motion and force at the wheel contact will be orthogonal. Further, the possible incertitudes in the application of force, due to non-modelized friction in the mechanical transmissions, have a smaller impact on the quality of the impedance based control, compared to the hybrid force/position control.

The actuation torques at joint level are computed following this classical impedance control law:

$$\boldsymbol{\tau} = \mathbf{K}_q(\dot{\boldsymbol{q}}^d - \dot{\boldsymbol{q}}) + \mathbf{J}^T\boldsymbol{f} \tag{13}$$

where $\dot{\boldsymbol{q}}$ are the desired joint velocities and $\boldsymbol{f}$ is the force to be applied at the wheel-soil contacts. The contact forces are obtained from the solution $\boldsymbol{x}^*$ of the optimization (12) and the equation (11). Whilst the desired joint velocities $\dot{\boldsymbol{q}}$ are computed with the robot kinematic model, in order to maintain the posture to the desired value. More details concerning the posture control can be find in [5].

3 Experimental Results

The experimental setup consists in an obstacle made with wooden planks, with a first inclined plane and a second horizontal plane. The angle of the first plan with respect to the horizontal ground is 60° and its height is 19 cm. The friction coefficient between the wheels of the robot and the wood used for this setup has been evaluated experimentally. The estimated value is around $\mu \approx 0.8$.

The resulting obstacle represents a high difficulty with respect to the robot geometry, we indeed observed during our experiments that the joint positions were close to the mechanical limits. The evolution of the robot is represented with some snapshots of our experiments on the Fig. 6.

The global control law, including optimization loop, is implemented on the robot. The robot uses DC motors for each actuated joint (16 dof). The motors torque is controlled using a digital motion controllers from AMC company (Advanced Motion Control) that implement current loop control. The motor controllers are connected to an embedded computer through a CAN bus running at 1MBits/S and implementing the CANopen protocol. The PC-104 embedded computer is based on a Intel Atom N270 (1.6GHz) processor. The robot includes encoders at each joint to measure the kinematic parameters and an IMU (XSens MTi) to evaluate the attitude of the platform. The embedded computer is running with a real-time Xenomai/Linux kernel and the control softwares are deployed using the ROS middleware [8]. The control is divided in two nodes: the first one runs as a 100 Hz real-time task to manage the low-level motion controllers distributed on the CAN bus and to compute the impedance control law (13); whilst the second one runs as non-RT task at 20 Hz to solve the optimization problem (12) and send target position of the CoM and the desired interation forces to the previous node.

One should note that this experimental prototype is not equipped with force sensors in the legs actuation chain. Thus, the required torque at joint level is directly obtained by controlling the current in the motors. Accordingly, the corresponding force that the robot is applying at the wheel-ground contact is not precisely known. Indeed, frictions in the mechanical transmission could affect these forces value. However, the choice of the stability margin used as criterion in the optimization procedure should be able to consider this effect as a perturbation, if we can check that the amount of friction force is less than the measured stability margin. To quantify the amount of friction force in the actuation system, we have applied a classical measurement protocol that consists to actuate each joint with out external contact force (no load motion) and to measure the resulting current in the motor. We have done two experiments: the first was to measure the needed torque to start the motion and that correspond to static friction (also called stiction force); the second was to measure the motor torque during the motion as a function of its position. The measurements are done during multiple trials on each actuators.

The Fig. 4 shows the results corresponding to the first type of tests (stiction value) that was obtained for one actuator whereas the Fig. 5 gives the results for the second

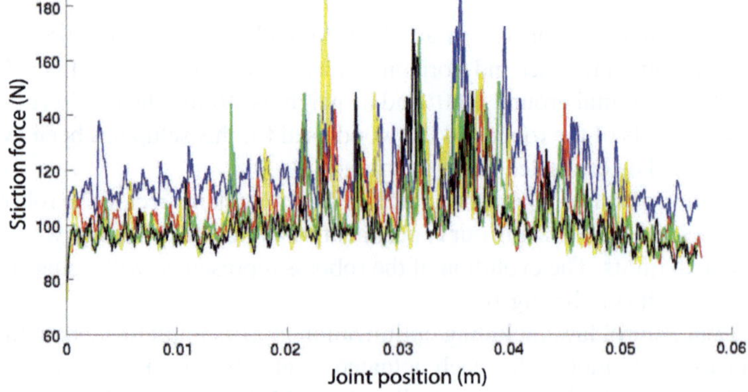

Fig. 4 Static friction

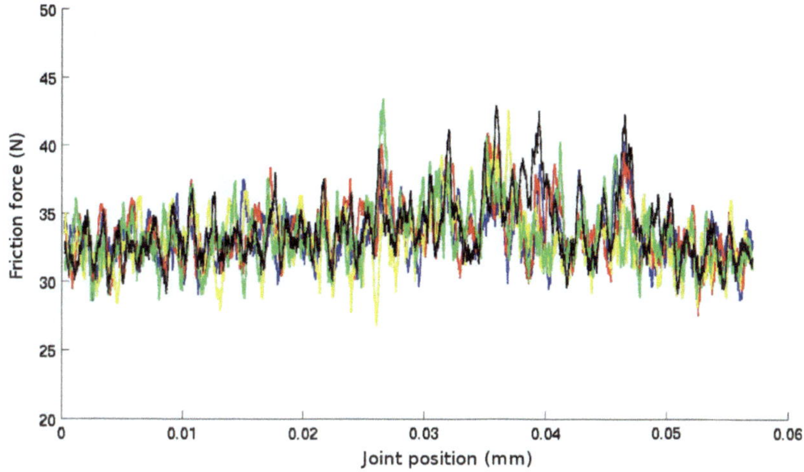

Fig. 5 Dynamic friction

type of tests (dynamic friction). As expected, we find that the stiction force is greater than the dynamic friction. Considering the geometry of the leg described by the jacobian matrix, the corresponding force at the wheel-ground contact point is, in the worst case, equal to 12 N.

The first result is that the robot is able to safely cross over a high step-like obstacle, despite the presence of important friction in the mechanical transmission that alter the precision of the impedance control. The Fig. 7 shows the obtained stability margin which represent the minimum residual tangential force of all the contact forces. This margin quantifies the distance to the unstable configuration arising from slippage. The minimum value is around 15 N that represents 10 % of the system weight. This margin seems to be large enough but it should be compared to the internal friction force. Unfortunately, this value is not measurable directly on the robot, but a measure of the actuation torques when no external forces are applied at the end of the wheel-legs, gives a maximum static friction of about 2 Nm that corresponds, in the worst case to an error of 12 N at the contact point (Fig. 6).

For sake of clarity, the running timeline is represented by three successive and repetitive phases. During phase 1, the robot wheels are in contact with a flat horizontal ground. Then the front wheels encounter the vertical part of the step and the system enter in phase 2 where adherence conditions at the front wheel become preponderant to insure the robot stability. Thus, we can see that the stability margin drops below zero, indicating that the controller can not produce the necessary forces to start the crossing motion. The desired position of the CoM is calculated and the robot reaches the optimal configuration. During this phase, the robot still uses the internal forces to keep the contact forces inside their friction cones and to sustain its weight. Once the stability margin has increased enough, the robot enters in phase 3 and starts to climb over the step. While climbing, the optimal position of the CoM is updated, avoiding

Fig. 6 Evaluation of the control algorithm during step crossing

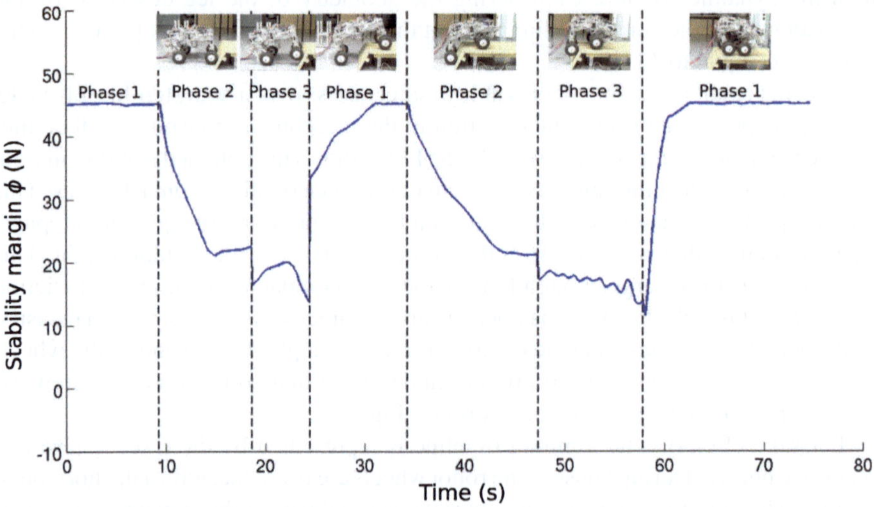

Fig. 7 Stability margin during a step crossing

a possible tip-over as the robot is rising. Once the front wheels have reached the top of the step, the robot is once again in a phase 1 where all the wheels are on horizontal planes. And, in a similar way, the same phases are following for the rear wheels of the robot.

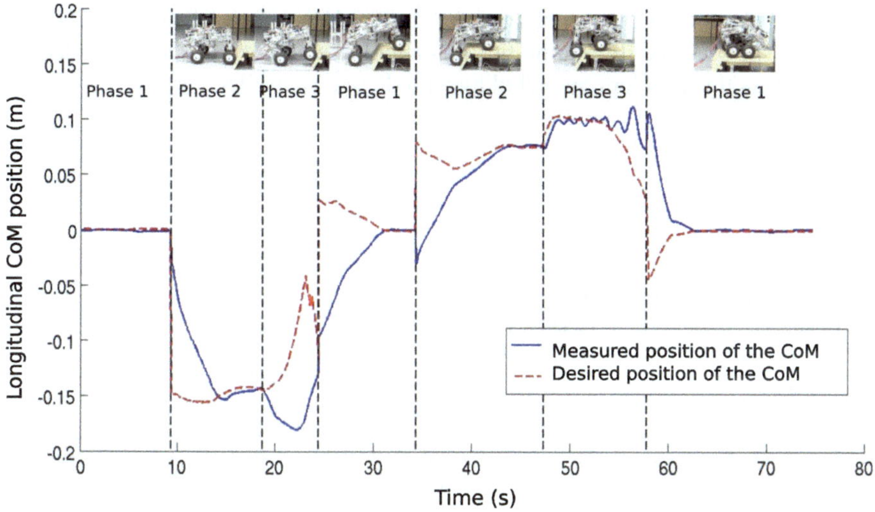

Fig. 8 Desired and actual horizontal position of the CoM

One can see in Fig. 7 between time 47–57 s some oscillations during the second "phase 3" that correspond to the crossing of the rear wheels. These perturbations result from the servoing of the CoM position (see Fig. 8) which is perturbed by the impedance control. The main reason is that in this phase some desired joint torques are close to zero which involves some variation of sign around zero and cause the irregular motion of the CoM. Also, we can see in the Fig. 8 that the position of the platform CoM position is quite good excepted again in the phase 3. This is due to the importance of internal forces during this phase which are critical to insure the obstacle clearance. Thus, the controller tries to move the legs in order to control the posture while trying to maintain a large contact forces needed to overcome the low friction at the wheel-ground contacts.

4 Conclusion

In this work, we have developed a control algorithm that enhances the climbing performance of a hybrid wheel-legged robot when it crossing a frontal step-like obstacle. The proposed methodology exploits the redundant actuation in order to improve the adhesion at the wheel-ground contacts. This algorithm is based on the optimization of both the robot posture and the distribution of the internal forces. The optimization criterion is based on the measure of the smallest force perturbation sustainable at the contacts level which represents the robustness of the contacts stability in term of traction. Experiments realized on our experimental platform Hylos2 validate the implementation of our algorithm on a real system. However, analysis of the posture trajectory shows some discontinuities that should be smoothed. Future works will try to address this problem by including a predictive controller.

References

1. Aoki, T., Murayama, Y., Hirose, S.: Development of a transformable three-wheeled lunar rover: Tri-Star IV. J. Field Robot. **31**(1), 206–223 (2014)
2. Cheng, F.T., Orin, D.E.: Efficient algorithm for optimal force distribution—the compact-dual LP method. IEEE Trans. Robot. Autom. **6**, 178–187 (1990)
3. Fu, Q., Zhou, X., Krovi, V.: The reconfigurable omnidirectional articulated mobile robot (roamer). In: Khatib, O., Kumar, V., Sukhatme, G. (eds.) Experimental Robotics, Springer Tracts in Advanced Robotics, vol. 79, pp. 871–882. Springer, New York (2014)
4. Grand, C., Ben Amar, F., Plumet, F., Bidaud, P.: Stability control of a wheel-legged mini-rover. In: Proceedings of CLAWAR'02: 5th International Conference on Climbing and Walking Robots, pp. 323–330. Paris, France (2002)
5. Grand, C., Amar, F.B., Plumet, F.: Motion kinematics analysis of wheeled-legged robot over 3d surface with posture adaptation. Mech. Mach. Theory **45**(3), 477–495 (2010)
6. Hogan, N.: Impedance control: an approach to manipulation. In: American Control Conference, pp. 304–313. IEEE (1984)
7. Kerr, J., Roth, B.: Analysis of multifingered hands. Int. J. Robot. Res. **4**, 3–17 (1986)
8. Quigley, M., Conley, K., Gerkey, B., Faust, J., Foote, T., Leibs, J., Wheeler, R., Ng, A.Y.: ROS: an open-source robot operating system. In: ICRA Workshop on Open Source Software (2009)
9. Smith, J., Sharf, I., Trentini, M.: Paw: a hybrid wheeled-leg robot. In: Proceedings of IEEE International Conference on Robotics and Automation (ICRA) (2006)
10. Wilcox, B.H., Litwin, T., Biesiadecki, J., Matthews, J., Heverly, M., Morrison, J., Townsend, J., Ahmad, N., Sirota, A., Cooper, B.: Athlete: a cargo handling and manipulation robot for the moon. J. Field Robot. **24**(5), 421–434 (2007)
11. Ylönen, S., Halme, A.: Further development and testing of the hybrid locomotion of work-partner robot. In: Proceedings of International Conference on Climbing on Walking Robots (CLAWAR) (2002)
12. Yoshioka, T., Takubo, T., Arai, T., Inoue, K.: Hybrid locomotion of leg-wheel ASTERISK H. J. Robot. Mechatron. **20**(3), 403 (2008)

Part II
Haptics

Robert J. Webster III
Vanderbilt University

What does a molecule feel like? How do our brains process haptic information? Can novel applications of haptics make surgery more effective? These questions and more were explored during this exciting session at the International Symposium on Experimental Robotics 2014. Presentations spanned the spectrum of haptics research from perception to high-fidelity interfaces to application-specific robot design. The first two papers addressed surgical robotics, describing two different paradigms: a master–slave approach and a hand-held approach. The third paper addressed a functional magnetic resonance imaging (fMRI) compatible robot designed to enable real-time monitoring of brain activity during haptic interaction. The last paper described the design of new high-fidelity haptic interfaces and their use in rendering scaled versions of the forces that occur at the micro and nano scales.

The first paper, *Haptic Control Implementation of a 3-RRR Spherical Parallel Manipulator for Medical Uses* by Houssem Saafi, Med Amine Laribi, and Said Zeghloul focused on a master–slave paradigm with a parallel robot master for haptic interaction in surgery. A key feature of the system is that the master presents the surgeon with the same interface he/she would have in standard laparoscopic surgery, with the master designed to constrain the surgeon's hand to a remote center of motion. The presentation described how the system was designed, addressing use of motion capture and an anastomosis task to define design specifications. The presentation also covered the kinematics, control, and experimental validation of the system.

The second paper *Experiments on the Simultaneous Hand-Held Control of Rigid Endoscopes and Robots Passing Through Them* by Richard J. Hendrick, S. Duke Herrell, Christopher R. Mitchell, and Robert J. Webster III described a new robotic approach for laser prostate surgery, in which the surgeon holds the entire robot in his or her hands and controls the motions of the endoscope and two miniature manipulators simultaneously. The surgeon can thus feel tissue reaction forces on the endoscope directly, obtaining some haptic information without the robot having to do any work to provide it. The presentation described the motivation for such a system in terms of bringing the HoLEP procedure to more patients, design considerations, prototype, and experiments exploring use of task space control versus joint space control. A culminating experiment with an anthropomorphic prostate phantom illustrated the potential of the system to make HoLEP surgery easier for surgeons.

The third paper *Using Haptic fMRI to Enable Interactive Motor Neuroimaging Experiments* by Samir Menon, Hari Ganti, and Oussama Khatib explored how the human brain process information, through the use of a fMRI compatible haptic interface. The presentation described how the incorporation of three degrees of freedom makes the interface unique among prior in-scanner robots of this type, and enables it to facilitate new neuroscience experiments on brain activity during coordinated multi-degree-of-freedom motions. The presentation addressed MRI compatibility experiments, and experiments to characterize force transmission inside the MRI scanner, as well as the results of initial human studies to map brain activity. This research illustrates the promise of haptics in characterizing the way the human brain accomplishes motion control.

The final paper in the session *Dual Stage Options for Interface Designs Suitable for Haptic Interaction at the Micro-Nano Scales* by Abdenbi Mohand Ousaid, Tianming Lu, Cecile Pacoret, Stephane Regnier, and Vincent Hayward delved into haptics at the micro and nano scale. Accurately rendering the non-intuitive physics that characterizes these size scales requires better haptic interfaces. The authors proposed two new dual stage design concepts that facilitate more accurate rendering of highly scaled versions of micro/nano scale forces. Experiments illustrated the excellent performance of both—in particular the ability to render the forces of a needle tip interacting with a tiny water droplet, and the ability to render Brownian motion of individual molecules.

Haptic Control Implementation of a 3-RRR Spherical Parallel Manipulator for Medical Uses

Houssem Saafi, Med Amine Laribi and Said Zeghloul

Abstract This paper presents a set of experiments which have been carried out in order to design a task oriented device, for teleoperated minimally invasive surgery. The workspace of the slave robot has been identified from the motion of an expert surgeon performing an anastomosis task thanks to the Nexus motion capture system. A system composed of master and slave robots as well as its controller has been designed, built and tested. This system provides to the surgeon a force feedback through a current loop of the actuators of the master device. Experiments are carried out and presented in order to show the feasibility of the control force feedback.

Keywords Haptic control · Spherical parallel manipulator · Tele-operation system

1 Introduction

Many researchers ask about the necessity of haptic feedback in robot-assisted surgical applications. The answer is: the haptic feed-back adds more security to the medical application. Indeed, haptic feedback provides to the surgeon a better perception of the environment of medical tasks. A number of tele-operated medical systems have been presented in the last decades. However, due to their complexity, most of those

H. Saafi · M.A. Laribi (✉) · S. Zeghloul
Department GMSC, Pprime Institute,
CNRS - University of Poitiers - ENSMA - UPR 3346,
Poitiers, France
e-mail: med.amine.laribi@univ-poitiers.fr
URL: http://www.pprime.fr

H. Saafi
e-mail: houssem.saafi@univ-poitiers.fr

S. Zeghloul
e-mail: said.zeghloul@univ-poitiers.fr

M.A. Hsieh et al. (eds.), *Experimental Robotics*, Springer Tracts
in Advanced Robotics 109, DOI 10.1007/978-3-319-23778-7_5

systems remain as prototypes [1, 2]. Few medical systems are commercialized. The Da Vinci medical station [3] is one of the more reliable and commercially available system. However it doesn't offer the possibility of a haptic feed-back.

This paper focuses on the implementation of haptic feed-back in the proposed tele-operation system. The motion control implementation has been presented in [4, 5]. The study of haptic feedback covers a wide range of features such as time-delay [6–8], optimal control [9], stability [10–12] and transparency [12–14].

The paper is organized as follows. In Sect. 2, the developed tele-operation system for surgical application is described. Section 3 deals with the control system architecture. In Sect. 4, the SPM kinematic model is studied. The implementation of the haptic control is presented in Sect. 5. An experimental validation is carried out in Sect. 6. Finally, Sect. 7 summarizes this paper.

2 Tele-Operation System

The tele-operation system is developed for minimally invasive tasks. It is composed of a slave surgical robot, a master device and a control system, as shown in Fig. 1. The role of the control system is to manage the surgical robot motions and to provide communication between master and slave with a haptic feedback strategy.

The instruments in MIS (Minimally Invasive Surgery) are designed to enter into the patient's abdominal cavity through tiny skin incisions using a trocar, see Fig. 2. The possible movements of the instruments can be defined by 3 rotations around the incision described by the Euler angles (ψ, θ and φ) and a translation along the instrument axis.

The architecture of master device is a 3-RRR Spherical Parallel Manipulator (SPM) with a decoupled translation attached to the end effector, see Fig. 3. The surgical robot is a 4 DoF serial robot. Its architecture is a RRR spherical serial manipulator with a translation along the last joint axis, see Fig. 4.

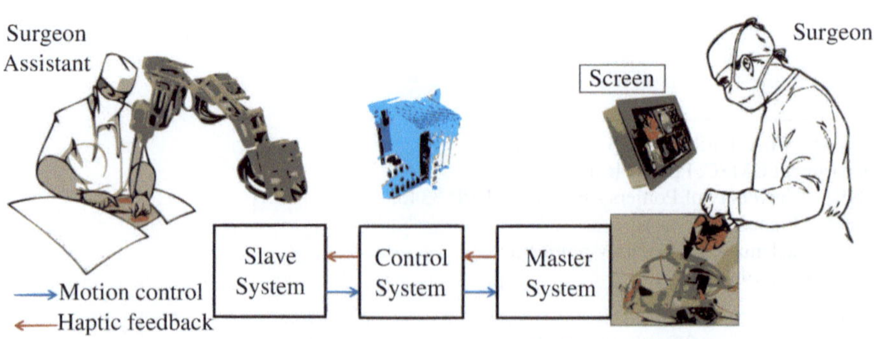

Fig. 1 Tele-operation system for MIS

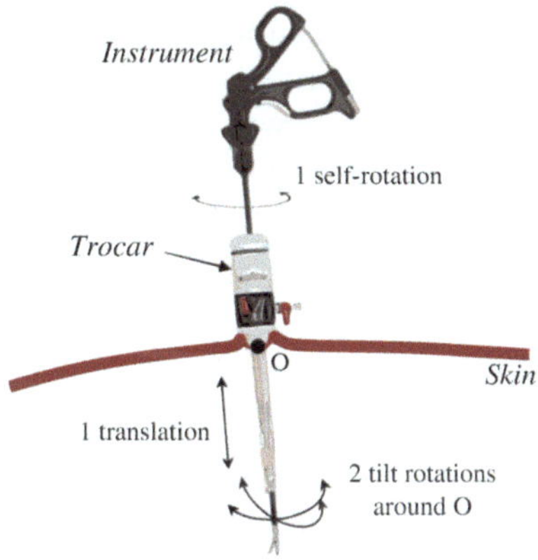

Fig. 2 MIS possible movements

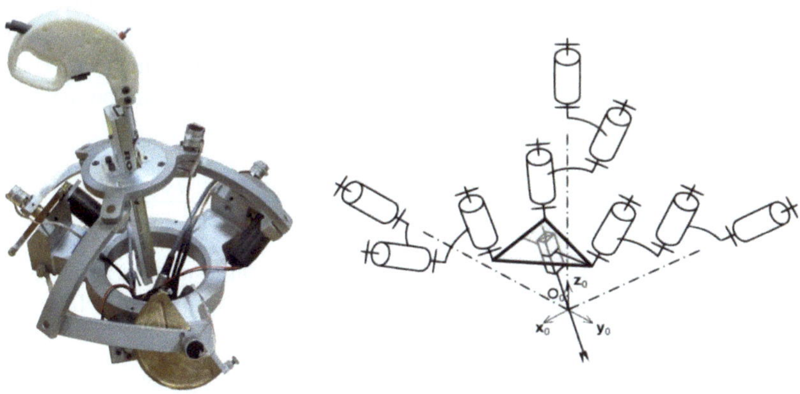

Fig. 3 The developed master haptic system and its kinematic diagram

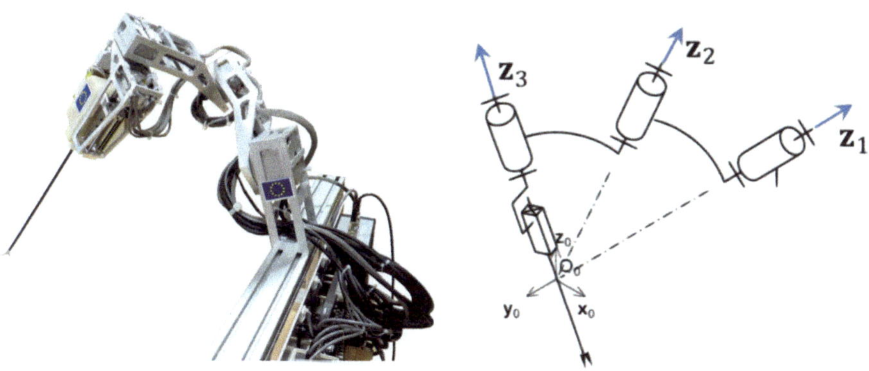

Fig. 4 The developed slave system and its kinematic diagram

An experimental study using the Vicon Nexus motion capture system was carried out to identify the workspace of an anastomosis surgical operation performed by an expert surgeon. The obtained results allowed us to identify the workspace, swept by the instrument, as a cone with an apex angle of 26°. The kinematic of the master and slave robot was optimized using the results of motion capture [15].

3 Control System Architecture

The control architecture of the tele-operation system can be summarized by the diagram of Fig. 5. It's based on the use of two PLC (Programmable Logic Controller), one for the slave and another for the master.

The PLC of the master device solves the Forward Kinematic Model (FKM). The Euler angles and translation parameters, obtained by the FKM, are then used to control the slave thanks to the seconds PLC. The PLC of the slave calculates the joints angular configurations and speeds using the Inverse Kinematic Model. The angular velocity of the slave joints are obtained by a numeric derivation method, see Eq. 1.

$$\dot{q}(t) \approx \frac{-q(t+2T) + 8q(t+T) - 8q(t-T) + q(t-2T)}{12T} \quad (1)$$

The force feedback is provided to the surgeon by three motors located on the master device. These motors are controlled by input signals given by a force sensor installed on the slave robot.

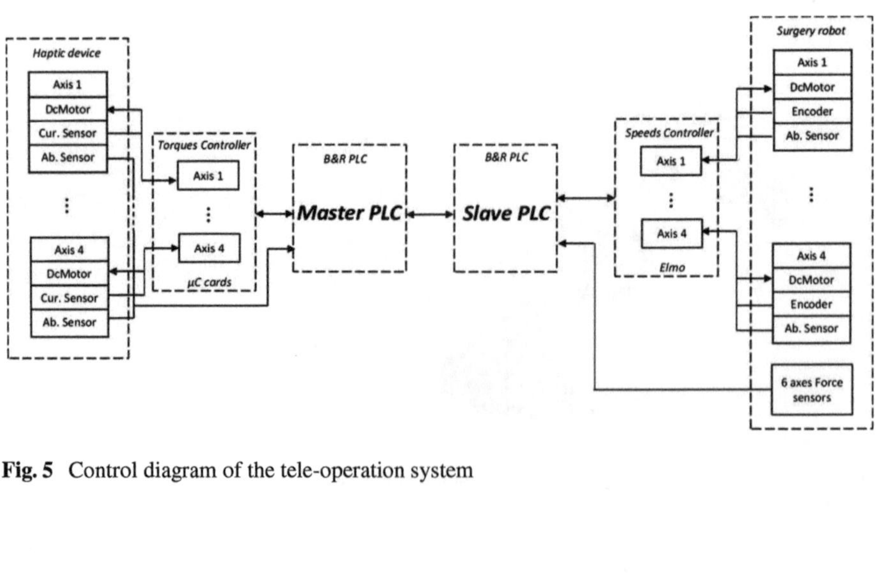

Fig. 5 Control diagram of the tele-operation system

4 Kinematic Model of the SPM

The SPM parallel robot has three identical legs. Each leg is made of two links and three revolute joints, Fig. 6. All axes of the revolute joints are intersecting in one common point, called CoR (Center of Rotation). Each link is characterized by the angle between its two revolute joints, as shown in Fig. 7. This angle is constant and it represents the dimension of the link. The angle, α, characterizes the first link and the angle, β, the second link. The angle, γ, defines the orientation of the axis Z_E normal to the moving platform with respect to the last joint. The axes of actuated joints are located along an orthogonal frame.

The kinematic model of the SPM can be solved by differentiating Eq. 2, defined as follows:

$$\mathbf{Z_{2k}} \cdot \mathbf{Z_{3k}} = \beta \tag{2}$$

For $k = A, B\ and\ C$, with:

$$\begin{cases} \mathbf{Z_{2k}} = \qquad\qquad\qquad\qquad Rot(\mathbf{Z_{1k}}, \theta_{1k}) \cdot Rot(\mathbf{X_{1k}}, \alpha) \cdot \mathbf{Z} \\ \mathbf{Z_{3k}} = Rot(\mathbf{Z}, \psi) \cdot Rot(\mathbf{X}', \theta) \cdot Rot(\mathbf{Z}'', \varphi) \cdot Rot(\mathbf{X}'', \gamma) \cdot \mathbf{Z} \end{cases} \tag{3}$$

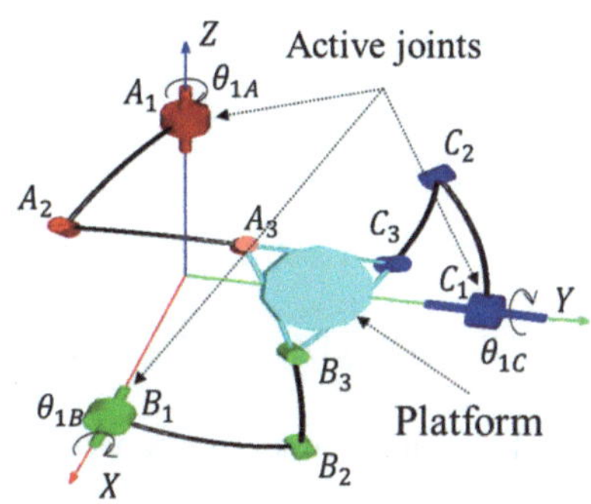

Fig. 6 Spherical parallel manipulator kinematic

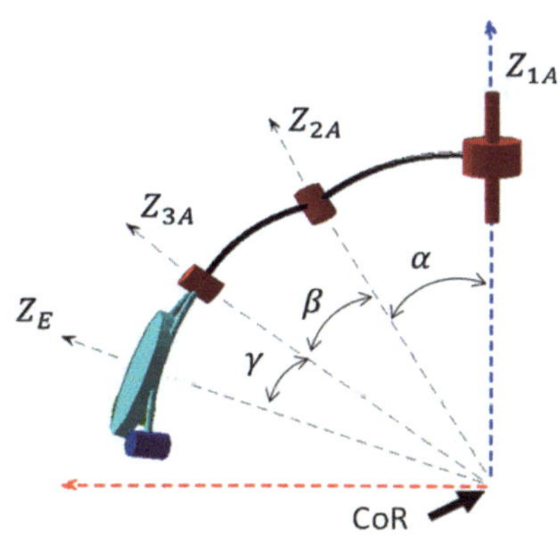

Fig. 7 Parameters of leg A

This leads to:

$$\dot{\mathbf{Z}}_{2k} \cdot \mathbf{Z_{3k}} + \mathbf{Z_{2k}} \cdot \dot{\mathbf{Z}}_{3k} = 0 \tag{4}$$

where

$$\begin{cases} \dot{\mathbf{Z}}_{2k} = \dot{\theta}_{1k}\, \mathbf{Z_{1k}} \times \mathbf{Z_{2k}} \\ \dot{\mathbf{Z}}_{3k} = \qquad \omega \times \mathbf{Z_{3k}} \end{cases} \tag{5}$$

ω is the angular velocity of the end-effector.

After substitution, Eq. 4 becomes:

$$\dot{\theta}_{1k}\, (\mathbf{Z_{1k}} \times \mathbf{Z_{2k}}) \cdot \mathbf{Z_{3k}} = \omega \cdot \mathbf{Z_{2k}} \times \mathbf{Z_{3k}} \tag{6}$$

In matrix form, Eq. 6 can be written as:

$$\mathbf{B} \cdot \dot{\theta} = \mathbf{A} \cdot \omega \tag{7}$$

where

$$\dot{\theta} = [\dot{\theta}_{1A}\ \ \dot{\theta}_{1B}\ \ \dot{\theta}_{1C}]^T \tag{8}$$

$$\omega = \begin{bmatrix} \dot{\theta}cos(\psi) + \dot{\varphi}sin(\theta)sin(\psi) \\ \dot{\theta}sin(\psi) + \dot{\varphi}sin(\theta)cos(\psi) \\ \dot{\psi} + cos(\theta) \end{bmatrix}. \tag{9}$$

$$\mathbf{A} = [\mathbf{Z}_{3A} \times \mathbf{Z}_{2A} \quad \mathbf{Z}_{3B} \times \mathbf{Z}_{2B} \quad \mathbf{Z}_{3C} \times \mathbf{Z}_{2C}]^T \tag{10}$$

$$\mathbf{B} = Diag[\mathbf{Z}_{1A} \times \mathbf{Z}_{2A} \cdot \mathbf{Z}_{3A} \quad \mathbf{Z}_{1B} \times \mathbf{Z}_{2B} \cdot \mathbf{Z}_{3B} \quad \mathbf{Z}_{1C} \times \mathbf{Z}_{2C} \cdot \mathbf{Z}_{3C}] \tag{11}$$

The kinematic model can be written as:

$$\omega = \mathbf{A}^{-1}\mathbf{B} \cdot \dot{\theta} = \mathbf{J}\dot{\theta} \tag{12}$$

J is the Jacobian matrix of the SPM. Since the translational motion is decoupled, only revolute joints are considered in the following kinematic model.

5 Implementation of the Haptic Control

This section deals with the haptic feedback implementation for the master device. The SPM is equipped with three DC motors, see Fig. 8. The torques control of the three actuated joints provides the haptic feedback.

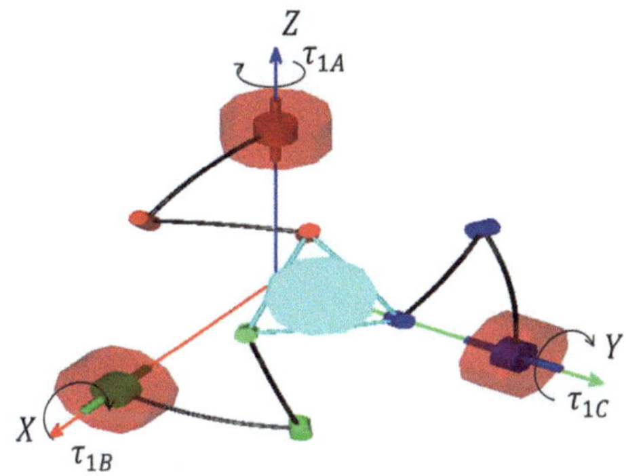

Fig. 8 Actuated joints (*red*)

Let **T** denotes the torque applied by the haptic device to the user hand and τ the torque supplied by the actuators. One can write:

$$\tau = \mathbf{J}^T\mathbf{T} + \tau_s + \tau_f + \tau_d \tag{13}$$

$$\tau = [\tau_{1A} \;\; \tau_{1B} \;\; \tau_{1C}]^T \tag{14}$$

where $\mathbf{J}^T$ is the transpose of the Jacobian matrix. τ_d is the dynamic torque provided by the actuators to overcome the dynamic behavior of the master device. τ_s is the static torque produced by the actuators to overcome the gravity effort applied on the master device. τ_f is the torque provided by the actuators to overcome the friction torque.

The dynamic behavior of the haptic device is neglected because of the low level of speed and acceleration of the application.

5.1 Static Compensation Torque

The static torque, τ_s, is calculated by solving the static equilibrium system of the SPM. The 3-RRR SPM is an over-constraint system. So the equilibrium is written for equivalent and non-overconstraint SPM which is 3-RRS SPM. To write the equilibrium we need to know the positions of the centers of gravity and the weight of all the parts of the SPM. Figure 9 shows the orientation of the gravity vector relative to the base frame. ψ is the angle between the vector **X** and the projection of the gravity vector **G** on the **XY** plane. It is equal to 45°. θ is the angle between the vector **Z** and **G**. It is equal to 54.7°.

The center of gravity of each link of the SPM is calculated using Eq. 15. It is assumed that each link is a circular arc.

$$\mathbf{OA} = 2R\frac{sin(\alpha/2)}{\alpha} \tag{15}$$

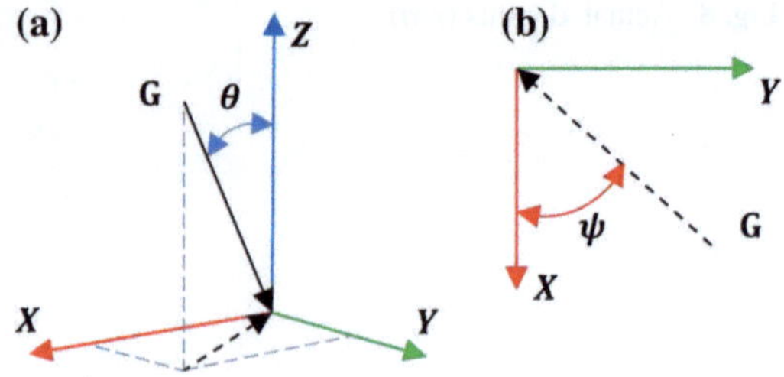

Fig. 9 Orientation of the gravity vector relative to the base frame. **a** 3D representation. **b** Projection of **G** on the **XY** plane

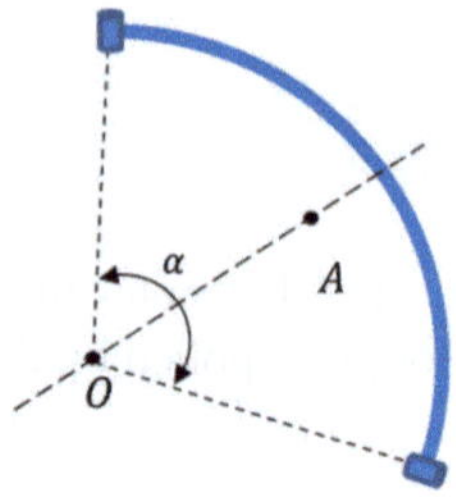

Fig. 10 The center of gravity of the link

where **A** is the center of gravity. Each link is described by the center denoted by the point O, the angle α and the radius **R**, as illustrated in Fig. 10. The end-effector center of gravity is calculated using a CAD software. The weight of each part is measured using a digital scale.

Then, the static equilibrium of force and moment using screw formulation can be written as follows:

Proximal link

$$\Im^k_{base\rightarrow proximal} + \Im^k_{gravity\rightarrow proximal} + \Im^k_{distal\rightarrow proximal} = \begin{Bmatrix} 0 \\ 0 \end{Bmatrix} \tag{16}$$

Distal link

$$\Im^k_{proximal\rightarrow distal} + \Im^k_{gravity\rightarrow distal} + \Im^k_{effector\rightarrow distal} = \begin{Bmatrix} 0 \\ 0 \end{Bmatrix} \tag{17}$$

End effector

$$\sum_{k}^{A,B,C} (\Im^k_{distal\rightarrow effector}) + \Im_{gravity\rightarrow effector} = \begin{Bmatrix} 0 \\ 0 \end{Bmatrix} \tag{18}$$

where

$$\Im^k_{base \to proximal} = \begin{Bmatrix} X_{1k} & L_{1k} \\ Y_{1k} & M_{1k} \\ Z_{1k} & \tau_{1k} \end{Bmatrix} \tag{19}$$

$$\Im^k_{distal \to proximal} = -\Im^k_{proximal \to distal} = \begin{Bmatrix} X_{2k} & L_{2k} \\ Y_{2k} & M_{2k} \\ Z_{2k} & 0 \end{Bmatrix} \tag{20}$$

$$\Im^k_{distal \to effector} = -\Im^k_{effector \to distal} = \begin{Bmatrix} X_{3k} & 0 \\ Y_{3k} & 0 \\ Z_{3k} & 0 \end{Bmatrix} \tag{21}$$

This model leads to a system of 42 equations and 42 unknowns. The execution time of the static model is about $66\mu s$ in master $B\&R$ PLC.

5.2 Friction Model

The friction model of the system can be simplified using the following model:

$$\begin{cases} \tau_{f,1k} = \tau_{fc} & if\ \dot{\theta}_{1k} > 0 \\ \tau_{f,1k} = -\tau_{fc} & if\ \dot{\theta}_{1k} < 0 \\ \tau_{f,1k} = 0 & if\ \dot{\theta}_{1k} = 0 \end{cases} \tag{22}$$

where τ_{fc} is the friction torque constant. A graphic representation of the considered friction model is presented in Fig. 11.

5.3 Torque Control of the Actuators

The current control diagram for each DC-motor installed in the haptic device is presented in Fig. 12. The PLC of the master device sends the reference torque to

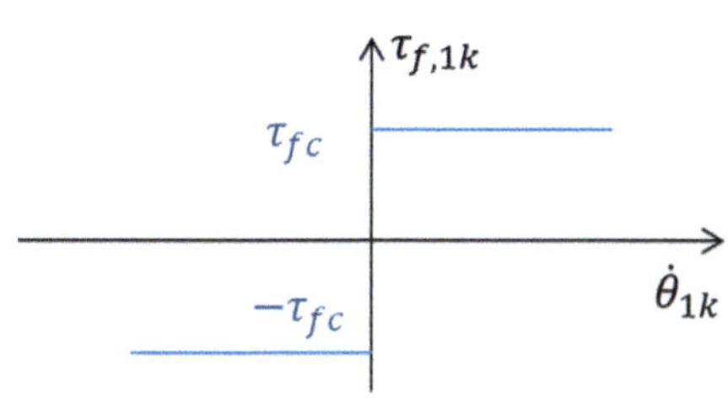

Fig. 11 Simplified friction model

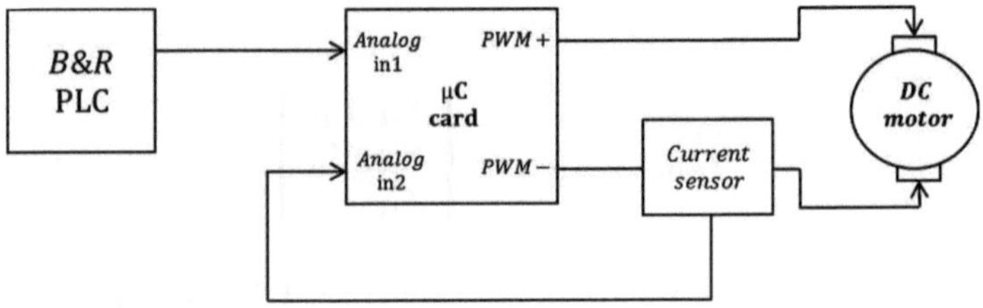

Fig. 12 Control diagram of the DC-motor current

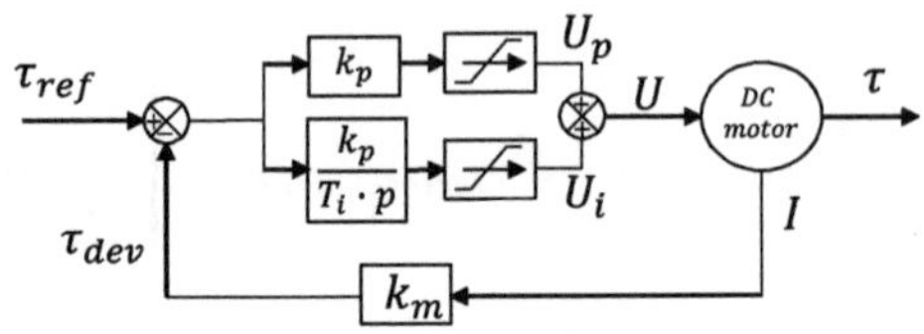

Fig. 13 Torque control scheme

a $\mu Controller$ card which regulates the torque of the DC-motors by varying the current intensity.

The regulation of the torque is made by PI regulator implemented in $\mu Controller$ card, see Fig. 13.

τ_{ref} is the reference torque calculated by the PLC using the model presented in Eq. 13. τ_{dev} is the torque developed by the DC-motor. k_m is coefficient that relate the DC-motor developed torque τ_{dev} to the DC-motor current I. The DC-motor voltage is varied using a PWM signal.

6 Experimental Validation

In order to prove the feasibility of the proposed haptic control scheme, experiments consisting to constrain a user to follow a circular trajectory are performed. One user applies a motion to the haptic device; the force feed-back control acts in such a way that the user is constrained to follow a circular path. Figure 14 shows the end-effector and reference trajectories in the plan (ψ, θ). The torques of the actuated joints are presented in Fig. 15.

The user follows the reference trajectory imposed by the haptic device with little fluctuations. These results validate the appropriate behavior of the haptic model, without using an effort sensor in the master device. To calibrate the force feedback a six components force sensor will be used in order to provide a more accurate control.

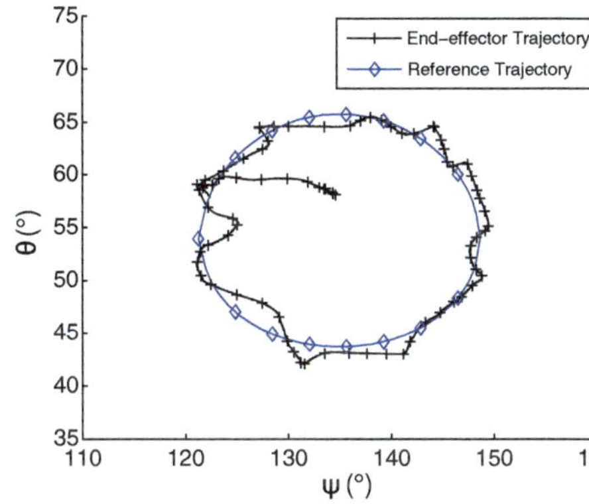

Fig. 14 Experimental results

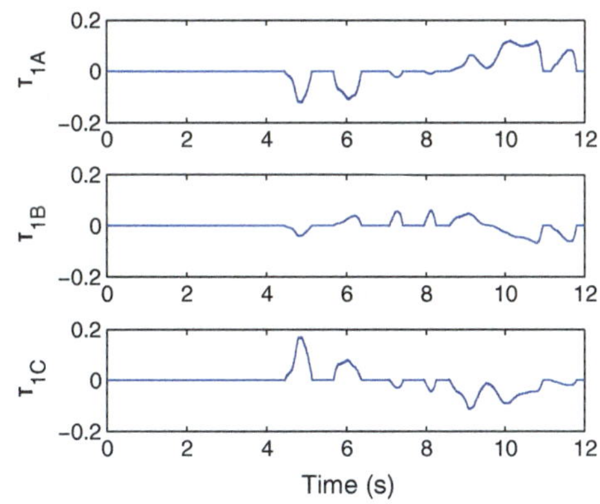

Fig. 15 Actuated joints torques in (Nm)

7 Conclusion

The implementation of the haptic feed-back for a master device was presented. The master device is used as a part of a tele-operation system for minimally invasive surgery tasks. The control architecture of the tele-operation system was presented and implemented. It consists of two PLC, one for the master and the other for the slave. The master PLC calculates the reference torque consisting of a static compensation torque, a friction compensation torque and the haptic feedback torque. The reference torque is then used by $\mu ControllerCard$ to control the actuated joint torques. The use of a force sensor in the master device is not required in order to ensure the haptic feedback thanks to the proposed control architecture. Experimentals were carried out successfully to prove the feasibility of the proposed haptic control.

Acknowledgments This research is supported by the Poitou-Charentes region 2007–2013 (program project 10 Images and interactivities), in partnership with the European Union (FEDER/ERDF, European Regional Development Fund). This research is supported by ROBOTEX, the French national network of robotics platforms (Project Number ANR-10-EQPX-44-01).

References

1. Andy, M., Shahram, P.: Analysis and experimentation of a 4-DOF haptic device. In: Symposium on Haptic Interfaces for Virtual Environments and Teleoperator Systems (2008)
2. Van den Bedem, L., Hendrix, R., Rosielle, N., Steinbuch, M., Nijmeijer, H.: Design of a minmally invasive surgical tele-operation master-slave system with haptic feedback. In: International Conference on Mechatronics and Automation, pp. 60–65, ICMA (2009)
3. Jaydeep, H.P.: Robotic assisted minimally invasive surgery. J. Minim. Access. Surg. **5**(1), 1–7 (2009)
4. Saafi, H., Laribi, M.A., Zeghloul, S., Ibrahim, M.Y.: Development of a spherical parallel manipulator as a haptic device for a tele-operation system: application to robotic surgery. In: Industrial Electronics Society, IECON 2013 - 39th Annual Conference of the IEEE, pp. 4097–4102 (2013)
5. Saafi, H., Laribi, M.A., Zeghloul, S., Ibrahim, M.Y.: Tele-operation control system for non-homothetic master/slave kinematics for minimally-invasive surgery. In: Industrial Electronics Society, IECON 2013 - 39th Annual Conference of the IEEE, pp. 3800–3805 (2013)
6. Haddadi, A., Hashtrudi-Zaad, K.: Robust stability of teleoperation systems with time delay: a new approach. IEEE Trans. Haptics **6**(2), 229–241 (2013)
7. Suzuki, A., Ohnishi, K.: Novel four-channel bilateral control design for haptic communication under time delay based on modal space analysis. IEEE Trans. Control Syst. Technol. **21**(3), 882–890 (2013)
8. Hashemzadeh, F., Tavakoli, M., Hassanzadeh, I.: Haptic teleoperation under variable delay and actuator saturation. In: World Haptics Conference (WHC), pp. 377–382 (2013)
9. Hulin, T., Camarero, R.G., Albu-Schaffer, A.: Optimal control for haptic rendering: fast energy dissipation and minimum overshoot. In: Intelligent Robots and Systems (IROS), 2013, International Conference on IEEE/RSJ, pp. 4505–4511 (2013)
10. Jazayeri, A., Dyck, M., Tavakoli, M.: Stability analysis of teleoperation systems under strictly passive and non-passive operator. In: World Haptics Conference (WHC), pp. 695–700 (2013)
11. Ciaurriz, P., Diaz, I., Gil, J.J.:Stable discrete-time impedances for haptic systems with vibration modes and delay. IEEE Trans. Control Syst. Technol. **99**, 1–10 (2013)
12. Bianchini, G., Prattichizzo, D.: Virtual coupling design for stability and transparency of multi-device haptic systems with delays. In: World Haptics Conference (WHC) 2013, pp. 223–228 (2013)
13. Wildenbeest, J.G.W., Abbink, D.A., Schorsch, J.F.: Haptic transparency increases the generalizability of motor learning during telemanipulation. In: World Haptics Conference (WHC) 2013, pp. 707–712 (2013)
14. Pacchierotti, C., Tirmizi, A., Bianchini, G., Prattichizzo, D.: Improving transparency in passive teleoperation by combining cutaneous and kinesthetic force feedback. In: Intelligent Robots and Systems (IROS) 2013, International Conference on IEEE/RSJ, pp. 4958–4963 (2013)
15. Chaker, A., Mlika, A., Laribi, M.A., Romdhane, L., Zeghoul, S.: Synthesis of a spherical parallel manipulator for a dexterous medical task. Front. Mech. Eng. **7**(2), 150–162 (2012)

Experiments on the Simultaneous Hand-Held Control of Rigid Endoscopes and Robots Passing Through Them

Richard J. Hendrick, S. Duke Herrell, Christopher R. Mitchell and Robert J. Webster III

Abstract Concentric tube manipulators have the diameter of needles and are consequently amenable to delivery into the human body through small ports in an endoscope. When this is done, the surgeon must manipulate both the endoscope and one or more concentric tube robots simultaneously. In this paper we explore a hand-held approach to this user interface challenge, in which the surgeon has direct physical control of endoscope pose and can use finger and thumb controls to specify the motion of two concentric tube robots that pass through the endoscope. In experiments with the system, we explore whether the endoscope-robot combination can reach locations in the prostate that are inaccessible to the endoscope alone. We also compare joint space and task space control for three-degree-of-freedom concentric tube robots, and demonstrate experimentally that laser resection of prostate tissue is possible using an anthropomorphic phantom.

1 Introduction

A great deal of interest in the surgical robotics community is currently focused on enabling less invasive access to the human body through natural orifices. Despite the relatively small diameter of the urethra, it is no exception. In fact, Transurethral Resection of the Prostate (TURP) was one of the earliest surgical robotics applica-

---

R.J. Hendrick (✉) · R.J. Webster III
Department of Mechanical Engineering,Vanderbilt University, VU Station B 351592, 2301 Vanderbilt Place, Nashville, TN 37235-1592, USA
e-mail: richard.j.hendrick@vanderbilt.edu

R.J. Webster III
e-mail: robert.webster@vanderbilt.edu

S.D. Herrell · C.R. Mitchell
Department of Urologic Surgery,Vanderbilt University Medical Center, A-1302 Medical Center, Nashville, TN 37232-2765, USA
e-mail: duke.herrell@vanderbilt.edu

C.R. Mitchell
e-mail: christopher.r.mitchell@vanderbilt.edu

M.A. Hsieh et al. (eds.), *Experimental Robotics*, Springer Tracts in Advanced Robotics 109, DOI 10.1007/978-3-319-23778-7_6

tions [1, 2]. Somewhat surprisingly, since that pioneering work, only a few research groups have developed robotic systems designed for transurethral deployment. In 2001, robotic transurethral laser resection of the prostate through a standard resectoscope was mentioned by Ho et al. [3], though the authors' main focus was on the Nd:YAG laser rather than on the robotic system. In 2002, Badajoz et al. proposed master-slave teleoperation of a robot holding a transurethral resectoscope for prostate resection [4]. In 2004, Hashimoto et al. described a 4 degree-of-freedom (DOF) manipulator intended for prostate resection that delivers a drill and cutter through an 8 mm rigid tube under ultrasound image guidance [5]. A major challenge was the removal of excised tissue from the body, since a morcellator was not incorporated into the intended workflow. In 2013, Goldman et al. demonstrated the feasibility of transurethral robotic bladder access, and simultaneously described the first use of a continuum robot in a transurethral application of which we are aware. Their robot delivered a laser fiber and biopsy forceps for bladder tumor resection [6].

We recently presented a new hand-held surgical robot aimed at improving transurethral prostate surgery by making it easier to perform [7]. Our system follows the continuum robot paradigm of Goldman et al. but uses a different, smaller type of continuum robot called a concentric tube robot, enabling it to deliver two manipulators through a 5 mm diameter port in a standard clinical resectoscope (see Fig. 1). One manipulator aims a holmium laser fiber, and the other provides tissue retraction [7].

This robotic system was motivated by the prospect of increasing the utilization of a procedure that is known to have excellent clinical outcomes, yet has failed to achieve widespread adoption due to its steep learning curve. The procedure is called Holmium Laser Enucleation of the Prostate (HoLEP), and is currently conducted using a straight, rigid endoscope. In this procedure, a holmium laser fiber is passed through the endoscope and used to cut prostate tissue that is obstructing urine flow. HoLEP has been clinically demonstrated to have significantly better outcomes for patients than traditional TURP which uses sharp dissection or electrocautery. These

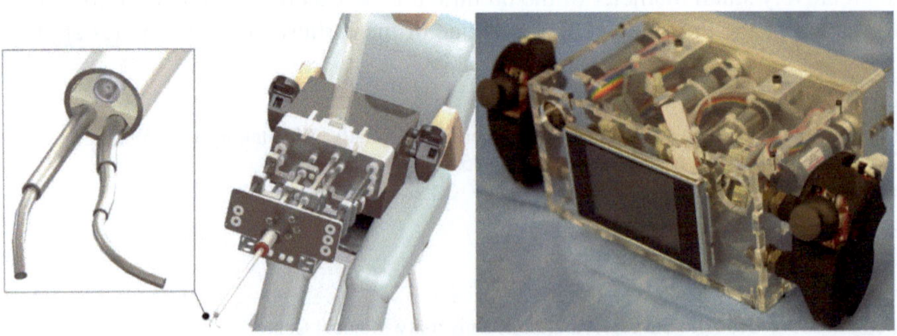

Fig. 1 The surgeon has hand-held control of the entire system, which is supported by a counterbalanced arm (see Fig. 4). The surgeon controls each concentric tube manipulator with joysticks and triggers located on the handles shown in the right image above, based on endoscopic video feedback

benefits include a 50 % reduction in catheterization time, a 33 % reduction in hospitalization duration, and the elimination of the need for blood transfusion [8]. Recently, long term follow-up data has shown that HoLEP requires fewer re-operations, which is leading many in the urology community to conclude that HoLEP should become the new gold standard treatment for enlarged prostate [9].

Despite its clinical advantages, HoLEP is conducted in only a few institutions because it is extremely challenging for the surgeon [10]. It is physically demanding because large forces are required to angle the endoscope, due to the soft tissues surrounding it. To make matters worse, the surgeon must simultaneously manipulate soft tissue using the endoscope itself, and bring the laser fiber (which has no articulation, aiming straight out from the endoscope at a fixed position with respect to the image) to bear on the desired surgical target, which requires immense surgical skill.

The robotic system described in this paper is designed to alleviate these challenges. It passes curved, flexible continuum manipulators through the same endoscope currently used in the manual procedure, to enable the surgeon to aim the laser with one manipulator and retract tissue with the other. The hand-held nature of our system and integration of a standard clinical endoscope enables it to fit seamlessly into the current clinical workflow, which we believe will facilitate adoption by the medical community.

Toward our overall goal of making HoLEP easier to perform, the specific objectives of this paper are (1) to experimentally illustrate that the concentric tube robots can enable the surgeon to reach a larger portion of the desired resection ellipsoid in HoLEP, (2) to experimentally compare the accuracy of joint space and task space mappings from the surgeon's thumb and finger controls to concentric manipulator motions, and (3) to experimentally demonstrate the system in a realistic setting by laser resecting an anatomically accurate prostate model. We accomplish these objectives via a set of user studies with two experienced urologic surgeons.

2 Technical Approach

2.1 Robot Design

Our robotic system consists of three main modules: the user interface, the transmission, and the endoscope. Detailed design information can be found in [7]. Briefly, at the back of the robot is the user interface module, as shown in Fig. 1. This module is designed to quick connect to the transmission module through spring-loaded shaft couplings and houses nine brushless motors. Fixed to the outside of the user interface module are handles, where the surgeon grasps the robot and can manually manipulate the entire robot to control endoscope pose. Each handle has an embedded joystick and trigger, which are used for controlling the concentric tube manipulators. A screen is also placed between the surgeon's hands, which can be used to display the endoscopic view. Originally, we thought the surgeons may prefer to have the visualization inline with the tools they are manipulating. In practice, however, we

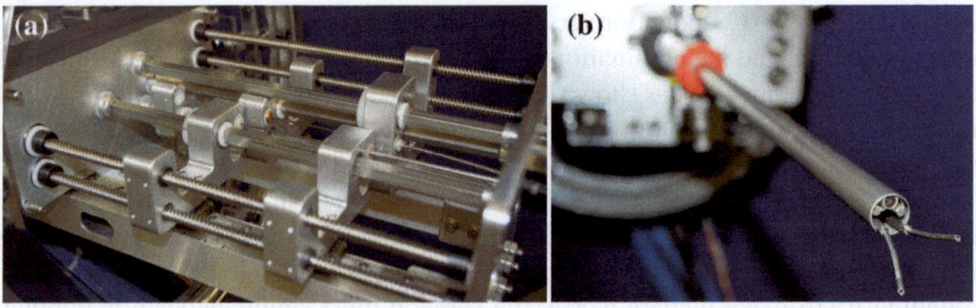

Fig. 2 **a** The transmission section: square shafts transmit the torque to rotate the concentric tubes, while lead screws drive carriers which translate the tubes. **b** The rigid endoscope can be seen mounted to the front of the robot, with two concentric tube robots passing through it. Two fiber optic bundles surround the lens to provide illumination. The outer diameter of the endoscope is 8.3 mm, and the two manipulators pass through a single 5 mm port in it

have found surgeons tend to prefer large high definition screens in the operating room.

The transmission section, shown in Fig. 2, converts the motion of the motors in the user interface module into translation and rotation of the tubes. The transmission section is capable of driving 9 DOF. One arm was designed as a three tube, 6 DOF manipulator, while the other was designed as a two tube, 3 DOF manipulator (in the set of experiments described in this paper, however, both arms behave as 3 DOF manipulators). Linear motion of the tubes is achieved via lead screws, which drive tube carriers that ride on ball bearing blocks on a guide rail. Rotation of the tubes is achieved via square shafts, which transmit torque through a gear train to the tube.

We leverage the surgeon's endoscopic manipulation skills and retain the current clinical workflow by designing this robot as a hand-held robot with a clinical endoscope (Storz, Inc. 27292 AMA, 26 Fr) mounted to the front of the transmission section, as shown in Fig. 2. Because this endoscope is rigidly mounted to the robot, manipulation of the robot via the user interface handles also manipulates the endoscope. This endoscope contains optics and light sources as well as a 5 mm tool channel, through which we pass our concentric tube robots. For a more detailed design description, see [7].

2.2 *Concentric Tube Robots*

Concentric tube robots are a class of continuum robots composed of concentrically nested, precurved, elastic tubes first proposed for use as robotic manipulators simultaneously in [11, 12]. They are typically made of superelastic nitinol, which is well suited for this application because it has 8 % recoverable strain and can be shape set into desired curves while maintaining its superelasticity. When these precurved tubes are translated and rotated at their bases, their elastic interaction creates "tentacle-like" motion (elongation and bending) of the device. Geometrically exact mechanics-based

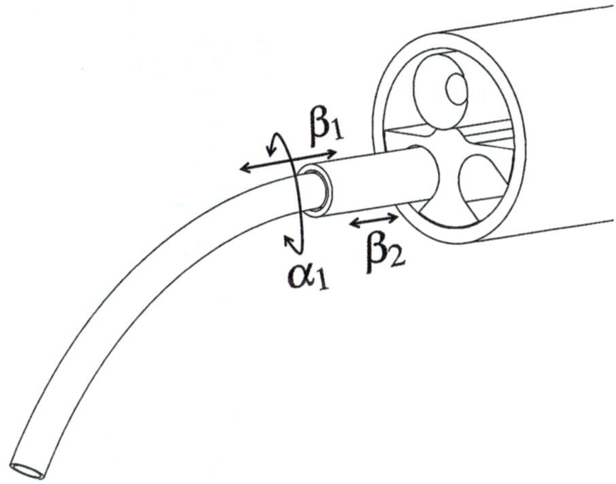

Fig. 3 The actuation variables α_1, β_1, and β_2 denote rotation of the inner tube, translation of the inner tube, and translation of the outer tube, respectively

models now exist for these manipulators [13, 14]. The manipulators used in the experiments in this paper are a special case concentric tube robot composed of two tubes: a nitinol inner tube with a constant curvature arc at its tip, and a straight, steel outer tube. Such a manipulator has three actuated degrees of freedom (see Fig. 3): rotation of the inner tube (α_1), translation of the inner tube (β_1), and translation of the outer tube (β_2). We approximate the outer tube as perfectly straight and rigid, which allows the kinematics to be computed purely geometrically, the details of which can be found in [7].

2.3 Counterbalance System

To assist the surgeon by supporting the weight of the the hand-held robot (while still permitting 6 DOF motion) a counterbalanced arm (Dectron, USA) is provided (Fig. 4). As the robot is used in transurethral endoscopic prostate surgery, the endoscope approximately operates through an anatomically constrained center of motion near the apex of the prostate. This constraint is created by the soft tissue pressure provided by the urogenital diaphragm. Thus, the surgeon must manually control 4 DOF (roll, pitch, yaw, and insertion).

2.4 User Interface Mappings

The introduction of concentric tube robots adds additional degrees-of-freedom for the surgeon. As illustrated in Fig. 1, the concentric tube robots add 9 DOF, three on the two-tube arm and six on three-tube arm. However, in the set of experiments described in this paper, for simplicity, both arms were configured as identical two-

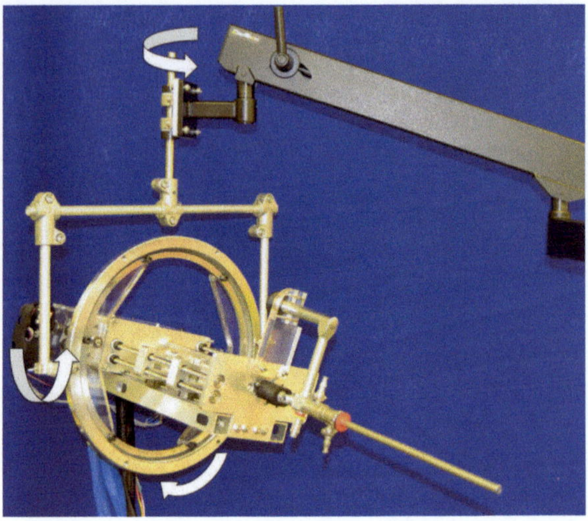

Fig. 4 A counterbalanced arm allows for manual 6-DOF spatial positioning of the robot by the surgeon

tube, 3 DOF manipulators. Thus, in our experiments, we assess the surgeon's ability to coordinate 10 DOF based on the endoscopic view. The surgeon manipulates the concentric tube robots via the embedded joystick (with pushbutton capability) under his/her thumb and an analog trigger under his/her index finger.

Initially, likely due to familiarity with manipulating manual tools through endoscope ports, surgeons expected to prefer direct joint space control of the rotation and axial extension of the tubes. Furthermore, in task space control, surgeons were initially surprised by rapid robot motions near singularities, and perceived these as a lack of control of the robot. Based on this, we implemented both joint and task space control and set out to experimentally compare the two.

In joint space control, the index finger trigger was mapped to rotational velocity of the inner tube ($\dot{\alpha}_1$), and upward motion of the joysticks (on each handle) were mapped to translational velocity of the tubes ($\dot{\beta}_1, \dot{\beta}_2$). The surgeon was able to reverse the direction of rotation by clicking the joystick and then again depressing the index finger trigger. All commanded velocities were linearly proportional to the deflection of the relevant analog input from thumb joysticks and index finger triggers.

In task space control, the tips of the manipulators move relative to the camera frame using a resolved rates algorithm. Thumb joystick deflections were mapped to end-effector velocity in the plane of the endoscopic view. The index finger trigger was mapped to end effector velocity perpendicular to the image plane, and clicking the joystick reversed the direction of motion perpendicular to the image plane (Fig. 5).

3 Experiments

Three distinct experiments were conducted: (1) to evaluate the ability of the concentric tube robots to access more of the prostate than a straight rigid endoscope, (2) to

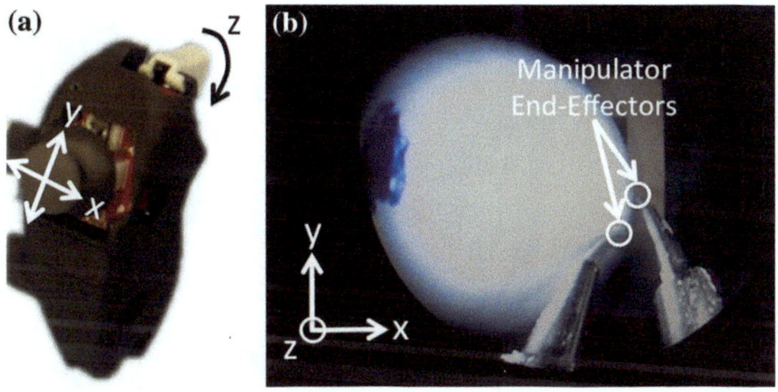

Fig. 5 **a** One of the handles that comprise the user interface, illustrating an analog joystick (with pushbutton) and trigger. The joystick provides for two bidirectional inputs, and the trigger gives a third unidirectional input. The pushbutton is used to reverse the direction of the trigger input. **b** The endoscopic view

compare the tracking ability of the hand-held system in task space and joint space, and (3) to show that this robotic system can perform the realistic laser resections required for this surgery.

3.1 Prostate Surface Access with Concentric Tube Robots

To approximate the center of motion (which in humans comes from anatomical constraints as mentioned earlier) in a benchtop setting, we used a wooden support with a hole through it, positioned at an anatomically accurate distance from the prostate model. To simulate the surgeon's desired surface for laser resection in HoLEP, we affixed an ellipsoid of anatomically correct dimensions (30 mm × 42 mm × 47 mm [15]) made of hard plastic behind this center of motion as shown in Fig. 6.

An Aurora Electromagnetic Measurement System (Northern Digital Inc.) was registered to the test stand, and an electromagnetic tracking coil was embedded in the tip of the concentric tube robot. Two experienced urologic surgeons then used the system to scan the surface of the desired resection ellipsoid. One two-tube, three DOF manipulator was used under both joint space and task space control and the surgeons scanned one quadrant of the model (the model is axially symmetric, so the ability to scan one quadrant demonstrates ability to scan all). The intent of this simple experiment was to compare the kinematic ability of this system against the straight, rigid endoscope which is currently used clinically, and to see if there were any differences between task space and joint space control in terms of resection surface accessibility when tissue deformation is not present. This experiment was completed prior to the design of the counterbalanced arm, so a passive lockable mechanical arm was used to support the robot. It was locked in place during the experiment, but the

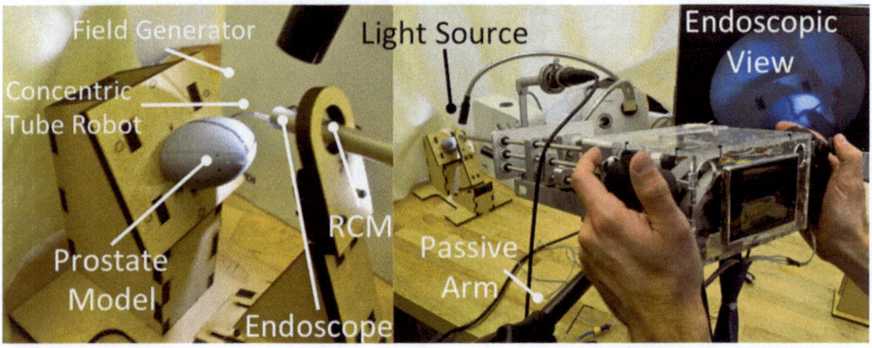

Fig. 6 The experimental setup constrains the endoscope to operate through a remote center of motion which would be anatomically constrained in a patient—we simulate the constraint here with a hole through a piece of wood. The surgeon scans the desired resection surface (the surface of the plastic model) with the instrument tip using endoscopic video feedback. The instrument tip is magnetically tracked

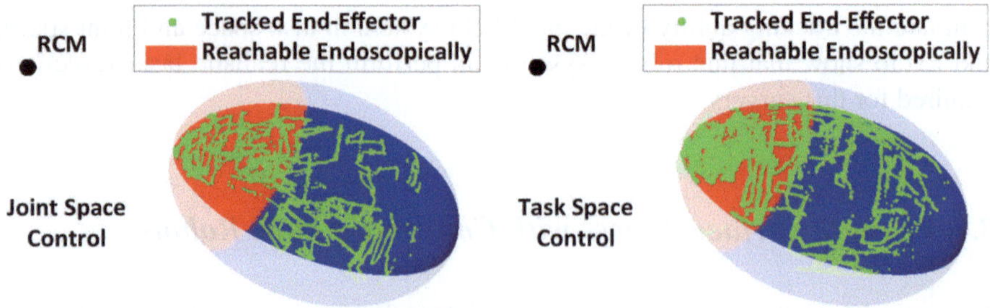

Fig. 7 Magnetic tracking data showing the positions on the prostate surface accessed by two surgeons using the system in joint space mode (*left*) and task space mode (*right*). The *red region* indicates the best-case scenario for the surface reachable by a conventional endoscope (without tissue deformation), since it is only capable of a straight-line approach

surgeons were allowed to unlock it and reposition the entire robot whenever desired, before re-locking it and continuing with the experiment.

The experimental results are shown in Fig. 7. A cannula tip point was considered to be on the surface of the resection ellipsoid if the point was <2 mm from the surface. The figure shows the tracked points projected onto the closest point on the resection surface model. Figure 7 illustrates that surgeons were able to access nearly the entire available ellipsoidal surface of the prostate model with both control mappings, and a typical endoscope image during the experiment is shown in Fig. 8. The area reached by the robot represents an improvement of approximately 65 % versus the best-case geometrically reachable area of existing straight, rigid clinical endoscopes without tissue deformation. This experiment shows that our robot is capable of reaching points relevant to prostate resection in a static case without tissue deformation, which is conservative, since tissue deformation tends to help make more points accessible rather than fewer. However, what it did not explore were quantitative differences

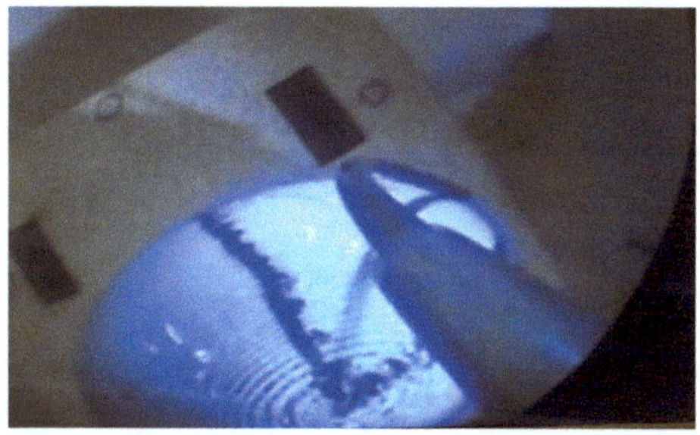

Fig. 8 Endoscopic view of a surgeon scanning the resection surface model as would be required for laser dissection

between joint and task space operation which inspired the experiment in the following section.

3.2 *Task Space Versus Joint Space Performance Comparison*

To explore the differences between task space and joint space operation, a second experiment was conducted using the same experimental setup described in Sect. 3.1. However, rather than scanning an entire quadrant of the prostate, the surgeons were asked to follow a specific curve along the surface of the prostate model. The surgeons were instructed to stay as close to the path as possible throughout the experiment, and task completion time was recorded. This experiment was conducted using the counterbalanced arm shown in Fig. 4. It was done once with the arm unlocked (so that the surgeons could position and orient the endoscope as well as the concentric tube manipulators) and once with the arm locked (forcing the surgeons to use only the concentric tube manipulators without changing the overall endoscope pose). The paths were slightly different for the two scenarios and can be seen in Fig. 9.

The path shown in Fig. 9a was traced with counterbalance assistance by two surgeons, each using both control modes on separate trials, each performing two trials per mode, yielding a total of eight experimental runs. The main experimental results are in Table 1. The mean accuracy for both control modes with counterbalance

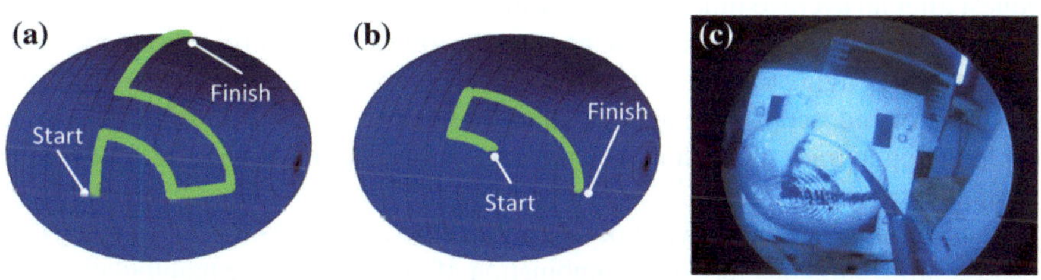

Fig. 9 **a** Desired path of concentric tube robot tip for the experiment with free endoscope manipulation. **b** Desired path of concentric tube robot tip for the experiment with a fixed endoscope pose. **c** Concentric tube robot following the desired path on the prostate resection surface model

Table 1 Counterbalanced tracing experiment (path shown in Fig. 9a)

	Joint space	Task space
Mean error (mm)	1.5	1.6
Max error (mm)	6.4	4.3
Time (s)	70.2	41.9

assistance was not significantly different for joint space control and task space control. The maximum tracked error (averaged over the eight runs) using joint space control was 6.4 mm, while it was only 4.3 mm when using task space control. Moreover, the total time to complete the task was almost twice as high using joint space control than task space control.

These results indicate that surgeons were capable of using both control modes to follow the prescribed path, but that while doing so in joint space mode, they occasionally made large errors. To explore this further, we examined the total time that the surgeons spent in a state of large error. In joint space mode, 5.4, 2.6, 1.9, and 1.0 % of task completion time was spent in excess of 3 mm, 4 mm, 5 mm, and 6 mm error, respectively. In task space mode, 5.4 % of task completion time was spent in excess of 3 mm, and error rarely exceeded 4 mm (0.2 % of the time).

Noting qualitatively that surgeons were using substantial endoscope manipulation —particularly when employing joint space control, we performed another similar experiment in which we fixed the endoscope position and orientation by locking the counterbalanced support arm. Surgeons were then asked to trace the path shown in Fig. 9b. Again, two surgeons attempted to complete this task in both control modes, twice each per mode, for a total of eight experimental runs.

The results for task space control were similar for this experiment compared to when the endoscope had free motion. The maximum error (averaged over the four runs) was 4.0 mm. The only significant difference without endoscope manipulation was that it took surgeons longer to trace a given path length. More specifically, it took them approximately the same amount of time to trace a shorter path in this experiment than it did to trace the longer path in the experiment with free endoscope manipulation.

In sharp contrast, completion of the task was not achievable using joint space control with a fixed endoscope. On all four runs, the experiment was ended at surgeon request after an average time of 2 min, with over 15 % of the time spent with >3 mm error (Table 2). We believe this indicates that the joint space control mapping was too mentally challenging for the surgeons to process rapidly or accurately without the assistance of endoscope pose manipulation, and that they were primarily using pose manipulation to trace the path in the corresponding experiment with free endoscope manipulation.

Considering all of the above experiments together, it appears that when using task space control, surgeons rely on a combination of endoscope pose manipulation and concentric tube manipulation to accomplish the task. In contrast, when using joint space control the surgeons appear to rely heavily on endoscope manipulation and use

Table 2 Fixed endoscope tracing experiment (path shown in Fig. 9b)

	Task space
Mean error (mm)	1.6
Max error (mm)	4.0
>3 mm error (%)	4.1
>4 mm error (%)	1.7
Time (s)	40.3

The task was not achievable in joint space mode. Surgeons attempted it for an average of 2 min per trial in joint space mode before requesting that the experiment be ended

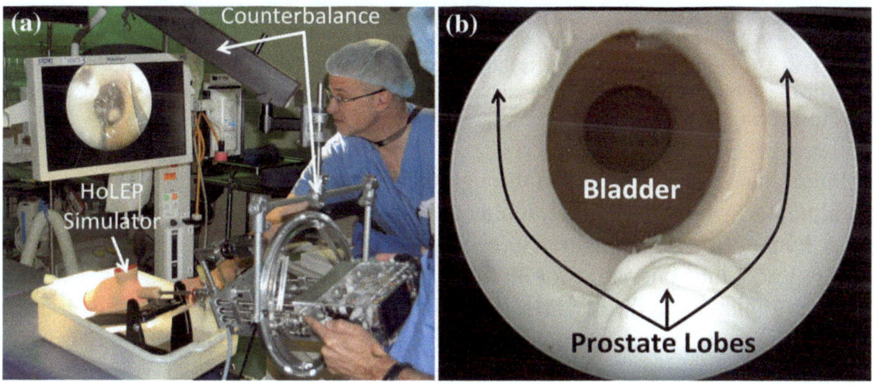

Fig. 10 **a** The counterbalanced robot operates transurethrally through the HoLEP simulator. The surgeons visually servo the concentric tube manipulator tips in task space with high definition endoscopic video feedback. **b** The endoscopic view shows the three lobes of the synthetic prostate. Each surgeon was tasked with laser resecting one lobe of the prostate phantom

little concentric tube robot manipulation to accomplish the task. It is also interesting to note that when using task space control, endoscope pose manipulation enables an increase in the speed of motion, without negatively impacting the accuracy of the manipulation. In future experiments we intend to encode the endoscope pose itself, to more deeply study these qualitative observations that appear to be consistent with the data collected.

3.3 Laser Resection of an Anthropomorphic Prostate Phantom

In this experiment we used a TruLase Prostate HoLEP simulator (TPR100, TruCorp Ltd.) designed for training surgeons in HoLEP (see Fig. 10). This simulator is anatomically accurate and allows for laser enucleation of a synthetic prostate specimen in a fluid-filled environment similar to that of HoLEP surgery. In this experiment, the surgeon used both concentric tube manipulators with task space control. A 500 μm

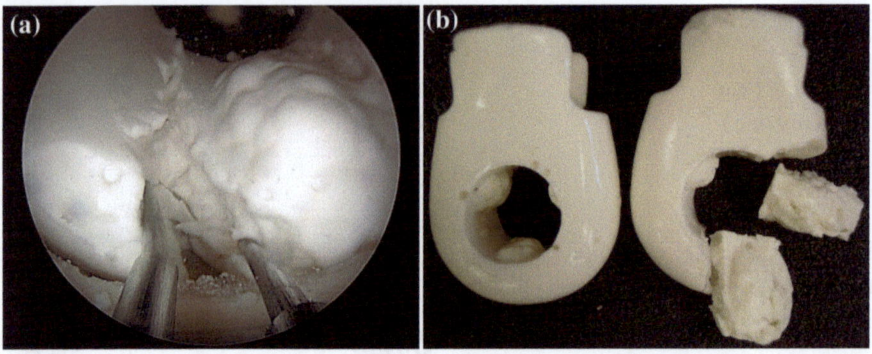

Fig. 11 **a** The left manipulator retracts the tissue, exposing the targets for the right manipulator to cut with the holmium laser. **b** A photograph of what the prostate model looked like before and after the experiment

holmium laser fiber was passed through the inner tube of the right concentric tube manipulator, while the left manipulator was used for tissue retraction with a clear inner lumen. Both manipulators were identical two-tube, 3-DOF devices, as shown in Fig. 3 with tube curvatures of 30 m^{-1}.

The simulator was filled with saline solution and endoscopic saline irrigation was used in the same manner as in a clinical HoLEP procedure. The synthetic prostate used in this experiment was the three lobe prostate insert (TPRO-03, TruCorp Ltd.) as shown in Fig. 10. A clinical 80 W holmium laser was used, and the surgeon could fire the laser on demand using a foot pedal, just as they would in a current clinical HoLEP case. The counterbalance arm was used and the endoscope (and robot) were free to be spatially oriented as desired. Surgeons were asked to laser resect and remove a single lobe of the prostate, and push it into the bladder, just as they would in a current clinical HoLEP case (the specimen then gets morcellated within the bladder in the current clinical HoLEP procedure, and our system is designed to follow the same protocol for specimen removal). There were no restrictions or instructions regarding use of robotic manipulators versus manual endoscopic placement; surgeons were free to use the device as they wished.

Two lobes were successfully laser resected from the prostate model and pushed into the bladder, one by each surgeon. One surgeon removed the median lobe (the lobe in the center of Fig. 10b), while the other removed one of the lateral lobes. The post-experiment synthetic prostate model can be seen in Fig. 11. After a short learning curve in which the surgeons focused primarily on the laser arm, surgeons began to use coordinated movement of both arms, increasingly relying on retraction from the retraction arm. The retraction arm was used to expose desired targets within tissue and the laser arm to cut the exposed surface with the laser.

4 Conclusions

Several important experimental insights have emerged from this series of experiments. First, we have shown that without tissue deformation our robotic system enables the surgeon to reach significantly more of the desired resection surface than is possible using a standard endoscope, regardless of the control mapping (see Fig. 7). These results indicate that our system may be able to reduce the amount of tissue deformation required in the procedure (both prostate tissue and surrounding tissues). If so, this would reduce the physical demands of the procedure on the surgeon, and reduce the technical challenge of the procedure associated with simultaneously deforming the prostate and aiming the laser with the same device.

We also sought to compare joint space control with task space control in terms of tracking accuracy. The primary motivation for this was the surgeon's initial inclination toward joint control due to their familiarity inserting other tools through endoscope ports manually, combined with their reluctance to see the robot move rapidly near singularities. Our primary observations were:

(1) With free endoscope movements, surgeons are likely to be able to complete the procedure, regardless of control mapping. This makes sense, since highly skilled surgeons can complete the procedure (albeit with significant difficulty) with a straight endoscope and no laser articulation. With free endoscope manipulation, joint and task space control yielded similar tip accuracy, with task space control enabling more rapid completion of the experiment.
(2) When surgeons were required to use the concentric tube manipulators exclusively (endoscope pose fixed), surgeons were able to follow complex curves accurately using task space control, but were unable to do so (or even complete the experiment) with joint space control.

These results lead us to believe that surgeons tend to primarily use endoscope motion rather than concentric tube robot motion when performing the task with free endoscope motion while using joint space control. Anecdotally, surgeons also found that their comfort level with task space control increased rapidly through performing these experiments, and they observed that robot motions, while sometimes fast, were predictable.

The prostate phantom resection experiment illustrated our robotic system functioning in a realistic model of clinical conditions. It also showed the surgeons could simultaneously coordinate many degrees of freedom, including controlling the pose of the endoscope while simultaneously using two concentric tube manipulators. We also observed that the learning curve for doing so was short for surgeons experienced in endoscopic surgery. Qualitatively, we observed that coordinated use of two arms made the procedure significantly easier, since one arm could be used to retract tissue and expose sites for the other to cut with the laser.

There remain many ways to improve the system in the future. Perhaps most importantly, surgeons observed that they would like to work as close as possible to the endoscope tip with the concentric tube manipulators. We explored this problem in [16], where we showed that it is possible to significantly expand the portion of the endoscope view that can be accessed by the concentric tube robots by enabling them to emerge and begin to curve a short distance behind the tip of the endoscope. We intend to implement this change and continue to iteratively refine the user interface and robotic hardware as we move forward toward cadaver experiments.

Acknowledgments This work was funded in part by the National Science Foundation (NSF) under IIS-105433, in part by the National Institutes of Health (NIH) under R01 EB017467, and in part by the Vanderbilt Initiative in Surgery and Engineering. The content is solely the responsibility of the authors and does not necessarily represent the official views of the NSF or NIH.

References

1. Davies, B.L., Hibberd, R.D., Ng, W.S., Timoney, A.G., Wickham, J.E.A.: The development of a surgeon robot for prostatectomies. Proc. Inst. Mech. Eng Part H: J. Eng. Med. **205**, 35–38 (1991)
2. Davies, B.: Medical robotics—a bright future. Lancet **368**, S53–S54 (2006)
3. Ho, G., Ng, W.S., Teo, M.Y., Kwoh, C.K., Cheng, W.S.C.: Experimental study of transurethral robotic laser resection of the prostate using the LaserTrode lightguide. J. Biomed. Opt. 244–251 (2001)
4. de Badajoz, E.S., Garrido, A.J., Vacas, F.G., Martinez, V.F.M., de Gabriel, J.G., Lozano, J.F., Cerezo, A.G.: New master arm for transurethral resection with a robot. Archivos Espanoles de Urologia **55**, 1.247–1.250 (1998)
5. Hashimoto, R., Kim, D., Hata, N., Dohi, T.: A tubular organ resection manipulator for transurethral resection of the prostate. In: IEEE/RSJ International Conference on Intelligent Robots and Systems, pp. 3954–3959 (2004)
6. Goldman, R.E., Bajo, A., MacLachlan, L.S., Pickens, R., Herrell, S.D., Simaan, N.: Design and performance evaluation of a minimally invasive telerobotic platform for transurethral surveillance and intervention. IEEE Trans. Biomed. Eng. **60**(4), 918–925 (2013)
7. Hendrick, R.J., Herrell, S.D., Webster III, R.J.: A Multi-arm hand-held robotic system for transurethral laser prostate surgery. In: IEEE International Conference on Robotics and Automation, pp. 2850–2855 (2014)
8. Ahyai, S.A., Lehrich, K., Kuntz, R.M.: Holmium laser enucleation versus transurethral resection of the prostate: 3-Year follow-up results of a randomized clinical trial. Eur. Urol. **52**(5), 1456–1463 (2007)
9. van Rij, S.,Gilling, P.J.: In 2013 Holmium laser enucleation of the prostate (HoLEP) may be the new 'Gold Standard'. Curr. Urol. Rep. **13**, 427–432 (2012)
10. Lingeman, J.E.: Holmium laser enucleation of the prostate-if not now, when? J. Urol. **186**(5), 1762–1763 (2011)
11. Webster III, R.J., Okamura, A., Cowan, N.J.: Toward active cannulas: miniature snake-like surgical robots. In: IEEE/RSJ International Conference on Intelligent Robots and Systems, pp. 2857–2863 (2006)
12. Sears, P., Dupont, P.: A Steerable needle technology using curved concentric tubes. In: IEEE/RSJ International Conference on Intelligent Robots and Systems, pp. 2850–2856 (2006)
13. Rucker, D.C., Jones, B.A., Webster III, R.J.: A geometrically exact model for externally loaded concentric-tube continuum robots. IEEE Trans. Robot. **26**(5), 769–780 (2010)
14. Dupont, P.E., Lock, J., Itkowitz, B., Butler, E.: Design and control of concentric-tube robots. IEEE Trans. Robot. **26**(2), 209–225 (2010)

15. Leenstra, J.L., Davis, B.J., Wilson, T.M., Mynderse, L.A., Herman, M.G., Hillman, D., Allen, K., Cheville, J., Holmes, D., King, B.: Prostate dimensions and volume in 700 patients undergoing primary surgical or radiotherapeutic management of localized adenocarcinoma: implications for design of minimally invasive prostate cancer devices. Int. J. Radiat. Oncol. **69**(3), S380–S381 (2007)
16. Hendrick, R.J., Mitchell, C.R., Herrell, S.D., Webster III, R.J.: Concentric tube robots for transurethral prostate surgery: matching the workspace to the endoscopic field of view. In: The Hamlyn Symposium on Medical Robotics (2014)

Using Haptic fMRI to Enable Interactive Motor Neuroimaging Experiments

Samir Menon, Hari Ganti and Oussama Khatib

Abstract Combining haptics with functional magnetic resonance imaging (Haptic fMRI) has enabled complex motor neuroimaging experiments that non-invasively map real-world motor tasks on to the human brain. The technique's resolution, fidelity and susceptibility to scanning artifacts, however, have not yet been estimated in a quantitative manner. Here, we demonstrate that unconstrained three degree-of-freedom Haptic fMRI experiments can reliably activate brain regions involved in planning, motor control, haptic perception, and vision. We show that associated neural measurements are reliable, heterogeneous at the millimeter scale, and free from measurable artifacts, and that their anatomical localization is consistent with past neuroscience experiments. In addition, we demonstrate the feasibility of using electromagnetic actuation in Haptic fMRI interfaces to apply high fidelity open-loop three-axis haptic forces (0.5–2N; square or 0.1–65Hz sine waveforms) while maintaining negligible temporal noise in pre-motor, motor, somatosensory, and visual cortex (<1 % of signal). Our results show that Haptic fMRI is a robust and reliable technique for characterizing the human brain's motor controller.

Keywords Haptic fMRI · MRI-compatible robotics · Motor control · Neuroscience · Neuroimaging

1 Introduction

Human motor neuroimaging experiments presently involve either uncontrolled limb motions or simple visuo-motor tasks within a constrained workspace, which has limited our understanding of motion and force control in the brain. Haptic fMRI promises to overcome this limitation by imaging the brain with fMRI [1, 2] while

S. Menon (✉) · H. Ganti · O. Khatib
A.I. Lab, Stanford University, Stanford, CA 94305, USA
e-mail: smenon@stanford.edu

H. Ganti
e-mail: hganti@stanford.edu

O. Khatib
e-mail: ok@cs.stanford.edu

M.A. Hsieh et al. (eds.), *Experimental Robotics*, Springer Tracts in Advanced Robotics 109, DOI 10.1007/978-3-319-23778-7_7

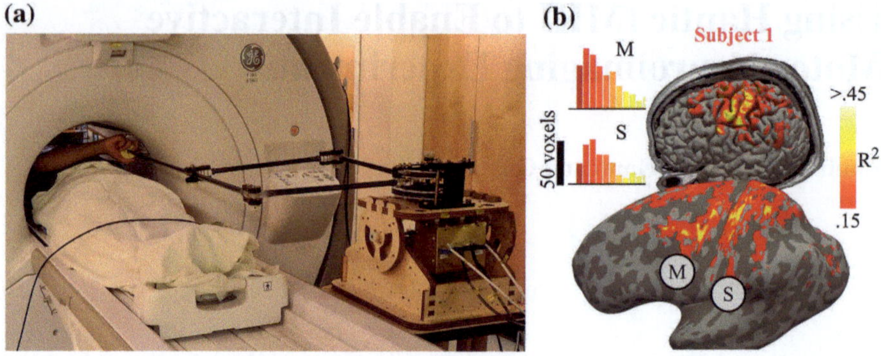

Fig. 1 Mapping motor responses in cortex: **a** HFI, a three degree-of-freedom fMRI-compatible haptic interface. **b** Reliable responses in motor (*M*) and somatosensory (*S*) cortex during a task that involves motor planning, reaching, and feedback trajectory control. Voxels with >15 % variance explained ($R^2 > 0.15$; see Appendix) are shown on the brain's surface and on an inflated mesh. Distribution histograms for motor and somatosensory cortex are inset

subjects simultaneously perform unconstrained three dimensional tasks in a virtual haptic simulation environment [3]. Engineering fMRI-compatible haptic interfaces for neuroscience experiments, however, requires achieving high-fidelity force control, backdrivability, natural operator-motion statistics [4], and uniform inertial properties [5] across a large three-dimensional workspace. Moreover, the devices must operate in a robust manner in large magnetic fields, elicit reliable subject motions and neural activation, and avoid injecting noise into fMRI measurements.

In this paper, we demonstrate that Haptic fMRI experiments elicit reliable neural activation for motor planning, unconstrained three-dimensional motions, and visual and force perception (Fig. 1). All neural activation patterns are localized to expected anatomical regions based on past experiments [6–10]. Our results use two experiments that independently contrast either planning and motion, or vision and force perception. To ensure accurate force rendering, we demonstrate that our electromagnetically actuated haptic interface can achieve high fidelity open-loop three-axis force control inside the MRI scanner while limiting performance decreases (<2–3 %) to levels below human perception [11]. Finally, we show that our experiments achieve temporal noise levels similar to the scanner's baseline (median noise-to-signal = 0.85 %) in pre-motor, motor, somatosensory, parietal, and visual cortex. Combined with past results that demonstrate HFI's haptic transparency, high fidelity motion monitoring, uniform inertial properties, reliable subject motions, and lack of confounding artifacts [12–14], this paper establishes Haptic fMRI as an effective motor neuroscience technique for three dimensional manipulation tasks.

2 Related Work and Technical Approach

Past research on fMRI-compatible haptic interfaces for motor control experiments has primarily focused on avoiding electromagnetic actuation, which simplified fMRI compatibility but limited device transparency, backdrivability, or degrees-of-freedom. MRI-compatible actuation mechanisms developed in the past include electro-active polymers [15], pneumatics [16, 17], hydraulics [18, 19], and cables driven by remote actuators [20]. An alternative approach modified an electromagnetically actuated PHANTOM device [21, 22] for MRI by using RF shields and a long carbon fiber extension rod attached to a linear rail (to improve stiffness); the device, however, can not span the entire MRI workspace.

To overcome the limitations of earlier designs, we designed and built an fMRI-compatible haptic interface, HFI (see Fig. 1a), with three degrees-of-freedom that span the entire MRI workspace (also see [12]). We used electromagnetic actuators for high frequency force control, low gear ratios (20 in-plane, 30 vertical axis) to ensure backdrivability, lightweight composites to ensure haptic transparency, and a mechanical structure that places actuators beyond the 400 Gauss static-field to limit magnetic interference in the motors. Furthermore, we mitigated eddy currents in motors induced by time-varying magnetic fields during fMRI scans by limiting motor displacement to <2 cm. The device's low and uniform inertia ($x = 0.81$, $y = 0.22$, $z = 0.27$ kg; condition number 3.81) help minimize friction (0.01–0.3 N), make it transparent, and elicit stereotypical motion patterns across subjects [13].

HFI's ability to support Haptic fMRI experiments that involve unconstrained arm motions promises to enable complex motor studies capable of testing how generative motor coordination models [23–25] map on to the human brain. Moreover, allowing non-invasive human motor neuroscience studies promises to help cross-validate decades of past animal neuroscience experiments, which have studied how the brain represents forces [26], movement direction [27–30], spatial gradients [31, 32], muscular activation [33–35], arm orientation [34, 36], limb joints [37], reach distance [38], motor planning [39], and hand position, velocity and acceleration [40].

3 Experimental Results

We focused on testing whether Haptic fMRI experiments conducted with our haptic interface, HFI, could enable controlled neuroscience experiments that localize neural activation patterns for planning, motion, vision, and force perception to expected brain regions. Our goal was to demonstrate that neural activation patterns obtained with Haptic fMRI match past research. This is important because the technique's novelty and reliance on indirect fMRI neural measurements (blood oxygenation response; see Appendix for details) have raised questions about the limits of spatio-temporal resolution and the potential for inducing complex noise patterns. Moreover, past attempts to use electromagnetic actuation with fMRI have resulted in unacceptable noise [18]—potentially because of inadequate radiofrequency shielding or unsuitable scanning hardware and protocols. In contrast, our earlier experiments found that HFI's shields limited noise due to electromagnetic actuation during uncontrolled haptic experiments [12]. Our past work, however, did not test the shields' effectiveness while applying arbitrary force magnitudes, directions, or frequencies. It also did not test HFI's ability to generate forces at the end-effector while operating in the MRI scanner room's large magnetic fields.

3.1 Neural Activation for Planning and Movement

We demonstrated HFI's ability to elicit reliable neural activation in the brain using a motor task that involves planning and reaching to different locations of space, followed by visuo-motor trajectory tracking in two directions centered around each reach destination (Figs. 1b, and 2; see video for protocol). The experiment's goal was to simultaneously activate as many motor-related regions as possible and test whether the resulting anatomical localization matched past research findings (see [6] for a review). As expected, the experiment evoked large reliable neural activation in

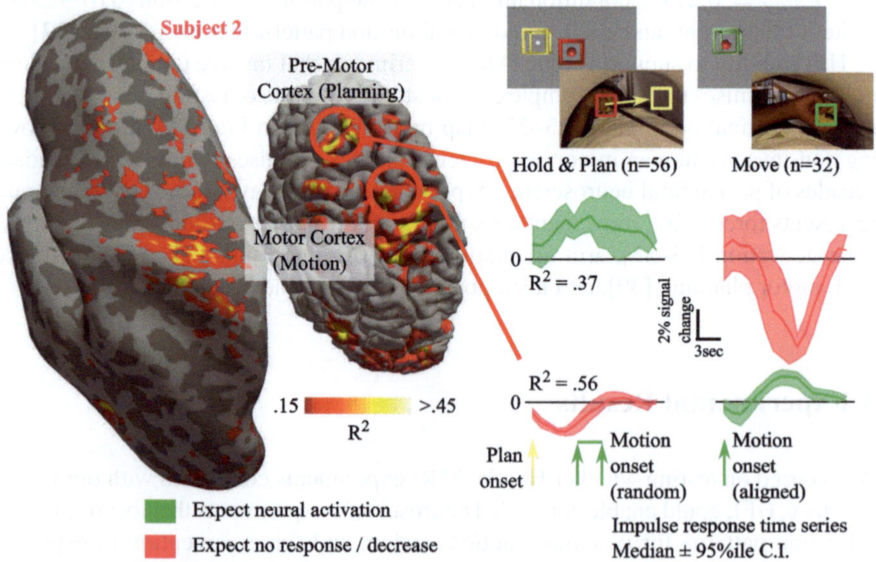

Fig. 2 Neural activation for planning and movement: Motion planning and reaching elicit reliable responses in pre-motor and motor cortex respectively. An impulse response model helped segregate neural activation during planning and movement into separate time series components (see Appendix for details). Time series fits are shown for two exemplar voxels for the left reaching condition; responses to the *center* and *right* conditions were similar. Heterogeneous time series in voxels that are a few centimeters apart indicate that neural activation signals dominate any potential motion-induced low spatial-frequency artifacts

primary motor and somatosensory cortex, which have been anatomically connected to low-level motor control and sensory perception. We also found activation in pre-motor, supplementary motor, and parietal cortex, which are involved in higher-level motor and visual processing.

In addition to mapping neural activation reliability across the brain's regions, we used a finite impulse response model (see Appendix for details) to estimate time series responses in individual $2.5 \times 2.5 \times 2.5\,\mathrm{mm}^3$ voxels for each task condition (see Fig. 2). As expected, neural activation in the pre-motor region increased during the motor planning period before actual movement, and decreased after movement onset. In contrast, primary motor cortex voxels typically showed a decrease in activation during planning and an increase during movement. Movement-related activation for visually-guided trajectory tracking (not shown) were also localized to the motor regions and were associated with a longer time series activation than reaching, as expected (trajectory tracking time was 8 s; reaching was 5 s).

This experiment demonstrates that Haptic fMRI's spatial resolution can reliably segregate time series responses across brain regions that are a few centimeters apart. To obtain our results, we avoided spatial smoothing as it reduces noise but correlates nearby neural activation patterns. As an added advantage, our results also implicitly indicate the lack of motion or haptic artifacts, which create low spatial-frequency task-correlated noise and can make time series responses and variability look similar across nearby brain regions. We noted that the number of reliable voxels in high-level cortical regions was lower than in the low-level regions for subject two when compared to subject one. As such, we plotted time series estimates for the less reliable subject.

3.2 *Neural Activation for Vision and Haptic Perception*

Having tested neural activation for planning and movement, we proceeded to determine whether we could localize activation for visual and force perception in visual and somatosensory cortex, respectively. We did so using an experiment design that exposed subjects to a high contrast checkerboard visual stimulus, or a haptic force perception stimulus (see Appendix for details). Both stimuli were designed to be simple, and were each expected to reliably activate their own corresponding (low-level) sensory region without activating the other's. As such, monitoring changes in visual cortex activation while our haptic interface applied forces would reveal any radiofrequency noise due to motor operation.

Our experiment elicited reliable neural activation in visual, somatosensory, and motor cortex (Fig. 3). While we found little pre-motor activation (there was no planning task), we did find activation in parietal (between vision and motor) and supplementary (inner brain) cortex, which have been associated with object perception and motor control, respectively.

As expected, somatosensory cortex did not respond to the visual stimulus, and early visual cortex did not respond to the force stimulus. In addition, the presence

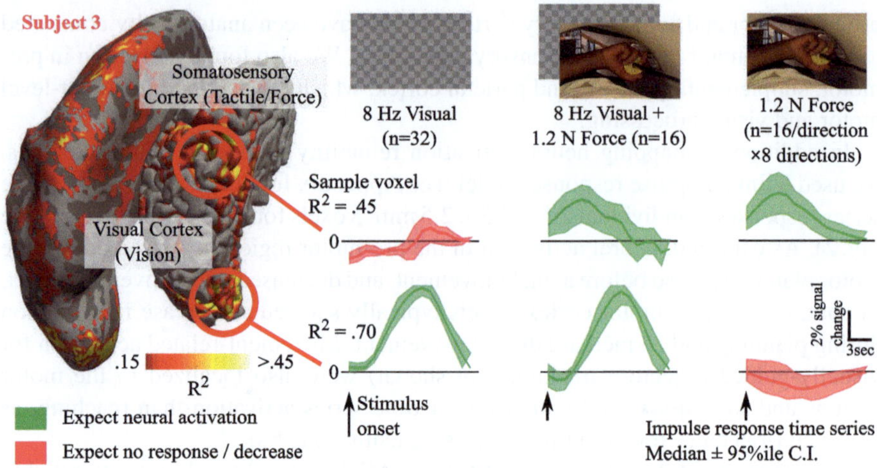

Fig. 3 Neural activation for visual and tactile perception: Vision and haptic-force perception elicit reliable responses in visual and somatosensory cortex, respectively. Impulse response time series for exemplar voxels are similar for simultaneous vision and force stimuli, as well as for each individual stimulus. The lack of spurious activation in visual regions during the force-only stimulus indicate that neural signals dominate any potential low temporal-frequency task-correlated noise due to HFI's electromagnetic actuation

of reliable activation in visual cortex (high R^2; see exemplar voxel's time series) indicates low radiofrequency interference when motors actively apply haptic forces.

It is noteworthy that visual cortex responses to the checkerboard are more reliable than somatosensory cortex responses to force perception. There are a few reasons for this. First, the high-contrast checkerboard is a well tested visual stimulus that is known to strongly activate early visual cortex [41]. Since no past fMRI studies have estimated time series activation for multi-axis force perception while the arm is unconstrained, our experiment was likely suboptimal and may be greatly improved in the future. Second, our force perception stimulus was oriented along one of eight random directions (see Appendix for details). We did so to probe visual cortex while actuating multiple motors in HFI with a variety of currents. While this increased the chance that HFI would generate interfering radiofrequency noise, it also led to less-reliable neural activation patterns for force perception.

3.3 Force Generation in the MRI Scanner

Having elicited expected neural activation for eight force directions in our perception experiment, we proceeded to test HFI's ability to apply forces in any direction (Fig. 4). Our force sensing rig, however, was not MRI-compatible and had to be placed at the 10 Gauss line, far away from the MRI machine. As such, we placed HFI at the 400 Gauss magnetic field line—its typical position—but inverted its direction to point

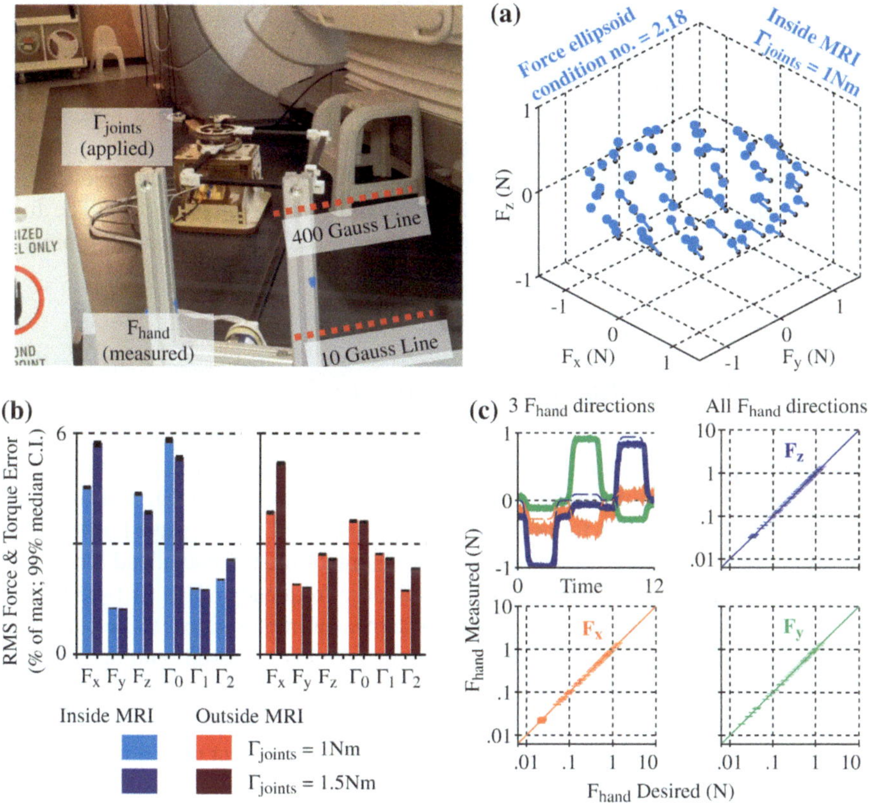

Fig. 4 Force generation in an MRI scanner: HFI reliably produced desired forces at the end-effector while operating at an MRI scanner's 400 Gauss line—its typical position during Haptic fMRI experiments. **a** Measured end-effector forces (*light blue*) closely track desired forces (*black*). Directions were uniformly sampled on a sphere in joint-torque space. **b** Root mean square force tracking errors for end-effector forces and joint torques for two applied joint torque levels (1, 1.5 nm; *light* and *dark bars*) shown for measurements made inside (*blue*) and outside (*red*) the MRI scan room. Noise levels increase marginally in the MRI scan room. **c** Raw end-effector force time series shown for three exemplar directions (*top-left*). Measured forces (*solid lines*) closely track desired forces (*dashed*). Breaking forces along all directions into three individual axes (x:bl, y:br, z:tr) demonstrates reliable force generation across three log decades. Error-bars are *vertical* (and small)

away from the scanner bore instead of into it. We then attached the end-effector to the force sensor and directly measured generated forces.

To test HFI's force generation, we uniformly sampled joint torque vectors on a unit sphere, applied each torque vector sequentially to the motors, and measured the resulting end-effector forces. HFI's kinematics transform the (unit) joint torque sphere to an end-effector force ellipsoid characterized by the Jacobian's eigenvectors. Our force measurements (see Fig. 4a) matched the theoretical Jacobian [12], and show that HFI's force generation is close to isotropic (condition number, 2.18).

The applied forces and torques have a negligible root mean square error (0.06–0.3 N; see Fig. 4b), which is below the human detection threshold [11]. Moreover, the force and torque errors are similar inside and outside the MRI machine. This indicates that HFI's motors are sufficiently far from the MRI machine and their performance is not unaffected by magnetic interference.

Finally, the measured end-effector forces match theoretical predictions over three decades on a log scale, indicating a high dynamic range (see Fig. 4c). The force generation becomes less reliable at very low forces (<0.08 N) due to a combination of sensor noise and device friction.

3.4 Noise During Haptic Force Transmission

Our final experiment quantitatively characterized the temporal noise in our fMRI measurements for the planning and moving tasks, as well as visual and force perception (Fig. 5). Since our force perception task involved square wave forces, which might not represent all possible haptic force interactions, we also added a noise test where HFI applied sine wave forces across a large frequency range (0.1–65 Hz). As a reference, we also compared temporal noise measurements across these conditions against noise levels with HFI's radiofrequency shields removed, and against MRI scanner baseline noise levels with no device and passive subjects.

We found limited noise in the haptic planning and movement task (Fig. 5a) across our regions of interest in the cortex. Temporal noise levels increased marginally in visual cortex for the vision and force perception task, potentially due to a random combination of higher head motion (up to 1.25 mm), scanner calibration drift, or interfering motor radiofrequency noise (Fig. 5b). This, however, was one of the worst runs of the experiment; there are usually scanner-calibration dependent changes across runs. The noise levels were much lower in other runs (see a second run in Fig. 5c1; head motion <0.4 mm). We did note a large systematic increase in noise near the center of the brain, which could be attributed to partial volume artifacts caused by a proximity to ventricles and other non-brain regions. Moreover, we used a 32 channel head coil, which surrounds the head and provides higher signal-to-noise near the surface of the skull.

Our final noise testing condition involved continuously applying a mixture of sine wave forces with HFI. While potentially unrealistic—applying forces non-stop will limit any statistical analysis—it does quantify worst-case noise patterns (Fig. 5c2). Temporal noise levels even in this extreme case were not significantly different from baseline, and are likely to be dominated by common fMRI artifacts related to head- and hand-motion, partial voluming, poor scanner calibration, or very high spatial resolution scanning. With shields removed, however, HFI's radiofrequency interference reaches unacceptable levels (Fig. 5c3).

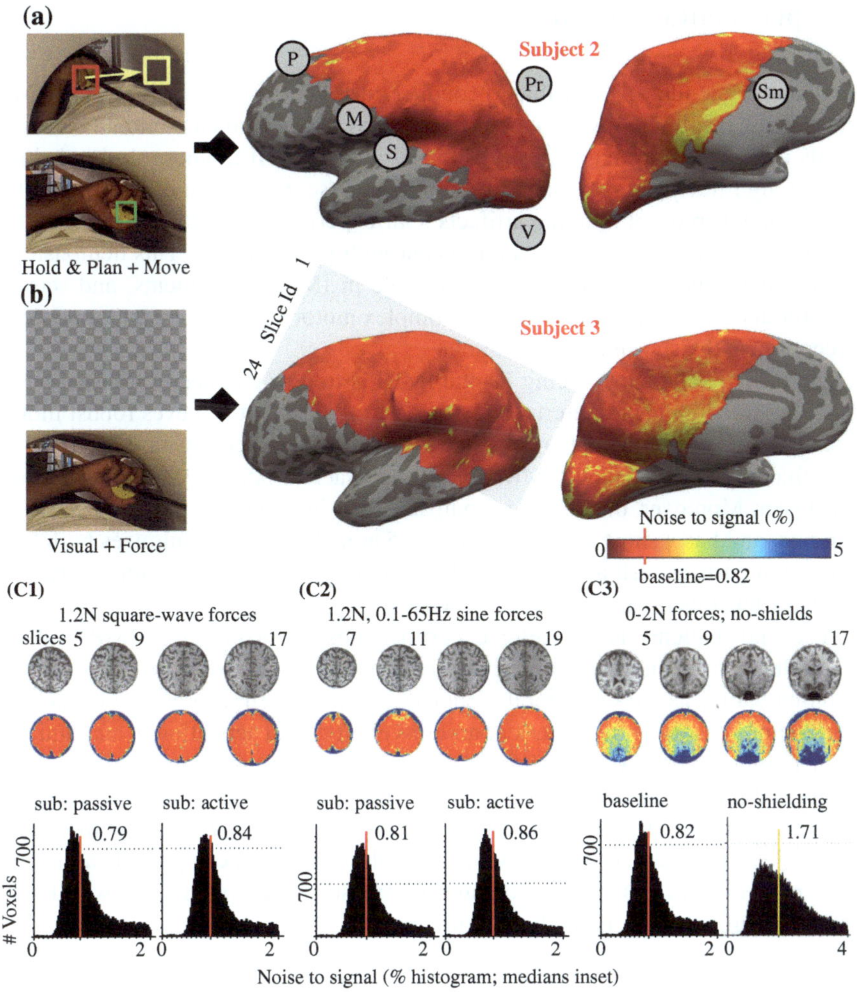

Fig. 5 Low noise during HFI operation: Temporal noise (high-frequency) for brain regions measured as a percent of the median time series signal. **a** fMRI measurements for the planning and movement task had little noise (<1 %) in pre-motor (*P*), motor (*M*), somatosensory (*S*), visual (*V*), parietal (*Pr*), and supplementary motor (*Sm*) cortex. **b** The force perception task induced greater temporal noise in the inner brain regions (slices match C.1). **c** Temporal noise histograms for three task types while the subject actively operated HFI or was passive (an operator held HFI). **c.1** Brain slices from a second run of the vision and force perception task. **c.2** HFI applied sine wave forces of 0.1–35 Hz at the end-effector instead of square waves. **c.3** Subjects randomly interacted with a haptic simulation with HFI's radiofrequency shields removed. Note the dramatic increase in temporal noise (*bottom right*)

4 Experimental Insights

We demonstrate that it is feasible to conduct three degree-of-freedom Haptic fMRI experiments involving force perception as well as unconstrained motion. Our experiments used an electromagnetically actuated interface, HFI, whose design helps it achieve high fidelity force generation in the MRI scan room, and whose radiofrequency shields prevent imaging artifacts. Our experiments used simple stimuli and elicited neural activation patterns consistent with past research. This demonstrates that we avoid the numerous artifacts possible in fMRI experiments, and sets the stage for detailed haptic studies to map complex motor coordination patterns, tactile perception, and visuo-motor integration.

The primary challenge facing Haptic fMRI is to now engineer a transparent six degree-of-freedom haptic interface that is cost-effective yet achieves robust motion tracking and force generation over the long-term. For instance, we have used HFI for more than eighty-five Haptic fMRI scans spread across sixteen sessions over more than 1 year. Moreover, the device is compact, costs less than ten thousand dollars to fabricate, and has a short setup time of 10–15 min. Thus, HFI minimizes both fixed fabrication costs as well as recurring MRI scanning costs. A six degree-of-freedom successor must meet or exceed HFI's metrics.

A second challenge is to develop experiment designs that leverage fMRI's ability to simultaneously image multiple brain regions at a high spatial resolution, while accommodating its slow temporal responses. Ideal experiments would ensure that overlapping sensory or motor task conditions elicit neural activation in anatomically distinct brain regions. Achieving this, however, requires improving upon past experiments that predict a fractured somatotopic organization where functionally related limbs are anatomically co-localized [7, 42]. Whole body control frameworks [24, 25] applied to subject-specific musculoskeletal models [43] can provide a theoretical basis to explain this complex organization; testing their predictions with Haptic fMRI experiments is an immediate future goal.

Appendix

MRI Protocol

All fMRI scans were conducted at Stanford University's Center for Cognitive and Neurobiological Imaging on a GE Discovery MR750 3T MRI scanner, with a thirty-two channel Nova Medical head coil. The scan protocol was gradient echo EPI with a 16 cm field of view sampled at a 64×64 resolution ($2.5 \times 2.5 \times 2.5\,mm^3$ voxels), a 1.57 s repetition time, a 28 ms echo time, and a 72° flip angle. Each scan run was preceded by 2nd-order polynomial shimming and was sandwiched by spiral fieldmap scans ($2.5 \times 2.5 \times 5\,mm^3$ voxels). Fieldmap scans were conducted within 10 s of each scan run's start and end. After scanning, the fMRI images were slice time corrected, motion corrected (SPM [44]), spatially undistorted using fieldmaps, and analyzed to compute temporal noise-to-signal.

fMRI Analysis

Temporal noise-to-signal computations used the median neural response distribution obtained by regressing out a line from each voxel's time series, computing the absolute value of the difference between successive time points, computing the median of these absolute differences, dividing the result by the mean of the original time series, and then multiplying by 100. Cortex segmentation used Freesurfer's Desikan-Killiany atlas [45]. Surface registration was done using Freesurfer, and all surface images were plotted using Freeview. Freeview smoothed the surface plots while rendering (2 steps).

Estimating fMRI Impulse Response Time Series and R^2

fMRI measures changes in blood oxygenation induced by neural metabolic activity [1, 2], which have a slower time course than neural computation and persist long after sensory stimuli and motor tasks terminate. Such persistent responses cause raw fMRI measurements to overlap in experiments where consecutive task conditions are not be separated by large time-intervals. Separating task conditions by large time-intervals, however, makes fMRI runs very long, which can induce a variety of unwanted artifacts related to MRI scanner calibration drift, neural adaptation, or subject attention lapses, microsleep and exhaustion. Instead, we optimized our experiments to ensure reliable motor task execution [13], which caused fMRI measurements for different task conditions to overlap.

We segregated neural activation for individual tasks using a finite impulse response (FIR) model (implemented using GLMdenoise [46]). The FIR model works by associating each task type with a unique time course and segregates time courses while assuming that overlapping responses sum linearly. fMRI signal linearity, however, is an active area of research [1, 41, 47]. As such, we randomized inter-task delays and randomly ordered tasks, which made the model's time series match anatomical expectations based on past research (see Figs. 2 and 3; read [6] for an overview). When tasks were closely spaced in time, as with planning and motion, this method was noisy. The parts of planning that overlap with motion are thus less reliable and the confidence interval for the planning time series estimates is larger after motion starts (but still above zero; see Fig. 1).

We computed 95 % confidence intervals by bootstrapping [48] runs (400 bootstraps), fitting FIR models to each, and taking the median percentile estimates across the estimated bootstrap time series. Finally, we computed R^2 values for each voxel by comparing the time series variance with the variance after regressing out median FIR model estimates.

Data Collection Protocols

See [13] for precise specifications of the motion protocol. Subjects executed one practice run inside the MRI scanner, and then executed at least eight scan runs (S1, 10; S2, 8). Each run was 630 s long.

The force and visual perception experiment protocol involved fixed duration stimuli instances with visual, motor, or visual and motor sensory input. The experiment was divided into runs, and each run was divided into blocks. During each block, the subject started with their hand at rest. Next, they were instructed to move their hand into free space. After a randomized delay period of 3–5 s, the subject experienced two randomly selected stimuli instances. Each stimulus instance was 3–5 s long and was separated from the other by a randomized delay 3–5 s. Finally, the subject was required to rest their hands for a random time interval (4–20 s), and then restarted the process. The subject executed four scan runs with multiple blocks. Each run was 459 s long.

Force magnitudes were set to evenly spaced directions along the x-y plane, with a magnitude of 1.2 N. The force vectors used were (1.2, 0.0), (0.0, 1.2), (−1.2, 0.0), (0.0, −1.2), (0.85, 0.0), (0.0, 0.85), (−0.85, 0.0), and (0.0, −0.85).

Haptic and Force Measurement Details

Haptic experiments were conducted with Haptic fMRI Interface [12], a three degree-of-freedom fMRI-compatible device. All motions were right handed, and the haptic control rate was 350 Hz. The reaching task spanned the entire workspace (see [13] for more details), but avoided arm motion artifacts [14].

Visual stimuli were displayed on a 30 in. diagonal (76 cm, 16 : 10 aspect ratio) flat panel display custom built by Resonance Technology. Subjects viewed visual stimuli through a dual-mirror setup. The visual distance from screen to mirror-2 is 184.4 cm, from mirror-2 to mirror-1 is 6 cm, and from mirror-1 to the eye is about 15 cm, for a total viewing distance of about 205 cm. The visual field of view is about 30°, making each visual checkerboard square span about one and a half degrees of the visual field. The display has a native resolution of 2560 × 1600 but stimuli were displayed at 1280 × 800. The display has a 7 ms temporal response, and 10-bit color rendering. The maximum luminance of the display is 329 cd/m^2 (red is 88, green is 117, and blue is 124 cd/m^2).

Forces were measured using a JR3 85M35A-U560 63N4S force sensor. The raw sensor data was sampled at 1 KHz, resampled to match HFI's control rate, and was finally filtered using a 75 Hz low pass filter to remove high frequency sensor noise.

Human Subjects

Subjects were healthy right-handed males with no history of motor disorders: S1, 29y, 185lb, 5′9″; S2, 19y, 170lb, 6′2″; S3, 21y, 160lb, 5′8″. Informed consent was obtained in advance on a protocol approved by the Institutional Review Board (IRB) at Stanford University.

References

1. Logothetis, N.K., Wandell, B.A.: Interpreting the bold signal. Annu. Rev. Physiol. **66**, 735–769 (2004)
2. Logothetis, N.K.: What we can do and what we cannot do with fMRI. Nature **453**, 869–878 (2008)
3. Hale, K.S., Stanney, K.M.: Deriving haptic design guidelines from human physiological, psychophysical, and neurological foundations. IEEE Comput. Graph. Appl. **24**, 33–39 (2004)
4. Kostic, M., Popovic, D., Popovic, M.: Influence of planar manipulandum to the hand trajectory during point to point movement. In: IEEE International Conference on Rehabilitation Robotics, pp. 1–4 (2011)
5. Khatib, O.: Inertial properties in robotic manipulation: an object-level framework. Int. J. Robot. Res. **14**, 19–36 (1995)
6. Rosenbaum, D.: Human Motor Control. Academic Press, San Diego (2009)
7. Meier, J.D., Aflalo, T.N., Kastner, S., Graziano, M.S.A.: Complex organization of human primary motor cortex: a high-resolution fMRI study. J. Neurophysiol. **100**, 1800–1812 (2008)
8. Churchland, M.M., Yu, B.M., Ryu, S.I., Santhanam, G., Shenoy, K.V.: Neural variability in premotor cortex provides a signature of motor preparation. J. Neurosci. **26**, 3697–3712 (2006)
9. Aflalo, T.N., Graziano, M.S.A.: Relationship between unconstrained arm movements and single-neuron firing in the macaque motor cortex. J. Neurosci. **27**, 2760–2780 (2007)
10. Kay, K.N., Winawer, J., Mezer, A., Wandell, B.A.: Compressive spatial summation in human visual cortex. J. Neurophysiol. **110**, 481–494 (2013)
11. Mesa-Munera, E., Ramirez-Salazar, J., Boulanger, P., Bischof, W.F., Branch, J.W.: Estimation of vibration and force stimulus thresholds for haptic guidance in MIS training. Rev. Ing. Biomed. **5**, 17–22 (2012)
12. Menon, S., Brantner, G., Aholt, C., Kay, K., Khatib, O.: Haptic fMRI: combining functional neuroimaging with haptics for studying the brain's motor control representation. In: Proceedings of the 13th Annual Conference of the IEEE Engineering in Medicine and Biology Society, pp. 4137–4142 (2013)
13. Menon, S., Yu, M., Kay, K., Khatib, O.: Haptic fMRI: accurately estimating neural responses in motor, pre-motor, and somatosensory cortex during complex motor tasks. In: Proceedings of the 14th Annual Conference of the IEEE Engineering in Medicine and Biology Society (2014)
14. Menon, S., Quigley, P., Yu, M., Khatib, O.: Haptic fMRI: using classification to quantify task-correlated noise during goal-directed reaching motions. In: Proceedings of the 14th Annual Conference of the IEEE Engineering in Medicine and Biology Society (2014)

15. Vogan, J., Wingert, A., Plante, J., Dubowsky, S., Hafez, M., Kacher, D.: Manipulation in MRI devices using electrostrictive polymer actuators: with an application to reconfigurable imaging coils. In: IEEE International Conference on Robotics and Automation, pp. 2498–2504 (2004)
16. Diedrichsen, J., Hashambhoy, Y., Rane, T., Shadmehr, R.: Neural correlates of reach errors. J. Neurosci. **25**, 9919–9931 (2005)
17. Menon, S., Stanley, A., Zhu, J., Okamura, A., Khatib, O.: Mapping stiffness perception in the brain with an fMRI-compatible particle-jamming haptic interface. In: Proceedings of the 14th Annual Conference of the IEEE Engineering in Medicine and Biology Society (2014)
18. Burdet, E., Gassert, R., Gowrishankar, G., Bleuler, H.: fMRI compatible haptic interfaces to investigate human motor control. Exp. Robot. IX **21**, 25–34 (2006)
19. Klare, S., Peer, A., Buss, M.: Development of a 3 DoF MR-compatible Haptic Interface for Pointing and Reaching Movements. Lecture Notes in Computer Science, vol. 6192, pp. 211–218. Springer, Berlin (2010)
20. Chapuis, D., Gassert, R., Gowrishankar, G., Burdet, E., Bleuler, H.: Investigation of a cable transmission for the actuation of MR compatible haptic interfaces. In: Proceedings of Biomedical Robotics and Biomechatronics, pp. 426–431 (2006)
21. Hribar, A., Koritnik, B., Munih, M.: Phantom haptic device upgrade for use in fMRI. Med. Biol. Eng. Comput. **47**, 677–684 (2009)
22. Massie, T., Salisbury, J.: The phantom haptic interface: a device for probing virtual objects. In: Proceedings of the ASME Winter Annual Meeting, Symposium on Haptic Interfaces for Virtual Environment and Teleoperator Systems, vol. 55(1), pp. 295–300. IOS Press (1994)
23. Khatib, O., Warren, J., Sapio, V.D., Sentis, L.: Human-like motion from physiologically-based potential energies. In: Advances in Robot Kinematics, pp. 149–163. Springer, The Netherlands (2004)
24. Sapio, V.D., Warren, J., Khatib, O., Delp, S.: Simulating the task-level control of human motion: a methodology and framework for implementation. Vis. Comput. **25**, 289–302 (2006)
25. Demircan, E., Besier, T., Menon, S., Khatib, O.: Human motion reconstruction and synthesis of human skills. In: Lenarčič, J., Stanisic, M. (eds.) Advances in Robot Kinematics, pp. 283–292. Springer, Berlin (2010)
26. Evarts, E.V.: Relation of pyramidal tract activity to force exerted during voluntary movement. J. Neurophysiol. **31**, 14–27 (1968)
27. Georgopoulos, A.P., Kalaska, J.F., Caminiti, R., Massey, J.T.: On the relations between the direction of two-dimensional arm movements and cell discharge in primate motor cortex. J. Neurosci. **2**, 1527–1537 (1982)
28. Georgopoulos, A.P., Schwartz, A.B., Kettner, R.E.: Neuronal population coding of movement direction. Science **233**, 1416–1419 (1986)
29. Georgopoulos, A.P., Ashe, J., Smyrnis, N., Taira, M.: The motor cortex and the coding of force. Science **256**, 1692–1695 (1992)
30. Schwartz, A.B., Kettner, R.E., Georgopoulos, A.P.: Primate motor cortex and free arm movements to visual targets in three-dimensional space. I. relations between single cell discharge and direction of movement. J. Neurosci. **8**, 2913–2927 (1988)
31. Georgopoulos, A.P., Caminiti, R., Kalaska, J.F.: Static spatial effects in motor cortex and area 5: quantitative relations in a two-dimensional space. Exp. Brain Res. **54**, 446–454 (1984)
32. Kettner, R.E., Schwartz, A.B., Georgopoulos, A.P.: Primate motor cortex and free arm movements to visual targets in three- dimensional space. iii. positional gradients and population coding of movement direction from various movement origins. J. Neurosci. **8**, 2938–2947 (1988)
33. Cheney, P.D., Fetz, E.E., Palmer, S.S.: Patterns of facilitation and suppression of antagonist forelimb muscles from motor cortex sites in the awake monkey. J. Neurophysiol. **53**, 805–820 (1985)
34. Kakei, S., Hoffman, D.S., Strick, P.L.: Muscle and movement representations in the primary motor cortex. Science **285**, 2136–2139 (1999)
35. Holdefer, R., Miller, L.: Primary motor cortical neurons encode functional muscle synergies. Exp. Brain Res. **146**, 233–243 (2002)

36. Caminiti, R., Johnson, P.B., Urbano, A.: Making arm movements within different parts of space: dynamic aspects in the primate motor cortex. J. Neurosci. **10**, 2039–2058 (1990)
37. Reina, G.A., Moran, D.W., Schwartz, A.B.: On the relationship between joint angular velocity and motor cortical discharge during reaching. J. Neurophysiol. **85**, 2576–2589 (2001)
38. Fu, Q.G., Suarez, J.I., Ebner, T.J.: Neuronal specification of direction and distance during reaching movements in the superior precentral premotor area and primary motor cortex of monkeys. J. Neurophysiol. **70**, 2097–2116 (1993)
39. Churchland, M.M., Santhanam, G., Shenoy, K.V.: Preparatory activity in premotor and motor cortex reflects the speed of the upcoming reach. J. Neurophysiol. **96**, 3130–3146 (2006)
40. Ashe, J., Georgopoulos, A.P.: Movement parameters and neural activity in motor cortex and area 5. Cereb. Cortex **4**, 590–600 (1994)
41. Boynton, G.M., Engel, S.A., Glover, G.H., Heeger, D.J.: Linear systems analysis of functional magnetic resonance imaging in human V1. J. Neurosci. **16**, 4207–4221 (1996)
42. Graziano, M.S.A., Aflalo, T.N.: Mapping behavioral repertoire onto the cortex. Neuron **56**, 239–51 (2007)
43. Delp, S.L., Anderson, F.C., Arnold, A.S.: OpenSim: open-source software to create and analyze dynamic simulations of movement. IEEE Trans. Biomed. Eng. **54**, 1940–1950 (2007)
44. Friston, K.J., Ashburner, J.T., Kiebel, S.J., Nichols, T.E., Penny, W.D.: Statistical Parametric Mapping: The Analysis of Functional Brain Images. Elsevier, London (2006)
45. Desikan, R.S., Killiany, R.J., et al.: An automated labeling system for subdividing the human cerebral cortex on MRI scans into gyral based regions of interest. Neuroimage **31**, 968–980 (2006)
46. Kay, K., Rokem, A., Winawer, J., Dougherty, R., Wandell, B.: GLMdenoise: a fast, automated technique for denoising task-based fMRI data. Front. Neurosci. **7** (2013)
47. Lin, A.L., Fox, P.T., Hardies, J., Duong, T.Q., Gao, J.H.: Nonlinear coupling between cerebral blood flow, oxygen consumption, and ATP production in human visual cortex. Proc. Natl. Acad. Sci. **107**, 8446–8451 (2010)
48. Efron, B., Tibshirani, R.: An Introduction to the Bootstrap, vol. 57. Chapman & Hall, New York (1994)

Dual Stage Options for Interface Designs Suitable for Haptic Interaction at the Micro-Nano Scales

Abdenbi Mohand Ousaid, Tianming Lu, Cécile Pacoret, Stéphane Régnier and Vincent Hayward

Abstract Direct, manual interaction with the micro/nano scales is not straightforward because the objects at this scale obey unituitive physics. For instance, in ambient conditions at the micro-scale, capillary forces dominate over many other forces. When the scale becomes smaller, Brownian motion becomes pervasive. Haptic interfaces give us the option to bring the experience of this physics with the direct reach of the human sensorimotor capabilities. To cope with the limitations of conventional force feedback devices, we present here two alternative dual-stage designs suitable to address the needs of the interaction with the micro/nano scales. The first one features very low apparent inertia, a large dynamics range and a wide bandwidth. This properties are obtained by coupling a large actuator to a small one via a viscous coupler. Feedback can then be used to achieve nearly perfect transparency. The second is a conventional force feedback device augmented with a tactile transducer. The two channels are frequency compensated to achieved a flat response from DC to 1 kHz.

Keywords Haptic interface · Dual-stage architecture · Pantograph · Transparency · Tactile transducer · Signal crossover

---

Contributed equally to this work.

A. Mohand Ousaid · T. Lu · C. Pacoret · S. Régnier · V. Hayward (✉)
Sorbonne Universités, UPMC University Paris 06, UMR 7222, ISIR,
75005 Paris, France
e-mail: hayward@upmc.fr

A. Mohand Ousaid
e-mail: mohand_ousaid@isir.upmc.fr

T. Lu
e-mail: lu@isir.upmc.fr

C. Pacoret
e-mail: pacoret@isir.upmc.fr

S. Régnier
e-mail: regnier@upmc.fr

M.A. Hsieh et al. (eds.), *Experimental Robotics*, Springer Tracts
in Advanced Robotics 109, DOI 10.1007/978-3-319-23778-7_8

1 Introduction

At the nano and micro scales, the mechanical behavior of objects no longer is dominated by gravity and friction. At short range, forces including electrostatic, capillary, and van der Waals forces are preeminent [1]. As a result, the physics of the micro/nano scales differs completely from that of the macro scale and are not accessible to humans sensory and motor capabilities.

Like a microscope magnifies viewed objects, haptic devices can be employed to magnify interaction between objects at a scale at which humans are naturally effective. Most haptic devices described in the past fail to match the human sensorimotor capacity and thus act as an obstacle between the hand and the phenomenon that could be accessed. Conventional interfaces are subjected to inherent friction and/or high inertia that affect the interface transparency, dynamic range, and bandwidth.

At the micro/nano scales, three issues arise in term of interface design [2]. Effective scaling systems must provide high transparency to convey the most of what can be sensed by humans and not taint it with noise; they must have a large dynamic range since human can motorically and sensorially deal with 4–5 orders of magnitude; and they must have wide bandwidth since the haptic capabilities cover from dc to about 1 kHz. Transparency thus quantifies a haptic interface's capacity to reproduce micro/nano interactions to a human operator. The dynamic range is related to the force resolution of the interface. It quantifies the ratio of the largest to the smallest forces that can be commanded, and thus sensed, through the interface. Bandwidth refers to the capacity of the interface to convey mechanical signals over a large frequency range, in order to render complex phenomena, such as Brownian motion.

In the past 25 years, several haptic interfaces have been proposed to deal with interaction with small objects. They enhance human interaction capabilities at the microscale using either electrodynamic transducers [3, 4], or single stage systems based on conventional robotic architectures, see [5], among several other examples. Such interfaces are subjected to inherent friction and high inertia that contribute spurious forces that mask the mechanical signals to be felt by the users. With single stage systems, sacrifices must be made regarding transparency, dynamic range, or the bandwidth in favor of other factors such as maximum force. To cope with these limitations, we describe here two alternative dual-stage haptic device designs (see Fig. 1) suitable to address the needs of interactions at the micro/nano scales.

2 Viscously Coupled Dual Stage Design

The first design, see Fig. 1a, focuses on transparency and dynamic range. It is based on a dual-stage architecture [6, 7]. The large, proximal motor provides power and the small, distal motor reproduces the force transients. The two stages are connected to each other by a passive viscous coupler based on Foucault (eddy) currents. Such

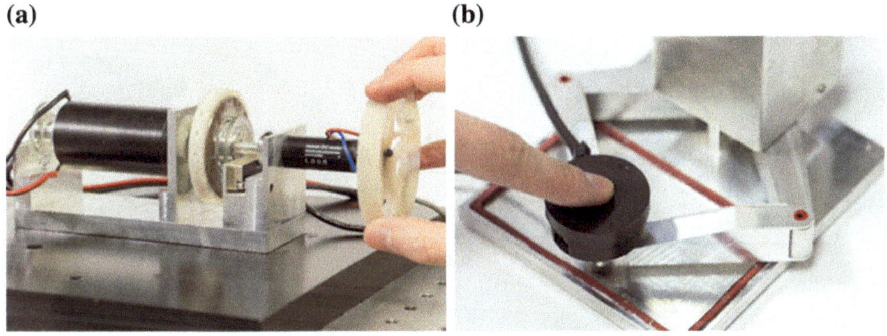

Fig. 1 **a** Dual-stage haptic interface with viscous coupling. **b** Direct drive haptic interface augmented with high-frequency tactile transducer

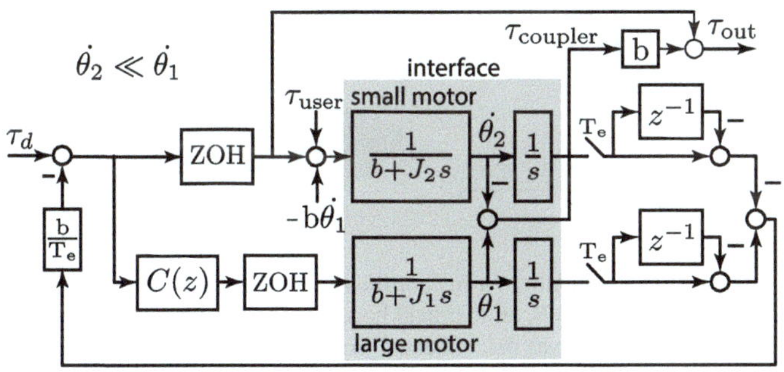

Fig. 2 The reference torque, τ_d, is compared to the torque produced by the coupler, τ_{coupler}. The large motor is enslaved by the compensator $C(z)$ to zero the torque error. The small motor compensates for the slower response of the large motor

coupler transforms slip velocity into torque with exacting accuracy and exhibits a linear behavior over a large velocity range. The viscous torque is proportional to the relative velocity between proximal and distal motor, $\tau_{\text{coupler}} = b(\dot{\theta}_1 - \dot{\theta}_2)$. The handle is driven by the sum of two torques, that of the coupler and that of the distal motor. A fundamental advantage is the possibility to decouple the output torque from the inertia of the motor. Referring to Fig. 2, a pole-placement controller, $C(z)$, designed by polynomial methods was shown to be able to achieve a near perfect transparency [7].

Figure 3a shows the response to a null reference torque. The torque felt by the operator, τ_{out}, is nearly zero despite rapid movements of the handle. The inertia of the proximal motor is entirely masked by the feedback control. Mechanically speaking, the power required to move the large motor is entirely supplied by the power amplifier and not by the operator. Only the inertia of the small motor is felt (Fig. 3b).

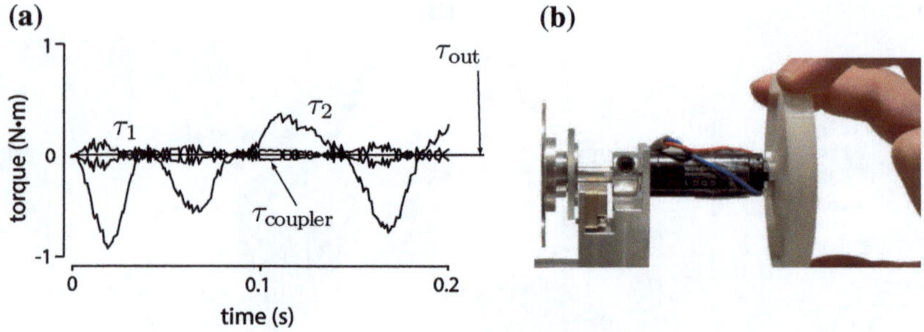

Fig. 3 **a** Interface response to a null reference. **b** What is felt by the user

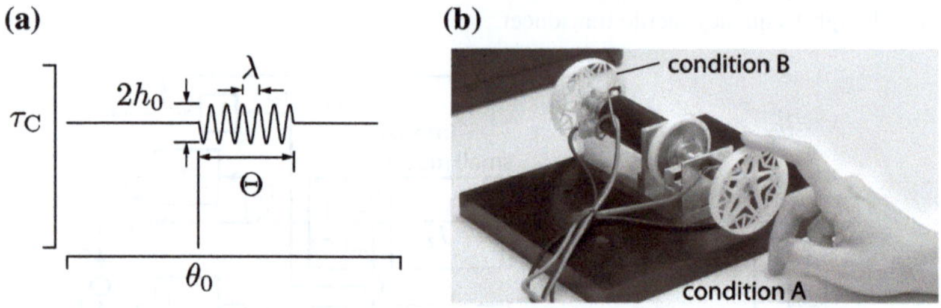

Fig. 4 Dual-stage interface validation. **a** In *condition A*, the handle was connected to the output of the dual-stage drive. In *condition B*, an identical handle was connected directly to the shaft of the proximal motor and the coupler was disconnected. **b** Stimulus. A small burst of oscillation of size Θ, amplitude h_0, and spatial period λ, was superposed onto a larger friction force, τ_C

2.1 Interface Evaluation

The objective was to verify that the high-level degree of transparency of the dual-stage had a measurable effect on human performance when using the interface. Human observers to performed a signal detection task using the stimulus depicted Fig. 4a. One component had a fixed magnitude and the other varied from imperceptible values to easily detectable values. To create an experimental condition that resembled actual use, observers detected the presence of weak force perturbations at random locations in the workspace. The stimuli was felt through two identical handles, one connected to the large motor, and other to the small motor, see Fig. 4b.

The observers were not aware of the condition under which they were working, that is with the dual-stage interface (condition A) or with the single motor (condition B), similar to a conventional interface. All subjects produced typical psychometric curves as shown in Fig. 5. Despite inter-subject variations, it could be concluded that the performance detection of the observers was improved by almost an order of magnitude. As a whole, the interface achieves a large dynamic range which goes from 5.2 mN to 5.7 N that is three orders of magnitude. Reader can find in [7] more details about the evaluation procedure.

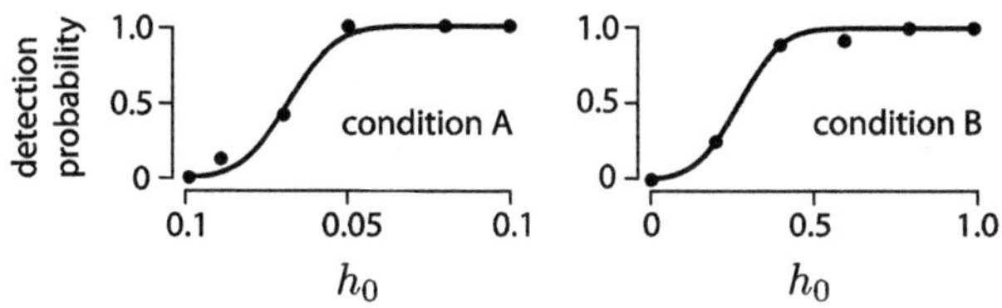

Fig. 5 Results and fits for conditions A and B a typical observer

2.2 Haptic Interaction with a Water Droplet

To test the ability of the haptic interface to bring the physics of micro scales to a human scale perception, we experimented with a actual interaction at micro scales. The test was carried out on a complex case for microscales force sensing, which entails measuring and feeling the interaction of a thin glass probe with a water droplet. The droplet was first approached towards the probe using the position of the interface handle as a set point for the droplet micro-positioner. Then, the droplet was contacted and retracted in approach-retraction cycle. The interaction force was measured by using a micro force sensor [8, 9] and fed back to the user through the dual-stage haptic interface. The dual-stage interface conveys the amplified interaction forces between the glass probe and water droplet at a scale where the operator could interact through his/her natural touch perception. Figure 6a, b represent respectively the force measured by the sensor over a cycle of approach-retraction and the force transmitted to the operator by a probe with a diameter of 80 μm. The force felt

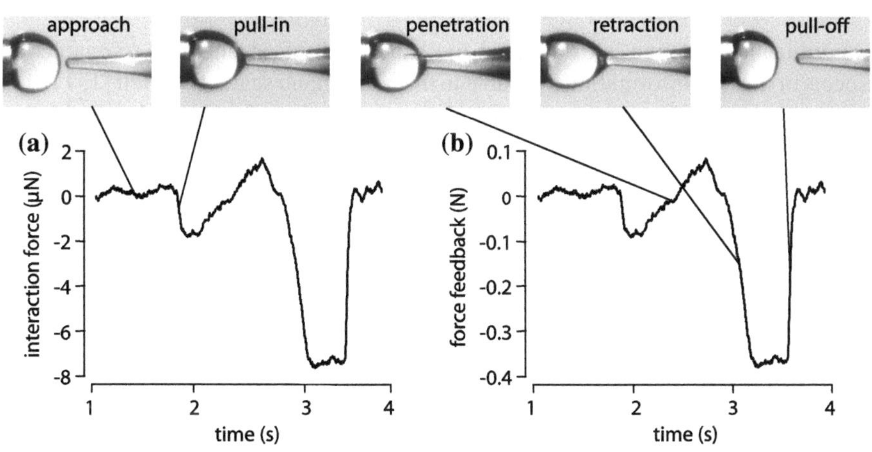

Fig. 6 Illustration of the main phases during the interaction, approach, pull-in, penetration, retraction and pull-off. Measuring and feeling the interaction between a *glass probe* and *water droplet*. **a** Interaction force measured by the sensor over a cycle of approach-retraction according to time. **b** Haptic force feedback felt by the operator through the dual-stage haptic interface

by the user through the interface handle was exactly the force measured by the sensor amplified by a coefficient $\alpha_f = 0.05 \times 10^6$. His system shows the capacity of the haptic interface to let humans to go beyond their sensory limitations. The system was also tested by several users with no experience in microscale phenomena. Various interaction forces, like pull-in and pull-off forces, are correctly rendered to the operators. Readers can refer to [10] for more details about the set-up and the experiment.

2.3 Discussion

We described a new type of electromagnetic drive for use in haptic interfaces and other applications. The interface increases its dynamic range and its transparency by decoupling the delivered maximum torque from its effective inertia. As a whole, the interface achieves a large dynamic range which goes from 5.2 mN to 5.7 N. In terms of transparency for a human user, the interface achieves two orders of magnitude of improvement over existing designs. However, the capacity of the haptic interface to convey large frequency signals remains limited, particularly tactile sensations that can be employed to feel fast phenomena like Brownian motion. To address this need, augmenting a conventional force feedback device with a vibrotactile transducer is an interesting option. Such as approach is described in the next section.

3 Haptic Interface Augmented with Tactile Transducer

A significant challenge is to increase the bandwidth of an interface to enable key aspects of the rich information available in the micro/nano scales to be felt [1, 11–13]. The problem arises from factors such as the presence of resonances in the mechanical structure of articulated systems of a scale sufficient to interact with a moving hand. Because of these resonances, typical haptic interfaces must be limited in bandwidth using appropriate low-pass filters [14]. To fix ideas, the best performing Sensable Phantom 1.0 A device has a useful mechanical bandwidth of only 30 Hz. As a result, high frequency phenomena such as Brownian motion, van der Waals and electrostatic forces cannot be reproduced by the interface. The mechanics of cantilevered structures of appropriate scales preclude any device to achieved a bandwidth able to transmit the signals arising from the micro/nano scales. In order to address this problem, we constructed a dual-stage design based on channel separation using signal crossover to separate haptic signals into low and high frequency channels that recombine at the tip of the device.

The main stage was a 'Pantograph' haptic interface, which is a planar parallel mechanism of two actuated degrees of freedom. It provided a stiff and light set of linkages driven without transmission [15, 16]. It has the desirable features of light inertia, high stiffness, compact design but large operating workspace. The mechan-

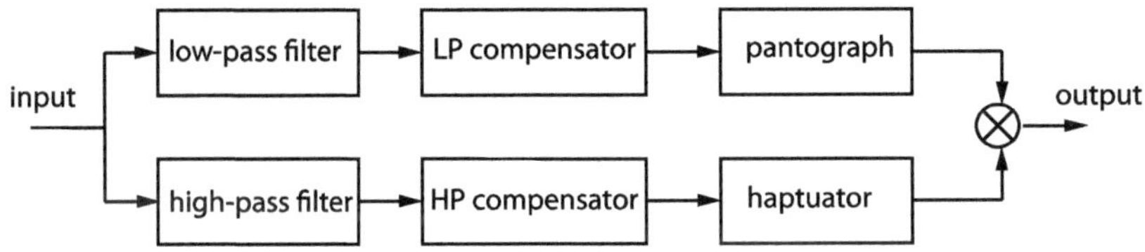

Fig. 7 System block diagram with compensation and crossover, L-P and H-P stand for low-pass and high-pass respectively

ical properties give rise to uniform magnitude performance on the workspace. A vibrotactile transducer, called a 'Haptuator' [17], was embedded inside the finger interface to reproduce fast oscillations. The Haptuator, based on the principle of conservation of momentum, is designed to work in the high frequency regions and is limited in the low frequency range because of displacement saturation. The resulting system can be viewed in Fig. 1b. Each component operated in its optimal range of frequencies. From the knowledge of the identified system's dynamics, a compensation/crossover scheme could achieve nearly ideal, uniform magnitude performance over the whole frequency band, from DC up to 1.0 kHz and up. The overall system block diagram is shown in Fig. 7.

3.1 Interface Evaluation

The magnitude performance of each subsystem was first experimentally determined using identification by the third of octave band method. Then, the compensation scheme and the crossover filter were implemented in a real-time computational system. The magnitude response over the required frequency bands, from DC to 1.0 kHz, is reported in Fig. 8. It can be seen that the interface exhibited a flat response in the desired frequency region.

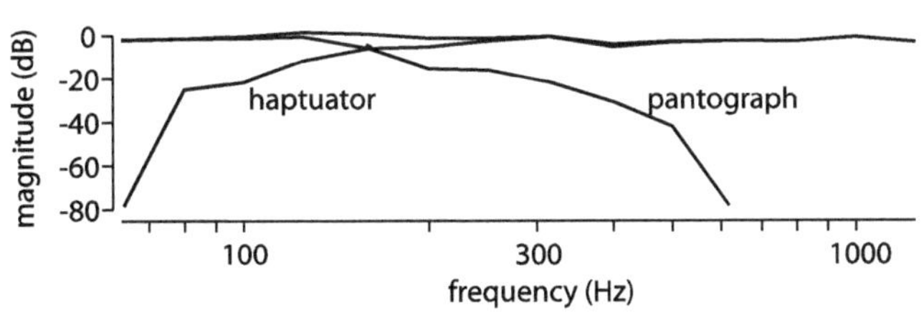

Fig. 8 Acceleration magnitude response of the interface with tactile feedback. The result shows a flat response up to 1.0 kHz

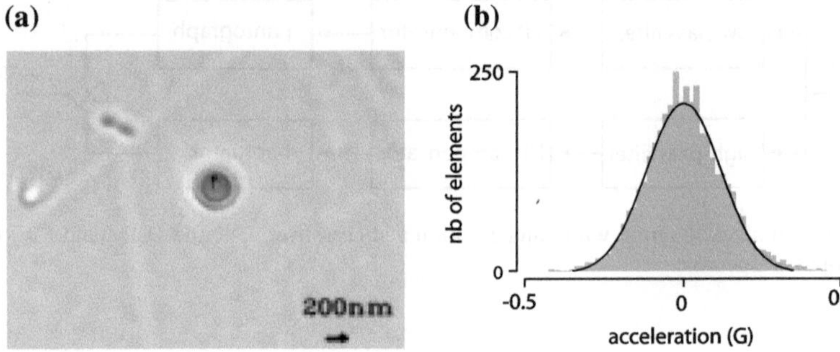

Fig. 9 **a** A virtual scenario created from records of an optical tweezers. The 200 nm spherical object was placed in a liquid environment and thus was subjected to Brownian motion. **b** Histogram of measured acceleration had a quasi-normal distribution. The results suggest the measurements conformed to a Gaussian distribution and the interface was able to render Brownian motion

3.2 Optical Tweezers Application

The interface was employed in a simulation scenario for an optical tweezers application to demonstrate its capabilities [18, 19], see Fig. 9a. The scenario involved a nano-scale spherical object placed in a liquid environment which was subjected to nothing else but the Brownian motion. Its movements were recorded but a novel high-speed image capture system [18]. During operation, operators were able to feel high frequency oscillations. In order to verify that the interface was capable of rendering the Brownian motion sensed by the optical tweezers simulation, a histogram was plotted to show the distribution of the signals recorded by an accelerometer mounted on the finger interface, see Fig. 9b. It can be concluded that the measurements reflected the expected normal distribution and that the haptic interface was capable of rendering Brownian motion. Observers invited to test the interface reported that the haptic sensations where novel and provided new intuition about what Brownian motion is.

3.3 Discussion

A haptic teleoperation interface was developed by combining a Pantograph and an Haptuator, which has uniform performance over a large frequency band from DC to 1.0 kHz. By applying the interface to data collected from optical tweezers, the observers were able to experience the Brownian motion present in the micro/nano world. The presented system demonstrates high bandwidth capability and reliability over determined frequency band. It gives rise to potential possibilities for complex applications which require uniform performance over large frequency regions. Further research aims at combining this concept with the viscously coupled dual stage design so that wide bandwidth, high transparency and dynamic range could be achieved all in one single system.

4 Conclusion

Dual-stage architectures were proposed to overcome the performance limitations of conventional haptic interfaces. A knob-type high fidelity haptic interface with one degree of freedom based on viscous coupling is advantageous when high torques are required since inertia of large motors are decoupled from operator experience. A dual stage design based on signal crossover and compensation filters could extend conventional interface bandwidth considerably. In future works, the complementary between the first and second design could be exploited to design high-performance and compact haptic interfaces suitable for dealing with micro/nano scales interactions.

Acknowledgments This work was supported in part by the eurostar project **REMIQUA**.

References

1. Gauthier, M., Régnier, S.: Robotic Microassembly. Wiley-IEEE press, publisher (2010)
2. Hayward, V., Astley, O.R.: Performance measures for haptic interface. In: Giralt, G., Hirzinger, G. (eds.) Robotics Research: The 7th International Symposium, pp. 195–207. Springer, Heidelberg (1996)
3. Hollis, R., Salcudean, S., Abraham, D.; Toward a tele-nanorobotic manipulation system with atomic scale force feedback and motion resolution. In: Micro Electro Mechanical Systems Conference, pp. 115–119 (1990)
4. Hunter, I.W., Lafontaine, S., Nielsen, P.M.F., Hunter, P.J., Hollerbach, J.M.: Manipulation and dynamic mechanical testing of microscopic objects using a tele-micro-robot system. IEEE Control Syst. Mag. **10**(2), 3–9 (1990)
5. Gaponov, I., Ryu, J-H., Choi, S.-J., Cho, H.-C., Poduraev Y.: Telerobotic system for cell manipulation. In: IEEE/ASME International Conference on Advanced Intelligent Mechatronics, pp. 165–169 (2008)
6. Millet, G., Haliyo, S., Régnier, S., Hayward, V.: The ultimate haptic device: first step. In: Eurohaptics Conference and Symposium on Haptic Interfaces for Virtual Environment and Teleoperator Systems. World Haptics, pp. 273–278 (2009)
7. Mohand Ousaid, A., Millet, G., Régnier, S., Haliyo, S., Hayward, V.: Haptic interface transparency achieved through viscous coupling. Int. J. Robot. Res. **31**(3), 319–329 (2011)
8. Mohand Ousaid, A., Haliyo, S., Régnier, S., Hayward, V.: Active electrostatic force-sensing probe in the microNewton range. In: IEEE/ASME International Conference on Advanced Intelligent Mechatronics, pp. 612–617 (2013)
9. Mohand Ousaid, A., Haliyo, S., Régnier, S., Hayward, V.: H-infinity optimal control enabled micro-force sensing. In: 3rd International Conference on Systems and Control, pp. 490–495 (2013)
10. Mohand Ousaid, A., Bolopion, A., Haliyo, S., Régnier S., Hayward, V.: Stability and transparency analysis of a teleoperation chain for microscale interaction. In: IEEE International Conference on Robotics and Automation (2014) in press
11. Romano, J.M., Kuchenbecker, K.J.: Creating realistic virtual textures from contact acceleration data. IEEE Trans. Haptics **5**(2), 109–119 (2012)
12. Sitti, M.: Micro and nano-scale robotics. Am. Control Conf. **1**, 1–8 (2004)
13. Finio, B., Galloway, K., Wood, R.: An ultra-high precision, high bandwidth torque sensor for microrobotics applications. In: IEEE/RSJ International Conference on Intelligent Robots and Systems, pp. 31–38 (2011)

14. Campion, G., Hayward, V.: Fundamental limits in the rendering of virtual haptic textures. In: Eurohaptics Conference and Symposium on Haptic Interfaces for Virtual Environment and Teleoperator Systems, pp. 263–270 (2005)
15. Hayward, V., Choksi, J. Lanvin, G., Ramstein, C.: Design and multi-objective optimization of a linkage for a haptic interface. In: Lenarcic, J., Ravani, B. (eds.) Advances in Robot Kinematics. Kluver Academic, Dordrecht, pp. 352–359 (1994)
16. Campion, G., Wang, Q., Hayward, V.: The pantograph Mk-II: a haptic instrument. In: IEEE/RSJ International Conference on Intelligent Robots and Systems, pp. 723–728 (2005)
17. Yao, H.-Y., Hayward, V.: Design and analysis of a recoil-type vibrotactile transducer. J. Acoust. Soc. Am. **128**(2), 619–627 (2010). http://link.aip.org/link/?JAS/128/619/1
18. Ni, Z., Pacoret, C., Benosman, R., Régnier, S.: 2D high speed force feedback teleoperation of optical tweezers. In: IEEE International Conference on Robotics and Automation, pp. 1700–1705 (2013)
19. Ni, Z., Pacoret, C., Benosman, R., Ieng, S., Régnier, S.: Asynchronous event based high speed vision for micro-particles tracking. J. Microsc. **245**(3), 236–244 (2011)

Part III
Manipulation

Jaydev P. Desai
University of Maryland, College Park

The session on "Manipulation" consisted of four papers from several different areas, which included manipulating flexible objects, passive grasping and manipulation of objects, robots capable of working in a disaster environment, and human–robot cooperative task execution where the human actively controls the robot by direct physical interaction. Caldwell et al. in the paper titled "Robotic Manipulation for Identification of Flexible Objects" describe their initial work on the identification of the stiffness of flexible objects via robotic manipulation. They chose to address this problem by both simulation and experimental approach. In this study they manipulated a Y-shaped object resembling a living plant. While the results presented were not consistent with the object's physical properties, it nonetheless provided useful insights into the problem of estimating the stiffness of flexible objects using an automated procedure. In the paper by Kessens and Desai, titled "Compact Hand with Passive Grasping," the idea of grasping objects based on the object and grasper interaction forces was presented. The advantage of this approach is that it is not necessary to have a specific grasper orientation since the grasper elements (suction cups) can be potentially, arbitrarily distributed on a flexible substrate. The results demonstrated grasping a wide variety of objects ranging from a dime to a soccer ball using a three-finger hand composed of an array of suction cups. Dellin et al. in the paper titled "Guided Manipulation Planning at the DARPA Robotics Challenge Trials" presented the results from Carnegie Mellon University's participation in the DARPA Robotics Challenge (DRC). They used virtual fixtures as a common language between the operator and the motion planner to solve a guided manipulation problem. The paper shows the operation of the Carnegie Mellon University Highly Intelligent Mobile Platform (CHIMP) involved in a variety of manipulation tasks (Debris, Door, Wall, Valve, and Hose) at the DRC trials. Finally, in the paper by Ficuciello et al. titled "Redundancy Resolution in Human-Robot Co-Manipulation with Cartesian Impedance Control," the results of controlling the robot arm that is cooperatively performing a task with a human via direct physical interaction of the human with the robot is presented. It is shown that the redundant degrees-of-freedom of the robot plays an important role during task performance. The proposed approach of inertial decoupling was tested on a 7-DOF KUKA LWR4 arm.

Compact Hand with Passive Grasping

Chad C. Kessens and Jaydev P. Desai

Abstract As robots are tasked with increasingly diverse requirements, the grasping and manipulation of unknown objects will become ever more important. Thus, end effectors should be able to grasp the widest possible range of object shapes and sizes. Previously, we showed the grasping potential of a self-sealing suction cup array for solving this problem. In this work, we show an improved design for the cup, which is both smaller and more robust. Seal quality and force-displacement data are presented for this new design. We also show the design of a sample three-finger hand in which embedded shape memory alloy (SMA) wires actuate the individual fingers comprising the self-sealing suction cup arrays. Force and range of motion data are also presented. Finally, we demonstrate the ability to grasp and release a soccer ball using the three-finger hand attached to the PA-10 robot manipulator.

Keywords Grasping · Unknown object · Suction · Shape memory alloy

1 Introduction

As robots continue to move from well-controlled factory floors into the dynamic and unstructured world in which we live, the multitude of tasks we would like robots to perform in our world continues to increase. Robotic abilities must be especially robust

C.C. Kessens (✉) · J.P. Desai
Robotics, Automation, and Medical Systems (RAMS) Laboratory,
University of Maryland, College Park, MD, USA
e-mail: ckessens@umd.edu
URL: http://rams.umd.edu

J.P. Desai
jaydev@umd.edu

C.C. Kessens
US Army Research Laboratory, Aberdeen Proving Ground, Aberdeen, MD, USA

M.A. Hsieh et al. (eds.), *Experimental Robotics*, Springer Tracts
in Advanced Robotics 109, DOI 10.1007/978-3-319-23778-7_9

and flexible to handle our demands of them. To achieve this, biological systems have proven to be very useful for inspiring robotic design in many ways [1–3]. The octopus has proven to be particularly inspirational, giving rise to new ideas on underwater propulsion [4], arm design [5], wall climbing [6–8], and grasping [9–13].

Unstructured grasping is a particularly useful but difficult challenge given the enormous range of potential object shapes and sizes of interest. As such, this has been the subject of much research [9, 14–16]. While dual arm manipulators provide many grasping advantages, many of today's robots are limited to a single manipulator due to cost, weight, and size constraints. Therefore, we chose to focus on the development of technologies that would maximize the range of object shapes and sizes graspable by a single manipulator. In addition, our long-term goal is to improve the speed, robustness, and stability with which such objects can be grasped, while maintaining a small, easy to operate package.

Considering the capabilities of the octopus, suction seems to be a technology that has been under-utilized in terms of unstructured grasping. While suction has been widely used, traditional applications have been for a very limited range of objects, such as featureless panels [11], limp sheets [12], or citrus fruits [13]. One reason for this is that passive pneumatic cups are rarely as reliable as active ones [6]. However, active pneumatic cups must either have an individual actuator for each cup [7] or must all be engaged on the object to prevent leaks and maintain a low pressure relative to the atmosphere. In [9], Tramacere et al. utilize the incompressibility of water for an effective passive sucker, but that seems impractical for everyday, on-land use. Therefore, in [17, 18], we proposed the use of self-sealing suction cup arrays for grasping. While similar in concept to the valve-regulated multiple (VM) suckers in [8], our design eliminated volume flow requirements by creating a nominally closed valve position, thereby improving scalability.

In this paper, we present a significantly re-designed self-sealing suction cup, which is smaller and more robust than the version in [18]. In addition, we present the design of a three-finger hand which is actuated by SMA wires embedded in the hand structure. SMAs were chosen for their ability to generate large forces in a compact space and have been used in various applications including surgical robotics [19], finger actuation [14], and even suction cup actuation [7]. We present the design of a release system, which provides a short burst of positively pressurized air to the system, similar to that found in [12, 20, 21]. However, to keep the design simple, we utilize the output of the vacuum pump as the source of the positive pressure in our system for object release. Finally, we present the results of our experiments on seal quality, force-displacement, blocked and unblocked force, range of motion, and demonstrate object grasping and release using our three-finger hand mounted on a PA-10 robot manipulator.

2 Technical Approach

2.1 Self-sealing Suction Cup Design

While we proved our self-sealing suction cup design to be quite capable of grasping a wide variety of objects in [18], our previous cup design had some limitations. Firstly, the number and thickness of the structures required a high overall cup height, thereby increasing the bulk of the device. Secondly, the design proved to be fragile with respect to hyper-extension as well as torsional and shear forces. Finally, because the cup is manufactured using the Objet® Connex®500 rapid prototyping machine, the support material must be removed. The geometry of the design made it difficult to remove this material in a repeatable fashion, which subsequently caused variation in the quality of the seal that could be achieved. We made several design revisions which address each of these issues.

To make the cup more compact, we consolidated the functions of several components, resulting in fewer overall components and a height reduction of nearly 50 percent to 0.72 cm in its uncompressed configuration. Note that the cup diameter was also reduced to 1.07 cm. In the new design, shown schematically in Fig. 1, the lever which pivots on the collar to lift the plug is built into the base of the cup, and it directly attaches to the plug and spring. Additionally, the tube connecting the cup to the hand now doubles as the spring, which restores the levers, and thus the plug, to their nominal closed position. This has the minor drawback of increasing the cross-section of spring material, thereby increasing the spring force that must be overcome to actuate the cup. However, the new geometry also allows for a greater mechanical advantage for the lever action.

To handle issues with hyper-extension, shear, and torsional forces, a plastic restraint surrounding the cup was added (see Fig. 2a). This restraint engages the tabs that extend from the lever pieces of the cup. To ensure that the restraint did not inadvertently cause cup activation or binding, radial and vertical clearances of 0.35 mm were maintained.

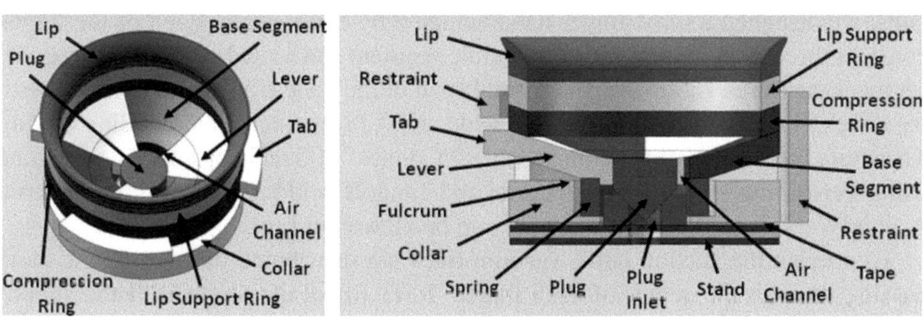

Fig. 1 New self-sealing cup design: *white parts* are plastic and *black parts* are rubber

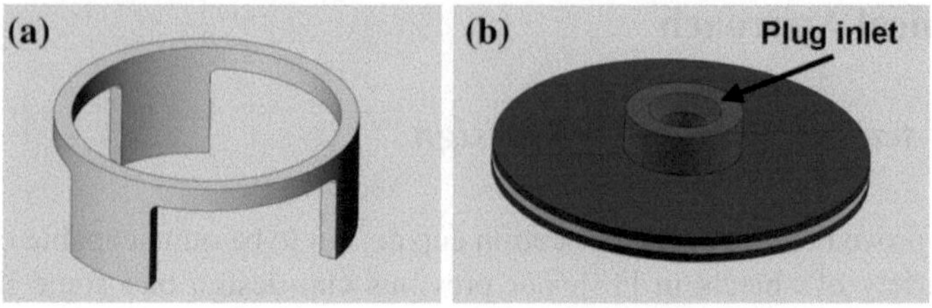

Fig. 2 **a** A plastic restraint prevents hyper-extension and torsion of the cup. **b** Stand with central suction port for plug insertion

To simplify the cleaning process and achieve better seal performance repeatability, the cup (containing the plug) was separated from the stand (containing the plug inlet—see Fig. 2b). This helped in two ways. Firstly, by printing them separately, the support material could easily be removed with good access and visibility. Secondly, to print the plug and plug inlet as separate parts in the same build, a small gap must necessarily exist between the components. However, if the parts are printed in separate builds, this gap can be closed at assembly.

As before, our cups are printed on the Objet® Connex®500 rapid prototyping machine, using VeroWhite for plastic components and TangoBlackPlus for rubber components. The openness of the design allows for reasonable visibility and access to areas needing support material removal. To achieve accurate placement of the plug relative to the stand, a piece of double-sided tape was used to adhere the cup to the stand. Once the cup and stand were assembled, Locktite was spread around the perimeter of the cup to solidify the bond, ensure a quality air seal, and to attach the restraint.

2.2 Hand Design

To demonstrate grasping with the use of the previously described self-sealing suction cups, we designed a three-finger hand actuated by SMA wires. Each of the fingers contains three digits composed of a flexible segment and a rigid segment, where each segment measures 1.2 cm long. The total span of the hand from fingertip to fingertip measures 21.8 cm. Each finger is 2 cm wide and 0.78 cm thick, not including the cup. The entire hand is constructed on an Objet® Connex®500 rapid prototyping machine, using VeroWhite as the plastic material and TangoBlackPlus as the rubber material. A SolidWorks® rendering of the hand can be viewed in Fig. 3.

To operate the suction cups, vacuum lines are distributed throughout the hand, passing through the center of each finger. To facilitate the removal of the support material, all lines are straight and fully through. After cleaning was completed, all holes that were not capped by a suction cup or inlet port were sealed using epoxy.

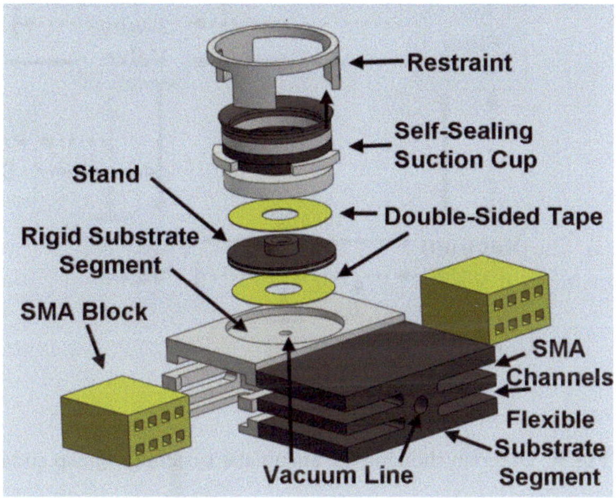

Fig. 5 Exploded view of a finger digit: *white* is plastic, *black* is rubber, and SMA blocks are Ultem®

2.5 Release System

Typically when one is manipulating an object, one wants to be able to grasp and release the object as desired during the manipulation process. Traditional release mechanisms have used a short puff or burst of compressed air to quickly reverse the pressure differential relative to the atmosphere [12, 20] and also overcome any adhesive effects of the gripper due to the soft rubber that is often used [21]. The system we designed uses the same concept, but to keep the system as simple as possible, we utilize the output of the vacuum pump as the source of the positive pressure. The key to this design is that the vacuum line connected to the hand must contain a normally closed valve that opens in release mode. This provides the pump with the volume of air necessary to feed back into the system. A flow diagram showing the valve system in both grasp and release modes is shown in Fig. 6.

3 Experimental Setup

3.1 Cup Force-Displacement

To determine the relationship between the force exerted on the cup and the cup's displacement, the test rig shown in Fig. 7 was assembled. Two pieces of information were of particular interest: (1) How much force is required to cause the cup to engage on an object? and (2) How much force will the cup exert prior to disengagement? A Transducer Techniques Model MLP-10 load cell with a capacity of 44.55 N was used to measure the force. This was attached to an MP-285 Sutter Instrument Company Micromanipulator. The force signal was passed through a Transducer Techniques

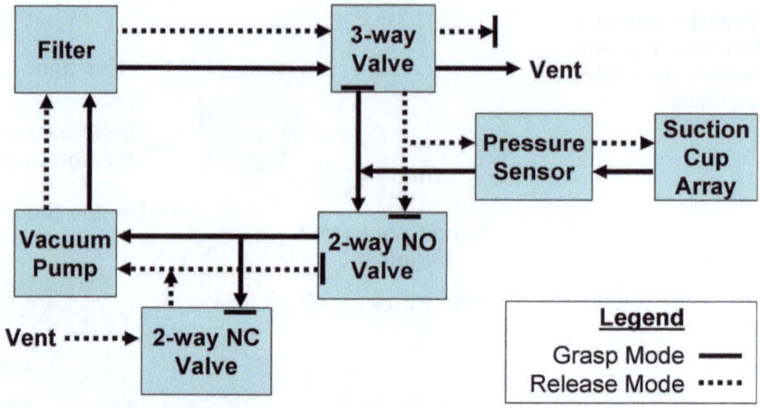

Fig. 6 Diagram showing the pneumatic system for grasp and release modes

Fig. 7 Picture of the cup force-displacement test setup

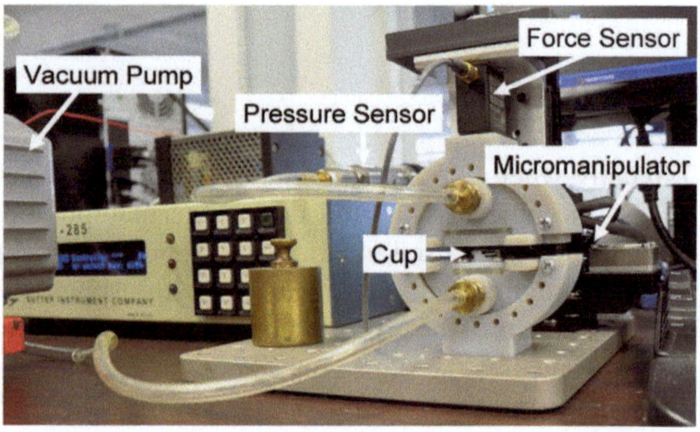

Model TM0-2 signal conditioner. It was calibrated in a vertical orientation using a 500 g weight. A cup was then rigidly attached to the base plate using a custom designed rig. The micromanipulator was positioned such that the contact plate was just above the cup without contacting it. The micromanipulator was then controlled such that the contact plate lowered to a final compression depth of 1.6 mm at a rate of 10 μm/s, approximating a quasi-static load. It was then raised at the same rate to a height of 1 mm above its starting position, sufficient to cause the cup to disengage. This was done for several cups both with and without vacuum pump power. The pressure on the object was continuously monitored using a Gems Sensors 2600SAG100EG3UB pressure transducer connected through a small hole in the contact plate, but contained within the bounds of the cup lip during contact.

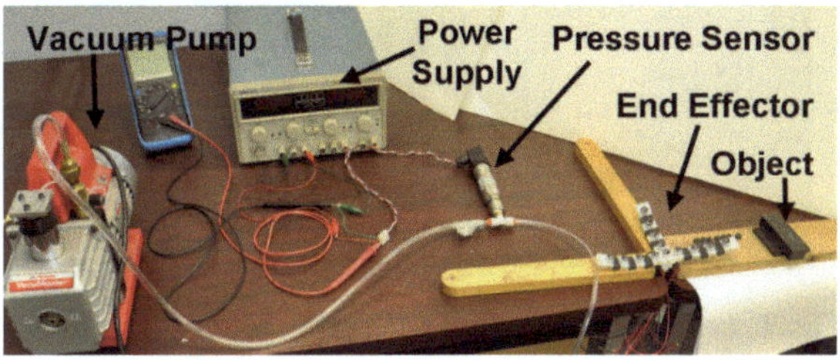

Fig. 8 Picture of the seal test setup

3.2 *Self-seal, Object Seal, and Reseal Quality*

Since we do not expect many situations in which all cups will be engaged, it is important for us to obtain quality seals to maximize the holding force of the cups on the object. Therefore, we were interested in three quantities. Firstly, we wanted to ensure that the seal between the plug and the plug inlet was of a high quality. This was an important aspect of the assembly procedure (see Sect. 2.4). To test this, we used a pressure transducer from Gems Sensors (Model number: 2600SAG100EG3UB) to measure the pressure in the vacuum line. A Robinair two-stage 1.5 cfm vacuum pump (Model: 15150) was used to supply the vacuum pressure. As each cup was added, any un-cupped ports were sealed with a custom fit plug, and the pressure in the line was measured and compared to the previous reading, prior to cup assembly.

Secondly, it is important that the cup is able to maintain a quality seal between the lip and the object being grasped to maximize the force on the object. Therefore, we placed a glass slide on top of the cup with sufficient force to cause engagement before measuring the pressure. This tested not only the lip seal, but also all seals within the cup, beyond the plug.

Thirdly, it is important that the cup be able to reliably reseal itself. Therefore, after the cup was engaged on the object, the object was removed, and the pressure was measured again. The object seal and reseal tests were repeated for a total of three tests per cup. Figure 8 shows the experimental setup used for each of these tests.

3.3 *SMA Unblocked and Blocked Force Tests*

Because we are using the SMA wires to engage the self-sealing suction cups, we wanted to be sure that the wires were capable of supplying sufficient force at various angles to cause engagement. However, while the blocked force test serves as the standard, in the real-world application, our hand would not be constrained in the manner simulated in the blocked force test. Therefore, we performed an unblocked force test of the actuator.

Using the same force sensor as described in Sect. 3.1, we used a traditional blocked force test for the distal link in the 0° orientation by clamping an aluminum block

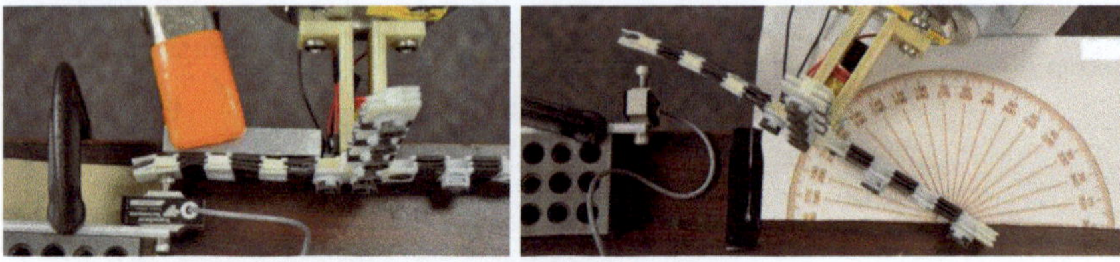

Fig. 9 *Left* blocked force test setup. *Right* unblocked force test setup

behind the finger, preventing bending. We then actuated the individual SMA finger with eight SMA wires (no antagonistic pair) by supplying 1.62 Amps and measured the peak force at the center of the distal digit, as shown in Fig. 9 (left). This test was conducted three times.

In addition, to determine the finger's ability to achieve the threshold force required to engage the cup, we tested the unblocked force at each rigid substrate segment at angles of 0°, 30°, 60°, and 90° relative to the base of the hand. We first aligned the force sensor at the desired test angle. Then, we actuated the finger to determine the location at which the digit of interest would contact the sensor flat-on and thus moved the force sensor to that location. We then verified the angle using a protractor, cooled the SMA, and restored the hand to its initial planar configuration. Finally, we actuated the individual SMA finger using the same conditions as in the blocked force test. Each test was conducted three times. Between each test, the SMA was completely cooled and then restored to the initial planar configuration by hand. Note that each proximal test ended when the distal end curled far enough to touch the force sensor. A photo of the experimental setup for the 30° test of the distal digit can be found in Fig. 9 (right).

3.4 SMA Range of Motion

Another area of interest was the range of motion achievable by the SMA wires. To test this, we placed markers at the center of the back of each rigid segment along the finger. We then placed the apparatus in a volume surrounded by Vicon® motion capture cameras. Calibration of the system showed an accuracy of 0.7 mm. Two fingers were tested, each with eight SMA wires and no opposing pair. Each finger was supplied with three Amperes of current, and the hand was held in an upright orientation to minimize gravitational effects. To eliminate any friction that might be caused by dragging along a table, the base was elevated slightly. Data for markers of interest were compared relative to stationary markered points.

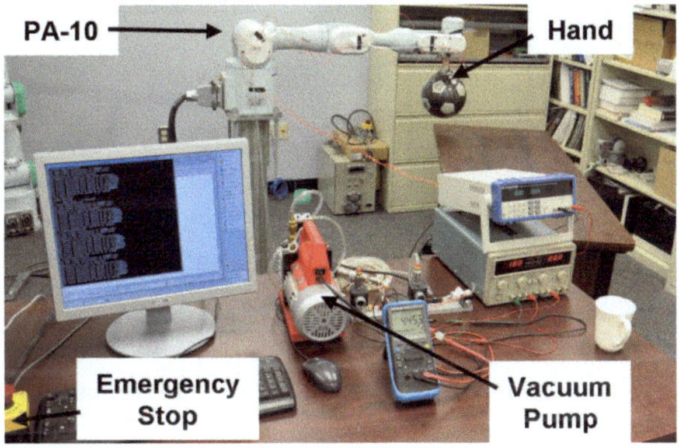

Fig. 10 Grasp and release setup with SMA actuated hand

3.5 Grasp and Release

Finally, the complete system was assembled to test the ability to grasp and release the object. The hand was mounted onto a Mitsubishi PA-10 robot manipulator, and the valve system shown in Sect. 2.5 was connected. The pressure in the system was measured in both grasp and release modes. The PA-10 was moved from an initial position to a secondary position, where it would make light contact with the object (a soccer ball) and then returned to its initial position.

For the actual test, the vacuum pump was first powered on. The robot was then actuated such that the hand contacted the ball. Next, 4 Amps of current were supplied to the hand until it was observed that the SMA wires were clearly actuated. Power to the SMAs was then cut, and the ball was taken off the table. Finally, after holding the ball off the table, the release mode was activated, causing the ball to drop. A figure showing the full system setup is seen in Fig. 10.

4 Results and Discussion

4.1 Cup Force-Displacement

Figure 11 shows the force-displacement curve for a representative cup with and without the applied vacuum, as the object moved downward (increasing cup compression), and subsequently upward (reducing compression and eventually increasing tension). The cup required 0.21 N of force to engage on the object. This force compressed the cup by 0.53 mm prior to engagement. Further, the cup demonstrated the ability to resist 6.94 N of force before disengagement at 0.76 mm extension beyond its neutral set-point. One experiment yielded a resistance force greater than 8 N.

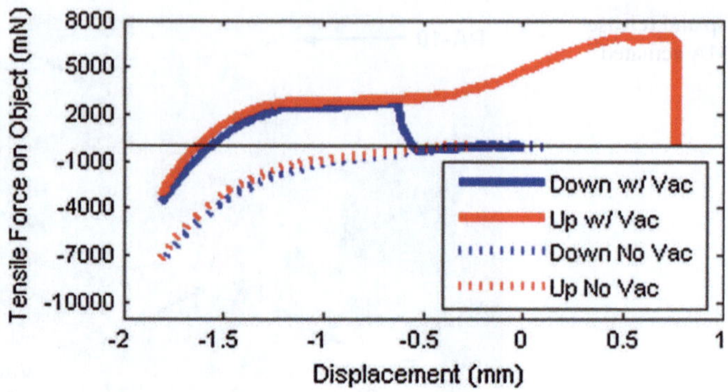

Fig. 11 Plot of force versus displacement of the cup when suction is (*solid line*) and is not (*dashed line*) applied

4.2 Self-seal, Object Seal, and Reseal Quality

We also conducted a self-seal quality test. As each cup was assembled on the hand, the in-line pressure was measured to ensure that the plug was achieving a quality seal. To highlight the advantage of the self-sealing design, the in-line pressure was compared between the self-sealing cups and an open port, as would be found with a traditional disengaged suction cup. The results are shown in Fig. 12 (left). The quality of the cup seals remained consistently high, maintaining a pressure differential of 98.2 % of atmospheric pressure when all 13 cups were in place. This represents a drop of only 1.04 % relative to the fully plugged case. The advantage of the self-sealing cup is obvious if even one cup is not in contact with the object.

The next test was the object seal and reseal quality test. Figure 12 (right) shows the results for all 13 cups. States D1, D2, and D3, which correspond to reseal after object disengagement, clearly show that all cups were able to reseal very reliably

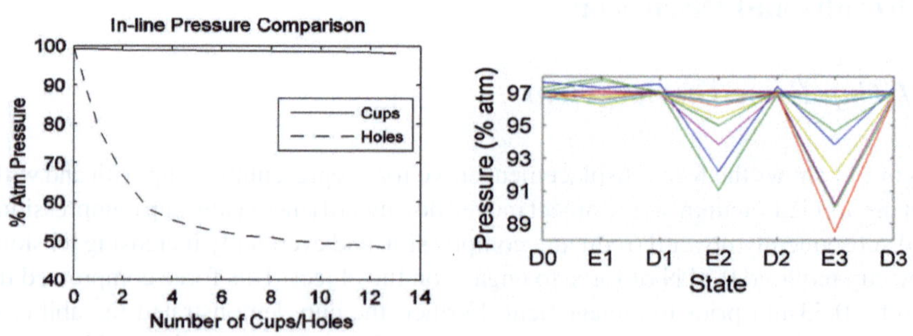

Fig. 12 *Left* Comparison of pressure differential between self-sealing cups and open penetrations. *Right* Pressure data for object seal and reseal tests: *D* states represent disengagement from object, and *E* states represent engagement on object

after contact. The worst reseal as compared to the initial state showed a loss of only 0.37 % of atmospheric pressure. Though the initial object seal was high quality, leaks seemed to develop in 6 of the cups after multiple object contacts. This is evident is state E3. We believe that the leaks are likely developing in the spring tube, which will need to be thickened to withstand the pressure differential in our future prototypes. Importantly, these leaks disappear upon disengagement from the object.

4.3 SMA Unblocked and Blocked Force Tests

SMA force test results are shown in Fig. 13. Based on the cup force-displacement test in Sect. 4.1, approximately 0.21 N of force is required to break the self-seal and cause the cup to engage. These tests show that the SMA is easily capable of providing sufficient force to cause cup engagement in all cases except for the distal link at 0°. This low value is likely explained by the significant backward deformation of the middle of the finger, causing the link to rotate significantly. Because the force sensor only measured a single axis, the majority of the force may have been missed due to the misalignment from the unblocked nature of the test. In a real-world application, the displacement of the finger may be reduced by the engagement of the central cups, and the force could be transferred in a direction that is closer to the local object surface normal.

4.4 SMA Range of Motion

To determine the ability of the SMA actuators to wrap around objects, we quantified the bending angles of each digit using a Vicon® motion capture system. Based on the output marker positions, the bending angles of the proximal, medial, and distal

Unblocked Force Test Results (N)						
	0a	0b	0c	30a	30b	30c
Distal	0.232	0.241	0.227	0.967	0.918	0.557
Middle	0.731	0.739	0.739	0.878	0.780	0.802
Proximal	0.606	0.655	0.855	-	-	-
	60a	60b	60c	90a	90b	90c
Distal	1.537	1.572	1.635	1.087	1.118	1.020
Middle	0.851	0.824	0.940	-	-	-
Proximal	-	-	-	-	-	-

Blocked Force Test (N):		
4.156	3.849	3.711

Fig. 13 Unblocked force results in N for each digit at 0°, 30°, 60°, and 90°. "–" indicates that the digit failed to achieve the required angle for the test. Also, blocked force test results are in N

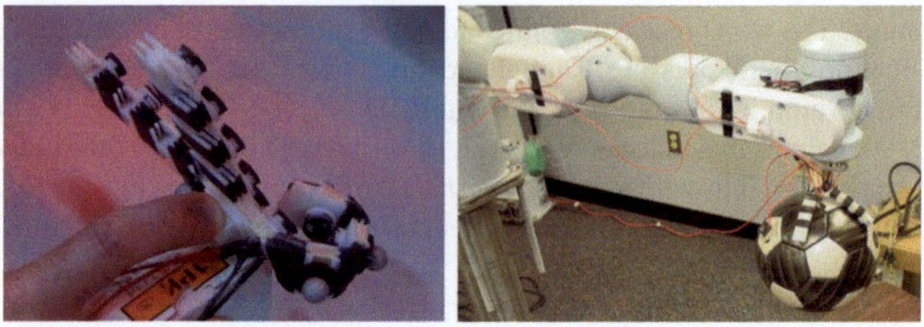

Fig. 14 *Left* Fully flexed finger. *Right* Hand grasping a soccer ball

segments were 33°, 83°, and 91°, respectively. A picture of the flexed finger is shown in Fig. 14 (left).

In addition, we desired to know how much the finger moved out-of-plane, an indication of the angular twist of the SMA actuators. The marker points in the fully flexed configuration were used to determine a best-fit plane, and then the maximum distances above and below this plane were determined. This analysis showed that the third marker was 0.96 mm below the plane, and the fourth marker was 0.86 mm above the plane.

4.5 Grasp and Release

Finally, the grasp and release abilities of the hand were tested on a soccer ball. Figure 14 (right) shows a successful grasp of the ball. Prior to grasping, the pressure differential of the system with the pump activated in grasp mode measured 94 % of atmospheric pressure. Upon release, a positive pressure differential of 12 % was generated in the line. However, this task was not highly repeatable and hence ensuring repeatability for grasp and release of a particular object as well as a wide variety of objects will be an active area of our future research.

5 Main Experimental Insights

In this paper, we presented a new type of three-finger hand which uses a combination of self-sealing suction cups and SMA wires for grasping. A brief burst of positive pressure generated by the vacuum pump output achieves object release. We presented experiments testing self-seal, object seal, and reseal quality; force-displacement behavior; range of motion; and blocked and unblocked force over a range of angles and digits. While we successfully demonstrated object grasping and release, several aspects of the design must be refined in our future work to improve and ensure repeatability of the grasping task. This includes modifying the design of the cup to prevent leaks generated when in prolonged contact with the object, enabling independent digit control, and addressing force distribution issues to improve grasp stability.

References

1. Beer, R., Quinn, R., Chiel, H., Ritzmann, R.: Biologically inspired approaches to robotics. What can we learn from insects?. Comm. ACM. 40, 31–38 (1997)
2. Hirose, S.: Biologically Inspired Robots—Snake-like Locomotors and Manipulators. Oxford University Press, Oxford (1993)
3. Autumn, K., Dittmore, A., Santos, D., Spenko, M., Cutkosky, M.: Frictional adhesion: a new angle on gecko attachment. J. Exp. Bio. **209**, 3569–3579 (2006)
4. Sfakiotakis, M., Kazakidi, A., Pateromichelakis, N., Ekaterinaris, J., Tsakiris, D.: Robotic underwater propulsion inspired by the octopus multi-arm swimming. In: IEEE International Conference on Robotics and Automation, pp. 3833–3839. Minneapolis (2012)
5. Margheri, L., Laschi, C., Mazzolai, B.: Soft robotic arm inspired by the octopus: I. from biological functions to artificial requirements. Bioinspiration Biomimetics 7 (2012)
6. Yoshida, Y., Shugen, M.: Design of a wall-climbing robot with passive suction cups. In: IEEE International Conference on Robotics and Biomimetics, pp. 1513–1518. (2010)
7. Bing-shan, H., Li-wen, W., Zhuang, F., Yan-zheng, Z.: Bioinspired miniature suction cups actuated by shape memory alloy. Int. J. Adv. Robot. Syst. 6 (2009)
8. Hirose, S., Nagakubo, A., Toyama, R.: Machine that can walk and climb on floors, walls and ceilings. In: Proceedings of the 5th International Conference on Advanced Robotics, Robots in Unstructured Environment, pp. 753–758 (1991)
9. Tramacere, F., Beccai, L., Mattioli, F., Sinibaldi, E., Mazzolai, B.: Artificial adhesion mechanisms inspired by octopus suckers. In: Proceedings of IEEE International Conference on Robotics and Automation, pp. 3846–3851 (2012)
10. Tsourveloudis, N., Kolluru, R., Valavanis, K., Gracanin, D.: Position and suction control of a reconfigurable robotic gripper. Mach. Intell. Rob. Control **1**, 53–62 (1999)
11. Bouchard, A.: Design and Control of a Manipulator for Autonomous Joining of Featureless Panels. Vanderbilt University, Nashville (2006)
12. Costo, S., Altamura, G., Bruzzone, L., Molfino, R., Zoppi, M.: Design of a reconfigurable gripper for the fast robotic picking and handling of limp sheets. In: Proceedings of the 33rd International Symposium on Robotics. Stockholm (2002)
13. Hannan, M., Burks, T.: Current Developments in Automated Citrus Harvesting. ASAE/CSAE Annual Int'l Mtg, Ottawa (2004)
14. Dilibal, S., Güner, E., Akturk, N.: Three-finger SMA robot hand and its practical analysis. Robotica **20**, 175–180 (2002)
15. Brown, E., Rodenberg, N., Amend, J., Mozeika, A., Steltz, E., Zakin, M., Lipson, H., Jaeger, H.: Universal robotic gripper based on the jamming of granular material. PNAS **107**, 18809–18814 (2010)
16. Xiong, C., Wang, M., Tang, Y., Xiong, Y.: Compliant grasping with passive forces. J. Robot. Syst. **22**, 271–285 (2005)
17. Kessens, C., Desai, J.: Design, fabrication, and implementation of self-sealing suction cup arrays for grasping. In: IEEE International Conference on Robotics and Automation, pp. 765–770 (2010)
18. Kessens, C., Desai, J.: A self-sealing suction cup array for grasping. J. Mech. Robot. **3**, 100–107 (2011)
19. Ho, M., Desai, J.: Towards a MRI compatible meso-scale SMA actuated robot using PWM control. In: 3rd IEEE/RAS-EMBS International Conference on Biomedical Robotics and Biomechatronics BioRob. Tokyo (2010)

20. Niemeyer, C.: Pick and place in a minifactory environment. Ph.D. Thesis: Swiss Federal Inst. of Tech. (2006)
21. Ruggeri, S., Fontana, G., Pagano, C., Fassi, I., Legnani, G.: Handling and manipulation of microcomponents: work-cell design and preliminary experiments. IFIP Adv. Inf. Comm. Tech. 371, 65–72 (2012)

Robotic Manipulation for Identification of Flexible Objects

T.M. Caldwell, D. Coleman and N. Correll

Abstract This paper provides preliminary insight into stiffness profile identification of a complex flexible object by robotic manipulation. The object is in the shape of the letter 'Y', chosen to resemble a living plant. The object is approximately modelled as a spring mass system. The robot manipulates the object with one or two arms grasped at the ends of the 'Y', and makes visual measurements which locates the object's position in space. Identification results from an optimization approach are compared for both one and two arm manipulation and sensing with and without vision. The results are not consistent with the expected physical object's properties due to a failure to observe the motion dependence between the object's connected segments. The result provides insight into the problem of assessing the minimal information needed to identify the stiffness of a flexible object, an issue of importance to automated approaches.

1 Introduction

The goal of this work is to use a robotic arm and an external vision system to identify the behavior of the flexible object shown in Fig. 1. This is an important step toward manipulation of flexible objects such as living plants, rubber tubes, and clothes [2, 6, 19, 22].

This work was supported by a NASA Early Career Faculty fellowship NNX12AQ47GS02. We are grateful for this support.

T.M. Caldwell · D. Coleman · N. Correll (✉)
Department of Computer Science, University of Colorado,
1111 Engineering Dr, Boulder, CO 80309, USA
e-mail: nikolaus.correll@colorado.edu

T.M. Caldwell
e-mail: caldwelt@colorado.edu

D. Coleman
e-mail: david.t.coleman@colorado.edu

M.A. Hsieh et al. (eds.), *Experimental Robotics*, Springer Tracts in Advanced Robotics 109, DOI 10.1007/978-3-319-23778-7_10

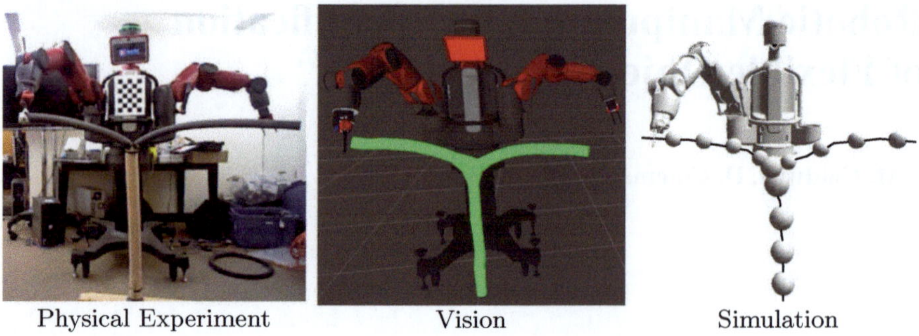

Fig. 1 The physical experiment, vision capture (in *green*), and simulation of baxter manipulating a flexible object

There are many methods to model and simulate flexible objects [11, 12]. A common approach is to model the object as a lattice or collection of links of masses and springs [11, 20, 21]. This approach has been used to simulate linear object like strings, hair, and electrical cables for which the model is a series of masses linked together with springs.

We also model the flexible object as a spring mass system. In [3] we modeled a rubber tyre in such a way, and identified the spring constants for its uniform stiffness with a novel identification method. In this paper, we explore a more complex structure in the form of a foam 'Y' made of tubes with differing stiffnesses. We observe that not only must we carefully plan where to manipulate an object in order to sufficiently excite all of its degrees of freedom for measurement—in this case at the end of each Y—but that this is insufficient to accurately identify its stiffness profile, even with additional measurements made by an external vision system. The failure in identification is due to the motion dependence between the object's segments.

Two types of sensing are investigated. The first is the joint angles and torques of Rethink Robotic's Baxter robot's arms. The second is an external vision system that produces a point cloud of the object and through filtering and fitting, can locate specific points on the object. We note that the vision cannot make any torque or force measurements and as such can not directly measure stiffness properties.

We model the object with the same underlying mechanics as the robot arm—i.e., as a collection of rigid bodies connected at joints by springs—allowing us to utilize the vast theory of rigid body mechanics [15]. Also, this enables planning and control to be done in the combined arm and object configuration space instead of only the end effector space or object space. We then use an optimal control approach in [3] for calculating model properties that best match the behavior of the flexible object. The physical experiment, the vision capture, and the model can be viewed in Fig. 1.

As in [3], we use variational integrators to simulate the robot and object. Variational integrators can be used to describe discrete-time equations of motion of a mechanical system. They are designed from the least action principle and have good properties that agree with known physical phenomenon like stable energy behavior [16]. All simulations were implemented in `trep` [8, 9], which is a tool to simulate articulated rigid bodies using midpoint variational integrators.

1.1 Organization of This Paper

This paper is organized as follows: Sect. 2 sets up the experiment with the robot and flexible object. In Sect. 3, the flexible object is modeled as a connection of springs and masses. This section also reviews variational integrators. Section 4 discusses the visual perception system and the techniques to filter and model fit the measured data. Section 5 reviews the identification algorithm from [3]. Section 6 conducts the identification on the flexible object and compares identification with and without vision, as well as discusses the results.

2 Example Experiment Setup

The goal of the example experiment is to identify stiffness properties of a flexible object. The flexible object has the shape of the letter 'Y'. It was chosen to resemble the basic geometry of a living plant. The base, or 'trunk' is attached to the ground and is labelled T_1. We investigate two scenarios. In the first, the robot has the point w^{T_2} of branch T_2 grasped, and the branch T_3 is free (refer to Fig. 2). In the second

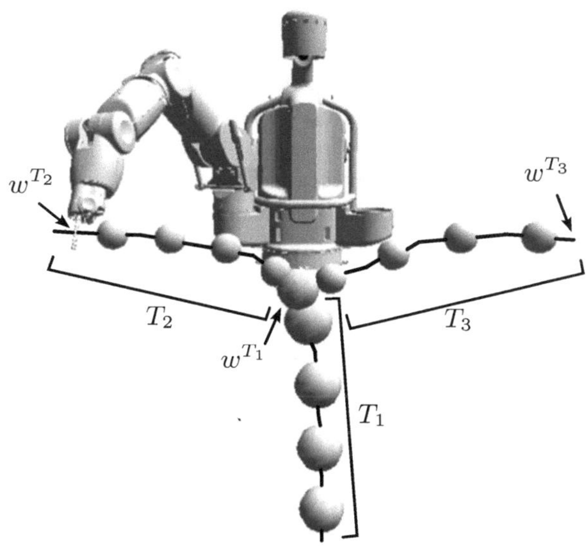

Fig. 2 Model of Baxter manipulating a flexible object. The object is composed of 3 *tubes*, each approximated with four rigid links. The *spheres* illustrate the location of the masses and their relative values. The tube segments are labelled T_1, T_2, and T_3, and the points at the ends of the tube are w^{T_1}, w^{T_2}, and w^{T_3}

scenario, the robot has both points w^{T_2} and w^{T_3} grasped using both arms. The goal is to manipulate the grasped branch so that the movement of the uncontrolled free branch can be predicted.

To improve prediction, we conduct the parameter identification optimization algorithm in [3] in order to identify the model's stiffness properties. The identification is made with physical contact data—i.e. joint angle and torques from one or two arms—and a vision system to capture the motion of the full flexible object. The identification process is the same as in [3] once the vision data has been fitted to the model.

The flexible object is foam tubing with differing widths for each of the T_1, T_2, and T_3 sections. The three tubes are connected with a 'Y' PVC joint and glue. The trunk tube has length 0.813 m and mass 0.064 kg. It is the widest tube with radius 0.032 m. The grasped tube has length 0.610 m and mass 0.015 kg and is the second widest tube with radius 0.025 m. The free tube has length 0.737 m and mass 0.016 kg. It is the thinnest with radius 0.019 m.

We use Rethink Robotics' Baxter robot [5] to both manipulate and measure the object. Baxter's arms each have 7 degrees of freedom. The arms are designed for compliance by means of series elastic actuators in each joint that allows for force sensing and control. Baxter publishes the joint angles and torques at 100 Hz. A picture of Baxter manipulating the flexible object is in Fig. 1.

3 Model and Simulation

We model the flexible object as a spring mass system by approximating the T_1, T_2, and T_3 segments each with four rigid links of uniform lengths and masses, connected by joints with torsional springs—see Fig. 2. Each joint is 3-dimensional allowing for bending and twisting motions of the flexible object. In total, the flexible object is 36 dimensional. The goal of the identification, Sect. 5, is to identify the torsional springs' spring constants. We label these parameters as ρ.

Due to Baxter's 7 degrees of freedom arms, the system of Baxter grasping one end of the flexible object (neglecting the other arm) has a total number of 43 configuration variables. When grasping both ends of the object, (using both arms) the model has a total number of 50 configuration variables. The dimensions, inertia, and other information concerning Baxter's arm can be obtained at https://github.com/RethinkRobotics.

3.1 Simulation

The model for both Baxter's arm and the flexible object are a series of rigid links connected by rotational joints. As such, the dynamics of both the manipulator and the object can be handled together. We use *variational integrators* to simulate the system dynamics. Variational integrators are a discrete-time representation of the equations of motion of a mechanical system. They are designed from the least action principle and have good properties that agree with known physical phenomenon like stable energy behavior [16].

Simulations are for a finite time interval $[0, t_f]$ with discrete times $t_0, t_1, \ldots, t_{k_f}$, where $t_0 = 0$, $t_{k_f} = t_f$ and $k_f + 1$ is the total number of discrete times in the interval. The simulation—i.e. the solving of the system dynamics—will result in a state $x_k := x(t_k)$ for each k. For variational integrators, the state is composed of the configuration, labelled q_k for time t_k, as well as a term labelled p_k, also for time t_k. For systems without external forcing, p_k is the conserved momentum. For the purposes of this paper, it can simply be thought of as analogous to the discrete velocity, which is often paired with q_k to make up the state. The state is $x_k := [q_k, p_k]^T$.

The literature on variational integrators [13] provides a one-step mapping to update the state at the previous time x_k to the next time x_{k+1}. We provide a short high-level review of variational integrators. We write the one-step mapping which constitutes the systems equations of motion as

$$x_{k+1} = f(x_k, \rho, t_k). \tag{1}$$

Here, f explicitly depends on the previous state, time, and the parameters which we wish to identify. While we write the equations of motion as an explicit equation, the equations are in fact implicit and rely on root solving to update the state.

The equations encapsulate the system's Lagrangian, any external forcing, holonomic constraints, as well as a choice of quadrature for approximating integrals. The function f can be linearized. We write

$$A_k = \frac{\partial}{\partial x_k} f(x_k, \rho, t_k) \text{ and } B_k = \frac{\partial}{\partial \rho} f(x_k, \rho, t_k). \tag{2}$$

The equations to calculate the linearization with respect to the state and parameters can be found in [3]. They are needed for calculating the gradient for parameter identification as part of an iterative optimization.

3.2 Simulation of Example

We use variational integrators to simulate Baxter manipulating the flexible object through the simulation tool `trep` [8]. The tool simulates articulated rigid bodies using midpoint variational integrators. It additionally provides partial derivative calculations that we need for the system linearization, Eq. (2).

The system of Baxter manipulating the flexible object with a single arm has a 43 dimensional configuration. Therefore, the system's state, $x_k = [q_k, p_k]^T$, is 86 dimensional. For the system of Baxter manipulating with both arms, the configuration is 50 dimensional and the state is 100 dimensional. The discrete dynamics f, Eq. (1), is given by the discrete system Lagrangian, discrete external forcing, and holonomoic constraints—see [3, 13]. The system Lagrangian is specified by the kinetic and potential energies of Baxter and the object. External forces enter the system through the torques applied by the motors at each of Baxter's joints. Additionally, holonomic constraints are needed to ensure that Baxter's end effectors remain in contact with the object. We chose a time step of 0.01 seconds, which matches the broadcast frequency of Baxter.

Nominally, the simulation will perfectly agree with Baxter's measured joint torque and angles for a given experiment. However, due to model and sensor disturbances, which are always an issue for real systems, this will not be the case. Furthermore, since the system is unstable—i.e. small disturbances can result in large changes to trajectory—directly feeding the measured torques into the model will not result in a meaningful simulation. Therefore, the measured joint torques, labelled $\overline{F}$, and measured joint angles, labelled $\overline{b}$, must be filtered through a feedback loop. We use a simple proportional control law with gain K:

$$F_k = \overline{F}_k - K_k(b_k - \overline{b}_k),$$

where F and b are the simulated joint torques and angles for the filtered control input. When K is large, the effect the parameters have on the simulation is dominated by the control and as such, the system cannot be identified. However, if K is too small, the system will remain unstable and not track the measured trajectory well enough to be meaningful. Correctly choosing K for the purposes of parameter identification of unstable systems is left for future work. For this paper, we chose K from a finite horizon LQR which results in an optimal feedback gain from the model linearized around b_{meas} and a quadratic cost functional (see [1] for LQR). The tradeoff between tracking the joint angles or joint torques is directly represented in the quadratic cost for specifying the size of K.

4 Vision

With the goal to improve parameter identification, the flexible object is visually tracked using an out of the box depth sensor, the Asus Xtion Pro Live. The depth sensor captures a point cloud, G, of the experiment. Processing G consists of three steps: filtering, segmentation, and graph creation.

A filtering component within the motion planning framework MoveIt! [4] performs the first step of self-filtering. It removes points detected on the robots body and arms. This is accomplished using the realtime joint states and calculated coordinate transforms to determine the robots configuration within the point cloud. A small amount of padding is included in the filtering to account for calibration error. The Point Cloud Library (PCL) [18] provides the second level of filtering, removing

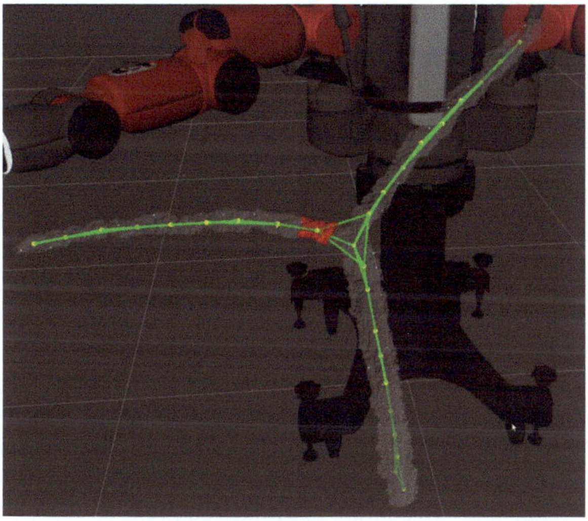

Fig. 3 Vision of flexible object

points detected behind the robot and on the floor. Finally, a statistical outlier removal filter removes remaining noise and measurement errors.

The segmentation of the filtered point cloud $G_{filtered}$ is accomplished using a custom algorithm built on top of PCL. This step converts the T_1, T_2, and T_3 sections (refer to Fig. 2) of the object into segmented components that can later be turned into a graph, as shown in Fig. 3. The lowest point in the point cloud (aligned with gravity) seeds the algorithm. The k_1 nearest neighbors to this point is then chosen using a Kd Tree to represent a segment? s of the plant model. The centroid of s is calculated and a second k_1 nearest neighbors search finds a centered segment $s_{centered}$ at the base of the plant.

Assuming the number of points in $s_{centered}$ are above a minimum threshold (to remove noise), the 3d centroid of $s_{centered}$ is added to a processed graph $G_{processed}$ and the points in $s_{centered}$ are removed from $G_{filtered}$. The next nearest neighbor to $s_{centered}$ is chosen as the new segment starting point and the algorithm repeats. Occasionally, there are insufficient nearest neighbors in a segment if, for example, the algorithm has reached the end of a plant branch. In this case, a random point is chosen in the remaining point cloud $G_{filtered}$ to continue the search, until no further points remain.

The final processing step takes the disconnected points in $G_{processed}$ and performs one final series of nearest neighbor searches to connect the nodes together to represent a flexible object modeled as a series of connected rigid bodies. These connected rigid bodies give a surprisingly accurate three dimensional reconstruction of a flexible object in soft-realtime, processing new point clouds at a rate averaging 3 Hz.

4.1 Fitting to Model

The processed points in $G_{processed}$ need to be fitted to the static model of the plant to be useful. The model is a discretization of the physical object where the flexible object's configuration specifies the location of the discretization points. Label the objects configuration as q_o.

The fitting is a calculation on q_o and is accomplished as follows: Let $G_{model}(q_o)$ be a graph specified for configuration q_o with the discrete points as its vertices and adjacent points in the model as its edges. Any two adjacent points in $G_{model}(q_o)$ can be connected by a line segment in space. Let $L_{model}(q_o)$ be the collection of these line segments. Further, let $d(p, \ell)$ be the shortest Euclidean distance between the point $p \in G_{processed}$ and line segment $\ell \in L_{model}(q_o)$. Define $d(p, L_{model}(q_o)) := \min_{\ell \in L_{model}(q_o)} d(p, \ell)$ as the least distance between p and any line segment in $L_{model}(q_o)$. This can be done for each $p \in G_{processed}$. The fitting is given by the q_o for which the points in $G_{processed}$ are nearest the line segments $L_{model}(q_o)$—i.e. by the optimization program

$$\arg\min_{q_0} \sum_{p \in G_{processed}} d(p, L_{model}(q_o)). \tag{3}$$

4.2 Vision Tracking Concerns

A number of tuned parameters make the vision filtering and fitting algorithms sensitive to object size, the distance between the camera and object, and variability of the object's thicknesses. However, it works well for our experimental goals.

One major shortcoming of the vision tracking pipeline developed for this experiment is occasional loss of data due to buffering issues and self occlusion. Because of the computational complexity of our flexible object manipulation pipeline, the experiment was run on 3 distributed commodity PCs using ROS [17]. A common issue is clock time synchronization between the three PCs, and ROS messages being dropped due to full buffers. This causes the robot transforms to be published with old time stamps and the robot self-filtering of the point clouds to stall, ultimately resulting in choppiness in the visual tracking of the plant model. This is an area of continued investigation and improvement.

5 Identification

The goal of the identification is to calculate the system model parameters that best agree with the physical behavior of the flexible object. For the example in the paper, the parameters we wish to identify are the spring constants associated with the flexible object spring mass model.

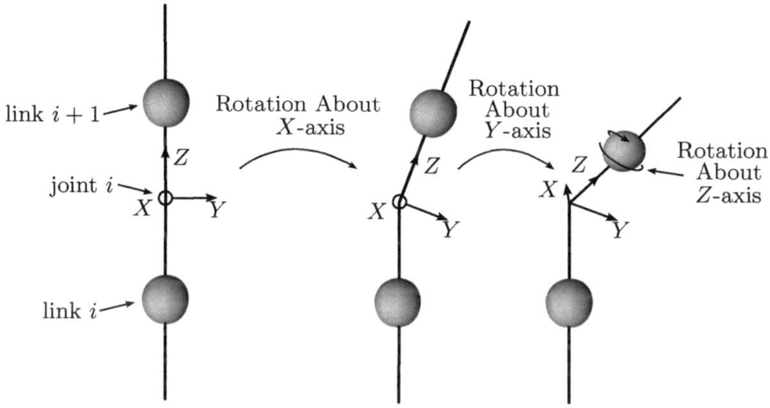

Fig. 4 Illustration of joint i connecting links i and $i+1$ of the flexible object. The joint is a rotation around the X, Y, and Z axes

Each joint of the flexible object is 3 dimensional; each can rotate around each axis. As seen in Fig. 4, each joint frame has the Z-axis aligned with the link. Therefore, a bend in the tube at a joint is realized by a rotation about the X- and Y-axes and a twist in the tube is a rotation about the Z-axis. The object's configuration specifies the amount each frame is rotated. Because the foam is uniformly distributed for each tube, we assume that the spring constants associated with bending—i.e. rotations about the X and Y axes—are the same for a single tube.

Label the torsional spring constant about the X-axis (alternatively Y or Z) for the ith tube as $\kappa_{T_i,X}$ (alt, $\kappa_{T_i,Y}$ or $\kappa_{T_i,Z}$). There are 6 total parameters $\rho = [\rho_1, \ldots, \rho_6]$ in our model, where

$$\begin{aligned}
\rho_1 &= \kappa_{T_1,X} = \kappa_{T_1,Y}, \\
\rho_2 &= \kappa_{T_1,Z}, \\
\rho_3 &= \kappa_{T_2,X} = \kappa_{T_2,Y}, \\
\rho_4 &= \kappa_{T_2,Z}, \\
\rho_5 &= \kappa_{T_3,X} = \kappa_{T_3,Y}, \text{ and} \\
\rho_6 &= \kappa_{T_3,Z}.
\end{aligned} \tag{4}$$

The goal of Sect. 6 is to identify ρ by calculating its corresponding simulation that best matches measured data. This parameter optimization is presented next.

5.1 *Optimal Parameter Identification*

The goal of parameter optimization is to calculate the model parameters ρ that minimize a cost functional. The cost functional is the integral of a running cost $\ell_d(x_k, \rho)$ plus a terminal cost $m(x_{k_f}, \rho)$:

$$\min_{\rho}\left[J_d(\rho) := \sum_{k=1}^{k_f} \ell_d(x_k, \rho) + m_d(x_{k_f}, \rho)\right]$$

constrained to the dynamics, $x_{k+1} = f(x_k, \rho, t_k)$. Since this is a nonlinear optimal controls problem, we turn to iterative methods like steepest descent to calculate a local minima. In order to apply steepest descent, we must have access to the gradient of the cost, which is given in the following Lemma from [3].

Lemma 1 *Suppose $f(x_k, \rho, t_k)$ is C^2 with respect to x_k and ρ. Let A_k and B_k form the linearization of f, Eq. (2), and assume f_k exists. Then,*

$$\nabla J_d(\rho) = \sum_{k=1}^{k_f} \lambda_k B_{k-1} + \frac{\partial}{\partial \rho}\ell_d(x_k, \rho) + \frac{\partial}{\partial \rho} m_d(x_{k_f}, \rho) \tag{5}$$

where λ_k is the solution to the backward one-step mapping

$$\lambda_k = \lambda_{k+1} A_k + \frac{\partial}{\partial x_k}\ell_d(x_k, \rho) \tag{6}$$

starting from $\lambda_{k_f} = \frac{\partial}{\partial x_{k_f}}\ell(x_{k_f}, \rho) + \frac{\partial}{\partial x_{k_f}} m_d(x_{k_f}, \rho)$.

It is worth noting that f_k is not guaranteed to exist, but its existence can be checked using the Implicit Function Theorem. In [7] a couple of scenarios are shown for which such singularities occur. Also, [14] reports the gradient and Hessian for optimal parameter identification in continuous time. The steepest descent direction is $-\nabla J_d(\rho)$ and the steepest descent algorithm can be applied [10].

6 Identification for Example

The parameters to be identified are the spring constants of the flexible object given by the six dimensional ρ. The identification calculates the value ρ with simulation that best matches the measured data, at least locally. We do the matching for two sets of measured data and for two scenarios—i.e. for a total of four experiments. The first measurement set is just Baxter's arm joint torque and joint angle measurements while the second also includes vision (see Sect. 4). The first scenario is Baxter manipulating one end of the object with a single arm and the second scenario is Baxter grasping both ends using both arms.

The identification is made by minimizing the error between simulated motion for a given ρ and measured motion at one to three points on the object, depending on the available measurements. The three points are the points at the end of the three tubes, labelled w^{T_1}, w^{T_2}, and w^{T^3} (refer to Fig. 2). The simulated points are labelled $w_k^{T_1}(\rho)$, $w_k^{T_2}(\rho)$, and $w_k^{T_3}(\rho)$ for parameters ρ at time t_k. The measured points at time t_k are labelled $\overline{w}_k^{T_1}$, $\overline{w}_k^{T_2}$, and $\overline{w}_k^{T_3}$. The point w^{T_1} can only be measured with vision, while the points w^{T_2} and w^{T_3} are measured by the robotic arm when grasped, otherwise they are measured by vision.

As discussed in Sect. 4.2, vision measurements arrive at irregular intervals and sometimes of poor quality due to missing data. The following process removes bad vision data and aligns the good data with the timing of the simulation: From Sect. 4, each frame of vision data is processed resulting in points $G_{processed}$ and fitted to the model with optimal fit of $q_o^\star$. Recall $q_o^\star$ is the object's configuration that best fits the data and is calculated from the program Eq. (3). The frame occurs at a time s and so we label that frame's fit as $q_o^\star(s)$. Furthermore, the quality of the fit is quantified by the value of $d(p, L_{model}(q_o^\star(s)))$, where lesser values correspond to better fits. As such, 'good' data is the configurations $q_o^\star(s)$ where $\sum_{p \in G_{processed}} d(p, L_{model}(q_o^\star(s))) < d_{max}$, a user specified tolerance. Data that does not meet this requirement is discarded. In order to align the data timing with the simulation, we first interpolate in time over the remaining data using a cubic spline and label the result $q_{o,interp}^\star$. Second, we calculate the simulation times t_k that are nearest the times s of the remaining data. Define $\sigma = \{\sigma_1, \ldots, \sigma_{k_f}\}$ as $\sigma_k = 1$ if t_k is the simulation time nearest a vision frame time s. In the identification, the cost function depends on the vision data $q_{o,interp}^\star(t_k)$ for which $\sigma_k = 1$, where the points $\overline{w}_k^{T_i}$, $i = 1, 2, 3$, are calculated using forward kinematics.

Aside: Since we have access to $q_{o,interp}^\star(t_k)$, we could alternatively choose to minimize the error in the simulated and measured configurations instead of at specific points. However, the position of any point on the object is not uniquely specified by a single configuration. In fact, since the vision system cannot measure a twist in a single tube, every configuration specifying the rotation around its local Z-axis (see Fig. 4) is arbitrarily set to 0. This results in a single measured object configuration, but it is unlikely that this choice results in the same configuration as the one simulated, even if the vision perfectly measures the position of all locations on the object.

The identification locally minimizes a cost function J_d given by running cost $\ell_d(q_k, \rho)$ and the terminal cost $m_d(q_{k_f}, \rho)$. Both depend on the error between simulated and measured. Label the errors for each of the three points as:

$$\epsilon_k^{T_1} = (w_k^{T_1}(\rho) - \overline{w}_k^{T_1}),\ \epsilon_k^{T_2} := w_k^{T_2}(\rho) - \overline{w}_k^{T_2},\text{ and } \epsilon_k^{T_3} = (w_k^{T_3}(\rho) - \overline{w}_k^{T_3}).$$

For the measurements that depend on vision, their corresponding error terms will be multiplied by σ_k in the running cost. The identification then locally minimizes J_d using the approach in [3] to calculate the locally optimal parameters $\rho^\star$ from an initial guess of $\rho = [3, 3, 3, 3, 3, 3]^T$ and inequality constraint $\rho_i \geq 0$—see Sect. 5.

The results follow:

6.1 Single Arm and No Vision

Without vision, the only measurements are Baxter's arm joint torques and angles. Baxter is only in contact with the object at the point w^{T_2}, and as such can only measure the flexible object's motion at that single point. This measurement is $\overline{w}_k^{T_2}$. The running cost, ℓ_d, and terminal cost, m_d, in the cost J_d are set as:

$$\ell_d(q_k, \rho) = \frac{1}{2}(\epsilon_k^{T_2})^T \epsilon_k^{T_2} \text{ and } m_d(q_{k_f}, \rho) = \frac{1}{2}(\epsilon_{k_f}^{T_2})^T \epsilon_{k_f}^{T_2}.$$

The locally optimal parameters for the single arm, no vision, case are $\rho^{\star}_{SA,NV} = [23.780, 0.000, 18.357, 11.103, 3.200, 3.059]^T$.

6.2 Single Arm and Vision

With vision, the motion of each point w^{T_1}, w^{T_2}, and w^{T_3} can be measured. Only $\overline{w}_k^{T_2}$ is measured from the robot arm, while $\overline{w}_k^{T_1}$ and $\overline{w}_k^{T_3}$ are measured from vision and are only valid at times t_k where $\sigma_k = 1$.

The running cost is

$$\ell_d(q_k, \rho) = \frac{1}{2}\sigma_k(\epsilon_k^{T_1})^T \epsilon_k^{T_1} + \frac{1}{2}(\epsilon_k^{T_2})^T \epsilon_k^{T_2} + \frac{1}{2}\sigma_k(\epsilon_k^{T_3})^T \epsilon_k^{T_3}.$$

We set the running cost to be $m_d(q_{k_f}, \rho) = \ell_d(q_{k_f}, \rho)$. Executing the identification results in locally optimal parameters $\rho^{\star}_{SA,V} = [22.270, 13.593, 10.735, 8.202, 11.5111, 9.692]^T$ for the single arm with vision measurements scenario.

6.3 Dual Arms and No Vision

With two arms and without vision, only the points w^{T_2} and w^{T_3} are measured. The running and terminal costs are

$$\ell_d(q_k, \rho) = \frac{1}{2}(\epsilon_k^{T_2})^T \epsilon_k^{T_2} + \frac{1}{2}(\epsilon_k^{T_3})^T \epsilon_k^{T_3}$$

and $m_d(q_{k_f}, \rho) = \ell_d(q_{k_f}, \rho)$. The identification results in the locally optimal parameters $\rho^{\star}_{DA,NV} = [19.559, 4.965, 8.575, 10.374, 0.000, 11.287]^T$.

6.4 Dual Arms and Vision

Finally, with both arms and vision, all points can be measured, but only the points w^{T_2} and w^{T_3} are measured from the arm and w^{T_3} is measured from vision. The running and terminal costs are

$$\ell_d(q_k, \rho) = \frac{1}{2}\sigma_k(\epsilon_k^{T_1})^T \epsilon_k^{T_1} + \frac{1}{2}(\epsilon_k^{T_2})^T \epsilon_k^{T_2} + \frac{1}{2}(\epsilon_k^{T_3})^T \epsilon_k^{T_3}.$$

and $m_d(q_{k_f}, \rho) = \ell_d(q_{k_f}, \rho)$. The optimal parameters are identified as $\rho^\star_{DA,V} = [24.840, 0.000, 12.373, 30.775, 1.406, 24.900]^T$.

6.5 Discussion of Results

The identified model parameters are not consistent with physical behavior. For example, it is unreasonable to expect that a physical tube does not have any stiffness associated with bending or twisting, which is reported by an identified spring constant of zero in all experiments except one. It is worth noting that even the experiment for which the robot manipulates both ends of the object and uses vision to locate the junction point, w^{T_1}, results in a spring constant with value 0.

We expect that this negative result is due to the motion dependence between the object's segments, which was not observable by the object's motion at the measurement locations $w^{T_1}, w^{T_2}, w^{T_3}$. In other words, we expect that with a perfect model and perfect measurements, the parameters ρ would not uniquely specify the motion of the measurement locations—or, at the least, that the parameters are highly sensitive to disturbances. This explanation is analogous to the observability gramian of linear control analysis being nonsingular.

To illustrate the issue, we look at a simple two spring system. Suppose both springs' spring constants differ and are unknown. One end of one spring is attached to one end of the other string. If the spring system is pulled apart and only the position and forces of the free ends are measured, then the spring constants cannot be identified.

Now, suppose external vision measurements are available and the attachment point can be measured. Then, the displacement of each spring caused by the external forces can be measured and the spring constants can be identified. We further complicate the system by connecting in series two torsional springs of differing, unknown spring constants. A known torque is applied to the system and the total angular displacement is measured. The goal is to identify the spring constants. However, assuming the vision system not able to measure the angular displacement of either torsional spring, there is once again an identifiability issue even with vision information.

This simple spring example and the identification results of the significantly higher dimensional system in this paper, provide insight into the problem of determining the minimal information needed to assess the stiffness of flexible objects. Solving such a problem would be invaluable to an automated identification routine for identification of other flexible objects with complex shapes or nonuniform stiffnesses. This problem will be addressed in future work.

Even though the identified parameters are not consistent with physical behavior, they are still useful depending on the desired task. For instance, the identified model is valid for planning a manipulation that moves a measurement point to a desired location. This problem is also planned future work.

7 Conclusion

The paper investigates the problem of identifying the stiffness profile of a flexible object shaped like the letter 'Y' through robotic manipulation. The robotic arms and a vision system measure the object's motion. Four experiments are run, once each for each of the following scenarios: manipulation with one or two arms and measurements with or without vision. The identification minimizes the error between the simulated and measured movement of up to three locations on the object depending on the available measurements. The identification results do not match the objects' expected physical properties, which we attribute to a failure to observe the motion dependence of the object's distinct segments. These results, while negative, provide insight into the problem of determining the minimal measurements needed to uniquely identify an object's stiffness.

References

1. Anderson, B.D.O., Moore, J.B.: Optimal Control: Linear Quadratic Methods. Dover Publications, INC (1990)
2. Bell, M.: Flexible object manipulation. PhD thesis, Dartmouth college, Hanover (2010)
3. Caldwell, T.M., Coleman, D., Correll, N.: Optimal parameter identification for discrete mechanical systems with application to flexible object manipulation. In: IEEE/RSJ International Conference on Intelligent Robots and Systems (2014)
4. Coleman, D., Sucan, I., Chitta, S., Correll, N.: Reducing the barrier to entry of complex robotic software. J. Softw. Eng. Robot. (Special issue on Best Practice in Robot Software Development) (2014)
5. Guizzo, E., Ackerman, E.: How rethink robotics built its new baxter robot worker. IEEE Spectrum (2011)
6. Jiménez, P.: Survey on model-based manipulation planning of deformable objects. Robot. Comput. Integr. Manufact. **28**(2), 154–163 (2012)
7. Johnson, E., Schultz, J., Murphey, T.D.: Structured linearization of discrete mechanical systems for analysis and optimal control. Trans. Autom. Sci. Eng. **12**(1), 140–152 (2014)
8. Johnson, E.R., Murphey, T.D.: Scalable variational integrators for constrained mechanical systems in generalized coordinates. IEEE Trans. Robot. **25**(6), 1249–1261 (2009)
9. Johnson, E.R., Murphey, T.D.: Linearizations for mechanical systems in generalized coordinates. In: American Control Conference, IEEE, pp. 629–633 (2010)
10. Kelley, C.T.: Iterative Methods for Optimization. Society for Industrial and Applied Mathematics, Philadelphia (1999)

11. Khalil, F.F., Payeur, P.: Dexterous robotic manipulation of deformable objects with multi-sensory feedback-a review. In: Robot Manipulators, Trends and Development, pp. 587–621. InTech (2010)
12. Lang, J., Pai, D.K., Woodham, R.J.: Acquisition of elastic models for interactive simulation. Int. J. Robot. Res. **21**(8), 713–733 (2002)
13. Marsden, J.E., West, M.: Discrete mechanics and variational integrators. Acta Numerica **10**(1), 357–514 (2001)
14. Miller, L.M., Murphey, T.D.: Simultaneous optimal estimation of mode transition times and parameters applied to simple traction models. IEEE Trans. Robot. **29**(6), 1496–1503 (2013)
15. Murray, R.M., Li, Z., Sastry, S.S.: A Mathematical Introduction to Robotic Manipulation. CRC press, Boca Raton (1994)
16. Pekarek, D., Murphey, T.D.: A backwards error analysis approach for simulation and control of nonsmooth mechanical systems. In: IEEE Conference on Decision and Control and European Control Conference (CDC-ECC), pp. 6942–6949 (2011)
17. Quigley, M., Conley, K., Gerkey, B., Faust, J., Foote, T., Leibs, J., Wheeler, R., Ng, A.Y.: ROS: an open-source robot operating system. In: ICRA workshop on open source software, vol. 3 (2009)
18. Rusu, R.B., Cousins, S.: 3d is here: point cloud library (pcl). In: IEEE International Conference on Robotics and Automation, IEEE, pp. 1–4 (2011)
19. Saha, Mitul, Isto, Pekka: Manipulation planning for deformable linear objects. IEEE Trans. Robot. **23**(6), 1141–1150 (2007)
20. Sahari, K.S.M., Min, C.H., Hou, Y.C.: Dynamic modeling of string for robotics application. In: Soft Computing and Intelligent Systems (SCIS) and 13th International Symposium on Advanced Intelligent Systems (ISIS), IEEE, pp. 774–779 (2012)
21. Wakamatsu, H, Takahashiand, K., Hirai, S.: Dynamic modeling of linear object deformation based on differential geometry coordinates. In: International Conference on Robotics and Automation, IEEE, pp. 1028–1033 (2005)
22. Wakamatsu, Hidefumi, Arai, Eiji, Hirai, Shinichi: Knotting/unknotting manipulation of deformable linear objects. Int. J. Robot. Res. **25**(4), 371–395 (2006)

Guided Manipulation Planning at the DARPA Robotics Challenge Trials

Christopher M. Dellin, Kyle Strabala, G. Clark Haynes, David Stager and Siddhartha S. Srinivasa

Abstract We document the empirical results from Carnegie Mellon University's entry into the DARPA Robotics Challenge Trials. Our system seamlessly and intelligently integrates recent advances in autonomous manipulation with the perspective and intuition of an expert human operator. Virtual fixtures are used as the common language between the operator and the motion planner. The planning system then solves a guided manipulation problem to perform disaster-response tasks.

1 Introduction

Motivation. Despite significant advancements in autonomous robotic manipulation, disasters such as the 2011 Fukashima Daiichi nuclear meltdown and the 2010 Deepwater Horizon oil spill have exposed severe limitations. Specifically, even when intact, these environments are engineered for human operation but not structured for robots. Worse, in a disaster, the environment changes unpredictably and significantly. Furthermore, errors in manipulation can be fatal not just for the robot, but can result in collateral damage.

When faced with such challenges, successful approaches often resort to direct teleoperation [16], as demonstrated in nuclear material handling [7], explosive ordnance disposal (EOD) [8], and assistive surgery [10]. Work to improve this interaction method has focused on providing assistance to the operator, e.g. graphical feedback on task orientations [5]. However, disaster scenarios have severe limitations in *bandwidth* and *latency*, making direct teleoperation challenging and time-consuming. We aim for a fast and robust approach to this problem which allows a robot directed by a human operator to efficiently perform task-relevant behaviors over a low-fidelity communications link.

C.M. Dellin (✉) · K. Strabala · G. C. Haynes · D. Stager · S. S. Srinivasa
National Robotics Engineering Center, The Robotics Institute,
Carnegie Mellon University, Pittsburgh, USA
e-mail: cdellin@cs.cmu.edu

M.A. Hsieh et al. (eds.), *Experimental Robotics*, Springer Tracts
in Advanced Robotics 109, DOI 10.1007/978-3-319-23778-7_11

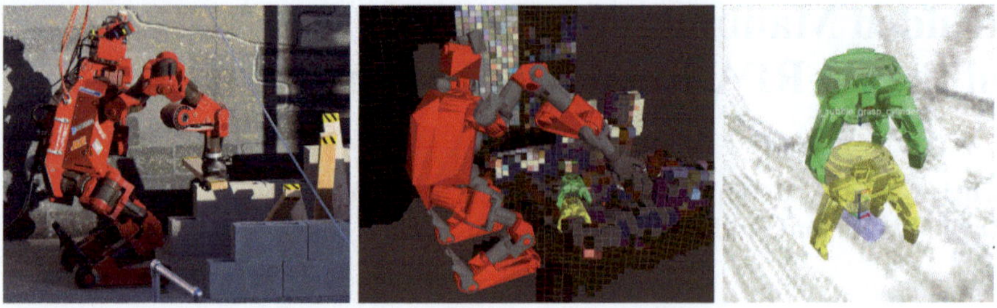

Fig. 1 CHIMP clearing ten pieces of wood during the Debris task at the DRC Trials, along with the operator's reconstructed 3D environment containing fixtures which comprise the common language between the operator and the manipulation planner

Problem Statement. The DARPA Robotics Challenge (DRC) program is a competition which aims to "develop ground robots capable of executing complex tasks in dangerous, degraded, human-engineered environments".[1] Carnegie Mellon's National Robotics Engineering Center entered the competition, and developed the CHIMP robot [2] [14] and its software system that demonstrates significant functionality in these tasks.

Contributions. Our key tenet is to efficiently integrate recent advances in autonomous manipulation (e.g. [1]) with the perspective and intuition of an expert human operator using *virtual fixtures* (Fig. 1). Instead of their typical use as an aid to the operator during teleoperation [13], we use virtual fixtures as a *common language* that enables the operator to specify constraints and annotations to the robot, which it then addresses autonomously. We expect that the system's manipulation capability under real-world task conditions will trigger further development in both theoretic research and system design.

Related Work. Our work is related to shared-autonomy manipulation planning systems such as the "human-in-the-loop" system developed by [9]; we extend their approach to general sets of constraints in a concurrent, multi-operator framework. We also note the relation to spacecraft telerobotics approaches (e.g. [17]) which must accommodate significant latency, albeit without the time pressure inherent in disaster response scenarios.

Outline. This paper presents an outline of our approach and analyzes our results from the DRC Trials, which took place in Homestead, FL in December 2013. We present our technical approach in Sect. 2. Section 3 details the application of the approach to each of the five manipulation tasks as the Trials, and Sect. 4 briefly describes our aggregated results and discusses experimental insights.

[1] http://www.darpa.mil/program/darpa-robotics-challenge.

[2] Carnegie Mellon University Highly Intelligent Mobile Platform.

2 Technical Approach

A comprehensive solution to this problem requires a wide range of intelligent components (e.g. high-performance hardware, compressed world modeling, etc.). In this paper, we focus on the foundational components used for describing the five manipulation tasks between the operator and the CHIMP planning and execution system. Details of the full system are available in [14].

Complementary Skills. When developing our approach to the manipulation tasks at the challenge, we wanted the system to exploit the complementary skills of the operator and the robot; see discussions of these skills e.g. in surgery [15] or space robotics [12]. The operator is generally endowed with *task awareness*, a high-level understanding of the task to be performed, along with the sequence of steps the robot must execute; however, direct teleoperation approaches often fail, especially under restricted networking fidelity or for constrained tasks. In contrast, the robot's motion planners are equipped with *kinematic awareness*, the capability to search and optimize over complex kinematic, collision, and stability constraints, but fully autonomous task planners suffer from brittleness and overfitting, particularly in unstructured environments. An efficient design of the interface between the two is therefore essential.

Fixtures as a Common Language. We chose to use a variant of *virtual fixtures* as this primary interface. In a traditional telepresence system, a virtual fixture [13] serves as a perceptual overlay or constraint, introduced to an operator's understanding of the remote scene, designed to reduce their requisite mental and sensory workload and improve task performance. Fixtures have been applied to fields such as medical [4] and space [17] robotics.

While our fixtures do serve this purpose for the operator, we also focus on the dual purpose: using fixtures as task-relevant specifications for the robot's planning and execution system. Fixtures are first-class objects, which can be defined by the operator in order to build a context for the robot to perform a task, or placed/adjusted automatically by the robot in response to perception data or its internal world model. Fixtures are spatial entities that are defined in *task space* relative to a world frame, robot link, or another fixture. To the operator, fixtures are presented and manipulated by overlaying them onto the voxelized world model.

Our key insight is that virtual fixtures, as first-class spatial entities, are an effective interface between the operator and the robot. They enable each to exercise their strengths, allowing the operator task-level control while abstracting away the complexities of the robot's kinematics, geometry, stability, etc. To the robot, fixtures impart task-rooted guidance, constraints, and intermediate goals which focus the planning problem.

Examples and Usage of Virtual Fixtures. In this paper, we focus primarily on the use of fixtures as task-space kinematic constraints. Each fixture *targets* either a robot link (e.g. gripper) or another fixture and advertises one or more named constraints for that target. Such constraints are then available to be selectively referenced by the planning and execution system, often in sequence for a particular task.

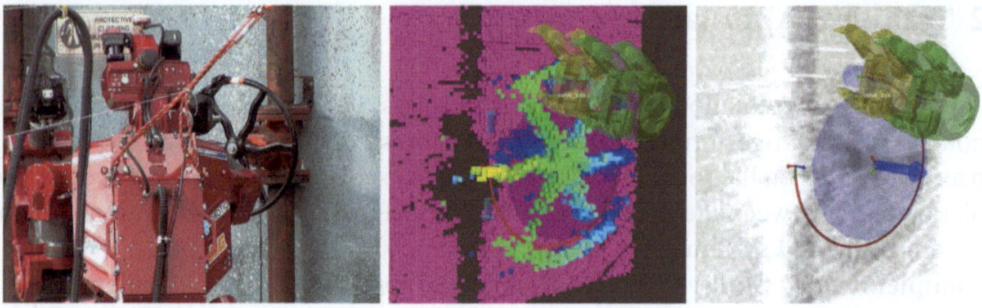

Fig. 2 CHIMP preparing to turn a valve at the DRC Trials, along with a `GraspCylinder` fixture (*upper* cylinder and grippers) and an `Axis` fixture (*lower* disk and axis *arrow*) configured for a 270° clockwise turn

For example, the `Axis` fixture codifies a generic rotational constraint about a fixed axis (Fig. 2). It is placed by the operator in the 3D world by first selecting a center point, and then selecting at least three points which define a plane normal to the axis. The fixture can be configured with labeled angles (e.g. `min`/`max`, `goal`, etc.). During the Door and Valve tasks, the handle was fixtured by a `GraspCylinder`, and the door/valve body by an `Axis` which targetted the `GraspCylinder` (i.e. constrained it to lie along a particular manifold).

The `GraspCylinder` fixture represents a graspable object using a simple grasping strategy that includes a configurable target gripper and grasp/pregrasp offsets. Figure 1 shows this fixture applied to a debris piece at the Trials. It advertises the `pregrasp`, `approach`, and `grasp` constraints, any of which may be active when the appropriate gripper satisfies the specification.

Fixtures may also be more task-specific when warranted; a `PlanarPath` fixture was created for the Wall task, which allowed the operator to define an arbitrary path of linear segments on a 2D surface (see Fig. 3). In preparation for the Trials tasks, we created instances of several other fixtures, including the `Vector`, `Box`, and `RotaryCutTool` which will not be discussed in detail.

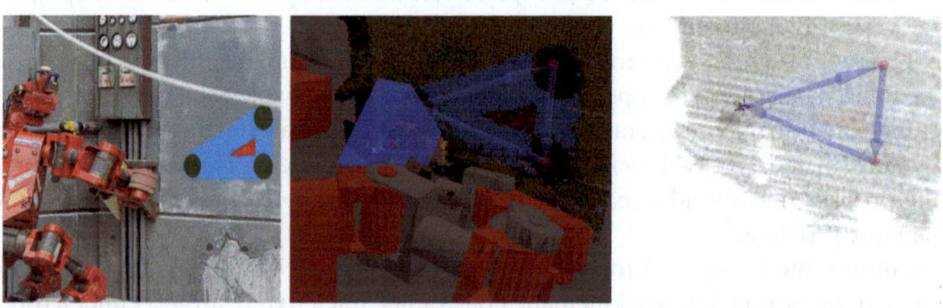

Fig. 3 CHIMP preparing to cut a triangular pattern from a wall in the DRC Trials, along with a `Box` fixture representing the volume of the drill in the hand, and a `PlanarPath` fixture showing the *trianglular shape* to be cut

Guided Manipulation Planning. To the robot's motion planning system, each fixture defining a Cartesian kinematic constraint induces a manifold in the robot's configuration space. We chose to represent these constraints on target objects in the scene via Task Space Regions (TSRs) [3]. Since all fixtures are first-class objects available to the planning system, multi-step planning requests can simply reference fixture identifiers. For example, this composite request represents a valve turn:

```
Plan(to=GraspCylinder1.pregrasp)
Plan(along=GraspCylinder1.approach, to=GraspCylinder1.grasp)
Plan(along=Axis1.initial_to_goal, to=Axis1.goal)
Plan(along=GraspCylinder1.approach, to=GraspCylinder1.pregrasp)
```

In this way, the role of the planner is simply to move the system between and along the constraint manifolds induced by the fixtures. The robot is tasked with what we term the *guided manipulation problem*, finding feasible paths guided by the given ordering of constraints. While the request above may be composed manually by the operator, the system provides shortcuts via "wizards" for composing common tasks (e.g. pick-and-place, operating valves/hinges, and completing a wall cut). During the Trials, we used the CBiRRT algorithm [2]) for planning with configuration-space constraint manifolds.

Trajectory Confirmation and Supervision. Once a candidate plan is computed, the resulting trajectory can be previewed by the operator before it is sent to the robot. Once executing, it can be supervised by the operator, who may choose to step through the trajectory one segment at a time for close inspection.

Trajectory Execution and Control. The robot executor maintains a queue of trajectory segments to be executed. Each segment produced by the planner is *tagged* with the fixture(s) that were asserted to be active; this allows the executor to validate that the fixtures' constraints are still met at the time of execution, and adjust the trajectory accordingly in some cases.

Trajectory segments which respect certain types of fixtures are tagged to be executed in particular ways; for example, segments which induce closed kinematic chains (e.g. valve turns) are executed using a workspace force controller which allows for looser gains in overconstrained directions.

3 Experimental Results

We designed a flexible system consisting of mobility actions, annotation tools, constrained motion planners, and tele-operation primitives that can be adapted to different task workflows. We leveraged prior work for the infrastructure (ROS [11]) and planning environment/algorithms (OpenRAVE [6], CBiRRT [2]), as well as significant technologies developed at the NREC. Here, we detail CHIMP's performance during the five manipulation tasks at the DRC Trials (see Figs. 4 and 5).

Fig. 4 CHIMP performing the five manipulation tasks at the DRC Trials: Debris, Door, Wall, Valve, and Hose

Fig. 5 During the Debris task, CHIMP clears piece 2, between 150 and 310 s

Network. During the competition, operators were physically isolated from the task area, with the network link alternating every minute between 1 Mbps bandwidth/100 ms latency and 100 kbps bandwidth/1 s latency.

Scoring. Teams were allowed 30 min to complete each task, and were judged by three metrics prioritized to break ties:

1. Points (Task Completion): teams were awarded 3 points for full completion of each task. Partial points were awarded for defined subtask completion.
2. Interventions: teams were permitted to (but penalized for) manually intervening during the task execution (e.g. falling on the safety belay). A bonus point was awarded for each task in which no interventions were needed.
3. Completion Time: aggregated time taken by the robot to perform the tasks.

The remainder of this section describes in detail how our technical approach was applied to each of the five manipulation tasks, and provides an analysis of time spent. We only provide a cursory description of the tasks themselves; we invite the reader to review the full task descriptions and detailed rules for the Trials competition at the DARPA archive website.[3]

Multiple Operators. For each task at the Trials, between two and four operators interacted simultaneously with the robot and its internal hardware, perception, fixturing, planning, execution, and teleoperation systems. Because fixtures are first-class

[3]`http://archive.darpa.mil/roboticschallengetrialsarchive/`.

objects, they can be defined by one operator, but seen, modified, and referenced by plans from other operators.

Timeline Figures. For each task, a figure is provided which details the fixtures, plans, network statistics, and robot execution mode throughout the 1800 s of allotted time (Figs. 6, 7, 8, 9, 10).

Fixture lifetimes are shown first. Fixtures are shared between operators once they are first created and saved. Black activity bars denote times when an operator is actively modifying fixture parameters. Planning durations are shown next, along with connections to the fixtures referenced in each planning request. Only operators who adjusted fixtures or requested plans are shown in these figures.

Estimates of network latency and aggregated bandwidth from ("Rx") and to ("Tx") the robot are shown. The robot executor was modal, either in trajectory execution ("Trj"), end-effector teleoperation ("Tel"), or driving ("Drv") mode; black activity bars denote individual trajectory segments or approximate motion request bandwidth. Last, estimates of torso speed and times of points awarded are illustrated.

3.1 Debris Task

Debris Setup. Each robot was allowed to `Setup` behind a start line. The task involved removing five pieces of wooden debris (1 pt.), removing an additional five pieces of debris (1 pt.), and driving through the open doorway (1 pt.). The approximate configuration of the debris was provided apriori.

Debris Approach. Fig. 6b outlines our approach for the debris task. The task was roughly organized into an initial `Setup` phase, followed by a four-step cycle of `Drive`, `Grasp`, `Liftoff`, and `Drop` phases. During the task, four operators operated the robot simultaneously, with loose roles of (a) world modeling, (b) fixturing/planning, (c) teleoperation, and (d) hardware monitoring.

The Debris task used the `GraspCylinder` fixture to define a grasp of each piece, along with a `Vector` fixture to define its liftoff direction and distance.

During `Setup`, the world modeler constructed approximate volumetric models of several static objects in the environment, in order to improve collision checking speed and accuracy. The `Grasp` and `Liftoff` phases are primarily managed by the fixturing/planning operator, who creates virtual fixture annotations for pregrasp, grasp, and piece liftoff constraints for the planner, invokes the planner, reviews the proposed trajectory, and supervises its execution. Once the piece is grasped and lifted, in the `Drop` phase, the third operator teleoperates the gripper towards a rough drop area to drop the piece.

Debris Results. We achieved four points in 1749 s; see Fig. 1 for a view of CHIMP mid-task. We prematurely dropped the ninth piece, and returned to it after completing the subsequent piece; the drop occurred at 1380 s, during its teleoperation move to the drop zone. We took advantage of pipelining (see Fig. 6c); for example, for many pieces, fixtures for subsequent pieces were created and positioned prior to the current piece being dropped.

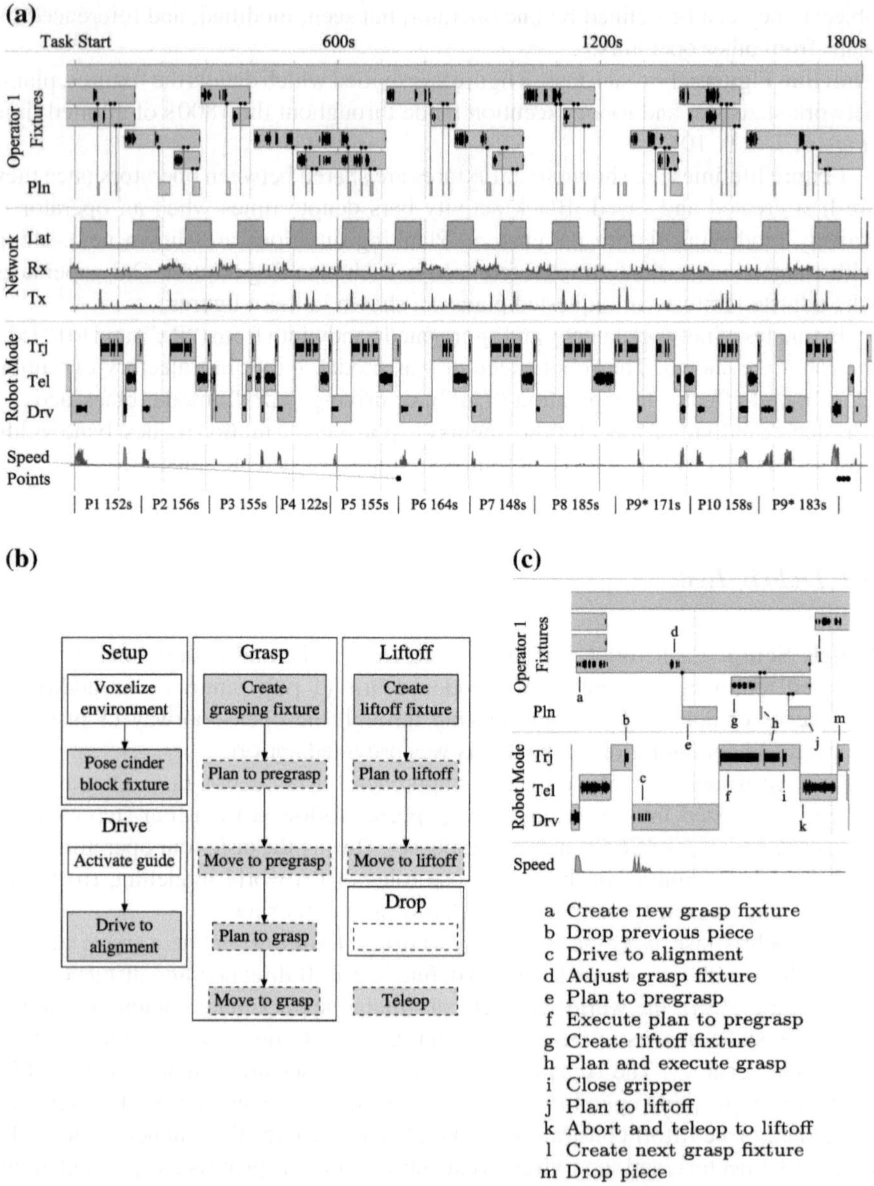

Fig. 6 Anatomy of the Debris task **a** Full debris task, moving ten pieces in 1749 s. Piece 9 (*) was dropped during its first attempted transfer, and was later retrieved to complete the task. **b** Flowchart of debris task. The cinder block fixture represents the environment; it is the first fixture created (*top row* in Fig. 6**a**) and exists for the entire task. *Dashed-border* steps transition to teleoperation on failure, **c** Annotated log of piece 2

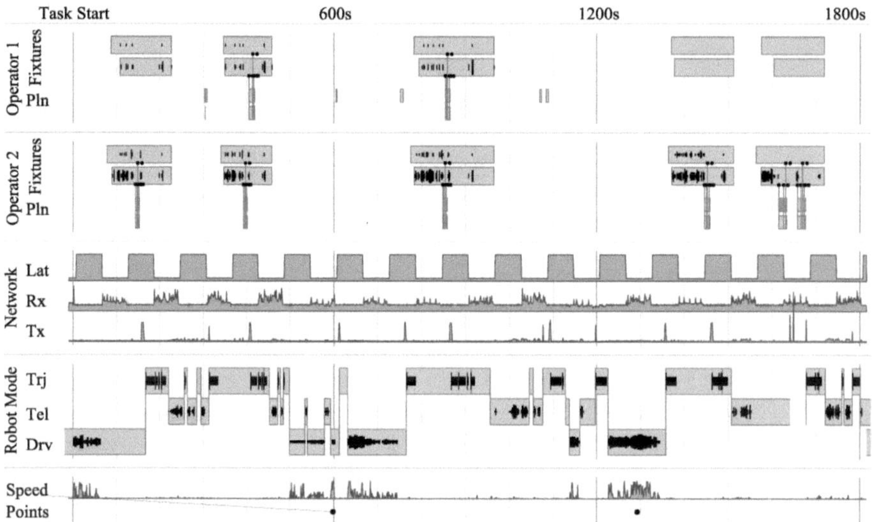

Fig. 7 Timeline of Door task. Each opening required two fixtures (a `GraspCylinder` for the handle, and an `Axis` for turning constraint); two operators cooperated to adjust fixtures and request plans. Two events resulted in premature door closings, necessitating repeated opening attempts: at 325 s, a strong wind gust blew the first closed, and at 1547 s, an errant teleoperation command caused the third to slip from CHIMP's control

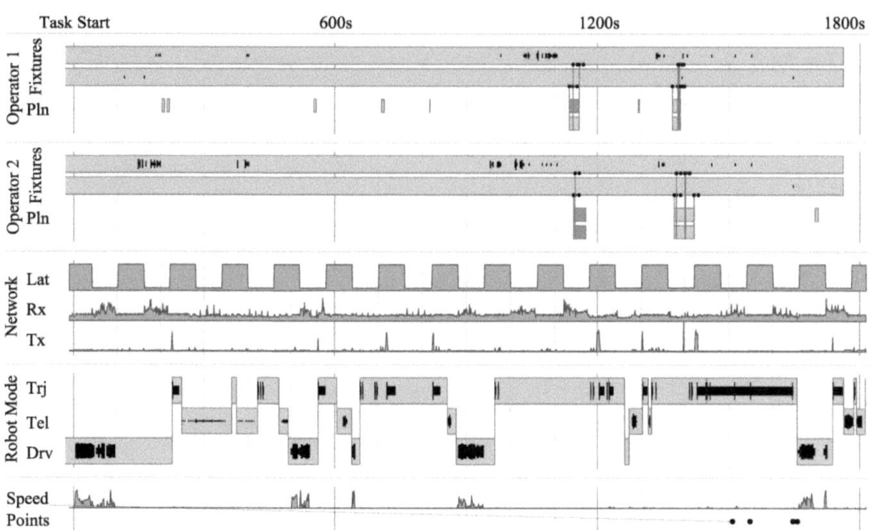

Fig. 8 Timeline of Wall task. The drill was successfully grasped and test-actuated around 550 s. The second operator made continual adjustments to the polygonal-path fixture, especially between 950–1050 s before the first plans. During supervision of the first execution at 1200 s, the operators visually detected that the cut was not as deep as desired, so the fixtures were re-adjusted. Cutting proceeded from 1425 to1650 s

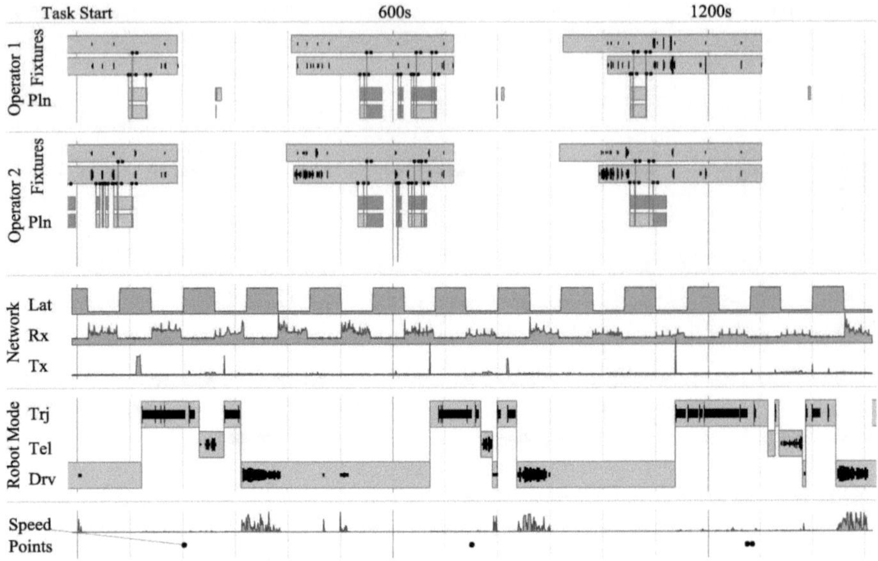

Fig. 9 Timeline of Valve task. Each valve was annotated using two fixtures. After the resulting motion was previewed, the trajectory was sent to the robot (at 110, 670, and 1140 s) for supervised execution. After each successful turn, the gripper was extracted using teleoperation (e.g. at 340 s) and the arm was reset to a driving posture (e.g. at 385 s). The robot was then driven to the next valve, and the process repeated

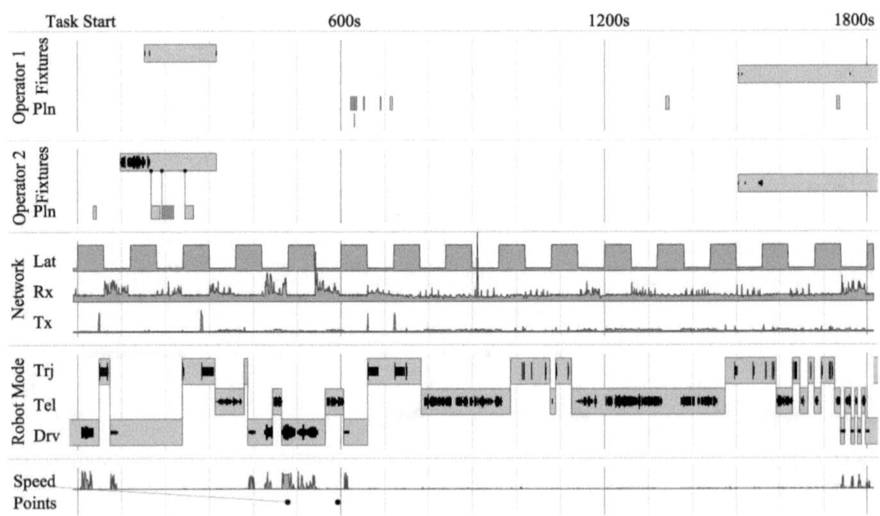

Fig. 10 Timeline of Hose task. The hose was grasped using a grasp strategy fixture at 380 s, and the first two points were achieved by 593. The remaining time was spent attempting to thread it onto the wye, with no success

A log of data collected at the debris task is shown in Fig. 6a. We moved all ten pieces in approximately 29 min. Figure 6c shows an annotated log of the process to clear the second of the ten pieces, and Fig. 5 shows the robot during its motion. The approach provides for several opportunities for pipelining. First, the fixture for the piece was created (a) before the previous piece was dropped (b), and before the robot was positioned appropriately (c). Second, the liftoff fixture was created and positioned (g) while the to-pregrasp motion was being executed. This piece also demonstrates a fail-over strategy that we used, whereby a long-running plan (j) or execution was interrupted and performed by teleoperation (k).

3.2 Door Task

Door Setup. The task required sequentially opening and traversing three doors: the first a "push" door (1 pt.), the second a "pull" door (1 pt.), and the third a "pull" door with spring closure (1 pt.). All doors had identical lever-style handles.
Door Approach. We used the `GraspCylinder` and `Axis` fixtures to approach, grasp, and turn each door handle. See Fig. 3 for examples of the axis fixture, which constrained the planner to move the valve handle body about a fixed axis. During execution, this trajectory segment was executed with a Cartesian force controller as described in Sect. 2. Subsequent manipulation of the doors (pulling and pushing) was performed via a combination of (a) pre-selected arm configurations and (b) gripper Cartesian-space teleoperation. Traversing the third door required positioning CHIMP in such a way that the door was actively held open while it was being traversed.
Door Results. See Fig. 7 for a time breakdown of the door task. We achieved two points in the alloted 1800 s. CHIMP successfully actuated the door handles in all 5 attempts, but suffered two events which lead to premature door closures on the first and third doors, requiring extra time.

3.3 Wall Task

Wall Setup. The task required grasping a cordless cutting instrument and using it to cut a prescribed triangle shape in drywall. Teams could choose between a drill loaded with a side-cutting bit, or a small circular reciprocating saw. Each of the three edges of the triangle successfully cut (without damaging the wall outside the lines) was worth one point.
Wall Approach. We used the `GraspCylinder` fixture to approach the drill, along with precision nudges to precisely grasp it so that the trigger was reachable. Actuation of the tool was visually inspected using a trigger-actuated light. We used the `Box` and `RotaryCutTool` fixtures to model the volume and cutting bit of the

drill, and used the `PlanarPath` and `Vector` fixtures to fully specify the location of the triangle shape on the wall, along with the approach direction and distance. Constrained planning was used to compute a full trajectory to perform all steps of the cut.

Wall Results. See Fig. 8 for a time breakdown of the Wall task. We achieved four points in 1647 s. After the first attempt, during supervision, the operator team determined that the bit may not have sufficiently punctured the wall; the puncture distance was adjusted, and after replanning, the trajectory was allowed to run to completion. Once the drill was grasped and the path annotated, the operators assumed only supervisory roles.

3.4 Valve Task

Valve Setup. The task required grasping and turning three valves (a 90-degree lever valve, a large circular valve, and a small circular valve). See Fig. 2. Each of the three valves completely turned (360° for the circular values) earned one point.

Valve Approach. We used the `GraspCylinder` fixture to describe the grasp strategy for each valve handle body. We then used the `Axis` fixture to label the axis of rotation of each valve. See Fig. 3 for examples of these fixtures for the valve task. Constrained planning was used to compute a full trajectory to turn each valve. During execution, this trajectory segment was executed with a Cartesian force controller as described in Sect. 2.

Valve Results. See Fig. 9 for a time breakdown of the Valve task. We achieved four points in 1275 s.

3.5 Hose Task

Hose Setup. The task required retrieving a hose from a wall-mounted spool, and transferring it for several meters (1 pt.), touching the hose nozzle to a wye (1 pt.) and threading it onto the wye (1 pt.).

Hose Approach. We used the generic grasp fixture to retrieve the hose, and quickly transferred and touched the nozzle to the wye. In limited testing, we hadn't found a robust way to accomplish the threading component.

Hose Results. See Fig. 10 for a time breakdown of the Hose task. We achieved two points in the first 593 s, and spent the remainder of the task time attempting to complete the third subtask.

4 Main Experimental Insights

Team Tartan Rescue placed third in the competition, and achieved 16 out of a possible 20 points on the five manipulation tasks (see Fig. 11). We were also the only team in the competition which was not penalized for an intervention.

The time spent for each task is shown in Fig. 12. Note that this allocation does not account for pipelining; when multiple operators were performing different actions simultaneously, the one deemed in the critical path was counted.

Our approach forms a strong foundation for human-guided manipulation applicable to disaster response scenarios.

Virtual Fixtures. We found the development and workflow of virtual fixtures to be at the appropriate level of generality and extensibility for the problem we addressed at the DRC Trials competition. As first-class spatial objects, operators found it straightforward to reason about their representation during the tasks.

Operator Experience. In contrast to a fully-autonomous system, we found performance was correlated with operator training. Over time, the operators learned heuristics for task parameters (e.g. base placements, grasp orientations, etc.) that lead to fast and robust solutions.

Multi-Step Planning Robustness. When solved naïvely, multi-step plans can often fail by committing to choices early that preclude efficient solutions to later steps. This happened occasionally during the trials (e.g. during debris piece 2 from Fig. 6c). A

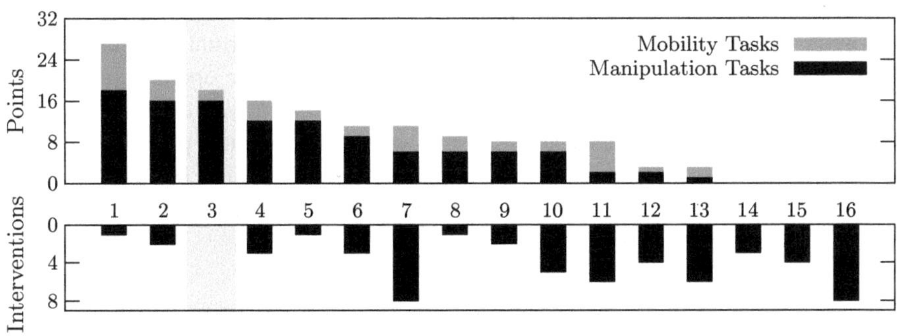

Fig. 11 Scores of each team at the DRC Trials, ranked by total points achieved. Team Tartan Rescue placed third with 18 points (tied for second with 16 points on the five manipulation tasks), and was the only team with zero interventions

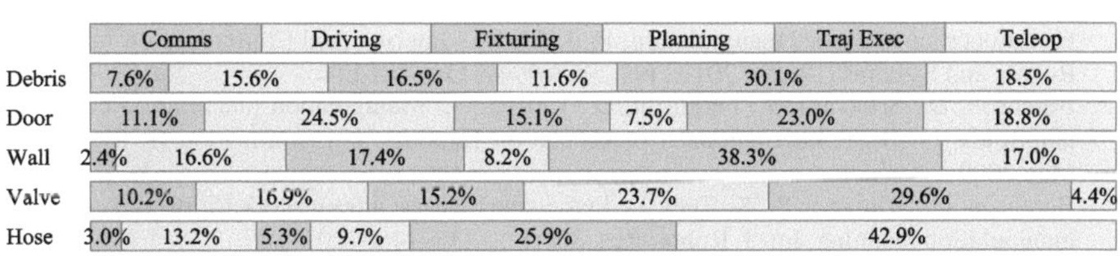

Fig. 12 Time breakdown of each task

solution to this problem may improve planning success rates, and is a promising area for future work.

Pipelining. We exploited pipelining between locomotion, fixturing, planning, and trajectory execution to improve our task completion times.

Failsafes. During trials, failed or long-running plans or executions were superseded by end-effector teleoperation or joint-level control. This strategy allowed for increased robustness and execution speed.

4.1 Future Directions

Our current approach and implementation is a first step towards developing a framework for guided manipulation. We are excited about two directions of future work: *autonomy* and *expressiveness*.

Towards greater autonomy. Our current framework relies completely on the operator for the deployment of virtual fixtues. By relying on the operator's spatial awareness, we are able to execute complex manipulation tasks with little perception: our system uses unstructured voxel worlds for collision avoidance and does not currently perform semantic perception.

Our framework does expose the scaffolding for semantic perception and learning. Given semantic information from a perception system, like objects, handles, and door kinematics, the system can automatically initialize fixtures that are cached or learned from previous demonstration.

Towards greater expressiveness. The expressiveness of virtual fixtures depends synergistically both on the capabilities and preferences of the operator, and on the capabilities of the underlying planning algorithms. Currently, we are restricted to virtual fixtures expressed as Cartesian task-space regions, and planning requests as fixed sequences of fixtures. In the future, we envision adding branching, giving the planner options to choose from, as well as more complex constraints related to stability and sensor visibility.

References

1. Bagnell, J., Cavalcanti, F., Cui, L., Galluzzo, T., Hebert, M., Kazemi, M. Klingensmith, M., Libby, J., Liu, T.Y., Pollard, N., Pivtoraiko, M., Valois, J.S., Zhu, R.: An integrated system for autonomous robotics manipulation. In: IEEE/RSJ International Conference on Intelligent Robots and Systems (IROS), 2012 , pp. 2955–2962, Oct. 2012
2. Berenson, D., Srinivasa, S., Ferguson, D., Kuffner, J.: Manipulation planning on constraint manifolds. In: IEEE International Conference on Robotics and Automation (ICRA), 2011, May 2009
3. Berenson, D., Srinivasa, S., Kuffner, J.: Task space regions: a framework for pose-constrained manipulation planning. Int. J. Robot. Res. **30**(12), 1435–1460 (2011)
4. Bettini, A., Marayong, P., Lang, S., Okamura, A., Hager, G.: Vision-assisted control for manipulation using virtual fixtures. IEEE Trans. Robot. **20**(6), 953–966 (2004)

5. Debus, T., Stoll, J., Howe, R.D., Dupont, P.: Cooperative human and machine perception in teleoperated assembly. In: Rus, D., Singh, S. (eds.) Experimental Robotics VII. Lecture Notes in Control and Information Sciences, vol. 271, pp. 51–60. Springer, Berlin Heidelberg (2001)
6. Diankov, R.: Automated construction of robotic manipulation programs. Ph.D. thesis, Carnegie Mellon University, Robotics Institute, Aug. 2010
7. Goertz, R.C.: Manipulators used for handling radioactive materials (Mcgraw-hill, New York, 1963)
8. Kron, A., Schmidt, G., Petzold, B., Zah, M., Hinterseer, P., Steinbach, E.: Disposal of explosive ordnances by use of a bimanual haptic telepresence system. In: Proceedings of Robotics and Automation, 2004. IEEE International Conference on ICRA '04. 2004, vol. 2, pp. 1968–1973, Apr. 2004
9. Leeper, A., Hsiao, K., Ciocarlie, M., Takayama, L., Gossow, D.: Strategies for human-in-the-loop robotic grasping. In: 7th ACM/IEEE International Conference on Human-Robot Interaction (HRI), 2012, pp. 1–8, Mar. 2012
10. Park, S., Howe, R.D., Torchiana, D.F.: Virtual fixtures for robotic cardiac surgery. In: Proceedings of the 4th International Conference on Medical Image Computing and Computer-Assisted Intervention, MICCAI '01, pp. 1419–1420, Springer, London, UK (2001)
11. Quigley, M., Gerkey, B., Williams, C.K., Faust, J., Foote, T., Leibs, J., Berger, E., Wheeler, R., A.Y. Ng. Ros: An open-source robot operating system. In: Proceedings of Open-Source Software workshop, IEEE International Conference on Robotics and Automation, (ICRA), 2009
12. Rodriguez, G., Weisbin, C.: A new method to evaluate human-robot system performance. Auton. Robot. **14**(2–3), 165–178 (2003)
13. Rosenberg, L.: Virtual fixtures: perceptual tools for telerobotic manipulation. In: Virtual Reality Annual International Symposium, 1993 IEEE, pp. 76–82, Sept. 1993
14. Stentz, T., Herman, H., Kelly, A., Meyhofer, E., Haynes, G., Stager, C.D., Zajac, B., Bagnell, J.A., Brindza, J., Dellin, C., George, M., Gonzalez-Mora, J., Hyde, S., Jones, M., Laverne, M., Likhachev, M., Lister, L., Powers, M., Ramos, O., Ray, J., Rice, D., Scheifflee, J., Sidki, R., Srinivasa, S., Strabala, K., Tardif, J.-P., Valois, J.-S., Vande Weghe, J.M., Wagner, M., Wellington, C.: CHIMP, the CMU highly intelligent mobile platform. J. Field Robot. Special Issue on the DARPA Robotics Challenge (2014)
15. Taylor, R.: Robots as surgical assistants: where we are, wither we are tending, and how to get there. In: Keravnou, E., Garbay, C., Baud, R., Wyatt, J. (eds.) Artificial Intelligence in Medicine. Lecture Notes in Computer Science, vol. 1211, pp. 1–11. Springer, Berlin Heidelberg (1997)
16. Trevelyan, J., Kang, S.-C., Hamel, W.: Robotics in hazardous applications. In: Siciliano, B., Khatib, O. (eds.) Springer Handbook of Robotics, pp. 1101–1126. Springer, Berlin Heidelberg (2008)
17. Xia, T., Leonard, S., Kandaswamy, I., Blank, A., Whitcomb, L., Kazanzides, P.: Model-based telerobotic control with virtual fixtures for satellite servicing tasks. In: IEEE International Conference on Robotics and Automation (ICRA) 2013, pp. 1479–1484, May 2013

Redundancy Resolution in Human-Robot Co-manipulation with Cartesian Impedance Control

Fanny Ficuciello, Luigi Villani and Bruno Siciliano

Abstract In this paper the role of redundancy in Cartesian impedance control of a robotic arm for the execution of tasks in co-manipulation with humans is considered. In particular, the problem of stability is experimentally investigated. When a human operator guides the robot through direct physical interaction, it is desirable to have a compliant behaviour at the end effector according to a decoupled impedance dynamics. In order to achieve a desired impedance behaviour, the robot's dynamics has to be suitably reshaped by the controller. Moreover, the stability of the coupled human-robot system should be guaranteed for any value of the impedance parameters within a prescribed region. If the robot is kinematically or functionally redundant, also the redundant degrees of freedom can be used to modify the robot dynamics. Through an extensive experimental study on a 7-DOF KUKA LWR4 arm, we compare two different strategies to solve redundancy and we show that, when redundancy is exploited to ensure a decoupled apparent inertia at the end effector, the stability region in the parameter space becomes larger. Thus, better performance can be achieved by using, e.g., variable impedance control laws tuned to human intentions.

Keywords Impedance control · Redundancy resolution · Human-robot interaction

This research has been partially funded by the EC Seventh Framework Programme within the SAPHARI project 287513 and RoDyMan project 320992.

F. Ficuciello (✉) · L. Villani · B. Siciliano
Dipartimento di Ingegneria Elettrica e Tecnologie dell'Informazione,
Università degli Studi di Napoli Federico II, Napoli, Italy
e-mail: fanny.ficuciello@unina.it

L. Villani
e-mail: luigi.villani@unina.it

B. Siciliano
e-mail: bruno.siciliano@unina.it

M.A. Hsieh et al. (eds.), *Experimental Robotics*, Springer Tracts in Advanced Robotics 109, DOI 10.1007/978-3-319-23778-7_12

1 Introduction

In all those applications where robots are used to cooperatively perform operations with humans, intentional human-robot physical interaction is required. Therefore, the stability of the interaction plays an essential role to ensure both safety and high performance. On the other hand, the performance level is related not only to the accuracy of the task execution but also to the comfort perceived by humans during manual guidance. Thus, the ability of the robot to adapt its dynamic behaviour to human intentions plays a crucial role. In the robotics literature, different impedance control strategies with modulation of the impedance parameters have been proposed [1–3].

In the application considered here, the robot must be free to move under the forces applied by the operator to the end effector for the execution of a cooperative writing task under human guidance. Hence, a Cartesian impedance strategy is used, where the desired stiffness and the desired position are set to zero while the damping and mass parameters can be suitably tuned. In a recent study [4], we have shown that, for the execution of this kind of co-manipulation task, the performance improves when the robot redundancy is suitably exploited. In detail, the Dynamic Conditioning Index (DCI) presented in [5] has been adopted to solve the redundancy of a 7-DOF KUKA LWR4 arm, with the aim of ensuring a decoupled apparent inertia at the end effector.

In this work, an extensive experimental study is presented, where the role of the redundancy on the stability of the coupled human-robot system is further investigated. In fact, during the interaction, stability depends not only on the implemented control strategy, but also on the robot's hardware (transmissions, sensors and actuators [6]). Moreover, the more far the desired impedance behaviour is from the natural robot dynamics, the more critical the stability tends to be, and the required control effort becomes more demanding. A theoretical stability analysis provides results that are often too conservative, especially in the presence of redundant degrees of freedom.

By using an experimental procedure, we first find the region in the parameter space (mass and damping) where stability is preserved, taking into account the non-isotropic features of the robot dynamics. Then, the performance of the Cartesian impedance control for different values of the parameters set within the stability region, by using different redundancy resolution strategies, are compared. In particular, the solution based on DCI is confronted to the classical solution based on the maximisation of the kinematic manipulability index. The results show that, using the first solution, the task can be completed for any choice of the impedance parameters (constant or variable) within the stability region. On the other hand, when the manipulability index is adopted, the performance is lower and, in some cases, for parameter values that are close to the frontier but inside the stability region, the task cannot be completed because instability occurs.

2 Impedance Control with Redundancy Resolution

The operator interacts with the robot by grasping the end effector and moving it along arbitrary trajectories. It is assumed that only forces can be applied. Since the robot's tip has to follow and adapt to the applied force, the end-effector dynamics can be set as a mass-damper system of equation

$$\boldsymbol{\Lambda}_d \ddot{\boldsymbol{x}} + \mathbf{D}_d \dot{\boldsymbol{x}} = \boldsymbol{F}_{ext}, \tag{1}$$

where $\boldsymbol{x} \in \mathbb{R}^3$ is the Cartesian position vector of the robot's tip, $\boldsymbol{F}_{ext} \in \mathbb{R}^3$ is the vector of the external forces and $\boldsymbol{\Lambda}_d$ and $\boldsymbol{D}_d$ are suitable inertia and damping matrices, that here are set as constant diagonal matrices.

The control law that imposes the impedance dynamics (1) can be implemented in the joint space in the form:

$$\boldsymbol{\tau}_{imp} = -\boldsymbol{J}^T(\boldsymbol{q})\boldsymbol{\Lambda}(\boldsymbol{q})[\dot{\boldsymbol{J}}\dot{\boldsymbol{q}} + \boldsymbol{\Lambda}_d^{-1}(\boldsymbol{D}_d\dot{\boldsymbol{x}} - \widehat{\boldsymbol{F}}_{ext})] + \boldsymbol{g}(\boldsymbol{q}) - \boldsymbol{r}, \tag{2}$$

where $\boldsymbol{q} \in \mathbb{R}^n$, with $n = 7$, is the vector of joint variables, $\boldsymbol{J}(\boldsymbol{q})$ is the robot Jacobian mapping $\dot{\boldsymbol{q}}$ to $\dot{\boldsymbol{x}}$ and $\boldsymbol{\Lambda}(\boldsymbol{q})$ is a (3×3) positive definite inertia matrix in the operational space [7].

In the above control law, an estimation of the external force $\widehat{\boldsymbol{F}}_{ext}$ is employed, assuming that force sensors on the end effector are not available. The estimation is computed using the residual technique developed in [8], as $\widehat{\boldsymbol{F}}_{ext} = \boldsymbol{J}^{\dagger T}(\boldsymbol{q})\boldsymbol{r}$, where the dynamically consistent generalised inverse of the Jacobian

$$\boldsymbol{J}^{\dagger}(\boldsymbol{q}) = \boldsymbol{M}^{-1}(\boldsymbol{q})\boldsymbol{J}^T(\boldsymbol{q})[\boldsymbol{J}(\boldsymbol{q})\boldsymbol{M}^{-1}(\boldsymbol{q})\boldsymbol{J}(\boldsymbol{q})^T]^{-1}$$

is used, being $\boldsymbol{M}(\boldsymbol{q})$ the positive definite inertia matrix of the robot in the joint space. Vector $\boldsymbol{r}$ is the residual providing an estimate of the external torques acting on the robot, that can be computed on the basis of the robot dynamic model and on the control torques.

In the application considered here, the human guidance of the end effector involves 3 of the 7 degrees of freedom of the KUKA LWR4 robot, thus there are 4 redundant degrees of freedom at disposal for a secondary task in the null space of the end-effector task. Our aim is to use these additional degrees of freedom to improve the effectiveness of the end-effector task, in terms of stability and performance.

The control law with redundancy resolution is:

$$\boldsymbol{\tau}_c = \boldsymbol{\tau}_{imp} + (\boldsymbol{I} - \boldsymbol{J}^T\boldsymbol{J}^{\dagger T})(\boldsymbol{u} - k_D\dot{\boldsymbol{q}}), \tag{3}$$

where $-k_D\dot{\boldsymbol{q}}$, with $k_D > 0$, is a suitable damping torque and $\boldsymbol{u}$ is a torque vector to be designed, corresponding to a secondary task, that can be set as:

$$\boldsymbol{u} = k_c \left(\frac{\partial \omega(\boldsymbol{q})}{\partial \boldsymbol{q}} \right)^T, \tag{4}$$

being $k_c > 0 (< 0)$ if the performance index has to be maximised (minimised).

2.1 Performance Indices

Different criteria can be pursued in order to choose the performance index corresponding to the secondary task.

One common choice is the kinematic manipulability index, proportional to the area of the velocity manipulability ellipsoid, which represents the capability of the robot to move the end effector along the Cartesian directions, with a given set of unit norm joints velocities. Hence, in a joint configuration where this index is (locally) maximised, it is possible to produce end-effector velocities in all possible directions with (locally) minimal joint velocities. The manipulability index (see, e.g., [9]) is defined as:

$$m(\boldsymbol{q}) = \sqrt{\det(\boldsymbol{J}\boldsymbol{J}^T)}, \tag{5}$$

i.e., by choosing $\boldsymbol{u}$ in (3) as

$$\boldsymbol{u} = k_m \nabla m(\boldsymbol{q}),$$

where the gradient of the manipulability measure $\nabla m(\boldsymbol{q})$ can be computed as [10]:

$$\nabla m(\boldsymbol{q}) = \frac{\partial m(\boldsymbol{q})}{\partial q_i} = m(\boldsymbol{q}) \text{trace} \left[\frac{\partial \boldsymbol{J}}{\partial q_i} \boldsymbol{J}^\dagger \right]. \tag{6}$$

Another possibility of exploiting redundancy is that of trying to optimise in some way the mapping between the forces applied to the end effector and the corresponding velocities or accelerations. As a matter of fact, in ideal conditions, the Cartesian impedance control law (2) allows cancelling out the robot dynamics as well as making the end-effector dynamics completely independent of the joint configuration. The most critical element is the equivalent inertia $\boldsymbol{\Lambda}(\boldsymbol{q})$, which is coupled and configuration dependent. At any given end-effector position, what can be done is to exploit the internal motion to move the robot towards configurations with maximally decoupled inertia. This can be achieved as in [4], by using a task function inspired to the dynamic conditioning index (DCI) introduced by [5] to measure the dynamic isotropy of robot manipulators in joint space.

In the operational space, the DCI index can be defined as the least-square difference between the generalized inertia matrix and an isotropic matrix, as:

$$\omega(\boldsymbol{q}) = -\frac{1}{2} \boldsymbol{E}(\boldsymbol{q})^T \boldsymbol{W} \boldsymbol{E}(\boldsymbol{q}) \tag{7}$$

where $\boldsymbol{W}$ is a diagonal weighting matrix and the error vector $\boldsymbol{E}(\boldsymbol{q})$ is defined as $\boldsymbol{E}(\boldsymbol{q}) = \left[\lambda_{11}(\boldsymbol{q}) - \sigma(\boldsymbol{q}), \lambda_{22}(\boldsymbol{q}) - \sigma(\boldsymbol{q}), \lambda_{33}(\boldsymbol{q}) - \sigma(\boldsymbol{q}), \lambda_{12}(\boldsymbol{q}), \lambda_{13}(\boldsymbol{q}), \lambda_{23}(\boldsymbol{q})\right]^T$, with σ defined as

$$\sigma(\boldsymbol{q}) = \frac{1}{3}\mathrm{Tr}(\boldsymbol{\Lambda}(\mathbf{q})).$$

The weighting matrix $\boldsymbol{W}$ has been chosen in order to give priority to the minimization of the norm of the off-diagonal elements of $\boldsymbol{\Lambda}(\boldsymbol{q})$, as $\boldsymbol{W} = \mathrm{diag}\{\boldsymbol{I}_3, 5\boldsymbol{I}_3\}$.

3 Variable Damping

A variable damping strategy has been selected to vary the damping properties of the end effector in order to accommodate the human movement during physical interaction. According to the results available in literature [1, 3, 6], the human perception is mainly influenced by the damping parameter, while, for a given damping, the desired mass determines the bandwidth of the system, and the stability. High damping values are desired when the operator performs fine movements while lower values should be used for fast movements.

In the next Section, the stability region in the parameter space will be evaluated experimentally for any damping value in the interval $[5, 60]$ Ns/m. The damping must be decreased if the operator wants to accelerate and increased if (s)he wants to decelerate and the desired mass have to be chosen accordingly.

In order to generate a law for the variable damping that is smooth also for sudden changes of the direction of motion, the damping can be related to the absolute value of the velocity.

When the operator wants to accelerate (decelerate) in a certain direction of the Cartesian space the velocity increases (decreases); thus, by decreasing (increasing) the damping accordingly, the impedance control strategy will accommodate the intention of the operator. The relationships used to vary the damping for each of the Cartesian principal directions is

$$D = a\,\mathrm{e}^{-b|\dot{x}|}. \tag{8}$$

whit $a = 60$ and $b = 4$. These parameters have been chosen empirically in order to have a variation of the damping within the interval $[5, 60]$ Ns/m for the possible range of velocities in the considered task.

4 Experimental Results

In this section, the two performance indices for redundancy resolution presented in Sect. 2 are evaluated and compared experimentally in terms of stability and performance.

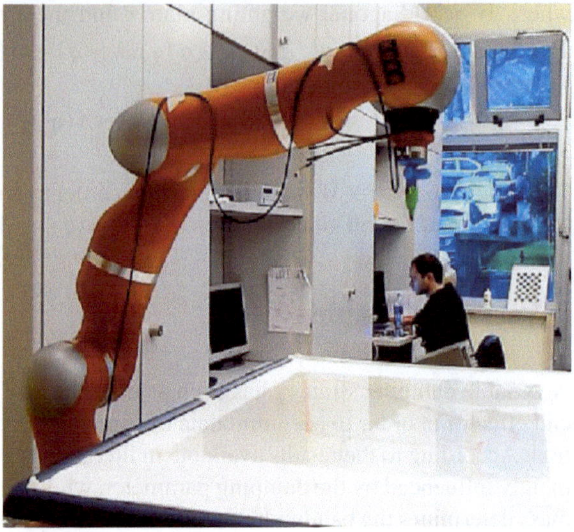

Fig. 1 Starting configuration for the writing task

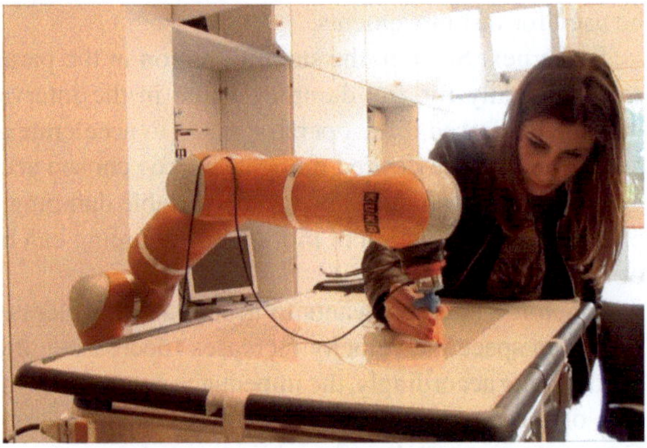

Fig. 2 Snapshot of the co-manipulation task

A case study has been selected, consisting in the execution of a writing task on a horizontal plane operated by a human: the operator guides a paint marker mounted on the robot's tip along a path drawn on a paper sheet. The initial configuration of the robot, shown in Fig. 1, has been chosen to facilitate the execution of the task. A snapshot of the task execution is reported in Fig. 2.

4.1 Stability

In our previous paper [4] the allowed range of variation of the impedance parameters of (1), so that stability is preserved, was evaluated experimentally. The corresponding stability region in the parameter space was found by setting the same damp-

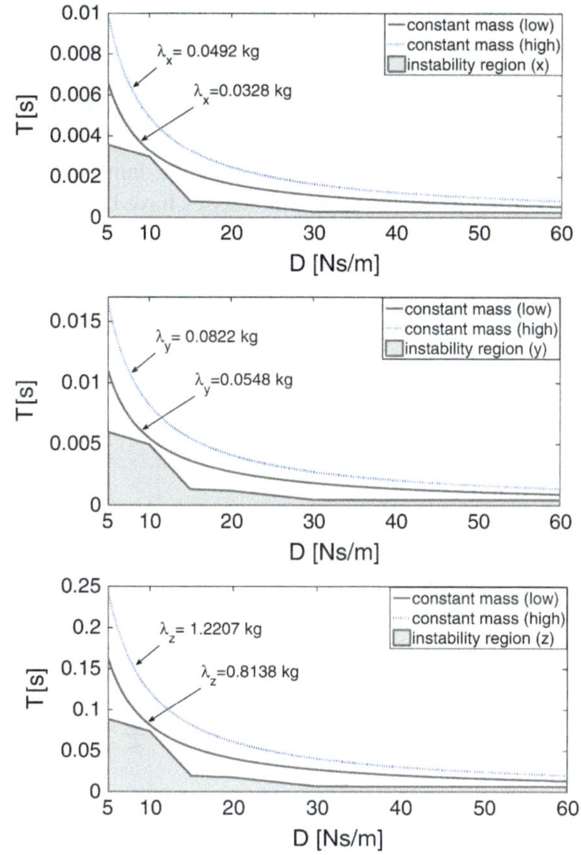

Fig. 3 Stability regions and constant mass curves for the three Cartesian directions

ing and mass parameters along all the Cartesian directions. This choice was rather conservative, considering that, e.g., the equivalent inertia at the end effector in the configuration of Fig. 1 is non homogenous being the inertia moment about the vertical direction almost ten times higher than the inertia moments about the horizontal axes. In this work, by using the same experimental procedure described in [4], a more accurate evaluation of the stability region has been carried out, but allowing the choice of different values of the parameters along the three Cartesian directions. The stability regions for the three Cartesian directions of the end effector, referred to the base frame, are shown in Fig. 3. The parameters are the desired damping D along a given Cartesian direction and the desired time constant T of the impedance equation (1), defined as D/λ, being λ the desired mass along the same Cartesian direction.

In the same figure, two set of curves corresponding to two different values of the mass for each Cartesian direction are represented. The curves with $\boldsymbol{\Lambda}_d = \mathrm{diag}\{0.0328, 0.0548, 0.8138\}$ kg constant mass are close to the instability frontiers, and can be assumed as minimal inertia curves. The dashed curves, with $\Lambda_d = \text{diag}\{0.0492, 0.0822, 1.2207\}$ kg, are safely within the stability regions.

In the first set of experiments, the comparison between the two redundancy resolution strategies is carried out by checking the stability, i.e., by verifying experimentally that the system remains stable during the task execution. The comparison have been performed for both constant and variable damping. The variable damping has been set according to Eq. (8) while the masses have been set constant, so that the impedance parameters vary on the constant mass curves of Fig. 3.

When the DC index is used for redundancy resolution, the task has been completed in both cases corresponding to low and high values of desired inertia, as it is shown in Figs. 4 and 5, where the paths of the paint marker are represented together with the reference path.

On the other hand, in the case of low inertia (Fig. 4), the task cannot be completed when the manipulability (Man) index is used for redundancy resolution, because the system tends to become unstable.

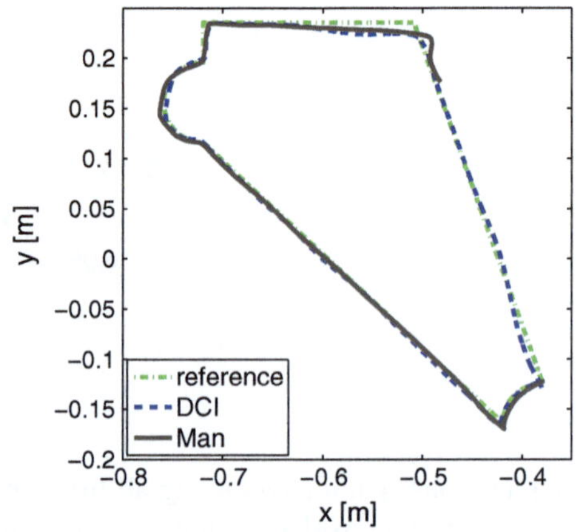

Fig. 4 Reference and actual paths for the writing task in the case of low desired inertia

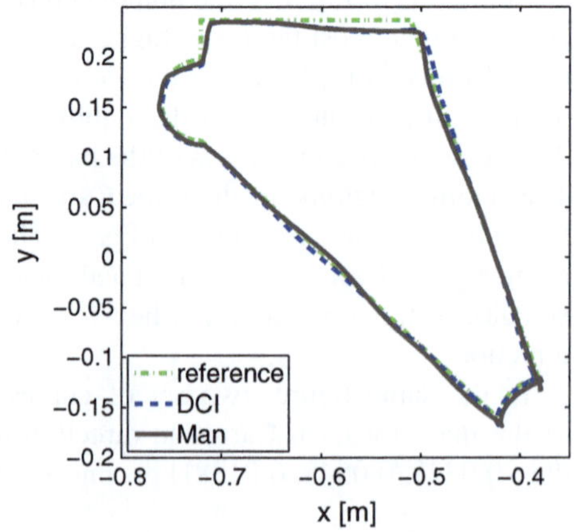

Fig. 5 Reference and actual paths for the writing task in the case of high desired inertia

Fig. 6 Time histories of the values of DC index in the case of low desired inertia and variable damping. The continuous (*dashed*) *line* represents the index when redundancy is used to increase manipulability (to minimise the DC index)

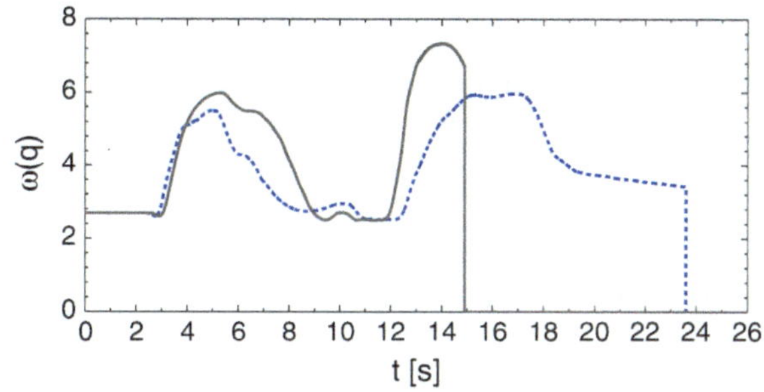

Fig. 7 Time histories of the values of DC index in the case of high desired inertia and variable damping. The continuous (*dashed*) *line* represents the index when redundancy is used to increase manipulability (to minimise the DC index)

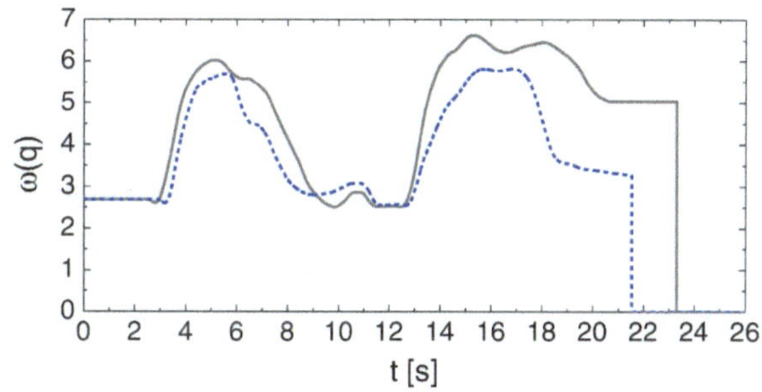

The corresponding time histories of the DC index during the execution of the experiment are reported in Figs. 6 and 7 respectively.

Notice that the values of the DC index are always lower when the minimization of the DC index is used as secondary task, as expected, with some exceptions in correspondence of abrupt changes of directions. Moreover, in the case of low inertia, the system tends to become unstable when the value of the DC index is too high, i.e., when the inertia of the robot at the end effector deviates significantly from the desired diagonal inertia imposed by the control.

Similar results can be observed in the case that the damping has been set constant (to the value of 60 Ns/m). The corresponding time histories of the DC index are reported in Figs. 8 and 9.

Fig. 8 Time histories of the values of DC index in the case of low desired inertia and constant damping. The continuous (*dashed*) *line* represents the index when redundancy is used to increase manipulability (to minimize the DC index)

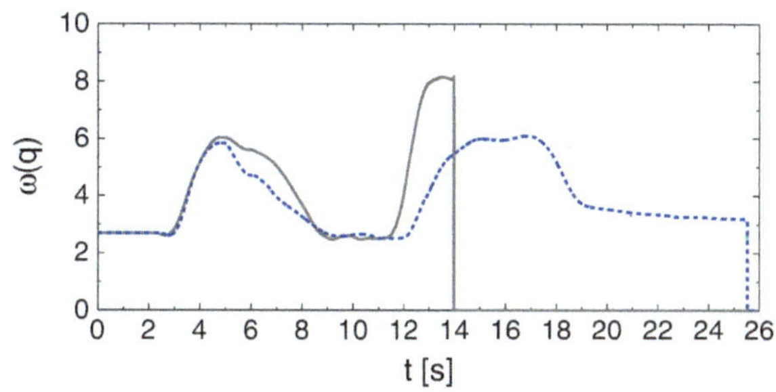

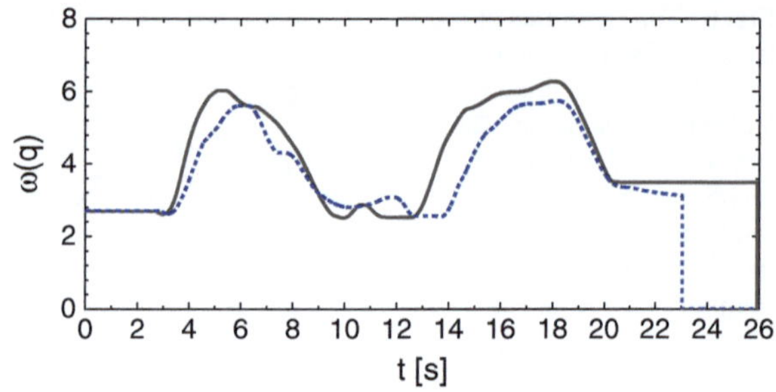

Fig. 9 Time histories of the values of DC index in the case of high desired inertia and constant damping. The continuous (*dashed*) *line* represents the index when redundancy is used to increase manipulability (to minimize the DC index)

4.2 Performance

The second set of experiments is aimed at evaluating the performance related to the two redundancy resolution strategies in terms of accuracy and execution time of the writing task. The accuracy has been measured by considering the error between the length of the path drawn in cooperation with the robot, l_e, and the ideal path length, l_d, namely:

$$e = |l_d - l_e|. \tag{9}$$

The execution time H is defined as the difference between the time when the entire path is completed and the time when the drawing tool touches the paper on the desk to start writing.

Two different impedance laws have been considered, one with constant parameters (set as $\lambda = 1.1\,\text{kg}$, $D = 60\,\text{Ns/m}$) and the other with variable damping (according to Eq. 8) and constant mass (set as $\lambda = 1.1\,\text{kg}$,). For simplicity, the impedance parameters have been set equal in all the Cartesian directions. The tests have been carried out on five different subjects that move the robot using their dominant hand. Each subject has been trained in advance, by executing the task with the different strategies to be tested. During the training phase, each subject was asked to look for the configuration which resulted the most comfortable, as well as for the best fitting starting point of the path, without any kind of conditioning. Moreover, they have been requested to pursue accuracy as a primary objective and execution time as a secondary objective. In addition, both during the training phase and the actual testing phase, the subjects have not been informed on the features of each control law, nor even which one of the four algorithms they were testing.

The results of the tests carried out both with constant and variable impedance are shown in Fig. 10, where the error on the length of the path e versus the execution time H is reported for all the subjects, as well as their mean values. In our previous work [4] the experimental results have shown that the error on the path is substantially reduced when a secondary task (exploiting DCI) is imposed. The experiments presented here confirm that the use of the DC index ensures better performance also with respect to other kind of secondary tasks, as that based on the kinematic manipulability index.

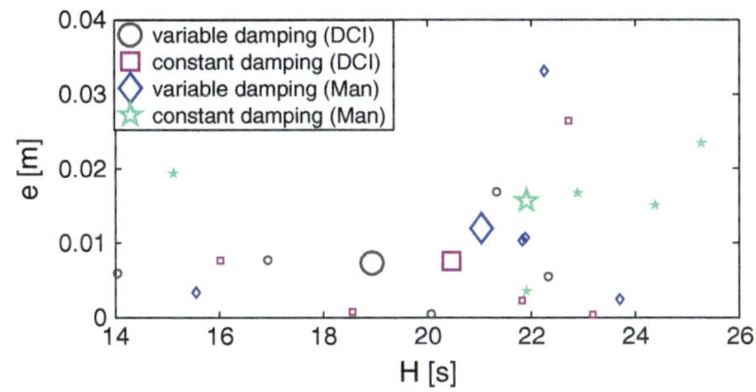

Fig. 10 Values of the length error e and execution time H in the experiments on five subjects using variable and constant impedance; both manipulability and DC index optimisation are used as secondary tasks. The bigger markers are the mean values on the five different subjects

Notice that, in the variable damping case, a significant reduction of the error on the path is obtained in spite of the strategy used to solve the redundancy.

Last but not least, all the subjects involved in the experiments have verified that the "feeling" of the manual guidance (in terms of intuitiveness and response of the robot) improves when the DC index is adopted, i.e., when redundancy is used to decouple the natural end-effector dynamics along the principal directions of the task and a further improvement is experienced when the damping parameters are tuned online according to the law (8).

5 Conclusions

A Cartesian impedance strategy with redundancy resolution has been proposed to control a 7-DOF KUKA LWR4 arm for a human-robot co-manipulation task. Experimental tests have shown the importance of exploiting the redundancy for both stability and performance. Two different strategies to solve redundancy, one based on the dynamic conditioning index and the other based on the kinematic manipulability index, have been compared and the corresponding stability regions have been found. The experiments have evidenced that, if redundancy is designed to minimise the coupling of the end-effector equivalent inertia, the stability region in the impedance parameters space is enlarged. Moreover, the use of the dynamic conditioning index allows to improve accuracy and execution time both for constant and variable impedance control.

References

1. Ikeura, R., Moriguchi, T., Mizutani, K.: Optimal variable impedance control for a robot and its application to lifting an object with a human. In: IEEE International Workshop on Robot and Human Interactive Communication, pp. 500–505. Berlin (2002)
2. Tsumugiwa, T., Yokogawa, R., Hara, K.: Variable impedance control based on estimation of human arm stiffness for human-robot cooperative calligraphic task. In: IEEE International Conference on Robotics and Automation, pp. 644–650. Washington (2002)

3. Lecours, A., Mayer-St-Onge, B., Gosselin, C.: Variable admittance control of a four-degree-of-freedom intelligent assist device. In: IEEE International Conference on Robotics and Automation, pp. 3903–3908. Saint Paul (2012)
4. Ficuciello, F., Romano, A., Villani, L., Siciliano, B.: Cartesian impedance control of redundant manipulators for human-robot co-manipulation. In: IEEE/RSJ International Conferance on Intelligent Robots and Systems, pp. 2120–2125. Chicago (2014)
5. Ma, O., Angeles, J.: The concept of dynamic isotropy and its applications to inverse kinematics and trajectory planning. In: IEEE International Conference on Robotics and Automation, pp. 10–15. San Francisco (1990)
6. Duchaine, V., Mayer-St-Onge, B., Gao, D., Gosselin, C.: Stable and intuitive control of an intelligent assist device. IEEE Trans. Haptics **5**(2), 1412–1939 (2012)
7. Khatib, O.: A unified approach for motion and force control of robot manipulators: the operational space formulation. IEEE J. Robot. Autom. **3**(1), 1115–1120 (1987)
8. Luca, A.D., Albu-Schäffer, A., Haddadin, S., Hirzinger, G.: Collision detection and safe reaction with the DLR-III lightweight robot arm. In: IEEE/RSJ International Conference on Intelligent Robots and Systems, pp. 1623–1630. Beijing (2006)
9. Sadeghian, H., Ficuciello, F., Villani, L., Keshmiri, M.: Global impedance control of dual-arm manipulation for safe human-robot interaction. In: 10th IFAC Symposium on Robot Control, pp. 767–773. Croatia (2012)
10. Park, J., Chung, W., Youm, Y.: Computation of gradient of manipulability for kinematically redundant manipulators including dual manipulators system. Trans. Control Autom. Syst. Eng. **1**(1), 8–15 (1999)

Part IV
Perception

Gaurav S. Sukhatme
University of Southern California

It has long been a dream of roboticists to be able to plug and play—attach sensors to a robot, power it on, and expect it to carry out the tasks for which it was programmed. This exciting session on Perception, brought together four papers that address different aspects of 'power on and go' autonomy in three settings (manipulator visual servoing, quadrotor flight control, and autonomous boat navigation). While these settings are diverse, the papers in the session share some common features. All four papers make use of small lightweight cameras. While this is natural on systems with payload limitations (e.g., quadrotors), the session illustrates that in a future where the semantics of the environment will likely matter more and more, cameras will inevitably be found on all kinds of robots. Sensor diversity is inevitable in unstructured environments (e.g., aerial flight or water navigation). It is also needed in situations where the environment is structured but task flexibility is needed (e.g., flexible manufacturing and assembly). Sensor diversity implies that inter-sensor calibration is inevitably needed. Three of the four papers in the session focused on calibration problems. Finally, two papers are 'data-driven' i.e., the robot uses data collected offline.

The first paper, *Online Camera Registration for Robot Manipulation* by N. Dantam, H. Ben Amor, H. Christensen, and M. Stilman focused on camera registration 'in the loop.' The approach converges in seconds, by visually tracking features on the robot and filtering the result. Online registration is useful in cases such as perturbed camera positions, wear and tear on camera mounts, and even a camera held by a human. The experimental results, on a Schunk LWA4 manipulator and Logitech C920 camera, servoing to target and pre-grasp configurations, demonstrate that millimeter-level manipulation accuracy can be achieved without the static camera registration typically required for visual servoing.

The next two papers in the session were on quadrotor navigation and quadrotor sensor calibration respectively. In *Collision Avoidance for Quadrotors with a Monocular Camera*, the authors H. Alvarez, L.M. Paz, J. Sturm, and D. Cremers, discuss an approach that allows a quadrotor with a single monocular camera to locally generate collision-free waypoints—an important ability to achieve automatic obstacle detection and avoidance. The approach is data-driven; a small set of images is acquired while the quadrotor is hovering. This is used to compute a dense depth map, which in turn is used to render a 2D scan and generate a suitable waypoint for navigation. A Parrot Ardrone quadrotor was successfully navigated through narrow passages including doors, boxes, and people.

In *Initialization-Free Monocular Visual-Inertial State Estimation with Application to Autonomous MAVs*, the authors S. Shen, Y. Mulgaonkar, N. Michael, and V. Kumar, present a monocular visual-inertial system for an autonomous quadrotor and describe a robust state estimator the allows the robot to execute trajectories at 2 m/s with roll and pitch angles of 20°, with accelerations over 4 m/s^2. The paper describes an approach to estimate the vehicle motion without initialization and a method to determine scale and metric state information without encountering any degeneracy in real time. In perhaps the most extreme version of power on and go autonomy, this research aims to throw an uncalibrated quadrotor into the air, have it calibrate in a split-second, and estimate its motion.

The final paper in the session, *Active Online Calibration of Multiple Sensors for Autonomous Surface Vessels* by H. Heidarsson and G. Sukhatme presented an approach to autonomous active calibration of the exteroceptive sensors of an Autonomous Surface Vessel (ASV). The approach consists of automatically locating suitable calibration sites in the environment from aerial imagery, navigating to them, gathering calibration data, and estimating the required parameters from data. The approach was successfully tested on a boat in a natural lake with no prior onshore calibration. It represents progress towards a system where sensors are affixed to the robot minutes before deployment and the calibration process enables the system to power on and go.

Online Camera Registration for Robot Manipulation

Neil Dantam, Heni Ben Amor, Henrik Christensen and Mike Stilman

Abstract We demonstrate that millimeter-level manipulation accuracy can be achieved without the static camera registration typically required for visual servoing. We register the camera online, converging in seconds, by visually tracking features on the robot and filtering the result. This online registration handles cases such as perturbed camera positions, wear and tear on camera mounts, and even a camera held by a human. We implement the approach on a Schunk LWA4 manipulator and Logitech C920 camera, servoing to target and pre-grasp configurations. Our filtering software is available under a permissive license (Software available at http://github.com/golems/reflex).

1 Introduction

Using visual feedback for robot manipulation requires registration between the camera and the manipulator. Typically, this is viewed as a static task: registration is computed offline and assumed to be constant. In reality, camera registration changes during operation due to external perturbations, wear and tear, or even human repositioning. For example, during the recent DARPA Robotics Challenge trials, impacts from falls resulted in camera issues which significantly affected the robot behavior for some teams [11]. Figure 1 shows additional use cases which may change the cam-

This work supported by a grant from Peugeot S.A.

N. Dantam (✉) · H. Ben Amor · H. Christensen · M. Stilman
Institute for Robotics and Intelligent Machines, Georgia Institute of Technology,
Atlanta, GA 30332, USA
e-mail: ntd@gatech.edu

H. Ben Amor
e-mail: hbenamor@cc.gatech.edu

H. Christensen
e-mail: hic@cc.gatech.edu

M. Stilman
e-mail: mstilman@cc.gatech.edu

M.A. Hsieh et al. (eds.), *Experimental Robotics*, Springer Tracts in Advanced Robotics 109, DOI 10.1007/978-3-319-23778-7_13

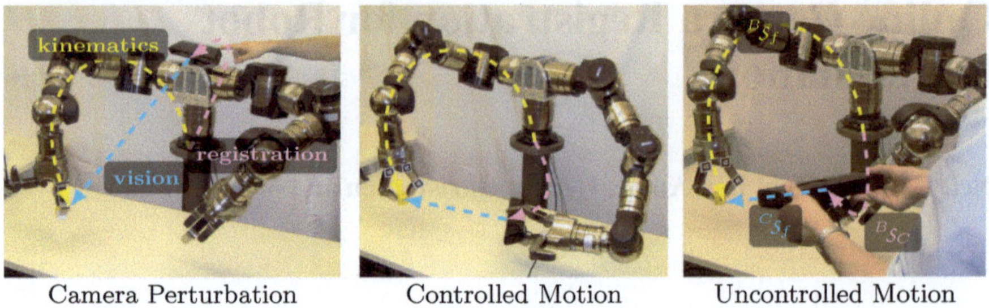

Fig. 1 Use cases for online camera registration. We combine the visual and kinematic pose estimates of end effector and filter the result to estimate the camera pose in robot body frame

era pose. The pose registration process should be treated as a dynamic task in which the involved parameters are continuously updated. Such an *online* approach to pose registration is challenging, since it requires the constant visibility of a calibration reference and sufficient accuracy to perform manipulation tasks.

To address changes in camera pose during operation, we propose an online camera registration method that combines (1) visual tracking of features on the manipulator, (2) a novel expectation-maximization inspired algorithm for pose filtering and tracking, and (3) a special Euclidean group constrained extended Kalman filter. Our key insight is to use the robot body as a reference for the registration process. By tracking known patterns or objects on the robot, we can continuously collect evidence for the current camera pose. However, naïve filtering of these pose estimates can lead to large variances in the calculated poses. The challenge is obtaining sufficient accuracy for manipulation through the online registration. To address this challenge, we combine pose filtering and manipulator control, incorporating camera registration into our manipulation feedback loop.

This paper presents a method for online registration and manipulation that combines object tracking, pose filtering, and visual servoing. First, we use perceptual information to identify the pose of specific features on the end-effector of the controlled robot (see Sect. 3.1). Then, we perform an initial fit to find offsets of the features on the robot, (see Sect. 3.2). A special Kalman filter is, then, used in conjunction with median filtering in order to perform online registration of the camera (Sect. 3.3). In our evaluation (see Sect. 4), we investigate the accuracy of the proposed method by applying it to robot grasping and manipulation tasks.

2 Related Work

Typical camera registration methods collect a set of calibration data using an external reference object, compute the calibration, then proceed assuming the calibration is static. OpenCV determines camera registration from point correspondences, typically using a chessboard [16]. Pradeep et. al. develop a camera and arm calibration approach based on bundle adjustment and demonstrate it on the PR2 robot [17]. This approach requires approximately 20 min to collect data and another 20 min for computation, a challenge for handling changing pose online.

Visual servo control incorporates camera feedback into robot motion control [2, 3]. The two main types of visual servoing are image-based visual servo control (IBVS), which operates on features in the 2D image, and position-based visual servo control, which operates on 3D parameters. Both of these methods assume a given camera registration. While IBVS is locally stable with regard to pose errors, under PBVS, even small pose errors can result in large tracking error [2]. Our proposed method addresses these challenges by correcting the camera registration online. In our experiments we show the importance of treating the registration process as a dynamic task. Furthermore, we show that our online registration achieves millimeter positioning accuracy of the manipulator. This is particularly important for grasping tasks performed using multi-fingered robot hands [1]. During such grasping tasks, inaccuracies in perception and forward kinematics often lead to premature contact between one finger and the object. As a result of the ensuing object movement, the intended grasp might not be satisfactorily executed or may fail altogether.

Other recent work has explored online visual parameter identification. [12] tracks a robot arm to identify encoder offsets. This method assumes a given camera registration, but is also tolerant of some registration error. In contrast, our work identifies the camera registration online, but does not explicitly consider encoder offsets. [9] considers bimanual arm and object tracking with vision and tactile feedback. Though the hardware and implementation differ from work presented in this paper, similar accuracy is obtained. [20] uses maps generated from a Simultaneous Localization and Mapping (SLAM) algorithm to calibrate a depth sensor. In our approach, unlike typical environments for SLAM, the object to which we are trying to register our camera—the manipulator—will necessarily be in motion.

3 Technical Approach

We determine the pose registration between the camera and the manipulator by visually tracking the 3D pose of the arm. We identify the pose of texture or shape features on the arm and fit a transformation based on the corresponding kinematic pose estimates of those features. To obtain sufficient accuracy for manipulation, we combine several methods to fit and filter the visual pose estimates before servoing to the target object. This estimation and control loop is summarized in Fig. 2.

For computational reasons, we used the *dual quaternion* representation for the special Euclidean group $\mathcal{SE}(3)$. Compared to matrices, the dual quaternion has lower dimensionality and is more easily normalized, both advantages for our filtering implementation. The relevant dual quaternion equations are summarized in appendix "A Dual Quaternion Computation." We represent the dual quaternion $\mathcal{S}$ for a transformation implicitly as a tuple of a rotation quaternion q and translation vector v: $\mathcal{S} = \left(q,\ v\right)$. This requires only seven elements. For Euclidean transformations, we

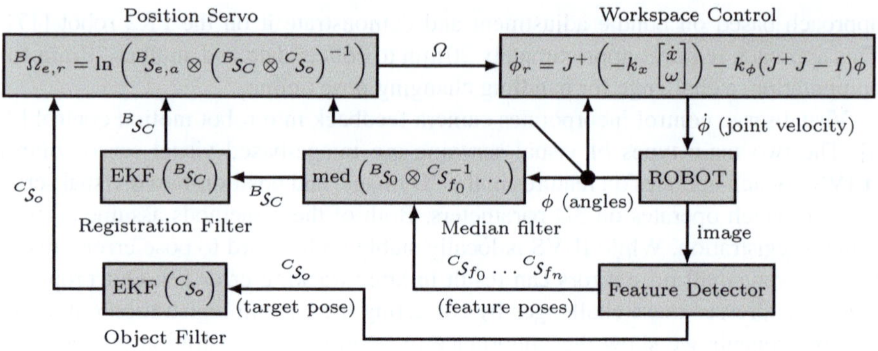

Fig. 2 Block diagram of control system. 3D poses for features are detected from visual data. The median camera transform is computed over all features and then Kalman filtered. With this registration, the robot servos in workspace to a target object location

use the typical coordinate notation where leading superscript denotes the parent frame and following subscript denotes the child frame, i.e., ${}^xS_y$ gives the origin of y relative to x. The transformation ${}^aS_b$ followed by ${}^bS_c$ is given as the dual quaternion multiplication ${}^aS_b \otimes {}^bS_c = {}^aS_c$.

3.1 Feature Estimation

To use the robot body as a reference for camera registration, it is important to identify and track body parts, e.g., the end-effector, in 3D. These 3D poses can be estimated with marker-based [18] and model-based approaches [4], see Fig. 3. Marker-based approaches require attaching fiducials to known locations on the robot, such as the fingers. Model-based tracking, on the other hand, requires accurate polygon meshes

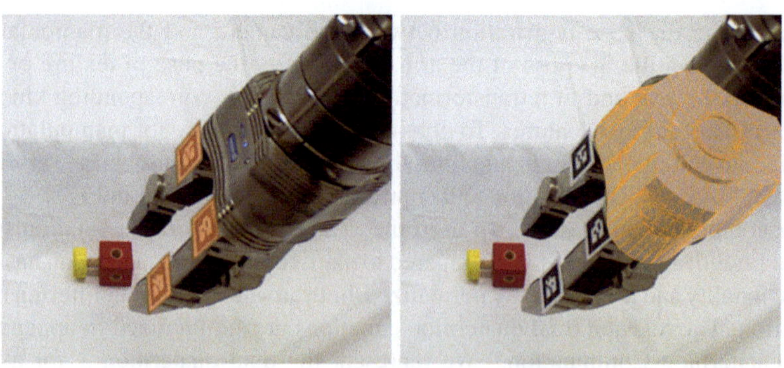

Fig. 3 Marker-based tracking (*left*) and model-based tracking (*right*)

of the tracked object. In our implementation, we use the ALVAR library [18] for marker-based tracking. For model-based tracking, we use the approach from [4]. In each frame, the 3D pose of the object is computed by projecting a 3D CAD model into the 2D image. After projection, we identify salient edges in the model and align them with edges in the 2D image. A particle filter is then used to filter the pose estimates over time. Both marker-based and model-based tracking provide 3D pose estimates of tracked features, but with frequent outliers and noise. Markers have the advantage of being easy to deploy, while model-based tracking can deal with partial occlusions of the scene.

3.2 Offset Identification

To improve the accuracy of kinematic pose estimates for features, we initially perform a static expectation-maximization-like [6] procedure, based on the following model:

$$^{B}\mathcal{S}_k \otimes {}^{k}\mathcal{S}_f = {}^{B}\mathcal{S}_C \otimes {}^{C}\mathcal{S}_f \tag{1}$$

where $^{B}\mathcal{S}_k$ is the measured nominal feature pose in the body frame determined from encoder positions and forward kinematics, $^{k}\mathcal{S}_f$ is the unknown static pose offset of the feature due to inaccuracy of manual placement, $^{B}\mathcal{S}_C$ is the unknown camera registration in the body frame, and $^{C}\mathcal{S}_f$ is the visually measured feature pose in the camera frame. These transforms are summarized in Fig. 1, with $^{B}\mathcal{S}_k \otimes {}^{k}\mathcal{S}_f$ combined as $^{B}\mathcal{S}_f$.

As an initialization step, we iteratively fix either $^{k}\mathcal{S}_f$ or $^{B}\mathcal{S}_C$ in (2) and solve for the other using Umeyama's algorithm [21]. This gives us the relative transforms for the features $^{k}\mathcal{S}_f$ which we assume are static.

3.3 Filtering

To compute the online registration, where $^{B}\mathcal{S}_C$ is changing, we combine median and Kalman filtering. The median filter is applied *independently* at each time step to reject major outliers in the estimated feature poses. Compared to weighted least squares methods, the median requires no parameter tuning and is especially resistant, tolerating outliers in up to 50 % of the data [8]. Given the median at each step, the Kalman filter is applied *over time* to generate an optimal registration estimate under a Gaussian noise assumption.

Based on (2), each observed feature on the robot gives on estimate for the camera registration $^{B}\mathcal{S}_C$:

$$^{B}\mathcal{S}_k \otimes {}^{k}\mathcal{S}_f \otimes ({}^{C}\mathcal{S}_f)^{-1} = {}^{B}\mathcal{S}_C \tag{2}$$

Median Filtering At each time step, we find the median registration over all observed features. Each observed feature gives a candidate registration ${}^{B}S_{C}$. First, we collect a set Q of the orientation candidates:

$$Q = \left\{({}^{B}q_{C})_i \mid ({}^{B}S_{k})_i \otimes {}^{k}S_{f} \otimes ({}^{C}S_{f})_i^{-1}\right\} \tag{3}$$

Then, we compute the median of the candidate orientation registrations Q. To find this median, the structure of rotations in $\mathcal{SO}(3)$ offers a convenient distance metric between two orientations: the angle between them. Using this geometric interpretation, the median orientation $\hat{q}$ is the orientation with minimum angular distance to all other orientations.

$$\widehat{{}^{B}q_{C}} = \underset{q_i \in Q}{\arg\min} \sum_{j=0}^{n} |\ln({q^*}_i \otimes q_j)| \tag{4}$$

The median translation $\hat{x}$ is the conventional geometric median, the translation with minimum Euclidean distance to all other translations. First, we find the set of candidate translations Z by rotating the feature translation in camera frame ${}^{C}v_f$ and subtracting from the body frame translation ${}^{B}v_f$:

$$Z = \left\{z_i \mid z_i = {}^{B}v_{f,i} - \widehat{{}^{B}q_{C}} \otimes {}^{C}v_{f,i} \otimes \widehat{{}^{B}q_{C}}^{*}\right\} \tag{5}$$

Then, we compute the geometric median of the candidate translations by finding the element with minimum distance to all other elements:

$$\widehat{{}^{B}v_{C}} = \underset{z_i \in Z}{\arg\min} \sum_{j=0}^{n} |z_i - z_j| \tag{6}$$

Then, the median transform is the combination of the orientation and translation parts:

$$\widehat{{}^{B}S_{C}} = \left(\widehat{{}^{B}q_{C}},\ \widehat{{}^{B}v_{C}}\right) \tag{7}$$

Kalman Filtering, we use an Extended Kalman filter (EKF) to attenuate noise over time, taking care to remain in the $\mathcal{SE}(3)$ manifold. Similar Kalman filters are discussed in [5, 14]. The quasi-linearity of quaternions means the EKF is suitable for orientation estimation in this application [13].

To filter $\mathcal{SE}(3)$ poses, we consider state x composed of a quaternion q, a translation vector v, and the translational and rotational velocities, $\dot{v}$ and ω:

$$x = (q, v) = [q_x, q_y, q_z, q_w, v_x, v_y, v_z, \dot{v}_x, \dot{v}_y, \dot{v}_z, \omega_x, \omega_y, \omega_z]$$

The measurement z is the pose:

$$z = (q, v) = [q_x, q_y, q_z, q_w, v_x, v_y, v_z]$$

The general EKF prediction step for time k is:

$$\hat{x}_{k|k-1} = f(\hat{x_{k-1}}) \tag{8}$$

$$F_{k-1} = \left.\frac{\partial f}{\partial x}\right|_{\hat{x}_{k-1|k-1}} \tag{9}$$

$$P_{k|k-1} = F_{k-1} P_{k-1|k-1} F_{k-1}^T + Q_{k-1} \tag{10}$$

where $\hat{x}$ is the estimated state, $f(x)$ is the process model, F is the Jacobian of f, P is the state covariance matrix, and Q is the process noise model.

The process model then integrates the translational and rotational velocity, staying in the $\mathcal{SE}(3)$ manifold using the dual quaternion exponential of the twist Ω:

$$\Omega(\omega, \dot{v}, v) = \left[\omega,\ v \times \omega + \dot{v}\right]$$

$$f(x) = \exp\left(\frac{\Delta t}{2}\Omega\right) \otimes (q,\ v) \tag{11}$$

Now, we find the process Jacobian F. The translation portion is a diagonal matrix of the translational velocity. For the orientation portion, we find the quaternion derivative $\dot{q}$ from the rotational velocity:

$$\dot{q} = \frac{1}{2}\omega \otimes q \tag{12}$$

This quaternion multiplication can be converted into the following matrix multiplication:

$$\frac{1}{2}\omega \otimes q = \frac{1}{2}M_r(q)\,\omega$$

$$M_r(q) = \begin{bmatrix} q_w & q_z & -q_y \\ -q_z & q_w & q_x \\ q_y & -q_x & q_w \\ -q_x & -q_y & -q_z \end{bmatrix} \tag{13}$$

Note that we omit the w column of the typical quaternion multiplication matrix because the w element of rotational velocity ω is zero.

This gives the following process 13×13 Jacobian F:

$$F = \begin{bmatrix} I_{4\times4} & 0 & 0 & \frac{1}{2}\Delta t M_r\left(q\right) \\ 0 & I_{3\times3} & \Delta t I_{3\times3} & 0 \\ 0 & 0 & I_{3\times3} & 0 \\ 0 & 0 & 0 & I_{3\times3} \end{bmatrix} \tag{14}$$

Now we consider the EKF correction step. The general form is:

$$\hat{z}_k = h(\hat{x}_{k|k-1}) \tag{15}$$

$$H_k = \left.\frac{\partial h}{\partial x}\right|_{\hat{x}_{k|k-1}} \tag{16}$$

$$y_k = v(z_k, \hat{z}) \tag{17}$$

$$S_k = H_k P_{k|k-1} H_k^T + R_k \tag{18}$$

$$H_k P_{k|k-1} = S_k K_k^T \tag{19}$$

$$\hat{x}_{k|k} = p(\hat{x}_{k|k-1}, K_k y_k) \tag{20}$$

$$P_{k|k} = (I - K_k H_k) P_{k|k-1} \tag{21}$$

where z is the measurement, h is the measurement model, H is the Jacobian of h, $\hat{z}$ is the estimated measurement, R is the measurement noise model, and K is the Kalman gain, v is a function to compute measurement residual, and p is a function to compute the state update.

We compute the EKF residuals and state updates using relative quaternions to remain in $\mathcal{SE}(3)$ without needing additional normalization. The observation $h(x)$ is a pose estimate:

$$h(x) = (q, v)$$
$$H = I_{7\times7} \tag{22}$$

We compute the measurement residual based on the relative rotation between the measured and estimated pose:

$$v(z, \hat{z}) = (y_q, y_v)$$
$$y_q = \ln\left(z_q \otimes \hat{z}_q^*\right) \otimes q$$
$$y_v = z_v - \hat{z}_v \tag{23}$$

where y_q is the orientation part of the residual and y_v the translation part. Note that $\ln\left(z_q \otimes \hat{z}_q^*\right)$ corresponds to a velocity in the direction of the relative transform between the actual and expected pose measurement and that we can consider y_q as a quaternion derivative. Then, the update function will integrate the pose portion of y, again using the exponential of the twist. First, we find the twist corresponding to the product of the Kalman gain K and the measurement residual y:

$$
\begin{aligned}
(Ky)_\phi &= (Ky)_q \otimes q^* \\
\Omega(Ky, v) &= \left((Ky)_\phi, v \times (Ky)_\phi + (Ky)_v\right)
\end{aligned}
\tag{24}
$$

Then, we integrate estimated pose using the exponential of this twist:

$$
(x_{(q,v)})_{k|k} = \exp\left(\frac{\Delta t}{2}\Omega\right) \otimes (q, v) \tag{25}
$$

Finally, the velocity component of innovation y is scaled and added:

$$
(x_{\omega,\dot{v}})_{k|k} = x_{\omega,\dot{v}} + (Ky)_{\omega,\dot{v}} \tag{26}
$$

3.4 Registered Visual Servoing

We use the computed camera registration ${}^B S_C$ to servo to a target object according to the control loop in Fig. 2. This is position-based visual servoing, incorporating the dynamically updated registration. First, we compute a reference twist ${}^B\Omega_{e,ref}$ from the position error using camera pose ${}^B S_C$ and object pose ${}^C S_o$:

$$
{}^B S_{e,ref} = {}^B S_C \otimes {}^C S_{obj} \tag{27}
$$

$$
{}^B\Omega_{e,ref} = \ln\left({}^B S_{e,act} \otimes {}^B S_{e,ref}^{-1}\right) \tag{28}
$$

Then, we find the reference velocity for twist ${}^B\Omega_{e,ref}$:

$$
\begin{bmatrix} \dot{x} \\ \omega \end{bmatrix} = \begin{bmatrix} \mathbb{D}({}^B\Omega_{e,ref}) - (2\mathbb{D}({}^B S_e) \otimes \mathbb{R}({}^B S_e)^{-1}) \times \mathbb{R}({}^B\Omega_{e,ref}) \\ \mathbb{R}({}^B\Omega_{e,ref}) \end{bmatrix} \tag{29}
$$

where $\mathbb{R}(X)$ is the real part of X and $\mathbb{D}(X)$ is the dual part of X.

Finally, we compute joint velocities using the Jacobian damped least squares, also using a nullspace projection to keep joints near the zero position:

$$
\dot{\phi}_r = J^+\left(-k_x \begin{bmatrix} \dot{x} \\ \omega \end{bmatrix}\right) - k_\phi (J^+ J - I)\phi \tag{30}
$$

where J is the manipulator Jacobian matrix, J^+ is its damped pseudoinverse, k_x is a gain for the position error, and k_ϕ is a gain for the joint error.

4 Experiments

We implement this approach on a Schunk LWA4 manipulator with SDH end-effector, see Fig. 1, and use a Logitech C920 webcam to track the robot and objects. The Schunk LWA4 has seven degrees of freedom and uses harmonic drives, which enable repeatable positioning *precision* of ± 0.15 mm [19]. However, absolute positioning *accuracy* is subject to encoder offset calibration and link rigidity. In practice, we achieve ± 1 cm accuracy when using only the joint encoders for feedback. The Logitech C920 provides a resolution of 1920×1080 at 15 frames per second. To measure ground-truth distances, we used a Bosch DLR165 laser rangefinder and a Craftsman 40181 vernier caliper.

We initially test the convergence and resistance of our approach while moving the camera. With the camera mounted on a tripod, we compute the filtered registration while the camera is perturbed, rotated, and translated.

The resulting registrations under moving camera are plotted in Fig. 4. The visual pose estimates contain frequent outliers in addition to a small amount of noise. The filtered registration removes the outliers and converges within 5 s.

To demonstrate the suitability of this approach for manipulation tasks, we test the positioning accuracy attainable with this online registration. As shown in Fig. 5, we place a marker on a table, measure linear distance to the marker with a laser ranger, servo the end-effector to the visually estimated marker position using the control loop in Fig. 2, and measure the distance to the end-effector which should be directly over the marker.

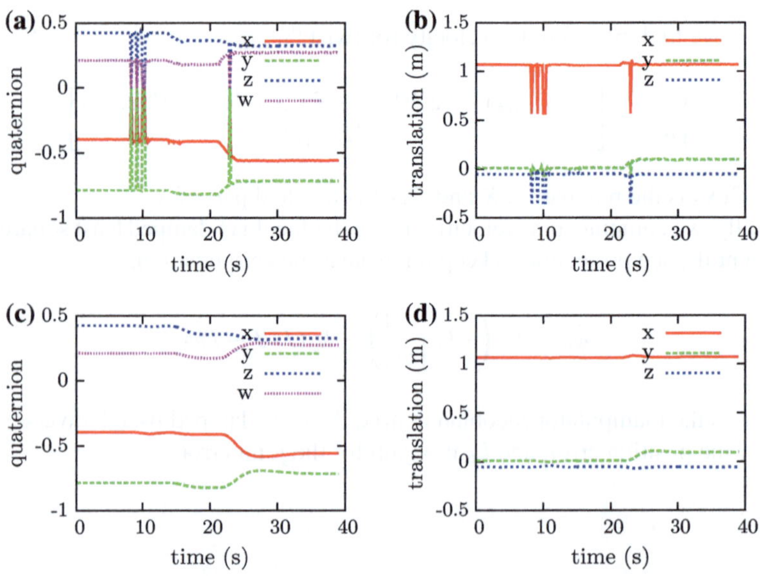

Fig. 4 Registration while camera is bumped (8 s), rotates (15 s) and translated (24 s). Camera is bumped. **a–b** Registration from raw visual pose estimates of one feature. Contains many outliers. **c–d** Filtered registration. Outliers and noise eliminated

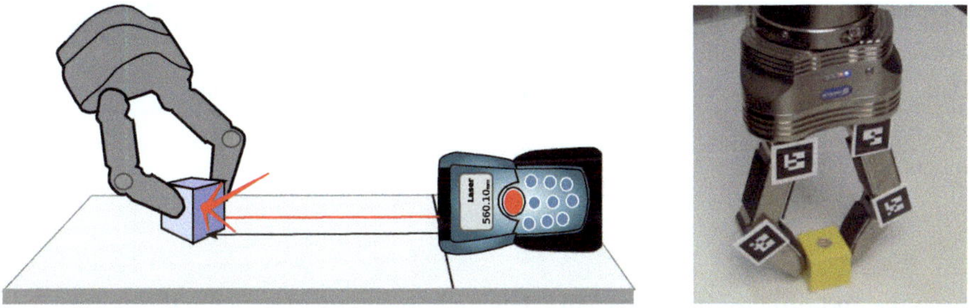

Fig. 5 Experimental setup for evaluating the positioning accuracy during camera registration. A cube was placed on a marker and the distance to a laser ranger was captured. Subsequently, the cube was placed in the hand of the robot, which, then, servoed to the position of the marker. Again, the distance was measured using the laser ranger

Table 1 Positioning experiment results

Setup	Average (mm)	Stdev (mm)
$\delta \leq 45°$	0.5	0.52
$\delta > 45°$	1.5	1.26

Average and standard deviation (mm) of measured difference between commanded position and object location

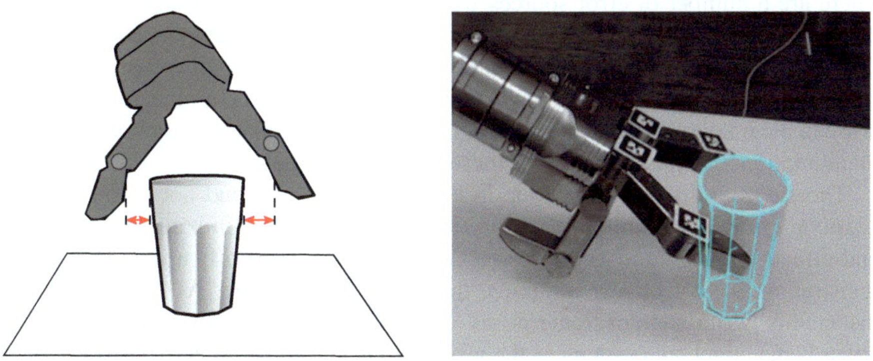

Fig. 6 Pre-grasp experiment: using the introduced camera registration, the open robot hand is servoed to the position of a glass. The distances between the fingers and the glass are then measured. Since the glass is rotationally symmetric, the distances of both used robot fingers should be identical in the ideal case

The resulting position accuracy achievable with online registration is summarized in Table 1. For an ideal camera placement with close, direct view of the end-effector (i.e. the angle δ between the camera and the markers is 45° or less), positioning accuracy is in the submillimiter range. Larger camera distances and angles, resulted in positioning error of 1–2 mm.

Finally, we test the pre-grasp positioning accuracy of this method as shown in Fig. 6. We place an object, in particular, a cup, at a variety of locations on the table,

Table 2 Pre-grasp experiment results. Average and standard deviation (mm) of measured difference between object and end-effector position

Data	Average (mm)	Stdev (mm)
All	5.8	8.5
Inliers	3.3	2.3

servo the end-effector to the visually detected object position using the control loop in Fig. 2, and then measure the distance of each finger to the object using a vernier caliper.

The results of the pre-grasp positioning are summarized in Table 2. A small number of trials resulted in centimeter-level error for objects placed near the edge of the image frame. Ommitting these outliers, the average positioning error of the pre-grasp configuration was 3.3 mm.

5 Experimental Insights

There are a number of error sources we must handle in this system. For the kinematics, error from encoder offsets in the arm, imprecise link lengths, and flexing of links all contribute inaccurate kinematic pose estimates. For perception, error from inaccurate camera intrinsics, imprecise fiducial sizes, offsets in object models, and noise in the image all contribute to error in visual pose estimates. To achieve accurate manipulation, we must account for these potential sources of error.

The key point of the servo loop in Fig. 2 is that we depend not on minimizing *absolute* error, but on minimizing *relative* error. We are minimizing error between end-effector pose $\mathcal{S}_e$ and target pose $\mathcal{S}_o$. Because we continually update the camera registration, we effectively minimize this error in the image. As long as there is distance between camera frame poses ${}^C\mathcal{S}_e$ and ${}^C\mathcal{S}_o$, we will move the end-effector towards the target, and as long as the visual distance estimate is zero when we reach the target, the arm will stop at the target. Thus, even if there is absolute registration error due to, e.g., unmodeled lens distortion, it is only necessary that relative error between visual estimates of the end-effector and target be small and converge to zero. The relative error between end-effector and target is crucial in manipulation, and our technique is well suited to minimizing this error.

The position of the tracked features on the robot has an important effect on error correction. Kinematic errors between the robot body origin and the tracked features, e.g., due to flex or encoder offsets, are incorporated into the camera registration and handled through the servo loop. Error between the observed features and the end-effector cannot be corrected. Thus, it is better to track features as close to the end-effector as possible. Consequently, we placed the fiducial markers on the fingers of the SDH end-effector.

The principal challenge in the implementation stems from observing the robot pose using small, ≈3 cm, markers. While marker translation is reliably detected, outliers in orientation are frequent. Ample lighting improves detection but does not eliminate outliers. The median pose, (4)–(6), was effective at eliminating outliers from visual estimates. Alternative methods for combining orientation estimates include Davenport's q-method [15] and the Huber loss function [10]. In contrast to these other methods, the median has no parameters such as thresholds which require adjustment. Thus, it is especially suited to this online registration application where outlier frequency may vary depending on camera placement, lighting, etc. A potential challenge is that the direct computation of (4) leads to an $O(n^2)$ algorithm in the number of orientations. However, for the small number of poses we consider at each step here, the computation time is negligible. On a Xeon E5-1620 CPU, computing the median of 32 orientations requires 30 μs.

6 Conclusion

We have presented an online method to identify the camera poses for robot manipulation tasks. This is useful for the typical case where camera registration is not static but changes due to model error, disturbances, or wear and tear. The key point is to track both the object and the robot in the image, and servo based on the visually estimated relative pose between the object and robot. By combining median and Kalman filtering of the registration pose, we are able to achieve millimeter-level manipulation accuracy. We have shown in our experiments that online registration can be used to improve positioning accuracy during grasping and manipulation tasks, thereby avoiding typical challenges such as premature contact between fingers and objects.

A useful extension to this work would be to handle online registration with multiple cameras. This could provide additional data to improve accuracy or permit greater field of view, e.g., observing both hands in bimanual tasks. We anticipate that considering median deviation and applying a similar extended Kalman filter to multiple simultaneous poses will extend this online approach to multi-camera setups.

Acknowledgments This work is dedicated to the memory of Mike Stilman, whose tireless guidance and support made these developments possible.

A Dual Quaternion Computation

Dual quaternions are a numerically convenient representation for Euclidean transformations, $\mathcal{SE}(3)$. Compared to ordinary quaternions which can represent rotation, dual quaternions can represent both rotation and translation. Mathematically, they are the extension of quaternions to the dual numbers [7]. Dual numbers are of the form $r + d\varepsilon$, where r is real part, d is the *dual* part, and ε is the dual element such that $\varepsilon^2 = 0$ and $\varepsilon \neq 0$. A dual quaternion $\mathcal{S}$ can be represented as a pair of quaternions, $\mathcal{S} = s_r + s_d\varepsilon$, which we represent with the tuple $\left[s_r,\ s_d\right]$.

The dual quaternion representing orientation q and translation v is:

$$\mathcal{S} = \left[s_r,\ s_d\right] = \left[q,\ \frac{1}{2} v \otimes q\right] \tag{31}$$

We represent the vector and scalar components of the ordinary quaternion parts of a dual quaternion as:

$$\begin{aligned} \mathcal{S} = \left[r,\ d\right] &= \left[\left(r_x i + r_y j + r_z k,\ r_w\right),\ \left(d_x i + d_y j + d_z k,\ d_w\right)\right] \\ &= \left[\left(r_v,\ r_w\right),\ \left(d_v,\ d_w\right)\right] \end{aligned} \tag{32}$$

where r_v and d_v are the vector parts and r_w and d_w are the scalar parts.

Dual quaternion Euclidean transforms are normalized by dividing by the real magnitude:

$$S' = \left[\frac{s_r}{|s_r|},\ \frac{s_d}{|s_r|}\right] \tag{33}$$

Operations on the dual quaternions can be derived from those of ordinary quaternions and the properties of dual numbers. However, this requires care to handle singularities. Generally, the values at these singularities can be computed by identifying singular factors with convergent Taylor series. While computer algebra systems, e.g., Maxima, Mathematica, can be used to compute the Taylor series, suitable singular factors must first be identified. We summarize the relevant functions and suitable Taylor series below.

Dual quaternion multiplication is:

$$\mathcal{A} \otimes \mathcal{B} = \left[a_r \otimes b_r,\quad a_r \otimes b_d + a_d \otimes b_r\right] \tag{34}$$

The dual quaternion exponential is:

$$\phi = |r_v| \tag{35}$$

$$k = r_v \cdot d_v \tag{36}$$

$$e^{\mathcal{S}} = e^{\tilde{w}} \left[\left(\frac{s}{\phi} r_v,\ c\right),\ \left(\frac{s}{\phi} d_v + \frac{c - \frac{s}{\phi}}{\phi^2} k r_v,\ -\frac{s}{\phi} k\right)\right] \tag{37}$$

where $s = \sin\phi$, $c = \cos\phi$, $\tilde{w} = r_w + d_w\varepsilon$, and $r_v \cdot d_v$ is the dot product of r_v and d_v.

Then, to handle the singularity at $\phi = 0$, we use the following Taylor expansions:

$$\frac{\sin\phi}{\phi} = 1 - \frac{\phi^2}{6} + \frac{\phi^4}{120} - \frac{\phi^6}{5040} + \dots \tag{38}$$

$$\frac{\cos\phi - \frac{\sin\phi}{\phi}}{\phi^2} = -\frac{1}{3} + \frac{\phi^2}{30} - \frac{\phi^4}{840} + \frac{\phi^6}{45360} + \dots \tag{39}$$

The dual quaternion logarithm is:

$$\phi = \operatorname{atan2}\left(|r_v|, r_w\right) \tag{40}$$

$$k = r_v \cdot d_v \tag{41}$$

$$\alpha = \frac{r_w - \frac{\phi}{|r_v|}|r|^2}{|r_v|^2} = \frac{1}{|r|}\left(\frac{\cos\phi}{\sin^2(\phi)} - \frac{\phi}{\sin^3(\phi)}\right) \tag{42}$$

$$(\ln S)_r = \left(\frac{\phi}{|r_v|}r_v, \quad \ln|r|\right) \tag{43}$$

$$(\ln S)_d = \left(\frac{k\alpha - d_w}{|r|^2}r_v + \frac{\phi}{|r_v|}d_v, \quad k + \frac{r_w d_w}{|r|^2}\right) \tag{44}$$

where $(\ln S)_r$ is the real part of the logarithm and $(\ln S)_d$ is the dual part of the logarithm. Note that ϕ represents the angle between the real and imaginary parts of unit quaternion r. Rewriting α in terms of sin and cos yields the convergent Taylor series below.

To handle the singularity at $|r_v| = 0$ and knowing $|r| = 1$:

$$\frac{\phi}{|r_v|} = \frac{\phi}{\frac{|r_v|}{|r|}} = \frac{\phi}{\sin\phi} \tag{45}$$

$$\frac{\phi}{\sin\phi} = 1 + \frac{\phi^2}{6} + \frac{7\phi^4}{360} + \frac{31\phi^6}{15120} + \dots \tag{46}$$

Then, for α in (42):

$$\frac{\cos\phi}{\sin^2(\phi)} - \frac{\phi}{\sin^3(\phi)} = -\frac{2}{3} - \frac{1}{5}\phi^2 - \frac{17}{420}\phi^4 - \frac{29}{4200}\phi^6 + \dots \tag{47}$$

References

1. Amor, H.B., Kroemer, O., Hillenbrand, U., Neumann, G., Peters, J.: Generalization of human grasping for multi-fingered robot hands. In: Proceedings of the International Conference on Robot Systems (IROS) (2012)
2. Chaumette, F., Hutchinson, S.: Visual servo control, part I: basic approaches. Robot. Autom. Mag. **13**(4), 82–90 (2006)
3. Chaumette, F., Hutchinson, S.: Visual servo control, part II: advanced approaches. Robot. Autom. Mag. **14**(1), 109–118 (2007)
4. Choi, C., Henrik, I.C.: Robust 3d visual tracking using particle filtering on the special euclidean group: a combined approach of keypoint and edge features. Int. J. Robot. Res. **31**(4), 498–519 (2012)
5. Choukroun, D., Bar-Itzhack, I.Y., Oshman, Y.: Novel quaternion Kalman filter. Trans. Aerosp. Electron. Syst. **42**(1), 174–190 (2006)
6. Dempster, A.P., Laird, N.M., Rubin, D.B.: Maximum likelihood from incomplete data via the em algorithm. J. Roy. Stat. Soc.: Ser. B **39**(1), 1–38 (1977)
7. Eduard Study. Geometrie der dynamen (1903)
8. Hampel, F.R., Ronchetti, E.M., Rousseeuw, P.J., Stahel, W.A.: Robust Statistics: The Approach Based on Influence Functions, vol. 114. Wiley (2011)
9. Hebert, P., Hudson,N., Ma, J., Burdick, J.W.: Dual arm estimation for coordinated bimanual manipulation. In: International Conference on Robotics and Automation (ICRA), IEEE, pp. 120–125 (2013)
10. Huber, P.J., et al.: Robust estimation of a location parameter. Ann. Math. Stat. **35**(1), 73–101 (1964)
11. Joo, S., Grey, M.: DRC-Hubo retrospective, Jan. 2014. Personal Communication
12. Klingensmith, M., Galluzzo, T., Dellin, C., Kazemi, M., Bagnell, J.A., Pollard, N.: Closed-loop servoing using real-time markerless arm tracking. In: International Conference on Robotics And Automation (Humanoids Workshop), May 2013
13. Laviola, J.J.: A comparison of unscented and extended Kalman filtering for estimating quaternion motion. In: American Control Conference, IEEE, vol. 3, pp. 2435–2440 (2003)
14. Lefferts, E.J., Markley, L.M., Malcolm, D.S: Kalman filtering for spacecraft attitude estimation. J. Guidance Control Dyn. **5**(5), 417–429 (1982)
15. Markley, L.F., Cheng, Y., Crassidis, J.L., Oshman, Y.: Averaging quaternions. J. Guidance Control Dyn. **30**(4), 1193–1197 (2007)
16. OpenCV API Reference. http://docs.opencv.org/master/modules/refman.html
17. Pradeep, V., Konolige, K., Berger, E.: Calibrating a multi-arm multi-sensor robot: a bundle adjustment approach. In: Experimental Robotics, pp. 211–225. Springer (2014)
18. Rainio, K., Boyer, A.: ALVAR – A Library for Virtual and Augmented Reality User's Manual. VTT Augmented Reality Team, December 2013
19. Schunk GmbH: Dextrous lightweight arm LWA 4D, technical data. http://mobile.schunk-microsite.com/en/produkte/produkte/dextrous-lightweight-arm-lwa-4d.html
20. Teichman, A., Miller, S., Thrun, S.: Unsupervised Intrinsic Calibration of Depth Sensors via Slam. In: Robotics Science and Systems (RSS) (2013)
21. Umeyama, S.: Least-squares estimation of transformation parameters between two point patterns. Pattern Anal. Mach. Intell. **13**(4), 376–380 (1991)

Collision Avoidance for Quadrotors with a Monocular Camera

H. Alvarez, L. M. Paz, J. Sturm and D. Cremers

Abstract Automatic obstacle detection and avoidance is a key component for the success of micro-aerial vehicles (MAVs) in the future. As the payload of MAVs is highly constrained, cameras are attractive sensors because they are both lightweight and provide rich information about the environment. In this paper, we present an approach that allows a quadrotor with a single monocular camera to locally generate collision-free waypoints. We acquire a small set of images while the quadrotor is hovering from which we compute a dense depth map. Based on this depth map, we render a 2D scan and generate a suitable waypoint for navigation. In our experiments, we found that the pose variation during hovering is already sufficient to obtain suitable depth maps. The computation takes less than one second which renders our approach applicable for obstacle avoidance in real-time. We demonstrate the validity of our approach in challenging environments where we navigate a Parrot Ardrone quadrotor successfully through narrow passages including doors, boxes, and people.

This work was partially supported by the German Academic Exchange Service (DAAD), the DFG under contract number FO 180/17-1 in the Mapping on Demand (MOD) project and the Ministerio de Economía y Competitividad under project DPI2012-36070: Semantic and Active SLAM for heterogeneous systems.

H. Alvarez · J. Sturm · D. Cremers
Department of Computer Science, Technische Universitt Mnchen,
Informatik 9, Boltzmannstrasse 3, 85748 Garching, Germany
e-mail: humalvarez@mytum.de

J. Sturm
e-mail: sturmju@in.tum.de

D. Cremers
e-mail: cremers@in.tum.de

L.M. Paz (✉)
Department of Engineering Science, University of Oxford, Parks Road,
Oxford OX1 3PJ, UK
e-mail: linapaz@robots.ox.ac.uk

L.M. Paz
I3A Instituto de Invertigación en Ingeniería de Aragón, University of Zaragoza,
C/Mariano Esquillor S/n, 50800 Zaragoza, Spain
e-mail: linapaz@unizar.es

M.A. Hsieh et al. (eds.), *Experimental Robotics*, Springer Tracts
in Advanced Robotics 109, DOI 10.1007/978-3-319-23778-7_14

Keywords Collision avoidance · Monocular depth-maps · Micro-aerial vehicles

1 Introduction

Micro-aerial vehicles (MAVs) have enormously gained in popularity over the past years. They are envisioned as versatile helpers for many different applications such as aerial surveillance, visual inspection, remote farming and filming.

All of these applications require that the quadrotors do not collide with the environment. While this can be achieved by a human pilot, it requires much training and puts a significant cognitive load onto the human operator. Therefore, it is desirable to enable the quadrotor to detect and avoid obstacles automatically. This can be achieved by adding distance sensors to the quadrotor, such as ultrasound sensors, laser-scanners [1, 2], stereo cameras [3, 4], Kinect [5], or combinations of multiple sensors [6, 7]. However, adding such sensors comes at the price of an increased weight and power consumption, so that they are not applicable to small-scale or nano-quadrotors. For example, platforms such as the Parrot Ardrone or the Bitcraze Crazyflie can only support a single monocular camera in terms of payload. Therefore, we are interested in methods for range mapping and obstacle avoidance for mobile robots that only have access to a single monocular camera.

Most structure-from-motion approaches (SfM) based on visual features such as PTAM [8] produce sparse maps that are not suited for collision-free navigation [9]: Typically the resulting point cloud of 3D features is highly noisy and, more importantly, the absence of visual features in regions with low texture does not necessarily imply free space. For example, although a cabinet might generate visual features around the edges or at the corners, its unicolored surface might not. In contrast, dense stereo methods [10, 11] are able to propagate depth information from the corners to texture-less regions through regularization, but are computationally intensive and difficult to apply on noisy quadrotor data. Other approaches implement reactive obstacle avoidance for quadrotors using optical flow [12], relative size change [13] or reinforcement learning [14].

The goal of this paper is to combine the advantages of dense mapping [11] with the robustness of feature-based SfM methods [8]. This enables collision-free navigation of a quadrotor during forward flight. Our approach consists of two steps: First, we compute a dense depth map from a small set of images (typically 30). Subsequently, we use this depth map to generate the next obstacle-free waypoint in forward direction (see Fig. 1). In contrast to previous work [9], this approach yields a dense depth map at approximately 1 Hz, so that we obtain absolute distance estimates for every pixel in the image in near real-time. In our experiments, we demonstrate that by using our approach, a quadrotor can safely navigate through and around obstacles such as boxes, doors, and persons. Furthermore, and not expected before experimentation, we found that no additional quadrotor motion is required during image acquisition: Small movements of the quadrotor during hovering are already sufficient to generate

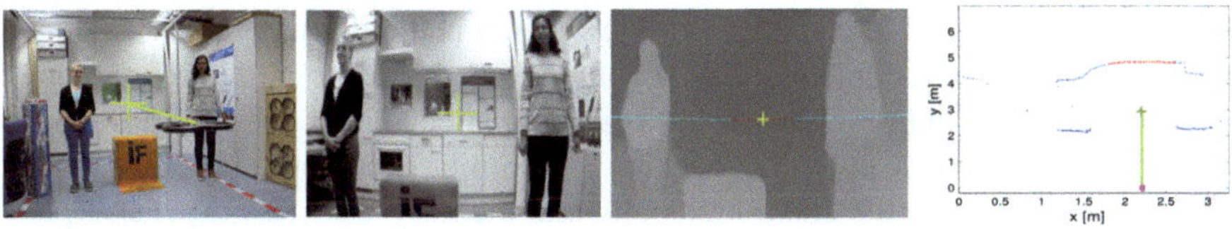

Fig. 1 Our goal is to enable a quadrotor with a monocular camera to avoid frontal obstacles and to navigate through narrow spaces. *Left* Quadrotor hovering in front of several obstacles, including persons. *Left middle* Camera view from the quadrotor. *Right middle* Estimated dense depth image. *Right* Top view of extracted 2D range scan and visualization of generated path

satisfactory depth maps. As a consequence, we expect that our approach applies to a large range of micro aerial vehicles.

In a series of experiments with a real quadrotor, we demonstrate the validity of our approach. The accompanying video `collision_avoidance.m4v` provide additional examples and more details. For all of our experiments, we used a Parrot Ardrone 2 quadrotor, but the proposed approach equally transfers to any mobile platform equipped with a monocular camera. Due to the limited computational resources on the quadrotor, we currently perform all computations on an external base station that communicates over wireless network with the quadrotor.

2 Related Work

Existing approaches to obstacle avoidance range from purely reactive approaches to full planning-based approaches [15, 16]. Reactive approaches aim at directly generating motion commands from sensor readings. As no intermediate representation has to be formed, reactive approaches are typically computationally lightweight and have low latency with well defined performance guarantees. Therefore, reactive approaches are well suited as safety mechanisms. Next to distance sensors, reactive approaches can also be applied to visual cues. For example, optical flow [12] can be used to determine the relative distance of an object to the quadrotor if the absolute speed of the quadrotor is known. However, optical flow can only be used with side-view cameras, as the optical flow of frontal obstacles is zero or close to zero. As an alternative, Mori and Scherer [13] recently proposed to detect relative size change in the image, which indicates approaching objects. Ross et al. [14] proposed an approach based on imitation learning, where multiple cues including optical flow and visual features are used for motion prediction.

When two frontal cameras are available, stereo methods can be used to generate a depth map [17]. If only a single camera is available (or, as in our case, affordable in terms of payload), stereo can also be computed between two or more consecutive images when the camera is moving. This is also called motion parallax. As images are inexpensive to acquire, Newcombe et al. [11] proposed in their DTAM approach to first accumulate a cost volume of 10's to 100's of images before generating a depth map. In contrast to classical two-view stereo methods, much more data from the environment is available in the cost volume which leads to more accurate depth maps. On the cost volume, regularization can be applied to infer missing values in the depth map and to reduce noise. Graber et al. [18] similarly compute a depth map from multiple images using the computationally and memory-wise less expensive plane sweep algorithm, but perform a more complex regularization during subsequent 3D reconstruction.

Both methods need accurate camera poses to construct the cost volume. While DTAM employs direct tracking on the currently estimated depth map, it implicitly assumes that the camera moves only slowly and with sufficient translational motion. On a quadrotor as the Ardrone, this is difficult to guarantee, as all flying maneuvers induce a change in attitude which presumably would often lead to a loss of tracking. Therefore, we decided to use the feature-based structure-from-motion approach of Klein and Murray [8], which contains an automatic recovery procedure and provides in our experience highly accurate camera poses. To integrate the visual information with the IMU data and to generate the control commands for the quadrotor, we make use of the tum_ardrone package of Engel et al. [9].

In this paper, we demonstrate that a single forward-facing camera on a low-cost quadrotor is enough to generate dense depth maps that are suitable for obstacle avoidance. We incrementally generate the next way point based on the current depth map. It should be noted that in this work, we do explicitly not deal with the problem of global path planning [15], as this requires a global (obstacle) map which is typically not available beforehand. Yet, it would be interesting in future work to integrate the individual depth maps into a global 3D map that could then be used for global path planning.

3 Dense Depth Map Estimation

The main challenge for our application is that although a sparse, feature-based representation provided by popular monocular SLAM/SfM approaches is sufficient for localization, it is generally not sufficient for autonomous path-planning. The reasons for this are that (1) large sparely-textured obstacles typically have no visual features on them and (2) outliers introduce unwanted obstacles in free space.

In Fig. 2, we illustrate this difficulty using the example of the PTAM system. The images in the top show the sparse keypoints (visualized as points) and the key frames (visualized with small axes in the top right image). Although PTAM detects many features in various parts of the scene, it does not contain any visual features for the upper cabinets, and therefore, a path planning system might generate a path that leads straight through the cabinet. In contrast, the dense depth map visualized in the bottom right image provides distance information for every pixel in the scene, in particular on the cabinet in the background as well as on the boxes in front.

In the following, we introduce our algorithm to estimate a regularized depth map from multiple input images. We assume that we obtain from a quadrotor a sequence

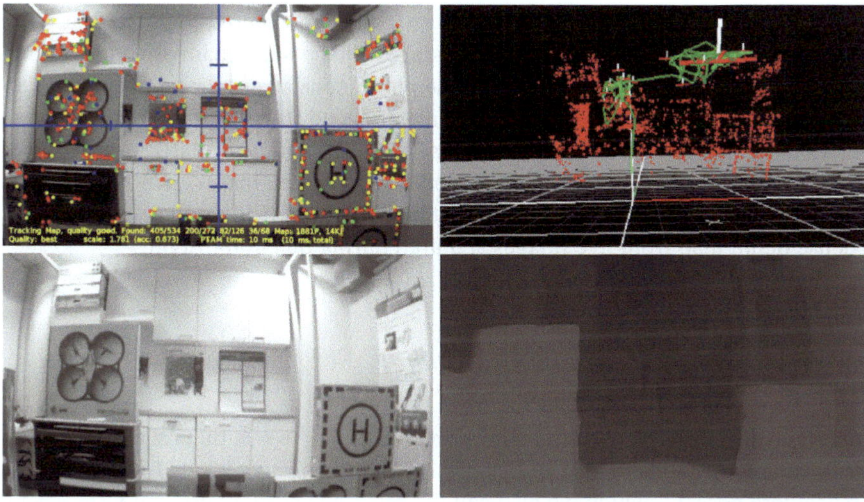

Fig. 2 *Top* Sparse feature maps, as for example generated by PTAM, cannot represent regions without texture (e.g., the *upper* kitchen cabinets). Therefore, they are generally not suitable for path planning or obstacle avoidance. *Bottom* A dense depth map using regularization provides distance estimates for every pixel

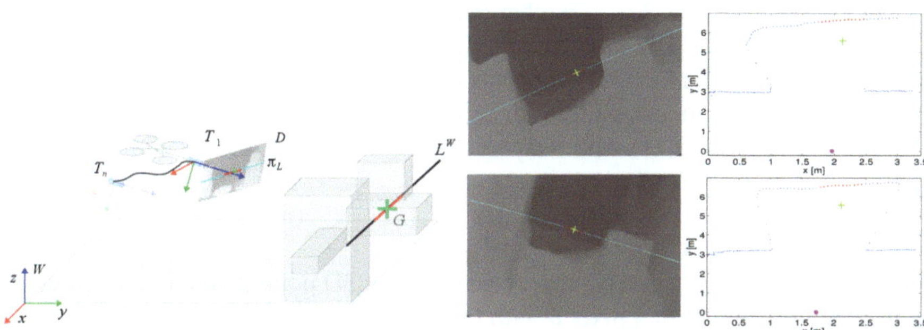

Fig. 3 Illustration of our approach and the involved coordinate systems: The quadrotor hovers in front of several obstacles, while it takes a series of images at camera poses $T_1, \dots, T_n$. From these images, it computes a depth map D and selects the next waypoint G. The quadrotor at first considers m distant candidate waypoints along a *horizontal line* L^W defined in world coordinates in front of its current position. We convert these points into the frame of the current reference camera, $\mathbf{u}_i := \pi(\mathbf{p}_i)$ and project them into the depth map to find the furthest 3D point. The *left column* shows two different depth maps with the projected *horizontal lines* (*cyan dots*), footprint points (*red*), and the selected waypoint as a *yelow cross*. We also show a *top* view of the same scene with the resemble 2D scan

of n gray-scale images $I_1, \dots, I_n : \Omega \subset \mathbb{R}^2 \mapsto \mathbb{R}$ with corresponding camera poses $T_1, \dots, T_n \in SE(3)$ that were gathered during hovering. In practice, we obtain the camera poses from PTAM as used by [9]. The geometry of our setup is visualized schematically in Fig. 3, where the quadrotor (left column) is looking at a simple scene (right column).

Analogous to DTAM [11], we follow a multi-view stereo approach based on energy minimization that consists of two stages: The first stage involves the calculation of a cost volume that accumulates the photo-consistency errors of the overlapping images for different inverse depths. In the second stage, we minimize an energy functional comprising the cost volume as data term and a regularization term that penalizes deviation from a spatially smooth inverse depth map.

In the following, we explain in more detail how we apply this approach to generate a dense depth map from live images acquired during quadrotor flight.

3.1 Cost Volume Creation and Update

Our algorithm starts by creating a cost volume $\mathbf{C} : \mathbb{R}^3 \mapsto \mathbb{R}$ that reflects our belief about the depth of every pixel. The cost volume is always located in front of the first camera pose T_1, which we also call the *reference* pose. Every cell $\mathbf{C}(u, v, \xi)$ in this volume indicates the cost (or negative loglikelihood) that the pixel $\mathbf{u} = (u, v)$ in the reference image I_1 has an inverse depth of ξ. We use a uniform discretization in the inverse depth range $[\xi_{min}, \xi_{max}]$ to assure an uniform sampling of the projected epipolar lines in the reference image space.

For every image $k = 2, \ldots, n$ other than the reference image, we now update the cost volume depending on how well the k image I_k matches the reference image I_1 at a certain inverse depth. This is done as follows: For every pixel $\mathbf{u}$ in the reference image and each hypothesized inverse depth ξ_i we update the accumulated average cost at cube voxel $\mathbf{C}(\mathbf{u}, \xi_i)$ with the photometric error $\rho(I_k, \mathbf{u}, \xi_i)$ defined by

$$\rho(I_k, \mathbf{u}, \xi_i) = I_1(\mathbf{u}) - I_k(\pi(T_k \pi^{-1}(\mathbf{u}, \xi_i)) \tag{1}$$

where $\pi(\mathbf{p})$ describes a perspective projection of a 3D point $\mathbf{p}$ and $\pi^{-1}(\mathbf{u}, \xi_i)$ refers to the back-projection of a pixel $\mathbf{u}$ with inverse depth ξ_i.

We adopt the camera model of [19], where the intrinsic calibration of the camera is given by the intrinsic parameters (f_u, f_v, c_u, c_v) and a single radial distortion parameter w. Since the images captured by the camera are of low resolution (in our case 640×360) and are transmitted to the ground-based desktop with lossy compression, we decided to not pre-rectify the images in order to not loose any more valuable information. Instead, we explicitly take the distortion model into account during projection and back-projection. The perspective projection $\mathbf{u} = \pi(\mathbf{p})$ is given by

$$\begin{bmatrix} u \\ v \end{bmatrix} = \begin{bmatrix} c_u \\ c_v \end{bmatrix} + \begin{bmatrix} f_u & 0 \\ 0 & c_v \end{bmatrix} \frac{r_d}{r} \begin{bmatrix} p_x/p_z \\ p_y/p_z \end{bmatrix} \tag{2}$$

where $r = \sqrt{\frac{p_x^2+p_y^2}{p_z^2}}$ and $r_d = \frac{1}{w} \arctan(2r \tan \frac{w}{2})$. On the contrary, to back-project $\pi^{-1}(\mathbf{u}, \xi_i)$ a pixel $\mathbf{u}$ with inverse depth ξ_i we use:

$$\mathbf{p} = \frac{1}{\xi_i}\begin{bmatrix} \frac{r}{r_d}\mathbf{u}_n \\ 1 \end{bmatrix} \tag{3}$$

$$\mathbf{u}_n = \begin{bmatrix} \frac{u-c_u}{f_u} \\ \frac{v-c_v}{f_v} \end{bmatrix} \tag{4}$$

where $\mathbf{u}_n$ is a normalized pixel, $r_d = \sqrt{u_n^2 + v_n^2}$ and $r = \frac{\tan(wr_d)}{2\tan(w/2)}$.

After all images have been integrated, the average photometric error $\mathbf{C}(\mathbf{u}, \xi_i)$ stored in the cost volume for pixel $\mathbf{u}$ at inverse depth ξ_i is

$$\mathbf{C}(\mathbf{u}, \xi_i) = \frac{1}{n}\sum_{k=2}^{n} ||\rho(I_k, \mathbf{u}, \xi_i)||_1. \tag{5}$$

Once the cost volume has been computed, an initial inverse depth map ξ_r can be extracted by searching for the inverse depth associated with the minimum cost for each pixel:

$$\xi_k(\mathbf{u}) = \arg\min_{\xi_i} \mathbf{C}_k(\mathbf{u}, \xi_i) \tag{6}$$

This solution is also called the *winner-takes-all* method. It should be noted that this initial inverse depth map will be very noisy, as so far no regularization has been applied. Moreover, it might happen that multiple disparities have equal cost, for example due to surfaces with little texture.

3.2 Primal-Dual Optimization of the Cost Volume

In order to find a better inverse depth map, the simple approach of (6) is used as a starting solution of a spatially regularized formulation based on total variation (TV). TV-regularization is known for its ability to preserve edges while smoothing homogeneous regions. The solution $\xi(\mathbf{u})$ is given by the minimization of the energy functional

$$E_\xi = \int_\Omega w(\mathbf{u})||\nabla\xi(\mathbf{u})||_\varepsilon + \lambda\mathbf{C}(\mathbf{u}, \xi(\mathbf{u}))d\mathbf{u}, \tag{7}$$

where Ω is the signal domain (in this case the image dimensions), $w(\mathbf{u})$ is a per pixel weight based on the image gradient that reduces the regularization across image edges, $|| \, ||_\varepsilon$ is the Huber norm and λ is a parameter used to define the tradeoff between the convex regularizer $w(\mathbf{u})||\nabla\xi(\mathbf{u})||_\varepsilon$ and the non-convex data term $\mathbf{C}_r(\mathbf{u}, \xi(\mathbf{u}))$. Since the data term is non-convex the energy functional is approximated by decoupling the data and regularization terms through an intermediate function $\alpha(\mathbf{u})$:

$$E_{\xi,\alpha} = \int_{\Omega} w(\mathbf{u})||\nabla\xi(\mathbf{u})||_{\varepsilon} + \frac{1}{2\theta}(\xi(\mathbf{u}) - \alpha(\mathbf{u}))^2 + \lambda \mathbf{C}(\mathbf{u}, \alpha(\mathbf{u}))d\mathbf{u}, \tag{8}$$

such that $E_{\xi,\alpha} \to E_{\xi}$ as $\theta \to 0$.

The new energy functional allows us to split the minimization into two different problems that are alternatively solved untill convergence:

- First, for a fixed $\alpha(\mathbf{u})$ solve:

$$\min_{\xi} \int_{\Omega} w(\mathbf{u})||\nabla\xi(\mathbf{u})||_{\varepsilon} + \frac{1}{2\theta}(\xi(\mathbf{u}) - \alpha(\mathbf{u}))^2 d\mathbf{u} \tag{9}$$

which corresponds to the well known Huber-ROF denoising problem that can be solved using a primal-dual algorithm [20]. In this case $\alpha(\mathbf{u})$ represents the noisy image whereas $\xi(\mathbf{u})$ is the searched denoised result.

- Second, for a fixed $\xi(\mathbf{u})$ solve:

$$\min_{\alpha} \int_{\Omega} \frac{1}{2\theta}(\xi(\mathbf{u}) - \alpha(\mathbf{u}))^2 + \lambda \mathbf{C}(\mathbf{u}, \alpha(\mathbf{u}))d\mathbf{u}. \tag{10}$$

This optimization is performed by a point-wise exhaustive search for each α in $\mathbf{C}$.

In practice, the update steps for $\xi(\mathbf{u})$ and $\alpha(\mathbf{u})$ can be performed efficiently and in parallel on modern GPU hardware for each independent pixel $\mathbf{u}$. In addition for the second step we implemented the accelerated and the improved sub-sample accuracy methods recommended in [11]. Finally, we convert the inverse depth map into a depth map D using

$$D(\mathbf{u}) := (\xi(\mathbf{u}))^{-1}. \tag{11}$$

In sum, this algorithm takes a sequence of n input images and camera poses, generates a cost volume, and then finds a regularized inverse depth map $\xi(\mathbf{u})$ with minimum total variation.

4 Obstacle Avoidance and Waypoint Selection

When a new depth map D becomes available, we use it to generate the next waypoint in forward direction. Our goal is to select the point where the quadrotor can fly the furthest at the current flying height. To this end, we calculate the most distant point in 3D space reachable by the quadrotor without collisions.

This selection process is illustrated in Fig. 3: The quadrotor at first considers m distant candidate waypoints $\mathbf{p}_1, \ldots, \mathbf{p}_m \in \mathbb{R}^3$ along a horizontal line defined in world coordinates in front of its current position. We convert these points into the frame of the current reference camera T_1 and project them into the depth map. Figure 3 right, shows two different depth maps with an overlay of the projected line. Note that in both cases, the quadrotor was tilted manually while it was recording the reference image to demonstrate that the projected line is in fact horizontal in world coordinates but not necessarily horizontal in the image.

The corresponding pixel coordinates $\mathbf{u}_i$ and depths d_i of these points are

$$\mathbf{u}_i := \pi(\mathbf{p}_i) \tag{12}$$

$$d_i := D(\mathbf{u}_i). \tag{13}$$

Using these virtual depth measurements, we generate a new set of points

$$\mathbf{q}_i := \pi^{-1}(\mathbf{u}_i, d_i^{-1}), \tag{14}$$

which are now located on the estimated surface. The resulting points are visualized from a top-down view in the right column of Fig. 3, and resemble a 2D laser scan.

In order to find the farthest admissible waypoint, we first maximize the minimum convolution given by

$$i^* = \arg \max_{i=1,\dots,m} \min_{j \in F_i} D(\mathbf{u}_j), \tag{15}$$

where $F_i \subset \{1, \dots, m\}$ refers to the footprint of the quadrotor on its way to point $\mathbf{p}_i$. We define the footprint of the quadrotor as all 3D points that the quadrotor would touch on its way from its current position to the respective point, under consideration of the quadrotor's shape. Subsequently, we determine the maximum distance the quadrotor can safely move forward as

$$d^* = \min_{j \in F_{i^*}} D(\mathbf{u}_j). \tag{16}$$

To generate the actual waypoint, we subtract a small safety boundary b (e.g., 1 m or 1.5 m) from this depth to prevent the quadrotor from crashing into the next obstacle. The selected waypoint thus becomes

$$\mathbf{p}^* := \pi^{-1}(\mathbf{u}_{i^*}, (d^* - b)^{-1}). \tag{17}$$

While it is in principle possible to evaluate the footprint for arbitrary trajectories, we found that the actual flying behavior of the quadrotor along two axes was somewhat undeterministic. Therefore, we decided to split the motion into two orthogonal segments (which we correctly consider during footprint generation): In the first segment, the quadrotor positions itself at the desired X position (left/right, see Fig. 5). In the second part, it then approaches the waypoint by a forward motion along the Y axis. Note that we keep the flying height fixed (Z axis) in all of our experiments, although it would be straight-forward to use it as a third degree-of-freedom during obstacle avoidance. Two examples of the selected waypoints as well as the corresponding footprint are visualized with a yellow cross and red dots, respectively, in the right column of Fig. 3.

To summarize this section, our waypoint generation method takes a dense depth map D as input and outputs a waypoint $\mathbf{p}^*$ that is expected to lead to the largest forward increment without collisions.

5 Experiments and Results

For all our experiments, we used a low-cost Parrot AR.Drone 2 quadrotor. It is equipped with an inertial measurement unit (IMU) consisting of a 3-axis gyroscopes and accelerometers operating at 200 Hz, an ultrasound altimeter with an update rate of 25 Hz and a frontal monocular camera covering a field of view of 73.5° × 58.5°, providing an image resolution of 640 × 360. The quadrotor communicates with a ground-based desktop over wireless LAN to perform the calculations of the all time demanding tasks. The video of the frontal camera is streamed in real-time to the desktop at 30 fps using a lossy compression to reduce bandwidth. The open-source implementation of [9] (tum_ardrone[1]) for quadrotor control and our path planning module run on the Intel Core i3-2120 CPU ×4 at 3.36 GHz, while the dense depth map estimation is executed on a NVIDIA GeForce GTX 560 TI graphic card with 384 CUDA cores and 1 GB of device memory. The tum_ardrone package continuously estimates the absolute scale of the visual map. As a consequence, the computed depth maps provide absolute distances, which is very useful for our purpose.

5.1 Dense Depth Map Evaluation

We first evaluated the performance of our depth map estimation approach using a synthetic dataset [21] which provides perfect camera poses and ground truth depth maps. This allowed us to analyze the impact on the accuracy with respect to the number of images integrated in the cost volume and to determine the required number of discrete depth layers. We calculated the Mean Absolute Error (MAE) on both the initial depth map and the final solution with respect to the ground truth. To the best

---

[1] http://wiki.ros.org/tum_ardrone.

We would like to thank Jakob Engel for his support and advice on using the tum_ardrone package.

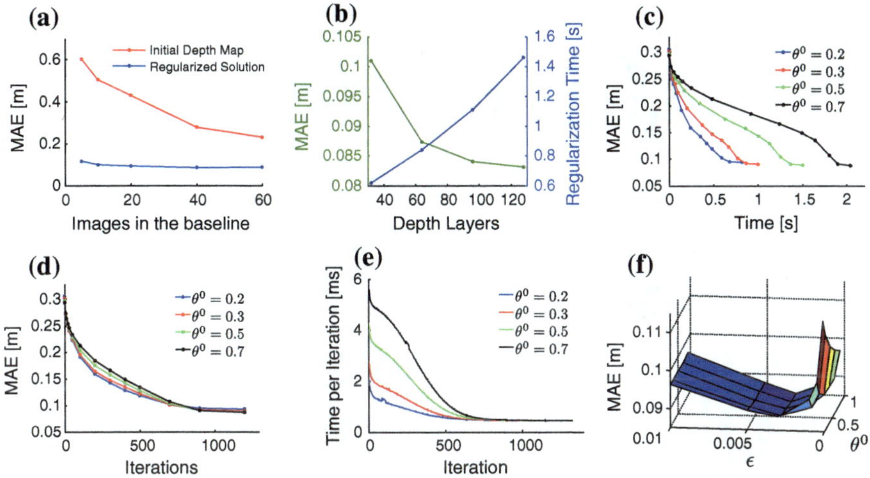

Fig. 4 Performance evaluation for the dense depth map estimation approach. **a** Error w.r.t. the number of images. Increasing the number of images within the same baseline significantly improves the accuracy of the initial depth map (*red curve*). In contrast, the quality of the regularized solution barely changes (*blue curve*). We selected 20 images to achieve the lower error. **b** Error (*green*) and computation time (*blue*) w.r.t. the number of layers of the cost volume. Increasing the number of depth layers improves the quality, but at the expense of increasing running time. A good compromise is achieved for 64 depth layers. **c** Error w.r.t. elapsed regularization time for different values of θ^0. Notice that for large values of θ^0, the regularization time increases substantially. A value of $\theta^0 = 0.2$ yields the least time for the same number of iterations without impairing precision. **d** Error w.r.t. the number of iterations. In all cases approx. 900 iterations are required to achieve convergence. **e** Time per iteration. Notice that the time per iteration decreases until convergence as the search interval–induced by the Quadratic Penalty term, decreases. **f** Error w.r.t. ε and θ^0. A valley is yielded for $\varepsilon = 0.003$, and it is independent of θ^0

of our knowledge, we are the first to provide such a quantitative evaluation of dense mapping approaches. Based on our findings, we then selected the parameters for the problem defined in Eq. 8 in terms of accuracy, computation time and convergence. The results are given in Fig. 4.

In the final experiments with the quadrotor we consider a depth range from 0.5 to 7 m, which we sampled into a cost volume of 64 inverse depth layers requiring approximately 84 MB ($640 \times 360 \times 64 \times 6$ bytes). In all experiments, we set $\theta = 0.2$, $\lambda = 0.9$ and 900 primal-dual iterations. Note that this process is carried out on the GPU and hence does not affect visual tracking and position control of the quadrotor. As expected, the energy optimization is the most expensive calculation, requiring approximately 500 ms in total to obtain the final depth map. However, integrating a single image into the cost volume requires only around 5 ms per frame. This means depth maps can be generated at 1.5 Hz.

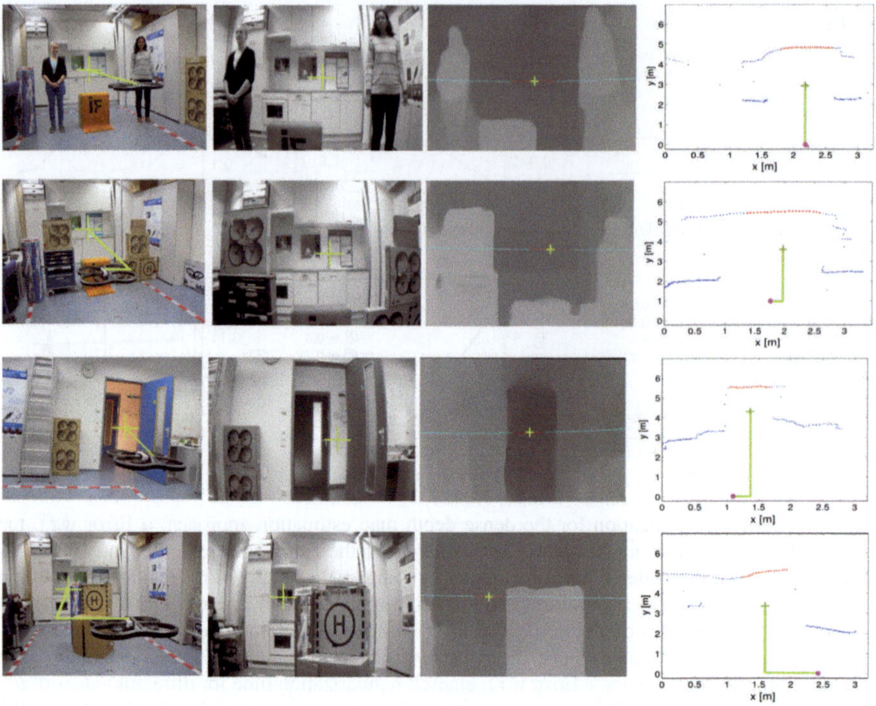

Fig. 5 Evaluation of the collision avoidance approach on four different environment configurations. From *top* to *bottom*: people, boxes on the outside, door, boxes on the inside. From *left* to *right*: External view, quadrotor view, depth map, top view of range scan and generated trajectory (*yellow lines*, added manually for visualization purposes)

5.2 Collision Avoidance Evaluation

To evaluate the reliability of our approach, we set up increasingly challenging environments through which the quadrotor had to navigate. The goal of our experiments was to measure the number of system successes and failures to pass the obstacle. Figure 5 shows the four different environments in which we tested our approach. In the "person" environment, two persons were standing approximately 1.5 m away from each other. In the "boxes on the outside" environment, we set up various boxes and a tool carriage in such a way that the quadrotor had to pass through the middle. In the "door" environment, the quadrotor had to navigate through an open door with a clearance of 90 cm. For the "boxes on the inside" environment, we put a stack of boxes in the middle of the room that the quadrotor had to circumvent, that is, either to the left or to the right.

In each environment, we carried out 10 trials with our collision avoidance approach. In each trial, the quadrotor was hovering for a couple of seconds in front of an obstacle, acquiring 30 images, computing the depth map, generating the way-

Table 1 Performance evaluation over 10 runs in each of the four experiments

Setting	Successes	Failures
People	10	0
Boxes on the outside	9	1
Boxes in the middle	9	1
Door	5	5
Sum	33	7
Overall performance	82.5 %	17.5 %

point and executing it. The overall time between starting the image capture and the generation of the next way point was less than 1.5 s. We deemed a trial successful when the quadrotor reached its desired waypoint without collision. Any other case, i.e., a failure to compute the depth map, to find a suitable waypoint, or a collision with the environment was counted as a failure. After every trial, the quadrotor was returned manually to its initial position, before it selected another random starting position for the next trial.

The results are summarized in Table 1. Over 40 trials, we achieved an overall success rate of 82.5 %, where the quadrotor reached the desired target position without collision. In 17.5 % of all trials, it failed. Most failures occurred in the "door" environment, where the quadrotor had to navigate through a very narrow space.

We consider our evaluation of the parameters a valuable contribution for future work on dense mapping approaches. We also demonstrated that obstacle avoidance and collision-free path planning is feasible from monocular images, which is in particular useful for upcoming nano-quadrotors with minimal payload. Interestingly, and opposed to our earlier assumption, we found that no additional quadrotor motion is required during image acquisition: Small movements of the quadrotor during hovering generate sufficient baselines leading to satisfactory depth maps.

6 Conclusion

In this paper, we presented a novel approach to obstacle avoidance for quadrotors that have only a single monocular camera. From a small set of consecutive images, we compute a regularized depth map that we subsequently use for waypoint generation. We demonstrated the applicability of our approach in four challenging environments using a Parrot Ardrone quadrotor. As our approach only requires an IMU and a monocular camera, it can be applied easily to many other lightweight aerial vehicles.

It would be interesting to augment our approach to full autonomous exploration, and to adapt it to CPU so that it becomes applicable to on-board computing. Furthermore, we believe that a collision avoidance system would clearly benefit from semantic scene understanding approaches (e.g., to recognize a door as a door), which we plan to look into in the near future.

Acknowledgments The authors would like to thank Pedro Piniés for the so many fruitful discussions about convex optimisation and variational methods. His insights about saddle point methods acquired during his stay in Graz University of Technology was crucial to understand and implement the core of the energy minimisation with TV regularisation.

References

1. Bachrach, A., de Winter, A., He, R., Hemann, G., Prentice, S., Roy, N.: RANGE—robust autonomous navigation in GPS-denied environments. In: ICRA (2010)
2. Grzonka, S., Grisetti, G., Burgard, W.: A fully autonomous indoor quadrotor. IEEE Trans. Robot. (T-RO) **8**(1), 90–100 (2012)
3. Fraundorfer, F., Heng, L., Honegger, D., Lee, G., Meier, L., Tanskanen, P., Pollefeys, M.: Vision-based autonomous mapping and exploration using a quadrotor MAV. In: IROS (2012)
4. Wagter, C.D., Tijmons, S., Remes, B., de Croon, G.: Autonomous flight of a20-gram flapping wing MAV with a 4-gram onboard stereo vision system. In: ICRA, Hong Kong (2014) [Accepted]
5. Huang, A.S., Bachrach, A., Henry, P., Krainin, M., Maturana, D., Fox, D., Roy, N.: Visual odometry and mapping for autonomous flight using an RGB-D camera. In: International Symposium on Robotics Research (ISRR) (2011)
6. Tomic, T., Schmid, K., Lutz, P., Domel, A., Kassecker, M., Mair, E., Grixa, I.L., Ruess, F., Suppa, M., Burschka, D.: Toward a fully autonomous UAV: research platform for indoor and outdoor urban search and rescue. Robot. Autom. Mag. IEEE **19**(3), 46–56 (2012)
7. Nieuwenhuisen, M., Droeschel, D., Holz, D., , Laebe, T., Behnke, S.: Multimodal obstacle detection and collision avoidance for micro aerial vehicles. In: ECMR (2013)
8. Klein, G., Murray, D.: Parallel Tracking and Mapping for Small AR Workspaces. In: Proceedings of the IEEE International Symposium on Mixed and Augmented Reality (ISMAR) (2007)
9. Engel, J., Sturm, J., Cremers, D.: Camera-based navigation of a low-cost quadrocopter. In: IROS (2012)
10. Hirschmüller, H.: Accurate and efficient stereo processing by semi-global matching and mutual information. In: CVPR (2005)
11. Newcombe, R.A., Lovegrove, S.J., Davison, A.J.: DTAM: dense tracking and mapping in real-time. In: Proceedings of the International Conference on Computer Vision ICCV (2011)
12. Green, W., Oh, P.: Optic-flow-based collision avoidance. Robot. Autom. Mag. IEEE **15**(1), 96–103 (2008)
13. Mori, T., Scherer, S.: First results in detecting and avoiding frontal obstacles from a monocular camera for micro unmanned aerial vehicles. In: ICRA (2013)
14. Ross, S., Melik-Barkhudarov, N., Shankar, K.S., Wendel, A., Dey, D., Bagnell, J.A.D., Hebert, M.: Learning monocular reactive uav control in cluttered natural environments. In: ICRA (2013)
15. Goerzen, C., Kong, Z., Mettler, B.: A survey of motion planning algorithms from the perspective of autonomous UAV guidance. J. Intell. Robot. Syst. **57**(1–4), 65–100 (2010)
16. Kendoul, F.: Survey of advances in guidance, navigation, and control of unmanned rotorcraft systems. J. Field Robot. **29**(2) (2012)
17. Hrabar, S., Sukhatme, G., Corke, P., Usher, K., Roberts, J.: Combined optic-flow and stereo-based navigation of urban canyons for a UAV. In: IROS (2005)
18. Graber, G., Pock, T., Bischof, H.: Online 3D reconstruction using convex optimization. In: ICCV LDRMC (2011)
19. Devernay, F., Faugeras, O.: Straight lines have to be straight: automatic calibration and removal of distortion from scenes of structured enviroments. Mach. Vis. Appl. **13**(1), 14–24 (2001)
20. Chambolle, A., Pock, T.: A first-order primal-dual algorithm for convex problems with applications to imaging. Math. Imaging Vis. **40**(1), 120–145 (2011)

21. Handa, A., Newcombe, R.A., Angeli, A., Davison, A.J.: Real-time camera tracking: when is high frame-rate best? In: ECCV (2012)

Initialization-Free Monocular Visual-Inertial State Estimation with Application to Autonomous MAVs

Shaojie Shen, Yash Mulgaonkar, Nathan Michael and Vijay Kumar

Abstract The quest to build smaller, more agile micro aerial vehicles has led the research community to address cameras and Inertial Measurement Units (IMUs) as the primary sensors for state estimation and autonomy. In this paper we present a monocular visual-inertial system (VINS) for an autonomous quadrotor which relies only on an inexpensive off-the-shelf camera and IMU, and describe a robust state estimator which allows the robot to execute trajectories at 2 m/s with roll and pitch angles of 20 degrees, with accelerations over 4 m/s^2. The main innovations in the paper are an approach to estimate the vehicle motion without initialization and a method to determine scale and metric state information without encountering any degeneracy in real time.

1 Introduction

Micro-aerial vehicles (MAVs) are ideal platforms for missions in complex confined environments due to its small size and superior mobility. Recently, there have been great successes in deploying sensor-equipped autonomous micro-aerial vehicles (MAVs) in complex GPS-denied environments. In particular, monocular visual-inertial systems (VINS) that consists of a camera and a low cost IMU are very attractive to MAVs with limited payload budget due to their small footprint, low

S. Shen (✉) · Y. Mulgaonkar · V. Kumar
GRASP Laboratory, University of Pennsylvania, Philadelphia, PA 19104, USA
e-mail: eeshaojie@ust.hk

Y. Mulgaonkar
e-mail: yashm@seas.upenn.edu

V. Kumar
e-mail: kumar@seas.upenn.edu

N. Michael
Robotics Institute, Carnegie Mellon University, Pittsburgh, PA 15213, USA
e-mail: nmichael@cmu.edu

M.A. Hsieh et al. (eds.), *Experimental Robotics*, Springer Tracts
in Advanced Robotics 109, DOI 10.1007/978-3-319-23778-7_15

Fig. 1 Our quadrotor experimental platform equipped with a Intel NUC computer, a MEMS IMU, and a camera. This quadrotor can be launched without initialization at a preferred state. It can autonomously maneuver through different trajectories, and recover from possible failure conditions on-the-fly

cost, and low maintenance. However, most existing monocular VINS approaches are inoperable without a good initial state estimate, nor are they able to recover/restart on the fly in case the estimator fails due to the lack of initial condition [1, 2]. Such autonomous MAVs are only capable of launching with known initial condition (e.g. stationary), and their operating envelop is limited in order to reduce the risk of estimator failure. This motivates the goal of this work as developing a dynamically-launchable and failure-recoverable MAV to autonomously fly through a wide variety of trajectories, including hovering, straight line, and aggressive maneuvers (Fig. 1).

Solutions to VINS has been proposed in a *filtering* setting [1–7] and in a *graph-based* optimization/bundle adjustment/smoothing setting [8–10]. Filtering approaches have the advantage of fast processing due to its continuous marginalization of past states, but their performance can be sub-optimal due to early fix of linearization points. Graph-based approaches benefit from iterative re-linearization of states but it requires more computation power. With proper marginalization, a constant complexity sliding window graph-based framework can be obtained [8]. A comparison between filtering and graph-based approaches is presented in [11]. The authors reported nearly identical results of two types of approaches. However, the platform for verification is only equipped with an optical flow sensor that is unable to perform long term feature tracking. This limits the power of graph-based approach, as a sufficiently connected graph is never constructed.

We can also categorize VINS solutions as *loosely coupled* [1] or *tightly coupled* [2–7, 9, 10]. Loosely coupled approaches usually utilize an independent vision processing module such as PTAM [12] for up-to-scale pose estimation, and integrating the vision pose with IMU using a filter for scale estimation. Tightly coupled approaches usually lead to better estimation results (up to linearization error) because they integrate camera measurement and noise models in a systematic manner.

However, all approaches mentioned above require good initializations due to the underlying linearized solver of the nonlinear VINS system.

Pioneering work on VINS *without* initialization is proposed in [13], where the authors proposed to perform the estimation in the body frame of the first pose in the sliding window. An IMU pre-integration technique is proposed to handle multi-rate sensor measurements. The authors show that the nonlinearity of the system mainly arises only from rotation drift.

Recent results suggests that by assuming the orientation is known or by estimating the rotation from short-term integrated gyroscope output, VINS may be solved in a linear *closed-form* [14–18]. It has been shown that both the initial gravity vector and the body frame velocity can be estimated linearly. These results have significant implications that a good initialization of the VINS problem may actually not be required. However, [14] is limited to use a fixed small number of IMU measurements, which makes it very sensitive to IMU noise. Approaches that utilize multiple IMU measurements in a sliding window [15–18] do not scale well to a large number of IMU measurements since they relie on double integration of accelerometer output over an extended period of time. Moreover, these closed-form approaches do not take the noise characteristic of the system into account, which lead to sub-optimal results.

Another issue of monocular VINS is the scale ambiguity due to degenerate motion. It is well known that in order to render the scale observable, accelerations in at least two axes are required [2, 4, 16]. However, for a MAV, degenerate motions such as hovering or constant velocity motions are unavoidable. The hover case is first addressed in [7] by proposing a last-in-first-out (LIFO) sliding window approach. However, [7] generates pessimistic covariance estimates due to the state-only measurement update scheme.

To address the problems of initialization, failure recovery, and degenerate motion altogether, we identify the contribution of this paper as threefold.

- We propose a linear sliding window formulation for monocular VINS that is able to estimate necessary navigation states (velocity and attitude) without any prior initial information (Sects. 2 and 4).
- We address the issues of degenerate motion and scale unobservability by proposing a two-way marginalization scheme (Sect. 3).
- We experimentally show that the proposed approach enables metric state estimation without initialization. We also show that our system is able to handle degenerate motion cases (Sect. 6).

2 Linear Sliding Window VINS Estimator

2.1 Notation

We consider (G) as the earth's inertial frame, (B) as the current IMU body frame, (B_k) as the camera frame while taking the kth image, Note that IMU usually runs at a higher rate than the camera, and that multiple IMU measurements may exist between

B_k and B_{k+1}. We assume that the camera and the IMU is pre-calibrated such that the camera optical axis is aligned with the z-axis of the IMU. $\mathbf{p}_Y^X$, $\mathbf{v}_Y^X$, and $\mathbf{R}_Y^X$ are 3D position , velocity, and rotation of frame X with respect to frame Y. $\mathbf{g}^G = [0,\ 0,\ g]^{\mathrm{T}}$ is the gravity vector in the world frame, and $\mathbf{g}^X$ is the earth's gravity vector expressed in frame X.

2.2 Formulation

Given two time instants (corresponds to two image frames), the IMU propagation model for position and velocity, expressed in the world frame, can be written as:

$$\begin{aligned}
\mathbf{p}_{B_{k+1}}^G &= \mathbf{p}_{B_k}^G + \mathbf{v}_{B_k}^G \Delta t + \iint_{B_k}^{B_{k+1}} \left(\mathbf{R}_B^G \mathbf{a}^B - \mathbf{g}^G\right) dt^2 \\
\mathbf{v}_{B_{k+1}}^G &= \mathbf{v}_{B_k}^G + \int_{B_k}^{B_{k+1}} \left(\mathbf{R}_B^G \mathbf{a}^B - \mathbf{g}^G\right) dt
\end{aligned} \tag{1}$$

where $\mathbf{a}^B$ is the accelerometer measurement in the body frame, Δt is the time difference between B_k and B_{k+1}. It can be seen that the rotation between the world frame and the body frame is required in order to propagate the states with IMU measurements. This rotation can only be determined if the initial attitude of the vehicle is known, which is not the case when the vehicle is dynamically launched or recovering from estimator failure. However, as suggested in [13], if the reference frame of the IMU propagation model is attached to the first pose of the VINS system (i.e. the first pose that we are trying to estimate), (1) can be rewritten as:

$$\begin{aligned}
\mathbf{p}_{B_{k+1}}^{B_0} &= \mathbf{p}_{B_k}^{B_0} + \mathbf{v}_{B_k}^{B_0} \Delta t - \mathbf{g}^{B_0} \Delta t^2/2 + \mathbf{R}_{B_k}^{B_0} \alpha_{B_{k+1}}^{B_k} \\
\mathbf{v}_{B_{k+1}}^{B_0} &= \mathbf{v}_{B_k}^{B_0} - \mathbf{g}^{B_0} \Delta t + \mathbf{R}_{B_k}^{B_0} \beta_{B_{k+1}}^{B_k} \\
\alpha_{B_{k+1}}^{B_k} &= \iint_{B_k}^{B_{k+1}} \mathbf{R}_B^{B_k} \mathbf{a}^B dt^2 \\
\beta_{B_{k+1}}^{B_k} &= \int_{B_k}^{B_{k+1}} \mathbf{R}_B^{B_k} \mathbf{a}^B dt
\end{aligned} \tag{2}$$

where $\mathbf{R}_{B_k}^{B_0}$ is the change in rotation since B_0, which can be obtained by combing the integral gyroscope measurements and relative epipolar constraints (Sect. 2.3). $\alpha_{B_{k+1}}^{B_k}$ and $\beta_{B_{k+1}}^{B_k}$ can be obtained solely with IMU measurements within Δt. We can see that the update equations for all the key quantities ($\mathbf{p}_{B_k}^{B_0}$, $\mathbf{v}_{B_k}^{B_0}$, $\mathbf{g}^{B_0}$) are now linear. It is thus expected that VINS system may be solved in a linear fashion, even without any knowledge of the initial condition.

2.3 Linear Rotation Estimation

The IMU propagation model in (2) can only be linear if good rotation estimates are provided. Although integrating gyroscope measurements will lead to reasonable rotation estimates, it still drifts over time. Therefore, we utilize additional epipolar constraints to eliminate rotation drift. We wish to estimate $R_{B_k}^{B_0}, k = 0, \ldots, N$, subject to following conditions:

$$\mathbf{R}_{B_0}^{B_0} = \mathbf{I}_3, \quad \mathbf{R}_{B_j}^{B_0} = \hat{\mathbf{R}}_{B_j}^{B_i} \mathbf{R}_{B_i}^{B_0} \tag{3}$$

where $\hat{\mathbf{R}}_{B_j}^{B_i}$ is a rotation that is obtained by either integrating gyroscope measurements between two consecutive images, or by finding the essential matrices between the current image and corresponding past images. For each incoming image, we try to compute the essential matrix between it and all other images within the sliding window.

As in [19], the above system can be solved linearly by relaxing orthonormality constraints of the rotations. Specifically, for a pair of rotation matrices $\mathbf{R}_{B_i}^{B_0}, \mathbf{R}_{B_j}^{B_0}$, and their relative constraint $\hat{\mathbf{R}}_{B_j}^{B_i}$, we have:

$$\begin{bmatrix} \mathbf{I}_3, & -\hat{\mathbf{R}}_{B_j}^{B_i} \end{bmatrix} \begin{bmatrix} \mathbf{r}_i^k \\ \mathbf{r}_j^k \end{bmatrix} = 0 \quad k = 1, 2, 3 \tag{4}$$

where $\mathbf{r}_i^k$ is the kth column of $\mathbf{R}_{B_i}^{B_0}$. The solution of the relaxed approximate rotation matrices can be found as the last three columns of the right singular matrix of the system (4). The true rotation matrices can then be obtained by enforcing unit singular values on the approximated rotation matrices: $\mathbf{R}_{B_i}^{B_0} = \mathbf{U}\mathbf{V}^{\mathrm{T}}$, where $\bar{\mathbf{R}}_{B_i}^{B_0} = \mathbf{U}\mathbf{S}\mathbf{V}^{\mathrm{T}}$ is an approximate rotation matrix. After this point, we assume that the rotation components within the sliding window is known and noise-free.

2.4 Linear Sliding Window Estimator

We apply a tightly-coupled, sliding window graph-based [8] formulation due to its constant computation complexity and its ability to incorporate constrains from multiple observations to refine its solution. The *full state* vector can be expressed as (the transpose is ignored for the simplicity of presentation):

$$\begin{aligned}
\mathcal{X} &= \left[\mathbf{x}_{B_0}^{B_0}, \ \mathbf{x}_{B_1}^{B_0}, \ \ldots \mathbf{x}_{B_N}^{B_0}, \ \lambda_0, \ \lambda_1, \ \ldots \lambda_M\right] \\
\mathbf{x}_{B_k}^{B_0} &= \left[\mathbf{p}_{B_k}^{B_0}, \ \mathbf{v}_{B_k}^{B_k}, \ \mathbf{g}^{B_k}\right] \ \text{for k} = 1, \ldots, \text{N} \\
\mathbf{p}_{B_0}^{B_0} &= [0, \ 0, \ 0]
\end{aligned} \tag{5}$$

where $\mathbf{x}_{B_k}^{B_0}$ is the kth *camera state*, N is the number of camera states in the sliding window, M is the number of all features that have been observed for at least twice within the sliding window. λ_l is the depth of the lth point feature from its first observation. We are able to use a one-dimensional representation for features due to the nature of the underlying image processing pipeline (Sect. 5.2). This saves significant amount of computation power. We keep the body frame velocity ($\mathbf{v}_{B_k}^{B_k}$) and gravity vector ($\mathbf{g}^{B_k}$) in the camera state. This helps reducing the impact of rotation error on the estimation results (Sect. 2.5).

Since the rotation is fixed as in Sect. 2.3, we can formulate the linear VINS by gathering all measurements from both the IMU and the monocular camera and solve for the maximum likelihood estimate by minimizing the sum of the Mahalanobis norm of all measurement errors:

$$\min_{\mathcal{X}} \left\{ (\mathbf{b}_p - \Lambda_p \mathcal{X}) + \sum_{k \in \mathcal{D}} \left\| \hat{\mathbf{z}}_{B_{k+1}}^{B_k} - \mathbf{H}_{B_{k+1}}^{B_k} \mathcal{X} \right\|_{\mathbf{P}_{B_{k+1}}^{B_k}}^2 + \sum_{(l,j) \in \mathcal{C}} \left\| \hat{\mathbf{z}}_l^{B_j} - \mathbf{H}_l^{B_j} \mathcal{X} \right\|_{\mathbf{P}_l^{B_j}}^2 \right\} \tag{6}$$

where the measurement triplets $\{\hat{\mathbf{z}}_{B_{k+1}}^{B_k}, \mathbf{H}_{B_{k+1}}^{B_k}, \mathbf{P}_{B_{k+1}}^{B_k}\}$ and $\{\hat{\mathbf{z}}_l^{B_j}, \mathbf{H}_l^{B_j}, \mathbf{P}_l^{B_j}\}$ are defined in Sects. 2.5 and 2.6 respectively. $\mathcal{D}$ is the set of all IMU measurements. $\mathcal{C}$ is the set of all observations between any features and any camera states within the sliding window. $\{\mathbf{b}_p, \Lambda_p\}$ is the *optional* prior for the system. This system can be solved by reorganizing in the following form:

$$\left(\Lambda_p + \Lambda_{imu} + \Lambda_{cam}\right) \mathcal{X} = \left(\mathbf{b}_p + \mathbf{b}_{imu} + \mathbf{b}_{cam}\right) \tag{7}$$

where $\{\Lambda_{imu}, \mathbf{b}_{imu}\}$ and $\{\Lambda_{cam}, \mathbf{b}_{cam}\}$ are information matrices and vectors for IMU and camera measurements respectively.

It should be noted that since the cost is linear with respect to the states, the system in (7) can have unique solution *without* the prior (initial condition):

$$(\Lambda_{imu} + \Lambda_{cam}) \mathcal{X} = (\mathbf{b}_{imu} + \mathbf{b}_{cam}) \tag{8}$$

This is the key to enable dynamic launching and failure recovery of MAVs. However, as will be shown in Sect. 3, there are degenerate motions for the monocular VINS setup, which will render the scale unobservable using only measurements within the sliding window. In such case, it is desirable to marginalize out states that are about to be removed from the window and convert them a prior to (implicitly) propagate the scale.

2.5 IMU Measurement Model

Given the locally drift-free rotation, we can rewrite (2) as a linear function of the state $\mathcal{X}$:

$$\begin{bmatrix} \alpha_{B_{k+1}}^{B_k} \\ \beta_{B_{k+1}}^{B_k} \\ \mathbf{0} \end{bmatrix} = \begin{bmatrix} \mathbf{R}_{B_0}^{B_k}\left(\mathbf{p}_{B_{k+1}}^{B_0} - \mathbf{p}_{B_k}^{B_0}\right) - \mathbf{v}_{B_k}^{B_k}\Delta t + \mathbf{g}^{B_k}\frac{\Delta t^2}{2} \\ \mathbf{R}_{B_{k+1}}^{B_k}\mathbf{v}_{B_{k+1}}^{B_{k+1}} - \mathbf{v}_{B_k}^{B_k} + \mathbf{g}^{B_k}\Delta t \\ \mathbf{R}_{B_{k+1}}^{B_k}\mathbf{g}^{B_{k+1}} - \mathbf{g}^{B_k} \end{bmatrix} = \mathbf{H}_{B_{k+1}}^{B_k}\mathcal{X} \tag{9}$$

The last block line in (9) represents prediction of the gravity vector. We estimate the gravity vector for each pose in order to avoid the negative effects due to possible accumulated rotation error. All variables except the position component are independent of the accumulated rotation $\mathbf{R}_{B_0}^{B_k}$, making them insensitive to rotation error. The linear IMU measurement model has the form:

$$\mathbf{z}_{B_{k+1}}^{B_k} \sim \mathcal{N}\left(\mathbf{H}_{B_{k+1}}^{B_k}\mathcal{X}, \begin{bmatrix} \mathbf{P}_{B_{k+1}}^{B_k{}^{\alpha\beta}} & \mathbf{0} \\ \mathbf{0} & \mathbf{P}_{B_{k+1}}^{B_k{}^{\mathbf{g}}} \end{bmatrix}\right) \tag{10}$$

Note that the terms $\alpha_{B_{k+1}}^{B_k}$ and $\beta_{B_{k+1}}^{B_k}$ and correlated since they both come from IMU measurements within Δt. Their joint covariance matrix $\mathbf{P}_{B_{k+1}}^{B_k{}^{\alpha\beta}}$ can be calculated using the pre-integration technique proposed in [13].

2.6 Camera Measurement Model

Let the lth feature be first detected in the ith frame. The observation of this feature in the jth normalized image plane $[u_l^{B_j}, v_l^{B_j}]^{\mathrm{T}}$ can be expressed as:

$$\lambda_l^{B_j}\begin{bmatrix} u_l^{B_j} \\ v_l^{B_j} \\ 1 \end{bmatrix} = \mathbf{R}_{B_0}^{B_j}\left(\mathbf{p}_{B_i}^{B_0} - \mathbf{p}_{B_j}^{B_0} + \lambda_l\mathbf{R}_{B_i}^{B_0}\begin{bmatrix} u_l^{B_i} \\ v_l^{B_i} \\ 1 \end{bmatrix}\right) \tag{11}$$

where $\lambda_l^{B_j}$ is the depth of the feature in the jth frame. We use tracking, instead of a descriptor-based method, as the tool for data association (Sect. 5.2). As such, the first observation *defines* the direction of a feature, and $[u_l^{B_i}, v_l^{B_i}]$ is noise-free. Note that (11) is now linear with respect to the state, but nonlinear to the image measurement since the depth is initially unknown. The unknown depth transforms into a unknown weighting factor to the measurement covariance. Still, we can rewrite (11) as:

$$\mathbf{0} = \begin{bmatrix} -1 & 0 & u_l^{B_j} \\ 0 & -1 & v_l^{B_j} \end{bmatrix} \mathbf{R}_{B_0}^{B_j} \left(\mathbf{p}_{B_i}^{B_0} - \mathbf{p}_{B_j}^{B_0} + \lambda_l \mathbf{R}_{B_i}^{B_0} \begin{bmatrix} u_l^{B_i} \\ v_l^{B_i} \\ 1 \end{bmatrix} \right) = \mathbf{H}_l^{B_j} \mathcal{X} \tag{12}$$

and the camera measurement model has the form:

$$\mathbf{z}_l^{B_j} \sim \mathcal{N}\left(\mathbf{H}_l^{B_j} \mathcal{X},\ \lambda_l^{B_j 2} \bar{\mathbf{P}}_l^{B_j}\right) \tag{13}$$

where $\bar{\mathbf{P}}_l^{B_j}$ is the feature observation noise in the normalized image plane. Note that although $\lambda_l^{B_j 2}$ is initially unknown, we can initialize it as the average depth of the scene. In practice, we found the solution very insensitive to the initial value of $\lambda_l^{B_j 2}$ as long as it is set to be *larger* than the actual depth. The reason for this is a possible direction for future research. Once the system is solved we can update the value of $\lambda_l^{B_k 2}$. The new value is used for subsequent optimization as long as both $\mathbf{p}_{B_k}^{B_0}$ and λ_l are still within the sliding window.

3 Handling Scale Ambiguity

It is well known that in order to render the scale of a monocular VINS observable, the IMU has to excite nonzero accelerations in at least two axes [2, 4, 16]. This is particularly critical for a initial-free sliding window estimator discussed in Sect. 2. In fact, when the vehicle is undergoing degenerate motions, such as constant velocity or hovering, and without any prior information, it can be verified that the position and velocity components of the solution of (8) in this situation can be scaled arbitrary without violating any constraints. Unfortunately, zero acceleration motion is unavoidable for a hover-capable MAV and it must be handled properly.

If the vehicle first undergoes generic motions with sufficient excitation in acceleration ($B_0, \ldots, B_n$), and then enters constant velocity motion ($B_{n+1}, \ldots, B_{N+n}$), the scale can only be observable if the camera states correspond to generic motion are included in the sliding window. This is unrealistic if available computation only allows N camera states in the sliding window. However, if we can provide an initial estimate of $\mathbf{x}_{B_{n+1}}^{B_0}$, we will be able to propagate (not observe) the scale from B_n to B_{n+1}. Naturally, this can be done by proper marginalization of $\mathbf{x}_{B_n}^{B_0}$ as it is removed from the window at the $(N+n)$th step.

For hovering, as proved in [7], if the vehicle first undergoes generic motions with sufficient acceleration excitation during $(B_0, \ldots, B_{N-1})$, and then enters a hover (B_N), the scale observability can be preserved by using a last-in-first-out (LIFO) sliding window scheme. [7] performs state-only measurement update during hovering, and covariance is updated only once as the vehicle exits hovering. This is due to the fact that features are not kept in the state, but instead marginalized out as covariance update is performed. However, this approach will lead to pessimistic covariance as observations obtained during hovering is not used to update the covariance.

3.1 Two-Way Marginalization

Based on the previous discussions, we propose a novel two-way marginalization scheme to handle both constant velocity and hovering cases. The pseudo code is shown in Algorithm 1. Consider the full state vector $\mathcal{X} = [\mathbf{x}_{B_0}^{B_0}, \ldots \mathbf{x}_{B_{N-1}}^{B_0} \mid \lambda_{\mathcal{L}}]$, where $\lambda_{\mathcal{L}}$ is the set of all features that have at least two observations within the sliding window. We add the next camera state $(\mathbf{x}_{B_N}^{B_0})$ to the sliding window if any of the following two criteria are satisfied:

1. The time between two images Δt is larger than δ.
2. After eliminating the relative rotation, the average parallax of all common features between the most recent two images is larger than ε.

The first condition ensures that the error in the integrated IMU measurement (Sect. 2.5) between two camera states is bounded, while second condition ensures that the new camera state is added when translation motion of the vehicle with respect to the scene is significant. We require that all newly added features $\lambda_{\mathcal{L}^+}$ to have at least two observations to ensure successful triangulation (Line 1). The system is then solved with all available measurements within the sliding window plus any available prior (Line 2).

We keep a variable s = *float*/*fix* to indicate whether we should marginalize out the second newest camera state $(\mathbf{x}_{B_{N-1}}^{B_0})$ or the oldest one $(\mathbf{x}_{B_0}^{B_0})$. To marginalize a chosen camera state $\mathbf{x}_{B_k}^{B_0}$, we first remove the camera state and all features $\lambda_{\mathcal{I}^-}$ that are first observed by it (Lines 5 and 13). We then construct a new prior based on all measurements related to the removed states (Lines 4 and 12):

$$\Lambda_p^+ = \Lambda_p + \sum_{k \in \mathcal{D}^-} \mathbf{H}_{B_{k+1}}^{B_k\,T} \mathbf{P}_{B_{k+1}}^{B_k\,-1} \mathbf{H}_{B_{k+1}}^{B_k} + \sum_{(l,j) \in \mathcal{C}^-} \mathbf{H}_l^{B_j\,T} \mathbf{P}_l^{B_j\,-1} \mathbf{H}_l^{B_j} \tag{14}$$

where $\mathcal{D}^-$ and $\mathcal{C}^-$ are sets of removed IMU and camera measurements respectively. The marginalization can the be carried out via Schur Complement [8].

The value of s is reevaluated after the marginalization based solely on the parallax between two most recent remaining images (Lines 6–9 and 14–17). Intuitively, our approach will keep removing the recent camera states if the vehicle has small or no motion. Keeping older camera states in this case will preserve acceleration information that is necessary to recover the scale, while still preserve all information provided by the marginalized states. On the other hand, if the vehicle is under fast constant speed motion, older camera states will be removed and converted into priors for the subsequent estimates. We do note that the scale in this case is subject to drifting. However, without global loop closure, marginalization is the best that can be done to propagate the scale information forward, while still maintaining constant computation complexity. Figure 2 demonstrates different working scenarios of the proposed marginalization approach.

We also note that due to the marginalization, the information matrix in (7) will eventually become dense. However, since our formulation is linear, no iterative method is required. In practice, we find that even a dense matrix solver is able to give real-time performance with tens of camera states and hundreds of features.

Algorithm 1 Two-Way Marginalization

Require:

$$\mathcal{X} \leftarrow \left[\mathbf{x}_{B_0}^{B_0}, \ldots \mathbf{x}_{B_{N-1}}^{B_0} \mid \lambda_{\mathcal{L}}\right]$$
$$s \leftarrow \textit{float} \textbf{ or } \textit{fix}$$
$$\{\Lambda_p, \mathbf{b}_p\} \leftarrow \text{Prior Information}$$

Ensure: $\Delta t > \delta$ **or** Parallex($\mathbf{x}_{B_{N-1}}^{B_0}$, $\mathbf{x}_{B_N}^{B_0}$) $> \varepsilon$

1: $\mathcal{X} \leftarrow \mathcal{X} \cup \left[\mathbf{x}_{B_N}^{B_0} \mid \lambda_{\mathcal{L}+}\right]$
2: Solve $\mathcal{X}$ using (6) and (7), optionally with $\{\Lambda_p, \mathbf{b}_p\}$
3: **if** $s = \textit{float}$ **then**
4: $\{\Lambda_p, \mathbf{b}_p, \lambda_{\mathcal{L}-}\} \leftarrow$ Marginalization($\mathbf{x}_{B_{N-1}}^{B_0}$)
5: $\mathcal{X} \leftarrow \mathcal{X} \setminus [\mathbf{x}_{B_{N-1}}^{B_0} \mid \lambda_{\mathcal{L}-}]$
6: **if** Parallex($\mathbf{x}_{B_{N-2}}^{B_0}$, $\mathbf{x}_{B_N}^{B_0}$) $< \varepsilon$ **then**
7: $s \leftarrow \textit{float}$
8: **else**
9: $s \leftarrow \textit{fix}$
10: **end if**
11: **else**
12: $\{\Lambda_p, \mathbf{b}_p, \lambda_{\mathcal{L}-}\} \leftarrow$ Marginalization($\mathbf{x}_{B_0}^{B_0}$)
13: $\mathcal{X} \leftarrow \mathcal{X} \setminus [\mathbf{x}_{B_0}^{B_0} \mid \lambda_{\mathcal{L}-}]$
14: **if** Parallex($\mathbf{x}_{B_{N-1}}^{B_0}$, $\mathbf{x}_{B_N}^{B_0}$) $< \varepsilon$ **then**
15: $s \leftarrow \textit{float}$
16: **else**
17: $s \leftarrow \textit{fix}$
18: **end if**
19: **end if**
20: **return** $\{\mathcal{X}, \Lambda_p, \mathbf{b}_p, s\}$

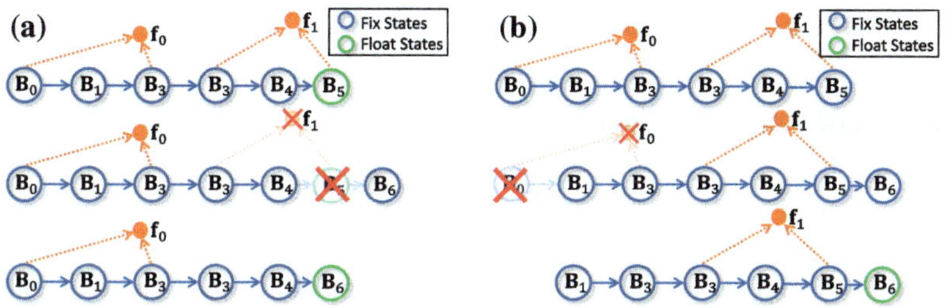

Fig. 2 **a** Structure of the full state before, during, and after marginalizing a recent camera state ($\mathbf{B}_5$) after a newer camera state $\mathbf{B}_6$ is added. Similar marginalization process of the oldest camera state is shown in (**b**)

4 Initialization and Failure Recovery

Our linear VINS formulation (Sect. 2) naturally allows on-the-fly initialization and failure recovery. In fact, these two tasks are identical, as both involve solving the linear system (7) with no priors after sufficient images are collected.

After initialization, in order to allow concurrently running the two-way marginalization algorithm for handling scale ambiguity (Sect. 3.1), we maintain two subsystems, which corresponds to two arrays of image/IMU measurements and camera/feature states. For the first subsystem, which is used for two-way marginalization, the newest two camera states may be separated arbitrarily far in time if the vehicle is hovering. On the other hand, the second subsystem refreshes independent of vehicle motion. It always maintains a queue structure where the oldest camera state and its corresponding measurements are removed as new image comes in. This way, we make sure that the IMU measurement in the queue always have bounded error.

When the system is in normal operation, only the first subsystem is solved and the second subsystem only collects data. A failure is indicated by insufficient features in the environment. When failure occurs, the first array, as well as all prior information, are discarded. Instead, we repeatedly try to find a valid solution using measurements within the second subsystem. Once a valid solution is found, we put all measurements back to the first array and resume normal operation. The block diagram of this failure recovery mechanism is shown in Fig. 3.

Fig. 3 Block diagram of the proposed failure recovery approach. Note the switching mechanism for two-way marginalization and the failure detector.

5 Implementation Details

5.1 Experimental Platform

The experimental platform shown in Fig. 1 is based on the Pelican quadrotor from Ascending Technologies, GmbH.[1] This platform is natively equipped with an AutoPilot board consisting of an IMU and a user-programmable ARM7 microcontroller. The main computation unit onboard is an Intel NUC with a 1.8 GHz Core i3 processor with 8 GB of RAM and a 120 GB SSD. The only addition to onboard sensing is a mvBlueFOX-MLC200w grayscale HDR camera with standard lens that capture 752×480 images at 25 Hz. The total mass of the platform is 1.39 kg, which leads to a thrust to weight ratio of approximately two. The entire algorithm is developed in C++ using ROS[2] as the interfacing robotics middleware.

5.2 Real-Time Implementation

Although the onboard camera captures images at 25 Hz, it is both computationally infeasible and unnecessary to perform the optimization (Sect. 2) at such a high rate. The system starts with one camera state in the sliding window, and a fixed number of corner features detected in that camera image. We utilize the KLT tracker to track features in the high-rate image sequence until the next camera state is added to the sliding window (Sect. 3.1). At this point, we apply RANSAC with epipolar constraints for outlier rejection, and then add extra features if computation permits. The pre-integrated IMU measurement (Sect. 2.5) is also computed as new camera state is added. All tracked features are used for rotation estimation (Sect. 2.3), however, only a subset of features with strong corner responses are used for position estimation (Sect. 2.4). We marginalize out additional features if the total number of features goes beyond a threshold.

We utilize multi-thread implementation to achieve real-time operation. Three threads run concurrently. The first thread is the image processing front end that we just described. The second thread is the main VINS optimizer (Sect. 2) and the marginalization module (Sect. 3.1). Finally, due to computation constraint, our VINS system runs at 10 Hz with approximate processing latency of 30 ms. This is not sufficient for autonomous control of MAVs. We therefore implement a third thread to propagate the latest VINS solution forward using the high-rate IMU measurements. The output of this thread is used directly as the feedback for the trajectory tracking controller. A breakdown of computation time of each components in our system is shown in Table 1. It suggests that our algorithm is able to run stably onboard.

---

[1] Ascending Technologies, GmbH, http://www.asctec.de/.

[2] Robot Operating System, http://www.ros.org/.

Table 1 Computation breakdown of major modules in our system for 30 camera states and 200 features

Module	Time (ms)	Rate (Hz)	Thread
Feature tracking	5	25	1
Add new camera state and features	14	10	1
Rotation estimation	2	10	2
Linear sliding window estimator	20	10	2
Marginalization	17	10	2
IMU forward propagation	1	100	3

6 Experimental Results

The experiments are conducted in a cluttered lab space with Vicon[3] motion capture system for ground truth comparison purpose. We highlight that in all experiments, the MAV has no prior knowledge of the environment, not even its own initial speed and attitude. With our linear formulation, the VINS estimator is initialized as the MAV takes off. This on-the-fly initialization enables rapid deployment of monocular visual-inertial systems.

6.1 Trajectory Tracking with Onboard State Estimation

We first test the performance of using the output from our VINS estimator for feedback control of an autonomous MAV. In this experiment, the MAV is set to autonomously fly through a figure eight pattern at different speeds. The time parameterized trajectory is generated using the minimum jerk cost function [20], which proved to be beneficial for vision-based approaches by reducing angular velocities.

In Fig. 4a, c and e, the MAV flies at average speed of 1 m/s with maximum acceleration of 1 m/s^2. It is well known that the position and the yaw angle of the platform in the world frame is unobservable without global reference sensors such as GPS. Indeed, we can see a drift in x, y, z at $\{-0.0584,\ 0.1191,\ 0.1229\}$ meters, and yaw at 1.563 degrees. On the other hand, the body frame velocity of the MAV, as well as the attitude which can be derived from the estimated gravity vector, remains observable throughout the flight. The standard deviation in three dimensional velocity compares close to the ground truth as $\{0.0734,\ 0.0410,\ 0.0229\}$, while the deviation in pitch and roll angles are $\{0.2487,\ 0.2477\}$, units in m/s and degrees, respectively. In Fig. 4b, d and f, the MAV is commanded a much higher speed of 2 m/s with maximum

[3]Vicon, http://www.vicon.com/.

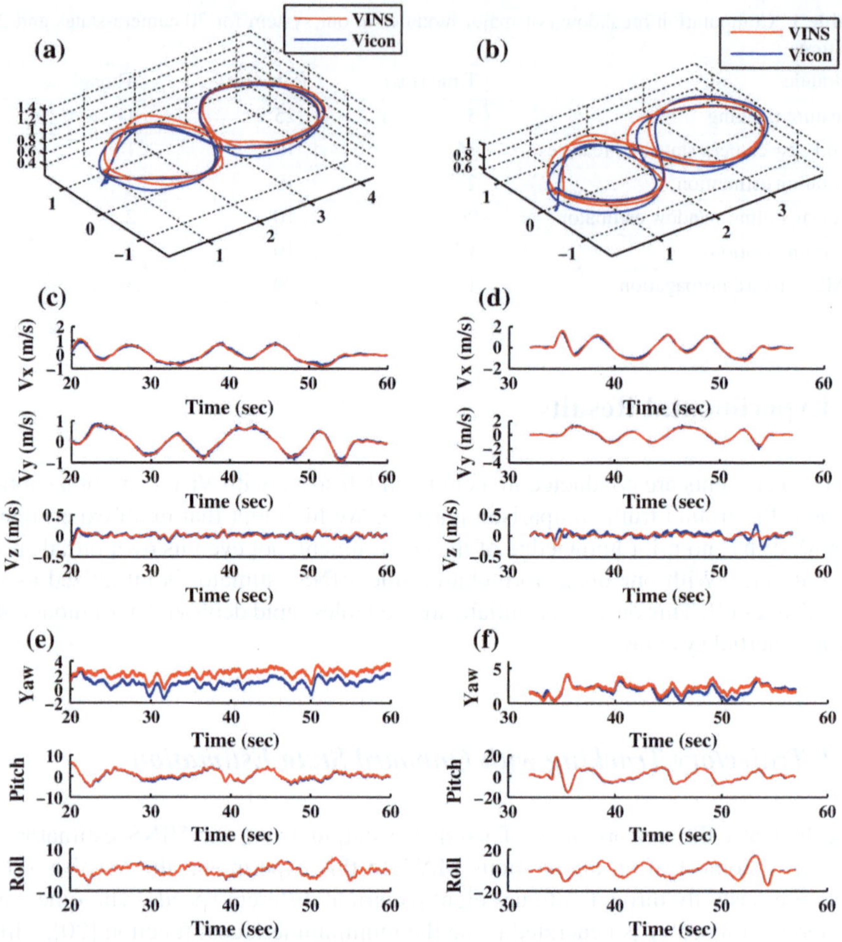

Fig. 4 The vehicle is set to autonomously fly through a figure eight pattern at the speed/acceleration of {1 m/s, 1 m/s^2} and {2 m/s, 4 m/s^2} respectively. Videos of the experiments are available at http://mrsl.grasp.upenn.edu/shaojie/ISER2014.mp4. **a** Slow velocity. **b** Fast velocity. **c** Slow orientation. **d** Fast orientation

acceleration of 4 m/s^2. The drift in position is now {0.0319, 0.0357, 0.2329} meters, while the yaw drift remains small as 0.409 degrees. The error statistics for velocity and attitude are {0.0921, 0.0678, 0.0731} and {0.4744, 0.4563} in meters and degrees respectively.

It can be seen that although the speed, acceleration, as well as attitude of the MAV increases significantly for the faster trajectory, the estimation quality largely remains the same. This highlights the robustness of our system for handling fast maneuvers.

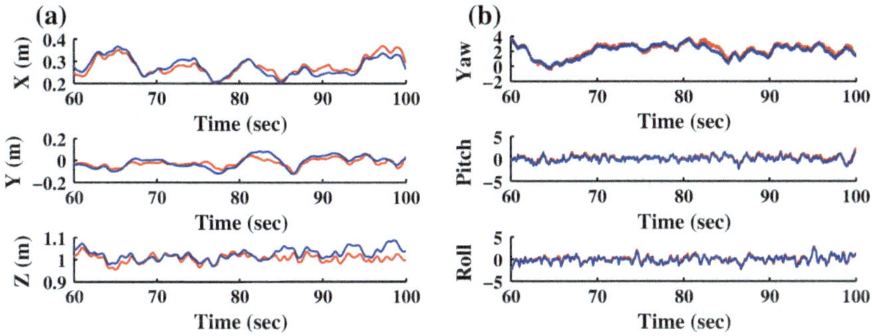

Fig. 5 Hover performance of the MAV using onboard state estimates for feedback control comparing with ground truth. Note that there is no drift in all directions. **a** Position, **b** Orientation

6.2 Hover Performance

One major issue of using a monocular VINS sensor suite for autonomous MAVs is the lack of direct measurement of metric scale. To this end, we use this experiment to highlight the effectiveness of our two-way marginalization (Sect. 3.1) scheme. The MAV takes off from the floor without any initial knowledge of its states. It is commanded to hover after the on-the-fly initialization is completed. As shown in Fig. 5, since the two-way marginalization always removes newer camera states while hovering, previous feature observations that have sufficient parallax, as well as IMU measurements that have sufficient accelerations are kept. As such, drift-free estimates in full 6 degree-of-freedom is obtained. During hovering, the onboard position estimate compares well with the ground truth with standard deviation of {0.018, 0.028, 0.019} meters. The hover performance with such onboard estimates has the standard deviation of {0.041, 0.053, 0.025}.

7 Conclusion and Future Work

In this paper, we presented what we believe to be the first autonomous quadrotor capable of fast navigation (2 m/s traversal of 0.9 m arcs with 0.5 g acceleration) that relies only on a single off-the-shelf camera and an IMU and does not require careful initialization. The main technical challenges we overcome stem from the fact that the state is not observable in general, and specifically, the scale is ambiguous for constant velocity or hover motions. The main practical challenge has to do with robustness to sudden changes in the number and quality of observed features and the dependence on robot/camera motions. Both of these challenges are addressed using a linear sliding window formulation for estimation and a two-way marginalization scheme that enable the estimation of the necessary navigation states without any prior initial information while being robust to degeneracies in the motion. We show results on our 1.4 kg quadrotor and show that the errors in an unstructured indoor environment are less than 5 cm using ground truth measurements from a motion capture system.

In the future, we would like to investigate into control and planning methodologies for generating motions that excite sufficient, but just enough accelerations for scale observability. We would also like to develop perception and mapping modules to enable autonomous obstacle avoidance in cluttered environments.

Acknowledgments We gratefully acknowledge support from ARL Micro Autonomous Systems and Technology Collaborative Technology Alliance Grant No. W911NF-08-2-0004.

References

1. Weiss, S., Achtelik, M.W., Lynen, S., Chli, M., Siegwart, R.: Real-time onboard visual-inertial state estimation and self-calibration of mavs in unknown environments. In: Proceedings of the IEEE International Conference on Robotics and Automation, pp. 957–964. Saint Paul, MN, May 2012
2. Jones, E.S., Soatto, S.: Visual-inertial navigation, mapping and localization: a scalable real-time causal approach. Intl. J. Robot. Res. **30**(4), 407–430 (2011)
3. Mourikis, A.I., Roumeliotis, S.I.: A multi-state constraint Kalman filter for vision-aided inertial navigation. In: Proceedings of the IEEE International Conference on Robotics and Automation, pp. 3565–3572. Roma, Italy, Apr 2007
4. Kelly, J., Sukhatme, G.S.: Visual-inertial sensor fusion: localization, mapping and sensor-to-sensor self-calibration. Intl. J. Robot. Res. **30**(1), 56–79 (2011). January
5. Kottas, D.G., Hesch, J.A., Bowman, S.L., Roumeliotis, S.I.: On the consistency of vision-aided inertial navigation. In: Proceedings of the International Symposium on Experimental Robotics, Quebec, Canada, June 2012
6. Li, M., Mourikis, A.: High-precision, consistent ekf-based visual-inertial odometry. Int. J. Robot. Res **32**(6), 690–711 (2013)
7. Kottas, D., Wu, K., Roumeliotis, S.: Detecting and dealing with hovering maneuvers in vision-aided inertial navigation systems. In: Proceedings of the IEEE/RSJ International Conference on Intelligent Robots and Systems, Tokyo, Japan, Nov 2013 (To appear)
8. Sibley, G., Matthies, L., Sukhatme, G.: Sliding window filter with application to planetary landing. J. Field Robot. **27**(5), 587–608 (2010)
9. Leutenegger, S., Furgale, P., Rabaud, V., Chli, M., Konolige, K., Siegwart, R.: Keyframe-based visual-inertial SLAM using nonlinear optimization. In: Proceedings of Robotics: Science and Systems, p. 0. Berlin, Germany June 2013
10. Indelman, V., Melim, A., Dellaert, F.: Incremental light bundle adjustment for robotics navigation. In: Proceedings of the IEEE/RSJ International Conference on Intelligent Robots and Systems, pp. 1952–1959. Tokyo, Japan, Nov 2013
11. Lange, S., Sunderhauf, N., Protzel, P.: Incremental smoothing vs. filtering for sensor fusion on an indoor UAV. In: Proceedings of the IEEE International Conference on Robotics and Automation, pp. 1773–1778. Karlsruhe, Germany, May 2013
12. Klein, G., Murray, D.: Parallel tracking and mapping for small AR workspaces. In: Proceedings of Sixth IEEE and ACM International Symposium on Mixed and Augmented Reality (ISMAR'07), Nara, Japan, Nov 2007
13. Lupton, T., Sukkarieh, S.: Visual-inertial-aided navigation for high-dynamic motion in built environments without initial conditions. IEEE Trans. Robot. **28**(1), 61–76 (2012)

14. Kneip, L., Weiss, S., Siegwart, R.: Deterministic initialization of metric state estimation filters for loosely-coupled monocular vision-inertial systems. In: Proceedings of the IEEE/RSJ International Conference on Intelligent Robots and Systems, pp. 2235–2241. San Francisco, CA, Sept 2011
15. Dong-Si, T., Mourikis, A.I.: Estimator initialization in vision-aided inertial navigation with unknown camera-IMU calibration. In: Proceedings of the IEEE/RSJ International Conference on Intelligent Robots and Systems, pp. 1064–1071. Vilamoura, Algarve, Portugal, Oct 2012
16. Martinelli, A.: Vision and imu data fusion: Closed-form solutions for attitude, speed, absolute scale, and bias determination. IEEE Trans. Robot. **28**(1), 44–60 (2012)
17. Martinelli, A.: Closed-form solution of visual-inertial structure from motion. IEEE Trans. Robot. (2013)
18. Lippiello, V., Mebarki, R.: Closed-form solution for absolute scale velocity estimation using visual and inertial data with a sliding least-squares estimation. In: Proceedings of Mediterranean Conference on Control and Automation, pp. 1261–1266. Platanias-Chania, Crete, Greece, June 2013
19. Martinec, D., Pajdla, T.: Robust rotation and translation estimation in multiview reconstruction. In: Proceedings of the IEEE International Conference on Pattern Recognition, pp. 1–8. Minneapolis, MN (2007)
20. Shen, S., Mulgaonkar, Y., Michael, N., Kumar, V.: Vision-based state estimation and trajectory control towards high-speed flight with a quadrotor. In: Proceedings of Robotics: Science and Systems. Berlin, Germany (2013)

Active Online Calibration of Multiple Sensors for Autonomous Surface Vessels

Hordur K. Heidarsson and Gaurav S. Sukhatme

Abstract We present an approach to actively calibrate the exteroceptive sensors of an Autonomous Surface Vessel (ASV) autonomously. The approach consists of locating suitable calibration sites in the environment from aerial imagery, navigating to them, gathering calibration data and estimating the required parameters from data. We have conducted experiments using an ASV in a lake to validate our approach.

Keywords Autonomous surface vessel · Calibration

1 Introduction

Autonomous Surface Vessels (ASVs) have the potential to become extremely useful tools for variety of tasks, such as environmental surveying, monitoring and mapping. However, at present, they are not ready to be operated at all times in shared environments such as marinas or lakes with boat traffic without human supervision. For practical long-term and cost-effective operation, an ASV needs to be able to operate without supervision whenever the need arises and should be adaptable to different situations through reconfiguration of sensors and payload. An ASV that is to be operated at any time, i.e. day or night, and in a variety of weather conditions (e.g. sun, rain, fog) needs to have a robust sensing system that is able to operate in these different conditions. Furthermore, the sensing system needs to be multimodal, since relying on a single type of sensor for detecting obstacles in this environment can result in serious errors. While most obstacles have expressions both below and above the water surface, there are some that have only one or the other. It is therefore

H.K. Heidarsson (✉) · G.S. Sukhatme
Robotic Embedded Systems Laboratory, University of Southern California,
Los Angeles, CA 90089, USA
e-mail: heidarss@usc.edu
URL: http://robotics.usc.edu/resl

G.S. Sukhatme
e-mail: gaurav@usc.edu

M.A. Hsieh et al. (eds.), *Experimental Robotics*, Springer Tracts
in Advanced Robotics 109, DOI 10.1007/978-3-319-23778-7_16

important that we are able to sense in both those domains. In order to do so we employ a laser and a camera above the surface and a sonar below.

For smaller ASVs, it can be useful to have a modular configuration, where sensors and equipment can be reconfigured easily, even in the field, to better suit the task at hand and as well as for easier transportation. By moving sensors around we have introduced another problem: now we might not know the transforms between the various sensors which are needed to be able to fuse data from them. This can be measured by hand, but it can be time consuming to do so accurately and especially so in the field. Furthermore, utilizing traditional data driven calibration methods using ready-made targets is non-trivial due to the different nature of some of the sensors, e.g. a laser and a camera need an above water target and a sonar needs an underwater target of a different type, which could be hard to deploy. This leads us to develop an active data driven method to obtain the calibration between the sensors from certain features we find in our environment.

2 ASV Description and Problem Statement

The ASV we consider in this paper is equipped with the following sensors: a forward-down facing monocular camera, a planar laser range finder that is mounted facing forward (with its measuring plane level with the water surface plane) and a mechanically scanning pencil-beam underwater sonar mounted facing forward (just below the surface of the water). Additionally, the ASV is equipped with a GPS receiver and an IMU for pose estimation. A diagram of the ASV can be seen in Fig. 1.

The monocular camera is utilized as a range sensor to points on the water surface plane. This is done by segmenting the water from non-water in the camera output [1, 17] and using the transform between the camera and water surface plane to project the camera image on to the plane. This means that all three sensors are sensing in different parallel planes: the laser slightly above the surface of the water, the camera directly on the surface of the water and the sonar just below the surface of the water. This enables us to detect a wide variety of obstacles, whether they have an above- or below- the surface expression or both.

The sonar we use is a single pencil beam mechanically scanning sonar. Roughly speaking it functions similar to a laser range finder, but since there's only a single beam it can only take one distance measurement at a time. The width of the sonar beam is also not as narrow as a laser (our unit has a beam width of 1.8°). The data received from the sonar is the echo strength binned evenly over the set range of the unit. This means that the sonar can detect more than one separate obstacle at a time if the obstacle doesn't cover the full width of the sonar beam. This can also increase noise due to reflections and other sources. The speed of sound in water varies significantly with temperature, requiring sonar data to be corrected (scaling) based on the water temperature. Our sonar does not do temperature compensation internally, requiring us either to scale the data based on a temperature reading, or, as we have done, estimate this scaling factor during our calibration.

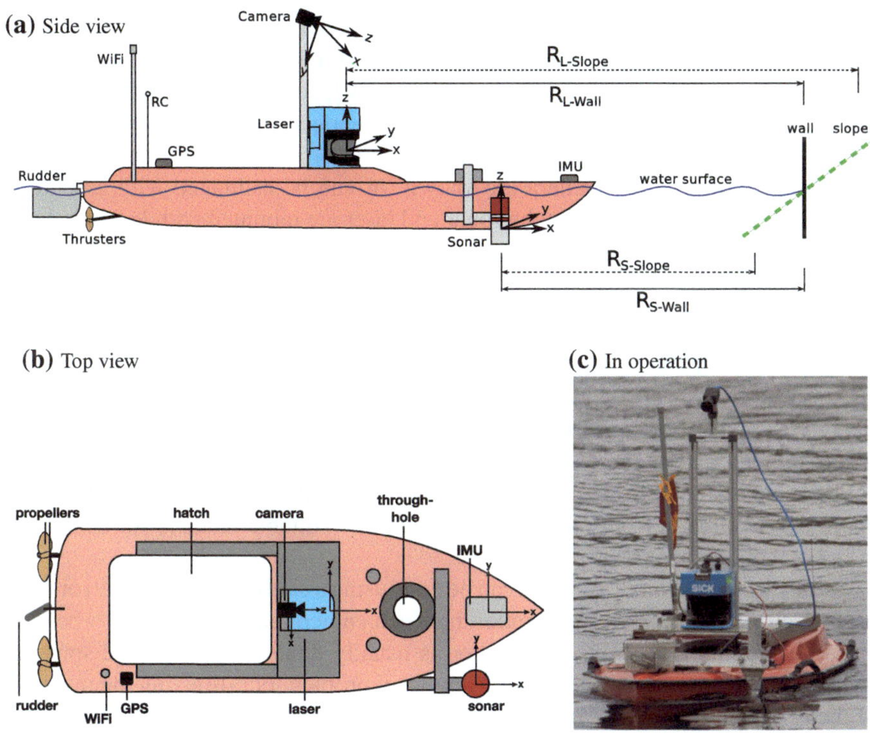

Fig. 1 The USC ASV. **a** Shows where the various sensors are mounted and a cross section of the sensing setup. The *black vertical line* and the angled *dashed line* to the *left* represent a vertical obstacle and a slanted one respectively. $R_{L-\{Wall,Slope\}}$ represent the detected ranges by the laser to the wall and sloped obstacle. $R_{S-\{Wall,Slope\}}$ similarly shows the detected ranges by the sonar. If the transform between laser and sonar is known, the detected ranges (converted to a common reference frame) are equal for the vertical obstacle, but differ for the sloped one, **b** shows a *top view*, **c** shows a photo of the ASV in operation

We assign a reference frame to each of the sensors. The camera frame $\{C\}$, with origin at the optical center of the camera and with z-axis aligned with the optical axis of the lens, the laser frame $\{L\}$, with origin at the center of the laser range finder and x-axis straight forward in the sensing plane, the sonar frame $\{S\}$, with origin at the center of the sonar with the x-axis straight forward in the sensing plane and the IMU and GPS frame $\{I\}$, with its origin at the center of the IMU body. Additionally, we define a fixed water surface frame $\{W\}$, that has the XY-plane on the surface of the water and origin fixed with respect to the ASV (for simplicity we define it to share its z-axis with $\{L\}$ so in 2D they can be used interchangeably).

Our problem is then as follows. With an ASV in the described sensor configuration and deployed in an unmapped environment, we want to estimate the planar rigid-body transform with added scaling between the laser and sonar, $\theta_{LS} = (a_S, R_{LS}, t_{LS})$ such that $X_L = a_S R_{LS} X_S + t_{LS}$ where X_L are points in the laser reference frame, $\{L\}$, corresponding to the points, X_S, in the sonar reference frame, $\{S\}$. We also want to estimate the homography transform, H_{WC}, between the (intrinsically calibrated) camera image plane and the water surface plane, such that $X_W = H_{CW}^{-1} p_C$ where X_W are points in the water surface reference frame, $\{W\}$, corresponding to the points p_C in the intrinsically calibrated camera image in homogeneous coordinates.

We start without any previous data from the sensors and assume that the intrinsic model parameters of the camera are known (or can be estimated with a different approach). We also assume that the sonar and laser are mounted with level pitch and roll on the ASV since otherwise their utility would be limited.

3 Related Work

There is a good body of work in cross-calibration of some types of sensors, especially involving cameras and lasers. One approach is to manually use a checkerboard camera calibration target and detect it simultaneously with both camera and laser [9]. Other approaches calibrate from outdoor scenes and find suitable features for calibration [12, 18], but use a 3D laser.

Intrinsic calibration of a camera can be done using a calibration target [19] or using online self-calibration [5]. Calibration between camera and IMU can also be done online from data [10]. Sonar calibration is usually done using deployed spherical targets [4, 6], but focuses on the more low-level problem of estimating the intrinsics of the sensor rather than its mounting location as we desire.

Work on sensing for small ASVs includes [2, 11]. Sonar sensing has mainly been done for AUVs, but some of the techniques [16] can be applied on ASVs as well.

4 Technical Approach

In this section we will begin by discussing the requirements for performing the calibration between the different sensors which will serve to motivate how we have approached the problem. We will follow that by outlining our process, step by step and finally provide a detailed explanation of each of the steps.

4.1 Calibration of Sensors in Different Domains

As previously described, the sensors we are considering are not sensing the exact same picture, but each in a different parallel plane. This has a great effect on what kind of data we can use to calibrate from. Looking at Fig. 1a one can see that when the different sensors look at a target that is not vertical, the different sensors will give conflicting range measurements to said target. Furthermore, since each sensor is in a different plane (and the transformation between them is currently not known), we won't even know that there is an inconsistency.

However, when looking at a vertical target that extends below the surface of the water, the sensors are measuring a common range, so this can be used for calibration. In the one dimensional case, this would be enough, but once we are in 2D, we have a set of lines that need to be aligned, but they could just slide along each other. Additionally, like previously mentioned, we are not able to tell if the target is slanted or not based on our sensor data. For these reasons, to have any hope of doing data association between the different sensors, we require our calibration target to have a corner and be vertical.

Since we want to calibrate from an existing feature in our environment, we would greatly benefit from some prior information on where to find those. To that effect we utilize overhead imagery to find potential calibration targets. With the overhead imagery we run into the same problem as before, it is not trivial to know if a potential target is vertical or sloped. However, most man made structures in water that have corners and appear straight edged (the main exception being ramps) tend to be vertical in the water. We will therefore build our approach on the assumption that a potential target from an overhead image will have a corner formed by straight edges, with the angle between them being close to 90°.

When calibrating the different sensors, we do not require to do all at once. It is sufficient to calibrate any two to each other at any time, but the additional sensor can provide additional redundancy against errors in case we are not looking at a suitable target.

4.2 Our Approach

Our calibration process is comprised of the following steps.

1. Collect overhead imagery.
2. Locate potential calibration targets.
3. Navigate to calibration site.
4. Collect calibration data.
5. Calibration.

Each step will now be described in detail.

Collect Overhead Imagery We start by gathering overhead image data. For up to date imagery an UAV with a downward facing camera can be used with simple auto generated way-point control to explore and gather overhead imagery. As an alternative, if available, existing aerial imagery can be used, such as from online mapping sources (at the risk of the imagery being outdated). We apply our previous approach [8] to detect water from the overhead imagery. We will refer to the resulting map as the obstacle map. This will facilitate knowing where the ASV can operate and provide input to our next step of locating potential sites for calibration.

Locate Potential Calibration Sites From the overhead imagery we want to locate potential calibration sites. To this end, we detect lines in the aerial imagery using a standard approach consisting of detecting edges in the image and then applying the probabilistic Hough transform [13]. Using the additional information about water/non-water from the previous step, we reject lines that appear either in mid-water (most likely ripples or noise) or lines that are on shore away from the water, leaving us with ones that close to the edge of the water. Now we cluster lines together based on location, where each cluster is a potential calibration site. Next we check all pairs of lines (first extending them by a small amount) for intersections to find corners and reject any that are outside some envelope of acceptable angles ($90° \pm n°$). Using this information allows us to assign a score based on number of corners, whether they are convex or concave, and length of the lines.

With the potential calibration sites found, we prioritize them by weighing the shortest path distance (estimated by planning a path on the obstacle map) from the ASV start location against the feature score. The site with the best combined score will be visited first. If there are many potential sites in the area, there might a benefit in finding the shortest expected path that leads to a suitable calibration target being found.

Navigate to Calibration Site Once we have picked a calibration site we need to navigate to it. In order to do so safely, we perform reactive obstacle avoidance along our planned path, based on a simple approach [7] using the sonar and laser individually. In case the site has multiple corners, we start with the closest one. We navigate to a point at a given distance out from the center line of the corner (see Fig. 2).

Collect Calibration Data Once in position, we keep station while performing a slow rotation to sweep the corner with our sensors to try to get good data points for the calibration. The need to keep relatively stationary comes from the fact that the mechanically scanning sonar has a single beam that sweeps back and forth in a arc. The scanning speed is limited by the speed of sound, so compared to the laser or camera, the sonar has a relatively slow update rate. Capturing a single frame (i.e. a full sweep from one side to the other), can take from 2 to 10 s, depending on the desired sector width and angular resolution. When a full frame has been captured, every measurement of it needs to be transformed to a common reference frame using pose estimates from the on-board pose estimate.

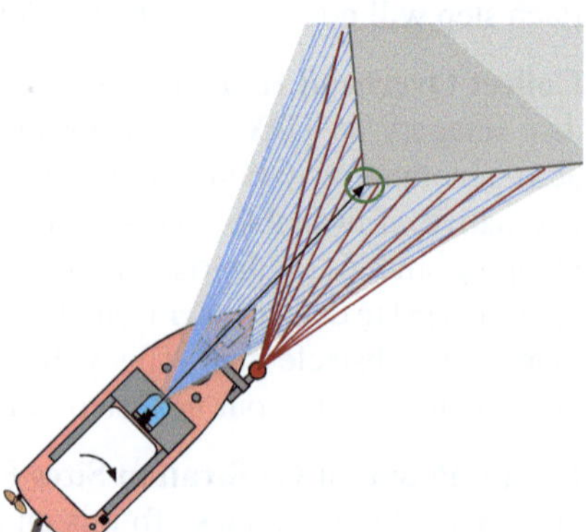

Fig. 2 ASV gathering calibration data

During the data collection phase, we need to verify that we are looking at a suitable calibration target before attempting to perform the calibration. Using the laser we can easily detect line and corner features with a fast update rate and low processing cost using a variety of algorithms [15]. After filtering out stray returns we use the probabilistic Hough transform on the laser data to extract lines. Next we merge close-by lines with similar angles and then elongate the remaining lines and check for intersections to locate corners.

Once a suitable laser frame is found, we check the corresponding sonar frame using a very similar process: filter away stray returns and extract lines to find corners, but since the sonar measurements are much noisier and less accurate than a laser, we use a variation of the Hough transform designed for underwater sonar [16] for improved accuracy. If corners are found in the laser and sonar images, we check if they are within a certain distance from each other (a rough upper bound on the translation between them) and add them to our calibration set. Example of the laser and sonar data can be seen in Fig. 5.

For the camera we use the fact that straight lines are preserved under projective transformation and attempt to find the corner(s) in the camera image. However, this is not as simple as for the overhead case or for the laser or sonar. Edge detection on the camera image generally results in many edges on the surface of the water in addition to those along our intended edges (refer to Fig. 3). These edges can be hard to filter out based on a single image. To remedy this, we use a technique based on [17] of employing optical flow [3] to do tracking over a few sequential frames. In our case, we place our trackers uniformly on the edges found in the image. Now we can compute the entropy of the tracked trajectories and also compute the dissimilarity between neighboring trajectories. Trackers with trajectories of high entropy or those highly dissimilar to their neighbors are very likely trying to track water and can be discarded along with their edges. We can now apply the Hough transform to the remaining edges in order to find lines which we then elongate and look for

Fig. 3 An example of a camera view during calibration. *Below* is a result from an edge extraction with multiple unwanted edges from water ripples

intersections. If we find a corner in the image which has no lines below it (i.e. larger y-coordinate), we save its coordinates along with the closest corner from the laser image as a pair for the camera calibration. Since we are tracking the corner we can get series of point correspondences through the sweeping maneuver. For the case of stereo cameras, this process is easier since we should be able to find the corner from the disparity information fairly easily and then calibrate in the same way from a set of point correspondences.

Calibration Once we have collected enough calibration data we can start estimating the calibration parameters. For the laser and sonar this is done by registering the two point sets to each other. For this we utilize the Coherent Point Drift (CPD) algorithm [14], a probabilistic registration algorithm capable of solving for scale as well as the rigid transform, with a slight modification where we force correspondence probabilities, $P\left(i|s_j\right)$, to zero for point pairs that should not be matched (i.e. points in separate frames). This allows us to jointly estimate the transform on several frames of data at once, resulting in a more robust estimation.

For the camera calibration we utilize the RANSAC version of findHomography in OpenCV, with which we supply our point correspondences between the image coordinates and water frame coordinates.

5 Experiments and Results

In this section we will show the results for our two main pieces of the calibration process, namely the locating of potential sites, and then the calibration itself.

5.1 Experimental Setup

For our experiments we have used the USC ASV (Fig. 1), which is approximately 1.8 m long and 0.8 m wide. It is actuated with two electric thrusters and a rudder, and is capable of speeds up to 1.6 m/s. The sensor package consists of a Microstrain 3DM-GX3 IMU, a U-Blox GPS unit, a Pt. Grey Grasshopper 3 USB3 camera, a SICK LMS-200 laser range finder and an Imagenex 881L sonar. On-board control and computation is done with a Core i7 computer running ROS.

Our field experiments have been conducted in Puddingstone Reservoir, which is an artificial lake in San Dimas, CA, USA.

5.2 Locating Calibration Sites

Using aerial imagery from Google Maps since we were unable to fly our UAV at the test site, we have performed the second step of our approach off-line where we

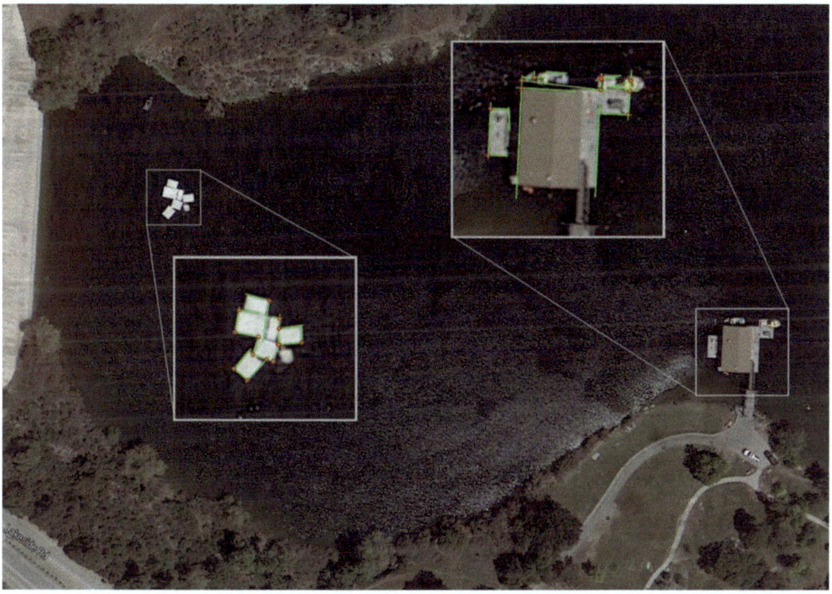

Fig. 4 Detected calibration sites in experimental area. *Green lines* indicate the detected lines and *red circles* indicate detected corners

locate potential calibration sites. We have identified two potential calibration sites in the area of operation and are able to detect both of them from the overhead imagery, see Fig. 4. The process resulted in several line edges along the shoreline of the lake, but they were rejected due the lack of suitable corners for the calibration. We also configured our process to use only convex corners as we did not want the ASV to navigate in to. This eliminated the dam on the left hand side of the image as a potential site.

Depending on where the ASV was launched the two calibration sites were ranked differently with the site in the middle of the lake being more valuable but the boat house site being closer to our launch spots.

5.3 Calibration

An example of the alignment between sonar and laser data can be seen in Fig. 5. This shows the alignment done on a single frame of data, which can work, but can suffer from scaling problem if there is only a single corner in the frame. To remedy that, we perform the alignment on multiple frames of data simultaneously, captured at different viewpoints so the scaling will stay fixed. Figure 6 shows a camera frame from the calibration, where the detected lines and corner are overlaid on the image along with the tracking points.

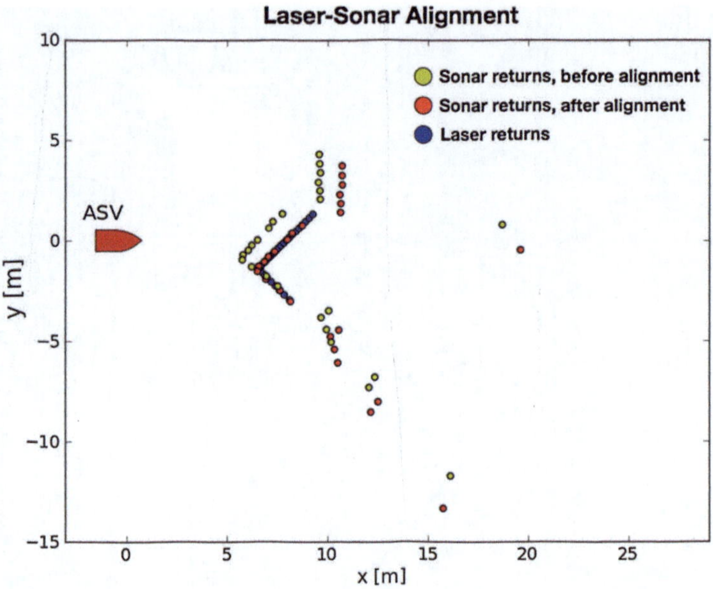

Fig. 5 Calibration between laser and sonar. The plot shows the alignment for a single frame from the calibration process. The *blue dots* are laser returns, the *yellow* are the uncalibrated sonar returns and the *red* are the calibrated sonar returns

Fig. 6 Camera view during calibration. The floating platform in front of the ASV is our calibration target that is being sensed by the sonar and laser in Fig. 5. The *blue lines* are *straight lines* detected in the image and the *green circle* indicates the corner that has been found. The *blue diamonds* are trackers that have been randomly placed on *lines* that have been found in previous frames

Fig. 7 Camera view with laser returns (*red dots*) and sonar returns (*dark blue dots* or *lines*) overlaid using transforms estimated by the calibration process. It is worth noting that since the floating platforms rise so low out of that water that the laser might not get a hit from the closest one. Also worth noting is that the overlaid sonar returns include the whole echo profile, not only the peak, which explains why they appear as *lines*

Finally, Fig. 7 shows the camera view with sonar and laser data overlaid using the estimated transforms.

6 Conclusions and Future Work

6.1 Conclusions

We have presented an approach to calibrate different sensors sensing in different planes on an ASV. Our method is based on finding physical corner features in the environment with the aid of overhead imagery, navigating to them and leveraging them to obtain a calibration between the different sensors. We have performed field experiments with our ASV in order to test and verify our approach.

6.2 Future Work

We plan to investigate the feasibility of getting a rough estimate of the slope of a target from the return data of the sonar. Since the sonar data contains information over the whole sensing range we have some additional information that we are currently mostly discarding. The spread and shape of the return could give some insight into the characteristics of the obstacle that generates the return. This would help improving the calibration process as well as general mapping and obstacle avoidance using the sonar and the other sensors which is our main focus. We are also working on applying self-supervised learning techniques for using one sensor to improve the sensor model of another. This allows us to use a sensor in which we have greater trust or that has a more established sensor model.

Acknowledgments Special thanks to RESL members Jnaneshwar Das, Stephanie Kemna, Jörg Müller and Carl Oberg for their help with conducting field experiments and the staff at Frank G. Bonelli Regional Park for being helpful and allowing us to conduct our experiments in the lake.

References

1. Achar, S., Sankaran, B., Nuske, S., Scherer, S., Singh, S.: Self-supervised segmentation of river scenes. In: 2011 IEEE International Conference on Robotics and Automation (ICRA), pp. 6227–6232, May 2011
2. Bandyophadyay, T., Sarcione, L., Hover, F.S.: A simple reactive obstacle avoidance algorithm and its application in singapore harbor. In: Field and Service, Robotics, pp. 455–465 (2009)
3. Bouguet, J.-Y.: Pyramidal implementation of the affine lucas kanade feature tracker description of the algorithm. Intel Corp. **2**, 3 (2001)
4. Chu, D., Baldwin, K.C., Foote, K.G., Li, Y., Mayer, L.A., Melvin, G.D.: Multibeam sonar calibration: target localization in azimuth. In: MTS/IEEE Conference and Exhibition OCEANS, vol. 4, pp. 2506–2510 (2001)
5. Civera, J., Bueno, D.R., Davison, A.J., Montiel J.M.M.: Camera self-calibration for sequential bayesian structure from motion. In: IEEE International Conference on Robotics and Automation, ICRA'09
6. Foote, K.G., Chu, D., Hammar, T.R., Baldwin, K.C., Mayer, L.A., Hufnagle L.C. Jr., Jech, J.M.: Protocols for calibrating multibeam sonar. J. Acoust. Soc. Am. **117**(4), 2013–2027 (2005)
7. Heidarsson, H.K., Sukhatme, G.S.: Obstacle detection and avoidance for an autonomous surface vehicle using a profiling sonar. In: 2011 IEEE International Conference on Robotics and Automation. Shanghai, May 2011
8. Heidarsson, H.K., Sukhatme, G.S.: Obstacle detection from overhead imagery using self-supervised learning for autonomous surface vehicles. In: IEEE/RSJ International Conference on Intelligent Robots and Systems. San Francisco, September 2011
9. Kassir, A., Peynot, T.: Reliable automatic camera-laser calibration. In: Australasian Conference on Robotics and Automation (2010)
10. Kelly, J., Sukhatme, G.S.: Visual-inertial sensor fusion: localization, mapping and sensor-to-sensor self-calibration. Int. J. Robot. Res. **30**(1), 56–79 (2011)
11. Leedekerken, J.C., Fallon, M.F., Leonard, J.J.: Mapping Complex Marine Environments with Autonomous Surface Craft. Delhi (2010)
12. Levinson, J., Thrun, S.: Automatic online calibration of cameras and lasers. In: Proceedings of Robotics Science and Systems (2013)
13. Matas, J., Galambos, C., Kittler, J.: Robust detection of lines using the progressive probabilistic Hough transform. Comput. Vis. Image Underst. **78**(1), 119–137 (2000)
14. Myronenko, A., Song, X.: Point set registration: coherent point drift. IEEE Trans. Pattern Anal. Mach. Intell. **32**(12), 2262–2275 (2010)
15. Nguyen, V., Martinelli, A., Tomatis, N., Siegwart, R.: A comparison of line extraction algorithms using 2D laser rangefinder for indoor mobile robotics. In: 2005 IEEE/RSJ International Conference on Intelligent Robots and Systems, (IROS 2005), pp. 1929–1934, August 2005

16. Ribas, D., Ridao, P., Tardos, J.D., Neira, J.: Underwater SLAM in man-made structured environments. J. Field Robot. **25**, 898–921 (2008)
17. Santana, P., Mendonca, R., Barata, J.: Water detection with segmentation guided dynamic texture recognition. In: 2012 IEEE International Conference on Robotics and Biomimetics (ROBIO), pp. 1836–1841 (2012)
18. Scaramuzza, D., Harati, A., Siegwart, R.: Extrinsic self calibration of a camera and a 3D laser range finder from natural scenes. In: IEEE/RSJ International Conference on Intelligent Robots and Systems, IROS, pp. 4164–4169 (2007)
19. Zhang, Z.: A flexible new technique for camera calibration. IEEE Trans. Pattern Anal. Mach. Intell. **22**(11), 1330–1334 (2000)

Part V
Human–Robot Interaction

Venkat Krovi
SUNY Buffalo

For robots to be truly accepted by humans, they need to be able to intermingle, interact, and blend within the surrounding crowd of other humans/robots. To achieve this goal, roboticists have long dreamed of robots that can observe humans, understand human intent from the sensed-data and exploit this intuited knowledge to enhance interactions. Understanding fellow man's intent can be quite challenging even for the most perceptive and capable of humans; realizing this with robots (with their significant sensing, computation and communication limitations in near-real-time) is nothing short of a robotic Herculean task. Yet the four papers in this session on human–robot Interactions attempt to realize this very capability in the robots in some very unique ways.

The first paper, *Experiments in Leader Classification and Following with an Autonomous Wheelchair*, by P. Stein, A. Spalanzani, V. Santos, and C. Laugier focuses on developing human-like motion capability for mobile-robots in highly-crowded environments. The premise is that by appropriately selecting human-leaders to follow, mobile-robots can exploit the human's advanced navigation and interaction skills to navigate cluttered environments. A machine-learning framework focuses on training of a leader-classifier capable of appropriate data-driven leader-selection, leader-following and leader-switching necessary for achieving the robust performance.

Authors R. Mead and M. Mataric present a very different slant on this problem—they propose to employ the social use of space and spatial-relations as engendered in the psychological subfield called proxemics. They build on the hypothesis that interpretation, manipulation, and dynamics of spatial behavior in face-to-face inter-human social encounters is significantly modulated by the distance between them. Hence, their paper entitled "Perceptual Models of Human-Robot Proxemics" takes an interesting two-pronged approach to studying socially situated inter-personal interaction between humans and/or robots. On the one hand, they seek to computationally capture and better understand the underlying human proxemic behaviors through carefully controlled and instrumented experimental studies; on the other hand they seek to use the developed insights for implementing robust proxemic controllers for sociable robots.

The third paper by authors D. Wollherr, S. Khan, L. Christian, and M. Buss is entitled *Interactive Urban Robot IURO: Towards Robot Action in Human Environments*. It examines a tantalizing scenario of a wheeled-mobile robot, navigating autonomously in an unstructured urban environment without maps, and localizing itself using sensory-data as well as by verbally obtaining directions from human passers-by. Key to realizing this capability is the merger of the probabilistic occupancy representation (SRTree for tightly interlinked hierarchical spatial and semantic representations) with reasoning using Markov Logic Networks. This framework is then systematically evaluated on a large-scale dataset (3D point cloud + RGB image data) from 2 km^2 region of downtown Munich—and shows enormous promise not only in terms of richness of relations but also for approaching real-time operations.

Rounding off the offerings in this session on Human Robot Interaction was the paper by M. Fiore, A. Clodic, and R. Alami from LAAS/CNRS entitled *On Planning and Task Achievement Modalities for Human-Robot Collaboration*. The paper presents a high-level global overview of the architecture and decisional components that allows seamless natural switching between the various interaction modalities such as joint attention, action observation, task-sharing, and action coordination. Such capabilities are critical for both for *creating* shared plans for human–robot interactions and for coordinated *execution*, taking into account the capacities of both agents. The example task of joint-cleaning of furniture (http://homepages.laas.fr/mfiore/iser2014.html) illustrates the ability to handle joint-goals, joint-actions, user-preferences, diverse agent-beliefs as well as to monitor human actions.

Experiments in Leader Classification and Following with an Autonomous Wheelchair

Procópio Stein, Anne Spalanzani, Vítor Santos and Christian Laugier

Abstract With decreasing costs in robotic platforms, mobile robots that provide assistance to humans are becoming a reality. A key requirement for these types of robots is the ability to efficiently and safely navigate in populated environments. This work proposes to address this issue by studying how robots can select and follow human leaders, to take advantage of their motion in complex situations. To accomplish this, a machine learning framework is proposed, comprising data acquisition with a real robot, data labeling, feature extraction and the training of a leader classifier. Preliminary experiments combined the classification system with a multi-mode navigation algorithm, to validate this approach using an autonomous wheelchair.

Keywords Leader selection · Leader following · Assistance robotics · Human-aware navigation

---

Procópio Stein is funded by INRIA's Large-scale initiative action *PAL (Personally Assisted Living)* `pal.inria.fr`.

P. Stein (✉) · A. Spalanzani · C. Laugier
INRIA Rhône-Alpes, 655 Avenue de L'Europe, 38334 Saint Ismier Cedex, France
e-mail: procopiostein@gmail.com

A. Spalanzani
e-mail: anne.spalanzani@inria.fr

C. Laugier
e-mail: christian.laugier@inria.fr

A. Spalanzani
Lab. LIG, Université Grenoble Alpes, Grenoble, France

V. Santos
Universidade de Aveiro, Campus Universitário de Santiago 3800 Aveiro, Aveiro, Portugal
e-mail: vitor@ua.pt

M.A. Hsieh et al. (eds.), *Experimental Robotics*, Springer Tracts in Advanced Robotics 109, DOI 10.1007/978-3-319-23778-7_17

1 Motivation, Problem Statement and Related Framework

The development of assistance and service robots requires special attention to the problem of navigation in populated environments, as how to interact with humans, respect social conventions and deal with unpredicted situations. The capacity of a robot to address these requirements directly impact its acceptance by humans.

Several successful techniques of motion planning address the problem of navigation in populated environments using probabilistic and predictive approaches [1, 2] and on models of social interactions [3]. The idea is that, by estimating areas that ought to be occupied in the future, these algorithms can create collision-free motion plans that may also respect social conventions.

Unfortunately, these algorithms usually do not incorporate the notion that humans will also adapt their motion to the environment, and it is usually assumed that the robot is the sole responsible for collision avoidance maneuvers. One consequence of these assumptions, is that, due to the uncertainty in predictions allied with situations of high human density, these algorithms may fail to find a path. This problem has been described in [4] and is known as the Freezing Robot Problem FRP.

The motivation for this work derives from the observation that humans can seamlessly move in highly populated environments, while addressing complex situations and interactions with other humans. Due to this complexity, the mechanisms that take place in this behavior are not yet fully understood. This is an active topic of research [5, 6] with some findings incorporated into motion planning algorithms [7, 8].

According to these observations, we study the novel hypothesis that a robot can that properly select persons as leaders in a complex environment, will benefit from their advanced navigation and interaction skills, by following them [9]. Figure 1 illustrates the advantages of this approach in comparison with a state of the art motion planning algorithm (RiskRRT), using a pedestrian simulator in a narrow corridor.

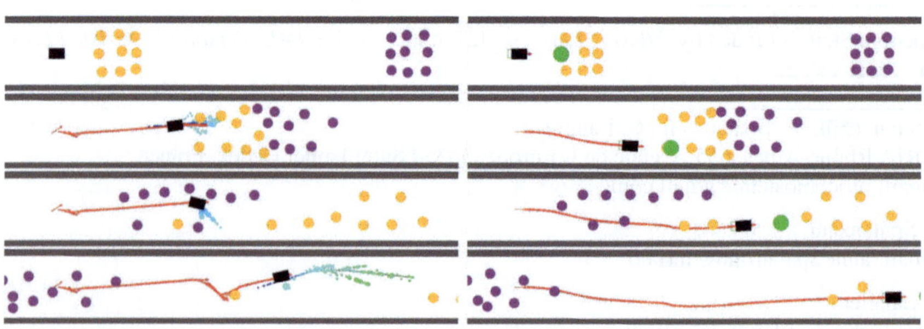

Fig. 1 Qualitative comparison of two navigation techniques in a narrow corridor. *Left column* RiskRRT, *right* leader following. Each row depicts the same instant during tests. Agents are represented by *circles*, the robot by the *black rectangle* and its path by a sequence of *red arrows*, showing its orientation in each instant

2 Technical Approach

Our experiments investigated how humans perform leader selection in populated environments. The approach was to associate LIDAR measurements with leader behaviors, in order to train a machine learning algorithm to classify persons as good or bad leaders. The general framework was built using ROS [10], while for machine learning we used the OpenCV library.

2.1 Data Acquisition

Data was acquired using a small car-like robot. Only two sensors were used: a laser scanning range finder and a camera fitted with a wide-angle lens. In this way, videos could be recorded and associated with laser scans during tests. Figure 2 shows the robotic platform, and a sample of the data collected from both sensors at a time instant. The purpose of this setup was to obtain information relative to the behavior and reactions of persons being followed by the robot.

2.2 Data Labeling and Feature Extraction

In many datasets, the labeling process is clear and objective, as marking faces in images, or pedestrians in laser scans. However, in this work the objective is to learn how humans decide when to start or stop following someone, or in other words, when someone is a good or bad leader. In order to do so, participants create labels based on their *feeling* about someone being a good leader or not.

A binary labeling system was used, with candidates identified either as good or bad leader. The process is depicted in . Volunteers should press a button whenever they *felt* a transition from good to bad leader occurred, while watching a video of

Fig. 2 *Left* platform used for data acquisition; *middle* sample image from a wide-angle camera; *right* sample of laser measurements showing three tracked subjects and their orientation as the dark vectors. The robot is represented as a *black rectangle*

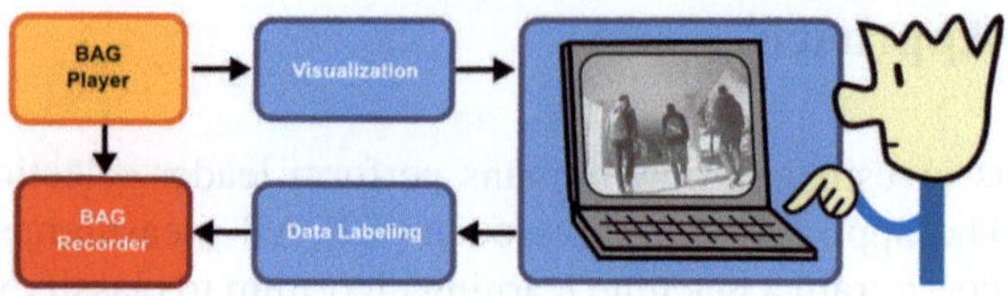

Fig. 3 Diagram of the data labeling process

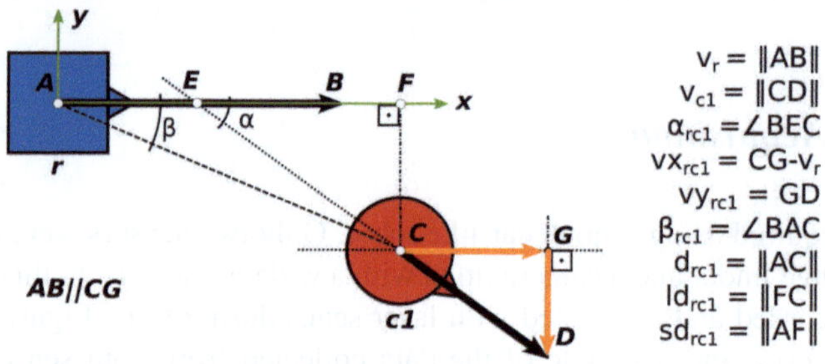

Fig. 4 Graphical representation of the features extracted while following. The robot is represented as a *blue square* while the human being followed as a *red circle*

persons being followed by the robot. This results in a mark that is stored along with all the original data recorded from experiments.

Videos of tests were shown to participants, who created labels based on their *feeling* about someone being a good leader or not. In the labeling process, the volunteers watched a video of persons being followed by the robot, and pressed a button whenever they *felt* a transition from good to bad leader occurred, as shown in Fig. 3.

As a result, a mark is stored along with all the original data recorded from experiments. The final label was computed as an average of individual inputs and then, LIDAR measurements were used to obtain the descriptors of each followed person. Figure 4 depicts the following features, which were calculated based on the robot's and subject's current state:

- subject's absolute velocity v_{c_i};
- relative heading between the robot and the subject α_{rc_i};
- subject's relative velocity w.r.t. the robot reference frame vx_{rc_i} and vy_{rc_i};
- angle between the robot heading and the subject's position β_{c_i};
- distance between the subject and the robot d_{c_i};
- lateral and sagittal displacement of the subject ld_{c_i} and sd_{c_i};

All the laser measurements were transformed to a fixed frame, to compensate for the robot's motion. Then, a motion tracker [11] was used to manage the identification and tracking of moving targets. The standard deviation and first derivative of each feature were also used, totaling 24 features.

2.3 *Classifier Training and Features Evaluation*

The chosen classifier is the adaptive Boosting (or AdaBoost) [12]. Among the advantages of this algorithm is the exponential convergence of the training error to zero and good generalization properties. But most important to the experiments conducted here, is the capacity of the algorithm to inherently select the features that contribute the most to the classification process.

This allows us to study of feature contribution, which is an important aspect of this work, enabling us to draw conclusions about why some features are more important than others, giving insights on what are the criteria used by humans when engaging or disengaging in following behaviors.

Learning and understanding what matters in following-leading engagements will be essential to the development of a more refined theory about human-following behaviors for robots.

3 Results

3.1 *Data Acquisiton*

In total 47 runs of the robot following persons or group of persons were recorded, with a mean duration of about 20 s each. Tests were conducted in an open corridor, about 3 m wide. The robot operator's goal was to move along this corridor back and forth, covering a distance of about 20 m in each pass.

Although each test had its own peculiarities, after the acquisition process, the data could be organized in four different classes, according to the type of transition from good to bad leader.

Before the labeling began, examples of different classes of good and bad leader situations were shown to the volunteers. Later they were asked to tell which of these situations occurred in each experiment. Although each test has its own peculiarities, the experiments can roughly fit one of the following classes:

1. **good leader (gd)**—leader(s) maintained their speed and orientation, without changing their behavior while being followed (Fig. 5);
2. **bad leader, moved aside (as)**—leaders gave room for the robot to pass, generally moving aside, while keeping their original motion direction (Fig. 6);
3. **bad leader, far or fast (fr)**—the distance between the leader and the robot grew to a point where it was not advantageous to keep following them (Fig. 7);
4. **bad leader, stopped (st)**—when the person being followed stops moving (Fig. 8).

Besides the four typical situations mentioned before, two other situations were recorded, which did not involve leader following, but could enhance the classifier training. These are the case when candidates were not moving, but rather standing

Fig. 5 Good leader behavior, although the robot is noticed, the leaders kept moving without changing their behaviors

Fig. 6 Bad leader: the person notices the robot and moves aside, giving room for the robot to pass

Fig. 7 Bad leader: the subject marked with a *white circle* moves to fast compared to the robot's speed and gets too far from it

Fig. 8 Bad leader: the person stops moving

close to the robot's path (nm), or when they were moving towards the opposite direction (od).

3.2 Classifier Training

After labeling, each resulting dataset was given a name to identify the situation they represented, together with a number to differentiate them. This organization allowed training and test datasets to encompass all the cases encountered. In total, 12911 samples were obtained, with the proportion of 37 % of bad leader labels and 63 %

Table 1 Composition of training dataset

Number of datasets used	Dataset situation
9	Good leader
7	Bad leader, leader moved aside
5	Bad leader, leader too far or fast
9	Bad leader, leader stopped
7	Bad leader, candidate not moving
10	Bad leader, candidate coming from opposite direction

of good leader labels. This is equivalent to 451 s of tests, divided in 47 experiments. Two datasets were created, one for the training comprising 8504 samples with 34 % cases of bad leader, and a test dataset with 2715 samples and 39 % cases of bad leader. The training dataset is composed of the following situations (Table 1):

An iterative process was used in the training of AdaBoost structures. In each iteration, the most important features were kept and then the training data was reprocessed and a new AdaBoost was trained. The objective of this procedure was to evaluate the influence played by different sets of features in the classifier performance and on the importance of features.

3.3 Feature Contribution

Figure 9 shows the contribution ratio of each feature used, after the first training the classifier using 24 features. The most important features are the lateral displacement and the distance between the robot and the leader. Besides those, the standard deviation of the distance, of the relative heading and of the target velocity have significant impact on the classification process. Finally, the derivatives features have almost no role at all in classification.

After an iterative process of picking the most important features and retraining the AdaBoost algorithm, results pointed to the **standard deviation of the distance between the robot and the leader** and the **lateral displacement of the leader** as the two most important features, as shown in Table 2. Together they contribution ratio was approximately 35 % when training the classifier using only the eight most important features.

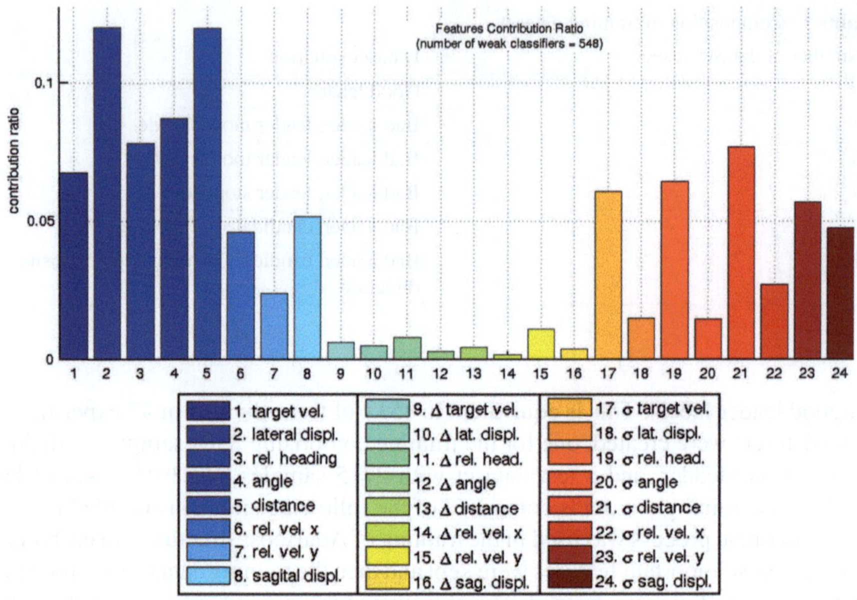

Fig. 9 Feature contribution ratio using 24 features, number of weak learners: 548

Table 2 Contribution ratio of most important features

Feature	Size of feature set			
	24	16	12	8
σ distance	0.077	**0.085**	**0.104**	**0.181**
Lateral displacement	**0.120**	**0.130**	**0.118**	**0.169**
Distance	**0.120**	**0.131**	**0.125**	**0.122**
Angle to robot	**0.087**	0.080	0.090	0.119
Sagittal displacement	0.052	0.053	0.078	0.118
Relative heading	0.078	0.083	0.081	0.113
Target velocity	0.068	0.068	0.082	0.093
σ relative velocity in y	0.057	0.062	0.075	0.085

3.4 Classifier Performance

To evaluate classification, two metrics were created. The first is the **false good leader**, where the classifier labeled a sample as *good leader* but the ground truth has a *bad leader* label. This is the most critical error, because it means that the algorithm would potentially select a bad leader to be followed, possibly disturbing the person. The second metrics is the **false bad leader**, that occurred when the classifier output a *bad leader* and the ground truth is labeled as *good leader*. This error is less critical because it translates only in a lost opportunity to classify someone as a good leader.

Regarding the performance of the classifiers, applied to the registered data, the *false good leader* relative error was particularly large at the *move aside* situations. However, the error was much smaller in the remaining situations, and the overall performance of the classifier was remarkable, with the total relative error across all the situations around 3 %.

4 Experiments

4.1 Platform and Setup

Although the aforementioned framework provided promising results, the classifier only operates over instantaneous measurements, and the ground truth was based on subjective impressions of volunteers. Therefore, to properly validate the proposed approach, interactive experiments are required.

Such experiments were conducted with an autonomous robotic wheelchair, equipped with two LIDAR sensors, the bottom one at 0.1 m and the top other at approximately 0.55 m in height. The platform also incorporated a RGBD camera, an on-board computer that takes care of low-level hardware control tasks and also a notebook computer that hosts the high-level algorithms, as the localization, planning and leader selection and following (Fig. 10).

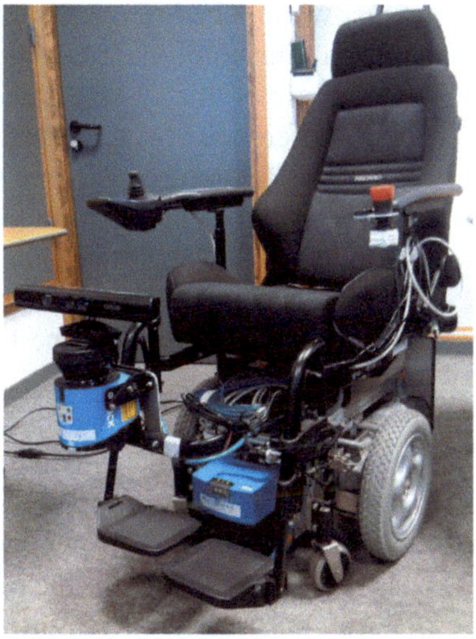

Fig. 10 Experimental platform, showing the two LIDARs and a RGBD camera

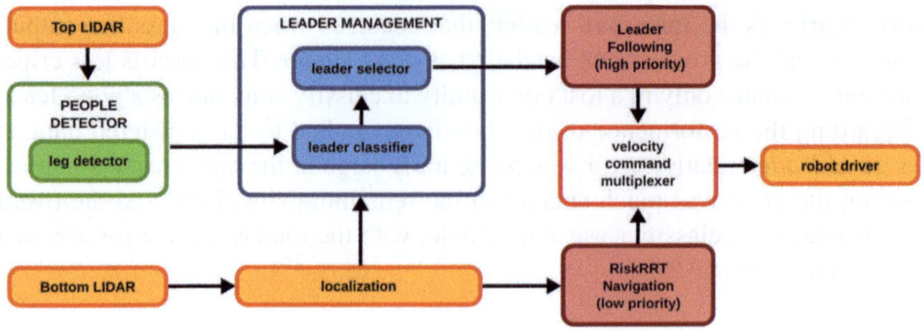

Fig. 11 Experimental architecture

Figure 11 illustrates the developed setup:

1. *People tracking*: the technique used for detecting and tracking persons uses the top LIDAR measurements in an AdaBoost classifier, which was previously trained with examples of legs patterns. If two legs are found and satisfy some requirements, it is assumed that a person has been detected.
2. *Leader classifier*: the information about tracked persons is passed to a module that implements the leader classification system presented in Sect. 2.3. This module provides a binary output (good/bad leader) based on a set of features extracted based on a tracked subject.
3. *Leader selector*: the previous module provides instantaneous classification of subjects based on the state of the robot and on their state. But to select a leader, a history of classifications must be considered. This module keeps a list of tracked subjects and associates a score to each one of them, which is a result of votes cast by different classifications. A good leader classification equals a vote of 0.01 points, while a bad leader vote is −0.1 points. Only after the score of a subject passes a threshold, that person can be considered a good leader to be followed. An example of this system is shown in Fig. 12.
   Such a difference of vote values means that someone needs several classifications of good leader to be considered to be followed, while only a couple of bad leader classifications is enough for a leader to be abandoned. Another advantage of this method is that a list with candidates scores allows for an easy switching of leaders. Once a leader is chosen, this module uses two proportional controllers, one for the distance and other for the heading, to compute velocity commands that are sent to the robot driver module, as shown in Fig. 13.
4. *RiskRRT Navigation*: if no feasible leader was found, the robot still needs to be able to navigate in dynamic environments. To accomplish this, this module uses the RiskRRT [1] algorithm to compute paths that avoid risky situations, based on the motion prediction of persons. This module is always active and computing possible paths, even when the robot is engaged in a leader following behavior. In this way it can keep navigating when a leader is abandoned.

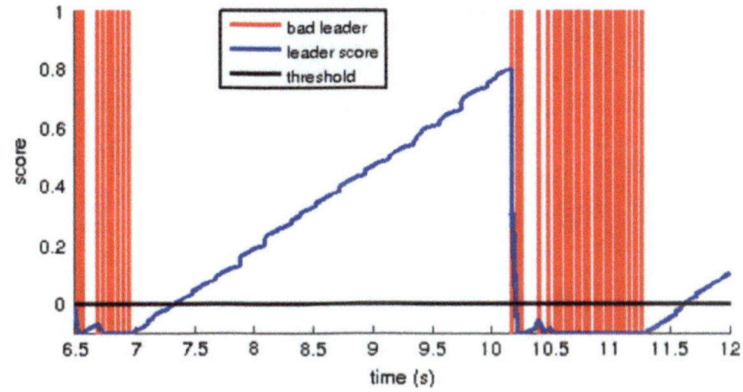

Fig. 12 Leader score example

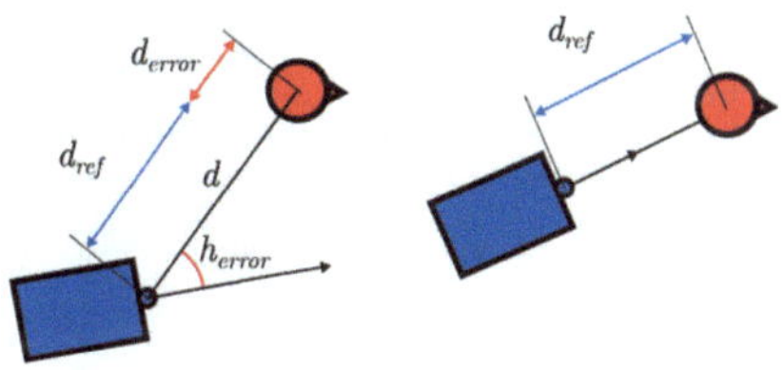

Fig. 13 Leader follower controllers

5. *Velocity command multiplexer*: in this setup, there are two modules that generate velocity commands to the robot. To manage which command will be sent to the robot, a simple multiplexer is used, which forwards the command with higher priority. The leader selector has the preference, but once a leader is lost or abandoned, it stops sending commands and them the RiskRRT module takes over.

4.2 Tests on Switching Navigation Method

The first experiment tests the performance of the proposed setup in switching between the leader following mode and the independent navigation, using the RiskRRT navigation algorithm. Figure 14 illustrates this experiment, where detected persons are represented by either a red circle (bad leaders) or a green circle (goo leaders). The chosen leader is represented by a light blue circle overlapping a green one. The small blue nodes are the exploration nodes of RiskRRT and the black line is the best path to navigate.

The first image of the sequence (1) shows the robot following a leader while, at the same time, the independent algorithm computes possible navigation routes. Once the leader stops moving (2), a sequence of bad leader classifications makes his score fall below an acceptable threshold, and the robot abandons that leader. As soon as this happens, the robot starts to follow the path computed by the independent navigation algorithm, avoiding a collision with the former leader and also with another person

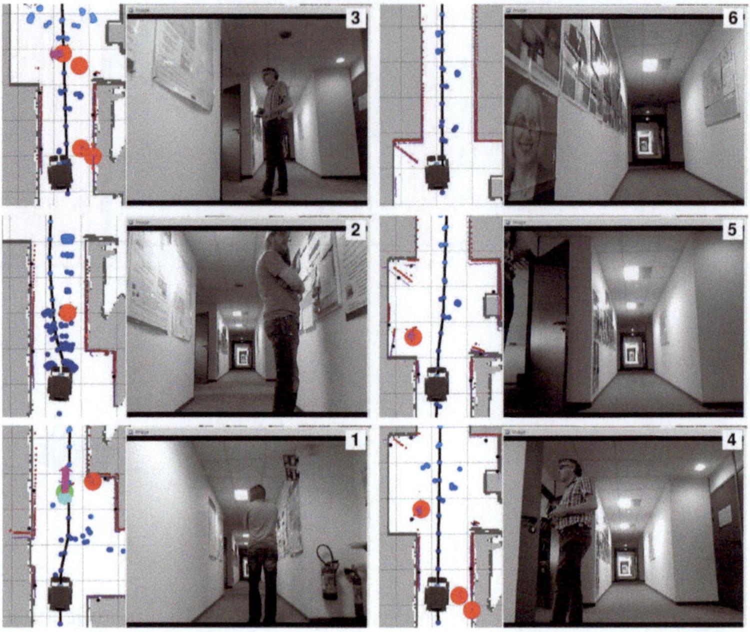

Fig. 14 Switching navigation method between leader following and independent navigation

crossing its path (3–5), which is also classified as a bad leader. The robot later continues to navigate using only the RiskRRT algorithm (6).

4.3 Tests on Leader Switching

This experiment tests the setup capacity of maintaining a list of feasible leaders and switching among them when the score of one surpass the score from another. Figure 15 shows two experiments (one in each column), and as in previous tests, a green circle means an instantaneous good leader classification, while a red one means the opposite. The blue circle marks the person with the highest leader score.

On the first examples (left column), the robot is behind two persons that are classified as good leaders and following the one on the right. As this person slows down to enter an office, he looses score points and the algorithm promptly switches to the other person, who continues to move along the corridor.

The second example (right column), shows the robot following a leader along a corridor. At a given moment, the leader slows down and another persons, which is moving faster passes the robot. According to the voting setup, the person that passes by the robot receives a higher score and then the robot changes the leader and starts to follow the new subject.

Fig. 15 Two tests on leader switching. On the *left column* the leader stops and the robot switches to follow another person. On the *right* a better leader appears on the scene

4.4 Tests on Leader Following Among Crowds

On the final experiment the robot must be able to select and follow leader in densely populated environments. The test occurs in a narrow corridor, where even small groups of people create difficult conditions for robot navigation. This test is shown in Fig. 16, and follow the same nomenclature from previous examples.

The experiment starts with the robot behind two persons moving along a corridor and following the one in the right side. Soon, a group of persons appear coming from the opposite direction and at the same time, the current leader stops to enter an office. The proposed setup manages to correctly identify a suitable leader among several subjects and successfully navigates through the crowd, taking advantage of the natural interactions among persons.

Fig. 16 Test on leader following in a densely populated corridor

5 Conclusions

This study and experiments provided insightful findings about the limitations and advantages of the proposed leader selection and following approach.

- *Autonomous wheelchairs are very interesting for experiments among persons.* This is mainly because people are used to see wheelchairs, and normally assume there is a human on the control of it. As a result, their reactions and behavior are more natural, which allows studies with minimal interference from the robot.
- *A robust person tracker is essential for a leader selection system.* When the robot followed someone, even short interruptions on the tracker resulted in erroneous leader classification and, therefore, in poor leader scoring, which hindered the navigation performance. Therefore, the identification of persons can be enhanced by using different sensors and detection methods, which must combined to provide a more robust detection and tracking system. Examples can be the fusion of images and laser scans to increase the reliability of people detection.
- *A balance must be found between taking advantage of a person's motion and disturbing that person.* Sometimes if the robotic wheelchair was too close to a leader, such person would stop to give room for the robot to pass, invalidating the advantages of the proposed method.

- *Distance-related features play an important role in leader selection.* Tests shown that the total and lateral distance to a candidate, together with their variance are crucial measurements to engage or disengage in a following behavior.
- *The feature space should be enriched with additional sensors and descriptors.* In some situations the selected features were not sufficient to properly classify a candidate. Different sensors and features must be used to improve the performance of machine learning classifiers. For example, a camera could be used detect gaze direction, and a RGBD camera could detect torso, head and arm patterns. Another interesting descriptor would be a distance from a subject to a precomputed path for the robot, which could avoid the situation where the robot drifts away from an objective.
- *Leader following has a great potential for enhancing robot navigations.* The presented framework and experiments validated the initial assumption that leader following can improve robotic navigation in dynamic environments, taking advantage of complex interactions among persons.

References

1. Fulgenzi, C., Tay, C., Spalanzani, A., Laugier, C.: Probabilistic navigation in dynamic environment using rapidly-exploring random trees and gaussian processes. In: 2008 IEEE/RSJ International Conference on Intelligent Robots and Systems (IROS), Sept 2008, pp. 1056–1062
2. Bennewitz, M., Burgard, W., Cielniak, G., Thrun, S.: Learning motion patterns of people for compliant robot motion. Int. J. Robot. Res. **24**(1), 31 (2005)
3. Rios-Martinez, J., Spalanzani, A., Laugier, C.: Understanding human interaction for probabilistic autonomous navigation using Risk-RRT approach. In: 2011 IEEE/RSJ International Conference on Intelligent Robots and Systems (IROS), Sept 2011, pp. 2014–2019
4. Trautman, P., Krause, A.: Unfreezing the robot: navigation in dense, interacting crowds. In: 2010 IEEE/RSJ International Conference on Intelligent Robots and Systems (IROS), Oct 2010, pp. 797–803
5. Mombaur, K., Truong, A., Laumond, J.: From human to humanoid locomotion—an inverse optimal control approach. Auton. Robots **28**, 369–383 (2009)
6. Dyer, J.R., Johansson, A., Helbing, D., Couzin, I.D., Krause, J.: Leadership, consensus decision making and collective behaviour in humans. Philos. Trans. R. Soc. B: Biol. Sci. **364**, 781–789 (2009)
7. Althoff, D., Wollherr, D., Buss, M.: Safety assessment of trajectories for navigation in uncertain and dynamic environments. In: 2011 IEEE International Conference on Robotics and Automation (ICRA), May 2011, pp. 5407–5412
8. Henry, P., Vollmer, C., Ferris, B., Fox, D.: Learning to navigate through crowded environments. In: 2010 IEEE International Conference on Robotics and Automation (ICRA), May 2010, pp. 981–986
9. Stein, P., Santos, V., Spalanzani, A., Laugier, C.: Navigating in populated environments by following a leader. In: International Symposium on Robot and Human Interactive Communication (RO-MAN) (2013)
10. Quigley, M., Conley, K., Gerkey, B., Faust, J., Foote, T., Leibs, J., Wheeler, R., Ng, A.Y.: ROS: an open-source robot operating system. In: ICRA Workshop on Open Source Software, vol. 3 (2009)

11. Almeida, J.: Target tracking using laser range finder with occlusion. Master's thesis, Universidade de Aveiro, Aveiro, Portugal (2010)
12. Freund, Y., Schapire, R.E.: A decision-theoretic generalization of on-line learning and an application to boosting. J. Comput. Syst. Sci. **55**(1), 119–139 (1997)

Perceptual Models of Human-Robot Proxemics

Ross Mead and Maja J. Matarić

Abstract To enable socially situated human-robot interaction, a robot must both understand and control *proxemics*—the social use of space—to employ communication mechanisms analogous to those used by humans. In this work, we considered how proxemic behavior is influenced by human speech and gesture production, and how this impacts robot speech and gesture recognition in face-to-face social interactions. We conducted a data collection to model these factors conditioned on distance. This resulting models of pose, speech, and gesture were consistent with related work in human-human interactions, but were inconsistent with related work in human-human interactions—participants in our data collection pos itioned themselves much farther away than has been observed in related work. These models have been integrated into a situated autonomous proxemic robot controller, in which the robot selects interagent pose parameters to maximize its expectation to recognize natural human speech and body gestures during an interaction. This work contributes to the understanding of the underlying per-cultural processes that govern human proxemic behavior, and has implications for the development of robust proxemic controllers for sociable and socially assistive robots situated in complex interactions (e.g., with multiple people or individuals with hearing/visual impairments) and environments (e.g., in which there is loud noise, reverberation, low lighting, or visual occlusion).

Keywords Human-robot interaction · Proxemics · Bayesian network

R. Mead (✉) · M.J. Matarić
Interaction Lab, Computer Science Department, University of Southern California,
3710 McClintock Avenue, RTH 423, Los Angeles, CA 90089-0781, USA
e-mail: rossmead@usc.edu
URL: http://robotics.usc.edu/interaction

M.J. Matarić
e-mail: mataric@usc.edu

M.A. Hsieh et al. (eds.), *Experimental Robotics*, Springer Tracts
in Advanced Robotics 109, DOI 10.1007/978-3-319-23778-7_18

1 Motivation and Problem Statement

To facilitate face-to-face human-robot interaction (HRI), a *sociable* or *socially assistive robot* [1] often employs multimodal communication mechanisms similar to those used by humans: speech production (via speakers), speech recognition (via microphones), gesture production (via physical embodiment), and gesture recognition (via cameras or motion trackers). Like signals in electrical systems, these *social signals* are attenuated by distance; this influences how the signals are produced by people (e.g., humans adapt to increased distance by talking louder [2]), and subsequently impacts how these signals are perceived by the robot.

This research focuses on answering the following questions: Where do people position themselves when interacting with a robot? How does this positioning influence how people produce speech and gestures? How do positioning and human social signal production impact robot perception through automated speech and gesture recognition systems? How can the robot dynamically adjust its position to maximize its performance during the social interaction?

These questions are related to the field of *proxemics*, which is concerned with the interpretation, manipulation, and dynamics of spatial behavior in face-to-face social encounters [3]. Human-robot proxemics typically considers either a *physical* representation (e.g., distance and orientation [4–6]) or a *psychological* representation (e.g., amount of eye contact or friendliness [7, 8]) of the interaction. In [9], we proposed a probabilistic framework for *psychophysical* proxemic representation to bridge the gap between these physical and psychological representations by considering the perceptual experience of each agent (human and robot) in the interaction. We now formally investigate our framework proposed in [9], modeling position and perception to inform human-robot proxemics.

2 Related Work

The anthropologist Hall [10] coined the term "proxemics", and proposed that *cultural* norms define zones of intimate, personal, social, and public space [3]. These zones are characterized by the pre-cultural and *psychophysical* visual, auditory (voice loudness), olfactory, thermal, touch, and kinesthetic experiences of each interacting participant [2, 11] (Fig. 1). These *psychophysical* proxemic dimensions serve as an alternative to the sole analysis of distance and orientation (cf. [12]), and provide a functional perceptual explanation of the human use of space in social interactions. Hall [11] seeks not only to answer questions of *where* a person will be, but, also, the question of *why* they are there.

Contemporary probabilistic modeling techniques have been applied to socially appropriate person-aware robot navigation in dynamic crowded environments [13, 14], to calculate a robot approach trajectory to initiate interaction with a walking

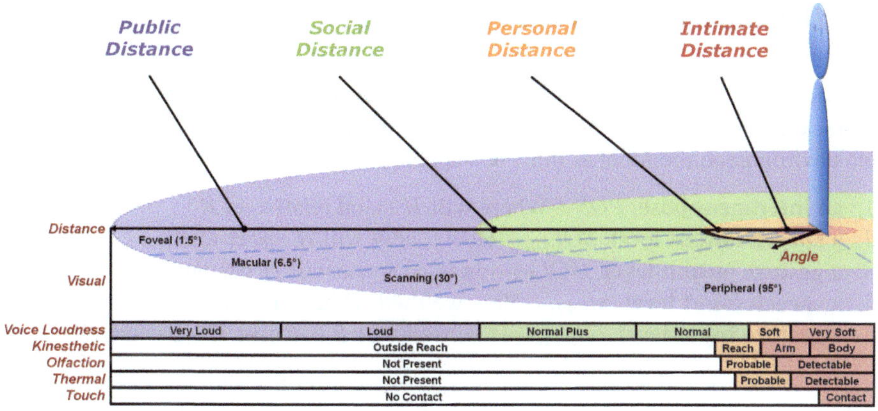

Fig. 1 Relationships between interpersonal pose and sensory experiences [2, 3, 11]

person [15], to recognize the averse and non-averse reactions of children with autism spectrum disorder to a socially assistive robot [16], and to position the robot for user comfort [4]. A lack of human-aware sensory systems has limited most efforts to coarse analyses [17, 18].

Our previous work utilized advancements in markerless motion capture (specifically, the Microsoft Kinect[1]) to automatically extract proxemic features based on metrics from the social sciences [19, 20]. These features were then used to recognize spatiotemporal interaction behaviors, such as the initiation, acceptance, aversion, and termination of an interaction [20–22]. These investigations offered insights into the development of spatially situated controllers for autonomous sociable robots, and suggested an alternative approach to the representation of proxemic behavior that goes beyond the physical [4–6] and the psychological [7, 8], considering psychophysical (perceptual) factors that contribute to both human-human and human-robot proxemic behavior [9, 23].

3 Framework for Human-Robot Proxemics

Our probabilistic proxemic framework considers how all represented sociable agents—both humans and robots—experience a co-present interaction [9]. We model the production (output) and perception (input) of speech and gesture conditioned on interagent pose (position and orientation).

[1] http://www.microsoft.com/en-us/kinectforwindows.

3.1 Definition of Framework Parameters

Consider two sociable agents—in this case, a human (H) and a robot (R)—that are co-located and would like to interact.[2] At any point in time and from any location in the environment, the robot R must be capable of estimating:

1. an **interagent pose**, POS—Where will H stand relative to R?
2. a **speech output level**, SOL_{HR}—How loudly will H speak to R?
3. a **gesture output level**, GOL_{HR}—In what space will H gesture to R?
4. a **speech input level**, SIL_{RH}—How well will R perceive H's speech?
5. a **gesture input level**, GIL_{RH}—How well will R perceive H's gestures?

These speech and gesture parameters are not concerned with the *meaning* of the behaviors, but, rather, the *manner* in which the behaviors are produced.

Interagent pose (POS) is expressed as a toe-to-toe distance (d) and two orientations—one from R to H (α), and one from H to R (β).

Speech output and **input levels** (SOL_{HR} and SIL_{RH}) are each represented as a sound pressure level, a logarithmic measure of sound pressure relative to a reference value, thus, serving as a signal-to-noise ratio. This relationship is particularly important when considering the impact of environmental auditory interference (detected as the same type of pressure signal) on SIL_{RH} and SOL_{HR} (our ongoing work).

Gesture output and **input levels** (GOL_{HR} and GIL_{RH}) are each represented as a 3D region of space called a gesture locus [24]. Related work in HRI suggests that nonverbal behaviors (e.g., gestures) should be parameterized based on proxemics [25]. For gesture production, we model the GOL_{HR} locus as the locations of H's body parts in physical space. The GIL_{RH} can then be modeled as the region of R's visual field occupied by the body parts associated with the gesture output of H (i.e., GOL_{HR}) [9].

3.2 Modeling Framework Parameters

We model distributions over these pose, speech, and gesture parameters and their relationships as a Bayesian network to represent: (1) how people position themselves relative to a robot; (2) how interagent spacing influences human speech and gesture production (output); and (3) how interagent spacing influences speech and gesture

[2]Our current work considers only dyadic interactions (i.e., interactions between two agents); however, our framework is extensible [9], providing a principled approach (supported by related work [2]) for spatially situated interactions between any number of sociable agents by maximizing how each of them will produce and perceive speech and gestures, as well as any other social signals not considered by this work.

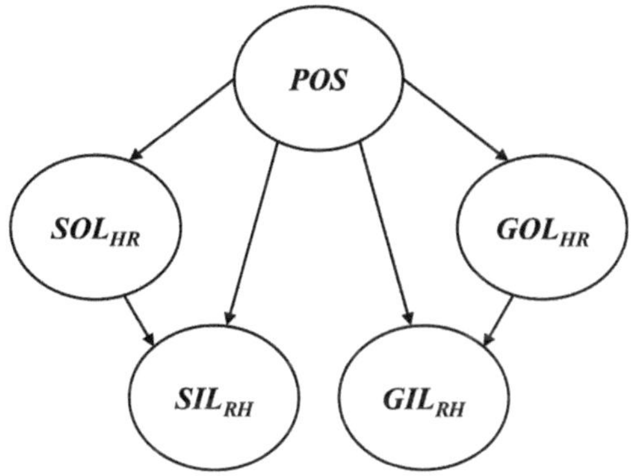

Fig. 2 A Bayesian network modeling relationships between pose, speech, and gesture

perception (input) for both humans and robots (Fig. 2). Formally, each component of the model can be written respectively as:

$$p(POS) \tag{1}$$

$$p(SOL_{HR}, GOL_{HR}|POS) \tag{2}$$

$$p(SIL_{RH}, GIL_{RH}|SOL_{HR}, GOL_{HR}, POS) \tag{3}$$

4 Data Collection

We designed a data collection to inform the parameters of the models represented by Eqs. 1–3. We performed this experimental procedure in the context of face-to-face human-human and human-robot interactions at controlled distances.

4.1 Procedure

Each participant watched a short (1–2 min) cartoon at separate stations (denoted $C1$ and $C2$ in Fig. 3), and then entered a shared space to discuss the cartoon with a social partner. Cartoons were chosen because they are commonly used in speech and gesture analysis studies, as people appear to gesture frequently when describing them [24]. Participants interacted for 4–6 min, and then separated into different rooms to watch another cartoon. Participants watched a total of six cartoons during the experimental session. For the first two (of six) interactions, participants were given no instructions as to where to stand in the room; this **“natural distance” within-participants condition** sought to elicit natural interagent poses, informing $p(POS)$ (i.e., Eq. 1), and

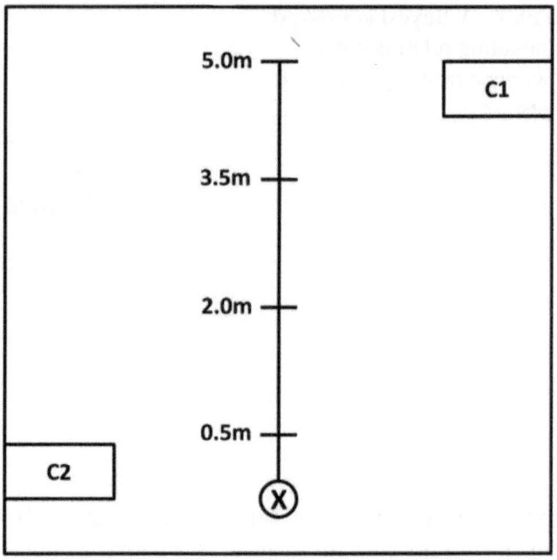

Fig. 3 The experimental setup for investigating proxemic behavior

also informing speech and gesture output/input parameters conditioned upon those natural poses (i.e., Eqs. 2 and 3). For the remaining four (of six) interactions, participants were instructed to stand at particular distances ($d = \{0.5, 2.0, 3.5, 5.0\}$ m) with respect to their social partner (who stood at floor mark X in Fig. 3), the order of which was randomly selected for each interaction. This "**controlled distance**" **within-participants condition** sought to expose how people modulate their speech and gestures to compensate for changes in interagent pose, exploring the state space for the perceptual models represented in Eqs. 2 and 3.

For each participant, the social partner was either another participant or a PR2 robot[3] ("**human versus robot**" **between-participants condition**). For each interactions, participants watched and discussed three different cartoons and three of the same cartoons to explore the space of interaction dynamics. For human-robot interactions, participants were informed that the robot was learning how to communicate about cartoons. The robot generated speech and gestures based on annotated data selected from the human-human interactions.

4.2 Materials

Six Kinects mounted around the perimeter of the room were used to monitor interagent pose (*POS*), as well as to extract the positions of human body parts, which served as our representation of human gesture output levels (GOL_{HR}) [20]. The field-of-view of a Kinect on-board the robot allowed for the extraction of its gesture input levels (GIL_{RH}) based on tracked human body features (e.g., head, shoulders, arms,

[3]https://www.willowgarage.com/pages/pr2/overview.

torso, hips, stance, etc.) [9]. Human speech output levels (SOL_{HR}) were recorded by calibrated microphone headsets; the corresponding robot speech input levels (SIL_{RH}) were recorded using the on-board Kinect microphone array.

4.3 Participants

Participants were recruited via mailing list, flyers, and word-of-mouth on the campus of the University of Southern California; thus, these results might not apply in other regions (though the procedure would). A total of 40 participants (20 male, 20 female) were recruited for the data collection. All participants were university students between the ages of 18 and 35, and most had technical backgrounds; 10 of these participants had prior experiences with robots. None of the participants had ever interacted with each other prior to the experiment (i.e., they were "strangers" or, at most, "acquaintances"). Ethnicity was recorded (varied; predominantly North American), but not discriminated in the model, as we wanted to model framework parameters over a general population.

4.4 Dataset

We collected pose, speech, and gesture data for 10 human-human interactions (20 participants) and 20 human-robot interactions (20 participants). Each data collection session lasted for 1 h, which included 20–45 min of interaction time. Recorded audio and visual data were independently annotated by two coders for occurrences of speech and gesture production; interrater reliability was high for both speech ($r_{speech} = 0.92$) and gesture ($r_{gesture} = 0.81$) annotations. For each of human-human and human-robot interactions, respectively, this dataset yielded 20 and 40 examples of natural distance selections, 4,914 and 4,464 continuous spoken phrases, and 2,284 and 1,804 continuous body gesture sequences; these numbers serve as the units of analysis in the following section.

5 Data Modeling and Analysis

The resulting dataset was used to model the relationships between speech and gesture output/input levels and interagent pose, as proposed in Eqs. 1–3.

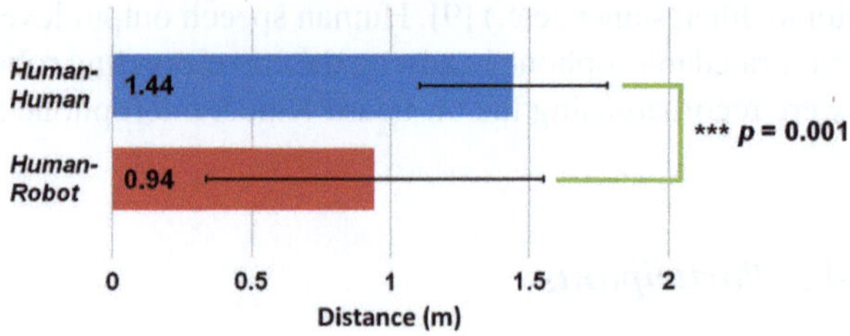

Fig. 4 Comparisons of human-human versus human-robot proxemics ($p = 0.001$)

5.1 Interagent Pose

To inform Eq. 1, interagent pose (*POS*) estimates from the Kinect during the **natural distance conditions** were used to generate the mean (μ) and standard deviation (σ) distance (in meters) in both human-human interactions ($\mu_{HH} = 1.44$, $\sigma_{HH} = 0.34$) and human-robot interactions ($\mu_{HR} = 0.94, \sigma_{HR} = 0.61$). An unpaired t-test revealed a very statistically significant difference ($p = 0.001$) in interagent pose under **human versus robot conditions** (Fig. 4).

The interagent poses in our human-human interactions are consistent with human-human proxemics literature—participants in our data collection positioned themselves at a common social distance [3] as predicted by their interpersonal relationship[4] [11].

However, the interagent poses in our human-robot interactions are inconsistent with human-robot proxemics literature—participants in our data collection positioned themselves much farther away than has been observed in related work. Walters et al. [6] consolidates and normalizes the results of many human-robot spacing studies, reporting mean distances of 0.49–0.71 m under a variety of conditions. Takayama and Pantofaru [8] investigated spacing between humans and the PR2 (the same robot used in our study), reporting mean distances of 0.25–0.52 m. We suspect that the apparent differences between our results and that of related work can be attributed to two differences in experimental procedure. First, in many of these studies, participants are explicitly told to respond to a distance or comfort cue; however, in our study, participants were more focused on the interaction itself, so the positioning might have been less conscious. Second, in many of these studies, the robot is not producing gestures; however, in our study, the robot was gesturing and has a long reach (0.92 m), so participants might have positioned themselves farther away from the robot to avoid physical contact. We will investigate this further in future work.

[4]Recall that participants were “strangers” or, at most, “acquaintances”.

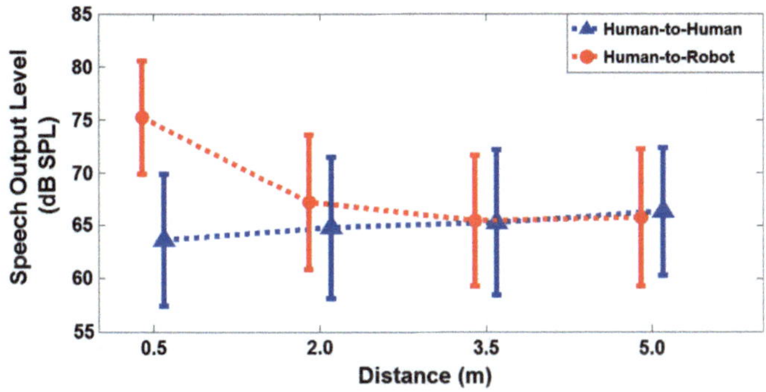

Fig. 5 Human speech output levels vary with distance in human versus robot conditions

5.2 Human Speech/Gesture Output Levels

To inform Eq. 2, we considered human speech and gesture output levels (SOL_{HR} and GOL_{HR}, respectively). In the **natural distance condition**, human-human and human-robot speech output levels were ($\mu_{HH} = 65.80$, $\sigma_{HH} = 6.68$) and ($\mu_{HR} = 67.23$, $\sigma_{HR} = 5.90$) dB SPL, respectively. An unpaired t-test revealed no statistically significant difference between speech output levels under **natural distance, human versus robot conditions**.

In the **controlled distance conditions**, an ANOVA F-test revealed significance ($p < 0.05$) in speech output levels at the nearest distance (0.5 m) across the **human versus robot conditions**. Furthermore, we found a trend towards significance ($p = 0.073$) for the statistical interaction of between distance and **human versus robot conditions** (Fig. 5) for speech output levels; this suggests that the way in which a person modulates speech to compensate for distance might be different when interacting with a robot rather than a person. These results are consistent with [26], which found that people tend to speak more loudly with a robot when they think it is trying to recognize what they are saying, often compensating for a perceived lack of linguistic understanding by the robot; our work expands upon this by illustrating the effect at multiple distances.

We detected no significant difference in gesture output levels in any condition, so each body part is modeled as a uniform distribution within a person's workspace (as measured by the Kinect). However, while our work suggests no significant relationship between gesture output level and *distance* (<5 m) in dyadic face-to-face interactions, related work suggests that the *orientation* between people (e.g., L-shaped, side-by-side, and circular; [27]) influences the way in which people produce gestures [28]. For example, during a face-to-face interaction, we might expect to see gestures made directly in front of a person [28], thus, likely in the center of the field-of-view of the robot; however, during a non-frontal interaction, we expect to see gestures located more laterally [28], potentially falling out of the field-of-view of the robot. We will investigate the relationships between gesture output level and orientation in future work.

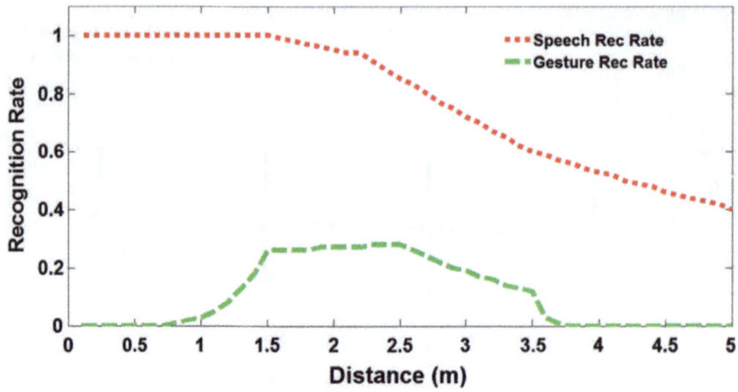

Fig. 6 Speech and gesture recognition rates as a function of distance

5.3 Robot Speech/Gesture Input Levels

To inform Eq. 3, we used the data recorded by the microphones and camera of the Kinect on-board the robot to estimate continuous performance rates of automated speech and gesture recognition systems to inform our perceptual models of *SIL* and *GIL*, respectively (Fig. 6).

For speech recognition as a function of human-robot *POS*, we trained the model on annotated spoken phrases from a subset of the data (2 male, 2 female; 25 phrase each; 100 phrase vocabulary). Figure 6 illustrates the impact of Eq. 2 (estimating SOL_{HR} based on *POS*) on speech recognition rates (SIL_{RH}). The estimation of SOL_{HR} is important for accurately predicting system performance, especially because it often indicates that increase in vocal effort improves recognition; the alternative is to assume constant speaker behavior, which would predict a performance rate inversely proportional to the distance squared [29]).

For gesture recognition as a function of human-robot *POS*, we used the annotated gesture frames to calculate the number of times a particular body part appeared in the Kinect visual field versus how many times the body part was actually tracked (Fig. 7). Figure 6 illustrates the impact of Eq. 2 (estimating GOL_{HR} based on *POS*) on gesture recognition rates (GIL_{RH}) by estimating the joint probability of recognizing three body parts commonly used for human gesturing (in this case, the head and both hands) at different distances.

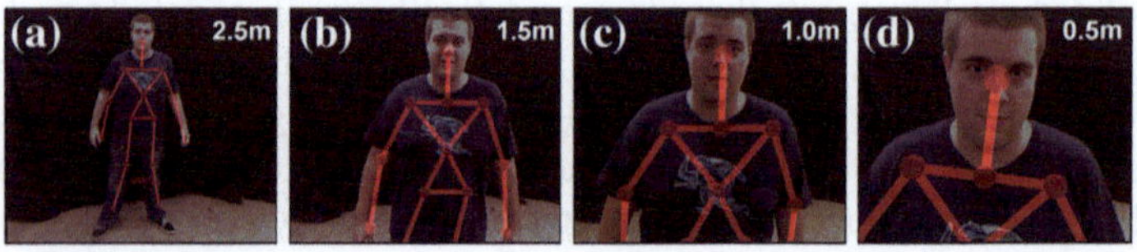

Fig. 7 Body features that fall into the Kinect field-of-view, depicted at four distances: **a** 2.5 m, **b** 1.5 m, **c** 1.0 m, and **d** 0.5 m [9]

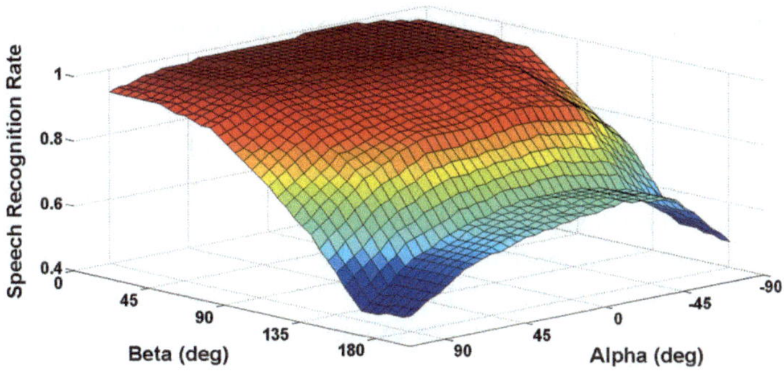

Fig. 8 Speech recognition rates as a function of speaker/listener orientations

Our data collection revealed very little about the impact of interagent *orientation* (as opposed to *position*) on robot speech and gesture input levels, as face-to-face interaction does not promote exploration in the orientation space. Thus, we exercised alternative techniques to construct models relating orientation to SIL_{RH} and GIL_{RH}. As noted in Sect. 3.1, two orientations were considered: robot-to-human orientation (α) and human-to-robot orientation (β).

The impact of α on SIL_{RH} is modeled as the *head-related transfer function* (HRTF) of the PR2 robot listener (determined using the technique described in [30]). The impact of β is based on existing validated models of human *speaker directivity* [31]. The use of an HRTF and speaker directivity is commonly used in the audio processing community. Figure 8 illustrates the relationship between orientation and SIL_{RH}.

A small data collection was performed to model the relationship between orientation and gesture recognition rates. Participants stood at specified positions and orientations, and moved their limbs around in their kinematic workspace. GIL_{RH} was modeled based on human body features tracked by the Kinect [9] (Fig. 7). Figure 9 illustrates the relationships between orientation and GIL_{RH}.

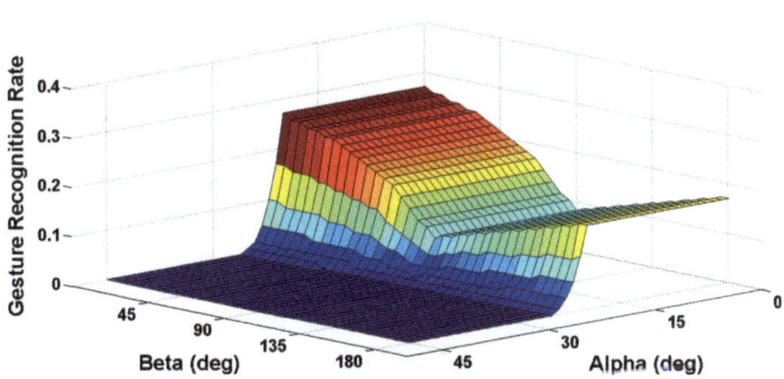

Fig. 9 Gesture recognition rates as a function of speaker/listener orientations

6 From Perceptual Models to Autonomous Controllers

The fundamental insight of this work is in the application of these perceptual models of human-robot proxemics to enable spatially situated communication in HRI. Autonomous sociable robots must utilize automated recognition systems to reliably recognize natural human speech and body gestures. The reported models of human communication enable the system to predict the manner with which a behavior will be produced by a person (Sect. 5.2, Fig. 5), which can then be used to predict system performance (Sect. 5.3, Fig. 6). For example, if the robot can detect interagent pose, then it could use its models to predict (1) how loudly the user will likely speak, (2) how loudly its sensors will likely detect the speech, and (3) how well its speech recognition system is likely to perform—all before a single word is spoken by the person. If the sociable robot is mobile, it could use its predictions to inform a decision-making mechanism to decide to move to a better position to maximize the potential for its performance in the interaction. This is a fundamental capability that autonomous sociable robots should have, and has the potential to improve autonomy, increase richness of interactions, and generally make robots more reliable, adaptable, and usable.

In our ongoing work, we have integrated these models into a **situated autonomous proxemic robot controller**, in which the robot selects interagent pose parameters to maximize its expectation to recognize natural human speech and body gestures (Eq. 4). The controller utilizes a sampling-based approach, wherein each sample represents interagent distance and orientation, as well as estimates of human speech and gesture output levels (production) and subsequent robot speech and gesture input levels (recognition); a particle filter uses these estimates to maximize the expected performance of the robot during the interaction (Fig. 10).

As an extension of this, the robot can dynamically adjust its own speech and gesture output levels (SOL_{RH} and GOL_{RH}, respectively) to be consistent with the models of human-human interaction behavior reported in Sect. 5.2 (Eq. 5) to maximize human recognition of its own speech and gestures.

$$\underset{POS}{\text{argmax}}\ E[SIL_{RH}, GIL_{RH} | SOL_{HR}, GOL_{HR}, POS] \tag{4}$$

$$\underset{SOL_{RH}, GOL_{RH}}{\text{argmax}}\ E[SIL_{HR}, GIL_{HR} | SOL_{RH}, GOL_{RH}, POS] \tag{5}$$

For interactions between two agents, this level of inference might be excessive; however, for interactions between three or more agents, such inference is necessary to determine appropriate pose, speech, and gesture parameters.

This controller has been implemented on the PR2 mobile robot (Fig. 11). The robot uses Eqs. 4 and 5 to estimate and select pose, speech, and gesture parameters to maximize its performance in the interaction, then uses the control equations described in our previous work [9] to realize these parameters.

We are in the process of validating this autonomous sociable robot controller.

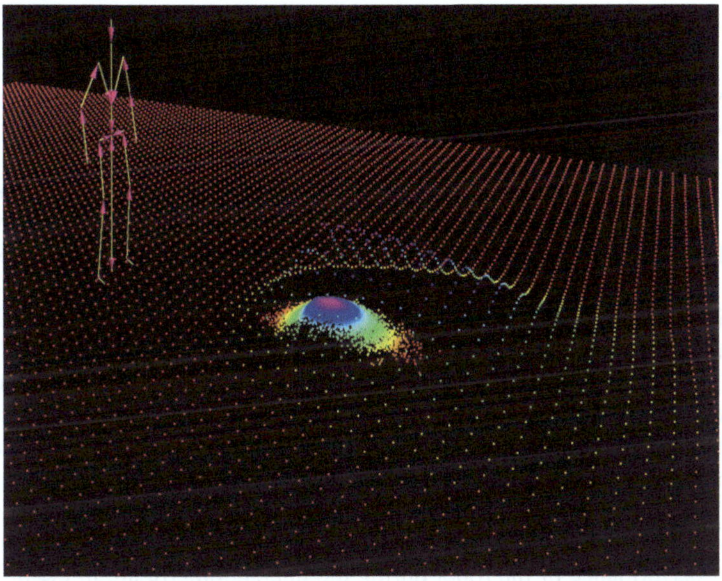

Fig. 10 Our autonomous sociable robot controller uses a sampling-based approach to estimate how the human produces social signals, and how the robot perceives them using models in our proxemic framework; a particle filter (with resampling) uses these estimates to maximize the expected performance of the robot during the interaction

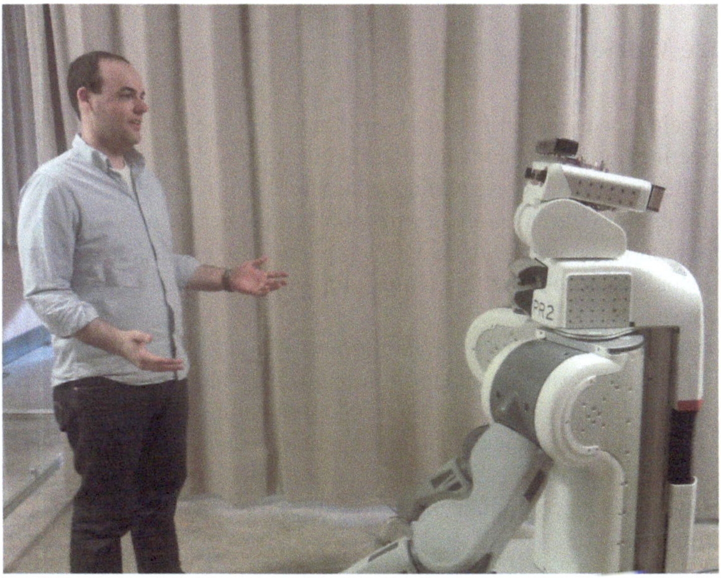

Fig. 11 The PR2 uses our controller to position itself in a conversation

7 Summary and Contributions

In this work, we formally investigated our framework proposed in [9], modeling how proxemic behavior is influenced by human speech and gesture production, and how this impacts robot speech and gesture recognition in face-to-face social interactions. This resulting models of pose, speech, and gesture were consistent with related work in human-human interactions [2], but were inconsistent with related work in human-human interactions [6, 8], warranting further investigation. This work contributes to the understanding of the underlying processes that govern human-human proxemic behavior [2], and provides an elegant extension into guiding principles of human-robot proxemic behavior. This research has implications for the development of robust spatially situated controllers for robots in complex interactions (e.g., with multiple people, or with individuals with hearing/visual impairments) and environments (e.g., in which there is loud noise, reverberation, low lighting, or visual occlusion).

The resulting models were used to implement a situated autonomous proxemic controller; the controller utilizes a sampling-based approach to maximize the expected performance of the robot during the interaction. We are in the process of validating the controller. This work further contributes to HRI community by providing data-driven models and software as part of the Social Behavior Library (SBL) in the USC ROS Packages Repository.[5]

8 Future Work

Proxemic behavior is *not* cross-cultural [3], and the desired sensory experience of each interacting human participant varies all over the world [2]. We did not constrain our data collection to focus on a particular culture, as the robot cannot currently make any assumptions or inferences about with whom it will be interacting. However, in the future, it would be beneficial to develope separate models conditioned on different cultures, which could be used in conjunction with contextual knowledge about the user and/or region.

Similarly, these models of proxemic behavior could be personalized to an individual. The use of a spatially situated communication has particular implications in socially assistive robotics domains [1], such as when interacting with a person with a visual or hearing impairment. The models of speech and gesture input would be conditioned on the sensory experience of such the individual, and the robot would dynamically adjust its communication modalities to maximize the perception of its social stimuli.

[5]https://code.google.com/p/usc-interaction-software.

The implementation of the system proposed in Sect. 6 has the robot producing speech and gesture based on models of how humans produce such behaviors (Sect. 5.2, Fig. 5). This is based on the assumption that humans actually want the robot to behave this way. Future work will investigate the production of these behaviors from the ground up by conducting a human psychometric study of robot speech and gesture output.

Complex environments introduce extrinsic sensory interference that requires socially situated proxemic behavior to be dynamic. For example, if one is speaking in a quiet room, listeners do not need to be nearby to hear; however, if one is speaking in a noisy room, listeners must be much closer to hear at the same volume and, thus, perceive the vocal cues contained in the utterance. Similarly, if one is speaking in a small, well-lit, and uncrowded or uncluttered room, observers may view the speaker from a number of different locations; however, if the room is large, dimly lit, or contains visual occlusions, observers must select locations strategically to properly perceive the speech and body language of the speaker. In future work, we will model and integrate extrinsic interference into our probabilistic framework of human-robot proxemics.

Acknowledgments This work is supported in part by an NSF Graduate Research Fellowship, the NSF National Robotics Initiative (IIS-1208500), NSF IIS-1117279 and CNS-0709296 grants, and the PR2 Beta Program.

References

1. Tapus, A., Matarić, M., Scassellati, B.: The grand challenges in socially assistive robotics. IEEE Robot. Autom. Mag. **14**(1), 35–42 (2007)
2. Hall, E.T.: The Hidden Dimension. Doubleday Company, Chicago (1966)
3. Hall, E.T.: American Anthropology Association. Handbook for Proxemic Research. Washington (1974)
4. Torta, E., Cuijpers, R.H., Juola, J.F., van der Pol, D.: Design of robust robotic proxemic behaviour. In: Proceedings of the Third International Conference on Social Robotics, ICSR'11, pp. 21–30 (2011)
5. Kuzuoka, H., Suzuki, Y., Yamashita, J., Yamazaki, K.: Reconfiguring spatial formation arrangement by robot body orientation. In: HRI. Osaka (2010)
6. Walters, M., Dautenhahn, K., Boekhorst, R., Koay, K., Syrdal, D., Nehaniv, C.: An empirical framework for human-robot proxemics. In: New Frontiers in Human-Robot Interaction, pp. 144–149. Edinburgh (2009)
7. Mumm, J., Mutlu, B.: Human-robot proxemics: Physical and psychological distancing in human-robot interaction. In: HRI, pp. 331–338. Lausanne (2011)
8. Takayama, L., Pantofaru, C.: Influences on proxemic behaviors in human-robot interaction. In: IEEE/RSJ International Conference on Intelligent Robots and Systems, IROS'09, pp. 5495–5502 (2009)
9. Mead, R., Matarić, M.J.: A probabilistic framework for autonomous proxemic control in situated and mobile human-robot interaction. In: 7th ACM/IEEE International Conference on Human-Robot Interaction, pp. 193–194. Boston (2012)
10. Hall, E.T.: The Silent Language. Doubleday Company, New York (1959)
11. Hall, E.: A system for notation of proxemic behavior. Am. Anthropol. **65**, 1003–1026 (1963)

12. Mehrabian, A.: Nonverbal Communication. Aldine Transcation, Piscataway (1972)
13. Vasquez, D., Stein, P., Rios-Martinez, J., Escobedo, A., Spalanzani, A., Laugier, C.: Human aware navigation for assistive robotics. In: Proceedings of the Thirteenth International Symposium on Experimental Robotics, ISER'12. Québec City (2012)
14. Trautman, P., Krause, A.: Unfreezing the robot: navigation in dense, interacting crowds. In: IROS. Taipei (2010)
15. Satake, S., Kanda, T., Glas, D.F., Imai, M., Ishiguro, H., Hagita, N.: How to approach humans?: strategies for social robots to initiate interaction. In: HRI, pp. 109–116 (2009)
16. Feil-Seifer, D., Matarić, M.: Automated detection and classification of positive versus negative robot interactions with children with autism using distance-based features. In: HRI'11, pp. 323–330. Lausanne (2011)
17. Oosterhout, T., Visser, A.: A visual method for robot proxemics measurements. In: HRI Workshop on Metrics for Human-Robot Interaction. Amsterdam (2008)
18. Jones, S., Aiello, J.: Proxemic behavior of black and white first-, third-, and fifth-grade children. J. Personal. Soc. Psychol. **25**(1), 21–27 (1973)
19. Mead, R., Atrash, A., Matarić, M.J.: Proxemic feature recognition for interactive robots: automating metrics from the social sciences. In: International Conference on Social Robotics, pp. 52–61. Amsterdam (2011)
20. Mead, R., Atrash, A., Matarić, M.J.: Automated proxemic feature extraction and behavior recognition: applications in human-robot interaction. Int. J. Soc. Robot. **5**(3), 367–378 (2013)
21. Mead, R., Atrash, A., Matarić, M.J.: Recognition of spatial dynamics for predicting social interaction. In: 6th ACM/IEEE International Conference on Human-Robot Interaction, pp. 201–202. Lausanne (2011)
22. Feil-Seifer, D., Matarić, M.: People-aware navigation for goal-oriented behavior involving a human partner. In: IEEE International Conference on Development and Learning, ICDL'11, vol. 2, pp. 1–6. Frankfurt Am Main, Germany (2011)
23. Mead, R., Atrash, A., Matarić, M.J.: Representations of proxemic behavior for human-machine interaction. In: NordiCHI 2012 Workshop on Proxemics in Human-Computer Interaction, Copenhagen (2012)
24. McNeill, D.: Hand and Mind: What Gestures Reveal about Thought. Chicago University Press, Chicago (1992)
25. Brooks, A.G., Arkin, R.C.: Behavioral overlays for non-verbal communication expression on a humanoid robot. Auton. Robot. **22**(1), 55–74 (2007)
26. Kriz, S., Anderson, G., Trafton, J.G.: Robot-directed speech: using language to assess first-time users' conceptualizations of a robot. In: Proceedings of the 5th ACM/IEEE International Conference on Human-Robot Interaction, HRI'10, pp. 267–274. IEEE Press (2010)
27. Kendon, A.: Conducting Interaction—Patterns of Behavior in Focused Encounters. Cambridge University Press, New York (1990)
28. Ozyurek, A.: Do speakers design their co-speech gestures for their addresees? the effects of addressee location on representational gestures. J. Mem. Lang. **46**(4), 688–704 (2002)
29. Wöelfel, M., McDonough, J.: Distant Speech Recognition. Wiley, West Sussex (2009)
30. Gardner, B., Martin, K.: HRTF measurements of a kemar dummy-head microphone. Technical Report 280. MIT Media Lab Perceptual Computing, Boston (1994)
31. Chu, W., Warnock, A.: Detailed Directivity of Sound Fields Around Human Talkers. Research Report. Institute for Research in Construction, Canada (2002)

The Interactive Urban Robot IURO: Towards Robot Action in Human Environments

Dirk Wollherr, Sheraz Khan, Christian Landsiedel and Martin Buss

Abstract Spatial reasoning and semantic environment understanding is a fundamental ability of robots navigating in unstructured dynamic environments. Since spatial and semantic reasoning is tightly linked to the sensor perception information, it is desirable that it is directly integrated into the environment model. This paper presents the interplay of a novel environment representation called *Semantic Rtree (SRTree)* and Markov Logic Networks for reasoning. The SRTree is a semantic occupancy grid based on the hierarchical Rtree data structure that models the probability of occupancy of each grid cell and additionally assigns a class label to it. The main advantages of the proposed approach are (1) a hierarchical representation of large scale outdoor urban environments, which (2) captures both quantitative (metric) and qualitative (semantic) aspects of the environment and allows reasoning in a single data structure, and (3) the capability of dealing with higher-order spatial relations. The proposed methods are experimentally evaluated on a large scale 3D point cloud dataset of downtown Munich enhanced by RGB image data. The dataset is being made publicly available.

Keywords Environment representation · Spatial reasoning · Markov logic networks

D. Wollherr · S. Khan · C. Landsiedel (✉) · M. Buss
Institute of Automatic Control Engineering, Technische Universität München,
Arcisstr. 21, 80290 Munich, Germany
e-mail: christian.landsiedel@tum.de
URL: http://www.lsr.ei.tum.de

D. Wollherr
e-mail: dw@tum.de

S. Khan
e-mail: sheraz.khan@tum.de

M. Buss
e-mail: mb@tum.de

M.A. Hsieh et al. (eds.), *Experimental Robotics*, Springer Tracts
in Advanced Robotics 109, DOI 10.1007/978-3-319-23778-7_19

1 Motivation, Problem Statement, Related Work

It has been a long-standing goal of researchers worldwide to bring robots into human environments and exploit their virtues in home applications as well as in the services sector [1]. To address some of the challenges towards the goal of robots as universal helpers being able to autonomously act in unstructured, dynamically changing environments, we developed the *Interactive Urban Robot (IURO)* robot, see Fig. 1. The goal of this project is to create a robot that must both navigate in an unknown urban environment and interact with human passers-by in order to retrieve information. The robot can be given a designated goal location in a city and successfully finds its way to this location without the use of map knowledge or GPS, obtaining and interpreting directions by asking pedestrians for the way. Building an autonomous outdoor interaction robot requires the design of a cognitive architecture that ties together functional modules for navigation, image processing and interaction. The interaction has to be natural and intuitive for the humans, as they are picked autonomously by the robot, have no prior contact with robotics technology, and are not instructed prior to the interaction.

In this scenario, a fundamental ability of the robot is to integrate environment information from multiple sources—particularly from sensors like cameras and laser range finders as well as verbal and gesture information from interactions with humans—into a concise environment representation. To this end, the robot must be capable

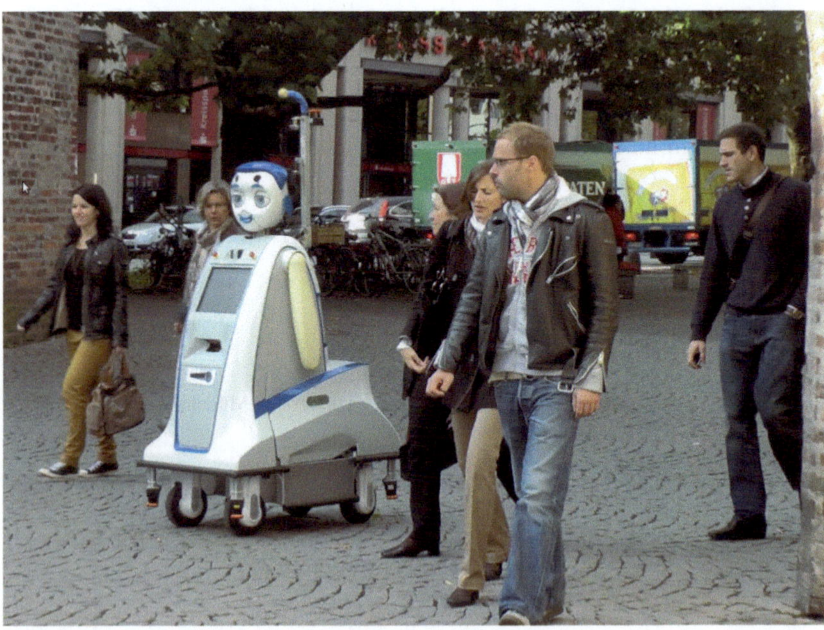

Fig. 1 The IURO robot in unstructured urban environment

of spatial reasoning, i.e. establishing geometric relations between objects, in terms understandable by humans. It is therefore desirable that metric and semantic environment representations are tightly interlinked.

The field of 3D robotic mapping has received a lot of attention in the scientific community. The most commonly used approach is an occupancy grid which divides the space into grid cells and estimates the probability of occupancy of each cell [2–5]. These representations can be useful for navigation and exploration; however, they do not capture symbol-level information about the environment. This is however, essential for reasoning, and it is considered to be the most common communication channel used and understood by humans. To develop autonomous interactive robots, robots must be capable of understanding the semantics and relationships between the objects in the environment. Recently, the focus of the robotics community has shifted towards semantic representations [6–8] and object relation modelling in semantic maps [9–14]. The majority of the works mentioned rely on point clouds or operate on the level of objects to represent the semantics of the environment, which is not suited for navigation. Occupancy grids are suitable for navigation; however, they do not capture the environment semantics, whereas standard semantic environment representations are ill suited for navigation purposes. *Hence, there is a requirement to generate a hybrid representation that combines the advantages of occupancy grids and semantic environment representations*. This paper presents the SRTree, which is capable of generating a probabilistic occupancy representation for the task of navigation and exploration and additionally captures symbolic information about the environment. In addition to capturing semantic information in the form of object and class information, the environment representation is capable of storing higher-order spatial relationships between objects. This representation can be useful in scenarios in which a robot is required to navigate in an environment while adhering to certain norms. In the context of IURO, the robot might be required to generate trajectories on sidewalks to avoid collisions with cars. Hence, given the SRTree environment representation, which contains occupancy probabilities and the semantic labels assigned to each grid cell, it is possible to generate trajectories in specific regions of the grid corresponding to the sidewalk. Additionally, the typical HRI scenarios encountered in the IURO project deal with higher-order spatial relationships. Hence it is important for the robot to be aware of the environment semantics and the relationships between objects present in the environment.

The paper presents the interplay of a novel environment representation, called *Semantic Rtree (SRTree)*, which is an extension of the authors' previous work [15], and a strategy to reason about spatial relations between objects using a Markov Logic Network. The main contributions of this paper are:

- A semantic occupancy grid (SRTree) that models the occupancy probabilities and assigns a class label to each grid cell.
- The capability of dealing with higher-order spatial relationships using Markov Logic Networks.
- A large scale colored pointcloud dataset of downtown Munich annotated with 10 different class labels.

The main advantage of the proposed approach is that it is capable of dealing with higher-order spatial relations and generates a hierarchical representation of *large scale outdoor urban environments*. SRTree provides the foundation for this higher-order spatial reasoning through its hierarchic structure which ensures fast access to the occupancy grid, the environment semantics and allows storing of spatial relations within the same structure. The dataset used for evaluation in this paper is being made publicly available.

2 Technical Approach

In this paper, an approach to efficiently perform semantic reasoning about spatial relations between typical objects in a urban environment is presented. This information can be deployed for example in a HRI scenario, where a human identifies an object by its position with respect to the environment; the robot needs to retrieve the most likely candidates for the object being referred to from its internal environment representation. This information can then be retrieved and put to use for example for semantic navigation or in an HRI scenario. An example for a task where these abilities are required is the reasoning on route descriptions given by humans [16].

The overall pipeline of operations performed by the approach proposed in this paper is shown in Fig. 2. The input of the processing is a multi-attribute pointcloud which defines the geometry, color, class label and object assignment for each point in the sensor field of view. The geometry and color information of each point in the pointcloud is obtained from the Z+F 5010C laser scanner which fuses data of the camera and the laser scanner. The proposed approach requires a pointcloud with pre-segmented objects and class labels assigned to each point in the pointcloud. The class labels are chosen based on categories which are commonly found in urban environments such as sidewalk, trees, building etc. In addition to the class labels, an object id is assigned to be able to define spatial relations (*LeftOf*, *On* etc.) between objects and to disambiguate between multiple instances of an object (multiple trees or cars etc.). The multi-attribute pointcloud is inserted into the SRTree (Semantic Rtree) which evaluates for each grid cell the occupancy probability, the most probable class label and object id attributes. This hybrid SRTree representation is then used to extract different meaningful geometric features which are employed in a Markov logic network framework to generate consistent spatial relations and determine higher

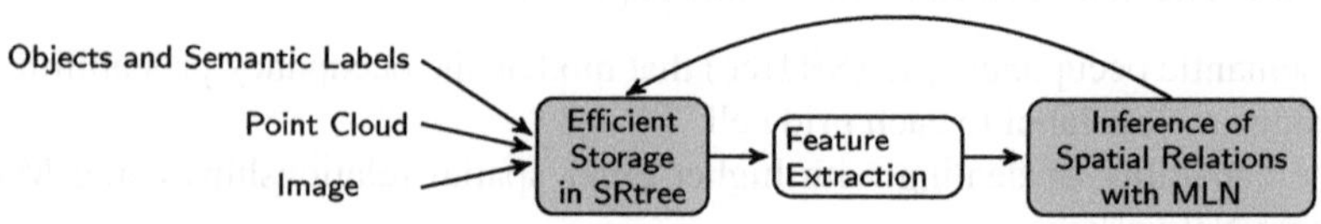

Fig. 2 The pipeline of operations performed in the proposed approach. The two components that are the focus of this paper appear *shaded*

order spatial relations between objects present in the urban environment. The pipeline becomes a closed chain once the determined spatial relations are inserted into the SRTree in form of a *spatial relation graph* between objects and used in typical HRI scenarios which require reasoning over route descriptions.

2.1 Semantic Rtree (SRTree)

The first component of the pipeline is the SRTree, which is used to generate a metric and semantic representation of the environment. The SRTree is based on the standard Rtree [17, 18] and is an extension of the authors' previous work [15], in which an occupancy grid based on the Rtree datastructure is proposed. The Rtree is composed of a hierarchy of axis-aligned rectangular cuboids and contains root, inner and leaf nodes. The root, inner node can have a maximum number M of children. The leaf and the inner nodes define a minimum bounding rectangular cuboid over its child branches.

The proposed SRTree approach generates an probabilistic occupancy grid and additionally assigns a class label to each grid cell. Consider a grid G_t consisting of m grid cells g_i, $i = 1, \dots, m$ of fixed resolution. The ith leaf branch of the Rtree hierarchy represents the ith grid cell g_i of the occupancy grid and contains the tuple $(o_i, l_i, \boldsymbol{x}_i, p_i, r_i)$. The term o_i represents the occupancy probability and l_i is the class label assigned to the grid cell based on the class occurrence probabilities $\boldsymbol{x}_i = [x_{i,1}, \dots, x_{i,N}]$ for the ith grid cell where N represents the total number of semantic categories. The element p_i in the tuple denotes the object id to which a specific cell belongs. The object id is assigned to a grid cell based on the mode of the histogram of object id's within a cell. The term r_i represents a pointer to the relation graph of the ith object to its neighbours. All grid cells belonging to the same object (for example an object *car #3*) have the same relation graph, and consequently contain the same instance of the variable r_i.

The sensor model employed in the SRTree is the same as used in the authors' previous work [15] and does not explicitly model free space. The occupancy probability of each grid cells is calculated as [5, 19, 20]

$$P(o_i|z_{1:t}) = \left[1 + \frac{1 - P(o_i|z_t)}{P(o_i|z_t)} \frac{1 - P(o_i|z_{1:t-1})}{P(o_i|z_{1:t-1})} \frac{P(o_i)}{1 - P(o_i)}\right]^{-1}, \quad (1)$$

which is a commonly used inverse sensor model in robotic mapping. $P(o_i|z_{1:t})$ represents the occupancy probability of the ith grid cell given all observations. $P(o_i)$ represents the occupancy probability of a grid cell prior to any observations. $P(o_i|z_t)$ and $P(o_i|z_{1:t-1})$ represent the probability given the most current observation z_t and observations since the beginning of time until time $t - 1$ respectively. In literature [5, 20], occupancy grids use a probability clamping threshold to prevent each cell from being overconfident about its state. Following the same pattern, the occupancy grid in the SRTree defines a minimum and maximum probability threshold $\alpha_{min}, \alpha_{max}$ respectively, the passing of which means that a grid cell is no longer updated.

In addition to the occupancy probabilities, the proposed SRTree assigns a class label to each grid cell based on the labels of the points inserted in it. Each grid cell maintains a list of the observed counts of each class label. Due to the noise resulting from the discretization effects of the occupancy grid or imperfect class label assignment, it is essential for each grid cell to take the uncertainty of the observed counts into account. To model this uncertainty, a Dirichlet distribution is used which generates a distribution over multinomial distributions for each object category.

Consider the probability density function of the Dirichlet distribution,

$$p^j(x_{i,1}, \ldots, x_{i,N-1}; \alpha_1^j, \ldots, \alpha_N^j) = \frac{1}{N(\alpha^j)} \prod_{k=1}^{N} x_{i,k}^{\alpha_k^j - 1},$$

where α_k^j represents the kth concentration parameter of the Dirichlet distribution that corresponds to the jth object category, where $k, j \leq N$. The concentration parameters $\forall k, \alpha_k^j$ are learnt offline using the moment matching method [21]. $x_{i,1}, \ldots, x_{i,N}$ represent the class occurrence probabilities ($\sum_{k=1}^{N} x_{i,k} = 1$) of the ith grid cell, which are calculated based on the normalized histogram of observed class counts. $N(\alpha)$ represents the normalization factor which can be expressed using the gamma function (γ) as follows

$$N(\alpha^j) = \frac{\prod_{k=1}^{N} \gamma(\alpha_k^j)}{\gamma(\sum_{k=1}^{N} \alpha_k^j)}.$$

The assignment of a specific label l_i to a grid cell g_i is based on

$$l_i = \underset{j}{\operatorname{argmax}}\, p^j(x_{i,1}, \ldots, x_{i,N-1}; \alpha_1^j, \ldots, \alpha_N^j), \forall j \leq N.$$

Hence the SRTree is capable of generating a probabilistic occupancy grid and additionally defines a class label for each grid cell. Figure 3 (left) shows the SRTree occupancy grid with the grid cells colored based on the running average of each RGB values of all points that falls within that cell. Figure 3 (right) shows the occupancy grid cells in different colors based on the class label assignment using the Dirichlet distribution. The relation graph between objects is obtained after inference on the Markov Logic Network as described in the following section.

2.2 Reasoning over Spatial Relations

This section describes how the metric environment information from the SRTree is augmented by estimates of qualitative spatial relationships between objects. This process starts by extracting numeric features from the metric representation of the

Fig. 3 The SRTree with grid cells colored based on RGB values and class labels (*green*: building, *dark red*: sidewalk, *blue*: street, *red*: car)

environment presented by the SRTree, which are then classified to a known set of relations. These estimates are subsequently postprocessed for a degree of global consistency with the application of rules in probabilistic logic, in the form of a Markov Logic Network [22]. The Maximum-a-Posteriori assignment of spatial relations between objects is then inferred. The MLN approach allows for an accessible specification of spatial relations, in particular ones of higher order (>2), such as the transitivity of the *On* relationship: $On(base, middle) \wedge On(middle, upper) \rightarrow On(base, upper)$. Similarly, the qualitative location of an object to another, with respect to an observer, is a higher-order function of the relations between the two objects and their relation to the observer [23].

The graph of objects and their relations built this way can then, in turn, be stored back into the SRTree representation. This section firstly details the model for spatial relations that was chosen with the application of reasoning over route directions in mind, and introduces related approaches. Then, the individual steps of the process—feature computation, baseline classfication and smoothing with a Markov Logic Network—are detailed.

We are concerned with relationships such as *left/right/behind/in front of* and relations of support between objects that can be used to locate objects in a scene. Thus, we restrict the set of qualitative spatial relations that are reasoned over in this work to the relations *On, LeftOf/RightOf* and *Behind/InFrontOf.*

This selection of relations is motivated by the typical terms used in route directions, the understanding of which is an important problem in collaborative robotic applications. The set of relations is inspired by more complex models used in qualitative spatial reasoning, such as the Region Connection Calculus [24], and in particular the Single Cross Calculus [25]. In the application context of processing route descriptions, the route graph [26] employs categories similar to the ones used here. The complexity of reasoning and the requirements to perception and environment understanding can make the application of these structures challenging in real-world robotics situations. For this reason, in this work we relax the constraints of logical consistency of the relations between objects. This enables the use of a probabilistic model with an approximate inference procedure.

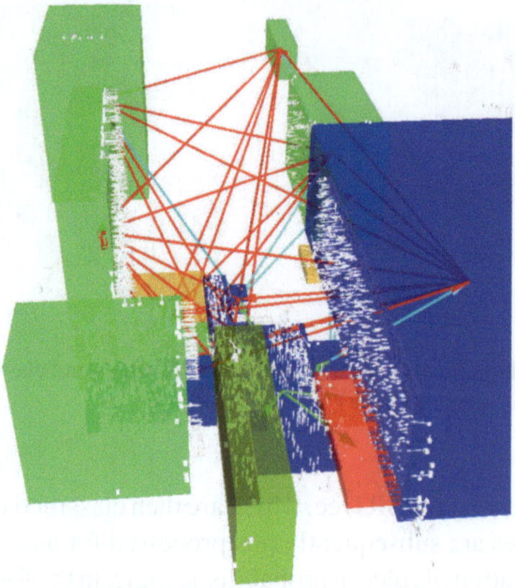

Fig. 4 Annotated spatial relations and bounding boxes of scene objects. *Red arrows* stand for *LeftOf* relationships, *blue* ones for *Behind*, and *green arrows* are *On* relationships

An approach related to the one presented here, which is also based on the specification of relations between objects in a logical language, has been proposed for a different set of relations in [27]. A methods to estimate support relationships between objects in indoor scenes is discussed in [28].

The approach presented here uses a basic set of features about the geometric relations of the objects in a scene as a first step. For every object in the scene, a bounding box is computed. The orientation of the bounding box is chosen such that it is aligned with the mode of the histogram of the normal vectors computed for each cuboid. The bounding boxes are shrunk along each of their axes to prune away a small percentile of points in order to obtain a tight fit. This step of computing features on bounding boxes assumes a scene that roughly adheres to a Manhattan model of the surroundings, which is a reasonable assumption for many objects and topologies encountered in an urban environment (e.g., houses and cars). A visualization of a scene with the point cloud as stored in the SRTree, the normal vectors and the bounding boxes, is shown in Fig. 4.

The features are computed for each pair of objects in a coordinate system that is aligned with the simulated viewpoint of a person describing the spatial relations present in the scene. The feature set consists of the distance between object centroids, as well as the maximum and minimum distances between the bounding boxes. These distances are also computed for projections of the bounding boxes along the axes of the coordinate system. So, for example the minimum vertical distance between the bounding boxes of two objects is a feature. For the projections of the objects along each of the coordinate axes, the percentage of points of an object that lie inside the convex hull of the projection of a second object is another feature. Further features are computed based on a cylindrical coordinate system. Here, the angles between the central axis of the field of view and the lateral boundary points as well as the centroid of the object. The differences between these coordinates for a pair of objects constitutes the set of angular features.

The features are then used for classification of the aforementioned set of spatial relations with a linear SVM classifier. The local estimates generated by this are combined in a MLN model, which is specified by a simple set of rules formulated in first-order logic. This way, consistency between the relations between objects can be increased. The knowledge base encoded by the MLN encompasses formulas describing the antisymmetry of support relations,

$$\begin{aligned} On(o_1, o_2) &\implies \neg On(o_2, o_1) \\ \mathit{LeftOf}(o_1, o_2) &\implies \neg \mathit{LeftOf}(o_2, o_1) \\ \mathit{Behind}(o_1, o_2) &\implies \neg \mathit{Behind}(o_2, o_1), \end{aligned}$$

their transitivity

$$\begin{aligned} On(o_1, o_2) \wedge On(o_2, o_3) &\implies On(o_1, o_3) \\ \mathit{LeftOf}(o_1, o_2) \wedge \mathit{LeftOf}(o_2, o_3) &\implies \mathit{LeftOf}(o_1, o_3) \\ \mathit{Behind}(o_1, o_2) \wedge \mathit{Behind}(o_2, o_3) &\implies \mathit{Behind}(o_1, o_3), \end{aligned}$$

and the exclusivity of the *On* relation

$$\begin{aligned} On(o_1, o_2) &\implies \neg \mathit{LeftOf}(o_2, o_1), & On(o_1, o_2) &\implies \neg \mathit{LeftOf}(o_1, o_2) \\ On(o_1, o_2) &\implies \neg \mathit{Behind}(o_2, o_1), & On(o_1, o_2) &\implies \neg \mathit{Behind}(o_1, o_2). \end{aligned}$$

The *RightOf* relation is defined with a hard rule complementary to the *LeftOf* relation according to

$$\mathit{LeftOf}(o_1, o_2) \iff \mathit{RightOf}(o_2, o_1),$$

and thus not labelled or reasoned over separately. An identical reasoning applies to the relationship between the predicates *Behind* and *InFrontOf*.

The information from the baseline classifier is entered into the model with formulas of the form $P_{SVM}(o_1, o_2) \iff P(o_1, o_2)$ for each predicate P, where the SVM output predicate is the binary decision of the classifier.

A Markov Logic Network built from this knowledge base defines a probability distribution over the application of the predicates in the knowledge base to all objects in the domain. An example for a *ground atom*, which is the result of this procedure, is the application of the predicate $On(o_1, o_2)$ in free logical variables o_1 and o_2 to the constants *car #1* and *street #3* to form the ground atom *On(car #1, street #3)*, which can take a truth value as its assignment. The probability distribution over the ground atoms $\mathbf{x}$ is defined by the sum of formulas satisfied by the current state of the variables, weighted by a weight w_i associated with each formula f_i, $i = 1, \ldots, N$ as

$$P(\mathbf{o}) = \frac{1}{Z} \sum_{i=1}^{N} w_i f_i(\mathbf{x}). \tag{2}$$

In this work, we are interested in an assignment to the variables that maximises this probability, the Maximum-a-priori (MAP) estimate. We formulate the MLN MAP inference problem as a discrete optimization problem in the binary random variables constituted by the ground atoms. This optimization problem can easily be formulated as a factor graph and solved approximately with Loopy Belief Propagation. The approximate MAP solution that is obtained in this way assigns a truth value to each of the ground predicates, and thus determines which pairs of objects are in a certain spatial relation. This information is represented in the *spatial relations graph*, which has a node for each object present in the scene, and a labelled edge for every pair of objects that has a *true* value for any relation in the MAP MLN solution. The edges are labelled with the predicate of the corresponding relations. Note that a pair oj objects can have multiple relations assigned to it, so an edge of the relation graph can have multiple labels. The spatial relations graph is part of the SRTree, where pairs of objects are annotated with the corresponding quantitative spatial information. Thus, the SRTree can be used to process queries like "List all objects that are to the left of a certain point on the sidewalk!".

3 3D Outdoor Urban Dataset

The dataset that is used for the experiments consists of 3D point cloud data enhanced by RGB image data using a Z+F 5010C laser range finder. The dataset consists of 62 scenes in downtown Munich, covering a total area of roughly 2 km^2. Images and point clouds for each scene have been manually segmented into objects and background, and objects are labelled with per-object class information as well spatial relations between objects. The nine classes used for the per-object annotation are *car*, *building*, *street*, *sidewalk*, *bicycle*, *other*, *tree*, *pole*, *sky*, *grass*. The segmentation of images and point clouds is based on an automatic segmentation [29] of the RGB images based on a graph based segmentation that was then manually corrected.

Additionally, the spatial relations *On*, *LeftOf* and *Behind* have been manually added for a half-space of each image, corresponding to the field of view of a person describing the scene.

The labelled data is used to learn the weights of the MLN and as a ground truth to evaluate the reasoning algorithm presented in this paper. This results in a 3D semantic dataset defining the object classes and additionally equipped with a relation graph between objects, including higher-order spatial relations.

Figure 4 shows an example scene with spatial relations represented by arrows connecting the objects.

4 Experiments

The integrated system, comprised of building the SRTree representation of dense point clouds along with its segmentation and classification, as well as the inference of spatial relations based on this data and their incorporation into the SRTree structure, is presented for the entire urban dataset.

Two different experiments are presented to outline the efficiency of the proposed approach. Firstly, the SRTree is evaluated for computational efficiency (insertion and access times) as well as the number of inner and leaf nodes required to represent the environment as a function of the grid cell size. Secondly, the approach at spatial reasoning is evaluated on the dataset described above.

4.1 Results

The evaluation of the SRtree is based on the normalized (per point) insertion and access times as well as the number of inner and leaf nodes required by the hierarchical data structure as the grid cell resolution is varied. The insertion time is defined as the time required to insert all the point cloud into the occupancy grid. The access time corresponds to the time required to access all the grid cells once all scans have been inserted. The sensor model $P(o_i|z_t)$ updates a grid cell by P_{occ} in case the beam is reflected within the volume whereas in case the beam traverses the grid cell an update (based on P_{free}) takes place if and only if a conflicting observation had been made during past observations (meaning a beam end point observation within this volume). Using such a sensor model the proposed approach avoids explicit modeling of a large amount of free space and updates only those regions of the grid that generate contradictory observations due to noise or dynamics. The values of P_{occ} and P_{free} have been set to 0.7 and 0.4 respectively. For all the results presented in this section the number of branches for all inner and leaf nodes in the Rtree datastructure is fixed to 16. All experiments were conducted on a Intel Core i5 3.2 GHz processor with 16 GB of RAM.

Figure 5a shows the normalized insertion time (per point) for a typical Z+F point-cloud scan containing around 20,000,000 points. It can be seen that the normalized insertion time is around 4 μs at 80 cm resolution grid, which is a reasonable grid cell size given the large size and scale of the point cloud (around 180 m). Thus, the SRtree occupancy grid takes a little more than a minute (at 80 cm resolution) to compute for a large point cloud. Figure 5b shows the time required to access all the grid cells of the occupancy grid. It can be seen that even at a high resolution (10 cm grid cell size), the SRtree can access the entire occupancy grid in 25 ms. Figure 5c, d shows the number of inner and leaf nodes present in the SRtree hierarchy. As the size of the grid cells is increased, the environment representation becomes coarser and the number of inner and leaf nodes reduce accordingly as can be seen in the figure.

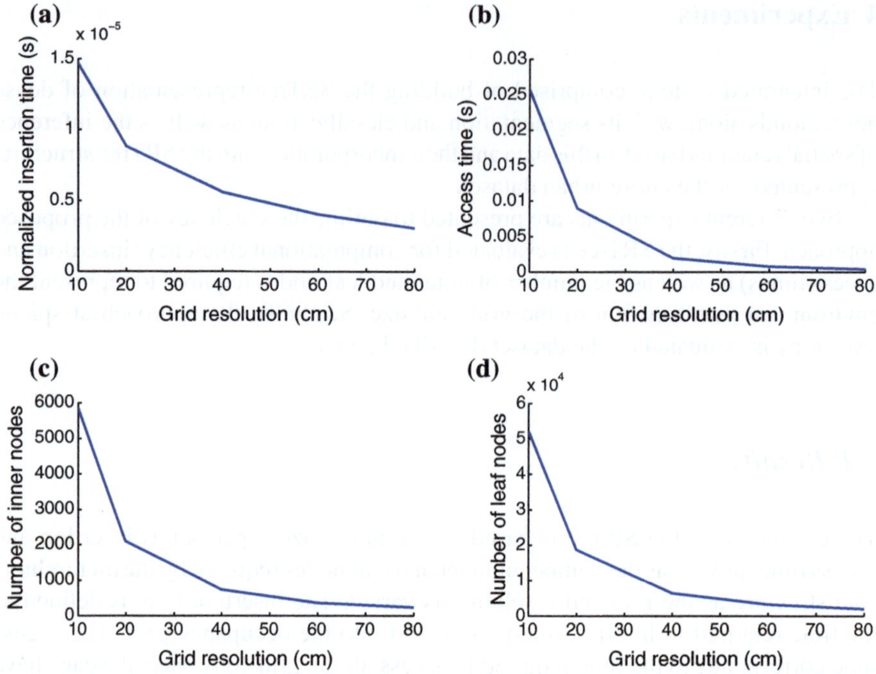

Fig. 5 Different aspects of the SRTree as a function of grid cell size. **a** Normalized insertion time of the SRtree occupancy grid. **b** Access time of all grid cells in the SRtree. **c** Number of inner nodes present in the hierarchy of the SRtree. **d** Number of leaf nodes present in the hierarchy of the SRtree

The MLN approach for spatial reasoning is evaluated on the dataset annotated with spatial relations as described above. The SVM and the MLN are trained on the same set of 48 labelled scenes; testing is performed on the remaining 14 scenes. For the SVM parameter learning, 5-fold cross validation is used. The formula weights of the spatial relations MLN are learned discriminatively using the Alchemy package.[1] Table 1 gives information metrics for the retrieval of spacial relations for all pairs of objects present in the test scenes, using the SVM formulation alone as well as the MLN model in addition to it. The main interest is in the correct identification of *true* values of relations, since these can be used for description of the environment. It can be seen that the SVM model slightly outperforms the MLN model on the *Behind* and *LeftOf* relations. The *On* relation however, which has the richest description in terms of rules in the MLN knowledge base, clearly profits from the added modelling effort in F_1 score.

[1]http://alchemy.cs.washington.edu/.

Table 1 Information retrieval metrics of the baseline SVM classifier and the added MLN inference for each of the three relations that were used

Relation	Model	Value	Precision	Recall	F_1-score
On	MLN	False	0.99	0.99	0.99
		True	0.57	0.57	0.57
	SVM	False	0.99	0.98	0.98
		True	0.45	0.60	0.51
Behind	MLN	False	0.95	0.93	0.94
		True	0.37	0.45	0.41
	SVM	False	0.94	0.95	0.95
		True	0.44	0.38	0.41
LeftOf	MLN	False	0.92	0.77	0.84
		True	0.56	0.82	0.66
	SVM	False	0.91	0.86	0.89
		True	0.65	0.75	0.69

5 Conclusion and Future Work

In this paper a novel environment representation titled SRTree is presented which generates a probabilistic 3D representation and captures the semantics of the environment. The environment representation provides the foundation for inference of higher-order spatial relationships between objects using a Markov Logic Network. The inferred spatial relations can furthermore be stored in the SRTree representation, which can useful in HRI interaction scenarios. The proposed framework is extensively evaluated on a large scale 3D dataset collected in downtown Munich and shows promising results.

Future work includes automatic segmentation of objects, taking advantage of the hierarchical nature of the SRTree data structure. The spatial relations model in the MLN framework can be extended to be aware of object classes to incorporate a more natural usage of these terms—e.g. an object *behind* a car might well actually be on the side of it with respect to the current position of the viewer, but the direction of the road and the car will still enable the use of the quantifier *behind*. Additionally, the approach can also be extended to arbitrary viewpoints.

Acknowledgments This work was supported in part by the ERC Advanced Grant project SHRINE (http://www.shrine-project.eu).

References

1. Gates, B.: A robot in every home. Sci. Am. **18**, 4–11 (2008)
2. Dryanovski, I., Morris, W., Xiao, J.: Multi-volume occupancy grids: an efficient probabilistic 3D mapping model for micro aerial vehicles. In: Proceedings of IEEE/RSJ International Conference Intelligent Robots and Systems, pp. 1553–1559 (2010)
3. Saarinen, J., Andreasson, H., Stoyanov, T., Ala-Luhtala, J., Lilienthal, A.J.: Normal distributions transform occupancy maps: application to large-scale online 3D mapping. In: Proceedings of IEEE International Conference on Robotics and Automation (2013)
4. Ryde, J., Corso, J.J.: Fast voxel maps with counting bloom filters. In: Proceedings of IEEE/RSJ International Conference on Intelligent Robots and Systems (2012)
5. Hornung, A., Wurm, K.M., Bennewitz, M., Stachniss, C., Burgard, W.: Octomap: an efficient probabilistic 3D mapping framework based on octrees. Auton. Robot. 1–18 (2013)
6. Pronobis, A., Jensfelt, P.: Large-scale semantic mapping and reasoning with heterogeneous modalities. In: Proceedings of IEEE International Conference on Robotics and Automation, pp. 3515–3522. IEEE (2012)
7. de Nijs, R., Ramos, S., Roig, G., Boix, X., van Gool, L., Kuhnlenz, K.: On-line semantic perception using uncertainty. In: Proceedings of IEEE International Conference Intelligent Robots and Systems, pp. 4185–4191. IEEE (2012)
8. Mitsou, N., de Nijs, R., Lenz, D., Frimberger, J., Wollherr, D., Kühnlenz, K., Tzafestas, C.: Online semantic mapping of urban environments. In: Spatial Cognition VIII, pp. 54–73. Springer (2012)
9. Limketkai, B., Liao, L., Fox, D.: Relational object maps for mobile robots. In: Proceedings of International Joint Conference on Artificial Intelligence, pp. 1471–1476 (2005)
10. Polastro, R., Corrêa, F., Cozman, F., Okamoto Jr., J.: Semantic mapping with a probabilistic description logic. Advances in Artificial Intelligence–SBIA 2010, pp. 62–71. Springer, Berlin (2011)
11. Anand, A., Koppula, H.S., Joachims, T., Saxena, A.: Contextually guided semantic labeling and search for three-dimensional point clouds. Int. J. Robot. Res. **32**(1), 19–34 (2013)
12. Tenorth, M., Kunze, L., Jain, D., Beetz, M.: KNOWROB-MAP—knowledge-linked semantic object maps. In: Proceedings of IEEE-RAS International Conference Humanoid Robots (Humanoids), pp. 430–435. IEEE (2010)
13. Pangercic, D., Pitzer, B., Tenorth, M., Beetz, M.: Semantic object maps for robotic housework-representation, acquisition and use. In: Proceedings of IEEE/RSJ International Conference Intelligent Robots and Systems, pp. 4644–4651. IEEE (2012)
14. Lemaignan, S., Ros, R., Sisbot, E.A., Alami, R., Beetz, M.: Grounding the interaction: anchoring situated discourse in everyday human-robot interaction. Int. J. Soc. Robot. **4**(2), 181–199 (2012)
15. Khan, S., Dometios, A., Verginis, C., Tzafestas, C., Wollherr, D., Buss, M.: Rmap: a rectangular cuboid approximation framework for 3D environment mapping. Auton. Robot. 1–17 (2014)
16. Landsiedel, C., de Nijs, R., Kühnlenz, K., Wollherr, D., Buss, M.: Route description interpretation on automatically labelled robot maps. In: Proceedings of the IEEE International Conference on Robotics and Automation, ICRA (2013)
17. Nanopoulos, A., Papadopoulos, A.N., Theodoridis, Y.: R-trees: Theory and Applications. Springer, London (2006)
18. Guttman, A.: R-trees: a dynamic index structure for spatial searching. Assoc. Comput. Mach. **14**(2) (1984)
19. Moravec, H., Elfes, A.: High resolution maps from wide angle sonar. In: Proceedings of IEEE International Conference Robotics and Automation, vol. 2, pp. 116–121 (1985)
20. Wurm, K., Hornung, A., Bennewitz, M., Stachniss, C., Burgard, W.: Octomap: A probabilistic, flexible, and compact 3D map representation for robotic systems. In: Proceedings of the ICRA workshop on best practice in 3D perception and modeling for mobile manipulation (2010)
21. Ronning, G.: Maximum likelihood estimation of Dirichlet distributions. J. Stat. Comput. Simul. **32**(4), 215–221 (1989)

22. Richardson, M., Domingos, P.: Markov logic networks. Mach. Learn. **62**(1), 107–136 (2006)
23. Moratz, R., Ragni, M.: Qualitative spatial reasoning about relative point position. J. Vis. Lang. Comput. **19**(1), 75–98 (2008)
24. Cohn, A.G., Bennett, B., Gooday, J., Gotts, N.M.: Qualitative spatial representation and reasoning with the region connection calculus. Geoinformatica **1**(3), 275–316 (1997)
25. Freksa, C.: Using orientation information for qualitative spatial reasoning. In: Frank, A., Campari, I., Formentini, U. (eds.) Theories and Methods of Spatio-Temporal Reasoning in Geographic Space. Lecture Notes in Computer Science, vol. 639, pp. 162–178. Springer, Berlin (1992)
26. Werner, S., Krieg-Brückner, B., Herrmann, T.: Modelling navigational knowledge by route graphs. Spatial Cognition II, pp. 295–316. Springer, Berlin (2000)
27. Sjöö, K., Pronobis, A., Jensfelt, P.: Functional topological relations for qualitative spatial representation. In: Proceedings of IEEE International Conference on Advanced Robotics, pp. 130–136. IEEE (2011)
28. Silberman, N., Hoiem, D., Kohli, P., Fergus, R.: Indoor segmentation and support inference from RGBD images. In: Proceedings of European Conference on Computer Vision, vol. 7576, pp. 746–760. Springer, Berlin (2012)
29. Felzenszwalb, P.F., Huttenlocher, D.P.: Efficient graph-based image segmentation. Int. J. Comput. Vis. **59**(2), 167–181 (2004)

On Planning and Task Achievement Modalities for Human-Robot Collaboration

Michelangelo Fiore, Aurélie Clodic and Rachid Alami

Abstract In this paper we present a robot supervision system designed to be able to execute collaborative tasks with humans in a flexible and robust way. Our system is designed to take into account the different preferences of the human partners, providing three operation modalities to interact with them. The robot is able to assume a leader role, planning and monitoring the execution of the task for itself and the human, to act as assistent of the human partner, following his orders, and also to adapt its plans to the human actions. We present several experiments that show that the robot can execute collaborative tasks with humans.

Keywords Human-robot interaction · Control architectures · Robot supervision

1 Introduction

In Human-Robot Interaction robots must be equipped with a complex set of skills, which allows them to reason about human agents intentions and statuses and to act accordingly. When interacting with a robot, different users will have different preferences. The robot must be able to take into account these preferences and provide different operation modalities to allow more natural interaction with the human. We have designed a system able to cooperate with humans to complete joint goals.

We present an object manipulation scenario, composed by tasks, such as fetching, giving objects and other collaborative operations, which are typical of domestic environments. In this scenario a human and a robot must cooperate to complete a

M. Fiore (✉) · A. Clodic · R. Alami
CNRS, LAAS, 7 Avenue du Colonel Roche, 31400 Toulouse, France
e-mail: mfiore@laas.fr

A. Clodic
e-mail: aclodic@laas.fr

R. Alami
e-mail: rachid.alami@laas.fr

M. Fiore
INSA, LAAS, Univ de Toulouse, 31400 Toulouse, France

M.A. Hsieh et al. (eds.), *Experimental Robotics*, Springer Tracts in Advanced Robotics 109, DOI 10.1007/978-3-319-23778-7_20

joint goal, which is a kind of goal that requires both agents to work together. This joint goal is known by both partners at the start of the scenario, perhaps because it has being agreed in a first interaction process. We have identified several ways—or modalities—to envisage how the robot planning and decisional abilities for task achievement can be used:

- Human Plans. The robot is not aware of the global plan or does not reason about the long-term global goal. The human decides when to ask the robot to perform individual tasks. The robot then acts by performing, in the context, the requested task. Decisional autonomy here is limited to how the robot refines and performs the task in the context. The robot is not completely passive when not executing an operation since it will continuously monitor and update the state of the environment.
- Robot plans. There is a joint goal between the human and the robot. The robot builds the 'best' plan to achieve this goal, taking into account the world status, the abilities of the two agents and the preferences of the human, then it verbalizes it and achieves it by doing its 'part of the job' and monitoring the human activity. This modality corresponds to a fully agreed upon plan that is built on-line or even predefined and known to both agents.
- Robot adapts. There is a joint goal between the human and the robot. The robot monitors what the human is doing and whenever possible, tries to achieve an action or a set of actions that advances the plan toward the goal. This could be seen as an intelligent robot reaction to the context including the possibility to have the robot proactively facilitating the action of the human whenever possible.

We think that the interaction process should be flexible, allowing eventually the robot to switch modality depending on the task status and of the human actions or simply upon request. To be able to interact with a human agent in a natural way, the robot needs to possess a number of different mechanisms, such as joint attention, action observation, task-sharing and action coordination. Using these mechanisms the robot can create shared plans, that take into account the capacities of both agents, and execute them in a coordinated way. We have developed an architecture and decisional components which allow us to run these different interaction modalities. The goal of the paper is not to provide a detailed description of any of these components but rather to provide a global view of the system and to show its flexibility.

2 Related Work

The topic of joint actions has been studied by different authors in the field of psychology. In [1] Bratman proposes a definition of the topic, giving several conditions deemed necessary to perform a joint action. Other researchers [17, 23] have studied a number of key mechanisms necessary to support joint actions between different partners: joint attention, action observation, task-sharing, action coordination and perception of agency.

Individual engaging in joint action use several ways, like language, gaze cues and gestures to establish a common ground. This mechanism appears to be crucial in successfully performing joint actions. Perspective taking, which is studied in psychology literature [5, 30], is a critical mechanism when interacting with people by allowing one to reason on others' understanding of the world in terms of visual perception, spatial descriptions, affordances and beliefs, etc. In [22] perspective taking is used to solve ambiguities in a human-robot interaction scenario by reasoning from the partner's point of view. In [29] perspective taking is integrated in a robotic system and applied to several kinds of problems.

Spatial reasoning [20], has been used for natural language processing for applications such as direction recognition [9, 15] or knowledge grounding [12]. Reference [26] presented a spatial reasoner integrated in a robot which computes symbolic positions of objects.

Understanding what other agents are doing is necessary to perform a joint action. In [7] the authors allow a robot to monitor users' actions by simulating their behaviors with the robot's motor, goal and perceptual levels. In [4] the authors present the HAMMER architecture, based on the idea of using inverse and forward models arranged in hierarchical and parallel manners. With this architecture the authors are able to use the same model to execute and recognize actions, an idea compatible with several biological evidences.

Other agents' actions can be predicted in different ways. By observing other agents we can predict the outcomes of their actions and understand what they're going to do next. Participants in a joint action form a shared representation of a task, used to predict other partners' actions. In [8] human intentions are estimated using a POMDP (Partially Observable Markov Decision Process) and a set of MDP (Markov Decision Process), that simulate human policies related to different intentions.

However, predicting other agents' actions is not enough to participate in a joint action, because the agent needs also to choose appropriate complementary actions, adapting them to the capability of the participants. The agents need to create a coordinated plan, giving particular care to timing information. In [19] the idea of cross-training is applied to shared-planning. A human and a robot iteratively switch roles to learn a shared plan for a collaborative task. This strategy is compared to standard reinforcement learning techniques, showing improvements in performances. These results support the idea of modeling practices for human teamwork in human robot interaction. In [24] a shared plan is executed using Chaski, a task-level executive which is used to adapt the robot's actions to the human partners. Plans can be executed in two different modalities: equal partners or leader and assistant. The authors show that this system reduces human idle time. In [10] an anticipatory temporal conditional random field is used to estimate possible users' actions, based on the calculation of object affordances and possible user trajectories. With this knowledge the robot can anticipate users' actions and react accordingly. In [2] the authors present a framework for human aware planning where the robot can observe and estimate human plans to coordinate its activites with those of the human partners. The framework doesn't support direct human-robot interaction.

Communication between partners in a joint task is crucial. The robot needs to be able to give information to its partners in a clear and socially acceptable way. Reference [27] proposes different strategies to modulate robot speech in order to produce more socially acceptable messages. In [13] the authors study how to generate socially appropriate deictic behaviors in a robot, balancing understandability and social appropriateness.

Few robotic architectures take humans into account to allow the execution of human-robot joint actions. Reference [28] presents ACT-R/E, a cognitive architecture, based on the ACT-R architecture, used for human robot interaction tasks. The architecture aims at simulating how humans think, perceive and act in the world. ACT-R/E has being tested in different scenarios, such as theory of mind and hide and seek, to show its capacity of modeling human behaviors and tought. In [6] the authors present HRI/OS, an agent-based system that allows humans and robots to work in teams. The system is able to produce and schedule tasks to different agents, based on their capacities, and allows the agents to interact mostly in a parallel and independent way, with loose coordination between them. Cooperation mainly takes place when one agent asks for help while dealing with a situation. In this case the HRI/OS will look for the best agent to help, based on their availability and capacities. In [3] the authors build SHARY, a supervision system for human robot interaction, tested in domestic environments to perform tasks such as serving a drink to a person. Our system is an evolution of Shary which includes new aspects, like spatial reasoning and modeling of joint actions.

3 Technical Approach

Our robot is controlled by an architecture composed of different components as shown in Fig. 1.

Supervision System. The component in charge of commanding the other components of the system in order to complete a task. After receiving a goal the supervision system will use the various planners of the system to obtain a list of actions for the robot and for the human. It will then be in charge of executing the robot actions and to monitor the human part of the plan. Specific treatment is reserved to joint actions (see Collaboration Planners). The supervision system aims at being flexible enough to be used in different robotic systems and robust so that it can recover from plan failures and adapt to human behaviors that are not expected, according to the current plan.

HATP. The Human-Aware Task Planner [11], based on a Hierarchical Task Network (HTN) refinement which performs an iterative task de-composition into sub-tasks until reaching atomic actions [18]. HATP is able to produce plans for the robot as well as for the other participants (humans or robots). By setting a different range of parameters the plans can be tuned to adapt the robot behavior to the desired level of cooperation. HATP is able to take into account the different beliefs of each

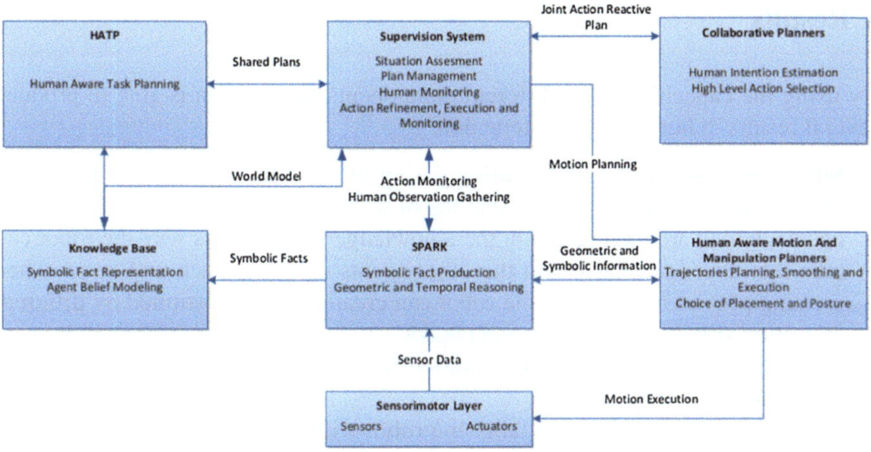

Fig. 1 System architecture

agents when producing a plan, eventually including actions that help creating joint attention [31].

Collaboration Planners. This set of planners are based on POMDP models used in joint actions, such as handovers, to estimate the user intentions and select an action to perform. In order to mantain a simple and flexible domain, the POMDP selects high level actions (like continue plan or wait for the user), which are adapted by the supervision system to the current situation. We can picture the interaction between HATP, the collaborative planners and the supervision system in the following way. HATP creates a plan composed by different actions to achieve a goal. The supervision system refines and executes each action in the plan, using the collaborative planners to adapt its actions to those of the other agents during a joint action.

SPARK. The Spatial Reasoning and Knowledge component, responsible for geometric information gathering [16]. SPARK embeds a number of decisional activities linked to abstraction (symbolic facts production) and inference based on geometric and temporal reasoning. SPARK maintains all geometric positions and configurations of agents, objects and furniture coming from perception and previous or a priori knowledge. SPARK computes perspective taking, allowing the system to reason on other agents' beliefs and capacities.

Knowledge Base. The facts produced by SPARK are stored in a central symbolic knowledge base. This base mantains a different model for each agent, allowing to represent divergent beliefs. For example, the fact representing the position of an object could point to a different location in two agent models in our knowledge base, representing the different information that the two agents possess.

A set of Human aware motion, placement and manipulation planners. These planners are in charge of choosing trajectories for the robot, taking into account the environment and the present agents [14, 21, 25].

4 Results

By using the capacities of its different components, our system is able to produce several results in human-robot interaction.

1. **Situation Assesment and Planning**. Using SPARK and its sensors, the robot is able to create different representations of the world for itself and for the other agents, which are then stored in the Knowledge Base. In this way the robot can take into account what it and the other agents can see, reach and know when creating plans. Using HATP the robot can create a plan constituted by different execution streams for every present agent.
   As said before, there are three operation modalities in the system: robot plans, user plans and robot adapts.
   Robot plans. In the first modality the robot will, using information present in the Knowledge Base and HATP, produce a plan to complete the joint goal. After that the robot will verbalize the plan to the user, explaining which actions will be performed by each agent and in which order. The robot will monitor the execution process, informing the human of which actions it's about to execute and also on when the human should execute its part of the plan. This modality, where the robot is the leader, can be helpful when interacting with naive users or in tasks where the robot has a better knowledge of the domain or of the environment than the other agents.
   Human plans. The human can also create plans, interacting with the robot by using a tablet application. This application allows the user to select different actions and parameters. The user can issue both high level goals (e.g. clean the table) and simpler actions (e.g. take the grey tape, give me the walle tape, stop your current action). The robot will simply observe the surroundings and wait for user inputs. This modality is always available and has a priority over the other two modalities. If the robot receives a command from the application while it is in another modality, it will abandon its current plan, stopping its actions at a safe point, and then execute the users' command. We feel that this interaction modality is important for two different reasons. First, some users will simply prefer to be in charge of the execution process, for a matter of personal preference or because they feel they have a deeper knowledge on how to realize the current task than the robot. We can picture, for example, industrial or medical scenarios, where the human is the leader and asks the robot to perform different tasks to help him, when needed. A second use of this modality is in situations where the robot doesn't have a clear estimation of the users' intentions and goals. For example, in a domestic environment, a user could decide to order a robot to bring him a drink, a need that the robot can't always anticipate.
   Robot adapts. In the last presented operation modality the robot will try to help the human to complete a task. At the start of the scenario, the robot will stand still and observe the environment. After the user takes an action the robot will

calculate a plan and try to help as it can, by performing actions related to that task and by giving helpful information to the user. In our implementation the robot will start this modality in a 'passive' role, simply observing the human until he takes the first action. We could also picture a more pro-active role for the robot, where the robot chooses a goal on its own and starts acting toward its completion, eventually asking for the help of the human when he can't complete an action.
This modality corresponds to what we feel is a very natural way of interaction between different partners, in particular in non-critical tasks, where defining an accurate plan between the partners is not fundamental. This situation relates quite easily to the scenario we will present in details in Sect. 5, where the two partners must clean a set of furnitures together. In this situation the two partners could simply choose to start executing the actions that they prefer, continuously adapting their plans to the other partners' actions.
The robot is able to switch from one modality to another during execution. For example, if the robot is in the 'robot plans' modality and the users' actions differ from the calculated plan the robot will interrupt its current action, create a new plan, and switch to the 'robot adapts' modality.

2. **Human Intention Estimation and Reactive Action Execution**. Using the Collaboration Planners we can execute joint actions in a reactive way. For example, in the case of an handover, a POMDP receives as input a set of observations, representing the distance between the human and the robot, the posture of the human arm and the user's orientation, used to estimate the current user intentions (i.e. user engaged, not engaged, not interested in the task). Using this information and the other variables that model the task status, the POMDP selects high level actions that are adapted by the supervisor. For example, the POMDP could decide to wait for an user that is not currently engaged in the joint action, but hasn't abandoned the task yet. The supervision system at this point will decide in which posture and for how long to wait for the user. If the user shows again interest in the task, the POMDP could select a 'continue plan' action, and the supervision system could choose to extend the robot arm as response.
3. **Human Action Monitor**. Using SPARK, the system is able to monitor human actions by creating Monitor Spheres associated to items considered interesting in a given context. A monitor sphere is a spheric area surrounding a point that can be associated to different events, like the hand of a human entering into it. Using this system and the vision capabilities of the robot we can monitor interesting activities, like the fact that a human takes an object or throws it into a trashbin. The monitor spheres for a human agent are created when he enters the scene, considering object affordances. If the human doesn't have items in his hands, the supervision system will use SPARK to create a monitor sphere associated to every pickable object. After a user takes an item, monitor spheres for pickable objects will be erased and the robot will create new spheres for containers, such as thrashbins, where the user can throw its items. For the moment we consider only these two kind of affordances, but we plan to include others in the future, allowing, for example, users to place objects on furnitures, such as tables.

4. **Robustness and Safety**. Our system incorporates different robustness and safety mechanisms. The robot can deal with failed actions by updating its knowledge of the environment and replanning accordingly. For example, if the robot tries to take an item and fails, it will update its knowledge introducing the information that the item is not reachable from the current position. The robot can then replan, for example by asking the user to take the item. The robot has the ability to stop its current action,,for example because of unexpected changes in the environment. The robot is also able to pause and resume the execution of an action, for example because the arm of the human is in its planned trajectory.
5. **Flexibility**. Our system is designed to be generic and easily expanded. New scenarios can be added by creating a new planning domain (for HATP and the Collaboration Planners), and eventually adding new actions to the system repertoire.

5 Experiments

For our experiments, we present a scenario where the robot and a human have a joint goal: cleaning a set of furniture. The two partners must place the tapes present on the furniture in a trashbin. We will use two pieces of furniture, identified as TABLE_4 and IKEA_SHELF and three tapes, identified as GREY_TAPE, LOTR_TAPE and WALLE_TAPE. We will use one trashbin, named PINK_TRASHBIN. We will place these items differently in each example, depending on our needs and won't necessary use all of them together.

We will present a set of runs of the system, which show its capacities[1]:

- **Robot adapts**: In this scenario (Fig. 2) the user is asked to clean the table, without agreeing before the start on a clear plan with the robot. The user is informed that the robot will try to help as it can. The user moves to the table and takes the WALLE_TAPE. At this point the robot notices that the user has completed an action and understands that he wants to clean the table.
  The robot creates a plan and executes its part of it while monitoring the human, which executes its part without deviating from the plan calculated by the robot.

- **Modality switch and user plans**: In this scenario (Fig. 3) the robot is the only agent able to reach both tapes, but it can't reach the trashbin, which can instead be reached by the human. We tested this scenario in two different runs. In the first one we start with the robot in 'robot plans' modality. After exploring the environment the robot produces a plan and starts its execution.
  While the robot is taking the LOTR_TAPE the human moves to take the WALLE_TAPE. This deviates from the robot plan, so it switches to the 'robot adapts' modality, communicating the change to the user. The user throws the WALLE_TAPE

[1] Videos from our experiments can be seen at http://homepages.laas.fr/mfiore/iser2014.html.

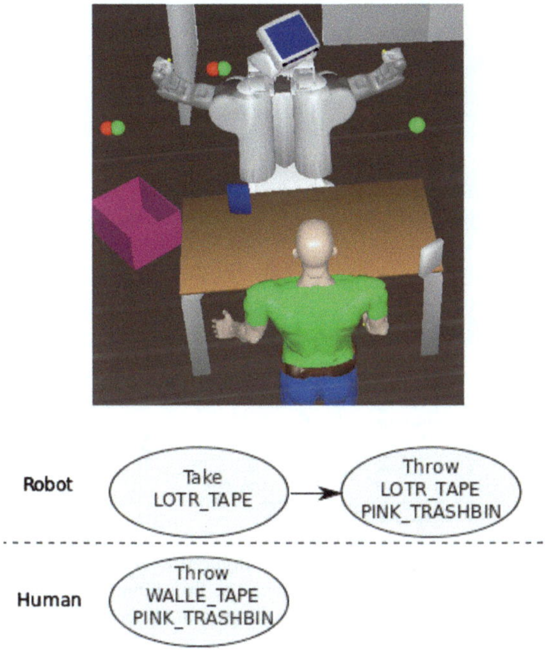

Fig. 2 Robot adapts.This figure shows the robot's representation of the scenario. The *white* tape is the WALLE_TAPE, while the *blue* one is the LOTR_TAPE. The *round shapes* represent the agents' rechabilities, with *red shapes* representing robot reachabilities and *green shapes* human reachabilities. In this case only the human can reach the WALLE_TAPE while both agents can reach the LOTR_TAPE and the PINK_TRASHBIN. After the human takes the WALLE_TAPE the robot produces a plan where the human must throw the tape in the thrashbin while the robot can take the LOTR_TAPE and throw it in the trashbin

in the PINK_TRASHBIN and meanwhile the robot takes the LOTR_TAPE and handles it to the user. The user takes the LOTR_TAPE and throws it in the PINK_TRASHBIN, completing the task.
In the second run the robot is in the 'user plans' mode. The user is asked to clean the table as he wishes. The user asks the robot to take each tape and give it to him, throwing them in the trashbin.

- **Replanning after failed action**: In this scenario (Fig. 4) the robot is the only agent able to reach the trashbin, while both agents can reach the two tapes. The robot is in 'robot plans' modality and, after examining the environment, produces a plan.
After taking and throwing the LOTR_TAPE, the robot tries to take the WALLE_TAPE, but fails because it's too far. The robot informs the user and replans. The agents execute the plan, completing the task.
- **Replanning after human inactivity**: In this run the robot computes that the GREY_TAPE and PINK_TRASHBIN are reachable only by the human, while the WALLE_TAPE is reachable only by the robot. The robot computes a plan and

Fig. 3 Modality switch and user plans. Another configuration of the environment, where the robot can reach the two tapes and the human can reach the thrashbin. The robot generates an initial plan from this situation. The block surrounding the give and receive actions means that they are considered a single joint action

starts executing it, observing the human reactions. After an initial stage when the human is commited to the task, he doesn't execute a part of the plan (taking the final tape and throwing it), so the robot looks for another plan. The only solution to the problem is the one already computed at the beginning, so the robot decides to ask the human to take the tape and throw it. A run of this scenario is shown in Fig. 5.

6 Conclusions

The studied experiment shows that our system is able to exhibit the capacities discussed in Sect. 3. Also, an interesting aspect of our system is that it's generic enough to be adapted to other manipulation scenarios, even involving more than one human. We review some of the main results of our experiments:

- **The system is able to handle joint goals**. The system is able to create shared plans with different users, taking into account the capabilities of each agent. When unexpected changes in the world or task status arise, the system is able to quickly replan, adapting to new scenarios. The system is able to execute this joint goal in a human aware way.

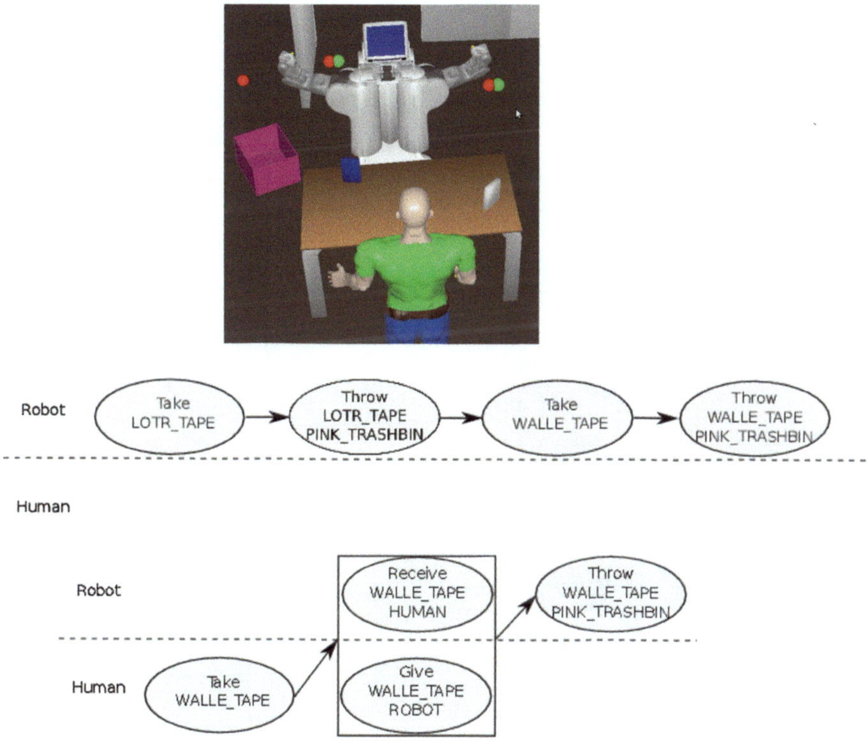

Fig. 4 Replanning after failed action. Here we can see a first plan, produced at the start of the scenario, and a second, produced after the robot fails to take the WALLE_TAPE

- **The system is able to handle joint actions**. The system is able to estimate user intentions in collaborative tasks and to choose appropriate actions, using a set of POMDP models.
- **The system is able to handle user preferences**. The system is able to adapt itself to user preferences, allowing the human partner to give commands or to be more passive in its role and switching from one modality to the other.
- **The system is able to handle each agent beliefs**. The system is able to represent different belief states for different agents and to take into accout what users can see, reach and know when creating a plan.
- **The system is able to monitor human actions**. The system is able to monitor human actions using a mechanism that is simple, but fast and efficient for the studied scenarios.

To further understand the advantages and disadvantages of these different modalities, and also in which conditions one or the others are pertinent, we need to conduct user studies, which will be done in the near future.

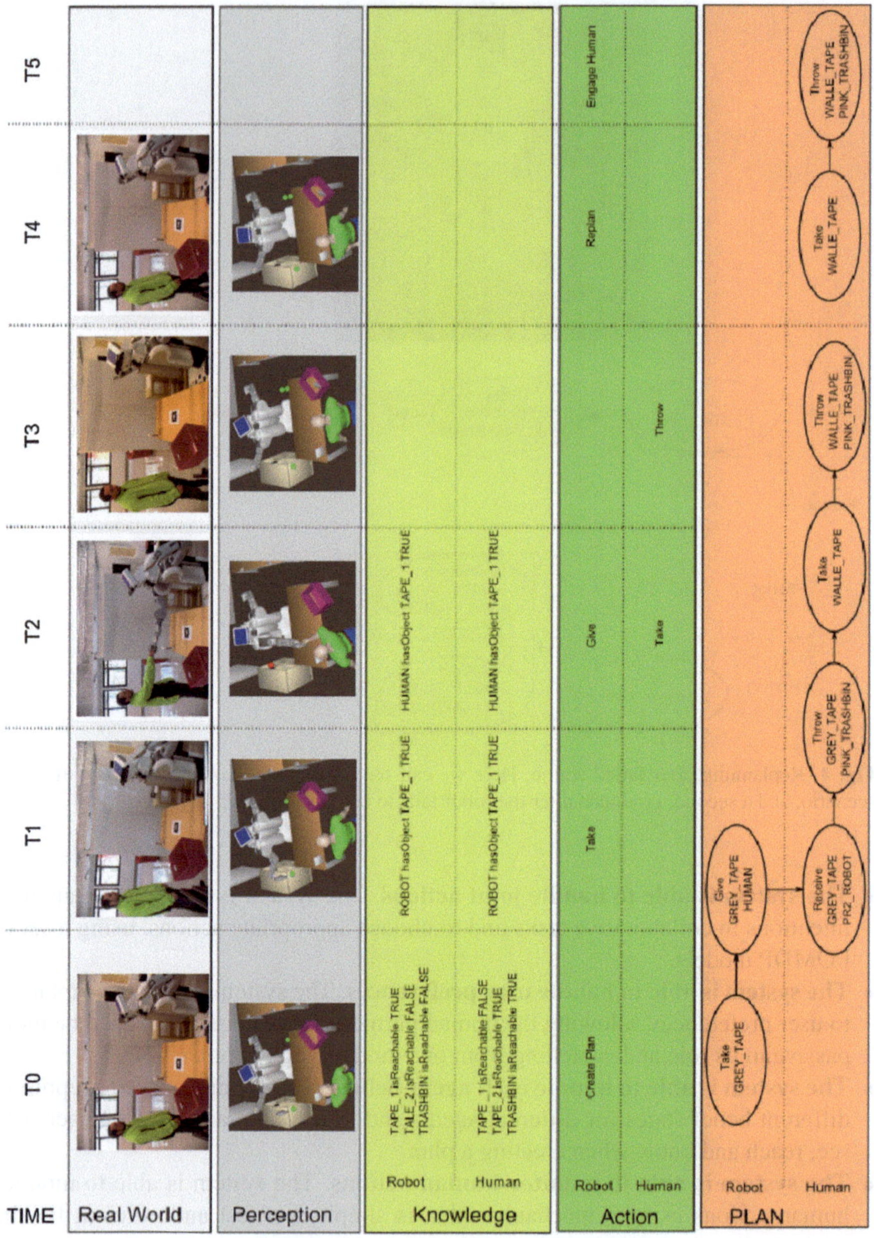

Fig. 5 The picture shows a run of our 'replanning after human inactivity scenario'. The different rows show, starting from top to bottom: the real world picture, the world state representation built by the robot, symbolic facts input in the knowledge base at each time step, action taken by each agent at each time step, the current plan calculated by the robot

Acknowledgments This work was conducted within the EU SAPHARI project (www.saphari.eu) funded by the E.C. division FP7-IST under contract ICT-287513.

References

1. Bratman, M.E.: Shared agency. In: Philosophy of the Social Sciences: Philosophical Theory and Scientific Practice, pp. 41–59 (2009)
2. Cirillo, M., Karlsson, L., Saffiotti, A.: A Framework for human-aware robot planning. In: Proceedings of the Scandinavian Conference on Artificial Intelligence (SCAI), Stockholm, SE (2008). http://www.aass.oru.se/~asaffio/
3. Clodic, A., Cao, H., Alili, S., Montreuil, V., Alami, R., Chatila, R.: Shary: a supervision system adapted to human-robot interaction. In: Experimental Robotics, pp. 229–238. Springer (2009)
4. Demiris, Y., Khadhouri, B.: Hierarchical attentive multiple models for execution and recognition of actions. Robot. Auton. Syst. **54**(5), 361–369 (2006)
5. Flavell, J.: Perspectives on Perspective Taking, pp. 107–139. L. Erlbaum Associates (1992)
6. Fong, T.W., Kunz, C., Hiatt, L., Bugajska, M.: The human-robot interaction operating system. In: 2006 Human-Robot Interaction Conference. ACM, March 2006
7. Gray, J., Breazeal, C., Berlin, M., Brooks, A., Lieberman, J.: Action parsing and goal inference using self as simulator. In: IEEE International Workshop on Robot and Human Interactive Communication, 2005. ROMAN 2005, pp. 202–209. IEEE (2005)
8. Karami, A.-B., Jeanpierre, L., Mouaddib, A.-I.: Human-robot collaboration for a shared mission. In: 5th ACM/IEEE International Conference on Human-Robot Interaction (HRI), 2010, pp. 155–156. IEEE (2010)
9. Kollar, T., Tellex, S., Roy, D., Roy, N.: Toward understanding natural language directions. In: HRI, pp. 259–266 (2010)
10. Koppula, H.S., Saxena, A.: Anticipating human activities using object affordances for reactive robotic response. In: Robotics: Science and Systems, Berlin (2013)
11. Lallement, R., de Silva, L., Alami, R.: HATP: an HTN planner for robotics. In: 2nd ICAPS Workshop on Planning and Robotics, PlanRob 2014 (2014)
12. Lemaignan, S., Sisbot, A., Alami, R.: Anchoring interaction through symbolic knowledge. In: Proceedings of the 2011 Human-Robot Interaction Pioneers Workshop (2011)
13. Liu, P., Glas, D.F., Kanda, T., Ishiguro, H., Hagita. N.: It's not polite to point: generating socially-appropriate deictic behaviors towards people. In: Proceedings of the 8th ACM/IEEE International Conference on Human-Robot Interaction, pp. 267–274. IEEE Press (2013)
14. Mainprice, J., Sisbot, E.A., Jaillet, L., Cortes, J., Alami, R., Simeon, T.: Planning human-aware motions using a sampling-based costmap planner. In: IEEE International Conference on Robotics and Automation (2011)
15. Matuszek, C., Fox, D., Koscher, K.: Following directions using statistical machine translation. In: International Conference on Human-Robot Interaction, ACM Press (2010)
16. Milliez, G., Warnier, M., Clodic, A., Alami, R.: A framework for endowing interactive robot with reasoning capabilities about perspective-taking and belief management. In: Proceedings of the 23rd IEEE International Symposium on Robot and Human Interactive Communication (2014)
17. Mutlu, B., Terrell, A., Huang, C.-M.: Coordination mechanisms in human-robot collaboration. In: ACM/IEEE International Conference on Human-Robot Interaction (HRI)-Workshop on Collaborative Manipulation, pp. 1–6 (2013)
18. Nau, D., Au, T.C., Ilghami, O., Kuter, U., Murdock, J.W., Wu, D, Yaman, F: SHOP2: an HTN planning system. J. Artif. Intell. Res. 379–404 (2003)
19. Nikolaidis, S., Shah, J.: Human-robot cross-training: computational formulation, modeling and evaluation of a human team training strategy. In: Proceedings of the 8th ACM/IEEE International Conference on Human-Robot Interaction, pp. 33–40. IEEE Press (2013)

20. O'Keefe, J.: The Spatial Prepositions. MIT Press, Cambridge (1999)
21. Pandey, A.K., Alami, R.: Mightability maps: a perceptual level decisional framework for co-operative and competitive human-robot interaction. In: IEEE/RSJ International Conference on Intelligent Robots and Systems (2010)
22. Ros, R., Lemaignan, S., Sisbot, E.A., Alami, R., Steinwender, J., Hamann, K., Warneken, F.: Which one? grounding the referent based on efficient human-robot interaction. In: 19th IEEE International Symposium in Robot and Human Interactive Communication (2010)
23. Sebanz, N., Bekkering, H., Knoblich, G.: Joint action: bodies and minds moving together. Trends Cogn. Sci. **10**(2), 70–76 (2006)
24. Shah, J., Wiken, J., Williams, B., Breazeal, C.: Improved human-robot team performance using chaski, a human-inspired plan execution system. In: Proceedings of the 6th International Conference on Human-Robot Interaction, pp. 29–36. ACM (2011)
25. Sisbot, E.A., Clodic, A, Alami, R., Ransan, M.: Supervision and Motion Planning for a Mobile Manipulator Interacting with Humans (2008)
26. Skubic, M., Perzanowski, D., Blisard, S., Schultz, A., Adams, W., Bugajska, M., Brock, D.: Spatial language for human-robot dialogs. IEEE Trans. Syst. Man Cybern. Part C: Appl. Rev. **34**(2), 154–167 (2004)
27. Torrey, C., Fussell, S.R., Kiesler, S.: How a robot should give advice. In: 8th ACM/IEEE International Conference on Human-Robot Interaction (HRI), 2013, pp. 275–282. IEEE (2013)
28. Trafton, G., Hiatt, L., Harrison, A., Tamborello, F., Khemlani, S., Schultz, A.: ACT-R/E: an embodied cognitive architecture for human-robot interaction. J. Hum. Rob. Interact. **2**(1), 30–55 (2013)
29. Trafton, J., Cassimatis, N., Bugajska, M., Brock, D., Mintz, F., Schultz, A.: Enabling effective human-robot interaction using perspective-taking in robots. IEEE Trans. Syst. Man Cybern. Part A, 460–470 (2005)
30. Tversky, B., Lee, P., Mainwaring, S.: Why do speakers mix perspectives? Spat. Cogn. Comput. **1**(4), 399–412 (1999)
31. Warnier, M., Guitton, J., Lemaignan, S., Alami, R.: When the robot puts itself in your shoes. Managing and exploiting human and robot beliefs. In: Proceedings of the 21th IEEE International Symposium in Robot and Human Interactive Communication (2012)

Part VI
Mapping and Localization

Gabe Sibley
University of Colorado, Boulder

The 2014 ISER section on localization and mapping presented five papers demonstrating robust real-world experiments for both perception and action. The breadth and depth of the proposed algorithms and systems showed capabilities that speak to the increasing maturity of the field.

In the first paper, Nima Keivan and Gabe Sibley from George Washington University presented work on visual-inertial simultaneous localization and mapping. Thanks to an adaptive and synchronous multi-threaded optimization strategy the work showed estimation with robust data-association even during degenerate and unobservable scenarios. The authors also showed how such capability leads directly to 'power-on-and-go' self-calibration for inertial monocular camera systems.

Next, Eric Nelson, Vadim Indelman, Nathan Michael and Frank Dellaert from Carnegie Mellon University and Georgia Tech presented their paper titled, "An Experimental Study of Robust Distributed Multi-Robot Data Association from Arbitrary Poses". The authors experimentally investigate the problem of computing the relative transformation between multiple vehicles from inter robot observations. They consider an EM-based methodology which evaluates sensory observations gathered over vehicle trajectories to establish robust relative pose transformations between robots. This work demonstrated robust multi-robot localization and mapping from quad-rotors equipped with 2D laser range finders, leading to convincing and useful 2D laser maps.

In the third paper, Daniele Nardi from Sapienza University of Rome presented, "Interactive Semantic Mapping: Experimental Evaluation", by Guglielmo Gemignani, Daniele Nardi, Domenico Daniele Bloisi, Roberto Capobianco Luca Iocchi. Prof. Nardi's presentation demonstrated robust interactive semantic mapping allowing for online model learning and object recognition. In this work, object models are learned after mutual human-robot data-association is established using laser pointers. Extensive experiments highlighted the effectiveness of the proposed method.

Next, Andrew Mittleider from University of Nebraska present work on Wirelessly charging sensor networks from UAV's. In his paper, "Experimental Analysis of a UAV-Based Wireless Power Transfer Localization System" with co-authors Brent Griffin and Carrick Detweiler, Andrew showed a robust magnetic localization system allowing quad-rotor to charge remote sensor network nodes robustly. Powering remote sensor networks over long periods of time is challenging. This line of research promises to allow long-term operation of sensor nodes embedded deep with structures which could otherwise not be charged.

In the final paper Felix Duvallet argued convincingly for how natural language can be used as an additional sensor for simultaneous localization and mapping. In this work the authors show how natural language directions, such as "move to the cone behind the barrier", can be used to infer plans that take into account the spatial information present in the language itself. The paper titled, "Inferring Maps and Behaviors from Natural Language Instructions" with co-authors Matt Walter, Thomas Howard, Sachithra Hemachandra, Jean Oh, Seth Teller, Nicholas Roy and Anthony Stentz uses natural language in a novel way for robot mapping and planning. With modern advances in symbol grounding the authors are able to leverage information implicit in linguistic instruction, thereby treating language as a new type of sensor for mapping, localization and planning.

Asynchronous Adaptive Conditioning for Visual-Inertial SLAM

Nima Keivan, Alonso Patron-Perez and Gabe Sibley

Abstract This paper is concerned with real-time monocular visual inertial simultaneous localization and mapping (VI-SLAM). In particular a tightly coupled nonlinear-optimization based solution that can match the global optimal result in real time is proposed. The methodology is motivated by the requirement to produce a scale-correct visual map, in an optimization framework that is able to incorporate relocalization and loop closure constraints. Special attention is paid to achieve robustness to many real world difficulties, including degenerate motions and unobservablity. A variety of helpful techniques are used, including: a relative manifold representation, a minimal-state inverse depth parameterization, and robust non-metric initialization and tracking. Importantly, to enable real-time operation and robustness, a novel numerical dog leg solver [16] is presented that employs multithreaded, asynchronous, adaptive conditioning. In this approach, the conditioning edges of the SLAM graph are adaptively identified and solved for both synchronously and asynchronously. In this way some threads focus on a small number of temporally immediate parameters and hence constitute a natural "front-end"; other threads adaptively focus on larger portions of the SLAM problem, and hence are able to capture functional constraints that are only observable over long periods of time—an ability which is useful for self-calibration, during degenerate motions, or when bias and gravity are poorly observed. Experiments with real and simulated data for both indoor and outdoor robots demonstrate that asynchronous adaptive conditioning is able to closely track the full-SLAM maximum likelihood solution in real-time, even during challenging non-observable and degenerate cases.

N. Keivan (✉) · A. Patron-Perez · G. Sibley
Autonomous Robotics and Perception Group, The George Washington University, Washington, USA
e-mail: nimski@gwu.edu, nima.keivan@colorado.edu

A. Patron-Perez
e-mail: apatron@gwu.edu

G. Sibley
e-mail: gsibley@gwu.edu

M.A. Hsieh et al. (eds.), *Experimental Robotics*, Springer Tracts in Advanced Robotics 109, DOI 10.1007/978-3-319-23778-7_21

1 Introduction

It is well known that the batch bundle-adjustment solution to monocular SLAM is the gold standard, in that it's form defines the Cramer-Rao lower bound and that it takes advantage of all measurements over all time to compute the maximum likelihood parameter estimate [1, 21]. Visual-inertial bundle adjustment is significantly more challenging than vision-only BA [7]. Vision-only monocular systems suffer from a well-studied scale ambiguity. Adding an IMU can makes scale observable, however inertial measurements complicate matters when it comes to computing the global MLE solution incrementally in real-time.

For bundle adjustment to be real-time for use on robots, a *local* approach is typically employed [12]. With an IMU this is difficult since the local adjustment region may need to be very large in order to ensure observability of certain parameters. Indeed, under certain degenerate motions such as constant velocity forward motion, some parameters may never be observable (though this rarely if ever happens in practice) [4, 5].

An alternative to local-bundle adjustment is to only keep a sliding window of the most recent poses and landmarks active, and marginalize the rest into a prior distribution [13, 18, 20]. This is equivalent to a fixed-lag Kalman smoother [2, 10] and recently such systems have shown remarkable results [3, 8, 9].

Marginalization into a prior distribution like this is predominantly employed for computational efficiency—if it were possible to compute the full MLE solution in real-time it would be preferable. Marginalization is also costly because it introduces conditional dependencies between the remaining parameters causing "fill-in". Fill-in can be addressed by cutting feature tracks and carefully marginalizing poses and landmarks simultaneously [14]. Marginalization is also potentially dangerous because it bakes in linearization errors which can lead to over-confident estimates or divergence unless one is careful to maintain consistency [3]. Carrying prior distributions induced from marginalization also necessitates an expensive global optimization at loop-closure to obtain the correct marginal. This paper attempts to remedy these issues by avoiding marginalization altogether.

Instead of relying on marginalization we take advantage of conditioning, which has shown surprisingly robust and accurate results in the computer vision community [1, 6] and avoids locking in incorrect parameter estimates when used adaptively [19]. Using a relative manifold is also important because optimal relative transformation estimates in SE3 are by definition near zero. This fact allows multiple threads to asynchronously optimize and update different overlapping subsets of the full problem without detriment.

Adaptive asynchronous conditioning has other benefits: it can (a) perform robust initialization even under degenerate motions, (b) allow constant-time loop closure without expensive loop-long re-linearization, (c) operate even during poor observability conditions, (e) avoid inconsistency associated with early marginalization and re-linearization, (f) track the relative-space maximum likelihood solution in constant time, (d) enable power-on-and-go self-calibration. We find that adaptive asynchronous conditioning closely tracks the global batch optimal solution, at a fraction of the computational cost, which enables real-time operation.

2 Sliding Window Optimization

2.1 Fixed Window Formulation

The state vector of the parameters inside the sliding window is defined as

$$\mathbf{x} = [\{\mathbf{T}_{wv}\mathbf{v}_w\mathbf{b}\}\,\{\rho_k\}]^T \tag{1}$$

where $\{\mathbf{T}_{wv}\mathbf{v}_w\mathbf{b}\}$ is the set of pose parameters: $\mathbf{T}_{wv} \in SE3$ is the transformation from vehicle to world coordinates, $\mathbf{v}_w \in \mathbb{R}^3$ is the velocity vector in world coordinates and $\mathbf{b} \in \mathbb{R}^6$ is the IMU bias vector for the gyroscope and accelerometer. As mentioned previously, *world* coordinates here refers to the lifted local coordinates on which the optimization takes place. Similarly $\{\rho_k\}$ is the set of 1-d inverse-depth parameters for each landmark [15]. Since landmarks are parameterized in inverse depth, each must be back projected from its reference frame before being projected into the measurement frame. The reprojection error of the kth landmark with reference frame j into the ith frame is defined by (Fig. 1):

$$\mathbf{r}_{\mathcal{P}_{ik}} = \mathbf{z}_{ik} - \pi(\mathbf{T}_{vs}^{-1}\mathbf{T}_{wv_i}^{-1}\mathbf{T}_{wv_j}\mathbf{T}_{vs}\mathbf{X}_{s_{jk}}) \tag{2}$$

where $\mathbf{z}_{ik}$ is the measurement in image coordinates, $\mathbf{X}_{s_{jk}} = [u\ v\ 1\ \rho]$ is the landmark inverse depth parameterization in the sensor frame obtained from back-projection, $\mathbf{T}_{vs}$ is the trasnformation from the sensor to the vehicle frame, and π is the non-linear projection function. Inertial constraints are formed between subsequent states by integrating IMU measurements. The constraint between two subsequent states i

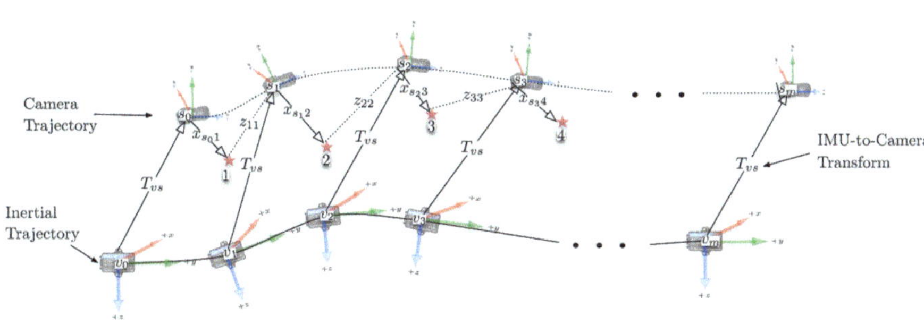

Fig. 1 Representation of the visual-inertial system

and j with poses $\mathbf{T}_{wv_i}$ and $\mathbf{T}_{wv_j}$ and velocities $\mathbf{v}_i$ and $\mathbf{v}_j$ is defined as $\mathbf{r}_{\mathcal{I}_{ij}} \in \mathbb{R}^{15}$ and is formulated as

$$\mathbf{r}_{\mathcal{I}_{ij}} = \begin{bmatrix} \log\left(\left(\mathbf{T}_{wv_i}\right)\mathbf{T}'_{v_iv_j}\left(\mathbf{T}_{wv_j}\right)^{-1}\right) \\ \mathbf{v}_i + \mathbf{v}'_{ij} - \mathbf{v}_j \\ \mathbf{b}_j - \mathbf{b}_i \end{bmatrix} \tag{3}$$

where $\mathbf{T}'_{v_iv_j}$ and $\mathbf{v}'_{ij}$ are the transformation and velocity deltas obtained by integrating the IMU measurements, and $\log(\cdot) \in \mathbb{R}^6$ is the SE3 logarithm function (with subsequent representation in minimal coordinates) as applied to an error state transformation. The residual on the gyroscope and accelerometer biases is derived from modeling them as random walk processes. The transformation delta due to the IMU integration is defined as

$$\mathbf{T}'_{v_iv_j} = \begin{bmatrix} \mathbf{R}'_{v_iv_j} & \mathbf{t}'_{ij} \\ 0 & 1 \end{bmatrix} \tag{4}$$

where $\mathbf{R}'_{v_iv_j} \in \mathbb{R}^{3\times3}$ is the rotation delta which is a function of the angular velocity measurements $\{\omega \in \mathbb{R}^3\}$ and the gyroscope biases $\mathbf{b}_g \in \mathbb{R}^3$. $\mathbf{t}'_{ij} \in \mathbb{R}^3$ is the translation delta which is a function of the acceleration measurements $\{\mathbf{a} \in \mathbb{R}^3\}$, the accelerometer biases $\mathbf{b}_a \in \mathbb{R}^3$, the gravity vector $\mathbf{g} \in \mathbb{R}^3$ and the initial velocity $\mathbf{v}_i \in \mathbb{R}^3$.

$\mathbf{R}'_{v_iv_j}$ is obtained by first integrating angular velocities in the world frame and then transforming the result to be relative to the starting orientation:

$$\mathbf{R}'_{v_iv_j} = \mathbf{R}^{-1}_{wv_i}\mathbf{R}'_{wv_j}$$

where $\mathbf{R}'_{wv_j}$ is the result of the discrete integration of angular velocities $\{\omega\}$ in the world frame, and also depends on the gyroscope biases $\mathbf{b}_g$. Each integration step is formulated as

$$\mathbf{R}'_{wt_{n+1}} = \exp\left(\mathbf{R}'_{wt_n}\left[\omega + \mathbf{b}_g\right]dt\right)\mathbf{R}'_{wt_n} \tag{5}$$

where $\mathbf{R}'_{wt_n}$ is the rotation matrix from world coordinates to the coordinate frame resulting from the integration up to time t_n, $\omega \in \mathbb{R}^3$ is the angular velocity vector obtained by the gyroscope at time t_n and $\mathbf{b}_g \in \mathbb{R}^3$ is the gyroscope bias vector. The angular velocity measurement, taken in the body frame represents a rotation in the tangent space of $\mathbf{R}'_{wt_n}$, however since the rotation is integrated in world coordinates, the angular velocities must be transformed from the body frame to the world frame. This transformation is undertaken by the adjunct, which for SO3 is simply a multiplication by the rotation $\mathbf{R}'_{wt_n}$. Once in the world frame, the angular velocities are integrated and a rotation delta is obtained via the SO3 exponential exp. Note that the transformation from/to minimal coordinates in exp has been omitted for brevity.

The translation vector $\mathbf{t}'_{ij} \in \mathbb{R}^3$ is obtained by integrating the body accelerations in the world frame, and removing the translation of the initial frame $\mathbf{t}_i$ as follows:

$$\mathbf{t}'_{ij} = \mathbf{t}'_{wj} - \mathbf{t}_{wi}$$

The discrete integration step for $\mathbf{t}'_{wj}$ is formulated as

$$\mathbf{t}'_{wt_{n+1}} = \mathbf{t}'_{wt_n} + \int_{t_n}^{t_{n+1}} \mathbf{v}_{t_n} dt \tag{6}$$

$$\mathbf{v}'_{t_n} = \mathbf{v}'_{t_{n-1}} + \int_{t_n}^{t_{n+1}} \left(\mathbf{R}'_{wt_n} [\mathbf{a} + \mathbf{b}_a] - \mathbf{g}\right) dt \tag{7}$$

where $\mathbf{v}'_{t_n} \in \mathbb{R}^3$ is the velocity integrated up to time t_n, $\mathbf{a} \in \mathbb{R}^3$ is the vector of accelerations measured in the body frame and $\mathbf{b}_a \in \mathbb{R}^3$ is the accelerometer bias vector. As the accelerometer measurements are integrated in the world frame, the measurements and biases in the body frame at time t_n must be transformed into the world frame, which is accomplished by multiplying by the body orientation at time t_n, $\mathbf{R}'_{wt_n}$.

Although the aforementioned derivation uses euler integration for simplication, all integrations, including the integration of angular velocities via the SO3 exponential exp are undertaken via fourth order Runge–Kutta. Considering (5) through (7), it can be observed that the rotation delta $\mathbf{R}'_{v_i v_j}$ is independent of the translation delta $\mathbf{t}'_{ij}$, and also that the contribution of the starting velocity $\mathbf{v}_i$ and the gravity vector $\mathbf{g}$ can be factored out of the translation delta $\mathbf{t}'_{ij}$ as follows

$$\mathbf{t}'_{ij} = \Delta t \mathbf{v}_i + \frac{1}{2} \Delta t^2 \mathbf{g} + \mathbf{t}^*{}_{ij}$$

where $\mathbf{t}^*{}_{ij}$ is integrated as in Eqs. 6 and 7, but with the starting velocity ($\mathbf{v}_i$) and gravity vector $\mathbf{g}$ set to zero, and Δt is the entire duration over which $\mathbf{t}^*{}_{ij}$ is integrated. Equation 4 can then be rewritten as

$$\mathbf{T}'_{v_i v_j} = \begin{bmatrix} \mathbf{I}_{3\times3} & \Delta t \mathbf{v}_i + \frac{1}{2} \Delta t^2 g \\ 0_{1\times3} & 1 \end{bmatrix} \begin{bmatrix} \mathbf{R}'_{v_i v_j} & \mathbf{t}^*{}_{ij} \\ 0_{1\times3} & 1 \end{bmatrix} \tag{8}$$

A similar factorization can be performed for $\mathbf{v}'_{ij}$ to separate out terms that depend on the initial rotation and translation. The factorization of $\mathbf{T}'_{v_i v_j}$ and $\mathbf{v}'_{ij}$ is undertaken in order to simplify the partial derivatives $\partial \mathbf{r}_{\mathcal{I}_{ij}} / \partial \mathbf{T}_{wv_i}$, $\partial \mathbf{r}_{\mathcal{I}_{ij}} / \partial \mathbf{T}_{wv_j}$, and $\partial \mathbf{r}_{\mathcal{I}_{ij}} / \partial v_i$ which are needed for the optimization. These would normally need to be propagated through the integration of inertial measurements via the chain rule. However, due to the factorization, these derivatives can be taken over the entire constraint by first integrating the inertial measurements as per (8) and then taking the aforementioned derivatives of (3). This both simplifies the process of calculating these derivatives, as well as avoidining the loss of accuracy due to the propagation of the derivatives through the Runge–Kutta integration. Unfortunately no such factorization can be made for $\partial \mathbf{r}_{\mathcal{I}_{ij}} / \partial \mathbf{b}$ which must be propagated through the Runge–Kutta integration via the chain rule.

2.2 Optimization Formulation

The cost function minimzed in the optimization consists of the aforementioned inertial and visual residuals and is formulated as

$$e = \sum_{i=1}^{n} \sum_{k=1}^{m} \|\mathbf{r}_{\mathcal{P}_{ik}}\|^2_{\Sigma_{P_{ik}}} + \sum_{k=1}^{n} \|\mathbf{r}_{\mathcal{I}_{ik}}\|^2_{\Sigma_{\mathcal{I}_k}} \tag{9}$$

where the notation $\|\mathbf{x}\|^2_{\Sigma}$ signifies the Mahalanobis distance given the measurement uncertainty Σ. In all cases residual uncertainties are calculated via Gaussian error-propagation from raw measurement uncertainties. In the case of visual measurements, a standard covariance of 1 pixel is used for both x and y image directions. For inertial measurements, the covariance of the final measurement must be propagated through the integration given the uncertainties in the accelerometer and gyroscope measurements supplied by the manufacturer. Since inertial measurements are integrated via the Runge–Kutta algorithm for accuracy, the uncertainties must also be propagated through each integration step as follows:

$$\mathbf{W}_{\mathcal{I}} = (\Sigma_{\mathcal{I}})^{-1} = \left(\frac{\partial \mathbf{r}_{\mathcal{I}}}{\partial \mathbf{x}_t} \left(\prod_{t=t_0}^{t_f} \frac{\partial \mathbf{x}_t}{\partial \mathbf{z}_t} \mathbf{C}_{\mathbf{z}_t} \frac{\partial \mathbf{x}_t}{\partial \mathbf{z}_t}^T \right) \frac{\partial \mathbf{r}_{\mathcal{I}}}{\partial \mathbf{x}_t}^T \right)^{-1} \tag{10}$$

where the weight for the particular residual $\mathbf{r}_{\mathcal{I}}$ is given by $\mathbf{W}_{\mathcal{I}} = (\Sigma_{\mathcal{I}})^{-1}$. To obtain the measurement covariance $\Sigma_{\mathcal{I}}$, the covariance of each inertial measurement at time t denoted by $\mathbf{C}_{\mathbf{z}_t}$ is propagated through the single-step state integration jacobian $\frac{\partial \mathbf{x}_t}{\partial \mathbf{z}_t}$. This step is then repeated for each inertial measurement and propagated via the chain rule to obtain the covariance for the final integration state. In order to obtain the covariance for the residual, the final integration state covariance is propagated through the residual jacobian $\frac{\partial \mathbf{r}_{\mathcal{I}}}{\partial \mathbf{x}_t}$, which is trivial with the exception of the derivatives for the SE3 logarithm.

Landmarks which are observed in the active window of frames but whose reference frame falls outside of it provide conditioning edges during the optimization. This is done by considering their reprojection error in poses outside of the active window where they were observed. A single IMU conditioning edge corresponding to the IMU residual between the last frame in the sliding window and its immediate parent frame is also used. The trajectory and map are represented in a relative graph [11]. This is shown in Fig. 2.

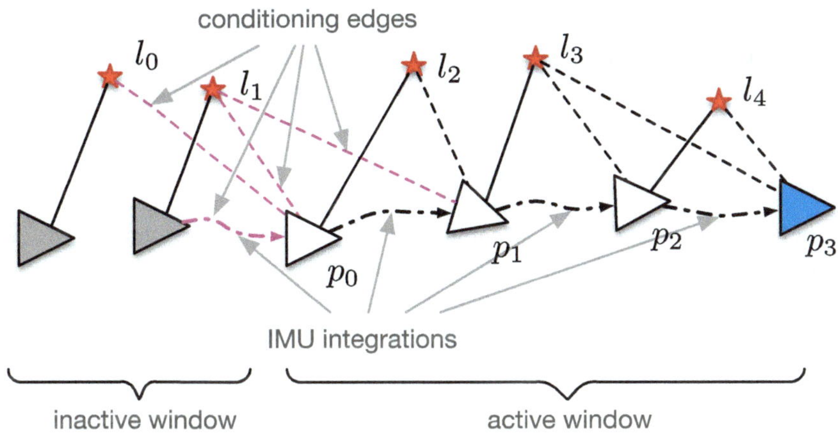

Fig. 2 Conditioning edges of the sliding window bundle adjustment

To reduce complexity, the optimization is performed on a *lifted* window of the relative chain in which all poses and landmarks are transformed into a consistent local coordinate system, referred to as the *world* coordinate system in the aforementioned methodology. This allows reprojection errors to form without requiring the traversal of intermediate poses between the reference and measurement frames. Once the optimization has finalized, the results are transformed back into the relative representation for map storage.

2.3 Adaptive Window Implementation

Since marginalization is forgoed, an adaptive local bundle adjustment is incorporated, which dynamically adjusts in size, in order to appropriately fold in parameters as needed. As is especially prominent when using an IMU, all parameters are not necessarily observable at any given time. Therefore, a fixed window cannot guarantee the optimization of a parameter at the time it becomes observable. Dynamically adjusting the window serves to allow the optimization to include parameters even if they are not immediately observable. Examples of these parameters are accelerometer and gyroscope biases and the direction of gravity, which is implicitly parameterized.

The condition used to assess whether the size of the window needs to be increased is based on the residuals observed in the conditioning edges shown in Fig. 2, after the optimization at iteration k has converged. The measurement covariances can then be used to assess whether the conditioning residuals are within expected bounds using a χ^2 test. The conditioning Mahalanobis distance is

$$e_c = \sum_{i \in C} \|\mathbf{r}_{\mathcal{P}_i}\|^2_{\Sigma_{P_i}} + \|\mathbf{r}_{\mathcal{I}_c}\|^2_{\Sigma_{\mathcal{I}_c}}$$

where the summation is over the set C comprising of all conditioning visual residuals, and $\mathbf{r}_{\mathcal{I}_c}$ is the single conditioning inertial residual connecting the active and inactive poses. Given e_c, an adaptive condition variable α_k can be defined as

$$\alpha_k = \frac{e_c}{\text{Inv}\,\chi^2\,(\beta, d)}, \quad d = 2|C| + 15$$

where Inv $\chi^2\,(\beta, d)$ is the inverse cumulative χ^2 distribution for d dimensions evaluated at probability β. The dimensionality d is derived from the 15 residuals of the single conditioning inertial residual plus 2 residuals for each visual conditioning residual. Initially if $\alpha_k > 1$, the conditioning residuals lie outside the βth percentile probability as expected from the residual covariance, so the window size is increased, and the optimization is run to convergence. While $\alpha_{k+1} > 1$ and $\alpha_{k+1} < \alpha_k$, the window size is continually increased and the optimziation is run once again to convergence, without adding additional frames. Otherwise, the window is resized to its default minimum length and new frames are added to the window.

The intuition behind this adaptive criterion is that when new residuals render past parameters observable, and they are not present in the active state, tension will be introduced into the conditioning edges. The residuals defining these edges would then fall outside their expected distributions. Increasing the window size until the conditioning edges are within expected bounds ensures that unobserved dimensions become part of the active window. In the case that the conditioning error is not decreasing but is still outside expected bounds, the window size is returned to its default minimal value, as the error is more likely explained by outlier measurements.

The size of the dynamic window could stretch far if a parameter does not become observable for an extended period of time. In this case obtaining a real-time solution will become infeasible. In order to obtain a real time solution, an asynchronous adaptive window BA is used in conjunction with a small fixed size window BA which runs synchronously in real time. The use of a relative map representation ensures that updates to the map remain small, allowing multiple BAs to update it asynchronously without clashing.

3 Experiments

To test the proposed method, experiments are run on two sensor platforms. Both platforms consist of a camera with wide-angle lens and a commercial grade MEMS accelerometer and gyroscope. The camera captures VGA images at 30 fps and the IMU sample rate is 120 Hz. In order to evaluate how closely the AAC method matches the global MLE solution, a batch solution for each recorded trajectory is estimated. Comparisons are also made with trajectories that were estimated by keeping a fixed-size sliding window to demonstrate the effects of the window size on the quality of the solution. For all experiments the synchronous part of the AAC method is run with a fixed window of 15 poses (Fig. 3).

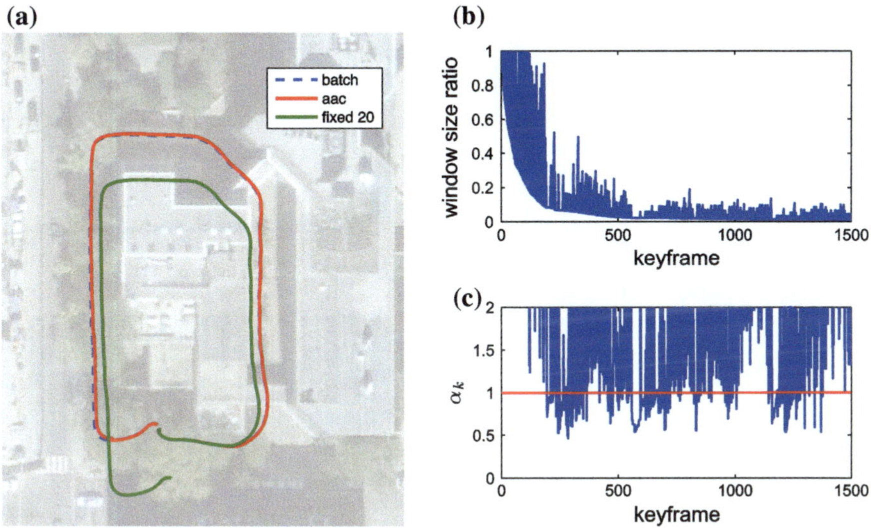

Fig. 3 A loop consisting of a 200 m dataset taken on foot around the GWU campus, superimposed over satellite imagery. The AAC optimization consists of a fixed window of 15 keyframes plus an asynchronous adaptive window with a minimum of 15 keyframes. The batch translation error as a percentage of traveled distance is 0.71 % while the AAC error is 0.72 %. **a** The resulting poses obtained by running AAC, batch and a fixed-window optimization over the data. **b** The ratio of the AAC window to the total number of keyframes. A ratio of 1.0 indicates a batch solve. **c** The condition variable α_k for every keyframe. Values larger than one indicate an expansion is necessary

The visual measurements are obtained by first extracting salient corners in the image where needed, to form landmarks. These landmarks are then tracked in subsequent images by minimizing the reprojective appearance error in a 9×9 pixel support area around the corner, between the two images similar to [17]. If the appearance error is between a predefined threshold and the reprojective constraints are not violated, the new position of the corner is added as a visual measurement of the landmark. All examples were run by attempting to track at most 128 landmarks. Keyframing was used as a means to increase the performance of the optimization by adding parameters only when sufficient motion was detected. This also alleviates problems arising from a stationary camera (Fig. 4).

The first experimental setup consists of a person walking in different indoor and outdoor environments. The first experiment within this setup was undertaken indoors and consisted of a closed loop sequence along a corridor. The trajectory length is approximately 80 m. The resulting images and IMU data were then processed with four different configurations of the solver: batch, two different fixed windows (15 and 25 poses) and AAC. The results from the four configurations run over the corridor sequence are shown in Fig. 5. From these results it can be observed that as the window size increases, the trajectory converges to the batch solution, as expected. However, it can also be seen that the adaptive mode, with an average window size of 33, matches the batch solution closely without a predetermined window size.

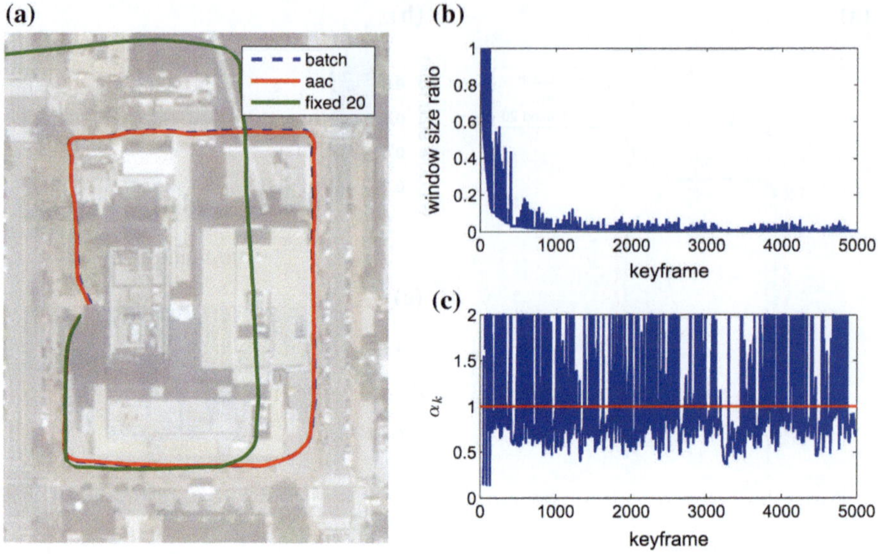

Fig. 4 A loop consisting of a 400 m dataset taken on foot around the GWU campus, superimposed over satellite imagery. The AAC optimization consists of a fixed window of 15 keyframes plus an asynchronous adaptive window with a minimum of 15 keyframes. The batch error as a percentage of traveled distance is 1.33 % while the AAC error is 1.42. For details on sub-figures (a), (b), and (c) refer to Fig. 3

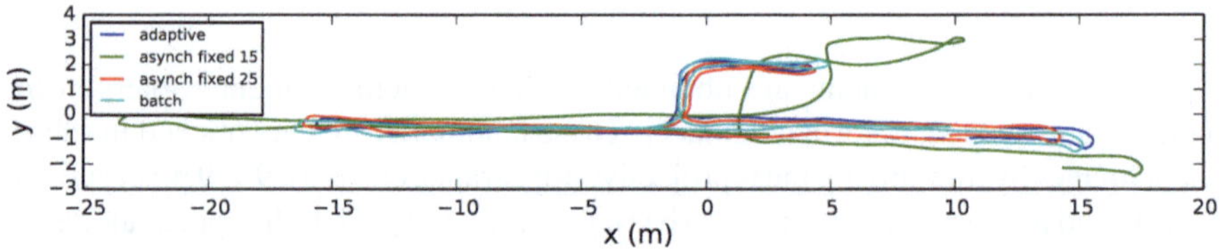

Fig. 5 Comparison of trajectories estimated by different bundle adjustment configurations. It can be seen that the 25 long asynchronous fixed window BA and the adaptive window BA both produce trajectories close to the batch solution, however the 15 long fixed asynchronous fixed window BA diverges substantially from the batch solution

Figures 3 and 4 show data obtained from running batch, AAC and fixed-window optimziations on two separate datasets taken on foot around the GWU campus. On both datasets, the AAC solution manages to adequately match the batch solution. Results using a fixed window size show that an adaptive solution is necessary to adequately approximate the batch solution. The subplots depicting the ratio of the active window to the total number of keyframes denote the initialization phase, where the AAC system pushes the optimization to batch, if the initial parameter estimates such as velocity and orientation with respect to gravity are mis-estimated. However after the initialization phase, the AAC window size reaches a constant-time phase where only minor expansions are required to keep an optimal estimate.

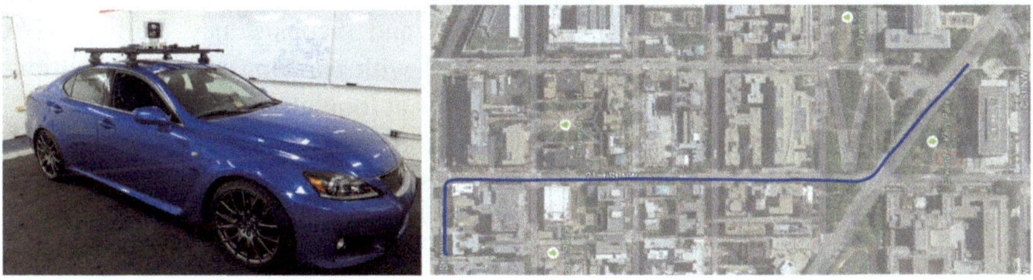

Fig. 6 a Autonomous vehicle used for data collection in this paper. b The 900 m trajectory estimated by the adaptive asynchronous system superimposed over aerial imagery

The experiments with the second platform were run using the autonomous car depicted in Fig. 6a. Images and IMU data were captured while driving around the GWU campus in Washington D.C. The trajectory generated by our adaptive method for a 900 m segment of the data is shown in Fig. 6b.

4 Experimental Insights

It was observed that in real-life situations, parameters such as velocity, gravity and bias are observable with adaptive conditioning. This is of course contingent upon sufficient excitation of the sensors. In the corridor dataset there is an ever present oscillatory acceleration which quickly renders the unknown parameters observable. Given this, we see a shorter required window size in order to closely estimate the MLE solution. As expected window growth is also seen in situations where scale and consequently velocity are ambiguous. An example of this is at the ends of the corridor where sharp turns introduce a slew of uninitialized new landmarks while simultaneously cutting tracks from established landmarks. The net result is a scale ambiguity that requires a larger window size to resolve, which is automatically discovered.

For data collected on the vehicle, scale was observable only over a large period of time where significant accelerations were imparted on the vehicle for example during turning. As such, a longer window was required in situations where scale became ambiguous and especially for initialization where prior estimates of velocity, biases and the gravity direction were not available.

When using asynchronous BA, care must be taken so as to ensure sufficient update frequency of the asynchronous solution in order to ensure overlap with the synchronous BA. This is required to keep the synchronous BA in the overall solution basin as solved by the asynchronous BA. As expected from the relative framework, the updates to the edges and inverse depth parameters for landmarks are small and no interference was observed between the two threads.

Adaptive asynchronous conditioning (AAC) is a novel solution to real-time visual-inertial SLAM. AAC is interesting because it automatically scales and focuses computation to capture the full MLE solution, and avoids the downsides associated with marginalization, such as incorrect linearization and inconsistency. Further, AAC avoids the computational difficulties associated with carrying prior distributions, such as the need to compute global optimizations at loop closure.

The proposed method offers a natural "front-end" while simultaneously allowing larger portions of the problem to influence the solution. It is thus able to produce estimates in real-time, and also capture functional-constraints that are only observable over long periods of time—an ability which is useful for self calibration, during degenerate motions, or when bias and gravity are poorly observed.

References

1. Engels, C., Stewenius, H., Nister, D.: Bundle adjustment rules. In: Photogrammetric Computer Vision (2006)
2. Gelb, A.: Applied Optimal Estimation. MIT Press, Cambridge (1974)
3. Hesch, J.A., Kottas, D.G., Bowman, S.L., Roumeliotis, S.I.: Camera-imu-based localization: observability analysis and consistency improvement. Int. J. Robot. Res. **33**, 182–201 (2013)
4. Jones, E., Vedaldi, A., Soatto, S.: Inertial structure from motion with autocalibration. In: ICCV Workshop on Dynamical Vision (2007)
5. Kelly, J., Sukhatme, G.S.: Visual-inertial sensor fusion: localization, mapping and sensor-to-sensor self-calibration. Int. J. Robot. Res. (2010)
6. Klein, G., Murray, D.: Parallel tracking and mapping for small ar workspaces. In: International Symposium on Mixed and Augmented Reality (2007)
7. Leutenegger, S., Furgale, P., Rabaud, V., Chli, M., Konolige, K., Siegwart, R.: Keyframe-based visual-inertial slam using nonlinear optimization. In: Robotics Science and Systems (2013)
8. Li, M., Kim, B., Mourikis. A.I.: Real-time cellphone localization using inertial sensing and a rolling-shutter camera. In: Proceedings of the IEEE International Conference on Robotics and Automation (2013)
9. Li, M., Mourikis, A.I.: High-precision, consistent ekf-based visual-inertial odometry. Int. J. Robot. Res. **33**, 690–711 (2014)
10. Maybeck, P.S.: Stochastic models, estimation, and control. In: Mathematics in Science and Engineering, vol. 141. Academic Press Inc, Boston (1979)
11. Mei, C., Sibley, G., Cummins, M., Newman, P., Reid, I.: RSLAM: a system for large-scale mapping in constant-time using stereo. Int. J. Comput. Vis. 1–17 (2010)
12. Mouragnon, E., Lhuillier, M., Dhome, M., Dekeyse, F., Sayd, P.: Real time localization and 3d reconstruction. In: Proceedings of Computer Vision and Pattern Recognition, New York, June 2006
13. Mourikis, A., Roumeliotis, S.: A multi-state constraint kalman filter for vision-aided inertial navigation. In: Proceedings of the IEEE International Conference on Robotics and Automation, pp. 3565–3572, Rome, April 2007
14. Nerurkar, E.D., Wu, K.J., Roumeliotis, S.I.: C-klam: constrained keyframe localization and mapping for long-term navigation. In: IEEE International Conference on Robotics and Automation Workshop on Long-term Autonomy (2013)
15. Pietzsch, T.: Efficient feature parameterisation for visual SLAM using inverse depth bundles. In: British Machine Vision Conference (2008)
16. Powell, M.J.D.: An efficient method for finding the minimum of a function of several variables without calculating derivatives. Comput. J. **7**(2), 155–162 (1964)
17. Scaramuzza, D., Forster, C., Pizzoli, M.: Svo: fast semi-direct monocular visual odometry. In: IEEE Conference on Robotics and Automation (2014)

18. Sibley, G., Matthies, L., Sukhatme, G.: Sliding window filter with applications to planetary landing. J. Field Robot. **27**(5), 587–608 (2010)
19. Sibley, G., Mei, C., Ried, I., Newman, P.: Adaptive relative bundle adjustment. In: Robotics Science and Systems (2009)
20. Sibley, G.: Sliding window filters for SLAM. Technical Report, University of Southern California, Center for Robotics and Embedded Systems, CRES-06-004 (2006)
21. Triggs, B., McLauchlan, P.F., Hartley, R.I., Fitzgibbon, A.W.: Bundle adjustment—a modern synthesis. In: ICCV '99: Proceedings of the International Workshop on Vision Algorithms, pp. 298–372. Springer-Verlag, London (2000)

An Experimental Study of Robust Distributed Multi-robot Data Association from Arbitrary Poses

Erik Nelson, Vadim Indelman, Nathan Michael and Frank Dellaert

Abstract In this work, we experimentally investigate the problem of computing the relative transformation between multiple vehicles from corresponding inter-robot observations during autonomous operation in a common unknown environment. Building on our prior work, we consider an EM-based methodology which evaluates sensory observations gathered over vehicle trajectories to establish robust relative pose transformations between robots. We focus on experimentally evaluating the performance of the approach as well as its computational complexity and shared data requirements using multiple autonomous vehicles (aerial robots). We describe an observation subsampling technique which utilizes laser scan autocovariance to reduce the total number of observations shared between robots. Employing this technique reduces run time of the algorithm significantly, while only slightly diminishing the accuracies of computed inter-robot transformations. Finally, we provide discussion on data transfer and the feasibility of implementing the approach on a mesh network.

1 Introduction and Related Work

In this work, we investigate the problem of computing the relative transformation between multiple vehicles based on corresponding inter-robot observations developed during autonomous operation in a common unknown environment. Applications that rely on distributed mapping and coordinated control must, in general, assume

E. Nelson (✉) · N. Michael
Carnegie Mellon University, Pittsburgh, PA 15213, USA
e-mail: enelson@cmu.edu

N. Michael
e-mail: nmichael@cmu.edu

V. Indelman · F. Dellaert
Georgia Institute of Technology, Atlanta, GA 30332, USA
e-mail: indelman@cc.gatech.edu

F. Dellaert
e-mail: dellaert@cc.gatech.edu

M.A. Hsieh et al. (eds.), *Experimental Robotics*, Springer Tracts in Advanced Robotics 109, DOI 10.1007/978-3-319-23778-7_22

the existence of a shared environment representation in order to establish a common reference frame for integration of distributed observations and joint cooperative control decisions. Therefore, a fundamental capability required by these applications is a robust strategy to establish the relative pose between individual vehicles.

Several approaches exist to address the problem of establishing a consistent reference frame in the multi-robot SLAM literature based on landmarks [1], direct relative inter-robot observations [2], and multi-robot data association [3]. Landmark-based strategies require additional prior knowledge of the environment or a means to instrument the environment (e.g., beacons [4]). Approaches that leverage direct inter-robot observations, such as when one robot detects the relative presence of another robot using onboard vision [5, 6] or RF ranging [7], assume that vehicles will proximally operate at the same time. In this work, we focus on environments and systems that do not admit prior or external instrumentation as well as temporal assumptions on spatial operation. Such scenarios can occur in complex or expansive environments where vehicles operate independently with the expectation of frequent and infrequent interactions (e.g., tunnel networks [8], large buildings [9]) or at disparate time schedules. Consequently, we assume that vehicles autonomously navigate the unknown environment, concurrently estimating their location and the map of the environment, and opportunistically coordinate with other vehicles toward furthering the application objective. For this reason, techniques most related to the emphasis of this work build on data association methods which seek to establish a consistent relative transform based on the existence of mutual environment observations made by each vehicle. Cunningham et al. [10] and Montijano et al. [11] propose robust methods for establishing a relative inter-robot transformation without requiring prior knowledge of the initial relative inter-robot pose based on variations of the RANSAC algorithm. Indelman et al. [12] suggest an alternative method based on the observation that vehicles will share common incremental observations in areas historically traversed by multiple vehicles, and pursue an EM-based methodology which evaluates present and historic observations developed along the trajectory transited by the vehicles.

In this work, we pursue an experimental sequel to the method presented by Indelman et al. [12]. The study focuses on analyzing the correctness of the resulting relative transformation as well as the relationship between the algorithm's computational complexity, shared data requirements, and team size, using trials of multiple autonomous vehicles (aerial robots). We briefly summarize the technical approach in Sect. 2. Section 3 details experiments designed to evaluate the technique using a multi-robot system. The experiments assess the correctness and robustness of the approach as well as key considerations toward reducing its computational complexity for real-time performance. Specifically, we describe an observation subsampling approach which utilizes laser scan autocovariance to select salient scans for sharing, and discuss considerations for implementing the approach over a capacity constrained network. Section 4 reports on the accuracy of inter-robot transforms resulting from three indoor and outdoor multi-robot trials, and analyzes the impact of laser scan saliency on both accuracy, and time consumed by individual algorithmic steps. Section 5 closes with a discussion and summarization of the experimental design and results.

2 Technical Approach

We now briefly review the formulation proposed in our prior work and defer to this work for a detailed discussion on the approach [12]. We consider a group of R robots deployed to collaboratively operate in some unknown environment and assume the robots start from different locations, without knowledge of the existence of other vehicles. Each robot r is assumed to be capable of estimating its trajectory X^r based on observations Z^r from its onboard sensors. We represent this estimation problem in a pose graph probabilistic formulation

$$p\left(X^r|Z^r\right) \propto p\left(x_0^r\right) \prod_i p\left(u_{i-1,i}^r|x_{i-1}^r, x_i^r\right) \tag{1}$$

where $x_i^r \in X^r$ is the robot's pose at time t_i, expressed relative to some reference frame, and $p\left(x_0^r\right)$ is a prior term. Since we assume no a priori knowledge about the environment and the initial pose of the robots, the reference frame of each robot is arbitrarily set to coincide with the initial pose.

The measurement likelihood term $p\left(u_{i-1,i}^r|x_{i-1}^r, x_i^r\right)$ in (1) involves the relative pose measurement $u_{i-1,i}^r$ that can be either directly obtained from odometry measurements or calculated from vision or laser sensor observations at the two time instances t_{i-1} and t_i. We follow the standard assumption in the SLAM community and model the measurement likelihood as a Gaussian:

$$p\left(u_{i-1,i}^r|x_{i-1}^r, x_i^r\right) \propto \exp\left(-\frac{1}{2}\left\|u_{i-1,i}^r \ominus h\left(x_{i-1}^r, x_i^r\right)\right\|_\Sigma^2\right)$$

with Σ being the measurement noise covariance and h the measurement model that, in the case of relative pose observations and robot poses expressed in the same reference frame is $h\left(x_{i-1}^r, x_i^r\right) \doteq x_{i-1}^r \ominus x_i^r$. We follow Lu and Milios [13] and use the notation $\ominus$ in $a \ominus b$ to express b locally in the frame of a for any two poses a, b.

The maximum a posteriori (MAP) estimate of the rth robot pose X^r using only local information is then given by

$$\hat{X}^r = \arg\max_{X^r} p\left(X^r|Z^r\right)$$

We denote by $\mathscr{F}$ the set of multi-robot data association, with each individual data association $(r_1, r_2, k, l) \in \mathscr{F}$ representing a relative pose constraint $u_{k,l}^{r_1,r_2}$ relating between the pose of robot r_1 at time t_k and the pose of robot r_2 at time t_l. This constraint can represent both direct observation of one robot pose relative to another robot, and also the estimated relative pose based on observation of a common scene by two robots. In the latter case, it is computed from the measurements of the two robots $z_k^{r_1} \in Z^{r_1}$ and $z_l^{r_2} \in Z^{r_2}$, that can represent, for example, laser scans or image observations.

Assuming multi-robot data association $\mathscr{F}$ has been established and appropriate constraints $u_{k,l}^{r_1,r_2}$ have been calculated, we can write a probabilistic formulation for the multi-robot joint pdf for the robots as follows:

$$p(X|Z) \propto \prod_{r} p(X^r|Z^r) \prod_{(r_1,r_2,k,l)\in\mathscr{F}} p\left(u_{k,l}^{r_1,r_2}|x_k^{r_1}, x_l^{r_2}\right) \tag{2}$$

where X and Z represent, respectively, the robot trajectories and the measurements.

As the robots express their local trajectories with respect to different reference systems, the measurement likelihood term in (2) is

$$p\left(u_{k,l}^{r_1,r_2}|x_k^{r_1}, x_l^{r_2}\right) \propto \exp\left(-\frac{1}{2}\left\|err\left(u_{k,l}^{r_1,r_2}, x_k^{r_1}, x_l^{r_2}\right)\right\|_{\Sigma}^{2}\right)$$

with

$$err\left(u_{k,l}^{r_1,r_2}, x_k^{r_1}, x_l^{r_2}\right) \doteq u_{k,l}^{r_1,r_2} \ominus h\left(x_k^{r_1}, x_l^{r_2}\right)$$

and

$$h\left(x_k^{r_1}, x_l^{r_2}\right) \doteq x_k^{r_1} \ominus \left(T_{r_2}^{r_1} \oplus x_l^{r_2}\right)$$

The notation $\oplus$ represents the compose operator [13], and $T_{r_2}^{r_1}$ is a transformation between the reference frames of robots r_1 and r_2. Since the robots start operating from different unknown locations, this transformation is initially unknown and arbitrary.

While the formulation (2) assumes multi-robot data association $\mathscr{F}$ is given, in practice it is unknown ahead of time and should therefore be established. In [12], we propose an expectation-maximization (EM) based framework to reliably infer the multi-robot data association $\mathscr{F}$ in a multi-robot pose SLAM framework, without assuming prior knowledge on initial relative poses between the robots, i.e., unknown $T_{r_j}^{r_i}$ for all pairs $r_i, r_j \in [1, \ldots, R]$. The remainder of this work experimentally evaluates the efficacy of this EM-based methodology to accurately estimate the unknown transformations, $T_{r_j}^{r_i}$, for all pairs of robots.

3 Experimental Study

3.1 Experimental Design and Approach

We consider the problem of developing accurate relative transforms between multiple aerial robots while operating in a common environment through shared laser sensor observations. As the system relies on a capacity constrained network, we focus on experimental questions relating computational complexity, shared data requirements,

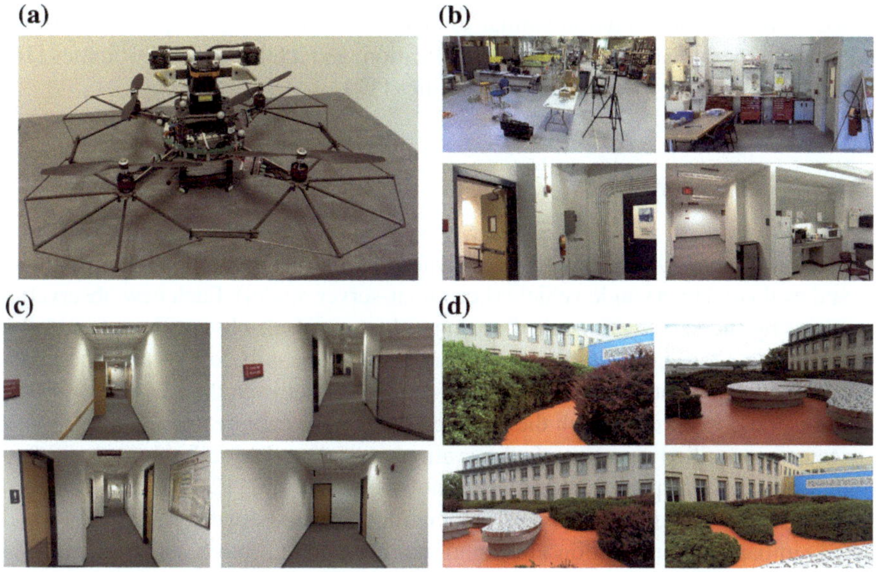

Fig. 1 **a** Quadrotor platform, equipped with an onboard computer (1.86 GHz Intel Core 2 Duo processor), IMU, laser, and beam deflector mirrors. Stereo cameras are not utilized in the described experiments. **b–d** Experiment environments for **T1**, **T2**, and **T3**

and team size. Toward studying this relationship, we propose a strategy that seeks to reduce data transfer and algorithmic complexity at the cost of reducing the accuracy of the computed relative transforms. We show the feasibility of the proposed technique experimentally and extend this discussion to larger teams where mild reductions in accuracy can permit real-time performance.

Trials of sensory information (laser scans) are captured from fleets of quadrotors operating in three different environments (Fig. 1). These environments consist of a set of paths transitioning from a wide open room with clutter to a hallway environment (trial **T1**), a series of connected corridors and hallways (trial **T2**), and an outdoor hedge maze (trial **T3**).

T1 Three robots navigating along a path from the same initial pose before diverging and traveling long distances in different directions.

T2 Three robots navigating through hallways from different starting poses. All robots meet, and travel in the same direction around a 10×16 m loop before diverging.

T3 Three robots navigating an outdoor structured environment. Robots navigate low to the ground to capture laser scans of bushes and walls. Robots 2 and 3 have no trajectory overlap, but capture laser scans which share features.

Pose estimates along the robot trajectories are generated through local instances of SLAM running on each quadrotor during operation. We use a laser and inertial based SLAM implementation similar to that of [14], which leverages ICP for laser odometry [15], a histogram filter for localization, and a UKF to fuse estimates [16]. Octomap is used for 3D mapping capabilities [17].

3.2 Implementation

We pursue a centralized implementation where laser scan observations are distributed to and evaluated by a single vehicle (i.e., client-server model). Each new observation received by the server is matched against a history of all observations from other robots using a variant of ICP [18]. If there is a significantly low covariance in any one ICP match, a data association $(r_i, r_j, k, l) \in \mathcal{F}$ between the two robots, r_i and r_j, links the poses from which the matched scans were captured.

After a specified number of observations are shared by all robots, transform hypotheses between pairs of robots are established using EM [12]. For each transform hypothesis generated this way, data associations formed between robot pairs which share similarity in translation and rotation are separated into a set of inliers. The transform hypothesis which contains the highest number of inlier data associations is chosen as the most probable transform. If the number of inliers is small for all transform hypotheses, none are selected, signifying that the robots are operating in disjoint environments. If a reliable transform has not yet been discovered, hypotheses are constantly generated and reassessed as new observations are shared between robots.

3.3 Computational Complexity and Saliency of Information

Given this implementation, we choose to evaluate three steps of the algorithm in an analysis of computational complexity (Table 1). SLAM is executed locally on each robot, and is therefore omitted. The server is responsible for managing all computations in these three steps, and executes them in sequence when presented with new pose and sensor information from a robot.

Step 1 Updates to a robot's pose graph to incorporate a new pose and any new data associations formed with other robots since the previous update.

Step 2 ICP and transform hypothesis generation between one robot's history of shared scans and any new scans recently shared by other robots.

Step 3 Sensor observation autocovariance calculation.

Table 1 Complexities of individual steps of the data association strategy

Algorithm step	1	2	3
Complexity	$O\left(E^2\right)$	$O\left(R^2\lvert\bar{Z}^r\rvert^2\right)$	$O\left(R\lvert Z^r\rvert\right)$

Step 1 Edges are added to the multi-robot pose graph upon individual robot pose updates as well as multi-robot pose correspondences generated through data association. The pose graph is implemented with the GTSAM optimization library [19], which computes individual updates (nearly) linearly in the number of edges in the pose graph. Continuing the notation introduced in Sect. 2, the shared pose graph has $E = R(|X^r| - 1) + |\mathscr{F}|$ edges if the poses of all robot are updated at a shared frequency. Since edges are added sequentially during run time, the total number of computations performed in Step 1 is roughly $\frac{1}{2}(E^2 + E)$, resulting in a complexity of $O(E^2)$.

Step 2 To identify data associations from arbitrary local robot coordinate frames in a temporally invariant manner, every laser scan shared by one robot must be compared against each other robot's history of laser scans. Assuming robot poses are updated with a shared frequency, every laser scan shared by an individual robot will be accompanied by $R - 1$ sensor observations shared by other robots. Growth in the laser scan histories of each robot results in a growth in the number of ICP comparisons performed between observations over time. Let each robot r gather a set of observations, Z^r, and share a subset, $\bar{Z}^r \subset Z^r$, with other robots. Then the total number of ICP comparisons performed over the duration of the trial run time is

$$
\begin{aligned}
&R(R-1)\sum_{i=1}^{|\bar{Z}^r|} i \\
=&\frac{1}{2}\left(R^2 - R\right)\left(|\bar{Z}^r|^2 + |\bar{Z}^r|\right)
\end{aligned}
$$

This result implies that the run time complexity of Step 2 is $O(R^2|\bar{Z}^r|^2)$.

Step 3 The dominant source of complexity in Step 2 arises from $R^2|\bar{Z}^r|^2$ laser scan ICP comparisons from potentially large initial offsets over the duration of the run time, as well as EM on a number of transform hypotheses quadratic in $|\bar{Z}^r|$. The algorithm can be made significantly more efficient through a reduction the number of observations in the set $\bar{Z}^r$ as $|\bar{Z}^r|$ is not restricted to be equal to $|X^r|$. We therefore propose a subsampling strategy that seeks to reduce $|\bar{Z}^r|$ based on laser scan autocovariance [20], which can be used as a scalar measure of laser scan saliency. Autocovariance is calculated by randomly perturbing the pose from which a laser scan was captured, performing ICP to match the perturbed scan against the original, and storing the resulting transformation mean and covariance. After N such iterations, autocovariance is computed by

$$
\delta = \frac{1}{\text{trace}\,(\Sigma)},
$$

where Σ is the covariance of the N-Gaussian mixture.

To subsample, we skip nine of ten sequential laser scans and maintain at least 0.1 m in normed (x, y) pose estimate between scans. Saliencies are calculated for the remaining set. Only scans with δ greater than a threshold, δ_s, are shared with

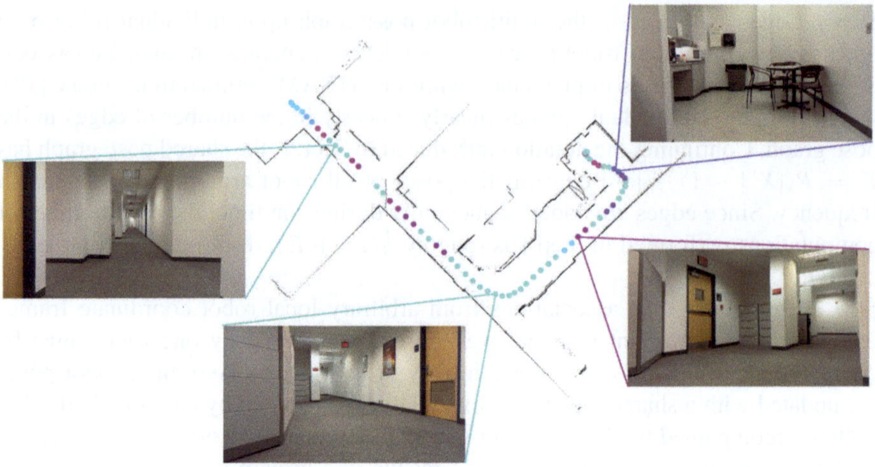

Fig. 2 Laser scan saliency along a robot's trajectory with inlaid images of the environment. Magenta corresponds to high saliency, while teal corresponds to low saliency

other robots. Because laser scan saliency is only computed once per laser scan, the run time complexity of Step 3 is $O(R|Z^r|)$. Laser scan saliency is plotted along a robot trajectory from trial **T2** in Fig. 2.

3.4 Network Complexity

Constrained network capacity is a concern when requiring the distribution of large amounts of shared data. Further, mesh networks exhibit a reduced capacity when distributing packets between multiple systems [21]. A conservative model of three robots sharing uncompressed scans (approximately 34 kB per scan) limits the data sharing rate of each robot to approximately 4 Hz. The rate reduces further with the number of robots where in practice an update rate of 1 Hz is expected for six vehicles [21].

4 Results and Discussion

The data association algorithm was used to calculate inter-robot relative transforms between robots in **T1**, **T2**, and **T3**. All reported transforms, $T_{r_i}^{r_1}$, are expressed as a rotation followed by a translation from the local coordinate frame of robot r_i to that of robot r_1.

We first report on the accuracy of transforms calculated without saliency thresholding on laser scans. Trajectories, inlier data associations, and outlier data associations from **T1** are shown in the local frame of each robot in Fig. 3. Robots begin from the same initial position and diverge after sharing a large number of laser scans in

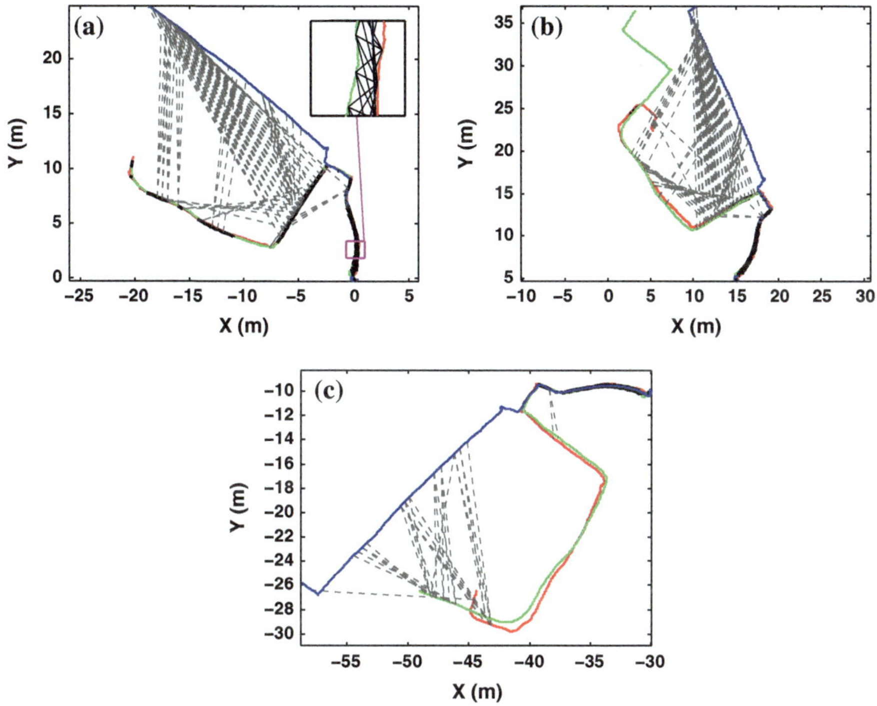

Fig. 3 Robot trajectories from **T1** in a common frame after applying transforms resulting from data association. Estimates are expressed in the local frame of each robot. Inlier (*black*) and outlier (*gray dashed*) correspondences after a common reference frame has been established are shown. **a** r_1 local frame (*red*). **b** r_2 local frame (*green*). **c** r_3 local frame (*blue*)

common. As such, strong hypotheses are formed in the initial stretch of the trajectory by the high number of inlier ICP correspondences. In the latter half of each robot's trajectory, incorrect data associations are made between robots due to the similarity in laser features throughout the hallway environment. However, given the strength of the transform established in the initial poses, these data associations are considered outliers.

Table 2 displays computed and measured transforms from all three datasets with no laser scan saliency thresholding. The transform with the largest error among all trials, $T_{r_3}^{r_1}$ from **T1**, has a translation error norm of 0.31 m. Rotational errors in all computed transforms are between zero and eight degrees.

Figure 4a displays saliency plotted across trajectories in **T2**. Structural symmetries in the hallway environment cause ICP comparisons to converge to local minima during autocovariance computation, resulting in low saliency in areas such as four-way crossroads and corridors. The set of salient laser scans is roughly a subset of the set of shared scans that form high numbers of ICP correspondences with other robots (Fig. 4b). This inclusion signifies that autocovariance is a suitable metric for laser scan subsampling.

Table 2 Computed and measured transformations T_{r2}^{r1} and T_{r3}^{r1}, with x, y in m and θ in rad

		Trial **T1**		Trial **T2**		Trial **T3**	
Transform		T_{r2}^{r1}	T_{r3}^{r1}	T_{r2}^{r1}	T_{r3}^{r1}	T_{r2}^{r1}	T_{r3}^{r1}
	x:	−0.12	0.15	2.62	−4.53	1.41	−13.59
Computed	y:	−0.03	−0.27	7.45	−4.09	−3.99	−1.24
	θ:	−0.02	0.03	−1.57	0.00	0.97	2.05
	x:	0.00	0.00	2.48	−4.60	1.42	−13.63
Measured	y:	0.00	0.00	7.50	−3.99	−3.90	−1.02
	θ:	0.00	0.00	−1.57	0.00	1.08	2.01

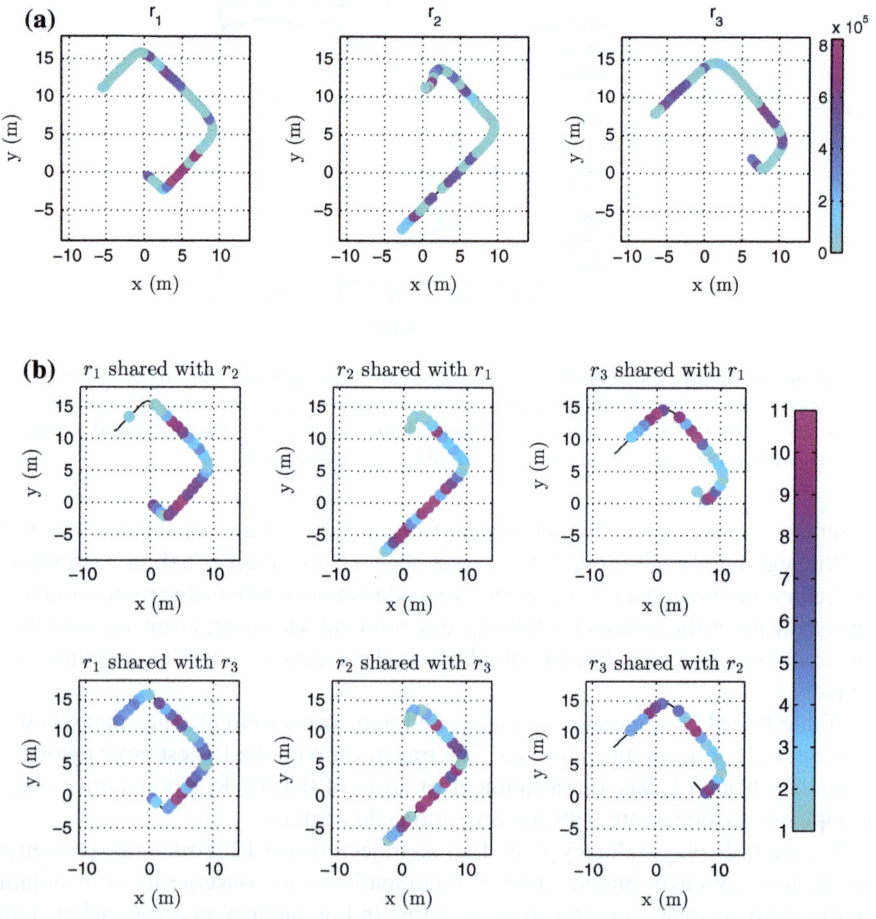

Fig. 4 **a** Saliency, δ, of shared scans in **T2**. **b** Number of ICP scan correspondences found between robot pairs in **T2**, plotted along robot trajectories. Counts mark the number of ICP correspondences made between the scan shared from the marked pose with all scans shared by the other robot

Table 3 Computed and measured transforms before and after saliency thresholding, with $\|x, y\|$ in m and θ in rad. $\sim$ signifies that a transform was not established

δ_s	Data set **D2**							Data set **D3**						
	Shared scans			$T_{r_2}^{r_1}$ error		$T_{r_3}^{r_1}$ error		Shared scans			$T_{r_2}^{r_1}$ error		$T_{r_3}^{r_1}$ error	
	r_1	r_2	r_3	$\|x, y\|$	θ	$\|x, y\|$	θ	r_1	r_2	r_3	$\|x, y\|$	θ	$\|x, y\|$	θ
0	75	77	65	0.15	0.00	0.20	0.00	74	55	71	0.09	0.10	0.22	0.05
2×10^5	22	26	23	0.19	0.00	0.24	0.00	26	18	36	0.22	0.08	0.59	0.13
4×10^5	22	24	23	0.19	0.00	0.24	0.00	24	16	35	$\sim$	$\sim$	0.59	0.13
6×10^5	16	18	19	0.18	0.01	0.29	0.02	22	15	31	$\sim$	$\sim$	0.67	0.13
8×10^5	8	6	4	$\sim$	$\sim$	$\sim$	$\sim$	8	1	15	$\sim$	$\sim$	$\sim$	$\sim$

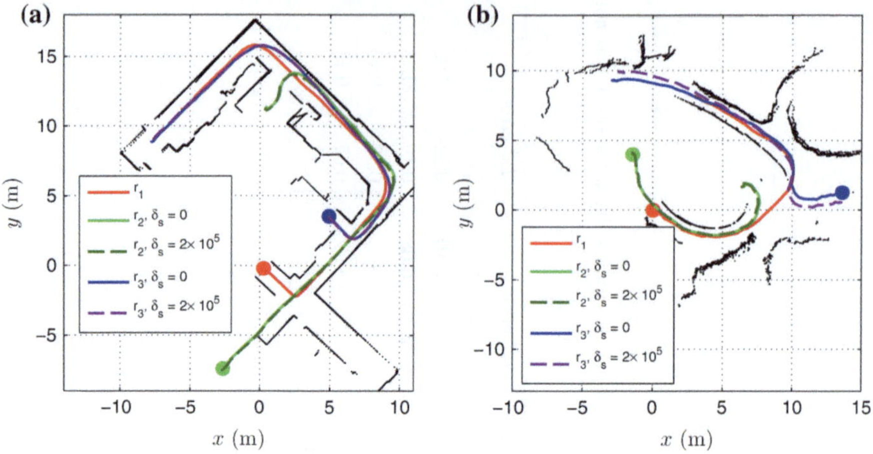

Fig. 5 **a**, **b** Robot trajectories from **T2** and **T3** in the frame of r_1 after applying transforms resulting from the data association strategy with and without saliency thresholding. Saliency thresholding makes little difference to transform estimates. Laser scans from r_1 are shown

Transform accuracies for **T2** and **T3** are shown in Table 3 with varying saliency thresholds, δ_s. Thresholding with $\delta_s = 2 \times 10^5$ reduces the average number of scans used to 36 percent of the total. By increasing the saliency threshold to $\delta_s = 4 \times 10^5$, there are no longer enough inlier ICP correspondences between robots r_1 and r_2 from **T2** to compute a transform. While thresholding scans by their saliency increases error in both translation and rotation, this inscrease, at worst, raises normed (x, y) error by 0.37 m, and rotation error by 0.08 radians over a trajectory roughly 20 m in length. At $\delta_s = 8 \times 10^5$, no robot pairs have enough inlier correspondences to form a transform.

Trajectories from **T2** and **T3** are shown in Fig. 5 with and without salient scans thresholded at $\delta_s = 2 \times 10^5$. Thresholding laser scans from robots in **T2** by saliency leads to a 0.04 m increase in translation error for both robots, with no difference in rotation error. In trial **T3**, thresholding increases translation error by 0.13 and 0.37 m, and rotation error by −0.02 and 0.08 radians for $T_{r_2}^{r_1}$ and $T_{r_3}^{r_1}$, respectively.

Without saliency thresholding, the total number of shared scans in all trials remains below the data sharing limit of 4 Hz described in Sect. 3.4 for three robots (Table 4). During trials, laser scans were captured on each robot at 20 Hz and uniformly subsampled to one tenth of the original amount. Mean sharing frequency across all robots with no laser scan subsampling remains within a small margin of 2 Hz, implying no robots were limited by the 0.1 m distance constraint between sequential laser scans. After thresholding shared scans by saliency with $\delta_s = 2 \times 10^5$, the mean sharing frequency decreases to below 1 Hz for all robots. The maximum sharing frequencies for both values of δ_s permit unconstrained data transfer on a mesh network with three robots. With a mesh network of six robots, the maximum sharing frequencies indicate that durations of increased sharing would exceed the network transfer limit of 1 Hz per robot by up to 208 percent for $\delta_s = 0$ and up to 131 percent for $\delta_s = 2 \times 10^5$.

Table 4 Trial durations, number of shared scans, and mean and maximum sharing frequencies for $\delta_s = 0$ and $\delta_s = 2 \times 10^5$

		$\delta_s = 0$			$\delta_s = 2 \times 10^5$		
Robot	Duration (s)	$\|Z^r\|$	Max (Hz)	Mean (Hz)	$\|Z^r\|$	Max (Hz)	Mean (Hz)
T2: r_1	37.4	75	2.08	2.00	22	1.01	0.59
T2: r_2	39.0	77	2.02	1.97	26	1.28	0.67
T2: r_3	32.5	65	2.02	2.00	23	0.95	0.71
T3: r_1	35.5	71	2.06	2.00	26	0.98	0.73
T3: r_2	27.6	55	1.99	1.99	18	1.31	0.65
T3: r_3	37.4	74	1.98	1.98	36	1.20	0.96

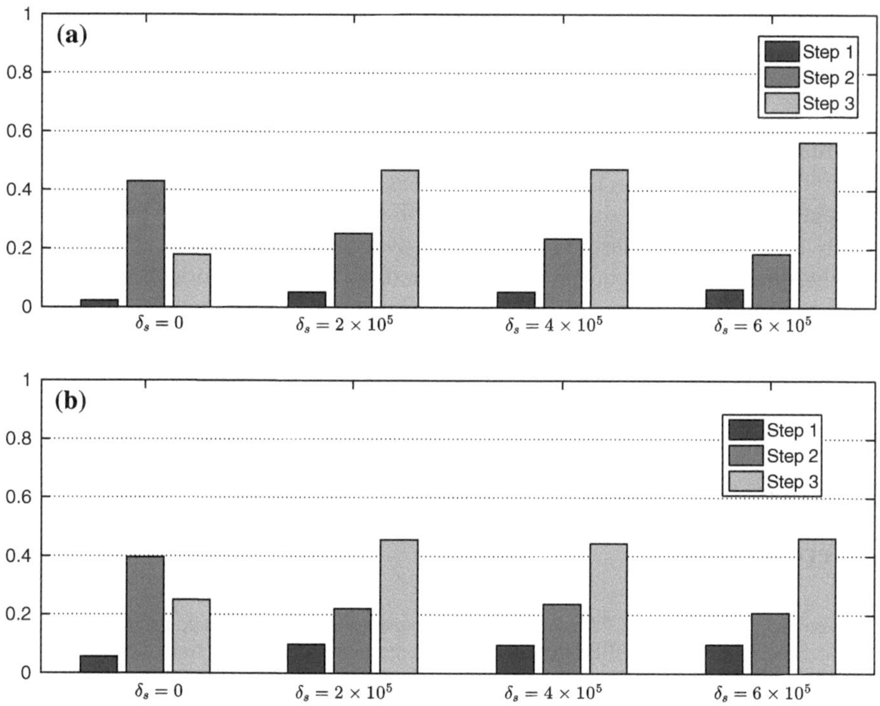

Fig. 6 Proportion of total run time consumed by each step of the algorithm for **a** trial **T2**, and **b** trial **T3** with varying values of δ_s. Time consumed by Step 3 is constant across values of δ_s, signifying that as δ_s increases the total run time of the algorithm decreases

Figure 6 shows the proportion of the process run time consumed by each step of the data association algorithm with $\delta_s = \{0, 2, 4, 6\} \times 10^5$ for trials **T2** and **T3**. After subsampling scans with $\delta_s = 2 \times 10^5$, Step 3 (autocovariance calculations for all laser scans) increases from 21.5 percent of the total computation time (averaged across both trials) to 46.3 percent. Because total time devoted to Step 3 is not a function of δ_s, thresholding on saliency by $\delta_s = 2 \times 10^5$ decreases the total run time of the algorithm by 46.4 percent. Note that due to the quadratic complexity of Step 2, this estimate is only reflective of trajectories of the same length as those in **T2** and **T3**.

5 Conclusion

In this work we investigated the problem of computing relative transformations between multiple vehicles from shared sensor observation correspondences. Experiments were developed to evaluate the accuracy, computational complexity, and network complexity of a data association strategy introduced by Indelman, et al. [12]. A sensory observation subsampling strategy based on laser scan autocovariance was introduced to reduce the number of laser scans shared between robots, therefore reducing the both the computational complexity of the algorithm as well as the rate of data sharing between robots.

Multi-robot trials were collected onboard quadrotors operating throughout three dissimilar environments. The trials were used to evaluate the accuracy of the data association algorithm in different domains of operation. We showed that by intelligently selecting which laser scans to share over the network, the total run time of the algorithm could be reduced by 46.4 % for three-robot trajectories on the order of 20 m in length. In addition, we demonstrated that the accuracy of transforms resulting from the algorithm does not suffer significantly from subsampling laser scans based on their autocovariance. Finally, we experimentally evaluated data sharing rates to show that this approach can be implemented using a mesh network for at least three robots.

References

1. Fenwick, J.W., Newman, P.M., Leonard, J.J.: Cooperative concurrent mapping and localization. In: Proceedings of the IEEE International Conference on Robotics and Automation, pp. 1810–1817, Washington (2002)
2. Bailey, T., Bryson, M., Hua, M., Vial, J., McCalman, L., Durrant-Whyte, H.: Decentralised cooperative localisation for heterogeneous teams of mobile robots. In: Proceedings of the IEEE International Conference on Robotics and Automation, pp. 2859–2865, Shanghai (2011)
3. Montijano, E., Aragues, R., Sagues, C.: Distributed data association in robotic networks with cameras and limited communications. IEEE Trans. Robot. **29**, 1408–1423 (2013)
4. Olson, E., Leonard, J., Teller, S.: Robust range-only beacon localization. In: Auton, I.E.E.E. (ed.) Underwater Vehicles, pp. 66–75. Maine, Sebasco (2004)
5. Howard, A., Parker, L.E., Sukhatme, G.S.: Experiments with a large heterogeneous mobile robot team: exploration, mapping, deployment and detection. Int. J. Robot. Res. **25**, 431–447 (2006)
6. Zhou, X.S., Roumeliotis, S.I.: Multi-robot SLAM with unknown initial correspondence: the robot rendezvous case. In: Proceedings of the IEEE/RSJ International Conference on Intelligent Robots and Systems, pp. 1785–1792, Beijing (2006)

7. Charrow, B., Michael, N., Kumar, V.: Cooperative multi-robot estimation and control for radio source localization. In: Proceedings of the International Symposium on Experimental Robotics, Quebec (2012)
8. Thrun, S., Thayer, S., Whittaker, W., Baker, C., Burgard, W., Ferguson, D., Hahnel, D., Montemerlo, M., Morris, A., Omohundro, Z., Reverte, C., Whittaker, W.: Autonomous exploration and mapping of abandoned mines. IEEE Robot. Autom. Mag. **11**, 79–91 (2004)
9. Michael, N., Shen, S., Mohta, K., Mulgaonkar, Y., Kumar, V., Nagatani, K., Okada, Y., Kiribayashi, S., Otake, K., Yoshida, K., Ohno, K., Takeuchi, E., Tadokoro, S.: Collaborative mapping of an earthquake-damaged building via ground and aerial robots. J. Field Robot. **29**, 832–841 (2012)
10. Cunningham, A., Wurm, K.M., Burgard, W., Dellaert, F.: Fully distributed scalable smoothing and mapping with robust multi-robot data association. In: Proceedings of the IEEE International Conference on Robotics and Automation, pp. 1093–1100, Saint Paul (2012)
11. Montijano, E., Martinez, S., Sagues, C.: Distributed robust data fusion based on dynamic voting. In: Proceedings of the IEEE International Conference on Robotics and Automation, pp. 5893–5898, Shanghai (2011)
12. Indelman, V., Keyes, D., Nelson, E., Michael, N., Dellaert, F.: Multi-robot pose graph localization and data association from unknown initial relative poses via expectation maximization. In: Proceedings of the IEEE International Conference on Robotics and Automation (To Appear), Hong Kong (2014)
13. Lu, F., Milios, E.: Robot pose estimation in unknown environments by matching 2D range scans. J. Intell. Robot. Syst. **18**, 249–275 (1997)
14. Shen, S., Michael, N., Kumar, V.: Autonomous multi-floor indoor navigation with a computationally constrained MAV. In: Proceedings of the IEEE International Conference on Robotics and Automation, pp. 20–25, Shanghai (2011)
15. Pomerleau, F., Colas, F., Siegwart, R., Magnenat, S.: Comparing ICP variants on real-world data sets. Auton. Robots **34**, 133–148 (2013)
16. Thrun, S., Burgard, W., Fox, D.: Probabilistic Robotics. The MIT Press, Cambridge (2005)
17. Hornung, A., Wurm, K.M., Bennewitz, M., Stachniss, C., Burgard, W.: OctoMap: an efficient probabilistic 3D mapping framework based on octrees. Auton. Robots **34**, 189–206 (2013)
18. Censi, A.: An accurate closed-form estimate of ICP's covariance. In: Proceedings of the IEEE International Conference on Robotics and Automation, IEEE, pp. 3167–3172 (2007)
19. Dellaert, F.: Factor graphs and GTSAM: a hands-on introduction. Technical Report GT-RIM-CP&R-2012-002 (2012)
20. Nieto, J., Bailey, T., Nebot, E.: Recursive scan-matching slam. Robot. Auton. Syst. **55**, 39–49 (2007)
21. Jun, J., Sichitiu, M.L.: The nominal capacity of wireless mesh networks. IEEE Wirel. Commun. **10**, 8–14 (2003)

Interactive Semantic Mapping: Experimental Evaluation

Guglielmo Gemignani, Daniele Nardi, Domenico Daniele Bloisi, Roberto Capobianco and Luca Iocchi

Abstract Robots that are launched in the consumer market need to provide more effective human robot interaction, and, in particular, spoken language interfaces. However, in order to support the execution of high level commands as they are specified in natural language, a semantic map is required. Such a map is a representation that enables the robot to ground the commands into the actual places and objects located in the environment. In this paper, we present the experimental evaluation of a system specifically designed to build semantically rich maps, through the interaction with the user. The results of the experiments not only provide the basis for a discussion of the features of the proposed approach, but also highlight the manifold issues that arise in the evaluation of semantic mapping.

Keywords Cognitive robotics · Human robot interaction · Knowledge representation and reasoning · Semantic mapping

1 Introduction

As robots are targeting the consumer market, the need for developing suitable interaction paradigms and interfaces for consumers is increasing. In this scenario, spoken language interaction plays a key role, as also demonstrated by other consumer

G. Gemignani (✉) · D. Nardi · D.D. Bloisi · R. Capobianco · L. Iocchi
Department of Computer, Control, and Management Engineering,
Sapienza University of Rome, Rome, Italy
e-mail: gemignani@dis.uniroma1.it

D. Nardi
e-mail: nardi@dis.uniroma1.it

D.D. Bloisi
e-mail: bloisi@dis.uniroma1.it

R. Capobianco
e-mail: capobianco@dis.uniroma1.it

L. Iocchi
e-mail: iocchi@dis.uniroma1.it

M.A. Hsieh et al. (eds.), *Experimental Robotics*, Springer Tracts in Advanced Robotics 109, DOI 10.1007/978-3-319-23778-7_23

products, such as cell phones and cars. However, in order to provide a system with the ability of interacting with the user using natural language, the robot must be able to interpret high level commands, such as "go to the printer near the secretary office". For executing such a command, the system must understand not only the meaning of the terms used by the user, but also to *ground* them into its world model (i.e. the representation of the operational environment).

To address this problem, several researchers have been developing *semantic maps*, that, according to the definition given in [1], should be able to integrate symbolic knowledge into the representation of the environment used by the robot. Although significant progress has been made in the last years, the semantic maps that robots can acquire and deploy are still limited. On the one hand, the acquisition of semantic knowledge by state-of-the-art approaches to perception is challenging, on the other hand, a systematic approach that exploits the interaction with the user, to build semantically rich representations of the environment has been only partially addressed.

The goal of our work is to rely on the interaction with the user, according to the paradigm of symbiotic autonomy [2] in order to build a representation of the environment that can allow a mobile robot to interpret and execute user commands that refer to places and objects in the environment. Specifically, we have developed a system that builds a layered semantic map through a multi modal interaction with the user that relies on the use of a simple pointer device [3]. The system has been deployed on four different robotic wheeled platforms and has been used to successfully build the semantic map of office and home environments. The system at an earlier stage has been presented in [4] and the present paper is specifically addressing the experimental evaluation of the proposed approach. To this end, we have reviewed the literature on semantic mapping to identify a proper methodology for a quantitative evaluation of the proposed approach. The outcome of our survey shows that there are no established methodologies for a quantitative evaluation of semantic mapping. In fact, several methods are adopted, each one covering a specific aspect of the proposed approach. Consequently, we have defined an experimental setting for each system component and evaluated their performance in isolation. Moreover, we have run several experiments aiming at the evaluation of the overall system. The results of this evaluation, that are discussed in detail in the paper, show that the proposed system has an overall very interesting performance. Moreover, since the representation of semantic knowledge requires several forms of approximations, our system shows a good trade-off between accuracy and ability to deal with high level semantic notions. This notwithstanding, several key issues remain to be addressed by the research on semantic mapping, to make possible the deployment of robots that are able to incrementally acquire and keep up-to-date the knowledge about the operational environment in the face of changes.

The paper is structured as follows. In the next section we review the state of the art on semantic mapping; then, we present a quick overview of our system (Sect. 3). The rest of the paper is devoted to discussing the experimental evaluation of the system (Sect. 4), by first analyzing the approaches found in the literature and then presenting a detailed evaluation of our system. A summary of the contributions of the proposed approach and hints for future work conclude the paper.

2 Related Work

The acquisition of the semantic knowledge needed to suitably interpret the commands given by a user to a robot is typically achieved through a process called *semantic mapping* [5]. The literature about such research topic can be divided into two main categories, by distinguishing automatic methods from the so called "human-in-the-loop" approaches, where a user is asked to help the robot in the acquisition process, as proposed also by [2].

As an example of automatic approaches, in Galindo et al. [6] environmental knowledge is represented by augmenting a topological map (extracted by means of fuzzy morphological operators) with semantic knowledge using anchoring. In Goerke et al. [7], in Brunskill et al. [8], and in Friedman et al. [9], instead, a set of techniques are used to automatically classify and cluster metric maps. Finally, in Mozos et al. [10] visual features are used for object recognition and place categorization. Although significant progress has been made in fully automated semantic mapping [11], even the most recent approaches still lack of robustness and generality.

Therefore researchers in the AI and Robotics community have started to enclose the human in the semantic acquisition process, trying to overcome the limitations that the current robotic systems have. As an example of "human-in-the-loop" approach, in [12] the authors describe a system for the creation of conceptual representations of indoor environments. In this work, a priori knowledge about spatial concepts is provided to the robotic platform, which produces an internal representation of the environment acquired through low-level sensors with the help of the user for place labeling. In [13], instead, an approach that uses heterogeneous modalities for a comprehensive multi-layered semantic mapping algorithm, aiming at place categorization and topological map construction, is presented. This system builds a probabilistic representation that includes information about the existence of objects and properties of space. Such a representation is used in order to estimate room labels. The user input, whenever provided, is integrated in the system as additional properties about existing objects. While in the latter described approach the support of the user does not play a central role, in [3] the authors propose a rich multi-modal interaction, including speech, gesture, and vision. Such an approach enables the system to perform a semantic labeling of the environment, without many pre-requisites on the features of the environment itself. In this system however, the authors do not attach any additional semantic information to the landmarks other than their position.

Compared with the related work, our approach, initially proposed in [4], improves the construction of semantic maps through the interaction with the user, aiming not only at representing objects as points in the metric map, but at creating a semantic map that holds manifold information of the objects (e.g., dimensions, colors, 3D models), which is needed by the robot for task execution and reasoning.

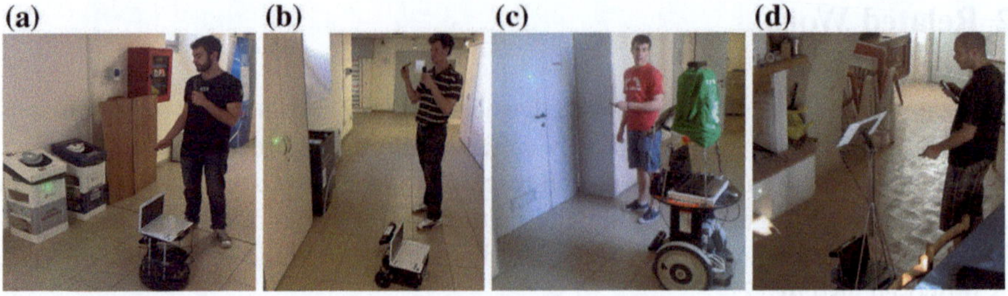

Fig. 1 Robots on which our system has been deployed. **a** Turtlebot. **b** MARRtino, a mobile base built by our students. **c** Mobile base derived from Segway. **d** Videre Design platform

3 System Overview

The proposed system is built for wheeled robots that are capable of mapping the environment through an off-line slam technique and, afterwords, can also navigate in it through the conventional ROS *movebase* module.[1] In our experiments we have used a Turtlebot (Fig. 1a), a MARRtino,[2] a mobile base built by our students (Fig. 1b), a mobile base derived from Segway (Fig. 1c), and a Videre Design platform (Fig. 1d). In addition to the navigation component, the robot is equipped with a kinect that can perform several functions, including the ability to detect a laser dot produced by a laser pointer, that the user exploits to point at the objects that the system should store in the map. The system can also acquire the image and the point cloud associated with the objects pointed by the user, later used to recognize previously seen objects.

The user can interact with the system using natural language through the use of a suitable human-robot interface. This component is implemented as a separate subsystem that can be deployed on different robotic platforms; it includes a speech processing component and a natural language processing chain that provide an interpretation of the user command in terms of frames, representing the commands executable by the robot. The knowledge acquired by the robot through the interaction with the user is stored in a multi-layered knowledge base, which contains the semantic knowledge about the environment, structured according to an abstract representation that is automatically built from the conventional 2D map.

The process of building the representation of the robot's knowledge is composed by the *Metric Map and Instance Signatures Construction Phase*, where a 2D metric map is generated through a SLAM module and the initial knowledge is extracted, and by the *Semantic Grid Map and Topological Graph Generation Phase*. In this latter phase, starting from the 2D metric map, a grid-based topological representation (*Semantic Grid Map*) is obtained, later used to produce the topological graph needed by the robot to perform high level behaviours.

---

[1] http://wiki.ros.org/move_base.

[2] http://www.dis.uniroma1.it/~spqr/MARRtino.

More in detail, in the first phase, the robot is used to navigate the environment in order to acquire the 2D map (using a Graph-based SLAM approach [14]) and to register the positions of the different objects of interest. During the robot exploration the user can in fact tag a specific object by using a commercial laser pointer. While the object is pointed through the laser, the user has to name it, so that a label can be assigned to it and its image and point cloud can be memorized. The registered object poses with the corresponding labels are processed to create the Semantic Grid Map and the Topological Graph. The Semantic Grid Map contains a high-level description about the regions, structural elements, and objects contained in the environment. The algorithm used to generate such a map, rasterizes the metric representation of the map into a grid-based topological representation, automatically labeling the areas of the environment (using contour closure and region filling techniques) and including representations of the objects described by the user. In the final step of the knowledge building process, a topological graph is created in order to represent the information needed by the robot for navigating and acting in the environment. The constituting nodes of this graph are locations associated to cells in the Semantic Grid Map, while the edges are connections between these locations (for a more detailed description about the representation and its building process we refer to [4]).

4 Experimental Approaches Analysis

Analyzing the literature on semantic mapping, no standard references for performing a correct evaluation of a system can be found. Due to this fact, in order to better explain why the evaluation part has been carried on as described in Sect. 5, this section will be dedicated to the problem of evaluating a semantic map.

The standard evaluation methods that can be found in the literature on Simultaneous Localization and Mapping (SLAM) consist in testing a particular system by processing a set of raw sensors data and then comparing the obtained output with a ground truth (see for example the Victoria Park Dataset [15]). Such a comparison is feasible thanks to the standard output generated by every SLAM algorithm. This is not the case for semantic mapping systems. In this particular research area, in fact, the output of each system is bound to subsets of the world model whose semantics is defined in an ad-hoc way; such an output is therefore hardly comparable with other systems. Due to this fact, research in semantic mapping has often focussed the evaluation on particular aspects of the proposed system.

An initial and probably the most simple evaluation approach adopted in the literature of semantic mapping consists in giving a qualitative evaluation of the output (usually a labelled metric map), by comparing it with a hand made ground truth (see for example [3]). While this method gives an idea of how well a system can perform and it is used to focus the evaluation process on the metric output, it can not be used to compare two different semantic mapping approaches. Moreover, while it is possible to compare at least qualitatively the metric output of the system, an evaluation of the semantic information stored in the map is usually not available.

Another testing method adopted to evaluate how well a system can acquire semantic information about an environment has been inspired by the literature on classifiers and consists in testing the system in a task of environment classification, by measuring the percentage of correctly classified places during a variable number of runs (see [7, 10]). This approach raises two issues: it implicitly assumes that it is possible to classify a place by the objects enclosed in it (e.g., it is not clear how to evaluate the classification of a room with a stove and a bed in it); it reduces the semantic mapping problem to a specific classification problem, not evaluating the system ability in acquiring other types of knowledge outside the ones needed for classification (e.g., spatial properties of the environment, objects' affordances, positions, and dimensions, etc.).

An alternative evaluation for the capability of acquiring semantic information consists in measuring the benefits gained from the addition of the acquired knowledge during a typical task executed by a robot. For example, semantic information is sometimes used to improve the performance of SLAM tasks [16]. Evaluating the improvements achieved by acquiring semantic information during a test run can indeed be used to get an idea of how well a system performs the semantic mapping task. This approach is typically the most complete evaluation approach; however, it is still not clear to the research community what are the types of tasks that should be considered to perform a full evaluation of an arbitrary semantic mapping technique.

An additional test that can be performed on systems that enclose the human in the mapping process consist in user evaluation studies. In this kind of tests, a system is tested to see how well it can interact with a user and how effectively it handles multiple, complex and dynamic interactions with a user.

As shown by the above analysis, no clear methodology is available to evaluate the performance of a semantic mapping approach. Consequently, we have chosen to analyze each system component separately in a quantitative way and, in addition, to test our system as a whole, both quantitatively and qualitatively during task execution. Such testing evaluation is described in the next section.

5 Experiments

In this section we discuss the experiments performed to validate the system and the results gathered from them over the last months, mainly focusing on the developments obtained after [4]. In order to validate the approach discussed in this abstract, we have tested both the single constituting components and the whole system. Specifically, we evaluated the spoken interaction, the Semantic Grid Map generation, the object segmentation and the spatial reasoning performed by the system. Since the spoken interaction and the Semantic Grid Map generation have already been tested in detail in [17] (the grammar based approach) and [18], respectively, in this section we will briefly report the results obtained for these two components, referring to the original articles for a more accurate evaluation and focussing our analysis on the object segmentation and the spatial reasoning components.

Table 1 Performance of the speech recognition component

Metric	P	R	F1
Action recognition	89.47	80.63	84.82
Full command recognition	75.43	67.98	71.51

5.1 System Component Evaluation

Spoken Interaction

The speech component has been designed mainly as a support for the *Augmented Mapping* task experiment described in this paper. For this part, we aimed at having a robust system, covering a controlled language with a low error rate in terms of transcription ability, instead of trying to deal with a wide range of linguistic phenomena. We therefore evaluated the performance of the Speech component with respect to the quality of the transcription of the user utterances and the command interpretation process. The former has been evaluated in terms of the Word Error Rate (*WER*) [19], obtaining a value of 0.258 on the transcription of commands uttered during the experiments. The second measurement has been carried out in terms of Precision (P), Recall (R) and F1-Measure (F1), as defined in [20]. The results obtained are reported in Table 1. Overall, the system satisfies the usability requirement, showing an acceptable performance during the interactions with the user, although covering a limited range of linguistic phenomena. Ongoing work is thus being carried out in order to improve this specific system component.

Semantic Grid Map Generation

In order to evaluate the Semantic Grid Map representation, a detailed set of experiments has been conducted. During these experiments, a set of 10 different metric maps has been processed by our system to get a qualitative evaluation of the capabilities of the system. An example of processed map and its output representation is shown in Fig. 2. When the objects are placed in the Semantic Grid Map, errors in their

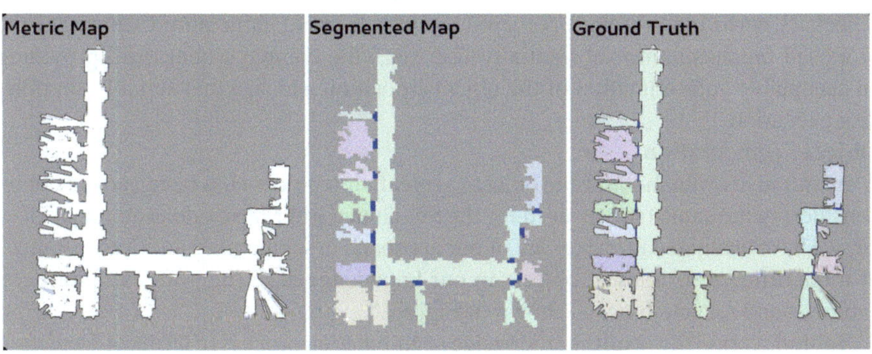

Fig. 2 Representation obtained for a metric map and respective ground truth

Table 2 Comparison between the pixels of each processed metric map and the cells of the corresponding Semantic Grid Map

Map	Pixels	Cells
BelgioiosoCastle	768,792	11,600
dis-B1	1,080,700	10,290
dis-B1-part	501,840	7,372
dis-Basement	992,785	13,455
FortAPHill	534,520	7,878
Freiburg	335,248	4,794
HospitalPart	30,000	285
Intel	336,399	4,473
scheggia	92,984	1,116
UBremen	831,264	10,962

Table 3 Average error evaluation for the width (W), depth (D) and area (A) of the objects in the Semantic Grid Map (SGM), normalized with the ground truth values

Object	Avg. SGM cells	Avg. Err. e_W	Avg. Err. e_D	Avg. Err. e_A
Cabinets	4.2	0.31	0.22	0.44
FireExtinguishers	1	1.13	0.67	2.6
RecycleBins	4	0.64	0.82	2.02

positions and dimensions are introduced because of the discretization of this map. To this end, we performed an additional evaluation by considering 11 instances of 3 different categories of objects in our department in order to measure their position and size errors with respect to a manually built ground truth. The results obtained are reported in Tables 2 and 3.

In general, even if the error for the object area can reach values around 3, loosing precision is still acceptable from the point of view of the task execution, since after reaching the desired location on the semantic map, an accurate localization of the objects is performed through perception. Overall, the data acquired show that the proposed representation substantially decreases the computational load, providing an acceptable approximation of the objects' position and size that suitably supports task execution.

Object Segmentation

A quantitative evaluation for the object segmentation process has been carried out by considering the same objects used for the Semantic Grid Map evaluation. In particular, we have evaluated the accuracy of our approach in segmenting multiple instances of three different classes of objects in our knowledge base (i.e., fire extinguishers, cabinets, and recycle bins) as shown in Fig. 3.

Table 4 reports the results of the image segmentation process in terms of Detection Rate (*DR*) and False Alarm Rate (*FAR*), computed as follows:

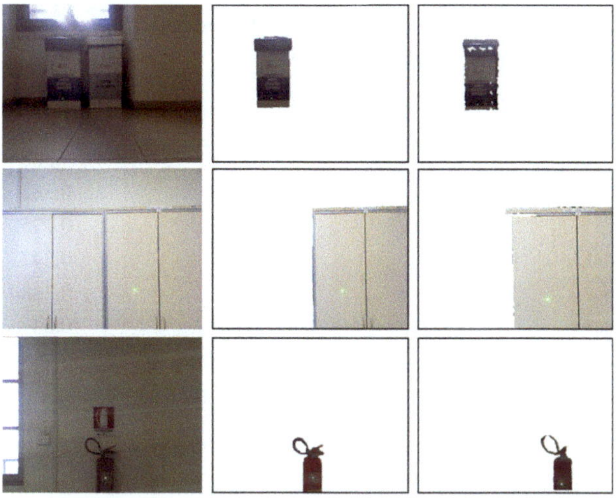

Fig. 3 Three classes of objects have been selected for the quantitative evaluation of the object segmentation module: fire extinguisher, cabinet, and recycle bin. In the first column the images of the objects are reported, while the manually obtained ground truth images for the silhouettes of the objects are shown in the second column. The third column contains the results of the segmentation process

Table 4 Error for the object segmentation module in terms of Detection Rate (*DR*) and False Alarm Rate (*FAR*)

Object	DR	FAR
Cabinet1	0.865	0.055
Cabinet2	0.946	0.010
Cabinet3	0.622	0.000
Cabinet4	0.841	0.037
Cabinet5	0.911	0.022
FireExtinguishis1	0.621	0.151
FireExtinguishis2	0.677	0.151
FireExtinguishis3	0.795	0.280
RecycleBin1	0.892	0.195
RecycleBin2	0.839	0.119
RecycleBin3	0.900	0.502
RecycleBin4	0.628	0.022

$$DR = \frac{TP}{TP + FN} \qquad FAR = \frac{FP}{TP + FP}$$

where *TP* are the true positives, i.e., correctly segmented pixels, *FN* are the false negatives, i.e., the number of object points detected as background, and *FP* are the

Table 5 Error in extracting the width (W size) of the tagged object

Object	Ground truth W (cm)	Detected W (cm)	Err. e_W
Cabinet1	100	96.56	0.034
Cabinet2		76.03	0.239
Cabinet3		79.16	0.208
Cabinet4		138.20	0.382
Cabinet5		80.50	0.195
FireExtinguishis1	15	11.29	0.247
FireExtinguishis2		11.72	0.218
FireExtinguishis3		15.71	0.047
RecycleBin1	38	44.30	0.165
RecycleBin2		29.25	0.230
RecycleBin3		79.30	1.086
RecycleBin4		34.85	0.082

false positives, i.e., the number of background points detected as object points. Low values for *DR* are mainly caused by holes in the depth data, especially along the borders of the objects. High values for *FAR* are mainly caused by a slight misalignment between the RGB image and the depth map provided by the sensor. The highest *FAR* value is obtained in the case of *RecycleBin3* since part of a cabinet alongside the tagged recycle bin is incorrectly segmented as part of it.

Since the final goal of our framework is to acquire knowledge for generating an accurate semantic map, we evaluate also the precision of our segmentation method in extracting the width (W size) of the tagged objects. The results are reported in Table 5. The error e_W is calculated as follows:

$$e_W = \frac{\left| detected_W - GT_W \right|}{GT_W}$$

where $detected_W$ is the width detected by our segmentation algorithm and GT_W is the ground truth width. The analysis of the results suggests that the proposed approach can recover the W size of the tagged objects with an acceptable error e_W. The highest e_W value is caused by the erroneously segmented *RecycleBin3*. It is worth noticing that in such a case the system memorizes the tagged object. However, since the W value for *RecycleBin3* is not coherent with the object properties stored in the conceptual KB, a clarification dialog has been implemented to flag this error.

Spatial Reasoning

Several tests have been conducted in order to demonstrate the improvements that qualitative spatial reasoning can determine in grounding the commands given by the users to a robot, as well as the efficacy of implementing such an approach on a real robot. Our validation work has been therefore focused on two different kinds of experiments.

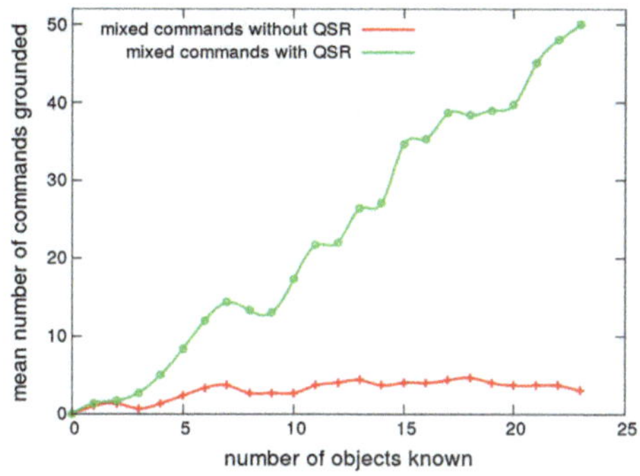

Fig. 4 Mean number of grounded commands with respect to the number of objects known in the environment, added in a random order

The purpose of the first experiment was to evaluate the impact of a qualitative spatial reasoner on an agent whose amount of knowledge continuously grows, as well as the influence of the already available knowledge on such a reasoning. Such an evaluation has been carried out by considering the number of unambiguous and ambiguous commands (i.e., commands referring to more than one object with a specific spatial property) grounded by the agent. Indeed, when full knowledge about the environment is available, grounding ambiguous commands would mostly lead to the execution of the wrong action with respect to the user expectation, while all the unambiguous commands are supposed to be correctly grounded. We therefore analyzed first the impact of the presence or absence of the qualitative spatial reasoner (QSR) and then the impact of the amount of knowledge available to the agent. In detail, we first asked to 26 students to provide a set of 3 commands containing spatial relations between objects, by looking at pictures of the test environment. Then, from the 78 acquired commands, we extracted two types of tasks: 28 ambiguous and 50 unambiguous. By gradually adding knowledge about the objects inside the knowledge base of the agent, we therefore measured how many commands were grounded. We repeated the experiment for both categories of commands, with or without the qualitative spatial reasoner. Since the curves depend on the order of the objects inserted in the knowledge base, the experiment has been performed five times in order to obtain its average trend (Fig. 4). In case the QSR was not present (red curve), only the objects in the environment, whose category has a unique member, were correctly identified. For example, since we had two cabinets in the test environment, there was no way of distinguish them without exploiting spatial relations. By comparing the two curves in the image, it can be noticed that the presence of the QSR does not greatly affect their trend when a little amount of knowledge is available, due to the absence of exploitable spatial relations between objects. On the contrary this is not true when substantial environmental information is accessible. Note that, when a complete knowledge about the relevant elements of the environment is known by the robot, the number of grounded commands, as expected, is equal to the number of unambiguous phrases (50 commands) present in the adopted set of commands.

Table 6 Number of correctly and wrongly grounded commands with respect to the expectations of the users

User	Correctly grounded commands	Wrongly grounded commands
1st	7	3
2nd	8	2
3rd	10	0
4th	6	4
5th	8	2
6th	8	2
7th	10	0
8th	7	3
9th	9	1
10th	8	2
Total	81	19

The second experiment performed aimed at understanding the limitations of the proposed approach. To this end, we measured the agreement between the user expectations and the grounding performed by the robot. In particular, we first produced a Semantic Grid Map by driving the robot on a tour of the environment and tagging 23 objects within an office environment, as well as the doors and the functional areas in it. Then, we asked 10 different non-expert users to assign 10 distinct tasks to the robot, additionally asking them to evaluate whether the robot correctly grounded their commands, meeting their expectations. The commands have been directly acquired through a Graphical User Interface, in order to avoid possible errors due to misunderstandings from the speech recognition system. In detail, the users had the possibility to choose the action to be executed by specifying the located object, the reference object and one of the 10 spatial relations implemented in our reasoner. Table 6 shows that approximately 80 % of the given commands have been correctly grounded. The remaining 20 % of wrongly grounded commands where due to two different phenomena: (i) the command given was ambiguous, requiring other properties, in addition to direction and distance, to identify the object; (ii) the users did not behave coherently during the interaction with the robot, by varying their concept of vicinity or by adopting different reference frames.

5.2 *Whole-System Evaluation*

For evaluating the system as a whole, three kinds of experiments have been performed, two qualitative and one quantitative. A first set of tests has been carried out to verify the mapping procedure and the automatic construction of the representation of different kinds of environments. The main focus of this first set of experiments has been on demonstrating how a robot, being deployed in an unknown environment, can be endowed with the ability of acquiring specific knowledge of the environment and later using it to accomplish motion tasks. For this type of qualitative validation, two different kinds of environments have been taken into consideration: homes and offices. More specifically, as described in Sect. 3, we have deployed our system on four different mobile bases in the office spaces of our department and in two different houses. During these tests, several non-expert users have been asked to guide the robot in discovering the environment and the objects in it. After having acquired the specific information about the environment, the users have also been asked to assign simple tasks to the robot through natural language, such as "move in front of the couch next to the tv-set", in order to test the consistency of the produced environmental representation. In particular our system has been tested in:

- the basement and the first floor of our department. In this environment we mapped four different laboratories and ten offices, as well as the corridors that connects them and asked several non-expert users to tag multiple objects during an open day of our lab.
- the ground floor of a house of one of the authors. With a couple of hours of work we were able to enter an unknown environment, extract a metric map of it and create a semantic map usable to fulfill the commands uttered by a user. In particular, a small environment composed by a kitchen and a living room was mapped and 41 different objects were successfully and easily added in the robot's knowledge with the aid of multiple users.
- a domestic environment used at Örebro University for domotic applications. During the First Örebro Winter School on "Artificial Intelligence and Robotics",[3] we created a representation of the apartment composed of a kitchen, a living room, a bed room and a dining-room. As part of their practical activity during the course, the students that participated in the school were invited to help the robot acquiring the knowledge about the objects in the environment. 15 different objects were tagged during this process. An image of the semantic map gathered during the school can be seen in Fig. 5.

The second set of tests was performed in order to validate the system in a long-run. We are in fact interested in understanding whether the developed approach is suitable for long-life learning and how well the produced representation can be consistently updated over time. To this end, we developed an on-line mapping experiment, where the segway and the Videre design robot were deployed for three weeks in our department. During this period, the robots interacted with multiple users in order to keep track of the objects that could change position over time. Twenty different object types that changed position over time were thus tagged and stored in the semantic map of the environment. Videos of some of the experiments and several data acquired during them can be found at http://www.dis.uniroma1.it/~gemignani/Articles/iser14.html.

[3]http://aass.oru.se/Agora/Lucia2013/.

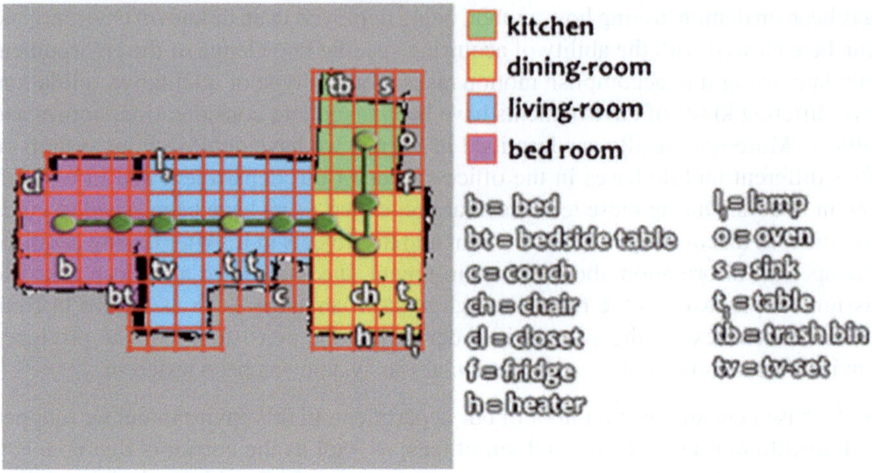

Fig. 5 Domestic environment mapped by the students during the First Örebro Winter School on "Artificial Intelligence and Robotics". The Topological Graph is depicted on *top* of the Semantic Grid Map and the objects in it. The metric map is also depicted in the background

The goal of the final quantitative experiment was to evaluate the whole system in a real environment during a typical task executed by the robot. For this reason we deployed our robot in an office environment and we asked both expert and non-expert users to drive the robot around using the vocal interface and to tag the various objects present in the environment. To test the robustness of our system in a noisy environment, we carried out a data collection during a public opening of our department asking 10 visitors, in addition to all of the authors of this paper (for a total of 16 users), to take part in the following experiment. The robot started with no knowledge about the objects enclosed in the environment and each user, after being explained for a minute the commands understood by the robot, had to drive, using the vocal interface, the mobile platform in front of a desired object and teach the robot its position and name. Having memorized different objects, the user had to ask the robot to move in front of them in order to demonstrate that the learning process had been carried out successfully. In this experiment all the users have been able to successfully memorize an object, thanks to the behaviors implemented on the robot that allowed to overcome the system components' limitations. After collecting the data needed, we calculated the distance between the position of the centroid of the learned objects with the one belonging to a ground truth manually created. The result of such a comparison is shown in Table 7. From the table it can be noticed that almost 90 % of the objects were placed with an error less than 50 cm. The remaining objects were placed at a distance between 50 cm and 1.5 m due to errors deriving from the object segmentation component, the Semantic Grid Map Generator and the robot pose localizer. It can also be noticed that the precision seems not to vary between expert and non-expert users, thus suggesting that this system does not require a specific training to be used. Overall, the evaluation of the performance shows that the

Table 7 Result obtained from the test performed on the whole system

Distance thresholds (m)	Average (%)	Experts (%)	Non-experts (%)
≤0.1	18	20	16
≤0.2	42	37	47
≤0.3	48	46	50
≤0.4	76	72	80
≤0.5	88	94	82

The position of the tagged objects is compared with the one obtained from a manually generated ground truth by calculating the distance between the two points

system can effectively acquire knowledge about the environment, allowing for the representation in the semantic map of a wide variety of elements. The evaluation also shows that several aspects of the system could be improved. In our view, the most critical improvement would arise from a tighter integration between state of the art techniques for object detection and categorization. Finally, the results of the final experiment with the users show that the approximations that have been introduced in the representation do not affect the execution of the task, thus providing some evidence of a good balance between abstraction and accuracy reached in our representation.

6 Conclusion

The experiments performed with our system show that our semantic mapping approach can be effectively deployed to build, represent and process environmental knowledge, acquired through the aid of the user. Indeed, this approach clearly supports the thesis that symbiotic autonomy [2] can help to make a step forward in the current robotic capabilities. Moreover, as it has been demonstrated by the deployment of different robotic platforms, the proposed approach is both independent from the chosen robotic platform and also independent from the user interacting with it. Such features allow for an easy deployment of various mobile bases over different experimental scenarios.

Summarizing, a simple, yet effective interaction with the user allows to build a semantic representation of the environment that is much richer and more accurate than existing automatic and user-guided approaches to semantic mapping. Indeed, the proposed approach can be substantially empowered by exploiting some of the state of the art approaches to automatically classify spaces, or to detect and classify objects. Specifically, the robot can take a more proactive role in handling knowledge that can be autonomously acquired through perception either by adding it in the semantic map or by querying the user about it, further developing the approach towards symbiotic autonomy. As a matter of fact, the proposed approach shows a different perspective on the implemented robot capabilities: the system performs intelligent behaviors (or it has an improved performance) not by fully relying on general knowledge, rather by acquiring specific knowledge about the operational environment. This shift of viewpoint, that is enabled by the interaction with the user, is applicable not only to the knowledge about the environment, but also in the knowledge about the tasks to be performed and also about the users of the system.

A second outcome of the proposed experimental setting is the notion of online semantic mapping. This should not be regarded just as a natural extension of the off-line procedure, that enables the robot to accumulate knowledge during operation; more generally, an online semantic mapping capability is needed to enable the robot to continuously adapt to the environment that changes over time. In this respect, our experiments on long-term performance of the robot brought up several interesting research challenges:

- update of the knowledge about objects in the face of new knowledge acquired either through perception or from the user (or different users);
- learn the spatio-temporal relations among the objects in the environment;

Our future research will focus on experiments that encompass the deployment[4] of the robot for long periods of time, thus allowing to investigate the above issues.

Acknowledgments We would like to thank Joachim Hertzberg for insightful discussions on the experimental evaluation of semantic mapping. Moreover, we acknowledge the contribution of Emanuele Bastianelli and Taigo M. Bonnani to the implementation of the system. This work is part of the activities in the RoCKIn Coordination Action[4], which is focussing on benchmarking of home robots through competitions.

References

1. Nüchter, A., Hertzberg, J.: Towards semantic maps for mobile robots. Robot. Auton. Syst. **56**(11), 915–926 (2008)
2. Rosenthal, S., Biswas, J., Veloso, M.: An effective personal mobile robot agent through symbiotic human-robot interaction. In: Proceedings of 9th International Joint Conference on Autonomous Agents and Multi-Agent Systems (AAMAS), pp. 915–922 (2010)
3. Randelli, G., Bonanni, T.M., Iocchi, L., Nardi, D.: Knowledge acquisition through human robot multimodal interaction. Intell. Serv. Robot. **6**, 19–31 (2013)
4. Bastianelli, E., Bloisi, D.D., Capobianco, R., Cossu, F., Gemignani, G., Iocchi, L., Nardi, D.: On-line semantic mapping. In: 16th International Conference on Advanced Robotics (ICAR), Nov 2013, pp. 1–6
5. Hertzberg, J., Saffiotti, A.: Using semantic knowledge in robotics. Robot. Auton. Syst. **56**(11), 875–877 (Semantic Knowledge in Robotics)
6. Galindo, C., Saffiotti, A., Coradeschi, S., Buschka, P., Fernández-Madrigal, J., González, J.: Multi-hierarchical semantic maps for mobile robotics. In: Proceedings of the IEEE/RSJ International Conference on Intelligent Robots and Systems (IROS), pp. 3492–3497 (2005)
7. Goerke, N., Braun, S.: Building semantic annotated maps by mobile robots. In: Proceedings of the Conference Towards Autonomous Robotic Systems, pp. 149–156 (2009)

[4] http://www.rockinrobotchallenge.eu.

8. Brunskill, E., Kollar, T., Roy, N.: Topological mapping using spectral clustering and classification. In: Proceedings of IEEE/RSJ Conference on Robots and Systems (IROS), pp. 3491–3496 (2007)
9. Friedman, S., Pasula, H., Fox, D.: Voronoi random fields: Extracting the topological structure of indoor environments via place labeling. In: Proceedings of 19th International Joint Conference on Artificial Intelligence (IJCAI), pp. 2109–2114 (2007)
10. Mozos, O.M., Mizutani, H., Kurazume, R., Hasegawa, T.: Categorization of indoor places using the kinect sensor. Sensors **12**(5), 6695–6711 (2012)
11. Gunther, M., Wiemann, T., Albrecht, S., Hertzberg, J.: Building semantic object maps from sparse and noisy 3D data. In: IEEE/RSJ International Conference on Intelligent Robots and Systems (IROS), IEEE, pp. 2228–2233 (2013)
12. Zender, H., Martínez, M.O., Jensfelt, P., Kruijff, G., Burgard, W.: Conceptual spatial representations for indoor mobile robots. Robot. Auton. Syst. **56**(6), 493–502 (2008)
13. Pronobis, A., Jensfelt, P.: Large-scale semantic mapping and reasoning with heterogeneous modalities. In: Proceedings of the 2012 IEEE International Conference on Robotics and Automation (ICRA'12), pp. 3515–3522 (2012)
14. Grisetti, G., Kuemmerle, R., Stachniss, C., Burgard, W.: A tutorial on graph-based SLAM. Intell. Transp. Syst. Mag. IEEE **2**(4), 31–43 (2010)
15. Guivant, J.E., Masson, F.R., Nebot, E.M.: Simultaneous localization and map building using natural features and absolute information. Robot. Auton. Syst. **40**(2), 79–90 (2002)
16. Nüchter, A., Wulf, O., Lingemann, K., Hertzberg, J., Wagner, B., Surmann, H.: 3D mapping with semantic knowledge. In: Robot Soccer World Cup IX RoboCup 2005, pp. 335–346 (2005)
17. Bastianelli, E., Castellucci, G., Croce, D., Basili, R., Nardi, D.: Effective and robust natural language understanding for human robot interaction. In: 21st European Conference on Artificial Intelligence, ECAI 2014 (2014)
18. Capobianco, R., Gemignani, G., Bloisi, D.D., Nardi, D., Iocchi, L.: Automatic extraction of structural representations of environmentss. In: Proceedings of the 13th Intelligent Autonomous System conference (2014)
19. Popović, M., Ney, H.: Word error rates: Decomposition over pos classes and applications for error analysis. In: Proceedings of the Second Workshop on Statistical Machine Translation, StatMT'07, Stroudsburg, PA, USA, Association for Computational Linguistics, pp. 48–55 (2007)
20. Croce, D., Castellucci, G., Bastianelli, E.: Structured learning for semantic role labeling. Intelligenza Artificiale **6**(2), 163–176 (2012)

Experimental Analysis of a UAV-Based Wireless Power Transfer Localization System

Andrew Mittleider, Brent Griffin and Carrick Detweiler

Abstract Sensors deployed in remote locations provide unprecedented amounts of data, but powering these sensors over long periods remains a challenge. In this paper, we develop and present a UAV-based wireless power transfer system. We discuss design considerations and present our system that allows a UAV to fly to remote locations to charge hard to access sensors. We analyze the impact of different materials on the wireless power transfer system. Since GPS does not provide sufficient accuracy, we develop and experimentally characterize a relative localization algorithm based on sensing the magnetic field of the power transfer system and optical flow that allows the UAV to localize the sensor with an average error of 15 cm to enable the transfer of on average 4.2 W. These results overcome some of the practical challenges associated with wirelessly charging sensors with a UAV and show that UAVs with wireless power transfer systems can greatly extend the life of remotely deployed sensors.

1 Introduction

Wireless sensor networks are used in a wide range of applications from large scale terrestrial habitat monitoring [1] to underground [2] and underwater [3] systems because of their ability to measure a multitude of environmental variables with high frequency over long periods of time. Their ability to monitor these locations over long

---

A. Mittleider · B. Griffin · C. Detweiler (✉)
Nebraska Intelligent MoBile Unmanned Systems (NIMBUS) Lab,
Department of Computer Science and Engineering, University of Nebraska-Lincoln,
Lincoln, NE 68588, USA
e-mail: carrick@cse.unl.edu
URL: http://nimbus.unl.edu

A. Mittleider
e-mail: amittlei@cse.unl.edu

B. Griffin
Department of Electrical Engineering and Computer Science, University of Michigan,
Ann Arbor, MI 48109, USA
e-mail: bagriffin@gmail.com

M.A. Hsieh et al. (eds.), *Experimental Robotics*, Springer Tracts
in Advanced Robotics 109, DOI 10.1007/978-3-319-23778-7_24

periods of time is significantly advancing science, however, powering these sensor networks remains a challenge despite advances in energy efficient sensor networks and battery technology. Current systems deployed for long periods either require additional infrastructure (e.g. power cables or solar panels) or periodic maintenance to replace batteries. We have developed a novel solution to this problem by using a UAV equipped with a resonant magnetic wireless power transfer system to charge sensors in hard-to-access locations [4]. The system can transfer over 10 Watts at close distances and a range of about 1 m. The system enables charging of sensors embedded in materials or in hard to access locations where physical charging connectors are impractical.

In this paper, we address the problem of getting the UAV close enough to the ground sensor to enable charging. We do this by developing a relative localization system based on sensing the magnetic field emitted by the wireless power transfer system and an optical flow camera. We use a weighted least-squares minimization approach based on the magnetic field model and empirical measurement estimations of the relative location. Precise relative localizing is critical since the positional error in the GPS that the UAV uses to navigate is too large to enable efficient charging. In addition, we want to avoid adding hardware to the UAV, as the additional payload would decrease flight time (e.g. adding a laser scanner to perform SLAM). The localization system we develop requires minimal additional hardware and we find experimentally that the average localization error in our system is 21 cm, which yields an average power transfer rate of 4.2 W. In addition, we investigate and show experimentally that the magnetic field sensing is not impacted by common materials non-magnetic materials such as dirt, woodchips, stone, and plastic. This means that sensors can be embedded, for instance in a field to monitor crops as shown in Fig. 1, and still be localized and charged.

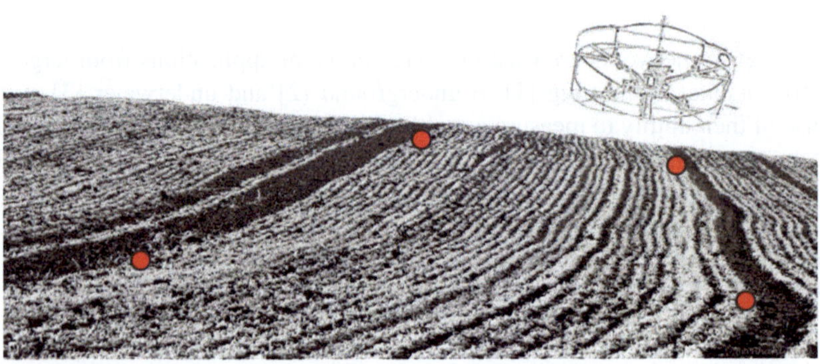

Fig. 1 Illustration showing a UAV charging WSN nodes embedded in a field

2 Related Work

There are a variety of techniques that can be used to wirelessly transmit energy ranging from directed microwave energy [5] to radio frequency power harvesting [6]. We use magnetic resonance wireless to transfer power from our UAV. This is related to systems developed by Tesla over a century ago [7] and more recently researchers have shown that it is possible to transfer tens of Watts over medium distances (few meters). For instance, Kurs et al. transferred 60 W over 2 m with 60 % efficiency to power a light bulb [8]. Duong and Lee [9] use a variable coupling technique to improve power transfer efficiency at different ranges, and Sample and Meyer [10] significantly reduce wireless power inefficiencies by developing an algorithm that automatically tunes the drive frequency during over-coupling of the system. Instead of focusing on improving the efficiency of magnetically resonant wireless power transfer systems, our research focuses on obtaining precise relative localization between the UAV carrying the power transfer system and the sensor node being charged. We also examine the impact of embedding the sensors we are localizing and charging in different materials. Prior work by Seo et al. examined the impact of building materials such as softwood lumber, concrete brick, and drywall with insulation on wireless power transfer [11]. We add to this by also examining common outdoor materials and looking specifically at how these impact our magnetic resonant localization sensor.

Our work is inspired by Moore and Tedrake [12] who use a magnetic field to localize a glider UAV to perch on a power line. They use a standard magnetometer to sense the magnetic field from low-frequency, high-current power lines (40 Amps). They use a Kalman filter and track the magnetic field and eliminate positional ambiguity in the magnetic field sensors. They conclude that their method of localization is sufficient for a small UAV to perch on a power line, however, their system does not fully close the loop to enable control based on the sensor readings. Our work differs from theirs in a number of ways. First, our system operates at higher frequencies and lower currents so standard magnetometers do not work. Instead we develop a sensor that measures the voltage in a small resonant coil (see Sect. 4.2). Second, we are primarily interested in having a final position close enough to transfer power and are less interested in the trajectory during approach, which is critical to enable perching. Thus, we do not need to consider the details of vehicle dynamics, which allows us to easily adapt our approach for new vehicles. Finally, we demonstrate our system and perform full closed-loop control.

Also related to our work is methods for localizing based on radio transmissions. For instance, Tokekar et al. used bearing and signal strength to locate radio tagged fish with a robotic boat [13]. They use many wireless sensor network (WSN) nodes equipped with rotatable antenna to locate the fish. First, they map a relationship between the signal strength and the distance from a ground truth measurement. They fit a linear regression model of this data. When the radio strength is at its maximum value, the bearing and radio strength information is sent to a centralized computer.

The position of the fish is then triangulated by creating an enclosed polygon with the bearing angles.

Finally, while we are one of the first to demonstrate charging of ground sensors from an aerial robot, the reverse problem of how to power an aerial vehicle was looked at in 1964 to wirelessly power a flying helicopter [14] and in 2011 was used to enable a 12 h, record-length quad-rotor flight [15].

3 Power Transfer Design Considerations

In this section we present background information on wireless magnetic resonant power transfer and discuss design considerations for UAV-based power transfer systems. Inductive power transfer involves at least two coils in close proximity sharing alternating magnetic fields. An alternating current (AC) in the transmitting coil produces a magnetic field that generates an alternating voltage in the receiving coil that can be applied to power or charge a device. Standard inductive power transfer is only efficient over short distances, but this limitation can be overcome with the use of strongly coupled magnetic resonances.

Power transfer is much more efficient over medium ranged distances by including two coupled resonant coils between the driven and loaded inductive coils, as seen in Fig. 2. In this configuration, the primary inductive coil, or Drive coil, is driven by an AC power supply. Due to the close proximity between the Drive coil and the first resonant coil, called the Tx resonant coil, oscillations occur and power is transferred to the Tx coil. The Tx coil causes the Rx coil to oscillate with a proportional degree of energy that is dependent on their coupling. The Tx and Rx coils do not have any direct load connected to them to interfere with the resonance. This allows them to couple and resonate over larger distances than is possible without resonant coils. The last coil, the Load coil, inductively receives power from the Rx coil in the same way

Fig. 2 Schematic for resonant power transfer

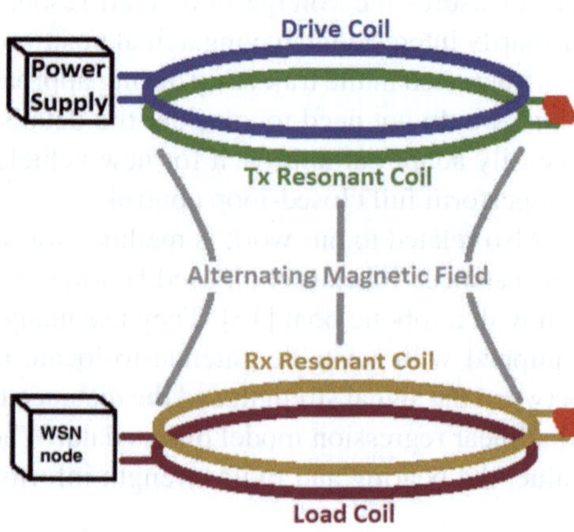

that the Drive coil transfers energy to the Tx coil, and it applies the voltage that it gains across a load to receive the power.

The two primary factors that impact resonant wireless power transfer performance are the quality factor (Q) of the coils and ensuring that all coils naturally resonate at a similar frequency. The quality factor represents how well a resonant coil can hold energy without losses to heat and is defined as [16]:

$$Q = \frac{1}{R}\sqrt{\frac{L}{C}} \tag{1}$$

where R is the resistance of the coil (Ω), L is the inductance of the coil (H), and C is capacitance (F). For details on how to compute these parameters see [17].

Intuitively, the quality factor can be thought of as how much energy a resonant system can hold compared to energy lost during a single cycle. However, as previously stated, while maximizing the quality factor is important when trying to minimize the losses within the coils, a high Q may have a detrimental effect. In practice, the resonant frequency of all of the coils will be slightly different from each other due to manufacturing imperfections and component tolerances. High Q factors cause a decrease in the bandwidth of the resonant coil, Δf in Hz, which is defined as: $\Delta f = \frac{f_r}{Q}$.

If the resonant frequency of one coil is not within the bandwidth range of the other coil, they will couple poorly as shown in Fig. 3. The figure represents two sets of resonant coils. The solid blue curves represent coils which have a resonant frequency f_1 and f_2, respectively, and have a high Q. The higher the curves intersect on the y-axis, the higher the power transfer. Notice that the high Q means that the peak power transfer is higher if they are aligned, but because of the difference in the coils natural resonant frequency (e.g. manufacturing tolerances), they are only able to transfer at about half of the optimal performance. The red, dashed curves show coils with the same resonate at the same frequencies f_1 and f_2, but lower Q.

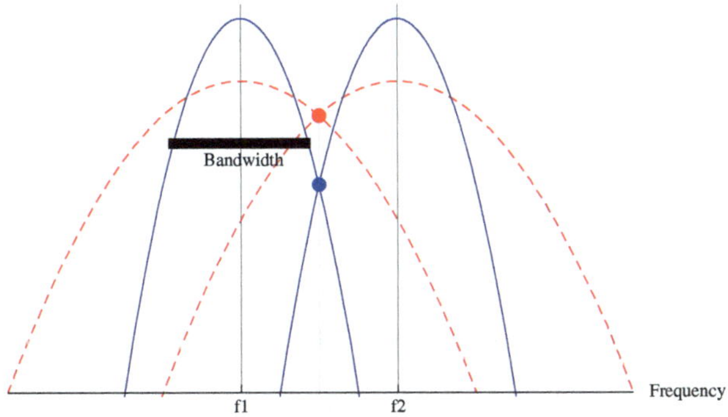

Fig. 3 Coils with high quality factor (*solid blue*) can have lower coupling than those with lower quality factors (*dashed red*) due to manufacturing tolerances

If they were precisely aligned they would have lower overall transfer, but since they have higher bandwidth they better tolerate misalignment that results in higher power transfer in this configuration. While fixed, ground based power transfer systems can be precisely tuned, this can be difficult for UAV-based systems or other field robotics applications. For instance to be lightweight, our coils flex significantly during flight and further after rough landings or crashes their shape becomes distorted, changing their resonant frequencies. So while having high Q is typically good, too high of a Q may reduce performance in many robot systems.

There are also a number of other considerations when designing resonant wireless power transfer systems. For low frequency systems energy losses are dominated by ohmic losses and high frequency losses by radiative losses [18]. The primary loss in our relatively low frequency system is ohmic, which causes the coils to heat when high currents move through the slightly resistive winds of the wire in each coil. In addition, we have switching losses in our AC power system. By decreasing the resistance, we can increase the Q and decrease the former losses. Unfortunately this may cause an increase in the switching losses due to higher currents moving through MOSFETs. Further, while using a thicker gauge wire is an easy way to decrease resistance and increase Q, it also adds weight to the UAV. Similar trade-offs must be made when adjusting the capacitance and inductance in the system to attempt to maximize the quality factor.

Now that we have briefly examined the theory behind magnetic resonant wireless power transfer and discussed design considerations, we go on to look at the specific components of our UAV-based wireless power transfer system and wireless sensor network node.

4 System Components

This section describes the hardware of the wireless power transfer system and the UAV. Figure 4 shows an overview of the wireless power transfer system on the UAV. The system consists of a UAV, the wireless power transmitter on the UAV, and the receiving wireless sensor node.

4.1 UAV Power Transfer System

Figure 4 shows an overview of the wireless power transfer system. On the UAV, the TX Drive Board sends an alternating current through the Drive Coil causing an alternating magnetic field that drives the neighboring Tx Resonant Coil. The Tx Resonant Coil serves to focus the field for transmission to the Rx Resonant Coil, which is placed on the WSN node along with the Load Coil. A magnetic resonant (MR) sensor is connected to the Rx Resonant Coil to detect the Tx system and enable localization. The Load Coil is connected to the receiving board, which draws energy

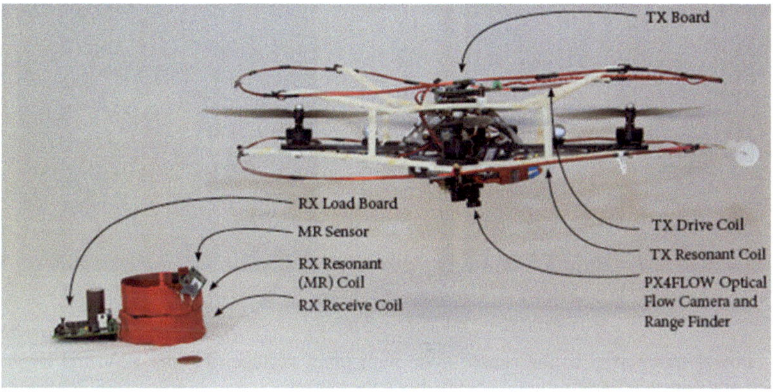

Fig. 4 The wireless power transfer system

Table 1 Power transfer system parameters for the UAV and WSN node coils

Description	Variable	Value
Coil length	l	1.47655 m
Coil radius	r	0.235 m
Resistance	R	0.0143 Ω
Number of wraps	N	2
Inductance	L	$5.20068 \times 10^{-6}\ \Omega$
Capacitance	C	1.5×10^{-7} F
Frequency	f_r	167 kHz
Bundle thickness	c	0.004 m
Quality factor	Q	411

from the Rx Resonant Coil. Finally, the energy from the Load Coil can be stored in the WSN node. Table 1 gives an overview of the specific parameters for the coils we use in our wireless power transfer system.

At the heart of the TX Drive Board is an AD9833 programmable waveform generator that can generate up to a 12.5 MHz signal. This signal is input into an H-Bridge that generates a high-power alternating current that is driven through the Drive Coil. Typically we operate with a 9–24 V input range with a current between 1–4 Amps, although the board is designed to support up to 45 V and 8 A with a maximum power rating of up to 100 W. In addition, the TX Board has a processor to control the frequency, enable or disable power transfer, monitor voltage and current, and communicate with the ground sensors and base station with a Zigbee radio.

Figure 5 shows the amount of power that can be received by the device attached to the load coil using an input of 12 V and 24 V. This figure shows that there is a region with about 30 cm radius which provides over 3 W of power transfer for 12 V input (our UAV operating voltage). As the distance from the transmitter to receiver

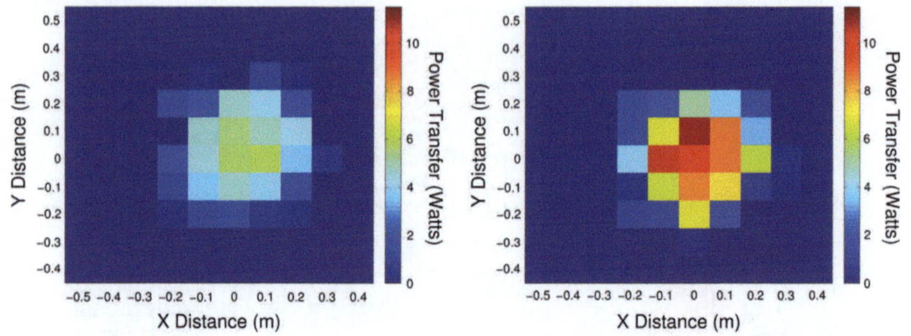

Fig. 5 Power transferred to the sensor node with input voltage of (*left*) 12 V and (*right*) 24 V recorded at a constant height roughly 5 cm above the sensor

increases past this radius, the amount of power transferred drops significantly. In Sect. 5 we show that the UAV can consistently localize to within 21 cm of the sensor. For our applications, 3 W in a 30 cm range suffices as this can charge a AAA battery with a single flight, which can power a WSN node for weeks or months. While we are less concerned with optimizing the power transfer system, it is important to note that the overall power transfer can be easily increased. Figure 5 (right) shows that we can double the power transfer, obtaining over 10 W by doubling the input voltage. This can be accomplished with an additional battery pack attached to the UAV or a UAV with higher operating voltage.

4.2 Wireless Power Receiver Sensor Node

A node in the wireless sensor network (WSN) consists of the wireless power receiver board, coils, a magnetic resonant sensor, and any other sensors that are specific to the nodes application, such as vibration, temperature, soil moisture, or pressure sensors. In this paper we omit any application specific sensing system and instead focus on the power transfer system and localizing the UAV onto the WSN node with the wireless power transfer system. A WSN node can receive about 6.1 W at peak efficiency. With 6.1 W power transfer for 5 min we can nearly charge a typical NiCd rechargable AAA battery, which can operate most types of low-power sensing systems for weeks. As with the transmitter, there is a Rx Resonant Coil in close proximity to the Load Coil. The receiver board draws energy from the Load Coil and may either use this energy directly or may charge batteries or super capacitors.

A Magnetic Resonant (MR) sensor, Fig. 6 (left), is connected to the Rx Resonant Coil and can detect the presence of the UAV power transfer system. When the Tx system approaches, the voltage in this resonant coil increases significantly and is

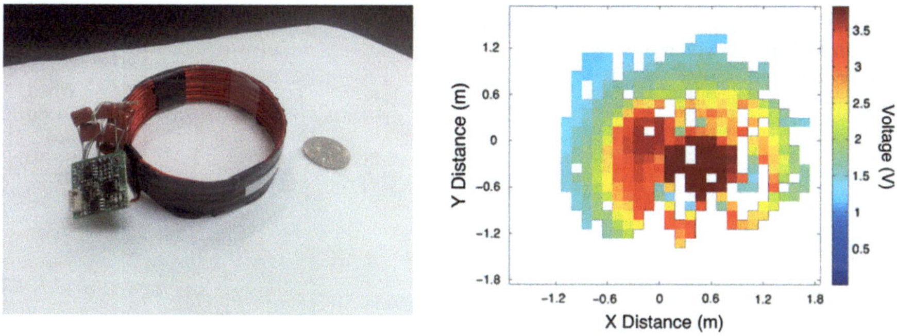

Fig. 6 (*left*) Magnetic resonant (MR) sensor, which reads the voltage through the resonant coil and (*right*) a heatmap of the values read 1.0 m above the sensor (*white* indicates areas of no data)

measured by the MR sensor. The advantage of having the MR sensor is that it can detect the power transfer system from three times farther away than the Rx Load Board and is thus the primary input to the localization algorithm. The MR sensor circuit consists of two opamps. The first controls the gain and the second is configured as a precision rectifier to transform the AC signal into a DC value that is then read by a microcontroller with a 10-bit analog to digital converter (ADC). Figure 6 (right) visualizes the data returned from the MR sensors. As the transmitter approaches the MR sensor, the voltage values approach a maximum value of about 3.7 V. As the distance between the MR sensor and the transmitter increase, the voltage values approach a minimum of around 1 V.

4.3 MR Sensing Through Different Materials

We imagine the WSN nodes charged by the UAV may be embedded in many different materials, for example, underground to monitor soil properties. After the UAV has localized the sensor, it can land on the node to maximize charging efficiency (removing the energy required for hovering). In this section, we test the effect of common outdoor materials on the MR sensor.

Figure 7 compares the voltage through the resonant coil that is embedded in plastic, wood, stone, soil, and steel compared with a baseline measurement of air. The x-axis is the horizontal distance, which is circularly symmetric around the MR sensor. The voltage on the y-axis is used for localization and is also directly correlated with the power transfer that the WSN node receives.

In this experiment, the WSN node was placed under at least 15 cm of each of the target materials (except for the steel, where it was covered with a single 1 mm sheet). The UAV flew a straight horizontal line directly over the WSN node at a constant

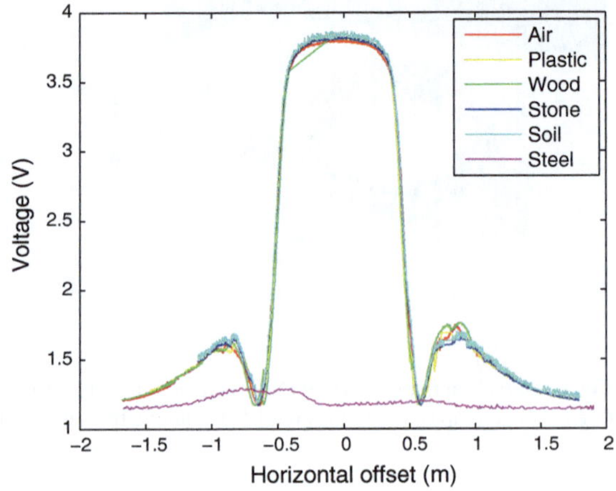

Fig. 7 Voltage readings from an MR sensor which is embedded in different materials

height of 1 m, and we record the values of the MR sensor. As Fig. 7 shows, the non-ferrous materials have no effect on the reception, however, the steel interferes with reception, as would other metals. This means that the sensors can be embedded in many common materials for long-term deployment.

5 Localization

In Sect. 4 we explored the details of the magnetic resonant power transfer system that is attached to the UAV and the WSN node. We showed that the system is able to transfer sufficient power to charge a low-powered WSN node. In this section we address the problem of getting the UAV close enough to the WSN node to transfer power. GPS can record the location of a WSN node when it is deployed, but GPS has up to 7.8 m error in a 95 % confidence range [19].

Since the UAV must be within 30 cm to efficiently transfer power, in this chapter we develop a localization algorithm that uses the sensed magnetic field information from the MR sensors to localize over the WSN node. We note that the information regarding intensity is measured by the MR sensor that is located with the WSN node. When the MR sensor first detects power transfer, it turns on its radio and starts broadcasting range information on a short-range radio to the UAV, so little energy is used when the UAV is not charging the node. In addition to using the information from the MR sensor, we also use information from an optical flow sensor on the UAV to provide better dead-reckoning capabilities. Using an optical flow camera for position and control of the UAV means that we add hardware to the UAV, but the WSN node can be significantly smaller. For the experiments presented in this section we simulate an optical flow camera by using a motion capture system with 0.2 m/s of Gaussian noise, which is the velocity error reported by the optical flow system developed by Honegger et al. [20].

5.1 Localization

In order to estimate the position of the UAV, we must map the voltage information received by the MR sensor to a position in space. First, we can get an estimation of the strength of the magnetic field based on the position of the UAV from an equation shown in [12]:

$$\bar{B}_i = \frac{\mu_0 N I_i a}{4\pi \hat{r_i}^3}[2cos(\hat{\theta}_i)\hat{r_i} + sin(\hat{\theta}_i)\hat{\theta}_i]. \tag{2}$$

where $\bar{B}_i$ is the strength of the magnetic field measured by the MR sensor in Tesla, μ_0 is the magnetic permeability of free space, N is the number of wraps of the coil, I_i is the current through the transmitter at time i, a is the area of the coil, $\hat{r}$ is the estimated radial distance from the UAV to the sensor, and $\hat{\theta}_i$ is the estimated angle from the UAV to the sensor.

The MR sensor provides an ADC value reading from 0 to 600 based on the voltage received by the surrounding magnetic field. These ADC values, α, are mapped to the field strength $\bar{B}$ by a lookup table function $\beta(\alpha)$. The lookup table is obtained using information similar to Fig. 7 that relates the relative position of the UAV and the MR sensor to the MR sensor reading.

Now that we have mapped the ADC values from the sensor to the field strength, we can develop a localization algorithm for the UAV using range readings from the MR sensor. In order to localize, we must have both the magnetic field strength and a position estimation of the UAV. However, since GPS error is too high, we use an optical flow camera, which can provide accurate motion estimates over short periods of time [21, 22] with higher accuracy than GPS.

The particular optical flow camera we attempted to use in our system (PX4Flow [21]) was very sensitive to its viewing surface. We tested many surfaces and found a minimum error of 0.2 m over a 10 s flight over a textured wood surface (similar to that reported by the developers), but very poor results with other materials. While the best error more than acceptable, we could not obtain enough material to cover the entire testing area. Therefore, we simulate the optical flow camera by introducing ten times as much noise (0.2 m/s) into the speed estimation of the UAV.

The MR sensor is placed directly on the receiving resonant coil of the WSN node. We then use a least squares approach to find the location of the WSN node. We do this by attempting to find a location $(\hat{x}_s, \hat{y}_s)$ which minimizes the difference between the position of the UAV and the range measurements from the MR sensor. Specifically, we minimize the function:

$$\underset{\hat{x}_s, \hat{y}_s, \hat{z}_s}{\arg\min} \sum_{i=1}^{n} \left(\frac{\mu_0 N I_i a}{4\pi \hat{r_i}^3}[2cos(\hat{\theta}_i)\hat{r_i} + sin(\hat{\theta}_i)\hat{\theta}_i] - \beta(\alpha_i)\right)^2 \tag{3}$$

where $\hat{r_i}^2 = (\hat{x}_{uav_i} - \hat{x}_s)^2 + (\hat{y}_{uav_i} - \hat{y}_s)^2$, $(\hat{x}_{uav_i}, \hat{y}_{uav_i})$ is the estimated position of the UAV given by the simulated optical flow, d_i is the distance measurement from the MR sensor that corresponds with that UAV position, and $\hat{z}_s = \hat{z}_{uav_i} - \hat{r}_i \sin \hat{\theta}_i$ is the height of the sensor.

Squaring the value means that the closer the UAV is to the receiver, the higher the weight is given to the function. This means that high voltages will penalize the function value much more for incorrect arguments.

Algorithm 1 Localization Algorithm

1: **procedure** LOCALIZE(x_{s_c}, y_{s_c}) ▷ Main localization method given the coarse position of the sensor
2: $GPSFlyTo(x_{s_c}, y_{s_c})$ ▷ Fly to the coarse position
3: ▷ Switch to Optical Flow + MR sensor control
4: $\boldsymbol{OptFlowFlyCircle(Radius = 2)}$
5: $OptFlowFlyTo(\hat{x}_s, \hat{y}_s)$
6: **end procedure**
7: **procedure** ON NON- ZERO MR READING($Volts$)
8: **while** $True$ **do** ▷ Continually refine estimate
9: $d \leftarrow Volts\_to\_range(Volts)$
10: $\boldsymbol{OptFlowFlyInCircle(Radius = d)}$
11: $OptFlowFlyTo(\hat{x}_s, \hat{y}_s)$
12: **end while**
13: **end procedure**
14: **procedure** ON NEW MR READING($Volts$)
15: $d \leftarrow Volts\_to\_range(Volts)$
16: $X_{uav} \leftarrow Append(X_{uav}, \hat{x}_{uav})$
17: $Y_{uav} \leftarrow Append(Y_{uav}, \hat{y}_{uav})$
18: $D \leftarrow Append(D, d)$
19: $\hat{x}_s, \hat{y}_s \leftarrow \sum_{i=1}^{n} (\frac{\mu_0 N I_i a}{4\pi \hat{r_i}^3}[2cos(\hat{\theta})\hat{r_i} + sin(\hat{\theta})\hat{\theta}] - \beta(\alpha_i))^2$ ▷ Eqn. 3
20: **return** $\hat{x}_s, \hat{y}_s$ ▷ The estimated position of the sensor.
21: **end procedure**

We then find the estimated position of the MR sensor, $(\hat{x}_s, \hat{y}_s)$, that minimizes this function over the n readings. The data used for input to the minimization function is in the structure of a circular queue that holds up to 10 s of data. Allowing this temporal property prevents the minimization function from becoming overweighted with areas with dense sampling, producing erroneous position estimates from too few samples, and problems from longer-term optical flow position estimation drift.

Minimizing Eq. 3 produces accurate position estimates as long as there are sufficient samples within 1 m of the MR sensor. Algorithm 1 shows the algorithm we use to ensure good sampling of the area. The localization algorithm works by first approaching the position of the sensor that was recorded during deployment using GPS alone. The UAV will be near the sensor when it arrives at its coarse position, but not close enough to be able to transfer power to the sensor. Immediately after a non-zero reading from the MR sensor is encoutered, the UAV will perform a circular trajectory with radius equal to the range estimation that it has received. Once the circle is complete, the UAV flies to the estimated position of the WSN node (line 5), which is computed in a separate thread continuously (lines 11 through 17). This

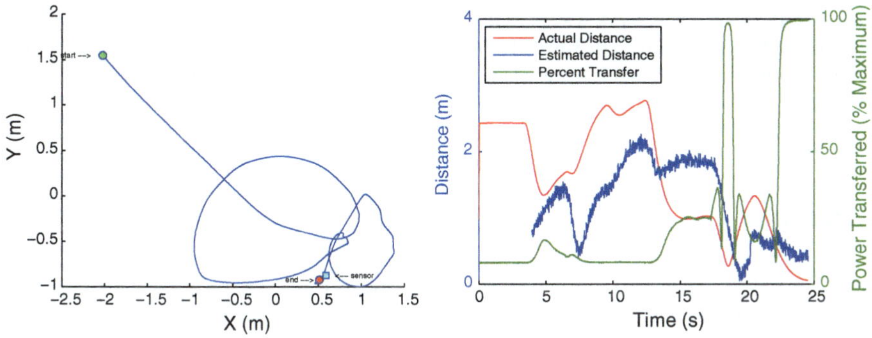

Fig. 8 (*left*) Localization path and (*right*) true distance, estimated distance, and power transfer from a localization experiment

procedure is performed in a loop to continually refine the estimate (lines 6 through 8) until some other action is taken (for example, land, fly home, fly to another WSN node).

We performed over 10 localization trials using this approach. On average, the UAV localized with an average error of 15 cm, with a 6 cm standard deviation and an average localization time of 46 s. Figure 8 (left) shows the UAV path localizing the sensor with this method. Figure 8 (right) shows the range readings compared to ground truth (obtained with a motion capture system) and the power transfer rate. During time $t = 0$ to $t = 5$, no valid range estimate had been found, thus there is no estimated distance. The UAV continued the scripted flight until $t = 15$, when it then flies to the current estimate of the sensor's position. In this particular experiment, the position was found within 6 cm of the true location after 24 s. At this distance, the WSN node receives 5.49 W and the MR sensor is at 100 % of its maximum value.

6 Conclusion

The UAV-based wireless power transfer system presented in this paper allows sensors in remote locations to be recharged autonomously. We discussed design considerations for UAV-based wireless power transfer systems and presented the details of our system. We showed that many common outdoor materials (wood, dirt, stone, soil) have little impact on the power transfer system We have presented a unique localization strategy that measures the magnetic field that is emitted from a resonant wireless power transfer system. The localization strategy allows the UAV to localize a WSN node with higher precision than is available with GPS. The UAV is able to localize to an average of 15 cm to land on the WSN node and charge it with an average of 4.2 W. This shows that it is possible for a UAV charge sensors deployed in remote and hard to access locations.

Acknowledgments This work was partially supported by NSF CSR-1217400 and USDA NIFA National Robotics Initiative 2013-67021-20947. Any opinions, findings, and conclusions or recommendations expressed in this material are those of the authors and do not reflect the views of the sponsors.

References

1. Szewczyk, R., Mainwaring, A., Polastre, J., Anderson, J., Culler, D.: An analysis of a large scale habitat monitoring application. In: Proceedings of the Conference on Embedded Networked Sensor Systems, pp. 214–226 (2004)
2. Akyildiz, I.F., Sun, Z., Vuran, M.C.: Signal propagation techniques for wireless underground communication networks. Phys. Commun. **2**(3), 167–183 (2009)
3. Detweiler, C., Banerjee, S., Doniec, M., Jiang, M., Peri, F., Chen, R.F., Rus, D.: Adaptive decentralized control of mobile underwater sensor networks and robots for modeling underwater phenomena. J. Sens. Actuator Netw. **3**(2), 113–149 (2014)
4. Griffin, B., Detweiler, C.: Resonant wireless power transfer to ground sensors from a uav. In: Robotics and Automation (ICRA), pp. 2660–2665 (2012)
5. McSpadden, J.O., Mankins, J.C.: Space solar power programs and microwave wireless power transmission technology. Microwave Mag. **3**(4), 46–57 (2002)
6. Sample, A., Smith, J.R.: Experimental results with two wireless power transfer systems. In: Radio and Wireless Symposium, pp. 16–18 (2009)
7. Tesla, N.: Apparatus for transmitting electrical energy (1914)
8. Kurs, A., Karalis, A., Moffatt, R., Joannopoulos, J.D., Fisher, P., Soljačić, M.: Wireless power transfer via strongly coupled magnetic resonances. Science **317**(5834), 83–86 (2007)
9. Duong, T., Lee, J.: Experimental results of high-efficiency resonant coupling wireless power transfer using a variable coupling method. Microwave Wirel. Compon. Lett. **21**(8), 442–444 (2011)
10. Sample, A.P., Meyer, D.A., Smith, J.R.: Analysis, experimental results, and range adaptation of magnetically coupled resonators for wireless power transfer. Trans. Industr. Electron. **58**(2), 544–554 (2011)
11. Seo, Y.-S., Hughes, Z., Hoang, M., Isom, D., Nguyen, M., Rao, S., Chiao, J.-C.: Investigation of wireless power transfer in through-wall applications. In: Microwave Conference Proceedings (APMC), pp. 403–405 (2012)
12. Moore, J., Tedrake, R.: Magnetic localization for perching uavs on powerlines. In: IEEE/RSJ International Conference on Intelligent Robots and Systems (IROS), IEEE, 2011, pp. 2700–2707
13. Tokekar, P., Bhadauria, D., Studenski, A., Isler, V.: A robotic system for monitoring carp in minnesota lakes. J. Field Robot. **27**(6), 779789 (2010)
14. Brown, W.C.: The history of power transmission by radio waves. Trans. Microw. Theory Tech. **32**(9), 1230–1242 (1984)
15. Achtelik, M.C., Stumpf, J., Gurdan, D., Doth, K.-M.: Design of a flexible high performance quadcopter platform breaking the mav endurance record with laser power beaming. In: International Conference on Intelligent Robots and Systems (IROS), pp. 5166–5172 (2011)
16. Bishop, O.: Electronics - Circuits and Systems. Taylor & Francis (2012)
17. Griffin, B.: Automated resonant wireless power transfer to remote sensors from an unmanned aerial vehicle. Master's thesis, University of Nebraska - Lincoln, Lincoln, NE (2012)
18. Karalis, A., Joannopoulos, J.D., Soljačić, M.: Efficient wireless non-radiative mid-range energy transfer. Ann. Phys. **323**(1), 34–48 (2008)
19. N. C. O. for Space-Based Positioning Navigation and Timing. In: Global positioning system standard positioning service performance standard 4e, Science

20. Honegger, D., Greisen, P., Meier, L., Tanskanen, P., Pollefeys, M.: Real-time velocity estimation based on optical flow and disparity matching. In: International Conference on Intelligent Robots and Systems (IROS), pp. 5177–5182 (2012)
21. Kendoul, F., Fantoni, I., Nonami, K.: Optic flow-based vision system for autonomous 3d localization and control of small aerial vehicles. Robot. Auton. Syst. **57**(6), 591–602 (2009)
22. Weiss, S., Scaramuzza, D., Siegwart, R.: Monocular-slambased navigation for autonomous micro helicopters in gps-denied environments. J. Field Robot. **28**(6), 854–874 (2011)

Inferring Maps and Behaviors from Natural Language Instructions

Felix Duvallet, Matthew R. Walter, Thomas Howard, Sachithra Hemachandra, Jean Oh, Seth Teller, Nicholas Roy and Anthony Stentz

Abstract Natural language provides a flexible, intuitive way for people to command robots, which is becoming increasingly important as robots transition to working alongside people in our homes and workplaces. To follow instructions in unknown environments, robots will be expected to reason about parts of the environments that were described in the instruction, but that the robot has no direct knowledge about. However, most existing approaches to natural language understanding require that the robot's environment be known a priori. This paper proposes a probabilistic framework that enables robots to follow commands given in natural language, without any prior knowledge of the environment. The novelty lies in exploiting environment information implicit in the instruction, thereby treating language as a type of sensor that is used to formulate a prior distribution over the unknown parts of the environment. The algorithm then uses this learned distribution to infer a sequence of actions that are most consistent with the command, updating our belief as we gather

The first four authors contributed equally to this paper.

F. Duvallet (✉) · J. Oh · A. Stentz
Robotics Institute, Carnegie Mellon University, Pittsburgh, PA, USA
e-mail: felixd@cmu.edu

J. Oh
e-mail: jeanoh@cmu.edu

A. Stentz
e-mail: tony@cmu.edu

M.R. Walter · T. Howard · S. Hemachandra · S. Teller · N. Roy
CS & AI Lab, Massachusetts Institute of Technology, Cambridge, MA, USA
e-mail: mwalter@csail.mit.edu

T. Howard
e-mail: tmhoward@csail.mit.edu

S. Hemachandra
e-mail: sachih@csail.mit.edu

S. Teller
e-mail: teller@csail.mit.edu

N. Roy
e-mail: nickroy@csail.mit.edu

M.A. Hsieh et al. (eds.), *Experimental Robotics*, Springer Tracts in Advanced Robotics 109, DOI 10.1007/978-3-319-23778-7_25

more metric information. We evaluate our approach through simulation as well as experiments on two mobile robots; our results demonstrate the algorithm's ability to follow navigation commands with performance comparable to that of a fully-known environment.

1 Introduction

Robots are increasingly performing collaborative tasks with people at home, in the workplace, and outdoors, and with this comes a need for efficient communication between human and robot teammates. Natural language offers an effective means for untrained users to control complex robots, without requiring specialized interfaces or extensive user training. Enabling robots to understand natural language instructions would facilitate seamless coordination in human-robot teams. However, interpreting instructions is a challenge, particularly when the robot has little or no prior knowledge of its environment. In such cases, the robot should be capable of reasoning over the parts of the environment that are relevant to understanding the instruction, but may not yet have been observed.

Oftentimes, the command itself provides information about the environment that can be used to hypothesize suitable world models, which can then be used to generate the correct robot actions. For example, suppose a first responder instructs a robot to "navigate to the car behind the building," where the car and building are outside the robot's field-of-view and their locations are not known. While the robot has no a priori information about the environment, the instruction conveys the knowledge that there is likely one or more buildings and cars in the environment, with at least one car being "behind" one of the buildings. The robot should be able to reason about the car's possible location, and refine its prior as it carries out the command (by updating the car's possible location when it observes a building).

This paper proposes a method that enables robots to interpret and execute natural language commands that refer to unknown regions and objects in the robot's environment. We exploit the information implicit in the user's command to learn an environment model from the natural language instruction, and then solve for the policy that is consistent with the command under this world model. The robot updates its internal representation of the world as it makes new metric observations (such as the location of perceived landmarks) and updates its policy appropriately. By reasoning and planning in the space of beliefs over object locations and groundings, we are able to reason about elements that are not initially observed, and robustly follow natural language instructions given by a human operator.

More specifically, we describe in our approach (Sect. 3) a probabilistic framework that first extracts annotations from a natural language instruction, consisting of the objects and regions described in the command and the given relations between them (Fig. 1a). We then treat these annotations as noisy sensor observations in a mapping framework, and use them to generate a distribution over a semantic

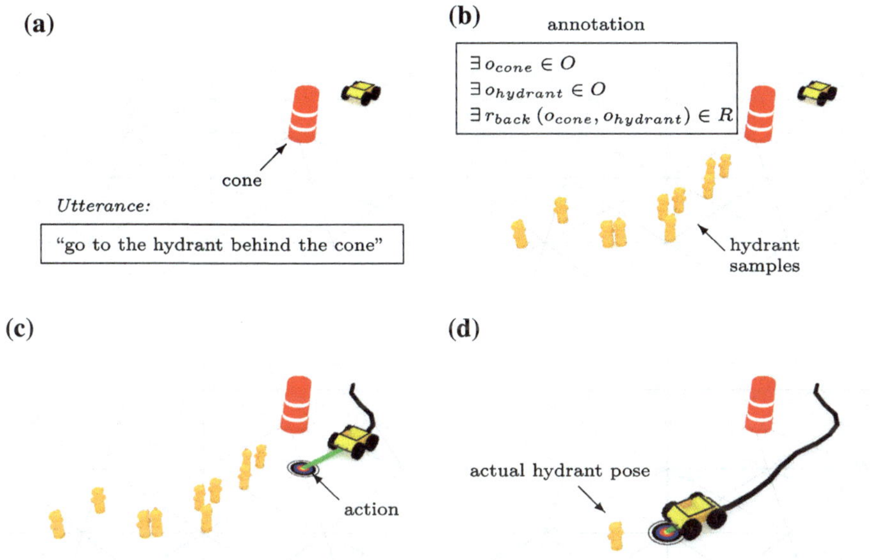

Fig. 1 Visualization of one run for the command "go to the hydrant behind the cone," showing the evolution of our beliefs (the possible locations of the hydrant). The robot begins with the cone in its field of view, but does not know the hydrant's location. **a** First, we receive a verbal instruction from the operator. **b** Next, we infer the map distribution from the utterance and prior observations. **c** We then take an action (green), using the map and behavior distributions. **d** This process repeats as the robot acquires new observations, refining its belief

model of the environment that also incorporates observations from the robot's sensor streams (Fig. 1b). This prior is used to ground the actions and goals from the command, resulting in a distribution over desired behaviors. This is then used to solve for a policy that yields an action that is most consistent with the command, under the map distribution so far (Fig. 1c). As the robot travels and senses new metric information, it updates its map prior and inferred behavior distribution, and continues to plan until it reaches its destination (Fig. 1d). This framework in outlined in Fig. 2.

We evaluate our algorithm (Sect. 4) through a series of simulation-based and physical experiments on two mobile robots that demonstrate its effectiveness at carrying out navigation commands, as well as highlight the conditions under which it fails. Our results indicate that exploiting the environment knowledge implicit in the instruction enables us to predict a world model upon which we can successfully estimate the action sequence most consistent with the command, approaching performance levels of complete a priori environment knowledge. These results suggest that utilizing information implicitly contained in natural language instructions can improve collaboration in human-robot teams.

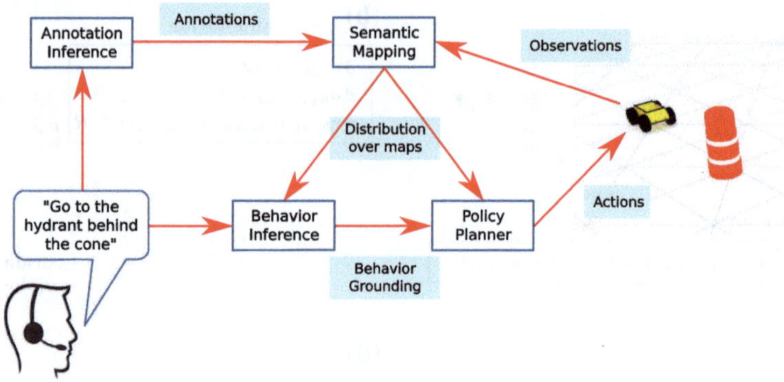

Fig. 2 Framework outline

2 Related Work

Natural language has proven to be effective for commanding robots to follow route directions [1–5] and manipulate objects [6]. The majority of prior approaches require a complete semantically-labeled environment model that captures the geometry, location, type, and label of objects and regions in the environment [2, 5, 6]. Understanding instructions in unknown environments is often more challenging. Previous approaches have either used a parser that maps language directly to plans [1, 3, 4], or trained a policy that reasons about uncertainty and can backtrack when needed [7]. However, none of these approaches directly use the information contained in the instruction to inform their environment representation or reason about its uncertainty. We instead treat language as a sensor that can be used to generate a prior over the possible locations of landmarks by exploiting the information implicitly contained in a given instruction.

State-of-the-art semantic mapping frameworks focus on using the robot's sensor observations to update its representation of the world [8–10]. Some approaches [10] integrate language descriptions to improve the representation but do not extend the maps based on natural language. Our approach treats natural language as another sensor and uses it to extend the spatial representation by adding both topological and metric information, which is then used for planning. Williams et al. [11] use a cognitive architecture to add unvisited locations to a partial map. However, they only reason about topological relationships to unknown places, do not maintain multiple hypotheses, and make strong assumptions about the environment limiting the applicability to real robot systems. In contrast, our approach reasons both topologically and metrically about objects and regions, and can deal with ambiguity which allows us to operate in challenging environments.

As we reason in the space of distributions over possible environments, we draw from strategies in the belief-space planning literature. Most importantly, we represent our belief using samples from the distribution, similar to work by Platt et al. [12]. Instead of solving the complete Partially-Observable Markov Decision Process (POMDP), we instead seek efficient approximate solutions [13, 14].

3 Technical Approach

Our goal is to infer the most likely future robot trajectory $x_{t+1:T}$ up to time horizon T, given the history of natural language utterances Λ^t, sensor observations z^t, and odometry u^t (we denote the history of a variable up to time t with a superscript):

$$\underset{x_{t+1:T} \in \Re^n}{\arg\max}\; p\left(x_{t+1:T} | \Lambda^t, z^t, u^t\right). \tag{1}$$

Inferring the maximum a posteriori trajectory (1) for a given natural language utterance is challenging without knowledge of the environment for all but trivial applications. To overcome this challenge, we introduce a latent random variable S_t that represents the world model as a *semantic map* that encodes the location, geometry, and type of the objects within the environment. This allows us to factor the distribution as:

$$\underset{x_{t+1:T} \in \Re^n}{\arg\max} \int_{S_t} p(x_{t+1:T} | S_t, \Lambda^t, z^t, u^t)\, p(S_t | \Lambda^t, z^t, u^t)\, dS_t. \tag{2}$$

As we maintain the distribution in the form of samples $S_t^{(i)}$, this simplifies to:

$$\underset{x_{t+1:T} \in \Re^n}{\arg\max} \sum_i p(x_{t+1:T} | S_t^{(i)}, \Lambda^t, z^t, u^t)\, p(S_t^{(i)} | \Lambda^t, z^t, u^t). \tag{3}$$

Based upon the robot's sensor and odometry streams and the user's natural language input, our algorithm learns this distribution online. We accomplish this through a filtering process whereby we first infer the distribution over the world model S_t based upon annotations identified from the utterance Λ^t (second term in the integral in (2)), upon which we then infer the constraints on the robot's action that are most consistent with the command given the current map distribution. At this point, the algorithm solves for the most likely policy under the learned distribution over trajectories (first term in the integral in (2)). During execution, we continuously update the semantic map S_t as sensor data arrives and refine the policy according to the re-grounded language.

To efficiently convert unstructured natural language to symbols that represent the spaces of annotations and behaviors, we use the Distributed Correspondence Graph (DCG) model [5]. The DCG model is a probabilistic graphical model composed of random variables that represent language λ, groundings γ, and correspondences between language and groundings ϕ and factors f. Each factor f_{i_j} in the DCG model is influenced by the current phrase λ_i, correspondence variable ϕ_{i_j}, grounding γ_{i_j}, and child phrase groundings $\gamma_{c_{i_j}}$. The parameters in each log-linear model υ are trained from a parallel corpus of labeled examples for annotations and behaviors in

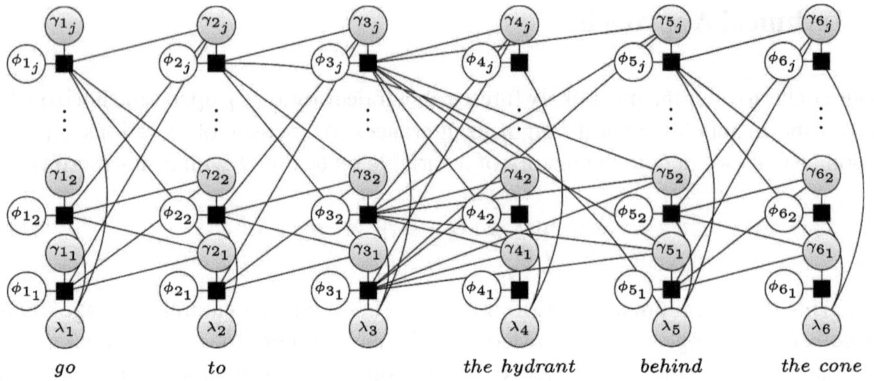

Fig. 3 A DCG used to infer annotations or behaviors from the utterance "go to the hydrant behind the cone." The factors f_{i_j}, groundings γ_{i_j}, and correspondence variables ϕ_{i_j} are functions of the symbols used to represent annotations and behaviors

the context of a world model Υ. In each, we search for the unknown correspondence variables that maximize the product of factors:

$$\arg\max_{\phi \in \Phi} \prod_{i} \prod_{j} f_{i_j} \left(\phi_{i_j}, \gamma_{i_j}, \gamma_{c_{i_j}}, \lambda_i, \Upsilon, \upsilon \right). \quad (4)$$

An illustration of the graphical model used to represent Eq. 4 is shown in Fig. 3. In this figure, the black squares, white circles, and gray circles represent factors, unknown random variables, and known random variables respectively. It is important to note that each phrase can have a different number of vertically aligned factors if the symbols used to ground particular phrases differ. In this paper we use a binary correspondence variable to indicate the expression or rejection of a particular grounding for a phrase. We construct the symbols used to represent each phrase using only the groundings with a true correspondence and take the meaning of a utterance as the symbol inferred at the root of parse tree.

Figure 2 illustrates the architecture of the integrated system that we consider for evaluation. First, the natural language understanding module infers a distribution over annotations conveyed by the utterance (Annotation Inference). The semantic map learning method then uses this information in conjunction with the prior annotations and sensor measurements to build a probabilistic model of objects and their relationships in the environment (Semantic Mapping). We then formulate a distribution over robot behaviors using the utterance and the semantic map distribution (Behavior Inference). Next, the planner computes a policy from this distribution over behaviors and maps (Policy Planner). As the robot makes more observations or receives additional human input, we repeat the last three steps to continuously update our understanding of the most recent utterance. We now describe in more detail each of these components.

3.1 Annotation Inference

The space of symbols used to represent the meaning of phrases in map inference is composed of objects, regions, and relations. Since no world model is assumed when inferring linguistic annotations from the utterance, the space of objects is equal to the number of possible object types that could exist in the scene. Regions are some portion of state-space that is typically associated with a relationship to some object. Relations are a particular type of association between a pair of objects or regions (e.g., front, back, near, far). Since any set of objects, regions, and relations may be inferred as part of the symbol grounding, the size of the space of groundings for map inference grows as the power set of the sum of these symbols. We use the trained DCG model to infer a set of annotations α_t from the positively expressed groundings at the root of the parse tree.

3.2 Semantic Mapping

We treat the annotations inferred from the utterance as noisy observations α that specify the existence and spatial relations between labeled objects in the environment. We use these observations along with those from the robot's sensors to learn the distribution over the semantic map $S_t = \{G_t, X_t\}$:

$$
\begin{aligned}
p(S_t|\Lambda^t, z^t, u^t) &\approx p(S_t|\alpha^t, z^t, u^t) && \text{(5a)}\\
&= p(G_t, X_t|\alpha^t, z^t, u^t) && \text{(5b)}\\
&= p(X_t|G_t, \alpha^t, z^t, u^t)\,p(G_t|\alpha^t, z^t, u^t), && \text{(5c)}
\end{aligned}
$$

where the last line expresses the factorization into a distribution over the environment topology (graph G_t) and a conditional distribution over the metric map (X_t). Owing to the combinatorial number of candidate topologies [10], we employ a sample-based approximation to the latter distribution and model the conditional posterior over poses with a Gaussian, parametrized in the canonical form. In this manner, each particle $S_t^{(i)} = \{G_t^{(i)}, X_t^{(i)}, w_t^{(i)}\}$ consists of a sampled topology $G_t^{(i)}$, a Gaussian distribution over the poses $X_t^{(i)}$, and a weight $w_t^{(i)}$. We note that this model is similar to that of Walter et al. [10], though in this work we don't treat the labels as being uncertain.

To efficiently maintain the semantic map distribution over time as the robot receives new annotations and observations during execution, we use a Rao-Blackwellized particle filter [15]. This filtering process has two key steps: First, the algorithm proposes updates to each sampled topology that express object observations and annotations inferred from the utterance. Next, the algorithm uses the proposed topology to perform a Bayesian update to the Gaussian distribution over the node (object) poses, and updates the particle weights so as to approximate the target distribution. We perform this process for each particle and repeat these steps at each time instance. The following paragraphs describe each operation in more detail.

During the proposal step, we first augment each sample topology with an additional node and edge that model the robot's motion u_t, resulting in a new topology $S_t^{(i)-}$. We then sample modifications to the graph $\Delta_t^{(i)} = \{\Delta_{\alpha_t}^{(i)}, \Delta_{z_t}^{(i)}\}$ based upon the most recent annotations (α_t) and sensor observations (z_t):

$$p(S_t^{(i)}|S_{t-1}^{(i)}, \alpha_t, z_t, u_t) = p(\Delta_{\alpha_t}^{(i)}|S_t^{(i)-}, \alpha_t)\, p(\Delta_{z_t}^{(i)}|S_t^{(i)-}, z_t)\, p(S_t^{(i)-}|S_{t-1}^{(i)}, u_t). \quad (6)$$

This updates the proposed graph topology $S_t^{(i)-}$ with the graph modifications $\Delta_t^{(i)}$ to yield the new semantic map $S_t^{(i)}$. The updates can include the addition of nodes to the graph representing newly hypothesized or observed objects. They also may include the addition of edges between nodes to express spatial relations inferred from observations or annotations.

The graph modifications are sampled from two similar but independent proposals for annotations and observations in a multi-stage process:

$$p(\Delta_{\alpha_t}^{(i)}|S_t^{(i)-}, \alpha_t) = \prod_j p(\Delta_{\alpha_{t,j}}^{(i)}|S_t^{(i)-}, \alpha_{t,j}) \quad (7a)$$

$$p(\Delta_{z_t}^{(i)}|S_t^{(i)-}, z_t) = \prod_j p(\Delta_{z_{t,j}}^{(i)}|S_t^{(i)-}, z_{t,j}). \quad (7b)$$

For each language annotation component $\alpha_{t,j}$, we use a likelihood model over the spatial relation to sample landmark and figure pairs for the grounding (7a). This model employs a Dirichlet process prior that accounts for the fact that the annotation may refer to existing or new objects. If the landmark and/or the figure are sampled as new objects, we add these objects to the particle, and create an edge between them. We also sample the metric constraint associated with this edge, based on the spatial relation. Similarly, for each object $z_{t,j}$ observed by the robot, we sample a grounding from the existing model of the world (7b). We add a new constraint to the object when the grounding is valid, and create a new object and constraint when it is not.

After proposing modifications to each particle, we perform a Bayesian update to their Gaussian distribution. We then re-weight each particle by taking into account the likelihood of generating language annotations, as well as positive and negative observations of objects:

$$w_t^{(i)} = p(z_t, \alpha_t|S_{t-1})\, w_{t-1}^{(i)} = p(\alpha_t|S_{t-1})\, p(z_t|S_{t-1})\, w_{t-1}^{(i)}. \quad (8)$$

For annotations, we use the natural language grounding likelihood under the map at the previous time step. For object observations, we use the likelihood that the observations were (or were not) generated based upon the previous map. This has the effect of down-weighting particles for which the observations are unexpected. We normalize the weights and re-sample if their entropy exceeds a threshold [15].

3.3 Behavior Inference

Given the utterance and the semantic map distribution, we now infer a distribution over robot behaviors. The space of symbols used to represent the meaning of phrases in behavior inference is composed of objects, regions, actions, and goals. Objects and regions are defined in the same manner as in map inference, though the presence of objects is a function of the inferred map. Actions and goals specify how the robot should perform a behavior to the planner. Since any set of actions and goals can be expressed to the planner, the space of groundings also grows as the power set of the sum of these symbols. For the experiments discussed later in Sect. 4 we assume a number of objects, regions, actions, and goals that are proportional to the number of objects in the hypothesized world model. We use the trained DCG model to infer a distribution of behaviors β from the positively expressed groundings at the root of the parse tree.

3.4 Policy Planner

Since it is difficult to both represent and search the continuum for a trajectory that best reflects the entire instruction in the context of the semantic map, we instead learn a policy that predicts a single action that maximizes the one-step expected value of taking the action a_t from the robot's current pose x_t. This process is repeated until the policy declares it is done following the command using a separate action a_{stop}.

As the robot moves in the environment, it builds and updates a graph of locations it has previously visited, as well as frontiers that lie at the edge of explored space. This graph is used to generate a candidate set of actions that consists of all frontier nodes $\mathcal{F}$ as well as previously-visited nodes $\mathcal{V}$ that the robot can travel to next:

$$A_t = \mathcal{F} \cup \mathcal{V} \cup \{a_{\text{stop}}\}. \tag{9}$$

The policy selects the action with the maximum value under our value function:

$$\pi(x_t) = \underset{a_t \in A_t}{\arg\max}\; V(x_t, a_t). \tag{10}$$

The value of a particular action is a function of the behavior and the semantic map, which are not observable. Instead, we solve this using the QMDP algorithm [13] by taking the expected value under the distributions of the semantic map S_t and inferred behavior β_j:

$$V(x_t, a_t) \approx \sum_{S_t^{(i)}} \sum_{\beta_j} V\left(x_t, a_t; S_t^{(i)}, \beta_j\right) p\left(\beta_j | S_t^{(i)}\right) p\left(S_t^{(i)}\right). \tag{11}$$

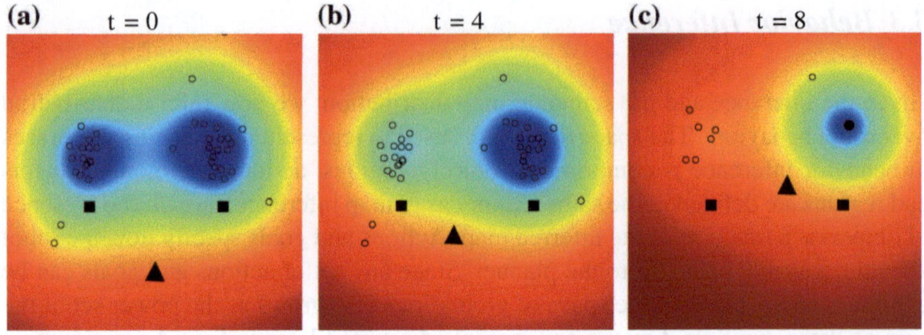

Fig. 4 Visualization of the value function over time for the command "go to the hydrant behind the cone," where the *triangle* denotes the robot, *squares* denote observed cones, and *circles* denote hydrants that are sampled (empty) and observed (filled). The robot starts off having observed the two cones, and hypothesizes possible hydrants that are consistent with the command (**a**). The robot first moves towards the left cluster, but after not observing the hydrant, the map distribution peaks at the right cluster (**b**). The robot then moves right and observes the actual hydrant (**c**)

There are many choices for the particular value function to use, in this work we define the value for a semantic map particle and behavior as an analogue of the MDP cost-to-go:

$$V\left(x_t, a_t; S_t^{(i)}, \beta_j\right) = \gamma^{d(a_t, g_s)}, \tag{12}$$

where γ is the MDP discount factor and d is the Euclidean distance between the action node and the behavior's goal position g_s. Our belief space policy π then picks the maximum value action. We re-evaluate this value function as the semantic map and behavior distributions improve with new observations. Figure 4 demonstrates the evolution of the value function over time.

4 Results

To analyze our approach, we first evaluate the ability of our natural language understanding module to independently infer the correct annotations and behaviors for given utterances. Next, we analyze the effectiveness of our end-to-end framework through simulations that consider environments and commands of varying complexity, and different amounts of prior knowledge. We then demonstrate the utility of our approach in practice using experiments run on two mobile robot platforms. These experiments provide insights into our algorithm's ability to infer the correct behavior in the presence of unknown and ambiguous environments.

Table 1 Natural language understanding results with 95 % confidence intervals

Model	Accuracy (%)	Training Time (sec)	Inference Time (sec)
Annotation	62.50 (10.83)	145.11 (7.55)	0.44 (0.03)
Behavior	55.77 (6.83)	18.30 (1.02)	0.05 (0.00)

4.1 Natural Language Understanding

We evaluate the performance of our natural language understanding component in terms of the accuracy and computational complexity of inference using holdout validation. In each experiment, the corpus was randomly divided into separate training and test sets to evaluate whether the model can recover the correct groundings from the utterance and the world model. Each model used 13,716 features that checked for the presence of words, properties of groundings and correspondence variables, and relationships between current and child groundings and searched the model with a beam width of 4. We conducted 8 experiments for each model type using a corpus of 39 labeled examples of instructions and groundings. For annotation inference we assumed that the space of groundings for every phrase is represented by 8 object types, 54 regions, and 432 relations. For behavior inference we assumed that noun and prepositions ground to hypothesized objects or regions while verbs ground to 2 possible actions, 3 possible modes, goal regions, and constraint regions. In the example illustrated in Fig. 3 with a world model composed of seven hypothesized objects the annotation inference DCG model contained 5,934 random variables and 2,964 factors while the behavior inference DCG model contained 772 random variables and 383 factors. In each experiment 33 % of the labeled examples in the corpus were randomly selected for the holdout. The mean number of log-linear model training examples extracted from the 26 randomly selected labeled examples for annotation and behavior inference was 83,547 and 9,224 respectively. Table 1 illustrates the statistics for the annotation and behavior models.

This experiment demonstrates that we are able to learn many of the relationships between phrases, groundings, and correspondences with a limited number of labeled instructions, and infer a distribution of symbols quickly enough for the proposed architecture. As expected the training and inference time for the annotation model is much higher because of the difference in the complexity of symbols. This is acceptable for our framework since the annotation model is only used once to infer a set of observations, while the behavior model is used continuously to process the map distributions as new observations are integrated.

4.2 Monte Carlo Simulations

Next, we evaluate the entire framework through an extended set of simulations in order to understand how the performance varies with the environment configuration

Table 2 Monte Carlo simulation results with 1σ confidence intervals (Hydrant, Cone)

			Success Rate (%)		Distance (m)	
World	Range (m)	Relation	Known	Ours	Known	Ours
1H, 1C	3.0	null	100.0	93.9	8.75 (1.69)	16.78 (7.90)
1H, 1C	3.0	"behind"	100.0	98.3	8.75 (1.69)	13.43 (7.02)
1H, 2C	3.0	null	100.0	100.0	11.18 (1.38)	32.54 (18.50)
1H, 2C	3.0	"behind"	100.0	99.5	11.18 (1.38)	40.02 (29.66)
2H, 1C	3.0	null	100.0	54.4	10.49 (1.81)	21.56 (10.32)
2H, 1C	3.0	"behind"	100.0	67.4	10.38 (1.86)	18.72 (10.23)
2H, 1C	5.0	"nearest"	100.0	46.2	9.19 (1.54)	12.05 (5.76)

and the command. We consider four environment templates, with different numbers of figures (hydrants) and landmarks (cones). For each configuration, we sample ten environments, each with different object poses. For these environments, we issued three natural language instructions "go to the hydrant," "go to the hydrant behind the cone," and "go to the hydrant nearest to the cone." We note that these commands were not part of the corpus that we used to train the DCG model. Additionally, we considered six different settings for the robot's sensing range (2, 3, 5, 10, 15, and 20 m) and performed approximately 100 simulations for each combination of environment, command, and range. As a ground-truth baseline, we performed ten runs of each configuration with a completely known world model.

Table 2 presents the success rate and distance traveled by the robot for these 100 simulation configurations. We considered a run to be successful if the planner stops within 1.5 m of the intended goal. Comparing against commands that do not provide a relation (i.e., "go to the hydrant"), the results demonstrate that our algorithm achieves greater success and yields more efficient paths by taking advantage of relations in the command (i.e., "go to the hydrant behind the cone"). This is apparent in environments consisting of a single figure (hydrant) as well as more ambiguous environments that consist of two figures. Particularly telling is the variation in performance as a result of different sensing range. Figure 5 shows how success rate increases and distance traveled decreases as the robot's sensing range increases, quickly approaching the performance of the system when it begins with a completely known map of the environment.

One interesting failure case is when the robot is instructed to "go to the hydrant nearest to the cone" in an environment with two hydrants. In instances where the robot sees a hydrant first, it hypothesizes the location of the cone, and then identifies the observed hydrants and hypothesized cones as being consistent with the command. Since the robot never actually confirms the existence of the cone in the real world, this results in the incorrect hydrant being labeled as the goal.

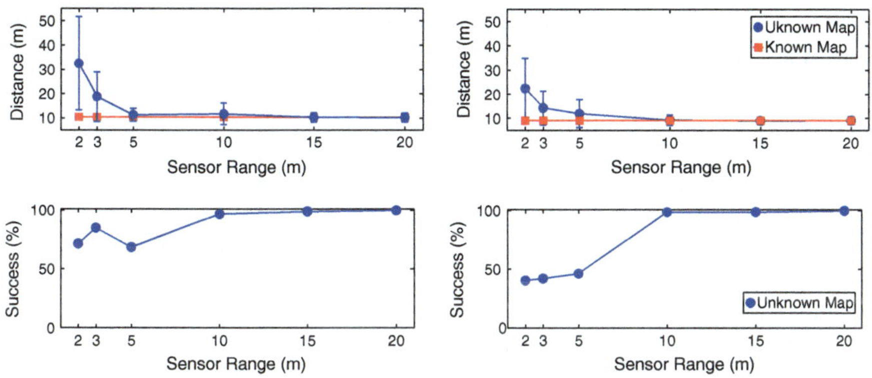

Fig. 5 Distance traveled (*top*) and success rate (*bottom*) as a function of the sensor range for the commands "go to the hydrant behind the cone" (*left*) and "go to the hydrant nearest to the cone" (*right*) in simulation

4.3 Physical Experiments

We applied our approach to two mobile robots, a Husky A200 mobile robot (Fig. 6a) and an autonomous robotic wheelchair [16] (Fig. 6b). The use of both platforms demonstrates the application of our algorithm to mobile robots with different vehicle configurations, underlying motion planners, and sensor configurations. The actions determined by the planner are translated into lists of waypoints that are handled by each robot's motion planner. We used AprilTag fiducials [17] to detect and estimate the relative pose of objects in the environment, subject to self-imposed angular and range restrictions.

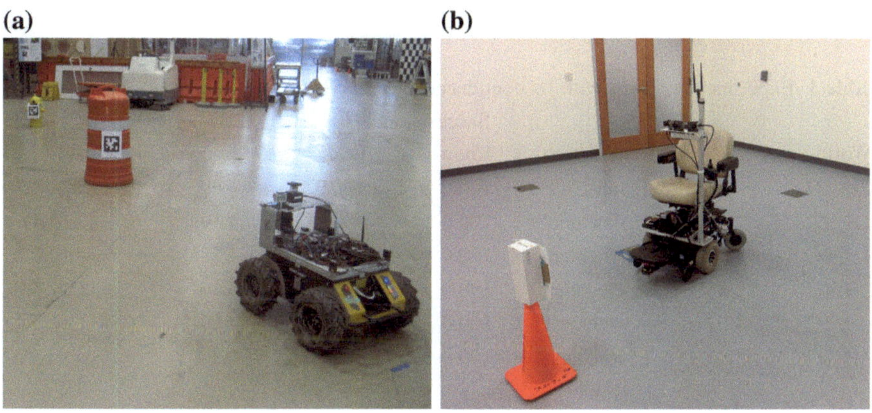

Fig. 6 The setup for the experiments with the **a** Husky and **b** wheelchair platform.s

In each experiment, a human operator issues natural language commands in the form of text that involve (possibly null) spatial relations between one or two objects. The results that follow involve the commands "go to the hydrant," "go to the hydrant behind the cone," and "go to the hydrant nearest to the cone." As with the simulation-based experiments, these instructions did not match those from our training set. For each of these commands, we consider different environments by varying the number and position of the cones and hydrants and by changing the robot's sensing range. For each configuration of the environment, command, and sensing range, we perform ten trials with our algorithm. For a ground-truth baseline, we perform an additional run with a completely known world model. We consider a run to be a success when the robot's final destination is within 1.5 m of the intended goal.

Table 3 presents the success rate and distance traveled by the wheelchair for these experiments. Compared to the scenario in which the command does not provide a relation (i.e., "go to the hydrant"), we find that our algorithm is able to take advantage of available relations ("go to the hydrant behind the cone") to yield behaviors closer to that of ground truth. The results are similar for the Husky platform, which resulted in an 83.3 % success rate when commanded to "go to the hydrant behind the cone" in an environment with one cone and one hydrant. These results demonstrate the usefulness of utilizing all of the information contained in the instruction, such as the relation between various landmarks in the environment that can be helpful during navigation.

The robot trials exhibited a similar failure mode as the simulated experiments: if the environment contains two figures (hydrants) and the robot only detects one, the semantic map distribution then hypothesizes the existence of cones in front of the hydrant, which leads to a behavior distribution peaked around this goal and plans that do not look for the possibility of another hydrant in the environment. As expected, this effect is most pronounced with shorter sensing ranges (e.g., a 3 m sensing range for the command "go to the hydrant nearest to the cone" resulted in the robot reaching the goal in only half of the trials compared to a 4 m sensing range).

Table 3 Experimental results with 1σ confidence intervals (Hydrant, Cone)

			Success Rate (%)		Distance (m)	
World	Range (m)	Relation	Known	Ours	Known	Ours
1H, 1C	2.5	null	100.0	100.0	4.69	16.56 (7.20)
1H, 1C	2.5	"behind"	100.0	100.0	4.69	9.91 (3.41)
1H, 2C	3.0	"behind"	100.0	100.0	4.58	7.64 (2.08)
2H, 1C	2.5	"behind"	100.0	80.0	5.29	6.00 (1.38)
2H, 1C	4.0	"nearest"	100.0	100.0	4.09	4.95 (0.39)
2H, 1C	3.0	"nearest"	100.0	50.0	6.30	7.05 (0.58)

5 Conclusions

Enabling robots to reason about parts of the environment that have not yet been visited solely from a natural language description serves as one step towards effective and natural collaboration in human-robot teams. By treating language as a sensor, we are able to paint a rough picture of what the unvisited parts of the environment could look like. We utilize this information during planning, and update our belief with actual sensor information during task execution.

Our approach exploits the information implicitly contained in the language to infer the relationship between objects that may not be initially observable, without having to consider those annotations as a separate utterance. By learning a distribution over the map, we generate a useful prior that enables the robot to sample possible hypotheses, representing different environment possibilities that are consistent with both the language and the available sensor data. Learning a policy that reasons in the belief space of these samples achieves a level of performance that approaches full knowledge of the world ahead of time.

We have evaluated our approach in simulation and on two robot platforms. These evaluations provide a preliminary validation of our framework. Future work will test the algorithm's ability to scale to larger environments (e.g., rooms and hallways), and handle utterances that present complex relations and more detailed behaviors than those considered so far. Additionally, we will focus on handling streams of commands, including those that are given *during* execution (e.g., "go to the other cone" uttered as the robot is moving towards the wrong cone). An additional direction for following work is to explicitly reason over exploratory behaviors that take information gathering actions to resolve uncertainty in the map. Currently, any exploration on the part of the algorithm is opportunistic, which might not be sufficient in more challenging scenarios.

Acknowledgments The authors would like to thank Bob Dean for his help with the Husky platform. This work was supported in part by the Robotics Consortium of the U.S. Army Research Laboratory under the Collaborative Technology Alliance Program, Cooperative Agreement W911NF-10-2-0016.

References

1. MacMahon, M., Stankiewicz, B., Kuipers, B.: Walk the talk: Connecting language, knowledge, and action in route instructions. In: Proceedings of National Conference on Artificial Intelligence (AAAI) (2006)
2. Kollar, T., Tellex, S., Roy, D., Roy, N.: Toward understanding natural language directions. In: Proceedings of International Conference on Human-Robot Interaction (2010)
3. Chen, D.L., Mooney, R.J.: Learning to interpret natural language navigation instructions from observations. In: Proceedings of National Conference on Artificial Intelligence (AAAI) (2011)
4. Matuszek, C., Herbst, E., Zettlemoyer, L., Fox, D.: Learning to parse natural language commands to a robot control system. In: Proceedings of International Symposium on Experimental Robotics (ISER) (2012)

5. Howard, T., Tellex, S., Roy, N.: A natural language planner interface for mobile manipulators. In: Proceedings of IEEE International Conference on Robotics and Automation (ICRA) (2014)
6. Tellex, S., Kollar, T., Dickerson, S., Walter, M.R., Banerjee, A.G., Teller, S., Roy, N.: Understanding natural language commands for robotic navigation and mobile manipulation. In: Proceedings of National Conference on Artificial Intelligence (AAAI) (2011)
7. Duvallet, F., Kollar, T., Stentz, A.: Imitation learning for natural language direction following through unknown environments. In: Proceedings of IEEE International Conference on Robotics and Automation (ICRA) (2013)
8. Zender, H., Martínez Mozos, O., Jensfelt, P., Kruijff, G., Burgard, W.: Conceptual spatial representations for indoor mobile robots. Robotics and Autonomous Systems (2008)
9. Pronobis, A., Martínez Mozos, O., Caputo, B., Jensfelt, P.: Multi-modal semantic place classification. International Journal of Robotics Research (2010)
10. Walter, M.R., Hemachandra, S., Homberg, B., Tellex, S., Teller, S.: Learning semantic maps from natural language descriptions. In: Proceedings of Robotics: Science and Systems (RSS) (2013)
11. Williams, T., Cantrell, R., Briggs, G., Schermerhorn, P., Scheutz, M.: Grounding natural language references to unvisited and hypothetical locations. In: Proceedings of National Conference on Artificial Intelligence (AAAI) (2013)
12. Platt, R., Kaelbling, L., Lozano-Perez, T., Tedrake, R.: Simultaneous localization and grasping as a belief space control problem. In: Proceedings of International Symposium of Robotics Research (ISRR) (2011)
13. Littman, M.L., Cassandra, A.R., Kaelbling, L.P.: Learning policies for partially observable environments: Scaling up. In: Proceedings of International Conference on Machine Learning (ICML) (1995)
14. Roy, N., Burgard, W., Fox, D., Thrun, S.: Coastal navigation-mobile robot navigation with uncertainty in dynamic environments. In: Proceedings of IEEE International Conference on Robotics and Automation (ICRA) (1999)
15. Doucet, A., de Freitas, N., Murphy, K., Russell, S.: Rao-Blackwellised particle filtering for dynamic Bayesian networks. In: Proceedings of the Conference on Uncertainty in Artificial Intelligence (UAI) (2000)
16. Hemachandra, S., Kollar, T., Roy, N., Teller, S.: Following and interpreting narrated guided tours. In: Proceedings of IEEE International Conference on Robotics and Automation (ICRA) (2011)
17. Olson, E.: AprilTag: A robust and flexible visual fiducial system. In: Proceedings of IEEE International Conference on Robotics and Automation (ICRA) (2011)

Part VII
Mechanisms

Nicola Bezzo

Recent advancements in technology and fabrication are enabling researchers to rapidly prototype and study new mechanisms for robotic applications.

Novel fabrication techniques such as print-and-fold and 3D printing are becoming more and more popular and accessible to the masses nowadays, allowing not only experts but also casual users to quickly prototype and create complex and low-cost mechanisms. Similarly, new actuation techniques that leverage soft materials and variable stiffness components are increasingly being used because they extend and prolong the range of motion of a robotic system. These exciting topics are the center of attention of this session in which the reader can find four papers describing: co-design techniques for fabrication, actuation prototypes, and validations through extensive experimental results.

The first paper, *Planar Cable Robot with Variable Stiffness* by X. Zhou, S. Jun, and V. Krovi presents a planar cable robot with variable stiffness. Thanks to the added active stiffness adjustment motor, the robot is able to independently achieve a large range of Cartesian stiffness and low tension. The authors propose an optimization problem with the objective of minimal tension to consider real world constraints like linear spring extension limits and validate the proposed system via tracking experiments.

The next three papers in the session deal with fabrication techniques using soft materials, and printable and foldable mechanisms. Specifically, in *Hydraulic Autonomous Soft Robotic Fish for 3D Swimming*, the authors R.K. Katzschmann, A.D. Marchese, and D. Rus present a soft robotic fish capable of prolonged and consistent 3D swimming by means of a novel hydraulic actuation system. A detailed description of the fabrication and actuation of the soft tail and the pitch control planes is presented together with experimental results of its locomotion capabilities in a swimming pool. In comparison to traditional pneumatically powered soft robots, it is demonstrated that hydraulic actuation allows for a higher force when high frequency change of direction is needed.

In *Foldable Joints for Foldable Robots* by C. Sung and D. Rus, the focus is on print-and-fold techniques for the construction and composition of basic joints.

Hinge, prismatic, and pivot joints are presented followed by the composition of folded structures to create more complex joints with higher degrees of freedom. The proposed system allows users to specify a type of joint and the desired parameters and automatically produces the corresponding fold pattern. The paper showcase examples of foldable robots such as a row-boat, a crane, and a camera mount, created by composing joints with rigid bodies using the technique in this paper.

The final paper of the session *A Design Environment for the Rapid Specification and Fabrication of Printable Robots* by A. Mehta, N. Bezzo, P. Gebhard, B. An, V. Kumar, I. Lee, and D. Rus presents a design environment for the rapid specification and generation of mechanical designs of robotic systems. Parametrized components are defined by code objects and then composed into higher order components using a graph-based representation to generate structurally specified robot designs. To further simplify the design flow for casual users, the authors propose an intuitive drag-and-drop graphical programming environment called ROSLab to generate the hardware specification of printable and foldable robots. Experimental results demonstrate the rapid design of a wheeled robot and an hexapod. Also cut-and-fold, 3D printing, and self-folding fabrication techniques are presented and compared for the same designed robots.

Planar Cable Robot with Variable Stiffness

Xiaobo Zhou, Seung-kook Jun and Venkat Krovi

Abstract Variable stiffness modules add significant robustness to mechanical systems during forceful interactions with uncertain environments. Traditionally, most existing variable stiffness modules tend to be bulky by virtue of their use of solid components making them less suitable for mobile applications. In recent times, pre-tensioned cable-based modules have been proposed to reduce weight. While passive, these modules depend on significant internal tension to provide the desired stiffness and their stiffness modulation capability tends to be limited. In this paper, we present a planar 2DOF cable robot formed by three active variable stiffness modules that we developed which decouples tension from stiffness. Controlled changes in structural parameters (independent of cable actuation) now permits independent modulation of the perceived stiffness. By varying each module's stiffness, the overall Cartesian stiffness of the robot can be modulated. We show that this approach is more effective than by increasing internal tension only. It is also easier than varying configuration to achieve variable stiffness. Further, thanks to the added active stiffness adjustment motor, it is able to independently vary stiffness and internal tension. Therefore such active module can achieve same Cartesian stiffness as passive modules but with much lower internal tension, which is more efficient. We present the analysis of the system and verified via both simulation and experimental results for the effectiveness of the Cartesian stiffness varying capability of the planar cable robot.

Keywords Cable robot · Variable stiffness · Decoupled tension

X. Zhou · S. Jun · V. Krovi (✉)
Department of Mechanical and Aerospace Engineering,
State University of New York at Buffalo,Buffalo, NY 14260, USA
e-mail: xzhou9@buffalo.edu

S. Jun
e-mail: seungjun@buffalo.edu

V. kriovi
e-mail: vkrovi@buffalo.edu

M.A. Hsieh et al. (eds.), *Experimental Robotics*, Springer Tracts
in Advanced Robotics 109, DOI 10.1007/978-3-319-23778-7_26

1 Motivation

Stiff actuation has the benefit of higher speed and positioning accuracy, while soft actuators are safer in forceful interactions with environment. To accommodate various operating conditions, variable stiffness actuators gain increasing popularity in recent developments. Variable stiffness implementation in hardware may be classified into two categories based on the types of components used. (1) Examples of VSAs using solid components include using a secondary motor to move spring location [1] or pivot point [2], using customized cam profile to change transmission angle [3], or using four bar linkages [4]; (2) VSAs using cable mechanisms usually involve nonlinear springs put in antagonistic configuration [5], or use customized cam profiles with linear springs [6, 7] to emulate nonlinear springs. For more comprehensive reviews of different VSA designs, please refer to [8, 9].

Variable stiffness in cable robots has recently been explored in the following three directions: (1) utilizing actuation redundancy, i.e. increasing internal tension, which was studied in our previous works [10, 11]. However, this is not very effective; (2) utilizing kinematic redundancy, i.e. changing the geometric configuration of the system with end-effector fixed at desired location to alter the pulling map and cause overall Cartesian stiffness to vary, which was explored in our recent work [12]. However, this is not straight forward; and (3) using a variable stiffness module to change joint space stiffness to shape task space stiffness, which include using triangular relationship similar to [13] but with cable pulleys [14] or directly route cable onto edge to reduce size [15], a variant of this using torsional springs is presented in [16]. Such triangular configuration using pulleys allow for a large range of stiffness change using a linear spring and is relatively compact and light weight. However, these are all passive variable stiffness module which relies on increasing internal tension to achieve larger stiffness. The main disadvantage with such passive variable stiffness modules is that the stiffness change is coupled with internal tension, the larger the stiffness the higher the tension. Therefore requiring large holding current of the drive motor for high stiffness and consumes more energy. This is our motivation to allow the spring attachment point to move using a secondary stiffness motor (we call it active variable stiffness). This stiffness adjustment motor mainly moves the spring such that the equilibrium point changes, therefore the force can be relatively small.

2 The Planar Variable Stiffness Cable Robot

2.1 Variable Stiffness Module

In [17], we developed an active cable-based variable stiffness module which has the advantage of decoupled tension from stiffness. Here, we briefly outline its working principle. Figure 1 illustrates the idea of the active variable stiffness module. Black line represents the cable, routed through three pulleys (red circle). Pulley C is con-

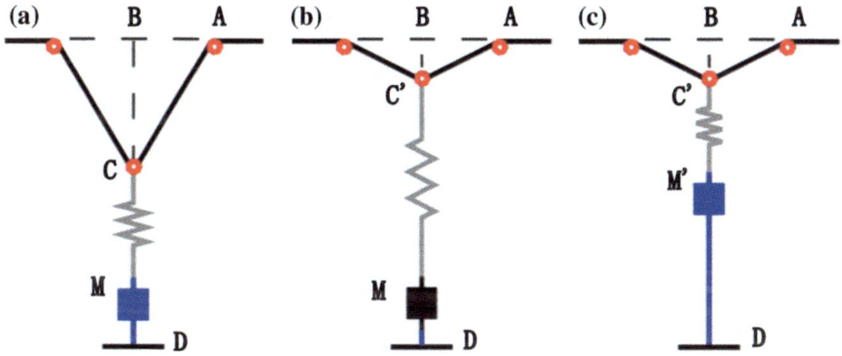

Fig. 1 Variable stiffness module: **a** Low stiffness configuration; **b** conventional passive module to achieve higher stiffness; **c** Active module to achieve higher stiffness with lower tension

nected with a linear spring, the other two pulleys are fixed to ground. The other end of the spring is fixed on M. For conventional passive variable stiffness modules Fig. 1b, M is directly fixed to ground (black). Our active module (Fig. 1a, c) uses a motor (in this case a slider) that can move the spring to change the equilibrium position.

Let $|AB| = b$, $|AC| = c$, $|BC| = a$, $|BD| = d$, $|CM| = e$, $|MD| = x_s$, $\angle ACB = \gamma$. Ignore diameter of pulley here. The output cable stiffness K_c is related to the constant spring stiffness K_s by:

$$K_c = \frac{dF_c}{dl} = \frac{K_s c^2}{4a^2}. \tag{1}$$

After pulling cable for δL length and a stiffness motor motion x_s', the configuration changes to Fig. 1c, let $|AC'| = c'$, $|BC'| = a'$, $|C'M| = e'$, $\angle AC'B = \gamma'$. Then the spring force becomes

$$F_s = K_s(e' - e_0), \tag{2}$$

where e_0 is the free length of spring. The cable tension becomes

$$F_c = \frac{F_s}{2\cos\gamma'}. \tag{3}$$

As can be seen, the stiffness is related to the included angle α, while the tension is proportional to the extension of the spring. In Fig. 1a, a low stiffness configuration is shown. The conventional approach to achieve a higher stiffness entails pulling at the cable to form a larger angle α', as shown in Fig. 1b. However, such an increase in stiffness is also coupled by an increased tension due to spring extension as seen in Fig. 1b. Our active module allows the change of equilibrium position therefore the spring force and thus cable tension can be controlled independently (Fig. 1c).

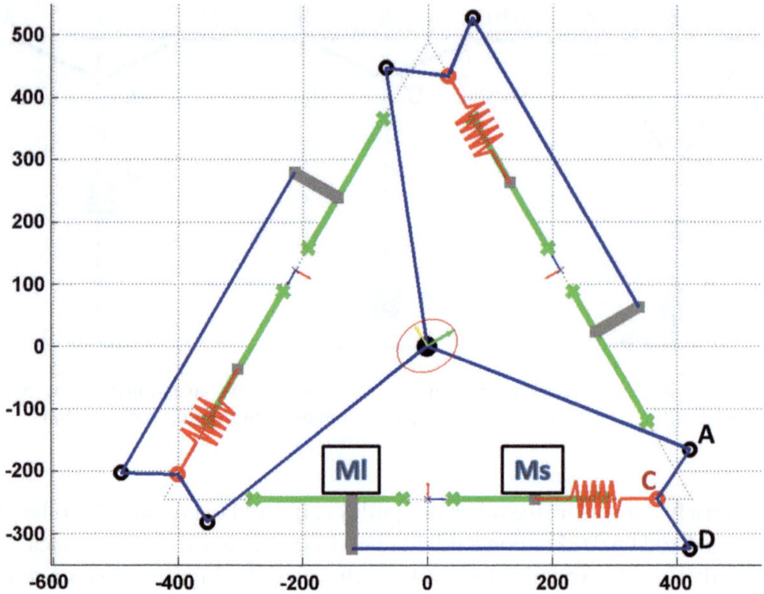

Fig. 2 Planar 2DOF cable robot with variable stiffness

2.2 Planar 2DOF Cable Robot

Now we put three of these active variable stiffness modules (Fig. 1c) together to form a planar 2DOF cable robot, shown in Fig. 2, units on the axis are in mm. Take a look at the marked bottom module: The length motor M_l moves along the left side green linear slider, it is mainly for controlling effective cable length (functions as a winch). The stiffness motor M_s connects to one end of the red linear spring and moves along the right side green linear slider. The other end of the spring is attached to the movable pulley C. The blue fixed length inextensible cable routes through black fixed pulley D, red movable pulley C, black fixed pulley A, and connects to the point mass end-effector. The red ellipse around the end-effector is the Cartesian stiffness ellipse (whose magnitude is scaled) at the current configuration.

2.2.1 Kinetostatics

The kinematics of the system needs to be solved together with statics problem, due to the added elasticity. Force equilibrium:

$$\mathbf{Pf} = -\underset{\sim}{w}, \tag{4}$$

where $\underset{\sim}{w}$ is the external wrench exerted on the payload, $\mathbf{f} = [F_{c_1}, F_{c_2}, F_{c_3}]^T$ is the cable tension vector, and P is the pulling map derived following [18] to be

$$\mathbf{P} = \begin{bmatrix} \frac{x_1-x_e}{\sqrt{(x_1-x_e)^2+(y_1-y_e)^2}} & \frac{x_2-x_e}{\sqrt{(x_2-x_e)^2+(y_2-y_e)^2}} & \frac{x_3-x_e}{\sqrt{(x_3-x_e)^2+(y_3-y_e)^2}} \\ \frac{y_1-y_e}{\sqrt{(x_1-x_e)^2+(y_1-y_e)^2}} & \frac{y_2-y_e}{\sqrt{(x_2-x_e)^2+(y_2-y_e)^2}} & \frac{y_3-y_e}{\sqrt{(x_3-x_e)^2+(y_3-y_e)^2}} \end{bmatrix}, \quad (5)$$

where (x_i, y_i) is the position of the fixed pulley A_i in world coordinate frame, (x_e, y_e) is the end-effector payload position. In this 2DOF case, $\underset{\sim}{w}$ is two dimensional therefore leaving one degree of redundancy in $\mathbf{f}$ which can be used to optimize internal tension, $\mathbf{f} = \mathbf{f_p} + \alpha \mathbf{f_n}$, where $\mathbf{f_p} = -\mathbf{P}^{\#} \underset{\sim}{w}$ is the particular solution and $\mathbf{f_n} = \ker(\mathbf{P})$ is the homogeneous solution.

3 Analysis

The Cartesian stiffness matrix can be written as

$$\mathbf{K}_x = -\frac{\partial \mathbf{P}}{\partial \mathbf{x}} \mathbf{f} + \mathbf{P} \mathbf{K}_q \mathbf{P}^T \quad (6)$$

where the second component is a similarity transform of the joint space stiffness matrix $\mathbf{K}_q$, and the first term comes from loading. Chen and Kao [19] termed this mapping as the Conservative Congruence Transformation (CCT). Examining Eq. 6, we see that the Cartesian stiffness may be controlled in three ways: (1) *internal tension* $\mathbf{f_p}$, which we explored in [10] where active stiffness control schemes were presented to take advantage of actuation redundancy and in [11] where lower bound of stiffness is guaranteed by controlling the smallest eigenvalue of the stiffness matrix; (2) *pulling map* $\mathbf{P}$, which we explored in [12] where Cartesian stiffness was modulated utilizing configuration redundancy; and (3) *joint stiffness* $\mathbf{K_q}$, which was explored by Yeo et.al [20] who designed a variable stiffness add-on module to enhance the overall stiffness of a planar cable robot, compared to their passive module, whose stiffness increases with internal tension, our module is active and has the advantage of decoupling internal tension from stiffness.

3.1 Range of Achievable Stiffness

The benefit of changing module stiffness compared to changing tension is its effectiveness in changing Cartesian stiffness. We compare the range of stiffness achieved by increasing internal tension only and by varying joint stiffness only. For illustration, consider the payload at the center of platform, with initial module stiffness set as $K_q = diag([1, 1, 1])$, and cable tensions $f = [1, 1, 1]^T$, which satisfies Eq. 4 with 0 external wrench due to symmetry. Figure 3 shows the resulting Cartesian stiffness by increasing internal tension only (Fig. 3a) and module stiffness only (Fig. 3b). Notice

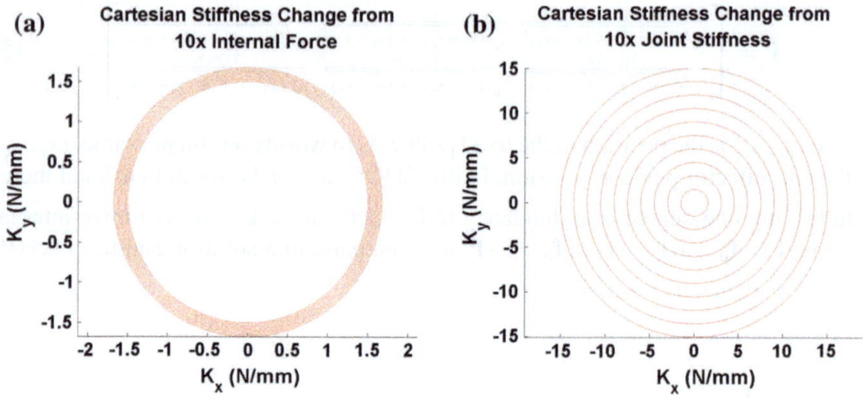

Fig. 3 Comparison of stiffness change (notice the scale): **a** increase cable tension only, **b** increase module stiffness only

the magnitude, it is evident that adjusting module stiffness is much more effective than adjusting internal tension in changing Cartesian stiffness.

3.2 Independence of Tension and Stiffness

The benefit of using active variable stiffness modules compared to passive modules is the capability of decoupling tension from stiffness. The passive module introduced in Yeo et.al [20] operates by increasing cable tension to pull the module to change geometry and thus increase its stiffness. Therefore, the higher the stiffness desired, the larger the internal tension results, which is bad for both safety as well as power consumption. Our active module has the additional degree of freedom to adjust the equilibrium position of the linear spring, therefore altering cable tension. This point is illustrated in Fig. 4. Notice that for both cases, the Cartesian stiffness (red ellipse) is the same, which can also be seen from the same angles formed by the modules. But it is clear that the tension in Fig. 4a case is much larger than in Fig. 4b, which is evident from both the spring length as well as stiffness slider position. As shown, with our active variable stiffness module, it is possible to set internal tension to be independent of desired Cartesian stiffness, with lower tension, smaller actuators may be used.

3.3 Achieving Desired Cartesian Stiffness with Lower Tension

Now that internal tension is decoupled from stiffness, theoretically, arbitrarily low positive internal tension may be set for desired Cartesian stiffness. The resulting

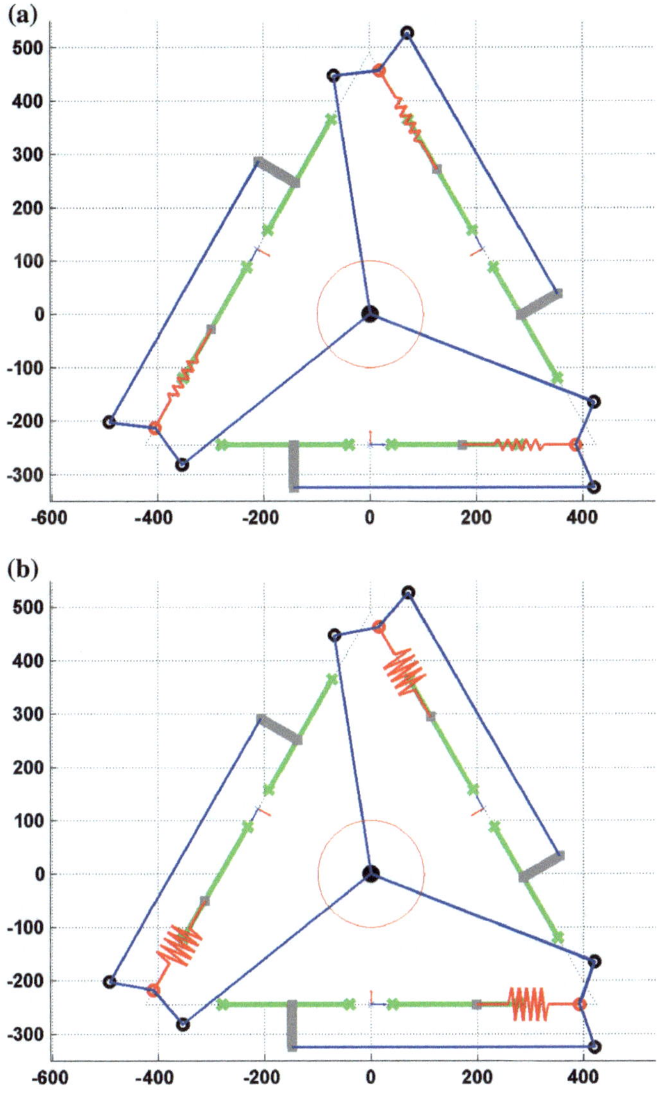

Fig. 4 Different tension for same desired cartesian stiffness: **a** larger internal tension, **b** smaller internal tension

module stiffness and its corresponding slider positions may be obtained by solving the following Eq. 6, with desired Cartesian stiffness $K_x^{\prime d}$, desired tension f^d, end-effector location x and the corresponding pulling map P all known. Note the symmetry of stiffness matrix, this results in a set of three linear equations with three module stiffness as unknowns and are straight forward to solve.

However, due to real world constraints such as slider travel limits and linear spring extension limits, it may not be easy to prescribe a desired tension that will work for all cases, therefore it is more convenient to pose it as the following optimization problem with the objective of minimal tension:

$$
\begin{aligned}
\text{minimize:} \quad & \alpha \\
\text{subject to:} \quad & K_x^d = -\frac{\partial \mathbf{P}}{\partial \mathbf{x}} \mathbf{f}^d + P K_q P^T \\
& \mathbf{f} > 0 \\
& x_{s_i} \in [l_{i1_{min}}, l_{i1_{max}}] \\
& x_{l_i} \in [l_{i2_{min}}, l_{i2_{max}}] \\
& \Delta\rho_i \in (\Delta\rho_{i_{min}}, \Delta\rho_{i_{max}}), (i \in 1, 2, 3) \\
& \dot{x}_{s_i} \leq V_{max} \\
& \dot{x}_{l_i} \leq V_{max}
\end{aligned}
\tag{7}
$$

where $\mathbf{f} = -\mathbf{P}^{\#} \underset{\sim}{w} + \alpha \, \ker(\mathbf{P})$, $\Delta\rho_i$ is the linear spring extension length, and the last two constraints are velocity limits to make trajectory smooth for hardware implementation, following [12]. The problem is solved using Fmincon in Matlab optimization toolbox, we use interior-point solver for possibility of stopping at a suboptimal which satisfies the constraints within step time limits.

3.4 Alternative Objectives of Cartesian Stiffness

Consider that the linear sliders have limited speed, they may not be able to go to desired position instantaneously to realize desired stiffness. Therefore, it may not be easy to prescribe desired Cartesian stiffness that is achievable during an entire trajectory. In this case, we can relax the equality constraint on stiffness and put the stiffness error into objective function to be minimized, or more generally, instead of specifying desired stiffness matrix, we may optimize objectives such as maximum stiffness, directional stiffness, etc. as explored in [12]. An example of orienting a stiffness ellipse to be horizontal and vertical is shown in Fig. 5.

4 Experiments

The experimental setup is shown in Fig. 6. OptiTrack cameras mounted on top of the frame are used to provide ground truth. The experimental setup corresponds to the illustrative diagram shown in Fig. 2. The $2DOF$ cable robot is formed by three legs (modules), each with two sliders that are position controlled. One slider M_l controls the effective cable length. The other slider M_s moves one end of the spring

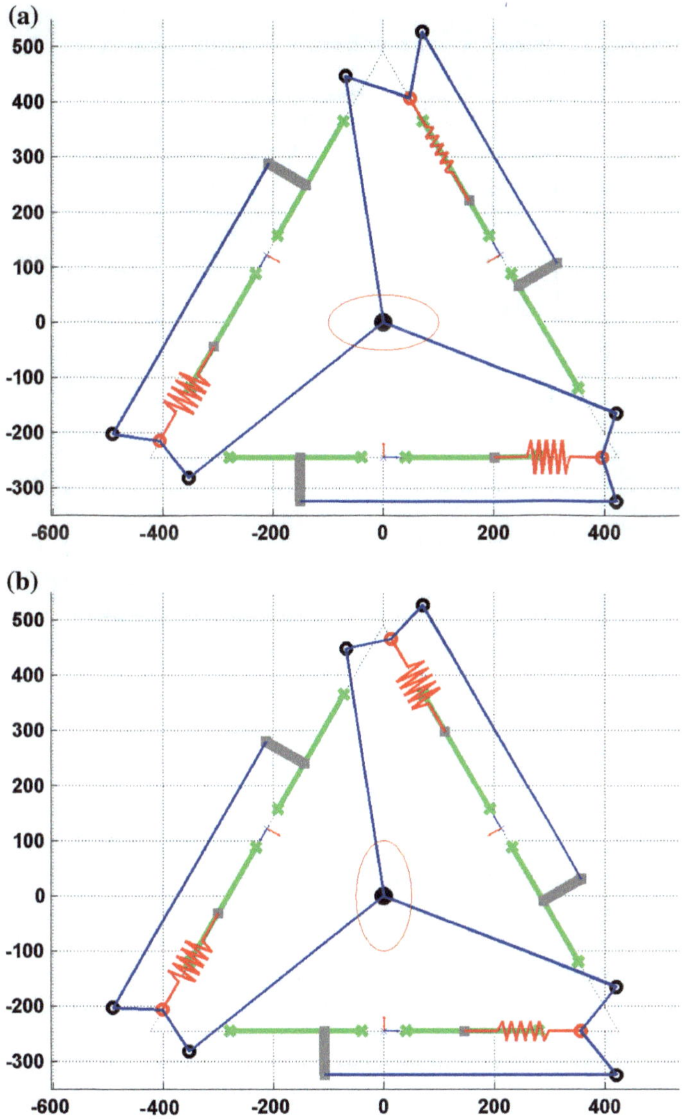

Fig. 5 Orientation of stiffness ellipse: **a** horizontal, **b** vertical

and thus changes the equilibrium position. The coordinated motions of the sliders control cable length as well as effective stiffness. The range of travel of the sliders are limited to [40, 320] mm. The constant stiffness of the linear spring is obtained by linear fitting of the calibration data and is found to be $K_s = 0.36$ (N/mm), relatively soft. The maximum extension length of the spring is limited to 40 mm for safety concerns.

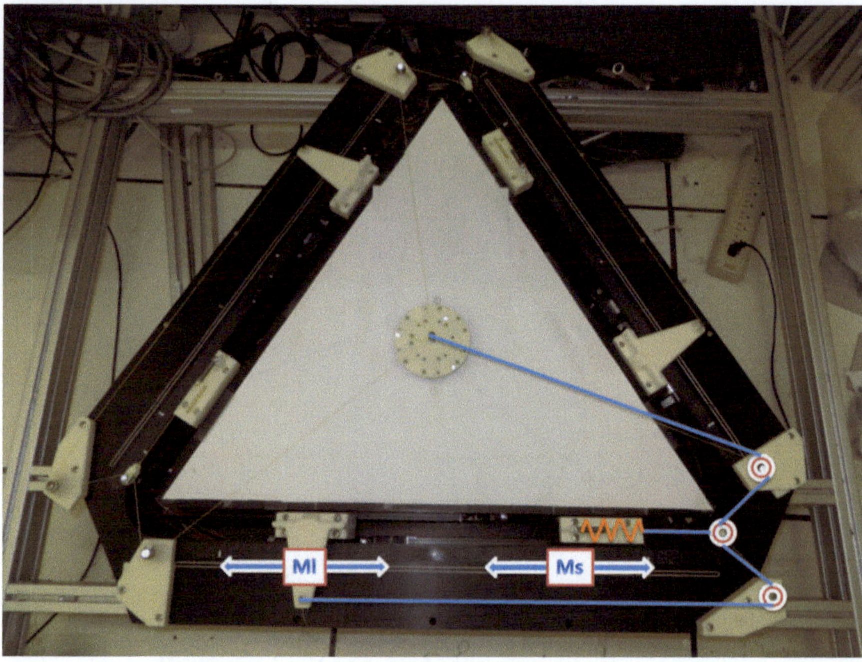

Fig. 6 Experimental setup

5 Results

Several experiments on different test conditions were performed. Here we show two sets of illustrative experimental results of tracking line and circle with disturbance using different desired Cartesian stiffness. The disturbance is introduced by manually releasing a rubber ball from the same spot on a slope to hit the end-effector.

Figure 7a shows the line tracking case, the end-effector starts at (0, 0) and moves in $-X$ direction. The dashed black line is commanded trajectory; red line with circular markers is result of lower stiffness (soft) setting; and the blue line with x markers is result of higher stiffness (stiff, 10 times that of soft case). It is clear from the figures that the stiff case has better disturbance rejection than the soft case. The magnitude is only about half lower, not 10 times reduction. This is mostly due to friction. As the ball hits, it first overcomes static friction of the end-effector with the ground (as well as friction in pulleys), then the stiffness takes effect. Qualitative testing by hand to move the end-effector while lifting it a bit off the table feels significant difference in the perceived stiffness. The quantitative disturbance rejection effect is not very significant as seen from the figure. However, if we look at the trajectory itself, it is clear that stiffer mode tracks better than soft mode. It is also seen that the vertical error is about 2 mm and is increasing. This is mainly due to geometric uncertainties of the setup as well as modeling error that currently ignores the pulley radius of 6 mm. More work on improving the test condition and modeling accuracy to better showcase the effectiveness of stiffness regulation is currently underway.

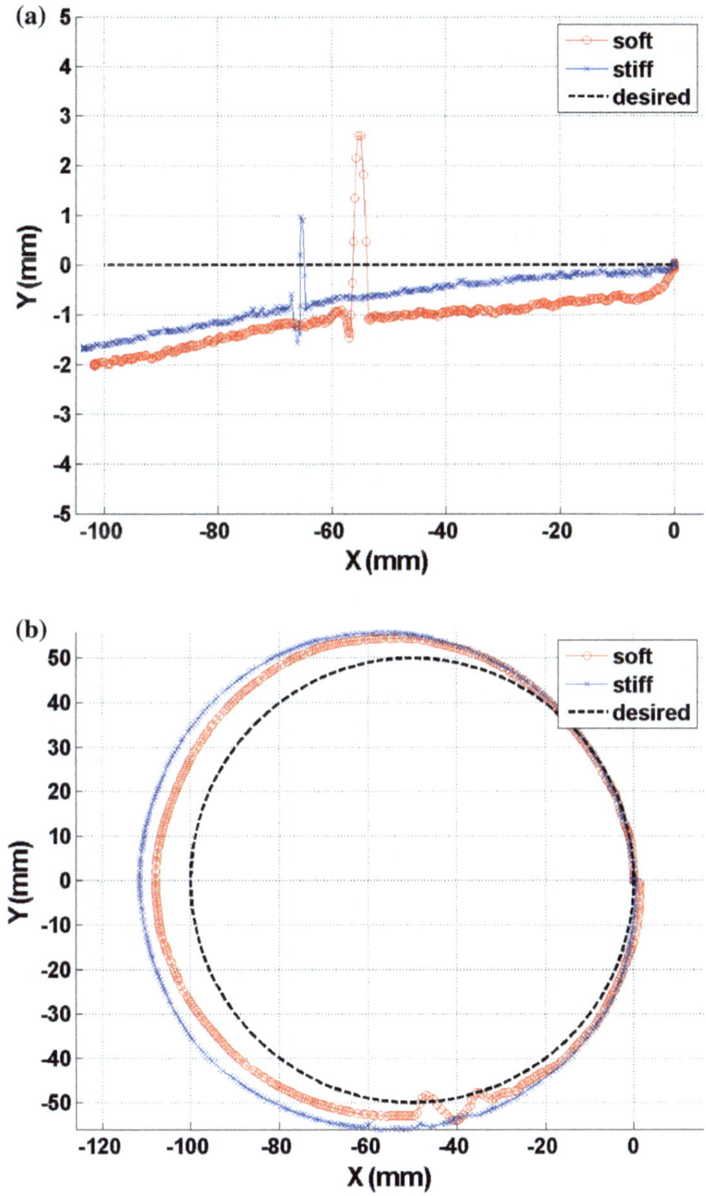

Fig. 7 Disturbance rejection performance for low and high stiffness settings. **a** Line tracking. **b** Circle tracking

Figure 7b is circle tracking. The end-effector starts from (0, 0) and traces a circle of radius 50 mm counter clockwise. The desired Cartesian stiffness settings are the same as the line tracking for both soft and stiff cases. The ball hits when the end-effector comes near the bottom the trajectory. And the results are consistent with that of line tracking case. The stiff case has better rejection than soft case. There is also tracking error present that relates to parameter uncertainty. A video of the experiment is available at http://youtu.be/pXH7rwrzh6s.

6 Main Experimental Insights

In this paper, we have presented a planar $2DOF$ cable robot that uses three active variable stiffness modules. Controlled changes in structural parameters (independent of cable length actuation) now permits independent modulation of the perceived stiffness from internal tension. Careful modulation of the three individual joint stiffness allows the cable robot to achieve a large range of Cartesian stiffness with low tension. A proof of concept experimental platform was setup to test its effectiveness. Compared with other ways of stiffness modulation, such as utilizing configuration redundancy to reconfigure the stiffness mapping (i.e. changing $\mathbf{G}$) [12] , or utilizing tension redundancy to regulate internal tension (i.e. changing $\mathbf{f_p}$) [11], such way of using a dedicated variable stiffness mechanism to set the joint stiffness $\mathbf{K_q}$ is more effective.

High internal tension can cause high friction in the bearings. The addition of mobility on changing the spring equilibrium location provides effective control to lower the internal tension, making cable robots more effective. Such decoupling of output stiffness from internal tension and the decoupling of cable length and tension provides great flexibility in cable robot controls. The vibration induced by the springs, coupled with geometric parameter uncertainties require better control for improved performance. The current proof of concept system uses linear actuators which appears to be bulky, we are working on a miniaturized system using rotary motors.

Acknowledgments This work was supported in part by the National Science Foundation Grants IIS-1319084 and CNS-1314484.

References

1. Jafari, A., Tsagarakis, N.G., Vanderborght, B., Caldwell, D.G.: A novel actuator with adjustable stiffness (awas). In: IEEE/RSJ International Conference on Intelligent Robots and Systems (IROS), 2010. 4201–4206
2. Jafari, A., Tsagarakis, N.G., Caldwell, D.G.: Awas-ii: A new actuator with adjustable stiffness based on the novel principle of adaptable pivot point and variable lever ratio. In: IEEE International Conference on Robotics and Automation (ICRA), pp. 4638–4643 (2011)
3. Hung Vu, Q., Aryananda, L., Sheikh, F.I., Casanova, F., Pfeifer, R.: A novel mechanism for varying stiffness via changing transmission angle. In: IEEE International Conference on Robotics and Automation (ICRA), pp. 5076–5081 (2011)

4. Schiavi, R., Grioli, G., Sen, S., Bicchi, A.: Vsa-ii: A novel prototype of variable stiffness actuator for safe and performing robots interacting with humans. In: IEEE International Conference on Robotics and Automation, ICRA 2008. pp. 2171–2176 (2008)
5. Laffranchi, M., Tsagarakis, N.G., Cannella, F., Caldwell, D.: Antagonistic and series elastic actuators: a comparative analysis on the energy consumption. In: IEEE/RSJ International Conference on Intelligent Robots and Systems, IROS 2009 (Oct 2009), pp. 5678–5684 (2009)
6. Hurst, J.W., Chestnutt, J.E., Rizzi, A.A.: The actuator with mechanically adjustable series compliance. IEEE Trans. Robot. **26**(4), 597–606 (2010)
7. Vanderborght, B., Tsagarakis, N.G., Semini, C., Van Ham, R., Caldwell, D.G.: Maccepa 2.0: Adjustable compliant actuator with stiffening characteristic for energy efficient hopping. In: IEEE International Conference on Robotics and Automation, ICRA '09, pp. 544–549 (2009)
8. Ham, R., Sugar, T., Vanderborght, B., Hollander, K., Lefeber, D.: Compliant actuator designs. IEEE Robot. Autom. Mag. **16**(3), 81–94 (2009)
9. Vanderborght, B., Albu-Schaeffer, A., Bicchi, A., Burdet, E., Caldwell, D.G., Carloni, R., Catalano, M., Eiberger, O., Friedl, W., Ganesh, G., Garabini, M., Grebenstein, M., Grioli, G., Haddadin, S., Hoppner, H., Jafari, A., Laffranchi, M., Lefeber, D., Petit, F., Stramigioli, S., Tsagarakis, N., Van Damme, M., Van Ham, R., Visser, L.C., Wolf, S.: Variable impedance actuators: a review. Robot. Auton. Syst. **61**(12), 1601–1614 (2013)
10. Yu, K., Lee, L.F., Krovi, V.: Simultaneous trajectory tracking and stiffness control of cable actuated parallel manipulator. In: Proceedings of the ASME Design Engineering Technical Conferences and Computer and Information in Engineering Conferences, vol. 7, San Diego, California, USA, pp. 55–63 (2009)
11. Yu, K., Lee, L.F., Tang, C.P., Krovi, V.: Enhanced trajectory tracking control with active lower bounded stiffness control for cable robot. In: IEEE International Conference on Robotics and Automation (ICRA), Anchorage, AK, pp. 669–674 (2010)
12. Zhou, X., Jun, S., Krovi, V.: Stiffness modulation exploiting configuration redundancy in mobile cable robots. In: IEEE International Conference on Robotics and Automation (ICRA), Hong Kong, China (May 29- June 6, 2014)
13. Tonietti, G., Schiavi, R., Bicchi, A.: Design and control of a variable stiffness actuator for safe and fast physical human/robot interaction. In: Proceedings of the IEEE International Conference on Robotics and Automation, ICRA 2005, pp. 526–531 (2005)
14. Friedl, W., Chalon, M., Reinecke, J., Grebenstein, M.: Fas a flexible antagonistic spring element for a high performance over actuated hand. In: IEEE/RSJ International Conference on Intelligent Robots and Systems (IROS), pp. 4201–4206 (2011)
15. Osada, M., Ito, N., Nakanishi, Y., Inaba, M.: Realization of flexible motion by musculoskeletal humanoid with add-on nonlinear spring units. In: IEEE-RAS International Conference on Humanoid Robots (Humanoids), 10th, pp. 174–179 (2010)
16. Yeo, S.H., Yang, G., Lim, W.B.: Design and analysis of cable-driven manipulators with variable stiffness. Mech. Mach. Theor. **69**, 230–244 (2013)
17. Zhou, X., Jun, S., Krovi, V.: A cable based active variable stiffness module with decoupled tension. In: ASME 2014 International Design Engineering Technical Conferences and Computers in Engineering Conference (IDETC 2014), Buffalo, NY (Aug. 17-20, 2014)
18. Zhou, X., Tang, C., Krovi, V.: 7. In: cooperating mobile cable robots: screw theoretic analysis. Lecture Notes in Electrical Engineering, vol. 57, pp. 109–123. Springer, Berlin Heidelberg (2013)
19. Chen, S.F., Kao, I.: Conservative congruence transformation for joint and cartesian stiffness matrices of robotic hands and fingers. Int. J. Robot. Res. **19**(9), 835–847 (2000)
20. Yeo, S., Yang, G., Lim, W.: Design and analysis of cable-driven manipulators with variable stiffness. Mech. Mach. Theor. **69**, 230–244 (2013)

Hydraulic Autonomous Soft Robotic Fish for 3D Swimming

Robert K. Katzschmann, Andrew D. Marchese and Daniela Rus

Abstract This work presents an autonomous soft-bodied robotic fish that is hydraulically actuated and capable of sustained swimming in three dimensions. The design of a fish-like soft body has been extended to deform under hydraulic instead of pneumatic power. Moreover, a new closed-circuit drive system that uses water as a transmission fluid is used to actuate the soft body. Circulation of water through internal body channels provides control over the fish's caudal fin propulsion and yaw motion. A new fabrication technique for the soft body is described, which allows for arbitrary internal fluidic channels, enabling a wide-range of continuous body deformations. Furthermore, dynamic diving capabilities are introduced through pectoral fins as dive planes. These innovations enable prolonged fish-like locomotion in three dimensions.

Keywords Soft robotics · Robotic fish · Hydraulic actuation · Underwater locomotion · Lost-wax silicone casting · Soft actuator fabrication · Fluidic elastomer actuator

---

This work was done in the Distributed Robotics Laboratory within CSAIL at MIT with support from the National Science Foundation, grant numbers NSF 1117178, NSF EAGER 1133224, NSF IIS1226883 and NSF CCF1138967, and National Science Foundation Graduate Research Fellowship Program, primary award number 1122374. We are grateful for this support. The authors declare no competing financial interests.

R.K. Katzschmann (✉) · A.D. Marchese · D. Rus
Computer Science and Artificial Intelligence Laboratory, Massachusetts Institute of Technology, 32 Vassar Street, 32-376, Cambridge, MA 02139, USA
e-mail: rkk@csail.mit.edu
URL: http://www.csail.mit.edu

A.D. Marchese
e-mail: andy@csail.mit.edu

D. Rus
e-mail: rus@csail.mit.edu

M.A. Hsieh et al. (eds.), *Experimental Robotics*, Springer Tracts in Advanced Robotics 109, DOI 10.1007/978-3-319-23778-7_27

1 Introduction

Many natural system have body compliance as an integral functional feature. Compliant bodies easily adapt to changes in the environment and compensate for its uncertainties. This adaptability reduces the complexity in modeling, planning, and control. During a collision, a compliant body can deform and absorb energy [1], which allows for safer interaction between humans and robots. The aim is to exploit the principle of body compliance and design softness into robots.

In this work, an approach to create and control an autonomous soft-bodied system for prolonged motion underwater is presented. It is demonstrated that a soft actuator can be used untethered for controlled continuous propulsive locomotion in a fluidic medium. This allows for use in underwater inspection tasks of complex shaped environments where body compliance and adaptability is key in navigating through branched structures like pipe systems. It also has the potential to be used for close studies of fish schools. Deploying a soft robotic fish which mimics the locomotion of a biological fish will more likely be accepted to swim along fish schools without disrupting their natural behavior.

1.1 Overview

This paper presents a hydraulically actuated soft-bodied robotic fish that can swim for long durations of time and move in three dimensions to control depth and planar trajectories. In order to enable these new capabilities, innovations in the design, fabrication, and control of the biologically inspired soft body and its drive system are shown. The soft body, fundamentally composed of distributed fluidic elastomer actuators [2–4], provides continuous undulatory motion as in [5], but is using hydraulic instead of pneumatic actuation. This modified body is fabricated through an innovative wax molding and casting process. The propulsive and steering actuation of the deformable body is driven by a novel closed-circulation water system. An open-loop controller for this new drive system is described and the integration into a closed-loop controller using an inertial measurement unit is laid out. Lastly, the entire system is experimentally validated by demonstrating forward swimming and yaw motions using the soft tail, and pitch control using dive planes.

1.2 Previous Work

Previous work has shown many approaches to building fish-like robots. Traditionally, robotic fish were hard, meaning they have bodies composed of rigid links and a finite number of joints [6–10]. A motor-less and gear-less approach was to use shape memory alloys to build robotic fish with a hard skeleton shell but with a deformable backbone for fish tail actuation [11].

Alternatively, soft-bodied robots have continuously deformable backbones and theoretically infinite degrees of freedom [12]. There are several examples of soft-bodied fish-like robots that use centralized actuation. A soft-bodied octopus-like arm developed by Laschi et al. demonstrated shortening, elongation and bending [13]. The robot fish FILOSE [14, 15] has a compliant posterior and demonstrated fishlike locomotion. Valdivia y Alvarado and Youcef-Toumi used a soft and compliant body in the design of a robotic fish to mimic the forward swimming kinematics of a real fish [16]. All three of these systems are cable-driven and actuated with an onboard servomotor, but require an external power supply and lack autonomy. Long et al. have developed a flexible biomimetic vertebral column used to propel an autonomous surface-swimming robot [17, 18]. Again, a single servomotor is used to actuate the compliant spine. It is an autonomous surface swimming system with only the posterior part of the tail being flexible. The above-mentioned compliant-bodied robotic fish operate on the principle of a passive, flexible mechanism driven by a traditional electromechanical actuator. They are primarily designed to study the hydrodynamics of the flexible body.

There are also examples of compliant active-bodied robots that achieve fish-like locomotion using distributed actuation. Shen et al. have used an oscillating strip of ionic polymer-metal composite as the posterior trunk of a dolphin-like robot [19]. Suzumori et al. developed a soft-bodied manta using pneumatic actuation [20]. Both systems are free swimming robots, but are also limited by an external tether.

The Airacuda fish developed by Festo [21] uses tube-shaped pneumatic muscles to actuate a flexible posterior body. The posterior body is composed of a rigid plastic skeleton covered by flexible skin and has the two actuating muscles at its center axis. In comparison, the posterior body presented in this work is composed almost entirely of soft rubber with many fluidic elastomer actuators embedded along both sides. The fluidic and electronic components of Airacuda are located in the fishs rigid anterior. The Airacuda uses its pneumatic actuation system not only for forward swimming and turning motions, but also for static diving, whereas the hydraulic system presented in this work uses dive planes for dynamic diving.

The previous work by Marchese et al. [5] has a fluidic actuation system that is embedded within the compliant and flexible body. This completely self-contained and autonomous system is capable of rapidly achieving continuum-body motion, emulating forward swimming and planar escape maneuvers of biological fish.

1.3 Extensions of Previous Work

This work extends in several ways the autonomous soft-bodied fish by Marchese et al. [5]. In this previous work, it was demonstrated that soft robots can be both self-contained and capable of rapid body motion. However, there are several prohibitive shortcomings of this previous implementation. (1) The fish used an open-circulation pneumatic actuation system that made prolonged operation difficult; after energy was delivered to the distributed body actuators, it was exhausted to the environment. This limitation is addressed in this paper by designing a closed-circulation drive system. (2) The previous implementation was pneumatically actuated without using an air bladder. The center of buoyancy was therefore uncontrollably changing throughout the actuation cycle. Designing the system to use water as a transmission fluid and being neutrally buoyant addresses this use. (3) The previous implementation was constrained to move on a pre-defined trajectory in two dimensions, which is addressed by introducing dive planes to allow for dynamic diving.

1.4 Major Contributions

This work differs from previous work in that distributed hydraulic actuation, a new drive system for this kind of actuation, and new fabrication techniques to cast the distributed fluid elastomer actuator are used. Specifically, the following contributions are made:

1. Design of a water-driven soft-bodied actuator and a closed-circulation hydraulic drive system;
2. Fabrication technique for a soft-bodied actuator that allows arbitrarily formed fluidic channels, enabling a wide-range of continuous bending profiles;
3. Biologically inspired fish-like gait and dynamic diving, enabling forward-swimming in 3D.

2 Technical Approach

The technical approach to develop an autonomously propulsing soft robotic fish includes the mechanical design of individual functional units, a new fabrication technique for fluidic elastomer actuators, and a control approach for the locomotion and steering of the fish.

2.1 Mechanical Design

The robotic fish consists of three major functional components. These are:

1. the fish's soft tail for forward propulsion and yaw motions,
2. a waterproof gear pump unit as actuation source for the tail, and
3. a pair of dive planes actuated by a waterproof servo to enable pitch control.

Furthermore, a dorsal and a ventral fin were added to protect the fish against rolling. The control electronics are placed in a watertight nose compartment. Figure 1 shows the robotic fish with all its major components.

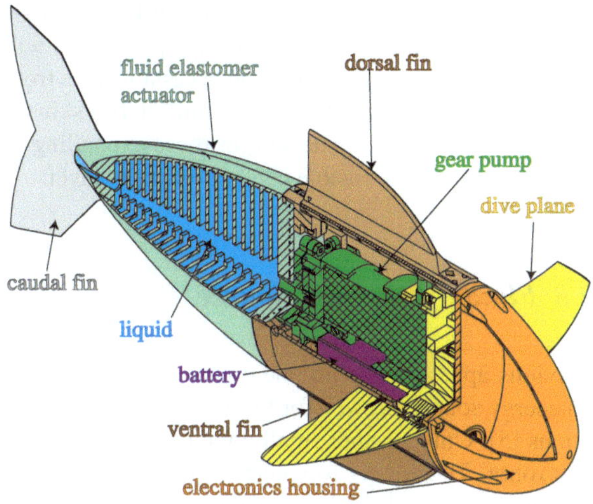

Fig. 1 Isometric cut of fish assembly

2.2 Fish Tail

The soft fish tail shown in Fig. 2 belongs to the group of fluidic elastomer actuators [2, 3]. The design mimics the rear portion of a fish, encompassing the posterior peduncle and the caudal fin. This tail can continuously bend along its vertical center constraint layer by fluidic actuation of two lateral cavity structures on each side. The

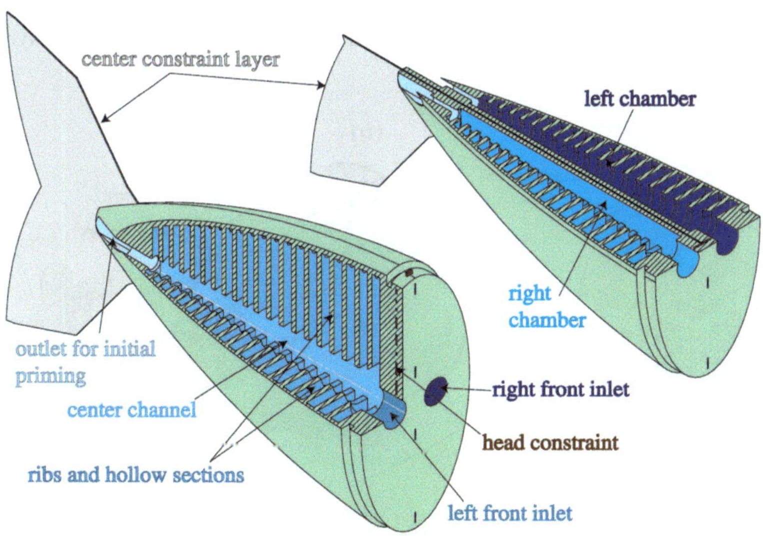

Fig. 2 Fish's soft tail as a fluidic elastomer actuator consisting of two fluidic chambers, visualized with two different cut views

inextensible and stiffer center constraint layer splits the tail evenly along a vertical plane. An actuator consists of evenly spaced ribs with hollow sections in between, connected by a center channel and accessible by a front inlet. The rib structure allows for expansion or contraction of the thin exterior skin under positive or negative fluidic pressure, respectively. The sum of these expanding or contracting motions leads to bending of the inextensible center constraint layer.

2.3 Fluidic Actuation Source

Previous approaches with pneumatic actuation using compressed gas cartridges as an energy source allowed for fast propulsion, but it was depleted after about 50 actuations [5]. Only a rather small compressed gas cartridge fits into the self-contained fish robot. The energy to weight ratio of a compressed gas cartridge is lower than modern batteries. In order to deflate an actuator, all the gas inside it is exhausted to the environment without reusing any of its kinetic energy. Furthermore, constantly releasing gas causes non-neglible changes in the overall buoyancy of the robotic fish, which either requires compensation through an air bladder or choosing a different actuation approach. Alternately transporting fluid from one chamber to the other does not require a storage unit in between, and the fluid does not need to be exhausted in order to deflate the actuator. This motivated a closed circulation actuation approach using an incompressible fluid like water and a gear pump to move it back and forth. The gear pump including its actuating DC motor are shown in Part b of Fig. 3.

The desired flapping frequency and curvature of the soft tail are determined based on previous studies on self-propelling foils driven by an external robotic actuator [22, 23]. The frequency and amplitude is applied onto a soft fish tail using a hydraulic

Fig. 3 **a** Dive plane attached to waterproof servo and clamp mount, and **b** gear pump with dc motor

cylinder pump. The measurements on the displaced volume and maximum pressures of the fish tail prototype combined with the desired flapping frequency result in the desired volumetric flow rate. Based on those results, a gear pump and its attached waterproof DC motor are designed and specified. The volumetric displacement per shaft revolution of an external gear pump is estimated with $Q = \frac{\pi}{4}\,(D^2 - d^2)\,w$, where w is the gear width, D is the gear's outer diameter, and d is the gear's inner diameter [24].

2.4 Dive Planes for Pitch Control

In order to allow for pitch motion and therefore dynamic diving, a pair of servo actuated dive planes are designed and added to the design at a place, where a fish's pectoral fins are usually located. Pectoral fins are responsible for the creation of a dynamic lifting forces to allow for depth control. One dive plane including its clamp mount to the back of the DC motor is shown in Part a of Fig. 3. The dive plane profile is designed using a loft limited by two symmetric air foil wing sections. A National Advisory Committee for Aeronautics (NACA) 0013 profile with a cord length of 0.06 m and a NACA 0010 profile with a cord length of 0.025 m was used. The cord length was determined by setting it to approximately 13 % of the entire length of the body of the fish. The thickness is defined by the size of the lever arm of the waterproof servo, which in return defines the size of the mounting plate needed. A symmetric profile is chosen so that no lift is produced when held in a horizontal position.

2.5 Fish Tail Fabrication

The actuated cavities of the fish tail are achieved by lost-wax casting. The fabrication process is depicted in Fig. 4.

In step (A), the rubber mold is poured and cured inside an assembly consisting of an outer mold with lid and a model for the core inside of it. In preparation for step (B), the lid and the model core are removed and the rubber mold is left inside the outer mold. The rubber mold receives a small carbon fiber tube as an inlay in its center cavity. This ensures that the wax core does not break when being removed from the rubber mold. Mold release spray is applied to the silicone rubber mold to ease the wax core removal process. The wax is heated up until it becomes fully liquefied. The assembly of rubber mold and outer mold is heated up for a few minutes to the same temperature as the wax. Using a syringe, the liquid wax is injected into the assembly. Within a few minutes, the injected wax will start to solidify and significantly shrink in volume; this is counteracted by injecting more hot wax into the solidifying wax core during the cool down. In step (B), the wax core is first allowed to completely

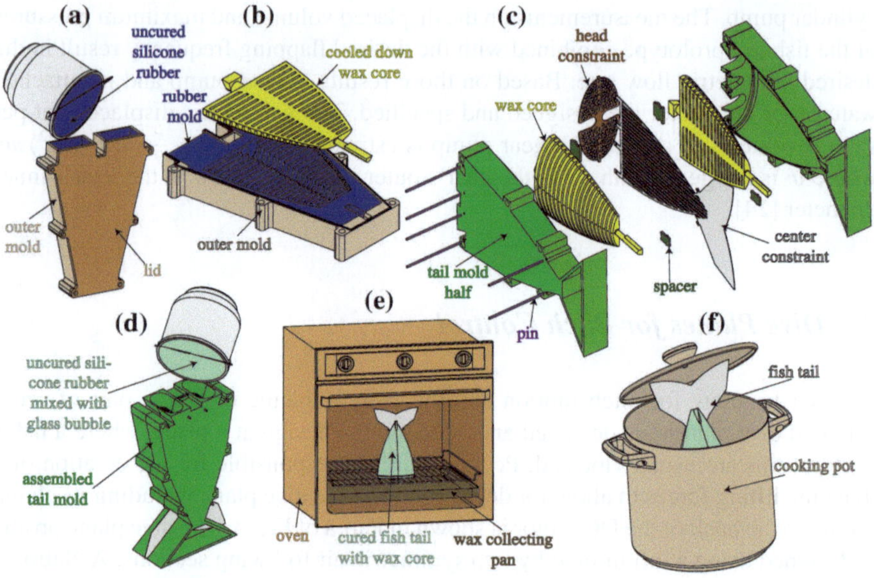

Fig. 4 Fish tail fabrication process: **a** Pour and cure a rubber mold, **b** pour wax cores, **c** combine head constraint, center constraint and wax cores with tail mold halves, **d** pour rubber mixed with glass bubbles into assembled tail mold, **e** using an oven melt out wax core from the cured fish tail, and **f** cook out remaining wax to create desired actuator cavities

cool down, then it is released from the mold. In step (C), a head constraint, a center constraint, and two wax cores are assembled together inside the tail mold halves using spacers, positioning pins and screws. In step (D), a mix of silicone rubber with glass bubbles is poured into the tail assembly and allowed to cure. In step (E), most of the wax core is melted out by placing the fish tail in an upright position into an oven. Finally, in step (F) the remaining wax residues are cooked out in a boiling water bath.

2.6 Control

The motion control of the robotic fish is mainly determined by the propulsive forward swimming motion of the soft tail. Yaw control is achieved through keeping, on average, more fluidic volume in one fish half than the other. Pitch control is done through adjusting the attack angle of the dive planes. Both yaw and pitch control effort depend on the forward swimming speed by the tail. An on-board 9 degrees of freedom inertial measurement unit provides absolute attitude measurements, which will be used in a future version of the fish for closed-loop attitude control.

2.6.1 Forward Swimming

Controlling the speed of the gear pump determines the volumetric flow from one side of the fin to the other. Alternating actuation at approximately 1 Hz results in a flapping motion of the posterior peduncle and phase-shifted flapping of the caudal fin. The voltage profile of the motor is shaped with an alternating trapezoidal profile to avoid impulsive switching, high peak currents, and high angular momentum around the roll axis. To adjust the forward swimming speed, the frequency and motor velocity are controlled.

2.6.2 Yaw Control

Controlling the heading requires that the fish tail has an adjustable average offset from the neutral tail position. This can be achieved by adjusting the alternating actuation so that on average more fluidic volume flows into one fish half than the other. This can be done by (a) adjusting the amplitude ratio between the positive and negative actuation trapezoid while keeping the time duration of both trapezoids the same, or (b) adjusting the ratio of time duration of both trapezoids so that one half is filled up longer than the other. The latter approach was chosen for the fish presented in this work. A combination of both methods is also possible.

2.6.3 Pitch Control

Controlling the pitch of the fish is achieved through adjusting the attack angle of the dive planes in a range of $-\frac{\pi}{4}$ to $\frac{\pi}{4}$ rad. The forward swimming speed directly determines how fast the pitch of the fish is changed and can be maintained.

3 Results

The implemented fabrication process for the wax is depicted in Fig. 5, showing each major step from silicone rubber mold creation to de-molding of the wax core. The outer mold, the lid, and the model core are 3d printed. The silicone rubber mold has an A30 durometer. Beeswax with a melting point of 63 °C is heated up to 95 °C for pouring into the rubber mold.

The titanium head constraint was water jetted out of a 0.9 mm thick highly corrosion-resistant grade 2 titanium plate. The center constraint was laser cut out of a 0.5 mm thick flexible acetal sheet. Both fabrication steps are depicted in Fig. 6.

In the final fabrication steps, two wax cores and both constraint layers are combined with the fish tail molds. The steps are depicted in Fig. 7.

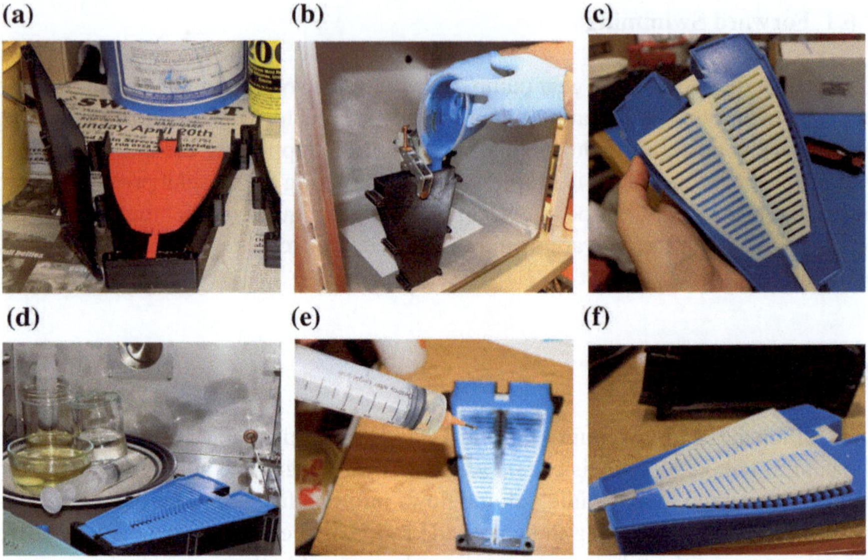

Fig. 5 Fabrication steps of the wax core. **a** 3d printed model core. **b** Create silicone mold. **c** Release model core. **d** Heating wax and mold. **e** Pour wax with syringe. **f** Demolding of wax core

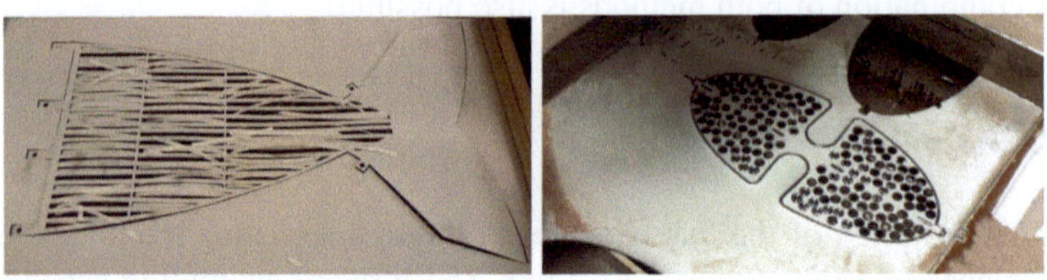

Fig. 6 Laser cut center constraint layer and water jet head constraint

The tail consists of A15 durometer silicone rubber mixed with low density crush resistant glass bubbles to achieve a mixed density just below the density of water. The mixing ratio, n_b, between the bubble mass, m_b, and the silicone rubber mass, m_s, is: $n_b = \frac{m_b}{m_s} = (1 - \frac{\rho_d}{\rho_s})/(\frac{\rho_d}{\rho_b} - 1)$, where ρ_b stands for the density of the glass bubbles, ρ_s stands for the density of the silicone rubber, and ρ_d stands for the desired mixed density. The density of the used silicone rubber is $\rho_s = 1.18\,\text{g/cm}^3$ and the density of the glass bubbles is $\rho_b = 0.125\,\text{g/cm}^3$. In order to make the fish tail slightly lighter than water, a desired mixed density of $\rho_d = 0.991\,\text{g/cm}^3$ with a mixing ratio of $n_b = 2.3\,\%$ is used. This mixing step is important for achieving overall neutral buoyancy of the robotic fish. Otherwise, the weight of the fish tail has to be compensated with thick styrofoam flotation attachments around the center of the fish, which introduces undesired drag.

The nose of the fish is a waterproof housing for the microcontroller, motor driver, and wireless communication electronics. The housing is 3D printed and waterproofed

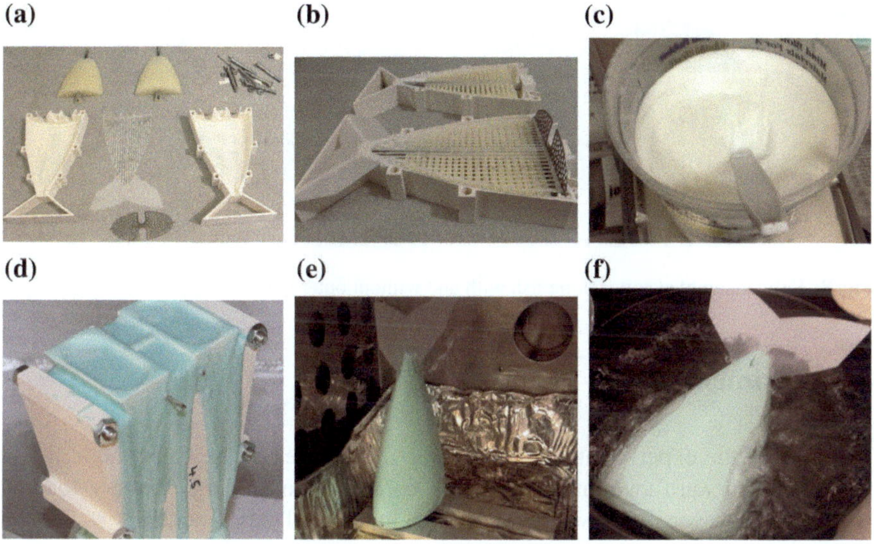

Fig. 7 Final fabrication steps of the silicone rubber fish tail. **a** Individual parts of tail mold assembly. **b** Create silicone mold. **c** Release model core. **d** Heating wax and mold. **e** Pour wax with syringe. **f** Demolding of wax core

by brush-coating it with a polyurethane paint and subsequent degassing [25]. Behind the nose is the dive plane assembly, consisting of two individually controllable dive plane units. Each unit consists of a dive plane, which is directly mounted onto the lever arm of a waterproof servo motor. The dive plane assembly is mounted to the end of the brushed DC motor of the gear pump. The motor and gear pump unit is directly attached to the soft fish tail. Underneath the gear pump motor sits a lithium polymer battery to power all components.

Each actuator has a removable plug at the caudal fin: the initial dive is started with plugs removed so water can fill the actuation chambers by running the self-priming gear pump at a low frequency for a short duration. After all air has been removed, the plugs are inserted to seal the chambers.

The 1.65 kg mass of the complete assembly was slightly adjusted to make it almost neutrally buoyant using foam attachments and additional weights placed outside and inside the 3D printed center hull of the fish. The fish has the dimensions: 0.45 m $\times$ 0.19 m $\times$ 0.13 m.

The fish receives 72 MHz wireless communication commands to move forward, move up and down, or turn left and right. The on-board micro-controller translates these high-level commands into the control law for the tail actuation and the angular positions of the dive planes.

The assembled fish with and without outer hulls is shown in Fig. 8.

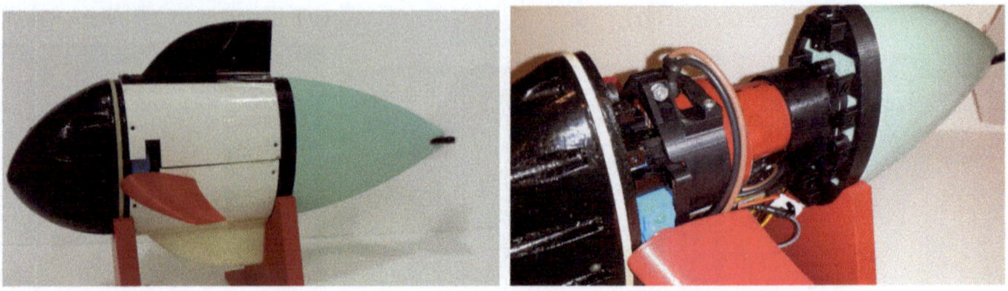

Fig. 8 Fully assembled soft robotic fish with and without outer hulls

4 Experiments

The goal of the experiments is to show the capabilities of the soft robotic fish in continuous forward swimming and pitch control, using a single battery loading. Additionally, yaw control was demonstrated in the pool experiments.

4.1 Fish Tank Experiments

The fish swam in a tank of a length of 1.22 m in a straight, horizontal line from wall to wall and repeated this 25 times. It was manually placed back to the start after each completion of one lap.

For an average horizontal distance of 0.74 m, the horizontal swimming speed was 0.10 m/s, which is equivalent to 0.15 body lengths per second. One repetition of the horizontal forward swimming experiment is shown in Fig. 9.

To test the pitch control, the fish's submergence was adjusted as it moved forward over the span of the tank by setting the pitch angle of the dive planes to $\frac{\pi}{4}$ rad. For 13 repetitions of this experiment, the average diving speed was 0.015 m/s over an average vertical distance of 0.13 m. The horizontal swimming speed was 0.08 m/s

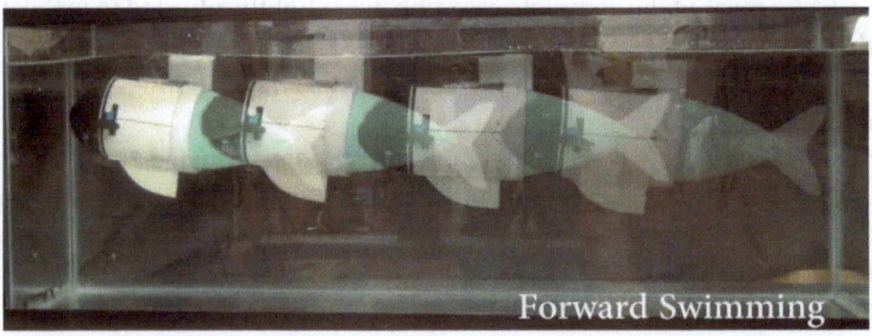

Fig. 9 Fish tank experiment: forward swimming

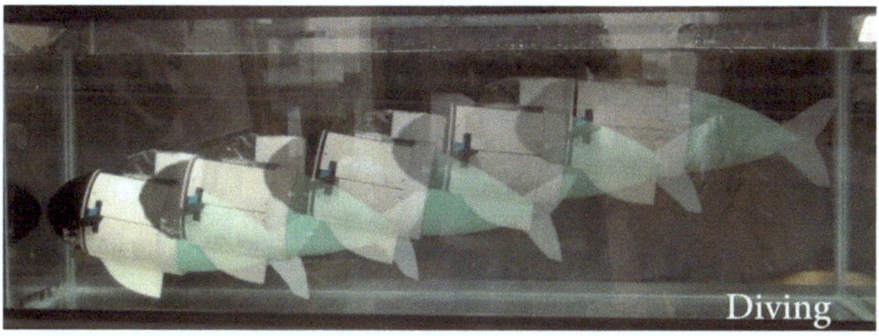

Fig. 10 Fish tank experiment: diving

over an average horizontal distance of 0.74 m. One repetition of the diving experiment is shown in Fig. 10.

After 35 min of continuous wireless underwater operation, the 1.3 Ah lithium polymer battery was almost depleted. This corresponded to 52 repeated runs through the tank, resulting in a total distance of approximately 40 m.

4.2 Pool Experiments

In a 13.25 m × 7 m × 1.12 m pool, the maximum horizontal swimming speed was 0.23 body lengths per second. One trial of the horizontal forward swimming experiment in the pool is shown in Fig. 9. The figure shows overlaid fish's poses for every time it had moved for one full body length.

To test the pitch control, the fish's submergence was adjusted as it moved near the side wall of the pool. The pitch angle of the dive planes was set to approximately $\frac{\pi}{4}$ rad. One trial of the diving experiment in the pool is shown in Fig. 11b.

The heading or yaw control was also tested. The fish was able to turn in the pool, as can be seen in Fig. 11c.

After about 40 min of continuous operation in the pool and swimming an accumulative distance of approximately 130 m, the 1.3 Ah lithium polymer battery was almost depleted.

5 Main Experimental Insights

The approach proposed in this work for creating an autonomous soft-bodied robotic fish resulted in a robot that demonstrated prolonged and consistent underwater operation. While most soft robots are pneumatically powered, it was shown that hydraulic power increases the capabilities for a given range of applications. For example, when high frequency change of actuation direction is needed, exemplified by the flapping

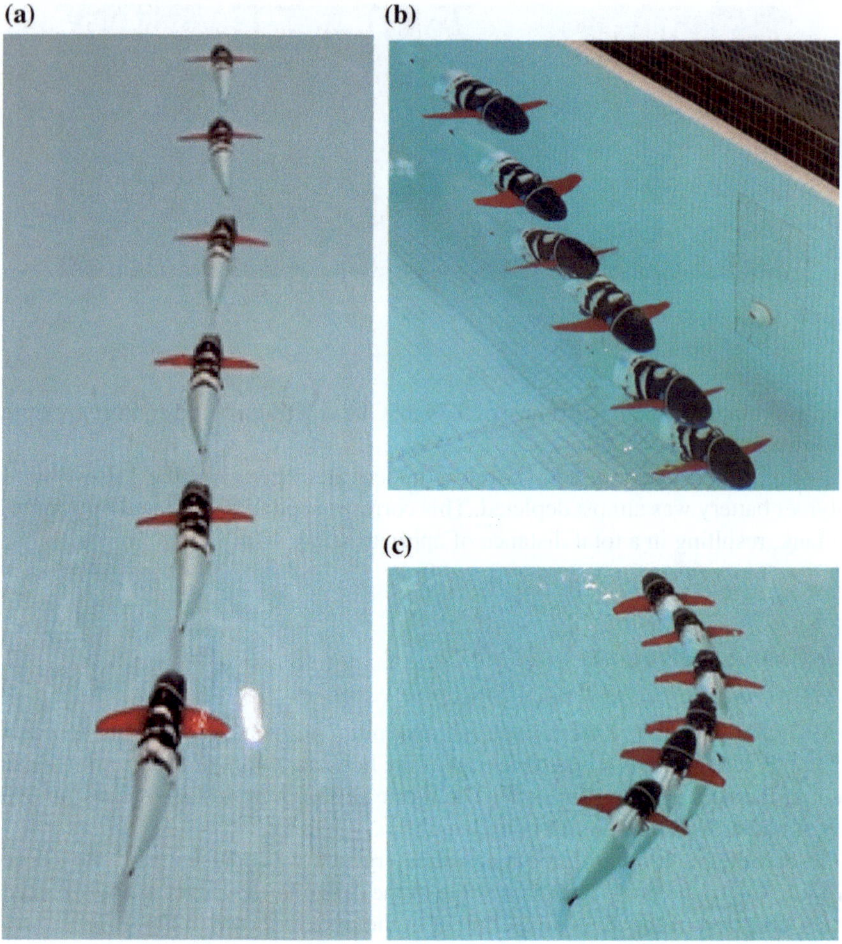

Fig. 11 Pool experiments demonstrating locomotion capabilities. **a** Forward swimming. **b** Turning

foil, or prolonged actuation of an autonomous soft robot is required. Hydraulic actuation allows for a higher force as shown by the overcoming of hydrodynamic resistance used for the fish-like locomotion [26]. The presented prototype is a step towards creating a closed-loop controlled biomimetic robotic fish which is inherently soft, performs continuous body deformations from nose to tail, and allows for safe interaction with other living beings. Based on the experimental results, the novel actuation system of the robot prototype will influence future work in the field, both in terms of fish robots and soft robots in general.

References

1. Albu-Schaffer, A., Eiberger, O., Grebenstein, M., Haddadin, S., Ott, C., Wimbock, T., Wolf, S., Hirzinger, G.: Soft robotics. IEEE Robot. Autom. Mag. **15**(3), 20–30 (2008)
2. Correll, N., Onal, C.D., Liang, H., Schoenfeld, E., Rus, D.: Soft autonomous materials - using active elasticity and embedded distributed computation. In: 12th International Symposium on Experimental Robotics (ISER), New Delhi, India (2010)
3. Onal, C.D., Chen, X., Whitesides, G.M., Rus, D.: Soft mobile robots with on-board chemical pressure generation. In: International Symposium on Robotics Research (ISRR) (2011)
4. Marchese, A.D., Onal, C.D., Rus, D.: Soft robot actuators using energy-efficient valves controlled by electropermanent magnets. In: 2011 IEEE/RSJ International Conference on Intelligent Robots and Systems (IROS), pp. 756–761. IEEE (2011)
5. Marchese, A.D., Onal, C.D., Rus, D.: Autonomous soft robotic fish capable of escape maneuvers using fluidic elastomer actuators. Soft Robot. **1**(1) (2014)
6. Barrett, D.S.: Propulsive efficiency of Robotuna. Ph.D. thesis, Massachusetts Institute of Technology (1988)
7. Triantafyllou, M.S., Triantafyllou, G.S.: An efficient swimming machine. Sci. Am. **272**(3), 64–71 (1995)
8. Zhong, Y., Chong, C., Zhou, C., Seet, G.G., Low, K.: Performance predict model for a body and caudal fin (bcf) biomimetics fish robot. In: International Conference on Advanced Intelligent Mechatronics, IEEE/ASME, pp. 1230–1235 (2009)
9. Liu, J., Hu, H.: Biological inspiration: from carangiform fish to multi-joint robotic fish. J. Bionic Eng. **7**(1), 35–48 (2010)
10. Wen, L., Wang, T., Wu, G., Liang, J.: Hydrodynamic investigation of a self-propelled robotic fish based on a force-feedback control method. Bioinspir. Biomim. **7**(3), 036012 (2012)
11. Rossi, C., Colorado, J., Coral, W., Barrientos, A.: Bending continuous structures with smas: a novel robotic fish design. Bioinspir. Biomim. **6**(4), 045005 (2011)
12. Trivedi, D., Rahn, C.D., Kier, W.M., Walker, I.D.: Soft robotics: biological inspiration, state of the art, and future research. Appl. Bionics Biomech. **5**(3), 99–117 (2008)
13. Laschi, C., Cianchetti, M., Mazzolai, B., Margheri, L., Follador, M., Dario, P.: Soft robot arm inspired by the octopus. Adv. Robot. **26**(7), 709–727 (2012)
14. El Daou, H., Salumae, T., Ristolainen, A., Toming, G., Listak, M., Kruusmaa, M.: A biomimetic design and control of a fish-like robot using compliant structures. In: 15th International Conference on Advanced Robotics (ICRA), pp. 563–568. IEEE (2011)
15. El Daou, H., Salumae, T., Toming, G., Kruusmaa, M.: A bio-inspired compliant robotic fish: Design and experiments. In: IEEE International Conference on Robotics and Automation (ICRA), pp. 5340–5345. IEEE (2012)
16. Valdivia y Alvarado, P., Youcef-Toumi, K.: Design of machines with compliant bodies for biomimetic locomotion in liquid environments. ASME J. Dyn. Syst. Meas. Control **128**, 3–13 (2006)
17. Long, J.H., Koob, T., Schaefer, J., Summers, A., Bantilan, K., Grotmol, S., Porter, M.: Inspired by sharks: a biomimetic skeleton for the flapping, propulsive tail of an aquatic robot. Mar. Technol. Soc. J. **45**(4), 119–129 (2011)
18. Long, J.H., Krenitsky, N.M., Roberts, S.F., Hirokawa, J., de Leeuw, J., Porter, M.E.: Testing biomimetic structures in bioinspired robots: how vertebrae control the stiffness of the body and the behavior of fish-like swimmers. Integr. Comp. Biol. **51**(1), 158–175 (2011)
19. Shen, Q., Wang, T., Liang, J., Wen, L.: Hydrodynamic performance of a biomimetic robotic swimmer actuated by ionic polymer-metal composite. Smart Mater. Struct. **22**(7), 075035 (2013)
20. Suzumori, K., Endo, S., Kanda, T., Kato, N., Suzuki, H.: A bending pneumatic rubber actuator realizing soft-bodied manta swimming robot. In: International Conference on Robotics and Automation, pp. 4975–4980. IEEE (2007)
21. Festo: Airacuda. http://www.festo.com/cms/en_corp/9761.htm (2006). Accessed 24 May 2014

22. Lauder, G.V., Flammang, B., Alben, S.: Passive robotic models of propulsion by the bodies and caudal fins of fish. Integr. Comp. Biol. **52**(5), 576–587 (2012)
23. Alben, S., Witt, C., Baker, T.V., Anderson, E., Lauder, G.V.: Dynamics of freely swimming flexible foils. Phys. Fluids (1994-present) **24**(5) 051901 (2012)
24. Karassik, I.J., Messina, J.P., Cooper, P., Heald, C.C.: Pump Handbook, vol. 4. McGraw-Hill, New York (2008)
25. Mireles, J., Adame, A., Espalin, D., Medina, F., Winker, R., Hoppe, T., Zinniel, B., Wicker, R.: Analysis of sealing methods for fdm-fabricated parts. Technical Report, W.M. Keck Center for 3D Innovation. The University of Texas, El Paso (2011)
26. Drucker, E.G., Lauder, G.V.: Locomotor forces on a swimming fish: three-dimensional vortex wake dynamics quantified using digital particle image velocimetry. J. Exp. Biol. **202**(18), 2393–2412 (1999)

Foldable Joints for Foldable Robots

Cynthia Sung and Daniela Rus

Abstract Print-and-fold approaches to robot fabrication allow entire robots to be produced using a single uniform process: fabricating them in-plane and then folding them into their 3-D forms. Current efforts to design print-and-fold robots have been limited by a lack of understanding of what motions can be achieved by folding. In this paper, we introduce fold patterns for three basic joints commonly used in robots, and we show how the patterns can be changed to accommodate user-specified ranges of motion. The joints are composed with each other to produce joints with higher degrees of freedom and with rigid bodies to produce entire foldable linkage mechanisms. We have folded our basic joints and composed mechanisms, and they achieve the expected kinematics. We have also printed control circuitry on and attached actuators directly to three of our designs, demonstrating that it possible to print and fold robots with many different kinematics.

Keywords Foldable joints · Linkages · Origami-inspired design · Print-and-fold robots

1 Introduction

Rapid design and customization of robots is limited by the practicalities of their fabrication. Current robot designs are often composed of multiple subcomponents that must be manufactured using distinct processes before assembly. In contrast, recent work in origami-inspired robot designs [1, 2] suggests that a print-and-fold approach to robot fabrication allows entire robots, including movable and rigid parts alike, to be produced quickly using a single uniform process.

C. Sung (✉) · D. Rus
Computer Science and Artificial Intelligence Laboratory,
Massachusetts Institute of Technology, 32 Vassar Street, 32-374,
Cambridge, MA 02139, USA
e-mail: crsung@csail.mit.edu

D. Rus
e-mail: rus@csail.mit.edu

M.A. Hsieh et al. (eds.), *Experimental Robotics*, Springer Tracts
in Advanced Robotics 109, DOI 10.1007/978-3-319-23778-7_28

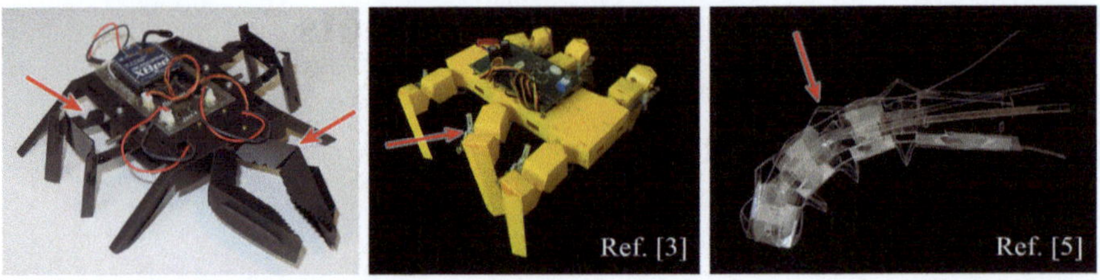

Fig. 1 Previous foldable robots, joints indicated by *arrows*. They all use hinge joints

Efforts to design foldable robots have been complicated by a lack of understanding of what types of motions can result from folding. Although many designs have been developed and tested [3–8], most designs are restricted to contain single degree-of-freedom (DOF) hinge-like joints achievable by a single fold (ref. Fig. 1). Those patterns that can achieve more complex motion [6–8] do not translate well to general robot design. Theoretical work on designing folded structures (see [9] for a review) mostly focuses on producing rigid structures rather than transformable structures, and although transformable folded structures were analyzed theoretically in [10–12], the resulting fold patterns are again application-specific.

In mechanism design not restricted to folding, the joints and links that make up the desired mechanisms are readily available and can be connected together straightforwardly, making methods for automated mechanism design [13, 14] and analysis [15] possible. Our approach is to design foldable joints that can be combined with rigid bodies so that designing foldable robots can be as easy as using preexisting mechanism design tools with folded structures instead. In this paper, we contribute fold patterns for revolute and prismatic joints that:

- are parameterized to achieve a user-specified size and range of motion,
- can be composed with each other into joints with higher degrees of freedom,
- can be composed with rigid bodies to produce foldable linkage mechanisms, and
- have been used to design foldable robots that have been experimentally validated.

Our designs open the way to making robots with any desired kinematics out of one sheet using a single print-and-fold process.

2 Parameterized Joint Patterns

Our mechanisms and robots are based on three joint types: a hinge joint, a prismatic joint, and a pivot joint. In this section, we present the parameterized fold patterns for each.

2.1 Definitions

We begin with informal definitions for the terms used in the following descriptions. A more formal treatment of folding theory can be found in [9]. Consider a non-self-intersecting 2-D polygon P (possibly with holes). A *fold* on P is a line segment such that both endpoints are on the boundary of P and the segment itself lies on the interior of P. Every fold is associated with a fold angle range that is a subset of $(-\pi, \pi)$. A *fold pattern* consists of the polygon P and the set of folds on P. A *folded state* of a fold pattern is a non-self-intersecting 3-D structure formed when all folds in the fold pattern are folded at an angle in their associated fold angle range. In this paper, all figures of fold patterns will display the boundary of P, or the cut lines, in solid black and fold lines in dotted gray.

The folds in a fold pattern divide the original polygon P into a set of smaller polygons that overlap only at the fold lines. These polygons are called *faces*. Since the purpose of joints is to connect other structures to each other, our joints have faces that exist specifically to allow attachment. We call these faces the *bases* of the joint. All of our joints are designed to connect two structures to each other and so each have two bases.

2.2 Hinge Joint

Hinge joints allow rotation about an axis parallel to a base. They are the easiest joint to create since they can be implemented as a single unconstrained fold on the base itself. However, when hinge joints are created in this way, the joint limits depend on the geometry of the bodies being connected and cannot be independently specified.

We have designed a hinge joint of a more general form, as shown in Fig. 2a. The joint consists of a base in the shape of a regular polygon (a hexagon in the example).

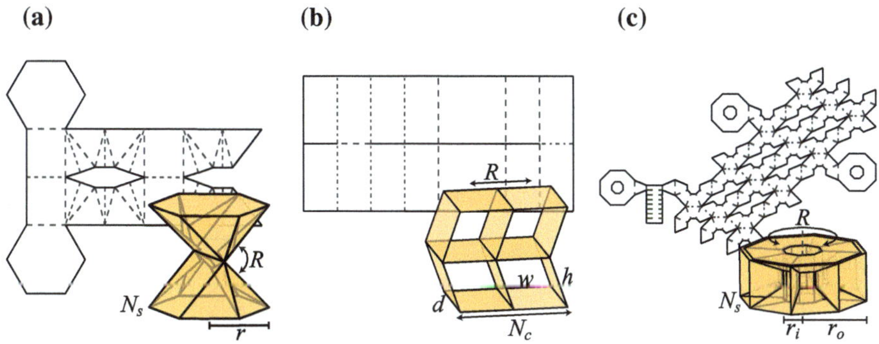

Fig. 2 Sample fold patterns and folded states for three basic joint types with input parameters indicated. **a** Hinge joint. **b** Prismatic joint. **c** Pivot joint

From two opposite sides, sloped faces angle to meet at the axis of rotation. The angle between these faces and the base is determined by the joint limits as $\frac{R}{4}$, where R is the total range of motion. On all other sides, triangular faces are attached to provide structural support.

The associated fold pattern is a strip that attaches to the outer edges of the base polygon and contains the rectangular and triangular faces. Additional folds that tuck away extra material help form the hinge shape. Input parameters to the pattern are the number of sides N_s, the radius r of the base, and the total range of motion R. The hinge joint is always symmetric. If asymmetric joint limits are desired, users can attach the joint at an angle by attaching a sloped polyhedron to the base first, as we do later for the mechanism in Fig. 8. Note that the joint angle is restricted to be between $-\pi$ and π radians, and that the length of the joint increases with the range of motion.

2.3 Prismatic Joint

Our prismatic joint allows translation either parallel to or normal to the base. To produce this motion, we use a grid of parallelogram linkages. In a single parallelogram linkage, horizontal and vertical translation are coupled. By connecting linkages in a grid, these two degrees of freedom can be decoupled. Figure 2b shows a two-by-two grid of linkages of height h. By restricting horizontal motion, the joint enables vertical translation by as much as $2h$. By restricting the vertical distance between the bases to be h, the joint enables horizontal translation by a distance h in either direction.

The prismatic joint fold pattern is a grid of rectangular faces, built by repeating and connecting units as shown in Fig. 3. On the leftmost side is a unit consisting of four faces that fold into a parallelogram linkage. To add columns, three-face add-on units are attached to the right, with the fold angles of each unit opposite in sign to the one before. For each layer, the entire row of units is duplicated and attached above the previous layer to the faces corresponding to the top link of the linkage below. Input parameters are the dimensions h, w, and d of one linkage in the grid, the joint's range of motion R, and the number of columns N_c in the grid. The number of layers needed to achieve the desired range of motion is is $N_\ell = \lceil \frac{R}{2h} \rceil + 1$. The range of motion can be increased by changing either h or N_ℓ.

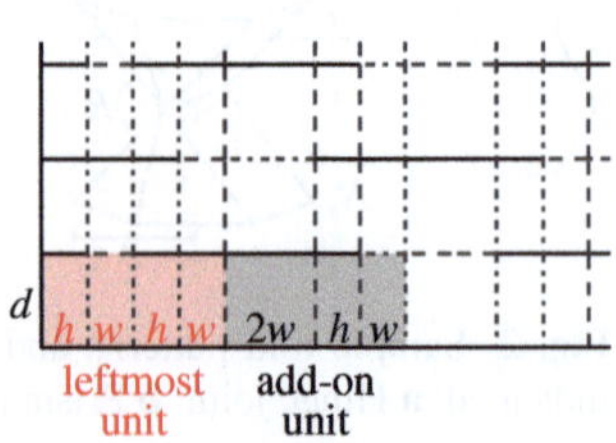

Fig. 3 Prismatic joint construction

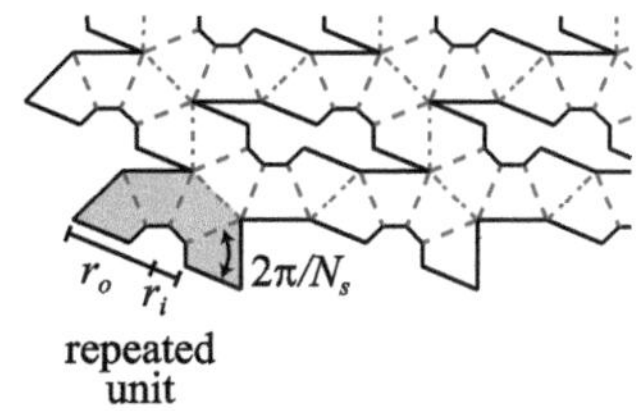

Fig. 4 Pivot joint construction

2.4 Pivot Joint

Pivot joints allow rotation about an axis perpendicular to the base. To achieve this twisting motion, we use a spherical parallelogram linkage. Like the hinge joint the base of the pivot joint is a regular polygon, and each side corresponds to one parallelogram linkage. Figure 2c shows an example of a pivot joint with an octagonal base. As in the prismatic joint, we use multiple layers of linkages in series to decouple vertical translation from twisting motion. Restricting the distance between the top and bottom faces to be the height of one layer enables pure twisting motion.

For simpler construction, we use square linkages. The pattern is built similarly to the prismatic joint, by repeating and connecting identical units. The unit, shown in Fig. 4, consists of four isosceles trapezoids, each with an angle equal to $\frac{2\pi}{N_s}$ between the legs of the trapezoid, connected along the legs. N_s units are attached to each other at the faces corresponding to the side links of the linkages in order to produce the N_s linkages that form one layer of the joint. The resulting strip of units is duplicated once for each layer and attached to the adjacent layers using the faces corresponding to the top and bottom links of the linkages. Input parameters to this design are the number of sides $N_s > 4$ of the base, the range of motion R, and the inner and outer radii r_i and r_o of the joint. The number of layers N_ℓ can be computed as $N_\ell = \lceil \frac{RN_s}{4\pi} \rceil + 1$. In Fig. 2c, the joint has 8 sides and 3 layers and can twist π radians.

2.5 Physical Models

We have built a system that allows users to specify a type of joint and the desired parameters and that automatically produces the corresponding fold pattern. We generated fold patterns for our basic joints and constructed them out of 0.13 mm thick polyester film, cutting them using a laser cutter and perforating the folds for easier assembly. Before printing, we added tabs and slots to the pattern to attach edges that should remain coincident in the folded state.

Figure 5 shows two of our folded joints. Figure 5a shows a 4-by-4 prismatic joint with every link 10 mm long. It is capable of 60 mm horizontal motion or 40 mm vertical motion. Figure 5b shows a 6-sided pivot joint with a $\frac{10}{3}\pi$ radian range of motion. The joint has 6 layers. Since plastic film has thickness, adding layers to increase a joint's range of motion increases the size of the joint: each layer adds the

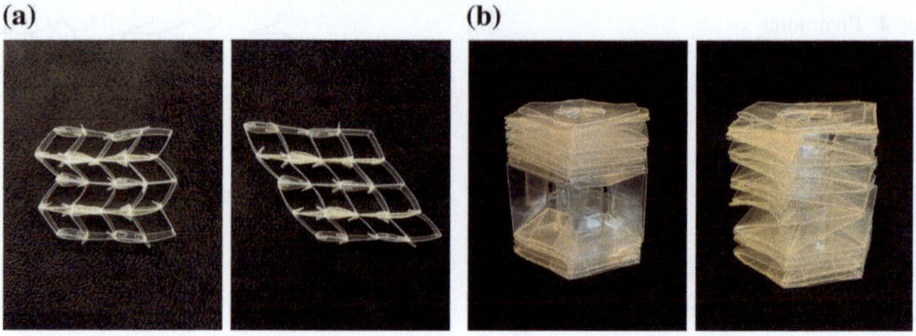

Fig. 5 Joints folded from polyester film in two different positions. **a** Prismatic joint. **b** Pivot joint

thickness of five sheets of plastic in the case of the prismatic joint, and four sheets of plastic for the pivot joint. For the joint in Fig. 5b, the additional thickness corresponds to almost half the joint length. This is not a concern for the hinge joint, which does not rely on layers to control the range of motion.

3 Composition of Folded Structures

As shown in [16], folded structures and their fold patterns can be composed to produce more complex designs. Our system supports such composition. In addition to generating joints, users can also input custom patterns for folded structures as a vector file. Users specify the edges or faces on separate folded structures that they wish to connect, and the system generates a single-sheet fold pattern for the composed structure. The system provides views of both the flat fold pattern and its folded state in 3-D so that users can visually verify that the composition is correct. We have tested this system for various joints, joint combinations, and linkage mechanisms.

3.1 Joints with Higher Degrees of Freedom

More complex joints with higher degrees of freedom can be created from our basic joints. In some cases, extra degrees of freedom come for free and composition is not necessary. For example, vertical translation can be added to the pivot joint to make a cylindrical joint by removing the distance constraint between the outer faces. In other cases, joints can be combined. For example, a universal joint can be made by connecting in series two hinge joints with orthogonal axes of rotation. A spherical joint is a pivot joint combined with a hinge. Since our designs are parameterized, basic joints can be adjusted for simpler joining (e.g., by having the same base) without restricting the joint limits.

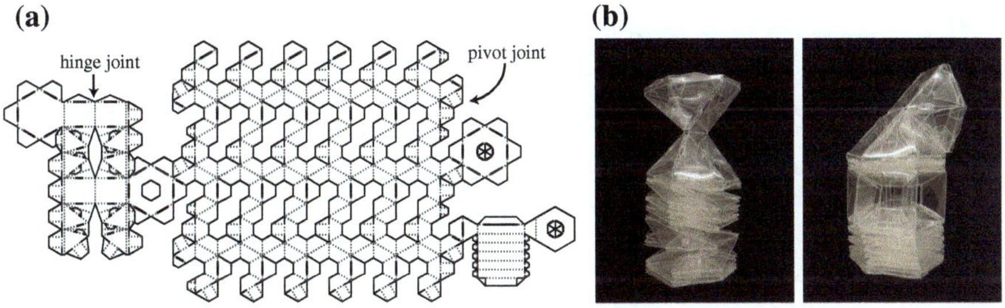

Fig. 6 Spherical joint composed from 6-sided pivot and hinge joints. **a** Composed fold pattern (with tabs and slots). **b** Folded joint in two positions

We tested composition of joints by generating a spherical joint, shown in Fig. 6. The composed 6-sided joint consists of a pivot joint with a $\frac{10}{3}\pi$ radian range of motion attached to a 6-sided hinge joint with a π radian range of motion. The resulting fold pattern is shown in Fig. 6a. Since the axes of rotation of the pivot and hinge joints intersect at the center of the hinge joint, the resulting joint approximates well the behavior of a spherical joint, despite the pivot and hinge joint being two separate entities.

3.2 Mechanisms

Entire linkages can be created by composing joints with rigid bodies. All of our joints have flat bases so that they can easily be attached to rigid bodies. To test the compatibility of our joints with rigid bodies, we used our system to compose a four-bar linkage. This linkage consists of four rectangular prisms, drawn manually, connected in a cycle using four pivot joints. In addition, we manually designed and added a motor mount to one joint to actuate the linkage. The resulting fold pattern is shown in Fig. 7a. The folded linkage behaves as expected. Frames of the resulting motion are shown in Fig. 7b. Note that since the joint angles are limited, care must be taken at the folding stage to ensure the joints are all connected at the correct position to take full advantage of the available range of motion.

Similarly, we composed two prismatic joints, two hinge joints, and a rectangular body to form a rowboat (Fig. 8). Each prismatic joint connects the rectangular body of the boat to one of the side paddles and can move independently. Paddles are made of a hinge joint with a $\frac{\pi}{2}$ radian range of motion mounted at a $\frac{\pi}{4}$ angle relative to the body of the boat. In this way, the paddles can lie either horizontally over the water or extend vertically down into the water to provide thrust. As shown in Fig. 8b, c, the folded rowboat is able to produce the intended movements.

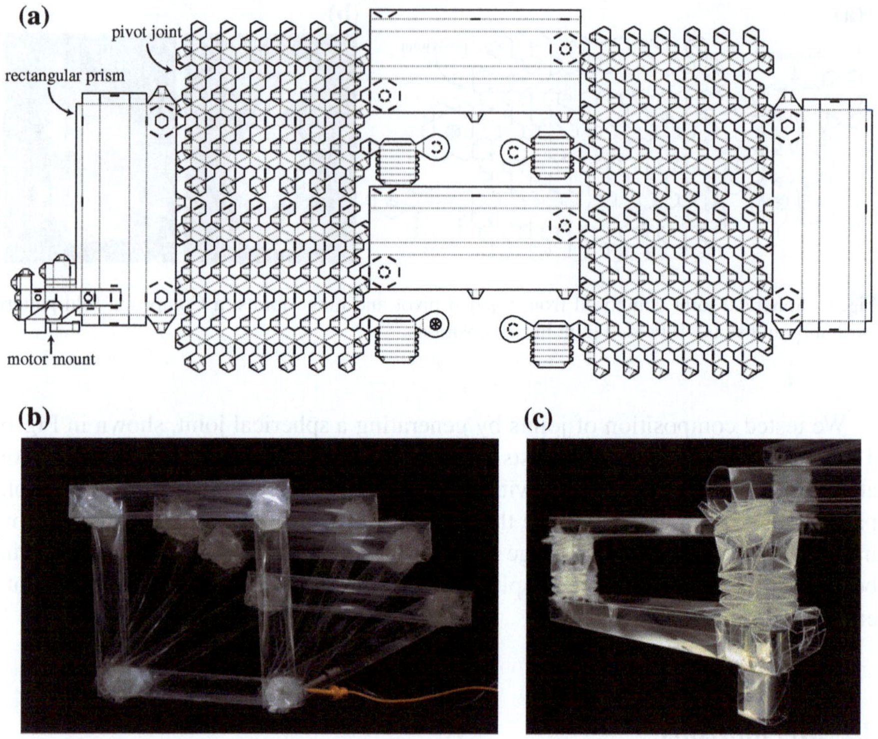

Fig. 7 Foldable four-bar linkage. **a** Composed fold pattern. **b** Movement of actuated four-bar. **c** Close-up of actuated joint

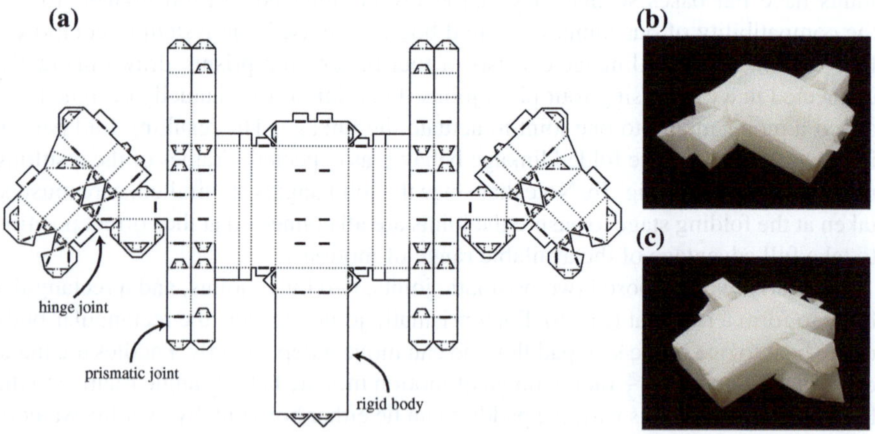

Fig. 8 Foldable rowboat. **a** Composed fold pattern. **b** Paddles up, **c** Paddles down

4 Foldable Robots

Print-and-fold manufacturing provides a natural method for incorporating actuation, sensing, and computation into a robot, specifically by printing circuitry and mounting components directly onto the fold pattern before folding. All of our joints are designed with this goal in mind and so have the space and the geometry for actuators and circuitry to be integrated directly into the folded structure, obviating the need for a post-folding stage of attaching circuit boards, actuators, and additional wires. In this section, we demonstrate joints and mechanisms with incorporated electronics.

4.1 Hinge Joint

We outfitted a 4-sided π radian hinge joint with a motor and potentiometer, as well as the control circuitry of a standard servo, to produce a hinge joint with position control. The motor was placed at the center of the hinge joint, with the output shaft aligned with the axis of rotation. It was kept in place using a manually designed motor mount (also used in the four bar linkage in Fig. 7) that was attached to the hinge joint. The control circuit was designed by hand and line the faces in the bottom half of the joint.

To fabricate the actuated joint, copper tape was first affixed to a sheet of polyester film and the circuit traces etched from the surface. The fold pattern was then cut and perforated on the film using a laser cutter. A DC motor, a potentiometer, and other circuit components were soldered to the circuit traces by hand. Lastly, the hinge joint was folded into shape.

The final hinge is pictured in Fig. 9. We used an external PCB to send a PWM signal to the joint to control its angle and were able to achieve the entire π radian range. This joint demonstrates that not only is the hinge joint able to be actuated, but

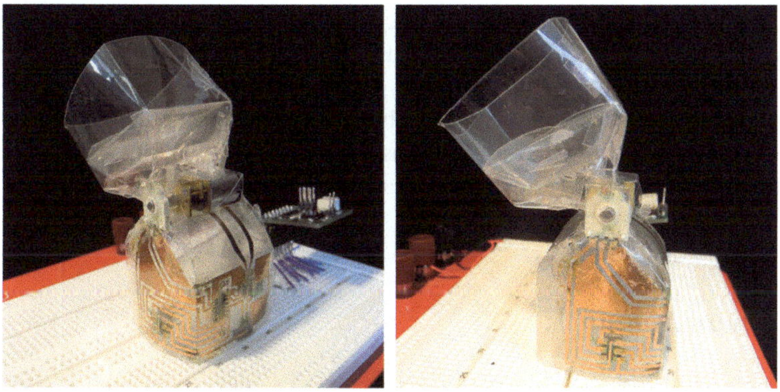

Fig. 9 Hinge joint with integrated electronics

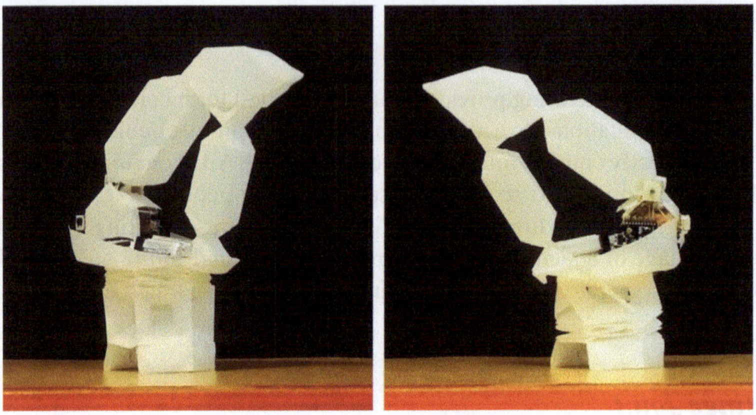

Fig. 10 Actuated four-bar linkage atop actuated pivot mount

also that sensors and actuators integrated directly into a folded circuit can emulate the behavior of an off-the-shelf servo.

4.2 Crane

When our system composes folded structures, the original fold patterns are preserved in their entirety. This feature enables us to also reuse circuits in their entirety. Taking the actuated hinge from Sect. 4.1, we composed with it bars of square cross-section and three more 4-sided π radian hinge joints to produce a four-bar linkage, which we then composed with a 6-sided 2π radian pivot mount to produce a machine with kinematics similar to those of a manufacturing crane (Fig. 10). In order to actuate the crane, we copied the circuit components and placement exactly from the hinge joint and added additional control circuitry for a DC motor mounted at the center of the pivot joint. Finally, we added circuitry to enable serial communication via a Digi XBee radio module so that the robot could be controlled wirelessly. Commands were sent from a nearby laptop, which controlled the direction of rotation of the pivot mount and the position of the hinge joint.

During testing, the hinge joint on the robot performed exactly as the original hinge joint did, although its range of motion was constrained due to the additional hinges and links attached to it. The pivot joint was able to achieve its full 2π radian range of motion.

4.3 Camera Mount

Actuation is not limited to one type. As a final test, we composed a mount for a smartphone using a spherical joint, yielding a camera with pan-tilt capabilities

Fig. 11 Smartphone mount attached to actuated spherical joint to allow pan and tilt

(Fig. 11), and we actuated both degrees of freedom independently using off-the shelf servos. Again circuitry was designed by hand and etched directly onto the robot body, except for the servos, whose wires were plugged into headers in the circuit. A laptop sent commands to tilt forward or backward or to pan left or right via serial communication through a Digi XBee radio module.

The spherical joint in the camera mount was designed for $\frac{3}{2}\pi$ radians of pan and $\frac{2}{3}\pi$ radians of tilt. During testing, the camera was able to achieve the full $\frac{3}{2}\pi$ radians of pan. However, since small servos are typically not designed to sustain large loads such as the weight of a smartphone, tilt had to be limited to $\pm\frac{\pi}{4}$ radians in order to maintain controllability when a smartphone was in place. In addition, since over half of the weight of the device lay above the hinge joint (camera mount: 91 g, smartphone: 116 g), the mount would bend or even topple when large tilt angles were attempted.

5 Discussion and Future Work

Our work demonstrates the feasibility of manufacturing an entire robot in one uniform process via print-and-fold. In this paper, we have presented designs for joints that are not only foldable but also parameterized to deliver a user-specified range of motion. We have shown that these fold patterns can be combined into fold patterns for entire robot bodies, which can be actuated using electronic components integrated during the printing process.

Further work is required before robots can reliably be fabricated using print-and-fold. First, although linkages are an important component of many robot designs, other common mechanisms should be investigated to see if folding can truly achieve any robot. Second, when folding robots from thin materials, strength, stiffness, and an actual ability to transfer and withstand high forces and torques becomes a concern. The camera mount, although stable for small angles, was top heavy when a smartphone was inserted, and it would become unstable when large displacements were attempted. Future work includes characterizing the mechanical properties of folded structures, as well as the dynamics of folded joints, so that print-and-fold robots can achieve the functionality, not just the movement, that they need.

Acknowledgments This work was funded in part by NSF Grant Nos. 1240383 and 1138967, and by the Department of Defense through the National Defense Science and Engineering Graduate Fellowship Program. We are grateful. We also thank Erik Demaine, Martin Demaine, and Cagdas Onal for helpful discussions.

References

1. Hoover, A.M., Steltz, E., Fearing, R.S.: RoACH: an autonomous 2.4g crawling hexapod robot. In: IEEE/RSJ International Conference on Intelligent Robots and Systems, pp. 26–33 (2008)
2. Onal, C.D., Wood, R.J., Rus, D.: An origami-inspired approach to worm robots. IEEE/ASME Trans. Mechatron. **18**(2), 430–438 (2013)
3. Soltero, D.E., Julian, B.J., Onal, C.D., Rus, D.: A lightweight modular 12-dof print-and-fold hexapod. In: Proceedings of IEEE/RSJ International Conference on Intelligent Robots and Systems, pp.1465–1471 (2013)
4. Niiyama, R., Rus, D., Kim, S.: Pouch motors: Printable/inflatable soft actuators for robotics. In: Proceedings of IEEE International Conference on Robotics and Automation, pp. 6332–6337 (2014)
5. Mehta, A., Rus, D.: An end-to-end system for designing mechanical structures for print-and-fold robots. In: Proceedings of IEEE International Conference on Robotics and Automation, pp. 1460–1465 (2014)
6. Lee, D., Kim, J., Kim, S., Koh, J., Cho, K.: The deformable wheel robot using magic-ball origami structure. In: Proceedings of ASME International Design Engineering Technical Conferences and Computers and Information in Engineering Conference, DETC2013–13016 (2013)
7. Gao, W., Ramani, K., Cipra, R.J., Siegmund, T.: Kinetogami: a reconfigurable, combinatorial, and printable sheet folding. J. Mech. Des. **135**(11), 111009 (2013)
8. Whitney, J.P., Sreetharan, P.S., Ma, K.Y., Wood, R.J.: Pop-up book MEMS. J. Micromech. Microeng. **21**(11), 115021 (2011)
9. Demaine, E.D., O'Rourke, J.: Geometric Folding Algorithms: Linkages, Origami, Polyhedra. Cambridge University Press (2008)
10. Abel, Z., Demaine, E.D., Demaine, M.L., Eisenstat, S., Lubiw, A., Schulz, A., Souvaine, D.L., Viglietta, G., Winslow, A.: Algorithms for designing pop-up cards. In: Proceedings of International Symposium on Theoretical Aspects of Computer Science, pp. 269–280 (2013)
11. Tachi, T., Miura, K.: Rigid-foldable cylinders and cells. J. Int. Asso. Shell Spat. Struct. (IASS) **53**(4), 217–226 (2012)
12. Mitani, J., Suzuki, H.: Computer aided design for origamic architecture models with polygonal representation. In: Proceedings of Computer Graphics International, pp. 93–99 (2004)

13. Zhu, L., Xu, W., Snyder, J., Liu, Y., Wang, G., Guo, B.: Motion-guided mechanical toy modeling. ACM Trans. Graph. **31**(6), 127 (2012)
14. Coros, S., Thomaszewski, B., Noris, G., Sueda, S., Forberg, M., Sumner, R.W., Matusik, W., Bickel, B.: Computational design of mechanical characters. ACM Trans. Graph. **32**(4), 83 (2013)
15. Mitra, N.J., Yang, Y.L., Yan, D.M., Li, W., Agrawala, M.: Illustrating how mechanical assemblies work. ACM Trans. Graph. **29**(4), 58 (2010)
16. Sung, C., Demaine, E.D., Demaine, M.L., Rus, D.: Edge-compositions of 3D surfaces. J. Mech. Des. **135**(11), 111001 (2013)

A Design Environment for the Rapid Specification and Fabrication of Printable Robots

Ankur Mehta, Nicola Bezzo, Peter Gebhard, Byoungkwon An, Vijay Kumar, Insup Lee and Daniela Rus

Abstract In this work, we have developed a design environment to allow casual users to quickly and easily create custom robots. A drag-and-drop graphical interface allows users to intuitively assemble electromechanical systems from a library of predesigned parametrized components. A script-based infrastructure encapsulates and automatically composes mechanical, electrical, and software subsystems based on the user input. The generated design can be passed through output plugins to produce fabrication drawings for a range of rapid manufacturing processes, along with the necessary firmware and software to control the device. From an intuitive description of the desired specification, this system generates ready-to-use printable robots on demand.

Keywords Design co-generation · Personal robots · Printable robots

1 Introduction

Creating a robotic system generally requires broad engineering expertise to successfully manage the interplay between constituent electromechanical subsystems. For an average user to be able to create their own robots for personal use, a system is needed to abstract away technical details in favor of easy to understand functionality specifications. To parallel a typical solution methodology in which a problem is broken into atomic tasks, we have developed a design environment in which robots are assembled by simply connecting self-contained components. These modular building blocks encapsulate mechanical, electrical, and software subsystems without relying on a specific manufacturing process. A full robot is then defined by the component hierarchy required to achieve each necessary behavior, and can be directly fabricated

A. Mehta (✉) · B. An · D. Rus
Massachusetts Institute of Technology, Cambridge, USA
e-mail: mehtank@csail.mit.edu

N. Bezzo · P. Gebhard · V. Kumar · I. Lee
University of Pennsylvania, Philadelphia, USA

M.A. Hsieh et al. (eds.), *Experimental Robotics*, Springer Tracts in Advanced Robotics 109, DOI 10.1007/978-3-319-23778-7_29

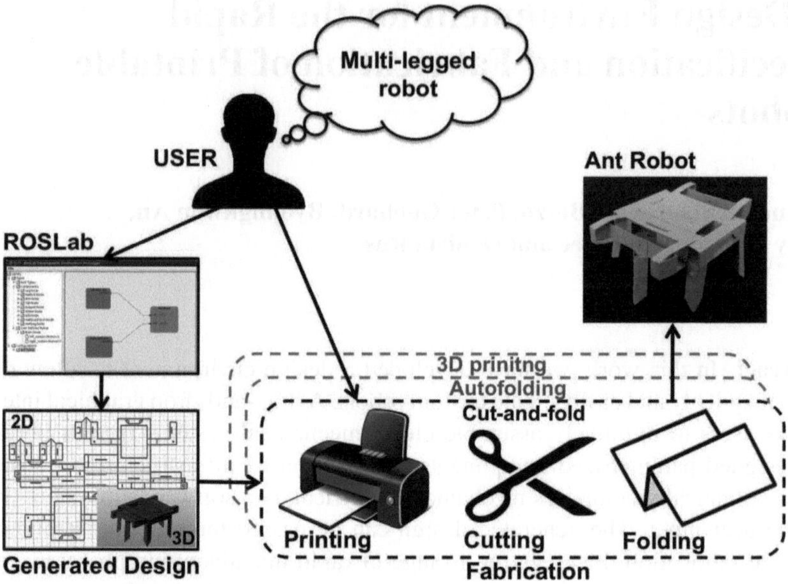

Fig. 1 Workflow diagram showing the steps necessary for a user to design, create, and fabricate a robot

from design files automatically co-generated after passing the specification through a suitable output plugin.

This definition is specified by the user in an intuitive and simplified high-level graphical programming language. Blocks describing distinct parts of a robot are graphically connected together on a workspace to describe the robot design. From this, Python code is automatically generated which, when executed, creates the hardware design description of printable robots. This graphical frontend provides intuitive user interfacing while providing versatile customizability, while the backend is compatible with a variety of 2D and 3D manufacturing processes suitable for automated home fabrication. The overall system, outlined in Fig. 1 takes as input a schematic outlining a high-level breakdown of the required components, and outputs a fabricated ready-to-use robot, with minimal required user intervention. This system paves the way towards realizing smart programmable cyber-physical systems on demand towards a future of pervasive personal robots.

2 Related Work

2.1 Fabrication

There are a variety of fabrication methods to create printable devices. Arbitrary 3D structures are generally achievable by additive manufacturing using 3D printers;

advances in printer technology have made desktop printers available to the general public. However, while complex solid geometries are easily manufactured with 3D printing, achieving the required compliance and mobility necessary for general robotic systems is nevertheless difficult to achieve using most common techniques [1]. Limited workarounds do exist [2, 3]; these often lack robustness or reliability, though current technology has been improving.

Alternatively, mechanical structures can be realized by patterning then folding 2D sheets to define the shell of the desired geometry. A variety of substrates are possible, including cardboard laminates [4], single layer plastic film [5], or more exotic materials [6, 7]. These designs can be manually folded by hand, folded by embedded or external active stimuli, or passively folded by controlled environmental conditions [8, 9].

2.2 Design

The fabrication processes listed in the previous section require a multitude of computer-aided design (CAD) tools to specify mechanical structures or electromechanical assemblies. Though some automated design tools have been developed, especially to translate 3D geometries into 2D unfoldings [10, 11], these are often limited in scope, resulting in custom designs needing to be manually drawn by experienced designers. Instead, the work presented in this paper builds off of the system presented in [12, 13], wherein mechanical designs for a particular cut-and-fold fabrication process are abstracted into code objects.

The use of graphical languages is common in the engineering and academic community. Simulink [14] and LabView [15] are two well known examples widely used for general purpose engineering computing. Authors in [16] use a formal graphical language to program medical operation between different medical devices. Standing more from a robotic perspective, only a few graphical programming environments are available and usually limited and constrained to specific platforms [17, 18]. To overcome this limitation, in [19] a high level programming language called ROSLab was proposed to generate C++/ Python code for general robotic applications involving different type of platforms such as aerial, wheeled, and multi-legged robots.

3 Technical Approach

3.1 Scripted Hardware Programming

The system presented in [12, 13] provides a scripted programming language to specify designs for printable robots realized using a cut-and-fold fabrication process. To enable greater versatility, that system was overhauled in this work to abstract designs into a process-agnostic representation of the component hierarchy.

In this system, parametrized components are defined by code objects, with scripted functions representing physical manipulations and design steps. Basic building blocks can define mechanical, electrical, or software elements, while composite blocks can be integrated across subsystems. Parameters can be used to customize variable geometric measurements, alternate electric components, or other design-time configurable quantities.

The code objects expose an interface allowing them to be hierarchically composed: component modules specify predetermined connections along which other components can be attached. Attaching modules via their connections forms a new higher-order module, establishing constraints on their free parameters, merging their mechanical structures, and wiring together their electrical components.

Complex designs, from electromechanical mechanisms up to full robots, can then be represented as simple software scripts implementing the above steps. Executing a script co-generates fabricable design files for the complete device, including mechanical drawings to be sent to a fabrication tool, electrical component requirements and wiring information, and firmware and software libraries and application code.

The composition of building blocks into higher order components is carried out using an internal graph-based representation to generate structurally specified robot designs. This graph specifies the connections and parameter constraints between constituent modules. To generate fabricable drawings, then, an interpreter plugin realizes the geometries along the design graph into a format required by the manufacturing process of choice. Plugins have been written for a number of fabrication methods as described in Sect. 4.2 below.

Mechanical geometries are stored using a face-edge graph that can be resolved to both 2D and 3D shapes as required by specific fabrication processes. A basic example of this is shown in by the beam in Fig. 2, generated from the code in listing 1.1. The blue squares in the graph represent the rectangular faces of the beam, connected to each other along folded edges represented by red circles. The unconnected dashed lines represent connections along which future components can be attached. A cut-and-fold pattern can be generated from the face graph, requiring the dotted edge to be replaced by a tab-and-slot connector. A 3D solid model can also be generated to display the structure resulting from folding the 2D pattern, or to directly generate a 3D object via 3D printing.

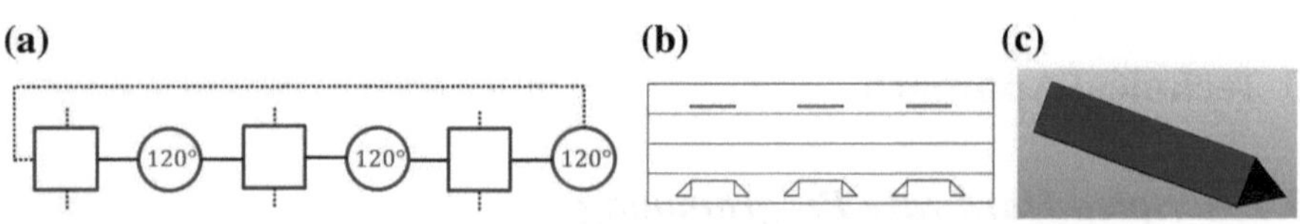

Fig. 2 Outputs generated from the code in listing 1.1. **a** Face-edge graph representation of a beam geometry. **b** Generated drawing to be sent to a 2D cutter. **c** Generated 3D solid model

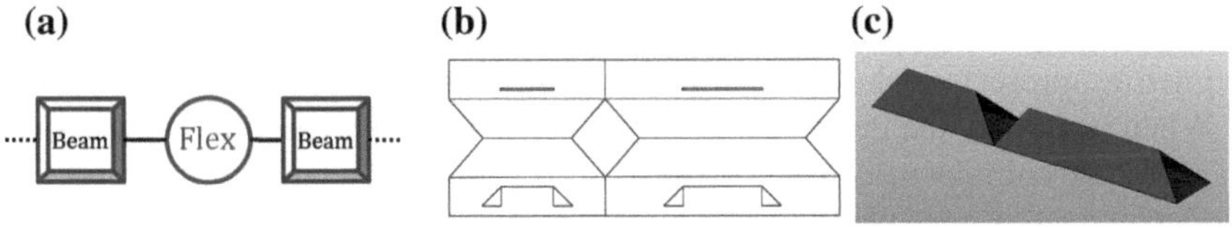

Fig. 3 Outputs generated from the code in listing 1.2. **a** Component-connection graph representation of a finger design hierarchy. **b** Generated drawing to be sent to a 2D cutter. **c** Generated 3D solid model

```
import Beam
b = Beam.Beam()
b.setParameter("length", 100)
b.setParameter("beamwidth", 10)
b.setParameter("shape", 3)
```

Listing 1 Scripted design of a mechanical beam

A simple composite structure is demonstrated in Fig. 3 from the code in listing 1.2. As above, constituent objects are instantiated and their parameters are set. However, these building blocks are attached along exposed connections in a new higher-order component. The connection type is set to be a flexible joint to allow for compliant motion, and geometric constraints are imposed on objects' parameters.

```
from api.Component import Component
from api.Edge import Flex
from Beam import Beam

finger = Component()

finger.addSubComponent("beam1", Beam)
finger.addSubComponent("beam2", Beam)

finger.setSubParameter("beam1", "length", 60)
finger.setSubParameter("beam1", "beamwidth", 10)
finger.setSubParameter("beam1", "shape", 3)
finger.setSubParameter("beam1", "angle", 45)

finger.setSubParameter("beam2", "length", 40)
finger.setSubParameter("beam2", "angle", 45)

finger.add("beam1")
finger.connect(("beam1", "topedge"),
               ("beam2", "botedge"),
               Flex())
```

Listing 2 Scripted design of a composite finger

As described in [13], electrical components, including sensors, actuators, and processors, can also be encapsulated in the same component framework of the mechanical designs described above. Purely electrical devices can be combined with mechanical structures to form integrated electromechanical mechanisms. Software drivers and UI elements can similarly be included within a component to define a fully self-contained robotic subsystem. The connections for such integrated components algorithmically combine electrical wiring and software blocks as well as mechanical geometries, preserving the modular design abstraction across robotic subsystems.

When compiled, the design scripts produce device specifications and wiring diagrams for the complete electronic subsystem, as well as integrated software packages to operate the designed electromechanical system. In this way, complete designs are automatically co-generated for the design hierarchy, thus enabling a user to design a print-and-play robot by scripting the composition of elements.

3.2 ROSLab for Hardware Design Generation

To further simplify the design flow for casual users, a graphical programming tool was adapted to generate the hardware specification scripts. ROSLab [19] provides an intuitive and simplified high-level development environment that builds software through a drag-and-drop interface. Code components are represented as blocks which can be imported from a library into a workspace, then connected along available interfaces to generate new designs.

In this work, we extend ROSLab to provide a design environment for creating printable robots with the scripted infrastructure described above. A predesigned collection of Python scripts representing parametrized robotic building blocks are available to a user as a component library in the ROSLab environment. Desired blocks can be dragged into a workspace, and parameters can be set by the user based on target specifications. Exposed interfaces on each robot component are represented by ports on the ROSLab block; these ports can be wired together to specify electromechanical connections. Compositions can themselves be saved as components in the library to be used in future higher order designs. In this way, a full robot can be hierarchically composed from its constituent blocks.

Code Generation and User Interface (UI) The programming workflow in ROSLab follows the intuitive logic of how one would construct a robot. Generally speaking, a robot is made of a main body part, which we call *brain* because it contains the processing unit (e.g., microcontroller, microprocessor, etc.), locomotion components (e.g., legs, wheels, propellers, etc.), and other extremities that hold sensors and actuators (e.g., gripper). A typical ROSLab program for fabrication starts by defining the brain node followed by the desired locomotion technique. For instance, a user that is interested in designing an ant-like robot would need to specify the *<brain>* component and two *<motion>* type *<leg-pair>* blocks. Similarly for a two wheeled robot (e.g., a Segway), the *<brain>* component will be connected to two *<wheel>* modules of type *<motion>*. The user will then place blocks in the ROSLab workspace to symbolize the brain and leg-pairs (or wheels for the Segway) and will connect these blocks together. Once the design is finalized, it will be automatically generated by ROSLab.

ROSLab is written in Java and is designed to follow a template logic. The templates are characterized by fields which are filled by specific code snippets according to

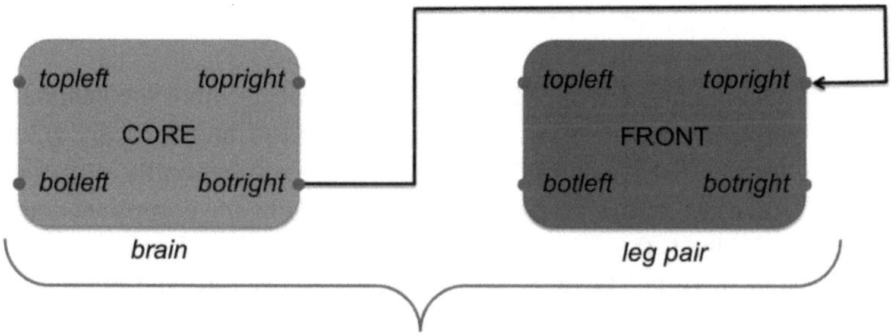


```
self.attach(("brain", "core", "botright"),
           ("legpair", "front", "topright"), Fold(-180))
```

Fig. 4 Pictorial representation of ROSLab code generation. The *bottom right* (botright) of the <brain> component named *core* by the user is connected to the *top right* (topright) of the <half> component named *front*. The last field inside the parenthesis is used to define how the the half will be folded during the assembling

how a component is connected to another component. Prior to use by a casual user, a ROSLab developer must first carefully construct the templates by finding common code primitives among the possible output files. Each component is described by (i) a set of ports which for a printable robot consist of foldable tabs; and (ii) a set of parameters which define the dimension and interface with the electronic hardware. These parameters can be easily edited within the ROSLab UI, thus giving the user the ability to create unlimited designs for the robot under development.

The code generation process is initiated by calling a function which parses the UI workspace, checks the connections and components in the design, and fills the template holes with code associated with each used component. Specifically, the code generation creates a Python script that contains details about the assembly of the different components.

Figure 4 shows an example of the blocks and code associated with the assembly of the brain to a leg-pair.

The user can create different designs however not all configurations are guaranteed to generate a stable mechanical system. The user can easily change connections and adapt the design from the ROSLab UI.

3.3 Robot Fabrication

Once a robot has been designed, it can be compiled to generate manufacturing specifications. Because the system can generate both a 2D representation of the surface of the robot body as well as the 3D volume, a number of rapid fabrication processes, as described in Sect. 2.1 above, can be used to create the specified mechanical structures. Output scripts in the system are used to translate the internal representation of the robot design into fabrication files suitable for manufacture. When compiling the design, the user can select from the available fabrication processes to create the robot body.

Since the system can generate the 3D geometry of the final mechanical body, a solid model output by the system can be sent to a 3D printer for fully autonomous fabrication of the desired structure. Though this process is generally versatile enough to make even the most complicated shapes, it often lacks the compliance needed for robot mobility. Instead, the 3D volume can also be realized by folding its surface from a patterned flexible 2D sheet, providing both structure and compliance. The underlying geometric data stores the face and edge geometries of the 2D unfolding, and so the body can be automatically self-folded by uniform heating of a patterned 3 layer laminate, as presented in [20]. Alternately, the cuts and folds can be patterned onto a plastic sheet using a laser cutter [21] or desktop vinyl cutter [12], and then manually folded to the final 3D geometry. Finally, it is possible to realize the mechanical body without any custom tools by printing the fold pattern onto a sheet of paper. A user can cut the design out with scissors, and fold the structure according to the printed instructions.

Once the body has been fabricated, the specified electronic components and electromechanical transducers must be mounted onto the body and wired together as specified by the system, with auto-generated firmware loaded onto the core microcontroller. The device then simply needs to be powered on and paired with a user interface for the process to be complete, delivering a fully functioning custom printed robot on demand.

4 Experiments and Results

4.1 Design

We use the ROSLab interface to create robots. A library of basic components was imported from [13], forming the building blocks from which new systems were designed. These systems were then manufactured and operated.

A simple wheeled robot can be specified as two motors attached to a central core. To add stability, a third point of contact, such as a tail, can be added to a free end. Symbolically, this is represented by the following relation:

$$Seg = left\ wheel + core + right\ wheel + tail. \tag{1}$$

The core, motors, and tail were all basic components from the library, and so the design of this simple robot consisted simply of dragging blocks into the workspace and connecting them as per Eq. 1. The ROSLab source along with the resulting printed robot are shown in Fig. 5. Similarly designed and fabricated wheels, also from the library, are added to complete the device.

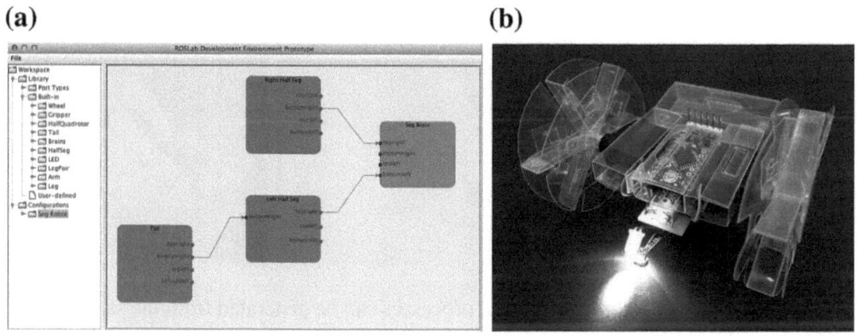

Fig. 5 A Seg robot designed within the ROSLab programming environment and fabricated in a cut-and-fold process. **a** A simple two wheeled robot design. **b** The resulting robot

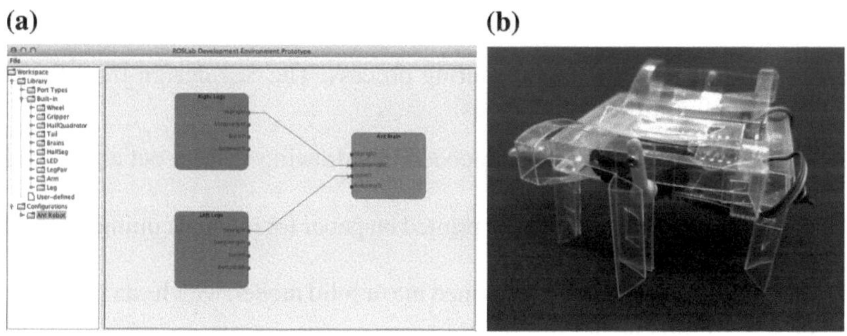

Fig. 6 The Ant robot generated by adapting the earlier Seg design. **a** Making an insect-like crawler with ROSLab. **b** A cut-and-fold robotic ant

Within the programming environment, labels can be added to the blocks: the central core provides a *brain* containing the low level computational electronics (e.g. processing, communications, and control), while the motor blocks provide *motion*, in particular mobility. These tags can be used to quickly generate alternate, functionally similar designs.

Instead of generating *motion* through wheels, an insect-like crawler can use a leg-pair block from the library to generate a walking motion. Since a tail is no longer needed to provide stability, the new design can be simply expressed as:

$$Ant = left\ leg\text{-}pair + core + right\ leg\text{-}pair. \tag{2}$$

The ROSLab source and generated robot are shown in Fig. 6.

The base components, and therefore the derived designs, are all parametrized; by using this framework, a user has the freedom to adjust design geometries in a variety of ways by tweaking the exposed parameters. As an example, some of the free user-defined parameters for a multi-legged robot are the leg length, the body length, the servo motor type, and the desired microcontroller in the brain.

(a) (b) (c)

Fig. 7 Fabrication files for three additional processes can be generated from the same source as in Fig. 5 above. **a** Cut and folded by hand. **b** 3D printed. **c** Self-folded in an oven

4.2 Fabrication

The user can also select a manufacturing process. The Seg design from Eq. 1 was fabricated with four different methods:

(I) a 2D edge unfolding can be encoded in a drawing to laser cut a plastic sheet, as in Fig. 5b;
(II) the edge unfolding can also be printed on paper for manual cutting and folding, as in Fig. 7a;
(III) the geometry can instead be formed into a solid model, which can get fabricated on a 3D printer as in Fig. 7b;
(IV) the geometric information can be used to preprogram an active patterned laminate for self-folding through uniform heating, as in Fig. 7c.

These four methods generated robots with the same geometry as specified by the robot design, but with significantly varying tradeoffs.

The traditional cut-and-fold process (I), wherein the structure was folded out of a patterned compliant plastic sheet, was overall a middle-of-the-road process. Though it took the shortest time to fabricate, needing only a single layer 2D cut, its required post-process folding step took a fair amount of user time and skill to assemble. The flexible source material was very useful in creating compliant degrees of freedom for moving parts such as in the legs of the crawler, but needed designed structural reinforcement when stiffness was desired as in the wheels of the Seg. It required an expensive laser cutter for fabrication.

The manual cut-and-fold process (II) was by far the cheapest process, requiring only a printed sheet of card stock and a pair of scissors or a knife. The generated design as shown in Fig. 8 can be printed using any standard color printer, fully defining the geometry. The lack of tooling was made up for by the labor necessary for fabrication and assembly as the user had to cut then fold the 2D design by hand, as shown in Fig. 9. The stiffer card stock made the structural elements simpler to design than the plastic substrate above, though the compliant folds were more prone to fatigue.

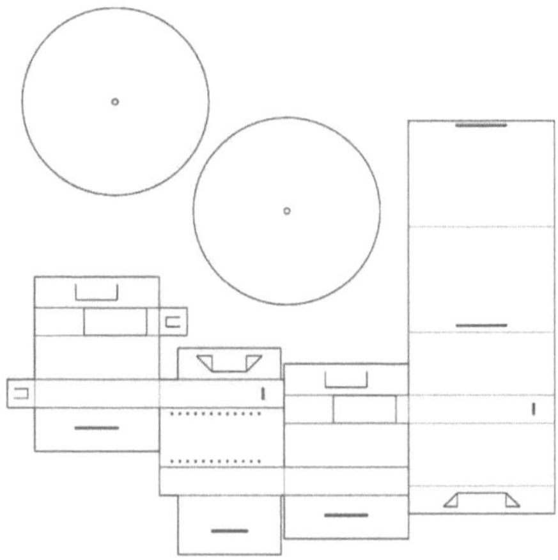

Fig. 8 The design file for a manual cut-and-fold process consists of a color diagram to be printed onto a sheet of cardstock. The user is responsible for cutting along the blue lines; red and green lines represent mountain and valley folds, respectively

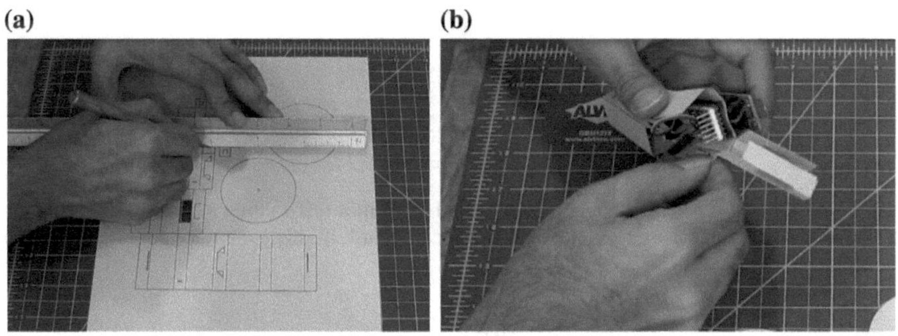

Fig. 9 The manual cut-and-fold fabrication process is labor intensive, but requires cheap tools and minimal infrastructure. **a** The designs are cut using a knife or scissors. **b** Electronic components are assembled while folding the body

The 3D printed process (III) took the least user input, directly building the generated 3D geometry as shown in Fig. 10, but took the longest overall time to fabricate. It produced the most rigid structural body, but could not generate flexible hinges necessary for compliant joints. Though flexible materials are slowly becoming available for 3D printers [22], they are still new and not well characterized.

Finally, the self-folding process (IV) combined the benefits of minimal user intervention from the 3D printing with the quicker fabrication of 2D layers. The generated layer designs are shown in Fig. 11. Though user labor was necessary to laminate the three laser-cut layers, this process was the quickest to fabricate the mechanical body, seen in Fig. 12. However, the final structure generated by the self-folded laminate

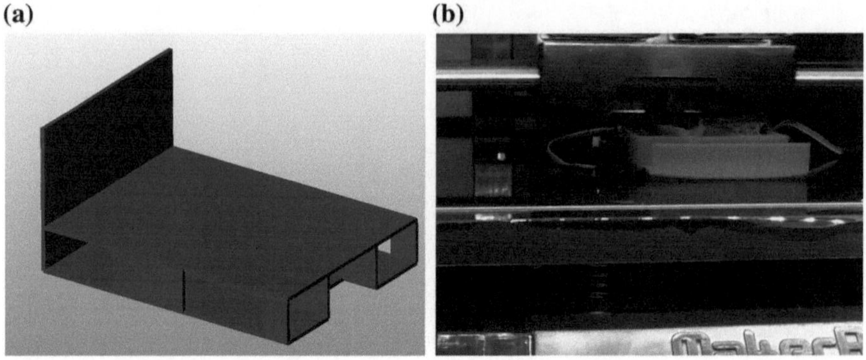

Fig. 10 The designed structure can be directly fabricated on a desktop 3D printer. **a** The generated solid model is sent to 3D printer software to manufacture the desired geometry. **b** Printing can be paused to incorporate embedded electronics, or they can be mounted afterward

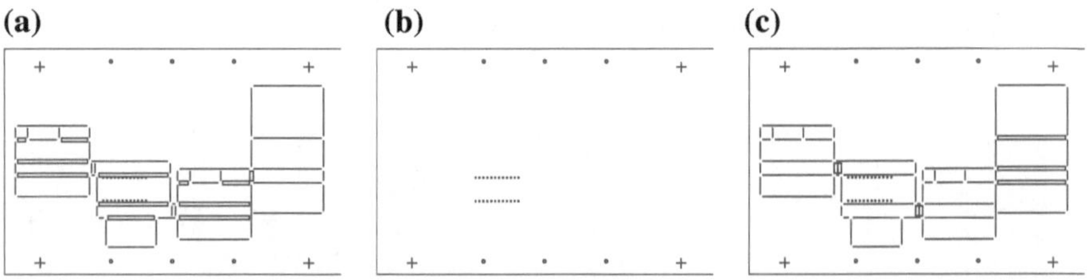

Fig. 11 A three layer laminate can be assembled to form a self-folding structure that forms the designed geometry under uniform heating. **a** Top cardboard. **b** Shape-memory polymer. **c** Bottom cardboard

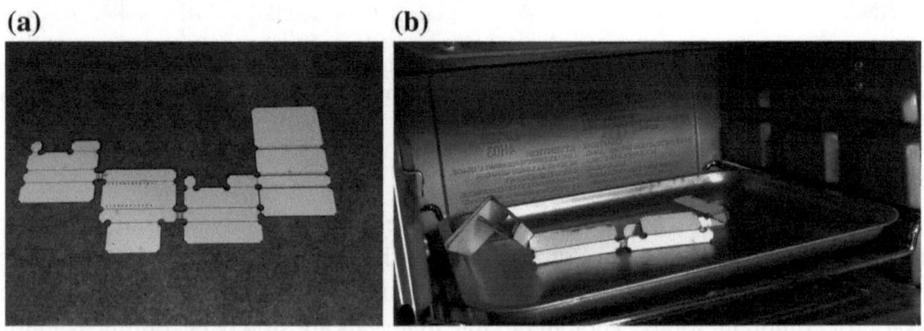

Fig. 12 Self-folding can automatically generate 3D geometry from a 2D unfolding. **a** Laser cut layers get assembled into a self-folding laminate. **b** Uniform heating in a toaster oven generates the desired folded geometry

Table 1 Analysis of various rapid fabrication methods

Process	Fabrication time (min)	Assembly time (min)	Body weight (g)	Process Strength	Process Weakness
(I)	2	15	15.2	Controlled compliance	Flexible structure
(II)	10	15	3.6	Low cost	Labor intensive
(III)	90	1	14.9	Structural integrity	Minimal compliance
(IV)	5	5	12.2	Construction speed	Large tolerances

occasionally had notable deviations from the designed geometry as the self-folding process halted before achieving to it's final configuration. The structural elements provided excellent rigidity, coming at the expense of less flexible compliant structures.

The comparison of these fabrication methods is summarized in Table 1.

5 Conclusions

The key development presented in this work was a design infrastructure which encapsulated process-independent definitions of robotic building blocks, allowing custom electromechanical systems to be hierarchically designed once for a range of fabrication methods. This was overlaid with the graphical interface of ROSLab allowing for direct transcription of design ideas into mechanism definitions. Together, these form a process flow for personal robot creation from vision to operation that is simple and intuitive. By abstracting and encapsulating the various stages of design, this modular pick-and-place design environment brings custom robot design to the realm of non-expert users. Simple robots such as the Seg and the Ant require only minutes to design; more complex robots can be hierarchically designed in similarly easy stages.

The unified design process highlighted important differences, summarized above, between the various rapid fabrication methods used for printable robotics. As user specifications for custom robots vary widely, so too do desired optimization targets, and thus the design paradigm presented in this work provides a valuable asset for personal robot creation.

This work suggests an important next step towards more autonomous robot design: developing the system to guide design decisions at both ends of the pipeline. Information about the tradeoffs among the fabrication processes can be incorporated into the system, and a recommended fabrication method can be then presented based on optimization goals on a user specified design. Conversely, given optimization goals and fabrication constraints, components can be suggested to generate a design based on a functional specification.

A full robot compiler can be built upon this framework, allowing casual users to design desired robots from a very high-level functional specification of the problems to be solved. The system presented in this work thus represents a major step forwards towards programmable cyber-physical systems on demand.

Acknowledgments This material is based on research sponsored by the National Science Foundation awards EFRI-1240383 and CCF-1138967, for which the authors express thanks.

References

1. Mavroidis, C., DeLaurentis, K.J., Won, J., Alam, M.: Fabrication of non-assembly mechanisms and robotic systems using rapid prototyping. J. Mech. Des. **123**(4), 516–524 (2001)
2. Richter, C., Lipson, H.: Untethered hovering flapping flight of a 3d-printed mechanical insect. Artif. life **17**(2), 73–86 (2011)
3. Rossiter, J., Walters, P., Stoimenov, B.: Printing 3d dielectric elastomer actuators for soft robotics. In SPIE Smart Structures and Materials+ Nondestructive Evaluation and Health Monitoring, International Society for Optics and Photonics, pp. 72870H–72870H (2009)
4. Hoover, A.M., Fearing, R.S.: Fast scale prototyping for folded millirobots. In Robotics and Automation (ICRA), pp. 886–892. IEEE (2008)
5. Liu, Y., Boyles, J., Genzer, J., Dickey, M.: Self-folding of polymer sheets using local light absorption. Soft Matter **8**, 1764–1769 (2012)
6. Shimoyama, I., Miura, H., Suzuki, K., Ezura, Y.: Insect-like microrobots with external skeletons. IEEE Control Syst. **13**(1), 37–41 (1993)
7. Brittain, S., et al.: Microorigami: Fabrication of small, three-dimensional, metallic structures. J. Phys. Chem. B **105**(2), 347–350 (2001)
8. Hawkes, E., et al.: Programmable matter by folding. Proc. Natl. Acad. Sci. **107**(28), 12441–12445 (2010)
9. Tolley, M., Felton, S., Miyashita, S., Xu, L., Shin, B., Zhou, M., Rus, D., Wood, R.: Self-folding shape memory laminates for automated fabrication. In IEEE/RSJ International Conference on Intelligent Robots and Systems (IROS), IEEE (2013)
10. Lang, R.: Origami Design Secrets : Mathematical Methods for an Ancient Art. A K Peters/CRC Press, Boca Raton (2012)
11. Demaine, E., Fekete, S., Lang, R.: Circle packing for origami design is hard. In: Proceddings of the 5th International Conference on Origami in Science, Mathematics and Education, pp. 609–626 (2010)
12. Mehta, A.M., et al.: A scripted printable quadrotor: rapid design and fabrication of a folded MAV. In 16th International Symposium on Robotics Research (2013)
13. Mehta, A.M., DelPreto, J., Rus, D.: Cogeneration of mechanical, electrical, and software designs for printable robots from structural specifications. In Intelligent Robots and Systems (IROS), (2014) (to appear)
14. Mathworks Simulink. http://www.mathworks.com/products/simulink/ Accessed 26 May 2014
15. NI Labview. http://www.ni.com/trylabview/ Accessed 26 May 2014
16. King, A.L., Feng, L., Sokolsky, O., Lee, I.: Assuring the safety of on-demand medical cyber-physical systems. In 1st International Conference on Cyber-Physical Systems, Networks, and Applications (CPSNA), pp. 1–6. IEEE (2013)
17. Lego mindstorms. http://mindstorms.lego.com Accessed 26 May 2014
18. VEX Robotics. http://www.vexrobotics.com Accessed 26 May 2014
19. Bezzo, N., Park, J., King, A., Geghard, P., Ivanov, R., Lee, I.: Demo abstract: Roslab a modular programming environment for robotic applications. In ACM/IEEE International Conference on Cyber-Physical Systems (ICCPS), IEEE (2014) (to appear)

20. An, B., et al.: An end-to-end approach to making self-folded 3d surface shapes by uniform heating. In IEEE International Conference on Robotics and Automation. IEEE (2014) (accepted)
21. Onal, C.D., Wood, R.J., Rus, D.: An origami-inspired approach to worm robots. IEEE/ASME Trans. Mechatron. **18**(2), 430–438 (2013)
22. Makerbot flexible filament. https://store.makerbot.com/flexible-filament Accessed 26 May 2014

Part VIII
Perception and Planning

For real-world systems such as robots, the intersection of perception and planning poses many challenges. This session brought together three research papers that considered perception and planning for different domains in robots: mobile manipulator, simple devices, and indoor navigation. While these settings are diverse, they presented end-to-end system with new algorithms, new concepts, and a new evaluation system respectively for the three papers.

The first paper "Anticipatory Planning for Human-Robot Teams" by Hema Koppula, Ashesh Jain, and Ashutosh Saxena focused on robots working alongside humans for performing collaborative tasks. Here the key was to using RGB-D sensor data for anticipating human's future actions, and then plan appropriate actions for the robot. Since the plan depends not only on the perceived state of the human and the environment, but also on the uncertainties in perception, one needs to jointly consider perception and planning. Furthermore, since it is a collaborative task, one needs to consider a multi-agent model. This paper presented a good learning model that considers all these properties. They evaluated their model on a large dataset, as well as on robotic experiments.

The second paper "Autonomous realization of simple machines" by Can Erdogan and Mike Stilman focused on robots utilizing the objects present in the environment for performing tasks. Specifically, they presented how a robot can create simple machines such as a lever-fulcrum assembly. Their setup included object detectors whose output is used into the motion planner for performing the mobile manipulation operations. They tested their approach on Golem Krang on several robotic experiments.

The third paper "An Experimental Protocol for Benchmarking Robotic Indoor Navigation" by Christoph Sprunk, Jorg Rowekamper, Gershon Parent, Luciano Spinello, Gian Diego Tipaldi, Wolfram Burgard, and Mihai Jalobeanu, focused on an integrative tesing of robot navigation. While several works have considered such evaluation, those evaluations have been only of sub-components such as mapping, localization and planning using datasets or simulations. This paper presented experimental protocols to evaluate the end-to-end navigation system deployed in a real environment.

In summary, the papers in this session presented new algorithms for joint perception and planning, innovative working systems, and also a way to extensively evaluate end-to-end systems.

Anticipatory Planning for Human-Robot Teams

Hema S. Koppula, Ashesh Jain and Ashutosh Saxena

Abstract When robots work alongside humans for performing collaborative tasks, they need to be able to anticipate human's future actions and plan appropriate actions. The tasks we consider are performed in contextually-rich environments containing objects, and there is a large variation in the way humans perform these tasks. We use a graphical model to represent the state-space, where we model the humans through their low-level kinematics as well as their high-level intent, and model their interactions with the objects through physically-grounded object affordances. This allows our model to anticipate a belief about possible future human actions, and we model the human's and robot's behavior through an MDP in this rich state-space. We further discuss that due to perception errors and the limitations of the model, the human may not take the optimal action and therefore we present robot's anticipatory planning with different behaviors of the human within the model's scope. In experiments on Cornell Activity Dataset, we show that our method performs better than various baselines for collaborative planning.

Keywords Collaborative task planning · Anticipation · Human activity perception · Object affordances · Human-robot interaction

1 Introduction

Currently, robots are being incorporated into human workspaces where they perform tasks with humans—assistive settings in nursing homes (e.g., [18]), collaborative assembly line manufacturing (e.g., [31]), or in other outdoor applications. The challenge here is two-fold: the robots often have to operate in *contextually-rich*

H.S. Koppula (✉) · A. Jain · A. Saxena
Department of Computer Science, Cornell University, Ithaca, USA
e-mail: hema@cs.cornell.edu

A. Jain
e-mail: ashesh@cs.cornell.edu

A. Saxena
e-mail: asaxena@cs.cornell.edu

M.A. Hsieh et al. (eds.), *Experimental Robotics*, Springer Tracts in Advanced Robotics 109, DOI 10.1007/978-3-319-23778-7_30

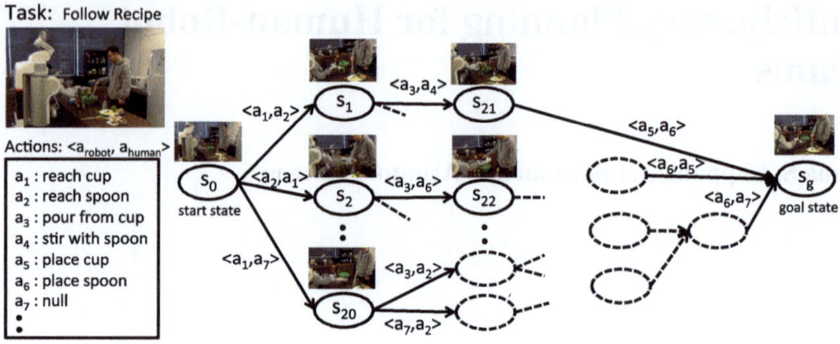

Fig. 1 **Robot performing collaborative task with human**: The human and the robot are asked to follow a recipe—pour the ingredients in to the bowl and stir. At every time step both the agents execute an action and change the state of the environment. The robot needs to plan its actions by taking in to account what actions the human would perform, where some human actions are more likely than the others based on the human's strategy

environments, where they have to perform tasks involving manipulation of objects, and they have to work closely *with humans* performing the same task (see Fig. 1).

Collaborative tasks are more challenging as compared to both reactive and role-based tasks. In collaborative tasks, the goal of the robot is to perform actions along side humans in order to achieve the goal of the task. For example, if the task is to set the dinner table, the various actions involved are reaching for the objects (e.g., plates, cups and spoons), and moving them to appropriate locations on the table. The robot can perform any action in order to achieve the goal as opposed to a role-based scenario where the robot has a pre-assigned role of setting plates or cups, etc. It needs to plan its actions by taking into account the actions of the human. In order to achieve this, there are three aspects we need to address: (i) model the contextually-rich environment to reason about what can be done and how, (ii) perceiving the human's actions and anticipating their future moves, and (iii) plan robot's actions taking into account the inherent uncertainty in the human actions.

In our previous works [17, 18, 20], we presented a perception algorithm for modeling the spatio-temporal relations of activities which allows us to detect the past actions and anticipate the future actions. However, the robotic responses were only reactive and hand-designed. In related works, Nikolaidis and Shah [32] consider collaboration for assembling tasks with pre-assigned roles for human and robot, where they do not explicitly model anticipation. Mainprice and Berenson [26] anticipate human actions to minimize penetration of robot in human workspace and Uyanik et al. [41] introduced social-affordances for planning. In comparison, we look at a more generic collaborative task planning problem, where the role of robot and human are indistinguishable.

In this paper, we formulate the collaborative task completion problem as a two-agent planning problem, where we model the ambiguities in perception as well as in the human's choice of actions. Unlike planning for multi-robot scenarios, where one has control over all agent behaviors [3, 39], the human does not perform his actions according to a fixed strategy. Humans tend to follow their habits when possible in a familiar environment, but will also try to adapt in response to the other agents in the environment. Therefore, our problem of robot-human collaborative planning can be viewed as a two-agent cooperative Markov game, where the goal of each game is to complete a pre-specified activity in a given environment. We aim to learn the optimal policy for the robot while taking into account the various human behaviors or strategies.

In detail, we represent the contextually rich environment in terms of the object affordances and incorporate them as the states of our collaborative Markov decision process. We propose a distributed Q-learning algorithm to learn the policies for both the agents. We model the human's actions in several ways—taking the ϵ-optimal action according to the MDP model, taking actions based on past habits as seen in a RGB-D video dataset, and taking appropriate actions by adapting to the environment and robot actions. Each human behavior results in exploring a different subspace of states by the robot, resulting in a different robot policy as shown in Fig. 1. Therefore, during learning we first estimate how adaptive the human is in the given environment and then jointly estimate the robot and human policies.

We evaluate our approach on five high-level activities in 60 environments from the CAD-250 dataset as well as in a user study. We predict the current object affordances from RGB-D videos and use our algorithm to plan appropriate actions by the robot to be performed along with the human. We compare our approach against the baselines on several metrics, and find that our approach performs better collaborative planning. Specifically, our robot policy learnt with an adaptive human model completes the tasks 36.5 % faster as compared to 13.8 % when the human agent is not modeled explicitly.

2 Related Work

Our approach of anticipatory planning has three main aspects: human-robot interaction, perception in contextually-rich environments, and planning algorithms. We now review the relevant works specific to these aspects.

Human-robot collaboration. Many tasks are parallelizable or involve complex interactions with objects in the environment, and can be more efficiently completed if human and robot collaborate. Some recent works have addressed this problem of collaboration in human-robot teams. Nikolaidis and Shah [32] consider collaboration for assembling tasks with pre-assigned roles for human and robot. Mainprice and Berenson [26] anticipates human actions to minimize penetration of robot in human workspace. Uyanık et al. [41] introduced social-affordances where robot's action depends on help from human. As opposed to them, in our work the role of robot and human are indistinguishable, and for task completion they interact with multiple objects performing different activities.

Another aspect of human-robot collaboration is the interaction between the agents and their compatibility. Some works [27, 28, 35] encode the compatibility in the form of constraints on the distance of robot from user, the visibility of robot and user arm comfort. Strabala et al. [36] and Cakmak et al. [5] consider handover tasks where the robot reasons about its location w.r.t. human and handover configuration. We differ from these in that, in our tasks both human and robot are active participants and collaborate towards a common goal.

Affordances. The concept of affordances was described by J.J. Gibson [8] as the "Action possibilities in the environment in relation to the action capabilities of an actor". Affordances have been widely used in robotics for obtaining a functional understanding of the scene as well as enabling robots to interact and manipulate objects. These works range from predicting opportunities for interaction with an object by using only visual cues [2, 9, 37] to observing effects of exploratory behaviors [10, 14, 29, 30]. For instance, Sun et al. [37] proposed a probabilistic graphical model that leverages visual object categorization for learning affordances. Katz et al. [14] propose a framework for learning to manipulate objects in clutter by choosing robot actions based on object affordances.

There is some recent work in interpreting human actions and interaction with objects [1, 12, 16, 19, 24] in context of learning actions from demonstrations. Lopes and Santos-Victor [24] use context from objects in terms of possible grasp affordances to focus the attention for recognition. Aksoy et al. [1] propose a dynamic graph sequence representation, constructed from human demonstrations, for robot manipulation. Affordances have also been used in planning (e.g., [25, 40]). In this work, we use object affordances to represent the state of the environment, and these affordances evolve as the objects are used in an activity [20].

Multi-agent Reinforcement Learning. The multi-agent reinforcement learning (MARL) literature focus on multiple autonomous agents learning how to solve dynamic tasks online. Besides single-agent reinforcement learning, MARL has strong connections with game theory, evolutionary computation, and optimization theory. We refer the reader to [4] for a survey of the works in this area and discuss some relevant ideas here. Many multi-agent algorithms exist for different tasks which range from fully cooperative setting [15, 21] to fully competitive setting [22]. When collaborating with humans, the robot needs to be aware of the human's behavior, which might not always be fully cooperative.

Adaptation of agents has been studied previously [33, 38, 43], where an agent's adaptation depends on the degree of awareness of other agent's behavior maintained by the learning algorithms. These algorithms use some form of opponent modeling to keep track of the other agent's policies [6, 11]. There is a tradeoff between the stability (convergence) of the algorithms and the degree of adaptability. We build upon some of these ideas and propose a two-agent reinforcement learning algorithm, which models the various human behaviors allowing the robot to learn an adaptive policy.

Fig. 2 **System overview**

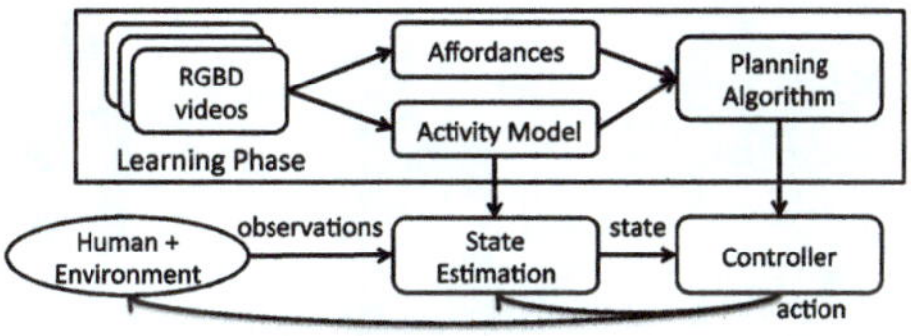

3 Approach

Our goal is to learn which actions a robot can perform in order to collaborate with the human and assist in the task. As illustrated in Fig. 2, we first learn the spatio-temporal structure of activities using a conditional random field (CRF) from RGB-D videos of people performing these activities. We model the sub-activities and affordances of the objects, how they change over time, and how they relate to each other (for details see [20]). We then learn a Q-value function in simulation using the learnt activity and affordance models. When working with the human, the robot first estimates the state of the environment by detecting the object affordances and human actions, and then chooses an appropriate action and executes it.

In detail, we consider a robot r working with a human h in an environment having objects O. The goal is to learn a policy for the robot, π^r, which maps the current environment to an action. We formulate the collaborative task planning problem as a Markov decision process (MDP) with two agents—the human and the robot. We define the following:

- State Space $\mathcal{S}$: Let $\mathbf{s_t} = \{s_t^1, \ldots, s_t^n\}$ denote the state of the environment, where s_t^i denotes the state of the ith object at time t and n denotes the number of objects.
- Action Space $\mathcal{A}$: Let $a_t = \langle a_t^h, a_t^r \rangle$ denote the joint action at time t, where a_t^h and a_t^r denote the human and robot actions respectively.
- Robot's policy π^r: $\mathcal{S} \times \mathcal{A}^r \rightarrow [0, 1]$, where $\mathcal{A}^r$ denotes the set of possible robot actions. $\pi^r(\mathbf{s}, a^r)$ specifies the probability of choosing action a^r in state $\mathbf{s}$.

We address the following challenging aspects of this problem: (i) Defining an efficient state-action representation that captures the contextually rich environments for performing complex activities, (ii) Learning task models which specify the effect of actions on the environment and the actions required for completing the task, and (iii) Modeling human's actions for learning the robot's policy.

3.1 Collaborative Markov Decision Processes

We use RGB-D videos of a single human performing the activities to define the state-action representation and learn the task model of the activities.[1] Once we have

[1] Such data is easier to collect for a wide variety of activities in a variety of environments [20, 42] as compared to collecting data of humans working with robots.

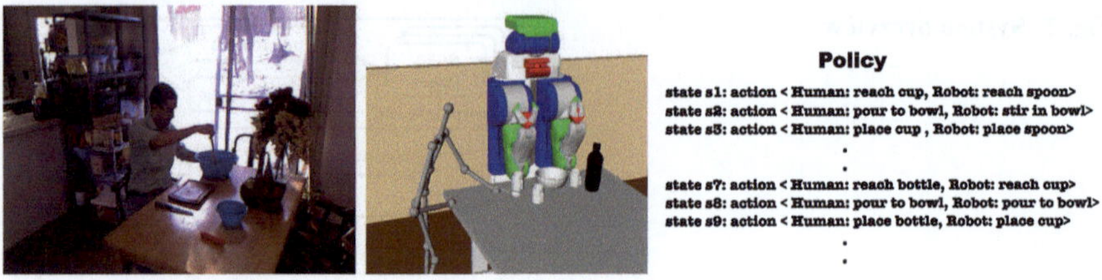

Fig. 3 Collaborative planning by the robot. In order to collaborate with the human on a *recipe following* task, the robot learns the activity model from RGB-D videos of human preparing a recipe (*left*), represents the environment via affordances and uses our planning algorithm (*middle*) to generate a policy for jointly performing the activity with the human (*right*)

the set of states, set of actions and the task model, we can solve the MDP using dynamic programming techniques [34]. However, with large joint state-action space, computing the optimal policy is computationally very expensive and therefore, we take the model-free approach of Q-learning and learn the Q-functions offline with the help of the learnt task models. When collaborating with humans, the robot choses actions greedily with respect to its learnt Q-function. We fix the robot's policy after the offline learning, however, one can also further refine the Q-functions on-the-fly while working with humans in the real world. We now describe the details of our collaborative MDP algorithm.

State-Action Representation: We represent the environment in terms of the object affordances, which leads to an efficient state action space for planning. For example in Fig. 3, the state of the environment is represented in terms of the affordance labels of the objects in the scene, i.e., the bowl is *stirrable*, the spoon is the *stirrer* and the rest of the objects are *stationary*. The *stir* action corresponds to the temporal motion trajectory of the spoon from the grounded stir affordance. On performing the stir action, the spoon becomes *placeable*, thus changing the state of the environment.

Task Model: State Transitions and Rewards. The affordance-based representation of the environment allows for factored representation of the transition and reward functions. That is, it is sufficient to specify the state transitions with respect to only a subset of affordances that are effected by an action. For example, a move action would change only the state of the movable object where as a pour action would change the state of the pourable and the pour-to objects. We assume that each action can be completed in one time step and hence given the nature of activities and affordances, the state transitions are deterministic. That is, on performing a valid action, the affordance of the object changes to another fixed affordance. The reward function allows us to specify valid actions at any given affordance state, where all valid actions receive a fixed positive reward and non-valid actions will incur a negative cost. We compute these task models (i.e., transition and the reward functions) for each high-level activity from the labeled RGB-D videos of a human performing the activity in various environments.

Algorithm 1 RUN-EPISODE (Q, π^h, π^r)

INPUT: State space S, Action space A

Initialize environment to start state; $R \leftarrow 0$; $i \leftarrow 0$;
loop
    if goal state **then**
        return Q-functions, R
    end if
    Sample a_h from π^h and a_r from π^r
    Take action (a_h, a_r) and observe $r, \mathbf{s}'$
    Update Q-functions as in Eq. 2
    $R \leftarrow R + \gamma^i * r$
    $i \leftarrow i + 1$
end loop

Learning Robot Policy. Given the deterministic nature of the state transitions, we use the distributed Q-learning [21] algorithm to learn the local value functions $q_t^h(s, a)$ and $q_t^r(s, a)$ for the human and the robot respectively. Each agent assumes that other agents are acting optimally and only updates their local Q-functions when it results in an increase. This ensures that the local Q-value always captures the maximum of the joint-action Q-values. Therefore at each iteration, the local Q-functions are updated as in Eq. 1 while maintaining the invariants in Eq. 2.

$$q_{t+1}^j(\mathbf{s}_t, a_t^j) = \max\{q_t^j(\mathbf{s}_t, a_t^j),\ R(\mathbf{s}_t, a_t^h, a_t^r) + \gamma \max_{a \in \mathcal{A}^j} q_t^j(\mathbf{s}_{t+1}, a)\},\ j \in \{r, h\} \quad (1)$$

$$q_t^r(\mathbf{s}, a) = \max_{a^h \in \mathcal{A}^h} Q_t(\mathbf{s}, a^h, a^r = a); \quad q_t^h(\mathbf{s}, a) = \max_{a^r \in \mathcal{A}^r} Q_t(\mathbf{s}, a^h = a, a^r) \quad (2)$$

Our collaborative distributed Q-learning algorithm is summarized in Algorithm 1. Here, an episode is defined as the sequence of actions performed by the robot and human from the initial configuration to the goal configuration.

4 Models of Human Behavior

Many studies on human behavior have shown that there are primarily two systems which drive the way humans think—the first being fast, intuitive and emotional; and the second system which is slower, more deliberative and logical [13]. This also applies to our problem of performing collaborative tasks, where humans can either think fast and perform activities following their habits or think more carefully about collaborating by taking into account what the robot can do. Therefore, the actions chosen by the human can range from fully cooperative, when humans are thinking for collaboration, to somewhat adversarial when their habits conflict with the robot's actions. Modeling these various types of human behavior becomes extremely important for collaboration.

Previous works in game theory literature study such scenarios in the setting of general sum Markov games, where the types of opponent behaviors have been roughly classified into fixed strategies or best-response strategies [23]. In the fixed strategy case, the opponent always executes a fixed unknown policy and Q-learning finds the best response with respect to the fixed opponent. In the second case, it is assumed that the opponent adapts and chooses the best response so that it is mutually beneficial to both agents. Following these ideas, we model the following behaviors of a human agent:

- ***Habit-following human***. In this model, we consider the perceptual data of the human from RGB-D videos, and assume that the human follows close to what he has done in the training videos. This is a fixed strategy behavior, where the human has a preferred way of performing activities and follows the same approach even when working with a robot. Let D be the set of activity videos and let $c(\mathbf{s}, a)$ be the number of times the human performed action a when in state $\mathbf{s}$ in D. The policy followed by a habit-following human, $\pi_d^h(\mathbf{s}, a)$, is defined as

$$\pi_d^h(\mathbf{s}, a_i) = \begin{cases} c(\mathbf{s}, a_i)/\sum_a c(\mathbf{s}, a) & \text{if } \mathbf{s} \in D \\ 1/n & \text{if } \mathbf{s} \notin D \\ 0 & \text{otherwise} \end{cases}$$

  where n is the number of possible actions in state $\mathbf{s}$.
- ***ϵ-optimal human***. In this model, we assume that the human takes the best action according to the value function most of the time, but makes a random choice ϵ fraction of the time. Here, human chooses a response that is mutually beneficial most of the time, according to the value function learnt so far.[2] This is equivalent to the ϵ-greedy exploration strategy which was shown to have better convergence properties [44] compared to always choosing the action greedily. The human policy is defined as

$$\pi_\epsilon^h(\mathbf{s}, a_i) = \begin{cases} (1-\epsilon) + (\epsilon/n) & \text{if } a_i = argmax_a(q^h(\mathbf{s}, a)) \\ \epsilon/n & \text{otherwise} \end{cases}$$

[2]The ϵ-*optimal human* behavior can differ from the *habit-following human* behavior, even when the reward model for learning the ϵ-*optimal human's* value function is extracted from the same data as the actions of *habit-following human*. There are two reasons for this: (i) The test environment is not present in the training data, and therefore, the reward function learnt from the training environments might not capture all valid ways of performing the activity in the test environment. This would lead to differences in what a human might do and the policy learnt from an incomplete reward function; (ii) Humans can follow a different reward model when working alone as compared to when collaborating with others. Since we have adapted the reward function learnt from a single-agent scenario to a two-agent scenario, it is possible for the optimal-human policy to deviate from the habit-following human.

Algorithm 2 Learn Robot Policy

INPUT: State space S, Action space A, Data D

Initialize π^h and π^r uniformly
$Q \leftarrow 0; \eta \leftarrow 0.5$
while *burn-in period* **do**
  Sample $s \sim Bern(\eta)$; $s \in \{d, \epsilon\}$
  Update π_s^h
  $Q, R \leftarrow$RUN-EPISODE(Q, π_s^h, π^r)
end while
Update η
loop
  Update π_a^h using Eq. 3
  $Q, R \leftarrow$RUN-EPISODE(Q, π_a^h, π^r)
end loop
return Robot's Policy π_r

- ***Adaptive human***. In the real world, when collaborating, humans usually adapt to other agents while trying to maintain their preferences or habits. That is, they follow their habits when possible in familiar situations, but when faced with new situations while working with the robot, they adapt and try to perform the action that is beneficial to both for completing the activity. We model this behavior by computing the probability of the human choosing one of the above two behaviors and define the human policy as

$$\pi_a^h(\mathbf{s}, a) = \eta * \pi_d^h(\mathbf{s}, a) + (1 - \eta) * \pi_\epsilon^h(\mathbf{s}, a); \quad \forall \mathbf{s}, a \tag{3}$$

where η denotes the probability of the human to follow habits.

During test time, when the robot is collaborating with the human on a new task, it should choose actions from the policy which is learnt with matching human behavior. One approach is to assume that the opponent type is known and fixed, and use the policy learnt with that type when executing the activities. Some works try to identify the opponents strategy on the fly and adapt accordingly. Such an approach requires the robot to perform activities with humans for long time durations, which is not very practical in most scenarios. In contrast to these approaches, we present an algorithm (Algorithm 2) which adaptively selects the human's actions for exploration during the learning phase.

We need to estimate the Q-values along with the value of η, which is the probability with which the human follows his habits. This probability depends on the human's familiarity of the environment as well as the cost of deviating from the optimal policy. Therefore, we model this probability as a function of the joint reward obtained when the human follows one of the two extreme behaviors, *habit-following* and *ϵ-optimal*, throughout the activity. Therefore, during an initial burn-in period, we sample the behavior uniformly and fix the behavior throughout an episode and learn the Q-values. We maintain a score for each of the behaviors, denoted by w_d and w_ϵ for the *habit-following* and *ϵ-optimal* behaviors respectively. At the end of Q-learning episode, we compute the normalized joint reward and update the corresponding score value as in Eq. 4.

$$\hat{R} \leftarrow \frac{R - \frac{r_{min}}{1-\gamma}}{\frac{r_{max}}{1-\gamma} - \frac{r_{min}}{1-\gamma}} \; ; \quad w \leftarrow w + \alpha(\hat{R} - w) \tag{4}$$

where r_{max} and r_{min} are the maximum and minimum reward at any given time, respectively, γ is the discount factor and α is the learning rate. At the end of the burn-in period we compute the value of η as:

$$\eta \leftarrow \frac{e^{(w_d)}}{e^{(w_d)} + e^{(w_\epsilon)}} \tag{5}$$

We then continue learning the Q-values for the adaptive human by updating the human policy according to Eq. 3 using the estimated value of η.

Effect of the human behavior on the learned robot policies. Figure 4 illustrates different human and robot policies corresponding to the different possible human behaviors we consider. Here, we consider an environment in which there are two cups, a bowl and a spoon, and the robot can only reach the cups and the bowl whereas the human can reach all objects. The goal of the activity is to follow a recipe involving transfer of the ingredients from the cups to the bowl and mix them with the spoon. In the training videos, at the beginning of the activity, the human reaches the first cup more often than the second cup as shown by the policy for the start state in Fig. 4a. The corresponding learned robot policy is to not do any action as reaching for a cup could result in a conflict.

Instead of following habits, if the human tries to optimizes for the joint reward, he would reach for the spoon and let the robot use the cups, which allows them to perform the activity together and complete it sooner. Figure 4b shows the policies corresponding to the *ϵ-optimal* human behavior. Given this environment, following habits turns out to be less rewarding and therefore a human would try to adapt more. This is reflected in the estimated value of η which is low for this particular scenario.

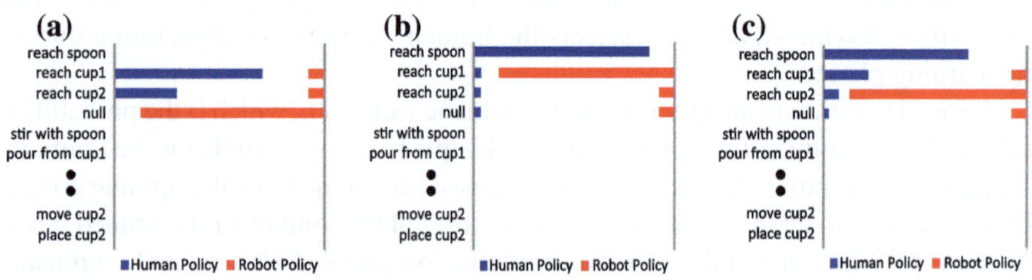

Fig. 4 **Illustration of policies learnt with different human behaviors**: Each figure shows the learnt probability distributions of the various possible actions at the start state of the *following recipe* activity. *Blue* and *red bars* represent the probability of choosing the corresponding actions by the human and robot respectively. **a** Habit-following human. **b** ϵ-optimal human. **c** Adaptive human

Figure 4c shows our learnt *adaptive human* policy and the corresponding robot policy. Note that even small changes in human behavior can result in significant changes in the robot's actions.

5 Experiments

We test our proposed algorithm and other baseline methods for generating collaborative plans for several household activities. We evaluate the learnt robot policies on both an activity dataset as well as in interaction with real humans. In this section we describe the data, experimental setup and the results.
Data: In order to evaluate our affordance and anticipatory planning models we expanded the CAD-120 dataset [20] to CAD-250 dataset, which has 130 additional RGB-D activity videos which contain more interesting object affordances and activities which allow human-robot collaboration. The sub-activities in the CAD-250 dataset include *{moving, stirring, pouring, drinking, cutting, eating, cleaning, reading, answering phone, wearing, exercising, hammering, measuring}* and the corresponding affordances are *{movable, stirrable, pourable, pourto, drinkable, cuttable, edible, cleanable, cleaner, readable, hearable, wearable, exercisable, hammer, hammerable, measurer, measurable}*.

We evaluated our planning algorithm on 60 RGB-D videos from the CAD-250 dataset which allow for collaboration. These activities include *two recipe making tasks*, *setting dinner table*, *cleaning house*, and *loading shelves*. These activities were performed by four subjects where each high-level activity is performed three times by each subject in a different environment. For each activity video, we labeled the sub-activities and the object affordances.

Baselines: We compare our method against the following baselines:

- **Human Expert**: A human expert manually designed collaborative plans for each activity in the dataset.
- **Chance**: This algorithm chooses actions uniformly at random from the set of possible actions.
- **Mental-model MDP** [32]: We follow Nikolaidis et al. and define a MDP to model the robot's mental model [32]. In this approach, the human actions are incorporated into the state transition function and the policy specifies only the robot's actions. Therefore, we use the same state and action spaces and reward function as described in our approach with only one agent and compute the transition function from the state action sequences from the training data. Note that in our adaptation of [32] we fix the transition function learned from the data and do not perform any cross training iterations as the roles are fully exchangeable in our collaborative setting.

5.1 Evaluation on Data

We evaluate the generated collaborative plans on the following two metrics: (i) Percentage time saving: The percentage of savings in time for task completion is computed as $\frac{n_h - n_c}{n_h} * 100$, where n_h denotes the number of time steps taken if only human performs the task and n_c denote the number of steps to task completion following the collaborative plan. (ii) Percentage conflicts: The percentage of time steps robot's chosen action conflicted with that of the human.

For each activity video in the dataset, we give the environment extracted from the first frame and the goal state as input to the planning algorithm. We perform leave-one-out cross-validation and use the rest of the activity videos for learning the task model and the robot policies as described in Sects. 3.1 and 4. The sequence of human actions are taken from the test video and executed together with the robot actions specified by the learnt robot policy. Table 1 shows the results averaged for each high-level activity as well as for all activities in the dataset.

As can be seen in Table 1, our algorithm allows for more collaboration between human and the robot, resulting in higher savings in the time required for task completion compared to the baseline algorithms. When the robot chooses actions uniformly at random (Chance baseline), it sometimes chooses action sequences that help in achieving the goal sooner, but can also perform undesirable actions requiring additional time to complete the activity. Therefore, on average it does not result in any savings in the execution time. These results show that modeling the human actions along with the contextually rich environments is very important for collaborative planning.

Table 1 Collaborative planning evaluation

Model	% time saving					% conflicts
	Recipe	Setting	Cleaning	Loading	Overall	
Human expert plans	36.8	53.1	16.4	42.4	37.2	0
Chance	3.3	10.5	−33.1	23.7	1.1	3.7
Mental-model MDP [32]	−2.6	30.4	−5.1	32.3	13.8	6.4
Our model—ϵ-optimal human	27.5	45.6	18.3	30.8	31.2	13.5
Our model—habit following human	28.4	48.1	18.6	41.4	33.4	11.9
Our model—adaptive human	**32.8**	**48.5**	**22.9**	**41.9**	**36.5**	13.7

Metrics computed for the collaborative plans generated on our RGB-D dataset

5.2 *User Study*

Experiment setup. We performed an user study with five subjects to evaluate the learned robot policies. We considered two high-level activities—*setting table* and *making recipe*, and four different environments for each activity. The subjects were asked to work with the robot to complete the tasks in a simulator. We re-created the environments in OpenRAVE [7] and provided an interface to the subjects to select an action they wish to execute. At every time instant, the users were shown the current state of the environment, and were asked to choose an action. The robot also selects an action based on the current state using its learned policy. Both the human and robot actions are then executed in the simulator. After completing each task, the users were asked to rate the following statements on Likert scale from 1 (strongly disagree) to 5 (strongly agree).

(a) The robot was collaborative and helped in the activity.
(b) The robot did the right thing at the right time.
(c) I am satisfied working with the robot.
(d) I will work with this robot again in future.

In this study, we compared the robot policies generated by the mental-model MDP [32] and our method learned with the three human behaviors. Therefore, every user performed each task four times, resulting in a total of 640 ratings.

Results. Figure 5 shows the comparison of the user ratings for the four different robot policies on the four criteria mentioned above. Users rated the robot trained with our collaborative MDP model significantly higher ($p < 0.001$) than the robot using the mental-model MDP on all four criteria. For the robot policies learnt with our collaborative MDP using different human behaviors, when asked if they thought the robot did the right action at the right time, the users rated the robot trained with *adaptive human* higher than others ($p = 0.08$). For other criteria, there is no significant difference in the user ratings, however, as can be seen in Fig. 5, there is a slight preference for the robot trained with *adaptive human*. Table 2 summarizes the two metrics on the collaborative plans generated in the user study. The users completed the tasks faster when working with robot trained with the *adaptive human* as compared to others.

5.3 *Robot Experiment*

We have also used the learned robot policy on our Kodiak (PR2) robot to work with a human on a *following recipe* task. Figure 6 shows the robot collaborating with human to prepare a recipe, where the robot is executing the pour action as the human is stirring, based on its learnt policy. Videos showing the human and robot collaborating are available at: http://pr.cs.cornell.edu/collaborativeplanning/.

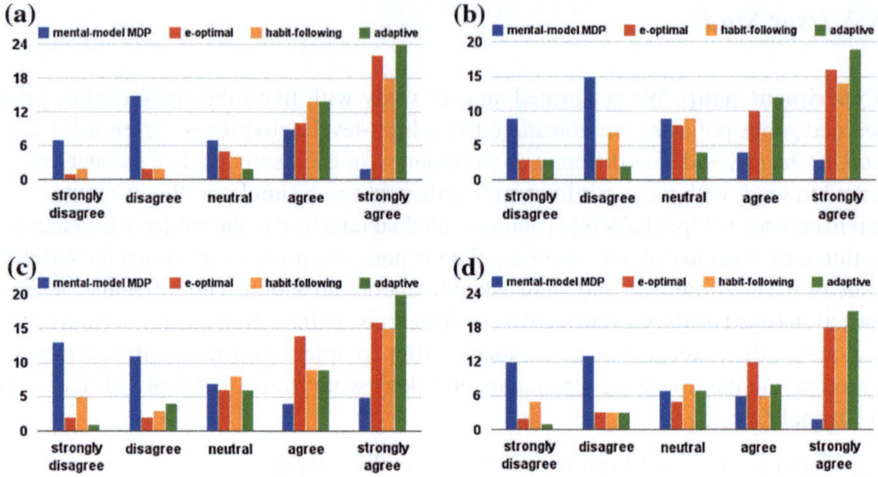

Fig. 5 **User study results**: The subjects collaborated with the robot on two different tasks in a total of eight different activities. They rated their experience based on four different criteria. The plots (a)–(d) show the comparison of the user ratings for four different robot policies—mental-model MDP [32], and our collaborative MDP trained with ϵ*-optimal human*, *habit-following human* and *adaptive human*. **a** The robot was collaborative and helped. **b** The robot did the right things. **c** I am satisfied working with the robot. **d** I will work with this robot again

Table 2 **Collaborative planning evaluation for user study**

Model	% Time saving			% Conflicts
	Recipe	Setting	Overall	
Mental-model MDP [32]	−0.9	13.9	6.5	2.7
Our model—ϵ-optimal human	34.3	46.5	40.4	4.6
Our model—habit following human	16.5	48.9	32.7	4.4
Our model—adaptive human	**38.2**	**52.7**	**45.5**	4.6

Metrics computed for the collaborative plans generated when working with humans during the user study

5.4 Discussion

We discuss the results of the evaluation on our dataset as well as the user study in the light of the following questions.

What is the advantage of our collaborative MDP model over a single-agent MDP model? The mental-model MDP doesn't model the human explicitly as an agent but incorporates the effect of human actions in to the state transition probabilities. We observe a large variation in the performance of the mental-model MDP [32] baseline across the types of activities—it performs well on the setting table and loading shelf

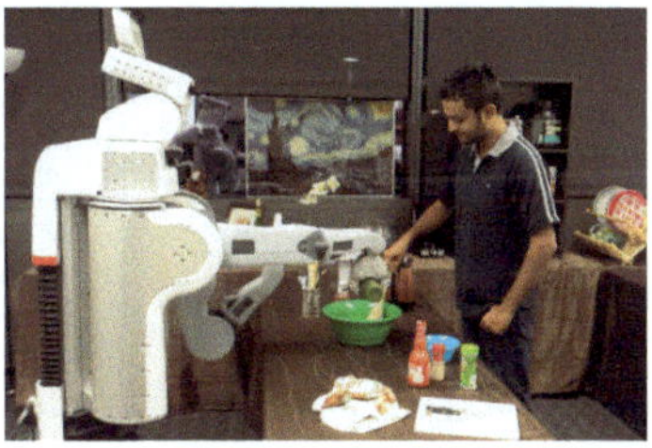

Fig. 6 Robot and human collaborating to prepare a recipe

tasks, but takes longer to complete the recipe and cleaning tasks. It is interesting to note that the setting table and loading shelf tasks have a smaller action space as compared to the cleaning and recipe tasks. Therefore, given limited training data, the mental-model MDP is sensitive to the estimated state transition probabilities and fails when the action space is large. On the other hand, our collaborative MDP approach, which models the actions of human explicitly, overcomes this problem and performs significantly better on all tasks.

How important is modeling human behavior for collaboration? As humans tend to have specific preferences for executing tasks, the robot policy learnt with *habit-following human* strategy, which incorporates these preferences into planning, achieves an additional 2.2 % saving in time compared to the *ϵ-optimal human*. However, when tested with new humans, whose habits were never seen in the training data, the robot policy learnt with *ϵ-optimal* human performs better (see Table 2). Modeling human as an adaptive agent always performs better and results in more collaboration—increasing the savings in the task completion time by 3.1 % when working with a familiar human (seen in the training data) and by 5.1 % when working with a new human.

We also study how the joint reward evolves over the Q-learning episodes during training. At the end of each episode, we use the learned robot policy to perform the activity with a human following the policy π_d^h corresponding to the test environment. Figure 7 shows the joint reward received by the human and the robot as a function of the number of training episodes. We see that the policy learnt with the *adaptive human* converges to the highest joint reward much faster in most cases. However, incorrect estimation of adaptation probability causes the *adaptive human* to perform sub-optimally in some cases (Fig. 7 right).

How often does the robot conflict with human? The savings in task completion time increase as a result of the robot's increased participation in the task. This also leads to an increase in the % of conflicts between the robot's and the human's actions. However, our model learnt with the *habit-following human* strategy reduces the % of conflicts compared to other baselines as it models the human's preferences. The number of conflicts again increase in case of *adaptive human* due to increased participation of the robot in the activity. When a conflict occurs, the preference is given to the human and the robot stops executing the action and chooses a new

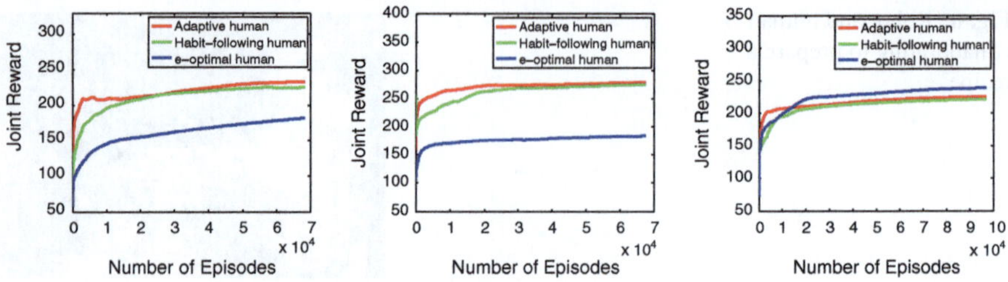

Fig. 7 Joint reward received during Q-learning. Plots showing the joint reward as a function of the number of Q-learning episodes for three test environments

action in the next time step. In our current model we prefer plans with increased collaboration and do not penalize conflicts heavily. However, it is possible to modify the reward function to incorporate this, and we plan to explore this in future work.

6 Conclusion

In this work, we considered the problem of anticipatory planning for human robot teams, for enabling robots to work along side humans in contextually rich environments to accomplish complex tasks. We proposed a two agent collaborative MDP model and learn robot policies by taking into account the actions that can be performed by the human. We represented the contextually rich environments in terms of the object affordances and learn the activity model from RGB-D videos of a human performing the activities. We used this learned task model in a distributed Q-learning algorithm to learn the robot policy for a given new environment. We model the different possible human behaviors—taking the ϵ-optimal action according to the MDP model, taking actions based on past habits, and taking appropriate actions by adapting to the environment and robot actions. We tested our collaborative MDP model on the activity dataset as well as while directly interacting with humans in a user study. We show that explicitly modeling the human actions in the MDP formulation results in learning better robot policies. We also showed that changes in the human behavior can lead to significant changes in desirable robot actions. Therefore, modeling human behavior is essential for collaborative planning.

References

1. Aksoy, E.E., Abramov, A., Dörr, J., Ning, K., Dellen, B., Wörgötter, F.: Learning the semantics of object-action relations by observation. IJRR **30**(10) (2011)
2. Aldoma, A., Tombari, F., Vincze, M.: Supervised learning of hidden and non-hidden 0-order affordances and detection in real scenes. In: ICRA (2012)

3. Bento, J., Derbinsky, N., Alonso-Mora, J., Yedidia, J.S.: A message-passing algorithm for multi-agent trajectory planning. In: NIPS (2013)
4. Busoniu, L., Babuska, R., De Schutter, B.: A comprehensive survey of multiagent reinforcement learning. IEEE SMC Part C (2008)
5. Cakmak, M., Srinivasa, S.S., Lee, M.K., Forlizzi, J.F., Kiesler, S.: Human preferences for robot-human hand-over configurations. In: IROS (2011)
6. Carmel, D., Markovitch, S.: Opponent modeling in multi-agent systems. In: Adaption and Learning In Multi-Agent Systems, vol. 1042 (1996)
7. Diankov, R.: Automated Construction of Robotic Manipulation Programs. Ph.D. thesis, CMU, RI, August 2010
8. Gibson, J.J.: The Ecological Approach to Visual Perception (1979)
9. Hermans, T., Rehg, J.M., Bobick, A.: Affordance prediction via learned object attributes. In: ICRA Workshop on SPME (2011)
10. Hermans, T., Rehg, J.M., Bobick, A.: Decoupling behavior, perception, and control for autonomous learning of affordances. In: ICRA (2013)
11. Hu, J., Wellman, M.P.: Multiagent reinforcement learning: theoretical framework and an algorithm. In: ICML (1998)
12. Jiang, Y., Koppula, H.S., Saxena, A.: Hallucinated humans as the hidden context for labeling 3D scenes. In: CVPR (2013)
13. Kahneman, D.: Thinking, fast and slow (2011)
14. Katz, D., Venkatraman, A., Kazemi, M., Bagnell, J.A., Stentz, A.: Perceiving, learning, and exploiting object affordances for autonomous pile manipulation. In: RSS (2013)
15. Koenig, S., Keskinocak, P., Tovey, C.A.: Progress on agent coordination with cooperative auctions. In: AAAI (2010)
16. Konidaris, G., Kuindersma, S., Grupen, R., Barto, A.: Robot learning from demonstration by constructing skill trees. IJRR **31** (2012)
17. Koppula, H.S., Saxena, A.: Learning spatio-temporal structure from RGB-D videos for human activity detection and anticipation. In: ICML (2013)
18. Koppula, H.S., Saxena, A.: Anticipating human activities using object affordances for reactive robotic response. In: RSS (2013)
19. Koppula, H.S., Saxena, A.: Physically grounded spatio-temporal object affordances. In: ECCV (2014)
20. Koppula, H.S., Gupta, R., Saxena, A.: Learning human activities and object affordances from RGB-D videos. In: IJRR (2013)
21. Lauer, M., Riedmiller, M.: An algorithm for distributed reinforcement learning in cooperative multi-agent systems. In: ICML (2000)
22. Littman, M.L.: Markov games as a framework for multi-agent reinforcement learning. In: ICML (1994)
23. Littman, M.L.: Friend-or-FOE Q-learning in general-sum games. In: ICML (2001)
24. Lopes, M., Santos-Victor, J.: Visual learning by imitation with motor representations. IEEE SMC, Part B (2005)
25. Lorken, C., Hertzberg, J.: Grounding planning operators by affordances. In: International Conference on Cognitive Systems (2008)
26. Mainprice, J., Berenson, D.: Human-robot collaborative manipulation planning using early prediction of human motion. In: IROS (2013)
27. Mainprice, J., Sisbot, E.A., Jaillet, L., Cortés, J., Alami, R., Siméon, T.: Planning human-aware motions using a sampling-based costmap planner. In: ICRA (2011)
28. Meisner, E., Isler, V., Trinkle, J.: Controller design for human-robot interaction. Auton. Robot. **24**(2), 123–134 (2008)
29. Moldovan, B., van Otterlo, M., Moreno, P., Santos-Victor, J., De Raedt, L.: Statistical relational learning of object affordances for robotic manipulation. Latest Adv. Inductive Logic Prog. (2012)
30. Montesano, L., Lopes, M., Bernardino, A., Santos-Victor, J.: Learning object affordances: from sensory-motor coordination to imitation. Trans. Robot. **24**(1), 15–26 (2008)

31. Nikolaidis, S., Shah, J.: Human-robot teaming using shared mental models. In: HRI, Workshop on Human-Agent-Robot Teamwork (2012)
32. Nikolaidis, S., Shah, J.: Human-robot cross-training: computational formulation, modeling and evaluation of a human team training strategy. In: HRI (2013)
33. Powers, R., Shoham, Y.: New criteria and a new algorithm for learning in multi-agent systems. In: NIPS (2004)
34. Puterman, M.L.: Markov Decision Processes: Discrete Stochastic Dynamic Programming. Wiley, New York (1994)
35. Sisbot, E.A., Marin, L.F., Alami, R.: Spatial reasoning for human robot interaction. In: IROS (2007)
36. Strabala, K.W., Lee, M.K., Dragan, A.D., Forlizzi, J.L., Srinivasa, S.S., Cakmak, M., Micelli, V.: Towards seamless human-robot handovers. JHRI (2013)
37. Sun, J., Moore, J.L., Bobick, A., Rehg, J.M.: Learning visual object categories for robot affordance prediction. IJRR (2009)
38. Tesauro, G.: Extending Q-learning to general adaptive multi-agent systems. In: NIPS (2004)
39. Turpin, M., Michael, N., Kumar, V.: Trajectory planning and assignment in multirobot systems. In: WAFR (2012)
40. Ugur, E., Sachin, E., Oztop, E.: Affordance learning from range data for multi-step planning. In: Epirob (2009)
41. Uyanik, K.F., Caliskan, Y., Bozcuoglu, A.K., Kalkan, S., Yuruten, O., Sahin, E.: Learning social affordances and using them for planning. In: CogSys (2013)
42. Wang, J., Liu, Z., Wu, Y., Yuan, J.: Mining actionlet ensemble for action recognition with depth cameras. In: CVPR (2012)
43. Wongpiromsarn, T., Ulusoy, A., Belta, C., Frazzoli, E., Rus, D.: Incremental synthesis of control policies for heterogeneous multi-agent systems with linear temporal logic specifications. In: ICRA (2013)
44. Wunder, M., Littman, M.L., Babes, M.: Classes of multiagent q-learning dynamics with epsilon-greedy exploration. In: ICML (2010)

Autonomous Realization of Simple Machines

Can Erdogan and Mike Stilman

Abstract For robots to become integral parts of human daily experience, they need to be able to utilize the objects in their environment to accomplish any range of tasks. In this work, we focus particularly on physically challenging tasks that push the limits on the robot kinodynamic constraints such as joint limits, joint torques and etc. Previously, we demonstrated an autonomous planner that instructs a human collaborator where to place the available objects in the environment to form a simple machine such as a lever-fulcrum assembly. In this work, we report results on the autonomous realization of such a design by the humanoid robot Golem Krang, focusing on the challenges of autonomous perception, manipulation and control.

1 Introduction

The ability to use the available objects in the environment towards accomplishing goals is essential to thriving in challenging circumstances. Everyday examples of tool use include simple machines such as levers and pulleys. The challenge in autonomous design of such simple machines is the space of discrete choices for the component options and the related high-dimensional continuous configuration space of the chosen components.

In previous work [1, 2], we demonstrated the constraint satisfaction approach to assembly design, specifically for robotic manipulation and locomotion. The idea is to represent the constraints between the components of the design and on the robot kinodynamics as generic equality and inequality functions within an optimization framework and solve for the global minima. Operations research [3] and architecture [4] fields also use global optimization in design problems.

C. Erdogan (✉) · M. Stilman
Institute for Robotics and Intelligent Machines, Georgia Institute of Technology, Atlanta, USA
e-mail: cerdogan3@gatech.edu

M.A. Hsieh et al. (eds.), *Experimental Robotics*, Springer Tracts in Advanced Robotics 109, DOI 10.1007/978-3-319-23778-7_31

Fig. 1 The mechanical advantage in forces

In this work, we take the next step towards full autonomy where the humanoid robot, Golem Krang, autonomously manipulates the objects in its environment to construct a simple machine. The robot perceives the available objects, specifically 15 kg cinder blocks and 10 kg wooden blocks (e.g. potential levers), relocates them to the desired configurations output by the constraint planner, and actuates them to flip a 50 kg load. Figure 1 demonstrates key scenes from this scenario such as (a) detection of a cinder block, (b) locomotion with a heavy load, (c) manipulating a lever while subject to multiple constraints and (d) application of force to the lever leading to a successful load motion.

Significant effort has been demonstrated by [5–7] to incorporate autonomous agents in human environments. Our work stands out in multiple aspects from the established state of the art. First, Golem Krang is a two-wheeled balancing robot, similar to a segway with two 7-dof robotic arms installed. The challenge with such a platform is the dynamic stability constraint where the robot has to ensure its center of mass is close to the wheel axis at all times as opposed to legged or multi-wheeled platforms. Secondly, to the best of our knowledge, Golem Krang is the tallest and heaviest two-wheeled robot with 150 kg at 1.9 m, a unique property among similar designs [8]. At this scale, the weight can help with heavy-duty manipulation but also complicates the autonomous locomotion. Lastly, Golem Krang perceives its environment with an onboard RGBD sensor with two degrees of freedom that can be manipulated for gaze control. Autonomous perception and scene recognition has only recently started to gather interest in the humanoid robotics field [9, 10] as opposed to the established motion capture methods [11].

2 Technical Approach

2.1 Constraint Satisfaction for Simple Machine Designs

The manipulation of multiple objects to achieve a goal can be readily represented in a constraint satisfaction paradigm where the constraints represent the relationships between the design components. In this work, we focus on simple machine designs such as lever-fulcrum assemblies or inclined planes that need to be structurally stable and provide mechanical leverage to their users. Reasoning about such design criteria requires analysis of more detailed concepts such as center of masses, robot kinodynamic constraints and physics principles.

The design process is composed of three steps. First, from the set of available objects in the environment, the planner needs to choose a subset that will be incorporated in the structure. Second, the structure components are assigned roles that designate how they should be put together—specifically, the constraints that *bind* one to another. Lastly, the planner needs to configure the objects such that the role constraints and the general design criteria are satisfied.

Component Choices A completeness property for a structural design planner is a crucial advantage for deployment in real-world applications (e.g. military or search-and-rescue operations). In emergency situations, when physical challenges require creative reasoning, the ability to consider all possible solutions and determine if one succeeds is a critical advantage.

The planner needs to exhaustively search the entire *finite* space of discrete assignments. In comparison to continuous choices, such as object configurations, the discrete nature of the component choices (i.e. in or out) makes such a search feasible. Despite the finite space, it is challenging to evaluate every option since there is still a combinational number of roles and infinite space of configurations to reason about. Note that we assume every object is used only once in the structure as opposed to the robot changing the structure throughout its use.

To remedy the computational challenge, pruning strategies and heuristics are significant tools in cutting back the search at the top level. For instance, in construction a lever-fulcrum design, two wooden blocks of the same size (or approximately to a degree of confidence) can be categorized under one class. Similarly, for loads that are known to be heavier than the maximum load a robot can handle, the longer lever candidates might be prioritized in the search.

Object Roles Imagine that a two-step stairs is needed to enable a swarm of rough-terrain vehicles, such as PackBots or RHexs [12, 13], to climb a window and survey a building, and some of the vehicles have robotic arms that can stack box-like objects. The goal for a planner would be to choose three boxes and stack two of them (e.g. box B on C) such that the swarm can first climb one step and then move on to the two stack, until it reaches the window sill.

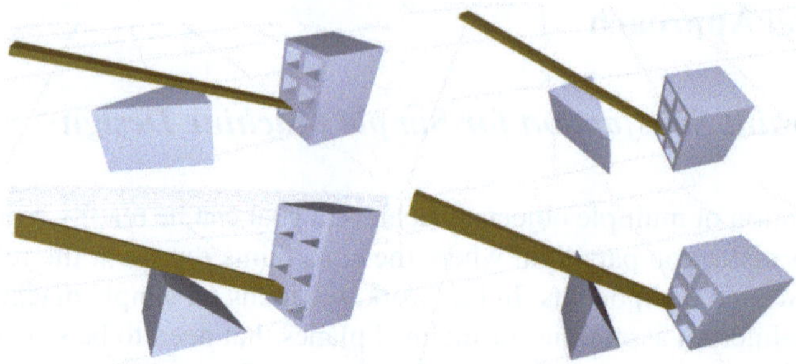

Fig. 2 Different face and edge assignments for a lever-fulcrum assembly. Some assignments, such as using the shortest edge of the lever to connect the fulcrum and the load, may not be feasible due to design constraints and collisions

In coming up with this solution, two types of choices are made. First, among the three objects, say A, B, and C, which object will be used as the first step and which one will be at the top in the second step needs to be decided. Secondly, in order to place an object on a surface, a base face needs to be chosen—in fact, doing so partitions the configuration space of the objects even before the design constraints are considered to prune infeasible assignments.

Hence, the goal of assigning a role to an object is to specify (1) the relationship of a component with respect to others, and (2) which of its face and edges are used in these connections. Figure 2 [2] depicts some outcome assemblies with a triangle prism acting as fulcrum positioned on different base faces and in contact with a lever on various edges. We also observe that the lever can contact the fulcrum on two faces (discounting symmetries) that lead to feasible assemblies.

Role assignment to available resources has been a thoroughly studied area, starting from classical planning [14, 15], and evolving into operations research [16, 17]. In this work, we adopt the STRIPS representation [18] as a method to represent the domain knowledge of possible actions that can be taken on the available objects in the environment. For instance, a lever can be *placed* on a fulcrum or a ramp can be *rested* against another object. A minor difference in our framework is that every action induces additional constraints to the assembly configurations and an action can be taken if and only if there exists some configuration that satisfies the accumulated constraints.

Continuous Configuration Space and Constraints The proposed framework accumulates design constraints via a classical planning framework and a feasibility process determines whether the constraints can be satisfied by some configuration of the components. Different types of constraints can all be generically expressed as equality or inequality functions on the space of object and robot configurations. If all the constraints are convex, for instance in a stair or bridge design with simplified robot assumptions, the feasibility can be determined by an efficient simplex algorithm implementation [1]. For nonconvex domains, especially when robot kinodynamics

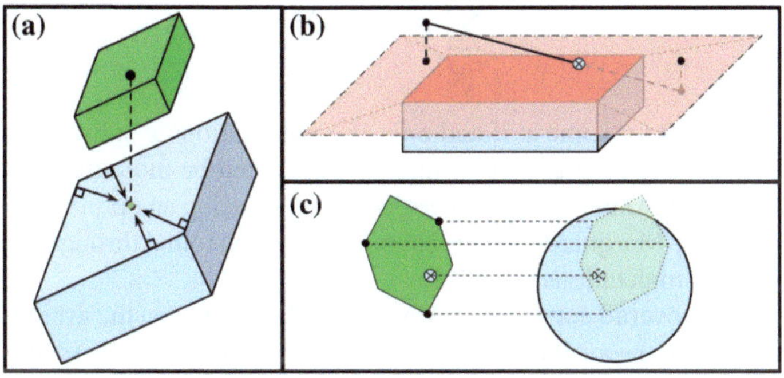

Fig. 3 Visualization of different types of geometric contact constraints

are considered, we propose a nonlinear optimization process which minimizes the violation of the constraints and attempts to find an assignment of configurations that satisfies all of them.

A number of design constraints such as stacking a box on top of another one or placing a lever at the edge of a fulcrum can be expressed with geometric projections. Figure 3 demonstrates three types of connections: (1) center of mass-face, (2) edge-face, and (3) face-face. The general idea is that points of interest such as center of mass, endpoints of an edge or vertices of a face are projected to the plane of another face and limits are imposed on its location. For instance, for two objects to be stacked successfully (A), the center of mass of the top object has to lie within the supporting face of the bottom object. Similarly, to ensure an edge is on a face (B), it is sufficient to confine the endpoints of the edge onto the face plane and guarantee there exists a shared point (e.g. cross) between the edge and the face. Lastly, for contact between two faces (C), three points of one of the faces has to lie on the other one and again, a shared point should exist. Observe that geometric contact concepts are easily expressed through equality and inequality expressions on the projections of significant points on the meshes.

Once the object choices and roles are determined, the equality and the inequality constraints between the assembly components can be gathered into two sets $\mathcal{F}$ and $\mathcal{G}$ respectively. The idea behind using optimization to find feasible samples in the configuration space is based on creating error functions by the violation of the constraints for a sample design $\boldsymbol{x}$:

$$f(\boldsymbol{x}) = 0 \quad \Rightarrow \quad E_f(\boldsymbol{x}) = f^2(\boldsymbol{x})$$

$$g(\boldsymbol{x}) \leq 0 \quad \Rightarrow \quad E_g(\boldsymbol{x}) = \begin{cases} g^2(\boldsymbol{x}) & \text{if } g(\boldsymbol{x}) > 0, \\ 0 & \text{otherwise.} \end{cases}$$

where $E_f(\boldsymbol{x})$ and $E_g(\boldsymbol{x})$ are the proposed squared error functions. Now, given the constraint sets $\mathcal{F}$ and $\mathcal{G}$, we define the total error:

$$\mathcal{E}(\boldsymbol{x}) = \sum_{f \in \mathcal{F}} E_f(\boldsymbol{x}) + \sum_{g \in \mathcal{G}} E_g(\boldsymbol{x}). \tag{1}$$

Note that $\mathcal{E}(\boldsymbol{x})$ is 0 for some $\boldsymbol{x}$ if and only if configurations $\boldsymbol{x}$ satisfy all the design constraints. Moreover, the global minima is guaranteed be more than or equal to 0 since the function is a sum of squared errors. Then, by using an optimization method, such as Levenberg-Marquardt, the global minima can be found through sampling the space for good initializations.

We outline the overall approach in Algorithm 1 below. Given the available objects, the goal criteria and the available actions, the planner searches in the space of discrete object roles (line 3) and attempts to take actions as long as they lead to feasible configurations (line 9). The forward search accumulates constraints until a feasible design is reached or backtracks.

Algorithm 1: ConstraintPlanner()

Input: *domain*: objects properties and generic actions; *goals*: list of goal literals to be fulfilled; *initialState*: discrete literals and no constraints;
Result: configurations: a feasible value in goal subspace;

1 *stateStack* ←createStack(*initialState*);
2 **while** *state* ← *stateStack.pop*() **do**
3   *actions* ←stateActions(*domain*);
4   **foreach** *action in the set actions* **do**
5   **if** *action.pres* ⊂ *state.literals* **then**
6   newConsts ← state.consts ∪ action.consts;
7   **for** *counter* = 1 ... *MAX_COUNT* **do**
8   {*localMin, confs*} ←optimize(*newConsts*, newSeed(*domain*));
9   **if** *abs*(*localMin*) ≤ $1e^{-4}$ **then**
10   **if** *goals* ⊂ *action.afters* **then return** *confs*;
11   **else**
12   *child* = {*state.literals* ∪ *action.afters*, *newConsts*}
13   *stateStack*.push(*child*);
14   **break**;
15 **return** ∅;

2.2 Humanoid Robot Platform: Golem Krang

Designed and built in the Humanoid Robotics Laboratory, Golem Krang is a 150 kg, 6.2 m segway-like humanoid robot with two wheels that uses a balancing strategy for locomotion and manipulation [19]. Given its unique design, the proposed planner needs to address several constraints. First, the contact point on the lever needs to be reachable by the robot. Second, the robot needs to apply sufficient force to overcome the opposing weight or friction, but we still need to maintain a maximum force limit

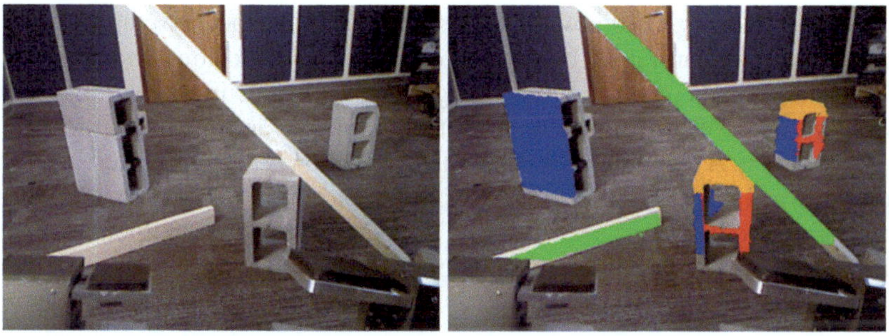

Fig. 4 Cinder block and wooden plates detected as fulcrum and lever objects

to protect the motors. Lastly, Golem Krang needs to maintain a stable posture before contact is made with the structure.

2.3 Perception

Equipped with a Microsoft Kinect that can pan and tilt, Golem Krang can inspect and reason about its environment with visual data. In perceiving the environment, we propose using a light-weight feature-based recognition approach as opposed to full 3D based approaches that use the entire mesh data such as the iterative closest point algorithm or over-segmentation methods. An assumption is that the planner knows the meshes of the available objects and with minimal additional feature knowledge, such as the top of a cinder block is at 44 cm from the ground or a lever is at least 2 m in one dimension, we can speed up the detection. The proposed approach, detects individual and/or assembly of cinder blocks, walls and wooden blocks as shown in Fig. 4.

2.4 Locomotion

The primary locomotion strategy for Golem Krang is to balance on its two wheels, keeping its center of mass on the vertical plane through its wheel axis. Modeled as an inverted pendulum, locomotion via balancing has a few advantages over running on both the wheels and the back caster. First, the footprint of the robot is smaller, 54 cm in width due to the wheel diameter while balancing, as opposed to 86 cm when the robot is grounded. Second, the locomotion is simpler to model since the fixed caster without omnidirectional wheels sustains different ground reaction forces as the robot spins, moves forward and backward.

The position and posture control is implemented using a proportional derivative controller based on the inertial readings that indicate the robot angle from the vertical and the wheel encoders. In this work, we assume the environment is setup such that the locomotion can be carried out by turning towards the goal position, moving forward and adjusting for the goal orientation—ignoring collisions in the world. To move forward, we use a velocity profile with limits on minimum and maximum acceleration and deceleration. A significant aspect of the locomotion is the manipulation of heavy objects such as 15 kg cinder blocks and 10 kg wooden plates. To enable stable dynamic balancing, the force-torque sensors at the grippers are used to incorporate the mass of the carried objects in the computation of the center of mass position.

2.5 Manipulation

The manipulation of multiple objects under motor and perception uncertainty requires a series of robust strategies both algorithmically and in practical implementation. In this work, a wide range of motion planning tools are adopted such as rapidly-expanding random trees (RRTs) [20], analytical inverse kinematics [21] and Jacobian control [22]. Additionally, we propose using guarded moves [23] that control the manipulator behavior until a predetermined tactile feedback is received. Moreover, "conformant motions" are used where the robot forces itself and its environment to a desired state without sensory feedback [24].

Motion planning Figure 5 depicts the analytical inverse kinematics and the steps during the manipulation of a cinder block to be used as a fulcrum. We use a predetermined grasp location, the top surface with the holes at the sides (see Fig. 5a). The planner first uses analytical inverse kinematics to find configurations close to the object that are collision-free. Figure 5a displays three configurations out of which the left most, semi-transparent one collides with a wheel. Once a goal in arm jointspace is determined, bidirectional RRTs with path shortening and smoothing are used to move the arm from its initial pose to the goal. Figure 5b demonstrates the keyframes as the

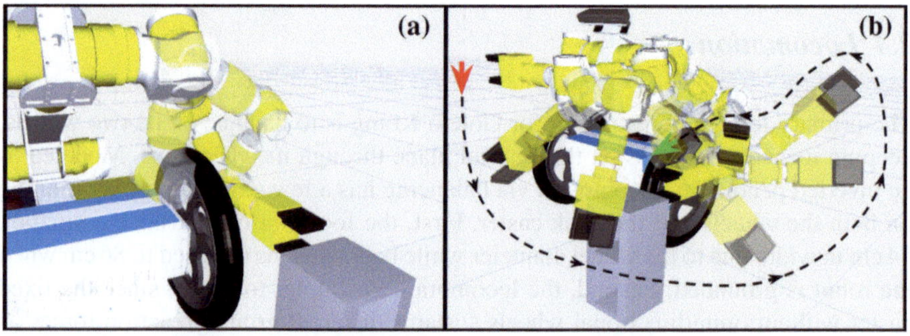

Fig. 5 *Left* Candidate grasp poses for the block—left most in collision with wheel. *Right* RRT trajectory to goal grasp pose, moving around the block to avoid collisions

arm moves from its initial pose (red), around the cinder block to avoid collisions, until it reaches the goal pose (green) in front of the grasp point.

Guarded and conformant motions Once in position for grasping, Golem Krang uses force/torque feedback at the end-effectors to reach out to the cinder block until contact and ensures its grippers can grasp it. Such guarded moves have proven to be simple, heuristic alternatives to visual servoing as the robot picks up levers and positions them on the fulcrum and inside the load. We also utilize conformant motions where, to localize the lever more precisely, the robot runs its wheels against the lever. When one of the wheels hit the obstacle first, the other wheel comes around until both wheels are in contact and any visual position error is removed. Such motions are used to eliminate uncertainty in the initial pose of objects in assembly tasks by pushing them into known poses [25].

3 Experiments

Golem Krang is tasked with overturning a 50 kg load using a lever-fulcrum assembly with a limit of 300 Nm on the force it can apply to the environment. Given the dimensions of the available objects, the robot has to design a structure, locate the components, position them and actuate the simple machine.

Placed in a random configuration in the room, Golem Krang begins by scanning the room for the available objects and finds the closest cinder block that would be used as a fulcrum (see Fig. 6). The scanning process is composed of a set of atomic behaviors which move its arms out of its sight to avoid occlusions. Once the fulcrum is located, the robot approaches until it positions itself in a predetermined distance to grasp the object. Using the motion planning tools, such as RRTs and guarded moves, the robot grasps the cinder block at its top.

An interesting observation is about how the location of a manipulated object and the uneven distribution of its weight over the wheels affect the locomotion accuracy. To minimize such an artifact, in Fig. 7, Golem Krang moves the grasped cinder block to the middle of its torso before turning around and localizing the load. Having

Fig. 6 Once Golem Krang detects the closest cinder block (*left*), it approaches (*middle*) and grasps the objects (*right*). Scene continues in Fig. 7

Fig. 7 Having grasped the fulcrum, the robot localizes the load and places the fulcrum in the initial design configuration. Scene continues in Fig. 8

Fig. 8 The lever is picked up by first using vision and then running the wheels against the object to make physical contact before manipulation. Scene continues in Fig. 9

detected the load, the final configuration of the fulcrum is deduced from the assembly design and the robot places it appropriately.

In the third part of the experiment, Golem Krang needs to detect and localize a candidate lever object and grasp it, as shown in Fig. 8. Given the size of the lever and the noisy perception data, we propose using the wheels to localize the lever object more accurately once the robot approaches it. Figure 8b displays the conformant behavior where the robot moves forward slowly to collide with the lever and have its localization error fixed. The left wheel first makes contact and the contact overcomes the input torque, while the right wheel keeps moving until the robot is parallel and directly in front of the lever.

To simplify the locomotion, we have assumed collision-free paths and when Golem Krang carries the lever, we ensure that the lever is carried high enough that it does not collide with other objects (see Fig. 9). Once the robot repositions itself in front of the fulcrum, using guarded moves, the robot first pushes the lever against

Fig. 9 Golem Krang places the lever in the planned pose and overturns the 50 kg load

the load horizontally to ensure it is at the correct distance and then tilts it until the design angle. When the lever reaches the goal pose, it is released so that it slides on the fulcrum into the load. Finally, the robot pushes the lever at the desired contact point and overturns the load.

4 Results

In this section, we provide experimental results on the functionality of the realized simple machines under different conditions, study the sources of inaccuracies in the assembly process and provide some insights on the autonomous manipulation of heavy objects.

Designs with Different Load Weights The proposed framework is tested with two load sizes, 50 and 100 kg, where the planner chooses from two levers 1.7 and 2.5 m long. Once the design is made, Golem Krang autonomously constructs it, as described in Sect. 3, and we measure the force/torque sensor readings at the actuating arm gripper. Figure 10a demonstrates the maximum applied forces at 242.27 and 368.63 Nm for 50 and 100 kg experiments respectively. We conclude that the mechanical advantage is 2.02:1 and 2.66:1 for each case, and the maximum force limit is preserved; and thus, the planned design, its assembly and actuation are successful.

Fulcrum Base Choice and Unmodeled Implications The proposed planner focuses on the initial moment of force application and ensures that the force induces enough torque to overturn the load. Although constraints are used to guarantee that the motion is collision-free, the possibility of the the contact between the lever and the cinder block moving is not considered. Figure 11a demonstrates a case where the planner *autonomously* adapts the fulcrum base choice to accommodate the short

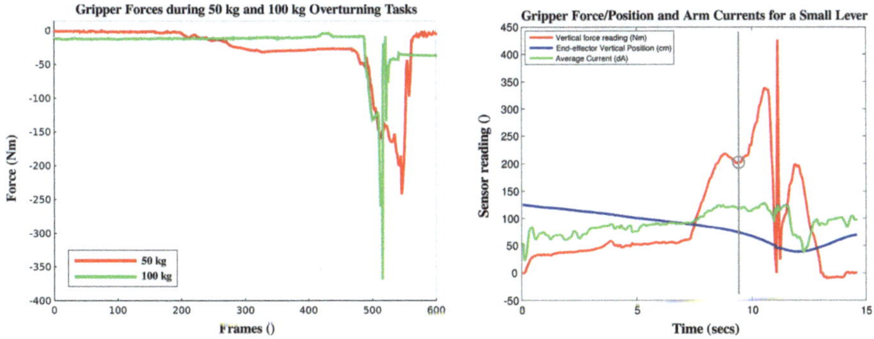

Fig. 10 *Left* Input forces during the 50 kg (*red*) and 100 kg (*green*) overturning tasks. *Right* Fulcrum shift while using a small lever leads to increased input force (*grey line*)

Fig. 11 The movement of the fulcrum-lever contact point for short lever objects

lever. However, the fulcrum point moves (Fig. 11b, c) because the contact height is shorter than the center of mass of the load. Figure 10b displays the increase in the required force as the fulcrum shifts.

Evaluating Accuracy A challenge in long sequential tasks is the accumulation of error over the course of the executions. To remedy such an error build-up, at the current state of our work, the authors intervene and instruct the robot to minimize its error if necessary. A cause of this inaccuracy is the reliability of the RGBD data. We observe a mean error of 2–3 cm within a 1 m bound increases up to 10 cm with 4 m. The increase in error with distance suggests a camera calibration problem which can be addressed with bundle adjustment [26].

Secondly, locomotion for heavy balancing robots is an active research area and our controller can be improved with [27]. A major challenge is the system modeling, specifically accurate center of mass and inertia information. Figure 12 demonstrates the behavior of the robot as it follows a velocity profile, moving straight. The sinusoidal behavior in the output position trajectory (red) is the robot regaining its

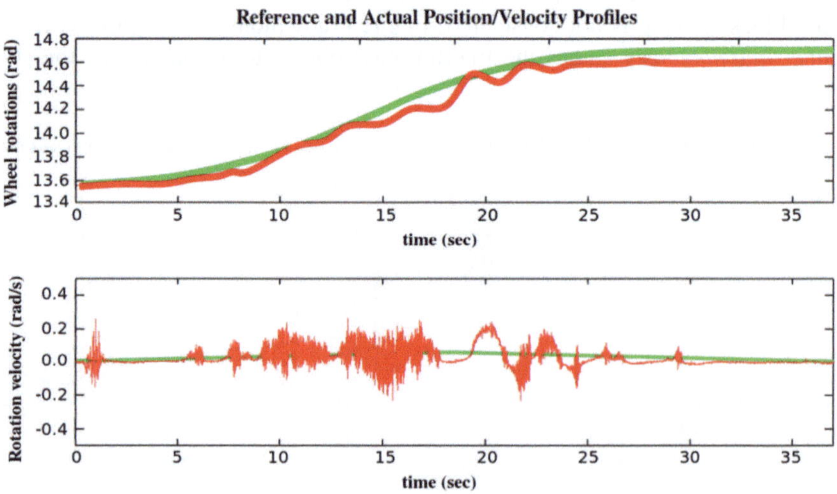

Fig. 12 The oscillations and steady state error in following a reference velocity profile

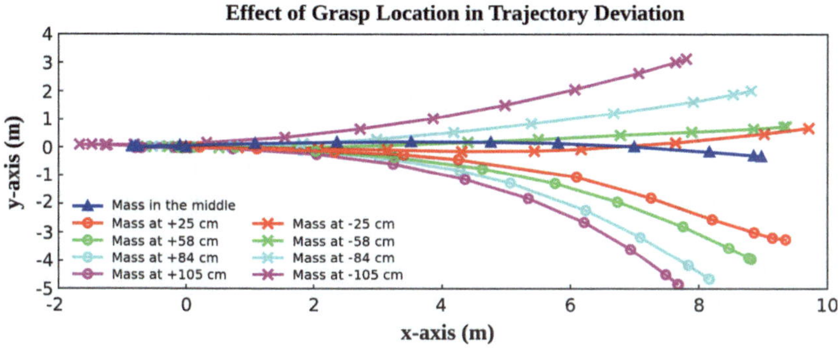

Fig. 13 The effect of the grasping location on the accuracy of locomotion

balance and the steady state error may be accommodated with an integral term in the controller. A concern is the high-frequency velocity sinusoids which may cause instability when heavy objects are manipulated.

Manipulation of Heavy Objects The manipulation of heavy objects brings about interesting challenges in manipulation, locomotion and control fields. First of all, even though the mass and the center of mass information of a carried load is sufficient in maintaining quasi-static balance, the inertias are needed when motion becomes faster. Although most of the time, the dynamic parameters of manipulated objects are not readily available, there are methods to estimate them with experiments [28] or with assumptions based on geometry and mass.

Secondly, we observe that the configuration of the grasped object with respect to the body frame can affect the accuracy of locomotion. For instance, carrying a 15 kg load in the middle of the two wheels as opposed to a meter away induces different trajectories for the same velocity profiles. Figure 13 shows the trajectories as the robot moves a cinder block from one side to another and attempts to follow a straight line. The results show a proportional relationship where trajectory errors increase with the distances between the load and the robot center. We speculate that the cause is that the wheel with more weight has to apply more torque to keep up. Note that the distribution of the load and its effect on the joint torques is a general issue that also affects legged humanoid robots.

Lastly, the mass and the sheer size of the manipulated objects can cause problems with robust grasping during locomotion. For instance, for Golem Krang's grippers with two prismatic finger joints, it is important to guarantee that an object would not slip from its grasp as the robot turns around swiftly. To handle such cases, we have adopted a two-hand grasp for the cinder block and ensured that the lever is held mostly from its middle region and at an upwards angle.

5 Discussion and Future Work

In this work, we present a framework where an autonomous agent, faced with a physically challenging task can perceive its environment, designs a multi-object manipulation solution and implements it. In addition to confirming the theory of autonomous design, we have the following lessons and new research directions.

Lessons learned First, the incorporation of visual servoing in manipulation and locomotion tasks is a crucial step towards more robust and accurate autonomous behavior. Although haptics-based and conformant actions are clearly effective in understanding and shaping the environment, visual data would result in more accurate and efficient executions.

Secondly, for locomotion with heavy loads, the compensation of the load weight distribution may be a significant motion planning and control problem as humanoid robots are involved in more physically straining tasks.

Third, the proposed constraint planner outputs design configurations to an accuracy of millimeters although it is clear that mobile manipulators can barely work reliably in centimeter or decimeter scale. The incorporation of the manipulator precision in the constraint optimization framework, and perhaps relaxing some of the design constraints, can lead to more efficient approaches.

Positive results To ensure the output designs are realistic, the authors had to incorporate a few buffer constraints in the planner that would allow room for error in manipulation. The results show that such leniency in the design process allows for replanning as the error builds-up in construction. In sum, the *planning in the constraint space* of the structures leads to functional simple machines.

Future work The results so far has been on the initial design where the focus is on the placement of the design components to overcome the physical constraints, with limited concern in the behavior of the structure as it is actuated. As the small lever example shows, motion planning is needed to guarantee that a design is buildable and functional, that is a planner should simulate its assembly and behavior until the successful completion of the task.

Figure 14 demonstrates a ramp structure composed of two inclined planes built for a ground vehicle to climb—a canonical simple machine example. The output is a

Fig. 14 Initial results on buildability and usability for ramp designs for ground vehicles

preliminary result where an initial design is computed with the proposed constraint planner and its buildability is checked with motion planning.

Lastly, we observe that the assembly of functional structures is closely related to the grasp problem where a manipulator chooses how to grasp the objects. In this work, we use fixed grasps for the fulcrums and the robot is limited to a single grasp orientation for the lever. Addressing the grasping problem is an important step towards generalizing the approach to a wider range of domains.

The videos are presented at: http://www.cc.gatech.edu/~cerdogan/iser2014.

References

1. Erdogan, C., Stilman, M.: Planning in constraint space: automated design of functional structures. In: ICRA (2013)
2. Erdogan, C., Stilman, M.: Incorporating kinodynamic constraints in automated design of simple machines. In: IROS (2014)
3. Vincent, V., Geffner, H.: Branching and pruning: an optimal temporal pocl planner based on constraint programming. In: AI (2006)
4. Yu, L.F., Yeung, S.K., Tang, C.K., Terzopoulos, D., Chan, T.F., Osher, S.: Make it home: automatic optimization of furniture arrangement. In: Siggraph (2011)
5. Beetz, M., Mosenlechner, L., Tenorth, M.: Crama cognitive robot abstract machine for everyday manipulation in human environments. In: IROS (2010)
6. Stilman, M., Kuffner, J.J.: Navigation among movable obstacles: real-time reasoning in complex environments. In: IJHR (2005)
7. Kemp, C.C., Edsinger, A., Torres-Jara, E.: Challenges for robot manipulation in human environments. IEEE Robot. Autom. Mag. **14**(1), 20 (2007)
8. Kuindersma, S.R., Hannigan, E., Ruiken, D., Grupen, R.A.: Dexterous mobility with the ubot-5 mobile manipulator. In: ICAR (2009)
9. Srinivasa, S.S., Ferguson, D., Helfrich, C.J., Berenson, D., Collet, A., Diankov, R., Gallagher, R., Hollinger, G., Kuffner, J., Weghe, M.V.: Herb: a home exploring robotic butler. Auton. Robots **28**(1), 5–20 (2010)
10. Nishiwaki, K., Sugihara, T., Kagami, S., Kanehiro, F., Inaba, M., Inoue, H.: Design and development of research platform for perception-action integration in humanoid robot: H6. In: IROS (2000)
11. Dasgupta, A., Nakamura, Y.: Making feasible walking motion of humanoid robots from human motion capture data. In: ICRA (1999)
12. Yamauchi, M.B.: Packbot: a versatile platform for military robotics. In: Defense and Security, pp. 228–237. International Society for Optics and Photonics (2004)
13. Saranli, U., Buehler, M., Koditschek, D.E.: Rhex: a simple and highly mobile hexapod robot. In: IJRR (2001)
14. Newell, A., Simon, H.: GPS, a program that simulates human thought. In: DTIC (1961)
15. McCarthy, J.: Programs with common sense. In: DTIC (1963)
16. Fulkerson, D.R.: A network flow computation for project cost curves. Manag. Sci. **7**(2), 167–178 (1961)
17. Taha, H.A.: Integer Programming: Theory, Applications, and Computations, vol. 975. Academic Press, New York (1975)
18. Fikes, R.E., Nilsson, N.J.: Strips: a new approach to the application of theorem proving to problem solving. Artif. Intell. **2**(3), 189–208 (1972)
19. Stilman, M., Olson, J., Gloss, W.: Golem krang: dynamically stable humanoid robot for mobile manipulation. In: ICRA (2010)

20. Kuffner, J.J., LaValle, S.M.: RRT-connect: an efficient approach to single-query path planning. In: ICRA. IEEE (2000)
21. Tolani, D., Goswami, A., Badler, N.I.: Real-time inverse kinematics techniques for anthropomorphic limbs. Graph. Models **62**(5), 353–388 (2000)
22. Whitney, D.E.: Resolved motion rate control of manipulators and human pros-theses. IEEE Trans. Man-Mach. Syst. (1969)
23. Bejczy, A.K.: Effect of hand-based sensors on manipulator control performance. Mech. Mach. Theory **12**(5), 547–567 (1977)
24. Goldman, R., Boddy, M.: Expressive planning and explicit knowledge. In: AIPS (1996)
25. Mason, M.T.: Mechanics and planning of manipulator pushing operations. Int. J. Robot. Res. **5**(3), 53–71 (1986)
26. Pradeep, V., Konolige, K., Berger, E.: Calibrating a multi-arm multi-sensor robot: a bundle adjustment approach. In: Experimental Robotics (1999)
27. Ha, Y., Yuta, S.: Trajectory tracking control for navigation of the inverse pendulum type self-contained mobile robot. Robot. Auton. Syst. (1996)
28. Saal, H.P., Ting, J.A., Vijayakumar, S.: Active estimation of object dynamics parameters with tactile sensors. In: IROS (2010)

An Experimental Protocol for Benchmarking Robotic Indoor Navigation

Christoph Sprunk, Jörg Röwekämper, Gershon Parent, Luciano Spinello, Gian Diego Tipaldi, Wolfram Burgard and Mihai Jalobeanu

Abstract Robot navigation is one of the most studied problems in robotics and the key capability for robot autonomy. Navigation techniques have become more and more reliable, but evaluation mainly focused on individual navigation components (i.e., mapping, localization, and planning) using datasets or simulations. The goal of this paper is to define an experimental protocol to evaluate the whole navigation system, deployed in a real environment. To ensure repeatability and reproducibility of experiments, our benchmark protocol provides detailed definitions and controls the environment dynamics. We define standardized environments and introduce the concept of a reference robot to allow comparison between different navigation systems at different experimentation sites. We present applications of our protocol in experiments in two different research groups, showing the usefulness of the benchmark.

Keywords Benchmark · Autonomous navigation · Indoor robots · Dynamic environments

1 Introduction

Robot navigation is a widely studied topic in robotics due to its cornerstone function for robot autonomy. Prior work on benchmarking robot navigation primarily focused on simultaneous localization and mapping (SLAM) techniques, and in particular on assessing the accuracy of the generated maps [4, 20]. These evaluations are useful when the robot task is to compute a precise map, e.g., for architectural or other surveying purposes. However, when the map is built for autonomous navigation, its

C. Sprunk, J. Röwekämper and G. Parent contributed equally to this work.

C. Sprunk (✉) · J. Röwekämper · L. Spinello · G.D. Tipaldi · W. Burgard
Department of Computer Science, University of Freiburg, Freiburg, Germany
e-mail: sprunkc@informatik.uni-freiburg.de

G. Parent · M. Jalobeanu
Microsoft Robotics, Microsoft Corporation, Washington, USA

M.A. Hsieh et al. (eds.), *Experimental Robotics*, Springer Tracts in Advanced Robotics 109, DOI 10.1007/978-3-319-23778-7_32

metric accuracy does not necessarily relate to the performance of the robot. A robot navigating in a real-world environment must be able to localize and reach destinations in environments that are populated with dynamic objects and that are changed with respect to the initial conditions. This includes environments shared with people or environments where objects may be moved around.

In this paper, we formulate an experimental protocol for benchmarking robot navigation. This fills the void of a missing evaluation method for repeatable, reproducible and comparable tests for autonomous indoor navigation consisting of performance metrics, methodology and baseline. We aim at accommodating for hardware differences between comparable solutions and for differences in sensors. In particular, we aim at reproducing identical environments, including environment dynamics between multiple runs at an experimentation site.

This paper represents the first time that navigation is quantified in a fashion similar to other hard sciences where environmental conditions are key for reproducibility and fair comparison. In other computer science disciplines, such as computer vision and machine learning, benchmarks had a large impact to standardize and to uniform evaluation procedures [1, 11]. Differently from these sciences, robot navigation cannot be evaluated only with datasets. The robot is immersed in the environment and interacts with it. For this reason, we provide to the community ways of measuring ground truth and suggest a reference robot.

In our benchmark, we aim to compute statistics about *a simulated year* of continuous robot operation. For this, we provide detailed definitions for the experimental environment and conditions. The experimental setup consists of definitions about the size, the dynamics, the environmental conditions and the overall duration of an experiment. This includes the number and the size of the rooms, the number of people walking in the scene, the kinds and amounts of objects and furniture that are moved and the number of goals for each environment. As reference robot, we selected the widespread commercial platform Pioneer P3-DX. We applied the benchmarking protocol to conduct experiments in two different research groups by using two different kinds of robots, showing the usefulness of the benchmark. The complete benchmark protocol along with detailed instructions and our evaluation software is publicly available at http://research.microsoft.com/brin/.

2 Related Work

Benchmarking plays an important role for comparison and evaluation in science. In particular, there are many benchmarking works in several fields related to robotics, including machine learning, computer vision and artificial intelligence. Machine learning is probably the field that received most attention, thanks to the use of very large evaluation datasets for different tasks [1, 2, 16]. Similarly, computer vision has many procedures and benchmarks available [6, 9, 11, 17].

Despite being one of the most studied field in robotics, there is a relatively small amount of literature related to benchmarking robot navigation. This is probably caused by the fact that robot navigation cannot be evaluated on a dataset. The robot navigates in a dynamic environment that is constantly changing. In NaviGates [14], the authors present an early benchmark for robot navigation. Here, they concentrate on robot skills and architecture but they do not take into account how to systematically evaluate the robot performance in a changing environment. Gutmann et al. [12] presented a set of extensive experiments evaluating the accuracy and robustness of localization systems using datasets. Calisi et al. [5] propose a benchmark framework that concentrates only on the evaluation of vehicle motion algorithms. Borenstein and Feng [3] introduce a method for measuring odometry errors of mobile robots. Specifically, it focuses on quantitative evaluation of systematic and non-systematic errors. The work of Nowack et al. [15] presents an investigation for an evaluation of two specific robot tasks, namely path planning and obstacle avoidance. In that work, the environment is considered static. Del Pobil et al. [8] and Dillmann et al. [10] survey efforts in quantification for a set of robot tasks, including robot cleaning, robot rescue and autonomous driving. Another way of evaluating navigation systems is to let them compete in a challenge such as the DARPA urban challenge [7]. However, such challenges typically require to transport all robot systems to one location and their outcome is rather a ranking of systems than an analysis.

3 Experimental Protocol

In this section, we provide a detailed description of the proposed experimental protocol. Further details beyond the presentation here are available at http://research.microsoft.com/brin/. The goal of the protocol is to evaluate and compare the performance of navigation systems (hardware and software) in real environments over long periods of time. In order to allow comparison between different navigation systems at different physical locations, we devise means for normalizing the performance across environments and platforms and take measures towards standardization and repeatability of evaluations.

First, we define a standard environment composed of four *areas*. Second, we define a set of *challenges* that the robot has to face. These challenges include changes in environment appearance, geometrical configuration, and dynamic obstacles. Third, we introduce the concept of a *reference robot* and a *reference navigation system* that will be identical across evaluation sites. Expressing the performance of the tested system relative to this reference system ensures comparability of results across robots and evaluation sites. Finally, we employ a vision-based *ground-truth system* to evaluate the navigation performance of both the test and the reference robot.

We propose to simulate an entire year of robot operation, defining 12 *loops*, each corresponding to a virtual month of operation. The experimenter defines 8 *way-points*, two for each area, and creates a route that visits all way-points and always changes areas between way-points. The task of the robot is to travel along this route in each loop, facing a different set of challenges for each loop.

Tables 1, 2 show an overview of the experimental protocol. The rows indicate the challenges, while the columns indicate their category, frequency, and configuration/location with respect to each of the twelve loops. In the remainder of this section, we will explain each element of our protocol in more detail.

3.1 Areas

We devised a standardized test environment consisting of four distinct areas: *atrium*, *lounge*, *office* and *hallway*. These areas are shown in the leftmost column of Tables 1, 2, grouping the challenges. The environment should contain at least one doorway and at least two different surfaces (e.g., carpet, tile, wood, cement). Ideally, the environment should not be a dedicated testing facility but rather a real building. Where possible, the test areas should be equipped with artificial lighting and with blinds or drapes to modify the environmental illumination.

The atrium is supposed to be a predominately open space with 90 % or more of its surface area clear of furniture with a recommended size of above 15 m × 15 m. The lounge is a social seating/dining area with an intended size of at least 12 m × 12 m. The office is densely occupied by desks, office chairs and shelves and has a recommended minimum size of 10 m × 10 m. The hallway has an intended length of at least 15 m and should have a low number of geometric and visual features. The above dimensions are recommendations, the experimenter is encouraged to respect the relative size of the areas in case of space limitations. Figures 2 and 4 show the real environments used in our experiments.

3.2 Challenges

We define a set of common environment dynamics, called challenges, to standardize the comparison with the reference robot and with tests conducted in different environments. Each challenge is listed as a numbered row in Tables 1, 2. The challenges are representative of events and dynamics that are highly likely to occur at least once over a year-long deployment of a robot in a typical indoor environment. They are divided into three main categories that are shown in Tables 1, 2 next to the challenge description:

Appearance (A): This category comprises visual appearance changes in the environment such as changing art work, whiteboard contents and lighting conditions. The challenges in this category are meant to test and assess the robustness of vision-based approaches.

Geometry (G): Challenges of this category include movable objects like doors, boxes, chairs, and ladders. These challenges simulate the natural variation of object configurations in environments and the different states of articulated objects

Table 1 The benchmark test grid proposed in this work

		Challenge	Cat.[a]	Freq.[b]	Month 1	Month 2	Month 3	Month 4	Month 5	Month 6
All areas	1	Artificial lighting	A	D	Off	On	On	Off	Off	On
	2	Lamps on/off	A	D	On	On	On	On	On	On
	3	Blinds or drapes open/closed	A	D	All Closed	All Open	50/50	All Closed	All Open	50/50
	4	Wall art changes	A	Y	Wall Art 1	–	–	–	–	–
	5	Door open/closed	G	H	Constantly	Constantly	Constantly	Constantly	Constantly	Constantly
	6	Wall color changes	A	Y	Color 1	–	–	–	–	–
Atrium	7	Large display monitors change content	A	D	Image 1	Image 2	Image 3	Image 1	Image 2	Image 3
	8	Person in path (ample room to avoid)	O	D	–	Position 1	Position2	Position 3	Position 4	–
	9	Small group in path (ample room to avoid)	O	D	–	Position 1	Position 2	Position 3	–	Position 1
	10	Person pushing cart (ample room to avoid)	O	D	–	Position 1	Position 2	–	Position 1	Position 2
Hallway	11	Shipping boxes on floor	G	D	–	1 Box	2 Boxes	3 Boxes	2 Boxes	1 Box
	12	Cart moves	G	D	Position 1	Position 2	Position 3	Position 4	Position 1	Position 2
	13	Ladders, tools, cables	G	Y	–	–	Position 1	–	–	–
	14	Two people blocking path (no room to avoid)	O	D	Position 1	–	–	–	Position 2	–
	15	Path completely blocked (door) for 1 min	O	D	–	Position 1	–	–	–	Position 2
	16	Path completely blocked (people) for 1 min	O	D	–		Position 1	–	–	–
	17	Person pushing cart (no room to avoid)	O	D	–	–	–	Position 1	–	–

(continued)

Table 1 (continued)

		Challenge	Cat.[a]	Freq.[b]	Month 1	Month 2	Month 3	Month 4	Month 5	Month 6
Lounge	18	Dining chairs shift	G	H	Neat	25 % Messy	50 % Messy	75 % Messy	100 % Messy	Neat
	19	Coats/jackets on coat racks	G	D	–	1/2 Full	Full	–	1/2 Full	Full
	20	Cart moves	G	D	Position 1	Position 2	Position 3	Position 4	Position 1	Position 2
	21	Caution sign (Janitor)	G	D	–	Position 1	Position 2	–	Position 2	Position 1
	22	Garbage/recycling bags	G	D	–	Black	White	–		2 Black
	23	Reconfigure furniture	G	Y	Configuration 1	–	–	–	–	–
	24	Person vacuuming or mopping	O	D	–	Position 1	Position 2	–	Position 1	Position 2
	25	Large work/social gathering (20–30 people)	O	M	–	–	Position 1	Position 1	–	–
Office	26	Whiteboard contents change	A	D	Clean	5 %	10 %	20 %	30 %	40 %
	27	Desk chairs shift (<1.5 m)	G	H	Neat	25 % Messy	50 % Messy	75 % Messy	100 % Messy	Neat
	28	Coats/jackets on chairs	G	D	–		5 %	–	10 %	2 %
	29	Bags on floor near desks	G	D	–	20 %	40 %	60 %	–	20 %
	30	Loose paper on floor	G	D	0 Pieces	5 Pieces	0 Pieces	5 Pieces	0 Pieces	5 Pieces
	31	Shelves contents change	G	M	20 % Full	–	40 % Full	–	60 % Full	–
	32	Shelves move	G	Y	Position 1	–	–	–	–	–
	33	Small gathering in work area (4–8 people)	O	D	–	Position 1	Position 2	–	Position 1	Position 2
	34	Social gathering (10–15 people)	O	M	–	–	Position 1	–	–	–

The table lists the configuration of each challenge for every loop of the benchmark, see Table 2 for the second part covering months/loops 7–12
[a]Challenge category. *A* Appearance, *G* geometry, *O* (moving) obstacle
[b]Frequency of occurence. *H* Hourly, *D* daily, *M* monthly, *Y* yearly

Table 2 Continuation of Table 1, the benchmark test grid proposed in this work

		Challenge	Cat.[a]	Freq.[b]	Month 7	Month 8	Month 9	Month 10	Month 11	Month 12
All areas	1	Artificial lighting	A	D	On	Off	Off	On	On	Off
	2	Lamps on/off	A	D	Off	Off	Off	Off	Off	Off
	3	Blinds or drapes open/closed	A	D	All closed	All Open	50/50	All closed	All Open	50/50
	4	Wall art changes	A	Y	Wall Art 2	–	–	–	–	–
	5	Door open/closed	G	H	Constantly	Constantly	Constantly	Constantly	Constantly	Constantly
	6	Wall color changes	A	Y	Color 2	–	–	–	–	–
Atrium	7	Large display monitors change content	A	D	Image 1	Image 2	Image 3	Image 1	Image 2	Image 3
	8	Person in path (ample room to avoid)	O	D	Position 1	Position 2	Position 3	Position 4	–	Position 1
	9	Small group in path (ample room to avoid)	O	D	Position 2	Position 3	–	Position 1	Position 2	Position 3
	10	Person pushing cart (ample room to avoid)	O	D	–	Position 1	Position 2	–	Position 1	Position 2
Hallway	11	Shipping boxes on floor	G	D	–	1 Box	2 Boxes	3 Boxes	2 Boxes	1 Box
	12	Cart moves	G	D	Position 3	Position 4	Position 1	Position 2	Position 3	Position 4
	13	Ladders, tools, cables	G	Y	–	Position 2	–	–	–	–
	14	Two people blocking path (no room to avoid)	O	D	–	–	Position 3	–	–	–
	15	Path completely blocked (door) for 1 min	O	D	–	–	–	Position 3	–	–
	16	Path completely blocked (people) for 1 min	O	D	Position 2	–	–	–	Position 3	–
	17	Person pushing cart (no room to avoid)	O	D	–	Position 1	–	–	–	Position 3

(continued)

Table 2 (continued)

		Challenge	Cat.[a]	Freq.[b]	Month 7	Month 8	Month 9	Month 10	Month 11	Month 12
Lounge	18	Dining chairs shift	G	H	25 % Messy	50 % Messy	75 % Messy	100 % Messy	Neat	25 % Messy
	19	Coats/jackets on coat racks	G	D	–	1/2 Full	Full	–	1/2 Full	Full
	20	Cart moves	G	D	Position 3	Position 4	Position 1	Position 2	Position 3	Position 4
	21	Caution sign (Janitor)	G	D	–	Position 1	Position 2	–	Position 2	Position 1
	22	Garbage/recycling bags	G	D	2 White	–	–	2 Black	2 White	–
	23	Reconfigure furniture	G	Y	Configuration 2	–	–	–	–	–
	24	Person vacuuming or mopping	O	D	–	Position 1	Position 2	–	Position 1	Position 2
	25	Large work/social gathering (20–30 people)	O	M	–	–	Position 2	Position 2	–	–
Office	26	Whiteboard contents change	A	D	50 %	60 %	70 %	80 %	90 %	100 %
	27	Desk chairs shift (<1.5 m)	G	H	25 % Messy	50 % Messy	75 % Messy	100 % Messy	Neat	25 % Messy
	28	Coats/jackets on chairs	G	D	20 %	4 %	30 %	6 %	40 %	8 %
	29	Bags on floor near desks	G	D	40 %	60 %	–	20 %	40 %	60 %
	30	Loose paper on floor	G	D	0 Pieces	5 Pieces	0 Pieces	5 Pieces	0 Pieces	5 Pieces
	31	Shelves contents change	G	M	20 % Full	–	40 % Full	–	60 % Full	–
	32	Shelves move	G	Y	Position 2	–	–	–	–	–
	33	Small gathering in work area (4–8 people)	O	D	–	Position 1	Position 2	–	Position 1	Position 2
	34	Social gathering (10–15 people)	O	M	–	–	–	Position 2	–	–

[a]Challenge category. *A* Appearance, *G* geometry, *O* (moving) obstacle
[b]Frequency of occurence. *H* Hourly, *D* daily, *M* monthly, *Y* yearly

such as doors. They test the robustness of navigation systems against geometry changes with respect to the setup and mapping phase. In addition to vision sensors, challenges in this category also affect proximity sensors.

Moving Obstacles (O): This category includes dynamic objects such as moving people, people transporting objects or gathering in groups, potentially (completely) blocking the path of the robot for an extended period of time. These challenges test the capabilities of a navigation system to deal with replanning while moving and to negotiate stalling situations.

All dynamic and moving elements have a designated frequency of occurrence and a designated location. The frequency can be hourly (H), daily (D), monthly (M) or yearly (Y) and is shown in the column next to the challenge category. The designated location/configuration of a challenge is shown in the respective column for each loop of the benchmark. If the navigation system of the robot does not rely on visual appearance (e.g., laser-based) one can skip the environment variations in the protocol that only affect visual appearance (category A).

3.3 Benchmark Test Grid

To ensure that the robot faces its challenges and the environment variations in a standardized and reproducible fashion, we devise a benchmark test grid that regulates the experimental evaluation. While the robot is traveling along its designated route, the environment is constantly modified according to the test grid shown in Tables 1, 2. The test grid contains instructions that describe the challenges the robot has to face. For each challenge, the table lists the specific configuration for each of the 12 benchmark loops.

The experimenter has to devise positions for the way-points 1–8. Then, the experimenter defines the order in which the robot has to visit the way-points, taking care to avoid traveling between two way-points in the same area. One complete visit of all way-points counts as one loop, or a benchmarking month, for the evaluation. With the knowledge of the robot's default path the experimenter is then able to provide meaningful positions for the generic configurations of challenges like "Two People Blocking Path (no room to avoid)" (line 14), or "Person in Path" (line 8). It also falls into the responsibility of the experimenter to concretely define configurations for the qualitative settings of the environment dynamics, e.g., a configuration change from "Neat" to "Messy" in an experiment script, see also Sect. 4. Additionally, the experimenter records the lengths of the default path segments of a loop for the evaluation.

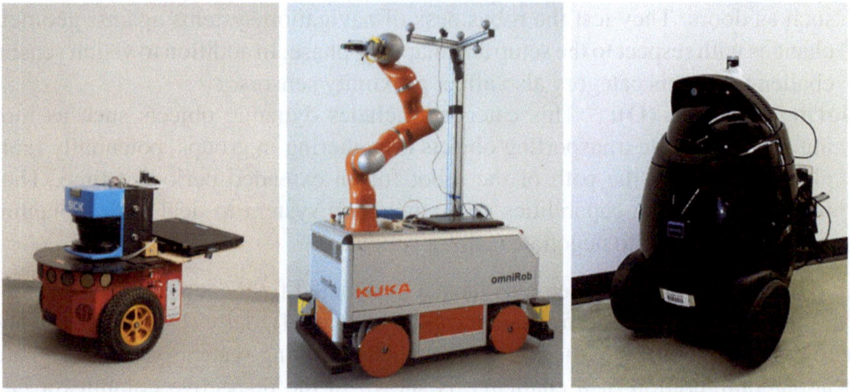

Fig. 1 Robots used in our experiments. All robots have an up-facing camera mounted for ground truth marker detection. *Left* The reference robot, a Pioneer P3-DX with a SICK LMS 200 laser scanner. *Middle* The omniRob used in the environment ALU-FR. *Right* The Microsoft Robotics Prototype 1 (MSR-P1), used in the environment MS

3.4 Reference Robot and Navigation System

For the baseline, we deploy the Pioneer P3-DX as *reference robot* in the same environment, running a *reference navigation software*. The software builds on the ARNL navigation stack shipped with the Pioneer, and is available at http://research.microsoft.com/brin/. We use ARNL 1.7.5.1 and BaseARNL 1.7.5.2 and change from the default values only the parameters *SecsToFail* to 90, *GoalOccupiedFailDistance* to 500 and *UseSonar* to "false".

The reference robot will visit the same way-points in the same order as the robot under evaluation. Thanks to the test grid introduced in the previous section it will also face the same challenges and configuration changes in a comparable manner. Figure 1 (left) shows one of the reference robots used in the experiments.

3.5 Ground-Truth Evaluation

We developed a cheap and affordable ground-truth system [13] to automatically detect if and when the robot has successful reached a way-point. The system consists of visual markers placed on the ceiling and an upward-pointing camera mounted on the robot. A dedicated software component, independent of the navigation system, is responsible for capturing the images from the camera at the way-points and for determining the positioning accuracy. It is available free of charge at http://research.microsoft.com/brin/. The system requires an initial calibration in which the user manually drives to the way-points and registers their position within the reference software. The visual markers are black-and-white checkerboards printed on foam-boards, and thus cheap and disposable, see Fig. 5. Whenever the robot reports an arrival at a way-point, the ground-truth system determines whether the way-point is reached, the accuracy with respect to the marker and the time elapsed from the last way-point.

We compute the following statistics: total number of failures, time to failure, distance to failure, average speed, accuracy at goal. The total number of failures is the number of segments in which the navigation system has been unable to arrive at a way-point. The time to failure is the operational time between consecutive failures, counted from the last restart to the last successfully visited way-point.

4 Experiments

We prepared two environments for the experiments. The first setup (environment ALU-FR) has been prepared in a large experimental area at the University of Freiburg, Germany. The second (environment MS) is a large, real office environment in the Microsoft Research building in Redmond, Washington, USA.

In the environment ALU-FR, we have benchmarked the navigation method proposed in [18, 19] installed on the omnidirectional robot omniRob shown in Fig. 1 (middle). In the environment MS, we evaluated an in-house experimental Microsoft navigation software, on the Microsoft Research Prototype 1 (MSR-P1) shown in Fig. 1 (right). The robot performs both SLAM and navigation by using only the Microsoft Kinect depth stream, gyroscope, and wheel odometry. In both environments, we have run the reference software on the reference platform Pioneer P3-DX, see Fig. 1 (left) and Sect. 3.4.

4.1 Environment ALU-FR

We furnished the environment to make each dedicated area verisimilar. This includes tables, cupboards, chairs, couches and computers. In particular, we have used wooden panels to subdivide the environment and fixed the fiducial markers at the way-points at a height of approximately 2.45 m. The complete environment measures 19 m × 12 m, the atrium 7.5 m × 11 m, the lounge 6 m × 9 m, the office 5.5 m × 12 m, and the hallway is 7 m long, see Fig. 2.

We instantiated the test grid from Tables 1, 2 into a concrete test script for our experiments. This is important to ensure that the test robot and the reference robot face the same challenges at the same time of each run. The laser-based occupancy grid map used for localization and navigation of the omniRob shown in Fig. 3 displays the eight way-points and some of the devised challenge positions. We specified a route by ordering the way-points as follows: 0→2→4→6→3→5→1→7→0. This order has succeeding way-points in different areas and the travel distance between

Fig. 2 Overall views of the ALU-FR environment: office (*top-left*), atrium (*top-right*) and detail views of the lounge (*bottom-left*) and the atrium (*bottom-right*)

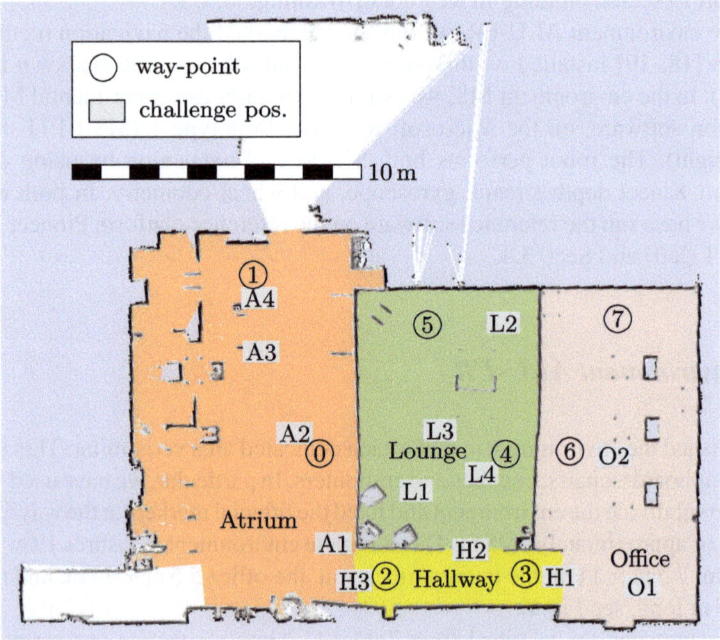

Fig. 3 The occupancy grid map used for the omniRob experiments in the ALU-FR environment. The four areas are marked by *color* and the map also shows the locations of way-points and some of the test grid challenges

way-points is varying from short to long. We devised positions for people to gather at and move to. Marking these positions on the floor is helpful for the participants during the experiments and to ensure repeatability.

Creating an experiment script from the test grid in Tables 1, 2 requires particular care on how to design the challenges and which of them can be omitted. The environment and the challenges have to be designed in a way that a path exists for the robot. As the omniRob is larger than the reference robot, we had to increase the size of doors and hallways. The navigation systems of the omniRob and the reference robot are not based on vision sensors but make only use of laser range finders. Therefore, we omitted challenges which have no or only minor effects on laser range finders such as changing artificial lighting, opening/closing blinds, wall art changes, wall color changes, and whiteboard content changes, i.e., lines 1–4, 6, 7, 26 from Tables 1, 2.

Furthermore, we did not put ladders, tools, cables and the cart in the hallway because of the omniRob's footprint and the particular manufacturing of its wheels (lines 12, 13). Due to the omniwheels of the omniRob, we skipped also the loose paper challenge (line 30). Moreover, we skipped the constant opening and closing of doors (line 5), the lounge coat racks (line 19), the janitor sign (line 21), modified the garbage bags to only be black (line 22) and limited the size of the biggest social gathering to 8 people (lines 25, 34).

The test grid only defines the challenges per loop but not at what time in the loop they occur. It is up to the experimenter to define when the robot faces the challenges in each loop. An excerpt of our experiment script is shown in Table 3. It shows all the travel segments for month/loop 3 of our test script that we derived from Table 1 and specifies which challenge configurations are applied for each loop segment. It is a detailed instruction procedure for the experimenter on how to modify the environment during the evaluation to ensure repeatability and reproducibility of the experiments: For example, while the robot travels between 2→4, it encounters two parcel boxes in the hallway and two people block the door H2 for 1 min. All chairs in the lounge are moved by 0.2 m with respect to their position while mapping the environment. The cart of the lounge is placed at L3 and one garbage bag was placed on the ground, see also Fig. 3.

4.2 Environment MS

The second environment consists of several areas of the Microsoft Research building 99 in Redmond, Washington, see Fig. 4. The atrium measures 25 m × 20 m, the lounge 20 m × 12 m, the office 10.5 m × 7.8 m and the hallway 17 m × 1.75 m. This environment includes an open floor plan in the atrium and lounge areas. It has substantial daylight coming in through the glass ceilings and the entrance. The lounge area includes a coffee shop, with multiple round tables and chairs, as well as tall rectangular tables with high chairs, couches and armchairs. The areas have carpet, linoleum, rough tile and hardwood as floor surface. Where practical, we chose the landmark locations close to interesting or meaningful locations when creating the test

Table 3 Excerpt of the instantiation of the test grid (see Tables 1, 2) to an evaluation script for loop 3 of environment ALU-FR

Loop segment	Area	Challenge/configuration	Row in Tables 1, 2
0→2	Atrium	Person with cart at A2	10
2→4	Hallway	2 boxes on the floor	11
	Hallway	2 people block at H2 for 1 min	16
	Lounge	move chairs by 0.2 m	18
	Lounge	cart at L3	20
	Lounge	1 garbage bag on the floor	22
4→6	Office	Move chairs by 0.2 m	27
	Office	1 jacket on chair	28
	Office	2 bags next to desks	29
	Office	Group of 4 people at O2	33
	Office	Shelves 40 % filled	31
6→3	–	–	–
3→5	Lounge	Group of 8 people at L1	25
	Lounge	Person vacuuming at L2	24
5→1	Atrium	Group of 4 people at A2	9
1→7	Atrium	Person at A2	8
	Office	Group of 8 people at O1	34
7→0	–	–	–

The specific challenges and their locations are shown for each segment of the loop, see also Fig. 3 for challenge locations

Fig. 4 The four areas of the environment MS: office (*top-left*), atrium (*top-right*), lounge (*bottom-left*) and hallway (*bottom-right*)

script for this environment, such as adjacent to the coffee stand, in front of the elevators and near the receptionist desk. The environment included a doorway between the hallway and the office as well as one additional doorway into an unmapped adjacent space that was alternately opened or closed for each loop. As we used a Microsoft Kinect depth sensor for mapping and navigation, we omitted the challenges involving lighting or appearance changes from the script, including lines 1–4, 6, 7, and 26 from Tables 1, 2. No shelf was available for the office, so we omitted challenges 31 and 32. Challenges 25 and 34 were omitted due to a lack of the required number of people. To avoid disturbances by direct sunlight or non-scripted interactions with people, we started the experiments in the evening.

4.3 Results

The performance of the different systems in the two environments is listed in Tables 4 and 5. The last column of each table shows the *relative* performance of a navigation system with respect to the reference one. Thanks to the benchmark protocol, it is now possible to say how accurate a system is with respect to a standardized baseline and environmental conditions. In environment ALU-FR, neither omniRob nor the

Table 4 Benchmark results in the environment ALU-FR

Performance	Freiburg	Reference	Ratio
Number of failures	0	0	–
Mean time to failure	–	–	–
Maximum time to failure	4343 s	5125 s	0.85
Mean distance to failure	–	–	–
Maximum distance to failure	1423 m	1349 m	1.05
Average speed	0.33 m/s	0.26 m/s	1.27
Positioning error	0.005 m ± 0.007 m	0.05 m ± 0.04 m	0.10

Table 5 Benchmark results in the environment MS

Performance	Microsoft	Reference	Ratio
Number of failures	5	9	0.56
Mean time to failure	2265 s	726 s	3.12
Maximum time to failure	5023 s	1971 s	2.55
Mean distance to failure	367 m	183 m	2.01
Maximum distance to failure	860 m	472 m	1.82
Average speed	0.16 m/s	0.25 m/s	0.64
Positioning error	0.23 m ± 0.2 m	0.22 m ± 0.1 m	1.05

reference system failed during the ∼1.5 km navigation length in circa 70 min. In environment ALU-FR the robots can always observe sufficient structure to properly localize.

In environment MS, the MSR-P1 and the reference system both encountered failures. The failures for the MSR-P1/reference robot were software problems (1/1), localization inaccuracies (3/1) and divergence (1/1), faulty obstacle perception (0/3), path oscillation for more than 5 min (0/1), not finding a path around a new obstacle (0/1) and not detecting a low obstacle (0/1). The benchmark revealed defects in several key areas of navigation including planning, localization, static and dynamic obstacle avoidance, reactive re-planning, remapping, and endurance, consistent with the limitations of each software. The experiments covered ∼2.1 km and took 6 h to conduct for each robot.

Three months prior to the experiments in environment MS, we conducted a stripped down version of the benchmark with older MSR-P1 software. We found that the MSR-P1 showed dramatic improvements (5 failures vs. 12) with respect to the pre-test, consistent with the improvements in navigation and mapping software done in the meantime. We also found that the reference system performed worse in the full benchmark (9 failures vs. 5). This before and after experiment confirms the benchmark's ability to expose the effects of both software and environmental changes.

We believe the results accurately reflect the capabilities and performance of all tested systems. In our observation this is primarily due to the wide coverage of possible failure modes. Moreover, the amount of challenges in our protocol seemed appropriate. The relatively small cumulative runtime seems sufficient to capture a good performance representation. However, as navigation systems get better, the total runtime might need to be increased.

5 Lessons Learned

Comparing autonomous navigation solutions according to their performance in real environments is an arduous task. During the process of setting up and performing the evaluation, we came across two aspects to be considered.

A first aspect is related to the comparison of different systems at different locations. The reference robot is instrumental in providing a sense of the complexity of each environment. However, one must consider that the shape and the size of the robot has a certain degree of influence on the results. The chosen benchmark targets navigation in office environments, thus slightly favoring small and circular robots. When the system under test differs from the reference robot in size, shape or even locomotion principles, the environment and the protocol should be slightly adapted to allow a fair comparison. This happened, for instance, when we evaluated the omniRob system, as described in Sect. 4.1.

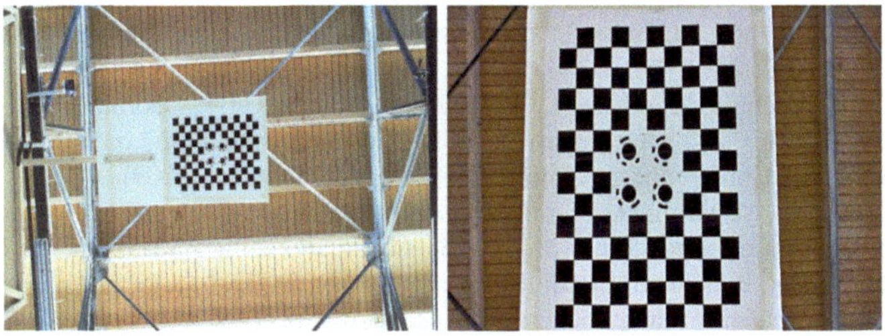

Fig. 5 Influence of camera mounting on marker detection tolerance. The pictures show the marker as seen from the camera for way-point 1 in environment ALU-FR. *Left* Reference robot P3-DX, the camera is mounted at a height of 0.45 m, see also Fig. 1 (left). *Right* Freiburg's omniRob, the camera is mounted 90° rotated with respect to the camera of the reference robot and at a height of 1.7 m, see also Fig. 1 (middle)

A second aspect lies in the fiducial system. The location of the camera on the robot is very important as the relative distance between the markers and the camera defines the *success* range for the failure detection system. A longer relative distance between them allows the marker to be detected from further away, see Fig. 5.

6 Conclusion

With this paper, for the first time, we have presented an experimental protocol to evaluate a robotic indoor navigation system as a whole. Differently from other scientific disciplines, robot navigation cannot be evaluated only with datasets. To ensure repeatability and reproducibility of experiments, our benchmark protocol provides detailed definitions for the environment dynamics. Additionally, we proposed the concept of a reference robot to allow comparison between different navigation systems at different experimentation sites. We applied our protocol and conducted experiments with different robots in two different research groups, showing the validity of the benchmark.

Acknowledgments This work has partly been supported by the EC under FP7-260026-TAPAS, FP7-610917-STAMINA, FP7-610603-EUROPA2, and FP7-267686-LIFENAV. The authors thank all members of the AIS Lab, the Microsoft Robotics Team, Studio99 and the Building 99 Hardware Lab for their patient help with the experiments.

References

1. Bache, K., Lichman, M.: UCI machine learning repository. University of California, Irvine (2013). http://archive.ics.uci.edu/ml
2. Bennett, J., Lanning, S.: The netflix prize. In: KDD Cup and Workshop at the 13th ACM SIGKDD International Conference on Knowledge Discovery and Data Minig (2007)

3. Borenstein, J., Feng, L.: Umbmark: a benchmark test for measuring odometry errors in mobile robots. Proc. SPIE **2591**, 113–124 (1995)
4. Burgard, W., Stachniss, C., Grisetti, G., Steder, B., Kümmerle, R., Dornhege, C., Ruhnke, M., Kleiner, A., Tardós, J.D.: A comparison of SLAM algorithms based on a graph of relations. In: International Conference on Intelligent Robots and Systems (2009)
5. Calisi, D., Iocchi, L., Nardi, D.: A unified benchmark framework for autonomous mobile robots and vehicles motion algorithms (MoVeMA benchmarks). In: RSS-Wksp. on Experimental Methodology and Benchmarking in Robotics Research (2008)
6. CAVIAR data sets. http://homepages.inf.ed.ac.uk/rbf/CAVIAR
7. DARPA urban challenge rules. http://archive.darpa.mil/grandchallenge/docs/Urban_Challenge_Rules_102707.pdf (2007)
8. Del Pobil, A.P., Madhavan, R., Messina, E.: Benchmarks in robotics research. In: IROS Workshop on Benchmarks in Robotics Research (2007)
9. Deng, J., Dong, W., Socher, R., Li, L.-J., Li, K., Fei-Fei, L.: Imagenet: a large-scale hierarchical image database. In: IEEE Conference on Computer Vision and Pattern Recognition (2009)
10. Dillmann, R.: Ka 1.10 benchmarks for robotics research (2004). http://www.cas.kth.se/euron/euron-deliverables/ka1-10-benchmarking.pdf
11. Everingham, M., Van Gool, L., Williams, C.K.I., Winn, J., Zisserman, A.: The pascal visual object classes (VOC) challenge. Int. J. Comput. Vis. **88**(2) (2010)
12. Gutmann, J.-S., Burgard, W., Fox, D., Konolige, K.: An experimental comparison of localization methods. In: International Conference on Robotics and Automation (1998)
13. Kikkeri, H., Parent, G., Jalobeanu, M., Birchfield, S.: An inexpensive methodology for evaluating the performance of a mobile robot navigation system. In: International Conference on Robotics and Automation (2014)
14. Knotts, R., Nourbakhsh, I., Morris, R.: Navigates: a benchmark for indoor navigation. In: International Conference and Experiments on Robotics for Challenging, Environments (1998)
15. Nowak, W., Zakharov, A., Blumenthal, S., Prassler, E.: Benchmarks for mobile manipulation and robust obstacle avoidance and navigation. BRICS Deliverable D3.1 (2010)
16. Pang, B., Lee, L.: Opinion mining and sentiment analysis. Found. Trends Inf. Retr. **2**(1–2), 1–135 (2008)
17. PETS 2009 data set. http://pets2009.net/
18. Röwekämper, J., Sprunk, C., Tipaldi, G., Stachniss, C., Pfaff, P., Burgard, W.: On the position accuracy of mobile robot localization based on particle filters combined with scan matching. In: International Conference on Intelligent Robots and Systems (2012)
19. Sprunk, C., Lau, B., Pfaff, P., Burgard, W.: Online generation of kinodynamic trajectories for non-circular omnidirectional robots. In: International Conference on Robotics and Automation (2011)
20. Sturm, J., Engelhard, N., Endres, F., Burgard, W., Cremers, D.: A benchmark for the evaluation of RGB-D SLAM systems. In: International Conference on Intelligent Robots and Systems (2012)

Part IX
Sensor Networks

Nikolaus Correll
University of Colorado, Boulder

Robotics enables sensor networks with mobility and autonomy. Mobility allows the sensor network to adapt to the environment and create rich representations using only a small number of—possibly expensive or bulky—sensors such as precision measurement equipment or high fidelity cameras. Autonomy enables applications in environments that thus far have been out of the reach of conventional sensor network deployments such as space, wilderness areas or water surfaces.

This chapter presents experimental advances on a diverse set of problems targeting applications from assembly of structures, entertainment, and environmental monitoring. In all cases robotics does not only contribute mobility and autonomy, but also leverages fundamental advances of the past decade such as Simultaneous Localization and Mapping (SLAM), stereo vision, and advanced manufacturing techniques for the sensor network community. These advances, together with an improved understanding of the challenges of long-term field deployment have the potential to make large scale deployment (and recuperation) of sensor networks economically viable.

The first paper "Precise Assembly of 3D Truss Structures Using EKF-based Error Prediction and Correction" by Erik Komendera and Nikolaus Correll combines high precision distance measurement systems that can be moved around within a truss structure during its construction with a SLAM approach to inform error correction in the system for the assembly of telescope optical benches in space. Extra information obtained during loop closure allows to significantly reduce the covariance matrix of the overall state estimate, which allows constructing the truss by adding to the nodes with the least uncertainty and correct for previous errors.

The second paper "Customized Sensing for Robot Swarms" by Dominic Jud, Javier Alonso-Mora, Jörn Rehder, Roland Siegwart and Paul Beardsley present a novel approach for the design of omnidirectional stereo camera pairs by 3D printing mirrors with application-specific shapes. They demonstrate their design approach on a swarm robotic platform geared toward entertainment and show how different areas of the 3D printed mirrors simultaneously provide up- and downward looking monocular as well as sideways stereo vision. While focusing on swarm robotics, the proposed approach has the potential to impact both static and flying sensor networks.

The third paper "Automatic Distribution of Disposable Self-Deploying Sensor Modules" by Paul Pounds, Timothy Potie, Farid Kendoul, Surya Singh, Raja Jurdak, and Jonathan Roberts reports on aerial deployment of a maple-seed inspired sensing platform that is made from a single flexible printed circuit board. The maple seed design allows this platform to land safely even when deployed from great heights, but poses major challenges in energy autonomy, localization and robustness to wind.

The last paper of this chapter "Towards Autonomous Lakeshore Monitoring" by Shane Griffith, Paul Drews and Cedric Pradalier overcomes a series of challenges in long-term autonomy and investigates algorithms for persistent visual monitoring by on-board SLAM and image stitching, enabling the acquisition of large-scale outdoor datasets. A key challenge here is to select features that are persistent despite daily lightning and seasonal foliage changes.

Precise Assembly of 3D Truss Structures Using EKF-Based Error Prediction and Correction

Erik Komendera and Nikolaus Correll

Abstract We describe a method to construct precise truss structures from non-precise commodity parts. Trusses with precision in the order of micrometers, such as the truss of a space telescope, can be accomplished with precisely machined truss connection systems. This approach is expensive, heavy, and prone to failure, e.g., when a single element is lost. In the past, we have proposed a novel concept in which non-precise commodity parts can be aligned using precise jigging robots and then welded in place. Even when using highly precise sensors and actuators, this approach can still lead to errors due to thermal expansion and structural deformation. In this paper, we describe and experimentally evaluate an EKF-based SLAM approach that allows a team of intelligent precision jigging robots (IPJR) to maintain a common estimate of the structure's pose, improve this estimate during loop closures in the construction process, and uses this estimate to correct for errors during construction. We also show that attaching a new node to the assembly site with the lowest uncertainty significantly increases accuracy.

1 Introduction

While high-precision assembly and welding have long been staples of factory automation, precise assembly in the field remains a difficult challenge. Industrial robots such as pick-and-place machines can achieve high precision with programmed motions, and are a critical component of modern industry, but do not function outside of their carefully controlled workspaces. Research into field assembly often focuses on self-correcting, interlocking components, freeing the robots from the task of ensuring accuracy, but this does not address the use of raw materials and on-line adjustments.

E. Komendera · N. Correll (✉)
Department of Computer Science, University of Colorado at Boulder, 430 UCB, Boulder, CO, USA
e-mail: ncorrell@colorado.edu

E. Komendera
e-mail: erik.komendera@colorado.edu

M.A. Hsieh et al. (eds.), *Experimental Robotics*, Springer Tracts in Advanced Robotics 109, DOI 10.1007/978-3-319-23778-7_33

Large scale construction projects do not rely solely on self-correcting, interlocking components, and rigid bodies are not valid assumptions. Instead, it is common to cut or bend parts from stock materials as needed, often incorporating the state of the structure (including any position errors). Likewise, welding and other thermal processes induce stresses on structures, which change the outcome slightly. If robots assembling the structure were to assume rigid bodies, eventually error would accumulate and lead to geometrical issues that cannot be overcome without significant repairs, including inducing forces on the structure to "jam" in parts, or disassembling and trying again.

Assembling structures in space [1] has the potential to overcome payload limitations of earth-based missions, thereby enabling the manufacturing of scalable structures. Space telescopes that could be assembled on orbit are a high priority for NASA [2], with proposed diameters of tens of meters up to hundreds of meters; in contrast, the James Webb Space Telescope, at 6.5 m, is at the upper limit for deployable telescopes [3]. Many of the benefits and techniques for assembling large space observatories are discussed in [4]. One recent robotic telescope assembly experiment [5] demonstrated the repeated assembly and disassembly of an 8 m telescope mirror, composed of 102 precisely machined truss members and 12 hexagonal panels, using an industrial manipulator arm and a rotating assembly platform. However, this approach for assembling large-scale telescope truss structures and systems in space have been perceived as very costly because they require high precision and custom components that rely on mechanical connections, increased launch mass, and the potential for critical failure if only one of the components is missing or defective [3].

This paper introduces a new prototype of the Intelligent Precision Jigging Robot (IPJR), for use with three dimensional truss structures such as shown in Fig. 1, extending on our previous work on 2D IPJRs [6, 7]. IPJRs work in groups to assemble trusses one cell at a time: each IPJR rests on a single strut between the structure and the new node, and adjusts the length of the strut until attachment is ready to be performed.

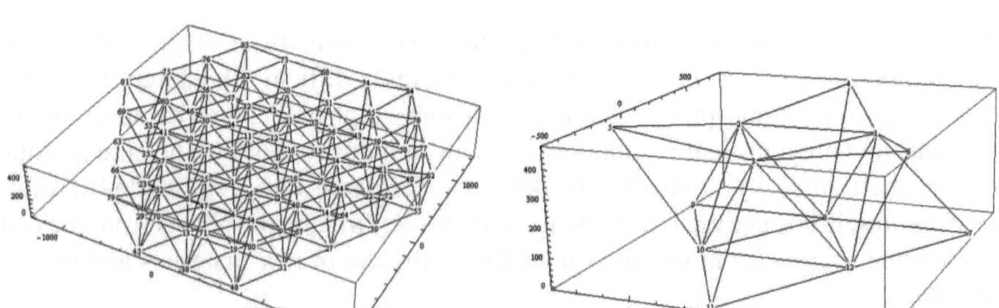

Fig. 1 *Left* The full telescope assembled in [2]. *Right* The innermost 12 nodes and 30 struts of the same telescope, with the top surface scaled by a factor of 4 to emphasize the curved surface. Both structures were considered in simulation, and the smaller one in physical experiments

Although the 2D IPJRs were able to assemble structures with accuracies that far exceeded that of the materials and processes used (glue gun and wooden dowels in [6] and spot welding of titanium rods facilitated by a large-scale robotic manipulator [7]), a fundamental problem in the IPJR approach are bias due to thermal expansion or structural deformation.

This paper also introduces an assembly algorithm based on the Extended Kalman Filter (EKF) and simultaneous localization and mapping (SLAM). Here, the EKF allows us to maintain and update a probabilistic state estimate (assuming a Gaussian distribution of the error), whereas SLAM allows us to take advantage of loop closures in the construction process to update the state of the entire structure. We can then use this improved estimate to change the build path during the construction process to minimize the expected variance as well as correct for errors by adjusting the length of future struts being placed.

To test the EKF-based algorithm, we perform several simulations and physical experiments on a telescope truss made of stock aluminum tubes, using IPJRs made from off-the-shelf and laser-cut parts. We show that the algorithm can detect and overcome hysteresis, bending and inconsistent IPJR robots by simulating the assembly process with an artificial bias. We compare the algorithm to a version without EKF in simulation, and assemble the structure with the EKF algorithm in a physical experiment. We show that the EKF assembly algorithm is a significant improvement over the assumption of zero-mean error, i.e., open-loop assembly of a structure.

2 Related Work

Algorithms for mechanical assembly planning (via disassembly sequences) for problems in well-known environments such as assembly lines, and with few assembly robots, were explored in the 80s and 90s, resulting in algorithms for finding fully-ordered sequences [8–11], or using opportunistic assembly planning [12] when necessary. All of these approaches attempted to find a suitable order for assembly that avoids construction deadlock, and assumed rigid bodies. In reality, structural forces and minor assembly errors often lead to situations where parts of a structure must be forced open to jam a new component in. Such algorithms do not consider these situations.

More recently, distributed field assembly by robots has been explored in large-scale assembly tasks. Three different robots are shown to successfully dock a part to an assembly [13]. An experiment performed at NASA's Jet Propulsion Laboratory demonstrated the precision assembly of beams by a pair of cooperative robots using highly rigid motions to ensure precision [14]. A robust algorithm is described in [15], which uses teams of robots to assemble structures while handling exceptions due to a wide range of failures, and relying only on a human operator when failures are beyond the scope of the assembly robots. Taking into account physical constraints such as structural stability and material properties into the build order has been shown in [16].

Recent robust and parallel assembly techniques include quadrotor teams that can assemble cubic truss structures [17], truss climbing and assembling robots [18], termite-inspired swarm assembly robots [19], and a robot team that can build IKEA furniture in cluttered environments [20]. Mobile assemblers [21] used visual feedback to estimate and correct errors in assembly. Many of these methods rely on self-correcting, interlocking mechanisms, which would add extra mass and expense due to machining requirements, and none consider welding or cutting.

3 Motivating Example and Experimental Setup

We are using the 84-node, 315-strut telescope truss described in [2] as the motivating example throughout the paper. We simulate its assembly, scaled by 1/6 and flattened on the bottom surface, as well as a 12-node, 30-strut subset of the structure with the top surface curvature scaled by a factor of 4, which we also assemble physically. Schematics are shown in Fig. 1. Our performance metric is the average error of the node positions in the coordinate frame defined by an origin node, an X-axis node, and an XY-plane node.

This experiment considers the assembly of truss structures one node at a time, starting from the predefined origin node, and working incrementally until the entire structure is finished. For each additional node added, at least as many struts as there are nodal degrees of freedom must be added simultaneously. In three dimensions, a minimum of three IPJRs are required to guarantee a unique node position, with the exception of grounded nodes, which require only two when the z-position is fixed.

In order to forgo premade accurate building materials, we chose one-size-fits-all telescoping aluminum rods (30″ long, 1/4″ and 7/32″ diameter) that can be locked in place with a shaft collar and that connect to the node balls via neodymium magnets. These design choices make assemblies temporary and allow us to reuse materials in different experiments. Each IPJR is designed to adjust the lengths of the telescoping struts by attaching to them during the assembly of a new part of the structure. In this paper, node balls are ball joints: the struts endpoints are free to slide around the surfaces of node balls.

The node-to-node nominal distances for the chosen structure range from 504.6 to 574.3 mm. A close-up image of a node ball with struts in the completed structure is shown in Fig. 2. We acquired 18 node balls and 51 struts, and used 12 node balls and 30 struts for the physical experiments.

We built five IPJRs for use in this experiment. Each IPJR, shown in Fig. 2, is an autonomous robot, consisting of a Raspberry Pi Model B for high level algorithmic control and communications, an Arduino for actuation and sensing, a motor driver board, and an Edimax WiFi dongle. The principal actuator is a Firgelli L-16 linear actuator with 140 mm extension, with advertised 0.5 mm accuracy, and potentiometer length feedback discretized to 140/1024 steps. To attach to both ends of a strut, each IPJR has two shaft collars. All components are fixed to a frame consisting of laser

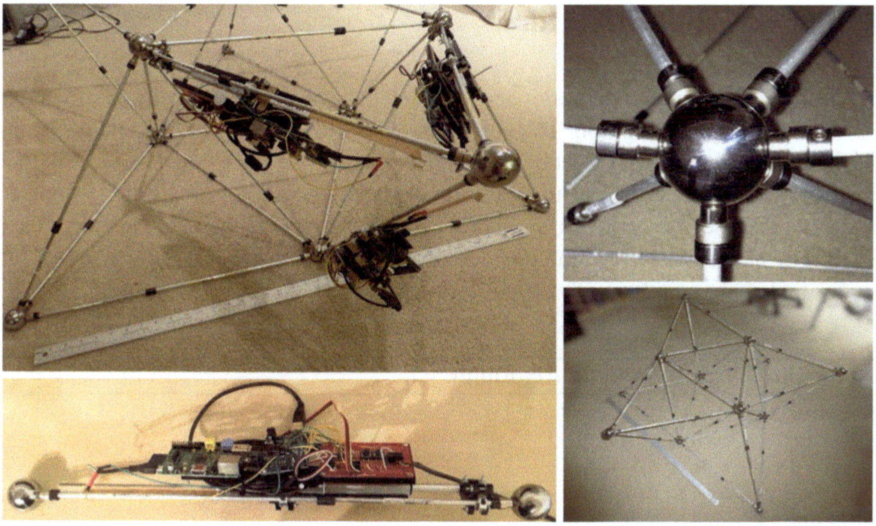

Fig. 2 *Top left* three IPJRs attached to a small truss. *Top right* close up of a node, showing the strut-node magnet interface. *Bottom left* an IPJR prototype attached to a strut. *Bottom right* a completed small truss

cut, $1/4''$ acrylic sheets. Each IPJR was powered by a power supply capable of 5 and 12 V output for the electronics and motors, respectively.

While each IPJR is fully capable of running our implementation of the EKF assembly algorithm as-is, we chose to run the algorithm from a central PC to avoid communication challenges. The Raspberry Pis instead run an HTTP server, which receives commands in the form of GET requests, and returns the data in response. Commands used in this experiment include checking the length potentiometer voltage, commanding the IPJR to move until it stops near a commanded location (without further control), and checking to see if the IPJR has finished moving. The control PC executes the experiment using a version of the EKF assembly algorithm implemented in Mathematica, which controls and monitors the IPJRs through GET requests.

For this experiment, a human external manipulator is required to attach the IPJRs to struts and to place each new cell in a coarse configuration. Once placed, the IPJRs refine the cell by taking measurements of the distances between node balls. The EKF algorithm requires direct measurements of the structure, but the IPJRs lacked that capability for these experiments. Instead, a ruler was used in order to obtain an unbiased measurement, not subject to bending due to the weight of the IPJR or play in the shaft collar attachment mechanism. The ruler was also used for verification of the physical experiments using a maximum likelihood estimator.

4 EKF Assembly Algorithm

A truss structure consists of nodes and struts that can be mathematically abstracted to vertices and edges of a graph. The geometry of the truss is fully defined by the graph topology and the length of each individual edge. As the truss consists of triangles, this information is sufficient to unambiguously define all angles in the system. In order to construct a specific truss, we therefore need to precisely set the length of each strut. The assembly algorithm considers the assembly of the structure by attaching nodes that share an edge with nodes already attached.

Algorithm 1 EKF Assembly Algorithm.

Require: $\bar{\mathbf{X}}, S, \Delta, \sigma_l, \sigma_c, ChooseSite \in \{MinTrace, MaxTrace, Random\}$
  $\hat{\mathbf{X}} = \mathbf{0}_{3nx3n}, \mathbf{P} = \mathbf{0}_{3nx3n}, \mathbf{A} = \{0\}$
  while $||\mathbf{A}|| < n$ **do**
    $f = ChooseSite(\hat{\mathbf{X}}, \mathbf{P}, S, \mathbf{A})$
    while $||\bar{\mathbf{x}}_f - \hat{\mathbf{x}}|| > \Delta$ **do**
      $\bar{\mathbf{c}}_f = ||\bar{\mathbf{x}_f} - \hat{\mathbf{x}}_i|| \forall i \in \mathbf{J}_f$
      Command all IPJRs with lengths $\bar{\mathbf{c}}_f$
      Collect measurements $\mathbf{y}$
      $\hat{\mathbf{X}}, \mathbf{P} = EKF(\hat{\mathbf{X}}, \mathbf{P}, \mathbf{y}, \bar{\mathbf{c}}_f, \sigma_l, \sigma_c)$
    end while
    $\mathbf{A} = \mathbf{A} \cup \{f\}$
  end while
  return $\hat{\mathbf{X}}, \mathbf{P}$

Algorithm 2 Extended Kalman Filter.

Require: $\hat{\mathbf{X}}, \mathbf{P}, \mathbf{y}, \bar{\mathbf{c}}_f, \sigma_l, \sigma_c$
  $\hat{\mathbf{X}}^P = f(\hat{\mathbf{X}}, \mathbf{0})$
  $\mathbf{F} = \left.\frac{\delta f(\mathbf{X},\mathbf{w})}{\delta \mathbf{X}}\right|_{\mathbf{X}=\hat{\mathbf{X}},\mathbf{w}=\mathbf{0}}$
  $\mathbf{G} = \left.\frac{\delta f(\mathbf{X},\mathbf{w})}{\delta \mathbf{w}}\right|_{\mathbf{X}=\hat{\mathbf{X}},\mathbf{w}=\mathbf{0}}$
  $\mathbf{P}^P = \mathbf{F}\mathbf{P}\mathbf{F}^T + \mathbf{G}(\mathbf{I}\sigma_c)\mathbf{G}^T$
  $\mathbf{H} = \left.\frac{\delta g(\mathbf{X},\mathbf{v})}{\delta \mathbf{X}}\right|_{\mathbf{X}=\hat{\mathbf{X}}^P,\mathbf{v}=\mathbf{0}}$
  $\mathbf{U} = \left.\frac{\delta g(\mathbf{X},\mathbf{v})}{\delta \mathbf{v}}\right|_{\mathbf{X}=\hat{\mathbf{X}}^P,\mathbf{v}=\mathbf{0}}$
  $\mathbf{K} = \mathbf{P}^P\mathbf{H}^T(\mathbf{H}\mathbf{P}^P\mathbf{H}^T + \mathbf{U}(\mathbf{I}\sigma_l)\mathbf{U}^T)^{-1}$
  $\hat{\mathbf{X}} = \hat{\mathbf{X}}^P + \mathbf{K}(\mathbf{y} - g(\hat{\mathbf{X}}^P, \mathbf{0}))$
  $\mathbf{P} = (\mathbf{I} - \mathbf{K}\mathbf{H})\mathbf{P}^P$
  return $\hat{\mathbf{X}}, \mathbf{P}$

The assembly algorithm (Algorithm 1) takes as input the nominal structure $\bar{\mathbf{X}}$ and maintains its estimated state in vector $\hat{\mathbf{X}}$. Both $\bar{\mathbf{X}}$ and $\hat{\mathbf{X}}$ are $3n$-length vectors of node positions $\mathbf{x_i} = \{x_i, y_i, z_i\}$. The state vector does not have an external reference for the coordinate system; instead, the coordinate frame is a function of the first three nodes $\mathbf{x_1} = \{0, 0, 0\}$, $\mathbf{x_2} = \{x_2, 0, 0\}$, and $\mathbf{x_3} = \{x_3, y_3, 0\}$, defining the xy-plane to be the triangle formed by these nodes, with node 1 at the origin and node 2 on the x-axis. The origin node is placed first. With the exception of the first three nodes, the ordering of the nodes in $\bar{\mathbf{X}}$ and $\hat{\mathbf{X}}$ is independent of the assembly order. The set of strut edges is defined by S, and indexes into the ordering of $\bar{\mathbf{X}}$ and $\hat{\mathbf{X}}$. Δ is the position distance tolerance. σ_l is a vector of edge measurement variances for each IPJR. σ_c is a vector of edge length actuation variances for each IPJR.

The function *ChooseSite* which chooses the next node to attach using a heuristic and the EKF algorithm are described in Sects. 4.1 and 4.2, respectively.

The algorithm initializes the structure state estimate $\hat{\mathbf{X}}$ and the $3n$-square covariance matrix $\mathbf{P}$ to zero, and the set of assembled nodes to be the origin node, $\mathbf{A} = \{0\}$. Then, until the size of $\mathbf{A}$ is equal to n:

- The algorithm chooses which node f to attach next, which is described in Sect. 4.1.
- The connecting nodes in A are mapped to the IPJR order in a vector $\mathbf{J}_f$; IPJR 1, for example, will connect f to the first node in $\mathbf{J}_f$. The commanded lengths $\bar{\mathbf{c}}_f$ are calculated to be the norm of the desired position for node f minus the estimated position of node i for all i in $\mathbf{J}_f$.
- The IPJRs are commanded to extend or contract to the lengths in $\bar{\mathbf{c}}_f$, with added (hidden) process noise $\mathbf{w} = N(0, \sigma_c)$
- Measurements of the IPJRs are collected in the same order, and assigned to measurement vector $\mathbf{y}$. Each measurement is of the form $\mathbf{y} = ||\mathbf{x}_i - \mathbf{x}_f||\forall i \in \mathbf{J}_f$ with added measurement noise $\mathbf{v} = N(0, \sigma_l)$.
- The Extended Kalman Filter is executed to update the current state estimate $\bar{\mathbf{X}}$ and covariance $\mathbf{P}$, with measurements $\mathbf{y}$, commands $\bar{\mathbf{c}}_f$, process variance σ_c, and measurement variance σ_l.
- If the estimated position of the new node $\hat{\mathbf{x}}_f$ is within Δ from the desired state $\bar{\mathbf{x}}_f$, command the IPJRs again, else place f into $\mathbf{A}$ and move on to the next node.
- When the assembly is complete, return $\bar{\mathbf{X}}$ and $\mathbf{P}$.

4.1 Choosing the Next Site

The function *ChooseSite* can be one of three choices: the minimum trace heuristic (2), the maximum trace heuristic (2), or the random heuristic (3):

$$
\begin{aligned}
f &= \underset{i}{\operatorname{argmin}}(Tr(\mathbf{P}_{\mathbf{J}_i})) && (1)\\
&\quad \forall i \in \{i|\exists a_1, a_2, a_3 \in \mathbf{A} \wedge (a_1, i) \in S \wedge (a_2, i) \in S \wedge (a_3, i) \in S\} \\
f &= \underset{i}{\operatorname{argmax}}(Tr(\mathbf{P}_{\mathbf{J}_i})) && (2)\\
&\quad \forall i \in \{i|\exists a_1, a_2, a_3 \in \mathbf{A} \wedge (a_1, i) \in S \wedge (a_2, i) \in S \wedge (a_3, i) \in S\} \\
f &= Random(\{i|\exists a_1, a_2, a_3 \in \mathbf{A} \wedge (a_1, i) \in S \wedge (a_2, i) \in S \wedge (a_3, i) \in S\}) && (3)
\end{aligned}
$$

For each heuristic, the first step is to collect a list of all of the possible next nodes, which requires that there are at least three struts in S between nodes in A and each possible node (the exceptions are nodes 2 and 3, which only require the origin and the first two nodes). If the heuristic is random choice, one of the possible nodes is randomly chosen. Otherwise, for each possible node, the trace of the submatrix of $\mathbf{P}$ corresponding to the connecting nodes in A is found. For the minimum trace heuristic, the possible node that minimizes the trace is returned; for the maximum trace heuristic, the node that maximizes the trace.

4.2 Extended Kalman Filter

The Extended Kalman Filter (Algorithm 2) is a standard sequential estimation technique that linearizes the state and measurement models around the estimated state so that the updated state is near-optimal given the new measurements. For brevity, we assume readership familiarity, and will define the additional terms without a tutorial. $\hat{\mathbf{X}}^P$ is the predicted update without noise. $\mathbf{F}$ and $\mathbf{G}$ are the Jacobians of the state function $f(\mathbf{X}, \mathbf{w})$ with respect to the state variables $\mathbf{X}$ and process noise variables $\mathbf{w}$, evaluated at the previous state $\hat{\mathbf{X}}$ without noise. The EKF considers the state transition function in the form $\mathbf{X}^{t+1} = f(\mathbf{X}^t, \mathbf{w})$ and $\mathbf{y} = g(\mathbf{X}^{t+1}, \mathbf{v})$. The state transition function is:

$$f(\mathbf{x}_i^{t+1}, \mathbf{w}) = \begin{cases} CalcNode(\bar{\mathbf{c}}_i, \mathbf{X}_{\mathbf{J}_i}, \mathbf{w}) & \text{if } i \text{ is currently being placed} \\ \mathbf{x}_i^t & \text{if } i \text{ is built or yet to be built} \end{cases}$$

The function *CalcNode*() is the solution for the attachment node position $\mathbf{x}_f$ to the nonlinear system of equations $\bar{c}_{if} + w_{if} = ||\mathbf{x}_i - \mathbf{x}_f||$ for all fixed nodes i and attachment node f: with desired lengths, known fixed nodes, and some noise, where does the attachment node go? With some algebraic manipulation, $\mathbf{x}_f$ can be found as a function of the states, lengths, and noises. The term $(\mathbf{I}\sigma_c)$ transforms the process variance vector into a diagonal matrix. The predicted covariance matrix estimate $\mathbf{P}^P$ is then updated with the sum of the state covariance propagation matrix and the process covariance matrix. Likewise, the measurement variance model $g(\mathbf{X}, \mathbf{v})$ is linearized with respect to the state variables $\mathbf{X}$ and measurement noise variables $\mathbf{v}$ to produce $\mathbf{H}$ and $\mathbf{U}$ at the predicted state $\hat{\mathbf{X}}^P$ without noise, and the term $(\mathbf{I}\sigma_l)$ transforms the measurement variance vector into a diagonal matrix. The Kalman gain $\mathbf{K}$ models the relative certainties of the state propagation and the measurement process. These terms are used to update $\hat{\mathbf{X}}$ and $\mathbf{P}$.

4.3 Verification

After each simulation was complete, the hidden state was directly compared to the final estimated states. For the physical experiments, we did not directly measure $\mathbf{x}_i$ and had to use a maximum likelihood estimator (MLE) on a set of ruler measurements. The MLE was used *independently* of the EKF algorithm. The MLE finds the structure that maximizes the joint probability distribution function of a structure, which is the product of the node position prior probabilities centered on $\bar{\mathbf{x}}$ and the measurement probabilities centered on the true distances $||\mathbf{x}_i - \mathbf{x}_j||$:

$$\tilde{\mathbf{X}} = \underset{\mathbf{X}}{\text{argmax}} \left(\sum_{i=1}^{n} Log \left(\frac{\exp\left(\frac{1}{2}\left(-\frac{(x_i-\bar{x}_i)^2}{\sigma_p} - \frac{(y_i-\bar{y}_i)^2}{\sigma_p} - \frac{(z_i-\bar{z}_i)^2}{\sigma_p}\right)\right)}{2\sqrt{2}\pi^{3/2}\sqrt{\sigma_p}^3} \right) + \sum_{i,j\in S_m} Log \left(\frac{\exp\left(-\frac{\left(\hat{l}_{ij}-\sqrt{(x_i-x_j)^2+(y_i-y_j)^2+(z_i-z_j)^2}\right)^2}{2\sigma_l^2}\right)}{\sqrt{2\pi}\sigma_l} \right) \right) \quad (4)$$

To get the best estimate, we measured as many node-node distances as our 1 m ruler would allow, which accounted for 57 pairs (S_m) of the possible 66 pairs in the 12-node structure, including 27 pairs not directly connected by struts. The ruler we used was 1 m long with a variance of $\sigma_l = 0.25\,\text{mm}^2$. To determine the accuracy of the MLE itself, we simulated 1000 structures with each node coordinate offset by a variance of $25\,\text{mm}^2$, and set the node prior variance $\sigma_p = 1000\,\text{mm}^2$ to allow the ruler to dominate. MLE is accurate to within 0.3 mm on average, and for the least certain node, 95 % of the estimates are more accurate than 1.2 mm (Fig. 3), making this our confidence interval for the physical measurements.

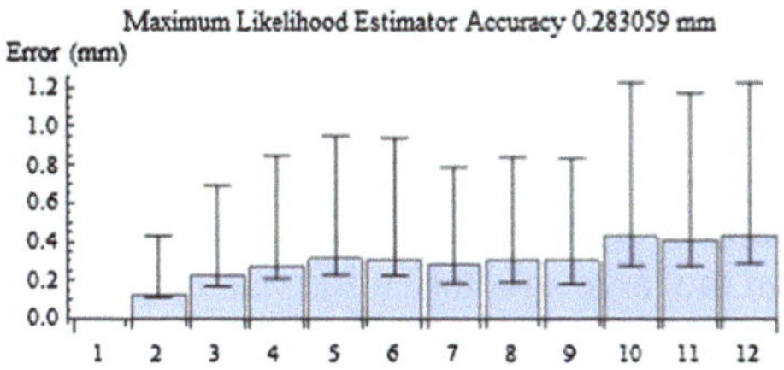

Fig. 3 The accuracy of the maximum likelihood estimator, used only for verification of the final physical assemblies, was simulated over 1000 random trials: the mean error of the maximum likelihood estimator with 5th and 95th percentiles is shown. The x-axis represents the node assembly order for the minimum trace case, and the further the node, the greater error in the MLE

5 Results

We first calibrated the IPJRs and found them have a minimum length of {496.5, 495.0, 500.7} mm, command variances $\sigma_c = \{1.15, 1.49, 0.16\}\,\text{mm}^2$, and step lengths $\{0.135, 0.138, 0.132\}$ mm. The ruler used to calculate node distances was $\sigma_l = 0.25\,\text{mm}^2$. These results are used both for conducting simulations and during the EKF on the real robot experiment.

5.1 Simulation

To come up with a realistic set of simulations and to explore the effects of inaccurate calibration, the EKF algorithm used the calibrated values, but the hidden, simulated IPJRs *did not* match the calibration (with the exception of a control case). Each time an IPJR is attached to a new strut, the hidden offset and step size are reset. The reason is threefold: when each IPJR is attached to a new strut, there will be a different offset each time; the attachment quality (and strut quality) varies; and the orientation of the strut and IPJR can lead to variable bending. For both the large and small truss structure, we used two control cases: open loop with perfect calibration, and open loop with hidden biases. In these cases, the Extended Kalman Filter uses only the IPJR command variances, and uses the *MinTrace* heuristic; thus, the control cases will attempt to place new nodes on what it thinks are the most certain sites. The control cases represent optimistic open loop results, the perfect calibration case more so than the hidden bias case.

The experimental cases test the EKF algorithm with measurements, using the minimum trace, random, and maximum trace heuristics. In the EKF assembly cases, the measurements were assumed to dominate the command variance, so we set the command variances to $\sigma_c = \{1000, 1000, 1000\}\,\text{mm}^2$ to treat the rulers as nearly-truth. To simulate hidden biases, each time an IPJR is placed on a strut, the hidden simulated IPJR offsets were randomly offset by up to 7.5 mm in either direction, and the step sizes were randomly offset by up to 5 μm. These values were chosen empirically based on observations in physical experiments with the IPJRs and the telescope trusses.

We ran 100 simulations of the two control cases and the three heuristics for both structures. We present the large structure results first.

The results of simulations of the 84-node, 315-strut structure are shown in Fig. 4. Each figure shows the errors of the nodes from the nominal in the order in which they were placed. The error bars represent the 5th and the 95th percentiles: 90 % of all results were between the bounds. The optimistic, unbiased IPJR case without measurements can be expected to produce average errors of 6.1 mm. The more realistic case with hidden biases should have average deviations of 22.6 mm. In both control cases, the error grows over time. These simulations show that the EKF assembly algorithm with measurements and heuristics performed considerably better than the

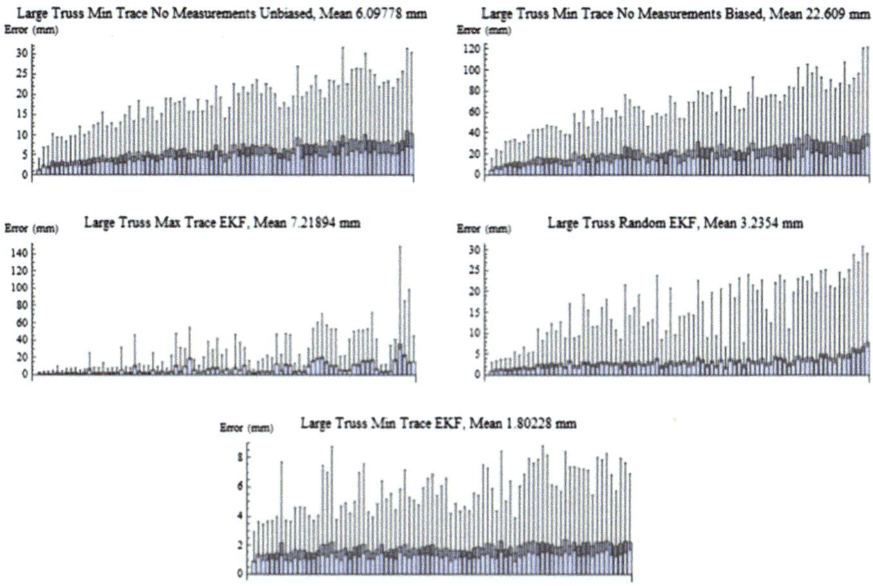

Fig. 4 Results of 100 simulated assemblies of the large truss. In reading order, starting at top left: the perfect calibration case without measurements, the hidden bias case without measurements, the maximum trace case with the EKF algorithm, the random case with the EKF algorithm, and the minimum trace case with the EKF algorithm. The x-axis indicates the node order placement. Error bars represent the 5th and 95th percentiles

biased case, and the quality of performance ranks as expected: the mean error of the minimum trace assembly sequences was 1.8 mm, the random case 3.2 mm, and the maximum trace case 7.2 mm.

The results of the smaller truss structure are shown in Fig. 5, following the same format as in Fig. 4. We observe two trends: the control cases are worse than the EKF assembly algorithm cases, as expected, but the differences between the assembly heuristics is not significant when the wide variability of simulation results is considered in the 5th and 95th percentiles. A perfectly calibrated IPJR had to contend with a growing error over time, and the mean node error is 3.3 mm. In the more realistic case with hidden biases, the mean node error is 11.9 mm, again with increasing errors. More importantly, when the 5th percentile errors of *each node* were averaged, the error was 4.0 mm: even when assuming low errors are correlated in such an ideal trial, at most 5 % of trials will have a better mean error.

The EKF assembly algorithm mean node errors followed the predicted trends with 100 trials: minimum trace (1.7 mm) is better than random choice (1.8 mm), which in turn is better than maximum trace (1.9 mm). No case violated the triangle inequality. However, all of these results fall within the 5th–95th percentile range, so these results are not significant. For a structure of only 12 struts and 30 nodes, with an insufficient number of opportunities to choose poorly, this is not surprising. Therefore, for the physical experiments, we chose to build only the minimum trace case.

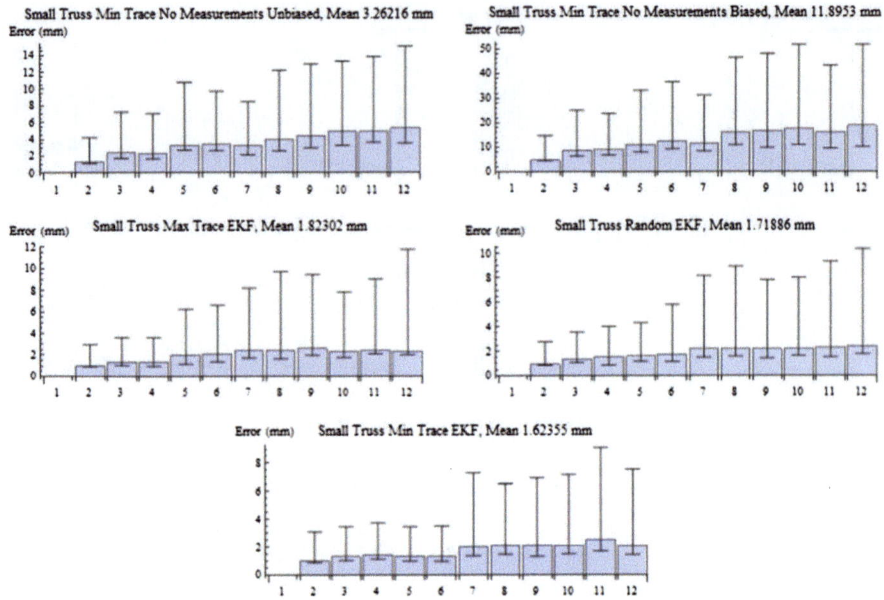

Fig. 5 Results of 100 simulated assemblies of the small truss. In reading order, starting at *top left* the perfect calibration case without measurements, the hidden bias case without measurements, the maximum trace case with the EKF algorithm, the random case with the EKF algorithm, and the minimum trace case with the EKF algorithm. The x-axis indicates the node order placement. *Error bars* represent the 5th and 95th percentiles

5.2 Physical Experiments

To test the validity of the EKF algorithm on real hardware, we performed two assemblies of the small truss with the IPJRs and the aluminum-strut, steel-node structure (Fig. 6), using the minimum trace heuristic. To collect measurements each step, in lieu of an on-board laser distance sensor planned for the future, the human operator measured the distance between node balls. Otherwise, the experiments progressed exactly in the same manner as the simulations.

Each assembled node required 1–3 attempts to converge to the desired tolerance. Each attempt partially corrected for the hidden biases due to the various factors affecting the parameters of the IPJRs: imprecise connections to the aluminum tubes, bent tubes, IPJR extension and contraction hysteresis, and deflection due to gravity.

The results of the two trials, bounded by the 95th percentile accuracy error of the maximum likelihood estimator, are shown in Fig. 7. The mean errors are 1.7 and 2.7 mm, with a confidence interval of 1.2 mm. This is a significant improvement over the realistic open loop assembly case with hidden biases. The second trial has the largest individual node error at 7 mm, which is better than the open loop case could achieve in less than 5 % of the trials.

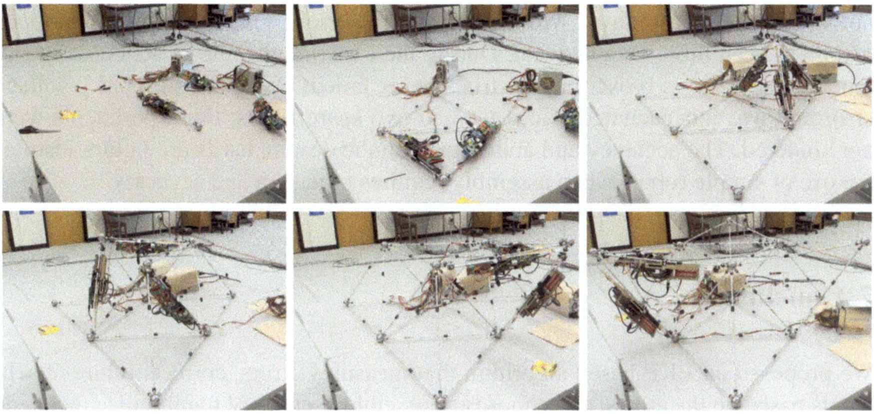

Fig. 6 In reading order, the first nodes placed requires only one IPJR, the other grounded nodes require 2 IPJRs, and the subsequent nodes require 3 IPJRs to be placed. The final frame shows the finished assembly, prior to the removal of the IPJRs

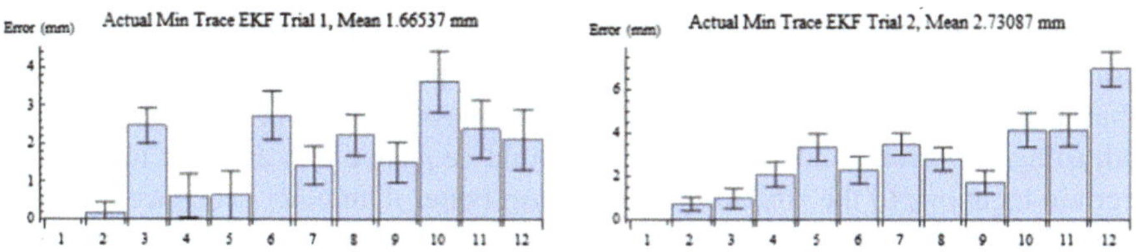

Fig. 7 The estimated errors of the two physical experiments, bounded by the MLE confidence interval

6 Discussion

The simulations show that the minimum trace assembly choice minimizes growing error in both truss structures, despite relying only on local measurements increasingly further from the origin. The data show that a suboptimal build sequence with EKF should do better on average than the open loop case: but only when the assembly trials finish. Half of the large truss maximum trace simulations failed, and three of the large truss random simulations failed. If the *CalcNode*() function failed to find a solution to the attachment node, the triangle inequality was violated: the IPJRs could not connect with the desired lengths. This algorithm did not implement backtracking, so such cases terimated.

The two physical assembly experiments with the EKF assembly algorithm performed far better than the open-loop control cases, even with the conservative MLE confidence interval of 1.2 mm. Considering that the struts and IPJRs were not made equally, this shows that precise assembly can be made possible by lower quality components through error detection and correction. The second physical trial shows a trend of growing error: this is attributed to two unlocked struts not sliding as expected, deforming the bottom layer. We believe that the error growth would have stopped had there been more nodes to add. During the experiments, the IPJRs occasionally rebooted, and struts broke, but the trials never failed. One IPJR suffered a voltage error and was incapacitated. Since we had two spare IPJRs, the experiments were not hindered. The accuracy and ability to continue despite hardware failure justifies the use of simple robots when assembly requires precision and accuracy.

7 Conclusion

We proposed an EKF-based algorithm that measures errors, corrects future attachments based on the errors, and chooses an assembly sequence to minimize the mean error. We argued that robotic assembly without active state monitoring, while common in the literature, is inappropriate for meeting precision requirements on structures made of stock materials. An EKF-based algorithm can be used to prevent errors from growing with each new addition, and can overcome biases due to imperfect calibration, hysteresis in the IPJRs, deflections from stresses, and inconsistent connections between the IPJRs and the struts on the structure. Simulations showed that using the EKF-algorithm is a significant improvement over not measuring and adjusting for errors, and that choosing a build sequence that adds on to the most certain locations to the structure will perform better than other sequences. Finally, we validated the proposed algorithm using three IPJRs to assemble a truss structure, consisting of 12 steel nodes and 30 aluminum struts, and showed that the EKF algorithm produced structures with 1.6 and 2.6 mm average node error.

We intend to continue using these IPJR prototypes and truss prototypes. We will thoroughly analyze the comparisons between the EKF algorithm and the open loop assembly algorithm, determine when and why 50 maximum trace and 3 random order trials failed, find the limit of the applicability of the Extended Kalman Filter, and analyze the effects of loop closure. We grouped together sources of error such as gravity, IPJR attachment error, imprecise struts, and thermal expansion under the label of "hidden biases", but we intend to model each of these explicitly, incorporating work done in [16]. These experiments chose assembly sites greedily, but offline planning, ranging from looking ahead several plies to finding a full sequence, may do better. We plan to implement backtracking if a failure occurs, permitting all maximum trace and random assemblies to finish. For struts with interchangeable locked and free-sliding states, we will explore allowing IPJRs to visit struts repeatedly. The MLE worked well for verification, and we will test whether it outperforms EKF in the assembly algorithm. Finally, to demonstrate the concept that a group of cheap IPJRs working in parallel can build suitably accurate truss structures, we intend to build several more IPJRs, modify the algorithm to allow concurrent assembly, and incorporate robotic manipulators.

Acknowledgments This work was supported by a NASA Office of the Chief Technologist's Space Technology Research Fellowship.

References

1. Zimpfer, D., Kachmar, P., Tuohy, S.: Autonomous rendezvous, capture and in-space assembly: past, present and future. In: 1st Space Exploration Conference: Continuing the Voyage of Discovery, vol. 1, pp. 234–245 (2005)
2. Watson, J.J., Collins, T.J., Bush, H.G.: Construction of large space structures at NASA Langley research center. In: IEEE Aerospace Conference (2002)
3. Dorsey, J.T., Doggett, W., Komendera, E., Correll, N., Hafley, R., King, B.D.: An efficient and versatile means for assembling and manufacturing systems in space. In: Proceedings of the AIAA SPACE Conference (2012)
4. Lillie, C.F.: On-orbit assembly and servicing of future space observatories. In: Proceedings of SPIE **6265**, 62652D-1 (2006)
5. Doggett, W.: Robotic assembly of truss structures for space systems and future research plans. In: IEEE Aerospace Conference Proceedings, vol. 7, IEEE (2002)
6. Komendera, E., Reishus, D., Dorsey, J.T., Doggett, W.R., Correll, N.: Precise truss assembly using commodity parts and low precision welding. Intel. Serv. Robot. **7**(2), 93–102 (2014)
7. Komendera, E., Dorsey, J.T., Doggett, W.R., Correll, N.: Truss assembly and welding by intelligent precision jigging robots. In: Proceedings of the 6th Annual IEEE International Conference on Technologies for Practical Robot Applications (TEPRA) (2014)
8. Homem de Mello, L.S., Lee, S.: Computer aided mechanical assembly planning. Springer, Berlin (1991)
9. DeFazio, T., Whitney, D.: Simplified generation of all mechanical assembly sequences. IEEE J. Rob. Autom. **RA-3**(6), 640–658 (1987)
10. Röhrdanz, F., Mosemann, H., Wahl, F.M.: A high level system for generating, representing and evaluating assembly sequences. In: In Proceedings of the International Joint Symposia on Intelligence and Systems (1996)
11. Wilson, R.H.: On Geometric assembly planning. PhD thesis, Stanford University (1992)
12. Fox, B.R., Kempf, K.G.: Opportunisitic scheduling for robotic assembly. In: Proceedings of the International Conference on Robotics and Automation (1985)
13. Simmons, R., Singh, S., Hershberger, D., Ramos, J., Smith, T.: First results in the coordination of heterogeneous robots for large-scale assembly. In: Experimental Robotics VII, pp. 323–332 Springer (2001)
14. Stroupe, A., Huntsberger, T., Okon, A., Aghazarian, H.: Precision manipulation with cooperative robots. In: Parker, L., Schneider, F., Schultz, A. (eds.) Multi-Robot Systems: From Swarms to Intelligent Automata (2005)
15. Heger, F.W.: Assembly planning in constrained environments: building structures with multiple mobile robots. PhD thesis, Carnegie Mellon University (2010)
16. McEvoy, M.A., Komendera, E., Correll, N.: Assembly path planning for stable robotic construction. In: Proceedings of the Sixth Annual IEEE International Conference on Technologies for Practical Robot Applications (2014)
17. Lindsey, Q., Kumar, V.: Distributed construction of truss structures. In: Algorithmic Foundations of Robotics X, pp. 209–225. Springer (2013)
18. Detweiler, C., Vona, M., Yoon, Y., Yun, S., Rus, D.: Self-assembling mobile linkages. IEEE Robot. Autom. Mag. **14**(4) (2007)
19. Werfel, J., Petersen, K., Nagpal, R.: Designing collective behavior in a termite-inspired robot construction team. Science **343**(6172), 754–758 (2014)
20. Knepper, R.A., Layton, T., Romanishin, J., Rus, D.: Ikeabot: An autonomous multi-robot coordinated furniture assembly system. In: IEEE International Conference on Robotics and Automation (ICRA), pp. 855–862. IEEE (2013)
21. Worcester, J., Lakaemper, R., Hsieh, M.y.A.: 3-dimensional tiling for distributed assembly by robot teams. In: Proceedings of the 13th International Symposium on Experimental Robotics (ISER2012), pp. 143–154. Springer (2012)

Customized Sensing for Robot Swarms

D. Jud, J. Alonso Mora, J. Rehder, R. Siegwart and P. Beardsley

Abstract This paper describes a novel and compact design for an omni-directional stereo camera. A key goal of the work is to investigate the use of rapid prototyping to make the mirrors for the device, by 3D printing the mirror shape and chroming the surface. The target application is in robot swarms, and we discuss how the ability to create a customized omni-camera enables sensing to become an integrated part of system design, avoiding the constraints that arise when using commercial sensors.

1 Introduction

Motivation Our previous works with ground [1] and aerial [2] robots represent two applications of robot swarms in entertainment where localization was performed with an external overhead tracking system (Fig. 1). The motivation of this work on readily customized sensing is to develop an on-board localization solution for a robot swarm (Fig. 2) that

(a) has a reusable core technology,
(b) is customizable for different robot platforms and
(c) is low-cost.

Related work External optical tracking systems[1] provide millimeter accuracy localization but have high cost, require a fixed infrastructure, and are constrained to line-of-sight to the robot swarm. Indoor wireless solutions include UWB[2] and RFID [3]. These solutions typically encounter interference in wirelessly active environments and multi-path issues in multi-robot setups. DGPS is restricted to outdoor use only

[1] www.vicon.com, www.optitrack.com.
[2] www.ubisense.com.

D. Jud (✉) · J. Alonso Mora · J. Rehder · R. Siegwart
Autonomous Systems Lab, ETH Zurich, Zurich, Switzerland
e-mail: djud@student.ethz.ch

J. Alonso Mora · J. Rehder · P. Beardsley
Disney Research, Zurich, Switzerland

M.A. Hsieh et al. (eds.), *Experimental Robotics*, Springer Tracts
in Advanced Robotics 109, DOI 10.1007/978-3-319-23778-7_34

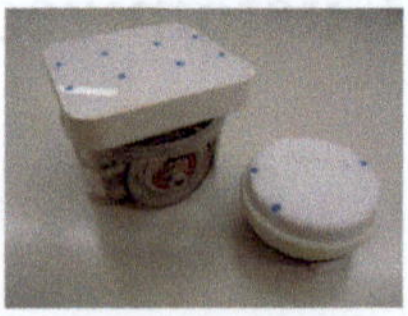

Fig. 1 Currently used setup with an overhead camera tracking a robot swarm [1]

Fig. 2 Robotic swarm of a central unit supporting the omnidirectional camera used to localize a satellite relative to the central unit

and might also be blocked by buildings. On-board localization can be used for localization relative to landmarks and neighbouring robots [4], or for relative localization only to enable collision avoidance. Although on-board localization imposes a cost per unit, an issue for swarms of tens to hundreds of robots, the cost of sensors and on-board computation is rapidly decreasing.

Different designs for omnidirectional stereo cameras already exist. The standard approach is to vertically align two omnidirectional sensors [5]. Other solutions only use one sensor, but still have stereo imaging. One approach uses a double lobed mirror [6] with a single camera for two different view points. A concave lens in between the mirror and the camera also enables stereo imaging [7]. Both of these two single camera omnidirectional stereo sensors have a very short baseline, but still a large overall size.

Problem statement The first goal of this paper is to develop a custom stereo omnidirectional camera (hereafter omni-cam) for on-board localization. The second goal is to demonstrate that rapid prototyping can be used to create mirrors specially designed for the application in hand. The approach is motivated by the Maker Movement and recent work on rapid prototyping for robots [8]. The contributions of this paper are:

- a novel and compact design of a stereo omni-cam.
- a demonstration that rapid prototyping can be used to make mirrors that are usable for computer vision.
- a method for calibrating the omni-cam setup, including the mirror shape.

2 Technical Approach

Stereo omni-cam The stereo omni-cam design has two cameras with opposing mirrors giving a stereo omnidirectional view, as shown in Fig. 4a. It is more compact than existing designs [5], because each camera is embedded inside one of the mirrors, with an aperture for viewing the opposing mirror. This enables the sensor to have any desired baseline. Apart from providing a stereo field-of-view (FOV) in the horizontal direction with the outer rim, the mirrors also have a smaller inner rim to provide a monocular field of view of the floor and ceiling in the vicinity of the robot, used for localizing the camera robot in its environment. The stereo view can be used to recover the position of any object in sight, i.e. our swarm robots. The different fields of view are shown in Fig. 3. Figure 4b shows the physical prototype with two synchronized Matrix Vision BlueFOX-MLC cameras running at a maximum of 5.8 Hz with a resolution of 2592 by 1944 pixels. The synchronized image acquisition is guaranteed by linking the cameras together by wire and only triggering the top camera whenever the bottom camera does so.
Robot platform The sensor is mounted on a mobile platform, Fig. 4c, which is deployed inside an acrylic sphere with a gravity-driven self-righting mechanism to maintain upright orientation, similar to the work of [9]. Figure 4d shows two sphere

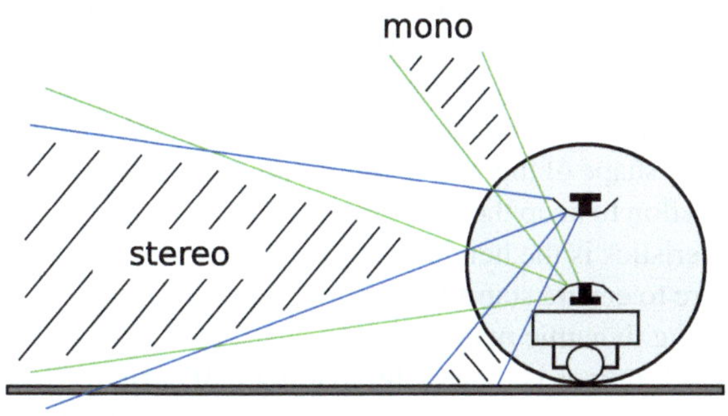

Fig. 3 The omni-cam has a large stereo view in the horizontal plane, but also monocular views of the ground and ceiling in the vicinity of the camera

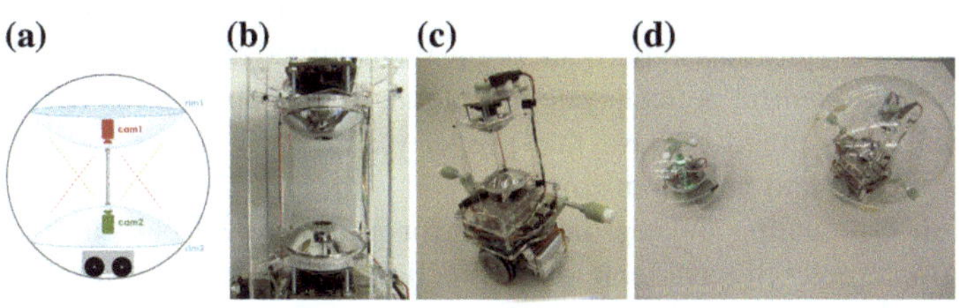

Fig. 4 **a** Schematic of the omni-cam design. There are two cameras with opposing mirrors giving a 360° stereo view. Each camera is embedded inside a mirror for compactness, with an aperture for viewing the opposing mirror. **b** Physical prototype. **c** Omni-cam mounted on robot platform. **d** Two robots inside acrylic spheres, the larger one containing the omni-cam

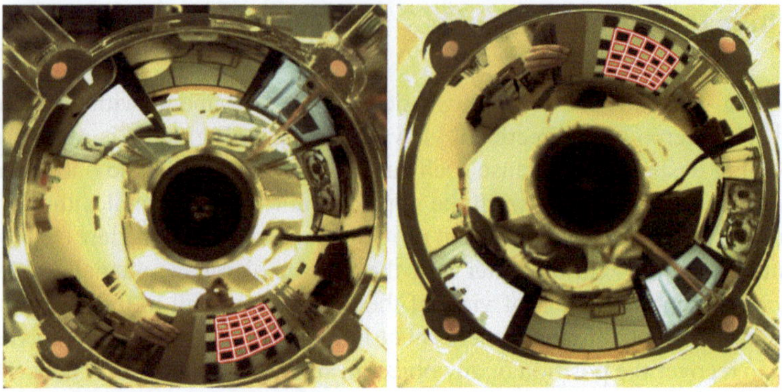

Fig. 5 Representative pair of images taken by the *top* (on the *left*) and *bottom* (on the *right*) cameras of the robot. Checkerboards were used for calibration and the images show example detections overlaid in *red*

robots, the larger one (diameter 250 mm) contains the omni-cam while the smaller one (diameter 160 mm) is a satellite robot without on-board sensing. The swarm consist of multiple satellite robots that are tracked by a single camera robot with the omni-cam. Coloured LEDs on the satellite robots are used to infer the position of the satellites relative to the camera robot. The images are processed on-board using an Odroid-XU Quad Core board. The motors and the belonging electronics are taken from an E-Puck robot [10]. All the other parts are specifically designed for the purpose of driving inside a sphere and are rapid-prototyped. The form of the wheels is made to match the shape of the sphere for maximum grip. On top is a spring loaded slider built from Teflon to keep the robot well positioned inside the sphere. The most important characteristics is the height of the center of gravity. It has to be below the center of the sphere to ensure stability. The lower the center of gravity, the better the robot behaves during dynamic manoeuvres.

Mirror design Each mirror is created by making a 3D print[3] of the desired shape, sanding, polishing the print and chroming it using a standard chroming service. The vapour deposition method only applies a thin chrome coat and does not cover up any irregularities of the surface, emphasising the importance of polishing out the layer steps from printing. Our mirror is designed using a second order polynomial function for the monocular rim and a hyperbola for the main curved part, see Fig. 6. This shape constitutes a non-central optical system (the majority of application specific mirror designs can be assumed to have this property). Figure 5 shows example camera images for the top and bottom mirrors.

Calibration Calculating the reprojection error is the crucial part in the optimization. Most commercially available omnidirectional cameras are central cameras as in Fig. 7, which means that all the outgoing rays from the mirror intersect in a virtual viewpoint. That point can be used to calculate the mirror intersection point. However

[3]We used the Objet Eden 350V, 16 micron layer resolution.

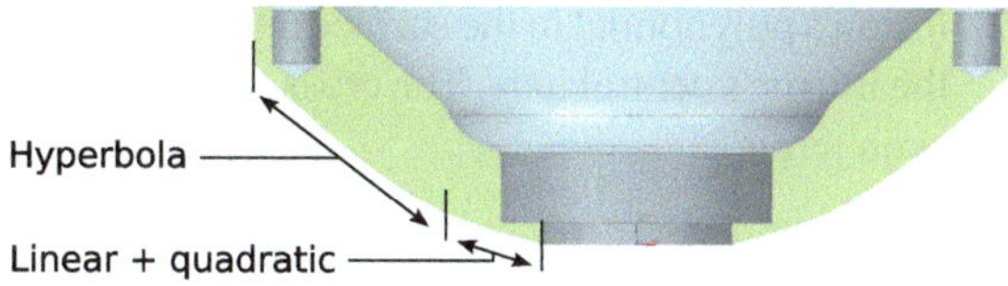

Fig. 6 Section view of the mirror CAD model showing the different functions resulting in different stereo and mono views

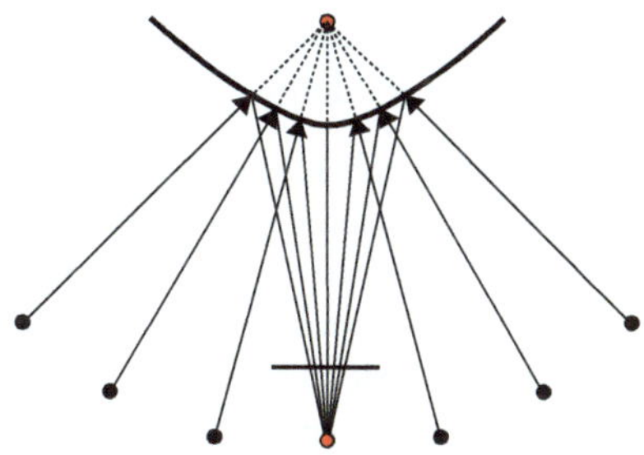

Fig. 7 Central camera system with a single viewpoint in the center of the mirror

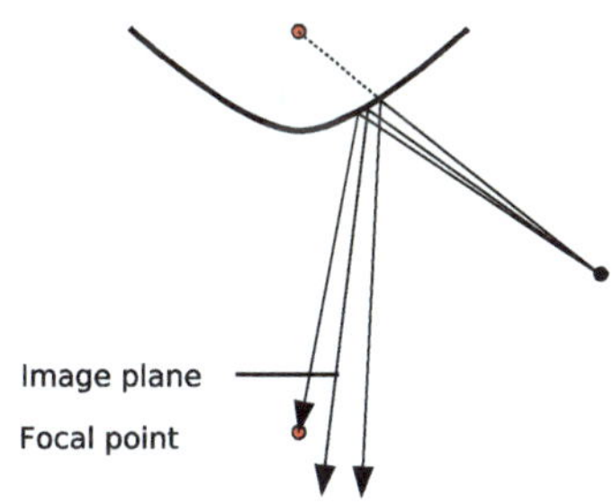

Fig. 8 Iteration process for the reprojection in non-central camera systems

our mirrors can have arbitrary shapes which are non-central and do not have such a closed form solution for the reprojection. Figure 8 shows the iterative procedure that is implemented to find the intersection point. First, an initial guess is calculated by assuming the camera system to be central. The intersection point is then iteratively found by minimizing the distance between the incident ray and the camera center.

The shape of the final mirror differs from the input CAD model due to printer resolution and effects of sanding. We investigate three calibration approaches to determine the final shape

1. fitting a discrete number of hyperbolas to the surface and interpolating

$$z(r,\phi) = b(\phi) - \frac{b(\phi)}{a(\phi)}\sqrt{(r - 10\,\text{mm})^2 + a(\phi)^2} \tag{1}$$

   The hyperbola parameters a and b are defined at discrete steps around the mirror, e.g. they would be defined at $\phi = 0$, $\phi = \frac{\pi}{2}$, $\phi = \pi$ and $\phi = \frac{3\pi}{2}$ when interpolating over four hyperbolas. These angle dependent parameters allow a rotationally unsymmetrical model of the mirror.

2. fitting a polynomial to the surface under the assumption of radial symmetry around the mirror's central axis. The cross-section is governed by a polynomial with only even terms

$$z(r) = a_0 r + a_1 r^2 + a_2 r^4 + \cdots + a_k r^{2k} \qquad k \in \mathbb{N} \tag{2}$$

3. fitting a polynomial to the surface in the radial direction as in Eq. 2 with additional terms for modelling the 2π-periodic asymmetries with Fourier series.

These three different models have completely different strengths and weaknesses. The first approach combines prior knowledge of the hyperbolic mirror surface with a simple model of the rotational asymmetries. Only 16 parameters are needed for both mirrors if four hyperbolas are used in the interpolation. The second approach includes the prior knowledge when initializing the polynomial parameters, but for higher order polynomials, it is less constraining than the hyperbolic model. It is simply a polynomial in radial direction and cannot model any rotational asymmetries. Although there might the drawback of needing a high order polynomial for a sufficiently good fit. The third approach is the most flexible of them all. It uses the same polynomial in radial direction as the second approach, but also includes an angular model for the rotational asymmetries based on Fourier series. Again the same problem might occur that there is a high order of degree needed for the polynomial and the Fourier series.

Stereo calibration is carried out using checkerboards. The camera intrinsics are calibrated beforehand using a standard algorithm from the OpenCV library.[4] An initial estimate of the mirror shape and the configuration of the cameras plus mirrors is taken from the CAD design. Figure 9 shows the configuration with the coordinate transformations being optimized in the calibration. With these initial parameters, an initial estimate of 3D pose for a set of calibration patterns (see examples of detected patterns in Fig. 5) is made from triangulating the corners of the checkerboards. For all approaches above, the mirror shape and the configuration is then recovered by optimizing over the shape parameters and coordinate transformations with the objective function being the discrepancy between the recorded checkerboard positions and the reprojection of the 3D reconstructions of the checkerboards. The optimization routine is a nonlinear least-squares solver.

3 Results

Two datasets were captured, each containing 50 images of checkerboard patterns. Checkerboard detection was done using the package in [11]. The first dataset of patterns is used to compute the mirror shape based on the method in Sect. 2. The second dataset is used to evaluate the calibration, doing a stereo reconstruction of the

[4] www.opencv.org.

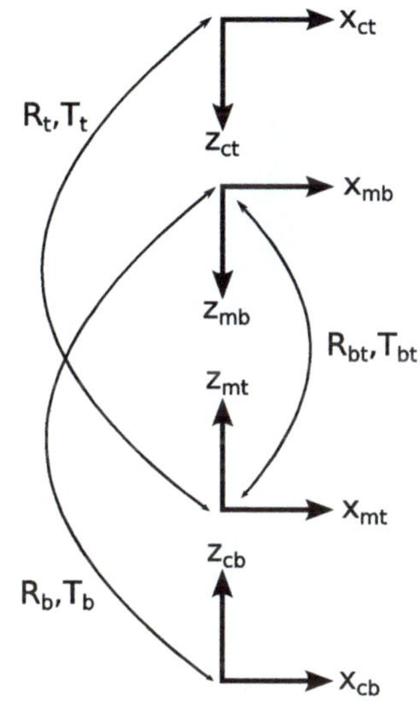

Fig. 9 Image of the omni cam showing the different coordinate frames and the according coordinate transformations as they are optimized in the calibration procedure

Table 1 Reprojection errors resulting from evaluating the calibration with the test dataset

(a) Reprojection error in pixels when mirror shape is modelled by discrete hyperbolas, equally-spaced with angle ϕ, with interpolation				(b) Reprojection error in pixels when mirror shape is modelled with a polynomial. Term k is the highest order of the polynomial expression in Eq. 2				
$\phi = 2\pi$	$\phi = \pi$	$\phi = \pi/2$	$\phi = \pi/4$	$k = 2$	$k = 4$	$k = 6$	$k = 8$	$k = 10$
2.72	2.66	2.2	2.36	3.93	3.37	3.25	3.23	3.21

calibration patterns, reprojecting to the image plane, and measuring the reprojection error between checkerboard detections and the projected 3D reconstructions.

Table 1a shows the results when the mirror shape is modelled using discrete hyperbolas, equally-spaced around the radial axis, with interpolation. The number of fitted hyperbolas is 1 (assumption of radial symmetry), 2, 4, and 8. The best result is obtained for 4 hyperbolas, while 8 hyperbolas is over-fitting. Table 1b shows the results when the mirror shape is modelled by a polynomial with assumption of radial symmetry, as in Eq. 2. The reprojection error decreases when the order of the polynomial expression increases. The best polynomial result (3.21) is not smaller than the lowest reprojection error with interpolated hyperbolas (2.2). This enforces the importance of a mirror model that includes the rotational asymmetries.

The third approach discussed in the calibration section is a mirror model that has the same polynomial in radial direction as Eq. 2 but also models the asymmetries with a Fourier series. A low order Fourier series did not make any difference to the result, whereas a higher order Fourier series renders the convergence to meaningful parameters difficult, because the cost surface exhibits multiple local minima. Consequently the experiments section shows data gathered with the interpolated hyperbolas

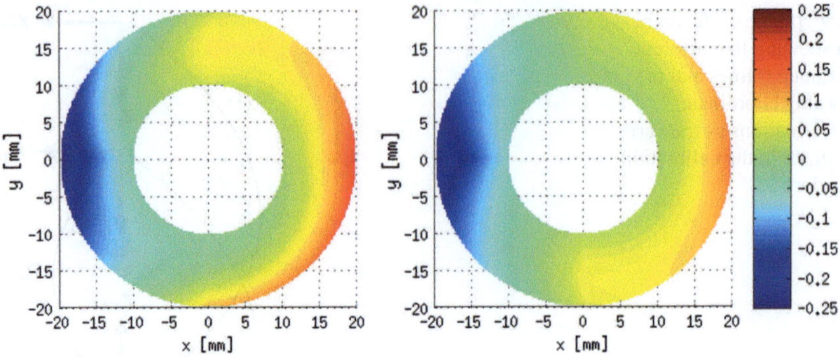

Fig. 10 Deviation of computed mirror shape from an averaged rotationally symmetric model for the *bottom* (on the *left*) and *top* (on the *right*) mirrors, for the case that the shape is modelled using four hyperbolas and interpolation. The height difference is displayed from −0.25 mm (*blue*) to +0.25 mm (*red*)

as mirror model. Figure 10 shows how the computed mirror shape diverges from an averaged rotationally symmetric representation.

The localization uncertainty [12] of satellite robots relative to the camera was not investigated within this work, but it is a part of future work on the project.

4 Experiments

This section focuses on the performance of localizing the satellite robot relative to the camera. Experiments were conducted using a central static robot equipped with the omni-cam and a satellite robot equipped with a color LED for detection (robots shown in Fig. 4d). The satellite robot was moved in a spiral with a range from 0.2 to 1.6 m from the central robot. Ground truth is obtained from an overhead tracking system. For this experiment, the mirror shape of the omni-cam was modelled using 4 hyperbolas with interpolation.

Figure 11a shows the robot trajectory. Figure 11b shows the radial component of the measured position, where the red line marks the ground truth distance and blue crosses correspond to individual localizations. While uncertainty grows with distance, the mean value is recovered correctly. A bias in the distribution around 1.3 m suggests imperfections which can most likely be attributed to the calibration. Figure 11c shows statistics for the difference in robot bearing angle between the measurement and the ground truth, with 50 % of the measurements within 0.02 rad error and 90 % within 0.04 rad error. The standard deviation of the angular error is 0.9 °.

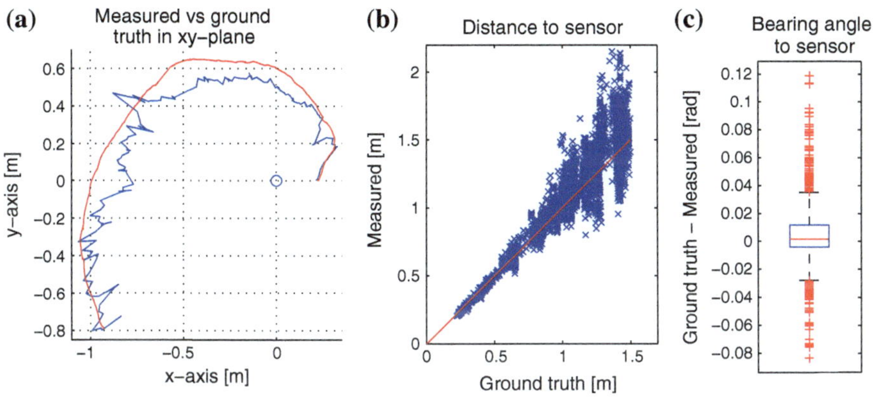

Fig. 11 **a** Ground-truth (*red*) and measured (*blue*) robot position using raw measurements with no filter on robot motion. **b** The radial distance of the satellite robot from the central robot. **c** The error of the angular measurement in a box plot. Both **b** and **c** are derived from measurements of the sensor

5 Application

The omni-cam is mounted onto a differential drive robot and used to track the satellite robot that is equipped with an LED. The triangulation of a single LED only provides a position and not the orientation of the robot. An extended Kalman filter is used to estimate the heading from the motion and the robots odometry. The measurement update of the Kalman filter is incorporating the position measurement at 5 Hz, whereas the state update is integrating the wheel odometry of the tracked robot at 40 Hz. This setup combined with a controller to drive the satellite robots enables the satellite to drive to waypoints relative to the camera robot. By including the wheel odometry of the camera robot into the state update, it is made possible to drive the camera robot while the satellite robots are moving to waypoints relative to the camera. Figure 12 shows the trajectory of a satellite robot with arrows indicating the estimated orientation along the trajectory.

The plot in Fig. 13 is showing the raw position measurement in x-direction from the omni-cam in blue and the according filter output in red. The increased noise at high distances from the sensor is successfully suppressed by including the odometry.

In another application, the camera robot can be driven by using a laser pointer. The robot sees the dot on the ground and follows it. This shows the utility of the additional monocular view of the camera. The bottom camera can detect the laser dot even if it is as close as 5 cm to the robot body where there is no stereo view anymore. The green laser pointer was implemented to pull the robot, whereas the red dot is pushing the robot away, as seen in Fig. 14.

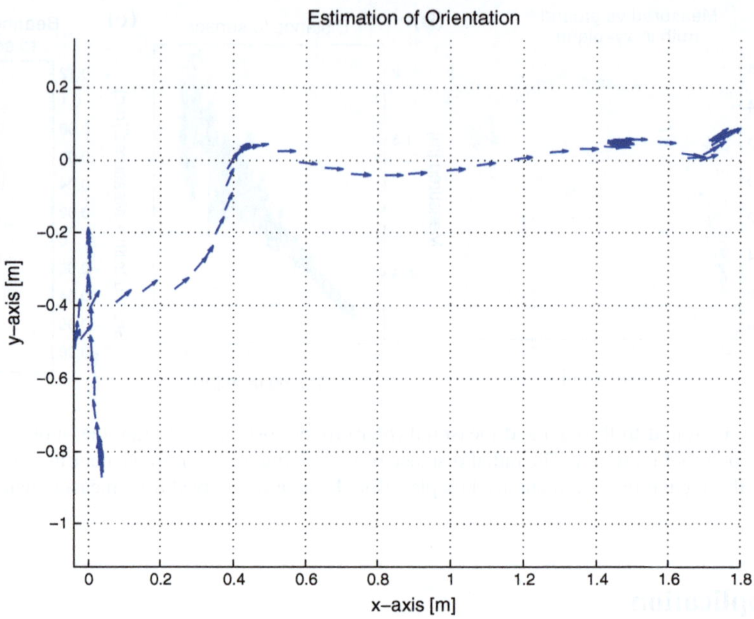

Fig. 12 Plot in the horizontal xy-plane showing the estimated orientation of the robot along a trajectory

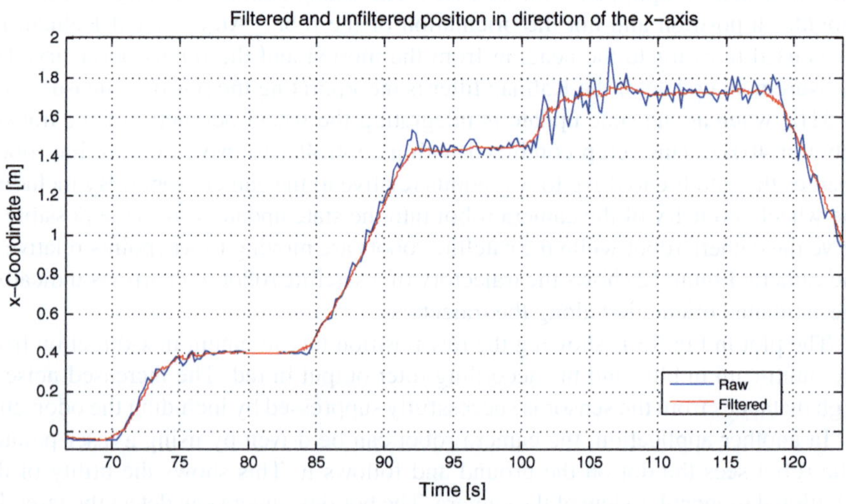

Fig. 13 A plot showing the raw measurement (*blue*) and the according Kalman filter output (*red*) in x-direction

Fig. 14 The camera robot follows the *green laser dot* while simultaneously driving away from the *red laser dot*

6 Discussion

Rapid prototyping for custom mirrors One approach to making a mirror is to mill and polish aluminium. This gives precise and high-quality results. Our focus has been on 3D printing followed by mirroring of the surface to make a custom mirror. The main motivation is ease-of-use, because 3D printing and associated design software are becoming standard hobbyist tools. In addition, the typical plastic from a 3D printer is about half the density of aluminium, making it more suitable for small, low power, ground and aerial robots. Furthermore, metal printing has now appeared at the high-end of the market, and can be expected to filter down to lower-cost solutions.

Methods to deposit a mirror on a surface include (a) traditional silvering using a silver spray, (b) chrome electroplating, and (c) vacuum deposition of chrome or aluminium. A lower-cost solution is mirror finish spray paint. Our approach was to print the 3D part, sand it by hand, and send it to a chroming service for vacuum deposition chroming. This method is relatively low cost (chroming services are widely available for the auto and home fittings markets) and can be used for large pieces.

The trade-off for this extra flexibility is that 3D printed parts are currently less precise than milled parts due to the coarse resolution of the 3D print, and the sanding and chroming process. We developed calibration methods to estimate the 3D shape, and experimental results demonstrate that the calibration is successful.

Adaptability to different robot platforms A motivation of the paper is to achieve reusable localization technology for use with different types of robot platform in varied settings. Rapid prototyping of omni-cams does not of course achieve this larger goal on its own. But by having greater control of the swarms sensing technology, the goal of a reusable localization system becomes more achievable. We are no longer limited to commercial omni-cam designs. As described in this paper, it was readily possible to prototype our own novel design of stereo omni-cam. And it is possible to adopt an iterative development cycle—for example to tune the stereo baseline, stereo field-of-view etc.—without excessive budgetary or time cost. Thus sensing becomes a controllable part of the design process, instead of an inflexible component which imposes undesirable constraints on the robot swarm system.

References

1. Alonso-Mora, J., Breitenmoser, A., Rufli, M., Siegwart, R., Beardsley, P.: Image and animation display with multiple robots. Int. J. Robot. Res. **31**(6), 753–773 (2012)
2. Alonso-Mora, J., Schoch, M., Breitenmoser, A., Siegwart, R., Beardsley, P.: Object and animation display with multiple aerial vehicles. In: IEEE/RSJ International Conference on Intelligent Robots and Systems (IROS), pp. 1078–1083 (2012)
3. Zhou, J., Shi, J.: RFID localization algorithms and applicationsa review. In: Journal of Intelligent Manufacturing (2009)
4. Leung, K.Y.K., Barfoot, T.D., Liu, H.H.: Distributed and decentralized cooperative simultaneous localization and mapping for dynamic and sparse robot networks, pp. 3841–3847 (2011)
5. Yi-ping, T., Qing, W., Ming-li, Z., Jun, J., Yi-hua, Z.: Design of vertically aligned binocular omnistereo vision sensor. EURASIP J. Image Video Proc. **2010** (2010)
6. Cabral, E.L., de Souza, J., Hunold, M.C.: Omnidirectional stereo vision with a hyperbolic double lobed mirror. In: Proceedings of the 17th International Conference on Pattern Recognition, ICPR 2004. vol. 1, pp. 1–9. IEEE (2004)
7. Yi, S., Ahuja, N.: An omnidirectional stereo vision system using a single camera. In: 18th International Conference on Pattern Recognition, 2006. ICPR 2006. vol. 4, pp. 861–865. IEEE (2006)
8. Soltero, D.E., Julian, B.J., Onal, C.D., Rus, D.: A lightweight modular 12-dof print-and-fold hexapod. In: IEEE/RSJ International Conference on Intelligent Robots and Systems (IROS), 2013, pp. 1465–1471. IEEE (2013)
9. Kawabata, K., Sato, H., Suzuki, T., Tobe, Y.: A wireless camera node with passive self-righting mechanism for capturing surrounding view. In: Gallegos-Funes, F. (ed.) Vision Sensors and Edge Detection, InTech (2010)
10. Mondada, F., Bonani, M., Raemy, X., Pugh, J., Cianci, C., Klaptocz, A., Magnenat, S., Zufferey, J.C., Floreano, D., Martinoli, A.: The e-puck, a robot designed for education in engineering. In: Proceedings of the 9th Conference on Autonomous Robot Systems and Competitions, vol. 1, pp. 59–65 (2009)
11. Geiger, A., Moosmann, F., Car, O., Schuster, B.: Automatic camera and range sensor calibration using a single shot. In: International Conference on Robotics and Automation (ICRA), St. Paul (2012)
12. Thrun, S., Burgard, W., Fox, D.: Probabilistic Robotics (Intelligent Robotics and Autonomous Agents). The MIT Press (2005)

Automatic Distribution of Disposable Self-Deploying Sensor Modules

Paul Pounds, Timothy Potie, Farid Kendoul, Surya Singh, Raja Jurdak and Jonathan Roberts

Abstract We present preliminary experiments of an aerial deployment system for environmental sensors with autorotative soft-landing. Key problems associated with this task are discussed, including design of the sensor modules, release mechanism and telemetry system. We evaluate performance of the initial proof-of-concept system and show that this approach to sensor deployment is viable.

Keywords Aerial robotics · Aerial sensor deployment · Disposable UAVs

1 Motivation, Problem Statement, Related Work

Real-time information about environmental conditions is valuable for the highly-dynamic and safety-critical activities of bushfire fighting. Rapidly shifting fire-fronts pose a danger to fire crews who may not have direct telemetry from observation posts or who suffer from degraded situational awareness due to smoke and wind. Obtaining measurements about fire development at ground level is extremely risky as sudden

P. Pounds (✉) · T. Potie · S. Singh
University of Queensland, St. Lucia, QLD 4072, Australia
e-mail: paul.pounds@uq.edu.au

T. Potie · F. Kendoul · R. Jurdak · J. Roberts
CSIRO, Pullenvale, QLD 4069, Australia

T. Potie
e-mail: timothy.potie@uqconnect.edu.au

F. Kendoul
e-mail: farid.kendoul@csiro.au

S. Singh
e-mail: spns@uq.edu.au

R. Jurdak
e-mail: raja.jurdak@csiro.au

J. Roberts
e-mail: jonathan.roberts@csiro.au

M.A. Hsieh et al. (eds.), *Experimental Robotics*, Springer Tracts in Advanced Robotics 109, DOI 10.1007/978-3-319-23778-7_35

changes in wind can cause workers to be overrun by the front. Even after a fire has passed through, an area may not be safe for crews to enter until the ground has cooled sufficiently; safety is difficult to ascertain without subjecting volunteers to potentially hazardous conditions.

We are developing an aerial deployment system to soft-land instruments into the proximity of a fire front or onto recently burned ground to take several atmospheric measurements of value in bushfire fighting. These sensors will survey the wide geographic expanse of large fire systems. The high likelihood that these sensors will be burnt and destroyed (or otherwise not recovered) during their mission makes the economy of device construction and assembly of great importance. Conventional construction of ruggardised sensors for hard-landing, landed deployment [1] or expensive arrest devices such as parachutes is thus impractical [2].

Our approach is to use a new integrated aero-electromechanical design technique that incorporates the aerodynamic structure of an autorotating monowing into the sensor printed circuit board [4]. The combined structure is easily fabricated with automated mass-production methods, requiring little manual assembly. We have previously described this approach and its application to both gliding and autorotating designs [3]; the aerodynamics of this style of aircraft is well-established [5–7]. Previous experiments in deployment and distribution of samara-like devices from a fixed-wing vehicle were carried out by Ulrich and Pines at U. Maryland in 2009 [8].

While preliminary static drop tests and single release trials have been promising, it was not known whether the combined sensor-antenna-autorotation system would function in practice. We have undertaken a small-scale trial at the CSIRO QCAT Pullenvale facility, using a quadrotor as a deployment vehicle (see Fig. 1). These tests demonstrated the successful operation of the autorotation landing system, integrated radio antenna, and automated deployment device. This paper does not consider the detailed integrated aerodynamic-electronic-material design of these modules or their underlying theory of construction, which have been previously published [3].

2 Technical Approach

The system is made up of three parts: the sensor modules, the deployment device and the transport vehicle. The monowings comprise a sensor circuit and integrated wing structure (see Fig. 2). Each module includes temperature, relative humidity, pressure and ambient light sensors. An Invensense MPU-9150 MEMS IMU detects descent spiral motion, as well as orientation on the ground. This allows the sensor to only begin taking measurements and transmitting data once landed. A daughter-board adds a Ublox GPS module with a chip-antenna.

Sensor measurements are transmitted to a receiving base station via a 434 MHz radio module. The antenna for the radio is itself an integrated part of the wing structure. An additional helical stub antenna is soldered at right-angles to the board—this ensures that a radiating element is always vertical, regardless of landing orientation.

Fig. 1 Quadrotor deploying a sensor module

The lift provided by the wing limits the size of the battery to a sub-1 g 20 mAh cell, giving up to 12 h of passive sensing time (less with continuous radio transmission or GPS measurement). The entire sensor, microprocessor, battery, GPS and wing assembly weighs 25 g.

The release mechanism consists of a container, auger screw, drive servo and control electronics. Sensors are loaded into the coils of the auger, separated by a fixed space; the wings may be orientated to the left or right (see Fig. 3). When activated,

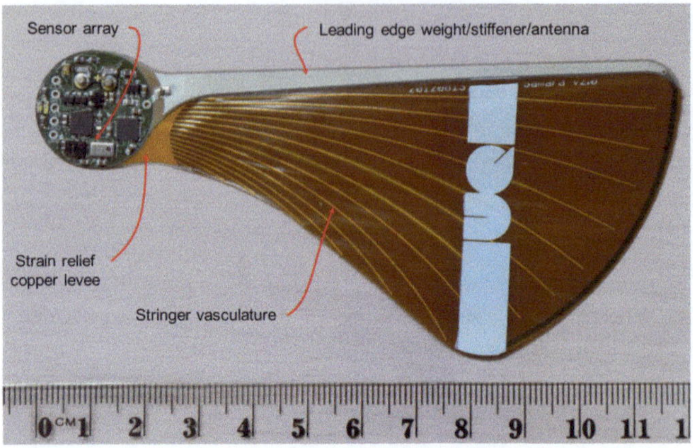

Fig. 2 Sensor module schematic

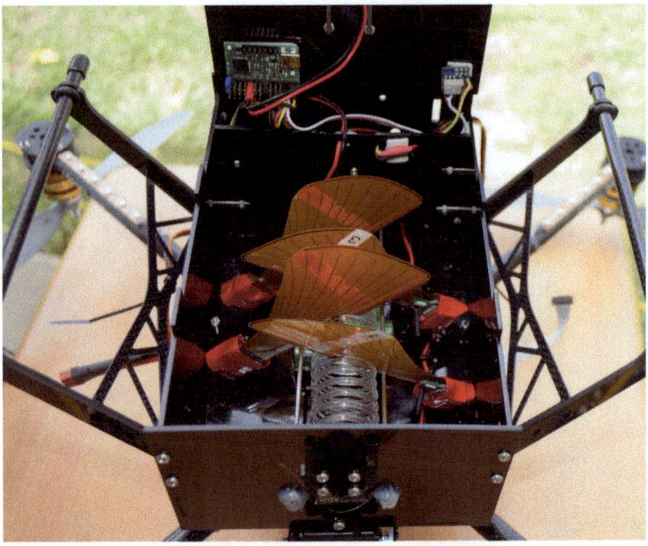

Fig. 3 Sensors loaded in release auger mechanism (*bottom view*)

the auger rotates, sliding the wings along guide rails until the front-most wing drops through an aperture. Once a wing has dropped, the auger reverses direction for a fixed number of turns to prevent the next sensor from accidentally vibrating free. The container was designed to carry up to 10 sensors. In this test, the increased thickness of the sensors due to the added GPS boards prevented more than four from being loaded at once.

The loaded container is mounted between the skids of a quadrotor platform, and connects to the avionics via a standard RC PWM interface. Software to automatically trigger a camera shutter can be used to command the release of a sensor when the UAV is at a target waypoint. The deployment platform is a 2 kg SkyBit Systems Eagle quadrotor, with autonomous waypoint navigation, altitude hold and camera activation functionality. The vehicle can fly for 16 min carrying the release mechanism and wings.

3 Experiments and Results

Experiments consisted of multiple flights to drop sensor modules at predefined waypoints. The release mechanism was triggered to deploy a sensor when the vehicle reached each waypoint.

Several sets of experiments were made. The first set was a dry run flight to test the release mechanism. Four wings were dropped from the same location at a height of 25 m. It was found that the mechanism was sensitive to loading orientation of the sensors: the first sensor deployed prematurely, the second worked normally, and the last two fouled in the release aperture. It was surmised that the asymmetrical centre of mass was causing the wing to impinge on the guide rails and catch. By turning all the wings to face the same orientation in the container, no further fouling occurred.

The second set of experiments involved releasing 16 sensors from four drop points over four flights (numbered one through four). This allowed the variability of drop location and orientation to be assessed, and in particular the influence of prevailing wind. The four drops were divided into two groups; the first two in close succession, a 2 h break, and then the final two—this allowed the varying wind direction to be measured.

For each drop, the landing GPS coordinates and orientation (right-side up, inverted upsidedown or sideways) were recorded (See Fig. 4). Flag numbers indicate the flight on which each sensor dropped; lines indicate the drop point each sensor was released from. On flight 4, a wing was unexpectedly deployed before the quadrotor reached the first waypoint; subsequent drops proceeded normally.

The soft-landing deployment system proved to be very reliable. While the time between release and the initiation of autorotation varied, every sensor completely entered autorotation and landed gently. It was found that the influence of the wind was significant, blowing sensors a significant distance from the drop point; the direction and distance traveled by each sensor varied. Contrary to expectations, the majority of sensors landed face-down. Once landed, environmental data was received from the sensors, but none achieved a GPS location fix: landing location had to be logged manually with a hand-held unit.

During the first drop deployment, a Go Pro camera recorded the descent of a sensor at 50 fps. This allows the relative motion of the sensor to be tracked as it descends (see Fig. 5). From the point of release, the wing's descent took approximately 6 s (297 frames), with the wing rotating at 10 Hz. During the first second of descent, the

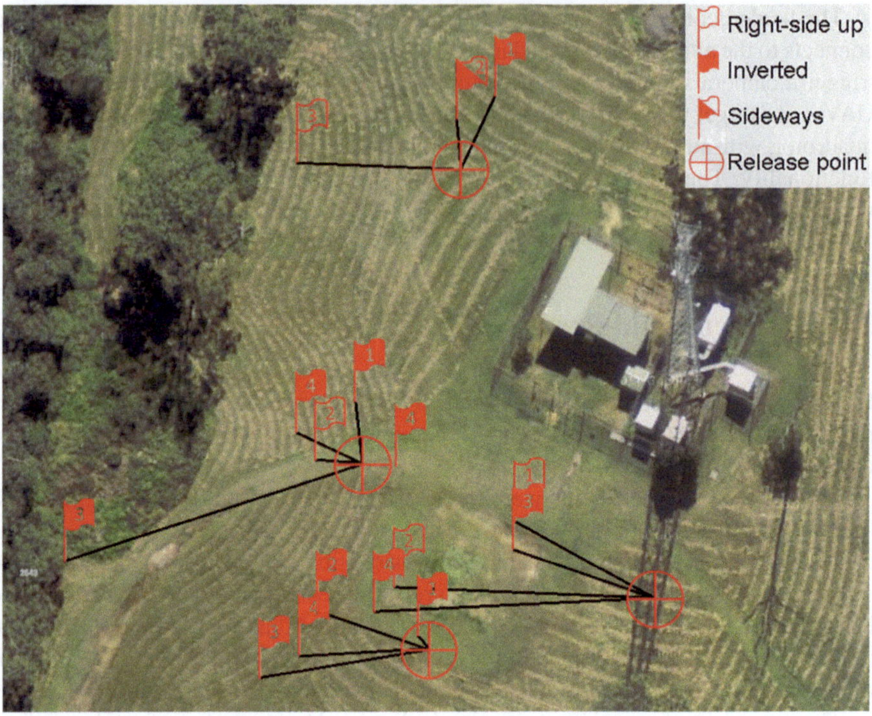

Fig. 4 Sensor variability scatterplot

dimensions of the wing are clearly visible, allowing a depth estimate to be extracted (as the total wing length is known). Beyond this point the wing could only be identified as a bright moving object. However, under day-time lighting conditions the sensor was readily tracked until it reached ground clutter.

A third set of experiments sought to map the wind velocity field around the testing area. Sensors were again released at 25 m, this time from twelve distinct drop points. The drops were performed over consecutive three flights. The corresponding landing location was logged for each drop, and plotted (see Fig. 6). Unlike the previous experiment, the more closely spaced samples show finer-grained detail about the wind pattern, and captures more noise. Unexpectedly, there are several measurements (most notably in flight 2) where the sensors landed in a direction opposing the prevailing wind conditions. It is thought that this is an indication the influence of local gusts and wind variability. During drop 2, it was also observed that gusts and updrafts close to scrub caused one of the sensors to briefly autorotate at constant altitude before coming to rest in a tree.

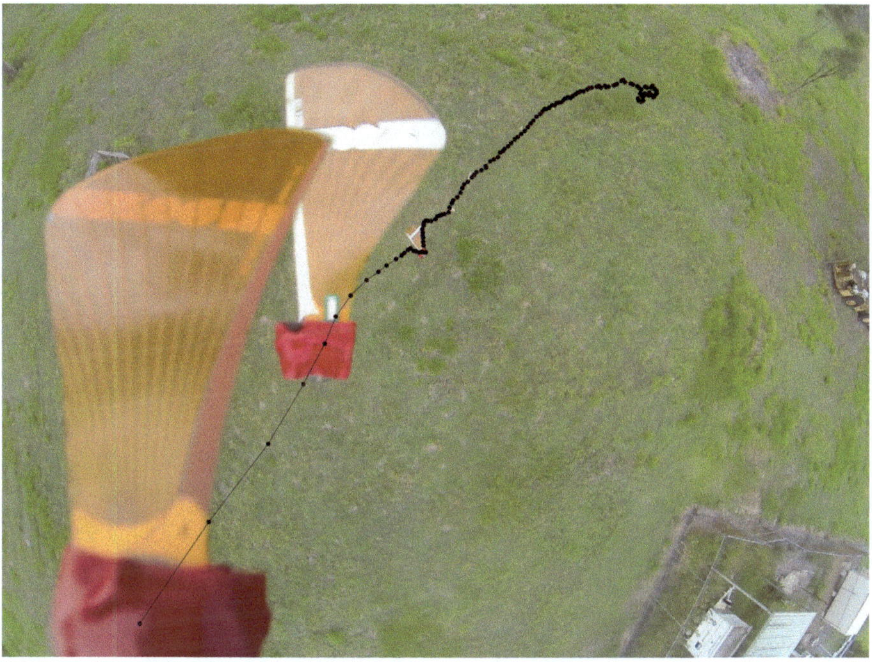

Fig. 5 Sensor descent trajectory montage. *Tracked point* is the centre of the sensor module in each frame, at 50 fps

4 Main Experimental Insights

The preliminary trials set out to identify hurdles to be overcome by practical deployments of monowing sensors—in particular, those that are not readily identified on the bench. Several potential complications were anticipated:

- Pathological aerodynamic modes (e.g. plunge or tumble)
- Unreliability of release mechanism
- Insufficient aerodynamic windmill braking
- Unpredictable landing orientation leading to poor antenna radiating angle
- Excessive drift in wind.

In practice, it was found that some of these potential problems did not arise, while others require further research.

The aerodynamics of the falling wings proved to be very reliable in triggering autorotation. Once the monowings begin to rotate, windmill braking rapidly slowed their descent. Even sensors that initially plummeted and began to rotate only 2 m above the ground slowed completely before touching down. However, in practice it will be essential to ensure the correct landing orientation, as long-term operation will require solar cells be exposed to the sky.

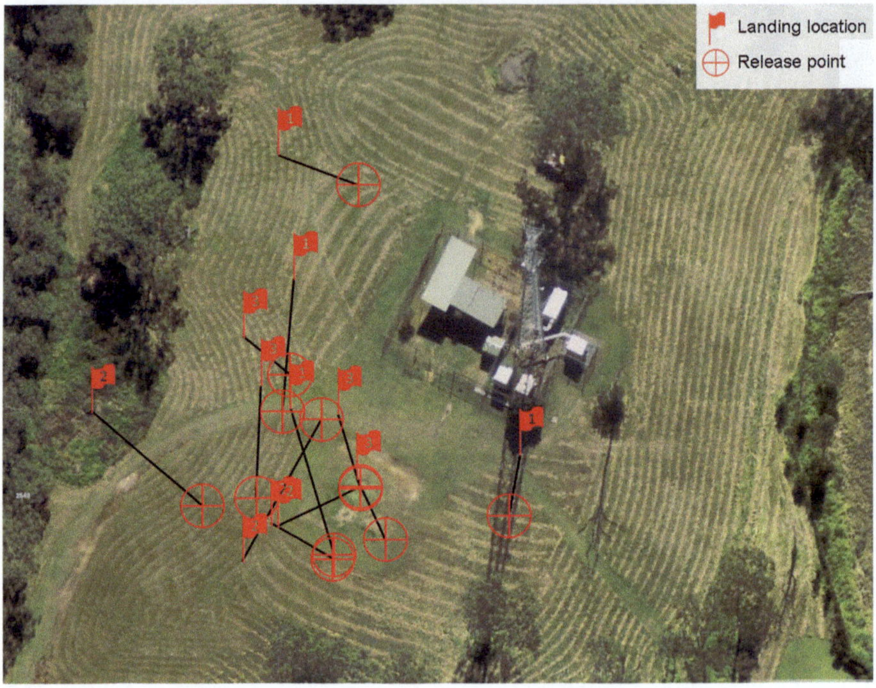

Fig. 6 Sensor distribution field

It is believed that the battery and added GPS module moved the centre of mass high on the wing, which caused it to invert in the first moments of release. This led the wing to be in an upsidedown pose as the wing started to enter autorotation, and maintaining the inverted attitude until the ground was reached. Future work will explore this influence and attempt to structure the aerodynamics to avoid undesirable autorotation orientation.

The inability of the GPS modules to quickly find lock indicates a key problem with the current design. It is likely that small antenna ground plane, face-down landings, long grass and surface clutter obstructing the antennas prevented the modules from finding satellites. In contrast, the multi-element antennas of the 434 MHz radios transmitted successfully. In future work, we will employ part of the flexible wing surface as a GPS antenna ground plane. Given the high probability of receiver obstructions, we may consider more sophisticated approaches where approximate high-altitude GPS fixes or optical tracking during descent are used in conjunction with radio communications to converge a global estimate of network node positions.

The prevailing wind direction is very clearly seen in the sensor scatter plot—the motion of the sensors is strongly dependent on air velocity. Its changing direction through the day can be readily identified. However, putting a sensor at a particular landing location with accuracy will require substantial new development. It is thought that by deploying 'scout' sensors and tracking their locations, a sparse wind map might be plotted so that future sensors can correct for local wind patterns. As more sensors are landed and tracked, greater precision of subsequent sensors may be achieved. However, as the wind variability plot indicates, there is substantial noise due to local time-varying gusts; it is expected that capturing comprehensive data will require a substantial increase in the number of sensors deployed.

Acknowledgments The authors would like to thank Craig Freakley and Stefan Harber for their assistance with this work.

References

1. Corke, P., Hrabar, S., Peterson, R., Rus, D., Saripalli, S., Sukhatme, G.: Deployment and connectivity repair of a sensor net with a flying robot, experimental robotics IX, pp. 333–343. Springer, Heidelberg (2006)
2. Song, W., Huang, R., Xu, M., Ma, A., Shirazi, B., LaHusen, R.: Air-dropped sensor network for real-time high-fidelity volcano monitoring. In: proceedings of 7th International Conference on Mobile Systems, Applications, and Services (2009)
3. Pounds, P., Singh, S.: Integrated electro-aeromechanical structures for low-cost, self-deploying environment sensors and disposable UAVs. In: proceedings of IEEE/RAS International Conference on Robotics and Automation (2013)
4. Valdes, S., Urza, I., Pounds, P., Singh, S.: Samara: low-cost deployment for environmental sensing using passive autorotation. In: Robotics: Science and Systems Workshop on Robotics for Environmental Monitoring (2012)
5. Ulrich, E., Pines, D.: Planform geometric variation, and its effect on the autorotation efficiency of a mechanical samara. In: AHS64th Annual Forum (2008)
6. Norberg, R.: Autorotation, self-stability, and structure of single-winged fruits and seeds (Samaras) with comparative remarks on animal flight. Biol. Rev. **48**(4), 561–596 (1973)
7. Varshney, K., Chang, S., Wang, Z.: The kinematics of falling maple seeds and the initial transition to a helical motion. Nonlinearity **25**(1), C1–C8 (2012)
8. Ulrich, E., Pines, D., Park, J.: Mechanical samara deployment and passive distribution by a fixed wing unmanned air vehicle. In: AHS International Specialist Meeting on Unmanned Rotorcraft (2009)

Towards Autonomous Lakeshore Monitoring

Shane Griffith, Paul Drews and Cédric Pradalier

Abstract This paper works towards autonomous lakeshore monitoring, which involves long-term operation over a large-scale, natural environment. Natural environments widely vary in appearance over time, which reduces the effectiveness of many appearance-based data association techniques. Rather than perform monitoring using appearance-based features, we are investigating whether the lakeshore geometry can provide a stable feature for this task. We have deployed an autonomous surface vessel 30 times over a duration of 8 months. This paper describes our initial analyses of this data, including our work towards a full simultaneous localization and mapping system and the shortcomings of using appearance-based features.

Keywords Lakeshore monitoring · SLAM · 3D reconstruction

1 Introduction

Efficiently monitoring a natural environment requires detecting and then exploring places that appear to be novel. With natural variation of appearance over short-, mid-, and long-term time scales, almost every location in an outdoor environment could be said to have changed, and thus, be a candidate for extensive exploration every time a robot is deployed there. The task of monitoring an expansive outdoor environment dictates a robot acquire an accurate model of the space. Yet, it is unclear how a robot could begin to acquire a model that would enable it to efficiently perform monitoring tasks.

S. Griffith (✉) · P. Drews · C. Pradalier
GeorgiaTech Lorraine-CNRS UMI 2958, Georgia Institute of Technology, Atlanta, USA
e-mail: sgriffith7@gatech.edu

P. Drews
e-mail: pdrews3@gatech.edu · C. Pradalier
e-mail: cedric.pradalier@gatech.edu

M.A. Hsieh et al. (eds.), *Experimental Robotics*, Springer Tracts in Advanced Robotics 109, DOI 10.1007/978-3-319-23778-7_36

Stationary monitoring tasks involve placing a camera in a predetermined location, which can simplify change detection because the camera is always pointing at the same spot and registering minor scene variation over consecutive frames. In this case, change detection is mostly unaffected by natural scene variation due to the high frame rate. Robotic monitoring, in contrast, involves a moving camera that captures intermittent snapshots of a scene. At first approximation, change detection following the stationary monitoring approach would require searching for images to compare and then finding an alignment between them. However, because the robot is capturing images of a non-planar surface, there usually is not a simple transform between images to align them. Furthermore, more variation accumulates between successive snapshots of a scene because the time interval between them is large.

Autonomous lakeshore monitoring calls for a representation that is stable across intermittent observations. The geometry of a lakeshore may be one such property. Some appearance-based features (e.g., SIFT, SURF) are invariant to many types of changes (e.g., scale, orientation, illumination), yet most are too unstable for comparing subsequent observations of a natural environment [7, 26]. However, it is the case that the appearance of many things in natural environments change with regularity, which may be possible to model in order to gain more predictive power. Thus, a robust representation of a lakeshore might use scene geometry as a basis for scene comparison, with spatiotemporal models of visual appearance supplementing it.

This paper starts to investigate the challenges of modeling a lakeshore environment using an autonomous surface vessel (ASV). Because a single sample of any outdoor environment is inadequate for capturing the distributions of its variations, we have deployed our robot for weekly data collection, so far 30 times. We are working towards applying SLAM techniques to extract an initial model of the lakeshore. This paper describes the challenges of using appearance-based features for data association between weekly surveys. This is ongoing work, in which we continue to collect data, improve our ASV system, and generate further analyses of our dataset.

2 Related Work

Autonomous lakeshore monitoring is a potential application of robotics for several reasons including the need to maintain water quality [11, 19], monitor the environmental effects of dams [25], identify adverse uses of a lakeshore [25], and survey rare plants [21]. Beyond these applications this problem is also interesting for its theoretical and practical challenges. Acquiring a 3D model of a lakeshore presents significant difficulties to the predominant 3D reconstruction and mapping techniques.

Existing work on 3D reconstruction, structure from motion, and simultaneous localization and mapping (SLAM) provide a well-established framework for addressing how to map an unchanging environment using camera images captured by a robot. These techniques usually involve extracting features from each image, performing feature matching between images, triangulating the 3D position of features using the estimated camera poses, and then refining these estimates using non-linear optimization techniques. Agarwal et al. [1] used a framework based on this approach to reconstruct a 3D depiction of some monuments of Rome in a day using images freely available on the internet. Davison et al. [6] showed how a robot can map an environment using a sequence of images from only a single camera.

Some papers have addressed the specific challenges of learning in outdoor environments: over a large area, with multiple observations over an extended period of time, accumulating several experiences of the same location, and in a variety of lighting and weather conditions, which make them especially relevant to our work. Churchill and Newman [4] present a system that avoids data association and instead accumulates "multiple experiences" of scenes, which consist of images that are localized using visual odometry and landmarks. This style of representation may be worth exploring for using with lakeshore monitoring. Glover et al. [7] combine two techniques for performing SLAM using images of outdoor scenes captured at different time intervals. Their approach can map an outdoor environment in a way that is somewhat robust to scene variation, but it undesirably generates a lot of new descriptors for re-visited locations. The authors speculate that their method's shortcomings are due to the fact that mapping is grounded in highly variable appearance-based keypoints. Nourani-Vatani and Pradalier [20] use optical flow to reduce feature matching time. The optimal flow indicates the direction the robot is headed. This information is saved to a topology, which indicates what set of visual features in the database to use for comparison. Ni et al. [18] propose a divide-and-conquer strategy for scalable optimization. A map is divided into submaps, which are optimized independently, and then combined later into a global map. Procopio et al. [22] show that an autonomous robot can use near-field stereo data to learn a classifier for identifying far-field obstacles in images. An ensemble of classifiers makes the system more robust to wide variations in the visual appearance of different outdoor scenes.

Images of a lakeshore consist primarily of land, water, and sky, yet research has found that different features may be more suitable for each one. Specifically, the predominant approach for modeling things on land involves extracting visual features (e.g., SIFT [16]); yet, mounting evidence suggests different techniques may be better suited for capturing information about water [8, 11, 12]. Iqbal et al. [12] reveal that a primary difficulty of trying to establish a fundamental vision-based feature for water detection is due to all the possible sources of variation in a scene. Furthermore, without the context that land provides, there is a lack of visual features in open and deep water [11]. Instead, other sensory modalities can be a good source of information about water, including laser [13], and audio and proprioception [8]. Currently, our research is more focused on representing the visual appearance of the lakeshore, rather than that of the water or the sky.

The use of a collection of image processing techniques for scene analysis is supported by successes modeling sources of scene variation. Sources of variation include, for example, shadows and artifacts, and methods have been developed to individually address each one. For example, explicitly representing the source of illumination (and then removing its effects, e.g., [5]) may allow us to more easily analyze how the appearance of certain plants correlates with changing levels of sunlight. Additionally, modeling scene variation may be easier if we eliminate effects that aren't likely to be repeated across surveys, like image artifacts (e.g., due to water droplets or dust on the camera protector) [9].

3 Experimental Setup

3.1 Robot

We used a Kingfisher from Clearpath Robotics for the experiments (see Fig. 1). The Kingfisher is an ASV propelled by a jet thruster in the end of each of its two pontoons. It is approximately 1.3 m long and 0.9 m wide, with space on top for optional sensors. Ours is equipped with an IMU, a compass, GPS, a forward-facing fish-eye camera, a top-mounted laser rangefinder, and a top-mounted pan-tilt camera. An onboard computer running ROS (see [23]) provides autonomous control, data logging, and communication for up to three hours on one charge.

3.2 Lake

We deployed the robot on Lac Symphonie in Metz, France, which is about 415 m long and spans roughly 220 m at its widest point (see Fig. 2). Its widest point is also the location of an island 131 m long and 87 m wide. Runoff water and a small tributary feed the lake while it drains into a nearby creek. Shrubs, bushes, trees, foliage, birds,

Fig. 1 The Kingfisher as it traversed the perimeter of Lac Symphonie

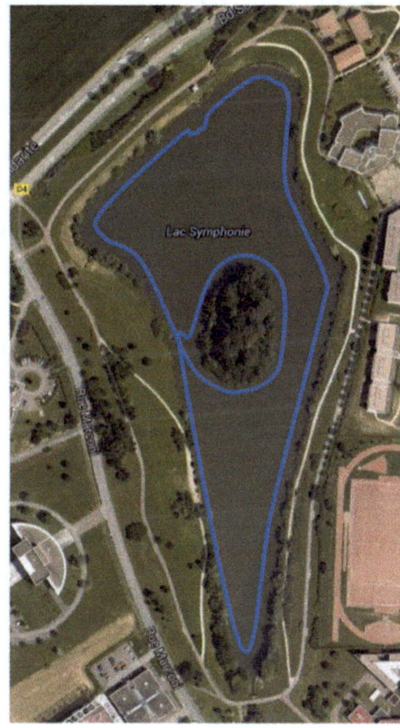

Fig. 2 Lac Symphonie in Metz, France from Google Maps. The *blue line* is approximately the path the Kingfisher traversed around the lake, roughly 10 m from the lakeshore

and pedestrians abound. A fitness path and a scenic trail encircle the lake. Collegiate and technology buildings loom in the background.

3.3 Behavior

The robot moves along the perimeter of Lac Symphonie and then the island with its pan-tilt camera pointed at the shore. It traverses them in a counter-clockwise direction at an average speed of 0.5 m/s. As it moves along the shore, the laser rangefinder captures a scan of distances to it, which a motion planner uses to optimize the boat's behavior for maintaining its 10 m distance. We chose 10 m to keep the kingfisher distant enough from the shore to avoid tree branches and shallow water, yet close enough to capture fairly high resolution snapshots. A human intervenes using an RC controller if the boat gets too close to tree branches or fishing lines. The boat performs one survey in approximately an hour.

A survey along the perimeter of the lakeshore was performed as often as once per week over a duration of ten months. From August 18, 2013 to June 13, 2014, the robot traversed the perimeter of the lakeshore a total of 30 separate times. Data could not be collected during weeks the lake was frozen, in rainy weather, or if we were traveling.

3.4 Natural Scene Variation

The robot captured significant scene variation across the entire observation period, with many instances of large scene variation between consecutive surveys. Much of the variation in the dataset comes from natural variation in appearance over three seasons. The trees and the bushes changed color and shed leaves, revealing buildings and other landmarks behind them. Moles continuously burrowed new mounds out of the grass. Changing water levels turned grass to mud and destroyed some plants. The cloud cover varied, the water rippled in different ways, and shadows appeared in different places.

Another source of scene variation in the dataset is due to the fact that the robot takes a slightly different path around the lake each time. Fluctuating water levels raise and lower the robot, but also change its distance to things on the shore. Swans occasionally affected the boat's path if they floated past its starboard side. The robot also sometimes got as close as 2–3 m away from shrubs near the shore that were too thin for the laser scan to consistently detect.

In the midst of natural scene variation, the most apparent changes in the scene were due to the activities of people. About half the surveys capture the construction of a new, shed-sized filtration building near the inlet to the lake. One survey captured canoes and kayaks on the shore for a water recreation event. Roughly a dozen fishermen held weekly competitions on Thursdays.

4 Methodology

4.1 Data Collection

This paper analyzes images primarily from two different surveys of the lakeshore. The boat's pose is estimated using GPS, IMU, and compass data. Images are captured as a sequence of 704×480 color images at 10 frames per second. A slight JPEG compression was applied to the images to increase the storage capacity of the boat.

4.2 Scene Reconstruction Using SIFT

We are working towards a monocular SLAM system for building a robust map of the environment, including dense scene reconstruction to capture the lakeshore geometry, data association across surveys, and optimization. This paper provides an analysis of scene reconstruction using SIFT.

Dense scene reconstruction is performed to precisely model the geometry of the lakeshore. The standard 3D estimation techniques provide a way to construct an estimate of the geometry from a sequence of images. Given matching points between two images, the 3D position is triangulated using Hartley and Thurm's iterative linear least-squares triangulation method [10]. Dense matches along a lakeshore enable the robot to more precisely estimate its geometry.

We initially used SIFT feature detection and matching [16] as provided by the OpenCV [2] library for acquiring a set of matches between pairs of images. Highly discriminative SIFT features (and other keypoint detection algorithms) are designed to be robust to large image transformations and changes in viewpoints. SIFT features are extracted from each image and then matches are found between pairs of images by comparing their feature sets. Because a lakeshore is, however, a mostly homogenous environment with very little change of viewpoint between frames, using SIFT can lead to a small number of matches, which are concentrated at high-contrast locations of each scene (e.g., buildings). Searching an entire feature set for each potential match can also make the process computationally inefficient. Our feature matching experiment illustrates these shortcomings.

5 Feature Matching Experiment

Our first experiment was useful for identifying the challenges of feature matching, both in within surveys and across different surveys. We applied our initial feature matching approach using SIFT to pairs of images of the same scene from the same survey, and to pairs images of the same scene from different surveys. Figure 3 shows the results. Many more matches are found between images from the same survey than from two different surveys. In the same survey condition, good matches are found on plants, the ground, and the building, but few are in the sky or in the water. Far fewer matches are found for images from two different surveys. The building is, however, the exception as its visual features are consistently matched across the surveys.

This result shows the difficulty of finding feature matches in images captured in natural environments at different times, even if they are of the same scene. With images of a lakeshore, the overall area of an image contributing to a successful match is small because SIFT features are only reliably detected on land. Additionally, the appearance of a natural scene in an outdoor environment can vary dramatically over any time scale, including abrupt changes in illumination, mid-term changes in water levels, and long-term seasonal variation in plants. Compared with flora, however, the contrast on buildings is more consistent across surveys, which is partly why many more features were found on the building.

Given that the number of feature matches on the lakeshore can be highly variable and that they are concentrated in high-contrast areas, we next identified the algorithm's coverage of the entire perimeter of the lakeshore. This analysis consisted of extracting features using SIFT, finding matches within the same survey, triangulating

Fig. 3 Snapshots of a scene along the lakeshore. The *colored dots* are the only features that could be matched. *Top left* the result of feature matching for two nearby images in the same survey. *Top right and bottom* the result of feature matching between this image and the *top left* image

the 3D locations of matched features, and then visualizing the points in a point cloud, for an entire survey. The point clouds for two different surveys are shown in Fig. 4. Each has enough points to show that their structures are very similar. The density of each point cloud shows where feature matches are found best around the perimeter of the lake. Noise is apparent in a good portion of the 3D position estimates due to the fact that optimization has not yet been performed on these results.

The point clouds capture much of the 3D structure of the lakeshore, but due to the sporadic coverage of SIFT feature matching, some locations are better represented than others. Many points are identified on high-contrast areas like buildings, trees, and unique terrain. The point cloud is a bit thinner in one survey due to the overcast weather. In general, areas with fewer points are either not illuminated well, are part of the featureless grassy bank, or are not viewed by the camera. In the top-left of both surveys, the low-setting sun caused sun glare, which reduced the performance of SIFT. Due to these shortcomings of SIFT, we are currently implementing further improvements to our feature extraction method (namely, KLT, and optimization) for improved coverage of natural environments.

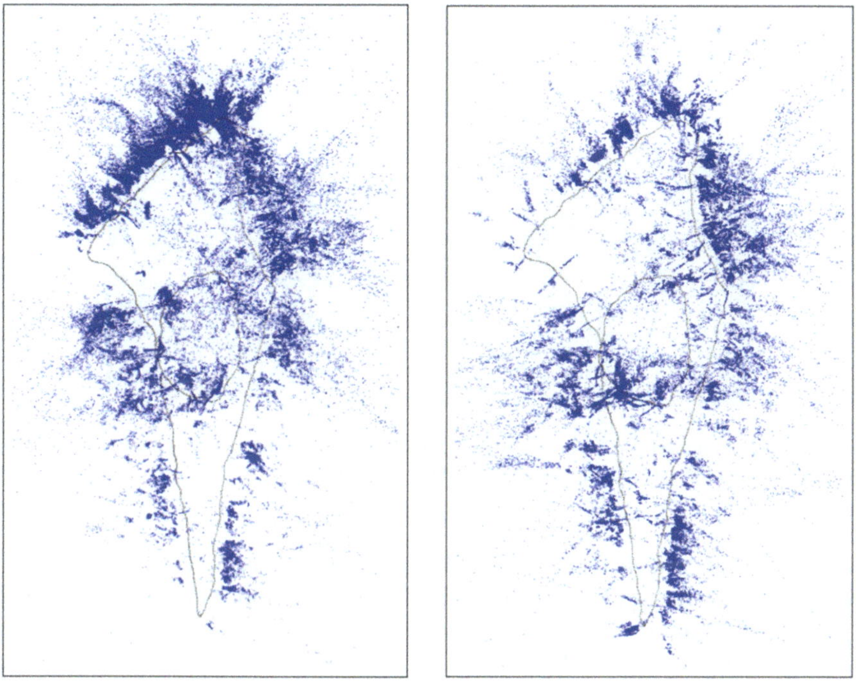

Fig. 4 Point clouds of *black* the boat's path around the lake, and *blue* the discriminative visual features in the scene, from two different surveys. Each image was generated using the Point Cloud Library [24]

6 Ongoing Work

6.1 Scene Reconstruction Using KLT

In light of the shortcomings of SIFT, we have identified that the pyramidal Lucas–Kanade tracker [17] (LKT or KLT, OpenCV) provides an excellent feature-tracking performance for our environment and experimental setup. Instead of matching feature descriptors, the Lucas–Kanade tracker uses the brightness constraint and a smoothness assumption to compute sparse optical flow on features detected by the Harris corner detector.

In practice, building reliable feature tracks over a large image sequence requires a few more steps, some generic and others specific to our experimental setup. We try to sustain a stable number (200–300) of features in each image, while keeping them well spread over the landscape. Toward this, new feature candidates are extracted for each image and then sorted into the cells of a grid, as illustrated in Fig. 5. Features are only added to cells in which there are currently no features being tracked, and feature

Fig. 5 Dense feature set and the grid used to enforce a relatively homogeneous feature distribution. The *red numbers* identify each feature and its ID

matches are only searched in a predicted neighborhood of cells. Thus, the number of features to track in each frame is limited, which ensures efficient computation time.

To limit the number of features ever tracked to a pre-specified maximum (300), a feature is removed from tracking when:

- OpenCV's Lucas–Kanade tracker cannot find a suitable match in the new image;
- its displacement between two frames is inconsistent with the robot's potential displacement (e.g., vertical displacement and large overall displacement);
- it moves into a cell in which multiple features (more than 5) have agglomerated (e.g., when features in the background are occluded);
- it is an outlier in a RANSAC-based fundamental matrix estimation.

The performance of feature tracking is depicted in Fig. 6. The number and the length of the black feature path shows that we can obtain reliable and stable features over fairly long sequences in a single survey. There are, however, several challenges associated with using this method for extracting dense coverage of a scene. A feature that is temporarily lost (e.g., if it is occluded or moves out of the field of view) and then reappears, for example, is not associated with its previous track in our implementation. Instead, a new track is created. Our current implementation does not address this limitation of KLT because it works well as-is.

6.2 Optimization

As the robot moves through the environment, it uses measurements of its motion and the projections of landmarks to compute estimates of the robot's poses and of the landmarks' positions, which are subsequently optimized using a non-linear optimization framework. Each measurement defines a constraint on the sequence of poses of

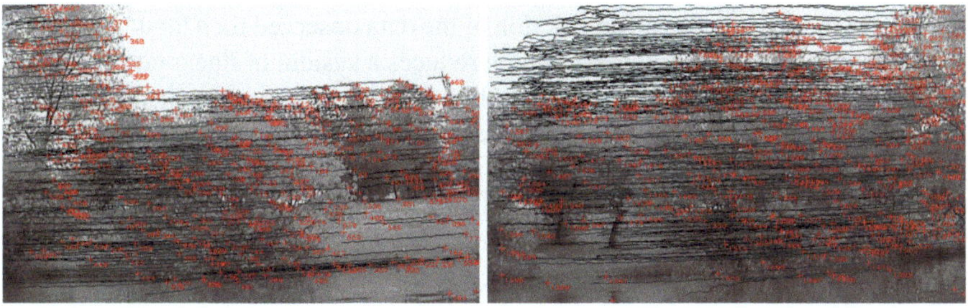

Fig. 6 Feature tracks (*black*) over two different sequences of 50 images. The *red text* identifies each feature and its ID. The length of each *black line* indicates the length of the feature track up that point

the robot and the positions of landmarks. The values of each pose and landmark position are optimized using the correlated data from the other measurements. As the robot explores, it acquires an increasing number of different poses and landmark positions to estimate.

This paper applies the iSAM2 framework to perform optimization [14]. iSAM2 first captures the dependencies among the sequence of robot poses and the landmark positions using a factor graph. Pose and projection measurements are represented as factors in the graph, which define the constraints on robot poses and landmark positions to be optimized. The robot pose and the landmarks' 3D positions are represented as hidden variables to be estimated. In the iSAM2 framework, the factor graph is converted into a Bayes tree, first by applying variable elimination to convert the factor graph into chordal Bayes net, and then by extracting a directed tree from the cliques of the Bayes net.

Representing the optimization problem as a Bayes tree allows for incremental updating without having to solve the entire optimization problem. The cliques of the Bayes tree affected by a new factor are removed and deconstructed into a factor graph. The variables of the new factor graph are reordered and converted back into a bayes tree, which is placed at the root of the unaffected portion of the deconstructed tree. For nonlinear factors, an additional relinearization step is performed to keep a valid linearization point for each variable.

To supplement iSAM2, we also utilize *smart factors* to reduce the computation time required for optimization [3]. Because we are applying SLAM in a long-term monitoring application, a very large number of factors accumulate to represent all the robot poses and the projection measurements of landmarks, and each directly increases the computation time required for optimization. It is possible, however, to reduce the number of factors used for optimization by taking advantage of conditional independence relations of landmark observations. Each landmark observation provides 'support' data, which is used for helping estimate a robot pose and a landmark's location. Fortunately, a set of support variables for one landmark is conditionally independent of a set of support variables for a different landmark given a *smart factor*.

A smart factor provides an abstraction of the data observed for a landmark. This is enabled by the Schur complement, which reduces a system of linear equations from one large problem into several smaller subproblems. As long as each subproblem is well-conditioned, solving these subproblems is equivalent to solving the large one. Because a smart factor abstracts all the projection data for a landmark, it can also seamlessly check that its subset of the data is well-conditioned. Degenerate cases like rotation-only movement, movement towards a landmark, and a single observation of a landmark are eliminated before they are used for optimization.

7 Conclusion and Future Work

Natural environments are challenging for the predominant SLAM data association techniques. Towards our long term goal, this initial paper found that SIFT, which is an appearance-based feature, provides spotty coverage of natural scenes and does not match well across surveys. These results are consistent with similar work on mapping in large natural environments over large time scales. In light of this, we have identified that KLT in addition to iSAM2 optimization may provide a more precise estimate of the geometry of a lakeshore.

This is ongoing work in which we are continually gathering data, improving our approach, and generating results. As we continue to get results, we are looking toward data fusion (using geometry for data association) in the optimization framework to help create a more comprehensive map of the lakeshore, as in [15]. We also plan to reduce the noise in robot pose estimates by applying the iterative closest point algorithm to the laser scan of the shore for improved odometry. Because this is still very preliminary work, we expect our methodology, analyses, and conclusions to substantially grow over time.

Acknowledgments Funding for this project is provided by the Lorraine Region, France.

References

1. Agarwal, S., Furukawa, Y., Snavely, N., Simon, I., Curless, B., Seitz, S.M., Szeliski, R.: Building rome in a day. Commun. ACM **54**(10), 105–112 (2011)
2. Bradski, G., Kaehler, A.: Learning OpenCV: computer vision with the OpenCV library. O'Reilly Media Inc. (2008)
3. Carlone, L., Kira, Z., Beall, C., Indelman, V., Dellaert, F.: Eliminating conditionally independent sets in factor graphs: a unifying perspective based on smart factors. In: 2014 IEEE International Conference on Robotics and Automation (ICRA) (2014)
4. Churchill, W., Newman, P.: Experience-based navigation for long-term localisation. Int. J. Robot. Res. **32**(14), 1645–1661 (2013)
5. Corke, P., Paul, R., Churchill, W., Newman, P.: Dealing with shadows: capturing intrinsic scene appearance for image-based outdoor localisation. In: 2013 IEEE/RSJ International Conference on Intelligent Robots and Systems (IROS), pp. 2085–2092. IEEE (2013)

6. Davison, A.J., Reid, I.D., Molton, N.D., Stasse, O.: MonoSLAM: real-time single camera SLAM. IEEE Trans. Pattern Anal. Mach. Intell. **29**(6), 1052–1067 (2007)
7. Glover, A.J., Maddern, W.P., Milford, M.J., Wyeth, G.F.: Fab-map+ ratslam: appearance-based slam for multiple times of day. In: 2010 IEEE International Conference on Robotics and Automation (ICRA), pp. 3507–3512. IEEE (2010)
8. Griffith, S., Sukhoy, V., Wegter, T., Stoytchev, A.: Object categorization in the sink: learning behavior–grounded object categories with water. In: Proceedings of the 2012 ICRA Workshop on Semantic Perception, Mapping and Exploration (2012)
9. Gu, J., Ramamoorthi, R., Belhumeur, P., Nayar, S.: Removing image artifacts due to dirty camera lenses and thin occluders. ACM Trans. Graph. (TOG) **28**(5), 144 (2009)
10. Hartley, R.I., Sturm, P.: Triangulation. Comput. Vis. image Underst. **68**(2), 146–157 (1997)
11. Hitz, G., Pomerleau, F., Garneau, M.E., Pradalier, C., Posch, T., Pernthaler, J., Siegwart, R.Y.: Autonomous inland water monitoring: design and application of a surface vessel. IEEE Robot. Autom. Mag. **19**(1), 62–72 (2012)
12. Iqbal, M., Morel, O., Meriaudeau, F., Komputer, F.I.: A survey on outdoor water hazard detection. In: The 5th International Conference on Information and Communication Technology and Systems (ICTS) (2009)
13. Jain, S., Nuske, S.T., Chambers, A.D., Yoder, L., Cover, H., Chamberlain, L.J., Scherer, S., Singh, S.: Autonomous river exploration. In: Field and Service Robotics, Brisbane (2013)
14. Kaess, M., Johannsson, H., Roberts, R., Ila, V., Leonard, J.J., Dellaert, F.: isam2: Incremental smoothing and mapping using the bayes tree. Int. J. Robot. Res. **31**(2), 216–235 (2012)
15. Kim, B., Kaess, M., Fletcher, L., Leonard, J., Bachrach, A., Roy, N., Teller, S.: Multiple relative pose graphs for robust cooperative mapping. In: 2010 IEEE International Conference on Robotics and Automation (ICRA), pp. 3185–3192 (2010)
16. Lowe, D.G.: Distinctive image features from scale-invariant keypoints. Int. J. Comput. Vis. **60**(2), 91–110 (2004)
17. Lucas, B.D., Kanade, T., et al.: An iterative image registration technique with an application to stereo vision. IJCAI **81**, 674–679 (1981)
18. Ni, K., Steedly, D., Dellaert, F.: Tectonic sam: exact, out-of-core, submap-based slam. In: 2007 IEEE International Conference on Robotics and Automation, pp. 1678–1685. IEEE (2007)
19. Niinioja, R., Holopainen, A.L., Lepistö, L., Rämö, A., Turkka, J.: Public participation in monitoring programmes as a tool for lakeshore monitoring: the example of lake pyhäjärvi, karelia, eastern finland. Limnol. Ecol. Manag. Inland Waters **34**(1), 154–159 (2004)
20. Nourani-Vatani, N., Pradalier, C.: Scene change detection for vision-based topological mapping and localization. In: 2010 IEEE/RSJ International Conference on Intelligent Robots and Systems (IROS), pp. 3792–3797. IEEE (2010)
21. Pavlovic, N.B., Bowles, M.L.: Rare plant monitoring at indiana dunes national lakeshore. Science and ecosystem management in the national parks. University of Arizona Press, Tucson, pp. 253–280 (1996)
22. Procopio, M.J., Mulligan, J., Grudic, G.: Learning terrain segmentation with classifier ensembles for autonomous robot navigation in unstructured environments. J. Field Robot. **26**(2), 145–175 (2009)
23. Quigley, M., Faust, J., Foote, T., Leibs, J., Wheeler, R., Ng, A.Y.: ROS: an open-source robot operating system. In: ICRA Workshop on Open Source Software (2009)
24. Rusu, R.B., Cousins, S.: 3d is here: point cloud library (pcl). In: 2011 IEEE International Conference on Robotics and Automation (ICRA), pp. 1–4. IEEE (2011)
25. Single, M.: Mokihinui hydro proposal consent applications review of assessment of effects at the coast. Technical Report (2008)
26. Valgren, C., Lilienthal, A.J.: SIFT, SURF and seasons: appearance-based long-term localization in outdoor environments. Robot. Auton. Syst. **58**(2), 149–156 (2010)

Part X
Many-Robot Systems

Alcherio Martinoli
EPFL

While the contributions of this session presented experimental work involving only a few robots (up to six), they leveraged radically different robotic platforms (quadrotors, miniature surface vessels and wheeled vehicles) confirming the recent trend of experimental diversification in distributed robotics. The classes of problem addressed were also varied: the work with aerial vehicles involved coverage missions, the contribution with surface vessels focused on loosely coordinated navigation across a stochastic field, and wheeled vehicles were used to investigate distributed learning policies. Although the approaches proposed took already into account realistic constraints of the robotic platforms (e.g., limited energetic autonomy, maneuverability, computational resources), only one contribution reported some initial field results outlining that it is still today quite difficult to gather systematic data of many-robot experiments outside the laboratory.

In the first paper, *Controlling Basin Breakout for Robots Operating in Uncertain Flow Environments*, C.R. Heckman, I.B. Schwartz and M. Ani Hsieh present their development and experimental validation of an autonomous surface/underwater vehicle (ASV/AUV) control strategy that leverages the environmental dynamics and noise to efficiently navigate in a stochastic fluidic environment. The environment is assumed to be tessellated in the so called Lagrangian Coherent Structures (LCS), structures that can be actually detected and leveraged in real oceans, typically for optimizing fuel consumption. The main contribution of the paper is to show how to formulate energy-efficient, minimal control laws able to leverage escape paths arising from noise-induced large fluctuations in order to navigate from one LCS-bounded region to another. The merit of the paper is to show how solid theoretical knowledge about the LCS can be translated into concrete control laws validated both in simulation and through an original laboratory test-bed.

The second and fourth papers in the session are concerned with distributed surveillance missions involving multiple quadrotors. In *QuadCloud: A Rapid Response Force with Quadrotor Teams*, the authors K. Mohta, M. Turpin, A. Kushleyev, D. Mellinger, N. Michael, and V. Kumar design their multi-robot system for achieving a fast response speed, for instance in case situational awareness must be rapidly acquired before the intervention of human rescue teams in a disaster scenario. The paper focuses on the integration of the various elements needed to achieve dependable operation of up to six quadrotors controlled by a single human operator. The merit of the paper is to report results related to an actual field deployment and to consider an end-to-end system. In *Provably Correct Persistent Surveillance for Unmanned Aerial Vehicles Subject to Charging Constraints*, the authors K. Leahy, D. Zhou, C. Vasile, K. Oikonomopoulos, M. Schwager, and C. Belta design their multi-robot system for achieving long-term patrolling of a given number of regions of interest. Their approach combine automata-based techniques (the constrained optimization problem is captured with Bounded Linear Temporal Logic) with a vector-field method assuming differentially flatness of the vehicle dynamics. In order to physically demonstrate the validity of the approach as well as the persistency of the surveillance, the contribution leverages original stations for automated charging of quadrotors. The merit of this paper is to show a concrete, provably correct solution for carrying out tasks lasting longer than the typical short battery life of current rotor-based aerial vehicles (30–60 min).

Finally, the third paper of the session, *Distributed Learning of Cooperative Robotic Behaviors using Particle Swarm Optimization* by E. Di Mario, I. Navarro, and A. Martinoli, proposes a method based on Particle Swarm Optimization for distributing the learning process across multiple robots. The method is robust to large parameter spaces (e.g., 26 real parameters per robot in the proposed case study) and achieves similar performances to those obtained using its centralized counterpart. The authors validate their method on a collaborative task (navigation in loose formation or flocking) involving a quantitative misalignment between local and global performance metrics used by the learning process in its distributed and centralized version, respectively. The merit of this paper is to show that distributed learning can deliver competitive performances even in collaborative tasks involving a large search space and to support this claim with a thorough experimental campaign based on high-fidelity simulations and real robots.

Controlling Basin Breakout for Robots Operating in Uncertain Flow Environments

Christoffer R. Heckman, M. Ani Hsieh and Ira B. Schwartz

Abstract We present the development and experimental validation of an autonomous surface/underwater vehicle (ASV/AUV) control strategy that leverages the environmental dynamics and noise to efficiently navigate in a stochastic fluidic environment. In this work, we assume the workspace is composed of the union of a collection of convex regions, each bounded by Lagrangian coherent structures (LCS). LCS are dynamical features in the flow field that function like invariant manifolds in general non-autonomous dynamical systems and they denote regions in the flow field where more escape events occur. We show through theory and simulation that a vehicle's likelihood of transition between adjacent LCS-bounded regions can be manipulated by the proposed control strategy, resulting in effective navigation strategies from one region to another. In addition, we show how optimal escape trajectories with respect to the transition probability between adjacent LCS-bounded regions can be determined. These trajectories correspond to energy-efficient trajectories since they leverage the inherent dynamics of the surrounding flow field. We experimentally show that the proposed control strategy exhibits a predictable exponential scaling of escape times and is effective even in situations where the structure of the flow is not fully known, there exist significant stochastic fluctuations, and control effort is costly.

Keywords Navigation · Underwater robots · Lagrangian coherent structures

---

C.R. Heckman (✉) · I.B. Schwartz
Nonlinear Systems Dynamics Section, Plasma Physics Division, Code 6792,
U.S. Naval Research Laboratory, Washington, DC 20375, USA
e-mail: christoffer.heckman@colorado.edu

I.B. Schwartz
e-mail: ira.schwartz@nrl.navy.mil

M.A. Hsieh
SAS Laboratory, Mechanical Engineering and Mechanics Department, Drexel University,
Philadelphia, PA19104, USA
e-mail: mhsieh1@drexel.edu

M.A. Hsieh et al. (eds.), *Experimental Robotics*, Springer Tracts
in Advanced Robotics 109, DOI 10.1007/978-3-319-23778-7_37

1 Introduction

An increase in ocean research over recent years has led to an explosion of interest in the deployment of robots to track various physical, chemical, and biological processes in the ocean, e.g., ocean temperature, the onset of harmful algae blooms, and dispersion of hazardous contaminants. However, geophysical fluid environments span large physical scales, are stochastic and have time-dependent dynamics and working in them presents significant challenges. For instance, as a result of the stochastic behavior of the ocean, sensors will tend to leave their monitoring region of interest, limiting the amount of useful data that can be collected. Despite these challenges, geophysical flows do exhibit coherent structure. Coherent structures are important because they give us insight into the dynamics of the fluidic environment. This knowledge, in turn, can be leveraged to improve the various deployment strategies for unmanned systems operating in them.

We are interested in a specific class of coherent structures called Lagrangian coherent structures (LCS) which are important for quantifying transport phenomena. LCS are the extensions of stable and unstable manifolds to general time-dependent flows [1], and divide the flow into dynamically distinct regions. For two-dimensional (2D) flows, LCS are analogous to ridges defined by local instability, and can be quantified by Finite-Time Lyapunov Exponents (FTLE) [2]. FTLEs provide a local measure of how fast a region in the flow field is expanding. Recent work has shown that fuel optimal paths in the ocean coincide with these LCS [3, 4]. Moreover, in stochastic systems it has been shown that optimal paths lie directly along local maxima of FTLEs [5]. Since LCS denote approximate boundaries between flow regions, they demarcate regions in the flow where escape events from one region to another occur with higher probability [6]. This suggests that using efficient controls to increase/reduce uncertainty near the LCS boundaries may increase/reduce the likelihood of the mobile sensor breaking out, or escaping, from its monitoring region and consequently provide an efficient navigation strategy or prolong the monitoring lifetime of the sensor.

Recently, new mathematical methods have been developed to elucidate and harness the effects of noise on dynamical switching behavior [6–9]. These methods accurately predict the expected switching time of robots between distinct basins of attraction. Moreover, the techniques predict the most probable (or optimal) escape path from a region resulting from a large fluctuation due to the underlying noise. Building on these techniques, we identify and analyze the most likely routes of transport between two adjacent LCS bounded regions for passive vehicles, i.e., vehicles with no actuation capabilities. We analyze the escape times via these most likely routes of transport and show that they correspond to fuel efficient paths since the environmental dynamics and noise is leveraged to achieve navigation. From these results, we design a simple control strategy and show how it can effectively manipulate a vehicle's escape times from an LCS-bounded region and achieve comparable or improved fuel efficiency. The result is a control strategy that can be implemented with limited knowledge of the flow field and leverages the surrounding fluid dynamics and the inherent environmental noise to minimize the overall control effort of the vehicle.

The structure of the paper is as follows: We formulate the problem and outline key assumptions in Sect. 2. The identification and analysis of the optimal paths for escape and the development of our control strategy is presented in Sect. 3. We briefly describe our experimental methodology in Sect. 4 and present our experimental results in Sect. 5. We conclude with a discussion of our results and experimental insights in Sect. 6 and directions for future work in Sect. 7.

2 Problem Statement

We consider the deployment of a planar autonomous underwater vehicle (AUV) within a two-dimensional (2D) obstacle free fluidic environment. We are interested in the deployment of mobile sensors with minimal actuation capabilities that can leverage the surrounding fluid dynamics to maneuver within the workspace. As such, we assume the following 2D kinematic model for the AUV:

$$\dot{\boldsymbol{q}} = \boldsymbol{u} + \boldsymbol{F}(\boldsymbol{q}) + \boldsymbol{\eta}, \tag{1}$$

where $\boldsymbol{q} = (x, y)$, $\boldsymbol{u}$ denotes the control input, $\boldsymbol{F}(\boldsymbol{q}) = [F_1(\boldsymbol{q}), F_2(\boldsymbol{q})]^T$ is a 2D planar vector field that describes the surrounding fluid dynamics, and $\boldsymbol{\eta}$ is a zero-mean stochastic white noise term with standard deviation $\sigma = \sqrt{2D}$ for a given noise intensity D. In this work $\boldsymbol{\eta}$ represents the inherent noise in the environment but can also reflect a vehicle's measurement and/or actuation uncertainty.

Let $\mathscr{W}$ denote the obstacle-free workspace. In this work, we assume $\mathscr{W} = \cup_{i=1}^{M} G_i$ where G_i denotes a convex Lagrangian coherent structure (LCS) bounded region. In 2D flows, LCS are one-dimensional (1D) boundaries that exhibit maximum Finite-Time Lyapunov Exponents (FTLE) measures [2]. Figure 1 shows an example 2D flow field and its corresponding FTLE field. A tessellation of the workspace along

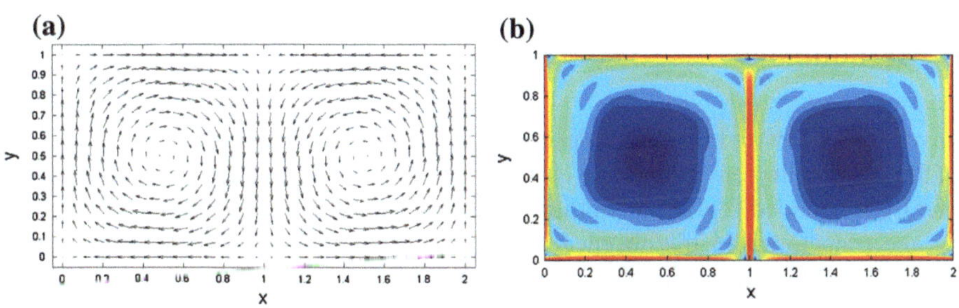

Fig. 1 Phase portrait **a** and corresponding FTLE field **b** of the model given by (9) with $A = 1$, $\mu = 0$, and $s = 1$

LCS, or boundaries characterized by maximum FTLE ridges, makes sense since: (1) they are dynamically distinct regions in a 2D flow field, and (2) they correspond to regions in the flow field where more escape events occur [6]. In the time-independent case, LCS correspond to stable and unstable manifolds of saddle points in the system where the manifolds can also be characterized by maximum FTLE ridges.[1] Since the manifolds demarcate the basin boundaries separating the distinct dynamical regions, these are also regions where uncertainty with respect to flow velocity vectors is high. Therefore, a passive particle's escape from G_i, or switching between two adjacent G_i's is influenced both by deterministic uncertainty as well as stochasticity due to external noise.

The objective is to synthesize a control strategy that enables the AUV or robot to navigate between physically adjacent G_i's by leveraging the surrounding fluid dynamics and the inherent environmental noise to minimize the vehicle's control effort. Specifically, we are interested in control strategies that can be implemented on mobile sensors with minimal actuation capabilities. In this work, we assume the vehicle has the ability to localize within $\mathscr{W}$ and the ability to measure the flow velocity at its current position. Additionally, we assume $\boldsymbol{F}(\boldsymbol{q})$ is known or can be inferred using knowledge of the LCS boundaries and that the tessellation of $\mathscr{W}$ into the individual G_i is given.

In general, it is unreasonable to assume $\boldsymbol{F}(\boldsymbol{q})$ is perfectly known for any given workspace $\mathscr{W}$. As such η in Eq. (1) not only encompasses both inherent environmental noise and noise in the vehicle's measurement and/or actuation, but also uncertainties in the model given by $\boldsymbol{F}(\boldsymbol{q})$. Furthermore, LCS delineate regions in the flow that exhibit distinct dynamics. This suggests that it is possible to provide a reduced-order representation of $\boldsymbol{F}(\boldsymbol{q})$ using only the locations of LCS boundaries. The locations of LCS boundaries can be determined in general using historical and ocean model data, however recent work has shown that LCS boundaries can also be tracked, potentially in real-time, using the strategy described in [10]. As such, a second objective of this work is to show that it is possible to achieve comparable performance when synthesizing control strategies for AUVs using either full knowledge of $\boldsymbol{F}(\boldsymbol{q})$ or only knowledge of the LCS boundary locations, i.e., partial knowledge of $\boldsymbol{F}(\boldsymbol{q})$. We describe our methodology in the following section.

3 Methodology

Given an AUV with kinematics described by Eq. (1), to better understand the impact of the surrounding environmental dynamics on the vehicle's ability to transition from one LCS bounded region to another, we first summarize the analysis of a vehicle's escape trajectories in the absence of controls from an LCS bounded region. The analysis will then motivate the proposed vehicle controller.

[1] The FTLE are computed based on a backward (attracting structures) or forward (repelling structures) integration in time.

3.1 Analysis

Under the influence of noise, the dynamical behavior of the system is determined by its stationary probability density. In particular, all equilibria are now peaks or troughs in a probability landscape describing where a particle is likely to be located. Although there exist many paths that transport a particle from one LCS bounded region to another, in the presence of small noise there are "most likely paths" in the sense that they will lie along a local peak in the probability density. We characterize this path to transition from one LCS bounded region to another by considering the most likely paths between the two states. We briefly summarize the analysis for the sake of completeness and refer the interested reader to [11] and [12] for the specific details.

The probability of escape from an attractor under the influence of small white noise scales exponentially [13] as

$$\mathscr{P}_{\eta}(\boldsymbol{q}) \asymp \exp(-\mathscr{R}(q))/D, \tag{2}$$

where $\mathscr{R}$ represents a dynamical quantity known as the *action*. Following the ideas in [14], we notice that for any given realization of noise, we have the approximate density scaling as $\exp(-\frac{1}{2}\int|\boldsymbol{\eta}|^2)$. If we continue the thinking of Feynman [13] and assume the noise is white, then for any realization of noise we get the action defined in [14]. We take a general Hamiltonian approach which also may be extended to escape induced by non-Gaussian noise [15].

To put the problem into Hamiltonian formulation, we characterize the paths that require the minimum action of the dynamics and noise to cause the transition. In this approach, the background flow field $\boldsymbol{F}(\boldsymbol{q})$ in Eq. (1) becomes a constraint. Assuming $\boldsymbol{u} \equiv \boldsymbol{0}$, the action functional for the noise is:

$$\mathscr{R}[\boldsymbol{q}, \boldsymbol{\eta}, \boldsymbol{\lambda}] = \frac{1}{2}\int \boldsymbol{\eta}(t) \cdot \boldsymbol{\eta}(t)dt + \int \boldsymbol{\lambda} \cdot (\dot{\boldsymbol{q}} - \boldsymbol{F}(\boldsymbol{q}) - \boldsymbol{\eta})dt. \tag{3}$$

When evaluating the action in Eq. (3), we compute $R = \min \mathscr{R}[\boldsymbol{q}, \boldsymbol{\eta}, \boldsymbol{\lambda}]$, where the minimum is taken over the functions $[\boldsymbol{q}, \boldsymbol{\eta}, \boldsymbol{\lambda}]$. Setting the first variation of the functional in Eq. (3) to zero will yield a system of differential equations that identify which solutions extremize the action in terms of the path and the minimum noise necessary to realize the path. This solution is the most probable path, even though it is rare and exists in the tail of the probability distribution for realizations of the noise. It is important to note that this procedure is analogous to the deterministic optimal trajectory generation problem where Eq. (3) is the objective function with $\boldsymbol{\eta}(t)$ as the control input. Different from the canonical optimal control problem [16], λ_i represent the conjugate variables to the state vectors.

The switching rate to transition from one LCS-bounded region to another is directly proportional to the probability of observing the most likely noise profile to induce such a switch [17]; all other noise realizations are exponentially less likely. Therefore, the mean switching time may be approximated as:

$$T_S = b \exp\left(\frac{R}{D}\right) \tag{4}$$

where b is a prefactor determined through numerical simulation or experiment. Note that even though this mean escape time may be approximated using Eq. (4), the fact that other noise realizations are exponentially less likely implies that the distribution of escape times due to noise is *exponential* in nature.

In order to calculate the switching time in Eq. (4), we must find the optimal trajectory and noise to induce a switch. Computing the variational derivatives of the action and setting them to zero results in the following differential equations:

$$\dot{x} = F_1(x, y) + \lambda_1 \tag{5}$$

$$\dot{y} = F_2(x, y) + \lambda_2 \tag{6}$$

$$\dot{\lambda}_1 = -\frac{\partial F_1}{\partial x}\lambda_1 - \frac{\partial F_2}{\partial x}\lambda_2 \tag{7}$$

$$\dot{\lambda}_2 = -\frac{\partial F_1}{\partial y}\lambda_1 - \frac{\partial F_2}{\partial y}\lambda_2 \tag{8}$$

where λ_i represent the conjugate momenta to x, y. Note that this is a conservative system which may be derived from a Hamiltonian.

The optimal path to transition between adjacent LCS-bounded regions is determined by solving the above system of equations with boundary conditions describing G_i and equilibria of the flow given by Eqs. (5)–(8). Let the equilibria $\boldsymbol{q}_A$ denote a location within G_i and $\boldsymbol{q}_B$ denote a location on the boundary of G_i. Then the boundary conditions for the most likely path are, for $\boldsymbol{F}(\boldsymbol{q}_A) = \boldsymbol{F}(\boldsymbol{q}_B) = \boldsymbol{0}$, $\boldsymbol{q}(t \to -\infty) = \boldsymbol{q}_A$, $\boldsymbol{q}(t \to \infty) = \boldsymbol{q}_B$, and $\boldsymbol{\lambda}(t \to -\infty) = \boldsymbol{\lambda}(t \to \infty) = \boldsymbol{0}$. Determining the optimal path connecting $\boldsymbol{q}_A$ and $\boldsymbol{q}_B$ is equivalent to solving a two point boundary value problem in four dimensions.

In this work, we model the external flow field $\boldsymbol{F}(\boldsymbol{q})$ using the wind-driven double gyre flow model which is often used to describe large scale recirculation in the ocean [18] and is given by:

$$F_1 = -\pi A \sin(\pi x)\cos(\pi y/s) - \mu x,$$
$$F_2 = \pi A \cos(\pi x)\sin(\pi y/s) - \mu y. \tag{9}$$

The parameter μ is a damping coefficient, s is a scaling dimension for the gyres, and A corresponds to the strength of the gyre flow. Figure 1a shows the phase portrait of Eq. (9) and Fig. 1b shows the corresponding FTLE ridges. In this model, the workspace consists of a grid of gyres where each gyre defines an LCS-bounded region. As shown in Fig. 1a, flows in adjacent gyres circulate in opposite directions. Each gyre has an attractor located in the center of G_i and, since the system is time-

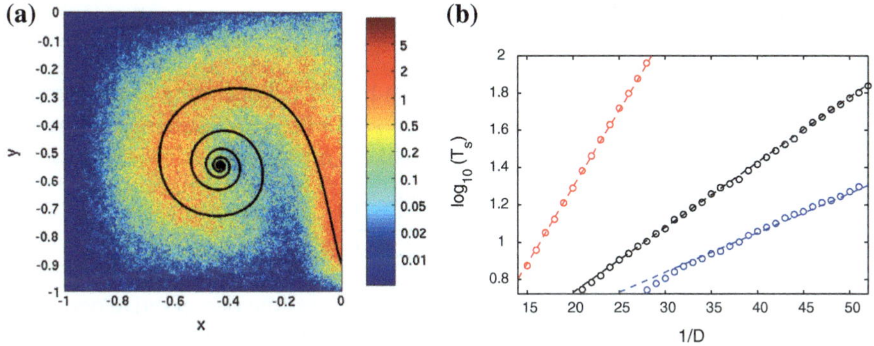

Fig. 2 **a** Comparison of the escape trajectories (*color*) and the optimal path to escape as calculated using theory (*line*) for the case of no control. *Red* indicates a higher incidence of occurrence while *blue* shows lower occurrence. The results are shown on a logarithmic scale to accentuate the ridge of higher probability. **b** Comparison of the switching time T_S as predicted by large-fluctuation theory (*dashed lines*) and computed from an average over many stochastic trials (points). Three regimes were examined: *black* represents no control, *red* represents $c \approx -0.31$ and *blue* represents $c \approx 0.1$. The parameters used to for both figures were $A = 1$, $s = 1$ and $\mu = 1$

invariant, the boundaries of each G_i consist of the stable and unstable manifolds of the saddle points located at the four corners of each G_i.

In our analysis, we set $\boldsymbol{q}_A$ to be the gyre attractor and $\boldsymbol{q}_B$ as one of the boundary saddle through which escape occurs.[2] Since both are saddle equilibria in the full set of the equations of motion, finding the most optimal path mathematically requires identifying a heteroclinic orbit connecting the two points. To solve for the path numerically, we implement an algorithm known as the IAMM [19]. We then employ continuation using AUTO's HomCont [20] to increase the accuracy of the approximation and to study the behavior of the path for different parameter values.

Figure 2a shows a simulation-derived probability density of paths to escape imposed over the optimal path as predicted by the theory. From this theoretical analysis, we note that "optimal" escape paths from G_i move the particle/passive vehicle along the direction of the flow field towards the boundaries of G_i. This suggests that it is possible to devise a simple control strategy that will enable an autonomous vehicle to efficiently navigate from one LCS-bounded region to another by leveraging the surrounding flows and inherent environmental noise without requiring explicit computation of "optimal" escape paths and subsequently full knowledge of the vector field. We present our controller synthesis in the following section.

[2] In the presence of noise, the likelihood of escape for any particle in G_i is dependent on the particle's proximity to the gyre boundaries and the noise intensity. Near the boundary saddle the vector field becomes very weak, resulting in noise dominating the dynamics. This results in high instability and corresponds to high escape likelihoods in the neighborhood of the boundary saddle.

3.2 Controller Synthesis

Consider a planar AUV with kinematics given by Eq. (1) operating in a workspace $\mathscr{W} = \cup_{i=1}^{M} G_i$ where each G_i denotes a convex region whose boundaries are defined by LCS such that the flow within each G_i is gyre-like. Consider the following controller:

$$\boldsymbol{u} = \boldsymbol{\omega} \times c \frac{[f_1, f_2, 0]}{||[f_1, f_2, 0]||}, \tag{10}$$

where c denotes the controller gain, $[f_1, f_2, 0]$ represents the conservative part of the vector field $\boldsymbol{F}$ with an augmented third dimension, and $\boldsymbol{\omega} = [0, 0, 1]^T$. The above control strategy was first introduced in [21] and further analyzed in [12]. When $c > 0$ the controller effectively pushes the vehicle towards the boundary of G_i in a direction that is perpendicular to the flow at the vehicle's location. Similarly, when $c < 0$ the controller pushes the vehicle towards the center of G_i.

A significant advantage of the proposed control strategy in Eq. (10) over the computation of optimal escape paths is that it does not require full knowledge of the flow field. Furthermore, when $c > 0$, a vehicle is more likely to escape a given G_i as c increases. Figure 2b demonstrates this relationship by comparing the theoretically predicted (based on the optimization framework described in Sect. 3.1) and numerically simulated (Monte-Carlo simulations) switching time from one G_i to any adjacent G_i as a function of the noise intensity for three values of control gain c. Note that escape times *exponentially* decrease when $c > 0$ while escape times increase when $c < 0$ for different noise intensities, as predicted by Eq. (4).

These results suggest that the controller given by Eq. (10) effectively aids or abates the probability to transition between basins and affects the switching time. Furthermore, similar to the "optimal" escape paths, the proposed control strategy requires minimal control effort expenditure of the vehicle by leveraging the surrounding environmental dynamics. As such, the proposed strategy should result in more energy-efficient trajectories compared to shortest distance paths that do not account for the external flow field. The experimental validation of our approach is presented in the following sections.

It is important to note that for robots to employ the proposed controller they must have knowledge of $\mathscr{W}$, the locations of the boundaries of G_i, and the ability to localize within each G_i. While it may be unreasonable to expect resource constrained AUVs to be able to determine the LCS locations in real-time, it is possible to compute the LCS boundary locations using historical and ocean model data obtained a priori. Furthermore, recent work has shown the possibility of tracking these boundaries using the strategy described in [10]. As such, these assumptions are no different than assuming that an autonomous ground or aerial vehicle has a map of the environment.

4 Experimental Methodology

To evaluate the control strategy using an experimental robotic platform, we will employ our multi-robot Coherent Structure Testbed (mCoSTe) [22]. The mCoSTe is an indoor laboratory experimental testbed that consists of several flow tanks and a fleet of micro-autonomous surface vehicles (mASVs). The mASVs are differential drive surface vehicles equipped with a micro-controller board, XBee radio module, and an inertial measurement unit (IMU). The vehicles are approximately 12 cm long and have a mass of about 45 g each (see Fig. 3b). Localization for the mASVs is provided by an external motion capture system.

The mCoSTe also includes two experimental flow tanks of relevance to this investigation: the High Reynolds number (HiRe) Tank and the Multi-Robot (MR) Tank, which are respectively $0.6 \times 0.6 \times 0.3\,\text{m}^3$ and $3 \times 3 \times 1\,\text{m}^3$ in size (see Fig. 3a), and the latter of which we used for our experiments. Both the HiRe and MR tanks were designed to be able to create time-independent and time-varying flow fields that exhibit kinematic and transport features similar to those observed in the ocean. The flows in the tanks are patterned after the gyre flow model given by Eq. (9). We refer the interested reader to [22] for a detailed validation and analysis of the flow fields created in the HiRe and MR tanks.

The objective of our experiments is to validate the exponential scaling for the noise-induced transitions between LCS-bounded regions as shown in Fig. 2b with and without the controller given by Eq. (10). Additionally, we show how "optimal escape paths" as predicted by minimizing the action functional given by Eq. (3) correspond to more energy efficient trajectories out of given G_i.

4.1 Stochastic Escape Times

To validate the exponential scaling for stochastically-induced transitions between LCS-bounded regions, we employ the mASVs and the MR tank. Using a simulated

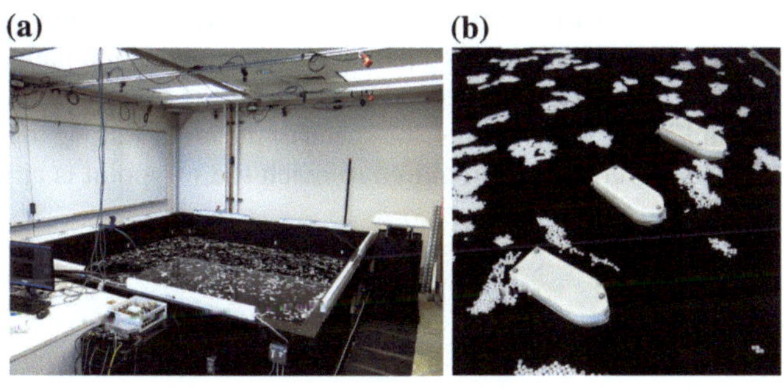

Fig. 3 Photos of a multi-robot (MR) tank and b three mASVs in the MR tank

time-invariant flow field given by Eq. (9), we operated the mASVs in still water and examine whether the vehicles eventually leave the specified region or gyre, G_i. The mASVs were controlled using a waypoint controller with the following waypoint-update scheme:

$$\boldsymbol{q}_{n+1} = \boldsymbol{q}_n + h(\boldsymbol{u}_n + \boldsymbol{F}(\boldsymbol{q}_n)) + \boldsymbol{\eta}_n$$

where we set $h = 0.1$ to create the desired number of waypoints. In our experiments, the parameters in Eq. (9) were chosen to be $A = 1$, $s = 1$, $\mu = 1$ such that no new waypoint is greater than 20 cm away from the vehicle's current position. Each G_i was a single $1.2 \times 1.2\,\text{m}^2$ gyre with surface flows and LCS boundaries similar to those shown Fig. 1b. We consider a stochastic flow to measure escape times for $c = \{-0.5, 0, 0.5\}$. In this case $\boldsymbol{\eta}_n = \sqrt{h}\sigma\mathcal{N}(0, 1)$ where $\mathcal{N}(0, 1)$ represents a vector of random numbers drawn from a Gaussian distribution with mean 0 and standard deviation 1. We set the noise intensity to be $D = 0.1$ resulting in $\sigma \approx 0.45$; this results in a random distribution of waypoints around the position $\boldsymbol{q}_n$ with a bias toward moving in the direction of the gyre flow.

4.2 Minimum Action Paths

To determine the energy efficiency of the minimum action paths, we consider the average control effort expenditure of a vehicle executing the optimal escape trajectory predicted by the minimization of the action functional given by Eq. (3). To create a gyre-like flow in the MR tank, four rotating cylinders with diameter approximately 9 cm are placed in the MR tank at the vertices of a $1 \times 1\,\text{m}^2$ square centered in the tank and set to rotate at approximately 20 Hz. The water depth is approximately 30 cm. The resulting flow in the MR tank is similar to that modeled by Eq. (9). In this environment, a single mASV is placed at a point along the calculated optimal path out of the flow and is instructed to exit the region by following two different paths. The first path is the minimum distance trajectory between the vehicle's starting position and the goal position, i.e., the boundary saddle. This is the baseline case where the minimum distance trajectory is the optimal path without considering the external flow field, e.g., $\boldsymbol{F}(\boldsymbol{q}) = \boldsymbol{0}$, which corresponds to a straight line between the start and goal positions. The second path is the minimal action path obtained by minimizing Eq. (3). The resulting desired trajectories are then approximated by a series of waypoints. In our experiments, the speed of the mASV along the trajectories was set such that the average time required to reach the waypoint is approximately 30 s.

5 Experimental Results

We begin with the results of the stochastic escape experiments. In the presence of noise, the mASV lingered in G_i for some time before stochastically escaping. We recorded the trajectories, waypoints and total time required for the mASV to leave the gyre for each of the three values of c. Figure 4 shows three histograms of measured escape times from the simulated gyre for the number of trials specified in each of the control regimes. Figure 6 shows the densities of the truncated prehistories for the mASVs as they escaped the simulated gyre for the three different gyroscopic control cases. Table 1 provides the calculated mean escape times for each value of c and the number of trials completed for each case.

Regarding the minimum action path experiments, we compare two separate metrics to determine the effort exerted by each vehicle. First note that the signal to the motors ranged from (−255, +255), so we use the root-mean-square of the signal to

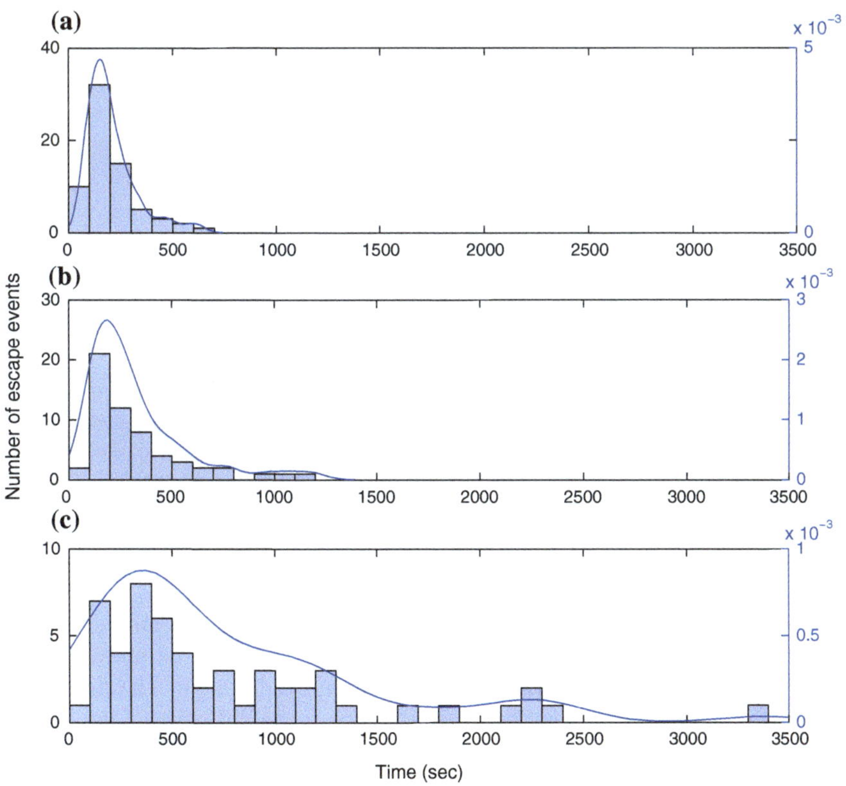

Fig. 4 Histograms of measured escape times from the simulated gyre with stochastic effects during the stochastic escape experiment, showing data for **a** $c = 0.5$, **b** $c = 0$ and **c** $c = -0.5$. The *overlaid curves* indicate the approximate pdf of escape times using a normal kernel for smoothening. These data are approximately an exponential distribution of escape times

Table 1 Comparison of escape times for the stochastic escape experiments for two different values of control and without control

c	Escape time (s)	Number of trials
−0.5	977	52
0	363	55
+0.5	223	65

$c = -0.5$ corresponds to the controller pointing inward, $c = 0$ no control and $c = +0.5$ the controller pointing outward

Table 2 Results from the path following experiments showing two metrics for average control effort to follow the path to escape

c	(Mean RMS signal)/(Travel time)	(Mean RMS signal)/(Path length)	Number of trials
Baseline	1857 ± 156	49538 ± 1773	31
+0.097	1750 ± 170	43223 ± 1932	34
0	1806 ± 139	45339 ± 2183	40

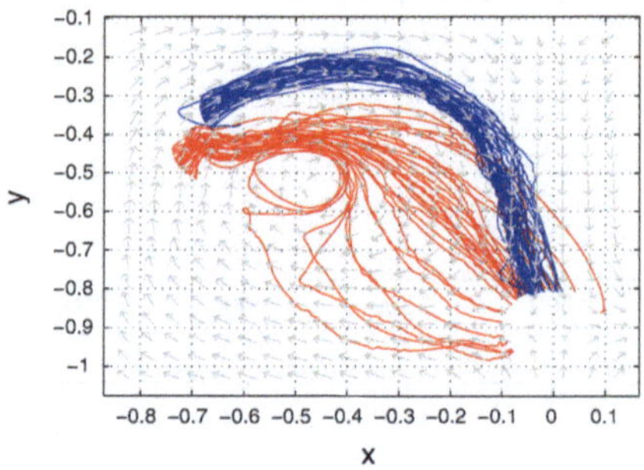

Fig. 5 The slope field of Eq. (9) overlaid with the actual paths followed by the mASV out of G_i. *Blue lines* correspond to the trials where the mASV was directed to follow the optimal path for $c = 0.097$ whereas *red lines* represent the straight-line path trials. The slope field provides only a theoretical representation of the flow lines and suggests that following the optimal path minimizes the current the mASV must overcome in order to reach its destination. The parameters used to generate the simulated flow field were $A = 1$, $s = 1$ and $\mu = 1$

both motors to represent the instantaneous control effort. The first metric we compare is the average RMS signal to the two motors for a given trial normalized by time-of-travel, and the second is normalized by total path length. The first metric gives a measurement of the total required actuation while normalizing the data for differences in total time in executing the path. The second provides a better comparison for whether the mASV was actually directed with the flow and thus leveraged the flow to reach its goal. Table 2 displays the data with 95 % confidence intervals. Figure 5 compares the paths that the mASV followed as it exited G_i following the two different trajectories.

6 Discussion and Experimental Insights

The escape trajectories taken by the mASV in the stochastic escape experiments and the optimal escape trajectories of a passive particle show good agreement. Both Fig. 2a and the left column of Fig. 6 show Monte Carlo simulations of a particle exiting its initial region and demonstrate a strong preference for the particle to approach the saddle point before exiting G_i. This is also exhibited in the stochastic escape experiment, shown in the right column of Fig. 6. Note that the results for stochastic escape are rotationally invariant, i.e. the escape paths were the same independent of which saddle was chosen as the exit point.

Furthermore, the scaling illustrated in Fig. 2b suggests that controllers which make use of the underlying dynamics and stochasticity may be rigorously developed to manipulate the switching time of a robot operating in a noisy flow. This result is confirmed experimentally in Fig. 4; by employing a simple control strategy, the mASV effectively manipulated its distribution of residence times in a simulated flow (see Table 1).

Finally, we have confirmed that the most likely path as predicted by the theory not only lies along the peak of the observed escape trajectories as shown in Fig. 2a, but also requires less control effort than following a straight-line path to exit the initial G_i. The last column in Table 2 demonstrates that the total control effort by the mASV following the optimal path is lower than the control effort exerted by the vehicle executing the straight-line path. This is not surprising since the mASV travels with the flow and is pushed along by the current and only actuates when necessary to correct course. More interestingly is that the proposed control strategy achieves comparable or slightly better performance from a control effort expenditure perspective. This is significant since the proposed controller can be implemented with only limited knowledge of the flow field.

This work demonstrates that the theory of large fluctuations applies to macroscopic systems of autonomous vehicles operating in noisy environments. The concepts about optimal paths and rare events explained in our methodology were originally developed for microscopic systems undergoing switching due to thermal fluctuations. However, our experiments suggest that a macroscopic agent operating in a massive stochastic environment is subject to the same phenomena as its microscopic counterparts. This has implications for the design of controllers in environments where the structure of the flow is not necessarily known, significant stochastic fluctuations are present, and control effort is costly.

7 Future Work

To the authors' knowledge, this is one of the first attempts to employ knowledge of coherent structures in designing more energy efficient navigation strategies for autonomous vehicles operating in uncertain fluidic environments. The main contri-

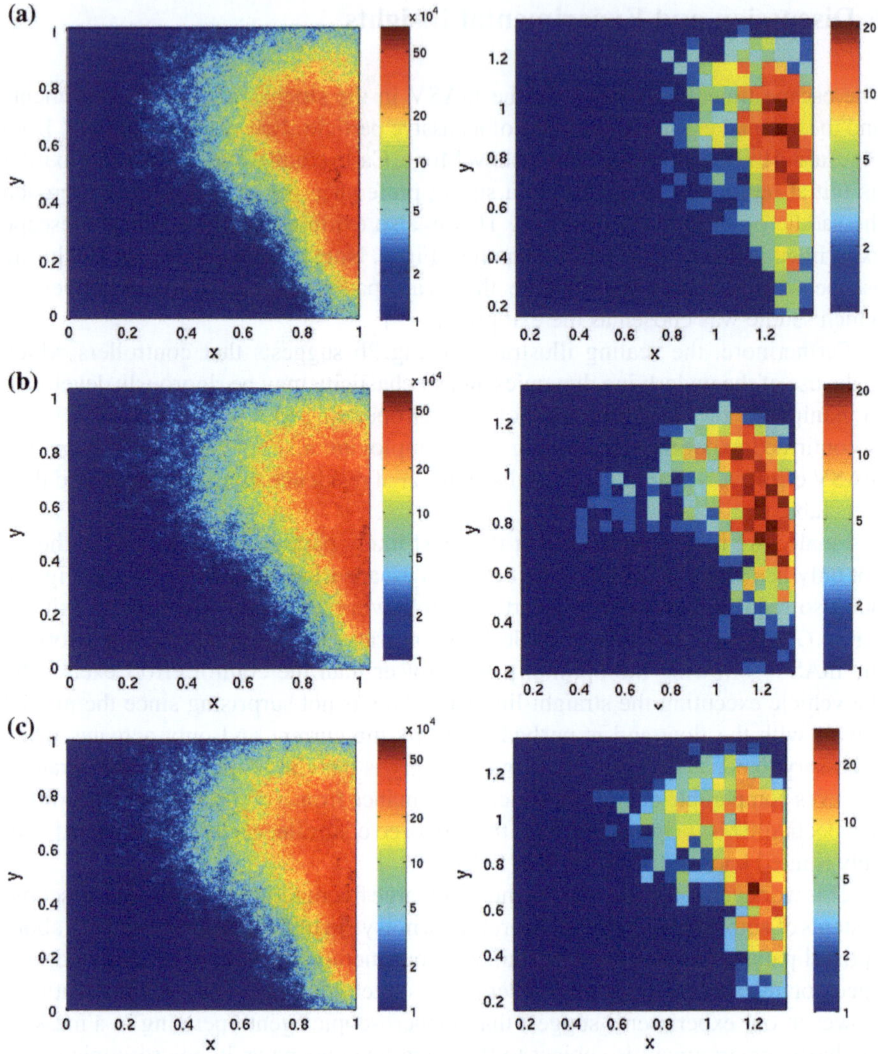

Fig. 6 Prehistories of stochastic escape from the initial region for controller gains **a** $c = -0.5$, **b** $c = 0$ and **c** $c = 0.5$. The *left column* shows Monte Carlo results for 1,000 trials at each controller gain, and the corresponding image in the *right column* shows the history of waypoints from the stochastic escape experiments for the same gain. The scaling for colors in both sets of figures is exponential and was truncated such that the approach to escape is accentuated compared with the time spent near the center of the region. The parameters used to generate the simulated flow fields were $A = 1$, $s = 1$ and $\mu = 1$

bution lies in the synthesis of minimal control effort strategies using knowledge of *most likely* escape paths arising from noise-induced large fluctuations. The result is a strategy that leverages the surrounding environmental dynamics and the inherent environmental noise to minimize the overall control effort of the to navigate from one LCS-bounded region to another. The proposed strategy was evaluated using a novel testbed capable of creating controlled Lagrangian coherent structures in a laboratory setting. This is the first experimental study of optimal escape trajectories and minimal control effort escape trajectories for unmanned underwater/surface vehicles operating in realistic complex flows.

One future application is the extension of the proposed AUV/ASV control strategy to domains consisting of geometrically complex LCS-bounded regions. For instance, the LCS may be time-dependent which would more closely resemble ocean flows or the LCS-bounded regions may be non-convex. We would also like to revisit the assumption of additive noise drawn from a Gaussian distribution and the use of the time-invariant gyre-driven flow model. Most noise is in fact drawn from a colored distribution and arises in modeling as multiplicative noise; both of these factors have an effect on the shape of the optimal paths and the distribution of rare events. Also, the time-invariant gyre-driven flow model used in Eq. (9) follows the position of a particle in a two-dimensional vector field whereas actual field deployments require taking into consideration the third dimension and the actual size and geometry of the vehicle. We would like to extend the existing framework to properly account for these factors.

Acknowledgments This research was performed while CRH held a National Research Council Research Associateship Award at the U.S. Naval Research Laboratory. This research was funded by Office of Naval Research (ONR) Award Nos. F1ATA01098G001 and N0001412WX-20083, and by Naval Research Base Program contract N0001412WX30002. MAH and the mCoSTe are also supported by ONR Award Nos. N000141211019 and N0001413-10731. CRH and MAH would like to thank Matt Michini from the SAS Lab for his assistance in conducting the experiments.

References

1. Haller, G., Yuan, G.: Lagrangian coherent structures and mixing in two-dimensional turbulence. Phys. D **147**, 352–370 (2000)
2. Shadden, S.C., Lekien, F., Marsden, J.E.: Definition and properties of lagrangian coherent structures from finite-time lyapunov exponents in two-dimensional aperiodicflows. Physica D: Nonlinear Phenomena **212**(3–4), 271–304 (2005)
3. Inanc, T., Shadden, S., Marsden, J.: Optimal trajectory generation in ocean flows. In: American Control Conference, 2005. Proceedings of the 2005, pp. 674–679 (2005)
4. Senatore, C., Ross, S.: Fuel-efficient navigation in complex flows. In: American Control Conference, 2008, pp. 1244–1248 (2008)
5. Schwartz, I.B., Forgoston, E., Bianco, S., Shaw, L.B.: Converging towards the optimal path to extinction. J. Royal Soc. Interf. **8**(65), 1699–1707 (2011)
6. Forgoston, E., Billings, L., Yecko, P., Schwartz, I.B.: Set-based corral control in stochastic dynamical systems: making almost invariant sets more invariant. Chaos **21**, 013116 (2011)
7. Bollt, E.M., Billings, L., Schwartz, I.B.: A manifold independent approach to understanding transport in stochastic dynamical systems. Physica D **173**, 153–177 (2002)
8. Billings, L., Schwartz, I.B.: Identifying almost invariant sets in stochastic dynamical systems. Chaos **18**, 023122 (2008)

9. Froyland, G., Padberg, K.: Almost-invariant sets and invariant manifolds: connecting probabilistic and geometric descriptions of coherent structures in flows. Physica D **238**, 1507–1523 (2009)
10. Michini, M., Hsieh, M.A., Forgoston, E., Schwartz, I.B.: Robotic tracking of coherent structures in flows. IEEE Trans. Robot. (2014)
11. Mallory, K., Hsieh, M.A., Forgoston, E., Schwartz, I.B.: Distributed allocation of mobile sensing swarms in gyre flows. Nonlinear Process. Geophys. **20**(5), 657–668 (2013)
12. Heckman, C.R., Hsieh, M.A., Schwartz, I.B.: Going with the flow: enhancing switching rates in multi-gyre systems. Accepted to the ASME Journal of Dynamic Systems, Measurement and Control (2014)
13. Feynman, R.P., Hibbs, A.R.: Quantum Mechanics and Path Integrals. McGraw-Hill Inc. (1965)
14. Freidlin, M.I., Wentzell, A.D.: Random Perturbations of Dynamical Systems. Springer, Berlin (1984)
15. Schwartz, I.B., Billings, L., Dykman, M., Landsman, A.: Predicting extinction rates in stochastic epidemic models. J. Stat. Mech. Theory E P01005 (2009)
16. Bryson, J.A.E., Ho, Y.C.: Applied Optimal Control: Optimization, Estimation and Control. Taylor and Francis, Routledge (1975)
17. Chan, H.B., Dykman, M.I., Stambaugh, C.: Switching-path distribution in multidimensional systems. Phys. Rev. E **78** (2008) Art. no. 051109
18. Veronis, G.: Wind-driven ocean circulation, part I and part II. Deep Sea Res. **13**(31) (1966)
19. Lindley, B.S., Schwartz, I.B.: An iterative action minimizing method for computing optimal paths in stochastic dynamical systems. Physica D **255**, 22–30 (2013)
20. Doedel, E.J., Champneys, A.R., Dercole, F., Fairgrieve, T., Kuznetsov, Y., Oldeman, B., Paffenroth, R., Sandstede, B., Wang, X., Zhang, C.: AUTO-07P: Continuation and Bifurcation Software for Ordinary Differential Equations (2008)
21. Mallory, K., Hsieh, M.A., Forgoston, E., Schwartz, I.B.: Distributed allocation of mobile sensing swarms in gyre flows (2013). arXiv:1303.0704
22. Larkin, D., Michini, M., Abad, A., Teleski, S., Hsieh, M.A.: Design of the multi-robot coherent structure testbed (mcoste) for distributed tracking of geophysical fluid dynamics. In: Submitted to ASME International Design Engineering Technical Conferences (IDETC), Buffalo (2014)

QuadCloud: A Rapid Response Force with Quadrotor Teams

Kartik Mohta, Matthew Turpin, Alex Kushleyev, Daniel Mellinger, Nathan Michael and Vijay Kumar

Abstract We describe the component technologies, the architecture and system design, and experimentation with a team of flying robots that can respond to emergencies or security threats where there is urgent need for situational awareness. We envision the team being launched either by high level commands from a dispatcher or automatically triggered by a threat detection system (for example, an alarm). Our first response team consists of autonomous quadrotors with downward-facing cameras that can navigate to a designated location in an urban environment and develop a integrated picture of areas around a building or a city block. We specifically address the design of the platform capable of autonomous navigation at speeds of over 30 mph, the control and estimation software, the algorithms for trajectory planning and allocation of robots to specific tasks, and a user interface that allows the specification of tasks with a situational awareness display.

Keywords Aerial robotics · Multi-robot systems · Field robotics

1 Introduction

Over the last decade, aerial robotics has received a lot of attention and there is extensive literature on both indoor [16, 18] and outdoor platforms. Indeed by some estimates,[1] the UAV market is estimated to exceed \$20 B in the next 3 years, and

[1] http://www.marketresearchmedia.com/?p=509.

K. Mohta (✉) · M. Turpin · V. Kumar
GRASP Laboratory, University of Pennsylvania,
Philadelphia, PA 19104, USA
e-mail: kmohta@seas.upenn.edu

A. Kushleyev · D. Mellinger
KMel Robotics, Philadelphia, PA 19146, USA

N. Michael
Robotics Institute, Carnegie Mellon University,
Pittsburgh, PA 15213, USA

M.A. Hsieh et al. (eds.), *Experimental Robotics*, Springer Tracts in Advanced Robotics 109, DOI 10.1007/978-3-319-23778-7_38

this forecast is conservative since it does not account for the thousands of micro-UAVs that are likely to be fielded in the near future. At the smaller (and lighter) end of the spectrum, quadrotors and hexrotors have become a standard platform for robotics research worldwide with the potential to support many indoor and outdoor applications [14]. In spite of the limitations of battery technologies and their inherent inefficiency, there are a number of short duration missions that make for very interesting applications. In this paper, we describe a rapid response force consisting of team of quadrotors that can quickly respond to disasters and emergencies by providing situational awareness before human responders can get to the scene. Our main goal in this paper is to describe the component technologies, our approach to the architecture and system design, and experimental data in support of an integrated system with applications to first response.

Our paper builds on previous work on designing outdoor quadrotor platforms. A tutorial on quadrotors is provided in [14]. In particular, the work by Huang et al. [7] is notable in its study of lift and drag in flight conditions. Commercial platforms for applications such as aerial photography are available from companies like DJI,[2] Ascending Technologies,[3] and Parrot. Multi vehicle demonstrations have been shown at conferences like ICRA 2013 and shows like the 50 robot outdoor event [2]. However, our main focus is on building a system of platforms that can function as a cohesive unit to perform a range of tasks within the broad scope of emergency response.

There is extensive literature on the control for quadrotors. Backstepping approaches to control system design based on linear control laws are discussed in [3]. A nonlinear controller that incorporates the curvature of the $SE(3)$ is described in [11, 15], while a closely related approach used in control showed significantly better performance over linear controllers [15]. Similarly the use of linear filters or extended Kalman Filters for state estimation in is quite standard [1]. However a UKF estimator yields significantly improved performance [17]. Our paper builds on these approaches as described later.

Finally, the study of quadrotor teams is also relevant to this work. Our own previous work is described in [9, 13]. Representative papers from other groups who have done similar work include [6]. The work in this paper primarily addresses a framework for coordination of aerial robots without assigning labels to the individual vehicles and without specificity of the number of robots in the team. Another related body of work is the design of user interfaces for multi robot control [4]. However, there is relatively less literature on the design of user interfaces for large teams of aerial robots where the three-dimensional environments and the short term duration of the missions impose new constraints.

[2]http://www.dji.com.

[3]http://www.asctec.de.

2 System Design

2.1 Architecture

The goal of QUADCLOUD is to monitor an area that is the size of a few city blocks, approximately a total area of 400 m × 400 m, while being responsive to the operator's commands and dynamic information collected by one or more robots. We want the response times to these requests of surveillance to be less than 30 s. This requires the robots to have speeds of around $10\,\mathrm{ms}^{-1}$, and acceleration of $3\,\mathrm{ms}^{-2}$. We want the size of the robots to be as small as possible for ease of handling and deployment (imagine deploying a team from the back of a pickup truck), and have enough payload capacity to carry all the components needed for autonomy and surveillance. Since the system runs outdoors, the robots also need to have enough thrust to be able to cope with moderate winds.

A key requirement for QUADCLOUD is that a single operator must be able to deploy, control and monitor the team of robots easily. This requires sufficient autonomy on each robot to able to handle the navigation task from trajectory generation to position stabilization on-board. Further, on each robot, we need downward-facing cameras to provide imagery, and the computational resources to allow processing of images at around 5 Hz from the on-board camera. Each robot must have the requisite on-board intelligence to look for salient information. In this paper, we do assume that the targets of interest are relatively simple and can be easily identified using regular cameras at heights of around 10–20 m. However, the algorithms must allow for false positives and a probabilistic representation of the belief state of the environment. Finally, in order to send data back from all parts of the monitored area and to respond to commands issued by the operator, each robot must be able to communicate with the base station within a 400 m distance.

Thus the architecture must incorporate some elements of decentralized planning, control, estimation and target detection and localization while allowing for a centralized, cloud computing model for command and control by the user and task planning. Further, we want a framework in which the user is agnostic to the number of robots, their identities and what exactly their individual states are. This attribute of anonymity increases the robustness of the system to failures and decreases the overhead on the human user.

The simplest architecture with these attributes is shown in Fig. 1. Robots are able to localize and control their motions to destinations, freeing the operator to work with high level task specifications such as target destinations or areas for surveillance. A centralized goal assignment and planning module decides which robot responds to what request and when. Individual robots independently decide how to follow these requests. Each robot periodically sends back its position estimates and images from the on-board camera to the base station. This information is presented to the operator on a simple user interface and through the user interface, the operator is also able to command goal positions which are sent to the planner.

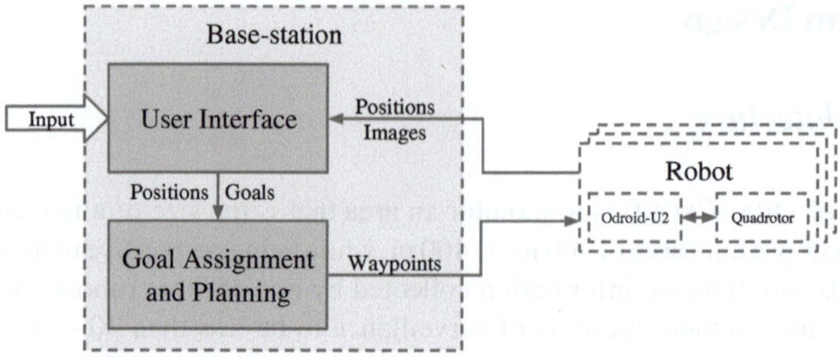

Fig. 1 Decentralized robot motion planning and control along with a centralized model for human interaction and task assignment

Fig. 2 The KMEL kQuad500 quadrotor equipped with a u-blox NEO-6T GPS module, a Matrix Vision mvBlueFOX camera, a Ubiquity Networks Bullet M2 and a Odroid-U2 quad-core ARM single board computer

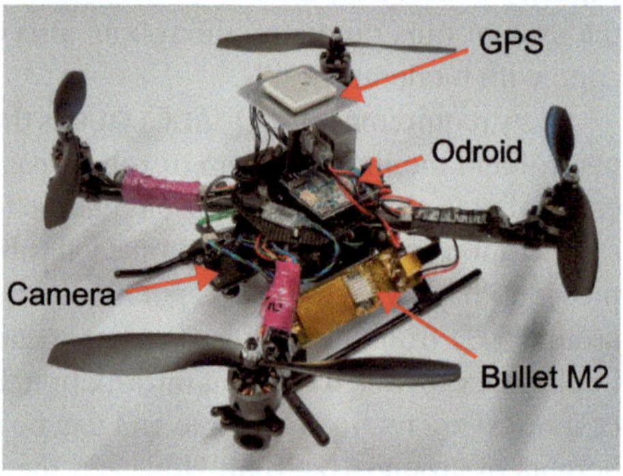

2.2 The Robots

The robots used for this project, shown in Fig. 2, are quadrotors designed and developed by KMEL Robotics. These quadrotors have a tip-to-tip diameter of about 0.54 m and weigh around 0.95 kg in the configuration used for this project. They are equipped with an ARM Cortex-M3 processor, 3-axis accelerometer, gyroscope, magnetometer and a pressure sensor while a u-blox NEO-6T GPS module was added in order to get position estimates. A control loop running at 500 Hz on the ARM processor stabilizes the attitude of the quadrotor. The communication with the on-board processor in order to receive sensor outputs and send thrust and attitude commands is done via a UART port.

All the high-level computations on the robot are performed on an Odroid-U2 quad-core ARM single board computer. This compact board has four Cortex-A9 cores each running at up to 1.7 GHz allowing us to have a powerful processor in a small form-factor. The Odroid-U2 comes with a big passive heat-sink which we replaced with a small active heat-sink cutting its weight from 130 g to around 50 g. Each quadrotor is also equipped with a Matrix Vision mvBlueFOX camera in order to capture images for target detection and surveillance. We wanted each robot to send images to the base station for surveillance purposes, thus we had a requirement of long-range communications with sufficient bandwidth. Since the Odroid-U2 does not have any built-in wifi, but has an ethernet port, we decided to use a Ubiquity Networks Bullet M2 which provides the required bandwidth with a range of more than 350 m. This allows us to stream compressed images at more than 40 fps from a single robot or around 5 fps from each robot when we have a team of 8 quadrotors sending images back. The Bullet M2 comes with a large and heavy antenna connector which we replaced with a smaller one, and we also removed the plastic shell for weight saving, reducing its weight from 180 to 50 g. The robots use a 3-cell 2.2 Ah LiPo battery which gives a flight time of around 8–10 min with the current configuration of the robot.

The software system on the robot consists of three main components: estimation, control and image processing. The image processing is simple and is composed of detecting specified features (that characterize the targets) in the image and image compression for transmitting the images from the camera to the base station. The estimation and control subsystems are described in more detail in the next section. We use ROS as the framework for all the software running on the Odroid-U2, because it provides a good inter-process communication framework allowing transparent relocation of the processes across the machines allowing us to run a particular set of nodes on the robot for testing and running others on a separate computer speeding up the development phase.

2.3 Estimation and Control

An overview of the estimation and control systems running on each robot is shown in Fig. 3. We use the GPS, IMU, magnetometer and pressure sensor for state estimation. First, the GPS latitude, longitude and height are transformed to a local cartesian frame using GeographicLib [8]. We ignore the height from the GPS measurement since it has a large drift and instead rely on the pressure sensor for getting the height. This processed output of the GPS along with the other sensor outputs, are then fed to a

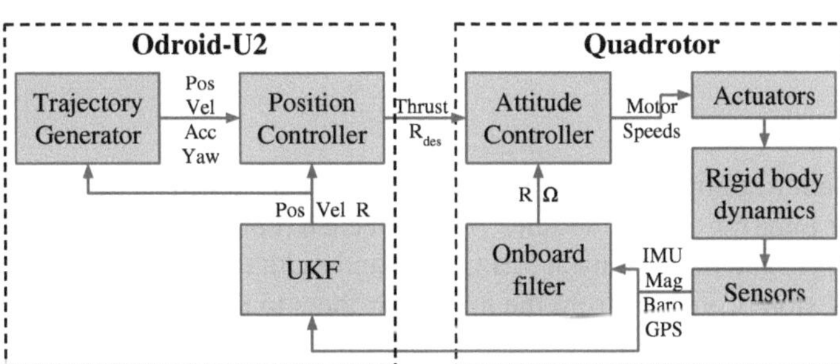

Fig. 3 The estimation and control systems running on each robot

Unscented Kalman Filter (UKF) in order to generate full 6-DoF pose estimates. The UKF state that we use is,

$$x = \left[p^T \; \dot{p}^T \; \psi \; \theta \; \phi \; a_b^T\right]^T$$

where p is the world-frame position of the robot, $\dot{p}$ is the world-frame velocity, ψ, θ and ϕ are the yaw, pitch and roll respectively and a_b is the accelerometer bias along the three axes.

Our control architecture has the common cascade structure of backstepping controllers [3, 16], with the attitude controller as the inner loop and the position controller as the outer loop around it. The controller is based on the non-linear controller developed in [11]. The attitude controller runs at a very high rate (600 Hz) on the on-board processor stabilizing the orientation of the quadrotor, allowing us to run the position controller at a much slower rate (50 Hz) on the Odroid-U2. This position controller takes position commands sent by a trajectory generator and, using the position estimates, converts them into thrust and attitude commands which are sent to the attitude controller running on the on-board processor. Finally, the attitude controller running on the quadrotor takes the thrust and attitude commands and converts them to commanded motor speeds.

2.4 Experimental Benchmarking

We have done extensive testing in order to test the performance of our estimation and control algorithms. The estimates of the UKF during a representative hover test are shown in Fig. 4. The plots also show the output from an OptiTrack system which was set up outdoors in order to provide ground truth reference. From the plots we can see that the errors have a standard deviation of around 16 cm in the horizontal plane and 39 cm in the vertical direction.

3 Communication and Supervision

Communication

The base station communicates with each of the robots via wifi through the Bullet M2 high-power long-range wifi modules. We want each robot to send back position estimates at 50 Hz and image data at 5 Hz. The bandwidth requirement is dominated by the transmission of images from the multiple robots to the base station. The camera on each robot has a resolution of 1280 × 960 which leads to a raw gray-scale image size of approximately 1.2 MB, so for raw image transmission at 5 Hz we require a bandwidth of about 48 Mbps. The Bullet M2 claims a maximum bandwidth of around 65 Mbps, but in real-world testing, we got a data rate of about 50 Mbps. Thus

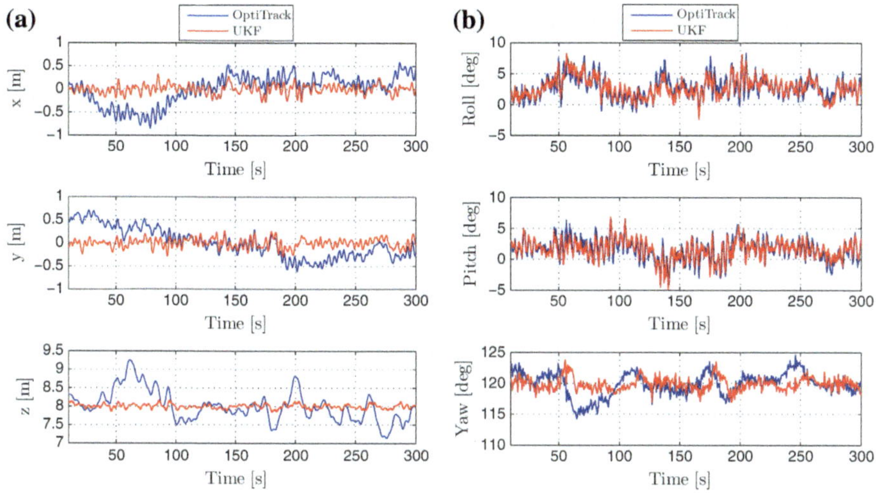

Fig. 4 UKF estimates during a representative hovering experiment in an open area. Ground truth from an OptiTrack motion capture system, which was set up specifically for this experiment, are shown for reference. The position tracking errors had a standard deviation of 0.158 m in the *horizontal direction* and 0.386 m in the *vertical direction*. **a** Position, **b** orientation

it is not possible to send raw images back from each of the robots at the desired rate. To reduce the bandwidth requirement, we decided to jpeg compress the images before sending. This brings down the size of the images from 1.2 MB to about 130 kB allowing us to stream images at 5 Hz from up to 10 robots. If we want to add more robots, we would need to decrease the frame rate of the image data being sent back from the robots. A frame rate of 2 Hz is sufficient for surveillance purposes and would allow us to scale to around 20 robots.

User Interface

Since all the computations for autonomy are being done on the robots themselves, the operator does not need a very powerful base station to control the team; the base station can just be a small laptop. As mentioned earlier, the robots send their position estimates to the base station. This information is presented to the operator in the form of markers on an overhead schematic map of the area. Figure 8 shows some screenshots of the user interface during an experiment. In addition to monitoring the system, the user is able to send goal positions to the system without needing to specify which robot is assigned which goal. Using the algorithm described in the next section, the system assigns the goals to the robots in order to minimize the maximum travel time and plans trajectories for each of them. This reduces the cognitive burden on the operator by allowing the operator to focus on the high-level tasks.

4 Combining Assignment and Trajectory Planning

To safely navigate the team of robots to goal locations, a motion planning algorithm is required that computes collision-free trajectories and respects the dynamics of the robots. It is well known that extending single robot motion planners to plan trajectories for a team of robots implies exponential computational complexity [5]. One attempt to solve this problem of computational intractability is to use a two step algorithm that decouples the path from the time parameterization of the trajectories [10]. These decoupled approaches first plan motions for each individual robot while disregarding collisions with other robots. The second step is to specify the time parameterization the robot follows its path. Unfortunately, these approaches are not complete and cannot guarantee they will find a solution if one exists.

Fortunately, our team of robots are identical and we can exploit this interchangeability to generate collision-free time parameterized trajectories in a computationally tractable manner using the GAP algorithm.

Goal Assignment and Planning (GAP)

In this paper, we leverage the authors' previous work Goal Assignment and Planning, or GAP [19, 20], to plan complete, collision-free trajectories with a computational complexity of $\mathcal{O}(N^3)$. This approach assumes full knowledge of obstacles present in the environment. The robots are modeled as second order systems with spherical extent. The radius of the robots is taken at a conservative 2 m to ensure that errors in localization will not cause a catastrophic collision.

The GAP algorithm is a decoupled motion planner that maintains completeness by leveraging the interchangeability of the robots. This algorithm begins by finding the cost associated with planning trajectories from the initial state of each robot to every goal location. We use Dijkstra's algorithm to quickly find these N^2 motion plans where robot i has cost C_{ij} to travel to goal j. The next step is the assignment of goals to robots where each robot is assigned to one goal. This assignment can be represented by a permutation matrix ϕ, where $\phi_{ij} = 1$ if and only if robot i is assigned to goal j. The Hungarian Algorithm is used to find the assignment which minimizes the maximum cost assignment:

$$\underset{\phi}{\text{minimize}} \sum_{i=1,2,\ldots,N} \sum_{j=1,2,\ldots,N} (\phi_{ij} C_{ij})^p$$

where p is a very large constant. In practice, $p = 50$ is used. Then, robots are prioritized using simple geometric considerations that are fully detailed in [20]. Finally, robots are assigned their full time parameterization to construct trajectories that guarantee collision avoidance.

For a team of 6 robots, these plans are generated in under 0.1 seconds. Additional details of this algorithm including boundary condition requirements for completeness are presented in [20].

5 Experimental Results

5.1 High Speed Tests

Since we are flying outdoors, we can fly along long trajectories, which provide sufficient distance to accelerate to high speeds. We ran some high speed tests with one of the quadcloud robots where we commanded the robot to fly approximately 80–100 m at speeds up to $15\,\text{ms}^{-1}$ and looked at the effect of drag.

For a quadrotor, if we ignore drag, since the thrust is only along the body Z-axis, we expect the accelerometer on the quadrotor to measure zero acceleration on the X and Y axes while the Z axis will measure the effect of thrust [12]. But, as shown in [12], the accelerometer on a quadrotor measures some acceleration in X and Y axes when moving due to the drag force acting on the robot.

From Fig. 5, we get,

$$a_x = \frac{-D\cos\phi}{m}$$
$$a_z = \frac{T - D\sin\phi}{m}$$

where D is the drag force and T is the thrust.

Thus, when the robot is flying fast and there is considerable drag, we expect it to have a significant measurement on the accelerometer X-axis. In our data (Fig. 6), we can see that the accelerometer measures approximately $4\,\text{ms}^{-2}$ acceleration along the X-axis when the quadrotor is flying at speeds of around $15\,\text{ms}^{-1}$. Using the above equations, from the accelerometer measurement and orientation estimates, we can estimate the drag force on the robot. The computation gives the magnitude of the drag force to be approximately 4.2 N which is significant considering that it is half of the gravitational force acting on the robot. This is a large external force that is not modeled in our estimator and can lead to errors in our position and orientation estimates. We are still looking into the effects of drag on the orientation estimates and also ways in which we can use the drag measurements for velocity estimates.

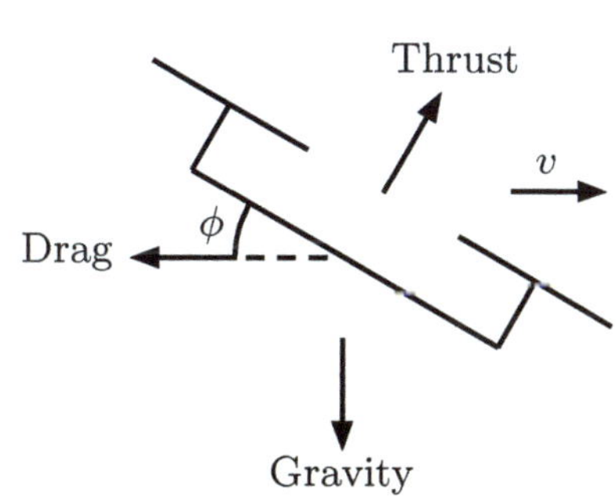

Fig. 5 Forces acting on a quadrotor moving towards the right with velocity v

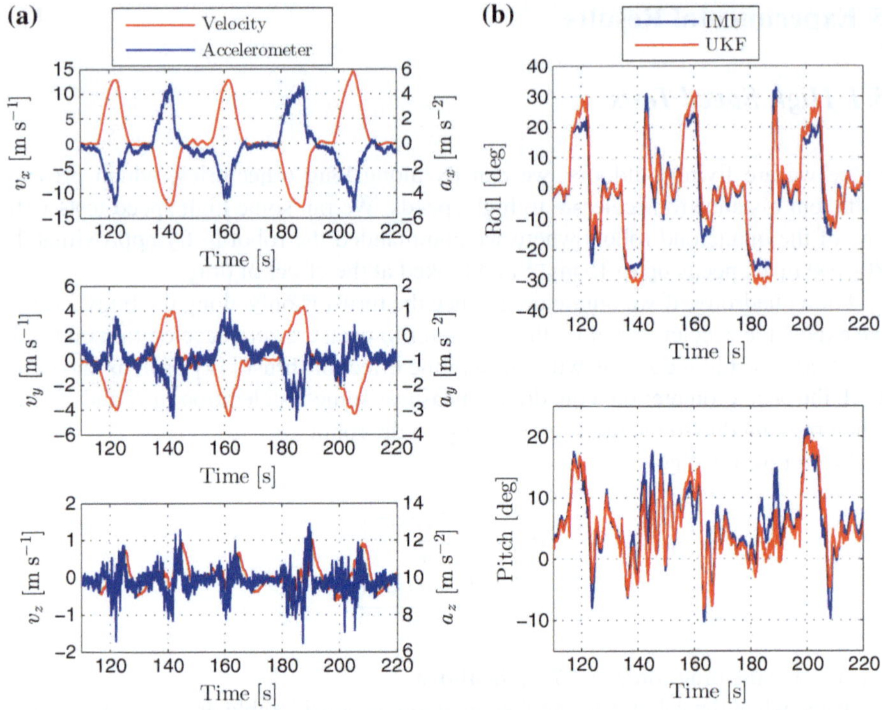

Fig. 6 Measurements during a high speed test. **a** Velocity and acceleration, **b** orientation

Fig. 7 An outdoor experiment with six robots

5.2 *Experimental Results with Multiple Robots in an Open Field*

Here, we describe an experiment we conducted with six robots in an open field (Fig. 7). Instead of flying around real-world obstacles, we provide virtual obstacles to the planner so that we can perform the experiment in a much safer manner. Figure 8

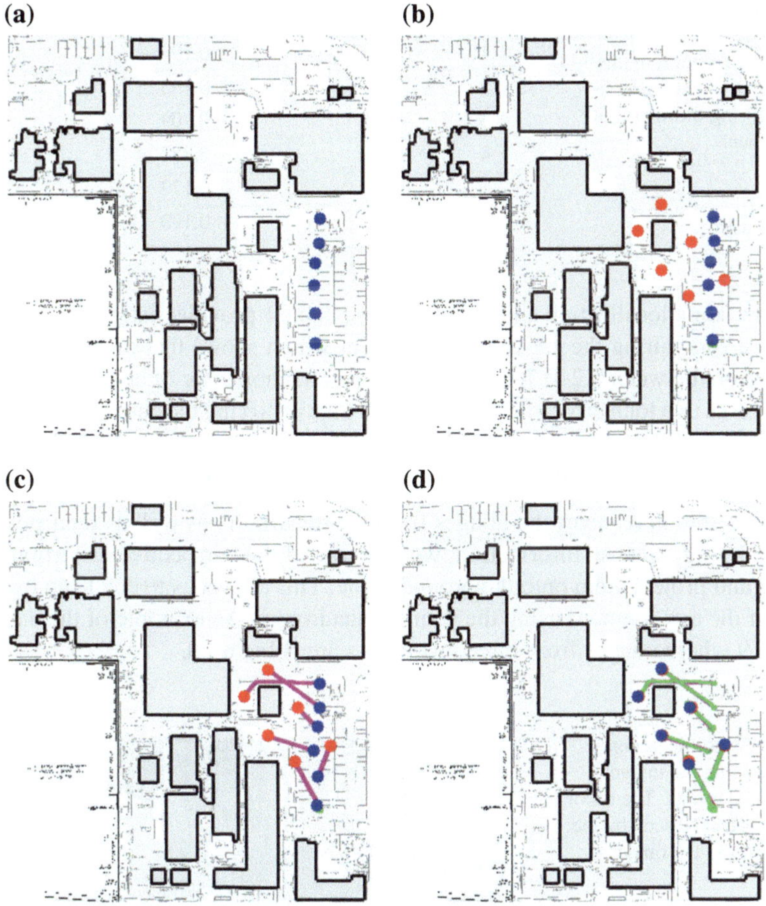

Fig. 8 A series of snapshots of the user interface while running an experiment with six robots. **a** Initial positions, **b** user species goal positions, **c** planner assigns the goals and generates trajectories, **d** actual trajectories followed by the robots

shows the various steps involved in the experiment. We start the robots from the ground with a separation of about 4 m between each other so that we can take-off without worrying about collisions between the robots. Once they take-off and reach a specified height we switch to the trajectory tracker which takes in inputs from the central planner and sends position commands to the position controller. Once this stage is reached (Fig. 8a), the operator can command the robots from the user interface and send the robots to the desired goal positions (Fig. 8b). Upon receiving the goal positions, the planner assigns the goals to the robots and plans trajectories for each of them (Fig. 8c) which are then followed by the robots (Fig. 8d).

As mentioned in the previous section, the planner models the robots as circles with a radius of 2 m, even though the actual robot radius is around 0.3 m, in order

Table 1 Mean error between the desired position and estimated position in the horizontal plane for each robot during the six robot experiment

Robot	XY error (m)
1	0.194
2	0.195
3	0.709
4	0.382
5	0.179
6	0.190

to allow some localization and control errors. Table 1 provides estimates of the controller errors during the six-robot experiment which shows that most of the robots have errors between 0.2–0.7 m. Adding the localization error of approximately 0.2–0.5 m gives us a total error of around 1–1.5 m thus justifying the choice in the planner.

Aggregation of Visual Imagery for Situational Awareness

The base station receives the images from the quadrotors as well as their pose estimates. Using the pose information, we can correct the perspective distortion of the image and project them onto the ground plane. This allows us to create an overhead map of the environment using the team of quadrotors. An example of this is shown in Fig. 9 where images from three quadrotors are being used.

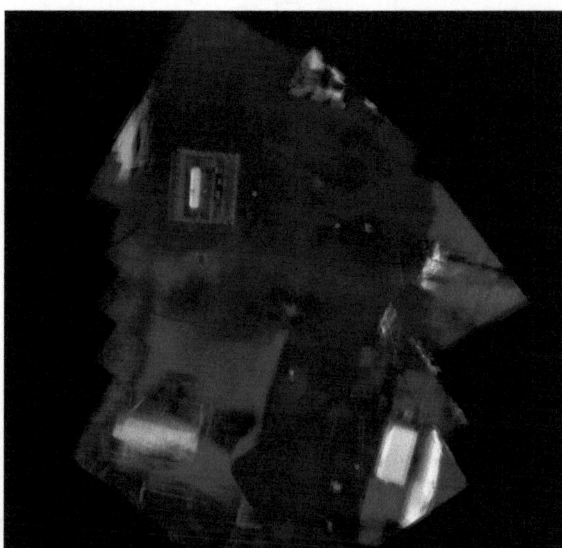

Fig. 9 A mosaic being created using the images from three robots. The robot positions can be seen by the *red circles* in the image

6 Conclusion and Future Work

We described the component technologies, the architecture and the software for QuadCloud, a prototype of a rapid response team for first response and disaster recovery. The key contributions are (a) the design of robust outdoor platforms with speeds that exceed 30 mph with an absolute positioning accuracy of well under 50 cm; (b) the architecture design that enables the rapid deployment of a small team of quadrotors for surveillance without specifying the roles of individuals; and (c) the design of algorithms for estimation, control and planning for multiple quadrotors. We showed experimental results with a team of six quadrotors. Our analysis suggests that the team size can be scaled up to 20 units without compromising system performance. Our future work would address vision-based stabilization to increase robustness to GPS drop out and experimentation with larger teams.

Acknowledgments We are grateful for the support of ARL grant W911NF-08-2-0004, ONR grants N00014-07-1-0829, N00014-09-1-1051 and N00014-09-1-103, NSF grants PFI-1113830 and IIS-1138847, and TerraSwarm, one of six centers of STARnet, a Semiconductor Research Corporation program sponsored by MARCO and DARPA.

References

1. Achtelik, M., Achtelik, M., Weiss, S., Siegwart, R.: Onboard IMU and monocular vision based control for MAVs in unknown in- and outdoor environments. In: IEEE International Conference on Robotics and Automation (ICRA) (2011)
2. Ars Electronica: Spaxels—Ars Electronica Quadcopter Swarm. http://www.aec.at/spaxels/en
3. Bouabdallah, S., Siegwart, R.: Full control of a quadrotor. In: 2007 IEEE/RSJ International Conference on Intelligent Robots and Systems, pp. 153–158 (2007)
4. Envarli, I.C., Adams, J.A.: Task lists for human-multiple robot interaction. In: IEEE International Workshop on Robot and Human Interactive Communication, 2005, pp. 119–124. IEEE (2005)
5. Erdmann, M., Lozano-Perez, T.: On multiple moving objects. In: 1986 IEEE International Conference on Robotics and Automation. vol. 3, pp. 1419–1424 (1986)
6. Hoffmann, G.M., Rajnarayan, D.G., Waslander, S.L., Dostal, D., Jang, J.S., Tomlin, C.J.: The Stanford testbed of autonomous rotorcraft for multi agent control (STARMAC). In: The 23rd Digital Avionics Systems Conference, pp. 12.E.4/1–10. IEEE (2004)
7. Huang, H., Hoffmann, G., Waslander, S., Tomlin, C.: Aerodynamics and control of autonomous quadrotor helicopters in aggressive maneuvering. In: 2009 IEEE International Conference on Robotics and Automation (ICRA), pp. 3277–3282 (2009)
8. Karney, C.F.F.: GeographicLib library. http://geographiclib.sf.net/
9. Kushleyev, A., Mellinger, D., Kumar, V.: Towards a swarm of agile micro quadrotors. In: Robotics: Science and Systems (2012)
10. Latombe, J.C.: Robot motion planning (1991)
11. Lee, T., Leok, M., McClamroch, N.H.: Geometric tracking control of a quadrotor UAV on SE(3). In: 2010 49th IEEE Conference on Decision and Control (CDC), pp. 5420–5425 (2010)
12. Leishman, R.C., Macdonald, J.C., Beard, R.W., McLain, T.W.: Quadrotors and accelerometers: state estimation with an improved dynamic model. IEEE Control Syst. Mag. **34**(1), 28–41 (2014)

13. Lindsey, Q., Mellinger, D., Kumar, V.: Construction with quadrotor teams. Auton. Robots **33**(3), 323–336 (2012)
14. Mahony, R., Kumar, V., Corke, P.: Multirotor aerial vehicles: modeling, estimation, and control of quadrotor. IEEE Robot. Autom. Mag. **19**(3), 20–32 (2012)
15. Mellinger, D., Kumar, V.: Minimum snap trajectory generation and control for quadrotors. In: 2011 IEEE International Conference on Robotics and Automation (ICRA), pp. 2520–2525. Shanghai (2011)
16. Michael, N., Mellinger, D., Lindsey, Q., Kumar, V.: The GRASP multiple micro-UAV testbed. IEEE Robot. Autom. Mag. **17**(3), 56–65 (2010)
17. Shen, S., Mulgaonkar, Y., Michael, N., Kumar, V.: Multi-sensor fusion for robust autonomous flight in indoor and outdoor environments with a rotorcraft MAV. In: IEEE International Conference on Robotics and Automation, Hong Kong (2014)
18. Shen, S., Michael, N., Kumar, V.: Autonomous multi-floor indoor navigation with a computationally constrained mav. In: 2011 IEEE International Conference on Robotics and Automation (ICRA), pp. 20–25 (2011)
19. Turpin, M., Michael, N., Kumar, V.: Concurrent assignment and planning of trajectories for large teams of interchangeable robots. In: 2013 IEEE International Conference on Robotics and Automation (ICRA), pp. 842–848. Karlsruhe (2013a)
20. Turpin, M., Mohta, K., Michael, N., Kumar, V.: Goal assignment and trajectory planning for large teams of aerial robots. In: Proceedings of Robotics: Science and Systems. Berlin (2013b)

Distributed Learning of Cooperative Robotic Behaviors Using Particle Swarm Optimization

Ezequiel Di Mario, Iñaki Navarro and Alcherio Martinoli

Abstract In this paper we study the automatic synthesis of robotic controllers for the coordinated movement of multiple mobile robots. The algorithm used to learn the controllers is a noise-resistant version of Particle Swarm Optimization, which is applied in two different settings: centralized and distributed learning. In centralized learning, every robot runs the same controller and the performance is evaluated with a global metric. In the distributed learning, robots run different controllers and the performance is evaluated independently on each robot with a local metric. Our results from learning in simulation show that it is possible to learn a cooperative task in a fully distributed way employing a local metric, and we validate the simulations with real robot experiments where the best solutions from distributed and centralized learning achieve similar performances.

1 Introduction

This paper deals with the synthesis of simple controllers for cooperative tasks performed by resource-constrained robots. Under these settings, evaluative machine learning techniques are an interesting tool for human designers that may be able to fully exploit the platforms' limited sensing capabilities as well as deal with noise in the performance evaluations [1–5].

E. Di Mario (✉) · I. Navarro · A. Martinoli
Distributed Intelligent Systems and Algorithms Laboratory, School of Architecture, Civil and Environmental Engineering, École Polytechnique Fédérale de Lausanne, Lausanne, Switzerland
e-mail: ezequiel.dimario@epfl.ch

I. Navarro
e-mail: inaki.navarro@epfl.ch

A. Martinoli
e-mail: alcherio.martinoli@epfl.ch

M.A. Hsieh et al. (eds.), *Experimental Robotics*, Springer Tracts in Advanced Robotics 109, DOI 10.1007/978-3-319-23778-7_39

The cooperative task chosen for this study is a loosely-coordinated collective movement or flocking [6–10], in which a set of robots move together as a group. Previous works have shown that it is feasible to use learning to generate cooperative behaviors [2, 3, 11, 12]. Mataric [11] and Parker [12] addressed the topic of learning in multi-robot teams using a small number of parameters per robot, as opposed to the large search space present in this paper. In the case of [2] and [3] learning has been done in a centralized manner, using homogeneous controllers and a global performance metric. The goal of this paper is to distribute the learning of a large parameter space, which increases robustness to failure of individual agents and may also speed up the learning process by testing several candidate solutions at the same time [13]. In order to achieve this goal, we aim to design a local or individual performance metric that can be evaluated by each robot but also leads to the desired cooperative behavior.

It should be noted that the task as implemented in this article is harder than those from previous work in that the robots are not physically connected to each other [2], they are required not only to aggregate but also move together [3], and there is no environmental template or goal to guide their movement [1].

Both the local and global performance metrics used in this article mimic in their components two of the flocking Reynolds' rules [14]: avoiding collisions and attraction to neighboring flock-mates. The alignment or velocity matching rule is not directly reflected in the performance metric in order to simplify the implementation on real robots. In [15], Q-learning is used to generate flocking behaviors of virtual agents (not robots) in the presence of a predator, where the agents individually learn discrete actions similar to Reynolds' rules.

Some researchers have used different optimization techniques to improve the performance of human designed flocking controllers [16–18]. Our approach in this article differs in that our behaviors are generated by a highly plastic feed-forward artificial neural network and not by a specific flocking controller.

The remainder of this article is organized as follows. Section 2 describes the learning algorithms, performance metrics and controller used. Section 3 describes the experimental setup and the different experiments performed. In Sect. 4 we present and discuss the results obtained both in simulation and with real robots. Finally, in Sect. 5 we summarize the findings of this article and discuss the limitations of the approach to be addressed in future work.

2 Methodology

In this article, a version of Particle Swarm Optimization (PSO) [19] is used in order to learn a coordinated collective movement behavior. The learning problem for PSO is choosing a set of parameters of an underlying robotic controller such that a given performance metric is maximized.

```
Intialize particles
for N_i iterations do
    for N_p particles do
        Update particle position
        Evaluate particle
        Re-evaluate personal best
        Aggregate with previous best
        Share personal best
    end for
end for
```

Fig. 1 Noise-resistant PSO algorithm

2.1 Learning Algorithm

The PSO algorithm used is a noise-resistant variation introduced by Pugh et al. [20], which operates by re-evaluating personal best positions and aggregating them with the previous evaluations (in our case a regular average performed at each iteration of the algorithm). The pseudocode for the algorithm is shown in Fig. 1.

The position of each particle represents a set of parameters of a controller. The movement of particle i in dimension j depends on three components: the velocity at the previous step weighted by an inertia coefficient w, a randomized attraction to its personal best $x^*_{i,j}$ weighted by w_p, and a randomized attraction to the neighborhood's best $x^*_{i',j}$ weighted by w_n (Eq. 1). $rand()$ is a random number drawn from a uniform distribution between 0 and 1.

$$v_{i,j} = w \cdot v_{i,j} + w_p \cdot rand() \cdot (x^*_{i,j} - x_{i,j}) + w_n \cdot rand() \cdot (x^*_{i',j} - x_{i,j}) \qquad (1)$$

The PSO neighborhood presents a ring topology with one neighbor on each side. Particles' positions and velocities are initialized randomly with a uniform distribution in the $[-20, 20]$ interval, and their maximum velocity is also limited to that interval. The PSO algorithmic parameters are set following the guidelines for limited-time adaptation we presented in our previous work [21] and are shown in Table 1. Since the dimension of the search space is 26, we round up to 28 particles in order to have exactly seven particles per robot in the distributed implementation involving four robots in total.

Using the PSO algorithm we explore two different learning schemes, in relation to how the particles are distributed among the robots and how the fitness function is defined. The first, global homogeneous, copies the same candidate solution (or set of weights) to every robot, and uses a global fitness function that evaluates the group behavior. The second, local heterogeneous, distributes a different candidate solution (or set of weights) to each robot, and uses a local fitness function that is evaluated independently and individually on each robot. The distributed version allows to speed up the evaluations by a factor equal to the number of robots, yet it makes the learning

Table 1 PSO parameter values

Parameter	Value
Number of robots N_{rob}	4
Population size N_p	28
Iterations N_i	50
Evaluation span t_e	4×45 s
Re-evaluations N_{re}	1
Personal weight pw	2.0
Neighborhood weight nw	2.0
Dimension D	26
Inertia w	0.8
V_{max}	20

harder, especially when the local and global performance metrics are not trivially aligned (e.g., the global performance cannot be represented by a linear combination of local performances).

2.2 Performance Functions

This section gives the mathematical definition of the performance metrics used for centralized and distributed learning. The way inputs are measured during the experiments in simulation and reality is described in Sect. 3.

Both global and local performance functions have three factors: movement, compactness, and collision avoidance. These factors reward robots that move as far as possible from their initial positions, stay close to each other, and avoid collisions between them. The factors are all normalized to the interval [0, 1].

The movement factor of the global performance metric (f_{1g}) is the normalized distance between the initial and the final positions of the center of mass of the group of robots. The normalization factor is the maximum distance that a robots can travel in one evaluation, i.e., the robot's maximum speed multiplied by the evaluation time.

$$f_{1g} = \frac{|\boldsymbol{x}_c(t_f) - \boldsymbol{x}_c(t_0)|}{D_{max}} \tag{2}$$

The global compactness factor (f_{2g}) is the average over the evaluation time and over each pair of robots of the inter-robot fitness. We define the inter-robot fitness between two robots as a function of the distance between them, as shown in Fig. 2. The fitness is maximum at 0 m, and it is zero when the robots are further apart than 0.7 m. At each time step, we calculate the inter-robot fitness for each pair of robots, and then average across all pairs:

$$f_{2g} = \frac{1}{N_{eval}} \sum_{k=1}^{N_{eval}} \left(\frac{1}{N_{pairs}} \sum_{j=1}^{N_{pairs}} fit\_inter_{j,k} \right) \quad (3)$$

where N_{eval} is the number of time steps in the evaluation period, N_{pairs} is number of inter-robot pairs and $fit\_inter_{j,k}$ is the inter-robot fitness for inter-robot pair j at time step k.

The global collision-avoidance factor (f_{3g}) is the average for every robot and over the evaluation time of the maximum value of the proximity sensors at each time step:

$$f_{3g} = \frac{1}{N_{robots}} \sum_{j=1}^{N_{robots}} \left(\frac{1}{N_{eval}} \sum_{k=1}^{N_{eval}} i_{max,j,k} \right) \quad (4)$$

where $i_{max,j,k}$ is the normalized proximity sensor activation value of the most active sensor at time step k for robot j, and N_{robots} is the number of robots.

The local performance metric is calculated individually by each robot, using exclusively on-board resources. The local movement factor (f_{1l}) is the normalized distance travelled by the robot, based on the final position, which is calculated with odometry using the wheel encoders.

$$f_{1l} = \frac{|\boldsymbol{x}_i(t_f) - \boldsymbol{x}_i(t_0)|}{D_{max}} \quad (5)$$

The local compactness factor (f_{2l}) is also based on the inter-robot fitness as defined in Fig. 2 and used in Eq. 3. However, in the local case the number of pairs N_{pairs} in Eq. 3 is modified so that each robot only measures the distance to the other three using an on-board range and bearing module, and then averages the inter-robot fitness only for those other three robots, as opposed to averaging across all pairs of robots. Another difference worth noting between the local and global compactness factors is that the local inter-robot distance measurements are affected by occlusion, while the global ones are not.

Finally, the local collision-avoidance factor (f_{3l}) is the single robot version of the global factor:

$$f_{3l} = \frac{1}{N_{eval}} \sum_{k=1}^{N_{eval}} i_{max,j,k} \quad (6)$$

Fig. 2 Inter-robot fitness as a function of the distance between two robots

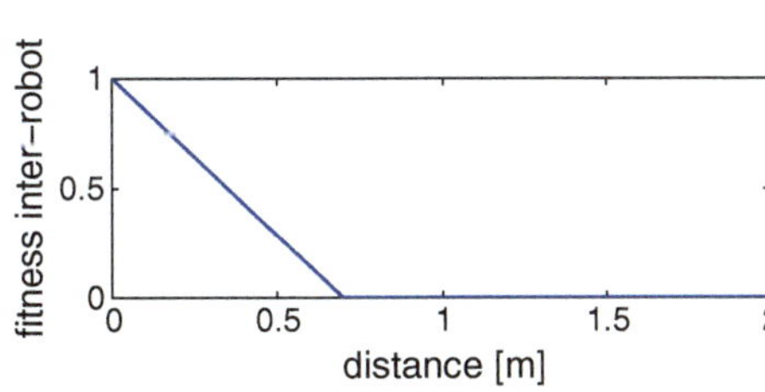

Both global and local fitness are obtained by aggregating the three corresponding factors using the generalized aggregation functions described by Zhang et al. [22]:

$$F = \left(\frac{\omega_1 f_1^s + \omega_2 f_2^s + \omega_3 f_3^s}{\omega_1 + \omega_2 + \omega_3} \right)^{\frac{1}{s}} \tag{7}$$

where f_i are the individual fitness factors, ω_i their corresponding aggregation weights, and s is the degree of compensation or trade-off strategy (higher s means that a high value for a certain factor can compensate for lower values in the others). For all experiments in this article we set the degree of compensation s equal to zero, simplifying Eq. 7 to:

$$F = (f_1^{\omega_1} f_2^{\omega_2} f_3^{\omega_3})^{\frac{1}{\omega_1+\omega_2+\omega_3}} \tag{8}$$

Since the three factors (f_i) are in the interval $[0, 1]$, the fitness function F will also be in the same range. The different combinations of aggregation weights explored in this article are as follows: $\{\omega_1 = 0.25, \omega_2 = 0.5, \omega_3 = 0.25\}$, $\{1/3, 1/3, 1/3\}$, and $\{0.1, 0.8, 0.1\}$.

In our previous work [23], we showed that the fitness evaluations for learning a simpler robotic task had a large standard deviation, and that performing re-evaluations was an effective way of dealing with this challenge in the learning. Given the more complex behavior to be learned in this article and the difficulties encountered while doing so, we decided to perform multiple internal evaluations of the fitness and average them in order to make the learning more robust. Concretely, each candidate solution is evaluated four times during 45 s and its performance averaged before consideration by the noise-resistant algorithm depicted in Fig. 1 ($F' = \frac{1}{4}\sum_{i=1}^{4} F_i$).

2.3 Controller Architecture

The controller is a feed-forward artificial neural network of two units which uses only local, on-board measurements regardless of the performance metric. Its inputs are the range and bearing measurements and the infrared proximity sensors, and it outputs the two wheel speeds. Each neuron has 13 input connections: 4 corresponding to the infrared proximity sensors, 8 corresponding to the range and bearing sensor, and one constant bias speed, resulting in 26 weight parameters (w_k) in total. The outputs of the neurons define the wheel speeds $\{v_l, v_r\}$ as given by Eqs. 9 and 10. $f(\cdot)$ represents the sigmoidal activation function.

$$v_l = f(w_1 + \sum_{k=1}^{4} i_k \cdot w_{k+1} + \sum_{k=1}^{8} rb_k \cdot w_{k+5}) \tag{9}$$

$$v_r = f(w_{14} + \sum_{k=1}^{4} i_k \cdot w_{k+14} + \sum_{k=1}^{8} rb_k \cdot w_{k+18}) \tag{10}$$

Instead of using the robot's nine proximity sensors as inputs, the neural network inputs use four virtual sensors ir_k (front-left, front-right, back-left and back-right) obtained from averaging in pairs and normalizing the sensor values of eight sensors and discarding the central sensor in the back part. This grouping allows us to reduce the number of weight parameters while still being able to detect and avoid obstacles [24].

The eight range and bearing inputs rb_k are obtained by dividing the bearing into eight sectors, and calculating the activation of each sector by taking the minimum range value measured in that sector and dividing it by the maximum possible range, which is approximately 3 m.

3 Experimental Setup

Our experimental platform is the Khepera III mobile robot, a differential wheeled vehicle with a diameter of 12 cm (Fig. 3a). It is equipped with nine infra-red sensors for short range obstacle detection, as well as a relative positioning system [25] which calculates range and bearing to nearby robots based on the strength of the infrared

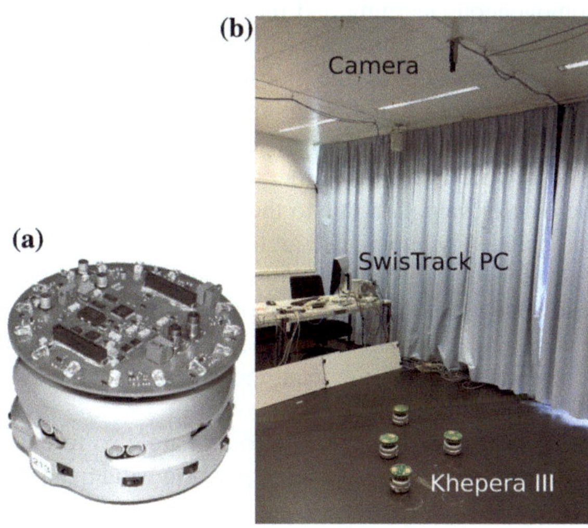

Fig. 3 **a** A Khepera III robot with a range and bearing module attached. **b** Experimental setup in the arena

signal. The Khepera III also has two wheel encoders, which are used to estimate the trajectory followed by the robots for the fitness movement factor calculations.

Since the response of the Khepera III proximity sensors is not a linear function of the distance to the obstacle, the proximity values are inverted and normalized using measurements of the real robot sensor's response as a function of distance. This inversion and normalization results in a proximity value of 1 when touching an obstacle, and a value of 0 when the distance to the obstacle is equal to or larger than 10 cm.

Simulations are performed in Webots [26], a high-fidelity submicroscopic simulator that models dynamical effects such as friction and inertia. In this context, by submicroscopic we mean that it provides a higher level of detail than usual microscopic models, faithfully reproducing intra-robot modules (e.g., individual sensors and actuators). The simulator has a built-in relative positioning system that gives information about the distance and direction to neighboring robots within line-of-sight, mimicking the one used in the real robots.

The learning process is performed completely in simulation. Each evaluation during the learning process has a duration of 45 s and takes place in an unbounded arena. Four robots are placed forming a square of side length equal to two robot diameters with random orientations. In order to calculate the local fitness function, robots only use their internal measurements (simulated range and bearing sensor, infra-red proximity sensors and wheel encoders, all with added noise). The global fitness function is calculated using the robots' global positions provided by the simulator, which have no errors or noise.

After the learning process is finished, the performance of the solution from each of the 20 learning runs is evaluated systematically in simulation, running 20 experiments of 45 s for each solution.

Based on the results of these tests in simulation, the best solution for global homogeneous learning and the best solution for local heterogeneous learning are selected for systematic tests with real robots. We run 20 experiments for each solution. In these experiments, the global positions is monitored using an overhead camera connected to a computer running SwisTrack [27] (see Fig. 3). The initial positions and number of robots are the same as used for learning in simulation, but the evaluation time is reduced to 10 s in order to be able to keep track of the robots' positions during the whole evaluation due to the limited field of view of the fixed overhead camera and the ideally unbounded arena.

Following the same scheme as done in simulation, the local fitness function is computed on each robot using only its on-board resources, while the global fitness is computed externally given the information provided by the overhead camera and complemented with the local measures for the avoidance factor obtained from the robots.

The two selected best controllers are also re-evaluated in simulation using the reduced time of 10 s in order to perform valid quantitative comparisons and validate our models.

4 Experiments and Results

As mentioned in Sect. 3 the learning is conducted in simulation with PSO, which is a stochastic optimization method. Therefore, for statistical significance, we perform 20 optimization runs for global homogeneous (centralized) learning and another 20 runs for local heterogeneous (distributed) learning. Figure 4 shows the progress of the best solution found at each iteration for the two different learning approaches. The curves show the average of the 20 runs, and the error bars represent one standard deviation.

Comparing Fig. 4b with 4a we can notice that the performance of the local heterogeneous learning measured with the global metric is not as high as the global homogeneous one measured with the same metric. However, it should be noted that in homogeneous learning each iteration uses four times the number of evaluations as heterogeneous learning, as in homogeneous learning each candidate solution must be copied to all robots while in heterogeneous learning each robot tests a different candidate solution at the same time.

In addition to the global metric, Fig. 4b shows the progress of the local performance metric for local heterogeneous learning. The global and local metrics are correlated, in the sense that learning with the local one leads to an improvement in the global one.

After the learning is finished, each of the 20 solutions found in the learning runs is tested in simulation for 20 evaluations of 45 s. Figure 5 shows the performance measured using the global metric obtained in this testing. The solutions from homogeneous learning outperform the ones from heterogeneous learning in average. However, the best solution from heterogeneous learning (number 15 in Fig. 5b) has the highest performance over all.

All solutions present a high variation between evaluations (note that in this case the boxplots represent the variation in the performance of each solution during the

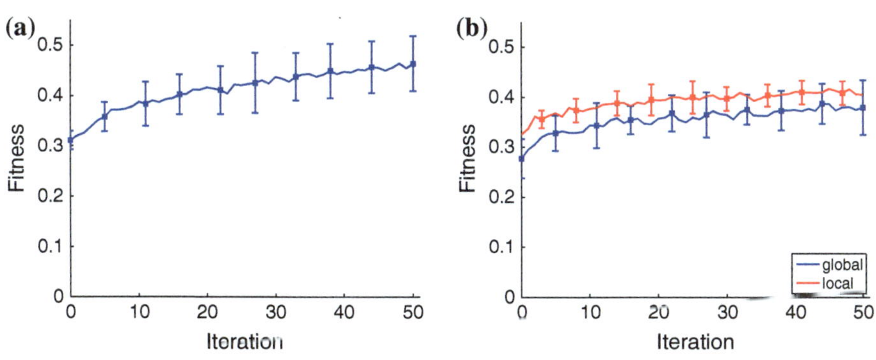

Fig. 4 **a** Learning progress measured using the global metric for global homogeneous (centralized) learning. **b** Learning progress measured using the global (*blue*) and local (*red*) metrics for local heterogeneous (distributed) learning

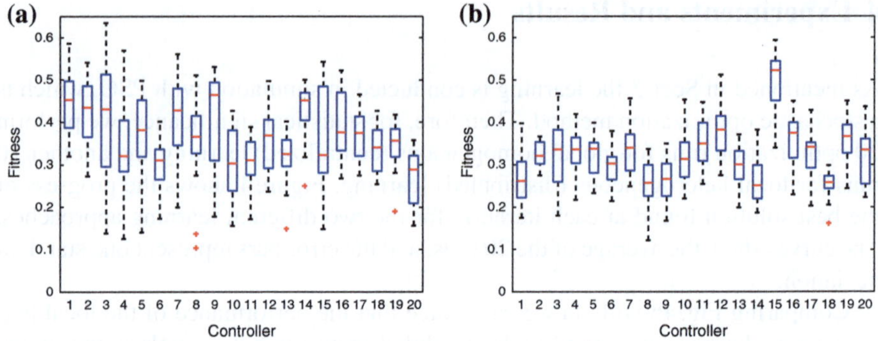

Fig. 5 Performance measured with the global metric in simulation for the 20 solutions found with **a** global homogeneous (centralized) learning and **b** local heterogeneous (distributed) learning

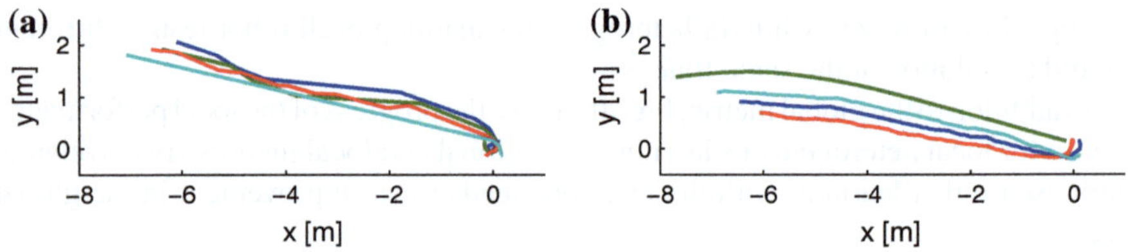

Fig. 6 Example of trajectories from simulation for successful runs with **a** local heterogeneous and **b** global homogeneous learning

20 evaluation runs, which is different from the variation in the learning shown in Fig. 4). This variation between individual evaluations implies that the controllers are sensitive to the initial conditions, i.e. the initial random orientations of the robots. The initial orientations affect the time it takes for the robots to find a common direction of movement, and therefore the total distance that the center of mass is able to travel in the 45 s. When robots fail to find a common direction of movement, they either aggregate close to their initial positions in a very compact group or split in smaller groups and go in separate directions. Figure 6 shows two example trajectories where a common direction of movement was found relatively quickly, allowing the robots to travel a large distance while staying close to each other.

It should be noted that the weights used in the fitness aggregation function also have a significant effect on the behavior of the resulting controllers. Before choosing the final values of {0.25, 0.5, 0.25} for movement, compactness, and avoidance respectively, preliminary tests were conducted in simulation with two other set of weights: {1/3, 1/3, 1/3} and {0.1, 0.8, 0.1}. Figure 7 shows the effect of these fitness aggregation weights on the resulting behaviors. Figure 7a has a low compactness weight, causing the robots to spread out, while Fig. 7b has a high compactness weight, causing robots to stay together without moving far. In order to keep the plots clear and avoid clutter, only the initial 10 s of the trajectories are shown.

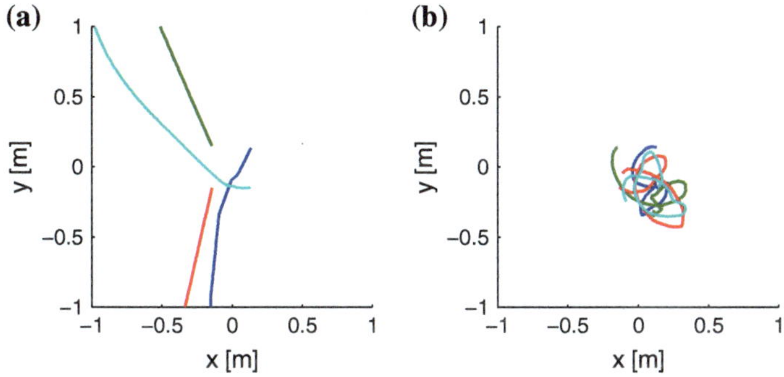

Fig. 7 Example of trajectories for different fitness aggregation weights: **a** {1/3, 1/3, 1/3} and **b** {0.1, 0.8, 0.1}, for compactness, movement, and avoidance, respectively

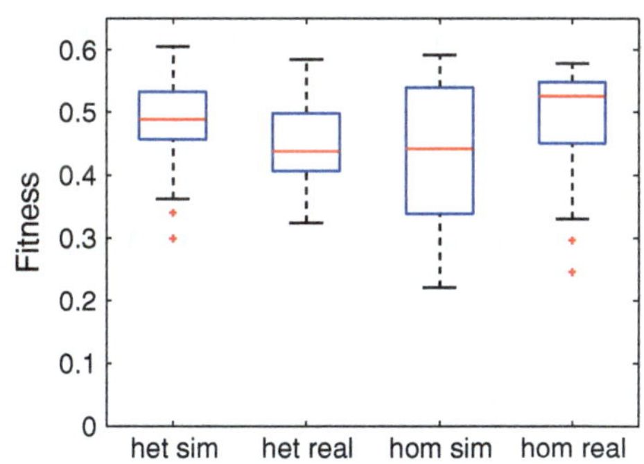

Fig. 8 Performance measured with the global metric in simulation and reality for the best controllers found with global homogeneous and local heterogeneous learning

Based on the results from the tests in simulation, the solutions with the highest medians were chosen to be tested on real robots (number 9 in Fig. 5a, homogeneous learning, and number 15 in Fig. 5b, heterogeneous learning). We conducted 20 evaluation runs of 10 s for each solution, both in simulation and reality. The evaluation time was reduced to 10 s to keep the robots in the overhead camera's field of view.

The quantitative performance measured in these evaluations is presented in Fig. 8. Overall, both controllers showed a satisfactory performance when tested in reality, even though there are differences between simulation and reality that are statistically significant according to the Mann-Whitney U test. The fact that the heterogeneous solution has higher performance in simulation than in reality while the homogeneous shows the inverse implies that there are unmodeled factors that affect the two controllers in a different way due to their different behaviors.

Qualitatively, the behaviors observed in reality were similar to the ones obtained in simulation. Figure 9 depicts two example trajectories from these 10 s evaluations. Note that the trajectories shown here for real robots last 10 s, representing only a fraction of the 45 s from those in Fig. 6, but the initial steps are very similar. During the real robot experiments, it became evident that for the heterogeneous solution the robot in front of the group was always the same, while the other three robots

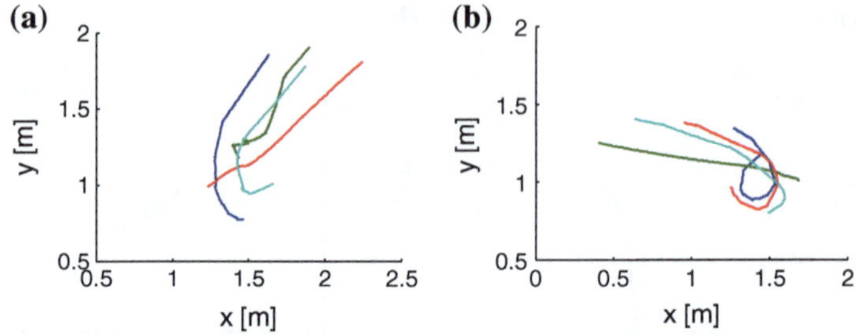

Fig. 9 Example of trajectories for successful runs with **a** local heterogeneous and **b** global homogeneous learning tested on the real robots

followed, meaning that heterogeneous learning led to specialized roles. On the other hand, for the homogeneous solution, the robot in front changed every time based on the initial random orientations.

5 Conclusion and Future Work

Our results have shown that it is feasible to learn a cooperative task in a fully distributed way with a local performance metric measured using local, on-board, noisy sensors. On average, the performance of the solutions found with the distributed approach measured with the global metric was not as high as the ones from centralized learning. This difference was not only due to the metric chosen for learning but also to the increased difficulty in coordination arising from heterogeneous controllers. However, the best solutions found for centralized and distributed learning performed similarly, both in simulation and in experiments with real robots. Additionally, the best solution from distributed learning exhibited specialized roles in which one robot consistently led the group while the others followed.

We have also seen that regardless of the learning method, the coordinated motion task was very sensitive to the initial configuration of the robots, and therefore the performance evaluations were noisy. We addressed this issue in the learning by using different initial orientations for each evaluation and averaging their performances.

As future work, we intend to make the solutions for the coordinated motion task more consistent and robust. In order to achieve this, we will explore increasing the complexity of the controller by using the relative velocity or the relative orientation to other robots as inputs, as well as a adding a corresponding alignment term in the local and global learning metrics. In addition, we would like to explore learning in the presence of obstacles in order to generate obstacle avoidance at the group level. Finally, we would like to test the learned controllers in larger flocks of robots by replicating the sets of controllers. We are specially interested in seeing how the heterogeneous solutions perform.

Acknowledgments This research was supported by the Swiss National Science Foundation through the National Center of Competence in Research Robotics.

References

1. Floreano, D., Mondada, F.: Evolution of homing navigation in a real mobile robot. IEEE Trans. Syst. Man Cybern. Part B Cybern. **26**(3), 396–407 (1996)
2. Baldassarre, G., Trianni, V., Bonani, M., Mondada, F., Dorigo, M., Nolfi, S.: Self-organized coordinated motion in groups of physically connected robots. IEEE Transactions on Systems, Man, and Cybernetics. Part B, Cybernetics: a publication of the IEEE Systems, Man, and Cybernetics Society **37**(1), 224–39 (2007)
3. Gauci, M., Chen, J., Dodd, T., Groß, R.: Evolving aggregation behaviors in multi-robot systems with binary sensors. In: Distributed Autonomous Robotic Systems, Springer Tracts in Advanced Robotics, pp. 355–367 (2014)
4. Jin, Y., Branke, J.: Evolutionary optimization in uncertain environments: a survey. IEEE Trans. Evol. Comput. **9**(3), 303–317 (2005)
5. Pugh, J., Martinoli, A.: Distributed scalable multi-robot learning using particle swarm optimization. Swarm Intell. **3**(3), 203–222 (2009)
6. Balch, T., Arkin, R.: Behavior-based formation control for multi-robot teams. IEEE Trans. Robot. Autom. **14**(6), 926–939 (1998)
7. Fredslund, J., Mataric, M.J.: A general algorithm for robot formations using local sensing and minimal communication. IEEE Transactions on Robotics and Automation, Special Issue on Multi Robot Systems **18**(5), 837–846 (2002)
8. Olfati-Saber, R.: Flocking for multi-agent dynamic systems: algorithms and theory. IEEE Trans. Autom. Control **51**, 401–420 (2006)
9. Antonelli, G., Arrichiello, F., Chiaverini, S.: Flocking for multi-robot systems via the null-space-based behavioral control. In: Proceedings of the IEEE/RSJ International Conference on Intelligent Robots and Systems, pp. 1409–1414 (2008)
10. Navarro, I., Matía, F.: A framework for collective movement of mobile robots based on distributed decisions. Robot. Auton. Syst. **59**(10), 685–697 (2011)
11. Mataric;, M.: Learning in behavior-based multi-robot systems: policies, models, and other agents. Cogn. Syst. Res. **2**, 81–93 (2001)
12. Parker, L.E.: L-ALLIANCE: task-oriented multi-robot learning in behavior-based systems. In: Advanced Robotics, Special Issue on Selected Papers from IROS'96, pp. 305–322 (1997)
13. Di Mario, E., Martinoli, A.: Distributed particle swarm optimization for limited time adaptation with real robots. Robotica **32**(02), 193–208 (2014)
14. Reynolds, C.W.: Flocks, herds, and schools: a distributed behavioral model. Comput. Graph. **21**(4), 25–34 (1987)
15. Morihiro, K., Isokawa, T., Nishimura, H., Matsui, N.: Emergence of flocking behavior based on reinforcement learning. Knowl. Based Intell. Inf. Eng. Syst. **4253**, 699–706 (2006)
16. Lee, S.M., Myung, H.: Particle swarm optimization-based distributed control scheme for flocking robots. Robot Intell. Technol. Appl. **208**, 517–524 (2013)
17. Vatankhah, R., Etemadi, S., Honarvar, M., Alasty, A., Boroushaki, M., Vossoughi, G.: Online velocity optimization of robotic swarm flocking using particle swarm optimization (pso) method. In: International Symposium on Mechatronics and its Applications (2009)
18. Celikkanat, H.: Optimization of self-organized flocking of a robot swarm via evolutionary strategies. In: International Symposium on Computer and Information Sciences (2008)
19. Kennedy, J., Eberhart, R.: Particle swarm optimization. In: IEEE International Conference on Neural Networks, pp. 1942–1948 (1995)
20. Pugh, J., Zhang, Y., Martinoli, A.: Particle swarm optimization for unsupervised robotic learning. In: IEEE Swarm Intelligence Symposium, pp. 92–99 (2005)

21. Di Mario, E., Martinoli, A.: Distributed particle swarm optimization for limited time adaptation in autonomous robots. In: Distributed Autonomous Robotic Systems, Springer Tracts in Advanced Robotics, pp. 383–396 (2014)
22. Zhang, Y., Antonsson, E., Martinoli, A.: Evolutionary engineering design synthesis of on-board traffic monitoring sensors. Res. Eng. Des. **19**(2), 113–125 (2008)
23. Di Mario, E., Navarro, I., Martinoli, A.: Analysis of fitness noise in particle swarm optimization: from robotic learning to benchmark functions. In: IEEE Congress on Evolutionary Computation, pp. 2785–2792 (2014)
24. Di Mario, E., Navarro, I.n., Martinoli, A.: The role of environmental and controller complexity in the distributed optimization of multi-robot obstacle avoidance. In: IEEE International Conference on Robotics and Automation, pp. 571–577 (2014)
25. Pugh, J., Raemy, X., Favre, C., Falconi, R., Martinoli, A.: A fast on-board relative positioning module for multi-robot systems. IEEE/ASME Transactions on Mechatronics Focused Section on Mechatronics in Multi Robot Systems, pp. 151–162 (2009)
26. Michel, O.: Webots: professional mobile robot simulation. Adv. Robot. Syst. **1**(1), 39–42 (2004)
27. Lochmatter, T., Roduit, P., Cianci, C., Correll, N., Jacot, J., Martinoli, A.: SwisTrack: a flexible open source tracking software for multi-agent systems. In: IEEE/RSJ International Conference on Intelligent Robots and Systems, pp. 4004–4010 (2008)

Provably Correct Persistent Surveillance for Unmanned Aerial Vehicles Subject to Charging Constraints

Kevin Leahy, Dingjiang Zhou, Cristian-Ioan Vasile, Konstantinos Oikonomopoulos, Mac Schwager and Calin Belta

Abstract In this work, we present a novel method for automating persistent surveillance missions involving multiple vehicles. Automata-based techniques were used to generate collision-free motion plans for a team of vehicles to satisfy a temporal logic specification. Vector fields were created for use with a differential flatness-based controller, allowing vehicle flight and deployment to be fully automated according to the motion plans. The use of charging platforms with the vehicles allows for truly persistent missions. Experiments were performed with two quadrotors over 50 runs to validate the theoretical results.

Keywords Persistent monitoring · Multi-robot systems · Aerial robotics · Formal methods

1 Introduction

In this paper, we investigate the automatic deployment of multiple quadrotors under resource constraints. The relatively short battery life in many unmanned aerial vehicles (UAVs) presents a significant barrier to their use in complex, long term surveillance missions. Moreover, the use of multiple vehicles allows for more complex behavior and longer mission horizons, but further complicates the task of deploying those vehicles given limited flight time. We present an algorithm that generates a feedback controller for multiple quadrotors with charging constraints to meet a complex temporal logic specification. The algorithm comprises a three-part tool chain that first plans a high level routing schedule for the quadrotors, then generates a vector field control input for the quadrotors to accomplish the schedule, and finally controls the quadrotors' nonlinear dynamics to follow the vector field with a feedback controller. The performance of the complete system, with its three interacting parts, is investigated in 50 experimental runs using two quadrotors and three charging stations in a motion capture environment.

K. Leahy (✉) · D. Zhou · C.-I. Vasile · K. Oikonomopoulos · M. Schwager · C. Belta
Boston University, Boston, MA 02215, USA
e-mail: kjleahy@bu.edu

M.A. Hsieh et al. (eds.), *Experimental Robotics*, Springer Tracts in Advanced Robotics 109, DOI 10.1007/978-3-319-23778-7_40

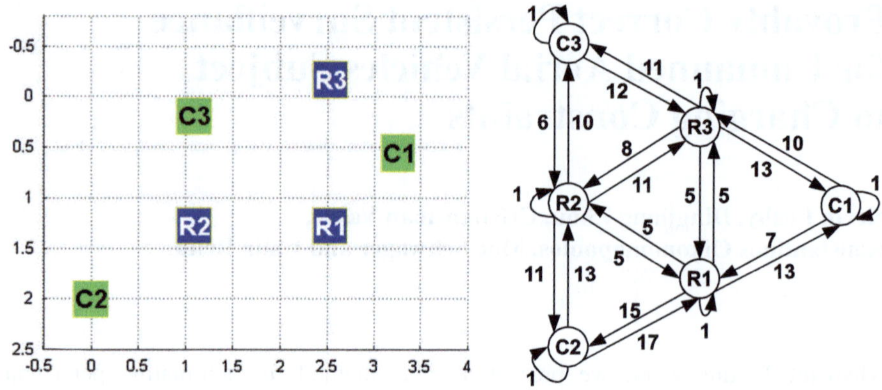

Fig. 1 Partitioned environment viewed from above and transition system. *Green squares* are charging stations, while *blue squares* are regions of interest. States in the transition system are charging stations and regions of interest. Weights on transitions are based on calculated time bounds

We consider the following problem: given an environment and a temporal logic mission specification with time deadlines that needs to be satisfied infinitely often, generate control policies for a team of quadrotors to complete the mission, while ensuring vehicles remain charged and collisions are avoided.

As a motivating example, we consider the environment shown in Fig. 1 consisting of three charging stations, three regions of interest, and two aerial vehicles. We assume vehicle battery life is 40 time units, and charging takes 120 time units, where time units are a generic unit that can be instantiated based on a particular implementation. Given this environment and these battery and charging constraints, the vehicles must perform a persistent surveillance mission defined by a rich linear temporal formula which imposes time bounds on each loop of the vehicle's (infinite) run. Thus, the specification is given as a bounded time formula which needs to be satisfied infinitely often. In this example, we wish the multi-robot system to satisfy the following mission specification *infinitely often*: "within 16 time units observe Region R3 for at least 3 time units; within 28 time units, observe Region R1 for at least 2 time units; and within 46 time units, observe Region R2 for at least 2 time units then within 8 time units observe Region R1 or Region R3 for at least 2 time units." We seek a method to generate a control policy ensuring that vehicles can be automatically deployed to successfully complete this mission in the specified environment.

The solution to this problem requires the use of several sophisticated systems, whose interaction both at a theoretical level and an experimental level produces many unique challenges. Our approach is related to the Vehicle Routing Problem (VRP) [1], which can be summarized as: given N identical vehicles at a depot and the distances among all sites and the depot, find a minimum distance tour for each vehicle such that it begins and ends at the depot and visits each site at least once. By placing time bounds on when each site must be visited, we obtain a version of

VRP called Time Window VRP (VRPTW) [2]. Multi-agent control for the VRPTW has also been considered without temporal logic constraints in [3, 4]. Our work uses temporal logic constraints for the VRPTW with richer specifications.

Temporal logic and formal methods [5] have been used for robot motion planning and control in persistent surveillance as in [6, 7]. These works, while considering optimal persistent surveillance with temporal logic constraints, do not consider resource constraints. These works also do not consider time windows, which we use in this paper. Resource constraints have been modeled in the routing problem for one vehicle without temporal logic constraints in [8]. Resource constraints have also been modeled for persistent monitoring in [9], in which the authors present a platform for autonomous charging of UAVs, including an algorithm for persistent surveillance for multiple vehicles without temporal logic constraints. Our work allows for richer mission specification while still modeling resource constraints.

The most closely related recent work includes [10] in which the authors propose a fragment of metric temporal logic, which restricts temporal operators to atomic propositions and their negation. In that work, each site may be visited only once, and bounds on transition duration are not allowed. Additionally, their work does not take into account resource constraints, and optimizes a weighted sum of distance traveled over a finite horizon. Our approach allows for a vehicle to visit a site multiple times during a tour if it is required, capturing resource constraints, and allowing bounds on transition durations.

A version of this work, involving formal methods for creating task plans, appears in [11]. Additionally, details on the differential flatness approach to vehicle control appears in [12].

2 Technical Approach

The solution is outlined as follows: first, motion plans are generated to satisfy the mission specification in Sect. 2.1. A vector field is constructed for navigating the quadrotors, from which the transition system is abstracted as explained in Sect. 2.2. Finally, in Sect. 2.3 a differential flatness-based approach is used to control the vehicles through the previously constructed vector field.

2.1 Control Policy Generation

Generating a control policy for our persistent surveillance problem first requires creating an abstraction of the environment and quadrotor behavior, including a model of the quadrotor battery charging and discharging. By specifying the mission using a temporal logic formula, we are able to use automata theoretic techniques in conjunction with theses abstractions to synthesize a control policy.

Finite Models of the Environment and Quadrotors For simplicity of presentation, we assume the team is made of N identical quadrotors. Consider a finite abstraction of the environment given as a graph $G = (V = \mathcal{S} \cup \mathcal{C}, E, w)$, where $\mathcal{S}$ is the set of sites and $\mathcal{C}$ is the set of charging stations or depots. An edge $e \in E \subseteq V \times V$ denotes that a vector field can be constructed such that a vehicle can move between the source and destination of the edge (see Sect. 2.2). Quadrotors can deterministically choose to traverse the edges of G, stay at a site for service, or stay docked in a charging station. A duration is associated with each edge, which represents the flight time and includes docking or undocking, if applicable, and is given by $w : E \to \mathbb{Z}_{\geq 1}$. The construction of the environment graph G is described in Sect. 2.2.

In this paper, we assume that the team has a mutual-exclusive (ME) operation mode, i.e. at any moment in time at most one quadrotor is flying. Thus, collision avoidance is conservatively guaranteed.

Each vehicle has a limited amount of battery life, specified as an integer value, and must regularly return to a charging station. The maximum operation time starting with a fully charged battery is denoted by t_{op}, while the maximum charging time starting with an empty battery is denoted by t_{ch}. The charge-discharge ratio, which denotes the amount of time required to charge the battery versus how long the vehicle may fly on a fully-charged battery, is $\gamma = \frac{t_{ch}}{t_{op}} \geq 1$ and may only take integer values. For simplicity, we assume that time is discretized, and all durations (e.g., $w(E)$, t_{op}, t_{ch}) are expressed as an integer multiple of a time interval Δt.

A battery is abstracted by a discrete *battery state* $b_t(i) \in \{0, \ldots, t_{ch}\}$, corresponding to quadrotor i at time $t \in \mathbb{Z}_{\geq 0}$, and an update rule, which specifies the change of charge after d time units:

$$b_{t+d}(i) = \begin{cases} \min\{b_t(i) + d, t_{ch}\} & \text{vehicle } i \text{ is docked} \\ b_t(i) - \gamma d & \text{otherwise} \end{cases} \tag{1}$$

It is assumed that the quadrotors are equipped with identical batteries. The batteries may be charged at any of the charging stations $\mathcal{C}$. Charging may start and stop at any battery state. Once a quadrotor is fully charged, it will remain fully charged until it leaves the charging station. We assume that at the start of the mission all quadrotors are fully charged and docked.

We will say that a quadrotor is *active* if it is flying, i.e. moving between sites and charging stations or servicing a request. A request at a site is said to be serviced if a quadrotor hovers above it. The time bounds in (2) represent the duration for which each site is to be serviced. A time interval in which all vehicles are docked and none are charging is called *idle time*.

Control Policy For $q \in V$, we use $\vec{q}$ to denote that a quadrotor is flying towards q. Let $\vec{V} = \{\vec{q} \mid q \in V\}$. A *control policy* for the team of quadrotors is a sequence $\mathbf{v} = v_1 v_2 \ldots$ where $v_t \in (V \cup \vec{V})^N$ specifies at each time $t \in \mathbb{Z}_{\geq 0}$ and for each quadrotor $i \in \{1, \ldots, N\}$ if quadrotor i is at a site or charging station or if it is moving. Let $v_t(i)$ and $v(i)$, $i \in \{1, \ldots, N\}$, denote the control value for quadrotor i

at time t and the control policy for quadrotor i (i.e., the sequence of control values), respectively. Then a transition $(q_1, q_2) \in E$ performed by quadrotor i starting at time t will correspond to $v_t(i) = q_1$, $v_{t+d}(i) = q_2$ and $v_{t+k}(i) = \vec{q}_2$, $k \in \{1, \ldots, d-1\}$, where $d = w(q_1, q_2)$ is the duration of the transition. Servicing or charging for one time interval (Δt time) by quadrotor i at time t corresponds to $v_t(i) = v_{t+1}(i) \in V$. A control policy $\mathbf{v} = v_1 v_2 \ldots$ determines an *output word* $\mathbf{o} = o_1 o_2 \ldots$ such that $o_t = \{v_t(i) | v_t(i) \in \mathcal{S}, i \in \{1, \ldots, N\}\}$ is the set of all sites occupied by the N quadrotors at time $t \in \mathbb{Z}_{\geq 0}$. We use ϵ to denote that no site is occupied. Note o_t is either ϵ or a singleton set, because of the ME operation mode assumption. Let $q^{[d]}$ and q^{ω} denote d and infinitely many repetitions of q, respectively.

Bounded Linear Temporal Logic To capture the richness of the specification, we use bounded linear temporal logic (BLTL) [13], a temporal logic with time bounds on each of its temporal operators. The mission specification presented in Sect. 1 can be expressed as $\mathbb{G}\phi$, where ϕ is given in (2) as a BLTL formula and the $\mathbb{G}$ operator indicates that ϕ should be satisfied infinitely often.

$$\phi = \mathbf{F}^{\leq 16}\mathbf{G}^{\leq 3} R3 \wedge \mathbf{F}^{\leq 28}\mathbf{G}^{\leq 2} R1 \wedge \mathbf{F}^{\leq 46}(\mathbf{G}^{\leq 2} R2 \wedge \mathbf{F}^{\leq 10}\mathbf{G}^{\leq 2}(R1 \vee R3)) \quad (2)$$

In (2), $\wedge$ and $\vee$ are the usual Boolean operators indicating conjunction and disjunction, while **F** and **G** are the temporal operators "eventually" and "always", respectively. Superscripts on the temporal operators are time bounds on those operators. Each Ri is a request associated with the region. A control policy is said to *satisfy* the persistent surveillance specification $\mathbb{G}\phi$, where ϕ is a BLTL formula, if the generated output word satisfies the BLTL formula ϕ infinitely often and there is no idle time between any two consecutive satisfactions of ϕ. Note that, between successive satisfactions of ϕ, the quadrotors may recharge their batteries, i.e. at least one may not be idle, because it is charging its battery.

Problem Formulation and Overview of the Approach Let $\mathbf{v}$ be a control policy. We say that $\mathbf{v}$ is *feasible* if at each moment in time all N quadrotors have non-negative battery states, i.e., $b_t(i) \geq 0$ for all $i \in \{1, \ldots, N\}$ and $t \in \mathbb{Z}_{\geq 0}$.

Problem 1 Given an environment $G = (V = \mathcal{S} \cup \mathcal{C}, E, w)$, N quadrotors with operation time t_{op} and charging time t_{ch}, and a BLTL formula ϕ over $\mathcal{S}$, find a feasible control policy that satisfies $\mathbb{G}\phi$ if one exists, otherwise report failure.

Let $\mathbf{v}$ be a feasible control policy satisfying $\mathbb{G}\phi$. We define a *loop* as a finite subsequence of $\mathbf{v}$ starting with the satisfaction of the formula ϕ and ending before the next satisfaction of ϕ.

The proposed approach to Problem 1 is based on automata techniques [5]. The motion model of the quadrotor team is represented as a product transition system between N copies of G which is pruned of any states and transitions which violate the ME operation mode. The product transition system is then composed with a finite state automaton which captures the charging constraints. The resulting product model is then composed with another finite state automaton which accepts the satisfying language corresponding to the given BLTL formula ϕ. The finite state automaton encoding ϕ is obtained by first translating it [14] to a syntactically co-safe Linear Temporal Logic formula [15] and then to an automaton using the *scheck* tool [16]. The satisfiability problem (Problem 1) is solved on the resulting product automaton by considering all possible states of the team at the start of a loop and paths between these states obtained with Dijkstra's algorithm. For more details about the procedure see [11], where the authors prove the completeness of the proposed approach. In [11], they also consider the fully-concurrent mode of operation and optimality.

2.2 Vector Field and Transition System Weights

We use a vector field for the implementation of the control policies synthesized as explained in Sect. 2.1, because it allows for the discrete environment model to be combined with the continuous dynamics necessary for vehicle navigation. Additionally, once the vector field has been created, upper limits on travel times through the vector field provide the weights w for the environment graph G such that a control policy can be synthesized.

Partition To generate the vector field, we first partition the environment into cubes. Each cube is defined by two vectors, $a = (a_1, a_2, a_3)$ and $b = (b_1, b_2, b_3)$ where $a_i < b_i$ for all $i = 1, 2, 3$. Thus, each cube may be written as

$$C(a, b) = \left\{x \in \mathbb{R}^3 | \forall i \in \{1, 2, 3\} : a_i \le x_i \le b_i\right\}. \tag{3}$$

Paths corresponding to edges in the environment are found as sequences of these cubes. The paths are constrained such that quadrotors fly to a fixed height from the charging stations and perform all observations from that fixed altitude. From these paths, we generate to ensure each sequence of cubes is followed.

Vector Field Construction A vector field everywhere inside a given cube can be created as a convex combination of a set vectors at its vertices [17], expressed as

$$h(x_1, \ldots, x_N) = \sum_{v \in \mathbf{V}(a,b)} \prod_{i=1}^{N} \left(\frac{x_i - a_i}{b_i - a_i}\right)^{\xi_i(v_i)} \left(\frac{b_i - x_i}{b_i - a_i}\right)^{1-\xi_i(v_i)} \cdot h(v), \tag{4}$$

where x_i is the coordinate in the ith dimension of a point in the cube, $\mathbf{V}(a, b)$ are the vertices of cube $C(a, b)$, $h(v)$ are the vectors at each vertex $v \in \mathbf{V}(a, b)$, $N = 3$, and $\xi_i(v_i)$ is an indicator function such that $\xi_i(a_i) = 0$ and $\xi_i(b_i) = 1$. Such a vector field can be used to keep the vehicle from leaving the cube (stay-in-cell) or to force it to leave through a given facet (control-to-facet), as displayed in Fig. 2.

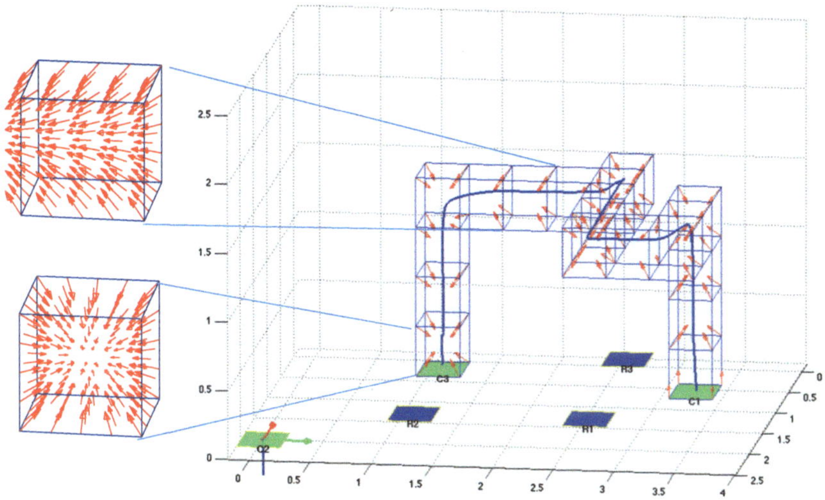

Fig. 2 Vector field detail and quadrotor flight data. The cube at the *top left* shows a control-to-facet vector field, and the cube at the *bottom left* shows a stay-in-cell vector field. One of these two kinds of fields is given to the quadrotor in each cell along its path to guide it through the desired trajectory

For each cube in any given path, we create a control-to-facet vector field to lead to the next cube in the path. Because discontinuities in the vector field could result in undesirable behavior of the quadrotors, we must ensure that velocity is continuous from one cube to the next. We ensure continuity by examining vectors at the facet where cubes meet. For each corner of such a facet, the vectors from the two cubes are compared to each other. Only the vector components that the two vectors have in common are kept. This process is illustrated in Fig. 3. In the figure, cells A, B, and C are joined together, and B then shares a facet with A and C. The vectors for cell B and C on their shared facet are identical, and continuity is ensured. But the vectors on A's shared facet with B are different (Fig. 3b). Thus the vertical components of these vectors are discarded, but the horizontal components, which are identical, are kept (Fig. 3c). Because of this process, there are limitations to the types of arrangements of cubes that can be constructed, because they would result in a vector of zero magnitude (see Fig. 4b), but in practical examples, such arrangements are unlikely to be desirable and can be avoided by using a finer partition of the environment if necessary.

Weights Because satisfaction of (2) depends on the time to travel among the regions of the environment, these times must be known. We can calculate the upper bound on the travel time between any two regions, which are captured as weights on the transition system described above. We model hovering over a region or charging as self-loop transitions of weight 1. Calculating the upper time bound for leaving a cube

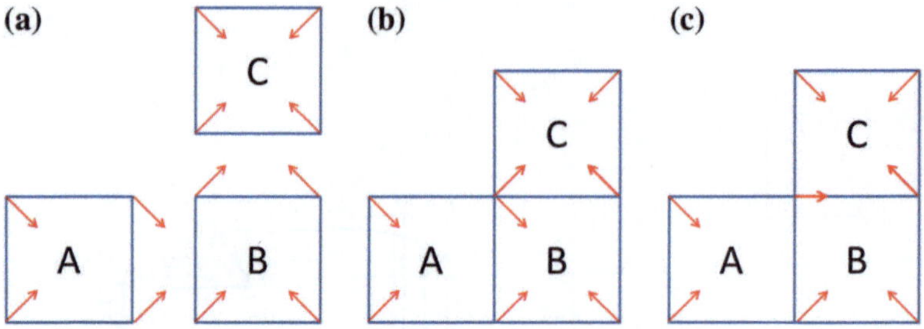

Fig. 3 Two-dimensional example of combining vectors. **a** Control-to-facet vector field from *A* to *B* and *B* to *C*, and stay-in-cell vector field for cell *C*. **b** Vector conflict where *A*, *B* and *C* meet. **c** Final vector field, keeping only non-conflicting vector components

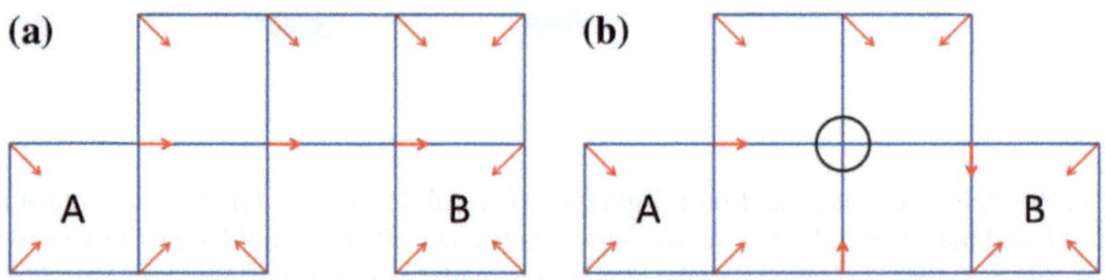

Fig. 4 Two-dimensional example of vector field configurations from *A* to *B*. **a** Allowable configuration results in vectors with some zero-magnitude components, while resulting in no vectors with zero-magnitude. **b** Not allowable configuration with an occurence of zero-magnitude for all components (*circled*)

depends on the vectors at the vertices. If none of these vectors has a component of magnitude zero, we calculate the time bound for exiting the cube through facet F as

$$T^F = ln\left(\frac{s_F}{s_{\bar{F}}}\right)\frac{b_i - a_i}{s_F - s_{\bar{F}}}, \tag{5}$$

where $\bar{F}$ is the facet opposite F, and $s_F, s_{\bar{F}}$ are the minimum vector components in the ith direction on facet F and $\bar{F}$, respectively [18]. In the event that s_F approaches $s_{\bar{F}}$, T^F approaches $(b_i - a_i)/s_{\bar{F}}$.

Because of the continuity requirements on the vector field, it is possible to have a vector with a component of magnitude zero (i.e. as seen in Fig. 4a). In this case, as long as there remains a non-zero component in another direction, there is a guaranteed upper bound on the time to leave the cell. This time bound, in the case of a zero-magnitude component in the ith direction and a non-zero component in the jth direction, while exiting in the ith direction through the facet containing the zero-magnitude component, can be expressed as

$$T^F = T_i^F + T_j^F = \left(\frac{b_i - a_i}{s_F\left(\frac{M}{2} - 1\right)}\right) ln\left(\frac{M}{2}\right) + \left(\frac{b_j - a_j}{-2s_{\overline{F}}}\right) ln\left(1 - M\right), \tag{6}$$

where $0 < M < 1$ is a measure of "conservatism." The closer M is to 1, the larger the time bound, and the higher the guarantee of the time bound being correct. This is due to the asymptotic nature of the solution approaching the zero-magnitude component in the ith direction.

2.3 Vector Field Following

Motion planning often involves the use of vector fields to be followed by a robot. This is easily accomplished with most ground robots as well as slow aerial robots. In our experiments however, we use quadrotors, which cannot easily follow a vector field because of their high dimensional, nonlinear dynamics. Thus, we exploit the differential flatness of quadrotor dynamics to design a controller which will allow the quadrotor to follow the vector field, compensating for the quadrotor's nonlinear dynamics [12].

Differential Flatness Quadrotor dynamics are given by the nonlinear system of equations

$$\dot{v} = ge_3 + \frac{1}{m}Rf_z e_3 \tag{7}$$
$$\dot{R} = R\Omega \tag{8}$$
$$\dot{\omega}_b = J^{-1}\tau - J^{-1}\Omega J\omega_b \tag{9}$$
$$\dot{h} = v, \tag{10}$$

where $v = [v_x, v_y, v_z]^T$ is the velocity in the world frame, g is the acceleration due to gravity, m is the mass, f_z is the total thrust force from the rotors, $e_3 = [0, 0, 1]^T$, and hence $f_z e_3$ is aligned with the negative vertical direction of the body frame, $-z_b$. R is the rotation matrix from the world frame to the body frame, defined in terms of Euler angles ψ, θ, and ϕ. $\omega_b = [p, q, r]^T$ is the angular velocity of the quadrotor expressed in the body frame, $\Omega = \omega_b^{\wedge}$ is the tensor form of ω_b. The torque on the quadrotor is given by τ in the body frame F_b. J is the inertia matrix of the quadrotor, and h is the position of the quadrotor in the world frame.

The system as defined in (7)–(10) has a 12-dimensional state, $\xi = [x, y, z, v_x, v_y, v_z, \psi, \theta, \phi, p, q, r]^T$, and input, $\mu = [f_z, \tau_x, \tau_y, \tau_z]^T$, which is the total thrust and three torques. The state and input are differentially flat. Their flat outputs

$$\sigma = [\sigma_1, \sigma_2, \sigma_3, \sigma_4]^T := [x, y, z, \psi]^T, \tag{11}$$

consisting of position and yaw, are such that the state, ξ is a function of these outputs and their derivatives. More precisely, $\xi = \beta(\sigma, \dot{\sigma}, \ddot{\sigma}, \dddot{\sigma})$, with

$$\begin{cases} [x, y, z, v_x, v_y, v_z, \psi]^T = \beta_{1:7}(\sigma, \dot{\sigma}) = [\sigma_1, \sigma_2, \sigma_3, \dot{\sigma}_1, \dot{\sigma}_2, \dot{\sigma}_3, \sigma_4]^T \\ \theta = \beta_8(\sigma, \dot{\sigma}, \ddot{\sigma}) = \text{atan2}(\beta_a, \beta_b) \\ \phi = \beta_9(\sigma, \dot{\sigma}, \ddot{\sigma}) = \text{atan2}(\beta_c, \sqrt{\beta_a^2 + \beta_b^2}) \\ [p, q, r]^T = \beta_{10:12}(\sigma, \dot{\sigma}, \ddot{\sigma}, \dddot{\sigma}) = (R^T \dot{R})^\vee, \end{cases} \tag{12}$$

where

$$\begin{cases} \beta_a = -\cos\sigma_4\ddot{\sigma}_1 - \sin\sigma_4\ddot{\sigma}_2 \\ \beta_b = -\ddot{\sigma}_3 + g \\ \beta_c = -\sin\sigma_4\ddot{\sigma}_1 + \cos\sigma_4\ddot{\sigma}_2, \end{cases} \tag{13}$$

and R is the rotation matrix with the Euler angles (ϕ, θ) defined in (12). Furthermore, the input, μ, is also a function of the flat outputs, expressed as $\mu = \gamma(\sigma, \dot{\sigma}, \ddot{\sigma}, \dddot{\sigma}, \ddddot{\sigma})$, with

$$\begin{cases} f_z = \gamma_1(\sigma, \dot{\sigma}, \ddot{\sigma}) = -m \parallel \ddot{\sigma}_{1:3} - ge_3 \parallel \\ [\tau_x, \tau_y, \tau_z]^T = \gamma_{2:4}(\sigma, \dot{\sigma}, \ddot{\sigma}, \dddot{\sigma}, \ddddot{\sigma}) \\ = J(\dot{R}^T \dot{R} + R^T \ddot{R})^\vee + R^T \dot{R} J (R^T \dot{R})^\vee, \end{cases} \tag{14}$$

where $\ddot{\sigma}_{1:3} = [\ddot{\sigma}_1, \ddot{\sigma}_2, \ddot{\sigma}_3]^T$ for short and the $^\vee$ map is the inverse operation of $^\wedge$. For details and a proof, please refer to [12].

With the flat outputs and their derivatives obtained as described below, the above equations can generate all the states and inputs. A standard $SE(3)$ controller [19] can be implemented to control the quadrotor flight along the vector field using the states and inputs as a control reference.

Vector Field Derivatives The inputs described in (14) require knowledge of velocity, acceleration, jerk, and snap. Hence it is necessary to find the time derivatives ($\dot{\sigma}$, $\ddot{\sigma}$, $\dddot{\sigma}$, $\ddddot{\sigma}$) by taking spatial derivatives of the vector field. We only consider vector fields which do not specify rotation, hence the yaw angle σ_4 is irrelevant. We arbitrarily set $\sigma_4(t) \equiv 0$. In general, the flat output derivatives $\dot{\sigma}_{1:3}$, $\ddot{\sigma}_{1:3}$, $\dddot{\sigma}_{1:3}$, $\ddddot{\sigma}_{1:3}$ at any point x in a vector field $h(x)$ can be recursively calculated by

$$\begin{cases} \dot{\sigma}_{1:3}(x) & = h(x) \\ \ddot{\sigma}_{1:3}(x) & = \mathcal{J}(\dot{\sigma}_{1:3}(x), x)\dot{\sigma}_{1:3}(x) \\ \dddot{\sigma}_{1:3}(x) & = \mathcal{J}(\ddot{\sigma}_{1:3}(x), x)\dot{\sigma}_{1:3}(x) \\ \ddddot{\sigma}_{1:3}(x) & = \mathcal{J}(\dddot{\sigma}_{1:3}(x), x)\dot{\sigma}_{1:3}(x), \end{cases} \tag{15}$$

where $\mathcal{J}(f(x), x)$ denotes the Jacobian matrix of the function $f(x)$.

The velocity is obtained directly from the vector field described by (4), from which the derivatives required for the differential flatness controller given in (15) can be derived analytically. First (4) is rewritten in matrix form as

$$h(x_1, \ldots, x_3) = [c_1, \ldots, c_8] \begin{bmatrix} h_{1x_1} & h_{1x_2} & h_{1x_3} \\ \vdots & \vdots & \vdots \\ h_{8x_1} & h_{8x_2} & h_{8x_3} \end{bmatrix}. \quad (16)$$

In this form, the coefficients c are functions of position, but the values of h are fixed for any given cube. This form is therefore convenient for computation of the acceleration and other vector field derivatives.

In general, the acceleration at x is given by

$$a(x) = \mathcal{J}(v(x), x)v(x), \quad (17)$$

where $\mathcal{J}(f(x), x)$ denotes the Jacobian matrix of the function $f(x)$, which is a 3×3 matrix with entries

$$\mathcal{J}_{ij} = \frac{\partial v_i}{\partial x_j} = h_{1x_i}\frac{\partial c_1}{\partial x_j} + \cdots + h_{8x_i}\frac{\partial c_8}{\partial x_j}. \quad (18)$$

Through straightforward calculation, acceleration is therefore given by

$$a_i = \sum_{j=1}^{3}\left(\sum_{k=1}^{8} h_{kx_i}\frac{\partial c_k}{\partial x_j}\right)v_j. \quad (19)$$

It should be noted that the vector fields for acceleration, jerk, and snap are continuous everywhere within a given cube but may be discontinuous at the facets between cubes. Similar calculations can be done for jerk and snap and are omitted due to space limitations.

3 Results and Experiments

The partitioned environment (Figs. 1 and 7) consists of 385 cubes each with edge length 0.36 m. Control policies for $\mathbb{G}\phi$ were calculated over the transition system displayed in Fig. 1. The computation time, excluding encoding of (2), was 301.7 s on a Linux system with a 2.1 GHz processor and 32 GB memory, and the final product automaton had 579,514 nodes and 2,079,208 edges. No solutions were found for quadrotors starting on Chargers C2 and C3, but all other combinations of starting positions yielded solutions.

Experiments were performed in the Boston University Multi-robot Systems Lab. The lab consists of a flight space with IR cameras to track reflective markers on the quadrotors using an OptiTrack system. This system allows for real-time localization of the quadrotors during experiments. Two K500 quadrotors from KMel robotics were used to execute the control policies described in Sect. 3.

Fig. 5 Quadrotor resting on charging station

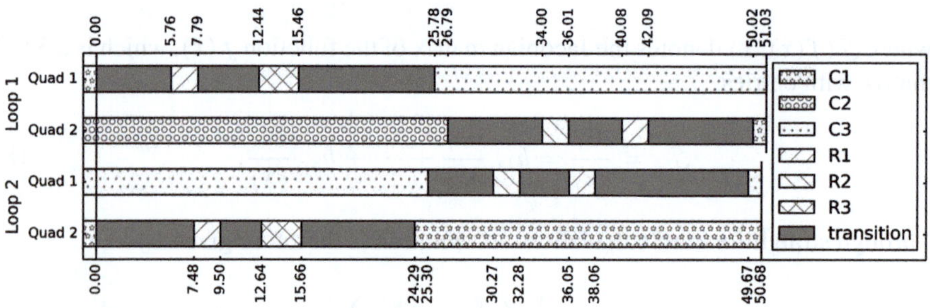

Fig. 6 Timeline of quadrotor flights for two loops. The *first two rows* display the first loop, with Quadrotor 1 flying before Quadrotor 2. The *next two rows* show the second loop, with Quadrotor 2 flying first

Charging stations (Fig. 5) were designed and built at Boston University for automatic docking and charging of quadrotors. These platforms allow a vehicle to land when its battery requires charging. When using multiple such platforms, another vehicle can then take off, ensuring continuous monitoring in situations where one vehicle may not be able to satisfy a persistent monitoring mission specification on its own.

The charging stations are made of laser cut acrylic parts connected with PLA plastic 3D printed parts. The electronics of the station consist of the Hyperion EOS0720i Net3AD charger, modified to enable control by MATLAB. To secure a robust connection with the stainless steel pads of the charging station, the quadrotors are equipped with stainless steel contacts mounted on springs with magnets. The platform is entirely controlled by MATLAB via USB connection, allowing for the detection of the presence of a quadrotor, real-time monitoring of battery and charging status, and control of the charging parameters including battery type, capacity, and charging rate. The maximum charging rate that can be achieved is 8 amps.

Figure 6 shows the results of a flight by two quadrotors. Seconds were used as the time units for these experiments so flights could be rapidly performed and analyzed.

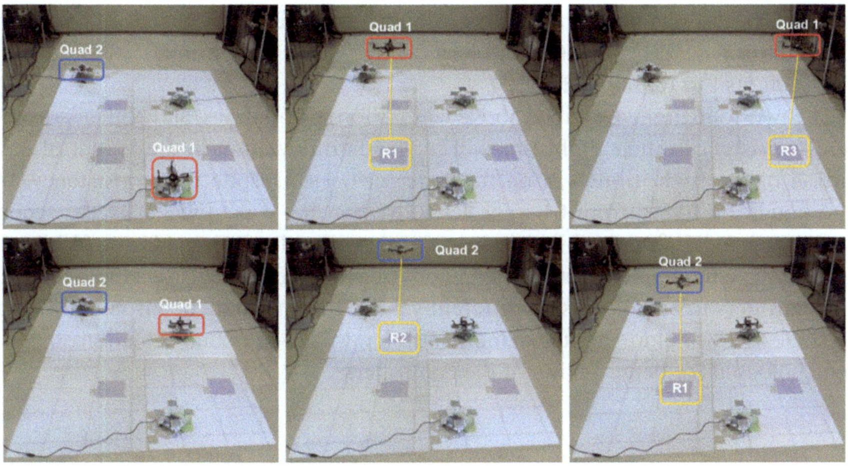

Fig. 7 Screencaps of the first flight loop

The quadrotors, shown in red (Quad 1) and blue (Quad 2) in Fig. 7, start fully charged from the charging stations $C1$ and $C2$, respectively. The control policy $\mathbf{v}$ for the two quadrotors, generated as described in Sect. 2.1, is the following:

$$
\begin{aligned}
v(1) = \quad & C1^{[1]}\vec{R}1^{[6]}R1^{[3]}\vec{R}3^{[4]}R3^{[4]}\vec{C}3^{[10]}C3^{[41]} \\
& \left(C3^{[31]}\vec{R}2^{[5]}R2^{[3]}\vec{R}3^{[10]}R3^{[3]}\vec{C}3^{[10]}\right)^{\omega} \\
v(2) = \quad & C2^{[29]}\vec{R}2^{[12]}R2^{[3]}\vec{R}1^{[10]}R1^{[3]}\vec{C}1^{[12]} \\
& \left(C1^{[1]}\vec{R}1^{[6]}R1^{[3]}\vec{R}3^{[4]}R3^{[4]}\vec{C}1^{[12]}C1^{[32]}\right)^{\omega} .
\end{aligned}
\tag{20}
$$

Under control strategy (20), in the first loop Quadrotor 1 (red) take-off first and services sites $R1$ and $R3$ and Quadrotor 2 (blue) completes the loop by servicing sites $R2$ and $R1$. In all subsequent loops, Quadrotor 2 (blue) takes-off first and services sites $R1$ and $R3$ and Quadrotor 1 complete the loop by servicing sites $R2$ and $R1$. After the first loop, Quadrotors 1 and 2 always return to $C3$ and $C1$, respectively. The corresponding output word is

$$
\begin{aligned}
o = \quad & \epsilon^{[7]}R1^{[3]}\epsilon^{[4]}R3^{[4]}\epsilon^{[23]}R2^{[3]}\epsilon^{[10]}R3^{[3]}\epsilon^{[12]} \\
& \left(\epsilon^{[7]}R1^{[3]}\epsilon^{[4]}R3^{[4]}\epsilon^{[18]}R2^{[3]}\epsilon^{[10]}R3^{[3]}\epsilon^{[10]}\right)^{\omega} .
\end{aligned}
$$

The flights presented in the experiments consist of the first two loops each satisfying ϕ. Any subsequent loop would be identical to the second loop. Since ϕ can be satisfied repeatedly, these flights can satisfy the mission specification, $\mathbb{G}\phi$.

Figure 6 shows that the specification was satisfied for both loops in the flight. Region R1 was visited in 5.76 s in Loop 1 and 7.48 s in Loop 2, ahead of the 28 s deadline. Likewise, Region R3 was visited in 12.44 and 12.64 s ahead of the 16 s deadline. In the second portion of each loop, Region R2 was visited in 34.00 and 30.27 s with a deadline of 46 s, and Region R1 was visited within the 8 s deadline after each visit to Region R2.

The two-loop flight described above was performed 50 times, and both quadrotors were consistent in their flight times. The standard deviation in the length of each portion of the flight time was on the order of 0.1 s. Despite this consistency, the time bound on flying from Charger C1 to Region R1 was violated by the second quadrotor in each flight, while not being violated by the first quadrotor. While the vehicles were nominally identical, small physical differences between them required the controllers to be tuned using different values. Because both quadrotors followed the same vector field using the same controller, this time bound violation suggests some potential for better tuning of the controllers.

4 Conclusion

The main insight gleaned from this experiment is how to automate a complex persistent surveillance mission specified as a temporal logic formula. The methodology explained herein allows for rapid experimentation following theoretic work using temporal logics. By using the environment partition and transition system generation with time bounds, minimal human input is required to establish an experimental framework for simulating and executing missions. Further, the inclusion of charging stations, whose performance can be modeled using automata, allows for long-term, truly persistent missions involving multiple vehicles not only to be modeled, but to actually be performed in the lab.

The implementation of the persistent surveillance framework required three systems to be integrated together: the BLTL control synthesis algorithm, the vector field generation algorithm, and the quadrotor differential flatness controller. Inevitably, limitations appear at the interfaces of such systems. For example, the use of multiple vehicles required tuning the controllers quite differently to ensure that the vector field was followed, even though the vehicles are of the same make and model. Regardless of any such complications, because a conservative approach was used, such as using upper bounds on travel time rather than expected travel time, the system met the specifications reliably and predictably. These experiments establish a framework that can be extended to a variety of future work. We are particularly interested in loosening restrictions on mutually exclusive operation so that multiple vehicles may be airborne simultaneously. This would also allow for more complex distributed tasks, such as simultaneously servicing several sites. We are also interested in extending this work to longer mission horizons with more vehicles.

Acknowledgments This work was supported in part by NSF grant number CNS-1035588, and ONR grant numbers N00014-12-1-1000, MURI N00014-10-10952 and MURI N00014-09-1051. The authors are grateful for this support.

References

1. Dantzig, G.B., Ramser, J.H.: The truck dispatching problem. Manag. Sci. **6**(1), 80–91 (1959)
2. Toth, P., Vigo, D.: The Vehicle Routing Problem. SIAM (2001)
3. Michael, N., Stump, E., Mohta, K.: Persistent surveillance with a team of mavs. In: Proceedings of the International Conference on Intelligent Robots and Systems (IROS 11), pp. 2708–2714. IEEE (2011)
4. Stump, E., Michael, N.: Multi-robot persistent surveillance planning as a vehicle routing problem. In: Proceedings of the IEEE Conference on Automation Science and Engineering (CASE), pp. 569–575. IEEE (2011)
5. Baier, C., Katoen, J.-P.: Principles of Model Checking. MIT Press (2008)
6. Smith, S., Tumova, J., Belta, C., Rus, D.: Optimal path planning for surveillance with temporal logic constraints. Int. J. Robot. Res. **30**(14), 1695–1708 (2011)
7. Ulusoy, A., Smith, S.L., Ding, X.C., Belta, C., Rus, D.: Optimality and robustness in multi-robot path planning with temporal logic constraints. Int. J. Robot. Res. **32**(8), 889–911 (2013)
8. Sundar, K., Rathinam, S.: Algorithms for routing an unmanned aerial vehicle in the presence of refueling depots. IEEE Trans. Autom. Sci. Eng. **11**(1), 287–294 (2014)
9. Mulgaonkar, Y., Kumar, V.: Autonomous charging to enable long-endurance missions for small aerial robots. Proc. SPIE-DSS 9083(64) (2014)
10. Karaman, S., Frazzoli, E.: Vehicle routing problem with metric temporal logic specifications. In: IEEE Conference on Decision and Control, pp. 3953–3958 (2008)
11. Vasile, C., Belta, C.: An automata-theoretic approach to the vehicle routing problem. In: Robotics: Science and Systems Conference (RSS), Berkeley (2014)
12. Zhou, D., Schwager, M.: Vector field following for quadrotors using differential flatness. In: Proceedings of the International Conference on Robotics and Automation (ICRA) (2014)
13. Jha, S., Clarke, E., Langmead, C., Legay, A., Platzer, A., Zuliani, P.: A bayesian approach to model checking biological systems. In: Proceedings of the 7th International Conference on Computational Methods in Systems Biology, CMSB '09, pp. 218–234. Springer, Berlin (2009)
14. Tkachev, I., Abate, A.: Formula-free finite abstractions for linear temporal verification of stochastic hybrid systems. In: Proceedings of the 16th International Conference on Hybrid Systems: Computation and Control, Philadelphia (2013)
15. Kupferman, O., Vardi, M.: Model checking of safety properties. Form. Methods Syst. Des. **19**(3), 291–314 (2001)
16. Latvala, T.: Effcient model checking of safety properties. In: 10th International SPIN Workshop, Model Checking Software, pp. 74–88. Springer (2003)
17. Belta, C., Habets, L.C.G.J.M.: Controlling a class of nonlinear systems on rectangles. IEEE Trans. Autom. Control **51**(11), 1749–1759 (2006)
18. Aydin Gol, E., Belta, C.: Time-constrained temporal logic control of multi-affine systems. Nonlinear Anal. Hybrid Syst. **10**, 21–33 (2013)
19. Mellinger, D., Kumar, V.: Minimum snap trajectory generation and control for quadrotors. In: 2011 IEEE International Conference on Robotics and Automation (ICRA), pp. 2520–2525. IEEE (2011)

Part XI
Interactive Presentations

Ross A. Knepper and Nora Ayanian
Cornell University and University of Southern California

Mirroring a trend in robotics conferences worldwide, ISER this year saw increased participation in interactive presentations. This format is well suited to experiments involving complex systems and mechanisms operating in real world environments. ISER 2014 included two interactive sessions in which a total of seventeen papers were presented, spanning the diverse robotic domains of manipulation, locomotion, vehicles traveling on water, land and air, and robotic smart materials.

Six papers addressed aspects of manipulation. ten Pas and Platt test a system capable of identifying graspable handles from point cloud data of novel objects. Using Baxter, the authors report 85 % accuracy in pick and place tasks on unfamiliar objects. Zhao et al. report human user study experiments that delve into the legibility of reaching motions where the target is ambiguous. It turns out that hand direction is a stronger indicator in signalling robot intent than arm motion. Dogar et al. present a system capable of marshalling a variety of sensors to achieve multi-scale assembly tasks. The system leverages robot coordination for transport of components too large for a single robot to handle. Kim et al. develop a model for teleoperated grasping that maps from human grasping behavior to a kinematically dissimilar robot grasp by composing synergies. The authors discovered a technique for personalizing the model to a given human user for greater performance. Demircan et al. examine the role of human body posture in forceful interactions such as pushing a large, heavy box. The authors collect data from humans in various pushing configurations to better understand muscular efficiency, which will inform robot behavior. Klingensmith et al. show how occupancy grid mapping techniques can be applied to object recognition and reconstruction. They argue that occupancy grids more naturally preserve the volumetric information provided by noisy RGB-D sensors compared to existing 3D point-cloud techniques. Experimental results comparing their occupancy grid based approach to traditional point-cloud methods show the effectiveness of the proposed strategy.

Two papers explored hexapod locomotion at different scales. St. Pierre et al. consider centimeter-scale magnetically actuated walkers. The authors seek to understand ground interaction forces at small scales, which are difficult to model at high frequency gaits. At a larger scale, Hoerger et al., examine the ability of marginally stable walkers to reject disturbances during walking. The robot, which is a stock hexapod robot extended with stilts, continually plans a path from unstable postures to stable states during walking.

Three interactive presentations involved vehicles traveling on or under the water's surface. Hitz et al. build a map using a nodding 2D LIDAR that is annotated with static and dynamic points of interest. Static points of reference help localize the vehicle, whereas dynamic points of interest are observed over the seasons for changes, such as the level of foliage in a willow tree. Smith et al. present a planner for finding and tracking underwater ocean fronts. Due to the dynamic and unpredictable nature of these fronts, an adaptive path planning approach is used. Ganesan et al. extend the traditional occupancy-grid method for mapping obstacles underwater—a context in which global localization is difficult—by placing the map in the robot frame.

Three papers focused on wheeled vehicles—three of them automotive. O'Callaghan and Ramos proposed a new technique for representing occupancy maps. The approach allows arbitrary resolution and can incorporate a temporal component for moving obstacles. Qin et al. present a method for obstacle recognition designed to work with low-cost sensors. The authors' key insight is that recognition ability is enhanced for noisy data by simultaneously considering a sequence of sensor readings within a time window. Rummelhard et al. evaluate a method for predicting risk of collision with obstacles without explicitly detecting or tracking the obstacles. Instead, the technique estimates collision risk of spatial cells through which the vehicle will imminently pass. Avoiding the issue of obstacle tracking avoids a common failure mode in collision avoidance algorithms.

Two interactive presentations consider problems in human-robot interactions. Tay et al. presents an algorithm to reason about a humanoid robot's motion that relies on single motions and interpolation between motion pairs. While the proposed strategy is focused on fall prediction for a humanoid robot, the work presents a framework towards enabling motion-based human-robot communication. Argall builds on a framework for a semi-autonomous wheelchair, which permits the user as much or as little control over the motion as desired through a modular architecture. This paper evaluates a module for the task of navigating doorways.

Two papers address problems of tracking and mapping using multiple quadrotors. Liu et al. demonstrate a technique for multiple quadrotors to perform frontier-based exploration of a space efficiently while building and merging disparate maps. Hausman et al. present a multi-quadrotor localization system for tracking moving objects without the benefit of motion capture. Instead, the robots form a visibility tree in which each robot can see one or more robots and other targets.

McEvoy and Correll coin the term robotic materials to describe smart materials with variable stiffness. The authors constructed a five-segment beam with embedded sensors, computation, and a cable drive system to control stiffness in each segment and achieve arbitrary pose.

Localizing Handle-Like Grasp Affordances in 3D Point Clouds

Andreas ten Pas and Robert Platt

Abstract We propose a new approach to localizing handle-like grasp affordances in 3-D point clouds. The main idea is to identify a set of sufficient geometric conditions for the existence of a grasp affordance and to search the point cloud for neighborhoods that satisfy these conditions. Our goal is not to find *all* possible grasp affordances, but instead to develop a method of localizing important types of grasp affordances quickly and reliably. The strength of this method relative to other current approaches is that it is very practical: it can have good precision/recall for the types of affordances under consideration, it runs in real-time, and it is easy to adapt to different robots and operating scenarios. We validate with a set of experiments where the approach is used to enable the Rethink Baxter robot to localize and grasp unmodelled objects.

Keywords Grasping · 3-D point clouds · Grasp affordances · Handle grasping

1 Introduction

Robust robot grasping in novel and unstructured environments is an important research problem that has many practical applications. A key sub-problem is localization of the objects or object parts to be grasped. Localization is challenging because it can be difficult to localize graspable surfaces on unmodelled objects. Moreover, even small localization errors can cause a grasp failure. In this paper, we develop an approach to localization-for-grasping based on localizing parts of objects rather than localizing the entire object. We refer to these graspable object parts as *grasp affordance geometries*: object geometries that can be grasped in a particular way by a particular robot hand. Although the idea of a grasp affordance has existed in the literature for a long time [5], the idea has new promise now because the availability of accurate range sensing information (i.e. the Microsoft Kinect) may make grasp affordance localization easier. In this paper, we develop an approach to searching a 3-D point cloud for grasp affordance geometries.

A. ten Pas (✉) · R. Platt
College of Computer and Information Science, Northeastern University, Boston, MA, USA
e-mail: atp@ccs.neu.edu

M.A. Hsieh et al. (eds.), *Experimental Robotics*, Springer Tracts in Advanced Robotics 109, DOI 10.1007/978-3-319-23778-7_41

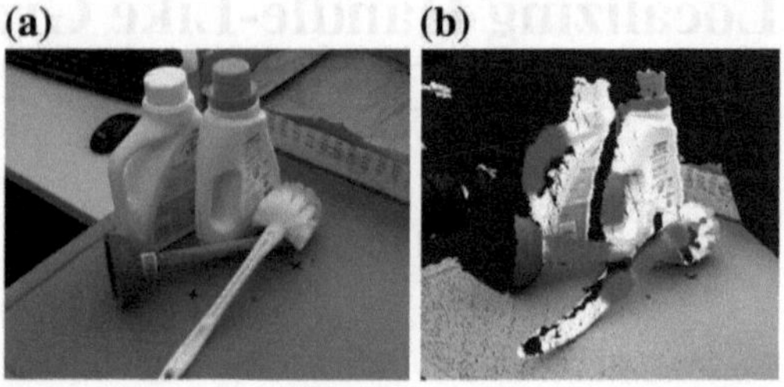

Fig. 1 **a** An RGB image of a typical scene. **b** Handle-like grasp affordances localized using our algorithm highlighted in cyan

The main idea is to identify a set of sufficient geometric conditions for the existence of a grasp affordance and to search the point cloud for neighborhoods that satisfy these conditions. Here, we concern ourselves with "handle-like" grasp affordance geometries. Our goal is not to find *all* possible grasp affordances, but instead to develop a method of localizing important types of grasp affordances quickly and reliably. Developing an efficient search is a key challenge. A complete handle configuration is determined by seven parameters and a brute force search of the point cloud would be infeasible in real time. We structure the search in two ways. First, we constrain the robot hand to grasp in a plane orthogonal to the minor principal curvature axis of the local object surface at the point where the grasp occurs. This constraint makes sense intuitively and ultimately enables us to reduce the search space down to three (spatial) dimensions. Second, we require a cylindrical gap to be present around an object surface to accommodate the grasping robot hand. This constraint enables us to eliminate many grasp candidates quickly. Figure 1 illustrates typical results of the overall process. The strength of this method relative to other current approaches is that it is very practical: it has good precision/recall for the types of affordances under consideration, it runs in real-time, and it is easy to adapt to different robots and operating scenarios. In addition, we have created an easy-to-use ROS package [15] that implements the algorithm and allows it to be used in most robotic manipulation operational scenarios.

2 Related Work

The problem of localizing graspable geometries has been the subject of extensive recent research. An important class of methods work by searching a height map or a range image for graspable regions. For example, Klingbeil et al. search for geometries in a range image that can be grasped by a parallel-jaw gripper [10]. A three-dimensional search (x, y, θ) is performed over the range image. The gripper is constrained to approach the object from a single direction. The work of Jiang et al. is related [8]. They search a registered RGBD image for regions that score high on a linear-in-the-features grasp score function, where feature weights were learned off-line. Closely related to the work of Jiang et al. is that of Fischinger and Vincze [4]. Rather than searching an RGBD image, they perform a 3-DOF search of a height map (calculated from a point cloud). They key element of this work is the introduction of a new type of feature used to develop a graspability score function. Our current work is distinguished from the above chiefly because we do not use a depth image or height map to structure our search for grasp affordances, but we operate directly on the point cloud instead. This brings several advantages including the ability to structure the search in different ways, and a looser coupling between how the affordance was perceived and the approach direction of the arm. Overall, our grasp success rates are at least as good as those of any of the work mentioned above. However, it is important to remember that this success rate assumes that objects can always be grasped by a handle that is within reach of the robot.

Other work loosely related to the above includes that of Herzog et al., who learn graspable height map "templates" based on user demonstrations of good and bad grasps [7]. Katz et al. develop a method that depends on physical interaction with the objects to be grasped [9]. The robot pushes the object under consideration and uses the resulting motion to perform segmentation accurately. The resulting system is very robust, but can require significant pushing interactions prior to grasping. Another line of current research approaches the problem of localization-for-grasping by searching for known modeled objects in a scene. Here, it is common to use feature-matching approaches. Appropriate 3-D features for use with point clouds include Fast Point Feature Histograms (FPFH) [12] and the SHOT feature [17]. It is typical to use RANSAC or Hough voting [13, 16] to align features found on an object model with features found in a scene. However, Glover and Popovic recently proposed a new method (loosely related to ICP [1]) that has demonstrated robustness advantages [6]. Often, the system may be ignorant of which object is present in a scene. Brook, Ciocarlie, and Hsiao develop a database-driven method that segments the point cloud into clusters and compare these clusters against 3D models in a database [2]. A Bayesian framework is used that incorporates uncertainty in object shape, object pose, and robot motion error.

3 Localizing Grasp Affordances

An enveloping grasp affordance is a handle-like object geometry that can be grasped by encircling it with the thumb and fingers of the robot hand. We locate these geometries in a 3D point cloud by searching for cylindrical shells that satisfy certain criteria with respect to local neighborhoods of the point cloud. A cylindrical shell is a pair of co-linear cylinders with different radii. We require the following conditions on the local point neighborhood to be satisfied:

1. Points near the center of the neighborhood must lie on a curved object surface (with respect to a parametrized threshold on curvature).
2. The axis of the cylindrical shell must be parallel to the secondary axis of curvature of the local object surface.

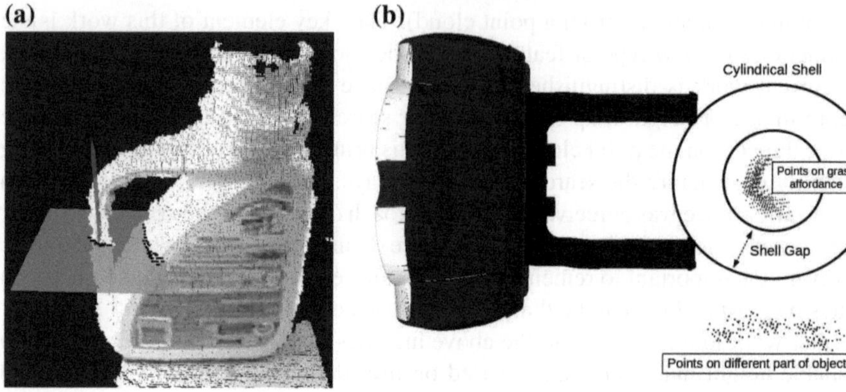

Fig. 2 Illustration of the affordance search. **a** Points in a local neighborhood are projected onto a plane orthogonal to the minor principal curvature axis of the object surface. **b** A shell is found that contains points within the inner circle but has a gap between the inner and outer circle

3. The gap between the inner and outer cylinders must contain zero points and be wide enough to contain the robot fingers.
4. The radius of the innermost cylinder must be no larger than the maximum hand aperture.

If the above conditions are satisfied, we say that an enveloping grasp affordance exists in the corresponding configuration. These are sufficient conditions for an enveloping grasp in the sense that if we assume they are satisfied and if we assume that points lie densely on all object surfaces in the neighborhood and if we assume the neighborhood can be reached by the robot hand, then we know that an object can be grasped using an enveloping grasp. This is illustrated in Fig. 2. In Fig. 2a, a locally curved surface has been found (at the root of the red arrow), and a plane has been drawn orthogonal to the secondary axis of curvature. Figure 2b shows the points after they have been projected onto the plane and a circular shell (a projection of the cylindrical shell) that satisfies the enveloping grasp affordance conditions.

Our overall algorithm has the following steps (see Algorithm 1). First, we randomly sample spherical point neighborhoods approximately 2 or 3 cm in radius. This is accomplished by sampling points uniformly at random from the cloud and then taking a point neighborhood about each sample (Step 3). Second, we fit an implicit quadratic function (in three variables) to each of these point neighborhoods using a least squares algebraic fit with Taubin normalization [14] (Step 4). As a result of fitting, we obtain an accurate measurement of the magnitudes and axes of principal surface curvature in the point neighborhood (Step 5). We eliminate from consideration all neighborhoods with an associated surface curvature below some parametrized threshold (Step 6), and project the point neighborhood onto the plane orthogonal to the axis of minor principal curvature (Step 7). Next, we fit a circle to the projected points (Step 8). We then fix the center of the shell to the center of the fitted circle and perform a 1-D search for cylindrical shells satisfying the enveloping grasp affordance

Algorithm 1 Handle Localization

1: $\mathcal{A} = \emptyset$
2: **for** $i = 1$ to I **do**
3: Sample x uniformly from cloud; calculate point neighborhood about x.
4: Fit a quadratic surface S to point neighborhood.
5: Estimate the median curvature $\hat{\kappa}$ of S.
6: **if** $\hat{\kappa} > K$ **then**
7: Project point neighborhood onto orthogonal plane
8: Fit a circle to points in plane; calculate cicle center, c.
9: Search for cylindrical shell, a, centered at c.
10: **if** a is found **then**
11: $\mathcal{A} = \mathcal{A} \cup a$
12: **end if**
13: **end if**
14: **end for**
15: $\mathcal{H} \leftarrow$ findHandles($\mathcal{A}$).

conditions (Step 9). Last, given the found enveloping grasp affordances, we search for sets of affordances that are roughly aligned and that exceed a minimum length (Step 15). Key elements of the algorithm are detailed in the sections below.

3.1 Estimating Object Surface Curvature by Fitting an Implicit Quadratic Surface

In order to find high-curvature regions of the point cloud and to estimate the axes of curvature accurately, we fit an implicit quadratic surface in three variables to points in the local neighborhood. A quadratic can be described by $f(\mathbf{c}, \mathbf{x}) = 0$, where

$$f(\mathbf{c}, \mathbf{x}) = c_1 x_1^2 + c_2 x_2^2 + c_3 x_3^2 + c_4 x_1 x_2 + c_5 x_2 x_3 + c_6 x_1 x_3 + c_7 x_1 + c_8 x_2 + c_9 x_3 + c_{10}, \tag{1}$$

and $\mathbf{c} \in \mathbb{R}^{10}$ denotes the parameters of the quadratic and $\mathbf{x} \in \mathbb{R}^3$ denotes the Cartesian coordinates of a point on the surface.

It turns out that there is no known fast (convex or closed form or etc.) method for finding the implicit quadratic surface that minimizes least squares geometric distances to a set of points (called the *geometric fit*). However, there do exist fast methods for solving for an *algebraic fit*, that is, a surface that solves the following optimization problem (Fig. 3):

$$\min_{c} \sum_{i=1}^{n} f(\mathbf{c}, \mathbf{x}^i)^2 = \mathbf{c}^T M \mathbf{c}, \tag{2}$$

Fig. 3 Two examples of implicit quadratic surfaces fit using Taubin normalization

where $M = \sum_{i=1}^{n} l(\mathbf{x}^i)l(\mathbf{x}^i)^T$, $\mathbf{x}^1, \ldots, \mathbf{x}^n \in R^3$ are the points to which the curve is fitted, and

$$l(\mathbf{x}) = (x_1^2, x_2^2, x_3^2, x_1x_2, x_1x_3, x_2x_3, x_1, x_2, x_3, 1)^T.$$

To avoid the trivial solution $\mathbf{c} = 0$, it is necessary to impose constraints on this problem. Different constraints produce different results. One that seems to produce fits that are intuitively close to the geometric fit is known as Taubin's method [14]. Taubin's method sets the constraint $\|\nabla_{\mathbf{x}} f(\mathbf{c}, \mathbf{x}^i)\|^2 = 1$. Equation 2 is reformulated as the generalized Eigen decomposition, $(M - \lambda N)\,\mathbf{c} = 0$, where

$$N = \sum_{i=0}^{n} l_x(\mathbf{x}^i)l_x(\mathbf{x}^i)^T + l_y(\mathbf{x}^i)l_y(\mathbf{x}^i)^T + l_z(\mathbf{x}^i)l_z(\mathbf{x}^i)^T.$$

Here, $l_x(\mathbf{x})$ denotes the derivative of $l(\mathbf{x})$ taken with respect to x_1 and the other derivatives are defined similarly. The eigenvector corresponding to the smallest eigenvalue provides the best-fit parameter vector.

To fix the axis of the cylindrical shell to lie along the axis of minor principal curvature, we need to estimate the magnitude and direction of the curvature of the quadratic surface. The curvature at a particular point can be calculated by evaluating the shape operator[1] on the plane tangent to the point of interest. The eigenvectors of the shape operator describe the principal directions of the surface and its eigenvalues describe the curvature in those directions. This can be calculated for a point, $\mathbf{x}$, on the surface by taking the Eigenvalues and Eigenvectors of:

$$\left(I - N(\mathbf{x})N(\mathbf{x})^T\right) \nabla N(\mathbf{x}),$$

where $N(\mathbf{x})$ denotes the surface normals of the quadratic surface. It is calculated by differentiating and normalizing the implicit surface:

$$N(\mathbf{x}) = \frac{\nabla f(\mathbf{c}, \mathbf{x})}{\|\nabla f(\mathbf{c}, \mathbf{x})\|},$$

[1]In general, the shape operator, S, can be calculated using the first and second fundamental forms of differential geometry: $S = \mathbf{I}^{-1}\mathbf{II}$.

where

$$\nabla f(\mathbf{c}, \mathbf{x}) = \begin{pmatrix} 2c_1x_1 + c_4x_2 + c_6x_3 + c_7 \\ 2c_2x_2 + c_4x_1 + c_5x_3 + c_8 \\ 2c_3x_3 + c_5x_2 + c_6x_1 + c_9 \end{pmatrix}.$$

Once a quadratic is fit to a point neighborhood, we evaluate the median curvature of the quadratic surface in the point neighborhood. This is accomplished by randomly sampling several points from the local quadratic surface and calculating the maximum curvature (maximum of the two principal curvatures) magnitude at each of them. Then, we take the median of these maximum curvature values and accept as grasp affordance candidates all quadrics where the median curvature is larger than that implied by the hand capture radius. On the assumption that all enveloping grasp affordances will be located in a high-curvature neighborhood, we eliminate from consideration all neighborhoods with an associated surface curvature below some parametrized threshold.

It is important to note that rather than fitting a quadratic surface in order to calculate local curvature magnitudes and axes, an alternative is to estimate curvature from surface normals associated with each point in the neighborhood. This works as follows. Each point is associated with a surface normal, $\mathbf{n}_i \in S^3$. Then, an Eigen decomposition is performed for the following matrix: $\sum_{i=1}^{n} \mathbf{n}_i \mathbf{n}_i^T$, $i \in [1, n]$. The major principal curvature axis is determined to lie in the direction of the Eigenvector associated with the minimum Eigenvalue. The curvature magnitudes are approximated by taking ratios between the eigenvectors. Although this type of approach is somewhat common in point cloud processing [11], our experience informally indicates that the method we present here is better: it seems to be more accurate, it is less noisy, and it can be computed faster than estimating surface normals for a (potentially large) set of points.

3.2 Cylindrical Shell Search

Once the directions and magnitudes of the axes of principal curvature are estimated and low-curvature regions are eliminated, we search for cylindrical shells in three steps. First, we project the points in the local neighborhood onto the plane orthogonal to the minor principal curvature axis (see Fig. 2a). Second, we calculate the center of the shell by fitting a circle to the points near the center of the neighborhood (i.e. points near the sampled point, x, in Step 3 of Algorithm 1). This is accomplished by minimizing algebraic distance as follows. Let x^i and y^i denote the two coordinates of the ith point in the plane. Let h_x, h_y, and r denote the coordinates of the center and radius of the circle. We calculate:

$$\mathbf{w} = -\left(\sum_{i=1}^{n} \mathbf{l}_i \mathbf{l}_i^T\right)^{-1} \sum_{i=1}^{n} \lambda_i \mathbf{l}_i, \tag{3}$$

where $\lambda_i = (x^i)^2 + (y^i)^2$ and $\mathbf{l}_i = (-x^i, -y^i, 1)^T$. Then calculate the center and radius using: $h_x = -0.5a$, $h_y = -0.5b$, and $r = \pm\sqrt{h_x^2 + h_y^2 - c}$.

Once the best-fit circle is calculated, the third step is to fix the center of the shell to the center of the circle and search (brute-force 1-D search) over different radii for a shell such that the gap contains no points and the radius of the inner cylinder is less than the diameter of the robot hand (conditions 3 and 4 for the existence of an enveloping grasp affordance).

3.3 Handle Search

The presence of an enveloping grasp affordance guarantees that a grasp is possible in that configuration as long as all object surfaces in the local area are densely covered with points. Unfortunately, this is not always the case. The assumption is particularly problematic for objects that are hard for the range sensor to perceive. For example, the PrimeSense device does very poorly measuring distances to highly reflective surfaces such as the body of the pot shown in Fig. 4a. One way to mitigate this problem is to search for sets of enveloping grasp affordances that form "handles", i.e. sets of affordances that are roughly aligned and that cover some minimum length. This helps reduce the number of false positives. True enveloping grasp affordances are typically found aligned along object handles. False positives (caused by sensor error) are typically found in arbitrary configurations. Figure 4b, c shows an example where the handle search eliminates all false positives.

We search for handles using brute-force search over all pairs of enveloping grasp affordances. For each pair of grasp affordances, i and j, with centroids h_i and h_j, major principal axes v_i and v_j, and radii r_i and r_j, we compute the following three distances: $d_o = \|(I - v_i v_i^T)v_j\|$, $d_c = \|(I - v_i v_i^T)(h_i - h_j)\|$, and $d_r = |r_i - r_j|$. An enveloping grasp affordance i is considered to be aligned with affordance j if

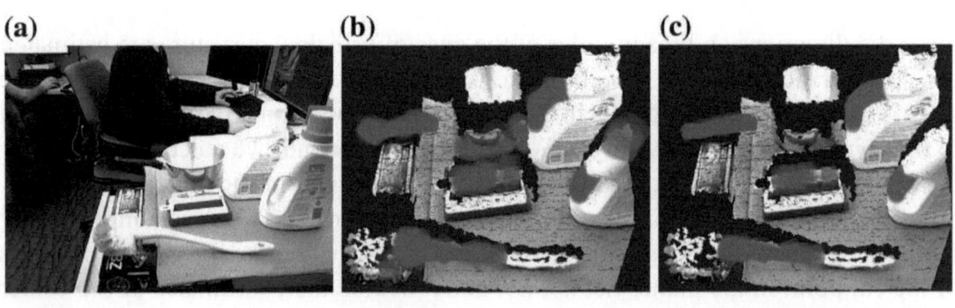

Fig. 4 **a** Illustration of handle search. **b** Shows all grasp affordances found in the point cloud. **c** Shows the handles found that satisfy alignment and minimum length constraints. The affordance search finds false positives on the surface of the pot and brush caused by measurement errors (the PrimeSense device fails to find accurate depths on reflective surfaces). However, they are eliminated in the handle search

d_o, d_c, and d_r are below parametrized thresholds. If an enveloping grasp affordance i is aligned with at least a minimum number of other grasp affordances, then it is considered to define a handle affordance. The handles found using this method constitute the output of our algorithm (Step 14, Algorithm 1).

3.4 Sampling Strategy

Sampling plays a key role in our algorithm. As shown in Algorithm 1 (Step 3), the basic approach is uniform random sampling. We sample a point uniformly randomly from the point cloud and operate on the neighborhood of points around that sample. Our experience indicates that in the manipulation scenarios outlined in Sect. 4, 20,000 samples are sufficient to localize all handles in a scene. With 20,000 samples, Algorithm 1 takes approximately 1.7 s to execute (see Sect. 4.3). A natural way to speed things up is to use a more effective sampling strategy. Here, we explore a sequential importance sampling method that can be viewed as an implementation of the Cross Entropy Method [3]. The method samples a fixed number of point neighborhoods in a series of rounds. In the first round, neighborhoods are chosen uniformly at random from the point cloud. After the first round, samples are drawn from a proposal distribution parametrized by the positions of the enveloping grasp affordances found in all prior rounds.

The form of the proposal distribution is a key choice that affects the performance of sampling. Here, we explore two variations on the Gaussian kernel density proposal distribution: a distribution expressed as a sum of Gaussians and a distribution expressed as a maximum over Gaussians. Let $x_i \in \mathbf{R}^3$, $i \in [1, n]$ denote the centroids of the n enveloping grasp affordances found in all prior rounds. The sum of Gaussians proposal distribution is:

$$g_{sum}(x) = \frac{1}{n} \sum_{i=1}^{n} \mathcal{N}(x|x_i, \Sigma),$$

where Σ is a constant parameter. The maximum of Gaussians proposal distribution is:

$$g_{max}(x) = \eta \max_{i \in [1,n]} \mathcal{N}(x|x_i, \Sigma),$$

where η is the normalization constant. It is relatively easy to sample from either of these proposal distributions. In order to draw k samples from g_{sum}, initialize $\mathcal{X} = \emptyset$ and do the following k times: choose an enveloping grasp affordance index, $j \in [1, n]$, uniformly randomly; draw one sample from $\mathcal{N}(x : x_i, \Sigma)$ and add it to $\mathcal{X}$. Sampling from g_{max} is slightly more complicated. A method based on rejection sampling is shown in Algorithm 2.

Algorithm 2 Sampling from a distribution expressed as a maximum over Gaussians

$\mathcal{X} = \emptyset$
for $i = 1$ to n **do**
  Choose i uniformly from $[1, n]$
  Sample $\hat{x} \sim \mathcal{N}(x|x_i, \Sigma)$.
  $m \leftarrow \max\{\mathcal{N}(\hat{x}|x_1, \Sigma), \mathcal{N}(\hat{x}|x_2, \Sigma), \ldots, \mathcal{N}(\hat{x}|x_n, \Sigma)\}$.
  if $\mathcal{N}(\hat{x}|x_i, \Sigma) \geq m$ **then**
    $\mathcal{X} \leftarrow \mathcal{X} \cup \hat{x}$.
  end if
end for

These two distributions, g_{sum} and g_{max}, differ in the way that they "allocate" samples to particular regions of space (i.e. to regions about potential handle locations). g_{sum} allocates samples to a region in direct proportion to the number of grasp affordances that have been found in that region. This can be a problem if there are multiple handles present in a scene, but one handle is more likely to be populated by enveloping grasp affordances than the others (perhaps it is larger, longer, or is more densely covered with points in the cloud). In this case, the handle where grasp affordances are more likely to be found is sampled even more densely on the next round. The result is that g_{sum} has a tendency to over-sample some handles in the scene and ignore others. g_{max} corrects for this effect somewhat by sampling from all handle regions with a more even probability.

This difference is illustrated in Fig. 5. Suppose that on a particular round of sampling, the algorithm has found all of the enveloping grasp affordances shown in cyan. Figure 5a shows a set of 100 samples drawn from g_{sum} and Fig. 5b shows the same number of samples drawn from g_{max}. Notice that the distribution drawn from Fig. 5a samples the object on the right more densely than the object on the left. This is because the object on the right was more densely covered with enveloping grasp affordances on prior rounds. Figure 5b shows that samples drawn from g_{max} cover both objects more evenly.

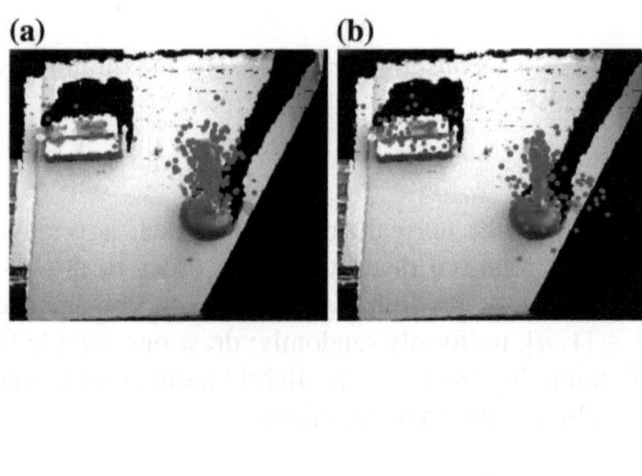

Fig. 5 Illustration of difference in sampling strategy. **a** Shows samples drawn from g_{sum}. **b** Shows samples drawn from g_{max}. Notice that the distribution in (**b**) covers the two handles more evenly

4 Experiments

4.1 Experimental Setup

We performed grasping experiments using the Rethink Robotics robot, Baxter. An Asus XTion Pro range sensor, mounted near the bottom of the robot's "chest", was used to sense a 3D point cloud containing the objects in front of the robot (see Fig. 6a). A typical grasp was performed as follows. First, the Asus range sensor captured a range image of the target objects that was immediately converted to a 3D point cloud. Second, our algorithm was run for this point cloud and handle affordances were localized. Third, the right arm reached toward the handle closest to the base of the right arm. The arm was moved to a position such that a point between the two gripper fingers was placed at the handle centroid. The target orientation was such that the gripper was perpendicular to the handle axis and an axis pointing outward from the gripper was co-linear with a line between the handle and the base of the right arm. After reaching the target pose, the gripper was closed, the object lifted into the air, and transported to a large box where the object was deposited. If a grasp failed on the first attempt, the robot continued to try to grasp by repeating this process. During each motion, the arm followed a straight line through configuration space.

We tested our localization and grasping algorithms in two ways. First, we performed a series of 12 single-object grasp trials for each of the 12 objects (shown in Fig. 6b) where each object was presented by itself. On each trial, the robot repeatedly attempted to grasp the presented object until either the object was grasped or it was pushed out of range. A grasp trial was run for each object in four different orientations at three different positions. Objects were placed such that a significant number of points on the handle were visible to the Asus range sensor and such that the handle was within the workspace of the robot's right arm. Second, we performed a series of 10 clear-the-table trials where we evaluated the capability for our approach to grasp a series of objects in the presence of clutter. On each clear-the-table trial, the robot attempted to clear five objects (selected from the set shown in Fig. 6b). Figure 7 shows a typical run of a clear-the-table experiment.

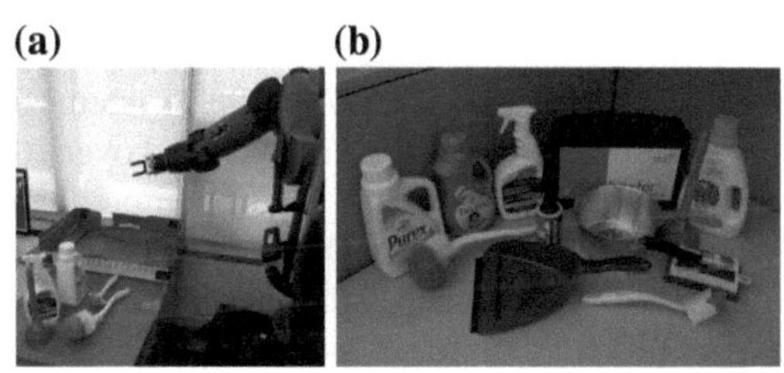

Fig. 6 **a** Typical grasping scenario. **b** The 12 objects used in our experiments. Notice that all objects have handles

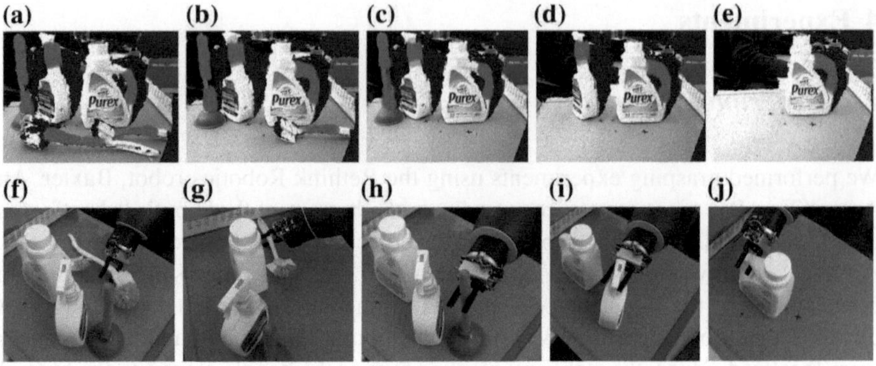

Fig. 7 Illustration of a typical clear-the-table experiment

4.2 Localization Results

The single-object experiments indicate that our approach is capable of robustly grasping objects with handles. Table 1 shows the results. Out of the 12 grasp trials for each object, the table shows the number of successful grasps performed on the first try (column 2), by the second try (column 3), and by the third try (column 4). Notice that our method successfully grasped each object on the first try approximately 85 % of the time. By the third try, it had nearly perfect grasp success. The only exception was for the Carrying Case where the object was pushed out of the workspace during a failed grasp attempt (collision between gripper and target object). Table 2 shows the results of ten clear-the-table experiments. The results show that our method sometimes failed to grasp one of the five presented objects. They also show that it sometimes took up to eight grasp attempts before all five objects were grasped.

4.3 Algorithm Runtime

This number is a conservative estimate of the maximum number of neighborhoods needed to localize all handles in our application scenarios. The algorithm was implemented in C++ on an Intel i7 3.5 GHz system (four physical CPU cores) with 8GB of system memory. Runtime was averaged over 10 runs. The results are shown in Fig. 8. As they show, total runtime is a little more than 0.5 Hz with the majority of the time taken by the brute-force 1-D shell search. We suspect that a closed-form approximation to the brute-force search exists that would reduce this time. Nevertheless, we expect this runtime to be fast enough for most application scenarios.

Table 1 Results for the single-object experiments

Object	Grasped on 1st attempt	Grasped on 2nd attempt	Grasped on 3rd attempt
Blue bottle	10/12	10/12	12/12
White purex bottle	11/12	12/12	12/12
White all bottle	9/12	12/12	12/12
Carrying case	11/12	11/12	11/12
Brush 1	10/12	11/12	12/12
Pot	11/12	12/12	12/12
Plunger	11/12	12/12	12/12
Sprayer	11/12	12/12	12/12
Dust pan	11/12	12/12	12/12
Brush 2	8/12	12/12	12/12
Sponge	8/12	12/12	12/12
Lint roller	11/12	12/12	12/12

Table 2 Results for the clear-the-table experiments

Trial num	1	2	3	4	5	6	7	8	9	10
Number of objects grasped out of total objects	5/5	4/5	5/5	4/5	5/5	4/5	5/5	5/5	4/5	5/5
Total grasp attempts	5	5	5	7	7	6	5	5	5	8

4.4 Comparison of Different Sampling Strategies

We also performed experiments to evaluate the number of handles in a scene missed by the algorithm as a function of the number of neighborhoods (I in Algorithm 1) and as a function of the sample strategy used. We tested with point clouds from seven scenes. The first five scenes contained exactly five different handles each. The last two scenes contained nine and ten handles, respectively. On each of these seven scenes, we tested the performance of our algorithm using three different sample strategies: uniform random Monte Carlo (MC), sequential importance sampling with g_{sum}, and sequential importance sampling with g_{max}. For each sample strategy, we performed experiments with 2000 and 5000 sampled neighborhoods. For uniform random MC we just sampled 2000 or 5000 samples in one batch. For sequential importance sampling with 2000 samples, we sampled 1000 neighborhoods in the first round and then 100 more neighborhoods in each of ten successive rounds. For sequential importance sampling with 5000 samples, we sampled 2000 samples in the first round and then 300 samples in each of 10 successive rounds.

Figure 9 show the results. Each bar shows the mean and standard deviation of 20 runs in the corresponding test scenario. The ground truth bar (yellow) shows the actual number of handles present in each scene. These results indicate the following.

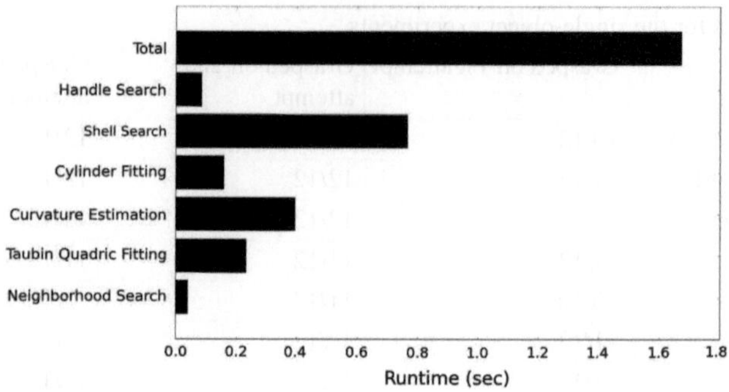

Fig. 8 Runtime of the localization algorithm for 20,000 samples averaged over 10 runs

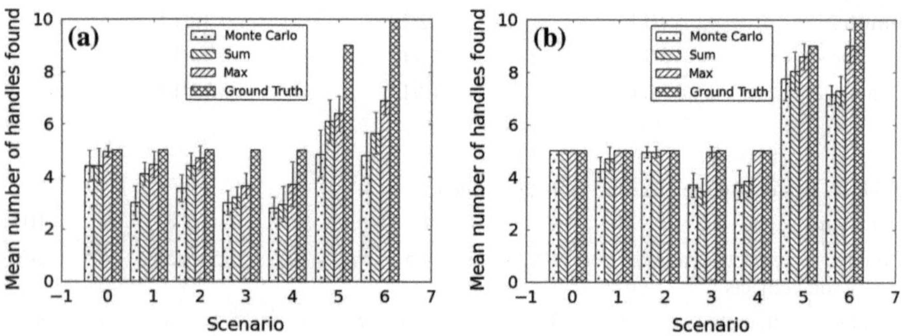

Fig. 9 Performance comparison between the three sampling strategies for 2000 (**a**) and 5000 (**b**) samples, averaged over 20 runs. The *error bars* show the standard deviation. **a** 2000 sampled neighborhoods, **b** 5000 sampled neighborhoods

First, our method can be expected to find two or three handles in any scene with as few as 2000 samples using any sampling method. This is sufficient for some tasks (such as table clearing), where it is only necessary to grasp one object at a time. However, even 5000 sampled neighborhoods might not be enough to find all handles in a complex scene, especially if uniform random Monte Carlo is used. We found that it was necessary to use as many as 20000 sampled neighborhoods in order to localize all handles using this method. The results also indicate that it is generally better to use a sequential sampling method. Moreover, the results show that sequential importance sampling using the g_{max} proposal distribution has the best performance. This strategy finds nearly all handles with 5000 sampled neighborhoods.

5 Conclusions

The paper proposes a new approach to localizing handle-like grasp affordances in 3-D point clouds. The core of the idea is to identify sufficient geometric conditions for the existence of a class of grasp affordances and to search the point cloud for point neighborhoods where these conditions are satisfied. Our work makes use of an approach to implicit quadratic curve fitting that (to our knowledge) has not been used in the robotics literature. Our reported results show high grasp success rates similar to those reported in Klingbeil et. al. [10] and Fischinger et. al. [4]. Moreover, our method has important advantages relative to other approaches including fast run time, the ability to operate on 3D point clouds rather than range images or height maps, and the ability to localize handles. In our single-object experiments, nearly all grasp failures were caused by attempting to grasp false positives found because of depth measurement errors or because of insufficient point density on object surfaces in the neighborhood of the false positive. For example, the grasp failures of BRUSH 2 were caused mainly by the algorithm localizing the brush part of the object because of significant measurements errors in that area. Our clear-the-table experiments also suffered from localization failures. However, there, the effects of localization errors were more serious because of the clutter. A failed attempt to grasp one object sometimes pushed other objects out of the workspace such that a complete clearing of the table became impossible. In general, we found the grasping process to be very robust as long as multiple re-grasp attempts were allowed. Overall, the results in Tables 1 and 2 indicate that our approach is practical for many real robot application scenarios. We have incorporated our work into a ROS package [15].

Acknowledgments This work was supported in part by NASA under Grant No. NNX13AQ85G and ONR under Grant No. N000141410047.

References

1. Besl, P., McKay, N.: A method for registration of 3d shapes. IEEE Trans. Pattern Anal. Mach. Intell. **14**(2), 239–256 (1992)
2. Brook, P., Ciocarlie, M., Hsiao, K.: Collaborative grasp planning with multiple object representations. In: IEEE International Conference on Robots and Automation (2011)
3. De Boer, P.-T., Kroese, D.P., Mannor, S., Rubinstein, R.Y.: A tutorial on the cross-entropy method. Ann. Oper. Res. **134**(1), 19–67 (2005)
4. Fischinger, D., Vincze, M.: Empty the basket—a shape based learning approach for grasping piles of unknown objects. In: IEEE International Conference on Intelligent Robot Systems (2012)
5. Gibson, J.: The Ecological Approach to Visual Perception. Psychology Press (1979)
6. Glover, J., Popovic, S.: Bingham procrustean alignment for object detection in clutter. In: IEEE International Conference on Intelligent Robot Systems (2013)
7. Herzog, A., Pastor, P., Kalakrishnan, M., Righetti, L., Asfour, T., Schaal, S.: Template-based learning of grasp selection. In: IEEE International Conference on Robotics and Automation (2012)
8. Jiang, Y., Moseson, S., Saxena, A.: Efficient grasping from rgbd images: learning using a new rectangle representation. In: IEEE International Conference on Robotics and Automation (2011)
9. Katz, D., Kazemi, M., Bagnell, D., Stentz, A.: Clearing a pile of unknown objects using interactive perception. In: IEEE International Conference on Robotics and Automation (2013)

10. Klingbeil, E., Rao, D., Carpenter, B., Ganapathi, B., Ng, A., Khatib, O.: Grasping with application to an autonomous checkout robot. In: IEEE International Conference on Robotics and Automation (2011)
11. Rusu, R., Cousins, S.: 3d is here: point cloud library (pcl). In: International Conference on Robotics and Automation (2011)
12. Rusu, R., Blodow, N., Beetz, M.: Fast point feature histograms (fpfh) for 3d registration. In: IEEE International Conference on Robots and Automation (2009)
13. Sun, M., Xu, B., Bradski, G., Savarese, S.: Depth-encoded hough voting for joint object detection and shape recovery. In: European Conference on Computer Vision (2010)
14. Taubin, G.: Estimation of planar curves, surfaces and nonplanar space curves defined by implicit equations, with applications to edge and range image segmentation. IEEE Trans. PAMI **13**, 1115–1138 (1991)
15. ten Pas, A., Platt, R.: Handle detector ROS package. http://wiki.ros.org/handle_detector
16. Tombari, F., Stefano, L.: Object recognition in 3d scenes with occlusions and clutter by hough voting. In: Pacific-Rim Symposium on Image and Video Technology (2010)
17. Tombari, F., Salti, S., Stefano, L.: Unique signatures of histograms for local surface description. In: European Conference on Computer Vision (2010)

An Experimental Study for Identifying Features of Legible Manipulator Paths

Min Zhao, Rahul Shome, Isaac Yochelson, Kostas Bekris and Eileen Kowler

Abstract This work performs an experimental study on the legibility of paths executed by a manipulation arm available on a *Baxter* robot. In this context, legibility is defined as the ability of people to effectively predict the target of the arm's motion. Paths that are legible can improve the collaboration of robots with humans since they allow people to intuitively understand the robot's intentions. Each experimental trial in this study reproduces manipulator motions to one of many targets in front of the robot. An appropriate experimental setup was developed in order to collect the responses of people in terms of the perceived robot's target during the execution of a trajectory by Baxter. The objective of the experimental setup was to minimize the cognitive load of the human subjects during the collection of data. The extensive experimental data provide insights into the features of motion that make certain paths more legible for humans than other paths. For instance, motions where the end-effector is oriented towards the intended target appear to be better in terms of legibility than alternatives.

Keywords Human-robot interaction · Legible paths · Manipulation · Co-robots

1 Introduction

The increasing availability of low-cost, compliant and human-friendly manipulators allows robots, such as Rethink Robotics' *Baxter* [1], to be placed in close proximity to human workers. Unlike traditional automation systems, which need to be kept in cages, these compliant robots can share a common workspace with human workers. A clear benefit of this close proximity is the opportunity for cooperation between a human worker and an assistive robot.

M. Zhao · R. Shome · I. Yochelson · K. Bekris (✉)
Department of Computer Science, Rutgers University, New Brunswick, USA
e-mail: kostas.bekris@rutgers.edu

M. Zhao · E. Kowler
Department of Psychology, Rutgers University, New Brunswick, USA

M.A. Hsieh et al. (eds.), *Experimental Robotics*, Springer Tracts in Advanced Robotics 109, DOI 10.1007/978-3-319-23778-7_42

In order for a robot to be effective in an assistive role, it is important that the human is able to easily and quickly understand the robot's intentions by observing its actions. Ideally, this understanding will come in an intuitive manner, similar to how humans are innately able to communicate with one another non-verbally when working in close quarters. Legible motion plans are an important part of making the robot understandable by human co-workers intuitively. In this context, the legibility of a motion corresponds to whether human subjects can realize early on which is the actual target of the moving arm out of many possible choices.

The goal of this work is to identify the key features of robotic motion for manipulators that contribute to their legibility. The motions are executed by two seven degree-of-freedom manipulation arms that are mounted on a *Baxter* robot. The arms move towards grasping multiple targets, which are positioned linearly in front of the robot. As the manipulator moves, human subjects observe the robot and report their belief regarding the intended target of the arm. An appropriate experimental setup was developed in order to collect these responses, so as to minimize the cognitive load of the human subjects and achieve good accuracy.

The five different types of trajectories that were considered during this experimental study cover a variety of discriminant legibility features. Some of the features correspond to arm policies, such as the shortest path in the configuration space, and other correspond to end-effector, i.e., "hand", policies, such as the orientation of the end-effector relative to the target. The experimental results show that the legibility of different trajectories is indeed different and consistent across different targets. Motions which allow the end-effector to point towards the intended target and move along a straight line in the workspace result in enhanced legibility.

The longer term objective of identifying these legibility characteristics is the design of motion planners that incorporate these features into the planning process so as to automatically generate legible motion. The hope is that co-robots, which can generate legible motion plans, can more effectively collaborate with humans.

1.1 Related Work

Previous work has emphasized the importance of anticipatory motion [2]. By identifying a symbol, or a socially representative element of the motion, and using it as early as possible during the motion, the robot's actions are easily and quickly interpreted by observers. It has also been indicated that legible, anticipatory motion greatly assists in collaborative tasks.

Research has also focused on exploiting the repeatability of common collaborative tasks to generate anthropomorphic motions [3]. There has been work on creating metrics that can reproduce motion plans to be more human-like [4]. Another philosophy in generating motion plans has been learning by demonstration. Motions, that are demonstrated by human teachers, are used to build the policy for the robot to map its state to an appropriate motion [5, 6]. This line of work leverages anthropomorphic motions. The legibility problem, however, does not necessarily correspond to the capability of a robot to reproduce human-like motion, but how a human perceives the robot's motion.

This crucial motivation has resulted in recent important efforts in identifying aspects of and generating legible robot motion [7, 8], which have inspired and influenced the current work. In particular, these efforts have resulted in a formalization of robot motion legibility, and approaches for autonomously generating legible robotic motion plans. Further work by the authors along this line has focused on distinguishing between predictability and legibility. In the corresponding experimental process the focus was on discriminating the legibility of motion using video recordings of a robot that can potentially reach two goals in an otherwise uncluttered workspace. Familiarization [9] has been shown to improve predictability when coupled with learning.

Human beings are good at interpreting actions and relative intentions of other moving agents in their environment. This ability is developed during the first 14 months of a person's life [10]. During daily life, there are usually two action interpretation processes [11]:

1. *Action-to-Goal* inference, in which people try to predict the result of the action based on the information accumulated during the action's execution.
2. *Goal-to-Action* inference, in which people try to predict a type of action that could achieve a determined goal.

The focus of legibility is on understanding action-to-goal inference, namely how humans interpret the observed actions and then discover the underlying intention [7]. Adults, young children, and even infants could selectively focus on the key components of the behavior of others, which is relative to their intention. In psychophysical experiments the human hand was discovered to play a crucial role during interpreting and sharing actions and intentions of people by others [12, 13]. Previous psychological studies show that between nine and twelve months, infants develop a perceptual link between pointing to the target object and the targeting mechanism itself. They understand that pointing is an object-oriented action [13]. These results motivate the focus of this study on features related to the robot's end-effector.

This work identifies the possible features of a manipulator's motion, which make it easier for a human to understand the intended target of this motion. The goal of the current work is to expand upon the existing experimental studies on this subject [7, 8] in two primary directions. This effort considers a workspace with many potential targets for the robot to interact with. Secondly, experiments are performed with human subjects placed in close proximity to the robot manipulator and do not make use of recordings. These experiments confirm aspects of previous work, such as the contradictory nature of shortest and legible paths, and they also reveal important features of legible paths in cluttered scenes. For instance, the direction and workspace path of the end effector are shown to significantly influence a human observer's capability to realize the robot's intended target.

2 Generating Different Manipulator Paths

In order to evaluate key features of legible paths for manipulators, this experimental study considers a variety of path types for a dual arm manipulator, i.e., a Baxter robot by Rethink Robotics. There are four *arm policies* considered in this study:

1. shortest path in configuration space (i.e., minimizing change in joint angles),
2. overhead motion frequently appearing in "pick and place" paths,
3. shortest, straight-line path for the end-effector in workspace, and
4. "curved" path for the end-effector in the workspace to exaggerate intent (see Fig. 1a).

Likewise, this study considers two potential *hand policies*:

1. the hand goes immediately to the final joint position (e.g., overhead grasp) and stays there for the duration of the motion, and
2. the hand points toward the goal in the workspace at all times. The pointing feature of these paths can be seen as a symbol generating anticipation of the motion [2].

Paths are generated both for the left and right arm of the robot. For each arm and for every type of trajectory, a fixed start position that is raised from *Baxter's* at-rest position is used. It helps in terms of target reachability. Different paths are generated for multiple targets. The targets are placed evenly along a line on a table in the manipulator's reachable workspace (see Fig. 1a, b). By combining the above mentioned policies and pruning incompatible combinations, five different classes of path are considered in the experimental study:

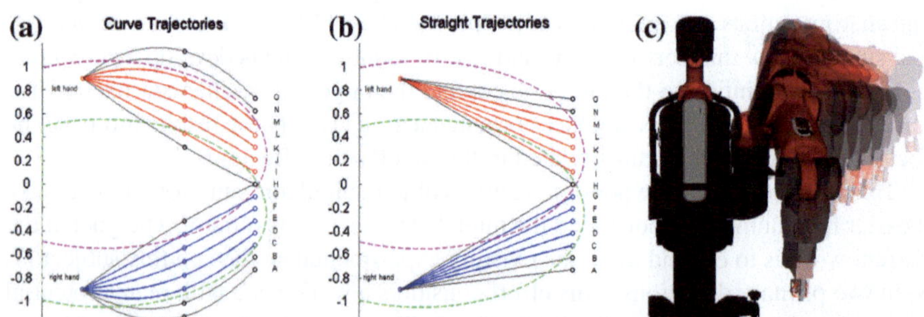

Fig. 1 *Left* (**a**) "curve" and *Center* (**b**) "straight" paths seen from above. The points on the *left side* of each plot represent the starting position for the *left* (*red*) and *right* (*blue*) end-effector. The *lines* show paths to reachable targets. Each hand has its own reachable region (*green curve for right*; *purple curve for left hand*). *Right* (**c**) one of the "overhead" paths in simulation [14]. The end-effector remains vertical and points downward

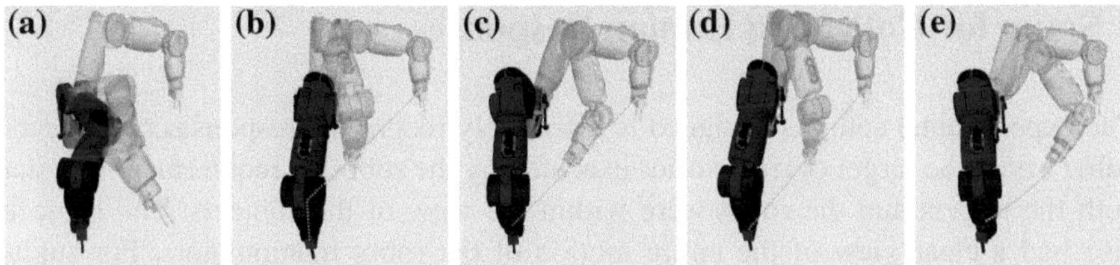

Fig. 2 *Left to right* **a** Shortest C-space path, **b** Overhead down, **c** Straight pointing to target, **d** Straight down, **e** Exaggerated "curved" motion pointing to target

1. "Shortest" path: This is the shortest path in the configuration space computed on an asymptotically near-optimal version [15] of a probabilistic roadmap [16] in the Open Motion Planning Library [17]. This class makes use of arm policy 1 (Fig. 2a) and immediately provides a path for the hand as well.
2. "Overhead Down" path: Similar to paths employed for pick-and-place tasks by Baxter robots in industrial settings, where the end-effector moves in a position over the target and points downwards throughout the motion (see Fig. 1c). This class makes use of arm policy 2 and hand policy 1 (Fig. 2b).
3. "Straight" path: The robot moves its end effector along a linear path from the initial position to the target object while the end effector points towards the target (see Fig. 1b). This class uses arm policy 3 and hand policy 2 (Fig. 2c).
4. "Straight Down" path: The robot moves its end effector along a linear path from the initial position to the target object while the end effector remains in a vertical orientation pointing down. This class makes use of arm policy 3 with hand policy 1 (Fig. 2d).
5. "Curved" path: The robot moves its end effector along an exaggerated curved path while pointing at the target. This class is inspired by ideas in previous work towards generating legible paths [7] (see Fig. 1a). This class makes use of arm policy 4 combined with hand policy 2 (Fig. 2e).

The above set of trajectories is designed to avoid confounding the effects of hand policies with the effects of arm policy, while keeping the total number of trajectories to a reasonable number so as to be able to extract useful conclusions. Note that there are two types of trajectories that are sharing the same arm policy (straight-line for the end effector in the workspace) but are different in terms of the employed hand policy. There are also two control classes, reflecting standard manipulation strategies ("shortest" and "overhead down" trajectories). In this way, the relative importance of these features can be discovered by comparing the time it takes for human subjects to realize the motion's target.

To ensure that for all classes there is ample time for subjects to give feedback about their belief of targets, all trajectories in this study are scaled to be performed in 8 s.

3 Setup for Collecting Human Responses

The experimental setup is designed to effectively record the responses of subjects' belief about the target of trajectories executed by the robot. A requirement was that both the targets and the robot were within the view of the subjects. The subjects also had a clear view of the entire motion of the robot manipulators. For studying legibility, the subject must be able to pay attention to the motion of the robot without distractions. Minimizing the cognitive load of the subject during the experiment involves minimizing distractions as well as making the data recording interface intuitive and effortless. In order to achieve this, an efficient recording mechanism is desired, which is both accurate in recording the responses and easy to assemble. The recording interface should also be resilient enough to withstand repeated experimental trials.

As shown in Fig. 3, the experimental setup consists of a Baxter robot, a workstation, a table with 15 colored cups, and a pointing device. Only 10 of the cups can be reached by the robot from its starting position with all 5 trajectory classes. Five for the left arm and five for the right one. These 10 cups were designated as targets. During a trial, the robot follows one of the trajectory classes from its starting position to a target. A human sitting behind the pointing device uses the pointer to indicate his or her belief of the robot motion's intended target. The position of the pointing device is then recorded in a log together with the target and the class of the trajectory followed.

The trajectories are stored and played back during the trials in order to ensure that artifacts from the random sampling in the motion planning process do not cause discrepancies between trials of the same trajectory class. Moreover, the overhead of planning for the execution for the trajectories is avoided by generating the trajectories once and replaying them. For each of the workspace constrained paths, MathWorks' MATLAB [18] is used to perform linear interpolation among a series of points in the workspace. Then, the MoveIt! package [19] with a KDL kinematics solver [20]

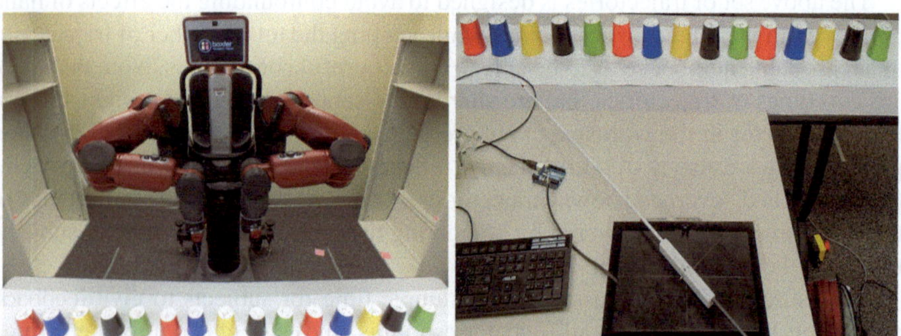

Fig. 3 (*left*) The start position of the trajectories on the Baxter robot during the experimental setup. (*right*) A view of the pointing device from the subject's perspective

and an OMPL [17] implementation of a PRM* variant is used to plan trajectories between the interpolated points. The final trajectories can be played on the robot using the Baxter RSDK [1].

The pointing device is fixed to the spindle of a linear potentiometer. The edges of the resistive track are then connected to the 5 volt and ground pins of an Arduino device and the wiper to an analog input pin. An Arduino sketch then performs the necessary calculations to extrapolate from the wiper voltage the position along the line of targets at which the ray of the pointer will intersect. This distance is then forwarded to the Arduino's USB port.

For each human subject, three random permutations of the 50 recorded trajectories are generated using a python script. The trajectories of each permutation are then executed in order, recording a log of the trajectory filename and pointer positions, with time-stamps, captured from the Arduino during the playback of each trajectory. In this manner, it is possible to ensure even coverage among the classes and targets while minimizing the chance of subjects guessing the target through means other than visual perception of the robot's motion.

The playback of a given trajectory is preceded by an auditory alert, a bell sound for the left arm or a buzz sound for the right arm. These sounds alert the subject regarding which arm they should direct their attention toward. The robot then plays the trajectory, which has been scaled to run in 8 s, as the subject's feedback is recorded. Following the playback, the robot returns to a starting position, which is common to all the trajectories. Then, the subject is shown the number of trials that have been completed and is prompted to press any key on the keyboard to continue.

After each block of 50 trials, which forms a permutation of the full set of trajectories, the subject is given a mandatory two minute break. These breaks allow the subject to rest, and to maintain attention on the perception task. Human subjects are asked to participate in this experiment only once. This is to ensure that the base legibility of the paths when first encountered, and also the rate at which learning about the paths takes place, can be accurately gauged.

4 Experimental Evaluation of Legibility

4.1 Reaction Time

In order to compare the legibility of different types of trajectories in detail, the study examined the reaction time of 30 subjects, who had the opportunity to observe 150 trials (i.e., 3 blocks of 50 unique trajectories randomly ordered). Reaction time is the time it takes a subject to converge to the correct target. This is measured as the last occurrence when the pointing device enters the target zone during a trial. Figure 4 plots three types of reaction times:

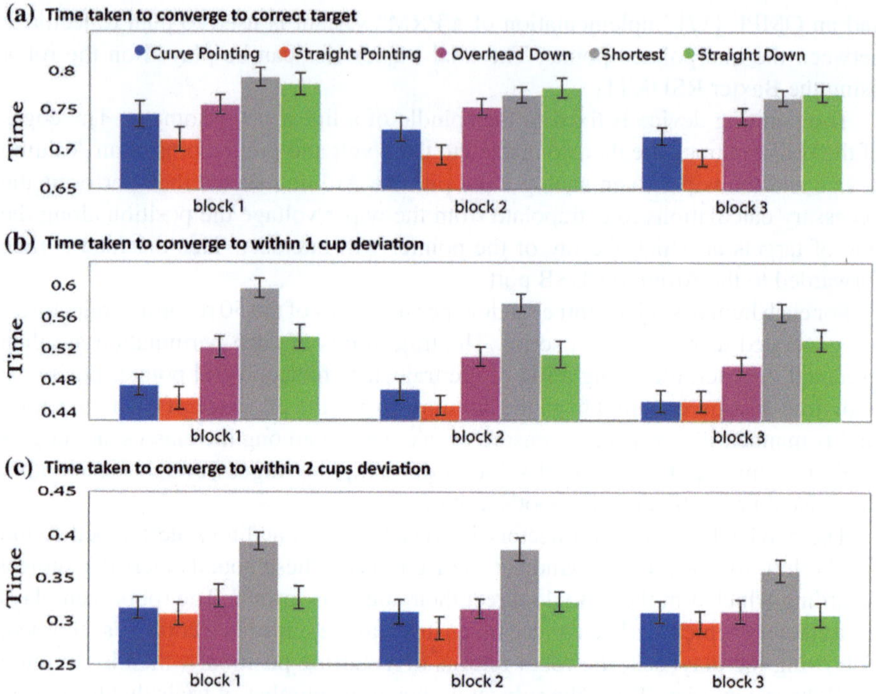

Fig. 4 Reaction time (RT): the time converging to (**a**) the correct target; (**b**) 1 cup away from the correct target; (**c**) 2 cups away from the correct target. There were five different types of trajectories: curve-pointing (*blue*), straight-pointing (*red*), overhead-down (*pink*), shortest (*gray*) and straight-down (*green*). There were three blocks of 50 trajectories in order: block 1 (*left*), block 2 (*middle*) and block 3 (*right*). The error bar represent ±1 standard deviation error

1. Time for convergence to the target itself (Fig. 4a), which happened at the late part of trials;
2. Time for convergence to the range within 1 cup away from the target (Fig. 4b), which happened at the middle range of trials; and
3. Time for convergence to the range within 2 cups away from the target (Fig. 4c), which happened at the beginning of trials.

Data from the three blocks of 50 trajectories provided to each human subject are presented in order from left to right. In general, the straight-pointing type (red bars) was always the best. The curve-pointing (blue bars) was the second best. The shortest type (gray bars) was the worst, especially regarding the early on convergence (2 cups away, or 1 cup away from the target). One-way ANOVA tests show that there were significant differences among the different types of trajectories for all groups (Table 1, F scores and p values).

Block 1 (plots on the leftmost column of each figure) provides the response times when each trajectory was first presented to subjects. The performance for the shortest (grey bars) type was always the worst when converging to all types of error range. The disadvantage of the shortest type was obvious when converging into relatively

Table 1 One-Way ANOVA analysis for RT of 2 cups away, 1 cup away and pointing to the target for each block

	Block 1	Block 2	Block 3
Target	5.2 (<0.01)	8.34 (<0.001)	11.37 (<0.001)
	2–4; 2–5	1–5; 2–3; 2–4; 2–5	1–4; 1–5; 2–3; 2–4; 2–5
1 cup	16.74 (<0.001)	14.38 (<0.001)	14.91 (<0.001)
	1–4; 1–5; 2–3; 2–4; 2–5; 3–4; 5–4	1–4; 2–3; 2–4; 2–5; 3–4; 5–4	1–4; 1–5; 2–4; 2–5; 3–4
2 cups	7.3 (<0.001)	7.61 (<0.001)	3.3 (=0.013)
	1–4; 2–4; 3–4; 5–4	1–4; 2–4; 3–4; 5–4	2–4; 5–4

In each cell, the values in the first row are the F-score (p-value). The second row lists all pairwise types which are significantly different from each other from post-hoc test (1-Curve-pointing; 2-Straight-pointing; 3-Overhead-down; 4-Shortest; 5-Straight-down)

large error ranges (2 cups or 1 cup away from the target). Pairwise comparison shows that it is significantly longer than the other four types (Table 1). This disadvantage decreased when approaching to the correct target finally. It suggests that the confusion of the shortest type usually appeared at the early stage of the trajectories.

The reaction time for the straight-pointing type (red bars) was consistently shorter than the others when converging to the range within 1 cup of the target and to the target, and it is marginally shorter than the others when converging to the range within 2 cups of the target. The curve-pointing was always longer than the straight-pointing but shorter than the remaining three. This implies the legibility of the straight-pointing path. The curve-pointing path is the second most legible among the five types. The ease of understanding of the straight-pointing and the curve-pointing trajectories could be due to the fact that the end-effector was always pointing to the target. The end effector (i.e., hand) was previously reported as an important cue in understanding people's intentions [10, 12, 13]. The advantages of the end-effector pointing to the target were strongest when converging to the range of 1 cup away from the target. It suggests that the characteristics of curve-pointing and straight-pointing helps people understand the intention of the robot by converging to the smaller error range more quickly.

The "curve-pointing" did not perform as well as "straight-pointing", which was surprising given the conclusions of previous studies [7]. Nevertheless, this could be due to the difference between the "two-targets" setting in previous studies and the "crowded-target" setting in the current experiment. With multiple targets in a crowded environment, the curve path was more likely to confuse people, rather than improve legibility.

Overhead-down was the third most legible trajectory and it was better than the straight-down. This also makes sense, because whenever the overhead-down trajectories reached to the top of the cup, subjects knew the answer for sure. The straight-down trajectories were still on the way to the top of the cup at the same time point. This leads to similar reaction times between these two types of trajectories when converging to 2 cups or 1 cup away from the target. It leads into shorter reaction time for the overhead-down trajectory when converging to the correct target.

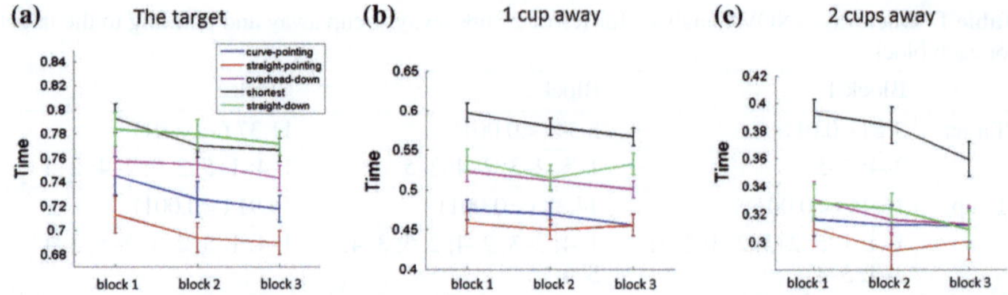

Fig. 5 Learning effects reflected in reaction time. The time converging to (**a**) the correct target; (**b**) within 1 cup deviation from the true target; (**c**) within 2 cups of the true target. There were five different types of trajectories: curve-pointing (*blue*), straight-pointing (*red*), overhead-down (*pink*), shortest (*black*) and straight-down (*green*). The error bars represent ±1 standard deviation error

Figure 5 provides a study of the effects of learning over the duration of the trajectory. It shows that the time converging to the target was decreasing across blocks, which means subjects did learn trajectories. The learning effect is larger from block 1 to block 2, than from block 2 to block 3. This could be because subjects were already well trained when entering into block 2 and might get tired in block 3. The learning effect also varies among different types of trajectories. The shortest type shows greater learning effect than the others in all convergence cases (Figure 5a, b, c). These results suggest that the shortest type was the hardest one to be interpreted at the early stage, but it can be learned by more training. Nevertheless, the learning does not allow it to reach the legibility level of alternatives such as the "straight-pointing" path. Additionally, the learning effect appears also as decreasing variance in later blocks (block 2 and 3).

4.2 Predicted Target Over Time

The root mean square of the distance between the pointed cup and the true target reflects the accuracy of the subject's prediction of the target over the course of the trial. It is averaged across trials for each subject, and then across all subjects. Figure 6 shows that the root mean square varied for different types of trajectories at the beginning and converged to the correct target location in the end over a normalized time scale. The convergence was fast during the middle range of trials (0.3–0.7) for all trajectories. A consistent result is that the predictions for the shortest trajectory were further away from the true target than any other type of trajectories.

The pointer velocities (Fig. 7), which was how fast subjects moved the pointer, were peaking during the middle range of trials (0.4–0.6). The graph shows that the shortest trajectory type resulted in different behavior than the other four. These results suggest that the subjects were not able to predict the target during the early parts of the shortest trajectories as well as for the other types. Frequently during the shortest paths, the end-effector was overshooting the target and then returning back to it. This complicated the interpretation of the motion even close to the completion of the path.

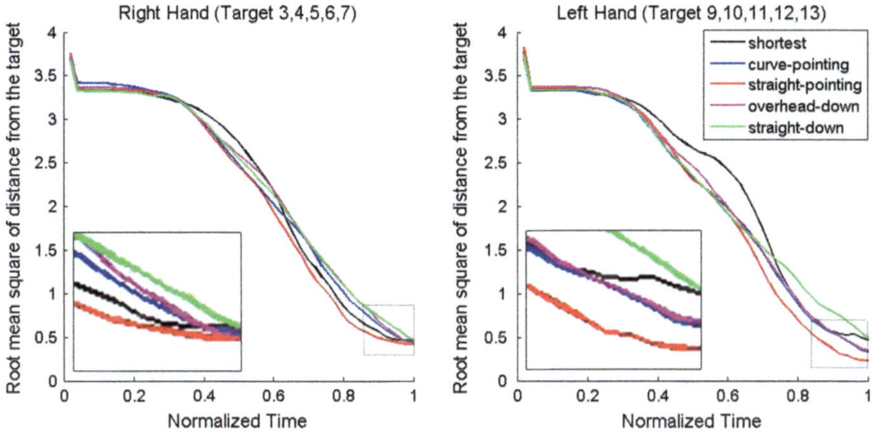

Fig. 6 Root mean square of distance from the target along the normalized time scale for five types of trajectories: shortest (*black*), curve-pointing (*blue*), straight-pointing (*red*), overhead-down (*pink*) and straight-down (*green*)

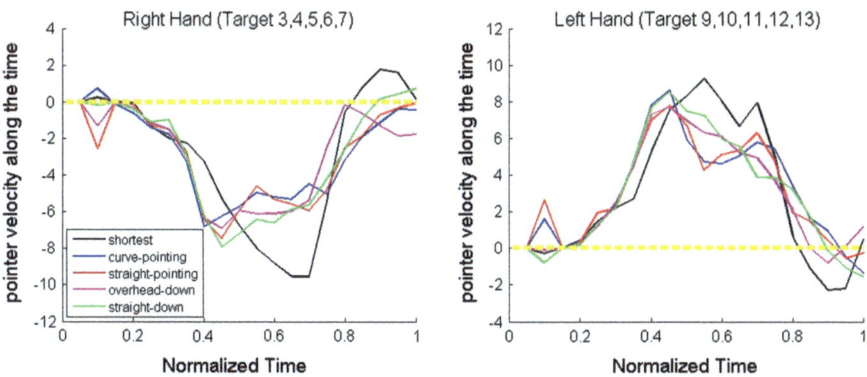

Fig. 7 Pointer velocity along the normalized time scale for five types of trajectories: shortest (*black*), curve-pointing (*blue*), straight-pointing (*red*), overhead-down (*pink*) and straight-down (*green*)

4.3 Performance by Cup

The understanding of different types of trajectories was also related to the location of the target. In order to better analyze different trajectories, it is useful to further examine performance (mean position and pointer velocity over time) for different cups. Figure 8 shows the mean distance from the target for each cup and Fig. 9 shows the pointer velocity for each cup.

By comparison to the cups located on the edge of the target set (i.e., 3, 4, 12, 13), for the cups located near the center (i.e., 6, 7, 9, 10), the mean deviation from the target was more likely to cross the 0 level (Fig. 8). As the subjects typically begin with

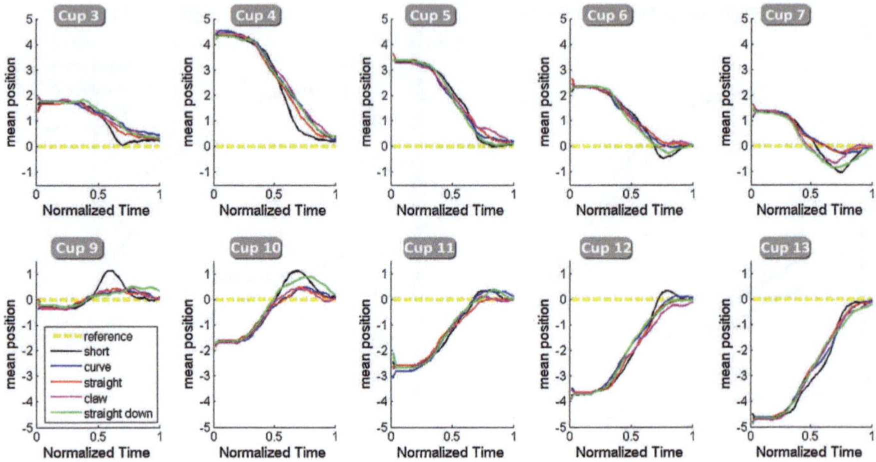

Fig. 8 Mean distance from target along normalized time scale for each cup (cup No. labeled on the top of each plot). Five types of trajectories: shortest (*black*), curve-pointing (*blue*), straight-pointing (*red*), overhead-down (*pink*) and straight-down (*green*)

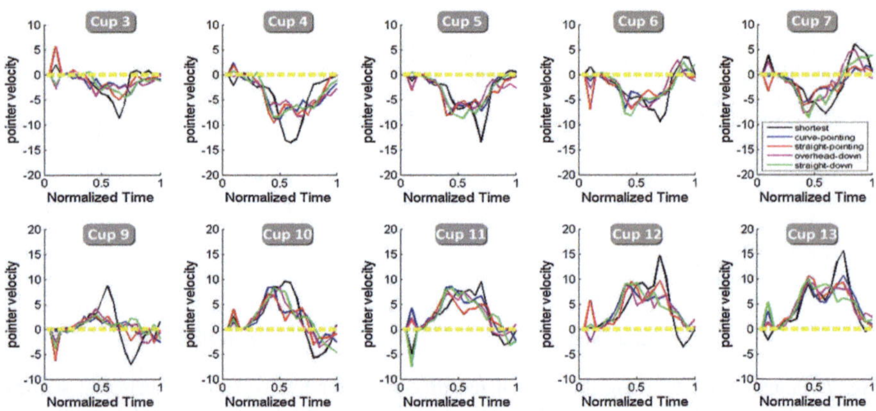

Fig. 9 Pointer velocity along the normalized time scale for each cup (cup No. labeled on the top of each plot). Five types of trajectories: shortest (*black*), curve-pointing (*blue*), straight-pointing (*red*), overhead-down (*pink*) and straight-down (*green*)

the pointing device centered, this suggests that subjects were more likely to overshoot the target. The starting position of the robot is nearer the edge cups than the center cups. The overshooting could indicate that they are following the arm rather than predicting the target accurately. Subjects were more likely to overshoot the target for the shortest type. The traces in Fig. 8 illustrate the subjects' reactions. The trajectories with the lowest reaction times also demonstrate the least overshooting.

4.4 Conclusions

Overall, the straight-pointing path was the easiest to understand, followed by the curve-pointing, and the shortest type was the least legible. The advantage of the straight-pointing path and the curve-pointing is likely due to the end-effector's orientation towards the target. Reaction times are slower with the straight-down than the straight-pointing, though both of them followed the same end-effector paths in the workspace.

The disadvantage of the shortest type appears as slow convergence; it takes more time to approach a certain error range than the other types. The overhead-down was better than straight-down when converging to the target. An obfuscating effect is that subjects could know the target for sure when the hand moved to the top of the cup, while at the same time point, the straight-down motion was still away from the target.

Even though the learning effect exists, it varies for different types of trajectories. The general learning benefit appears with less variance in blocks 2 and 3. Three types—shortest, curve-pointing and straight-pointing, could be learned across blocks, while the overhead-down and the straight-down show the least amount of learning. Even though the shortest type can be learned, it still doesn't become more legible than the other four trajectories. This implies, that optimizing a C-space metric may not be the best strategy for providing legible trajectories.

5 Discussion

This study partially supports previous findings regarding the legibility of robot motion [7], i.e., different types of paths can have highly divergent levels of legibility. Shortest C-space paths, which are frequently the focus of motion planning methods, can be poor choices in terms of legibility. Similarly, paths that are currently used for pick-and-place tasks in industrial setups (e.g., "overhead") also appear not to be intuitively interpreted. Paths that focus on the orientation of the end-effector seem to be advantageous in terms of legibility, since they exhibited the best performance in estimation (high accuracy and less convergence time).

The fact that "straight-pointing" paths were more legible relative to "curve-pointing" paths corresponds to a difference from previous findings. The idea behind the "curve-pointing" paths is that legibility may increase by exaggerating the arms' motion so that it moves away from unintended targets (Fig. 1a). The difference seems to be due to the presence of multiple targets in the current work. When only two targets are present, exaggerating the motion in one direction can significantly assist in identifying the target but can be confusing in the case of multiple targets, or in cluttered workspaces. This study is intended to inform legibility in such cluttered environments.

A significant observation corresponded to the importance of the end-effector's orientation relative to the target. It was hypothesized that humans might pay particular attention to the pose and orientation of a robotic end-effector, similar to the way they respond to a human's hands. The experimental results confirmed this hypothesis. It would be worthwhile to incorporate the maintenance of such end effector orientations into the cost functions of motion planners in the future.

A question that needs to be answered is whether it is worthwhile to consider legibility of robotic motion planning paths, as opposed to relying on learning to take place in human observers. There is a learning effect when the subjects observed the same trajectories. The benefit primarily appears as reduced variance during repetitions of the same trajectories and varies across types of trajectories. Three types of paths, shortest, curve-pointing and straight-pointing, could be learned across blocks, while the overhead-down and the straight-down did not exhibit significant learning behavior. Although the learning effect existed, the benefit of the learning might not be able to override the advantage of the legible information, which was supported by the fact that the performance of the shortest type was improved in later blocks, but it was still not as good as the performance of other types.

Note that in the experiments the subjects witnessed the same path to a target in 3 blocks of time rather than 3 variations of the same type of path to the same target. Certain planners, such as sampling-based ones, can vary in the repeatability of their solutions. It is not necessarily the case that similar degrees of learning would occur for the general case of repeated exposure to motion plans generated from such motion planners. Furthermore, in this study the initial condition was always the same. When a robot needs to plan on the fly and transition from one task to another, the human subject will not be exposed to the same exact trajectories repeatedly. It is interesting to consider the effects of legibility in the context of trajectories that have different initial conditions.

During preliminary experiments, there was a transition from a web-based UI in the pilot trials, to the physical pointer feedback device used in gathering the data included here. This change decreased the cognitive load placed on human subjects by the data collection interface and resulted in a reduction between the best-performing and worst-performing path classes relative to the pilot study. A human co-worker in a collaborative setting is likely to have additional mental demands beyond the robot interaction. While minimizing the cognitive load might clarify the effects of legible features, such distractions might exaggerate the legibility of robot motions. An interesting line of future research is to analyze the effect of cognitive load on legibility.

Initial pilot trials also used trajectories which varied in duration. Increasing duration of trajectory execution gives the subject more time to recognize the legible features of the motion. However, it is not clear whether the effect persists if the trajectory duration keeps on increasing. Unnaturally slow trajectories might obfuscate the features that contribute to legibility. A scope for future work would be to understand the effects of the duration and speed of trajectories on legibility consistent among different types of trajectory.

Future experiments could involve the random placement of targets over a two-dimensional subspace, the presence of obstacles, as well as stopping the motion of the arm half-way towards the target and asking the user to guess the intended target. The longer-term objective is the definition of appropriate motion planners that generate highly-legible paths. It appears that such planners and accompanying cost metrics need to be reasoning for the orientation of the end-effector and its workspace path. This line of work can eventually lead to robots that use time-efficient paths when they operate in a dark factory floor and automatically switch to humanly-legible but less efficient paths when people enter their workspace and collaborate.

References

1. Rethink Robotics: Baxter Research Robot Software Development Kit (SDK) Version 0.7.0 (2013)
2. Gielniak, M., Thomaz, A.: Generating anticipation in robot motion. In: IEEE RO-MAN, pp. 449–454 (2011)
3. Beetz, M., Stulp, F., Esden-Tempski, P., Fedrizzi, A., Klank, U., Kresse, I., Maldonado, A., Ruiz, F.: Generality and legibility in mobile manipulation. Auton. Robot. **28**(1), 21–44 (2010)
4. Gielniak, M.J.: Spatiotemporal correspondence as a metric for human-like robot motion. In: In ACM/IEEE HRI (2011)
5. Argall, B.D., Chernova, S., Veloso, M., Browning, B.: A survey of robot learning from demonstration. Robot. Auton. Syst. **57**(5), 469–483 (2009)
6. Billard, A., Calinon, S., Dillmann, R., Schaal, S.: Robot programming by demonstration. In: Siciliano, B., Khatib, O. (eds.) Springer Handbook of Robotics, pp. 1371–1394. Springer, Berlin (2008)
7. Dragan, A.D., Lee, K.T., Srinivasa, S.S.: Legibility and predictability of robot motion. In: International Conference on Human-Robot Interaction (HRI) (2013)
8. Dragan, A., Srinivasa, S.S.: Generating legible motion. In: Proceedings of Robotics: Science and Systems, Berlin, Germany (2013)
9. Dragan, A., Srinivasa, S.S.: Familiarization to robot motion. In: International Conference on Human-Robot Interaction (HRI) (2014)
10. Momasello, M., Carpenter, M., Call, J., Behne, T., Moll, H.: Understanding and sharing intentions: the origins of cultural cognition. In: Behavioral and Brain Science (2005)
11. Csibra, G., Gergely, G.: Obsessed with goals: functions and mechanisims of teleological interpretation of action in humans. Acta Psychol. **1**(124), 60–78 (2007)
12. Woodward, A.L.: Infants selectively encode the goal object of an actor's reach. In: Cognition (1998)
13. Woodward, A.L., Guajardo, J.J.: Infants' understanding of the point gesture as an object-directed action. In: Cognitive Development (2002)
14. Kimmel, A., Dobson, A., Littlefield, Z., Krontiris, A., Marble, J., Bekris, K.E.: Pracsys: an extensible architecture for composing motion controllers and planners. In: Simulation, Modeling and Programming for Autonomous Robots (SIMPAR), Tsukuba, Japan (2012)
15. Marble, J., Bekris, K.E.: Asymptotically near-optimal is good enough for motion planning. In: Proceedings of the 15th International Symposium on Robotics Research (ISRR-11), Flagstaff, AZ 28 Aug–1 Sept 2011
16. Kavraki, L.E., Svestka, P., Latombe, J.C., Overmars, M.: Probabilistic roadmaps for path planning in high-dimensional configuration spaces. IEEE TRA **12**(4), 566–580 (1996)
17. Şucan, I.A., Moll, M., Kavraki, L.E.: The open motion planning library. IEEE Rob. Autom. Mag. **19**(4), 72–82 (2012). http://ompl.kavrakilab.org
18. MATLAB: version 8.2.0.701 (R2013b). The MathWorks Inc., Natick, MA (2013)
19. Sucan, I.A., Chitta, S.: Moveit! Online
20. Smits, R.: KDL: kinematics and dynamics library. http://www.orocos.org/kdl

Towards Coordinated Precision Assembly with Robot Teams

Mehmet Dogar, Ross A. Knepper, Andrew Spielberg, Changhyun Choi, Henrik I. Christensen and Daniela Rus

Abstract We present a system in which a flexible team of robots coordinates to assemble large, complex, and diverse structures autonomously. Our system operates across a wide range of spatial scales and tolerances, using a hierarchical perception architecture. For the successful execution of very precise assembly operations under initial uncertainty, our system starts with high-field of view but low accuracy sensors, and gradually uses low field-of-view but high accuracy sensors. Our system also uses a failure detection and recovery system, integrated with this hierarchical perception architecture: upon losing track of a feature, our system retracts to using high-field of view systems to re-localize. Additionally, we contribute manipulation skills and tools necessary to assemble large structures with high precision. First, the team of robots coordinates to transport large assembly parts which are too heavy for a single robot to carry. Second, we develop a new tool which is capable of co-localizing holes and fasteners for robust insertion and fastening. We present real robot experiments where we measure the contribution of the hierarchical perception and failure recovery approach to the robustness of our system. We also present

M. Dogar, R.A. Knepper and A. Spielberg contributed equally to this paper.

M. Dogar (✉) · R.A. Knepper · A. Spielberg · D. Rus
Computer Science and Artificial Intelligence Lab, MIT, Cambridge, USA
e-mail: mdogar@csail.mit.edu

R.A. Knepper
e-mail: rak@csail.mit.edu

A. Spielberg
e-mail: aespielberg@csail.mit.edu

D. Rus
e-mail: rus@csail.mit.edu

C. Choi · H.I. Christensen
Institute for Robotics and Intelligent Machines, College of Computing,
Georgia Tech, Atlanta, Georgia
e-mail: cchoi@cc.gatech.edu

H.I. Christensen
e-mail: hic@cc.gatech.edu

M.A. Hsieh et al. (eds.), *Experimental Robotics*, Springer Tracts
in Advanced Robotics 109, DOI 10.1007/978-3-319-23778-7_43

an extensive set of experiments where our robots successfully insert all 80 of the attempted fastener insertion operations.

Keywords Robotic assembly · Robotic manufacturing · Robot teams · Distributed control · Precision assembly · Perception

1 Introduction

Manufacturing systems of today have very limited flexibility, often requiring months of fine-tuning before an industrial assembly line is ready for production. We envision the manufacturing systems of the future, in which agile, flexible teams of mobile robots coordinate to assemble complex and diverse structures autonomously. Here, we define flexibility as the ability for robots to change tasks, factory floors to be reconfigured, and similar parts to be interchanged without reprogramming the system.

This approach has the potential to meet the demands of modern production: ever-shortening product life-cycles, customized production, and efficiency [2]. In this paper we present a significant step in this direction through an exemplar task which requires a heterogeneous team of four robots with different skills to align and fasten a panel to a corresponding box. We outline the task below (see Fig. 1 for details):

- A robot specializing in fine perception and manipulation localizes a hole on the box.
- A fleet of two robots lift and rotate the panel which would be too heavy for a single robot to manipulate single-handedly.
- Using a robot specializing in coarse perception for guidance, the fleet aligns the panel to the box.
- Using the fine perception/manipulation robot for guidance, the fleet aligns a hole on the panel with a hole on the box.
- The fine perception/manipulation robot inserts the first fastener.
- Using the bounding geometry of the box, one of the fleet robots aligns the panel with the box.
- Finally, with all four holes aligned, the fine perception/manipulation robot inserts the three remaining fasteners into the remaining holes.

In particular, we present the following contributions:

1. A hierarchical perception system formalized in the context of integrated perception and manipulation over changing task scales and scopes.
2. A failure recovery approach which allows us to re-seed search and tracking procedures.
3. A simple but robust controller for the collaborative manipulation of objects whose manipulations are outside the physical limits of single individual robots.
4. The demonstration of a new rigid LIDAR and fastener tool which allows for simultaneous localization and fastener insertion within the same coordinate frame.

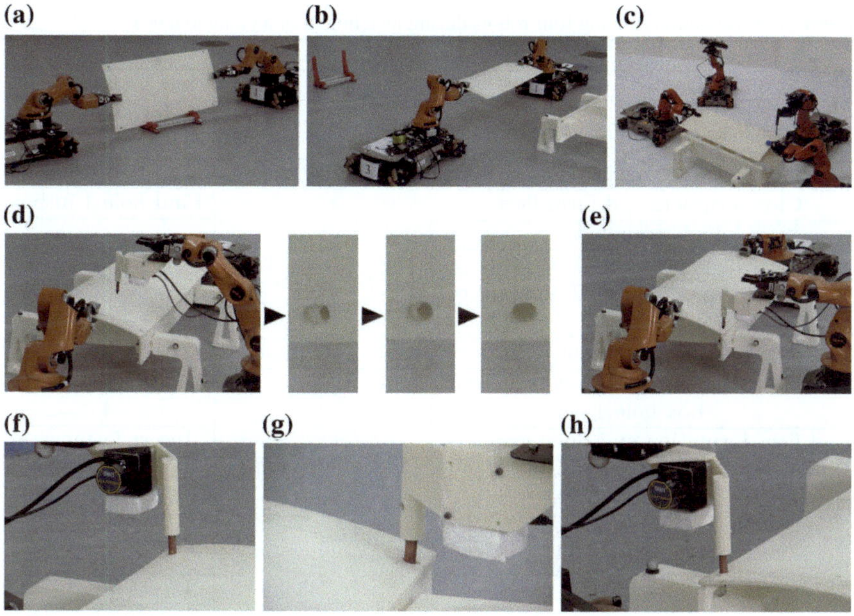

Fig. 1 Assembly tasks involve large-scale operations such as transport and fine manipulation operations such as hole alignment and fastener insertion. **a** Locate/grasp parts, **b** transport of parts, **c** part alignment, **d** hole alignment, **e** Fastener insertion, **f** Fastener 2, **g** Fastener 3, **h** Fastener 4

Hierarchical approach Autonomous manufacturing systems must operate across a wide range of spatial scales and tolerances. Consider the task of attaching a panel to a complex assembly. First, the robots move the panel from a storage rack to the assembly site (Fig. 1a–c). Second, the robots insert fasteners to attach the panel to the assembly (Fig. 1d–h). The first task requires perception and control at a spatial scale which captures the parts and sometimes the whole factory floor and tolerates relatively large errors in positioning. The second task requires fine perception and control with much tighter tolerances. With existing technologies, no monolithic perception and control approach solves both problems. In this work, we contribute a hierarchical approach in which different layers of localization and control systems interact to satisfy the continuously changing scale and precision requirements. See Table 1 for an example flow of control across the levels of the hierarchy.

Failure recovery By exploiting our system's hierarchical perception formalization we also introduce a failure recovery system. We present systems which can determine when insufficient precision has been obtained. Our system allows us to move freely between adjacent levels in the perception hierarchy, allowing us to re-seed failed searches and tracking procedures with better initial guesses. This allows us to avoid lengthy searches in the absence of useful feature information by falling back to estimates which are coarser but larger in scope. Such a system is applied to hole alignment but could also be applied to a number of other manipulation tasks in

Table 1 Flow of actions among four robots during attachment of a panel to a box

<table>
<tr><th colspan="4">Robot</th></tr>
<tr><th>R1</th><th>R2</th><th>R3</th><th>R4</th></tr>
<tr><td></td><td></td><td></td><td>Move to hole 1 neighborhood</td></tr>
<tr><td colspan="2">Navigate to and move gripper to panel</td><td>Localize box</td><td>Find hole 1 in box</td></tr>
<tr><td colspan="2">Close grippers and form fleet</td><td></td><td>Find hole 1 in box</td></tr>
<tr><td colspan="2">Pick up panel</td><td></td><td></td></tr>
<tr><td colspan="2">Orient panel to horizontal</td><td></td><td></td></tr>
<tr><td colspan="2">Transport panel into neighborhood of box</td><td></td><td></td></tr>
<tr><td colspan="2">Servo panel into alignment with box</td><td>Localize panel</td><td></td></tr>
<tr><td colspan="2">Servo panel hole 1 into alignment with box hole 1</td><td></td><td>Localize panel hole 1</td></tr>
<tr><td colspan="2">End fleet formation and open grippers</td><td></td><td>Insert fastener 1</td></tr>
<tr><td>Move out of the way</td><td>Align panel hole 2 to box hole 2</td><td>Move out of the way</td><td>Navigate to panel hole 2</td></tr>
<tr><td></td><td>Move out of the way</td><td></td><td>Localize hole 2</td></tr>
<tr><td></td><td></td><td></td><td>Insert fastener 2</td></tr>
<tr><td></td><td></td><td></td><td>Navigate to hole 3</td></tr>
<tr><td></td><td></td><td></td><td>Localize hole 3</td></tr>
<tr><td></td><td></td><td></td><td>Insert fastener 3</td></tr>
<tr><td></td><td></td><td></td><td>Navigate to hole 4</td></tr>
<tr><td></td><td></td><td></td><td>Localize hole 4</td></tr>
<tr><td></td><td></td><td></td><td>Insert fastener 4</td></tr>
</table>

Time flows from top to bottom. Box colors indicate the type of localization used in each action. Blue boxes indicate fiducial-based localization. Green boxes denote object-shape-based tracking. Pink boxes indicate functional-feature level localization. White boxes indicate sensorless operations

other systems which involve active perception and estimation, including precision grasping and collision-free navigation of cluttered factory environments.

Collaborative manipulation A team of robots working in a factory requires coordination and collaboration. The coordination can be loosely coupled, as in collision-free navigation, or tightly-coupled, as when carrying a large part (Fig. 1b) as a team. Our system displays coordination between robots at these various levels.

Specialized tools for robotic manipulation An important challenge in flexible factory automation is enabling fine manipulation skills, e.g. inserting a fastener or screwing a nut. Much like human workers, robots need specialized tools and skills (control algorithms) to perform these operations to specifications. We've developed such a tool (Fig. 3) to unify sensing and actuation in the tool frame, thus delivering high precision, as suggested in our second listed contribution.

Our approach is holistic: we are interested in the challenges and questions that a complete system raises. The literature has approached the underlying problems separately. Many methods have been proposed for collaborative manipulation/transport of objects by a team of robots [9, 12, 15, 23, 25]. Particularly, Desai and Kumar [9] propose a motion planning approach for a team of robots transporting an object among obstacles; and Khatib et al. [12] present a decentralized control framework for the manipulation of an object with a system of multiple manipulators. Similar approaches have been applied to the factory floor [14, 20] where a team of robots transports an object with the help of human input. We present a system where the team of robots transports an object in the context of a complex task. To do this, they must form a fleet, and maintain specified relative arm configurations while making progress toward goal positions. We develop control algorithms which treat fleets (connected by manipulated objects) as rigid movable bodies and are able to correct for erroneous deviations. Our control/perception environment is not structured specifically for a transport task, but is generic enough to accommodate other assembly tasks.

One generic and important assembly operation is fastening multiple parts together. In our system this is achieved by inserting fasteners through holes on the parts. This operation, sometimes called *peg-in-hole* in the literature, has been studied extensively. One approach to this problem is to use a hybrid force-position control [17, 19], which, through force sensing and compliant motion [11], enables a manipulator to slide along surfaces. Combined with a principled approach to dealing with uncertainty [16], a high-precision operation such as peg-in-hole can be accomplished through a set of *guarded-moves*. This approach, however, may not be feasible if the assembly parts are very sensitive and prone to scratching. In our implementation we avoid making forceful interaction with the surfaces of assembly parts. Instead of a series of guarded moves, we use extensive and high-accuracy sensor readings to localize the hole, and a compliant shape for the fastener tip to account for any remaining inaccuracy in localization.

Robotic perception literature and technology provide a rich set of tools [4, 6, 21] which can be used for certain tasks in the factory setting. While these systems work best when the object is closer than a few meters, the accuracy drops as the object gets too far or too close. In addition, visual perception is highly challenged in many cases: occlusions, cluttered backgrounds, and image blurring because of fast motions either in objects or camera. To overcome these limitations of visual perception, it is often combined with motion estimation [13] or tactile sensing [1, 10]. Skotheim et al. [22] use functional feature detection for low-level industrial manipulation. Although the literature provides these powerful tools, any single one is insufficient to overcome the challenges of flexible factory environments.

2 Hierarchical Localization and Control Approach

Various objects and features of a flexible factory environment require various perception and control technologies and a smooth integration among them. We present a three-tiered perception and control structure, comprising fiducial-based, object-shape-based, and functional-feature-based approaches.

Fiducial-based technology tracks non-production parts, part sources, and robots, using a motion capture system like Vicon.[1] Motion capture provides highly accurate, sub-centimeter localization accuracy, but it is restricted to tracking parts to which external fiducials may be affixed. For many production parts, attaching fiducials is undesirable and impractical. Furthermore, occlusion can be a problem. Thus, complementary localization methods are needed.

Object-shape-based tracking is implemented as a particle filtering approach using an RGB-D camera [5]. 3D mesh models of production parts are known a priori, and three visual features—colors, depth points, and normals—are used to calculate the likelihood of each particle hypothesis with respect to the current RGB-D scene. Our system localizes the box and panel from a single RGB-D camera. The robot carrying the camera can be seen in Fig. 1c. The system may exploit the freedom of the camera's point of view to avoid occlusion.

Functional-feature-based tracking for hole alignment and insertion is the most demanding part of our task as it requires very high-precision coordination among multiple robots. We use a coordinated control procedure along with a specialized tool, explained in the next section.

We hypothesize that without the use of all three levels in the sensing and control hierarchy, the system cannot achieve robust fastener insertion. In the rest of this section, we discuss the levels of the hierarchy and how the robots may smoothly transition up and down through them.

2.1 Sequential Composition of Sensors

The *funnel* analogy has long served in robotics literature to represent the act of reducing uncertainty or error in the configuration of an object. Mason [18] first introduced the concept in the context of performing sensorless manipulation actions that employ passive mechanics to reduce part uncertainty. Later, Burridge et al. [3] applied the funnel analogy to feedback control in the form of sequential composition of controllers, spawning much follow-on work [7, 8, 24]. This body of work is sensor-agnostic in that the type and quality of sensor data is assumed to be homogeneous throughout the configuration space.

A contribution of this paper is sequential composition of sensors used for localization. Each sensor operates over some capture volume, or scope, which is the top of the funnel. Within the scope, it delivers to the robot a pose estimate that reduces uncertainty with some accuracy, which is the bottom of the funnel. Each of the localization technologies we employ imposes errors that limit accuracy in three categories: (1) *sensor error*, (2) *indirection error* and (3) *semantic calibration error*. Sensor error, the accuracy claimed by the sensor manufacturer, is typically the smallest contribution to overall error in performing localization.

---

[1] http://www.vicon.com/.

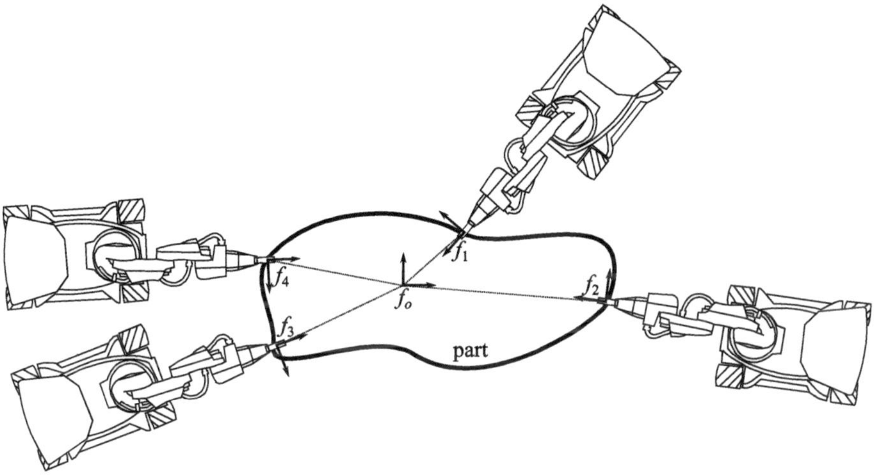

Fig. 2 Through fleet control, an arbitrary number of robots collaboratively carry a part in an arbitrary shape formation. Individual robot motions are computed with respect to a commanded twist at the fleet origin, o. Each robot n maintains the pose of the fleet origin in its own local coordinate frame, f_n, so there is no need for a global reference. The algorithm is fully distributed

Indirection error stems from the fact that sensors rarely localize the desired coordinate frame directly. Instead, they sense some set of features, each with some transform to the desired frame. This indirection leads to two sources of error: (1) small errors in orientation are magnified by translation, and (2) the feature poses may not be well calibrated to the desired frame. All three localization technologies exhibit indirection error. Since fiducials cannot be applied directly to the part being assembled, the robot hands must be tracked instead with fiducial-based technology. The position of each hand on the object may not be well known (as in Fig. 2). Since the hands grasp the perimeter of the object, the indirection error of fiducial-based methods is proportional to the size of the part, making them the coarsest level of the hierarchy. In the case of object-shape-based tracking, a point cloud over a smooth panel surface (as in Fig. 1) produces substantial ambiguity about the location of each point on the object and hence of the the object origin. It is the indirection error that the particle filter strives to minimize. Finally, the functional-feature-based hole detector tracks the hole's circumference, whereas the center of the hole is desired. In the case of a circular hole, the resulting indirection error is minimal due to symmetry.

Finally, semantic calibration error originates from the fact that a perception model used for localization must be calibrated against the semantic model used for manipulation. For example, fiducials placed on the robot for motion capture must be manually calibrated to the robot's pose. Similarly, for object-shape-based tracking, the origin and shape of the CAD model of the tracked object may not match the origin and shape of the physical object. The functional-feature-based hole tracker has no semantic calibration error because the sensor directly tracks a semantic feature.

Given a position estimate of the object with uncertainty, it may be within scope of several sensors, giving the system some flexibility in which technology to use (see

Table 2 Order of magnitude of sensor capabilities and of errors induced by the usage model

Approach	Sensor	Scope (m^3)	Error (m)			Net accuracy (m^3)
			Sensor	Indirection	Semantic calib.	
Fiducial-based	Vicon	10^2	10^{-3}	10^{-1}	10^{-2}	10^{-3}
Object-shape-based	Kinect	10^0	10^{-2}	10^{-2}	10^{-2}	10^{-6}
Functional-feature-based	Hokuyo	10^{-2}	10^{-3}	10^{-3}	0	10^{-9}

See the text for a description of error sources. Net accuracy is the volume resulting from the sum of the three distance errors

Table 2 for a summary of sensor capabilities). This flexibility allows the system to be tolerant to effects such as occlusion or communication drop-outs. The typical progression of the localized feedback control system is to servo the object into position at increasingly detailed scales.

2.2 Failure Recovery

Failures in execution can happen at any step of the assembly operation. To make sure that the assembly operation completes successfully, our system detects and tries to recover from failures.

The hierarchical perception/control structure provides the backbone of our failure recovery approach. During successful execution, the control is handed-off from higher levels to the lower levels: higher levels perform coarse localization and lower levels perform precise tasks. Failure recovery is implemented as the inverse process, where the control is handed off from lower levels to higher levels: lower levels of perception are precise in tracking objects/features but have limited scope, which may result in the tracked objects/features getting lost. In such a case the control is handed-off to the higher level for a coarse but larger scope localization.

A crucial example of the failure recovery process occurs during alignment of the panel-hole with the box-hole. To accomplish this task, the panel is first aligned with the box using the object-shape-based perception system, which has a large scope but low accuracy. Once the panel is coarsely aligned with the box, the functional-feature-based localizer takes over to track the panel-hole and align it with the box-hole. This localizer has high accuracy but a small scope. The scanner occasionally loses track of the hole due to the small scope and the noise in the arm and base motions of the robots during alignment. In such a case, the system reverts back to the previous level, the object-shape-based alignment. The larger scope re-aligns the panel with the box and hands over the control to the functional-feature-based tracker once more. This process continues until this sensor successfully tracks the panel-hole and aligns it with the box-hole.

This approach to detecting and recovering from failure provides significant robustness to our system. Even if the individual layers permit failure, the overall architecture displays very high robustness as long as failures are detected and the system is started from a recoverable state.

2.3 Fleet Control

For collaborative transport of large parts, the robots perform a distributed, collective behavior inspired by human group behavior using force feedback and observation of others. In fleet control mode, the robots maintain a fixed formation of arbitrary shape while holding an object, as in Fig. 2.

Initially, each robot separately moves into formation by grasping the object at an appropriate location. Robot n's pose, f_n is measured at this grasp point because the other robots can readily localize its hand. Formation control initializes via a synchronization broadcast message. Upon initialization, the robots compute a common reference origin f_o for the object. Robot n represents the fleet origin in its own frame as $T_{f_n}^{f_o}$. The position of the origin defaults to the mean of all robot hand positions, and its orientation initializes to that of the global coordinate frame (i.e. Vicon frame). Henceforth, the global frame is not needed as all coordinates are given in f_o or f_n. If desired, f_o can be moved with respect to the fleet.

Group motions are commanded as a twist w specified in frame f_o. Each robot computes its own hand motion in order to comply with w in six degrees of freedom (DoFs). Hand motions are achieved through base motion when possible (X, Y, yaw) and arm motion otherwise (Z, roll, pitch). It should be noted, however, that the KUKA youBot cannot achieve full six DoF motion due to their arm kinematics. Therefore, the task presented in this paper involves only five DoF object manipulation.

An important function of the fleet controller is to maintain a stable fleet formation. Any position error introduced by group motion will cause the fleet origin to drift away from its target pose in the frame of the robots. A PD controller introduces correction terms to the body and arm motions in order to maintain the correct fleet formation.

Similarly, force exchange among the robots through the object can indicate an error in desired position. The robots' arm joints perform PD velocity control on joint angle. In the steady state, an error derived from the joint torques can be attributed to a combination of gravity and an error in the fleet formation. Thus, the robot has detected a resultant force from the combined motion of the rest of the fleet. In response to this force, the fleet controller applies a correction term to $T_{f_n}^{f_o}$.

Since each robot computes a motion consistent with the fleet twist command, any residual force results from an error in the formation, which may have two causes. First, the robot may drift slightly out of formation while carrying a rigid object. Second, the object may be somewhat deformable. Although the fleet cannot deliberately exploit

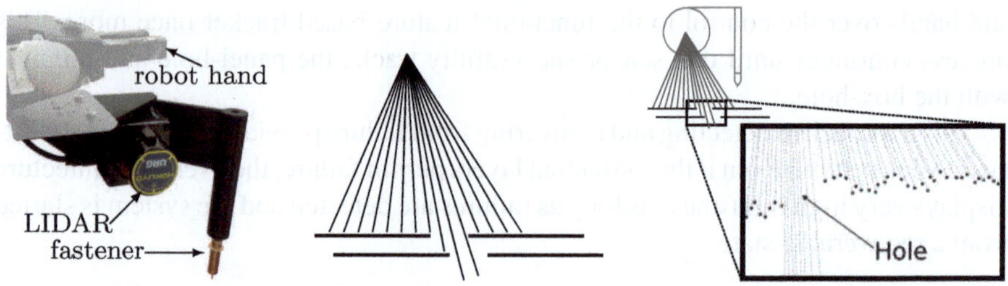

Fig. 3 *Left* Hole alignment and insertion tool. *Center* Alignment of two holes is achieved by estimating the width of the opening. *Right* Example real data used to estimate the width of the opening

deformability of material, it will accommodate deformations induced by the carrying operation by slightly varying the formation in response to these joint torques.

2.4 Coordinated Mating of Holes and Fastener Insertion

To achieve millimeter-scale accuracy, we employ a custom-built end-effector tool on which both a Hokuyo LIDAR and a fastener are rigidly affixed (Fig. 3-left). This sensor fulfills the functional-feature-based localization in the hierarchy.

Algorithm 1 Coordinated alignment of holes

```
1: function ALIGNHOLES
2:     while hole-width < threshold do
3:         twist ← FastenerRobot.DesiredPartMotion(history)
4:         Fleet.MovePart(twist)
5:         hole-width ← FastenerRobot.EstimateHoleWidth()
6:         history.Add(hole-width)
7: function FLEET.MOVEPART(twist)
8:     for each robot in fleet do
9:         pose ← robot.ComputePoseRelativeToPart()
10:        robot-twist ← Transform(twist,pose)
11:        Robot.Move(robot-twist)
```

We present the collaborative procedure by which our system aligns the holes of two different parts in Algorithm 1. This procedure is executed after the robot with the fastener locates the hole on one of the parts (the box, in our example) and the fleet of robots brings the panel into the vicinity using the object-level tracking.

The goal in Algorithm 1 is to achieve an alignment within the tolerance required by the fastener. At each step the robot with the tool estimates (line 5) the alignment of the two holes (Fig. 3-center) by measuring the width of the opening (Fig. 3-right). If the opening is not large enough (line 2), the fastener robot commands a new velocity twist for the moving part (lines 3–4). In computing this, the fastener robot can use the history of readings to maximize the alignment using gradient ascent. We implement this by making the fleet follow a series of waypoints.

A twist for the moving part commands the robots in the fleet to move (lines 7–11) using decentralized fleet control. After the holes are aligned, the fastener can be inserted. The fastener is placed directly in line with the LIDAR's laser scan, thus allowing the robot to know exactly where the fastener is with respect to a detected hole at all times, and to bring the fastener over the hole.

3 Experiments

We use a team of four KUKA Youbots for our experiments. These robots are tasked with assembling a panel (Fig. 4a) on a box (Fig. 4b) using fasteners (Fig. 4c). The panel and box are initially placed on supporting racks, which have markers for the fiducial-based Vicon tracking system. Two of the robots, R1 and R2, are responsible for the manipulation of the panel. Robot R3 carries a Kinect RGB-D camera which performs the object-shape-based tracking of the panel and the box. Robot R4 carries the insertion tool (Fig. 3-left). The insertion tool has an integrated Hokuyo laser scanner which performs the functional-feature-based alignment with the holes on the box and the panel.

We measure the effectiveness of different components of our perception and control hierarchy by running experiments with three different configurations of this system:

1. *Fiducial-based + Object-shape-based (FO):* In this case, the panel and box are aligned only using the object-shape-based tracking and control. The functional-feature-based tracking, i.e. the Hokuyo laser scanner is *not* used.

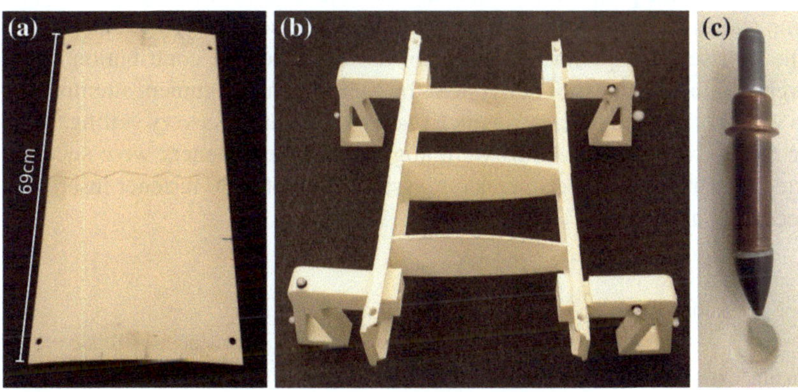

Fig. 4 Assembly parts used in our experiments. **a** Panel. **b** Box. **c** A fastener and hole (misaligned) as used in this task. The fastener is an adapted cleco. The holes were drilled to permit a cleco to fit up to the flange with a tolerance of 1.5 mm

2. *Fiducial-based + Functional-feature-based (FF):* In this case, the object-shape-based tracking of the panel and box is left out, i.e. the Kinect RGB-D sensor is *not* used. Instead, the robots remember their grasping configuration of the panel and assume it does not change relative to the robot hands during the course of the task.
3. *Fiducial-based + Object-shape-based + Functional-feature-based (FOF):* Our system as described in Sect. 2 where the objects are tracked using the Kinect RGB-D camera and the hole is aligned using the Hokuyo laser scanner.

With our system we performed two sets of experiments. First, we ran our system in the FOF configuration 22 times to measure the robustness, the contribution of our failure recovery system to the robustness, and the overall speed of our system. A video of one such run is available at: http://youtu.be/cmJTsyIgCRo.

Second, we performed experiments to measure the contribution of the hierarchical perception architecture to the robustness of our system. In this set of experiments we created perturbations to the pose of the panel as it was being carried. Under these perturbations we ran our system four times in each of the FO, FF, and FOF configurations, totaling to twelve more runs.

4 Results

We start with reporting the results of 22 experiments in the FOF configuration. Our system showed a remarkable robustness for such a complicated and long task. Aside from two hardware failures of unknown cause, the system succeeded 20 out of 20 times. Table 3 shows the average time of 20 successful runs along with the minimum and maximum durations. The first column shows the time spent for localizing the four holes on the assembly during each run. The second column shows the time spent during aligning the panel to the box using the object-based tracking system. The last column shows the execution time for the complete assembly operation.

The first set of experiments also showed the important contribution of failure recovery to the robustness of our system. In 20 % of panel alignment attempts the two holes were not aligned precisely, which resulted in failure recovery getting triggered. After failure recovery the holes were aligned and the fasteners were successfully inserted. During these experiments our system attempted 80 fastener insertions and succeeded in all of them.

Table 3 Execution times

	Hole localization	Ladder-panel alignment	Total
Mean time (s)	92	37	679
Min time (s)	27	17	569
Max time (s)	259	141	849

Table 4 Comparison of the performance of different configurations of our system

	Success	Notes
FO	1/4	Successful run scratched panel surface on 2 of the 4 holes
FF	2/4	Panel hole search timed out at 10 min
FOF	3/4	All succeeded for hole alignment but one failed during fastener insertion

We report the result of our second set of experiments in Table 4. Here we perturb the position of the grasped panel to measure the robustness of our system. The first two cases show the system running with certain layers of the hierarchical perception system removed. In these cases the system was not able to get precise alignment between the holes of the panel and the box. The full hierarchical perception system was able to get precise alignment between the holes in all four cases, but had trouble with the insertion of the fastener since the insertion routine was not adaptive to the changed height of the panel due to the perturbation. However our full system was robust in achieving the precise hole alignment.

5 Insights and Conclusion

The results show that intelligent use of a hierarchical perception system can greatly improve the robustness of a manufacturing system to be nearly perfect. The system is not only able to perform collaborative carrying, precise alignment, and collision-free insertion, but is also able to detect and fix the rare errors in alignment. Further, the only failures were in the cases of high-torque-driven arm failures, in which the system failed in the collaborative carry step. In addition, we have demonstrated that use of object-shape-based tracking makes the system robust to outside perturbations or other internal errors that could lead to poor grasps.

Robustness is a key attribute for maximizing productivity in manufacturing. Traditional factory robots are bolted to the floor, thus achieving sensorless high precision through kinematics. Modern factory automation processes eliminate uncertainty through careful, time-consuming human design. Product changes require re-engineering of the process, contributing to a lack of versatility. Instead, we present a flexible system which achieved alignment within tolerance in 100 % of trials, but the fastener motion caused a failure in one of the perturbed cases.

We have identified several avenues for improving fastener insertion. Most importantly, torque feedback at the fastener tip flags an insertion failure. Soft guard materials around the fastener might help to avoid damaging fragile parts.

Our experiments validated our hypothesis that a hierarchical sensing system improves robustness in assembly. Removal of either the functional-feature-based or the object-shape-based localization from the hierarchy substantially diminished the successful completion performance of the system. With all three sensing systems, any gaps in perception by one sensor can be filled by one of the other sensors, allowing for a smooth transition among operating scales.

As many assembly procedures are composed of successive individual steps each of which must succeed, identifying and recovering from failures is crucial. A single failed step either requires a method of recovery failure or requires a restart of the procedure. Potential failure modes of the system include: misalignment of the two holes and fastener, dropping the panel prematurely, and incorrectly tracking the panel or features. We implemented automatic failure detection and handling algorithms for many of these problems and have designed the system to minimize the incidence of failure.

Acknowledgments This work was supported by The Boeing Company. We are grateful for their support.

References

1. Allen, P.K.: Integrating vision and touch for object recognition tasks. Int. J. Robot. Res. **7**(6), 15–33 (1988)
2. Bourne, D.: My boss the robot. Sci. Am. **308**(5), 38–41 (2013)
3. Burridge, R.R., Rizzi, A.A., Koditschek, D.E.: Sequential composition of dynamically dexterous robot behaviors. Int. J. Robot. Res. **18**(6), 534–555 (1999)
4. Choi, C., Christensen, H.I.: Robust 3D visual tracking using particle filtering on the Special Euclidean group: a combined approach of keypoint and edge features. Int. J. Robot. Res. **31**(4), 498–519 (2012)
5. Choi, C., Christensen, H.I.: RGB-D object tracking: a particle filter approach on GPU. In: Proceedings of the IEEE International Conference on Intelligent Robots and Systems, pp. 1084–1091 (2013)
6. Collet, A., Martinez, M., Srinivasa, S.S.: The moped framework: object recognition and pose estimation for manipulation. Int. J. Robot. Res. **30**(10), 1284–1306 (2011)
7. Conner, D.C., Rizzi, A.A., Choset, H.: Composition of local potential functions for global robot control and navigation. In: Proceedings of the IEEE International Conference on Intelligent Robots and Systems, vol. 4, pp. 3546–3551. IEEE (2003)
8. Das, A.K., Fierro, R., Kumar, V., Ostrowski, J.P., Spletzer, J., Taylor, C.J.: A vision-based formation control framework. IEEE Trans. Robot. Autom. **18**(5), 813–825 (2002)
9. Desai, J.P., Kumar, V.: Motion planning for cooperating mobile manipulators. J. Robot. Syst. **16**(10), 557–579 (1999)
10. Ilonen, J., Bohg, J., Kyrki, V.: Fusing visual and tactile sensing for 3-D object reconstruction while grasping. In: Proceedings of the IEEE International Conference on Robotics and Automation, pp. 3547–3554 (2013)
11. Inoue, H.: Force feedback in precise assembly tasks. Technical report, DTIC Document (1974)
12. Khatib, O., Yokoi, K., Chang, K., Ruspini, D., Holmberg, R., Casal, A.: Coordination and decentralized cooperation of multiple mobile manipulators. J. Robot. Syst. **13**(11), 755–764 (1996)
13. Klein, G., Drummond, T.: Tightly integrated sensor fusion for robust visual tracking. Image Vis. Comput. **22**(10), 769–776 (2004)
14. Lenz, C., Nair, S., Rickert, M., Knoll, A., Rosel, W., Gast, J., Bannat, A., Wallhoff, F.: Joint-action for humans and industrial robots for assembly tasks. In: RO-MAN, pp. 130–135 (2008)
15. Li, Z., Ge, S.S., Wang, Z.: Robust adaptive control of coordinated multiple mobile manipulators. Mechatronics **18**(5–6), 239–250 (2008)

16. Lozano-Perez, T., Mason, M.T., Taylor, R.H.: Automatic synthesis of fine-motion strategies for robots. Int. J. Robot. Res. **3**(1), 3–24 (1984)
17. Mason, M.T.: Compliance and force control for computer controlled manipulators. IEEE Trans. Syst. Man Cybern. **11**(6), 418–432 (1981)
18. Mason, M.T.: The mechanics of manipulation. In: Proceedings of the IEEE International Conference on Robotics and Automation, vol. 2, pp. 544–548. IEEE (1985)
19. Raibert, M.H., Craig, J.J.: Hybrid position/force control of manipulators. J. Dyn. Syst. Meas. Control **103**(2), 126–133 (1981)
20. Reinhart, G., Zaidan, S.: A generic framework for workpiece-based programming of cooperating industrial robots. In: ICMA, pp. 37–42 (2009)
21. Rusu, R.B., Bradski, G., Thibaux, R., Hsu, J.: Fast 3d recognition and pose using the viewpoint feature histogram. In: IROS, pp. 2155–2162 (2010)
22. Skotheim, Ø., Nygaard, J.O., Thielemann, J., Vollset, T.: A flexible 3d vision system based on structured light for in-line product inspection. Electr. Imaging **2008**, 680505–680505 (2008)
23. Sugar, T.G., Kumar, V.: Control of cooperating mobile manipulators. IEEE Trans. Robot. Autom. **18**(1), 94–103 (2002)
24. Tedrake, R., Manchester, I.R., Tobenkin, M., Roberts, J.W.: Lqr-trees: feedback motion planning via sums-of-squares verification. Int. J. Robot. Res. **29**(8), 1038–1052 (2010)
25. Yamashita, A., Arai, T., Ota, J., Asama, H.: Motion planning of multiple mobile robots for cooperative manipulation and transportation. IEEE Trans. Robot. Autom. **19**(2), 223–237 (2003)

Robot Hand Synergy Mapping Using Multi-factor Model and EMG Signal

Sanghyun Kim, Mingon Kim, Jimin Lee and Jaeheung Park

Abstract In this paper, it is investigated how a robot hand can be controlled from a human motion and an EMG signal in a tele-operation system. The proposed method uses a tensor to represent a multi-factor model relevant to different individuals and motions in multiple dimensions. Therefore, the synergies extracted by the proposed algorithm can account for not only various grasping motions but also the different characteristics of different people. Moreover, a synergy-level controller which generates motion and force of the robot is developed with postural synergies and an EMG signal. The effectiveness of the proposed new mapping algorithm is verified through experiments, which demonstrate better representation of hand motions with synergies and greater performance on grasping tasks than those of conventional synergy-based algorithms.

Keywords Synergy · Robot hand · Mapping · Multi-factor model · EMG

1 Introduction

A robot hand can provide a great deal of manipulation capability to its user in a tele-manipulation system. The method of controlling the robot hand with human hand motion is one of the most important parts of such a system, as a human hand can perform many types of operations given its number of joints, whereas a robot hand is limited in terms of motion compared to that of human hand motion. Thus,

S. Kim · M. Kim · J. Lee · J. Park (✉)
Graduate School of Convergence Science and Technology, Seoul National University, Seoul, Korea
e-mail: park73@snu.ac.kr

S. Kim
e-mail: ggory15@snu.ac.kr

M. Kim
e-mail: mingonkim@snu.ac.kr

J. Lee
e-mail: jmpechem@snu.ac.kr

M.A. Hsieh et al. (eds.), *Experimental Robotics*, Springer Tracts in Advanced Robotics 109, DOI 10.1007/978-3-319-23778-7_44

the functionality and controllability of dexterous robot hands have been investigated in an effort to overcome the difficulty stemming from the kinematic dissimilarity between robot and human hands.

Various kinematic maps between a human hand and a robot hand have been proposed using the joint angles of the human hand [1], fingertips [2], and poses [3] or separate motion controllers using a supervisory control methods [4].

In particular, Santello et al. [5] demonstrated that there exist strong correlations between grasping postures of the hand. These correlation patterns, referred to as synergies, can be defined as a spatial configuration of the hand shape, as more than 80 % of all grasping postures can be described with only two synergies. Thus, how to extract synergies and how to use synergies have recently been studied in attempts to reduce the high-dimensional data in the matrix representation to a lower dimensional space for human-like control and grasping [6–9].

These approaches, however, still have limitations. First, these synergy mapping schemes cannot account for individual characteristics, while Santello et al. [5] found that synergies differ from person to person. Hence, reconstructed motion with synergies shows erroneous results for a user who does not undergo the training. Secondly, the grasping force cannot be represented using motion-based synergy-level controllers [7]. For example, it is difficult for a robot hand to grasp a thin object, as a robot hand only follows human hand motion but not the grasping force.

Hence the following questions are proposed: if the synergy is defined as the linear combination of the grasping type and individual characteristics, can we obtain synergies that represent human hand motion more accurately? Additionally, if this is possible, can we extract synergies for a new user by only extracting the factor of the new user? Finally, can we manipulate the robot hand dexterously, if the grasping force of the robot is generated from the intention of the user? These issues constitute the main focus of this paper.

In this paper, a new type of mapping algorithm is proposed to answer these questions. Our goal is to extract the synergies of each user and to generate motion and grasping force in a robot using these synergies. The proposed algorithm uses a tensor composed of data relevant to different individuals and various motions in multiple dimensions, as shown in Fig. 1. This tensor, which is regarded as a multi-factor model, is then decomposed into the two factors of the information of the grasping motions and the individual characteristics. Thus, we can extract synergies accounting for not only various grasping motions but also the different characteristics of different people. Furthermore, the EMG signal of a user is used to estimate the intention of the grasping force. This estimated grasping force is generated by the robot hand using synergies.

This paper is organized as follows. Section 2 introduces the method used to extract the postural synergies and control the robot hand using the multi-factor model and the synergy-level controller. Section 3 presents the experimental setup and data sets to extract synergies. Section 4 discusses the result of the proposed algorithm for extracting synergies compared with other methods, and the paper is concluded in Sect. 5.

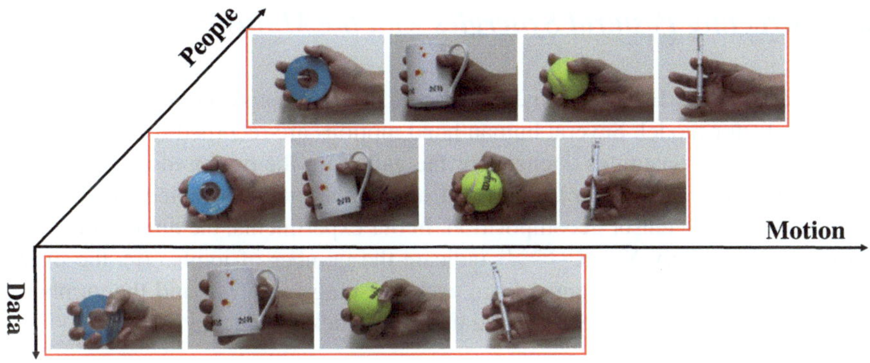

Fig. 1 Illustration of a tensor including the factors of grasping motion and the factors of individuals

2 Technical Approach

As shown in Fig. 2, the proposed algorithm consists of two parts: the extraction of the postural synergies with the multi-factor model, and the synergy-level controller part for a robot with postural synergies from the multi-factor model and the EMG signal.

In Sect. 2.1, we introduce the method used to extract the postural synergies of the user with the multi-factor model. The multi-factor model using tensor decomposition is applied to separate the factors of the individuals and the grasping hand motion. Therefore, when a new user performs several types of training motions, the factor for the new user is computed by optimization using the multi-factor model. Consequently, the postural synergies for the new user are extracted by combining the multi-factor model and the factor for the new user. In Sect. 2.2, we discuss the approaches used to overcome the kinematic dissimilarity between the robot hand and the user and to generate the grasping force of the robot hand with the synergy-level controller. We use the EMG signal to generate grasping force of a user, as there is a certain linear relationship between the EMG signal and the grasp force [10].

The sections below describe the details of the proposed algorithm. In our paper, bold lower case (**a**), bold upper case (**A**), and underlined characters ($\underline{A}$) are used to denote the vector, matrix, and tensor, respectively.

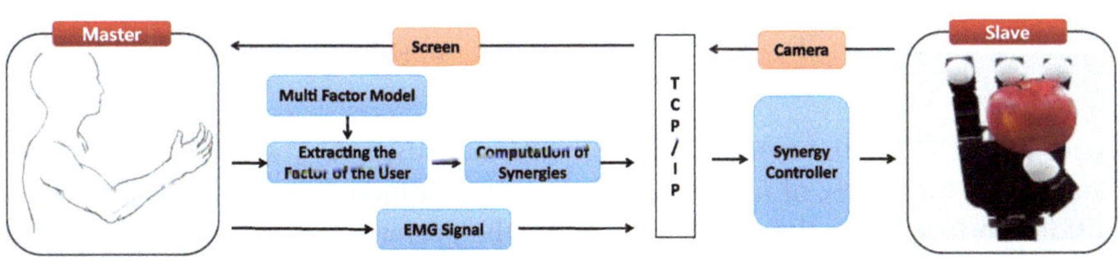

Fig. 2 Schematic diagram using the multi-factor model and EMG signal for hand-teleoperation

2.1 Extracting Postural Synergies with the Multi-factor Model

In this section, we discuss the method used to extract the synergies of a user with the multi-factor model. Several human hand motions are recorded for training. The training data set constitutes a tensor for the gallery in the proposed algorithm. The factors of individuals and grasping hand motions are then extracted by the multi-factor model using tensor decomposition. In our experiments, the dimensions of the tensor for the gallery $\underline{\mathrm{Y}} \in \mathbb{R}^{I_1 \times I_2 \times I_3}$ indicate the number of joints (I_1), the overall number of sample data obtained from each grasping type (I_2), and the number of people (I_3). The tensor is decomposed using the Tucker model [11, 12],

$$\underline{\mathrm{Y}} = \underline{\mathrm{G}} \times_1 \mathbf{A}_1 \times_2 \mathbf{A}_2 \times_3 \mathbf{A}_3 + \underline{\mathrm{E}}, \tag{1}$$

where $\underline{\mathrm{G}} \in \mathbb{R}^{J_1 \times J_2 \times J_3}$ is the core tensor, $\mathbf{A}_\mathrm{k} \in \mathbb{R}^{I_k \times J_k}$ denotes the factor matrices of mode-k, and $\underline{\mathrm{E}} \in \mathbb{R}^{I_1 \times I_2 \times I_3}$ is the error tensor.

Tensor $\underline{\mathrm{G}}$ and the matrices $\mathbf{A}_\mathrm{k}$ are calculated by the alternative least square (ALS) algorithm [12] to minimize the Frobenius norm F, as follows:

$$\min_{\underline{\mathrm{G}}, \mathbf{A}_1, \mathbf{A}_2, \mathbf{A}_3} \quad \| \underline{\mathrm{Y}} - \underline{\mathrm{G}} \times_1 \mathbf{A}_1 \times_2 \mathbf{A}_2 \times_3 \mathbf{A}_3 \|_F^2$$
$$\text{subject to } \underline{\mathrm{G}} \in \mathbb{R}^{J_1 \times J_2 \times J_3}, \ \mathbf{A}_\mathrm{k} \in \mathbb{R}^{I_\mathrm{k} \times J_\mathrm{k}} : \text{orthonormal}. \tag{2}$$

Thus, the factor of joint $\mathbf{A}_1$ spans the space of the joint angles and the factor of motion $\mathbf{A}_2$ represents the space of the grasping motion denoting the principal motion factor regardless of the individuals. The factor matrix of individuals $\mathbf{A}_3$ spans the space of the characteristics of an individual.

Finally, as shown in Fig. 3, Eq. (1) can be reshaped so that it represents the decomposition of only $\mathbf{A}_2$ and $\mathbf{A}_3$ by means of

$$\underline{\mathrm{Y}} \simeq \underline{\acute{\mathrm{G}}} \times_2 \mathbf{A}_2 \times_3 \mathbf{A}_3, \quad \text{where } \underline{\acute{\mathrm{G}}} = \underline{\mathbf{G}} \times_1 \mathbf{A}_1. \tag{3}$$

The tensor $\underline{\acute{\mathrm{G}}} \in \mathbb{R}^{I_1 \times J_2 \times J_3}$ is a new core tensor.

When a new user who has not participated in the multi-factor model performs a few of the same types of motions existing in the training data, the factor for the user, $\mathbf{p}$, which denotes the relationship between the characteristics of the new user and the individuals in the multi-factor model, is computed by

$$\min_{\mathbf{p}} \quad \| \mathbf{F} - \underline{\acute{\mathrm{G}}} \times_2 \mathbf{A}_{2,\mathrm{n}} \times_3 \mathbf{p}^\mathrm{T} \|_1 \quad \text{subject to } \mathbf{p} \in \mathbb{R}^{J_3}, \tag{4}$$

where $\mathbf{F} \in \mathbb{R}^{I_1 \times n}$ is the motion data set of the new user, $\mathbf{A}_{2,\mathrm{n}} \in \mathbb{R}^{I_1 \times n}$ is the corresponding factor matrix of motion in the multi-factor model, and n is the number of motions performed by the new user. Finally, the postural synergies $\mathbf{b}_i \in \mathbb{R}^{I_1}$ of the user and the corresponding coefficients $\mathbf{c}$ of each synergy are represented as

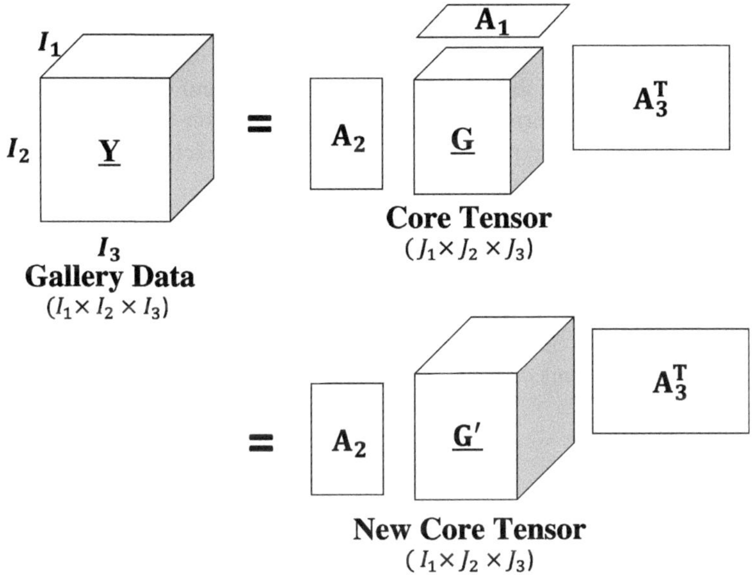

Fig. 3 Illustration of a tensor including the factors of grasping motion and the factors of individuals

$$\left(\mathbf{b}_1\ \mathbf{b}_2 \ldots \mathbf{b}_k\right) = \acute{\underline{G}} \times_3 \mathbf{p}^{\mathrm{T}}, \tag{5}$$

$$\mathbf{c} = \left(\mathbf{b}_1\ \mathbf{b}_2 \ldots \mathbf{b}_k\right)^+ \times \mathbf{q}_h, \tag{6}$$

where $\mathbf{q}_h$ is the vector of the joint angles of a human. As noted in the Appendix, the first synergy contains the greatest amount of information among the synergies because the multi-factor model provides the bases in the order of importance, like to the SVD algorithm [11, 12]. For example, if the first two synergies are selected to represent the posture, the approximated joint angles of the human hand $\hat{\mathbf{q}}_\mathbf{h}$ are represented as

$$\hat{\mathbf{q}}_\mathbf{h} = \left(\mathbf{b}_1\ \mathbf{b}_2\ 0 \ldots 0\right) \times \mathbf{c}. \tag{7}$$

2.2 Synergy-Level Controller for Grasping Motion and Force

The synergy-level controller generates the motion and the force of the robot by matching the synergies of the human to those of the robot. The synergies of the user and the robot differ due to the level of kinematic dissimilarity. However, the corresponding synergies of the robot hand can be computed by the assumption that the coefficients of the synergies of the human hand are identical to those of the robot hand. Thus, the synergies of the robot $\mathbf{b}_{r,i}$ are computed by

$$\left(\mathbf{b}_{r,1}\ \mathbf{b}_{r,2}\ \ldots\ \mathbf{b}_{r,k}\right) = \left(\mathbf{q}_{r,1}\ \mathbf{q}_{r,2}\ \ldots\ \mathbf{q}_{r,k}\right) \times \mathbf{c}^{+}, \tag{8}$$

where, $\mathbf{q}_{r,i}$ is the vector of the joint angles of the robot hand when the robot hand is configured to match the corresponding human hand motion for each grasping type, and where (+) refers to the pseudo-inverse. The size of each synergy vector $\mathbf{b}_{r,i}$ is equal to the number of joints of the robot hand.

Next, when the robot grasps an object, the grasping force of the robot can be generated by increasing the coefficients of the synergies by the amount of the preprocessed EMG signal $V(t)$ which is assumed to be proportional to the intention of the user with regard to their grasping force. Also, it should be noted that the coefficients of the synergies vary at approximately the same ratio during one type of grasping motion. Therefore, the coefficients of the robot hand synergies, $\mathbf{c}_r$, are computed by

$$\mathbf{c}_r = \mathbf{c} + K \times V(t) \times \frac{\dot{\mathbf{c}}}{|\dot{\mathbf{c}}|}, \tag{9}$$

where K is a scalar component for gain and the vector $\dot{\mathbf{c}}$ is the rate of change for each coefficient. The grasping posture is maintained while the grasping force is created using (9) and (10).

$$\mathbf{q}_r = \left(\mathbf{b}_{r,1}\ \mathbf{b}_{r,2}\ \ldots\ \mathbf{b}_{r,k}\right) \times \mathbf{c}_r \tag{10}$$

3 Experiment

The proposed algorithm is validated through experiments. On the master side, a motion capture system is used to track a human hand motion. The physical hardware of the robot hand is used as a slave system. Finally, between the master system and the slave system, there is a communication line (TCP/IP) to transfer the human hand motion information. The subsections below describe the details of the system configuration and the data sets.

3.1 System Overview

The human hand motion is tracked by a motion-capture system which uses the software *NEXUS (Vicon, Co. USA)*, twenty-four 5 mm markers (Fig. 4a), and fourteen cameras *(Vicon T160 Camera)*. The frame rate was 100 frames per second. On the other hand, the EMG sensor is placed on the extensor digitorum communis (the forearm) of the user. This surface EMG system, *Trigo (Delsys, Co. USA)* is used to obtain the EMG signal. The EMG signal is sampled at 1000 Hz. During the preprocessing step, we used a linear envelope method to rectify the amplitude of the signal [13]. To do this, we used a fourth-order Butterworth IIR filter. The cutoff frequency of the high-pass filter was 30 Hz and the cutoff frequency of the low-pass

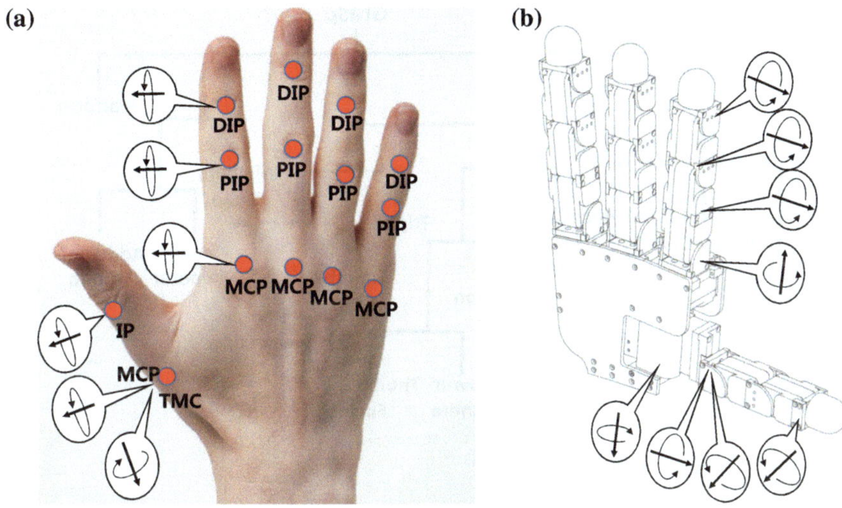

Fig. 4 The comparison of kinematic structure between the human hand and the robot hand: **a** the joint location of hand; **b** the joint location of robot hand

filter was 1 Hz. Finally, we experimentally determined the threshold of the EMG signal so that the noise caused by the movement of the hand could be ignored.

In the slave system, a fully actuated four-fingered robot hand, the *Allegro hand* [14], is used to achieve dexterous manipulation. The appearance and joint location of the *Allegro hand* are described in Fig. 4b.

3.2 Data Sets of Human Grasping

According to the grasp taxonomy, there are fifteen grasping types [15] (Fig. 5). In our experiments, each grasping type (except for the platform) was collected twice. Thus, our data sets for human grasping were collected from twenty-eight different grasping motions. In order to generate the multi-factor model, the data sets of five subjects (five men) are collected by the motion-capture system. The average age and hand length, defined as the distance from the tip of the middle finger to the midpoint interstylon line, of these participants were 27.3 ± 1.45 years and 18.1 ± 2.14 cm, respectively. Finally, each grasping motion consisted of 100 frames through Dynamic Time Warping (DTW) [16]. Thus, the dimensions of the tensor for the gallery were $\mathbb{R}^{15\times2800\times5}$ in our experiments.

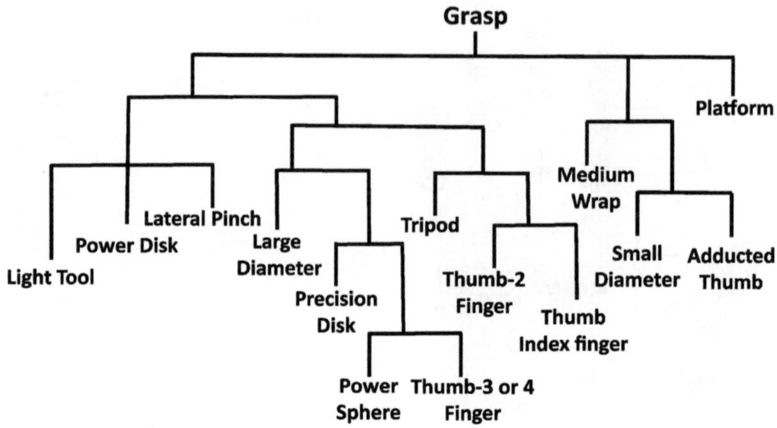

Fig. 5 Grasp taxonomy according to Dai [15]

4 Results

In this section, we experimentally validate the proposed algorithms for extracting postural synergies and for generating the grasping force of the robot hand.

4.1 Extracting Postural Synergies with the Multi-factor Model

We validate the proposed algorithm for synergy compared to other synergy-extraction methods, in this case Principal Component Analysis (PCA) and Robust Principal Component Analysis (RPCA) [17, 18]. PCA is one of the most widely used algorithm to extract synergies [5–9], and RPCA is a modified version which is robust with respect to the different tendencies of people when used for training. Also, it is fair to compare with linear methods and our proposed algorithm, as these both algorithms are linear methods which reduce the dimensionality.

We measured the angular difference of each joint between the actual posture and the reconstructed posture to compare how well each algorithm reconstructs the human hand motion using two or three synergies.

Figure 6 shows the mean absolute error (MAE) between the hand posture and the reconstructed posture with two or three synergies for people who participated in the training for the gallery data. Thus, the gallery data set for each algorithm contained the motion data of the users. As shown in Fig. 6, the proposed algorithm produced a much closer posture to the hand motion than the other methods. This is because the proposed algorithm extracts the postural synergies for each training user with tensor representation.

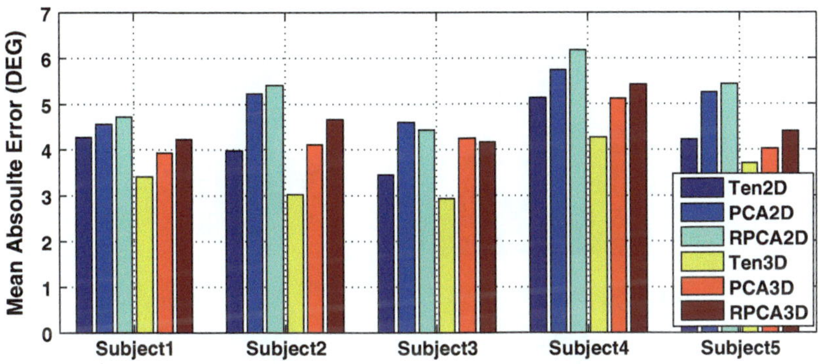

Fig. 6 The mean absolute error between the hand posture and the reconstructed posture with two and three synergies of the subjects who participated in the training

Fig. 7 The mean absolute error between the hand posture and the reconstructed posture with two and three synergies of the subjects who did not participate in the training

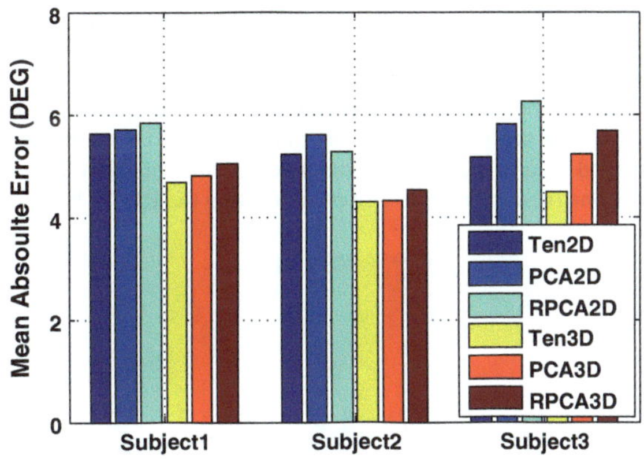

Figure 7 shows the MAE result of the extracted postural synergies for people who did not participate in the training. Although the gallery data did not contain the motion data of the users, better hand motions were represented by the proposed algorithm. This is because the proposed algorithm accounts for the characteristic of the users. The synergies for the user are calculated by the linear combination of the synergies of all training of the people in the multi-factor model and using the factor of the user $\mathbf{p}$.

4.2 Synergy-Level Controller for Grasping Motion and Force

In this section, the result of the hand motion mapping to the robot hand is presented by means of a simulation and experiments. First, we demonstrated that the robot can imitate human motions during a simulation using the physics-based simulation software *RoboticsLab* [19]. Figure 8 shows the configurations of the robot hand when

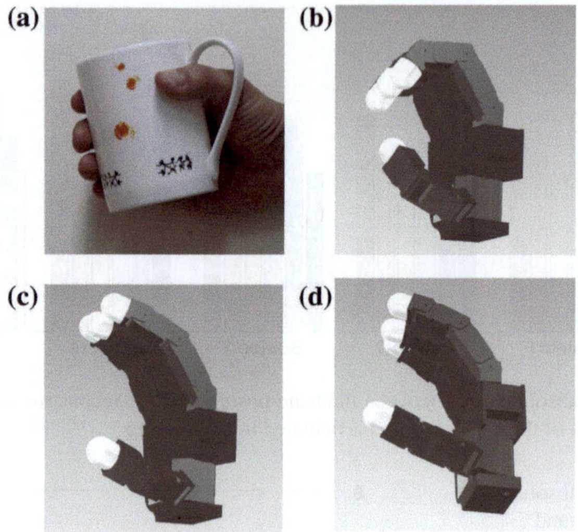

Fig. 8 The comparison of postures when the operator grasps the drinking cup (The power disk grasp); **a** the posture of human hand, **b** the reconstructed posture of robot hand using the proposed algorithm, **c** the reconstructed posture of robot hand using the PCA-based algorithm, **d** the reconstructed posture of robot hand using the RPCA-based algorithm

Fig. 9 The comparison of postures when the operator grasps pen (thumb-index grasp); **a** the proposed algorithm, **b** PCA, **c** RPCA

the operator grasped the drinking cup using the proposed algorithm, the PCA-based algorithm, and the RPCA-based algorithm. The proposed algorithm showed a grasping motion that was more similar to the original hand motion used in grasping the drinking cup than the other algorithms.

We also demonstrated the performance of the proposed algorithm when the robot hand tracks a human hand motion in real-time using the synergy-level controller. Figure 9 shows an image taken during the grasping of the tennis ball using the first two synergies by the proposed algorithm, by the PCA algorithm, and by the RPCA algorithm. First, the posture of the robot hand using the proposed algorithm was much more similar to that of the operator than the posture of the robot using other algorithms when a thumb-index grasp was performed. Also, the ball was successfully grasped using only the proposed algorithm.

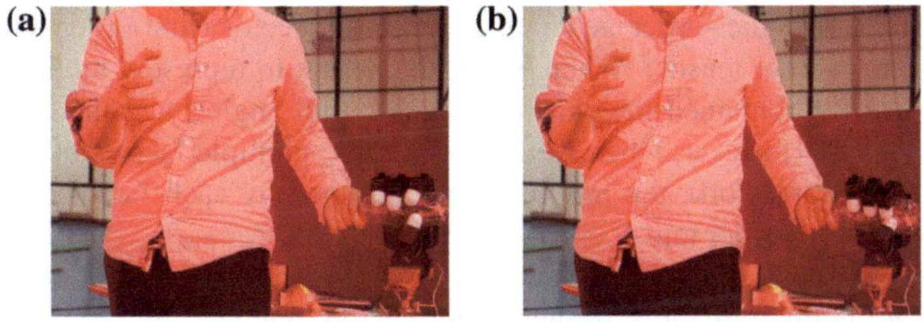

Fig. 10 The comparison of the robot grasping; **a** without the EMG signal, **b** with the EMG signal

Second, the performance of the proposed algorithm with an EMG signal was also demonstrated through experiments. As shown in Fig. 10, the robot hand produced grasping force due to the EMG signal while the posture of the human hand did not change.

5 Conclusion

In this paper, a novel synergy-based mapping algorithm for a robot hand which uses a multi-factor model and an EMG signal is proposed. The main experimental points of insight are summarized as follows. First, we extracted synergies that represent human hand motion more accurately by accounting for the characteristics of individuals. Second, we demonstrated that the synergies of a new user could be obtained simply by extracting the factor pertaining to the new user. Third, the posture of the robot hand using the proposed algorithm was more similar to the human hand than that of using other algorithms. This therefore enabled dexterous manipulation through tele-operation. Finally, the grasping force of the robot can be generated by using an EMG sensor. Future works will involve extending the workspace by integrating the robot hand with an arm, and simplifying the system by replacing the motion-capture system with other compact and practical solutions. Also, an algorithm to extract the grasping force of a human by means of EMG will be investigated.

Acknowledgments This work was supported by the Global Frontier R&D Program on Human-centered Interaction for Coexistence through the National Research Foundation of Korea (NRFM1 AXA003-2011-0032014).

Appendix: Tensor Representation

This section is a condensation of [11, 12]. For details of tensor, please read the references.

A tensor is a higher-order generalization of the vector (first-order tensor) and matrix (second-order tensor). When we used a matrix to represent motion data set, the rows usually contained the channels for the joint angles, and the columns for motion samples. However, when we considered the multiple factors of human and used a tensor framework, the motions were grouped by each factor so that they constituted an Nth-order tensor.

The order of tensor $\underline{Y} \in \mathbb{R}^{I_1 \times I_2 \times \ldots \times I_N}$ is N. The mode-n vectors of an Nth-order tensor $\underline{Y}$ are defined as the I_n-dimensional vectors obtained by varying index I_n while keeping the other indices fixed. All mode-n vectors can be arranged together as column vectors to compose a mode-n flattening matrix $\mathbf{Y}_\mathrm{n} \in \mathbb{R}^{I_n \times (I_1 I_2, \ldots, I_{n-1} I_{n+1}, \ldots, I_n)}$. The I_n-dimensional vectors of $\mathbf{Y}_\mathrm{n}$ are obtained from tensor $\underline{Y}$ by varying index I_n while keeping other indices fixed.

The multiplication of a high-order tensor $\underline{Y} \in \mathbb{R}^{I_1 \times I_2 \times \ldots \times I_N}$ by a matrix $\mathbf{A} \in \mathbb{R}^{J_n \times I_n}$ is a mode-n product of tensor $\underline{Y}$ by $\mathbf{A}$, which is denoted as $\underline{Y} \times_n \mathbf{A}$. It can also be expressed in terms of flattened matrices. The entries of the product are given as

$$(\underline{Y} \times_n \mathbf{A})_{i_1 \ldots i_{n-1} j_n i_{n+1} \ldots i_n} = \sum_{i_n} d_{i_1 \ldots i_{n-1} i_n i_{n+1} \ldots i_n} a_{j_n i_n} \tag{11}$$

The tensor decomposition of $\underline{Y}$ seeks for N orthonormal mode matrices as Eq. (12), which is obtained by HOSVD.

$$\underline{Y} = \underline{G} \times_1 \mathbf{U}_1 \times_2 \mathbf{U}_2 \ldots \times_n \mathbf{U}_\mathrm{n} \tag{12}$$

The column vectors of $\mathbf{A}_\mathrm{n}$ are the orthonormal basis vectors of the mode-n unfolding matrix $\mathbf{Y}_\mathrm{n}$. Core tensor $\underline{G}$ governs the relationship among mode matrices $\mathbf{U}_\mathrm{n}$.

Consequently, tensor representation is helpful to treat multi-factorization problem, as matrix is able to decompose factor using non-negative matrix factorization such as PCA and NMF.

References

1. Hong, J., Hollerbach, J., Jacobsen, S.: Calibrating a VPL dataglove for teleoperating the Utah/MIT hand. In: IEEE International Conference on Robotics and Automation, pp. 1752–1755 (1989)
2. Rohling, R., Hollerbach, J.: Optimized fingertip mapping for teleoperation of dexterous robot hands. In: IEEE International Conference on Robotics and Automation, pp. 796–775 (1993)
3. Griffin, W., Findley, R., Turner, M., Cutkosky, M.: Calibration and mapping of a human hand for dexterous telemanipulation. In: ASME IMECE, Conference on Haptic Interfaces for Virtual Environments and Teleoperator System Symposium, pp. 1–8 (2000)
4. Ozawa, R., Ueda, N.: Supervisory control of a multi-fingered robotic hand system with data glove. In: IEEE/RSJ International Conference Intelligent Robots and System, pp. 1606–1611 (2007)
5. Santello, M., Flanders, M., Soechting, J.: Postural hand synergies for tool use. J. Neurosci. 10105–10115 (1998)

6. Ciocarlie, M.T., Allen, P.K.: Hand posture subspaces for dexterous robotic grasping. Int. J. Robot. Res. **28**(7), 851–867 (2009)
7. Matrone, G.C., Cipriani, C., Secco, E.L., Magenes, G., Carrozza, M.C.: Principal components analysis based control of a multi-dof underactuated prosthetic hand. J. NeuroEng. Rehabil. **7**(1) (2010)
8. Ajoudani, A., Godfrey, S.B., Catalano, M., Grioli, G., Tsagarakis, N.G., Bicchi, A.: Teleimpedance control of a synergy-driven anthropomorphic hand. In: IEEE/RSJ Intelligent Robots and System, pp. 1985–1991 (2013)
9. Romero, J., Feix, T., EK, C.H., Kjeelstrom, H.: Extracting postural synergies for robotic grasping. IEEE Trans. Robot. **29**(6), 1342–1352 (2013)
10. Pedersen, D.R., Arora, J.S., Brand, R.A., Cheng, C.: Direct comparison of muscle force predictionjs using linear and nonlinear programming. J. Biomed. Eng. **109**(3), 192–199 (1987)
11. Cichocki, A., Zdunek, R., Phan, A.H., Amari, S.: Nonnegative Matrix and Tensor Factorizations: Applications to Exploratory Multi-way Data Analysis and Blind Source Separation. Wiley, New York (2009)
12. Kolda, T.G., Brett, W.: Tensor decompositions and applications. SIAM Rev. **51**,455–500 (2009)
13. Konrad, P.: The abc of EMG. A practical introduction to kinesiological electromyography, 1 (2005)
14. SimLab.: Allegro Hand is a low-cost and highly adaptive robotic hand. http://www.simlab.co.kr
15. Dai, W., Sun, Y., Qian, X.: Functional analysis of grasping motion. In: IEEE/RSJ Intelligent Robots and System, pp. 3507–3513 (2013)
16. Berndt, D.J., Clifford, J.: Using dynamic time warping to find patterns in time series. KDD Workshop **10**(16), 359–370 (1994)
17. Jolliffe, I.: Principal Component Analysis. Wiley Online Library (2005)
18. Candes, E.J., Li, X.: Robust principal component analysis? J. ACM **58**(3) (2011)
19. Simlab. http://www.rlab.co.kr

Muscular Effort for the Characterization of Human Postural Behaviors

Emel Demircan, Akihiko Murai, Oussama Khatib and Yoshihiko Nakamura

Abstract The human selection of specific postures among the infinity of possibilities is the result of a long and complex process of learning. Through learning, humans seem to come to discover the properties of their bodies and how best to put them to use when performing a task. Exploiting the body's kinematic characteristics, humans effectively use the body's mechanical advantage to improve the transmission of the muscles' tension into the forces the task requires. However, the efficiency of this transmission is also affected by the human muscle actuation physiology and dynamics. By also adjusting the body configurations to maximize this transmission of muscle tensions to resulting task forces, humans are in fact exploiting what can be termed the physiomechanical advantage of their musculoskeletal system. Here, we investigate the physiomechanical advantage of humans through several experimental validations. Based on the results of the analysis, we conclude that in learned tasks the optimization of the physiomechanical advantage corresponds to the overall minimization of the human muscular effort. The approach presented here can be applied for the motion control of human musculoskeletal models where the control is task-driven and the task consistent postures are driven by the muscular criteria.

Keywords Muscular effort · Motion analysis · Postural behaviors

1 Introduction

1.1 Related Work

The ability of humans to move and coordinate their limbs in the performance of common tasks is remarkable. When holding a heavy object or applying a force to the environment through a tool, the arms and body of a skillful human are configured

E. Demircan (✉) · A. Murai · O. Khatib · Y. Nakamura
Department of Mechano-Informatics, The University of Tokyo, Tokyo, Japan
e-mail: emeld@stanford.edu

E. Demircan · A. Murai · O. Khatib · Y. Nakamura
Department of Computer Science, Stanford University, Stanford, USA

M.A. Hsieh et al. (eds.), *Experimental Robotics*, Springer Tracts in Advanced Robotics 109, DOI 10.1007/978-3-319-23778-7_45

in the most effective fashion for the task. Mathematical models have proven to be valuable tools for motor control prediction [1, 2] and for predicting the kinematically redundant body motion [3]. These models frequently characterize some element of musculoskeletal effort. Robotics-based effort models [4] frequently utilize quantities that are derivable purely from skeletal kinematics and that are not specific to muscle actuation and dynamics. It is thus useful to consider an analogous measure that encodes information about the overall musculoskeletal system to account for muscle actuation and its redundancy. Activation, which represents the normalized exertion of muscles, provides a natural starting point for constructing such a measure. The magnitude of muscle activation vector has been used as an optimization criterion in both static and dynamic optimizations [5].

1.2 Motivation

Recently, our effort on human motion analysis have resulted in the robotics-based synthesis of human motion [6]. Our hypothesis is that by exploiting the body's kinematic characteristics, humans effectively use the body's mechanical advantage to improve the transmission of the muscles' tension into the forces the task requires. However, the efficiency of this transmission is also affected by the human muscle actuation physiology and dynamics. By also adjusting the body configurations to maximize this transmission of muscle tensions to resulting task forces, humans are in fact exploiting what can be termed the physiomechanical advantage of their musculoskeletal system. Here, we investigate the physiomechanical advantage of human postural behaviors through several experimental validations. We speculate that in learned tasks the optimization of the physiomechanical advantage corresponds simply to the overall minimization of the human muscular effort. By validating this criteria through natural human motions, we aim at using it for the real-time posture control of human and humanoid models.

2 Muscular Effort Criteria

Robotics-based effort models frequently utilize quantities that are derivable purely from skeletal kinematics and that are not specific to muscle actuation and dynamics. It is thus useful to consider an analogous measure that encodes information about the overall musculoskeletal system to account for muscle actuation and its redundancy. Since human motions are frequently linked by physiological capacities, the use of a model-based characterization of muscle systems that accounts for muscle kinematic and strength properties is critical to simulating human motion. A strategy that employs mechanical advantage to produce a posture and task force corresponds to a minimization of muscular effort (Fig. 1). While in static poses, the muscles produce joint torques to act against the gravitational torques, in dynamic skills, inertial forces are part of the effort and are taken into account accordingly.

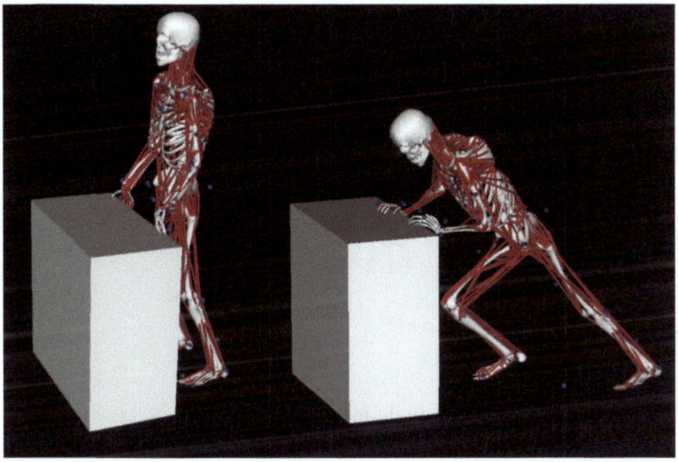

Fig. 1 The hypothesis is that in learned tasks humans minimize a criterion associated with the muscular effort. In order to push a heavy object, a human tries to best position his body to maximize the *physiomechanical advantage*

2.1 Muscular Effort Formulation

Musculoskeletal kinematics provide us the basics for understanding the effect of the musculoskeletal geometry on muscle function and for establishing the relationship between muscle forces and resulting joint moments. Here, we define the muscle Jacobian of humen musculoskeletel system and present the muscle/task relationship for task-oriented characterization of human motion. For a human musculoskeletal system of n degrees of freedom and r muscles, a set of muscle forces, m, arises based on muscle activations, a, as well as the skeletal configuration, q and $\dot{q}$. In this model, all musculo-tendon lengths, l, can be uniquely determined from the joint angle, q, and differential changes, dl, are given by:

$$dl = L(q)dq. \tag{1}$$

Here, L is the muscle Jacobian representing the muscle moment arms (Fig. 2). The muscle-induced joint torques, Γ, can be given using the muscle forces, m, and the task forces, F, by the relationship:

$$\Gamma = J^T(q)F = L^T(q)m. \tag{2}$$

In order to investigate the muscular effort in terms of the musculoskeletal parameters, we introduce the function, $\Phi(q)$, to represent the physiomechanics of the human musculoskeletal system including the Jacobian, $J(q)$, the muscle Jacobian,

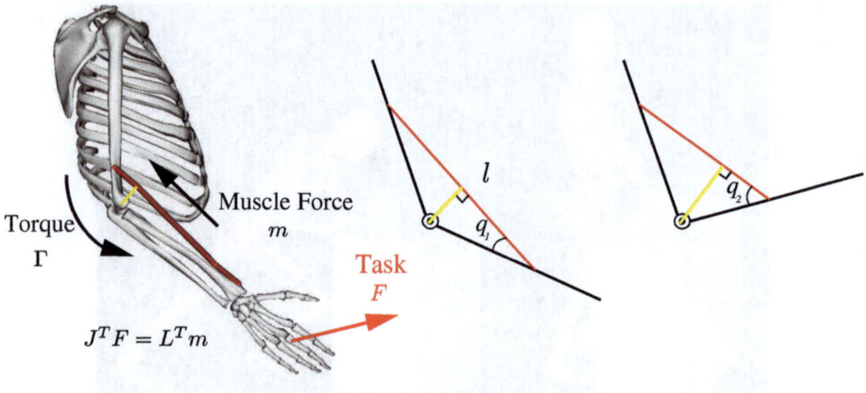

Fig. 2 Muscle/task relationship: the muscle-induced joint torques, Γ, are dependent on the force production, m, as well as the moment arm of the muscle, defined as the perpendicular distance (*yellow line*) from point of force application to the axis of rotation. Different skeletal configurations (i.e., q_1 and q_2) would result in different muscle Jacobians, $L(q)$ (figure adapted from [7])

$L(q)$, and the muscle force-generating capacity, C (i.e., peak isometric forces). The physiomechanical advantage function, $\Phi(q)$, is defined as:

$$\Phi = J(q)(L^T(q)C^2L(q))^{-1}J^T(q), \tag{3}$$

and captures the spacial characterization of the muscular effort measure by connecting the muscle physiology to the resulting task through the Jacobian.

Using the physiomechanical advantage function, $\Phi(q)$, for task-based effort criteria, and the generalized operational space forces, F, for the resulting task requirements, the overall effort function can be written in the form:

$$E = F^T \Phi(q) F. \tag{4}$$

2.2 Dynamic Consistency with Contacts

For k contacts with the environment, we form the contact Jacobian by concatenating all the Jacobians for each contact:

$$J_c(q) = \begin{bmatrix} J_{c_1} \\ J_{c_2} \\ \cdot \\ \cdot \\ \cdot \\ J_{c_k} \end{bmatrix}. \tag{5}$$

Similarly, for m tasks to be performed by the musculoskeletal system, we form the task Jacobian by concatenating all the Jacobians for each task:

$$J_t(q) = \begin{bmatrix} J_{t_1} \\ J_{t_2} \\ \cdot \\ \cdot \\ \cdot \\ J_{t_m} \end{bmatrix}. \tag{6}$$

In the presence of supporting contacts, we use the contact consistent Jacobian of the task,

$$J_{t|c}(q) = J_t(q)N_c(q), \tag{7}$$

as defined in [8] and the dynamically consistent null-space matrix associated with $J_c(q)$,

$$N_c(q) \triangleq I - \bar{J}_c(q)J_c(q), \tag{8}$$

established by [9].

Here, the subscript $t|c$ indicates that the task point is consistent with the contacts. The range of the Jacobian, $J_{t|c}(q)$, is the instantaneous space of task motion that is consistent with the contacts. $\bar{J}_c(q)$ is the generalized inverse [9] that ensures the dynamic consistency between the contacts and the tasks, is unique and is given by,

$$\bar{J}_c(q) = A_c^{-1}(q)J_c^T(q)\Lambda(q), \tag{9}$$

where

$$\Lambda(q) = (J_c(q)A_c^{-1}(q)J_c^T(q))^{-1}, \tag{10}$$

is the operational space kinetic energy matrix [10].

Thus, Eq. (3) becomes:

$$\Phi = J_{t|c}(q)(L^T(q)C^2L(q))^{-1}J_{t|c}^T(q). \tag{11}$$

In our case, the left/right foot and the left/right hand represent the contact and the task points, respectively. The task Jacobian and the contact Jacobian can be given, respectively, by:

$$J_t(q) = \begin{bmatrix} J_{RightHand} \\ J_{LeftHand} \end{bmatrix}, J_c(q) = \begin{bmatrix} J_{RightFoot} \\ J_{LeftFoot} \end{bmatrix}. \tag{12}$$

3 Musculoskeletal Models and Experiments

3.1 Musculoskeletal Models

The musculoskeletal model used in this work is derived from upper [11] and lower [12] body models. The skeletal part of the model is represented by rigid bodies or bone segments. A reference frame is attached to each body segment and inertial parameters of the body segment are expressed in this reference frame. The upper body's kinematics contain 14 degrees-of-freedom (DOFs), which represent the shoulder, elbow, forearm, wrist, and hand. The lower body's kinematics contain 17 DOFs, which represent the hip, knee, ankle, subtalar, and metatarsophalangeal joints. The arms-torso, torso-pelvis, and pelvis-leg joints are represented by ball-and-socket joints. The remaining joints are revolute. The hip is modeled as a ball-and-socket joint, the knee is modeled as a custom joint with one DOF [13], and the foot and ankle are modeled as a custom joint with two DOFs (i.e., ankle dorsi-/plantar flexion, tarsal eversion/inversion). Lumbar motion is modeled as a ball-and-socket joint [14]. The shoulder is modeled as a ball-and-socket joint, the elbow is modeled with a revolute joint and the wrist is modeled with a custom joint with three DOFs (i.e., flexion/extension, ulnar/radial deviations, pronation/supination). Muscles span the joints and generate forces and movement. Muscle line of action is defined by the origin, insertion, and via points connected to the rigid bodies. The contraction dynamics of each musculo-tendon units is modeled using a conventional Hill-type phenomenological model of muscle [15]. In our musculoskeletal model, the upper extremity, lower extremity, and back joints are actuated by 118 musculotendon actuators [12, 14]. The functional groups of muscles used in the whole-body effort characterization of pushing are given in Table 1. The generic model was scaled based on the measurement-based scaling [16] to match subjects anthropometry based on experimentally measured markers placed on anatomical landmarks. A virtual marker set is placed on the unscaled model based on these anatomical landmarks. The scale factors for a body segment are determined by comparing distance measurements between the virtual markers and the corresponding experimental marker positions. The marker locations are obtained using motion capture equipment. Before each motion capture experiment, a static trial is performed on the subject to assist scaling the musculoskeletal model (for example, with markers attached to the medial and lateral femoral epicondyles and medial and lateral malleoli). Then, each body segment of the model is scaled by measuring the positions of two experimental markers on that body segment on the subject, and the size of the body segment is uniformly adjusted so the virtual markers are co-located with the experimental markers. The mass of the model is scaled by proportionally adjusting the mass of each body segment so the total mass of the model equals the measured mass of the subject.

Table 1 Functional groups of muscles [12, 14, 16] used in the whole-body muscular effort analysis

Shoulder adduction	Shoulder abduction
Coracobrachialis	Deltoid
Infraspinatus	Subscapularis
Latissimus dorsi	Trapezius
Pectoralis major	
Teres major	
Scapular retraction	Scapular elevation/depression
Trapezius	Levator scapulae
	Latissimus dorsi
	Trapezius
Arm flexion	Arm extension
Biceps brachii	Latissimus dorsi
Coracobrachialis	Teres major
Pectoralis major	Triceps minor
Hip adduction	Hip abduction
Adductor magnus	Gluteus medius
Hip flexion	Hip extension
Gluteus medius	Adductor magnus
	Biceps femoris
	Gluteus maximus
	Semimembranosus
Knee flexion	Knee extension
Biceps femoris	Vasti
Semimembranosus	Rectus femoris
Ankle dorsi-/plantar flexion	Ankle eversion/inversion
Extensor digitorum	Extensor digitorum
Extensor hallucis	Peroneus brevis
Peroneus tertius	Peroneus longus
Tibialis anterior/posterior	Peroneus tertius
Flexor digitorum	Extensor hallucis
Flexor hallucis	Flexor digitorum
Gastrocnemius	Flexor hallucis
Peroneus brevis/longus	Tibialis anterior
Soleus	Tibialis posterior
Trunk flexion/extension/rotation	
External oblique	
Internal oblique	
Erector spinae	

3.2 Experiments

Motion capture experiments were conducted on a healthy man pushing against a heavy object. The subject was instructed to push against the object with the highest possible force he could exert in a horizontal direction. After each pushing trial, the subject was asked to change his posture in order to push the object more comfortably. This was repeated for seven different configurations. The subject's motion was captured at 100 Hz using an eight-camera Motion Analysis [17] system. The ground reaction forces were captured at 4000 Hz with two Bertec force plates [18]. The pushing force was measured from the force sensors [19] placed between the object and the hands of the subject.

4 Musculoskeletal Simulations and Results

4.1 Belted Ellipsoid Representation of Effort Function

Robotics brought efficient algorithms and tools for the analysis and control of multi-degree of freedom redundant manipulators. The belted ellipsoid established by the author [20] is a geometric representation that characterizes the inertial properties perceived at a given position and an orientation of an end effector. Ellipsoid representations only provide a description of the square roots of effective mass (inertia) in (or about) a direction. A belted ellipsoid is defined as a geometric representation that characterizes the actual magnitude of these properties. A point on the ellipsoid represented by a vector v is transformed into a point on the belted ellipsoid represented by a vector w, where the vector w is collinear to v and is of magnitude equal to $v^T v$ [20]. In robotics, the belted ellipsoid representation [20] is used to assess the performance of a multi-degree of freedom manipulator by describing its effective mass along a desired direction of the end-effector motion. In illustrating human muscular effort geometrically, the belted ellipsoid representation can be used to efficiently describe the actual value of the effort along a desired direction. It provides an understanding of the muscular effort required to position the body or to perform a desired task.

4.2 Simulation and Analysis Results

Figure 3 shows the three-dimensional simulations of two extreme postures and five intermediate postures of the subject while pushing an object. Figure 4 illustrates two different postures used while pushing the object. The ground contact forces, F_c, shown with green arrows, are measured from the force plates and used for the calculation of the muscular effort. In Fig. 4, the musculoskeletal simulation of pushing while adopting an unnatural posture is shown in the left image with the belted ellipsoid

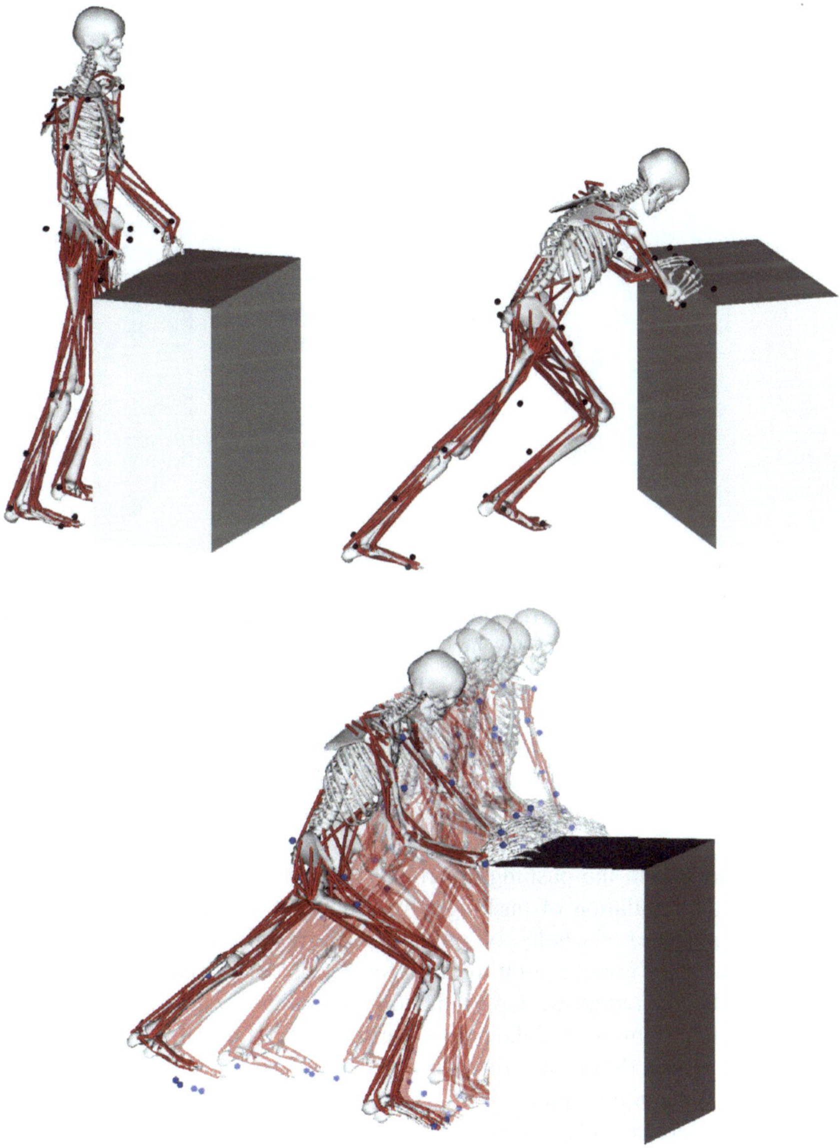

Fig. 3 Three-dimensional simulations of (*top*) two extreme postures (i.e., the most and the least effort), and (*bottom*) intermediate postures (i.e., decreasing effort from the *right* to the *left* postures) while pushing an object. Motion capture and force plate data were collected with a healthy male subject pushing against an object and the scaled full-body model was used to generate the pushing simulations

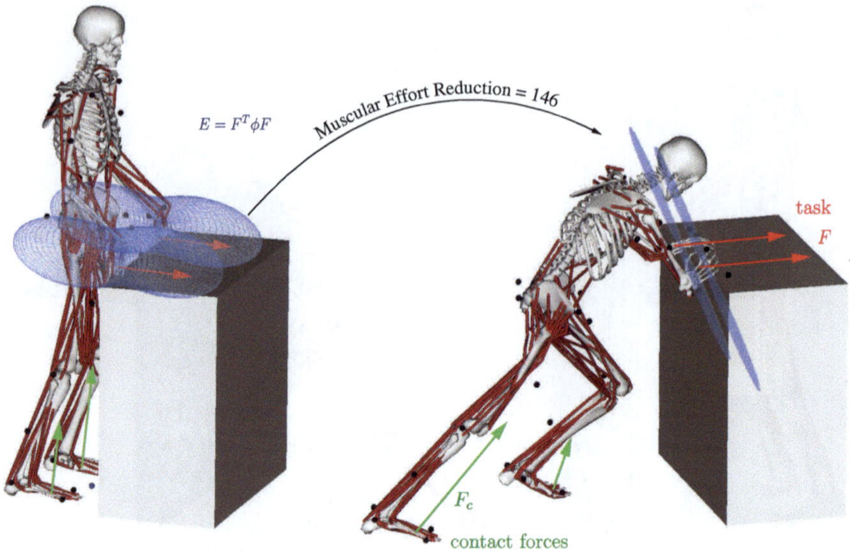

Fig. 4 Whole-body effort during pushing: the musculoskeletal simulations of pushing while adopting the least (*left*) and the most (*right*) comfortable pushing postures. The belted ellipsoid [20] (*blue*) of whole-body muscular effort is given with the directions of the task forces (*red arrows*), F, and the ground reaction forces (*green arrows*), F_c

(blue) of whole-body muscular effort reflected at both hands. The measured force at the contact point from the force sensor is 50.4 N in total for the left and right hands. The red arrow depicts the direction of the applied force, F, aligned with the maximum semi-axis of the belted ellipsoid. This shows that the whole-body effort is maximized in the pushing direction. The right image in Fig. 4 shows the musculoskeletal simulation of pushing while adopting a natural posture with the belted ellipsoid (blue) of whole-body muscular effort reflected at both hands. The measured force at the contact point from the force sensor is 379 N in total for the left and right hands. The red arrow depicts the direction of the applied force, F, aligned with the minimum semi-axis of the belted ellipsoid. This shows that the whole-body effort is minimized in the pushing direction. Between the two pushing postures, there is a 146 times reduction in muscular effort in the direction of pushing (i.e. task).

Figure 5a illustrates the change in whole-body effort of pushing versus the total pushing force (i.e., summation of horizontal forces applied from the left and right hands) measured from the force sensors and force plates, for seven different configurations. Figure 5b illustrates the total pushing force versus the pushing configuration. The total force of pushing is 50.4, 82.6, 106, 111, 121, 239.6, and 379 N for the configurations from one (i.e., most effort) to seven (i.e., least effort).

As more comfortable postures were adopted, the muscular effort decreased significantly with maximum reduction of 146, while the maximum force exerted on the object increased from 82.6 to 239.6 N at intermediate postures to reach its maximum, 379 N, for the posture with the least effort production.

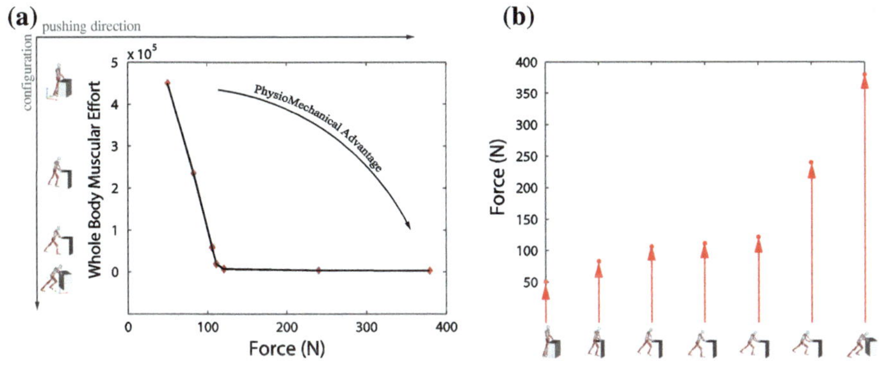

Fig. 5 **a** Whole-body muscular effort versus the pushing force **b** total pushing force versus the pushing configuration

5 Conclusion

A strategy that employs physiomechanical advantage to produce a posture and task force corresponds to a minimization of muscular effort. The whole-body muscle effort criterion was implemented to analyze several pushing postures. While in static poses, the muscles produce joint torques to act against the gravitational torques, in dynamic skills, inertial forces are part of the effort and are taken into account accordingly. In order to achieve dynamic consistency between the task and the contacts, the contact consistent Jacobian of the task was calculated and was included in the computation of the whole-body muscular effort. The results of the analysis showed that as the subject adopted more comfortable postures to push against an object, the transmission from the ground reaction forces and the weight of the subject to the resulting force increased significantly. This optimal transmission resulted in a whole-body effort reduction of 146, benefiting from the skeletal configuration as well as from the muscle kinematics. The minimization of the whole-body effort is a criterion associated with natural human motion and can be used to contitute the desired musculoskeletal posture to be achieved by the controller. Our approach for the characterization of human muscular effort involves scaling a musculoskeletal model to match an individual's anthropometry. It provides subject-specific muscle and joint parameters, such as musculotendon lengths, moment arms, lines of action, and joint topology. As such, this technique inherently accounts for differences between individuals stemming from differences in body size. The technique would, therefore, predict that subjects of different stature would perform the same task (i.e., pushing with maximum force) with slightly different joint kinematics.

Acknowledgments The authors thank Jessica Rose, Kristine Bruce and Gerald Brantner for their help with the motion capture experiments. The data collection was done at Lucile Packard Children Gait Hospital at Stanford University. This work was supported by the Global Creative Leaders (GCL) program of the University of Tokyo Graduate School of Engineering.

References

1. Hermens, F., Gielen, S.: Posture-based or trajectory-based movement planning: a comparison of direct and indirect pointing movements. Exp. Brain Res. **159**(3), 304–348 (2004)
2. Vetter, P., Flash, T., Wolpert, D.M.: Planning movements in a simple redundant task. Curr. Biol. **12**(6), 488–491 (2002)
3. DeSapio, V., Warren, J., Khatib, O.: Predicting reaching postures using a kinematically constrained shoulder model. In: Proceedings of the Tenth International Symposium Advances in Robot Kinematics, pp. 209–218. Heidelberg, Germany (2006)
4. Khatib, O., Warren, J., DeSapio, V., Sentis, L.: Human-like motion from physiologically-based potential energies. In: Lenarčič, J., Galletti, C. (eds.) On Advances in Robot Kinematics, pp. 149–163 (2004)
5. Thelen, D.G., Anderson, F.C., Delp, S.L.: Generating dynamic simulations of movement using computed muscle control. J. Biomech. **36**, 321–332 (2003)
6. Khatib, O., Demircan, E., DeSapio, V., Sentis, L., Besier, T., Delp, S.: Robotics-based synthesis of human motion. J. Physiol. Paris. **103**(3–5), 211–219 (2009)
7. Lieber, R.L.: Skeletal Muscle Structure, Function, and Plasticity, 2nd edn. Lippincott Williams and Wilkins (2002)
8. Park, J., Khatib, O.: Multi-link multi-vontact force control for manipulators. In: IEEE International Conference on Robotics and Automation, pp. 3613–3618 (2005)
9. Khatib, O.: Motion/force redundancy of manipulators. In: Proceedings of the Japan-USA Symposium on Flexible Automation, pp. 337–342, Kyoto, Japan, July 1990
10. Khatib, O.: A unified approach for motion and force control of robot manipulators: the operational space formulation. Int. J. Robot. Autom. **3**(1), 43–53 (1987)
11. Holzbaur, K.R., Murray, W.M., Delp, S.L.: A model of the upper extremity for simulating musculoskeletal surgery and analyzing neuromuscular control. Exp. Brain Res. **33**(6), 829–840 (2005)
12. Delp, S.L., Loan, P., Hoy, M.G., Zajac, F.E., Topp, E.L., Rosen, J.M.: An interactive graphics-based model of the lower extremity to study orthopaedic surgical procedures. IEEE Trans. Biomed. Eng. **37**(8), 757–767 (1990)
13. Seth, A., Sherman, M., Eastman, P., Delp, S.L.: Minimal formulation of joint motion for biomechanisms. Nonlinear Dyn. **62**(1), 291–303 (2010)
14. Anderson, F.C., Pandy, M.G.: A dynamic optimization solution for vertical jumping in three dimensions. Comput. Methods Biomech. Biomed. Eng. **2**(3), 201–231 (1999)
15. Hill, A.V.: The heat of shortening and dynamics constants of muscles. Proc. R. Soc. Lond. B (London: Royal Society) **126**(843), 136–195 (1938)
16. Delp, S.L., Anderson, F.C., Arnold, A.S., Loan, P., Habib, A., John, C.T., Guendelman, E., Thelen, D.G.: OpenSim: open-source software to create and analyze dynamic simulations of movement. IEEE Trans. Biomed. Eng. **55**, 1940–1950 (2007)
17. Motion Analysis, Santa Rosa, CA, USA. http://motionanalysis.com/
18. Bertec, Columbus, OH, USA. http://bertec.com/
19. Tekscan Inc., South Boston, MA, USA. http://www.tekscan.com/flexible-force-sensors/
20. Khatib, O.: Inertial properties in robotics manipulation: an object-level framework. Int. J. Robot. Res. **14**(1), 19–36 (1995)

Object Modeling and Recognition from Sparse, Noisy Data via Voxel Depth Carving

Matthew Klingensmith, Martin Herrmann and Siddhartha S. Srinivasa

Abstract In this work, we make the case for using volumetric information for shape reconstruction and recognition from noisy depth images for robotic manipulation. We provide an efficient algorithm, Voxel Depth Carving (a variant of Occupancy Grid Mapping) which accomplishes this goal. Real-world experiments with lasers, RGB-D cameras, and simulated sensors in both 2D and 3D verify the effectiveness of our algorithm in comparison to traditional point-cloud based methods.

1 Introduction

3D sensors are cheaper and more readily available than ever before. Commercial depth sensors, such as Microsoft's *Kinect* or the Asus *Xtion Pro* provide inexpensive, low-latency, colored depth data. However, this comes at the expense of significant noise and missing data [21]. The rise of cheap 3D sensing presents the challenge of using noisy, incomplete depth data for robotic manipulation. Here, we focus on two key perception challenges: *shape reconstruction* and *object recognition*, which are necessary for robotic manipulation of everyday objects.

Many previous works on object recognition [13, 16, 25] and reconstruction [11, 18, 20, 26] use a point cloud, a set of 3D points which encode data from the sensor. Unfortunately, point clouds throw away important information encoded in depth images: locality of adjacent depth pixels, the implicit *ray* from the focal point of the depth camera to each point in the cloud, and, importantly, the implicit *volumetric* information implied by the ray passing through empty space from the camera to the scene. Consequently, corruption from noise and missing data can severely distort a point cloud, reducing its usefulness.

In contrast, researchers in the mobile robot navigation community have long used occupancy grids [5] and ray clouds rather than point clouds to overcome the problem of noisy, sparse data. Occupancy grids provide a natural way to reason about the uncertainty of noisy sensors by incorporating ray noise models into the occupancy

M. Klingensmith (✉) · M. Herrmann · S.S. Srinivasa
The Robotics Institute, Carnegie Mellon University, Pittsburgh, USA
e-mail: mklingen89@gmail.com

M.A. Hsieh et al. (eds.), *Experimental Robotics*, Springer Tracts in Advanced Robotics 109, DOI 10.1007/978-3-319-23778-7_46

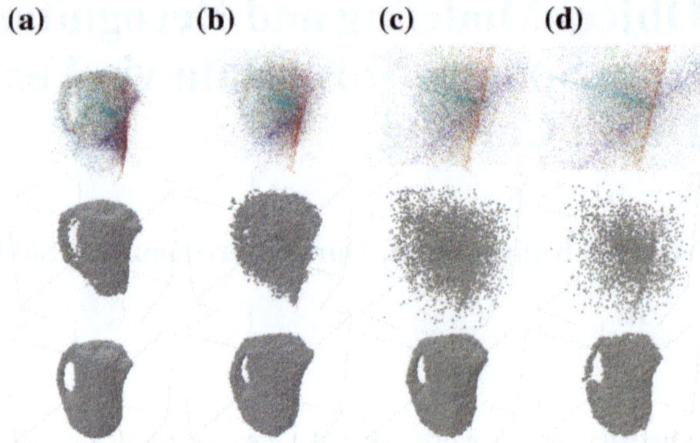

Fig. 1 The effect of noisy data: the *top row* shows the raw point clouds. The *middle row* shows the result of the reconstruction using hit data only. The *bottom row* shows the result of our method. We vary σ_z, the standard deviation of noise on the depth sampled from $\mathcal{N}(0, \sigma_z)$ from 0 to 15 cm. **a** 0 cm, **b** 5 cm, **c** 10 cm, **d** 15 cm

update probabilities. By applying these techniques from occupancy grid mapping to object recognition and reconstruction, our work aims to exploit and recover the implicit volumetric information encoded in noisy, incomplete depth images; while at the same time being faster than typical occupancy grid mapping techniques in our domain. As a result, we are able to reconstruct and recognize objects even in extremely noisy conditions—where the corresponding point cloud is so distorted as to be unrecognizable (Fig. 1).

However, directly applying occupancy grid mapping to the domain of recognizing and reconstructing common objects presents many challenges. Traditional occupancy grid mapping techniques [5] assume laser-like or sonar-like sensors carving out grid cells in spaces much larger than the robot, wheras we are concerned with very dense depth scans covering a much smaller space at higher resolution. In our domain, ray rasterization is slowed by the extreme number of rays, and results in artifacts around the fine details of objects. To solve this problem, we iterate over a fixed set of voxels, rather than over rays. We use projection and interpolation rather than rasterization to carve out space much more quickly and conservatively. We call this technique *Voxel Depth Carving*. Recognizing objects in a noisy occupancy grid also presents difficulties. We are unable to use typical surface descriptors or point descriptors used in 2D images or point clouds, and instead we must use volumetric descriptors to recognize objects. In this work, we show that simple affine-invariant geometric moments can sufficiently recognize objects from a database of hundreds of candidates when only noisy volumetric information is given. Further, in our domain, the majority of the space around objects may be dominated by occlusion, and remains "unknown." To deal with unknown space from occlusions, we construct a Markov Random Field with strong structural priors to make assumptions about the space behind objects. Doing so gives us much better object recognition performance when the number of views of the object is small.

2 Related Works

Despite extensive research, object recognition and reconstruction remains challenging in both the 2D [24] and 3D [10] cases. We are interested in a more specific subset of the problem where multiple, registered viewpoints are considered.

Shape Reconstruction When only 2D data is given, silhouette information may be used in the form of a *visual hull* [17]. However, generating silhouettes is not always feasible. Algorithms that incorporate color information from Lambertian scenes by evaluating *photo consistency* [15] produce high quality, high resolution geometry reconstruction, but fail for lower resolutions.

Another body of work concerns constructing geometry from high resolution laser scans [12, 20, 26]. Such scans typically contain very little noise, and are extremely dense. Most algorithms use only hit information and discard passthrough information. In the absence of noise, these approaches are well suited, but with real-world data from commercial sensors, their prerequisite of having dense, watertight point clouds falls short. Our approach begins by assuming large chunks of missing data, occlusion, and noise are all present.

In contrast to the methods used in 3D object recognition and reconstruction, works in the robotic navigation and mapping community have long made extensive use of volumetric information in the form of Occupancy Grid Maps [5] for 2D navigation, and in the case of 3D navigation, as Octree mapping [12] and related algorithms. Unlike point cloud based methods, which only consider the endpoints of rays, these methods integrate passthrough information to construct a probabilistic representation of the space around the robot. Our work can be seen as an extension of Occupancy Grid Mapping for object modeling which uses a voxel-centric approach rather than a ray-centric approach to more efficiently compute the occupancy probability of the space around the object.

This paper builds on one of our earlier unpublished reports in which we first proposed the method of Voxel Depth Carving [8]. Since then, another work by Pajarola et al. [23] independently introduced the same method to the computer graphics community, where it is known by the same name. While their method is very similar to ours, they have not verified the method using real-world sensor data, as we have—-and only consider the problem of high-resolution surface reconstruction, whereas we are also interested in matching and probabilistic object models.

Object Recognition Template matching approaches such as LINEMOD [9] have been extended to noisy 3D data successfully, but such techniques cannot efficiently select models from thousands of candidates in a database, which is our goal. Hsiao et al. [13] explore the problem of recognizing houshold objects for grasping from extremely noisy and incomplete depth data by considering alignment to a number of 3D objects from a very small database—but their approach (which relies only on simple ICP [1]) only scales to tens of objects. The Point Cloud Library [25] includes several methods of 3D object detection and recognition based on point features, but (as the name of the library suggests) all such methods rely purely on point cloud hits, and do not consider volumetric data.

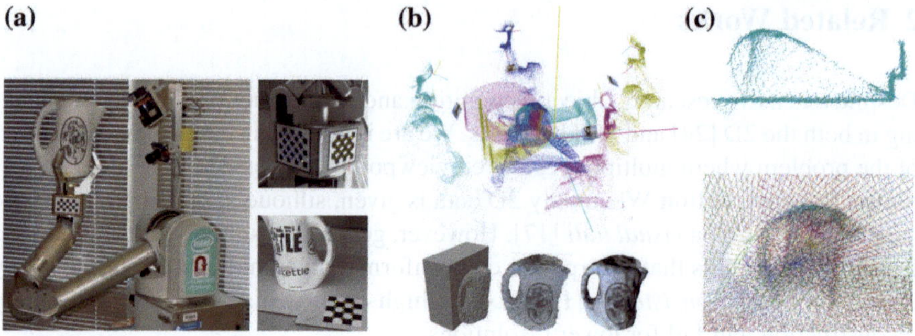

Fig. 2 Kettle reconstruction: **a** Experimental setup. **b** (*Top*) Multiple registered views, (*Bottom*) Voxel carving and mesh generation. **c** (*Top*) Grazing incidence from a single view, (*Bottom*) aggregated views

Our work complements other object recognition techniques by using volumetric descriptors (such as moments) to match against a database of thousands of objects. We build on the work of Goldfeder [7], who recognize everyday objects by the Canterakis [4] moments of their point clouds; we simply compute the moments of their depth hulls instead (Fig. 2).

3 Technical Approach

Problem Assume that the robot takes N scans of a scene. We have $H_1, \ldots, H_N$ globally registered rigid poses of the robot's sensor. For each scan, assume we have M rays emanating from the sensor $R_k = \{r_1, \ldots, r_M\}$; where a ray $r_i = \{o_i, p_i\}$ has an origin o_i, and an endpoint "hit" p_i. We may further assume that for each R_k, all $o_i \in R_k$ are the same (that is, all the rays pass through a focal point). This is often the case for RGB-D sensors and laser scanners. We will use o_k to mean the focal point of scan R_k.

In the absence of noise, each $p_i \in R_k$ is a point on the surface of an object. However, we will assume that every ray in the scan has a length which is corrupted by noise, i.e

$$\tilde{d}_i = \|p_i - o_i\| = d_i + n_i$$

where d_i is the true length of the ray, and a random variable n_i denotes noise, drawn from a probability distribution $f_{n_i}(x)$. In our experiments, we use a simple depth-dependant Gaussian model of the noise i.e.: $f_{n_i}(x, d_i) = \mathcal{N}\big(d_i, \sigma_z(d_i)\big)$, where σ_z is a funciton which varies with depth.

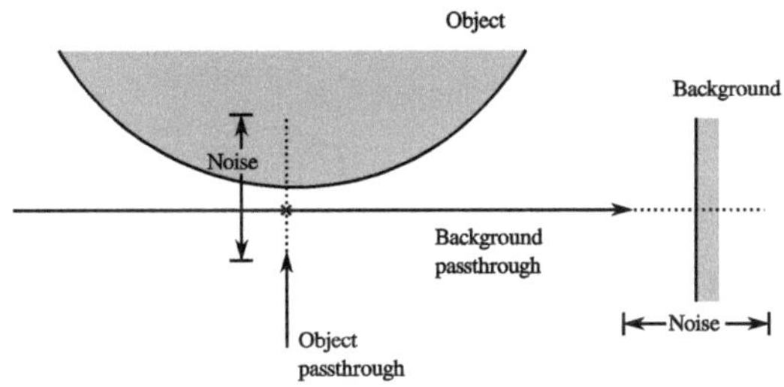

Fig. 3 With noise, the hit information becomes meaningless for evaluating the occupancy of the marked point. The passthrough information from the ray passing by the object remains useful

The goal is to find a function $S(x) : \mathbf{R}^3 \rightarrow \{-1, 1\}$, the *shape function*, which is -1 whenever the space is *free* (i.e. it does not contain an object), and 1 whenever the space is *filled.*

Specifcially, we may consider the probabilistic shape function $P_s(x|R_1, \ldots R_N)$ We will assume that a discrete representation (in the form of voxels) of the probabilistic shape function is sufficient. We construct a voxel grid representation of the space $V \in \mathbf{R}^{X \times Y \times Z}$, where X, Y, and Z are the number of cells along each axis of the workspace. And, $V[x] = P_s(x|R_1 \ldots R_N)$. We can similarly write the probability that a cell is free, $\bar{P}_s(x|\ldots)$. The joint distribution over all the cells, $P_s(x_1, \ldots, x_{XYZ}|R_1 \ldots R_N)$ is labeled $P_s(V|R_1, \ldots, R_N)$ for convenience.

Passthrough Information To find the shape function, it is necessary not only to consider the end-points of the rays in each scan (called the *point cloud*), but also the presence of rays passing through space between the scan origin and the end-point. To see why this is important, consider the physical process behind a laser scan. Rays of light (which have non-zero thickness) emenate from a central point through the scene. Some of the rays strike objects in the scene directly (we will call these Type I rays), others will not hit any objects (Type II rays), and still others will *graze* objects (Type III rays). Grazing ray hits [22] are the most interesting of these, as in practice, the sensor will randomly return depths intermediate between the surface of the object and the background (Fig. 3).

If we only consider the endpoints of rays, Type II and III rays immediately become useless, since for Type II rays, there is no endpoint, and for Type III rays, the endpoint is wildly corrupted by noise. In contrast, by using the entirety of the ray, we are able to use *all three* kinds of rays to determine which parts of the space are *free*, even if we can't say anything about which parts of the space are *occupied.*

Consider the simple case where $f_{n_i}(x)$ is Gaussian, centered at the true depth with some standard deviation σ_n (Fig. 4). As we increase σ_n, the occupancy probability of space becomes more and more blurred around the hit, while the probability of space around the sensor being free remains high. This fact remains true even for non-Gaussian noise models—notably, it is true for Type III (grazing) rays.

Depth Images In addition to storing the rays from each scan, we can also construct the *depth image* of the scan. We define a depth image of sensor j as a function $D_j(u, v) : \mathbf{R}^2 \rightarrow \mathbf{R}$, which takes in two real parameters u, and v, and returns the

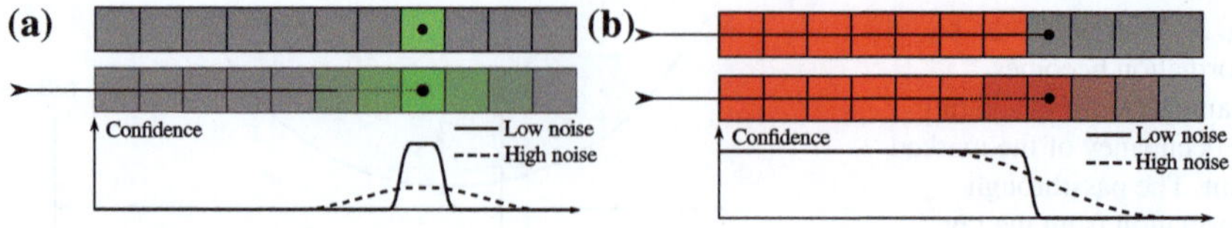

Fig. 4 The relative confidence of hit and passthrough information under low and high noise. While the usefulness of hit information declines rapidly with noise (**a**), most of the passthrough information (**b**) remains intact

length of a stored ray at that location. For RGB-D images, the depth image is merely a 2-dimensional grid of depth values, where each grid cell is the length of the ray passing through that grid cell on the image plane of the camera to the scene. For values of u and v which do not fall in the center of a grid cell, we use bilinear interpolation to determine the depth value. On the other hand, for laser scanners, the depth image must be synthesized.

We will also assume there exists an *projection mapping* of the depth image $D_j^p(x) : \mathbf{R^3} \to \mathbf{R^2}$, which, given a point x in the scene, projects that point back onto the depth image. In the case of RGB-D sensors, the projection mapping is simply a perspective projection of the scene onto the image plane. For laser scanners, we can use the pinhole camera model:

$$D_j^p(x) = \left[\frac{(x - o_j) \cdot e_2}{(x - o_j) \cdot e_1}, \frac{(x - o_j) \cdot e_3}{(x - o_j) \cdot e_1}\right]^T$$

where (e_1, e_2, e_3) form an arbitrary orthonormal basis, with e_1 being the viewing direction. This model is only suitable for scans with a field of view significantly less than 180°.

Occupancy Grid Mapping Consider the typical occupancy grid mapping problem. If we first assume that each of the voxels is conditionally independent of one another given measurements, then what we wish to find is the probability of a cell being occupied given all of the measurements:

$$P_s(V|R_1, \ldots, R_N) = \prod_{c_i \in V} P_s(c_i|R_1, \ldots, R_N)$$

Which, with the further assumption that each sensor measurement is independent becomes

$$\prod_i \prod_j \prod_{r_k \in R_j} P_s(c_i|r_k)$$

In other words, the probability of any instatiation of the voxel grid's MRF becomes a matter of simply computing the likelihood of each voxel being occupied given each ray in every scan independently. To improve numerical stability, we can use the log-odds of the occupancy probability rather than the probability distribution itself:

$$l_i = \sum_j \sum_{r_k \in R_j} \log \frac{P_s(c_i|r_k)}{1 - P_s(c_i|r_k)}$$

Traditional occupancy grid map approaches [5] solve the problem of finding the probabilistic shape function by *rasterizing* each ray using Brensham's algorithm [3], and lowering the occupancy probability for each cell that the ray passes through, while raising the occupancy probability for the cell that the ray ends in. Unfortunately, this approach has the disadvantage that as rays diverge from the sensor, they cover less space, and thus the reconstruction degrades as distance from the sensor increases. This problem is especially visible when the resolution of the voxel grid is high. Rasterizing each ray also takes a considerable amount of time when the resolution of the grid is high.

Voxel Depth Carving To speed up occupancy grid mapping in our domain, we make an approximation which uses projection instead of rasterization to carve away voxels which are likely to be free. Instead of iterating over rays and rasterizing them, we instead iterate over voxel cells, and determine whether a voxel cell should be marked as *free* from the collected sensor measurements. This method makes more sense in our domain, because typically the number of rays (from a collection of registered depth camera views) will be much higher than the number of voxels (which need only capture the shape of a small object). However, we can only approximate the true depth in areas that no rays actually pass through by interpolating between depth values of nearby rays. The approximation will be worse where there are rapid changes in depth smaller than the resolution of the sensor.

For each voxel cell in the scene, we project its center c_i onto the depth images of each sensor (Fig. 5a). We then compare the linearly-interpolated value from the

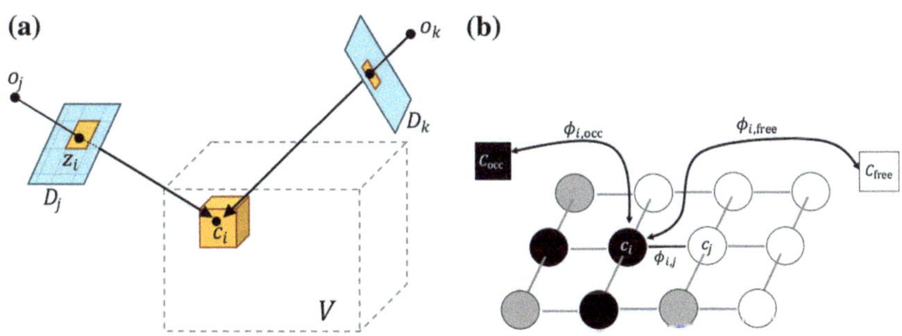

Fig. 5 **a** Two scans, R_i and R_k are shown, with one voxel c_i. Projections onto the depth images D_j and D_k, with the minimal projection shown as z_i. **b** A diagram of the Markov Random Field in 2D. $\phi_{i,j}$ is the edge with weight $\phi(c_i, c_j)$. The two special nodes, c_{free} and c_{occ} are shown as boxes. In 3D, cells are 6-connected. The darkness of a cell indicates its label

depth image $z_{i,j}$, with the Euclidean distance from the voxel cell to the sensor ($e_{i,j}$), which will tell us whether or not the cell should be free.

$$z_{i,j} = D_j\left[D_j^p(c_i)\right], \quad e_{i,j} = \|c_i - o_j\|$$

From here, we can compute the probability that a cell is free given a particular ray. When the cell is sufficiently close to the sensor, we have:

$$\bar{P}_s(c_i|r_j) = P(z_{i,j} > e_{i,j}) = \int_{e_{i,j}}^{z_{MAX}} P(z_{i,j} = z)dz$$

but as the distance to the cell greatly exceeds $z_{i,j}$, the probability a cell is free becomes unknown. For an ideal sensor (i.e. no noise), the probability distribution is a step function. The more noise is added to the sensor, the less steep the step function becomes.

Because of this, $P(z_{i,j} > e_{i,j})$ is very large near sensors, and very small near the vicinity of hits. If all we are interested in is whether or not a cell is free (and make no claims about the occupancy of voxels otherwise), we can make use of this fact by only considering the *minimum* distance to any sensor to decide whether a cell is free. That is,

$$z_i = \min_j z_{i,j}, \, e_i = \min_j e_{i,j}$$

We can then simply threshold the differences in the depth of the cell and measurements so as to only consider a cell free when it is sufficiently far away from a hit. This approach is a very conservative approximation of occupancy grid mapping: hit data can not be used at all, and only cells which are very likely to be free will be updated in each step. However, under conditions of very high noise, this conservative approach prevents us from carving important features of the object away.

$$\text{label}(c_i) = \begin{cases} -1 \text{ (free)} & \text{if } e_i < t_i \\ 0 \text{ (unknown)} & \text{else} \end{cases} \quad (1)$$

with some threshold t_i. All points which are closer to the origin than the threshold are classified as "free". We choose the threshold by subtracting a *margin* Δd from the measured depth, $t_i = z_i - \Delta d$. We generally want to use a low value for Δd because otherwise, cavities will look flatter than they actually are. Cavities with a depth less than Δd cannot be reconstructed at all. Using the noise model of the sensor, we can determine a suitable value for the margin by considering the probability that there are points which are misclassified as "free", that is:

$$\exists p \in [o_i, p_i] : (\|p - o_i\| > d_i) \text{ and } (\|p - o_i\| < t_i) \iff d_i < z_i - \Delta d$$

The probability for a misclassification is given by

$$\begin{aligned} P_{\text{mis}} &= P(d_i < z_i - \Delta d) \\ &= P(n_i > \Delta d) \\ &= 1 - \int_{-\infty}^{\Delta d} f_n(x)dx \\ &= 1 - F_n(\Delta d) \end{aligned}$$

where F_n is the probability distribution of the noise. Given a maximum acceptable misclassification probability $P_{\text{mis,max}}$, we can determine the smallest value of Δd that results in $P_{\text{mis}} \leq P_{\text{mis, max}}$.

In particular, for Guassian noise with mean μ and standard deviation σ, we have an optimal Δd of

$$\Delta d = \Phi^{-1}(1 - P_{\text{mis, max}})\sigma + \mu$$

By iterating over each voxel, we are able to "carve" large volumes of space which are likely to be free, leaving only the occupied space of the object as the number of views of the scene increase (Fig. 2b). Note that since we take the center of the voxel only, as cell resolution decreases, the accuracy of our method degrades. This effect can be mitigated by additionally projecting the vertices of the voxel to the depth image and taking the minimum depth over the convex hull of the projected vertices. We are left with an algorithm which has performance characteristics linear in the number of voxels (rather than the number of rays, as in the occupancy grid mapping case). With several dense RGB-D scans, this is a significant performance improvement; at the cost of throwing away data near ray hits.

Unknown Space and Shape Priors Until now, we have only been concerned with determining whether or not a cell is "free", and make no claims about whether a cell is "occupied", instead opting to call all cells which are not free "unknown."

However, it is possible to determine which cells are likely to be occupied based on the free and unknown cells using a prior on the structure of objects. We can capture this structure using a Markov Random Field that has pairwise energies between cells (in addition to energies associated with sensor data). In this sense, Voxel Depth Carving becomes a method of updating the MRF given sensor data, with strong structural priors determining which cells are actually labeled as "occupied."

Markov Random Field Using a Markov Random Field (MRF), we can encode prior assumptions about the structure of objects into the pairwise energies between adjacent cells. That is:

$$P_s(V) = \frac{1}{Z} \prod_{c_i, c_j \in V} \phi(c_i, c_j)$$

where ϕ is the joint probability of two adjacent cells being occupied, and Z is a normalizing constant. One choice of ϕ which results in local smoothing and preserves boundaries is an Ising-like local model

$$\phi(c_i, c_j) = \exp\big(-\gamma L_i L_j\big)$$

where γ is a smoothness parameter, and L_i, L_j are the labels of voxel i and j (for the MRF we are required to label voxels containing points from the point cloud as "occupied" ($L_i = 1$)). This model penalizes labelings which are not locally contiguous. If we use the Ising model to propagate occupancy probability from ray hits to the rest of the voxel grid, what we will be left with is a distribution which is smooth in "unknown" areas, but maintains the hard edges obtained from Voxel Carving. Additionally, we can apply a prior to the structure of the Markov Random Field by adding additional vertices representing the labels "free" and "occupied", called c_{free} and c_{occ} (Fig. 5b). Every voxel in the grid is then connected to these vertices with an energy computed from a prior and evidence from voxel carving. That is:

$$\phi(c_i, c_{\text{free}}) = \alpha I(L_i = -1) + \pi_{\text{free}}(c_i)$$
$$\phi(c_i, c_{\text{occ}}) = \alpha I(L_i = 1) + \pi_{\text{occ}}(c_i)$$

where π_{free}, π_{occ} are priors which weight "unknown" cells based on assumptions about the shape of the object.

Minimum Graph Cut We are left with a binary labeling problem on a Markov Random Field. The goal is to segment the field into two components: "occupied" and "free", in a way that minimizes the energy of the field. With a field of this type, it is true from the MaxFlow Mincut theorem [28] that the minimum-cut of the graph produces the optimal labeling. That is, we want to find a set of edges, $C = \{\phi_1, \ldots, \phi_M\}$, where ϕ_i is an edge between a cell that is "occupied" and "free"; and we want this set of edges to have minimal energy. The edges must divide the graph into two connected components, one containing c_{free}, the other containing c_{occ} The minimum graph cut can be found efficiently using Ford-Fulkerson algorithm [6]. Our problem is thus reduced to the binary image segmentation problem, which is efficiently solved by finding the minimum graph cut [2].

Shape Priors One prior we consider is to assume that the object is essentially a sphere of radius r_s and center c_r, called the "sphere prior" (Fig. 6b). In this case,

$$\pi_{\text{free}}(c_i) = \frac{\|c_i - c_r\|}{r_s}$$

That is, a cell is less likely to be free the closer it is to the center of the sphere. Notice however that if α is very high, the minimum cut of the MRF will only use the prior in areas that are "unknown", using evidence from the sensor in other areas

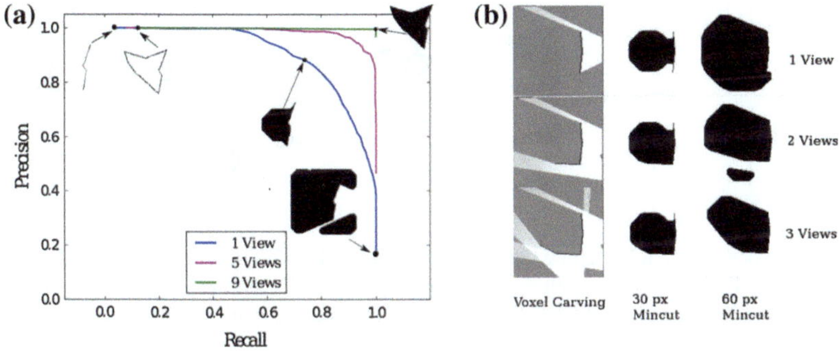

Fig. 6 **a** Precision versus recall graph from varying r_s from 0 to 100 for several views of a random polygon. The graph cut is shown for the bet and worst values of r_s for 1 and 9 views. **b** The minimum cut of an 8-connected 2D voxel grid obtained by voxel-carving a random polygon. Three views are shown, with two prior radii shown for the sphere prior

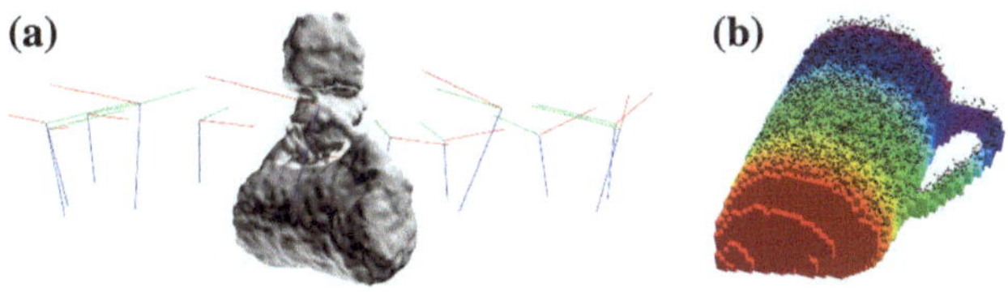

Fig. 7 **a** Thirteen registered views of a kettle held in the robot's hand are used to carve a voxel grid, which is then meshed using Marching Cubes. The robot's hand is reconstructed as well. **b** A coffee mug scanned by a simulated Kinect at a depth of 3 m in the intermediate stages of voxel carving. Voxels (*colorful*) are constructed from multiple noisy depth scans (*black points*)

(Fig. 7). We can also use a prior for the field which considers *each ray* independently using the following formula (which we call the "thin object prior"):

$$\pi_{\text{free}}(c_i) = \exp(-\beta h_x)$$

where $h_x = \min_j(\|x - o_j\| - z_j)$. This prior causes us to be less confident about voxels being occupied the further they are away from observed sensor points (Fig. 8b). The parameter β controls just how thin we believe objects to be. When $\beta = \infty$, for instance, we are left with a reconstruction of the surface of the object (which is less noisy than the point cloud); and when $\beta = 0$, we simply have the depth hull of the object. Varying β produces shapes which interpolate between the depth hull and the object surface.

Surface Reconstruction Based on the carved voxel shape, the surface of the object can be reconstructed for visualization or analysis. Many algorithms may be used for this task, including the simplest, Marching Cubes [19]. Additionally, color can be applied to the reconstructed surface by the projection of each vertex onto the color image of the sensor (Fig. 2b). Admittedly, the surfaces we have constructed are quite ugly in comparison to state-of-the-art techniques from computer graphics (as in

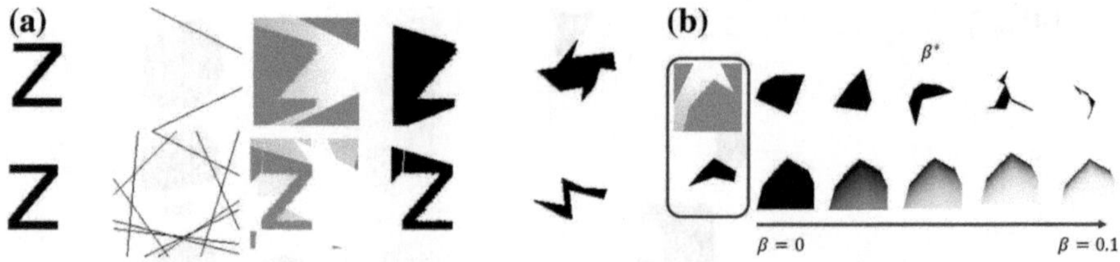

Fig. 8 2D experiments. **a** Matching the letter Z to 10k random 2D polygons. The number of views increases from 1 to 4, and the match becomes more semantically similar. The *leftmost* image is the input, followed by the camera FOVs, the carved voxels, the occupancy probability, segmentation, and finally the top match. **b** As β increases from 0 to 0.1, we increase the weight of the object surface. At β^*, the match to the nearest polygon in the database is semantically closest. The *leftmost* column represents the occupancy probability, and the *rightmost* is the top match. **a** Effect of increasing views, **b** single view thinness prior

[23]), but we are concerned with creating a probabilistically accurate representation of the occupied portions of the space for use in grasping and recognition—and not with constructing an aesthetically pleasing surface mesh of the object.

Object Recognition Once the shape of the object has been reconstructed using voxel depth carving, we may want to match the object to one of several in a database, or else classify the object. Recognizing objects from their volumetric representations is a very well-studied topic, known variously as Shape Retrieval or Shape Classification. The Princeton Shape Benchmark [27] lists several volumetric descriptors, including Shape Histograms, the Spherical Harmonic Descriptor, and others. The volumetric descriptor we chose for our experiments is the 3D Zernike [4] descriptor (as in [7]). The Zernike descriptor is constructed by projecting the object's voxelization onto a set of N basis polynomials defined on the unit sphere. It is affine invariant, compact, and descriptive.

Matching is performed by comparing the Zernike descriptor of the carved object with pre-computed Zernike descriptors in an object database, and taking the nearest neighbor in the Euclidean sense.

4 Experiments

Voxel Carving Experiments We validated our method in two real-world setups: the first one is an object standing on a table, and the second one is an object in the robot's hand. When recording ray clouds, we have determined the pose of the laser scanner. Pose registration is beyond the scope of this work. For our purposes, we found that determining the pose of checkerboards on the wrist of the robot or on the table, respectively, with a camera mounted on the robot, is sufficient. The camera was also used to colorize the resulting model, as seen in Fig. 2b.

Object on the Table For the object on the table, we recorded seven ray clouds at a low resolution, taking about 1.5 s per cloud. Each of the clouds contains about 64,000 points. After filtering out points lying outside the region of interest, about 10,000 points wer cloud remain. The point clouds after filtering are shown in Fig. 2c.

From each of the point clouds, we synthesize a depth image. After carving the voxel grid, we use the Marching Cubes algorithm and Laplacian smoothing to create a mesh from the voxel reconstruction. Finally, we use the camera images to colorize the mesh by re-projecting vertices onto RGB cameras co-registered with the laser scanner. More sophisticated methods for creating a mesh out of a voxel grid are available, but are beyond the scope of this work.

Creation of the depth images from the laser scan takes less than half a second. Reconstruction of the voxel grid takes about 14.7 s for a grid of 456×10^3 voxels. The time for generating and colorizing the mesh is less than one second. All of these results were obtained on an Intel Core 2 duo processor.

Object in the Hand We had the HERB hold a tea-kettle in its hand. HERB then rotated its wrist to take 13 scans of the object Fig. 2a. These scans were co-registered using a fiducial on HERB's wrist, and were then carved using the depth carving method. For this experiment, we did not colorize the mesh (Fig. 7a).

Robustness to Noise For evaluating the robustness of our method to noise, we used the acquired data and added zero-mean Gaussian noise to the depth. Note that neither of these properties are required for our method—the probability distribution can be arbitrary and have non-zero mean. We chose a maximum acceptable misclassification probability of $P_{\text{mis, max}} = 0.2$, resulting in a margin of $\Delta d \approx 1\sigma$. The results of the reconstruction can be seen in Fig. 1.

Figure 1a shows the original point cloud with no additional noise. Both the conventional method, using hit information, and our method, using depth images, approximate the shape well, but with the traditional method, part of the surface to the lower left is missing due to the low density of points. In Fig. 1b, noise with a standard deviation of $\sigma = 5$ cm was added. The conventional method results in a very jagged surface. With even more noise (Fig. 1c, d), the object can barely be recognized in the point cloud anymore. The conventional method does not produce any useful results, while our method still performs well. Note that in Fig. 1d, the standard deviation of the noise is more than *half* the size of the object.

Simulation Environment To explore the effect of noise, multiple views, and object priors, we created both a 2D and 3D simulation environment. In the 2D environment, laser scans are simulated by casting 2D rays from the focal point of a camera in an arc. The rays strike a binary image representing an object. Noise is then added to the length of the rays. In our 3D simulation environment (Fig. 7b), we simulate a *Kinect* with realistic resolution, noise parameters [21], and depth discretization using ray casting against 3D meshes.

Classification Accuracy with Shape Priors To explore the behavior of the object "thin-ness" prior, we varied the falloff parameter β over a number of views of the object. Voxels were said to be "occupied" if their occupancy probability was greater

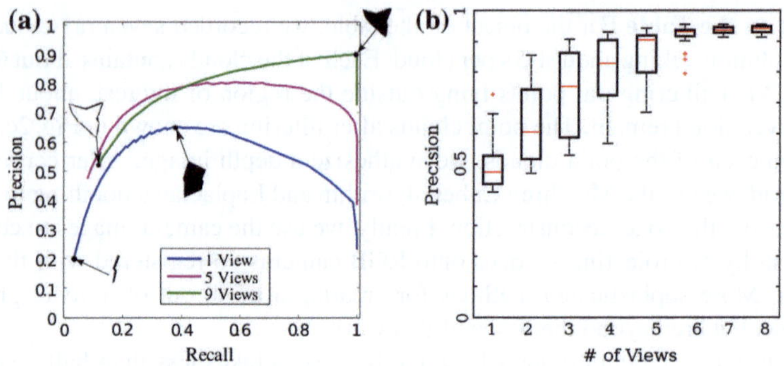

Fig. 9 Precision versus recall curves varying the thin-object prior's falloff parameter β. Each *curve* represents varying β for the same collection of scans. The most and least accurate reconstructions are shown for 1 and 9 views. **a** 2D precision versus recall, **b** 3D precision distribution

than 0.49, and were said to be "free" otherwise. After carving, the voxels were compared to the true object in simulation to see how accurately the space would be classified into "free" and "occupied" cells. We calculated Precision and Recall for each shape estimation where shape reconstruction is treated as a classification problem for each voxel independently.

In the 2D case, we took between 1 and 9 scans of a randomly generated polygon, and varied β for the thin object prior between 0 and 0.5; in another experiment we vary r_s in the sphere prior from 0 to 100. An example result is shown in Fig. 9a. In the 3D case, we took between 1 and 9 simulated scans of objects from the Willow Garage Household Objects Database, and varied β between 0 and 40. Since the resolution of our 3D voxel grid was smaller than in the 2D case, Recall was very high for all scans, so instead we plot a distribution of precision vs. β (Fig. 9b).

Moment Matching Experiments We additionally explored the possibility of using volumetric shape descriptors to match objects with a database of known object models in our simulation environment.

In the 2D case, we generated 10,000 random polygons and rasterized them to 256 × 256 images. We then computed their Hu [14] moments, and stored them in a database. Novel test objects (Latin letters) were scanned in the simulation, and the Hu moments of their probabilistic shape functions were computed. The test objects were then matched to their nearest counterparts in the database by comparing their Hu moments in the Euclidean sense.

In the 3D case, we voxelized 256 objects in the Willow Garage Household database, and matched their 20-dimensional Zernike moments against scans of novel test objects. We compared the matching performance (with realistic sensor noise) while using only the hits to using Voxel Depth Carving (Fig. 10).

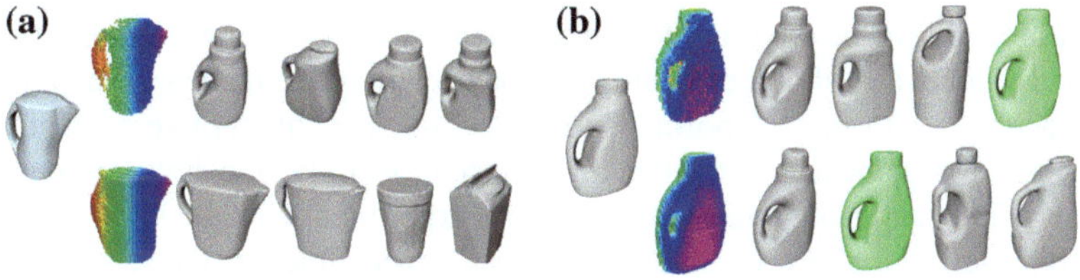

Fig. 10 The top 4 matches according to the L2 norm of Zernike Moments to a voxel-carved pitcher (*far left*) from 3 simulated views from a Microsoft *Kinect* sensor. The *top row* shows matches using hits only. The *bottom row* shows matches using passthrough information. **a** Matching a novel object, **b** matching a known object

5 Results and Main Experimental Insights

Voxel Carving Experiments Figure 2 shows an intermediate stage during carving and the completely carved and colorized voxel grid. Figure 2b shows the colorized and smoothed trimesh generated from the voxel grid. Figure 1 shows the effect of Gaussian noise on voxel carving versus reconstruction from the point cloud alone.

Even when the noise of the sensor approaches the scale of the object, our method is able to extract its shape. Our analysis indicates that our robustness to noise is greatly enhanced by integrating *near misses* (Fig. 2c) of the object, which discriminate the object from the background even when the noise is exceedingly high. In contrast, the pointcloud rapidly decays with noise, losing its descriptive power.

Classification Accuracy and Shape Priors Even when only one view of the object is given, using a simple prior on object thin-ness greatly affects occupancy classification accuracy (Fig. 9a). The prior becomes less important as the number of views increases (Fig. 9b). This is because regardless of the prior, as more views are taken the set of un-carved voxels approaches the true shape of the object. When $\beta = 0$, the estimated object shape is simply a representation fo the object's surface which is much less noisy than the input point cloud. A similar effect happens when we vary r_s for the sphere prior. When $r_s = 0$, we again get a representation of the object's surface, and as r_s increases toward infinity, the occupied space approaches the depth hull.

Moment Matching Experiments Figure 8a shows the effect of multiple views on voxel carving and the resulting top match from Hu moments in a database of 10k random polygons and the simple prior that voxels which are not carved are occupied. With 1 view, the match is not very semantically similar to the letter Z. With 4 views, the match is closer. When using the naive prior that un-carved voxels are occupied, our method fails to reconstruct the shape from a single view, instead leaving a long "shadow" behind the object. This result highlights the need for strong priors on the shape of objects in the scene given depth measurements. Figure 8b shows how the thin object prior can be used to find a good match via moments, even when there is only a single view of the object. As the weight on surface voxels increases, the occupancy probability tends to favor the thin shell of the object over the interior.

Figure 10 shows the top four matches from the Willow Garage Household Objects database to a novel test object (a pitcher, not already in the database), and a known object according to the L2 norm of their 20-dimensional Zernike moments. The top row shows matches using the hit information only, the bottom row shows matches using passthrough information. In the experiment with the known test object (Fig. 10b) viewed under favorable conditions, the true object is the second match in the database using our method, whereas it is fourth when using only the hits. In the experiment with the novel object (Fig. 10a), the handle's interior was not fully visible, and so during voxel carving it was left occupied—while the hit data leaves the hole unoccupied.

This result shows again the need for strong priors on occupancy when using voxel carving. Nevertheless, the first two object matches (both water filters) are more semantically similar to the test object than the first two matches using only the hits. We are left to wonder whether Zernike descriptors are strong enough descriptors to capture the kind of information we desire about manipulable objects. Indeed, the fact that Zernike descriptors are affine-invariant might actually be an undesirable property when considering household objects—whose uses are strongly correlated with scale.

6 Conclusion

We have shown that Occpuancy Grid Mapping techniques taken from robotic navigation is useful in a robotic manipulation setting. When only noisy, partial views of the object of interest are given, a volumetric approach to object modelling and recognition can be more useful than a surface-centric or point-based approach. Our experiments show that voxel carving leads to much more robust shape reconstruction under highly noisy, low-resolution conditions than methods which use point clouds alone. Using techniques from graph-based image segmentation, we introduced an efficient way to reconstruct 3D shapes with the minimum graph cut of a Markov Random Field. By estimating the occluded volumes of sensed objects, we are able to utilize implicit information from depth sensors which would otherwise be thrown away in surface or point-based models.

References

1. Besl, P.: A method for registration of 3-D shapes. IEEE Trans. Pattern Anal. Mach. Intell. (1992)
2. Boykov, Y., Funka-Lea, G.: Graph cuts and efficient n-d image segmentation. Int. J. Comput. Vision **70**(2), 109–131 (2006)
3. Bresenham, J.E.: Algorithm for computer control of a digital plotter. IBM Syst. J. **4**(1), 25–30 (1965)

4. Canterakis, N.: 3D Zernike moments and Zernike affine invariants for 3D image analysis and recognition. In: 11th ICSA (1999)
5. Elfes, A.: Using occupancy grids for mobile robot perception and navigation. Computer (1989) (Long Beach, CA)
6. Ford, L.R., Fulkerson, D.R., A simple algorithm for finding maximal network flows and an application to the hitchcock problem. Can. J. Math. (1957)
7. Goldfeder, C.: Data-driven grasping. Auton. Robots **31**(1), 1–20 (2011)
8. Herrmann, M.: Exploiting passthrough information for multi-view object reconstruction with sparse and noisy laser data (2010)
9. Hinterstoisser, S., Lepetit, V., Ilic, S.: Model based training, detection and pose estimation of texture-less 3d objects in heavily cluttered scenes. ACCV **7724**, 548–562 (2013)
10. Hoiem, D., Savarese, S.: Representations and techniques for 3D object recognition and scene interpretation. Synth. Lect. AI Mach. Learn. **5**(5), 1–169 (2011)
11. Hornung, A., Kobbelt, L.: Robust reconstruction of watertight 3D models from non-uniformly sampled point clouds without normal information. SGP (2006)
12. Hornung, A., Wurm, K.M., Bennewitz, M., Stachniss, C., Burgard, W.: OctoMap: an efficient probabilistic 3D mapping framework based on octrees. Auton. Robots **34**(3), 189–206 (2013)
13. Hsiao, K., Ciocarlie, M., Brook, P.: Bayesian grasp planning. ICRA **2011**, (2011)
14. Hu, M.K.: Visual pattern recognition by moment invariants. IRE Trans. Inf. Theory 66–70 (1962)
15. Kutulakos, K.N., Seitz, S.M.: A theory of shape by space carving. ICCV **1**, (1999)
16. Lai, K., Fox, D.: Object recognition in 3D point clouds using web data and domain adaptation. IJRR **29**(8), 1019–1037 (2010)
17. Laurentini, A.: The visual hull concept for silhouette-based image understanding. IEEE Trans. Pattern Anal. Mach. Intell. **16**(2) (1994)
18. Li, Y., Wu, X., Chrysathou, Y., Sharf, A.: GlobFit: consistently fitting primitives by discovering global relations. ACM Trans. Graph. (2011)
19. Lorensen, W.E., Cline, H.E.: Marching cubes: a high resolution 3D surface construction algorithm. ACM Siggraph **21**(4), 163–169 (1987)
20. Marton, Z.C., Rusu, R.B., Beetz, M.: On fast surface reconstruction methods for large and noisy point clouds. ICRA 3218–3223 (2009)
21. Nguyen, C.V., Izadi, S., Lovell, D.: Modeling kinect sensor noise for improved 3D reconstruction and tracking. In: 3D Imaging, Modeling, pp. 524–530 (2012)
22. Olsen, M.J.: Avoiding indicents with incidence. LIDAR Mag. **2**, 2 (2012)
23. Pajarola, R., Guggeri, F., Scateni, R.: Shape reconstruction from raw point clouds using depth carving. Eurographics (2012)
24. Prasad, D.K.: Survey of the problem of object detection in real images. IJIP **6**, 441–466 (2012)
25. Rusu, R.B., Cousins, S.: 3D is here: point cloud library (PCL). ICRA 1–4 (2011)
26. Shewchuk, J.R., Brien, J.F.O.: Spectral surface reconstruction from noisy point clouds. SGP **14** (2004)
27. Shilane, P., Min, P.: The princeton shape benchmark. Shape Model. **08540** (2004)
28. Staples-moore, A.: Network flows and the max-flow min-cut theorem

Model-Based Insights on the Design of a Hexapod Magnetic Walker

Ryan St. Pierre, Dana Vogtmann and Sarah Bergbreiter

Abstract We present a design for a 1 cm^3 magnetically actuated hexapedal walking robot fabricated using an all-polymer multi-material fabrication process capable of integrating sub-mm elastomeric joints into rigid polymer links. Modeling, optimization, and experimental insights from several different leg designs were used to improve the robot performance. The fabricated robots are capable of speeds up to 10.6 body lengths per second. This work focuses on the development of models for locomotion at small-scales as well as experimental validation and input back to the model.

Keywords Millirobot · Magnetic actuation · Optimization

1 Introduction

Legged robots on the millimeter scale present challenges in design, actuation, and fabrication. Models including SLIP (spring loaded inverted pendulum [1]) and LLS (lateral leg spring [2]) provide insight on how to design larger legged robots, but it is still unclear that these models make sense at smaller size scales. As robots shrink down in size, contact forces with the ground can surpass inertial forces in their effect on locomotion. The ultimate goal of this work is to understand how legged robots below 1 cm^3 in size can walk and run efficiently and effectively. In particular, this paper focuses on an experimental setup that can validate models for locomotion in robots sized 1 cm^3 and smaller.

R. St. Pierre (✉) · D. Vogtmann · S. Bergbreiter
Department of Mechanical Engineering, University of Maryland,
College Park 20742, USA
e-mail: rstpierr@umd.edu

D. Vogtmann
e-mail: dvogtman@umd.edu

S. Bergbreiter
e-mail: sarahb@umd.edu

M.A. Hsieh et al. (eds.), *Experimental Robotics*, Springer Tracts
in Advanced Robotics 109, DOI 10.1007/978-3-319-23778-7_47

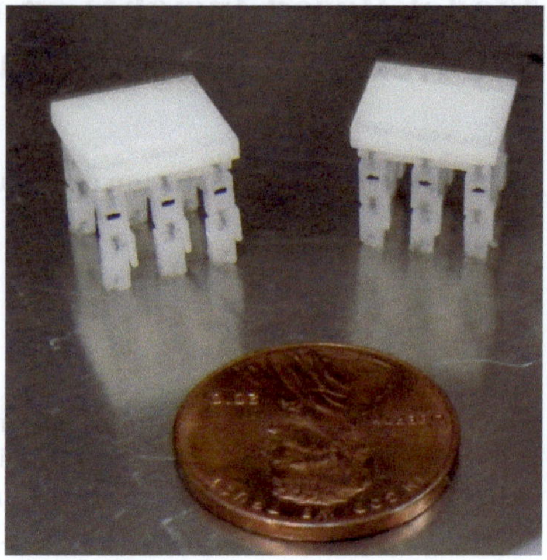

Fig. 1 Picture of two magnetic walkers

Previous work on legged robots at the sub-cm^3 scale has focused primarily on fabrication and includes [3, 4]. The robots in these works have been fabricated primarily from silicon in traditional micro manufacturing processes. Ebefors includes polyimide joints in this process, improving the robot's robustness. However, the actuation was slow-less than 1 body length per second. The robot in [4] demonstrated actuation, but did not walk forward. The robot shown in Fig. 1 uses polymer joints similar to Ebefors, but in this case, the polymer joints are used to provide additional passive degrees of freedom in the robot's legs.

One challenge in fabricating and testing robots at this scale is the integration of high power density actuation. The thermal actuators used by Ebefors [3] require a tether for power delivery and the actuators used by Hollar [4] only allow for very slow (<1 body length per second) locomotion. However, small-scale magnetically actuated robots have been frequently demonstrated in a variety of contexts including the MagMites for micromanipulation [5]; modular robots which remain connected through active magnetic interfaces [6]; flexible magnetic structures actuated by external fields [7]; and medical devices such as magnetically actuated endoscopic robots [8] and micro surgical robots [9]. In this work, actuation of the multi-material elastomeric leg mechanisms is done using an embedded permanent magnet on which a torque is exerted by an external magnetic field.

Multi-material robot legs actuated through an external magnetic field provide a novel experimental method with which to validate experimental models of robot locomotion at small size scales. Section 2.1 describes the basic dynamic model of the robot legs. Two different optimizations are used to ultimately improve robot speed over flat terrain; a strictly kinematic optimization maximizes stride length, and a leg geometry optimization improves walking speed at a constant driving frequency. Previous work in optimization of miniature robots has yielded improved results such as jump height [10] and payload capacity [11]. Work in optimization of legged walking robots on the larger scale has yielded improvements in results such as energy efficiency [12], strength to weight ratio [13], and torque distribution [14]. Section 3 describes the resulting fabricated robots and experimental setup, and Sect. 4 provides results from tests of each fabricated robot driven at a constant drive frequency and results from tests of one of the optimized robots at varying drive frequencies. Finally, Sect. 5 discusses insights obtained from the experiment along with future work.

2 Technical Approach

2.1 Design of Leg/Walker

The leg designs presented here, shown in Fig. 2, are simple, underactuated legs, having only two degrees of freedom: an active rotational hip joint and a passive rotational knee joint. The active hip joint is actuated by an applied torque (an embedded permanent magnet in the fabricated mechanism), while the knee joint is free to bend.

Additionally, asymmetries have been incorporated into the leg design: a hard mechanical stop on one side of the knee joint and a high-friction feature applied to only one side of the foot. The hard mechanical stop allows the knee joint to swing freely in one direction while stopping it at 180° in the other, similar to [15], which drives the robot preferentially in the direction of the stop. The high-friction feature creates a differential friction in the foot, similar to [16], which drives the robot preferentially in the direction opposite to the high friction feature.

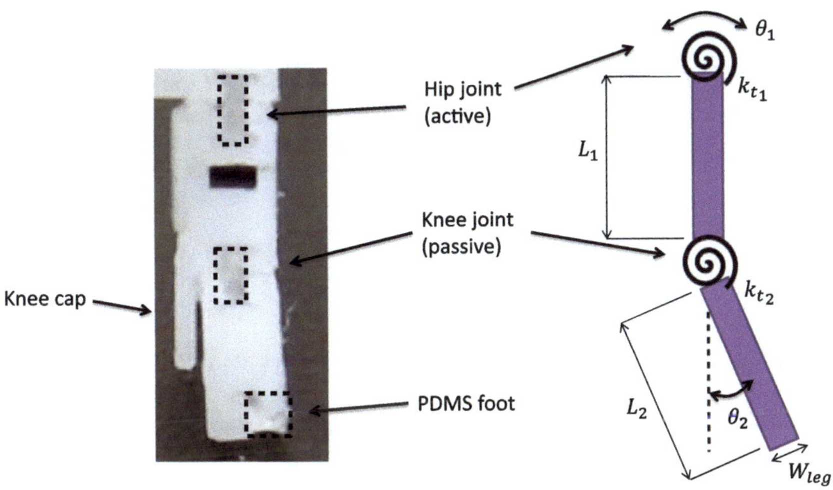

Fig. 2 Magnetic walker left (*left*) and schematic of model (*right*)

Six of these legs are arranged 90° to the flat walking surface, with three on either side of a central body. The legs are then driven from the active hip with alternating positive and negative applied torques. The sign of the applied torque is opposite on opposing tripods, resulting in an alternating tripod gait.

2.2 Model of Leg

The equations of motion for the hexapod leg, modeled as a set of coupled rigid bodies, are of the form

$$\mathbf{M}(\theta)\ddot{\theta} + \mathbf{C}(\theta, \dot{\theta})\dot{\theta} + \mathbf{N}(\theta) = \tau, \tag{1}$$

where $\mathbf{M}(\theta) \in \Re^{n \times n}$ is the mass matrix, $\mathbf{C}(\theta, \dot{\theta}) \in \Re^{n \times n}$ is the Coriolis matrix, and $\mathbf{N}(\theta) \in \Re^{n}$ includes gravity and spring forces. In this model, the leg is modeled as a two-link pendulum with two degrees of freedom ($n = 2$), θ_1 and θ_2 (θ_2 is defined relative to θ_1), which correspond to the motion of the hip joint and the knee joint. The flexures at the hip and knee joints are modeled as a torsional spring as in [17], and are located between the two links and between the body and the top leg length. A schematic of the model is shown in Fig. 2.

The mass matrix for the system is given by

$$\mathbf{M}(\theta) = \begin{bmatrix} M_{11} & M_{12} \\ M_{21} & M_{22} \end{bmatrix}, \tag{2}$$

where

$$\begin{aligned}
M_{11} &= J_1 + J_2 + \frac{{L_1}^2 m_1}{4} + {L_1}^2 m_2 + \frac{{L_2}^2 m_2}{4} + L_1 L_2 m_2 \cos(\theta_2) \\
M_{12} &= \frac{m_2 {L_2}^2}{4} + \frac{L_1 m_2 \cos(\theta_2) L_2}{2} + J_2 \\
M_{21} &= \frac{m_2 {L_2}^2}{4} + \frac{L_1 m_2 \cos(\theta_2) L_2}{2} + J_2 \\
M_{22} &= \frac{m_2 {L_2}^2}{4} + J_2.
\end{aligned}$$

The Coriolis matrix is given by

$$\mathbf{C}(\theta, \dot{\theta}) = \begin{bmatrix} C_{11} & C_{12} \\ C_{21} & C_{22} \end{bmatrix}, \tag{3}$$

where

$$
\begin{aligned}
C_{11} &= -\frac{\dot{\theta}_2\, L_1\, L_2\, m_2\, \sin(\theta_2)}{2}\\
C_{12} &= -\frac{L_1\, L_2\, m_2\, \sin(\theta_2)\, \left(\dot{\theta}_1 + \dot{\theta}_2\right)}{2}\\
C_{21} &= \frac{\dot{\theta}_1\, L_1\, L_2\, m_2\, \sin(\theta_2)}{2}\\
C_{22} &= 0.
\end{aligned}
$$

The potential vector is given by

$$
\mathbf{N}(\theta) = \begin{bmatrix} N_1 \\ N_2 \end{bmatrix}, \tag{4}
$$

where

$$
\begin{aligned}
N_1 &= k_{t_1}\, \theta_1 - g\, \sin(\theta_1) \left(\frac{L_1\, m_1}{2} + L_1\, m_2\right) - \frac{g\, L_2\, m_2\, \sin(\theta_1 + \theta_2)}{2}\\
N_2 &= k_{t_2}\, \theta_2 - \frac{g\, L_2\, m_2\, \sin(\theta_1 + \theta_2)}{2}.
\end{aligned}
$$

The input torque vector is given by

$$
\tau = \begin{bmatrix} \tau_1 \\ \tau_2 \end{bmatrix}, \tag{5}
$$

where

$$
\begin{aligned}
\tau_1 &= \tau_{magnetic}\\
\tau_2 &= 0.
\end{aligned}
$$

In Eqs. 2–4, the subscripts 1 and 2 refer to the top and bottom leg links respectively. m is the mass, L is the length, and J is the mass moment of inertia. k_{t_1} and k_{t_2} are the spring constants of the hip joint and knee joint respectively. g is the acceleration of gravity, and $\tau_{magnetic}$ is the torque put on the hip joint from the magnetic field.

The equations are solved using numerically implemented differential equations. The numerical solution is constrained to represent the physical stop (knee cap) in the leg design.

One of the challenges with modeling walking robots is incorporating contact with the ground. In this model, it is assumed that the hexapod walks with a stick-slip motion, meaning that the foot is always in contact with the ground. When the walker is actuated, one set of legs move forward due to their low friction interaction with the ground, as described in Sect. 2.1. When the direction of the magnetic field is

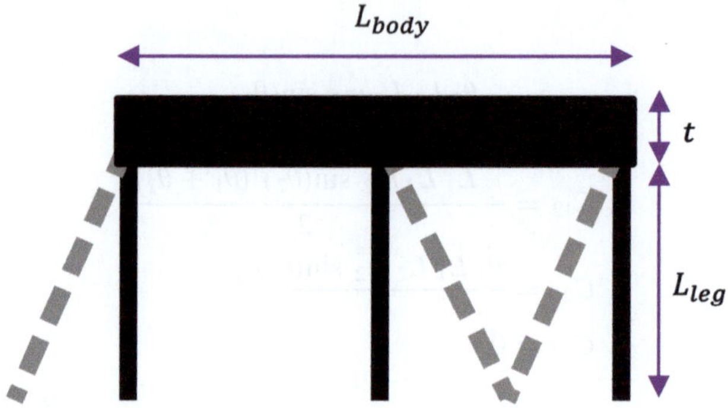

Fig. 3 A schematic of the kinematic optimization to maximize stride length. The actuated position of the legs (*dotted lines*) show that the legs touch, limiting the stride length

reversed, the high friction component of the foot comes into contact with the ground, causing the foot to stick. Other forms of contact with the ground are not captured in this model.

The motion of the foot is tracked to determine the appropriate ground friction, either low friction during the forward swing or high friction during the back swing. The normal force due to the weight of the robot itself, and tangential force from ground friction on the foot, are calculated based upon the foot motion. The forces on each foot are summed, and this summation determines the forces and torques on the body, which accelerate the robot forward.

2.3 Maximizing Stride Length

The model of the leg indicates that by maximizing the stride length, the walker can take larger steps, ultimately walking faster. To maximize the stride length, the body is elongated while the legs are shortened, keeping the volume of the walker constant. The lengthening of the body prevents the legs from touching during each stride, and allows the legs to use their full range of motion, offsetting the reduction in leg length. A schematic of this is shown in Fig. 3. A strictly kinematic optimization problem can be formulated by treating the leg as one link. The stride length maximization can be set up as an optimization problem as in Eq. 6,

$$\begin{aligned} \underset{L_{body}}{\text{maximize}} \quad & \sin^{-1}\left(\frac{L_{body}}{4L_{leg}}\right) \\ \text{subject to} \quad & (L_{leg}+t)WL_{body} \leq 1\,\text{cm}^3, \\ & L_{leg} \geq 0.4\,\text{cm}, \\ & \frac{L_{body}}{4L_{leg}} \leq 1. \end{aligned} \tag{6}$$

L_{body} and W are the length and width respectively of the robot body. L_{leg} is the length of the leg, considered as one link in this case, and t is the thickness of the body and attachment points of the leg. In this problem, W and t are held constant at 1 and 0.19 cm respectively. Using *fmincon* in MATLAB, L_{body} is maximized at 1.6 cm and L_{leg} is 0.4 cm, resulting in a walker volume of 0.94 cm^3.

2.4 Optimization of Leg

A genetic algorithm in MATLAB was used to maximize the simulated forward average velocity of the magnetic walker. The genetic algorithm allows for many different parameters to be varied in the optimization. In this case, the geometric parameters-leg link lengths, leg link widths-and knee joint stiffness, are varied to maximize the simulated average velocity. Driving frequency is not varied in the optimization and was held constant at 5 Hz.

The model indicates that varying the leg link lengths impacts the stride length of the walker, and that the width of the leg impacts the mass of the leg, influencing the dynamics. The width of the flexure changes the stiffness of the knee joint. A less stiff knee joint doesn't return the second link of the leg in time for the next step. However, if the knee joint is too stiff, the leg acts like one rigid link. Given these insights from the model, the dimensions varied in the genetic algorithm were the length of each of the leg links, L_1 and L_2, the width of the leg, W_{leg}, and the width of the knee joint flexure, $W_{flexure}$. The total length of the leg is constrained to be less than or equal to 7 mm, to keep the volume of the walkers constant. All of the parameters were constrained with a lower and upper bound, as shown in Table 1.

2.5 Fabrication

The legs were fabricated using a prototyping process, Laser Cut Elastomer Refill, (LaCER). This process was developed in [18] and consists of laser patterning rigid material, selectively removing portions of the patterned material, pouring uncured

Table 1 Lower bounds, upper bounds and optimized parameters in genetic algorithm optimization. All dimensions in mm

Parameter	Lower bound	Upper bound	Optimized parameters
L_1	1.75	5.50	4.5
L_2	1.75	5.50	2.5
W_{leg}	1.20	2.00	2.0
$W_{flexure}$	0.20	1.00	0.6

low-modulus polymeric material into the trenches left by the removed portions, photo- or heat-curing the material inside the trenches, and planarizing and removing the final pieces from the surrounding rigid material.

The rigid portions of the test legs were fabricated from Delrin and cut using a desktop laser cutter, a 60 W VersaLaser VLS3.50. The elastomeric hinges were fabricated from a pourable two-part elastomer, Dow Corning Sylgard 184 PDMS, cured at 100 °C for 30 min. Additionally, small areas of PDMS were formed in the foot of the legs to provide differential friction against the ground.

Small magnets, 1 mm length 0.75 mm diameter cylindrical N50 neodymium magnets with a parylene coating from SuperMagnetMan (Cyl0010 Parylene http://www.supermagnetman.net/), were embedded in each of the legs to provide the alternating actuation torque under influence of a rotating external magnetic field. To embed the magnets, the legs were oriented over a large magnet so that the small magnets placed on the surface of the leg were automatically oriented and held in place in the external magnetic field. The oriented small magnets were then pushed down with tweezers into a space cut into the legs intended to hold the embedded magnets. They are held in place by a surrounding thin ring of PDMS deposited during the fabrication of the hinges, as well as a dot of glue for added security.

3 Experiments

3.1 Fabricated Walkers

Four magnetic walkers were fabricated with different dimensions for their legs. The dimensions of the four walkers are summarized in Table 2.

Build O has dimensions that were identified as optimal for leg link lengths. Build E has even link lengths. Build R has link lengths reversed from the optimized geometry. Build S has shorter, even link lengths, and a longer body length as identified in Sect. 2.3. An additional build P was also created with "peg-legs"-without a knee joint (Fig. 4).

Table 2 Dimensions of fabricated walkers

Build	L_1	L_2	W_{leg}	$W_{flexure}$	L_{body}
O	4.5 ± 0.1	2.5 ± 0.1	1.8 ± 0.1	0.3 ± 0.1	10
E	3.5 ± 0.1	3.5 ± 0.1	1.8 ± 0.1	0.3 ± 0.1	10
R	2.5 ± 0.1	4.5 ± 0.1	1.8 ± 0.1	0.3 ± 0.1	10
S	2.5 ± 0.1	2.5 ± 0.1	1.8 ± 0.1	0.3 ± 0.1	14

All dimensions in mm

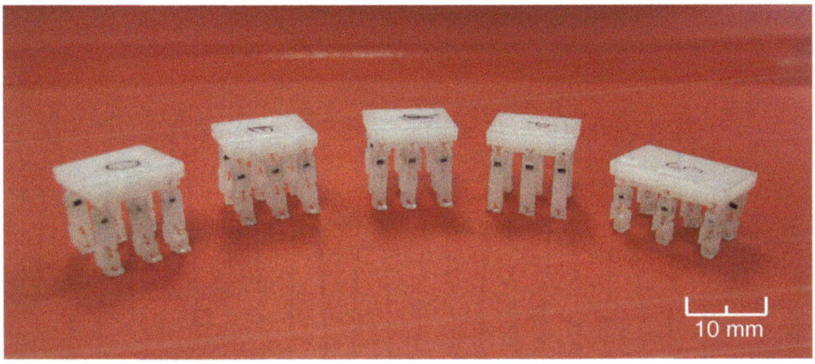

Fig. 4 Five walker builds. From *left* Optimized, Even, Reversed, Peg-leg, and Short

3.2 Experimental Setup

Each walker is actuated by an external magnetic field rotating at 5 Hz. In a separate experiment, frequency was varied. The rotating magnetic field is provided by mounting a 1 inch^3 N55 neodymium magnet on carbon fiber rods mounted on ball bearings to support the weight of the magnet. One end of the mounted magnet assembly is attached to a 30 mm diameter DC motor which was driven at 3.5 V to produce rotation at 5 Hz, and up to 12 V to produce rotation up to 26.7 Hz. The mounted magnet setup is shown in Fig. 5.

A thin sheet of transparent acrylic was supported just above the rotating magnet setup. Cardstock guides were mounted to the acrylic sheet to direct the path of the walkers. Because the distance at which the rotating magnet can actuate embedded

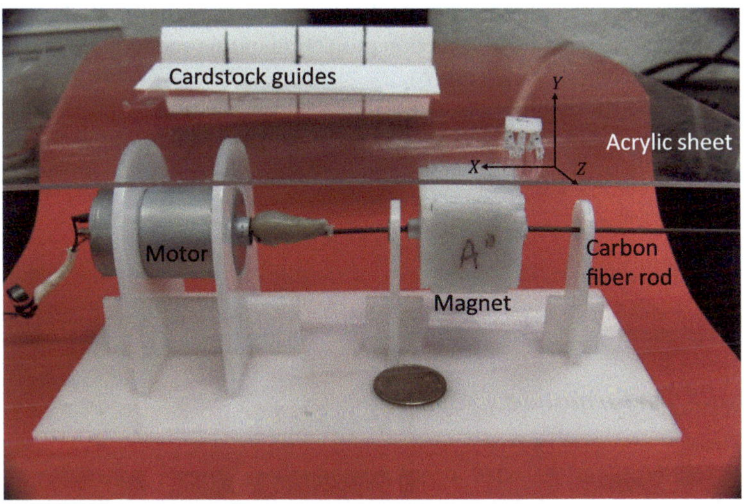

Fig. 5 Experimental setup showing the magnet mounted and attached to a DC motor

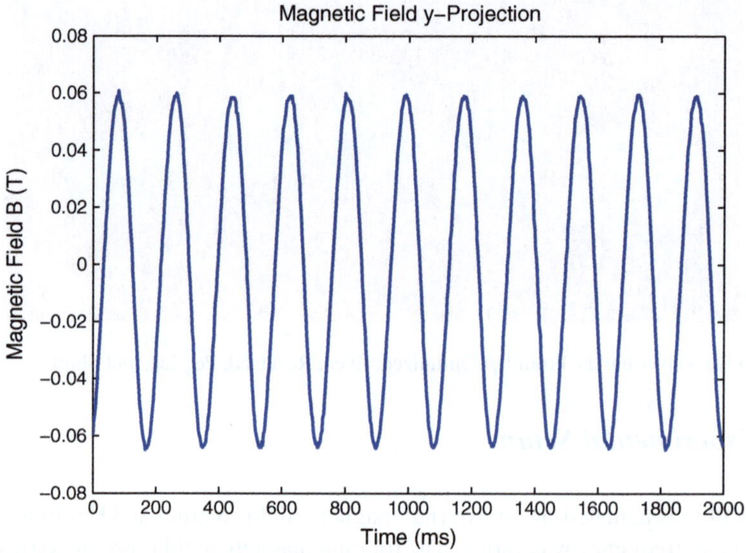

Fig. 6 The y-component of the magnetic field vector applied to the hexapod legs over several cycles at about 5 Hz

magnets in the walkers is only a few centimeters, the acrylic sheet was moved over the setup to keep the walker's current position centered over the rotating magnet during testing. When the walkers are placed on the acrylic sheeting, the magnets embedded in the legs are positioned 25–27 mm (depending on the build) from the surface of the rotating magnet. An AD22151 sensing chip from Analog Devices was used to measure a field intensity, B, of approximately 60 mT at this distance. The direction of the magnetic field vector at this location, $\overrightarrow{B}$, varies with the rotation of the mounted magnet, and the y-direction projection of this vector was collected over several cycles at roughly 5 Hz (Fig. 6).

A sample of each build was tested walking in this rotating magnetic field over a distance of 60 mm during several trials. Distance markers were drawn on the card-stock guides to determine the distance and the time was recorded using video of the trials.

4 Results

4.1 Build Performance

The different builds were tested as described in Sect. 3.2 over at least 7 trials each. The walkers were recorded during their motion. The distance traveled and travel time was evaluated from the videos. The experimental data and simulated average

Table 3 Experimental and simulated average velocities

Build	Experimental (mm/s)	Simulated (mm/s)
O	9.7 ± 1.4	9.1
E	8.3 ± 1.0	8.3
R	4.2 ± 0.6	3.9
S	9.4 ± 1.2	7.9

velocities are shown in Table 3. The peg-leg walker's performance was not simulated and is not reported in the table. However, it was tested and walked at an average velocity of 1.4 mm/s.

4.2 Frequency Testing

The optimized build (build O) was tested as described in Sect. 3.2, with a driving voltage applied to the motor between 3.0–12.0 V in 0.5 V increments, resulting in driving frequencies between 3.7–26.7 Hz. Average speeds over at least nine trials were found for each frequency tested, along with the standard deviation for each. Frequencies between 3–30 Hz were additionally simulated using the model described in Sect. 2.2. A comparison of the results are shown in Fig. 7.

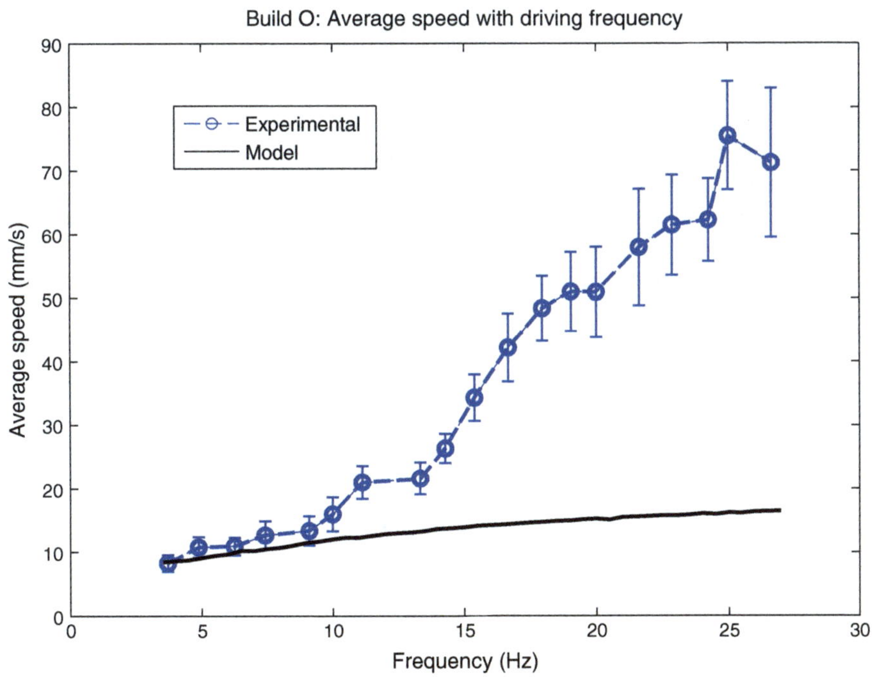

Fig. 7 Experimental and simulated performance of optimized design over a range of frequencies

The average speed of the walker was measured at up to 75.5 mm/s at a frequency of 25 Hz. Although the average speed at 26.7 Hz was somewhat lower, the maximum speed for a single trial at this frequency was measured at 106 mm/s, or 10.6 body lengths per second. This is the fastest speed (in body lengths per second) observed thus far in the magnetic walkers.

At higher frequencies, greater than 10 Hz, the modeling results do not match the behavior of the experimental results. Based on observations of the walker's gait at high frequencies, the stick-slip contact model used for lower frequencies is no longer the main mode of ground contact. A flight phase for the walker is observed at frequencies of 15 Hz and greater. A more dynamic approach is required to accurately capture the walker's behavior at higher speeds.

5 Main Experimental Insights

Based on the experimental results, the design and optimization processes were successful in identifying parameters which would result in improved speed performance, at a constant driving frequency of 5 Hz. At lower speeds, the simulations accurately predict the performance of the hexapedal walkers. However, when the walkers are driven at higher frequencies, the simulations do not accurately predict their performance. Similarly the stride-optimized build S moved faster experimentally than simulated. There appear to be parameters not yet accounted for in the model which allow the performance of the walkers to reach up to 10.6 body lengths per second when driven at 25 Hz. Close investigation of the behavior of the walkers during experimental testing will provide more insights into the current limitations of the model, particularly around contact. Using the experimental results, more parameters can be identified, improving both the model and future design iterations.

When the model is more able to more accurately predict the behavior of the walkers, it will be used to design walkers optimized over single or multiple objectives. Different behaviors could be optimized, such as obstacle clearance height, payload capacity, or energy efficiency. Additionally, factors beyond the stride length-currently captured well by the model-clearly contribute to improved speed. These factors may be dynamic material properties, complicated ground interactions, or other behaviors not currently well-captured by the model. In this case, the design optimized for speed may change with further model iterations, informed and improved by experimental results. Additionally, these magnetically actuated robots can be used to calculate actuation and power needed for robots with self contained actuation systems.

Acknowledgments This work is supported by NSF Award ECCS1055675. Thanks also to Eric Diller for early discussions around this idea.

References

1. Altendorfer, R., Koditschek, D.E., Holmes, P.: Stability analysis of a clock-driven rigid-body SLIP model for RHex. Int. J. Robot. Res. **23**, 1001–1012 (2004)
2. Seipel, J.E., Holmes, P.: Running in three dimensions: analysis of a point-mass sprung-leg model. Int. J. Robot. Res. **24**, 657–674 (2005)
3. Ebefors, T., Mattsson, J.U., Kälvesten, E., Stemme, G.: A walking silicon micro-robot. In: Proceedings of the Transducers 99, pp. 1202–1205. Citeseer (1999)
4. Hollar, S., Flynn, A., Bellew, C., Pister, K.: Solar powered 10 mg silicon robot. In: IEEE The Sixteenth Annual International Conference on Micro Electro Mechanical Systems (MEMS), pp. 706–711. IEEE, Kyoto (2003)
5. Frutiger, D.R., Vollmers, K., Kratochvil, B.E., Nelson, B.J.: Small, fast, and under control: wireless resonant magnetic micro-agents. Int. J. Robot. Res. **29**(5), 613–636 (2010)
6. Gilpin, K., Rus, D.: Self-disassembling robot pebbles: new results and ideas for self-assembly of 3d structures. In: IEEE International Conference on Robotics and Automation Workshop Modular Robots: The State of the Art, pp. 94–99 (2010)
7. Garstecki, P., Tierno, P., Weibel, D.B., Sagués, F., Whitesides, G.M.: Propulsion of flexible polymer structures in a rotating magnetic field. J. Phys.: Condens. Matter **21**(20), 204110 (2009)
8. Sendoh, M., Ishiyama, K., Arai, K.I.: Fabrication of magnetic actuator for use in a capsule endoscope. IEEE Trans. Magn. **39**(5), 3232–3234 (2003)
9. Yesin, K.B., Exner, P., Vollmers, K., Nelson, B.J.: Design and control of in-vivo magnetic micro-robots. In: Medical Image Computing and Computer-Assisted Intervention-MICCAI 2005, pp. 819–826. Springer (2005)
10. Kovac, M., Fuchs, M., Guignard, A., Zufferey, J.-C., Floreano, D.: A miniature 7g jumping robot. In: IEEE International Conference on Robotics and Automation (ICRA), pp. 373–378. IEEE (2008)
11. Song, Y.S., Suhr, S.H., Sitti, M.: Modeling of the supporting legs for designing biomimetic water strider robots. In: Proceedings of the 2006 IEEE International Conference on Robotics and Automation (ICRA), pp. 2303–2310. IEEE (2006)
12. Gregorio, P., Ahmadi, M., Buehler, M.: Design, control, and energetics of an electrically actuated legged robot. IEEE Trans. Syst., Man, Cybern., Part B: Cybern. **27**(4), 626–634 (1997)
13. Ananthanarayanan, A., Azadi, M., Kim, S.: Towards a bio-inspired leg design for high-speed running. Bioinspiration Biomimetics **7**(4), 046005 (2012)
14. Erden, M.S., Leblebicioglu, K.: Torque distribution in a six-legged robot. IEEE Trans. on Robot. **23**(1), 179–186 (2007)
15. Pratt, J.E., Pratt, G.A.: Exploiting natural dynamics in the control of a planar bipedal walking robot. In: Proceedings of the Annual Allerton Conference on Communication Control and Computing, vol. 36, pp. 739–748. University of Illinois (1998)
16. Senoo, T., Takano, M., Ishikawa, M.: Dynamic horizontal movement of a bipedal robot using frictional asymmetry. In: IEEE/RSJ International Conference on Intelligent Robots and Systems (IROS), pp. 1834–1839. IEEE (2012)
17. Howell, L.: Compliant Mechanisms. Wiley
18. Vogtmann, D.E., Gupta, S.K., Bergbreiter, S.: Multi-material compliant mechanisms for mobile millirobots. In: IEEE International Conference on Robotics and Automation (ICRA), pp. 3169–3174. IEEE (2011)

Real-Time Stabilisation for Hexapod Robots

Marcus Hörger, Navinda Kottege, Tirthankar Bandyopadhyay, Alberto Elfes and Peyman Moghadam

Abstract Legged robots such as hexapod robots are capable of navigating in rough and unstructured terrain. When the terrain model is either known a priori or is observed by on-board sensors, motion planners can be used to give desired motion and stability for the robot. However, unexpected leg disturbances could occur due to inaccuracies of the model or sensors or simply due to the dynamic nature of the terrain. We provide a state space based framework for stabilisation of a high dimensional multi-legged robot which detects and recovers from unexpected events such as leg slip. We experimentally evaluate our approach using a modified PhantomX hexapod robot with extended tibia segments which significantly reduces its stability. Our results show that roll and pitch stability is improved by $2\times$ when using the proposed method.

1 Introduction

Legged robots such as hexapods shown in Fig. 1 are well suited for navigating in rough and uneven terrain that can be challenging to conventional wheeled or tracked vehicles. These robots can adapt to the complex terrain by adjusting their gait patterns, footfall trajectory or footholds. For effective and stable navigation in such terrain, the robot requires a mechanism to detect and recover from unexpected events such as leg slippage. In the event of a slip, identifying that the slip has occurred and subsequently taking corrective actions to move to a stable configuration allows it to continue its navigation task. Often such stabilisation requires the control of the whole system rather than just the individual leg that has slipped. This implies that a simple servo control on the leg motors alone are incapable of achieving the desired outcome.

In this paper, we frame the problem of hexapod stabilisation as a path finding problem from a critical region to a stable region in the configuration space of the robot. Such a formulation allows us to take advantage of the whole body kinematics

M. Hörger · N. Kottege (✉) · T. Bandyopadhyay · A. Elfes · P. Moghadam
Autonomous Systems, CSIRO Computational Informatics, Queensland Centre for Advanced Technologies, Brisbane, QLD 4069, Australia
e-mail: Navinda.Kottege@csiro.au

M.A. Hsieh et al. (eds.), *Experimental Robotics*, Springer Tracts in Advanced Robotics 109, DOI 10.1007/978-3-319-23778-7_48

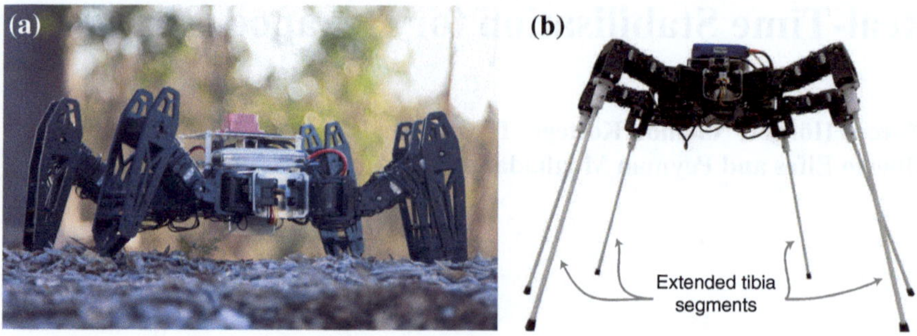

Fig. 1 Modified phantomX hexapods **a** with additional computing and sensing, **b** with extended tibia segments which reduces its stability during locomotion

in generating stable postures for future slip proof steps without explicit knowledge of the local terrain.

There are broadly two approaches in attaining hexapod locomotion: when terrain is unknown, the hexapod executes repeated pattern of coupled footfall or a gait with little feedback [1]; and when the terrain model is completely known or is observed by on-board sensors with sufficient accuracy, the footfalls are computed to give desired motion and stability [2–5]. The former approach relies on the stochastic nature of the hexapod's interaction with the terrain to recover from slips and trips. However, there is no guarantee of the approach working in very challenging environments like steep slopes or on slippery surfaces in the event of unexpected disturbances in the leg-terrain interaction such as leg slips or changes in the body orientation due to incomplete and uncertain terrain information. In order to navigate such challenging terrains, a fast reactive approach is necessary which is able to compensate these unforeseen disturbances. Another approach uses a set of behaviours to control a hexapod robot [6, 7]. In order for a robot to adapt its behaviours, it must have the ability to autonomously detect and classify terrain types [8]. An issue which is not adequately addressed in such an approach is the the effects due to terrain attributes such as loose soil or slippery surfaces causing the robot to slip, since the actions necessary to stabilise the system might not be defined in the set of behaviours.

Stable footfall generation for a known terrain is computationally expensive due to the high dimensionality of the planning space. When a slip occurs, the planner often has to recompute the footfall from a new post-slip configuration. In this paper, we propose an algorithm to detect and arrest the slip in real-time without knowledge of the local terrain before the body moves to an unstable configuration, potentially damaging itself, or to a configuration from where recovery is difficult.

2 Technical Approach

The complexity of the legged robot locomotion on uneven terrain comes from the high degrees of freedom that need to be controlled using imprecise knowledge of the environmental interaction during the robots gait. The challenge arises not only for planning footfalls on a high DoF robot but also while executing the footfalls in the presence of unexpected events. While unknown terrain poses a great challenge to legged locomotion, even with full knowledge of the terrain, the robot encounters inherent slips due to loose gravel, slippery vegetation, uneven surfaces etc. Many approaches [4, 6, 9, 10] have attempted to provide stability to the body by taking corrective actions to control the slipping joint or the leg. While individual controls can be added to each leg to prevent its slip, often the stability of the whole body is dependent on all leg positions in a coupled manner. Fixing the slip of one leg may not inherently provide stability. Due to this we need to provide control in the higher dimensional space.

In this preliminary study we take the first steps towards developing a framework for high-dimensional control of the hexapod to prevent slipping in planar terrains. Slipping on a simplified planar terrain model gives us an opportunity to study the high-dimensional slip control without the complexity of unknown terrain.

2.1 Formulation

In this study we focus on a hexapod robot with each leg having 3 actuators giving 18 joints to control. The state space C for the hexapod then lies in 18 dimensions. C consists configurations which satisfy the joint limits, without considering collisions of the vehicle with itself or the environment. Even with tight joint limits, not all configurations in the state space C are valid for stable navigation.

In this work we only look at quasi-static walking gaits, i.e., at any configuration of the robot during gait execution, the robot is inherently stable. While we ignore any dynamic gaits, we believe that our framework lends to an extension for dynamic gaits in a straightforward manner.

A main characteristic of the hexapod leg slip during gait execution is the foot tip position moving away from the planned footfall during the support phase of the gait cycle. When this happens, the support polygon gets skewed and often leads to instability of the whole robot. Another characteristic of leg slip is sudden change in body orientation. This can be detected in body roll, pitch or a combination of both depending on the number and position of the slipping legs. In this study we check for the position of the foot tips and the body orientation compared to a small tolerance around a desired foot tip and body orientation to satisfy the stable configuration. While this check could be extended to the rate of change of the tip position and orientation error, in this study we focus on the measure of error and not its rate of change.

To satisfy the footfall position and the body orientation, we use the 6-dimensional world task space T, in which we define constraints on the tip position and body orientations that limit the planning in C to desired tip and body orientations in T. These constraints generate a desired region T_{constr} inside T. We define constraining cuboids T_i with edges d_x, d_y and d_z for the desired tip position for each leg i, capturing the full span and an operational tolerance of the tip position during the gait cycle. Furthermore, we formally define two frames: a body frame B and a local frame L both with origins at the centre of the robot. The centre could be the centre of mass or any other suitable origin on the body. L is a 3-dimensional frame with its x, y plane being parallel to the ground plane. The body frame B is defined such that the x, y plane is congruent to the plane through the body.

Note that the tip positions are expressed in L. The orientation of the body is therefore defined as the orientation of B about L.

Let θ_r and θ_p be B's roll and pitch angle about L. By imposing constraints Θ on θ_r and θ_p, we define

$$T_{constr} = (T_n, \Theta) \tag{1}$$

with $T_n = (T_1, \ldots, T_6)$, where $(T_1, T_2, \ldots)$ are the space of allowed tip position for the corresponding legs.

Figure 2 illustrates the constraint cuboids around the initial tip positions and the body orientation constraints. Any configuration in C that does not satisfy the tip position or body orientation constraints are discarded during planning, thus, T_{constr} induces a manifold M_{constr} such that

$$M_{constr} = \{c \in C \mid f_C^T(c) \in T_{constr}\} \tag{2}$$

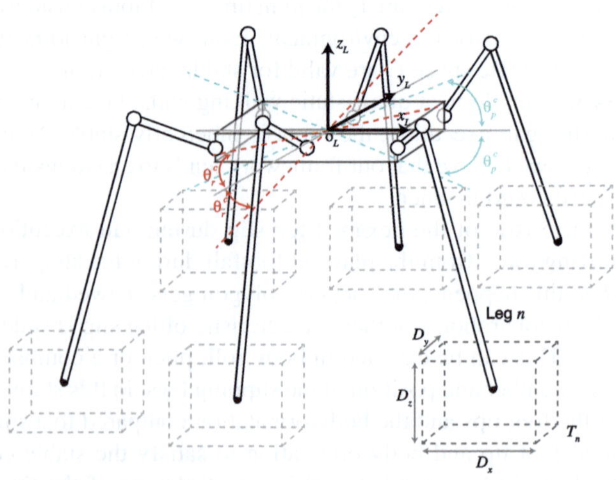

Fig. 2 Foot tip constraint cuboids T_{1-6} $(D_x \times D_y \times D_z)$ and body orientation constraints $\Theta = (\theta_r^c, \theta_p^c)$ shown with respect to configuration of the hexapod robot

with f_C^T being a mapping from C to T. We call a configuration c *stable* if $c \in M_{constr}$, *unstable* otherwise.

During a stable gait, the robot's foot tip positions follow a precomputed trajectory. For a tripod gait, 3 legs move at a time while for a wave gate one leg moves at a time. The stationary legs provide stable support for the body. In this study, we focus on the tripod gait, but the approach is applicable to any other gait with minimal modifications to our implementation.

2.2 Proposed Solution

We start with the assumption that any stable configuration has a neighbourhood in C that is stable. During the stable gait execution, the body moves through a periodic cycle of stable states $c_i \in M_{constr}$. During a slip event, as the leg tip moves closer to the boundaries of T_{constr}, the robot configuration $c_{current}$ moves closer to the boundary of M_{constr}. Our overall objective of stabilisation is to keep the body configuration inside M_{constr} and react, when the configuration $c_{current}$ travels outside M_{constr}. To do so, we cache known stable configurations in the neighbourhood of the gait cycle in M_{constr} offline and at run time, when the robot slips, quickly find a path from $c_{current}$ back into a stable configuration $c_s \in M_{constr}$.

One problem with that approach is that the configuration c_s might be too close to the boundaries of M_{constr} such that from c_s, even a slight disturbance might push the configuration outside M_{constr} again. In order to prevent such situations, we try to restrict c_s to a much smaller region within C which has a sufficient distance to the boundaries of M_{constr}. This is achieved by adding an operational tolerance to T_{constr} yielding $T_{stable} = (T_n^*, \Theta)$. T_n^* consists of a second set of tip constraint cuboids $(T_1^*, \ldots, T_6^*)$ with dimensions d_x, d_y, d_z, where $0 < d_x, d_y, d_z < D_x, D_y, D_z$. Figure 3 illustrates the second set of tip constraint cuboids. T_{stable} induces a smaller manifold $M_{stable} \subset M_{constr}$.

2.2.1 Stable Configuration Caching

We are interested in stable configurations which are close to the known stable configuration cycle. Thus, instead of randomly selecting a configuration inside M_{stable} to connect to, the algorithm selects a configuration $c \in M_{stable}$ within a certain neighbourhood of the current configuration cycle. However, sampling M_{stable} can be a difficult task. The shape and properties of M_{stable} depends on the dimensions of the tip position constraint cuboids defined in T. A naive sampling technique would be to use rejection sampling by uniformly picking a sample $c \in C$ and check whether

$$f_C^T(c) \in T_{stable} \tag{3}$$

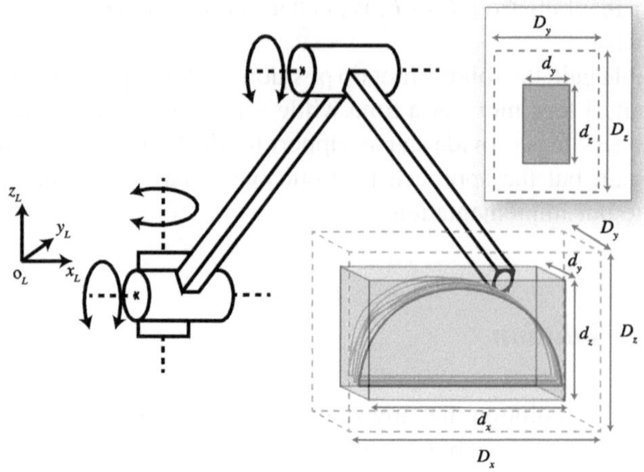

Fig. 3 Detailed illustration of foot tip constraints with cuboid centred at the initial position of foot tip

This sampling techniques samples the full state space C and rejects samples which lie outside M_{stable}. Since the volume of M_{stable} is unknown a priori, it could be the case that it is only a small fraction of the volume of C. In those cases, rejection sampling fails to sample a number configuration candidates inside M_{stable} within a reasonable amount of time. Therefore we generate a point cloud P of configurations inside M_{stable} offline, from which the algorithm can select a configuration to connect to during run-time.

Another approach is to grow a space filling tree, in our case an *RRT*, from a known stable configuration (possibly an initial state or a known home state) inside M_{stable}. The nodes generated by *RRT* are then used as M_{stable} samples. This techniques quickly generates n samples inside M_{stable}, however, it suffers from the same limitations of *RRT*. Despite its probabilistic completeness, *RRT* performs poor in complex environments with narrow passages. Several techniques have been proposed to improve sampling under the occurrence of narrow passages, such as filtering or adaptive sampling techniques [11–13] or retraction based approaches [14–18].

We are especially interested in regions inside M_{stable} which can be reached from a known path in M_{stable} defined by a gait cycle of the vehicle, not necessarily a single initial configuration. For this we pick n uniformly spaced samples from the configuration cycle while the robot performs a specific gait (e.g. tripod gait). These n samples are then used as seeds for an *RRT* forest with n trees. From each seed a tree is grown inside M_{stable} with a predetermined number of nodes, using the standard *RRT* algorithm [19].

The nodes of the generated *RRT* forest form a point cloud P within M_{stable}. These points are potential candidates the system can connect to when the task space constraints are violated. Figure 4 illustrates the offline generation of P using *RRT*.

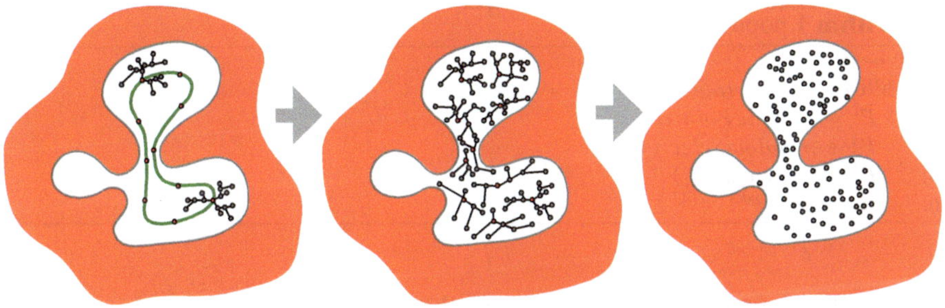

Fig. 4 Illustration of the sampling method based on the *RRT* forest approach to obtain a point cloud of stable configurations. The outer shading shows the area within C which are unstable regarding the task space constraints T_{stable}. The *green line* shows the path inside M_{stable} for a specific gait cycle. The *red dots* are the seeds from which the *RRT* trees are grown inside M_{stable}

2.2.2 Run Time Stabilisation

During run-time, when the current configuration $c_{current}$ is inside M_{constr}, the algorithm populates a list A with sets of stable configurations located in the neighbourhood of the current configuration $c_{current}$ (Fig. 5). This is done by sampling k points from P that are within a hypersphere with radius r around $c_{current}$. Recall from Sect. 2.2 that the P entirely lies within M_{stable}. The sampled set of points $P^k \subset P$ is then appended to the list. Algorithm 1 shows the population of the list during run-time. The function *samplePointcloud*(r) selects $P^i \subseteq P$ such that

$$P^i = \{p \in P \ s.t \quad ||p - c_{current}|| \leq r\} \tag{4}$$

where $||.||$ is a standard Euclidean distance metric. Out of P^i, *randomSelect(set, k)* randomly selects k points. This set $P^k \subset P^i$ is appended to the list of stable sets.

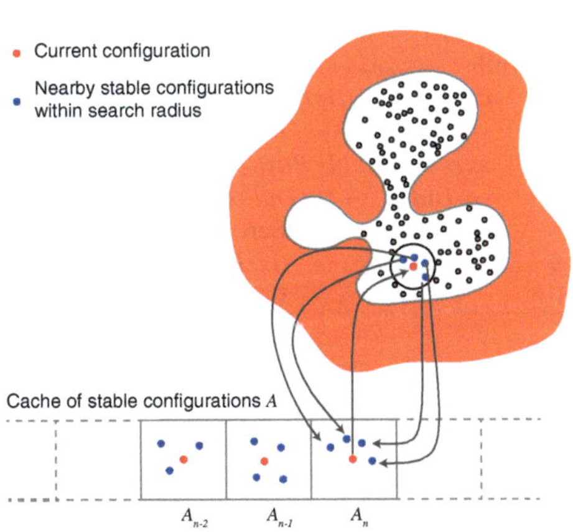

Fig. 5 The cache of stable configurations is populated at run-time with points from the point cloud P. The current configuration (*red dot*) does not necessarily have to be inside M_{stable}. Caching of stable configurations happens as long as the current configuration is inside M_{constr}

Algorithm 1 populate_list

```
A ← []
while c_current is stable do
    P^i ← samplePointcloud(r)
    P^k ← randomSelect(P^i, k)
    P^k.append(c_current)
    A.append(P^k)
```

Note that the population of the list only happens when $c_{current} \in M_{constr}$. If this is not the case, the *populate_list* algorithm is interrupted until the $c_{current}$ becomes stable again.

This list serves as a source for stable configurations the stabilisation algorithm tries to connect to when the configuration becomes unstable. If this is the case, the algorithm selects the last element A_n of the list (the last set of known stable configurations). Within A_n, the algorithm selects the first configuration as c_{stable} (line 4 in Algorithm 2). Since the structure and connectivity of M_{constr} is unknown, instead of joining the states with a direct path, we use *RRTConnect* [20] to find a path from $c_{current}$ to c_{stable} (function *findStablePath* in Algorithm 2). For details of *RRTConnect*, the reader is referred to [20].

After finding a path p connecting $c_{current}$ and c_{stable} the system executes p (function *execute(stablePath)* in Algorithm 2). In some cases, executing p doesn't result in the system to obtain a configuration c such that $c \in M_{stable}$, due to unconsidered interactions of the robot with the ground. Instead c might still be outside M_{constr} after executing p. If this is the case, the algorithm successively selects $c_i \in A_n$ as c_{stable}, calculates p from $c_{current}$ to c_{stable}, and executes p. In case c is still critical after connecting to each $c_i \in A_n$, the next set A_{n-1} in A is chosen and the algorithm tries to connect to the c_i's within that set. This is repeated until the system is in a stable configuration again.

Algorithm 2 connect_to_c_stable

```
1: listIndex ← A.last()
2: setIndex ← 0
3: while c_current is critical do
4:     c_stable ← A[listIndex][setIndex]
5:     stablePath ← findStablePath(c_current, c_stable)
6:     execute(stablePath)
7:     setIndex ← setIndex + 1
8:     if setIndex = k then
9:         listIndex ← listIndex − 1
10:        setIndex ← 0
```

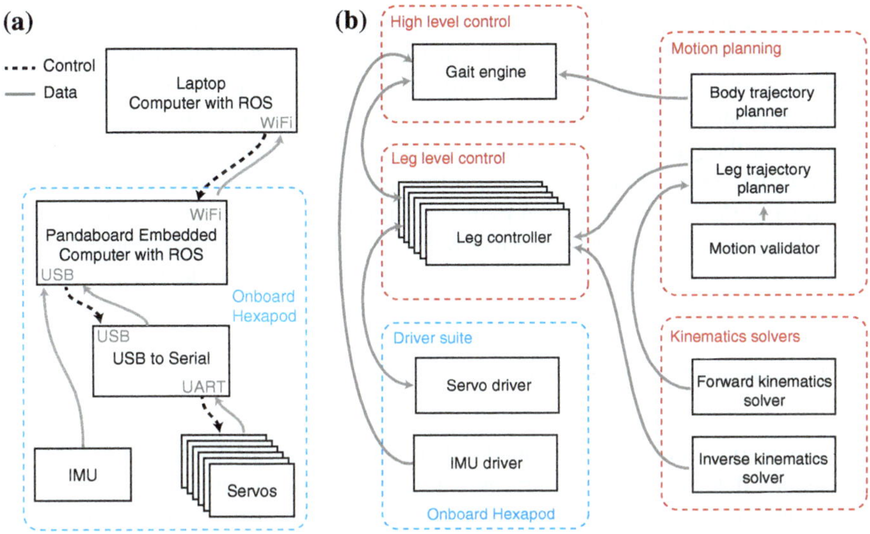

Fig. 6 Hardware (**a**) and software (**b**) block diagrams showing control and data flows of the hexapod system

3 Experimental Setup

3.1 Hardware

Our experimental setup consists of a modified PhantomX hexapod kit by Trossen Robotics[1] using Dynamixel AX-18 servosmotors (3 per leg, 18 in total). We have added a Pandaboard embedded computer running Robot Operating System (ROS) and an SBG IG-500N attitude and heading reference system (AHRS) on to our hexapod platform (Fig. 1b). The length of the tibia links have been extended (140 mm → 405 mm) in order to gain a broader range of unstable situations. The robot's body orientation is measured using the IMU of the AHRS. The joint angles of the servomotors are received at a rate of 35 Hz and the IMU data is received at a rate of 100 Hz. Figure 6a shows the control and data flow between the hardware components.

3.2 Software Architecture

The software architecture in which the proposed stabilisation method is embedded is shown in Fig. 6b. It consists of a high level gait engine which generates the body

[1] http://www.trossenrobotics.com.

locomotion during run-time, by defining desired tip positions for a given gait pattern. These tip position goals are sent to the leg controllers which calculate a trajectory from the current tip position of leg i to the commanded tip position, using a leg trajectory planner and an inverse kinematics solver. For the proposed stabilisation mechanism, the gait engine module monitors the current state of the system (joint angles from the servomotors and body orientation obtained from the on-board IMU) and utilises OMPL (Open Motion Planning Library) to determine if a state is stable or not. It also uses OMPL and its *RRTConnect* implementation to compute a path in M_{constr} from a configuration outside M_{constr} to a stable one inside M_{stable}.

3.3 Constraint Dimensions

The dimensions of the foot tip constraint cuboids ‘ on the performance, since these dimensions determine the time it takes for the system to detect a leg slip. If the constraints are too narrow, even slight deviations of the tip positions will lead to a slip detection. On the other hand, if the dimensions of the cuboids are too large, a slip gets detected too late for the system to stabilise the robot. The dimensions of the foot tip constraint cuboids T^*_{1-6} and T_{1-6} used during our experiments are $d_x = 100$ mm, $d_y = 60$ mm, $d_z = 140$ mm and $D_x = 120$ mm, $D_y = 80$ mm, $D_z = 160$ mm respectively. Body orientation constraints Θ used during our experiments are $\theta^c_r = \pm 5°$ and $\theta^c_p = \pm 6°$.

3.4 Performance Metrics

In order to be able to define metrics which measure the quality of the proposed stabilisation approach, we need to gain insight about what happens with the robot

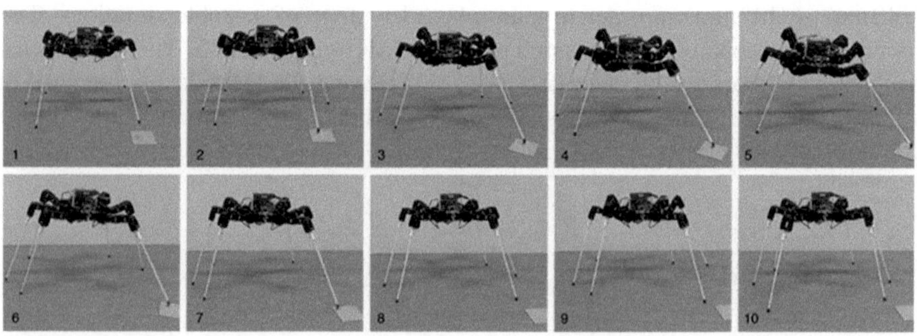

Fig. 7 A sequence of frames showing an experiment where the front left foot of the robot steps on an aluminium plate causing the leg to slip away during a gait (*top row*). Our stabilisation system allows the robot to recover and continue the gait (*bottom row*)

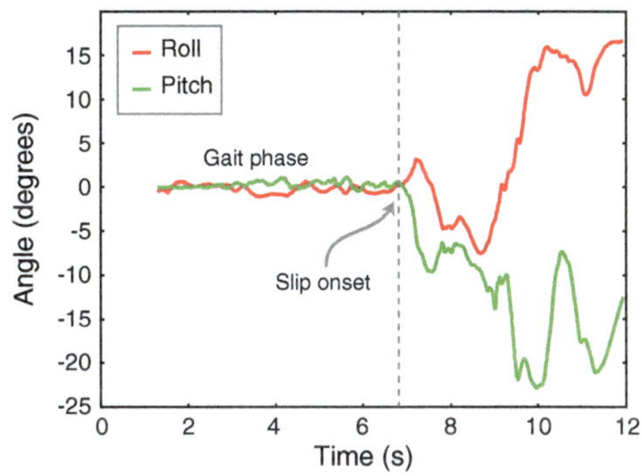

Fig. 8 This figure shows the body's roll (*red*) and pitch (*green*) orientation in degree during a gait phase and a leg slipping event without stabilisation

during a leg slip when no stabilisation is performed. A leg slip induces unexpected movement of the body, resulting in significant variances of the body orientation, as it can be seen in Fig. 7.

These uncontrolled variances in the body's orientation angles can significantly impact the stability of the robot. Therefore we aspire to limit sudden changes in the orientation of the body. In other words, the standard deviation of the orientation of the body frame B about the local frame L gives us a metric regarding the stability of the robot.

Another performance metric is the time t_{rec} it takes the system to stabilise the robot after a leg slip has been detected. This is the time interval between a stability criterion being violated to the time the robot is able to continue its locomotion with the specified gait pattern. The longer the robot continues to slip, more difficult it is to recover to a stable state and it becomes more vulnerable to damage either by toppling over or stripping gears. t_{rec} is calculated as

$$t_{rec} = t_c - t_s \tag{5}$$

where t_c is the time when the robot continues its locomotion, and t_s the time when a leg slip gets detected. Note that t_{rec} includes all stabilisation attempts since the system may perform multiple stabilisation attempts when its still in an unstable state after a particular recovery phase.

3.5 Experimental Procedure

In our experiments the hexapod walks straight using an alternating tripod gait. During its locomotion we let it step on a slipping obstacle (a small aluminium plate with a low friction coefficient) with a specific leg, causing the interacting leg to slip away from its intended foothold position (Fig. 8). This causes the whole robot to become unstable and, in certain cases, without our stabilisation system, the body to topple over. We repeat that experiment three times for each leg. This experimental

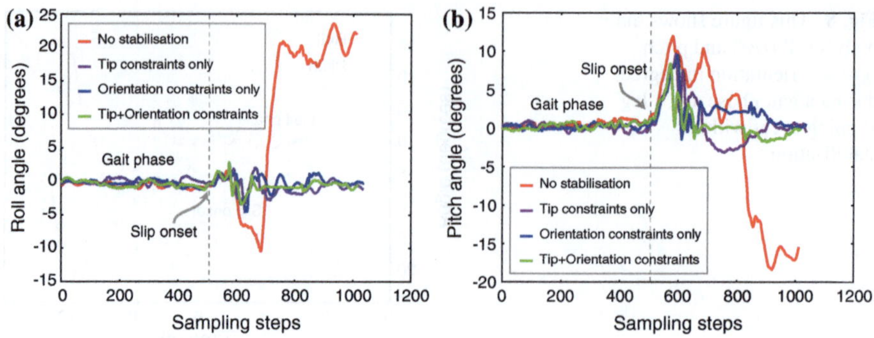

Fig. 9 Body roll (**a**) and pitch (**b**) during a leg slipping event without stabilisation and with stabilisation using tip, orientation and both constraints

design is repeated for each of the two task-space constraints (tip position cuboids and orientation constraints) and a combination of the two constraints. We also conducted a set of experiments where the stabilisation system was turned off. The standard deviations of body roll and pitch angles are recorded for each of the experimental runs and presented in the following section along with a comparison of recovery times for different stability constraint combinations.

4 Results and Insights

Tables 1, 2, 3 and 4 show the standard deviation of the body pitch and roll angles with respect to the local frame L during a leg slip and the subsequent stabilisation phase. The first column refers to the leg which was slipping. The second and third column refers to the standard deviation of the respective angular direction.

The results in Table 1 were obtained without any stabilisation, whereas for the Tables 2, 3 and 4 the proposed stabilisation approach used different task space

Table 1 Standard deviation of body roll and pitch (θ_r and θ_p) with no stabilisation for a series a leg slipping experiments

Slipping leg	σ_{θ_r}	σ_{θ_p}
Front left	3.65°	4.02°
Front right	4.75°	5.21°
Middle left	3.45°	4.12°
Middle right	4.89°	4.76°
Rear left	7.58°	8.93°
Rear right	5.51°	6.96°

Table 2 Standard deviation of body roll and pitch (θ_r and θ_p) with stabilisation using orientation constraints only for a series a leg slipping experiments

Slipping leg	$\sigma\theta_r$	$\sigma\theta_p$
Front left	1.03°	3.19°
Front right	1.58°	2.36°
Middle left	1.55°	3.09°
Middle right	1.00°	2.65°
Rear left	2.26°	2.26°
Rear right	3.01°	2.80°

Table 3 Standard deviation of body roll and pitch (θ_r and θ_p) with stabilisation using tip constraints only for a series a leg slipping experiments

Slipping leg	$\sigma\theta_r$	$\sigma\theta_p$
Front left	1.76°	3.17°
Front right	1.30°	2.20°
Middle left	1.92°	1.59°
Middle right	1.70°	2.65°
Rear left	1.28°	2.08°
Rear right	1.75°	1.38°

Table 4 Standard deviation of body roll and pitch (θ_r and θ_p) with stabilisation using tip constraints and orientation constraints for a series a leg slipping experiments

Slipping leg	$\sigma\theta_r$	$\sigma\theta_p$
Front left	1.74°	2.73°
Front right	1.02°	1.65°
Middle left	1.27°	2.07°
Middle right	1.57°	1.82°
Rear left	1.72°	2.63°
Rear right	1.37°	1.48°

constraints (orientation constraints only, tip position constraints only, orientation and tip position constraints).

It can be seen that the standard deviation for the roll and pitch angles are significantly reduced using the proposed stabilisation mechanism. Using the orientation constraints only had a higher overall standard deviation in roll and pitch angles, followed by the tip position constraints. The best results were obtained by using both types of task space constraints (Fig. 9).

Table 5 shows the time t_{rec} it takes for the system to get back to a stable state after a leg slip occurred and the number of stabilisation attempts, using different constraints. This includes the time $t_{connect}$ it takes for *RRTConnect* to find a path from a state

Table 5 Mean recovery times t_{rec} for leg slip recovery for each leg with different constraints along with the mean number of stabilisation attempts given within brackets

Slipping leg	t_{rec} for tip and orientation constraints	t_{rec} for tip constraints only	t_{rec} for orientation constraints only
Front left	0.85 s (1.0)	1.06 s (1.3)	1.53 s (1.6)
Front right	0.89 s (1.0)	1.16 s (1.0)	1.93 s (1.6)
Middle left	1.26 s (1.0)	0.96 s (1.3)	1.92 s (1.6)
Middle right	0.52 s (1.6)	1.66 s (2.6)	1.79 s (2.0)
Rear left	0.72 s (1.6)	4.34 s (3.6)	1.87 s (2.0)
Rear right	1.49 s (1.3)	2.99 s (3.0)	1.96 s (1.6)

that violates the task space constraints to a state inside the stable manifold M_{stable}, and the time t_{exec} the system takes to execute that path. On average, *RRTConnect* took 0.11 s until the algorithm found a path back to M_{stable}. The table also gives the mean number of stabilisation attempts after a leg slip for each leg. The recovery time using only the tip constraint cuboids was much higher compared to using orientation constraints alone when a rear leg was slipping. We believe that this was due to the direction of motion of the rear foot tips during the support phase in the gait cycle being in the same direction as the slip. This makes the overall slip much severe compared to the front or middle legs. Therefore, slips of the rear legs took a greater number of stabilisation attempts before the robot was able to continue its locomotion resulting in an overall longer recovery time for slips of the rear legs when using the tip constraint cuboids only. Using orientation constraints appeared detect and arrest the slip event much quicker in such instances. The stabilisation system showed best overall performance when both the tip constraint cuboids and the orientation constraints were used as seen from the results.

5 Conclusions

We presented a framework for real-time stabilisation of a high dimensional multi-legged robot. An experimental evaluation of this framework was performed using a hexapod robot and the results demonstrated the method effectively detects and recovers from unexpected events such as leg slip. The standard deviation of roll and pitch of the robot's body was used as a metric for stability. The stability of the robot improved by 2× when the proposed method was used, with a reduction of $3.27° \rightarrow 1.45°$ standard deviation for roll angle and a reduction of $3.94° \rightarrow 2.06°$ standard deviation for pitch angle. We also presented results for reaction time of the system when using different stability constraints based on foot tip positions as well as the body roll and pitch angles and demonstrated that the best results are achieved when both types of constrains are used for detecting instability at run-time. With the proposed real-time stabilisation system, the hexapod robot was capable of successfully recovering from unexpected leg-slip events and re-commence locomotion without explicit knowledge of the local terrain. Our current study is limited to planar terrains of zero elevation. We believe that our approach can easily extend to elevated planar slopes with the help of an onboard IMU. We are currently extending our approach to work on elevated planar surfaces. We plan to incorporate system dynamics with a better model of the hexapod platform and knowledge of the terrain from an onboard perception module to make our approach robust to challenging environments in the near future.

References

1. Lee, T.T., Liao, C.M., Chen, T.: On the stability properties of hexapod tripod gait. IEEE J. Robot. Autom. **4**(4), 427–434 (1988)
2. Hauser, K.K., Bretl, T., Latombe, J.C., Harada, K., Wilcox, B.: Motion planning for legged robots on varied terrain. Int. J. Robot. Res. **27**(11–12), 1325–1349 (2008)
3. Bretl, T., Rock, S.M., Latombe, J.C.: Motion planning for a three-limbed climbing robot in vertical natural terrain. In: Proceedings of the IEEE International Conference on Robotics and Automation (ICRA), pp. 2946–2953 (2003)
4. Belter, D., Skrzypczynski, P.: Integrated motion planning for a hexapod robot walking on rough terrain. In: Proceedings of the IFAC World Congress, vol. 18, pp. 6918–6923 (2011)
5. Belter, D., Skrzypczynski, P.: Posture optimization strategy for a statically stable robot traversing rough terrain. In: Proceedings of the IEEE/RSJ International Conference on Intelligent Robots and Systems (IROS), pp. 2204–2209 (2012)
6. Kerscher, T., Rönnau, A., Ziegenmeyer, M., Gassmann, B., Zoellner, J., Dillmann, R.: Behaviour-based control of a six-legged walking machine LAURON IVc. In: Proceedings of the 11th International Conference on Climbing and Walking Robots (CLAWAR), pp. 8–10 (2008)
7. Rönnau, A., Kerscher, T., Ziegenmeyer, M., Zollner, J., Dillmann, R.: Adaptation of a six-legged walking robot to its local environment. In: Kozowski, K. (ed.) Robot Motion and Control 2009. Lecture Notes in Control and Information Sciences, vol. 396, pp. 155–164. Springer, London (2009)
8. Best, G., Moghadam, P., Kottege, N., Kleeman, L.: Terrain classification using a hexapod robot. In: proceedings of the Australasian Conference on Robotics and Automation (ACRA) (2013)
9. Wettergreen, D., Thorpe, C.: Developing planning and reactive control for a hexapod robot. In: Proceedings of the IEEE International Conference on Robotics and Automation (ICRA), pp. 2718–2723 (1996)
10. Lewinger, W.A., Quinn, R.D.: A hexapod walks over irregular terrain using a controller adapted from an insect's nervous system. In: Proceedings of the IEEE/RSJ International Conference on Intelligent Robots and Systems (IROS), pp. 3386–3391 (2010)
11. Boor, V., Overmars, M.H., Van der Stappen, A.F.: The Gaussian sampling strategy for probabilistic roadmap planners. In: Proceeding of the IEEE International Conference on Robotics and Automation (ICRA), pp. 1018–1023 (1999)
12. Simon, T., Laumond, J.P., Nissoux, C.: Visibility-based probabilistic roadmaps for motion planning. Adv. Robot. **14**(6), 477–493 (2000)
13. Sun, Z., Hsu, D., Jiang, T., Kurniawati, H., Reif, J.H.: Narrow passage sampling for probabilistic roadmap planning. IEEE Trans. Robot. **21**(6), 1105–1115 (2005)
14. Lee, J., Kwon, O., Zhang, L., Yoon, S.E.: SR-RRT: Selective retraction-based RRT planner. In: Proceeding of the IEEE International Conference on Robotics and Automation (ICRA), pp. 2543–2550 (2012)

15. Zhang, L., Manocha, D.: An efficient retraction-based RRT planner. In: Proceeding of the IEEE International Conference on Robotics and Automation (ICRA), pp. 3743–3750 (2008)
16. Rodrguez, S., Tang, X., Lien, J.M., Amato, N.M.: An obstacle-based rapidly-exploring random tree. In: Proceeding of the IEEE International Conference on Robotics and Automation (ICRA), pp. 895–900 (2006)
17. Hsu, D., Sãnchez-Ante, G., Cheng, H.L., Latombe, J.C.: Multi-level free-space dilation for sampling narrow passages in PRM planning. In: Proceeding of the IEEE International Conference on Robotics and Automation (ICRA), pp. 1255–1260 (2006)
18. Saha, M., Latombe, J.C.: Finding narrow passages with probabilistic roadmaps: the small step retraction method. In: Proceedings of the IEEE/RSJ International Conference on Intelligent Robots and Systems (IROS), pp. 622–627 (2005)
19. Lavalle, S.M., Kuffner, J.J., Jr.: Rapidly-exploring random trees: progress and prospects. In: Algorithmic and Computational Robotics: New Directions, pp. 293–308 (2000)
20. Kuffner Jr., J.J., Lavalle, S.M.: RRT-Connect: an efficient approach to single-query path planning. In: Proceeding of the IEEE International Conference on Robotics and Automation (ICRA), pp. 995–1001 (2000)

State Estimation for Shore Monitoring Using an Autonomous Surface Vessel

Gregory Hitz, François Pomerlesau, Francis Colas and Roland Siegwart

Abstract Although many applications of small Autonomous Surface Vessels rely on two-dimensional state estimation, inspection tasks based on long-range sensors require more accurate attitude estimates. In the context of shoreline monitoring relying on a nodding laser scanner, we evaluate three different extended Kalman filter approaches with respect to an accurate ground truth in the range of millimeters. Our experimental setup allowed us to track the impact of sensors noise, including GPS non-Gaussian error, a phenomenon often underestimated. Extensive field experiments demonstrate that the use of a complementary filter in combination with a model-based extended Kalman filter performed best and reduced velocity errors by 73 % compared to GPS. Finally, following our state estimation observations, we present a long-term shore monitoring result highlighting changes in the environment over a period of 6 months.

Keywords State estimation · ASV · ICP · Shore monitoring · 3D point clouds

1 Introduction

Most of the small and medium Autonomous Surface Vessels (ASVs), like the one presented in Fig. 1, operate in open environments and handle many tasks, such as station keeping [17], path following [1], or environmental monitoring [7, 8]. Those researches demonstrated that a low precision two-dimensional (2D) state estimation based on GPS and compass measurements is sufficient for such applications. Unfortunately, other applications, such as mapping shorelines [20] or bathymetry [3], rely on long-range sensors. Such applications are sensitive to poor attitude estimation leading to noise often referred to as motion blur. For example, many three-dimensional (3D) scanners used in robotics are assembled from a 2D laser rangefinder and a motor.

G. Hitz (✉) · F. Pomerleau · F. Colas · R. Siegwart
Autonomous Systems Lab, ETH Zurich, Zurich 8092, Switzerland
e-mail: hitzg@ethz.ch

F. Pomerleau
Département d'informatique et de génie logiciel, Université Laval, QC, Canada

M.A. Hsieh et al. (eds.), *Experimental Robotics*, Springer Tracts in Advanced Robotics 109, DOI 10.1007/978-3-319-23778-7_49

This type of scanner takes approximately 2 seconds to complete a swipe. During this time, the ASV rocks with the waves and, even at low speed, can change position significantly. To give an idea of the distortion, a typical 5° oscillation of the boat causes more than a meter vertical uncertainty on the localization of a feature at a distance of 10 m, leading to significant motion blur in the resulting point clouds. There are ways to circumvent the problem of attitude estimation. Increasing the acquisition rate by ensuring a fast shutter speed of cameras can be achieved by using the platform in bright daylight [4]. Also, high-end laser rangefinders allowing high scanning frequency can be used [16], increasing significantly the cost of the platform. Finally, the attitude problem can be directly addressed at the design level by implementing passive damping systems [10] or by increasing the size and inertia of the ASV. Such systems can reduce the impact of the main motion frequency, but can hardly be built generic enough to handle the large spectrum of events that an ASV can encounter during a mission. To overcome those limitations, we investigate how attitude estimation can be added in a 3D state estimation algorithm fulfilling properly the constraints of an ASV.

Solutions for 3D state estimation were widely investigated in the recent years. Developed in the context of large ocean vehicles, Fossen [6] proposed a more traditional approach using a dynamic system model to feed as prediction to an Extended Kalman Filter (EKF), while all the sensors are used in the correction step. Motion models linking the commands to states can be difficult to build or computationally expensive to compute. This is particularly true for the the Unmanned Aerial Vehicle (UAV) field. To overcome this problem, EKFs using the Inertia Measurement Unit (IMU) directly in the prediction step were proposed [12]. Focussing only on the attitude estimation based on an IMU, Mahony et al. [13] introduced the concept of the Complementary filter (CF) and demonstrated its performance on a micro aerial vehicle. This variety of solutions require a deeper investigation on how the IMU should be used in an EKF. Moreover, those solutions were not developed for ASVs, which have their own motion characteristics. To adopt the proper solution, we compare the current state-of-the-art EKF algorithms using an external tracking system which provides us with highly accurate ground truth measurements and discuss the impact of the ASV particularities on those filters.

Finally, range measurements can be used to monitor shoreline vegetation, but can also facilitate an additional localization method for the ASV. The registration of subsequent point clouds provides relative position updates and, a step further, the comparison against a known map links the position information to a global frame. Rectifying laser scan measurements via state prediction during the swiping has been done for other types of robots (e.g. ground robots [11] and quadrotors [15]). Another approach is to include the laser information directly in the state estimation [2]. This solution works well when the surrounding environment constrains properly the scan registration, but sensor maximum range coupled with water reflectivity can reduce the returned points to an unstable number. Shoreline monitoring has been shown to be a delicate operation for the integrity of the platform when using a quadrotor [20]. ASVs can support large payloads and repeatedly navigate in a given area. As reported by Kelly et al. [9], experiments aiming at long-term autonomy require a higher level

Fig. 1 *Left* Our ASV, equipped with a reflective prism and a nodding laser scanner. *Right* The theodolite (Leica Total Station TS15), used for ground truth measurements, positioned on shore

of integration than short term experiments. This might explain why, to the knowledge of the authors, no prior work on shore monitoring was presented. Thus, we conclude our paper by presenting 3D mapping results highlighting environmental changes that occur over a period of 6 months. Map management techniques allow us to maintain a clean representation of the environment, while being able to adapt to changes. The ASV used for the paper is shown in Fig. 1, with the up and down nodding Hokuyo laser in front. The other sensors are a GPS receiver, a compass and an IMU. The boat measures $2.5 \times 1.8 \times 0.9$ m and weighs approximately 155 kg. It is used as a toxic cyanobacteria monitoring tool and is fully described in [8].

2 State Estimation for ASVs

The goal of using state estimation procedures is to estimate the state of a system more accurately, given measurements from a set of sensors. In our application, we want to estimate the position and attitude of the body fixed frame $\{B\}$ with respect to an inertial world frame $\{W\}$. For this purpose, we use an EKF, which makes use of a prediction model to assess how probable measurements are when they are fed to the filter. The basis of Kalman filter (KF) will be considered as textbook knowledge for the purpose of this article, so we will only describe the specificities of the filter variants we investigated in this section.

In our application, there are three sensors which provide data to the filter. The GPS receiver measures the position of the boat with respect to the world frame. The compass (in our case, a one-dimensional one) provides the north vector in the body frame. Furthermore, the IMU measures the accelerations and the angular velocities of the body in the body frame. We assume that all our sensor measurements are corrupted by additive white Gaussian noise. We also assume that the accelerometers and gyroscopes of the IMU are distorted by time varying, additive biases, which need

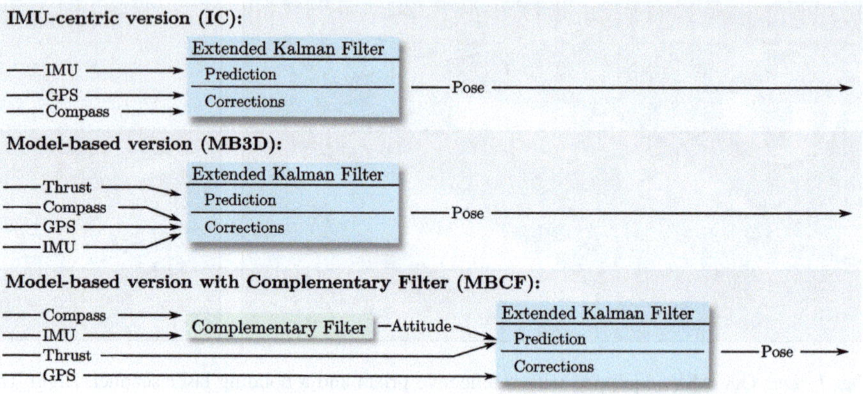

Fig. 2 The three filtering schemes used for state estimation. From *top* to *bottom*: IC, MB3D and MBCF

to be estimated along with the states of the system. Assuming that GPS measurements are only affected by white Gaussian noise is problematic, yet greatly simplifies the mathematical description. We will discuss this issue in the results section.

The world frame $\{W\}$ is defined as a metric North-East-Down frame (NED). In our specific case, we use the Swiss grid [5], which is based on a Mercator projection. However, any other metric coordinates could be used (e.g. UTM). We directly convert GPS measurements from WGS84 to Swiss grid coordinates before feeding them to the state estimator. The body frame of the boat is defined similarly to the world frame with the x axis pointing forwards, the y axis to the right and z downwards. In the following, we describe three different formulations of EKFs to perform state estimation on our ASV. Figure 2 gives an overview of the three filter versions.

2.1 IMU-centric Extended Kalman Filter

The IMU-centric filter implements the state-of-the-art EKF formulation that is mostly used in UAV applications. It is favored for complex systems since it circumvents the use of dynamical models by directly integrating the IMU measurements in the prediction steps, treating them as a system input rather than actual sensor measurements. This simplifies the prediction model to simple kinematic equations. The measurements from the GPS receiver and the compass are treated as regular updates in the EKF. Our formulation of the IMU-centric EKF follows closely the one of Leutenegger and Siegwart [12]. The state vector is defined as $x = [p, q, v, b_g, b_a]^T$ where $p \in \mathbb{R}^3$ defines the position in the world frame, $q \in SO(3)$ is a unit quaternion describing the attitude, $v \in \mathbb{R}^3$ is the linear velocity of the boat in the world frame, $b_g \in \mathbb{R}^3$ and $b_a \in \mathbb{R}^3$ describe the biases of the gyroscopes and accelerometers respectively. The covariance matrix of the quaternion representing the attitude is not well defined due to its unit length constraint [21]. Therefore, the filter is defined to operate on the error state. For the attitude it is defined multiplicative: $\hat{q} = \delta q \otimes q$, whereas for the other states it is additive: $\hat{x}_{\backslash q} = x_{\backslash q} + \Delta x_{\backslash q}$. The resulting error quaternion δq can be reduced to a 3D representation using small angle approximations [21], which renders the covariance matrix non-singular again. The full derivation of the update equations and Jacobian matrix are not provided here, but follow closely the ones provided by Leutenegger and Siegwart [12].

2.2 Model-Based Extended Kalman Filter

The model-based EKF implementation uses a dynamic model, which describes the dynamics of the system based on the commands sent to the motors. For the derivation of the model, we follow the work of Fossen [6]. The model is defined by a kinematic part (Eq. 1) and a dynamic part (Eq. 2):

$$\dot{\eta} = J(\eta)\nu \tag{1}$$

$$M\dot{\nu} + C(\nu)\nu + D(\nu)\nu + g(\eta) = \tau_m + \tau_e \tag{2}$$

where $\eta \in \mathbb{R}^6$ is the position and attitude of the boat defined in the world frame $\{W\}$, $\nu \in \mathbb{R}^6$ its linear and rotational velocities in the body frame $\{B\}$, M the inertia matrix, C the Coriolis matrix, D the damping matrix, g a vector of hydrostatic restoring forces, τ_m the thruster forces as input and τ_e unknown external perturbations. For this model, we use a set of Euler angles to represent the attitude of the boat, which simplifies the modeling of the hydrodynamic restoring forces. Note that the well-known singularity at 90° pitch is not problematic for a surface vessel. We formulate the model (Eq. 2) around the center of gravity of the body. Therefore, the inertia matrix can be formulated as follows:

$$M = \begin{bmatrix} m\,\mathbb{I}_{3\times3} & 0_{3\times3} \\ 0_{3\times3} & I_g \end{bmatrix}$$

where m is the mass of the boat, $\mathbb{I}_{3\times3}$ is the 3×3 identity matrix, and $I_g \in \mathbb{R}^{3\times3}$ is the rotational inertia matrix. To further simplify the model and reduce the number of parameters, we assume that $I_g = \text{diag}(I_x, I_y, I_z)$.

The Coriolis matrix is defined as follows:

$$C(\nu) = \begin{bmatrix} mS(\nu_2) & 0_{3\times3} \\ 0_{3\times3} & -S(I_g\nu_2) \end{bmatrix}, \quad \text{where} \quad S(x) = \begin{bmatrix} 0 & -x_3 & x_2 \\ x_3 & 0 & -x_1 \\ -x_2 & x_1 & 0 \end{bmatrix} \tag{3}$$

$\nu_2 \in \mathbb{R}^3$ denotes the rotational part of ν and $S(x)$ denotes the cross-product operator.

Simple linear damping forces have proven to yield good results, at least in the velocity regimes that our ASV operates in (i.e. up to 1.5 m/s). Therefore, we use the following damping matrix:

$$D(\nu) = \text{diag}(D_x, D_y, D_z, D_\phi, D_\theta, D_\psi)$$

The hydrostatic restoring forces are most difficult to model. The buoyancy of a vessel is defined by the mass of water that it displaces, which depends on the (unknown) shape of the the local water surface surrounding the boat. Even when assuming a flat water surface, the buoyancy depends on the geometry of the boat. While the geometry is usually known, it is less straightforward to define the derivatives thereof, which are necessary to formulate the Jacobian matrix of the entire system. Fossen [6] has derived a simplified formulation under the assumptions of small roll and pitch angles and box-shaped vessels. It decouples the different dimensions and results in a linear equations:

$$g(\eta) = G\eta \quad \text{where} \quad G = \text{diag}(0, 0, G_z, G_\phi, G_\theta, 0) \tag{4}$$

Using this simplification the resulting dynamics are described by linear second order systems.

The state vector for the filter is defined as $x = [\eta^T, \nu^T, b_g]^T$ and is discretized at constant time intervals Δt. The discretized state transition and the Jacobian matrix are then defined as follows:

$$x_{k+1} = x_k + \Delta t\, \dot{x}_k \qquad\qquad F = \frac{\partial x_{k+1}}{\partial x_k} = I + \Delta t \frac{\partial \dot{x}_k}{\partial x_k}$$

The Jacobian matrix can be computed analytically. Due to space restrictions we have to omit this derivation here. While the updates of compass, GPS and gyroscope measurements are straightforward, the measured accelerations can not be used directly. In the general case, the IMU is not situated in the center of gravity and thus the measured accelerations need to be translated to the center of gravity. Since this is done in a moving frame $\{B\}$, the translation depends not only on the rotational velocities, but also on the rotational accelerations. As this is unpractical, the other option would be to shift the center of the model equations to the IMU. This however, makes the model itself far more complicated. For these reasons we do not use the measured accelerations in the model-based version.

2.3 Attitude Estimation with the Complementary Filter

In our third approach we use a separate estimator for the attitude of the ASV. For this purpose we use the Complementary filter (CF) [13], which integrates the gyroscopic measurements ω and corrects them with measurements of known directions v_i. The state of the filter consists of the attitude represented by a unit quaternion $\hat{q}$ and a vector of additive gyroscope biases $\hat{b}_g$. The CF is then defined as follows:

$$
\begin{aligned}
\omega_c &= -\text{vex}\left(\sum_{i=1}^{N} \frac{k_i}{2}\left(v_i \hat{v}_i^T - \hat{v}_i v_i^T\right)\right) \\
\dot{\hat{q}} &= \frac{1}{2}\hat{q} \otimes \left(\omega - \hat{b}_g + k_P \omega_c\right) \\
\dot{\hat{b}}_g &= -k_I \omega_c
\end{aligned}
$$

where vex$(\cdot)$ defines the inverse operator of the cross-product operator (Eq. 3): $\text{vex}(S(a)) = a$ for $a \in \mathbb{R}^3$. The correction term ω_c depends on the measured directions v, on the estimates thereof $\hat{v}$ and on their respective weights k_i. k_P and k_I denote filter parameters. In our case, the measured directions are provided by the compass and the accelerometers. The compass measures the north vector and the accelerometers provide the direction of gravity. However, the gravity vector is affected by the accelerations, which are induced by the motors. Such distortions can be avoided by having the model described above estimate the forces caused by the motors and correct the measurements. Given the attitude estimate from the CF, we use a simplified version of the model-based state estimator to estimate the remaining states (positions p and linear velocities v).

2.4 Point Cloud Registration

Building upon the state estimation procedures presented above, we can assemble undistorted point clouds from the range measurements of the laser scanner. Using an optimized version of the Iterative Closest Point algorithm based on the observation of Pomerleau et al. [19], we register the point clouds to a global map. Furthermore, state-of-the-art map management methods [18] allow us to distinguish static and dynamic points. This brings several benefits. First, shorelines are usually not entirely static, for instance boats that are moored at a single buoy have different positions depending on wind conditions. Second, noisy points which are introduced by state estimation errors, are classified as dynamic points and can be removed. Finally, it allows to detect seasonal changes in the shoreline vegetation, which can be of interest for biological studies, one of the scientific goals of our ASV.

3 Experiment Setup

To evaluate the presented state estimation approaches, we conducted several series of field tests. We used the collected data to estimate model parameters and assess the noise level of the GPS receiver. We also collected shoreline mapping data sets with the laser rangefinder over the course of 6 months.

3.1 Ground Truth Positioning Information

To go beyond comparing the results of the state estimation against its sensor input (GPS and compass measurements), we recorded ground truth positioning information with an external positioning device. For this purpose, we mounted a reflective prism on the boat (see Fig. 1), which was then tracked by a theodolite (Leica Total Station TS15). This device is able to track the prism in a range of up to 2 km with 2 mm accuracy. However, this only provides positioning ground truth and no information about the attitude of the boat. As Fig. 1 shows, the prism had to be mounted 116 cm above the center of gravity of the boat to ensure visibility at all times. Rocking motions of 5° lead thus to a horizontal offset of up to 0.1 m. Such errors can either be corrected by the attitude estimate (which might however also introduce additional errors), or be ignored since the error is still an order of magnitude smaller when compared to GPS noise.

3.2 Point Cloud Distortion Measure

Measuring the attitude of the ASV in a very precise manner is very difficult. One approach, which was used for work on rovers for lunar missions [14], would be to track three or more reference points on the boat simultaneously, creating a setup similar to indoor tracking systems. Having only one theodolite available rendered the use of external observation methods infeasible. The use of visual markers on shore and a camera on the boat could be a good solution, but the distance between the shore and the boat would reduce greatly the accuracy of those measurements.

To circumvent this issue, we chose not to quantify the absolute attitude error, but rather the distortion of the resulting point clouds. For this purpose, we set up a specific experiment, referred to as *pole experiment* henceforth. A pole covered by reflective material was mounted on a pontoon (see Fig. 3) and was scanned repeatedly using

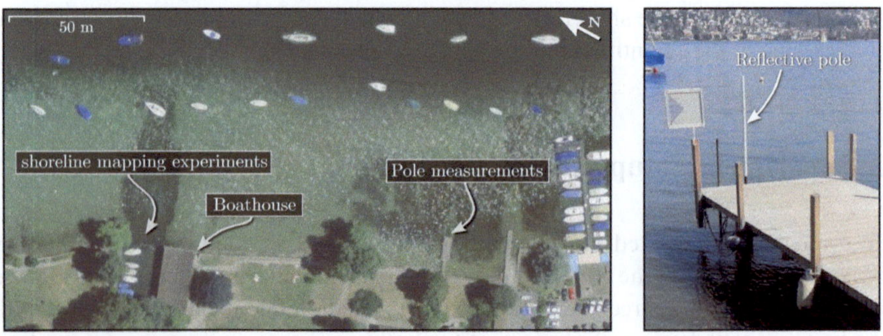

Fig. 3 *Left* Overview of the testing area. *Image source* Google Earth, 47.32023°N, 8.553017°E, Feb. 6, 2014. *Right* The reflective pole used to assess the distortion of point clouds

the laser mounted on the boat. As the measured points on the pole have significantly higher return intensities, they can be extracted from the point cloud simply by applying a constant threshold. Fitting a line to the extracted points provides two measures which we have used to assess the distortion of the point cloud. First, the root mean squared error (RMSE) of the line fit provides a measure of distortion. Second, the deviation of the inclination of the fitted line from the vertical provides an additional performance measure. During this test, the boat was not actuated, i.e. only affected by waves. To eliminate distortion caused by linear motion of the boat, we used the theodolite to track the motion of the ASV during the test.

3.3 State Estimation Evaluation

In addition to the *pole experiment* described above, we have defined the following error metrics to evaluate the results of the three state estimation approaches.

$$e_{\text{pos},k} = \left\| p_k - p_{l,k} \right\|_2 \tag{5}$$
$$e_{\text{vel},k} = \left\| v_k - v_{l,k} \right\|_2 \tag{6}$$

$$d_{\text{pos},k} = \left\| p_{k+1} - p_k \right\|_2 \tag{7}$$
$$d_{\text{vel},k} = \left\| v_{k+1} - v_k \right\|_2 \tag{8}$$
$$d_{\text{att},k} = \left| \ln \left(q_{k+1} \otimes q_k^{-1} \right) \right| \tag{9}$$

where p_l and v_l refer to position and derived velocity measurements from the theodolite. The index k refers to the value at time step t_k. Equations 5 and 6 describe the error in position (e_{pos}) and velocity (e_{vel}). These are measured by computing the estimation error with respect to ground truth measurements and give thus an absolute error. To measure the smoothness of an estimated state trajectory, we use a measure of discontinuity d which is defined for position (d_{pos}), velocity (d_{vel}) and attitude (d_{att}) by the Eqs. 7–9, respectively. Measuring smoothness reveals drastic update step which can occur whenever predicted measurements and actual sensor values do not match well in the update steps of the EKF.

4 Results

We have conducted a series of field experiments, during which we have collected data sets consisting of GPS, compass and IMU measurements, system specific information (such as thrust values of the motors) and external position information from a theodolite. We collected data in 5 one-day campaigns, during which the boat trav-

Table 1 Parameters for the model of the ASV

Damping		Inertia ($kg\,m^2$)	Restoring forces ($kg\,m^2\,s^{-2}$)
$kg\,s^{-1}$	$kg\,m^2\,s^{-1}$		
D_x: 80	D_ϕ: 0.5	I_ϕ: 5	G_ϕ: 70
D_y: 1000	D_θ: 0.5	I_θ: 2	G_θ: 70
D_z: 900	D_ψ: 200	I_ψ: 190	G_z: 600

eled an overall distance of 10.3 km. We encountered varying conditions ranging from strong winds and rain to very calm days, resulting in data sets with different environmental influences.

4.1 Model Parameters

The proposed model-based state estimator relies on the underlying model of the ASV, which then is highly dependent on its parameters. Besides the weight of the boat, which we have measured with a spring scale (155 kg), there are several other parameters that can not be measured directly. To estimate them we have manually conducted an iterative optimization with respect to measured data. Table 1 provides an overview of those parameters.

4.2 State Estimation

Figure 4 shows both the raw GPS data points and the theodolite measurements (GT) along an example of a short round trip path. The graph clearly shows the limitations of the GPS measurements, with an average error of 1.9 m. However, the graph also shows that the error is not Gaussian, but rather shows constant offsets in a particular direction, which is highlighted in the first inset (1), where the GPS measurements start diverging severely from the ground truth. The third EKF version (MBCF) is not shown in Fig. 4, because it bases on the same motion model as MB3D and thus the resulting trajectory is almost identical. While the overview on the right shows that both versions follow the ground truth nicely, the two insets (1) and (2) show in more detail the errors caused by the GPS offsets: as the EKFs assume Gaussian noise on the GPS signal, they eventually drift to reduce the relative error to the GPS signal (see inset (1)). As long as the offset on the GPS signal remains relatively constant (see inset (2)), the estimates from both EKFs cannot reduce the error with respect to the ground truth. This is a common problem when using GPS signals and it emphasizes why we are interested in feeding exteroceptive measurements into the state estimators. As a result of this the overall performance of all EKF implementations in terms of the position error (Eq. 5) is only slightly better than the raw GPS positioning.

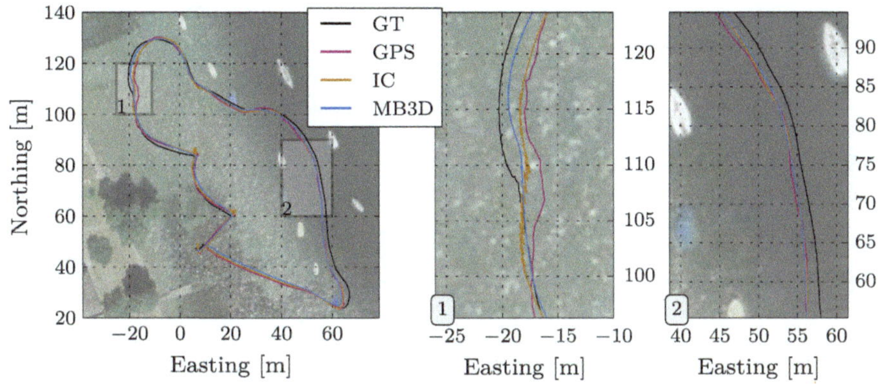

Fig. 4 *Left* Top view of resulting trajectories. *GT* refers to the ground truth data from the theodolite (*black line*), *IC* denotes the IMU-centric approach (*brown line*) and *MB3D* the model-based approach (*blue line*). The *GPS* measurement are represented with a *purple line*. *Middle and Right* Zoomed results showing GPS offset

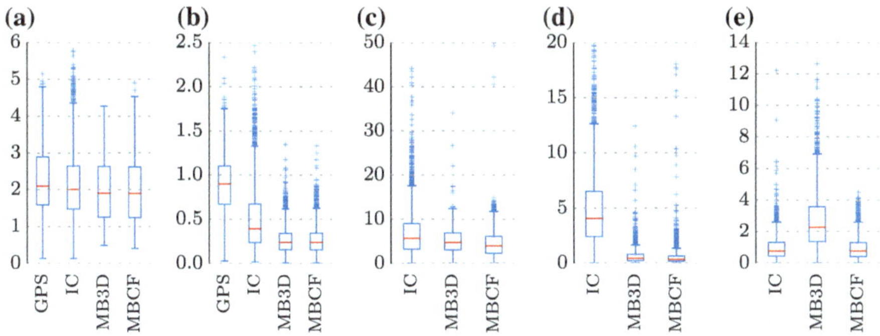

Fig. 5 Results of the evaluation. The metrics are: position error (e_{pos}), velocity error (e_{vel}), position discontinuity (d_{pos}), velocity discontinuity (d_{vel}), attitude discontinuity (d_{att}). For the first two, the raw GPS is shown as a *baseline*. **a** e_{pos} (m), **b** e_{vel} (m/s), **c** d_{pos} (mm), **d** d_{vel} (mm/s), **e** d_{att} (mrad)

Figure 5 shows the results of the evaluation according to the metrics defined in Eqs. 5–9 of combined data from all 5 test days. For the absolute errors in position (a) and velocity (b) the raw GPS is provided as a comparison. In the case of the velocity, this refers to the differentiation of the GPS points. Since the discontinuity measures are not absolute, we can not compare them to the raw GPS input.

As already observed in the top view in Fig. 4, there is only very little improvement in terms of the absolute position error e_{pos} with respect to the raw GPS measurements. The medians reduced by 4.1, 9.1 and 9.4 % for the IC, MB3D and MBCF versions. While these differences are statistically significant,[1] the difference between MB3D and MBCF is not ($p = 0.24$). This correlates with the fact that these two versions

[1] We use the Mann-Whitney U test, with $p < 0.001$ as significance threshold.

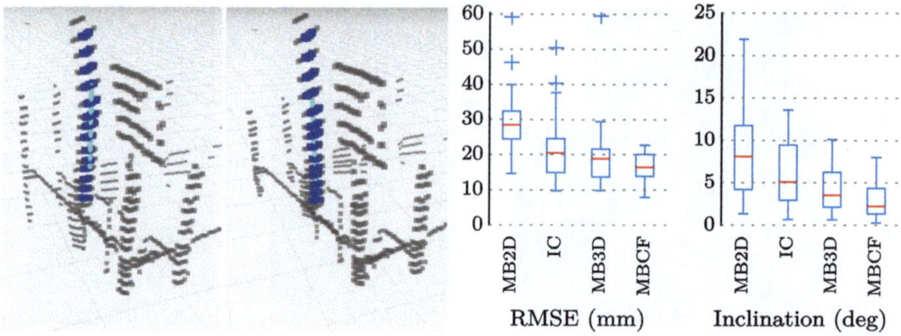

Fig. 6 Results of the pole experiment. *Left* An example of an assembled point cloud without and with attitude estimation. The *blue points* indicate the pole. The *light blue line* shows a 1 m segment of the fitted line. *Right* The fitting errors and the resulting inclination of the fitted line. 46 point clouds were used

are based on the same model for linear motion and thus have very similar errors. The high position errors for all EKF versions highlight the problem of non-Gaussian GPS offsets. Whereas the positioning information can not be improved significantly, the estimates of the velocity clearly improve with respect to the GPS baseline. The model-based versions (MB3D and MBCF) show the lowest velocity errors (73.5 and 73.2 % better than GPS). The IMU-centric EKF has still quite some outliers and only improves by 56.3 % with respect to the baseline, which is due to wrong estimates of the accelerometer biases. Since the GPS measurements clearly have non-Gaussian components, the corresponding position updates map the error to the accelerometer biases and velocity estimates. This effect also leads to less smooth state trajectories in the position and velocity space. Figures 5c and 4d show this in terms of the discontinuity measure (cf. Eq. 7–9). A high discontinuity measure indicates that the corresponding state was frequently corrected, leading to large differences between two subsequent states. Wrong estimates of the accelerometer biases distort the velocity estimates during the predictions, which then need to be corrected by sensor updates. Especially the discontinuity of the velocity estimates shows that the model-based versions (MB3D and MBCF) result in smoother trajectories than the IC version.

The discontinuity of the attitude estimates suggests that MB3D has higher inconsistencies, which relates to the fact that it relies on a parameter-based, strongly simplified model. Even though we estimated these parameters with respect to field data, the linear formulation of the hydrostatic restoring forces (cf. Eq. 4) might be too simple to accurately describe the motion. In relation to the MB3D version, the IC and MBCF implementations have lower discontinuities (improvements by 66.4 and 66.7 %). Both versions achieve relatively smooth attitude trajectories as they directly integrate the gyroscope measurements. Even though there are quite some difference in their implementations, the resulting distributions of discontinuity measures do not differ significantly ($p = 0.08$).

Figure 6 presents the results of the pole experiment. On the left hand side, an example of a distorted point cloud and its rectified version is shown. The pole that was used as a vertical and straight reference is highlighted in blue in both pictures. To assemble the distorted point clouds as a baseline, the model-based state estimator was used in 2D, thus ignoring any roll and pitch. The images also show a 1 m segment of the fitted line in light blue. A total of 46 point clouds were processed, i. e. the pole was extracted in each point cloud and a line was fitted to it. The resulting statistics are shown on the right in Fig. 6. The RMSE of the fitted lines show that there is clearly most distortion in the baseline, where no attitude estimation is used. The MBCF version performs best, owing to its good model-based translational components and the direct integration of the gyroscope measurements in the complementary filter. A possible explanation for the poor performance of the IMU-centric implementation is that, also here, erroneous bias estimates distort the gravity vector. Those biases influence the final attitude estimates leading to a larger error. The inclination of the line also shows that the MBCF performs best. The inclination angles have the lowest median and the smallest spread, meaning that the fitted lines were most upright. One could note that the lack of a ground truth measurement of the poles inclination causes a few degrees of uncertainty in the inclination measure in an absolute sense, but the relative performances between solutions remain the same.

The experimental comparison of the three proposed state estimators showed that the model-based version in combination with a complementary filter (MBCF), performed best for the system at hand. It combined the good performance of the linear components of the model-based approach, with the simplicity of the IMU-centric version for the attitude estimation. And thus, circumvented the need for formulating an accurate model of the hydrostatic restoring forces, which describe the rocking motion of the boat. Our results also show that the IMU-centric version does not perform as well as one might expect, which is mostly due to the error on the GPS signal. In comparison to the work of [12], the motion of an ASV is in the same order of magnitude as the error level on the GPS readings. This fact and the non-Gaussian characteristics of the noise lead to a lower performance when applied on an ASV.

4.3 Point Cloud Mapping

As it was our goal to collect point cloud information of shoreline areas, we have also collected a data set consisting of laser scans, GPS, compass and IMU measurements during a period of four months from October 10, 2013 to April 16, 2014. Different weather conditions were encountered, such as rain, bright sunlight, moderate wind and consequently waves. Figure 7 shows the final map on the left and gives a qualitative assessment of the achieved accuracy. The boathouse has straight and sharp walls and consistent rectangular corners. The map also shows large trees in the back of the small harbor, which can cause severe GPS shading. At the water front, there is a large willow tree which had only hanging branches, but no leafs during winter (depicted in the right of Fig. 7). The corresponding point cloud map is shown in the middle

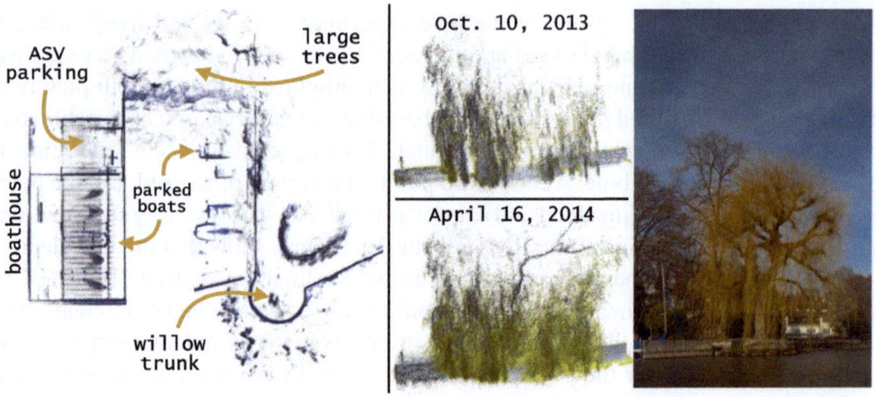

Fig. 7 Results of the shoreline mapping. *Left* Top view of the resulting 3D point cloud map from combined data over 6 months. *Middle* Side view of the willow tree, in October (Autumn) and April (Spring), respectively. *Right* The willow tree in October

(Oct. 10, 2013). Over the course of spring the willow tree grew more leaves and the corresponding point clouds became denser, which would have led to a very cluttered map. By detecting dynamic points, we are able to only use points for the final map, of which we are certain enough that they belong to a static structure. In Fig. 7, the points, which were classified as dynamic, are highlighted in yellow. At the example of the willow tree, we demonstrated both the importance of map maintenance, and the ability to readily use 3D mapping techniques to track environmental changes on shore.

5 Conclusion

In this article, we evaluated different types of filters aiming at a more accurate 3D state estimation in a context of shoreline monitoring with an ASV. Although building an accurate motion model of forces applied to a floating platform is not trivial, we demonstrated that filters using an approximated model provide better performance than filters without. Unfortunately, the simplicity of the model selected doesn't recover properly roll and pitch angles. Thus, we showed that adding the directly using IMU measurements and a complementary filter produces more accurate state estimates. Also, acceleration forces applied to a small or medium ASV are smaller than an UAV. In this application context, the combination of low accelerations and non-Gaussian noises on the GPS results in a wrong estimation of the IMU biases. This renders the use of IMU-centric filter less attractive in the case of an ASV. Moreover, GPS positions can have static offsets depending of the visible satellite constellations. This static bias can only be estimated by feeding another position sensor in the EKF. As we could highlight with our experimental setup, this offset causes a considerable error on all the solutions evaluated. Further research on how to handle this type of non-Gaussian GPS noise should be investigated in the future.

Acknowledgments This work was funded by the Swiss National Science Fund (CR22I2–130023) and the EU FP7 projects NIFTi (ICT-247870) and Tradr (ICT-609763). We also thank the Natural Sciences and Engineering Research Council of Canada (NSERC) and the *Fonds Québécois de la Recherche sur la Nature et les Technologies* (FQRNT) for the funds granted to Prof. P. Giguère.

References

1. Bibuli, M., Bruzzone, G., Caccia, M., Lapierre, L.: Path-following algorithms and experiments for an unmanned surface vehicle. J. Field Robot. **26**(8), 669–688 (2009)
2. Bosse, M., Zlot, R.: Continuous 3D scan-matching with a spinning 2D laser. In: IEEE International Conference on Robotics and Automation, pp. 4312–4319 (2009)
3. Chen, V.L., Batalin, M.A., Kaiser, W.J., Sukhatme, G.: Towards spatial and semantic mapping in aquatic environments. In: IEEE International Conference on Robotics and Automation (ICRA), pp. 629–636 (2008)
4. Dunbabin, M., Lang, B., Wood, B.: Vision-based docking using an autonomous surface vehicle. In: IEEE International Conference on Robotics and Automation (ICRA), pp. 26–32 (2008)
5. Federal Office of Topography (swisstopo): Swiss map projections, http://www.swisstopo.admin.ch/internet/swisstopo/en/home/topics/survey/sys/refsys/projections.html (2014). Accessed 18 May 2014
6. Fossen, T.I.: Handbook of marine craft hydrodynamics and motion control. Wiley, Chichester (2011)
7. Grinham, A., Dunbabin, M., Gale, D., Udy, J.: Quantification of ebullitive and diffusive methane release to atmosphere from a water storage. Atmos. Environ. **45**(39), 7166–7173 (2011)
8. Hitz, G., Pomerleau, F., Garneau, M.E., Pradalier, C., Posch, T., Pernthaler, J., Siegwart, R.Y.: Autonomous Inland water monitoring: design and application of a surface vessel. Robot. Autom. Mag, IEEE **19**(1), 62–72 (2012)
9. Kelly, J., Sibley, G., Barfoot, T., Newman, P.: Taking the long view: a report on two recent workshops on long-term autonomy. Robot. Autom. Mag, IEEE **19**(1), 109–111 (2012)
10. Kitts, C., Mahacek, P., Adamek, T., Rasal, K., Howard, V., Li, S., Badaoui, A., Kirkwood, W., Wheat, G., Hulme, S.: Field operation of a robotic small waterplane area twin hull boat for shallow-water bathymetric characterization. J. Field Robot. **29**(6), 924–938 (2012)
11. Kubelka, V., Oswald, L., Pomerleau, F., Colas, F., Svoboda, T., Reinštein, M.: Robust data fusion of multi-modal sensory information for mobile robots. J. Field Robot. (2014) (to appear)
12. Leutenegger, S., Siegwart, R.Y.: A low-cost and fail-safe inertial navigation system for airplanes. In: IEEE international conference on Robotics and Automation (ICRA), pp. 612–618 (2012)
13. Mahony, R., Hamel, T., Pflimlin, J.: Nonlinear complementary filters on the special orthogonal group. IEEE Trans. Automa.Control **53**(5), 1203–1218 (2008)
14. Maimone, M., Cheng, Y., Matthies, L.: Two years of visual odometry on the mars exploration rovers. J. Field Robot. **24**(3), 169–186 (2007)
15. Michael, N., Shen, S., Mohta, K., Mulgaonkar, Y., Kumar, V., Nagatani, K., Okada, Y., Kiribayashi, S., Otake, K., Yoshida, K., Ohno, K., Takeuchi, E., Tadokoro, S.: Collaborative mapping of an earthquake-damaged building via ground and aerial robots. J. Field Robot. **29**(5), (2012)
16. Papadopoulos, G., Kurniawati, H., Sharif, A.S.B.M., Wong, L.J., Patrikalakis, N.M.: 3D-surface reconstruction for partially submerged marine structures using an autonomous surface vehicle. In: IEEE/RSJ International Conference on Intelligent Robots and Systems (IROS), pp. 3551–3557 (2011)

17. Pereira, A., Das, J., Sukhatme, G.: An experimental study of station keeping on an underactuated ASV. In: IEEE/RSJ International Conference on Intelligent Robots and Systems (IROS), pp. 3164–3171 (2008)
18. Pomerleau, F., Krüsi, P., Colas, F., Furgale, P., Siegwart, R.Y.: Long-term 3D map maintenance in dynamic environments. In: IEEE International Conference on Robotics and Automation (ICRA), (2014) (accepted)
19. Pomerleau, F., Colas, F., Siegwart, R., Magnenat, S.: Comparing ICP variants on real-world data sets. Auton. Robot. **34**(3), 133–148 (2013)
20. Scherer, S., Rehder, J., Achar, S., Cover, H., Chambers, A., Nuske, S., Singh, S.: River mapping from a flying robot: state estimation, river detection, and obstacle mapping. Auton. Robot. **33**(1–2), 189–214 (2012)
21. Trawny, N., Roumeliotis, S.I.: Indirect Kalman filter for 3D attitude estimation a tutorial for quaternion algebra. Tech. Rep. 612, University of Minnesota, Departement of Computing Science and Engineering (2005)

Adaptive Path Planning for Tracking Ocean Fronts with an Autonomous Underwater Vehicle

Ryan N. Smith, Philip Cooksey, Frederic Py, Gaurav S. Sukhatme and Kanna Rajan

Abstract Ocean fronts are productivity *hot spots*, supporting marine life from plankton to whales. These dynamic systems contain a vast amount of information, and have the potential to significantly expand our knowledge of aquatic ecosystems in relation to climate change. However, ocean fronts and other dynamic features cannot be studied through conventional oceanographic techniques. In this paper, we address the problem of sampling and tracking an ocean front with an Autonomous Underwater Vehicle based on predictions and/or priors provided by a heterogeneous team of assets and ocean models. Specifically, given a prior (that may not be accurate or up-to-date) we present a method for an underwater vehicle to plan a mission and adapt this mission on-the-fly to track a dynamic feature. Results from field trials are presented, and demonstrate that the vehicle is able to adapt its path to follow a desired contour.

R.N. Smith (✉)
Physics and Engineering Department, Fort Lewis College, Durango, CO, USA
e-mail: rnsmith@fortlewis.edu

P. Cooksey
California State University, Monterey Bay, Monterey, CO, USA
e-mail: pcooksey@csumb.edu

F. Py · K. Rajan
Monterey Bay Aquarium Research Institute, Moss Landing, CA, USA
e-mail: fpy@mbari.org

K. Rajan
e-mail: kanna.rajan@mbari.org

G.S. Sukhatme
Robotic Embedded Systems Laboratory and the Department of Computer Science,
University of Southern California, Los Angeles, CA 90089, USA
e-mail: gaurav@usc.edu

M.A. Hsieh et al. (eds.), *Experimental Robotics*, Springer Tracts
in Advanced Robotics 109, DOI 10.1007/978-3-319-23778-7_50

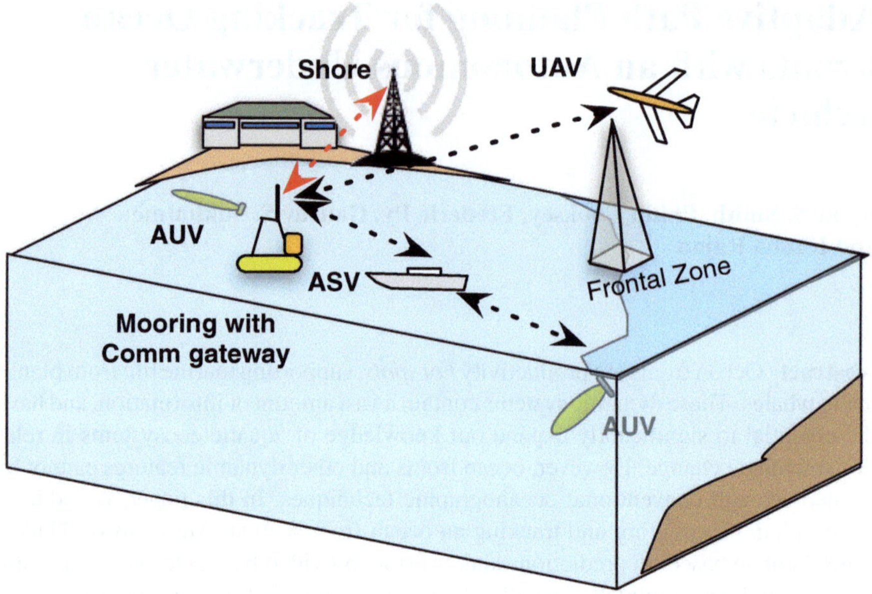

Fig. 1 An envisioned scenario in the near future, with the use of AUVs, autonomous surface and aerial platforms finding, tracking and sampling frontal zones

1 Introduction

An oceanic front is a narrow band of enhanced horizontal gradients of water properties (temperature, salinity, nutrients, etc.) dividing broader areas with different water masses or vertical structure. Fronts occur across a variety of spatial scales; along-frontal scales of 1–10,000 km, cross-frontal scales of 10 m–100 km, and vertical scales of 10–1000 m. Temporal scales vary from days for transient fronts, e.g., upwelling front in Monterey Bay, to millions of years for quasi-stationary, large-scale transoceanic fronts, e.g., Kuroshio Front off the eastern coast of Japan. Ocean fronts have been linked to elevated primary production and enhanced diversity of species; *hot spots* for marine life, across an astonishing spectrum of scales from plankton to whales. These frontal *hot spots* hold a wealth of information to improve our understanding of aquatic ecosystems in relation to climate change, however ocean fronts cannot be studied through conventional oceanographic techniques [1].

It is of interest to determine whether or not it is possible to develop a biological model for activity at or within an ocean front for ecological purposes. To develop such a model, we must first gather enough information to begin to characterize these dynamic systems at multiple scales, both temporally and spatially.

At a high level, we envision a scenario similar to that shown in Fig. 1. The objective is that coordinated observation between aerial, surface and underwater platforms is critical in observing dynamic biological phenomenon with varying spatial and temporal scales (from minutes to weeks and tens of km^2 to hundreds of km^2). The frontal signature is initially detected by an Unmanned Aerial Vehicle (UAV) or remotely sensed by satellite. This information is communicated to shore or a relay for analysis and response. A team of Autonomous Surface Vehicles (ASVs) and Autonomous Underwater Vehicles (AUVs) is then deployed to locate, sample and track the front, with the goal to characterize the physical and biological processes occurring.

In this paper, we address the problem of sampling and tracking an ocean front with an AUV based on the predictions and/or priors provided by other members of the heterogeneous team of assets, as well as ocean models. Specifically, given a prior (that may not be accurate or up-to-date) can the AUV adapt its mission to find and track a dynamic feature? From the initial prior, a static plan can be created for optimal sampling, if we assume the feature is static. However, an ocean front is expected to evolve and move throughout the duration of sampling, and there is latency in the communication from the aerial sensor to the aquatic sampling assets. Hence, we must develop a technique for the vehicle to adapt to an ever-changing frontal signature.

2 Related Work

Robotic platforms hold the promise of a revolution in ocean sampling. Considerable study has been reported on control design for AUVs for adaptive ocean sampling and coordinated control of multi-vehicle systems, e.g., [2–8]. Applications of ocean sampling techniques for AUVs are discussed in [5, 6, 9–12], with ocean front perception and detection specifically addressed in [2, 13–15]. Ocean front detection and characterization has been extensively studied without in situ robotics through satellite remote sensing [16, 17]. These algorithms provide the foundation for the priors supplied by the aerial sensor platforms seen in Fig. 1. Along with steering an AUV to the right locations, research exists in the area of static sensor placement to maximize knowledge return [18].

Existing sampling methods are currently all based on a geographical coordinate system, i.e., latitude and longitude. However, the definition of geographical space is ill-defined in the ocean and complex ocean dynamics make geographic-relative navigation difficult, specifically when tracking dynamic and episodic events. Sampling at uniformly distributed geographic coordinates can generate a suboptimal distribution of samples given the dynamic nature of the oceans' water masses. Adapting to the changing environment is crucial to characterizing and eventually understanding these dynamic frontal systems.

To address adaptation, researchers have implemented human-in-the-loop solutions; static paths are created and alternative static paths are generated by domain experts after analyzing collected data [19–21]. These methods have their advantages, however we are interested in enabling decision making for path adaptation on-board the vehicle. Some results have implemented information-based metrics and machine learning to optimally determine a path or sensor placement based on reduction of overall covariance of the scalar field in question. Recent work in [2] begins to address the issue of on-board decision making and adaptation, but approaches the problem from a multiple underwater vehicle point-of-view, with constraints on communication among the fleet. Here, we enable in situ robotic adaptation to environmental conditions for a single vehicle based on real-time measurements for targeted sampling of ocean fronts.

3 Algorithm Design

The goal of this paper is to plan a path for an AUV to execute that enables sampling of a dynamically-evolving ocean front. The first step we take is to generate an initial path based on given priors which assumes that the front is static in space and time. This initial path provides the basis from which we will adapt to track the evolving feature. The initial path is generated as a regular, zig-zag pattern, crossing the frontal boundary with a predefined swath width, Fig. 2. Then, we will adapt on-the-fly to track the front based on in situ measurements. Since the desired sampling strategy inherently requires repeated traversal through the ocean front, the adapted path will also have a zig-zag structure.

3.1 Initial Path

As our focus is on the mission adaptation of the AUV, we provide the vehicle with a prior estimation characterizing the location of an ocean front from an Unmanned

Fig. 2 Representation of the initially computed survey path

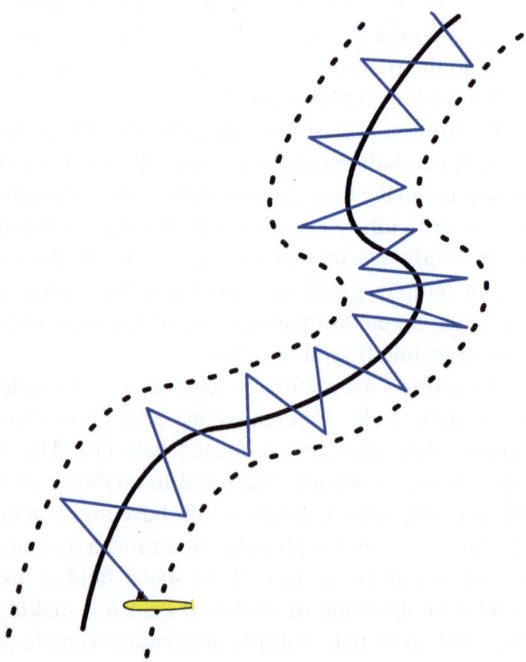

Aerial Vehicle (UAV) or satellite, see Fig. 1. This prior can be provided as a KML file and is assumed to be temporally latent and inaccurate. The KML file characterizing the prior for the observed front is the input to the proposed algorithm to plan an initial path to survey the *observed* ocean front. As it is assumed that the front has moved since the observation, the initial path plan is very conservative with regard to the spatial footprint of the front. From the provided prior observation, we parameterize the boundary of the front to a curvilinear line in 2-D. We assume that a front can be represented as a single,[1] continuously-differentiable function, i.e., we assume the parameterization of the front is $\mathcal{C}^1$. The initial path plan, computed on-board the AUV, is a regular, zig-zag pattern, crossing the frontal boundary with a predefined swath width, see Fig. 2. Assuming that the speed over ground of the AUV is constant, we vary the speed along the parameterized frontal boundary, and hence the horizontal spatial resolution, by expanding or contracting the periodicity of the zig-zags. The technique applied is similar to what was done in [5]. In this approach, we rotate the planning method 90° to generate zig-zag paths in the horizontal plane. The candidate waypoints for this zig-zag path are the points defined by the two dashed lines in Fig. 2; one on either side of the parameterized prior (the solid balck line in Fig. 2). The lines are located a distance one-half the predefined swath width from the parameterized prior, with their tangent vector at each point matching that of the respective point of the parameterized prior. The degree of compression or expansion of each zig-zag period is proportional to the inverse of the derivative (curvature) of the parameterized boundary at each location. Specifically, in areas of low curvature (straight line) we prescribe a zig-zag path that covers equal distance in the along and cross-track directions for each period. In areas of high curvature, we perform higher density sampling by reducing the distance covered in the along-track direction for each period. The proposed method acts to resolve the non-linear frontal boundary.

3.2 Adaptation Algorithm

The basis for path adaptation to track and sample an ocean front is governed by two assumptions. First, the frontal boundary can be represented as a continuously-differentiable function in 2-D, i.e., the front is $\mathcal{C}^1$, over short horizons. Secondly, optimal sampling of the front is carried out by repeatedly crossing through the frontal boundary with the AUV. Given these assumptions, the adaptation algorithm takes the relevant science parameter(s) as an input, analyzes the collected data, determines the location of the frontal boundary, predicts the location of this frontal boundary, and assigns a new waypoint that steers the vehicle through the front. For each path segment (T_i), data is recorded in a ROS bag file and sent to the T-REX. The T-REX then perceives if and when the AUV has crossed through the frontal boundary, and records this location (l_i). Determination that the AUV has crossed through the frontal boundary is based on the technique developed in [14].

[1] A discontinuity is assumed to represent two different fronts, and ocean physics negate the possibility of non-differentiable locations along a front.

Algorithm 1. NEWWAYPOINTS([Step, Lon, Lat, Param][1...*n*])

```
CubicSpline spline
comment: CubicSpline fixed 2nd derivative
{ set<pair<double,double>> waypoints
{ int splineCount = 0, nextStep = 1
{ bool inFront = false
{ beginBoundaryTracking = false
for iterator ← 1 to n (in data[Step][])
  do
if not beginBoundaryTracking AND data[Step][iterator] ≥ nextStep
  then beginBoundaryTracking = true
if beginBoundaryTracking
        { double depth = data[Depth][iterator]
        { if (not inFront AND Param ≥ VALUE)
        { OR (inFront AND Param ≤ VALUE)
  then  {          { spline.addPoint(data[Lon][iterator])
        {          { spline.addPoint(data[Lat][iterator])
        {    then  { inFront = not inFront
        {          { splineCount += 1
        {          { beginBoundaryTracking = false
if splineCount ≥ 3 AND data[Step][iterator] ≥ nextStep
        { double longitude = data[Lon][iterator] − 0.0002
        { comment: Get next longitude value.
        { comment: Based on location and fixed interval.
  then  { double latitude = spline(longitude)
        { comment: Spline generates latitude value.
        { comment: Based on longitude input and previous spline points.
        { waypoints.insert(longitude, latitude)
        { nextStep = data[Step][iterator] + 1
```

To begin, the vehicle executes the first three path segments of the initially computed path. During this phase, the only adaptation we allow T-REX to make is that the vehicle can continue a fixed distance past a prescribed waypoint if it perceives that the vehicle has not reached or is still within the front, or it allows the vehicle to stop early if it perceives that the vehicle has already crossed through the front. This initialization process is implemented because we expect the front to move spatially between the time of acquiring the initial delineation and the time the AUV beings its mission. Additionally, we require some data on the location of the frontal boundary to base our predictions for adaptation. We remark here that the proposed algorithm is highly sensitive to the initial conditions of the deployment. Specifically, this method is not intended to *find* or *locate* ocean fronts, but rather to track and sample them adaptively. If the initial prior is incorrect, or the front has significantly moved from the original detection location, the proposed method will not work.

The initially detected frontal boundary locations, l_i for $i = 1, 2, 3$ are used by T-REX to predict the frontal boundary location over a short horizon. Since we assume the front to be $\mathcal{C}^1$, we use a cubic spline to fit the previously gathered data and predict the most likely location of the frontal boundary. After the initial three path segments (four waypoints) are executed, we are only interested in computing the *next* waypoint for the AUV. T-REX generates this single waypoint based on the predicted location of the front. This process is iterated until the front is no longer detected or the AUV reaches the end of the desired sampling region. At any time, the prediction algorithm relies on at most, the last three observed locations of the frontal boundary. Thus, we assume that locally the front is $\mathcal{C}^1$, but that globally this may not be the case. In particular, we do not expect that the proposed short horizon prediction will effectively describe a front over one million kilometers.

Once the adaptation phase is initiated after the execution of the initial three path segments, the path execution is simply applying a specified heading computed by T-REX based on the predicted location of the frontal boundary. While T-REX does output a waypoint to steer the AUV through the front, the length of the adapted path segment is ultimately determined by data from in situ sampling along the path. The pseudocode for the adaptation of the vehicle path is presented in Algorithm 1.

4 Algorithm Implementation

For this research, field experiments were carried out in Monterey Bay, California with a YSI EcoMapper AUV [22], as shown in Fig. 3. The adaptation of the sampling path occurs incrementally as the vehicle follows and samples within and along the front.

Fig. 3 The YSI Ecomapper Vehicle (*top*) and the beach deployment of the vehicle (*bottom*)

The low-level functionality of navigation to a specified location is handled through a ROS interface on-board the vehicle [23, 24]. ROS was also used to gather and distribute the sensor data, which informed the high-level, decision-making process for adaptation. The high-level decision of plan synthesis is executed on-board by T-REX [25, 26]. The T-REX framework receives the sensor data published by ROS, and deliberates about future states, plans for actions and executes generated activities while monitoring plans for anomalous conditions. Hence, plans are not scripted a priori but synthesized on-board with high-level directives instead of low-level commands. The adaptation algorithm presented in Algorithm 1 is implemented as a decision tool within T-REX for the intended planning purposes.

5 Experiments

Field experiments were carried out near the harbor channel mouth in Moss Landing, CA, directly in front of the Monterey Bay Aquarium Research Institute (MBARI). The EcoMapper AUV was deployed and recovered from the beach, enabling ease of multiple tests without the need for significant support infrastructure. To validate the utility of the algorithm based on an ability to ground truth results and eliminate relying on finding an actual ocean front, we chose to survey a pseudo-front. The pseudo-front was defined to be a fixed depth contour of 30 feet along the edge of the Monterey Canyon. This provided a non-dynamic feature to examine and track for algorithm development and validation. Monterey Canyon, located at the center of Monterey Bay, is the largest submarine canyon along the coast of North America, and begins at the mouth of the harbor channel, see Fig. 4. By choosing a fixed depth, and zig-zagging across the canyon edge, we were able to conduct repeated experiments over a known feature for validation of our adaptation methodology.

Fig. 4 Relief of the Monterey Canyon

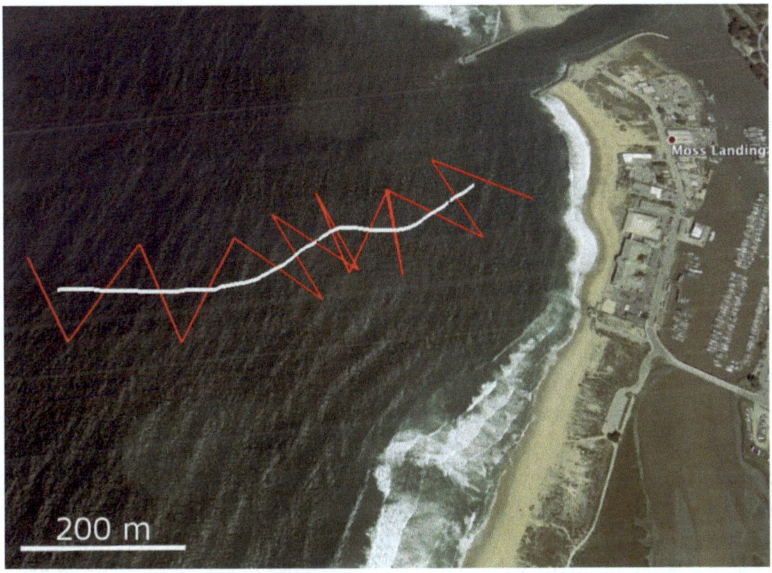

Fig. 5 A delineation of the chosen Monterey Canyon depth contour as the initial prior for the pseudo-front (*white*). The initially computed path for the AUV is the zig-zag path shown in *red*

The initial prior for the pseudo-front (canyon edge) was delineated by hand from existing bathymetry maps and from analyzing data collected from previous vehicle deployments within the region. This initial prior is shown as the curved, white line in Fig. 5. The initially computed path based on this prior as described in Sect. 3.1 is given by the zig-zag, red line in Fig. 5.

For all field deployments, the AUV started its mission at a fixed deployment location, then navigated on the surface of the water to the first waypoint of the path. Considering Fig. 5, the first waypoint of the path is the one located the furthest East. The prescribed paths were traversed from this location heading westerly. During path execution, the AUV maintained a constant depth of 2 m from the surface for the entire mission. During the experiments presented in this paper, T-REX was operating and making decisions for plan adaptation, however these adaptations were not being passed down to ROS for execution during the mission. Thus, in the following section, we present the revised path as computed by T-REX for the mission as compared to ground truth data. We remark that the paths computed by T-REX that adapt the initial path to follow the pseudo-front were not executed by the vehicle in an in situ adaptive fashion. Hence, the adapted waypoints shown are predicted based on the previous three path segments implemented by the AUV. Had the on-board decisions been acted upon in situ the location of adapted waypoints 2, 3, ... would be slightly different, as the executed path segments would have been different from those actually executed.

6 Results

During the experimental campaign, multiple paths were executed with the AUV. Initially, bathymetry data were gathered to generate the initial prior upon which to base the sampling mission. Seven initial delineations for the initial prior were tested, and for each of these the initial swath width for the initial zig-zag path was varied from 50 to 300 m. For the experimental results presented, the pseudo-front was set to coincide with a depth contour of 30 feet and the swath width for the initial path was set to 200 m. An example of the initially planned path and a sample execution of this path is shown in Fig. 6.

As previously noted, it was found that the proposed method is very sensitive to the initial conditions. This was evidenced through multiple experiments where the initial prior was positioned such that the initial path never crossed the prescribed depth contour. Since the vehicle never encountered the pseudo-front in this scenario, the adaptation process was never invoked. This was particularly the case for the narrower swath widths; those less than 100 m. Additionally, the choice of a depth contour as a pseudo-front was intended to eliminate variability and focus experiments on algorithm development and validation. However, since the average tidal flux in Monterey Bay is approximately 2 m, the location of the 30 m depth contour was a dynamically evolving boundary feature, even throughout a given day. For the bathymetric relief at the testing location, the location of the 30′ contour was shown to move on the order

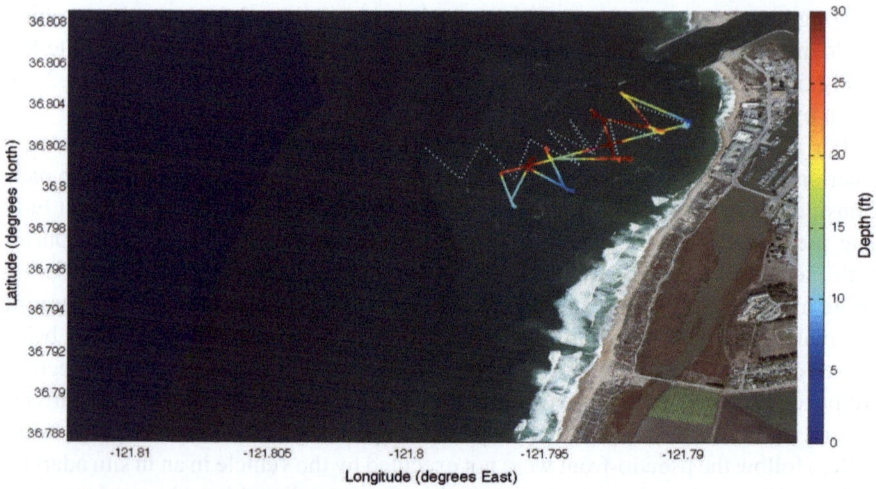

Fig. 6 Initial planned path (*dashed white line*) and the depth recorded along the path executed by the AUV (solid, multi-colored path) for the mission in Monterey Bay

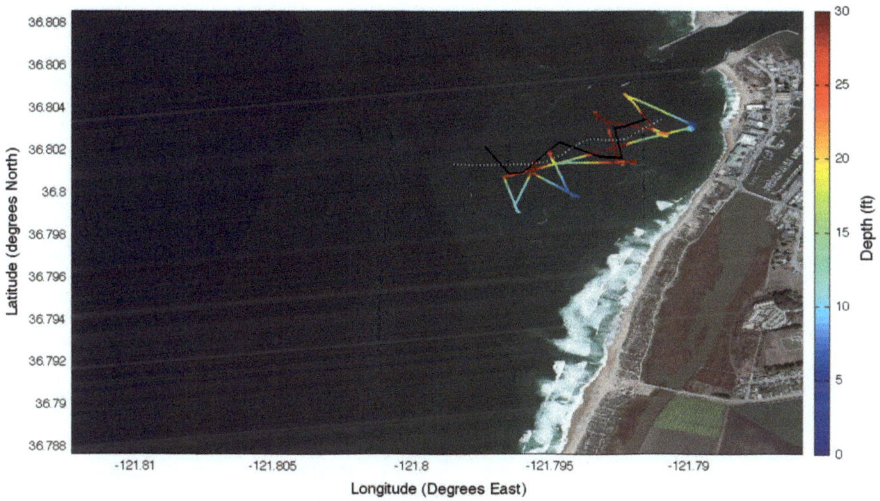

Fig. 7 The depth data collected along the path executed by the AUV (multicolored, zig-zag line) with the initial prior (*white dashed line*) and the location of the pseudofront (*solid black line*), equal to the 30′ depth contour, estimated from in situ data

of 10 m between the MLLW and MHHW[2] levels. Thus, if the prior was delineated at one of these instances, and the experiment conducted during the other extreme we experienced a significant difference in frontal boundary location for a *seemingly* static feature. This is evidenced by noticing the difference in the observed/measured 30′ depth contour in Figs. 8 and 9.

A representative set of depth data collected during a mission execution is presented in Fig. 7 as the multi-colored solid line. The initial prior used for this mission is overlaid as the white dashed line and the location of the 30′ depth contour estimated from in situ data is given by the black line. Here we see that even with a *static* boundary, the initial prior is different from the detected location of the pseudo-front. Additionally, even if the frontal boundary and the initial prior coincided, the execution of the initially designed path is significantly different from what was prescribed. This is a typical result for an AUV dead-reckoning between prescribed waypoints [27, 28]. The combination of these two artifacts further motivates the necessity for in situ adaptation for AUVs to follow a static contour or a dynamically evolving feature.

In Figs. 8 and 9, we present the results from two separate field trials. Here, the initial prior and initially design path are the same for both experiments. The initially designed path is denoted by the white dashed line, with the estimated location of the pseudo-front given by the solid black line. The computed waypoints are given by the cyan stars. For the adapted path (thin cyan line), we choose the initial waypoint to

[2]Mean Low Low Water (MLLW) and Mean High High Water (MHHW) levels are the averages of the lower low water height and higher high water height of each tidal day observed over the National Tidal Datum Epoch.

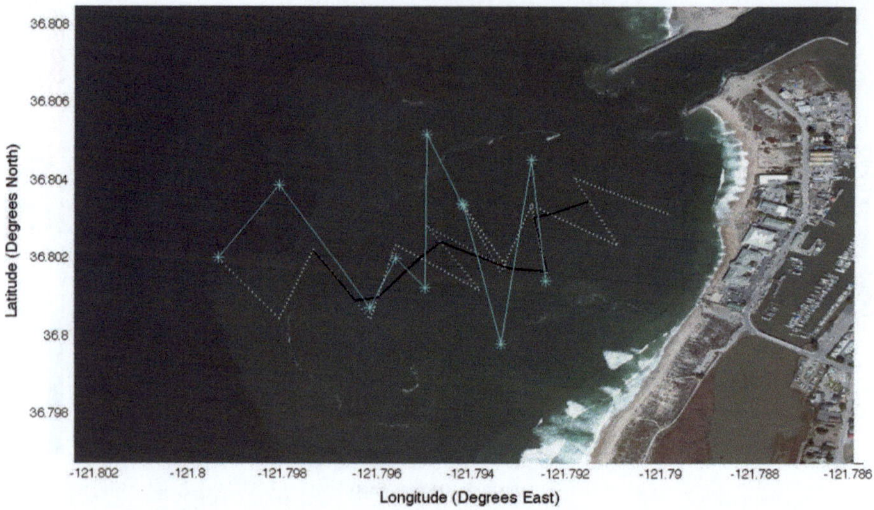

Fig. 8 Adapted waypoints predicted by the proposed algorithm (*cyan stars*) overlaid on the initially planned path (*white dashed line*) and the estimated pseudofront contour (*solid black line*)

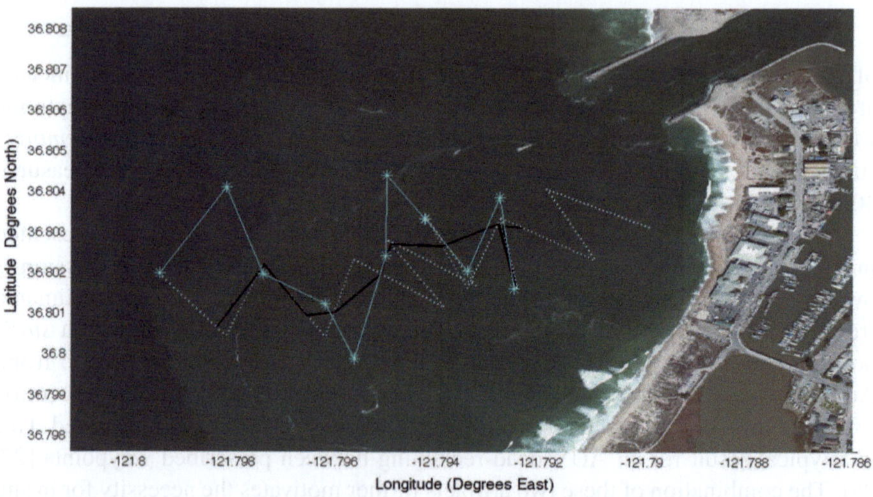

Fig. 9 Adapted waypoints predicted by the proposed algorithm (*cyan stars*) overlaid on the initially planned path (*white dashed line*) and the estimated pseudofront contour (*solid black line*)

coincide with the current location of the vehicle at the first point the adaptation is invoked. The final waypoint of this path is fixed to the final waypoint of the initially designed path to ensure the vehicle a safe return home.

From the presented results, the initially planned path appears to achieve better results for tracking the prescribed contour. As the proposed algorithm is designed to be very conservative, assuming the frontal boundary is moving, and since this feature was relatively static, we expect that the initially designed path will perform well. Additionally, since the adaptation and computation of waypoints was done offline, and the vehicle was not adapting during the deployment, we must look at the previous three path segments *executed*; these form the basis for the subsequently predicted waypoint. Given this, we see that the predicted waypoints are a good guide for steering the AUV along a path that will definitely cross the frontal boundary. In each instance of adaptation, the frontal boundary would have been crossed crossed if the AUV were to drive from its current location to the computed waypoint. This is not the case for the initially designed path, which does not cross the pseudo-front along each prescribed path segment. This is expected for static paths that attempt to track dynamic features or features that have uncertain evolution.

In Fig. 8 the 4th computed waypoint, and in Fig. 9, the 4th and 6th computed waypoints were revised and the path segment extended based on data collected along the path after the initial computation. Here, T-REX decided that the AUV either had not encountered the front (prescribed 30′ contour) or was not entirely through the boundary.

Comparing the initially designed path with the path defined by the computed waypoints, we notice that the adapted path traverses a longer distance. This is a result of the proposed algorithm acting very conservatively, to ensure that the frontal boundary is crossed. Additionally, the use of a cubic spine is likely overestimating the curvature of the pseudo-front, forcing T-REX to compute long path segments to cover all potential locations for the front.

7 Conclusions

In this paper, we have demonstrated a successful integration of ROS and T-REX on-board an EcoMapper AUV for adaptive, goal-oriented mission execution. This system was utilized to demonstrate an algorithm for tracking a frontal boundary with adaptive path planning based only on a prior delineation from remotely sensed data. Initial path plans were developed from priors and implemented during field trials. Data collected from field trials were acted upon by T-REX to adapt and re-plan, enabling the vehicle to follow a desired contour.

From the analysis of the performed experiments and algorithm implementation, the choice of a cubic spline to estimate the evolution of the frontal boundary needs to be re-evaluated. For the proposed pseudo-front and actual ocean fronts, a more gradual along-track gradient may be expected in practice. There exists a trade-off in the selection of the prediction tool and the conservative nature of the proposed algorithm. Proper selection is critical and depends on the actual front under study.

8 Future Work

The fist step for future work is to allow T-REX to supply ROS with updated and adapted plans during field deployments. This study has shown that the developed algorithms work well to adapt to in situ data and enable the vehicle to follow a desired contour. An investigation into alternative spline tools or other prediction methods for front evolution will be the focus of forthcoming work. Currently, we are investigating the integration of ocean model predictions to inform the vehicle of how the frontal system has moved prior to the start of the survey, or how it is moving during the execution of the mission. Combining the ocean model predictions with the cubic spline estimations will provide a longer horizon for path planning and hopefully place the AUV in the *right place at the right time* to sample the frontal boundary. In future deployments, we aim to evaluate the utility of this method for sampling an actual ocean front. The difficulty here lies in the location and initial delineation of such a feature. As presented in [2] the use of ocean model data for validation through simulations will be an initial step prior to field deployment.

References

1. Olson, D., Hitchcock, G., Mariano, A., Ashjan, C., Peng, G., Nero, R., Podesta, G.: Life on the edge: marine life and fronts. Oceanography **7**, 52–60 (1994)
2. Reed, B., Hover, F.: Tracking ocean fronts with multiple vehicles and mixed communication losses. In: International Conference on Intelligent Robots and Systems (IROS), IEEE/RSJ, pp. 3374–3381 (2013)
3. Yuh, J.: Design and control of autonomous underwater robots: a survey. Auton. Robots **8**, 7–24 (2000)
4. Paley, D., Zhang, F., Leonard, N.: Cooperative control for ocean sampling: the glider coordinated control system. IEEE Trans. Control Syst. Technol. **16**, 735–744 (2008)
5. Smith, R.N., Schwager, M., Smith, S.L., Jones, B.H., Rus, D., Sukhatme, G.S.: Persistent ocean monitoring with underwater gliders: adapting sampling resolution. J. Field Robot. **28**, 714–741 (2011)
6. Smith, R.N., Das, J., Heidarsson, H., Pereira, A., Cetinić, I., Darjany, L., Ève Garneau, M., Howard, M.D., Oberg, C., Ragan, M., Schnetzer, A., Seubert, E., Smith, E.C., Stauffer, B.A., Toro-Farmer, G., Caron, D.A., Jones, B.H., Sukhatme, G.S.: USC CINAPS builds bridges: observing and monitoring the southern California bight. IEEE Robot. Autom. Mag. Spec. Issue Mar. Robot. Syst. **17**, 20–30 (2010)
7. Smith, R.N., Chao, Y., Li, P.P., Caron, D.A., Jones, B.H., Sukhatme, G.S.: Planning and implementing trajectories for autonomous underwater vehicles to track evolving ocean processes based on predictions from a regional ocean model. Int. J. Robot. Res. **29**, 1475–1497 (2010)
8. Smith, R.N., Das, J., Chao, Y., Caron, D.A., Jones, B.H., Sukhatme, G.S.: Cooperative multi-auv tracking of phytoplankton blooms based on ocean model predictions. In: Proceedings of Oceans '10—IEEE Sydney, Sydney, Australia, pp. 1–10 (2010)
9. Creed, E.L., Mudgal, C., Glenn, S., Schofield, O., Jones, C., Webb, D.C.: Using a fleet of slocum battery gliders in a regional scale coastal ocean observatory. In: Oceans '02 MTS/IEEE (2002)
10. Garcia-Olaya, A., Py, F., Das, J., Rajan, K.: An online utility-based approach for sampling dynamic ocean fields. IEEE J. Oceanic Eng. **37**, 185–203 (2012)

11. Singh, H., Yoerger, D., Bradley, A.: Issues in auv design and deployment for oceanographic research. In: Proceedings of the 1997 IEEE International Conference on Robotics and Automation, vol. 3, pp. 1857–1862 (1997) (Invited paper)
12. Fiorelli, E., Leonard, N., Bhatta, P., Paley, D., Bachmayer, R., Fratantoni, D.: Multi-auv control and adaptive sampling in monterey bay. IEEE J. Oceanic Eng. **31**, 935–948 (2006)
13. Zhang, Y., Godin, M., Bellingham, J., Ryan, J.: Ocean front detection and tracking by an autonomous underwater vehicle. In: MTS/IEEE Oceans, pp. 1–4 (2011)
14. Zhang, Y., Godin, M.A., Bellingham, J.G., Ryan, J.P.: Using an autonomous underwater vehicle to track a coastal upwelling front. IEEE J. Oceanic Eng. **37**, 338–347 (2012)
15. Zhang, Y., Bellingham, J., Ryan, J., Kieft, B., Stanway, M.: Two-dimensional mapping and tracking of a coastal upwelling front by an autonomous underwater vehicle. In: MTS/IEEE Oceans—San Diego, pp. 1–4 (2013)
16. Belkin, I.M., O'Reilly, J.E.: An algorithm for oceanic front detection in chlorophyll and SST satellite imagery. J. Mar. Syst. **78**, 319–326 (2009) (Special Issue on Observational Studies of Oceanic Fronts)
17. Belkin, I.M., Cornillon, P.C., Sherman, K.: Fronts in large marine ecosystems. Prog. Oceanogr. **81**, 223–236 (2009) (Comparative Marine Ecosystem Structure and Function: Descriptors and Characteristics)
18. Zhang, B., Sukhatme, G.S.: Adaptive sampling with multiple mobile robots. In: IEEE International Conference on Robotics and Automation (submitted) (2008)
19. Leonard, N.E., Paley, D.A., Davis, R.E., Fratantoni, D.M., Lekien, F., Zhang, F.: Coordinated control of an underwater glider fleet in an adaptive ocean sampling field experiment in monterey bay. J. Field Robot. **27**, 718–740 (2010)
20. Das, J., Maughan, T., McCann, M., Godin, M., O'Reilly, T., Messie, M., Bahr, F., Gomes, K., Py, F., Bellingham, J., Sukhatme, G., Rajan, K.: Towards mixed-initiative, multi-robot field experiments: design, deployment, and lessons learned. In: Proceedings of the Intelligent Robots and Systems (IROS) Conference, San Francisco (2011)
21. Gomes, K., Cline, D., Edgington, D., Godin, M., Maughan, T., McCann, M., O'Reilly, T., Bahr, F., Chavez, F., Messi, M., Das, J., Rajan, K.: ODSS: a decision support system for ocean exploration. In: Workshop on Data-Driven Decision Guidance and Support Systems (DGSS) at the 29th IEEE International Conference on Data Engineering, Brisbane, Australia (2013)
22. YSI Incorporated: YSI EcoMapper. http://www.ysi.com/productsdetail.php?EcoMapper-41 (2014). Accessed May 2014
23. Willow garage: robot operating system (ROS). http://www.ros.org (2007)
24. Quigley, M., Gerkey, B., Conley, K., Faust, J., Foote, T., Leibs, J., Berger, E., Wheeler, R., Ng, A.Y.: ROS: an open-source robot operating system. In: Proceedings of the Open-Source Software Workshop of the International Conference on Robotics and Automation (2009)
25. Rajan, K., Py, F.: T-REX: Partitioned inference for AUV mission control. In: Roberts, G.N., Sutton, R. (eds.) Further Advances in Unmanned Marine Vehicles. The Institution of Engineering and Technology (IET) (2012)
26. Rajan, K., Py, F., Barreiro, J.: Towards deliberative control in marine robotics. In: Seto, M. (ed) Autonomy in Marine Robots. Springer, New York (2012)
27. Smith, R.N., Kelly, J., Chao, Y., Jones, B.H., Sukhatme, G.S.: Towards improvement of autonomous glider navigation accuracy through the use of regional ocean models. In: Proceedings of the 29th International Conference on Offshore Mechanics and Arctic Engineering, pp. 597–606, Shanghai, China (2010)
28. Smith, R.N., Kelly, J., Sukhatme, G.S.: Towards improving mission execution for autonomous gliders with an ocean model and kalman filter. In: Proceedings of the IEEE International Conference on Robotics and Automation, Minneapolis, MN (2012)

Robust Underwater Obstacle Detection for Collision Avoidance

Varadarajan Ganesan, Mandar Chitre and Edmund Brekke

Abstract Underwater obstacle detection and avoidance is essential for safe deployment of autonomous underwater vehicles (AUVs). A forward-looking sonar is typically used to detect and localize potential obstacles. Such sensors tend to have a coarser sensor resolution and a lower signal-to-noise ratio (SNR) than electromagnetic sensors typically used for similar tasks in land-based robotics. Lack of access to GPS causes additional uncertainty in vehicle navigation, making it difficult to detect and localize potential obstacles relative to a world-fixed reference frame. In this paper, we propose an obstacle detection algorithm for AUVs which is based on occupancy grids. The proposed method differs from existing occupancy grid-techniques in two key aspects. First, we use an occupancy grid attached to the body frame of the AUV, and not to the world frame. Second, our technique takes detection probabilities and false alarm rates into account, in order to deal with the high amounts of noise present in the sonar data. The proposed algorithm is tested online during field trials at Pandan Reservoir in Singapore and in the sea at Selat Pauh off the coast of Singapore.

Keywords Underwater obstacle detection · Collision avoidance · Occupancy Grids

1 Motivation

In recent years, we have seen an increasing interest in autonomous underwater navigation and exploration. Although significant advances have been made in the development of autonomous underwater vehicles (AUVs), the technology for effective

V. Ganesan (✉) · M. Chitre · E. Brekke
ARL, Tropical Marine Science Institute and Department of Electrical and Computer Engineering, National University of Singapore, 119227 Singapore, Singapore
e-mail: varadarajan@arl.nus.edu.sg
URL: http://www.arl.nus.edu.sg

M. Chitre
e-mail: mandar@arl.nus.edu.sg

E. Brekke
e-mail: edmund@arl.nus.edu.sg

M.A. Hsieh et al. (eds.), *Experimental Robotics*, Springer Tracts in Advanced Robotics 109, DOI 10.1007/978-3-319-23778-7_51

obstacle avoidance remains relatively immature. Devices such as multibeam and sector-scanning forward looking sonars (FLS) are available for obstacle detection. Although multibeam FLS are commonly adopted as underwater obstacle avoidance sensors due to their superior performance, they are usually much costlier than sector scanning sonars. Our aim in this paper is to develop an algorithm for reliable obstacle detection that may be used with either type of FLS. We demonstrate our algorithm experimentally using data from the more challenging of the two, i.e., the sector-scanning sonar.

Accurate localization of obstacles is essential for collision avoidance. Due to lack of availability of GPS signals underwater, AUVs generally rely on on-board proprioceptive sensors such as compass, Doppler velocity log (DVL) and inertial navigation system (INS) for underwater navigation. Dead-reckoning using these sensors suffers from unbounded positioning error growth [1], which in turn leads to inaccurate localization of potential obstacles. This problem is even more acute in low-cost AUVs where the proprioceptive sensors have low accuracy.

The conventional approach to solving this problem is to improve the AUV's positioning accuracy. This may be achieved by using sensors of higher accuracy, or by deploying external aids such as acoustic beacons. Both solutions will incur additional costs. An interesting alternative is to use simultaneous localization and mapping (SLAM) where the detected obstacles are used as landmarks to improve positioning [2, 3]. SLAM holds great promise to solve the navigation and obstacle avoidance problems together, but issues such as feature representation, data association and consistency are still undergoing active research [4]. In our opinion, SLAM is therefore not yet mature enough for reliable underwater obstacle detection and avoidance.

We propose a method for detection and localization of obstacles which employs an occupancy grid attached to the AUV's body frame. This entails several novelties. Although occupancy grid formulations are popular in land-based robotics [2, 5–7], this approach does not appear to be common in the underwater domain. Feature-based solutions appear to be more popular [8–11]. Existing publications on occupancy grids for FLS, such as [12] and [13], present results from a controlled environment and under static conditions. In contrast, we present results from both lake trials and sea trials with the AUV in a dynamic state. We believe that the occupancy grid approach is particularly suitable for underwater robotics, since it often is very difficult to extract reliable features from FLS data, especially when a sector-scanning sonar is used.

Furthermore, we use a *local occupancy grid* in the AUV's frame of reference, as opposed to more conventional geo-referenced occupancy grid. This is somewhat similar to the concept of robocentric SLAM [14]. The key insight underlying this is that for the purpose of obstacle avoidance, as opposed to more comprehensive mapping, the obstacles only need to be accurately localized *relative* to the AUV. Accurate localization in a geo-referenced frame is not necessary. Adopting the AUV's body frame for obstacle localization makes the obstacle detection and avoidance performance less sensitive to the AUV's positioning error growth.

Finally, our formulation incorporates motion uncertainties and sensor parameters such as false alarm rate and detection probability in a Bayesian framework. When the

AUV moves, the obstacles "move" in the AUV's body frame in a predictable way. Our *motion model* updates the occupancy probabilities from the estimated translational and rotational motion. When a sonar measurement becomes available, the occupancy probabilities are updated using a Bayesian *measurement model* that integrates new information from the measurement into the belief represented by the occupancy grid. The occupancy grid is used to determine the location of nearby obstacles. If these obstacles pose a threat of collision, the AUV's command and control system takes evasive maneuvers.

2 Technical Approach

As briefly outlined above, we use a local occupancy grid to represent our belief of the location of nearby obstacles. To update the occupancy grid as the AUV moves and sonar measurements becomes available, we require a motion model and a measurement model. Finally, we require a detection procedure that operates on the occupancy grid to yield a set of potential obstacles. This set of potential obstacles is sent to the AUV's command and control system for consideration of possible avoidance maneuvers.

2.1 Occupancy Grid

The local occupancy grid is rectangular with $m \times n$ occupancy cells, each at a fixed location with respect to the AUV. An illustration of the local occupancy grid is shown in Fig. 1. We use $O_{x,y}$ to denote an occupancy cell with index (x, y). Each occupancy cell $O_{x,y}$ is associated with the events $\mathbb{O}_{x,y}$ that it is occupied and $\widehat{\mathbb{O}_{x,y}}$ that it is not occupied. Therefore, they would be related as $P(\mathbb{O}_{x,y}) + P(\widehat{\mathbb{O}_{x,y}}) = 1$. The $m \times n$ matrix of occupancy probabilities $[P(\mathbb{O}_{x,y}) \ \forall \ x, y]$ fully describes the belief held by the algorithm about obstacles in the vicinity of the AUV.

2.2 Measurement Model

An FLS sends out a sonar "ping" in a given direction and listens for echoes. The echo intensity profile returned from the environment is discretized into a set of bins (k, θ) where index k represents the range and index θ represents the bearing. Let the measurement observed in bin (k, θ) be $z_{k,\theta}$. Given a threshold value t_k for range bin k, we report a detection $S_{k,\theta} = 1$ if $z_{k,\theta} \geq t_k$ and $S_{k,\theta} = 0$ otherwise.

Let p_k be the probability of detection of an obstacle at a range corresponding to bin k, and f_k be the probability of false alarm which are necessary operational

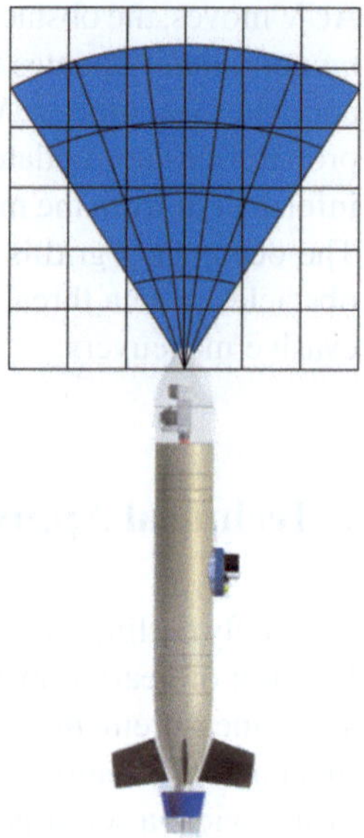

Fig. 1 Illustration of local occupancy grid attached to the AUV and its sensor frame (*blue color*)

parameters. A plot of p_k versus f_k (parametrized by t_k) is known as the receiver operating characteristic (ROC) curve. This ROC curve varies with signal-to-noise ratio (SNR) and environmental characteristics; we can experimentally measure this for a sonar in an operational environment of interest. We set a constant acceptable false alarm rate f (i.e., set $f_k = f$) and obtain the corresponding p_k and t_k for each range bin k.

The experimentally measured ROC curves matched existing models for detection of targets in the presence of noise as proposed in [15]. At Pandan reservoir, the ROC curves obtained matched that of detection of targets giving constant amplitude returns in Gaussian noise. The model for this case is as follows:

$$p_k = \frac{1}{2}\mathbf{erfc}\left\{\mathbf{erfc}^{-1}(2f_k) - \sqrt{\frac{\text{SNR}}{2}}\right\} \quad (1)$$

where SNR is the signal to noise ratio, **erfc** is the complementary error function. This can be explained by the enclosed nature of the reservoir resulting in the presence of Gaussian noise and targets like lake walls with surfaces which would give returns of constant amplitude.

At the sea in Selat Pauh, the background noise did not particularly match any of the existing distribution for background noise models in literature like the Gaussian or Rayleigh distribution. Hence, there is no model for the detection of targets in literature to verify the experimentally obtained ROC curves.

When a measurement becomes available, the occupancy grid serves as a Bayesian prior. Depending on whether $S_{k,\theta} = 1$ ($z_{k,\theta} \geq t_k$) or $S_{k,\theta} = 0$ ($z_{k,\theta} < t_k$), the

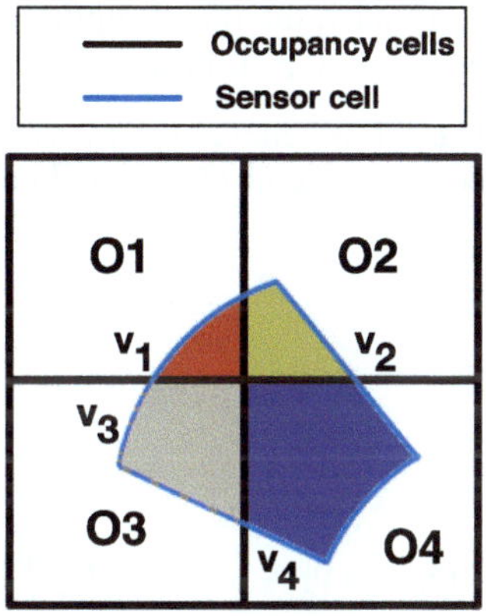

Fig. 2 Illustration of overlap between occupancy cells and a sensor cell. The area of overlap between a range bin and O$\{i\}$, is $v_{\{i\}}$ where $i \in \{1, \ldots, 4\}$

occupancy cells are updated to the posterior probabilities using Bayes' rule and the probabilities p_k and f obtained above.

Figure 2 shows the overlap between occupancy cells and a particular range bin. Let the region of overlap between any range bin (k, θ) and any occupancy cell $O_{x,y}$ be denoted by $O_{k,\theta}^{x,y}$. Also, let $\mathbb{O}_{k,\theta}^{x,y}$ denote the event that the region $O_{k,\theta}^{x,y}$ be occupied. We define our measurement model such that $S_{k,\theta} = 1$ will be observed when a target is present in any one of the overlapping regions $O_{k,\theta}^{x,y}$ with a probability equal to the probability of detection. This give rise to four possible combination of events as follows:

$$P(S_{k,\theta} = 1|\mathbb{O}_{k,\theta}^{x,y}) = p_k \tag{2}$$

$$P(S_{k,\theta} = 1|\widehat{\mathbb{O}_{k,\theta}^{x,y}}) = f \tag{3}$$

$$P(S_{k,\theta} = 0|\mathbb{O}_{k,\theta}^{x,y}) = 1 - p_k \tag{4}$$

$$P(S_{k,\theta} = 0|\widehat{\mathbb{O}_{k,\theta}^{x,y}}) = 1 - f \tag{5}$$

Let the area of overlap between range bin (k, θ) and occupancy cell $O_{x,y}$ be $v_{k,\theta}^{x,y}$ and the area of an occupancy cell be denoted by $A(O_{x,y})$. Now the events $\mathbb{O}_{k,\theta}^{x,y}$ and $\mathbb{O}_{x,y}$ are related as follows:

$$P(\mathbb{O}_{k,\theta}^{x,y}|\mathbb{O}_{x,y}) = \frac{v_{k,\theta}^{x,y}}{A(O_{x,y})} = a_{k,\theta}^{x,y} \tag{6}$$

$$P(\widehat{\mathbb{O}_{k,\theta}^{x,y}}|\mathbb{O}_{x,y}) = 1 - a_{k,\theta}^{x,y} \tag{7}$$

$$P(\widehat{\mathbb{O}_{k,\theta}^{x,y}}|\widehat{\mathbb{O}_{x,y}}) = 1 \tag{8}$$

$$P(\mathbb{O}_{k,\theta}^{x,y}|\widehat{\mathbb{O}_{x,y}}) = 0 \tag{9}$$

Finally, the map is updated for the two possible cases corresponding to $S_{k,\theta} = 1$ or $S_{k,\theta} = 0$ as follows:

Case 1 Whenever the measurement obtained is such that $S_{k,\theta} = 1$ ($z_{k,\theta} \geq t_k$), the occupancy cell $O_{x,y}$ is updated as follows:

$$P(\mathbb{O}_{x,y}|S_{k,\theta}=1) = \frac{P(S_{k,\theta}=1|\mathbb{O}_{x,y})P(\mathbb{O}_{x,y})}{P(S_{k,\theta}=1)} \tag{10}$$

$$P(S_{k,\theta}=1|\mathbb{O}_{x,y}) = 1 - P(S_{k,\theta}=0|\mathbb{O}_{x,y}) \tag{11}$$

$$\begin{aligned} P(S_{k,\theta}=0|\mathbb{O}_{x,y}) &= \prod_{i=1}^{m}\prod_{j=1}^{n}\left\{\sum_{\mathbb{O}_{i,j}}^{\widehat{\mathbb{O}_{i,j}}}\sum_{\mathbb{O}_{k,\theta}^{i,j}}^{\widehat{\mathbb{O}_{k,\theta}^{i,j}}} P(S_{k,\theta}=0|\mathbb{O}_{k,\theta}^{i,j})P(\mathbb{O}_{k,\theta}^{i,j}|\mathbb{O}_{i,j})P(\mathbb{O}_{i,j}|\mathbb{O}_{x,y})\right\} \\ &= \prod_{i=1}^{m}\prod_{j=1}^{n}\Big\{P(S_{k,\theta}=0|\mathbb{O}_{k,\theta}^{i,j})P(\mathbb{O}_{k,\theta}^{i,j}|\mathbb{O}_{i,j})P(\mathbb{O}_{i,j}|\mathbb{O}_{x,y}) \\ &\quad + P(S_{k,\theta}=0|\widehat{\mathbb{O}_{k,\theta}^{i,j}})P(\widehat{\mathbb{O}_{k,\theta}^{i,j}}|\mathbb{O}_{i,j})P(\mathbb{O}_{i,j}|\mathbb{O}_{x,y}) \\ &\quad + P(S_{k,\theta}=0|\widehat{\mathbb{O}_{k,\theta}^{i,j}})P(\widehat{\mathbb{O}_{k,\theta}^{i,j}}|\widehat{\mathbb{O}_{i,j}})P(\widehat{\mathbb{O}_{i,j}}|\mathbb{O}_{x,y}) \\ &\quad + P(S_{k,\theta}=0|\mathbb{O}_{k,\theta}^{i,j})P(\mathbb{O}_{k,\theta}^{i,j}|\widehat{\mathbb{O}_{i,j}})P(\widehat{\mathbb{O}_{i,j}}|\mathbb{O}_{x,y})\Big\} \end{aligned} \tag{12}$$

$$\begin{aligned} &= \left(1 - f + a_{k,\theta}^{x,y}(f - p_k)\right)\left\{\prod_{i=1}^{m}\prod_{j=1}^{n}\left\{\left(1 - f + a_{k,\theta}^{i,j}(f - p_k)\right)P(\mathbb{O}_{i,j})\right.\right. \\ &\quad \left.\left. + (1-f)P(\widehat{\mathbb{O}_{i,j}})\right\}\right\}\forall(i,j) \neq (x,y) \end{aligned} \tag{13}$$

$$P(S_{k,\theta}=1) = 1 - P(S_{k,\theta}=0) \tag{14}$$

$$\begin{aligned} P(S_{k,\theta}=0) &= \prod_{i=1}^{m}\prod_{j=1}^{n}\left\{\sum_{\mathbb{O}_{i,j}}^{\widehat{\mathbb{O}_{i,j}}}\sum_{\mathbb{O}_{k,\theta}^{i,j}}^{\widehat{\mathbb{O}_{k,\theta}^{i,j}}} P(S_{k,\theta}=0|\mathbb{O}_{k,\theta}^{i,j})P(\mathbb{O}_{k,\theta}^{i,j}|\mathbb{O}_{i,j})P(\mathbb{O}_{i,j})\right\} \\ P(S_{k,\theta}=0) &= \prod_{i=1}^{m}\prod_{j=1}^{n}\left\{\left(1 - f + a_{k,\theta}^{i,j}(f - p_k)\right)P(\mathbb{O}_{i,j})\right. \\ &\quad \left. + (1-f)P(\widehat{\mathbb{O}_{i,j}})\right\} \end{aligned} \tag{15}$$

where $P(S_{k,\theta} = 1|\mathbb{O}_{x,y})$ denotes the likelihood of getting a measurement $z_{k,\theta} \geq t_k$ from range bin (k, θ) given $O_{x,y}$ is already occupied and $P(S_{k,\theta} = 1)$ is the normalizing constant. $a_{k,\theta}^{i,j}$ becomes zero when the occupancy cell is far away from the range bin (k, θ). Hence, we only update the probabilities within the neighborhood of $r \times r$ occupancy cells that enclose range bin (k, θ). Also, while updating each occupancy cell $O_{x,y}$ in the $r \times r$ neighborhood, only the other occupancy cells $O_{i,j}$ in the same neighborhood will be involved.

It should be noted that for the case when $S_{k,\theta} = 1$, all possible combinations of detections and/or false alarms from all possible combinations of overlapping occupancy cells need to be considered. Hence calculating $P(S_{k,\theta} = 1)$ becomes rather involved. But $S_{k,\theta} = 0$ occurs only when a detection was missed or there was no target present in **all** the overlapping cells for which the probability can be calculated in a straightforward manner. Following which, $P(S_{k,\theta} = 1)$ can be calculated by taking the compliment of $P(S_{k,\theta} = 0)$.

Case 2 When the measurement obtained is such that $S_{k,\theta} = 0$ $(z_{k,\theta} < t_k)$, the occupancy cell $O_{x,y}$ is updated is a slightly different manner.

$$P(\mathbb{O}_{x,y}|S_{k,\theta} = 0) = \frac{P(S_{k,\theta} = 0|\mathbb{O}_{x,y})P(\mathbb{O}_{x,y})}{P(S_{k,\theta} = 0)} \tag{16}$$

where $P(S_{k,\theta} = 0|\mathbb{O}_{x,y})$ denotes the likelihood of getting a measurement $z_k < t_k$ from a range bin (k, θ) given $O_{x,y}$ is occupied. It can be obtained as per Eq. 12 and the normalizing constant, $P(S_{k,\theta} = 0)$, can be obtained from Eq. 15.

The implicit assumption made in the formulation is that the probabilities with which the cells are occupied are independent from one another.

2.3 Motion model

The motion model takes into account the translation and the rotational motion of the AUV and tracks the probabilities of the occupancy cells accordingly. It is defined such that the translational motion and rotational motion are decoupled from one another.

Translational Motion We model the translational motion as a convolution between the cell probabilities and an appropriate kernel **K**. The choice of kernel **K** depends on whether the AUV undergoes deterministic or probabilistic motion.

Deterministic Motion It is reasonable to model the AUV's motion as deterministic when GPS is available due to the high accuracy of GPS signals. For such a case, the occupancy grid is simply shifted by the amount of displacement. Figure 3 shows how the probability is updated through a convolution when the robot undergoes translational motion.

The kernel is a representation of the amount of displacement the robot has undergone. In our case, the kernel is two dimensional represented by an $N \times N$ matrix.

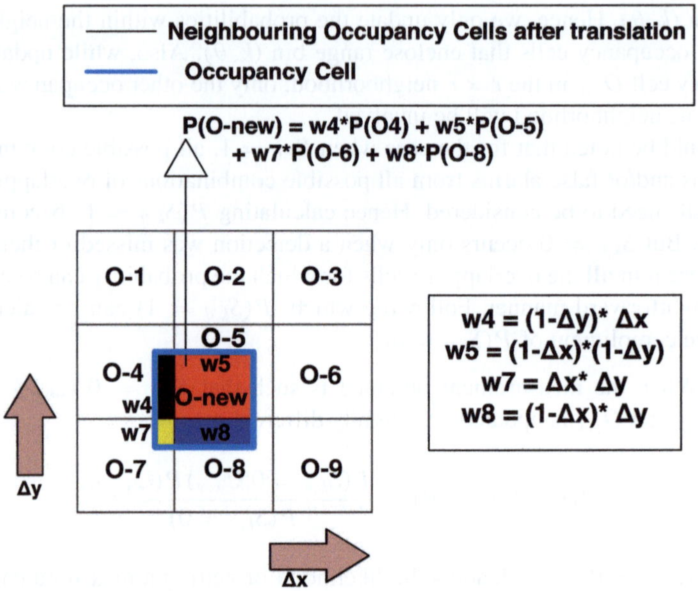

Fig. 3 Illustration of overlap of neighboring occupancy cells after undergoing translation with a particular occupancy cell. The area of overlap between O-new and O-{i}, is w-{i} where $i \in$ {4, 5, 7 and 8}

Elements of the kernel, which is the area of overlap, are shown in Fig. 3. The mathematical form of the motion update is as follows:

$$\mathbf{P} \otimes \mathbf{K} \tag{17}$$

where $\otimes$ is the convolution symbol and $\mathbf{P}$ is the matrix representation of the entire occupancy grid.

Probabilistic Motion When there is no GPS or DVL available, the displacement is unimodal with its peak representing the mean translational motion, and spread modelling the uncertainty associated with the motion estimate. The uncertainty is modeled as a Gaussian distribution, denoted by $\mathcal{N}(\boldsymbol{\mu}, \mathbf{R})$ where $\boldsymbol{\mu}$ is the mean displacement of the AUV and variance, $\mathbf{R}$, is the process noise of the thruster model. Hence the area under the distribution would give the desired kernel $\mathbf{K}$. A typical element for this type of kernel would be of the form:

$$\mathbf{K}_{ij} = \iint\limits_{A} \mathcal{N}(\boldsymbol{\mu}, \mathbf{R})dxdy \tag{18}$$

The integral is evaluated over the region of the distribution represented by the element $\mathbf{K}_{ij}$. The grid is updated using Eq. 17.

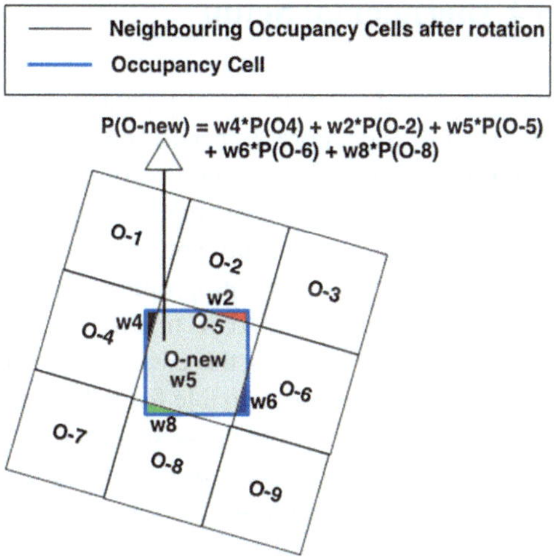

Fig. 4 Illustration of overlap of neighboring occupancy cells after undergoing rotation with a particular occupancy cell. The area of overlap between O-new and O-$\{i\}$, is w-$\{i\}$ where $i \in \{2, 4, 5, 6 \text{ and } 8\}$

Rotational Motion We model the rotational motion of the AUV as deterministic. To avoid rounding errors, we accumulate changes in heading until they reach $\pm 1°$. The area of overlap of rotated neighboring occupancy cells $O'_{x-i,y-j}$ $\forall\, i, j \in \{-1, 0, 1\}$ with a particular occupancy cell $O_{x,y}$ is calculated. Then the new probability of occupancy is updated as:

$$P(\mathbb{O}_{x,y}) = \sum_i \sum_j w_{x,y}^{x-i,y-j} P(\mathbb{O}'_{x-i,y-j}) \tag{19}$$

where $w_{x,y}^{x-i,y-j}$ is the ratio of the area of overlap between occupancy cell $O'_{x-i,y-j}$ and $O_{x,y}$ and the area of occupancy cell $O_{x,y}$. Figure 4 shows how the probability is updated in the presence of rotation.

2.4 *Obstacle Detection*

The expected number of obstacles $N_{x,y}$ in a neighborhood of a occupancy cell $O_{x,y}$ can be estimated from the occupancy grid:

$$N_{x,y} = \sum_i \sum_j P(\mathbb{O}_{x-i,y-j}) \;\forall\, i, j \in \{-1, 0, 1\}. \tag{20}$$

Here we have taken the neighborhood to be ± 1. We set a threshold P_{thresh} and declare a detected obstacle if $N_{x,y} \geq P_{\text{thresh}}$. At the end of every scan, the obstacles detected throughout the grid is sent to the Navigator of the AUV to carry out necessary avoidance maneuvers if necessary.

3 Experimental Setup

We conducted experiments at Pandan reservoir in Singapore and also in the sea off the coast of Singapore. For both sets of experiments, we used a Micron DST sector scanning sonar [16] integrated on our STARFISH AUV [17].

During the Pandan experiment, the mission was planned such that the AUV would be operating near some static buoys and the reservoir's embankments. The sonar was configured for 50 m operating range with 44 bins and 90° scan sector. The mission was executed with the AUV maintaining a constant depth of 0.5 m. The mission path and the obstacles in the environment are shown in Fig. 5a. Note that the lower embankment wall is not visible from the surface but marked in Fig. 5a using a dashed line. A illustration of the cross-section of the embankment is shown in Fig. 5c.

The experiment at sea was conducted at Selat Pauh, an anchorage area south of Singapore with a depth of 10–20 m. The AUV mission plan led the AUV to an area close to shallow coral reefs (<5 m). During this mission, the AUV swam at the surface. Figure 5b shows the AUV path and the location of the shallow reefs.

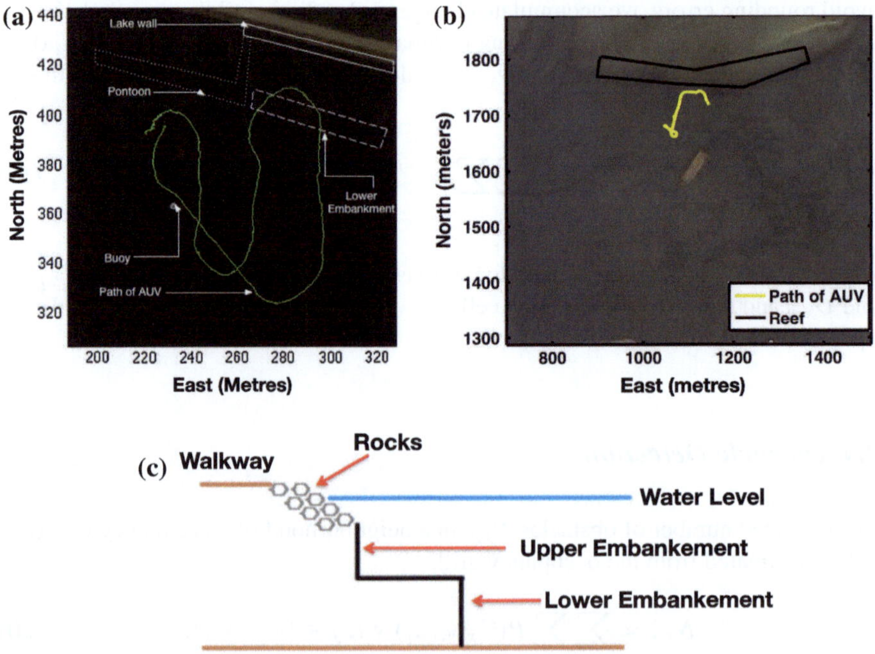

Fig. 5 Experiments at Pandan reservoir and at sea. **a** AUV path and obstacle locations at Pandan reservoir, **b** AUV path and reef location at sea, **c** illustration showing the structure of embankments at Pandan reservoir

4 Results

The experimentally obtained ROC curves at Pandan reservoir matched those obtained from a Gaussian noise model with an appropriate SNR at operational values of f_k (0.01–0.04) as shown in Fig. 6a. We set the desired false alarm rate $f = 0.02$ and obtained p_k and t_k values for all range bins. The scans from the FLS were processed

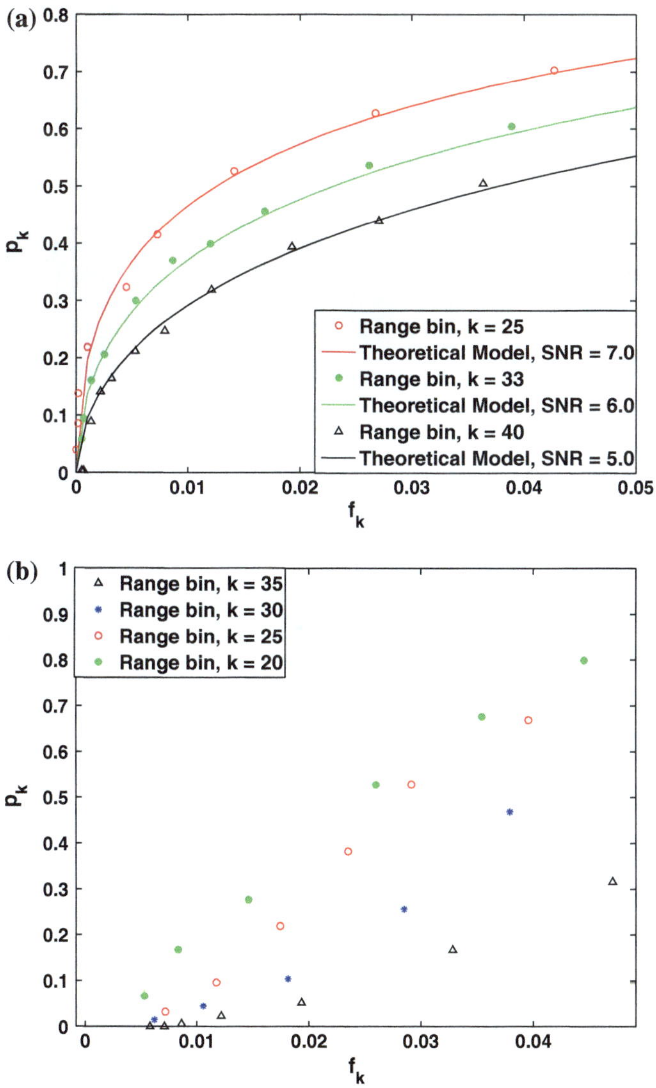

Fig. 6 Experimentally obtained ROC plots. **a** ROC plot at Pandan Reservoir and the corresponding theoretical curves, **b** ROC plot at Selat Pauh

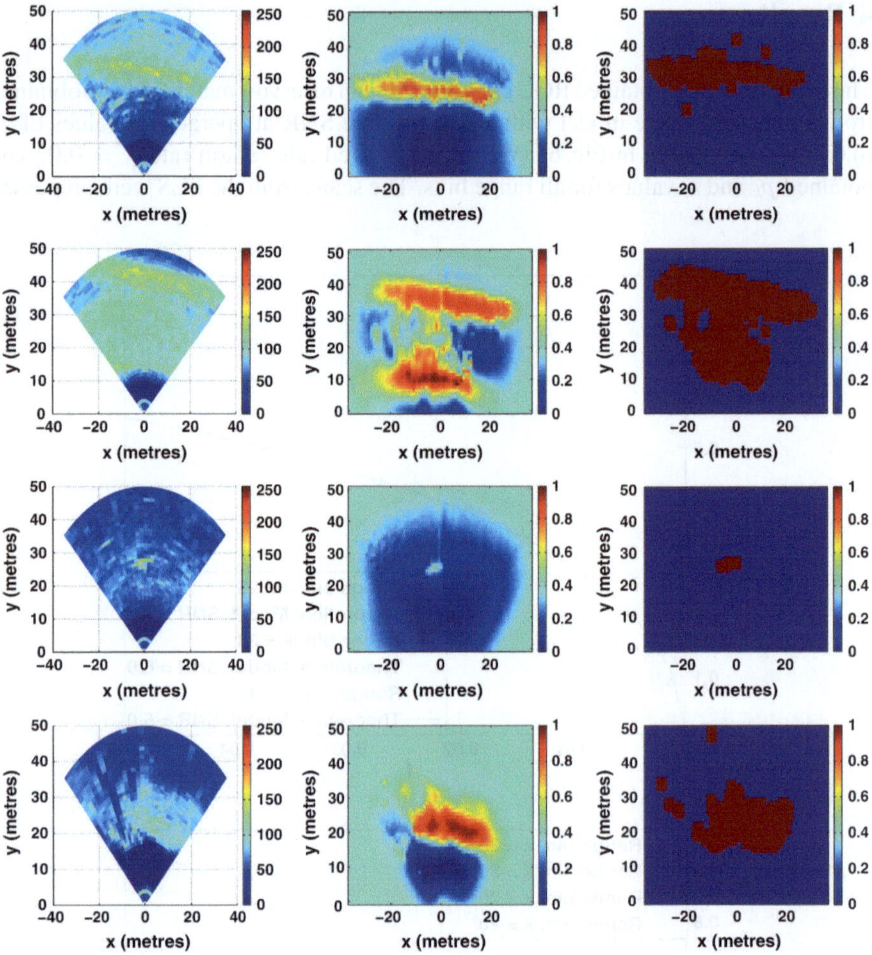

Fig. 7 Unprocessed scans (*left column*), occupancy grid (*middle column*) and obstacle detection (*right column*) of various targets. The *first two rows* show the reservior's embankments during the Pandan experiment, while the *third row* shows a buoy during the same experiment. The *bottom row* shows a patch of coral reef during the sea experiment

online and local occupancy grids were generated. Obstacles such as the reservoir embankments and buoys were clearly detected. Unprocessed scans, local occupancy grids and obstacle detections are show in Fig. 7 (first, second and third row).

The ROC curves obtained from the experiments held at the sea are shown in Fig. 6b for operational values of f_k. As the sea was much noisier than the reservoir, we set a slightly higher rate of false alarm $f = 0.03$ to ensure good detections. The FLS scans were processed in the same way as the Pandan experiment, and the results are shown in Fig. 7 (bottom row).

5 Experimental Insights

From the unprocessed sonar scans shown in Fig. 7 (left column), we see that the targets cannot be clearly distinguished from the background noise. Multiple scans are processed and assimilated into the local occupancy grid as the AUV moves. The results from this process are seen in Fig. 7 (middle column). We observe that the cells corresponding to obstacles show a high probability of occupancy. The improvement comes from combining information from multiple scans. The Bayesian update effectively weighs the information from multiple scans based on its reliability. Finally, a hard-decision detection procedure is used at the end of each scan to detect potential obstacles. Obstacles such as buoys, reservoir embankments and coral reefs are detected reliably as shown in Fig. 7 (right column). These obstacle detections are then sent to the AUV's command and control system.

While Bayesian updates of an occupancy grid can be implemented in a georeferenced frame, accumulation of errors in the AUV's position estimate can render this approach ineffective. By noting that obstacle avoidance only requires accurate knowledge of obstacle locations in an AUV's body frame, we are able to use a local occupancy grid in concert with a uncertainty-aware motion model. The result is an algorithm that accurately tracks and detects obstacles in the AUV's frame of reference. Although this approach is ideally suited to obstacle avoidance, it does not provide an absolute location for each detected obstacle and therefore is unsuitable for mapping applications. The approach limits the occupancy grid to a small region around the AUV; this limits memory requirements and computational load and makes the algorithm appropriate for real-time implementation. However it also results in the AUV "forgetting" obstacles that it might have seen during a previous visit to a given area. Since revisiting areas in not common during most AUV missions, and since obstacles can be reliably re-detected, we do not see this as a significant shortcoming. We therefore believe that our proposed algorithm is well suited to underwater obstacle detection and collision avoidance for AUVs, and can be not only used with expensive multibeam sonars, but also with cheaper sector scanning FLS.

6 Conclusion

We developed a novel method for underwater obstacle detection using a probabilistic local occupancy grid. We demonstrated its capability to detect obstacles robustly and localize them accurately in the AUV's frame of reference. Compared to previous published approaches, our approach deals more directly with positional uncertainty by adopting an occupancy grid in the AUV's frame of reference. Hence, the obstacles are accurately localized relative to the AUV. Finally, this method is computationally less intensive compared to other image processing techniques or SLAM techniques and can be implemented on board an AUV. Future work may explore the possibility of tackling the problem of the "forgetting" nature of the local occupancy grid.

References

1. Teck, T.Y., Chitre, M.: Direct policy search with variable-length genetic algorithm for single beacon cooperative path planning. In: International Symposium on Distributed Autonomous Robotic Systems (DARS) (2012)
2. Thrun, S., Burgard, W., Fox, D.: Probabilistic Robotics (Intelligent Robotics and Autonomous Agents). The MIT Press (2005)
3. Leedekerken, J.-C., Leonard, J.J, Bosse, M.-C., Balasuriya, A.: Real-time obstacle avoidance and mapping for AUVs operating in complex environments. In: Proceedings of the 7th International Mine Warfare Symposium, Monterey (2006)
4. Brekke, E., Chitre, M.: Bayesian multi-hypothesis scan matching. In: Proceedings of OCEANS'13, Bergen (2013)
5. Elfes, A.: Using occupancy grids for mobile robot perception and navigation. Computer **22**(6), 46–57 (1989)
6. Konolige, K.: Improved occupancy grids for map building. Auton. Robots **4**(4), 351–367 (1997)
7. Eliazar, A.: DP-SLAM. Ph.D. dissertation, Department of Computer Science, Duke University (2005)
8. Horner, D.P., Healey, A.-J., Kragelund, S.P.: AUV experiments in obstacle avoidance. In: Proceedings of MTS/IEEE OCEANS (2005)
9. Quidu, I., Hetet, A., Dupas, Y., Lefevre, S.: AUV (Redermor) obstacle detection and avoidance experimental evaluation. In: OCEANS 2007-Europe, Aberdeen (2007)
10. Teo, K., Ong, K.W., Lai, H.C.: Obstacle detection, avoidance and anti collision for MEREDITH AUV. In: MTS/IEEE Biloxi-Marine Technology for Our Future: Global and Local Challenges, OCEANS 2009, Biloxi (2009)
11. Heidarsson, H., Sukhatme, G.: Obstacle detection and avoidance for an Auonomous Surface Vehicle using a profiling sonar. In: IEEE International Conference on Robotics and Automation (ICRA) (2011)
12. Martin, A., An, E., Nelson, K., Smith, S.: Obstacle detection by a forward looking sonar integrated in an autonomous underwater vehicle. In: Oceans 2000, vol. 1. Providence RI (2000)
13. Chew, J.-L. Chitre, M.: Object detection with sector scanning sonar. In: Proceedings of OCEANS'13, San Diego (2013)
14. Robocentric map joining: Improving the consistency of EKF-SLAM. Robot. Auton. Syst. **55**(1), 21–29 (2007)
15. Richards, M.: Fundamentals of radar signal processing. pp. 304–316 (2005)
16. Tritech. Micron DST sector scanning sonar. http://www.tritech.co.uk/product/small-rov-mechanical-sector-scanning-sonar-tritech-micron
17. Koay, T., Tan, Y., Eng, Y., Gao, R., Chitre, M., Chew, J., Chandhavarkar, N., Khan, R., Taher, T., Koh, J.: STARFISH: a small team of autonomous robotic fish. Indian J. Geo-Marine Sci. **40**(2), 157–167 (2011)

Gaussian Process Occupancy Maps for Dynamic Environments

Simon T. O'Callaghan and Fabio T. Ramos

Abstract We present a continuous Bayesian occupancy representation for dynamic environments. The method builds on Gaussian processes classifiers and addresses the main limitations of occupancy grids such as the need to discretise the space, strong assumptions of independence between cells, and difficulty to represent occupancy in dynamic environments. We develop a novel covariance function (or kernel) to capture space and time statistical dependencies given a motion map of the environment. This enables the model to perform predictions on how the occupancy state of the environment will be in the future given past observations. We show results on a simulated environment with multiple dynamic objects, and on a busy urban intersection.

1 Introduction

The Gaussian process occupancy map (GPOM) [12] is a continuous occupancy representation of the environment that overcome some of the limitations with occupancy grids [4]. The method places a Gaussian process (GP) [15] prior over functions mapping the 2D or 3D space into the probability of occupancy. Both laser beams and laser returns are used as free-space and occupied observations respectively to train a GP classifier. The resulting model is not limited to a particular resolution and naturally captures spatial relationships between data points, offering a principled methodology to reason about occlusions, and informative exploration strategies [1, 7].

Formally affiliated with the Australian Centre for Field Robotics, The University of Sydney, Australia.

S.T. O'Callaghan (✉)
NICTA, Eveleigh, Australia
e-mail: simon.ocallaghan@nicta.com.au

F.T. Ramos
Australian Centre for Field Robotics, School of Information Technologies,
The University of Sydney, Sydney, Australia
e-mail: fabio.ramos@sydney.edu.au

M.A. Hsieh et al. (eds.), *Experimental Robotics*, Springer Tracts
in Advanced Robotics 109, DOI 10.1007/978-3-319-23778-7_52

Despite overcoming some of the issues present in discrete representations such as occupancy grids, GPOM assumed a static world and ignored the effects of time on the model. In reality, mobile robots frequently operate in dynamic environments where the motion of pedestrians, vehicles and other moving objects play an important role in affecting the state of the world and consequently in determining the outcomes of the automaton's actions. In this work, we extend the GPOM to incorporate a temporal dimension and predict the probability of occupancy at a point in time and space using information from past observations.

Commonly, dynamic entities are represented as a single point (generally the estimated centroid of the object [16]) and observations associated with moving objects are removed before the remaining laser hits are passed to a static-world mapper. Recently, a number of works have focused on an alternative approach to classical object tracking by modelling motion on a "sub-object" level within geometric maps. The work in [3, 11] augment the properties of the occupancy grid map so as to model predicted velocity in a cell as well as its occupancy. In this model, concepts such as objects or tracks do not exist. They are replaced by properties such as occupancy or risk, which are directly estimated for each cell of the grid using sensor observations from consecutive scans to propagate motion and hence the cells' probabilities of occupancy into the future. Gindele et al. [8] add prior information of the scene to this model in order to predict more complex mobile object motions such as vehicles turning at corners, etc.

Here, we adopt a similar approach by modelling motion on a sub-object level while conserving the continuous nature of our technique and the associated benefits of operating within a fully Bayesian setting. We adapt the framework of a Gaussian process classifier to account for the effects of motion and, consequently, to learn dependencies between consecutive observations to model occupancy in dynamic environments. Our proposed approach learns the dynamic regions of the map and expands on the GPOM, enabling it to propagate the hypothesis of occupancy temporally as well as spatially within a single Bayesian model.

2 Dynamic Gaussian Process Occupancy Maps

In GPOM we model the probability of occupancy at a query location $\mathbf{x}_*$ given a set of N observations $\{\mathbf{x}_i, y_i\}_{i=1}^N$, as $p(y|X, \mathbf{y}, \mathbf{x}_*) = \Phi(\frac{\alpha\mu_*+\beta}{1+\alpha^2\sigma_*^2})$, where $\Phi(\cdot)$ is the cumulative Gaussian distribution, μ_* and σ_*^2 are the predictive mean and variance for the posterior of $\mathbf{x}_*$ respectively, and α, β are parameters of the cumulative Gaussian estimated through leave-one-out cross validation [12]. The posterior is obtained from a $\mathcal{GP}(m(\mathbf{x}), k(\mathbf{x}, \mathbf{x}'))$, with mean function $m(\mathbf{x})$ and covariance function $k(\mathbf{x}, \mathbf{x}')$. In this work we generalise GPOM to handle dynamic environments by jointly learning occupancy properties with a motion model of the environment. A block diagram illustrating the proposed inference framework is presented in Fig. 1.

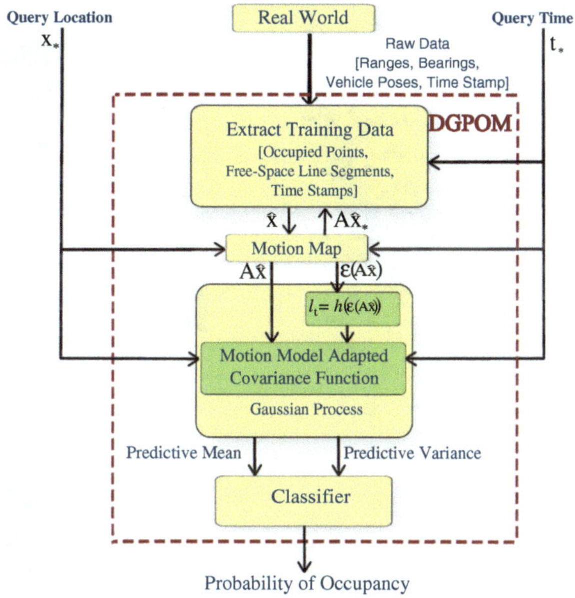

Fig. 1 Block diagram of the inference procedure for a GPOM with motion-map adapted covariance function

2.1 Motion-Model Adapted Gaussian Process

Defining $\Upsilon(\cdot)$ as a vector field estimate of the mean velocity or drift of an underlying function, we can compute the spatial displacement between instants $t_{\mathbf{x}}$ and $t'_{\mathbf{x}}$ of a particle originally at location $\mathbf{x}$ as,

$$\psi(t_{\mathbf{x}}, t_{\mathbf{x}'}) = \int_{t_{\mathbf{x}}}^{t_{\mathbf{x}'}} \mathbf{v}(t)dt, \tag{1}$$

where $\mathbf{v}(t)$ is obtained from the velocity vector field. We can then modify the covariance function to take into account displacements. For example, for the squared exponential covariance function, the dynamic version takes the form,

$$k(\hat{\mathbf{x}}, \hat{\mathbf{x}}') = \sigma_f^2 \exp\Big(-\frac{|\hat{\mathbf{x}} - A\hat{\mathbf{x}}'|^2}{2L^2}\Big), \tag{2}$$

where $L = \begin{bmatrix} L_s & 0 \\ 0 & l_t \end{bmatrix}$, $A = \begin{bmatrix} 1 & 0 & \psi(t_{\mathbf{x}}, t_{\mathbf{x}'})_1 \\ 0 & 1 & \psi(t_{\mathbf{x}}, t_{\mathbf{x}'})_2 \\ 0 & 0 & 1 \end{bmatrix}$, and $\hat{\mathbf{x}}$ is the concatenation of the observation's location input vector, $\mathbf{x}$ and $t_{\mathbf{x}}$. L_s is length-scale hyperparameter matrix pertaining to the spatial dimensions and l_t is the temporal length-scale hyperparameter.

Incorporating a motion model into the covariance function enables the GP to learn dependencies in the observations along the direction of motion rather than along the

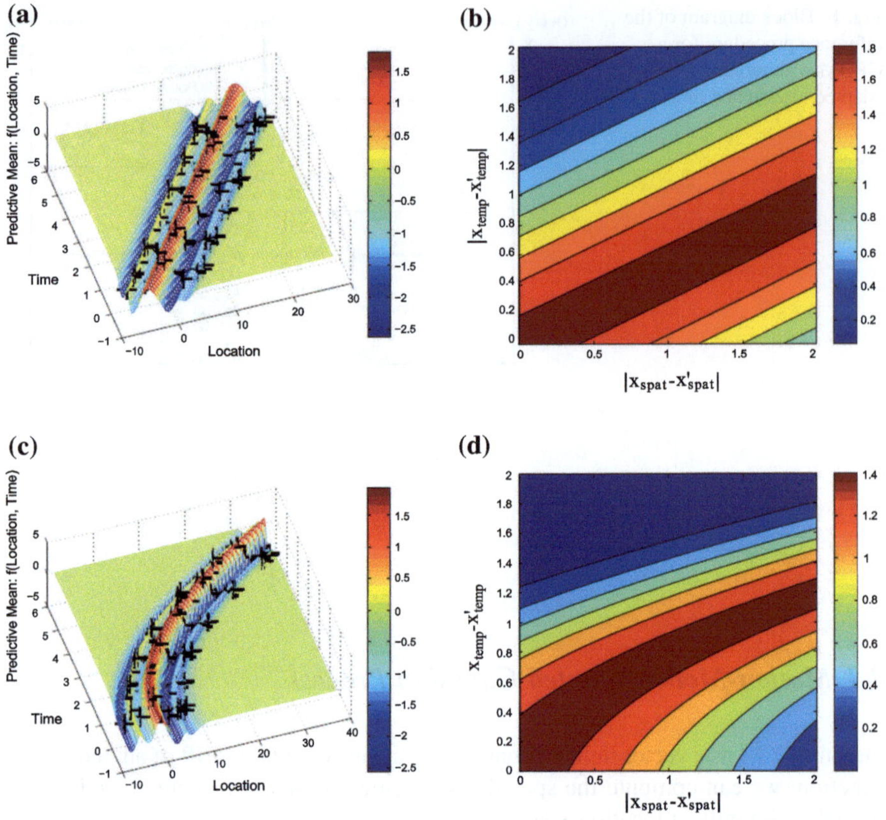

Fig. 2 Predictive mean functions (*left column*) and their associated covariance contours (*right column*) after training. The results shown are from a GP employing a constant velocity motion model (*top row*) and a constant acceleration model (*bottom row*) for $\boldsymbol{\psi}(t_{\mathbf{x}}, t_{\mathbf{x}'})$ in Eq. 2. Observations of the function are represented as *black crosses*

temporal axis as can be seen in Fig. 2a, c. During the training phase, both the spatial and dynamic elements of the model are learnt jointly. Improving the estimate of the underlying function's motion, increases the marginal likelihood of the Gaussian process. Similarly, optimising the GP's spatial representation of the function allows for a better alignment of consecutive sets of observations using the motion model.

Motion-model adapted covariance functions offer some useful capabilities in terms of training a Gaussian process to represent the spatial and dynamic behaviour of an underlying function. However, using such a GP to model occupancy in a map proves problematic due to the model's assumption that the entire underlying function is subject to the same motion.

A possible solution to the limitations of the technique is to modify the displacement function so that it is also dependent on location as well as time, $\boldsymbol{\psi}(t_{\mathbf{x}}, t_{\mathbf{x}'}, \mathbf{x})$. The notion of associating a location with information about the motion of objects at that point is not a new one. Gindele et al. [8] and Brechtel et al. [2] predict the hypothesis of occupancy based on past map states and on velocity values assigned to each cell using prior map knowledge such as a GPS nav-map to propagate occupancy in dynamic regions of the environment. Ellis et al. [5] and O'Callaghan et al. [13] use information from pedestrian trajectory traces to learn motion maps for the environment with applications in object tracking, anomaly detection and path planning. In our case, we seek to learn a motion map that will enable the GP to shape its covariance contours and to determine the extent of temporal dependencies.

2.2 Motion-Map Adapted Gaussian Process

Remodelling Eq. 2 to accommodate for spatial changes in velocities, for the 2-D case, we obtain:

$$A = \begin{bmatrix} 1 & 0 & \psi(t_{\mathbf{x}}, t_{\mathbf{x}'}, \mathbf{x})_1 \\ 0 & 1 & \psi(t_{\mathbf{x}}, t_{\mathbf{x}'}, \mathbf{x})_2 \\ 0 & 0 & 1 \end{bmatrix}. \tag{3}$$

As before, the input vector, $\mathbf{x}'$, is translated using A when used as training data in the Gaussian process. The displacement vector $\boldsymbol{\psi}(t_{\mathbf{x}}, t_{\mathbf{x}'}, \mathbf{x})$ represents the spatial displacement undergone by a point initialised at $\mathbf{x}$ over a time interval of $t_{\mathbf{x}'} - t_{\mathbf{x}}$. Crucially, its value now also depends on the initial location of the observation rather than simply the time stamps,

$$\boldsymbol{\psi}(t_{\mathbf{x}}, t_{\mathbf{x}'}, \mathbf{x}) = \int_{t_{\mathbf{x}}}^{t_{\mathbf{x}'}} \mathbf{v}(\mathbf{x}(t))dt - \mathbf{x}, \tag{4}$$

where the velocity, $\mathbf{v}(\mathbf{x})$, at any point is governed by the motion map.

The estimated motion map will inevitably have some degree of error in it and so it is important that this uncertainty is reflected in the GPOM's hypothesis of occupancy estimates. Consequently, we employ a non-stationary covariance function with the ability to locally alter its temporal length-scale, $l_t(\mathbf{x})$, based on the observed quality of the motion map in each region.

Melkumyan and Ramos [10] provides a comprehensive list of analytical solutions for many popular stationary covariance function's non-stationary form. Here, we use a summation of two non-stationary Matérn class covariance functions with ς values of $3/2$ (Eq. 5) to model the covariance in the temporal domain as it provides a good balance between capturing sudden changes in the function while also learning long-term trends of the data;

$$k(x, x'; l_t(x), l_t(x')) = \qquad (5)$$
$$\sigma^2\Big(l_t(x)\exp\Big(-\sqrt{3}\frac{|x-x'|}{l_t(x)}\Big) - l_t(x')\exp\Big(-\sqrt{3}\frac{|x-x'|}{l_t(x')}\Big)\Big),$$

where $\sigma^2 = 2\sqrt{l_t(x)l_t(x')}/(l_t(x)^2 - l_t(x')^2)$.

2.3 Learning the Motion Map

Motion-map adapted Gaussian processes enable the GPOM to handle dynamic objects by propagating the effects of the movement into the inference model. In this section, we discuss one possible method for deriving the motion map which we represent here as a mean velocity field $\Upsilon(\cdot)$ and an associated error field $\varepsilon(\cdot)$ based on previous observations of the environment.

The procedure initially builds an occupancy map for each scan. A large body of literature exists on various optical flow techniques, [6, 9], for extracting regions of motion between two images. Here we opt for a straightforward and fast agglomerative clustering of query points deemed occupied followed by data association to estimate changes between scans.

Comparing two consecutive maps the algorithm identifies query points in each that have a high probability of being occupied and also have a high probability of being either occupied or free space in the other map. The resulting maps, referred to as *Frame1* and *Frame2* in the figure, are then subtracted from each other and element-wise multiplied by one another to produce *Diff* and *Same*, respectively. The *Diff* map contains clusters of value -1 representing an area that an object has just vacated and clusters of value $+1$ in an area where an object has just moved into. *Same* highlights regions that have remained occupied in both scans and may possibly contain a stationary object. Sample outputs of these 3 stages are shown in Fig. 3.

The algorithm then clusters the positive valued cells as objects and the non-zero cells from *Diff* and *Same* as motion clusters. For each motion cluster in *Diff*, the algorithm searches for an adjacent object. If one is found, this object is assumed to have either just vacated or occupied that cluster depending on the value of that cluster ($+1$ or -1). The direction of motion is obtained by calculating the angle between the centroid of the motion cluster and the associated object cluster. The magnitude of the motion is simply the width of the cluster. Stationary objects are identified as clusters in *Same* that do not have a neighbouring motion cluster from *Diff*.

These clusters and their velocities are then stored in memory and the procedure is repeated for each pair of consecutive scans in the training set. A regressor is trained using a portion of these clusters to infer a velocity at any point in space, i.e. the mean velocity field $\Upsilon(\cdot)$. Finally, the error of this field is assessed using clusters withheld from the training and a second regressor is then trained to approximate our estimated accuracy of $\Upsilon(\cdot)$ over the entire region.

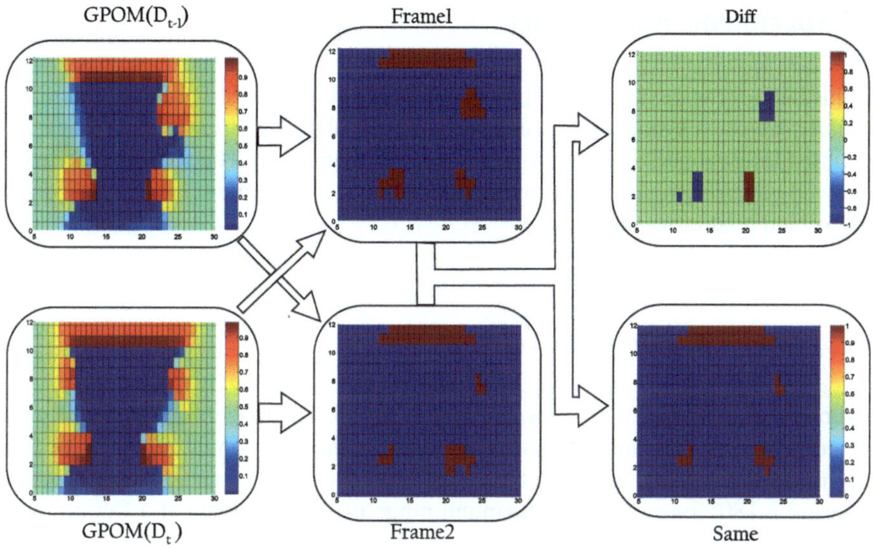

Fig. 3 The stages involved in generating the motion clusters. First stage: Occupancy maps for 2 consecutive scans. Second stage: *Frame1* and *Frame2* of the respective scans. Third stage: *Diff* and *Same* maps

2.4 Relating $\varepsilon(\cdot)$ to $l_t(\cdot)$

Theoretically, the latent temporal length-scale hyperparameter function, $l_{t(\mathbf{x})}$, can be learnt by maximising its marginal likelihood such as in [14]. Implementing such a scheme unfortunately is quite expensive due to the large search space for learning the parameters of the function combined with the relatively slow iteration time of the optimiser due to the requirement of inverting an $n \times n$ covariance matrix each time.

A convenient approximation is to learn a parametric mapping, $h(\cdot)$, from a point's estimated translation error, $\varepsilon(\mathbf{x})$ to the temporal length-scale,

$$l_t(\mathbf{x}) \approx h(\varepsilon(\mathbf{x})) = a * \varepsilon(\mathbf{x})^{-(\frac{1}{b})}. \tag{6}$$

The chosen form of the mapping is described by Eq. 6. The parameter b controls the rate of decay of l_t as $\varepsilon(\mathbf{x})$ increases while a serves as a scaling parameter.

This approach to learning the temporal length-scale results in two additional dimensions being added to the search space during training. However, it is a significantly more constrained problem than attempting to train a latent non-parametric function, $l_t(\cdot)$ during optimisation of the marginal likelihood.

3 Experiments and Results

In this section, we examine the performance of the proposed DGPOM algorithm. Initially, we use a synthetic dataset simulating the observations received from a car-mounted laser rangefinder sensor positioned at a T-junction. The second experiment involved gathering range data at a city intersection. Both experiments include quantitative and qualitative analysis.

3.1 Simulated Data Experiment

Figure 4 presents a summary of a synthetic scenario. The rangefinder sensor is positioned at $\mathbf{x} = (15; 0)$ and observes cars passing the junction at velocities of 2 or -1 m/s depending on which lane the vehicle occupies. Each scan contains 70 beams covering a 180° sweep with a maximum range of 20 m sampled at a frequency of 1 Hz.

The Motion Map $\Upsilon(\cdot)$ and $\varepsilon(\cdot)$ for the environment are learnt with data acquired a priori. The motion clusters and their velocities are shown in Fig. 5a. A subset consisting of 500 of these points was used to train a regressor to model the horizontal and vertical components of $\Upsilon(\cdot)$. Figure 5b presents the regressor's output for the horizontal component of the estimated mean velocity field. Both lanes are clearly distinguishable from the plot with the regressor also learning that vehicles observed in the lane traveling from left to right tended to move at approximately twice the speed of the vehicles in the opposite lane. The mean velocity field also estimates an average velocity of 0 m/s in the area occupied by the building at the top of the scene. A quiver plot of the resulting mean velocity field is superimposed onto the environment in Fig. 5c.

DGPOM Outputs Using the Dynamic GPOM framework the probability of occupancy can be inferred at various instances in time and space. The outputs from two

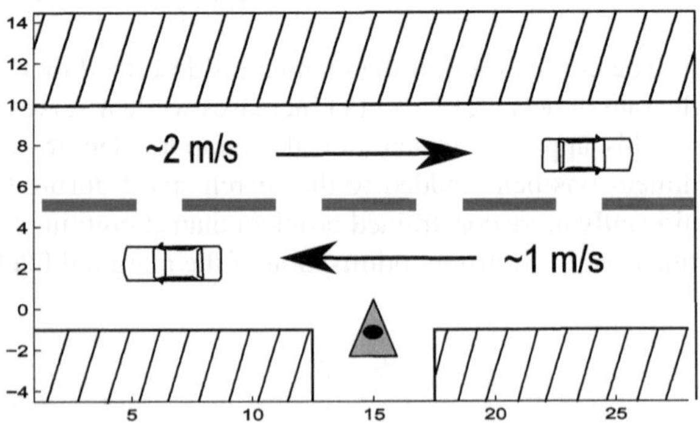

Fig. 4 Ground truth of simulated experiment

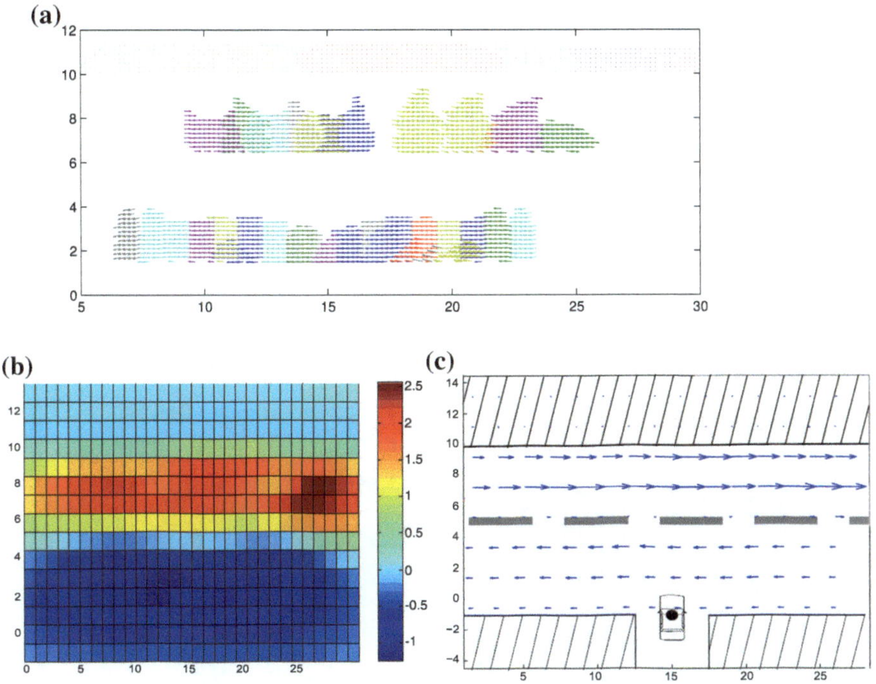

Fig. 5 **a** The resulting motion clusters with their associated velocity vectors (dynamic and static). **b** Output of regressor modelling the horizontal component of $\Upsilon(\cdot)$. **c** A quiver plot of the motion map's velocity superimposed on the environment

of those instances are displayed in Fig. 6. The ground truths at $t = 3$ and $t = 5$ (top row) show two vehicles in each lane (blue rectangles) with a large amount of occlusion created by the cars closer to the sensor. The red lines represent the range-finder sensor's observations at that time step. Incorporating observations from previous scans using the DGPOM framework enables the algorithm to accurately infer a large amount of the scene (middle row). Despite not observing the car at (17; 7.5) in the first image directly, the procedure uses the learnt motion map to infer a strong covariance between that location and previous observations of the vehicle resulting in a high probability of occupancy in the mapper's output at the aforementioned location. Similar behaviour can also be seen in the second column at (5; 7.5). The associated predictive variance maps (bottom row) offer an insight into how the scene was reconstructed. The dark blue regions of high confidence rely primarily on the most recent observations. Light blue and green areas indicate estimates in regions that have not been observed in the most recent scans and, due to a degree of inaccuracy in motion map, $\Upsilon(\cdot)$, are less confident in the hypothesis of occupancy.

The additional flexibility afforded to the algorithm by incorporating location as an extra parameter in determining the behaviour of the covariance function enables the GP to exploit strong temporal dependencies between scans containing multiple

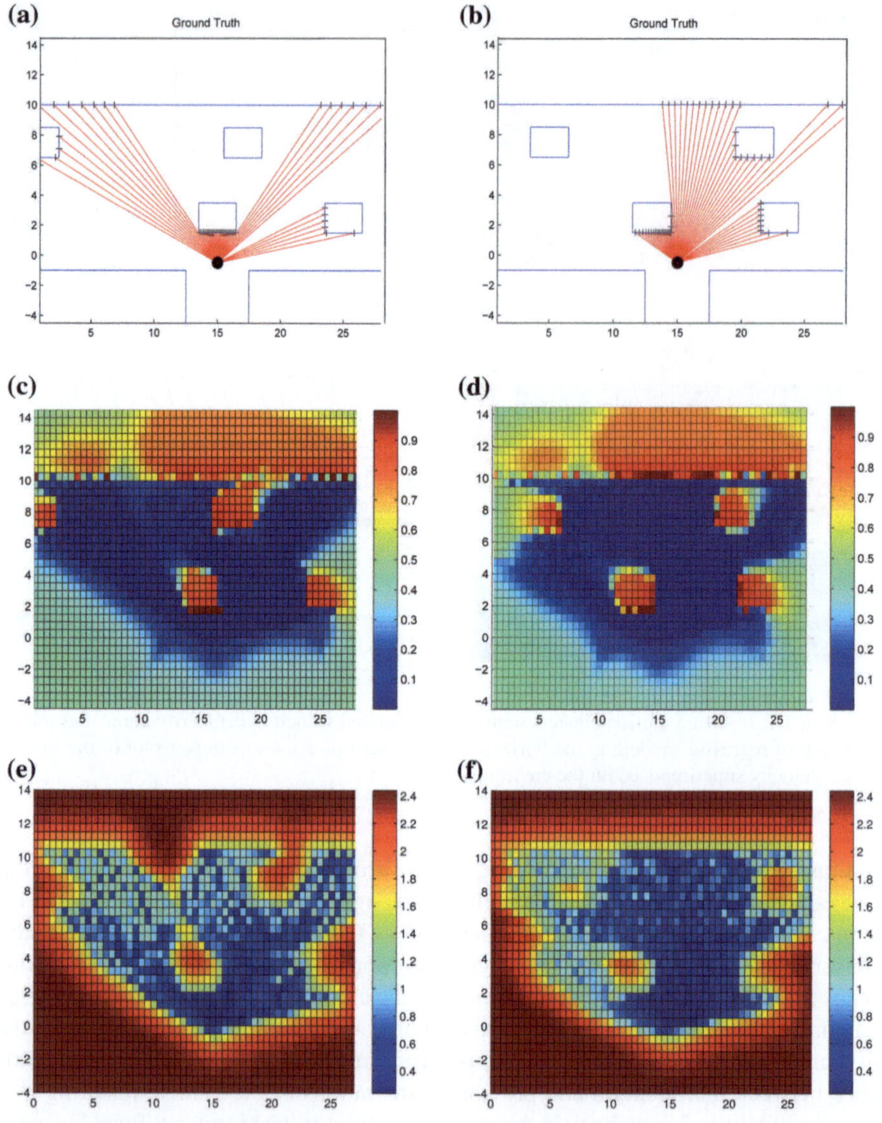

Fig. 6 Results for simulated experiment at two separate instances in time. $t = 3$ (*left column*) and $t = 5$ (*right column*). *Top* Ground truth of the environment's state. *Middle* Probability of occupancy versus location which incorporates observations from previous time steps. *Bottom* Predictive variance maps

dynamic and stationary objects. As a result, the probability of occupancy for the entire region can be handled within the same Bayesian setting without the need for filtering out dynamic objects and handling them in a separate procedure. Large covariances along traffic lanes learnt by the covariance function allows the GP to accurately infer

Table 1 Comparison of areas under the ROC curve at various instances in time (columns) using 3 different approaches

	Query time (t_{x_*})						
Algorithm	$t_{x_*}=3$	$t_{x_*}=4$	$t_{x_*}=5$	$t_{x_*}=6$	$t_{x_*}=7$	$t_{x_*}=8$	$t_{x_*}=9$
DGPOM $\Re^{D+1}$	0.9497	0.9301	0.9300	0.9288	0.9480	0.9372	0.9435
GPOM $\Re^{D}$	0.7736	0.8040	0.8248	0.8227	0.8364	0.8376	0.8850
GPOM $\Re^{D+1}$	0.8383	0.8773	0.8903	0.8954	0.9267	0.8922	0.8974

the location of the car by propagating the influence of past observations forward through time using the velocity vectors of the motion map.

ROC Tests With a known ground truth it is possible to determine the precision of the estimates using the ROC curves once more. Table 1 shows the results of a comparison between the inference algorithm's outputs and two variations of the standard GPOM; GPOM ($\Re^D$) employs a D-dimensional Gaussian process and hence ignores the observations' time stamps while GPOM ($\Re^{D+1}$) includes time as an additional feature in the classic GP architecture. The table lists the areas under the ROC curve produced by each algorithm over a series of time steps with the DGPOM consistently outperforming its static counterparts. All three algorithms utilise rangefinder data acquired at the query time (t_{x_*}) and three previous scans. While the hypothesis of occupancy for the GPOM ($\Re^D$) in motionless regions such as the building wall is comparable to the DGPOM, the static-world assumptions it makes result in an inability to reconstruct the vehicles accurately. The dynamic objects in the scene also negatively influence the performance of the GPOM ($\Re^{D+1}$) leading to a short temporal length-scale. Consequently, only the observations acquired at the query instances have any significant influence on the probability of occupancy estimate.

3.2 *Real Data Experiment*

A Pioneer 2-AT robot equipped with a SICK LMS291 laser rangefinder gathered observations from 3 min of traffic flow at a busy intersection. Figure 7 provides an aerial view of the area including the location of the robot during testing. Superimposed on the image are the laser returns from the static objects (manually classified) and the mean velocity field. The road lanes can be identified clearly in the plot as well as a number of zero-magnitude velocity vectors around the buildings.

A contour plot generated from sampling the $\varepsilon(\cdot)$ function has also been superimposed on the aerial image in Fig. 8. Comparing the $\Upsilon(\cdot)$ with motion clusters from a test set indicated a degree of error in the mean velocity field along the roadways as well as a large error in the region of the intersection. A certain level of error is to be expected due to variations in speed between vehicles however the considerable error over the junction is primarily due to the fact that the direction of motion in

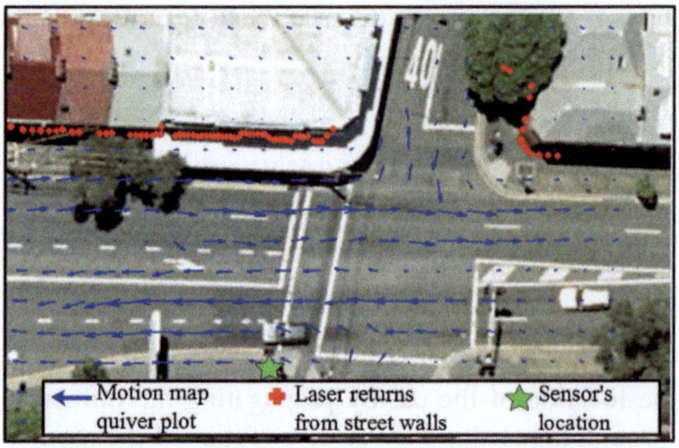

Fig. 7 Aerial image of the outdoor environment used in the experiment with the learned velocity field superimposed

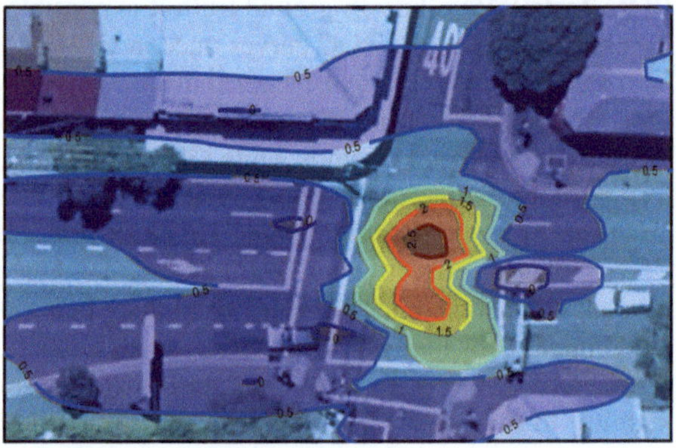

Fig. 8 Contours from the error map superimposed over the environment

this region can fluctuate greatly here. Consequently, the local temporal length-scale in this region will be affected, resulting in past observations having a comparatively small influence on the hypothesis of occupancy. The parameters of Eq. 6 converged to $a = 0.8625$ and $b = 1.242$ after training corresponding to an $l_t < 0.5$ across the centre of the intersection. A possible solution to address is discussed in Sect. 4.

The outputs of the DGPOM's inference algorithm at four time steps and the observations acquired at each instant are displayed in Fig. 9. To illustrate the sense of motion in the estimates, the outputs in the second and fourth row are at a query time of one time step after the first and third rows, respectively. Similar to its performance on the synthetic dataset, the algorithm's probability of occupancy map predicts high probabilities in locations occupied by vehicles despite being occluded from the sensor at the t_{x_*}. In the first set of outputs, the algorithm predicts the presence of two cars at the center of map, $\mathbf{x}_* \approx (0; 13)$, using data from previous scans. There is an increase in uncertainty in the second set of prediction as the time since both cars were last observed increases. The accuracy of the $\Upsilon(\cdot)$ in some areas enables information obtained from scans in the past to accurately predict the location and outline of both buildings on either side of the intersection wall.

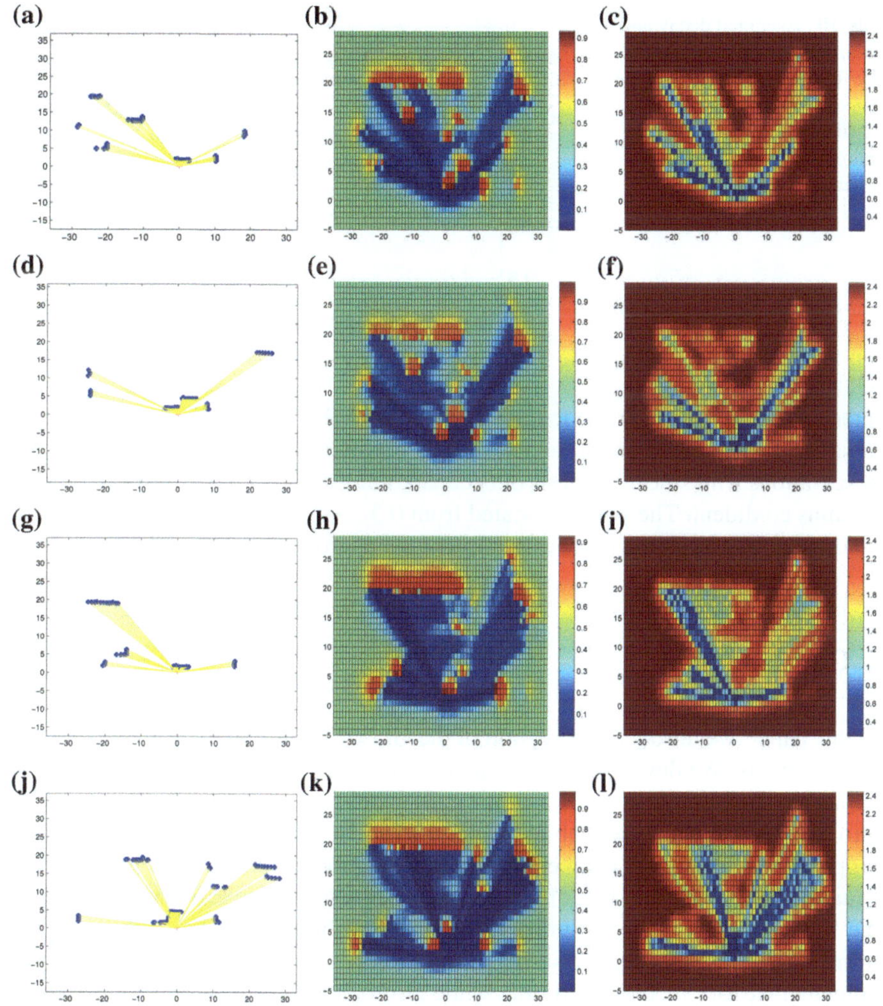

Fig. 9 Sequence of images illustrating the output of the DGPOM at different instances in time. From *top* to *bottom* $t = 4$, $t = 5$, $t = 29$, $t = 30$. *Left column* Range observations recorded at each time stamp. *Middle column* Probability of occupancy versus location. *Right column* Predictive variance verus location

To analyse the performance of the DGPOM's predictions as time since the last observation increases, the ground truth is required. Although the exact state of the environment at any given instant is unknown, it is approximated by an occupancy map generated using observations acquired exclusively at the given $t_{\mathbf{x}_*}$ and comparisons are made only in areas where this map has a high degree of confidence. Figure 10 shows the averaged behaviour of the area under the ROC curve as time since the last observation is increased. Initially, there is a shape falloff in performance mainly due to the predicted probability of occupancy in regions such as the centre of the

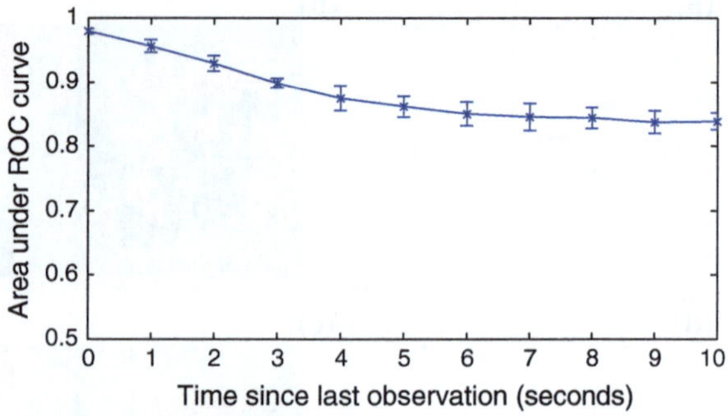

Fig. 10 Averaged decay and saturation of the area under the ROC curve as the time since the last observation increases. The *error bars* represent the 1 standard deviation boundary

intersection rapidly reverting to the global mean of 0.5. Eventually the curve begins to saturate once the dynamic regions return to a global mean due to a lack of new observations while the hypothesis of occupancy areas of the map believed to be static remains confident. The y-axis is scaled from $0.5 \rightarrow 1$ to represent the range from a random guess to a perfect reconstruction of the approximated ground truth.

4 Conclusions

In this paper, we introduced a version of the GPOM algorithm to deal with dynamic environments. We developed a continuous occupancy map capable of learning static and dynamic regions and integrating observations from multiple points in time into a single continuous probabilistic spatio-temporal model of the environment. The proposed motion-model adapted covariance function enables the parameters of an equation describing the drift of an underlying function to be learnt in tandem with the hyperparameters by maximising the marginal likelihood. The benefits of this framework could make interesting contributions to modelling environmental phenomena that vary in space and time while also being subject to drift such as wind, air pollution, and cloud movement.

However, to model spatially non-uniform drift (multiple dynamic objects), the search space for jointly optimising the GP and latent displacement function becomes infeasibly large. Consequently, we constrain the problem by assuming the function or motion map, $\Psi(\cdot)$, is temporally intransigent. Dependencies between observations are propagated temporally as well as spatially while employing a non-stationary covariance function to adjust their magnitude depending on the estimated accuracy of the translation $A\mathbf{x}$.

Results using the DGPOM are encouraging although the implementation will need to exploit its parallelisable properties using multiple cores to be used in an online setting effectively (similar to [2, 3, 8, 11]). Static obstacles such as buildings are represented clearly despite considerable occlusions. In dynamic regions, the motion of cars is inferred using past observations and the motion map. The hypothesis of occupancy is weighted by the estimated error of the mean velocity assumption to prevent inaccurate modelling of the motion leading to the creation of phantom obstacles.

References

1. Bohg, J., Johnson-Roberson, M., Björkman, M., Kragic, D.: Strategies for multi-modal scene exploration. In: Proceedings of the 2010 IEEE/RSJ International Conference on Intelligent Robots and Systems, October 2010
2. Brechtel, S., Gindele, T., Dillmann, R.: Recursive importance sampling for efficient grid-based occupancy filtering in dynamic environments. In: IEEE International Conference on Robotics and Automation (ICRA), pp. 3932–3938, May 2010
3. Chen, C., Tay, C., Laugier, C., Mekhnacha, K.: Dynamic environment modeling with gridmap: a multiple-object tracking application. In: 9th International Conference on Control, Automation, Robotics and Vision, 2006. ICARCV '06, pp. 1–6, December 2006
4. Elfes, A.: Occupancy grids: a probabilistic framework for robot perception and navigation. PhD thesis, Carnegie Mellon University (1989)
5. Ellis, D., Sommerlade, E., Reid, I.: Modelling pedestrian trajectories with gaussian processes. In: Ninth International Workshop on Visual Surveillance (2009)
6. Fleet, D., Weiss, Y.: Optical Flow Estimation. Springer, Berlin (2005)
7. Gan, S., Yang, K., Sukkarieh, S.: 3d path planning for a rotary wing UAV using a Gaussian process occupancy map. In: Proceedings of the Australasian Conference on Robotics and Automation (2009)
8. Gindele, T., Brechtel, S., Schroder, J., Dillmann, R.: Bayesian occupancy grid filter for dynamic environments using prior map knowledge. In: IEEE Intelligent Vehicles Symposium, 2009, pp. 669–676, June 2009
9. Lucas, B.D., Kanade, T.: An iterative image registration technique with an application to stereo vision. In: Proceedings of the 7th International Joint Conference on Artificial Intelligence—Volume 2, pp. 674–679. Morgan Kaufmann Publishers Inc., San Francisco, CA, USA (1981)
10. Melkumyan, A., Ramos, F.: Multi-kernel gaussian processes. In: IJCAI, pp. 1408–1413 (2011)
11. Moras, J., Cherfaoui, V., Bonnifait, P.: Credibilist occupancy grids for vehicle perception in dynamic environments. In: IEEE International Conference on Robotics and Automation (ICRA), 2011, pp. 84–89 (2011)
12. O'Callaghan, S.T., Ramos, F.T.: Gaussian process occupancy maps. Int. J. Robot. Res. **31**(1), 42–62 (2012)
13. O'Callaghan, S.T., Singh, S.P.N., Alempijevic, A., Ramos, F.T.: Learning navigational maps by observing human motion patterns. In: IEEE International Conference on Robotics and Automation (ICRA), 2011, pp. 4333–4340, May 2011
14. Plagemann, C., Kersting, K., Burgard, W.: Nonstationary Gaussian process regression using point estimates of local smoothness. In: European Conference on Machine Learning (ECML) (2008)
15. Rasmussen, C.E.: Gaussian Processes for Machine Learning. MIT Press, Cambridge (2006)
16. Wang, C.C.: Simultaneous localization, mapping and moving object tracking. Ph.D. thesis, Robotics Institute, Carnegie Mellon University, Pittsburgh, PA, April 2004

A Spatial-Temporal Approach for Moving Object Recognition with 2D LIDAR

B. Qin, Z.J. Chong, S.H. Soh, T. Bandyopadhyay, M.H. Ang, E. Frazzoli and D. Rus

Abstract Moving object recognition is one of the most fundamental functions for autonomous vehicles, which occupy an environment shared by other dynamic agents. We propose a spatial-temporal (ST) approach for moving object recognition using a 2D LIDAR. Our experiments show reliable performance. The contributions of this paper include: (i) the design of ST features for accumulated 2D LIDAR data; (ii) a real-time implementation for moving object recognition using the ST features.

1 Introduction

In this paper, we propose a spatial-temporal (ST) approach for moving object recognition using only modest sensory data. Compared to more elaborate and costly solutions (e.g., outdoor depth cameras and 3D ranger finders), our method works with range readings obtained from a planar 2D LIDAR on a mobile platform. Using only range readings complicates object recognition because information is sparse relative to richer modalities such as vision. Furthermore, noise introduced by ego-motion (and other sources) can make static objects appear dynamic. We show that it is possible to obtain highly accurate object classification via temporal accumulation and a coupled classification process.

B. Qin (✉) · Z.J. Chong · M.H. Ang
National University of Singapore, Kent Ridge, Singapore, Singapore
e-mail: baoxing.qin@u.nus.edu

S.H. Soh
Singapore-MIT Alliance for Research and Technology, Singapore, Singapore

T. Bandyopadhyay
The Commonwealth Scientific and Industrial Research Organization, Canberra, Australia

E. Frazzoli · D. Rus
Massachusetts Institute of Technology, Cambridge, MA, USA

M.A. Hsieh et al. (eds.), *Experimental Robotics*, Springer Tracts in Advanced Robotics 109, DOI 10.1007/978-3-319-23778-7_53

1.1 Related Work

Existing work in moving object recognition decomposes the problem into two distinct sub-tasks: detection and classification. The former aims to discern the existence of moving objects, while the latter aims to recognize the objects' identities. We can categorize existing methods of moving evidence detection into two types: tracking-based methods and SLAMMOT (Simultaneously Localization And Mapping, and Moving Object Tracking) methods.

Tracking-based methods (e.g. [4, 6]) work at the object level: they first segment a laser scan into multiple segments as the measurements of different objects. The segments are then fed into a tracker to estimate their positions and velocities, and objects exceeding a defined speed or displacement are reported as the moving objects. The accuracies of these approaches are mainly determined by the tracking process, which has to solve the notorious data association problem and may fail in cluttered environments.

SLAMMOT methods detect moving objects at the atomic level [7, 11, 12]. An occupancy grid map of the local environment is created through a SLAM process and the changes of the grids' occupancy indicate the existence of moving objects. Compared to the tracking-based methods, the SLAMMOT techniques have two major advantages: first, they are more robust to ego-motion estimation errors, which are compensated by the SLAM process. Second, they do not have to address the tracking problem. However, the computational cost associated with SLAM is usually high, leading to a low update frequency. This is undesirable for high-speed robots such as autonomous vehicles. Moreover, since SLAMMOT methods usually assume robots move on a flat surface, they are not applicable to the bumpy road environments.

In both types of methods discussed above, object classification is performed independently from motion detection, either before or after it. The classification process usually relies on the sparse geometric features of the 2D segments and hence, is vulnerable to similar-looking background noise. For better performance, classification results at different times can be fused to achieve continuous estimation when object tracking information is available.

1.2 A Spatial-Temporal Solution

A spatial-temporal approach for moving object recognition couples detection and classification into a single process. The basic idea of our approach derives from the observation that accumulated laser scans generally provide sufficient information for the task. For example, it is difficult to recognize a vehicle from a single scan segment, because of its simple shape contour. However, Fig. 1 illustrates that in the ST domain, the moving vehicle shows unique geometric features, i.e., a chain of shifted "L" shapes. The uniqueness of these features comes from not only the vehicle's appearance in the spatial domain, but also from its motion pattern in the

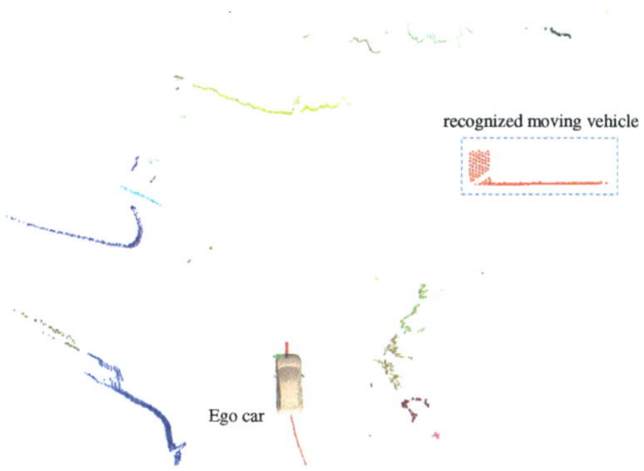

Fig. 1 One example to illustrate our idea. The *red-green* axes attached to the ego car represent the mounted 2D LIDAR. This image also captures a typical snapshot of a clustered campus environment

temporal domain. We show that these features can be exploited to create accurate classifiers.

Our method consists of three basic steps: (1) laser scans are first accumulated over a certain time window, (2) segmentation is then performed on the accumulated data to generate clusters, and (3) moving objects are finally recognized using the spatial-temporal features of these clusters. Compared to existing methods, our approach does not rely on object tracking nor local environment mapping and hence, it is more robust in cluttered environments and computationally lighter. Furthermore, since detection and classification are conducted in one single process, better recognition accuracy can be achieved. The rationale is that while motion patterns in the T-domain can aid object classification, appearance features in the S-domain can also help determine whether an object is moving (e.g., a bizarre-shaped cluster is more likely to be a static bush rather than a moving vehicle). A coupled process is able to fully utilize the ST information and benefit both sub-tasks.

2 Technical Approach

In brief, our method segments and clusters accumulated laser scans in a time-window, extracts relevant spatio-temporal features and then classifies each cluster. Segmentation is performed using a graph-based algorithm in the ST domain and classification is performed using the widely-used Support Vector Machine (SVM). The flowchart of our algorithm is illustrated by Fig. 2.

2.1 Data Accumulation in T-Domain

Laser scans are accumulated over a defined time window to collect N scans: $\mathcal{S} = \{s_{t_1}, s_{t_2}, \ldots, s_{t_N}\}$, where $\mathcal{S}$ denotes the collected scan set, and s_i each scan component.

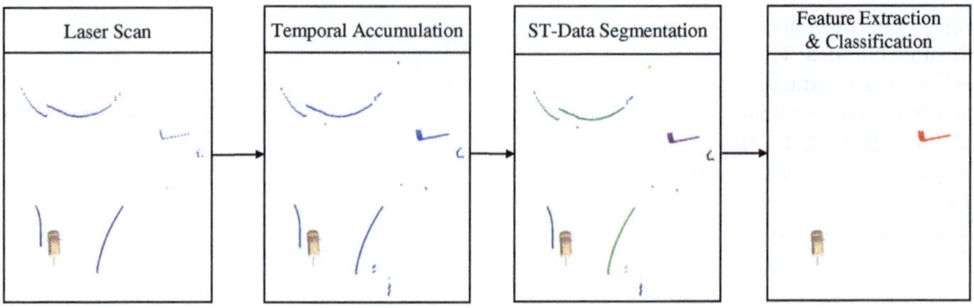

Fig. 2 Algorithm flowchart. Laser scans are received from the LIDAR sensor, and then accumulated in the temporal domain, visualized as *blue points*; the accumulated ST-data are then segmented into different ST-clusters, visualized in various *colors*; feature extraction and classification are performed on each ST-cluster, to recognize the moving objects, colored in *red*. The *clay-colored* vehicle model visualizes the ego car

To represent ego-motion, we record the LIDAR's pose (according to robot's odometry system) at each corresponding time stamp: $\mathcal{X} = \{x_{t_1}, x_{t_2}, \ldots, x_{t_N}\}$, where each x_i is the LIDAR pose corresponding to scan s_i. The accumulated laser scans $\mathcal{S}$ and associated ego poses $\mathcal{X}$ carry all the raw information required in our system.

2.2 Graph-Based ST Segmentation

To segment the accumulated scans, we first convert the scan set $\mathcal{S}$ into a point set $\mathcal{D}$, where each point d_i contains the position information x_i in the robot's fixed odometry coordinate system, and its collected time t_i. In addition to this information, we maintain the conversion relationship between the scans and the points, such that each point in $\mathcal{D}$ is mapped to its angle and range reading in $\mathcal{S}$.

We employ the graph-based region merging method [3] for segmentation of the transformed set $\mathcal{D}$. The advantage of this approach is that it is able to find a segmentation that is neither too coarse nor too fine. In brief, the data is treated as a graph, with points as nodes and edge weights indicating the dissimilarities between nodes. Initially each node is an individual component, and the algorithm performs pairwise region merging iteratively if the minimum edge weight connecting two components is less than the minimum internal difference (a scoring function); see [3] for more details. In our work, the edge weight (dissimilarity measure) between two points in the ST-domain is the weighted Euclidean distance:

$$\mathcal{E}_w(d_1, d_2) = \|x_{t_1} - x_{t_2}\| + \alpha \times |t_1 - t_2| \tag{1}$$

where α is a weight parameter. Intuitively, this metric ensures that points which are close in both spatial and temporal domains are placed in the same cluster.

After the segmentation process, clusters of points in the ST-domain are obtained. Preliminary results showed that if we used each cluster as unorganized data and simply extracted its statistical features as a whole, performance was degraded, presumably due to a loss of information. As such, we define a ST cluster, denoted as $\mathcal{ST}$, as a collection of scan segments in a sequence together with their LIDAR poses:

$$\mathcal{ST} = \{z_{t_1}, z_{t_2}, \ldots, z_{t_N}, x_{t_1}, x_{t_2}, \ldots, x_{t_N}\} \tag{2}$$

where z_{t_i} is the scan segment collected at time t_i, and x_{t_i} its corresponding LIDAR pose. Given the segmentation results of data $\mathcal{D}$, to construct $\mathcal{ST}$ is straightforward.

2.3 Spatial-Temporal (ST) Features

In this section, we discuss the design of our spatial-temporal features. Recall that $\mathcal{ST}$ not only contains the information about the object's shape, but also the information related to their motion patterns. We construct our feature vector $\mathcal{F}$ to maximize the amount of original information, while keeping its structure invariant to the scan number N:

$$\mathcal{F} = \{\{\hat{z}_{t_i}\}_N, \mathcal{M}, \mathcal{X}\} \tag{3}$$

where $\hat{z}_{t_i}$ is a compressed representation for raw scan segment z_{t_i}, $\mathcal{M}$ is a set of "shape moments" that captures the shape characteristics of the cluster, and $\mathcal{X}$ is the pose set.

The compressed segment ($\hat{z}$) The compressed segment (CS) approximates each scan segment by a fixed number of key points and selected statistical features. Figure 3 illustrates the idea of the compressed segment. Here, we have used the Douglas-Peucker algorithm [2] to find the relevant key points. In addition, the number of points in between each pair of neighboring key points, and the variance of their distances to the line formed by the pair are also incorporated in the feature vector. To represent positional information relative to the background, range differences

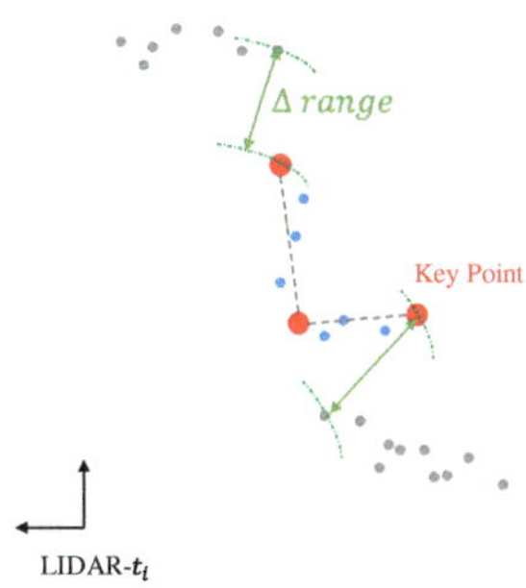

Fig. 3 One example of compressed segment

Table 1 Feature vector of the compressed segment

Feature name	Description
Key points	x, y position of the points in the LIDAR coordinate; intensity values of these points (if intensity values are provided)
Points in between key points	Number of points in between a pair of key points; variances of distances from these points to their key point lines
Range distances to background	Range differences between two extreme points to their neighboring background points

between the extreme points and their respective neighboring background points are also used. Table 1 summaries all the features in a CS feature vector.

The shape moments ($\mathcal{M}$) Although the scan segment information is incorporated into the feature vector by the way of the compressed segments $\hat{z}_{t_i}$, some geometric information may still be lost due to compression. To better preserve the information, we project the scan points into the global odometry coordinate, and then extract the Hu-Moments [5] of the contour to convey the shape information of the overall point set.

The pose set ($\mathcal{X}$) To take into account robot ego motion, LIDAR poses at different time are incorporated in the feature vector. However, rather than using their original pose values in the global odometry frame, we transfer all the LIDAR poses into the latest LIDAR coordinate. This helps remove the irrelevant information of absolute positions and concentrate the classification on the relative movements.

Pose-Variant and Pose-Invariant Feature Sets Figure 4a illustrates the spatial-temporal feature vector $\mathcal{F}$. Note that it captures not only object appearance and movement, but also the information relating to the sensing scenario, such as how far the object is and at what angle. The scenario information is important for multiple

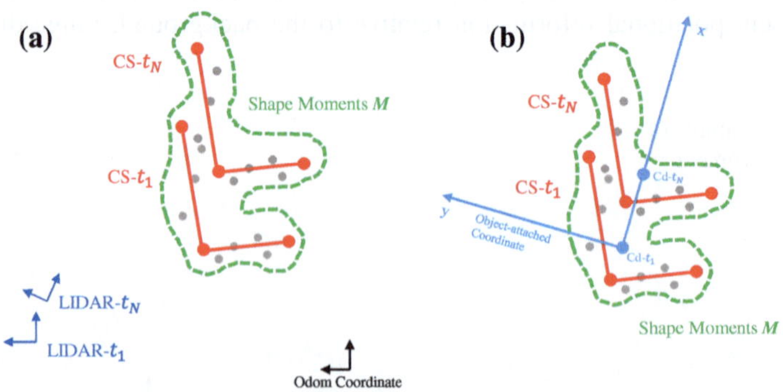

Fig. 4 **a** Pose-variant and **b** pose-invariant feature sets

reasons. First, the distance to the object affects the number of laser points cast on it, due to LIDAR's limited angular resolution and detection range. Second, the observation angle on the object determines the measurements, e.g., the side of an object may be occluded when observed from the front. The importance of scenario information for object recognition is demonstrated by the experiment results described in Sect. 4.

Because the compressed scan $\hat{z}_{t_i}$ is defined in the sensor coordinate LIDAR$-t_i$, $\mathcal{F}$ is pose-variant and a large number of training instances may be needed to cover different sensing situations. For this reason, this paper also proposes a pose-invariant feature vector, where the compressed scan $\hat{z}_{t_i}$ is transformed into an object-attached coordinate, as shown in Fig. 4b. Denoting the centroid of LIDAR segment z_{t_i} as Cd_{t_i}, the origin of the object-attached coordinate is defined to be Cd_{t_1}, with its x axis pointing from Cd_{t_1} to Cd_{t_N}. Compared to the pose-variant feature vector, the pose-invariant vector is more general in terms of object positions and orientations, but at the cost of losing scenario information.

3 Experiments

The objective of our experiments is threefold. First, we seek to validate that accumulated scans will result in higher accuracies compared to single scans. Second, we attempt to better understand the effect the length of the time-window had on classification accuracy. Third, we seek to analyze the performance of our designed spatial-temporal features.

Our test bed is a converted iMiev with a 2D LIDAR (SICK LMS 151) mounted on the front of the vehicle, as shown by Fig. 5. The LIDAR runs at 50 Hz, with 270° FOV. The entire system is developed using the Robot Operating System (ROS) [9]. To test the performance of our algorithm, we conduct experiments in two different environments: a university campus and a highway. The former is a cluttered environment with average vehicle speeds of 10–30 km/h, while the highway is a more "structured"

Fig. 5 iMiev test bed

environment with vehicle speeds of 60–100 km/h. In this experiment, we focus on recognizing moving vehicles, but our algorithm is applicable for general-purpose moving object recognition. Ground truths of moving vehicles are obtained via manual labelling for both environments, with 232 positive vehicles samples labelled for the campus environment and 1212 positive samples for the highway one. Note that negative samples are also manually labelled in the experiments, the numbers of which change with the temporal window lengths, as will be shown in the next section.

4 Results

We evaluate our algorithm using five different metrics: segmentation ratios, classification accuracy, spatial analysis, performance of different feature sets, and the computational cost. Note that all the analyses are performed with the pose-variant feature vector, except where performances of different feature sets are studied. Major insights of the experimental results will be summarized at the end of this section.

Segmentation ratios: Figure 6 shows the ratio of background clusters to vehicle clusters. Compared to the highway environment, the campus environment is far more cluttered, resulting in a larger number of background clusters. However, the number of background clusters decreases drastically as the number of accumulated scans is increased. This suggests that the temporal accumulation prevents the background from being over-segmented, which as we will see, leads to an improvement in classification accuracy.

Classification Accuracy: Figure 7 shows the classification results for moving vehicle recognition under fivefold cross validation. Note that the classification problem here is a unbalanced binary classification problem, and the number of background clusters varies with the accumulated scan number. For above reasons, while apparently good total accuracies (>97 %) are achieved in both environments, we analyze the precision and recall rates to better evaluate our algorithm: precision measures what fraction of

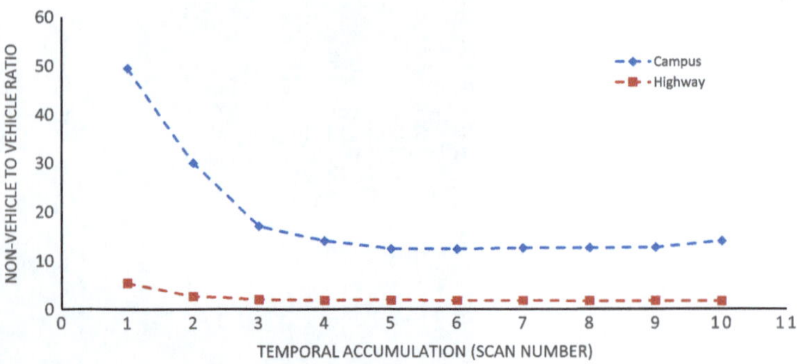

Fig. 6 Non-vehicle to vehicle ratio in different environments

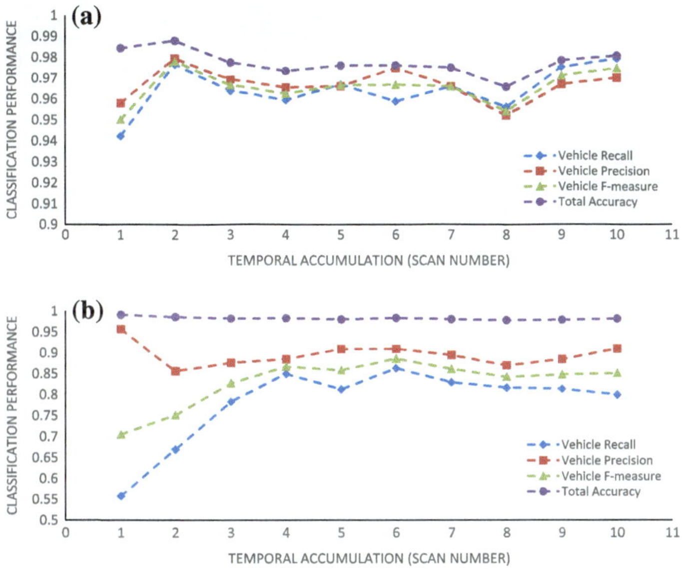

Fig. 7 Classification at different environments. **a** Highway environment, **b** campus environment

the detections are actually moving vehicles, and recall measures what fraction of the actual moving vehicles are detected [8].

In the clean highway environment, both the precision and recall rates are high (>94 %) even when using only single scan segments. With a temporal window length larger than two, the SVM attains performances above 97 %. In the cluttered campus environment, vehicle detection appears more challenging. However, we observe that it is in this environment that our approach yields the most positive effect. While the precision remains decent (>85 %), the recall rate using no temporal windowing is at a low 55 %. The recall rate rises rapidly with the temporal accumulation from 55 % to a high of 86 % (at $N = 6$). The F-measure (F_1-score) shows a weighted average of the precision and recall, where our algorithm achieves best performance at $N = 6$. As N continues to grow, the performance seems to tail off. We believe this occurs due to the "curse of dimensionality", which hampers the classification process as the feature vector length grows.

Figure 8 shows examples of moving vehicle detection in the two environments ($N = 2$ for highway, and $N = 6$ for campus). The top row shows the images captured from an on-board camera, which is calibrated with the LIDAR sensor, and the bottom row shows the recognition results from the accumulated LIDAR data. Temporal accumulation and ST segmentation are performed to extract individual ST-clusters (shown in different colors), which are then classified to extract the moving vehicles (shown as red blobs). The results are also projected into the camera image for visualization purpose. Since the camera has much smaller field of view (≈70°) compared to the LIDAR, some of the results are not shown in the image. Note that a

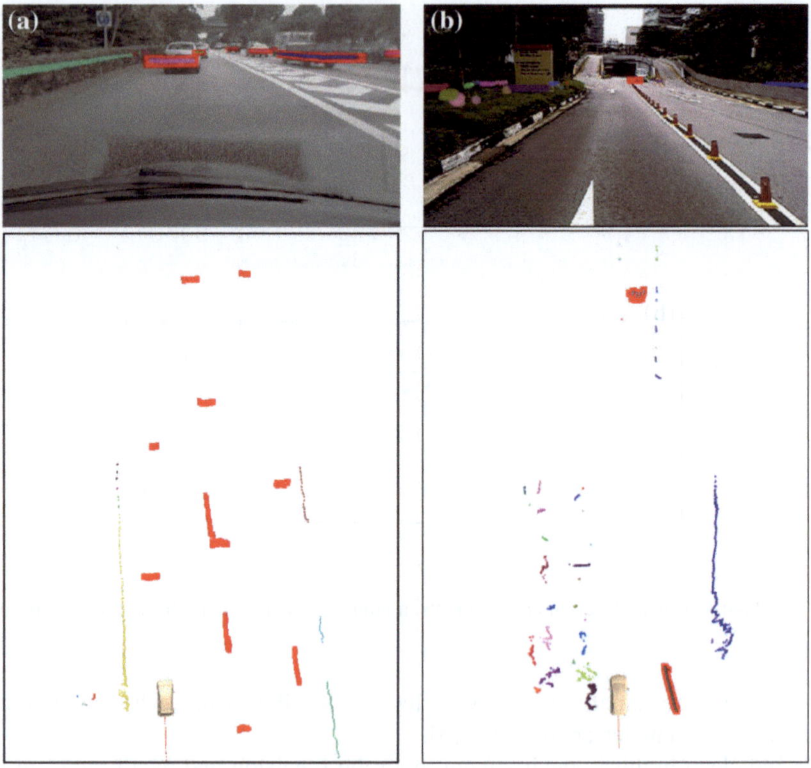

Fig. 8 Vehicle detection examples. **a** Vehicle detection at highway, **b** vehicle detection at campus

minor misalignment exists between the camera images and the LIDAR data, which is attributed to the time difference between the laser points (accumulated in the past) and the captured image.

From the two examples it is easily observed that the campus environment is much more cluttered than the highway. In the highway example, the road barrier and bushes at the two sides are generally neat and consistent, making it relatively easy to differentiate the foreground objects and the background noise. There are 20 clusters extracted in the shown case, with 10 of them recognized as moving vehicles. Compared to the clean highway scenario, the campus environment is much more "dirty": its background usually consists of various unconnected objects, and the bumpiness of the ground may also bring noise when the LIDAR scans strike on the road surface. Given all these challenges, our algorithm is still able to perform robust recognition: 2 moving objects are correctly identified from the 37 extracted clusters in the shown case.

Spatial analysis: Figure 9 presents us with more insights into the performance of our algorithm from a spatial perspective ($N = 2$ for highway, and $N = 6$ for campus). Since our test locations are left-hand drive, many of the vehicle samples in

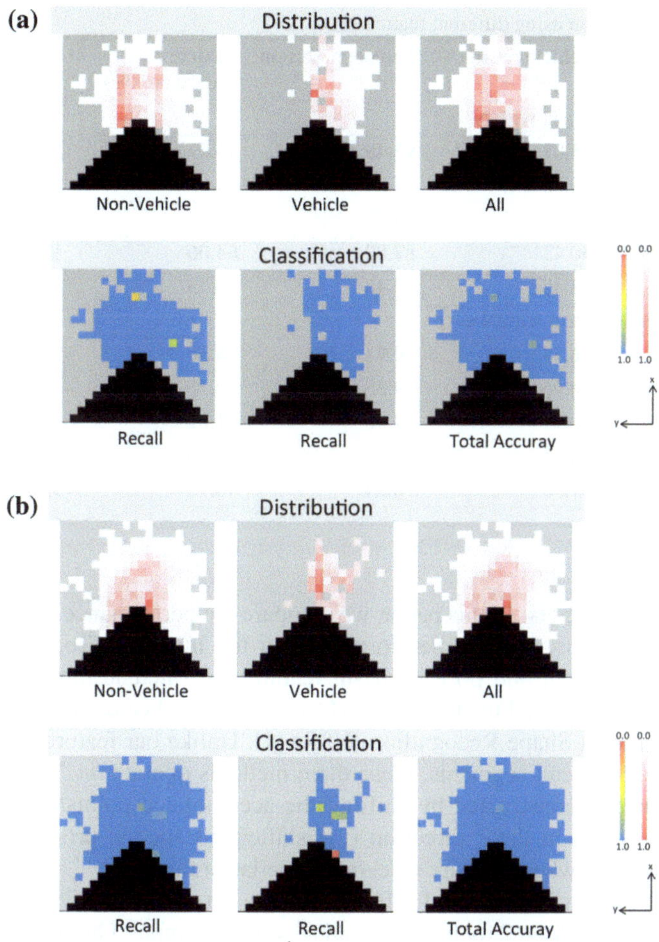

Fig. 9 Overall vehicle detection performance. The center of each plot is the LIDAR origin, with LIDAR orientation shown by the legend. Each pixel in the figures represents a 5×5 m grid place. The *grey* pixels are places where no sufficient samples collected, and the *dark areas* are places beyond LIDAR FOV. In the distribution plots, the density value of each grid represents the number of collected samples in this place, which is normalized by the largest value. **a** Highway, **b** campus

our collected training data are at the front and right sides of the iMiev. In Fig. 9b, we see that high vehicle detection errors occur at the boundary of LIDAR FOV, where only parts of the clusters are observed. Other errors take place over >20 m away from the LIDAR center. We posit that this is due to the fact that when observing objects from a distance, the LIDAR readings are occluded by other objects or missing due to low reflectivity. Importantly, the results show that in the vicinity of the LIDAR, the detection accuracy is nearly 100 %, which is essential for safe navigation.

Table 2 Classification using different feature sets

Feature sets	Vehicle recall (%)	Vehicle precision (%)	Vehicle F-measure (%)	Total accuracy (%)
(a) Highway environment with N = 2				
Pose-variant	**95.87**	**97.48**	**96.67**	**97.61**
Pose-invariant	94.68	94.50	94.59	96.87
ESF	73.86	94.39	82.87	91.17
VFH	50.42	87.80	64.06	83.63
USR	82.45	79.48	80.93	88.77
(b) Campus environment with N = 6				
Pose-variant	**86.21**	**90.91**	**88.50**	**98.32**
Pose-invariant	70.96	89.35	79.10	96.70
ESF	50.37	77.40	61.02	94.33
VFH	24.63	82.72	37.96	92.90
USR	37.50	70.34	48.92	93.10

Different Feature Sets: In this paper, we compare the performance of our designed features with existing feature sets proposed in the literature. Together with our designed pose-variant and pose-invariant features, we include three more feature sets: Ensemble of Shape Functions (ESF) [13], Viewpoint Feature Histogram (VFH) [10], and Ultrafast Shape Recognition (USR) [1]. Unlike our feature sets extracted from compressed scan segments, these three methods operate on 3-D spatial data. Here, 3-D data is constructed by shifting the accumulated points (point set D in Sect. refSTspsdata) in the z direction (the shifted distance is proportional to the elapsed time from when they are received to the latest time).

To assess the performances of different feature sets, the same temporal window length is used, with $N = 2$ for highway and $N = 6$ for campus. Our results are shown in Table 2. It is observed that the pose-variant and pose-invariant features outperform the 3D feature sets, which are designed for dense 3D data and appear not suitable for the ST data accumulated from the LIDAR. The better performance of pose-variant over pose-invariant features indicates the usefulness of the scenario information as discussed in Sect. 2.3.

Computational cost: On our experimental platform (computer equipped with a Core i7-4770 processor), the computational time required to process one new scan is 5–10 ms for the campus environment with $N = 6$. In the highway environment with $N = 2$, the processing time is only 1–4 ms. In short, computational costs are low, making our method suitable for real-time applications.

Summary: From the precceeding discussion, three major insights can be derived from our experimental results:

1. Accumulation in the temporal domain helps to prevent over-segmentation of sensor data in the cluttered environment (Fig. 6).
2. The spatial-temporal features enable more accurate classification compared to using only spatial features from a single measurement (Figs. 7 and 9);
3. The increase of the accumulation time window improves recognition accuracy in the cluttered environment up to a maximum time window (Fig. 7).

5 Conclusions

In this paper, we propose and investigate a novel spatial-temporal approach for moving object recognition with a single 2D LIDAR. By using crafted spatial-temporal features, we obtain promising classification results in two different experimental settings. Our results suggest that our approach is particularly applicable in cluttered environments, where temporal windowing prevents over-segmentation of the observations and the accumulation of sensor information makes moving object recognition more accurate. As future work, we plan to investigate the performance of our approach on other classes of moving objects (e.g., pedestrians and motorcycles).

Acknowledgments This research was supported by the National Research Foundation (NRF) Singapore through the Singapore-MIT Alliance for Research and Technology's (FM IRG) research programme, in addition to the partnership with the Defence Science Organisation (DSO). We are grateful for their support.

References

1. Ballester, P.J., Richards, W.G.: Ultrafast shape recognition for similarity search in molecular databases. Proc. R. Soc. A: Math. Phys. Eng. Sci. **463**(2081), 1307–1321 (2007)
2. Douglas, D.H., Peucker, T.K.: Algorithms for the reduction of the number of points required to represent a digitized line or its caricature. Cartogr. Int. J. Geogr. Inf. Geovis. **10**(2), 112–122 (1973)
3. Felzenszwalb, P.F., Huttenlocher, D.P.: Efficient graph-based image segmentation. Int. J. Comput. Vis. **59**(2), 167–181 (2004)
4. Gate, G., Nashashibi, F.: Fast algorithm for pedestrian and group of pedestrians detection using a laser scanner. In: 2009 IEEE Intelligent Vehicles Symposium, pp. 1322–1327. IEEE (2009)
5. Hu, M.-K.: Visual pattern recognition by moment invariants. IRE Trans. Inf. Theory **8**(2), 179–187 (1962)
6. Mertz, C., Navarro-Serment, L.E., MacLachlan, R., Rybski, P., Steinfeld, A., Suppé, A., Urmson, C., Vandapel, N., Hebert, M., Thorpe, C., et al.: Moving object detection with laser scanners. J. Field Robot. **30**(1), 17–43 (2013)
7. Miyasaka, T., Ohama, Y., Ninomiya, Y.: Ego-motion estimation and moving object tracking using multi-layer lidar. In: 2009 IEEE Intelligent Vehicles Symposium, pp. 151–156. IEEE (2009)
8. Murphy, K.P.: Machine Learning: A Probabilistic Perspective. MIT Press (2012)

9. Quigley, M., Conley, K., Gerkey, B.P., Faust, J., Foote, T., Leibs, J., Wheeler, R., Ng, A.Y.: ROS: an open-source Robot Operating System. In: ICRA Workshop on Open Source Software (2009)
10. Rusu, R.B., Bradski, G., Thibaux, R., Hsu, J.: Fast 3d recognition and pose using the viewpoint feature histogram. In: 2010 IEEE/RSJ International Conference on Intelligent Robots and Systems (IROS), pp. 2155–2162. IEEE (2010)
11. Vu, T.D., Burlet, J., Aycard, O.: Grid-based localization and local mapping with moving object detection and tracking. Inf. Fusion (2011)
12. Wang, C.C., Thorpe, C., Thrun, S.: Online simultaneous localization and mapping with detection and tracking of moving objects: theory and results from a ground vehicle in crowded urban areas. In: 2013 IEEE International Conference on Robotics and Automation, ICRA'03. IEEE (2013)
13. Wohlkinger, W., Vincze, M.: Ensemble of shape functions for 3d object classification. In: 2011 IEEE International Conference on Robotics and Biomimetics (ROBIO), pp. 2987–2992. IEEE (2011)

Probabilistic Grid-Based Collision Risk Prediction for Driving Application

Lukas Rummelhard, Amaury Nègre, Mathias Perrollaz and Christian Laugier

Abstract In the recent years, more and more modern cars have been equipped with perception capabilities. One of the key applications of such perception systems is the estimation of a risk of collision. This is necessary for both Advanced Driver Assistance Systems and Autonomous Navigation. Most approach for risk estimation propose to detect and track the dynamic objects in the scene. Then the risk is estimated as a Time To Collision (TTC) by projecting the object's trajectory in the future. In this paper, we propose a new grid-based approach for collision risk prediction, based on the Hybrid-Sampling Bayesian Occupancy Filter framework. The idea is to compute an estimation of the TTC for each cell of the grid, instead of reasoning on objects. This strategy avoids to solve the difficult problem of multi-objects detection and tracking and provides a probabilistic estimation of the risk associated to each TTC value. After promising initial results, we propose in this paper to evaluate the relevance of the method for real on-road applications, by using a real-time implementation of our method in an experimental vehicle.

Keywords Dynamic occupancy grid · Risk assessment · Time to collision · Intelligent vehicle

L. Rummelhard (✉) · M. Perrollaz · C. Laugier
e-Motion, Inria Grenoble Rhône Alpes, Montbonnot-Saint-Martin, France
e-mail: lukas.rummelhard@inria.fr

M. Perrollaz
e-mail: mathias.perrollaz@inria.fr

C. Laugier
e-mail: christian.laugier@inria.fr

A. Nègre
Laboratoire d'Informatique de Grenoble, CNRS, Paris, France
e-mail: amaury.negre@imag.fr

M.A. Hsieh et al. (eds.), *Experimental Robotics*, Springer Tracts in Advanced Robotics 109, DOI 10.1007/978-3-319-23778-7_54

1 Introduction

In the development of ADAS (Advanced Driver Assistance Systems) that can prevent collision to happen, a basic requirement is an assessment of the collision risk. This paper presents an original method for collision risk estimation. The idea is to compute an estimation of the TTC for each cell of a grid, instead of reasoning on objects. This strategy avoids to solve the difficult problem of multi-objects detection and tracking and provides a probabilistic estimation of the risk associated to each TTC value.

The method has been implemented for real time operation on GPU in our experimental vehicle. Then the vehicle is used for real on-road testing in order to evaluate the method in real conditions.

The paper is organized as follow: Sect. 2 presents related work. Section 3 describes the method. Section 4 provides experimental results. Section 5 concludes.

2 Related Work

In the literature, various approaches have been proposed in order to detect potential collisions in advance and trigger an assistance like a driver alarm or an automatic braking. Most of them rely on the detection and tracking of the moving objects in the scene. This tracking is used to estimate a risk metric, used for a decision layer. The simplest approach consists in computing a simple Time to Collision (TTC) by predicting the trajectories of both the ego-vehicle and the other objects using a simple linear motion model [1]. A more advanced approach proposes to generate a set of possible trajectories for all the objects, including the ego-vehicle, and compute the number of possible intersections between those trajectories [2]. The trajectory generation is performed by applying a set of possible controls to the objects. This approach has the advantage of providing a probabilistic estimation of the risk.

Considering that a control can change over time and is dependent from the intention of the driver, the authors in [3] propose to constrain the possible trajectories using the estimated maneuver intention of the driver. This allows to consider predictions over a longer time period.

All this methods rely on the detection and tracking of the moving objects [4]. This stage is a difficult problem to solve, which can be computationally costly and can generate errors in the process. Alternatively to methods based on object trajectory predictions, other methods propose to estimate time to collision based on visual features [5], but such methods are only designed to work with computer vision.

Another alternative to the object-based approaches are the grid-based approaches, like the Bayesian Occupancy Filter [6], or its extension Hybrid Sampling Bayesian Occupancy Filter (HSBOF) [7], which represents the environment as a probabilistic occupancy grid. This framework allows to model both the static and the dynamic environments, by estimating velocity probability distributions for each cell in the grid. It is efficient for short term prediction, but since there are no notions of objects or behaviors, longer term prediction requires to integrate additional prior knowledge like map information [8].

Grid based methods are designed to be very efficiently implemented on parallel architecture and avoid to deal with object-level data association for both tracking and sensor fusion. Consequently, they are able to provide in real time a robust representation of the environment. However, in order to use such approaches for driving applications, the authors generally add a subsequent clustering stage to retrieve the notion of objects before any application [3, 9].

An hybrid approach is presented in [10], with *velocity obstacles*. The authors combine a notion of objects with an estimation of velocities within a dynamic occupancy grid.

Based on the occupancy grid framework, we propose in this paper a method that allows to estimate both a probabilistic collision risk and a TTC for each cell of a grid, without any detection and tracking of objects. This dense representation of the risk is intended to be used to take decisions about the short term evolution of the scene. In this work, we present the approach and propose to evaluate how it can be used for automotive applications.

3 Description of the Method

3.1 HSBOF Presentation

The Hybrid Sampling Bayesian Occupancy Filter (HSBOF) [7] is a Bayesian filtering perception technique which models the environment at a sub-object level, in term of spatial occupancy and dynamics. The surrounding of the subject is divided into cells, to which are associated random variables, symbolizing their occupancies and velocities. The scene is interpreted through the estimated distributions of those variables, estimations that are recursively updated according to the observations. The motion field of the scene is inferred through the spatial occupancy evolution, and is described as the combination of a static part, depicted by likelihood values in a regular grid, and a dynamic part, sampled by moving particles which transfer occupancy between cells. All the distributions are jointly generated and updated. The distinction in dynamics allows to apply specific motion representations, and then to efficiently allocate memory and computation power. Indeed, while the static part of the scene does not require an important number of samples to symbolize its velocity distribution, the dynamic one necessitate a substantial precision and reactivity. The Fig. 1 summarizes the HSBOF scene representation. This data structure also allows to conveniently change the reference frame, which is needed in case of a moving subject. Using motion sensors embedded in the mobile robot (Inertial Measurement Unit, GPS, Wheel speed and steering sensor, visual odometry, etc.), the displacement

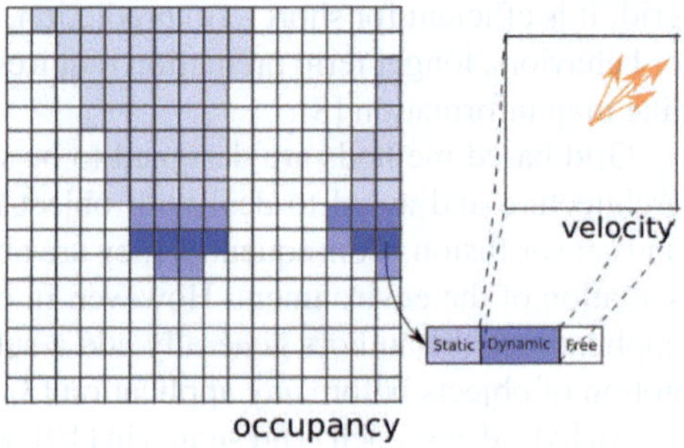

Fig. 1 HSBOF data representation. A dense grid is used to represent the occupancy while for each cell velocity is split among a static field and a set of particles

of the grid between two updates is estimated, then applied on the models. The global framework of the HSBOF is summarized on the Fig. 2.

3.2 Grid-Based Collision Assessment

As an output of the HSBOF, are available estimates of the robot surrounding occupancy and dynamics. Thanks to the already mentioned embedded motion sensors, the ego-motion is estimated. Using those and a robot model, which specifies its form, size and motion model, a time projection is achieved to assess collision risks in the future and localize them in the grid. Indeed, to each current occupancy component is associated a velocity and motion model, in such a way that its position over time can be predicted. Simultaneously, the space occupied by the robot over time can also be assessed. The risk evaluation consists in a succession of scene configuration and robot position predictions, and so potential overlaps, in other words collisions can be assessed in time and space (Fig. 3).

In practice, to each static cell and dynamic particle of the HSBOF representation are associated a "Time To Contact" (TTC) value and a risk weight. Until a time horizon, both vehicle and elementary components are iteratively moved according to their motion models. The time step used is chosen according to the grid resolution and the maximal reachable speed, to prevent any failed collision detection due to the time resolution. The intersection between the vehicle and each basic component is evaluated. If a collision is confirmed, the TTC value is set and a weight is associated to the element, according to its associated quantity of occupancy. Otherwise, the next time step is processed. Beyond the chosen time range, if no collision is observed, a null weight is assigned (since there is no predicted collision, the TTC value is meaningless, by convention it is set to the time limit). Those computations performed, for each cell in the current occupancy grid a risk value over the time range is calculated, corresponding to the likelihood of the cell to be occupied, and for its content to collide with the robot during the selected period. The risk calculation simply consists in a sum over this time range of the related weights.

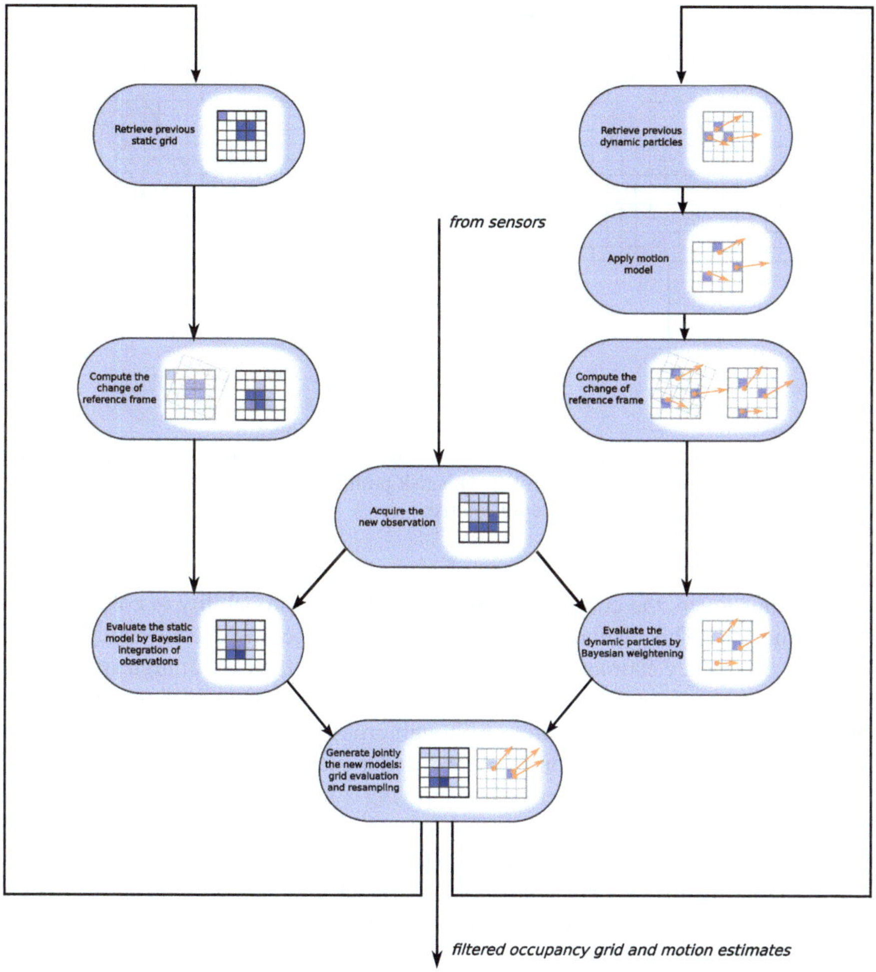

Fig. 2 HSBOF algorithm summary. From sensor data instantaneous occupancy grids are successively computed. Those observations are integrated in a Bayesian filter in which coexist and jointly adapt two models, a static grid and a dynamic set of moving particles. The result is obtained by their combination, which provides a filtered occupancy grid as well as inferred motion distributions for cells

As a result at this stage, a risk grid presents for each cell the probability to hit an obstacle coming from that cell over a given period if the scene dynamics remain unchanged (conservative model, relevant for short term prediction). Given that grid, a single global collision value is computed, and used to inform the system of any imminent danger. Two approaches are proposed to evaluate this general risk:

The first one proposes to evaluate the likelihood of colliding with a selected surface size. According to this area, and to the area of a cell, the number of n cells to be

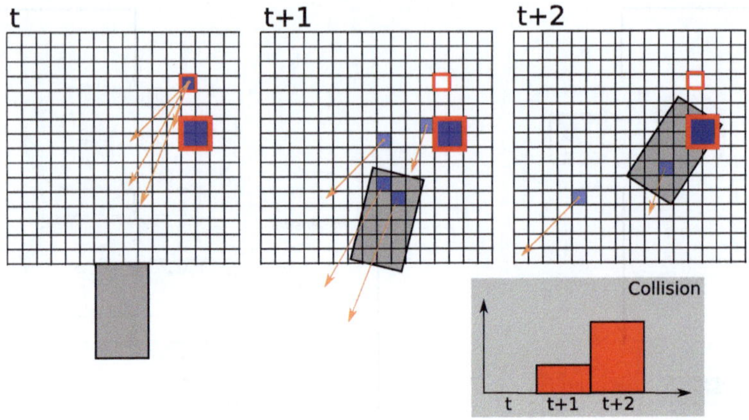

Fig. 3 Collision risk estimation over time for a specific cell. The cell position is predicted according to its velocity, along with the mobile robot. This risk profile is computed for every cell, and then used to integrate over time the global collision risk

struck is thus given, the computed risk value being the probability to collide with n cells of the grid. This evaluation presents the advantage of rightfully integrate risk when the occupancy estimation is diffuse. However, in the case of high speed robot displacement, the integration of occluded spaces, yet with unknown occupancy, can lead to systematic high risk scores. Although crossing unknown areas is rightfully dangerous, the estimation becomes useless under such circumstances.

The second evaluation selects, after applying a median filter on the risk grid to remove noisy data, the maximum value of collision over the cells. This simpler approach turns out to be more effective in practice.

By comparing this value to a threshold, a risk detector is thus defined. The system is then used to generate alerts, for the driver if the timing makes it relevant, and for a driving assistance device.

4 Experimental Results

4.1 Experimental Platform

For the experiments, we equipped a Lexus LS600h car with two Ibeo Lux multilayer lidars under the two front lights (see Fig. 4). The horizontal field of view covered by the two lidars is almost 160°. For the odometry calculation we use vehicle velocity and steering data collected from the CAN bus system. The input of our algorithm is a sequence of occupancy grids computed from the two lidars. To merge the eight laser scan layers acquired by the two lidars and to compute the input observation occupancy grid, we use a method similar to [11]. The vehicle is used to collect

Fig. 4 Experimental platform: Lexus LS600h car equipped with two Ibeo Lux Lidars and cameras

real on-road data for both downtown and highway scenarios, in order to assess the performances of the proposed approach.

The collected data are also processed in real time using a computer embedded in the car trunk. As the method is designed to be efficiently implemented on parallel, it is implemented on GPU, using nvidia CUDA. The grid resolution used is of 750×300, and the number of particles is 262,144 (0.86 particle/cell on average). The complete chain for perception (including lidar to grid mapping, HSBOF and TTC estimation) runs in real time at 15 FPS on a nVidia Quadro 2000 graphic board.

4.2 Methodology

The experiments focus on validating the approach in different contexts. On the one side, real on-road data is used to assess the realism of the approach in complex scenarios. On the other side, it is not possible to generate real risky situations on open road. Therefore, collisions are performed with soft objects in a closed environment, to simulate risky driving situations.

Beyond the various contexts, we propose to use our risk estimation strategy for various applications. Therefore, we consider 3 different time horizons corresponding to 3 different TTC values:

A1—pre-crash: a collision is likely to appear in less than $t_1 = 500$ ms. In this context, the driver is no longer able to react and the collision detection is used to prepare the vehicle to the impact;

A2—collision mitigation: a collision is likely to appear between t_1 and $t_2 = 1$ s. In this case, the vehicle can perform an automatic action (e.g. automatic braking) in order to mitigate the danger or avoid the impact;

A3—dangerous driving: a collision is likely to appear between t_2 and $t_3 = 1.5$ s. In this last case, the assistance system can warn the driver of a potential risk and the driver may react in a way or another.

4.3 First Experiment: Dynamic Collisions

The first set of experiments was performed in a mostly controlled environment. On a parking lot, the experimental vehicle was used to model its environment and predicts potential collisions. To simulate obstacles, 3 bouncy balls of 55, 65 and 75 cm of diameter were thrown around the vehicle and on its path. The main advantage of the use of balls is the immediate validation of the existence of a risk. Indeed, on real road data, as actual collisions are not conceivable, the ground truth for risk is a hard task as it implies a subjective assessment. In this experiment, the vehicle could hit the obstacles, the collisions being annotated, confirming a preceding risk without ambiguity. Many scenarios were staged, including:

- immobile or moving vehicle
- immobile or moving obstacles
- various number of obstacles
- various trajectory options
- impacts or close crossings
- limited visibility (late appearance of obstacles, occlusions, intersection during curves).

Estimated risk values Figure 5a shows an example of the estimated risk values while the balls are thrown toward the vehicle. The impacts are annotated as the vertical purple lines. It appears that for all the impacts, a high risk value is properly predicted in advance. Figure 5b shows in more detail an example of risk prediction before the impact. As expected, the risk corresponding to $TTC \leq 1.5$ s rises first (between 1.5 and 2 s before the collision), followed by the risk values corresponding to $TTC \leq 1.0$ s, and $TTC \leq 0.5$ s. Figure 5c shows an example where the ball crosses close to the car without hitting it. In this case, the risk corresponding to $TTC \leq 1.5$ s is high and then decreases as soon as the likelihood of a possible collision becomes very low.
Alerts generation The experiment with the balloons has been repeated about a hundred times, including cases where the car hits the balloons and other where the balls come close to the car without hitting it. This provides the opportunity to evaluate possible thresholds for generating alerts for the different values of TTC. Figure 6 shows how the precision and recall rates change with respect to the threshold.

It appears on the precision/recall curves that efficient detectors can be made by selecting the adequate threshold for every time horizon.

4.4 Second Experiment: The Mannequin

In order to go further with more realistic scenarios, we designed a fake pedestrian in the form of a tissue mannequin filled with bubble wrap. This allows to make collisions with a human-shaped object. Figure 7 shows the mannequin and the system

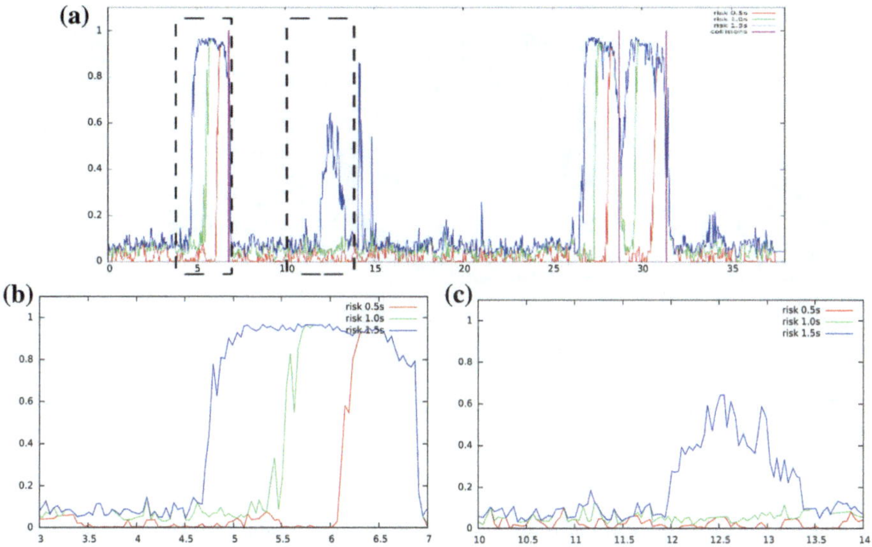

Fig. 5 **a** Evolution of the collision risk for TTC $\leq$ 0.5, 1.0, 1.5 s, in the case of multiple balls thrown toward the vehicle. **b** Details when a collision happens. **c** Details in case of the ball closely without collision. For 1.5 s, a significant risk is estimated, while for shorter TTC nothing is detected, as expected

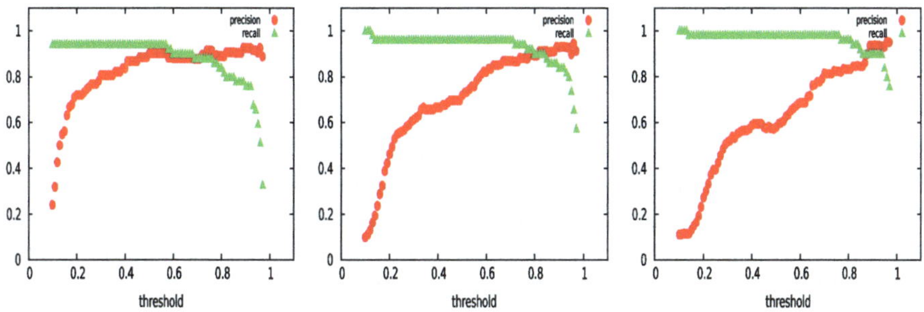

Fig. 6 Precision (*red dots*) and recall (*green triangles*) of risk detection according to the threshold selected. From *left* to *right*, those are given for a estimation over 0.5, 1 and 1.5 s

used for hanging it. This system is equipped with a runner, in order to permit lateral displacements of the pedestrian.

Over the different experimented scenarios of car velocity, pedestrian motion and occlusion, the system successfully generated risk alarms every time. Figure 8 shows examples of those experiments with the mannequin. The first examples correspond to the sudden appearance of a pedestrian which was hidden by the preceding car. The risk map then shows immediately a important risk at the position of the pedestrian. It means that the system is reactive enough to detect a risk appearing at the last moment before the collision. The second one corresponds to the road crossing of a moving

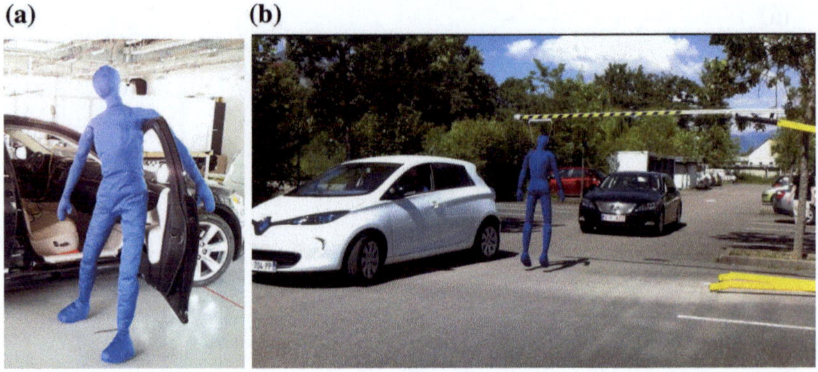

Fig. 7 **a** Fake pedestrian used for experiments. **b** The mannequin is attached to a system with a runner, in order to allow lateral displacements

pedestrian. As the system estimate its motion, a risk appears on the risk map even before the pedestrian is on the car trajectory.

4.5 Third Experiment: On-Road Evaluation

HSBOF and risk estimation We have performed experiments on various recorded sequences of road journey, in highway and downtown environments. Figure 9 shows the output of the HSBOF and the extraction of cells with a high short-term probability of collision. On the first example, the driver arrives too quickly at the traffic light, the car ahead then generates a collision risk corresponding to a A3 alert. On the second example, we can see that the system rightly detects the risk of collision with the pedestrian crossing the road, even through he is not already on the trajectory of the vehicle. Indeed, as the occupancy dynamics are inferred by the HSBOF, the method predicts the likely impact.

Figure 10 shows an example of the risk values estimated during a urban driving sequence. Since we are driving safely, the risk corresponding to $TTC \leq 1.0$ s remains very low. There are a few peaks for the risk corresponding to $TTC \leq 1.5$ s. They happen at the end of turning maneuvers where the predicted trajectory of the car is supposed to be circular, while the driver is about to go back to a straight trajectory. Thus, these cases do not generate high values of risk for shorter TTC values.

Alarms generation on-road In order to evaluate the relevance of the approach for pre-crash applications, we focus on the A1 alerts. The car is driven in various road environments (downtown, highway, national roads) to collect data. A dataset containing 27,000 frames, including complex situations is recorded and the algorithm is applied, with a variable alarm threshold. The curve on Fig. 11 shows the number of A1 alarms with respect to the risk threshold. With a risk threshold set to 0.8 (which is relevant with respect to the curves shown on Fig. 6) only 4 alarms were generated.

(a)
(b)
(c)
(d)

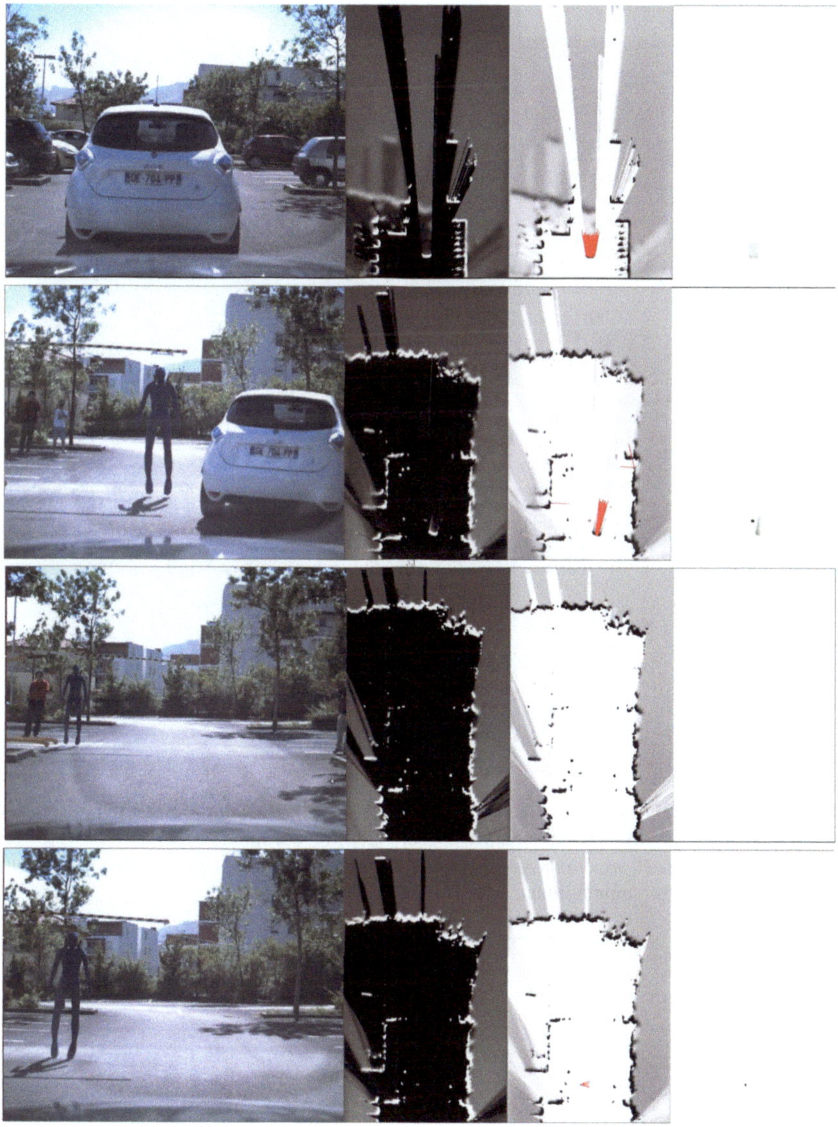

Fig. 8 Results of the system. Are presented each line a visual capture from the embedded camera, the estimated occupancy grid in front of the car (*white* for occupied, *grey* for unknown, *black* for empty), the estimated motion field (if a case is seen as dynamic, a *red* motion vector showing the average velocity in the cell is drawn on the map) and finally the estimated risk map for 0.5 s. The first sequence **a**, **b** presents the appearance of an occluded pedestrian, the second **c**, **d** a moving pedestrian heading towards the road

Fig. 9 First HSBOF and risk estimation results. *Left image* output of the HSBOF process with the filtered occupancy grid (*white* for empty space, *black* for occupied space) and the velocity field (from each moving cell are drawn *red* rays representing the velocities). *Right image* dangerous cells are reprojected in the camera image

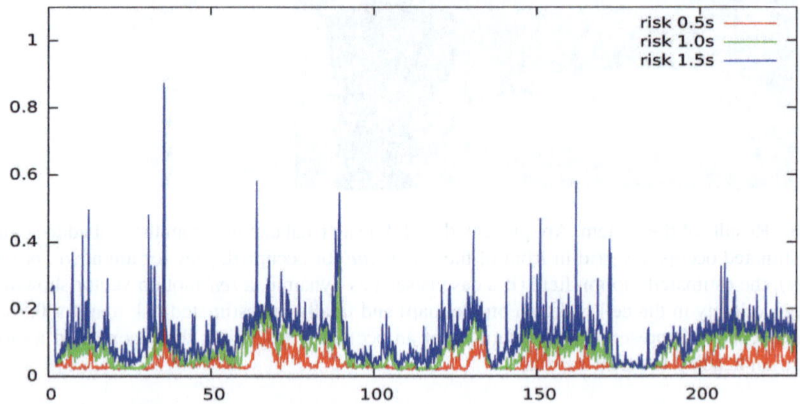

Fig. 10 Example of a sequence of driving downtown

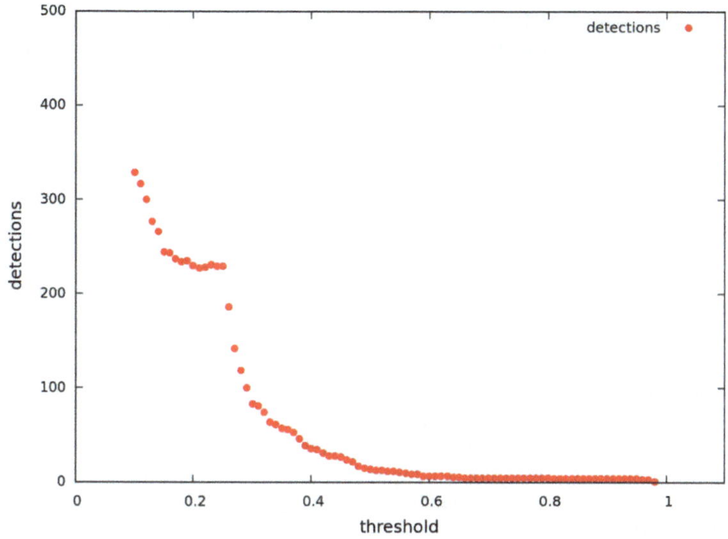

Fig. 11 Alarms generated for risky situations at $TTC \leq 0.5$ s while driving in crowded urban situations, with respect to the risk threshold (tested on 27,000 frames)

5 Conclusion

In this paper, we presented a method for short-term collision risk estimation. The method takes advantage of the ability of the HSBOF algorithm to provide both an occupancy grid and an accurate estimation of velocities over its cells. The use of those pieces of information, combined with the prediction of the vehicle trajectory, allows to estimate a Time To Collision for each cell of the grid. The method presents several advantages, like the ability to have probabilistic estimations of collision risk, the ability to be implemented efficiently in parallel, and also to estimate TTC in highly dynamic environment, without any object tracking algorithm.

A comprehensive evaluation of the algorithm has been proposed, based on both real road data and controlled situations. This evaluation showed that the method can perform with convincing results. In controlled environments the system is very reactive and every collision is properly detected while in difficult scenarios the system do not generate abusive alarms. There is still some work to do to deal with the remaining errors and to validate more extensively the algorithm on hundreds of kilometers, but the results obtained here from the experiments are promising. Therefore, we project to couple this perception layer with prototypes of Human-Machine-Interfaces for testing their use in Advanced Driver Assistance Systems.

Acknowledgments The authors would like to thank Toyota Motor Europe for their collaboration and their continuous support on the Lexus car. The authors acknowledge the contribution of the "Institut de Recherche Technologique NanoElec" which has been founded by the french program

"Investissement d'Avenirs" ANR-10-AIRT-05. Our thanks are also given to Nicolas Turro, Laurence Boissieux and Jean-François Cuniberto for their assistance in setting up the experiments.

References

1. Labayrade, R., Royere, C., Aubert, D.: Experimental assessment of the rescue collision-mitigation system. IEEE Trans. Veh. Technol. **56**(1), 89–102 (2007)
2. Kaempchen, N., Schiele, B., Dietmayer, K.: Situation assessment of an autonomous emergency brake for arbitrary vehicle-to-vehicle collision scenarios. IEEE Trans. Intell. Transp. Syst. **10**(4) (2009)
3. Laugier, C., Paromtchik, I., Perrollaz, M., Mao, Y., Yoder, J.-D., Tay, C., Mekhnacha, K., Nègre, A.: Probabilistic analysis of dynamic scenes and collision risk assessment to improve driving safety. Intell. Transp. Syst. J. **3**(4), 4–19 (2011)
4. Petrovskaya, A., Perrollaz, M., Oliveira, L., Spinello, L., Triebel, R., Makris, A. Yoder, J.-D., Nunes, U., Laugier, C., Bessière, P.: Awareness of road scene participants for autonomous driving. In: Eskandarian A. (ed.) Handbook of Intelligent Vehicles, pp. 1383–1432. Springer, New York (2012)
5. Alenya, G., Nègre, A., Crowley, J.L.: A comparison of three methods for measure of time to contact. In: IEEE Conference on Intelligent Robotics and Systems (IROS), St. Louis, États-Unis, October 2009
6. Tay, M.K., Mekhnacha, K., Chen, C., Yguel, M.: An efficient formulation of the bayesian occupation filter for target tracking in dynamic environments. Int. J. Veh. Autonom. Syst. **6**(1), 155–171 (2008)
7. Negre, A., Rummelhard, L., Laugier, C.: Hybrid sampling bayesian occupancy filter. In: IEEE International Symposium on Intelligent Vehicles (2014)
8. Gindele, T., Brechtel, S., Schroder, J., Dillmann, R.: Bayesian occupancy grid filter for dynamic environments using prior map knowledge. In: 2009 IEEE Intelligent Vehicles Symposium, pp. 669–676. IEEE (2009)
9. Danescu, R., Oniga, F., Nedevschi, S.: Modeling and tracking the driving environment with a particle-based occupancy grid. IEEE Trans. Intell. Transp. Syst. **12**(4), 1331–1342 (2011)
10. Laugier, C., Fulgenzi, C., Spalanzani, A.: Dynamic obstacle avoidance in uncertain environment combining pvos and occupancy grids. In: IEEE International Conference on Robotics and Automation (2007)
11. David Adarve, J., Perrollaz, M., Makris, A., Laugier, C.: Computing occupancy grids from multiple sensors using linear opinion pools. In: IEEE International Conference on Robotics and Automation, St Paul, Minnesota, US, May 2012

Modular and Adaptive Wheelchair Automation

Brenna D. Argall

Abstract We present in this paper a novel framework for the design of a modular and adaptive partial-autonomy wheelchair. Our design in particular aims to address hurdles to the adoption of partial-autonomy wheelchairs within general society. In this experimental work, a single assistance module (assisted doorway traversal) is evaluated, with arbitration between multiple goals (from multiple detected doors) and multiple control signals (from an autonomous path planner, and the human user). The experimental work provides the foundation and proof-of-concept for the technical components of our proposed modular and adaptive wheelchair robot. The system is evaluated within multiple environmental scenarios and shows good performance.

1 Introduction and Related Work

We envision a future where the partial-automation of powered wheelchairs will be the standard: that when a person is being fit for a wheelchair by a therapist, a variety of autonomy options will be available, just like today a variety of seating options are available.

While many individuals achieve sufficient mobility using manual and powered wheelchairs, a survey of 65 clinicians within the United States found that between 10 and 40 % could not be prescribed either [1], leaving those individuals reliant on a caretaker for mobility. The potential for "smart" wheelchairs—which incorporate robotics technologies—to aid the mobility of those with motor or cognitive impairments has been recognized for decades [2]. A survey of epidemiological data estimates that between 1.4 and 2.1 million individuals would benefit from a smart wheelchair at least some of the time [3]. Robotics autonomy can help with obstacle avoidance, navigation, route planning and spatially-constrained maneuvers. However, despite decades of development, and significant advances in capabilities [4, 5],

B.D. Argall (✉)
Northwestern University, Evanston, IL 60208, USA
e-mail: brenna.argall@northwestern.edu

B.D. Argall
Rehabilitation Institute of Chicago, Chicago, IL 60211, USA

M.A. Hsieh et al. (eds.), *Experimental Robotics*, Springer Tracts in Advanced Robotics 109, DOI 10.1007/978-3-319-23778-7_55

control [6–8] and interfaces [9, 10], very few smart wheelchair technologies have made the transition to the public and commercial sectors.[1]

One dominating confound to practical adoption is cost: at least in the short term and within the United States, these technologies will not be covered by Medicare/Medicaid or insurance plans, and so any system that is going to be of practical benefit to general society must be reasonable to finance out of pocket. The general trend for the majority of work in smart wheelchairs has been to offer a complete system: that is very capable, but also involves a fair amount of infrastructure, and components that are costly. Many are developed in their entirety from the ground up, including the wheelchair hardware and software systems [13–18]. While this historically has been the most common approach to smart wheelchair development, some more recent projects do take a more modular approach: to software, for example to accommodate multiple control interfaces [14, 15] or sensor modules [19]; or to hardware, to be able to interface with existing powered wheelchairs [15–17].

An important observation is that users of assistive devices overwhelmingly prefer to retain as much control as possible, and cede only a minimum amount of control authority to the machine [20, 21]. Thus, many smart wheelchairs offer a variety, often a hierarchy, of autonomous and semi-autonomous control modes within their shared control schemes [14, 22, 23]. Others explicitly target low-profile automation [10, 24], create new customized levels of autonomy [25], or blend the user's control commands with the automation's control commands [21, 26, 27]. Most commonly, shared-control smart wheelchair platforms place the high-level control (e.g. goal selection, route planning) with the user, and the low-level control (e.g. motion control commands, obstacle avoidance) with the machine [6, 14, 28–30].

In this paper, we introduce a system that prioritizes *customization, modularity* and the use of *commercial hardware*, to facilitate practical adoption by users. The result will be a complete system consisting of modular software and hardware components, easily added to a commercial wheelchair platform, and able to be customized to and by the user. We present in this paper our control framework, and first expermimental results.

2 Technical Approach

We introduce a system of modular software and hardware components—which scale with a user's physical needs, financial means and personal preferences. The framework introduced in this section will be grounded in Sect. 3 with concrete implementations of goal arbitration and control sharing, including example data.

[1] Exceptions include the Smart Wheelchair from Smile Rehab [11] and TAO-7 Intelligent Wheelchair Base from AAI Canada [12].

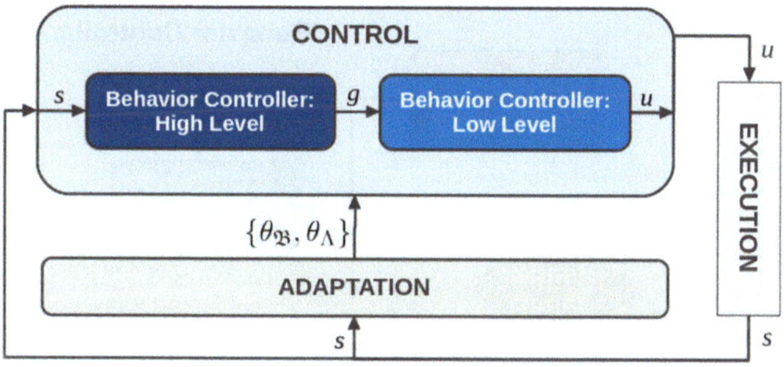

Fig. 1 Schematic of our full control architecture with adaptation. Behavior parameters $\theta_{\mathfrak{B}}$ and control sharing strategy parameters θ_{Λ} are adapted in response to cues from the user and metrics computed from data observed by the robot's sensors about world state s. The high-level behavior controller generates goals g, and the low-level behavior controller generates control commands $\mathbf{u}$

2.1 High and Low Level Behaviors

Our control framework (Fig. 1) assumes the existence of a set $\mathfrak{B}$ of autonomous robot behaviors, and a set Λ of control sharing strategies. The user is able to select a custom set $\mathfrak{B}_u \subseteq \mathfrak{B}$ of behaviors. The set $\mathfrak{B}$ furthermore is partitioned into high-level behaviors $\mathfrak{B}_h$ and low-level behaviors $\mathfrak{B}_\ell$. In our work with wheelchair navigation, behaviors in $\mathfrak{B}_h$ typically are associated with planning (e.g. path generation), while behaviors in $\mathfrak{B}_\ell$ are associated with motion generation (e.g. path driving).

A high-level behavior $b_h \in \mathfrak{B}_h$ outputs a goal g given state input $\mathbf{x}$,

$$g \leftarrow b_h(\mathbf{x}) \tag{1}$$

Each autonomy goal g is evaluated for confidence c_g that it is the user's goal. A goal g is passed to the low-level control module only if its confidence is both over threshold, $\tau_g < c_g$, and significantly higher than the second-highest confident goal, $\delta\tau_g < c_g - c_{g_\circ}$. (Grounded in Sect. 3.1).

A low-level behavior $b_\ell \in \mathfrak{B}_\ell$ outputs a control command $\mathbf{u}$ given state input $\mathbf{x}$ and goal g, (Fig. 2).

$$\mathbf{u} \leftarrow b_\ell(\mathbf{x}, g) \tag{2}$$

Within the low-level control module, commands generated by the autonomy are then reasoned about within the control sharing logic. (Grounded in Sect. 3.2).

The robot autonomy and control sharing components together enable a flexible and modular architecture, where any combination of autonomy behaviors in $\mathfrak{B}$ can be selected by the user for inclusion in $\mathfrak{B}_u$. The autonomy resolves conflicts between competing behaviors in $\mathfrak{B}_u$—via a resource controller that registers the data and control signal needs of each behavior $b \in \mathfrak{B}_u$, as well as what data that

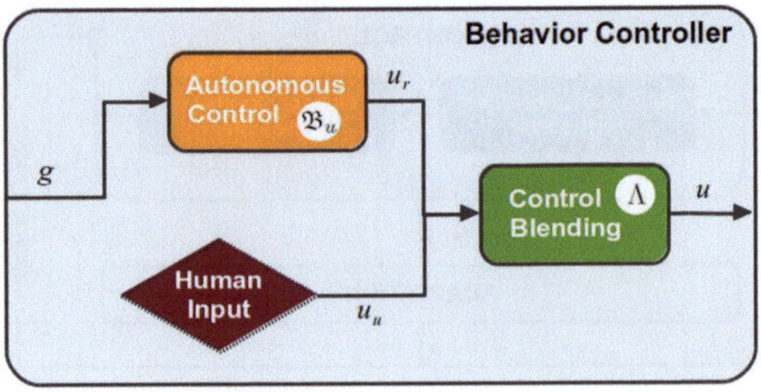

Fig. 2 Schematic of a behavior controller within our system (low-level; an equivalent architecture is used for the high-level behavior controller). The behavior controller considers multiple autonomous behaviors (in $\mathfrak{B}_u$) and considers user control signals ($\mathbf{u}_u$) to produce a single blended output signal ($\mathbf{u}$)

behavior provides. The robot autonomy and control sharing also reason about human input—which might play a role in behavior selection, be a cue for an already-running behavior, or be blended with autonomy-generated commands for reasons of safety.

2.2 Control Sharing

A control sharing strategy $\lambda \in \Lambda$ can take on one of three formulations: (i) All control to the user; (ii) All control to the automation; or (iii) A shared control formulation that blends the two control inputs.

Each behavior $b \in \mathfrak{B}$ has an associated control sharing strategy $\lambda_b \in \Lambda$. A strategy λ_b furthermore has an associated set of values $\boldsymbol{\theta}_{\lambda_b}$ that parameterize the function used to blend the user command $\mathbf{u}_u$ and autonomy command $\mathbf{u}_r$. Thus, each autonomy behavior b has associated with it a unique combination of control sharing strategy λ_b and paramterization $\boldsymbol{\theta}_{\lambda_b}$ for that strategy.

2.3 Adaptation

A defining feature of our architecture is the adaptation of the autonomy behaviors in $\mathfrak{B}_u$ and control strategies Λ_u associated with them.

Each behavior $b \in \mathfrak{B}$ available within our system has an associated set of parameters $\boldsymbol{\theta}_b$ which are available for modulation by the adaptation component of our framework. For example, the path planner [31] used on our development platform (Fig. 4) has parameters to modulate how much curvature there is in the generated trajectory, and how aggressively the robot will attempt to reach the goal position.

Similarly, each control sharing strategy $\lambda \in \Lambda$ has an associated set of parameters $\boldsymbol{\theta}_\lambda$ which are available for modulation by the adaptation component of our framework. For example, in a linear control sharing formulation, the parameter which dictates how much control is allocated to the user might be increased as the user becomes a more proficient driver.

Exactly what influence the parameters $\boldsymbol{\theta}_b$ have on associated behavior b (and parameters $\boldsymbol{\theta}_\lambda$ on strategy λ) varies across behaviors (and strategies). However, the approaches used to *modulate* the parameters will be common across behaviors and strategies, and any number of machine learning algorithms may be used to perform this modulation. One key factor to consider will be the *feedback signal* received by a machine learning algorithm—which will be computed autonomously from environmental cues as well as gathered from the user, who not only is a non-expert in the area of robotics but additionally has motor impairments and provides signals through a possibly limited control interface (e.g. used to drive the wheelchair).

2.4 Integration with Commercial Hardware

Lastly, each behavior has an associated specification ϑ of what form of input signals it expects to receive from the human user, which will change depending on the control device employed. For example, a traditional 2-axis joystick provides a 2-D continuous-valued control signal; while a Sip-N-Puff interface (Fig. 3) typically provides a 1-D *non-proportional* control signal, whose magnitude does not scale with the magnitude of the user's input (i.e. blowing or sucking). For a given input specification ϑ, the set $\mathfrak{B}$ is partitioned into a subset of behaviors $\mathfrak{B}_\vartheta$ which satisfy that specification.

Our system will prioritize good performance with non-proportional control interfaces (e.g. Fig. 3), which we believe provide a greater opportunity for autonomy to make an impact—since control with these devices is more difficult. More broadly, the interfaces used for human input will be restricted to those which are commercially available. Not only is this technology extensively validated (having been evaluated by thousands of users), but it also is covered by insurance—and thus, from the standpoint of financial feasibility, more readily accessible to users.

Fig. 3 *Left* Sip-N-Puff control interface [32], where commands are issued by blowing and sucking on a straw. *Right* Switch-based electronic head array with three proximity sensors [33]

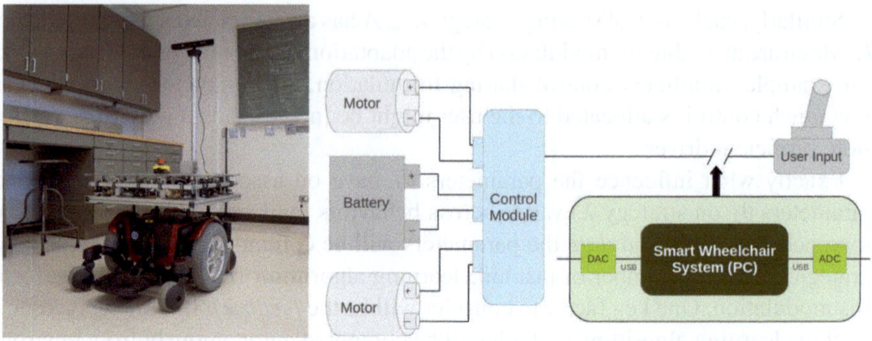

Fig. 4 *Left* Development platform. A differential drive mobile robot built on a wheelchair base, with a ring of IR and ultrasonic sensors and a top-mounted Kinect. *Right* Proposed integration with commercial wheelchair platforms. Signals from the user input device (e.g. joystick) are interrupted and processed, along with data from the onboard sensors, within our smart wheelchair PC system. Control signals—generated by the human (i.e. unmodified input signals), the automation, or a blend of the two—are then sent to the commercial control module of the wheelchair

To interface with multiple electronics packages from different wheelchair vendors, our proposed add-on system will be inserted between the input device and the control module of the commercial wheelchair system (Fig. 4, right). Several makes and models of wheelchairs are specifically designed to accept signals from expandable input controllers,[2] which present the user's signals to the proprietary control electronics. Importantly, presenting our control signals to an expandable input mechanism *should not void* the wheelchair warranty.

By restricting our system to integrate easily with commercial hardware and maintain a low price point, we are knowingly making trade-offs with respect to how "complete" the system is in the capabilities it offers to the user. The idea is that, in exchange, the system becomes more accessible, and transfer to the general public thus more feasible.

2.5 Development Platform

Our development platform (Fig. 4, left) consists of a Pride Mobility Quantum 600 base [36], modified to be drive-by-wire (including inverter and wheel encoders) by Sensible Machines [37]. To this we have added sensing and computing components.[3]

[2]Including: the Q-Logic system from Pride Mobility; PG Drive Systems' R-Net electronics system used by Permobil, Pride Mobility, and Sunrise Medical; and the MK6i electronics system used by Invacare [34, 35].

[3]Full specifications: *mini-PC* = Shuttle XH61 mini-PC with Intel i7-2600S processor, 16GB DDR3 SDRAM, 40GB solid state hard drive; *IR range sensor* = Sharp GP2Y0A02YK IR distance sensors (×10); *Ultra-sonic range sensor* = Maxbotix LV-MaxSonar-EZ1 Ultra-sonic range sensors (×4); *Sensor interface board* = Arduino Mega2560. *Total cost:* $1070.

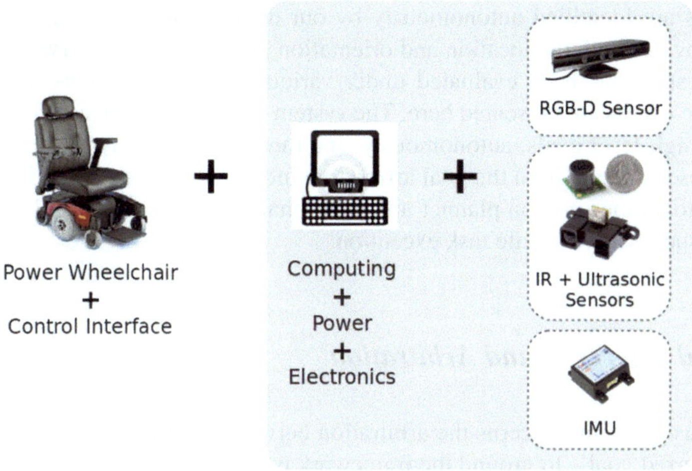

Fig. 5 Schematic of modular sensing components (*right*) within our add-on autonomy system (*gray box*) for commercial powered wheelchairs

The schematic in Fig. 5 overviews our add-on system with various example modular sensing components. The base system consists of an RGB-D sensor (e.g. Microsoft Kinect, Asus Xtion) and wheel encoders. Additional requirements include an inverter for power and an input device interface board, in order to interface with the expandable input mechanism of various commercial wheelchairs. This modular system will be interfaced with participants' own wheelchairs in future subject studies.

The base system, consisting of only a Kinect sensor and wheel encoders, is evaluated in the following section. All software has been developed within the Robot Operating System (ROS), with each high and low level behavior operating as an individual ROS node. Customization thus consists of bringing up only those nodes identified in $\mathfrak{B}_u$, and having the resource controller reason about and resolve any conflicts between those nodes.

3 Assessment

In this experimental work, a single assistance module (assisted doorway traversal) is evaluated, with arbitration between multiple goals (from multiple detected doors and the inferred user's goal) and multiple control signals (from an autonomous path planner and the human user).[4] Doorway navigation was chosen as a task frequently cited as challenging for powered wheelchair drivers [1], due to tight spatial constraints.

[4]Implementation of the adaptation components of our framework are currently under development, and are not included in this assessment.

Doorways are identified autonomously by our doorway detection algorithm [38], which provides both the location and orientation of an observed doorway.

Our system has been evaluated under various testing conditions, from which illustrative results are presented here. The system was found to successfully identify multiple high-level goals, autonomously, and then to reason between them and the inferred user's goal within the goal arbitration module. Speed commands generated by the autonomous motion planner and the human operator were blended, with the result of successful and safe task execution.

3.1 Goal Inference and Arbitration

Our first assessment concerns the arbitration between multiple goals, including the user's inferred goal—to ground the framework presented in Sect. 2.1.

We infer the user's goal from only those control signals used to teleoperate the wheelchair (rotational and translational speed commands). While there are undoubtedly many advantages to using custom interfaces like touch screens, that are tailored to the task or a user's particular needs, our intent here instead is to push the limits of existing commercial control interface technologies within an autonomy framework, using software solutions whenever possible. Small screens with menu-based interfaces are available with many control devices (especially ones like in Fig. 3), however menu navigation with these interfaces can be cumbersome and our aim is for the user to be able to indicate their goal or preference more intuitively.

To infer the user's immediate goal g_u, our system maintains a smoothed estimate $\tilde{\mathbf{u}}_u$ of the user's command, weighted by the time since the last update:

$$\tilde{\mathbf{u}}_u^t \leftarrow \kappa \cdot \mathbf{u}_u^t + (\kappa - 1) \cdot \tilde{\mathbf{u}}_u^{t-1} \tag{3}$$

$$\kappa = e^{-\Delta\tau}$$

where $\Delta\tau$ is the difference between the timestamps of $\tilde{\mathbf{u}}_u^{t-1}$ and $\mathbf{u}_u^t$. The smooth command is then forward projected to calculate the immediate user goal g_u. (In our implementation, the projection time is 3.0 s).

To determine which of the autonomy goals might be the user's final (high-level) goal, a confidence measure is computed for each autonomously detected goal. Associated with each autonomy goal g is a set of N navigation goals $\{g_{nav}^i\}_{i=0}^N$. Navigation goals are executed as a sequence, with one goal $g_{nav}^\star \in \{g_{nav}\}$ active at a given time. For example, during doorway assistance two navigation goals are set, along the normal of the pose of the identified door on opposite sides of the door frame; achieving the first aligns the robot for doorway traversal and achieving the second has the robot pass through the doorway.

In our implementation, the confidence $c_g \in [0, 1]$ associated with an observed goal g is calculated based on the distance d and heading ϕ (absolute value) to the current active navigation goal $g_{nav}^\star$:

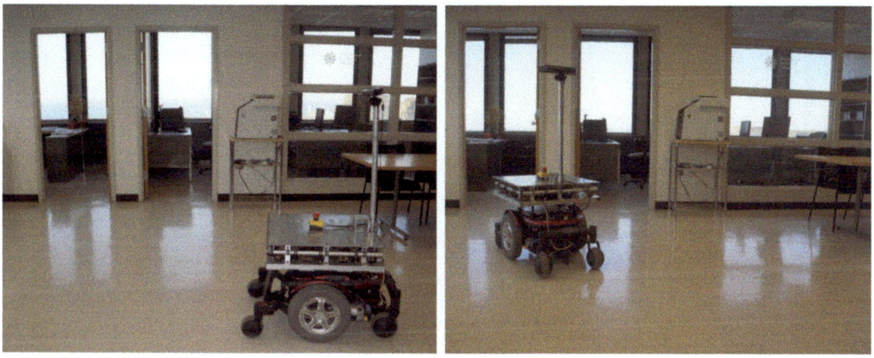

Fig. 6 Example environment scenario, with two side-by-side doors

$$c_g = c_p \cdot \left(\beta \cdot \left(\frac{2}{1+e^{\phi}}\right) + (1-\beta) \cdot \left(\frac{2}{1+e^{d}}\right)\right) \tag{4}$$

$$\beta = \texttt{min}\,(1, d)$$

where $c_p \in [0, 1]$ is the perception confidence from observing g.[5] The parameter β dictates that when the robot is far from $g^{\star}_{nav}$ ($d > 1$ m), aiming towards the navigation goal ($\phi \to 0$) during driving matters most; while closing the distance matters most when near to $g^{\star}_{nav}$.

If there exists a goal g whose confidence is both above threshold and significantly so ($\delta\tau_g < c_g - c_{g_\circ}$), this goal is considered active and passed to the low-level control module. If no autonomy goal is active, then the user-inferred goal is sent instead.

Figure 7 shows the goal confidence calculated over a sample run in an environment with two side-by-side doors (Fig. 6).

3.2 Command Arbitration and Safety Monitoring

Our second assessment concerns the sharing of control between the user and the autonomy—to ground the framework presented in Sect. 2.2.

Goals passed from the high-level behavior module, post-arbitration, are presented to the low-level control module to generate control commands. In the presented assessment, the low-level controller is a velocity-based path planner [31], which generates rotational and translational speeds for the robot. The autonomy command $\mathbf{u}_r$ is blended with the user's command $\mathbf{u}_u$, according to the control sharing strategy λ_b associated with the behavior b that generated the goal g.

[5]Since d and ϕ are both always positive, the logistic function (fractions in parentheses) range is [0, 0.5]. The factor of 2 in this equation compensates for this fact, so that the range of c_g becomes [0, 1].

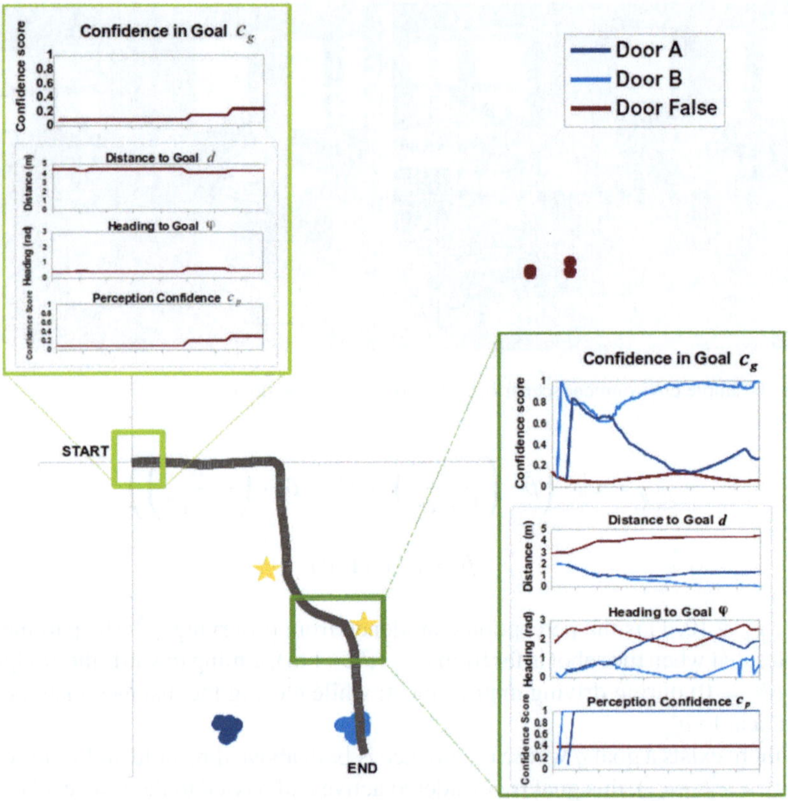

Fig. 7 Confidence associated with autonomously observed goals. Two doors (*light blue dots*, *dark blue dots*) are observed within the environment; a third (*dark red dots*) is falsely identified. Plot panels show (*top* → *bottom*) goal confidence c_g, distance to navigation goal d, heading to navigation goal ϕ and perception confidence c_p. At the start of the run (*light green inset*), the false positive door is observed, however the low perception confidence keeps the overall confidence c_g also low. As the robot turns towards the actual doors, both are identified with similarly high confidence. As the user issues commands that show preference for Door B (*first star*), its confidence rises until it is sufficiently greater than that of Door A for Door B to become the active goal (*dark green inset*). The user initially retains control however, as dictated by the associated control sharing strategy. When the user ceases issuing commands (*second star*), the autonomy takes over in full. Robot ground path shown in *dark gray*

The results presented in Figs. 7, 8 and 9 utilize a linear control blending formulation:

$$\mathbf{u} \leftarrow \alpha \cdot \mathbf{u}_u + (1 - \alpha) \cdot \mathbf{u}_r \tag{5}$$

$$\alpha \in [0, 1]$$

The automation command $\mathbf{u}_r$ is generated by the path planner, which takes the inferred user goal g_u as its target.

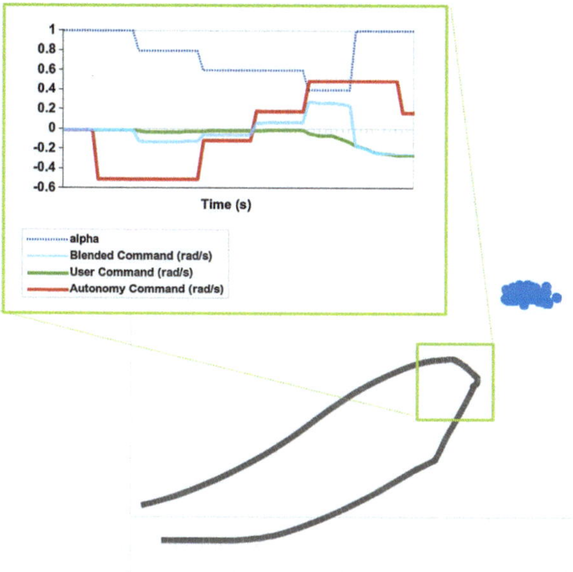

Fig. 8 Nimble transfers of control authority during control blending. The user never relinquishes control as s/he drives near to an autonomously detected goal (*blue dots*), which is also near an obstacle. The autonomy therefore never takes over control to achieve the detected goal. However, as the robot path (*dark gray line*) nears the obstacle, the autonomy gradually takes over some of the control (reduced α, *dashed line* in plot) to avoid collision. When the user turns away from the obstacle and goal however, the autonomy immediately transfers control back to the user ($\alpha = 1$)

The control sharing strategies implemented within our architecture to date include: *all-user* ($\alpha = 1$); *all-autonomy* ($\alpha = 0$); *blending-zero-relinquish* ($0 \leq \alpha \leq 1$), where a zero command from the user is interpreted relinquishing control to the autonomy; and *blending-zero-stop* ($0 \leq \alpha \leq 1$), where a zero command from the user is interpreted as a stop command.

Figure 8 presents the blending of control commands issued by the user and the autonomy during a sample run when the user never relinquishes control (under sharing strategy *blending-zero-relinquish*), in the presence of a detected autonomy goal and also an obstacle.

Before the command $\mathbf{u}$ is passed to the robot for execution, it is assessed for safety by forward projecting (3.0 s) the command and evaluating the resultant path for collisions. If the projected path collides with an obstacle, the control balance is iteratively shifted away from the user and to the autonomy, whose path planner is accounting for obstacles, until the resulting command projection no longer results in a collision.

Specifically, α is initialized to the value specified in $\boldsymbol{\theta}_\lambda$, but decremented according to $\alpha \leftarrow \alpha - \delta\alpha$ if the projected path collides with an obstacle. The decrementation is incremental, until either the projected command no longer collides with an obstacle or all of the control lies with the autonomy ($\alpha = 0$). This paradigm is a balance between

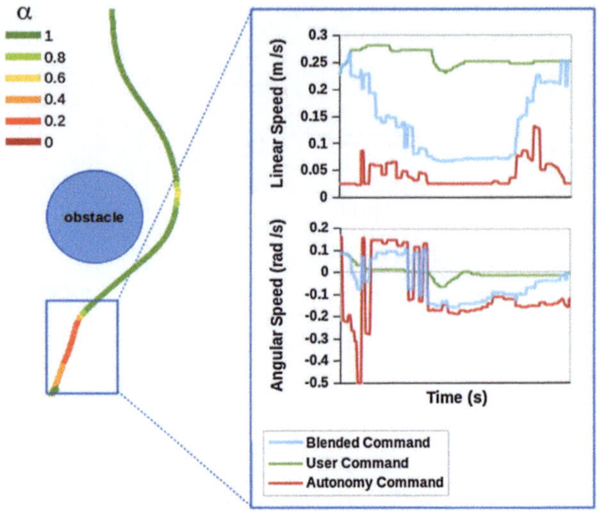

Fig. 9 Command blending to maintain safety. As the forward projection of the user's commands (*green* in graphs) generate a path which collides with an obstacle, control is iteratively shifted from the user to the autonomy (by reducing the value of α). The resultant blended command (*light blue* in graphs) prioritizes foremost safety, but also keeping as much control as possible (and within the constraints of $\boldsymbol{\theta}_\lambda$) with the user. Robot ground path shown with colors that reflect the value of the control blending parameter α at that time

competing aims: keeping control maximally with the user (i.e. α as large as possible), but limiting the number of forward projection roll-outs (to limit computational costs). In practice $\delta\alpha = \mathtt{min}(\frac{\alpha}{5}, 0.1)$, and so the upper limit on the number of roll-outs that might occur is 10.

Figure 9 presents the blending of control commands issued by the user and the autonomy during a sample run in the presence of an obstacle and user commands that would collide with that obstacle.

4 Conclusions and Future Work

The customization of exactly which behaviors are selected for inclusion in $\mathfrak{B}_u$ is one mechanism by which customization to the user's physical abilities and preferences is accomplished. The other mechanism will be the *adaptation* of the autonomous behaviors in $\mathfrak{B}_u$, and of the control strategies Λ_u associated with them (Fig. 1). The idea is to leverage machine learning to autonomously adapt the robot behaviors and control sharing strategies in order to customize to a user's physical abilities and personal preferences. The next step in the development of our software architecture thus will be to complete the development and evaluation of the adaptation modules.

The experimental work presented in this paper has provided the foundation and proof-of-concept for the technical components of our proposed modular and adaptive wheelchair robot. Our system prioritizes simple integration with existing commercial chairs and control interfaces, to mitigate costs not covered by insurance and thus accelerate adoption by users. Furthermore, our system is distinguished by its focus on customization to the user, via the selection and adaptation of a unique set of autonomy behaviors and control sharing strategies, and also modularity in the sensor add-ons. The system has been evaluated within multiple environmental scenarios and shown good performance. The technical components are of course only one half of the story, and our future work will evaluate the operation of this system by limited-mobility users at the Rehabilitation Institute of Chicago—the #1 ranked rehabilitation hospital in the United States.

Acknowledgments Many thanks to Matthew Derry for his significant contributions to the development of the software infrastructure for this system.

References

1. Fehr, L., Langbein, W.E., Skaar, S.B.: Adequacy of power wheelchair control interfaces for persons with severe disabilities: a clinical survey. J. Rehabil. Res. Dev. **37**(3), 353–360 (2000)
2. Simpson, R.: Smart wheelchairs: a literature review. J. Rehabil. Res. Dev. **42**(4), 423–438 (2005)
3. Simpson, R., LoPresti, E., Cooper, R.: How many people would benefit from a smart wheelchair? J. Rehabil. Res. Dev. **45**(1), 53–72 (2008)
4. Wang, Y., Chen, W.: Hybrid map-based navigation for intelligent wheelchair. In: Proceedings of ICRA (2011)
5. Demeester, E., Hüntemann, A., Vanhooydonck, D., Vanacker, G., Degeest, A., Brussel, H.V., Nuttin, M.: Bayesian estimation of wheelchair driver intents: modeling intents as geometric paths tracked by the driver. In: Proceedings of IROS (2006)
6. Röfer, T., Mandel, C., Laue, T.: Controlling an automated wheelchair via joystick/head-joystick supported by smart driving assistance. In: Proceedings of ICORR (2009)
7. Li, Q., Chen, W., Wang, J.: Dynamic shared control for human-wheelchair cooperation. In: Proceedings of ICRA (2011)
8. Philips, J., del R. Millán, J., Vanacker, G., Lew, E., Galan, F., Ferrez, P.W., Brussel, H.V., Nuttin, M.: Adaptive shared control of a brain-actuated simulated wheelchair. In: Proceedings of ICORR (2007)
9. Iturrate, I., Antelis, J.M., Kübler, A., Minguez, J.: A noninvasive brain-actuated wheelchair based on a p300 neurophysiological protocol and automated navigation. IEEE Trans. Ind. Electron. **25**(3) (2009)
10. Katsura, S., Ohnishi, K.: Human cooperative wheelchair for haptic interaction based on dual compliance control. IEEE Trans. Ind. Electron. **51**(1) (2004)
11. Katsura, S., Ohnishi, K.: The smart wheelchair from smile rehab ltd. http://smilerehab.com/smart-wheelchair.php
12. Katsura, S., Ohnishi, K.: The TAO-7 intelligent wheelchair base from AAI Canada, Inc. http://www.aai.ca/robots/tao_7.html
13. Borgolte, U., Hoyer, H., Bühler, C., Heck, H., Hoelper, R.: Architectural concepts of a semi-autonomous wheelchair. J. Intell. Robot. Syst. **22**(3–4), 233–253 (1998)
14. Mazo, M.: An integral system for assisted mobility. IEEE Robot. Autom. Mag. **8**(1), 46–56 (2001)

15. Braga, R.A., Petry, M., Reis, L.P., Moreira, A.P.: Intellwheels: modular development platform for intelligent wheelchairs. J. Rehabil. Res. Dev. **48**(9) (2011)
16. Urdiales, C.: Collaborative Assistive Robot for Mobility Enhancement (CARMEN): The bare necessities assisted wheelchair navigation and beyond. Springer, Berlin (2012)
17. Desmond, R., Dickerman, M., Fleming, J., Sinyukov, D., Schaufeld, J., Padir, T.: Development of modular sensors for semi-autonomous wheelchairs. In: Proceedings of TePRA (2013)
18. Civit-Balcells, A., Gonz, J.A.: Tetranauta: a wheelchair controller for users with very severe mobility restrictions. In: Proceedings of the 3rd TIDE Congress (1998)
19. Prassler, E., Scholz, J., Fiorini, P.: A robotics wheelchair for crowded public environment. IEEE Robot. Autom. Mag. **8**(1), 38–45 (2001)
20. Biddiss, E.A., Chau, T.T.: Upper limb prosthesis use and abandonment: a survey of the last 25 years. Prosthet. Orthot. Int. **31**(3), 236–257 (2007)
21. Lankenau, A., Röfer, T.: A versatile and safe mobility assistant. IEEE Robot. Autom. Mag. **8**(1), 29–37 (2001)
22. Bley, F., Rous, M., Canzler, U., Kraiss, K.F.: Supervised navigation and manipulation for impaired wheelchair users. In: Proceedings of SMC (2004)
23. Parikh, S.P., Rao, R., Jung, S.H., Kumar, V., Ostrowski, J.P., Taylor, C.J.: Human robot interaction and usability studies for a smart wheelchair. In: Proceedings of IROS (2003)
24. Zeng, Q., Teo, C.L., Rebsamen, B., Burdet, E.: A collaborative wheelchair system. IEEE Trans. Neural Syst. Rehabil. Eng. **16**(2), 161–170 (2008)
25. Desai, M., Yanco, H.A.: Blending human and robot inputs for sliding scale autonomy. In: Proceedings of RO-MAN (2005)
26. Simpson, R.C., Poirot, D., Baxter, F.: The hephaestus smart wheelchair system. IEEE Trans. Neural Syst. Rehabil. Eng. **10**(2) (2002)
27. System, T.N.A.W.N.: Simon p. levine and david a. bell and lincoln a. jaros and richard c. simpson and yoram koren and johann borenstein. IEEE Trans. Rehabil. Eng. **7**(4) (1999)
28. Bourhis, G., Horn, O., Habert, O., Pruski, A.: An autonomous vehicle for people with motor disabilities. IEEE Robot. Autom. Mag. **8**(1), 20–28 (2001)
29. Iturrate, I., Antelis, J.M., Kübler, A., Minguez, J.: A noninvasive brain-actuated wheelchair based on a p300 neurophysiological protocol and automated navigation. IEEE Trans. Ind. Electron. **25**(3) (2009)
30. Yanco, H.A.: Shared user-computer control of a robotic wheelchair system. PhD thesis, Department of Electrical Engineering and Computer Science, Massachusetts Institute of Technology, Boston (2000)
31. Park, J.J., Johnson, C., Kuipers, B.: Robot navigation with model predictive equilibrium point control. In: Proceedings of IROS (2012)
32. Park, J.J., Johnson, C., Kuipers, B.: Sip-N-Puff control interface from Therafin Corporation. http://www.therafin.com
33. Park, J.J., Johnson, C., Kuipers, B.: Electronic head array from Adaptive Switch Laboratories. http://www.asl-inc.com
34. Simpson, R., LoPresti, E., Hayashi, S., Nourbakhsh, I., Miller, D.: The smart wheelchair component system. J. Rehabil. Res. Dev. **41**(3B) (2004)
35. Sharma, V., Simpson, R., LoPresti, E., Schmeler, M.: Evaluation of semiautonomous navigation assistance system for power wheelchairs with blindfolded nondisabled individuals. J. Rehabil. Res. Dev. **47**(9) (2010)
36. Sharma, V., Simpson, R., LoPresti, E., Schmeler, M.: The Quantum 600 powered wheelchair from Pride Mobility. http://www.pridemobility.com
37. Sharma, V., Simpson, R., LoPresti, E., Schmeler, M.: The MR-100 drive-by-wire wheelchair-base robot from Sensible Machines. http://www.sensiblemachines.com/enabling-modules.php
38. Derry, M., Argall, B.: Automated doorway detection for assistive shared-control wheelchairs. In: Proceedings of ICRA (2013)

Fall Prediction for New Sequences of Motions

Junyun Tay, I-Ming Chen and Manuela Veloso

Abstract Motions reinforce meanings in human-robot communication, when they are relevant and initiated at the right times. Given a task of using motions for an autonomous humanoid robot to communicate, different sequences of relevant motions are generated from the motion library. Each motion in the motion library is stable, but a sequence may cause the robot to be unstable and fall. We are interested in predicting if a sequence of motions will result in a fall, without executing the sequence on the robot. We contribute a novel algorithm, ProFeaSM, that uses only body angles collected during the execution of single motions and interpolations between pairs of motions, to predict whether a sequence will cause the robot to fall. We demonstrate the efficacy of ProFeaSM on the NAO humanoid robot in a real-time simulator, Webots, and on a real NAO and explore the trade-off between precision and recall.

Keywords Fall prediction · Sequence of motions · Humanoid robot

1 Introduction

Research showed that as much as 70% of meanings in communication is derived from non-verbal behavior such as gestures and other body movements [3]. To enable effective human-robot interaction, we leverage on the use of motions for an autonomous

---

Junyun Tay is in the Nanyang Technological University-Carnegie Mellon University (NTU-CMU) Dual Degree Ph.D. Programme in Engineering (Robotics).

J. Tay (✉) · I. Chen
Nanyang Technological University Singapore, Singapore 639798, Singapore
e-mail: junyunt@andrew.cmu.edu

I. Chen
e-mail: michen@ntu.edu.sg

J. Tay · M. Veloso
Carnegie Mellon University, 15213 Pittsburgh, USA
e-mail: veloso@cs.cmu.edu

M.A. Hsieh et al. (eds.), *Experimental Robotics*, Springer Tracts in Advanced Robotics 109, DOI 10.1007/978-3-319-23778-7_56

task-performing humanoid robot to complement the traditional form of communication, e.g., text-to-speech. In our previous work, we contribute algorithms to enable humanoid robots to autonomously dance to the beats and emotions of music [14] and animate speech with gestures [12]. Motions are meaningful when they are relevant and initiated at the right times: a switch to a new motion is cued by a change in a task. For example, an autonomous dancing humanoid robot plans for a switch in the dance movement according to the change in emotions of the music [14], and a storytelling robot plans to switch to a pointing motion to refer to a specific object [12] given the cue of the object.

From our previous experiences, we observe that several sequences of motions are generated to express the same meaning but only one sequence of motion can be executed by the robot at a time. However, not all the sequences of motions are stable. We are interested in predicting the stability of a sequence of motions before executing it on the robot. As a motivating example, we consider the task of an autonomous humanoid robot playing a game of charades. In our library of motions, each motion is labeled with meanings expressed by the motion. For example, a motion where the robot waves its arms in the air is labeled with the word "happy." Each label can be mapped to multiple motions, e.g., two motions are labeled with the word "sad." Likewise, each motion can be mapped to multiple labels. There is a sequence of words to play in the game of charades, and for every word, several motions express its meaning. Hence, multiple sequences of motions that are relevant and synchronized to the task are generated. We need to determine the sequences that will cause the robot to fall. In particular, we consider the case where each motion in the library is stable by itself for execution on the robot, but a sequence of motions may result in a fall.

We contribute an algorithm, ProFeaSM, that predicts if a sequence of motions will fall. Our approach is novel as conventional methods to check for stability require an accurate model of the robot. We do not require any model of the robot to perform the prediction. Motion planning methods for stability modify the sequence of motions (e.g., timings and motion trajectories), but these methods are not applicable for our problem as the motions have to be synchronized with the task and reflect the meanings. Existing fall avoidance approaches also cannot be applied as they require training instances of the sequences for their algorithm, whereas ProFeaSM does not require execution of any sequence.

Our algorithm, ProFeaSM, is based on the concept of momentum and inertia. We predict how the body angles change as the sequence of motions is executed. The body angles are computed using the gyroscope and accelerometer sensors, which are typically found in humanoid robots. ProFeaSM uses the body angle values from the executions of single motions and interpolations between pairs of motions to derive the velocity and acceleration of the body angles. ProFeaSM then uses these velocities and accelerations to determine how the body angles will change during a sequence of motions. We design ProFeaSM to include a parameter we can vary to explore the trade-off between precision and recall. With a threshold of the body angle, we can determine if the robot will fall from executing the sequence of motions. We show the efficacy of ProFeaSM using a complex real-time simulator, Webots, and also on a real NAO humanoid robot.

We discuss related work and highlight the differences of this work in Sect. 2. We explain our algorithm in Sect. 3. We describe our experimental setup and analyze the results in Sect. 4. Lastly, we conclude our work in Sect. 5.

2 Related Work

We want to prevent falls for a humanoid robot as it takes time to get up and may cause the robot to break or cause more serious wear and tear. To predict a fall, we can determine the dynamic stability with a model of the robot and its environment and test if a sequence of motions is stable in simulation. However, it is often difficult to have an accurate model to predict stability reliably given that it is difficult to model variables such as friction, slippage and wear and tear.

Conventional motion planners for stability require an accurate model of the robot and its environment and generating dynamically balanced motion for humanoid robots is challenging with the high number of degrees of freedom and "the size of the space to explore is augmented with the robot velocity and footprint positions [2]." Geometric paths can be planned by approximating a dynamic trajectory [2]. However, the drawback of this method is that "some feasible dynamic motions are inherently impossible to compute with this approach [2]." Though there are planners that will compute dynamically stable motion trajectories offline [7, 8], these planners require an accurate robot model and change the desired motion in terms of the timings and the trajectory. Changing the timings of the desired motion causes the motion to be no longer synchronized to the task and changing the motion trajectory changes the meaning expressed.

Falls that can cause the robot to break or serious wear and tear can be predicted and is generally done through online monitoring methods. These online monitoring methods predict falls by thresholding relevant physical quantities (e.g., angular momenta) or determine stability by tracking the position of the center of pressure (CoP) such that the CoP stays within the support polygon. Others analytically model the robot's dynamics to determine if a fall will occur. However, these methods do not scale well to humanoid robots with complex geometries and high degrees of freedom. Since it is difficult to model real world variables such as friction and wear and tear etc., a data-driven approach is used to collect sensor data of stable and unstable trajectories and classified to determine if a fall will occur [5, 6]. The prediction is done during the execution before a fall occurs. For instance, the robot's internal sensors are monitored over time to detect the onset of a robot's failure by using supervised learning techniques and create a classifier to determine failures for a dynamically balancing Segway RMP [11]. Our approach is to use sensor values from previous executions of motions and predict if a sequence is stable offline, without execution or monitoring on the robot. We also do not require a model of the robot or its environment.

Falls can be avoided using reflex motions [4, 9] or to execute a controlled falling motion [5]. However, the desired motion trajectory is changed by executing reflex motions or a controlled falling motion, which also results in the change of the meanings expressed. We want to execute sequences of motions on the robot without falling. Fall avoidance methods are also triggered at a time where the motions are in the midst of execution and only slightly before the fall occurs. Though they can reduce the probability of the robot breaking, these methods cannot predict falls before execution and require training data of instances of the robot falling. Fall prediction methods can also be falsely triggered if the training data is insufficient or not executing controlled falling motions in time to prevent bad consequences. Hence, offline predictions of falls without execution will be better than trying to reduce the impact of a fall. Fall avoidance should only be used as a last resort to avoid significant damage to the robot when a fall occurs.

3 Technical Approach

We formalize the problem of predicting the stability of a sequence of motions in Sect. 3.1 and describe our algorithm in Sect. 3.2. We define a motion as a motion primitive and describe the components of a motion primitive and how a sequence of motion primitives is determined. We also explain how the list of possible combinations of sequences is derived. We use the task of an autonomous humanoid robot playing a game of charades as a motivating example.

3.1 Formalization

Let D denote the number of generalized coordinates or degrees of freedom (DOF) of the robot R and let C be the D-dimensional configuration space of R. A keyframe $k \in C$ is a vector of D real numbers specifying values for each of the joint angles for the respective degrees of freedom (joint) of R. A keyframe $k \in C$ is valid if it is collision-free and the joint angles stay within joint limits.

A motion primitive contains a series of keyframes (static poses) and durations [12]. Each duration is the time to interpolate (move) between two keyframes. A sequence of motion primitives starts with an initial pose and contains a list of motion primitives that expresses the meanings required of the task.

Definition 1 A **motion primitive** P is a tuple of N primitives and is defined as $P = (M_1, \ldots, M_N)$ and $N \in \mathbb{Z}^+$. The primitive M_n is a tuple of 2 keyframes, k_{n-1} and k_n, and the time to interpolate between these two keyframes, $t_{n-1,n}$, where $M_n = (k_{n-1}, t_{n-1,n}, k_n)$. k_0, the first keyframe in M_1 is the initial pose of the robot, R, which contains all the joint angles for D degrees of freedom. Let P be the set of all motion primitives in the motion primitive library.

To interpolate between two keyframes, the interpolation method, e.g., linear interpolation or bezier interpolation, is defined. We assume the trajectories to interpolate between the keyframe are generated by a motion planner and will fulfill the following conditions:

1. be collision-free
2. be within physical capabilities (joint angular and velocity limits)

The time to interpolate between two keyframes, k_n and k_{n+1}, is computed by the interpolation time computation function $T : C \times C \rightarrow \mathbb{R}^+$. $t_{n,n+1}$ specifies the minimum duration required to interpolate from the angles of the respective joints in k_n to the angles defined in k_{n+1} or is pre-defined by the motion choreographer. If the minimum duration required to interpolate from one keyframe to another is longer than the time pre-defined by the motion choreographer, the minimum duration will be used for $t_{n,n+1}$. The minimum duration depends on the interpolation method used and is determined using the maximum joint angular velocities. Therefore, to compute $t_{n,n+1}$, we use $t_{n,n+1} = T(k_n, k_{n+1})$.

Definition 2 A **sequence of motion primitives** $\mathbb{S}$ consists of L sequence primitives, where $\mathbb{S} = (\mathbb{P}_1, \ldots, \mathbb{P}_L)$, $L \in \mathbb{Z}^+$. Let S be the set of all possible sequences of motion primitives. $\mathbb{P}_l$ is a tuple containing 2 motion primitives, P_{l-1} and P_l; $t_{l-1,l}$ denotes the time to interpolate between P_{l-1} and P_l (time to interpolate between the last keyframe of P_{l-1} and the first keyframe of P_l).

Similarly, $t_{l-1,l}$, the minimum time to interpolate between P_{l-1} and P_l, is computed by the interpolation time computation function T. The joint angles of the initial pose of the robot R is defined by the first keyframe k_0 of the first primitive M_1, in the first motion primitive P_0, in the sequence primitive $\mathbb{P}_1$, in the **sequence of motion primitives** $\mathbb{S}$.

We aim to predict if a sequence of motion primitives $\mathbb{S}$ will cause the robot to fall without the robot executing the sequence, and without having a model of the robot and the environment. Every sequence in the list of possible combinations of sequences starts with the same initial pose.

3.2 Algorithm

Before the prediction of whether a sequence of motion primitives will fall, we record the body angles of the robot whilst executing each motion primitive in the library and the interpolations between pairs of motion primitives. We explain how our algorithm works using the body angles recorded to determine the stability of a sequence of motion primitives.

Body angles are recorded at a regular frequency f. For example, we execute a motion primitive and the b body angles are $(t_0, s_0), \ldots, (t_b, s_b)$, where t_i is the timestamp, s_i consists of the body angles at time t_i, and s_0 are the body angles of the initial pose. s_i comprises of the body angle X and Y readings per time step. b is calculated using the duration of the execution, d, using the equation $b = (d \times f) + 1$ as we also record the body angles for one time step before the motion primitive is executed.

We collect the body angles of three different groups of executions:

1. single: Each motion primitive is executed individually with the initial pose of the start of every sequence as the first keyframe, followed by the keyframes in the motion primitive. We denote single$_{P_i}$ as the body angles collected for the motion primitive P_i.
2. startSingle: Each motion primitive is executed individually with its first keyframe (typically different from the initial pose of Fig. 1a). We denote startSingle$_{P_i}$ as the body angles collected for the motion primitive P_i.
3. interpolation: We find all possible pairs of motion primitives, P_i and P_j, where $i \neq j$ and execute the interpolation between each pair of motion primitive. The interpolation between two motion primitives, P_i and P_j, is done with the execution of the last keyframe K_n of the first motion primitive P_i, and the first keyframe K_1 of the second motion primitive P_j. We denote interpolation$_{P_i,P_j}$ as the body angles collected for the interpolation between the two motion primitives, P_i and P_j.

We collect m iterations of each of the three groups of executions. With these three groups of body angles, we use our algorithm ProFeaSM to predict if a particular sequence of motion primitives, $\mathbb{S}$, will cause the robot to fall. With m iterations of each of the three groups and $|\mathsf{P}|$ number of motion primitives, we collect a total of $m \times (|\mathsf{P}| + |\mathsf{P}| + |\mathsf{P}|(|\mathsf{P}| - 1)) = m \times |\mathsf{P}|(|\mathsf{P}| + 1)$ executions.

Our algorithm ProFeaSM is made up of four algorithms, namely Process, Feasibility, Stitch and Multiplier (Algorithms 2–5). We process the body angles of the three groups of executions if we collect more than 1 iteration of the three groups of executions using Algorithm 2.

Algorithm 1 ProFeaSM: Process-Feasibility-Stitch-Multiplier

```
ProFeaSM(𝕊, inertialMultiplier, single, startSingle, interpolation)
(m, n) ← size(single)
if m = 1 then
    hasFallen ← Feasibility(𝕊, inertialMultiplier, single, startSingle, interpolation)
else
    single ← Process(single)
    startSingle ← Process(startSingle)
    interpolation ← Process(interpolation)
    hasFallen ← Feasibility(𝕊, inertialMultiplier)
end if
return hasFallen
```

Next, we process the m iterations collected of each of the three different groups of executions by determining the median for body angle X, $\mathcal{M}_{\mathrm{bax}}$, and body angle Y, $\mathcal{M}_{\mathrm{bay}}$. Process (Algorithm 2) returns the median of a trajectory of body angles (X or

Y), given ba, a list of m body angle trajectories. For example, $\mathcal{M}_{\text{bax}} = \text{Process(bax)}$, where bax contains the m trajectories of body angle X. We use these median body angles to predict if a sequence of motion primitives, $\mathbb{S}$, will fall using Algorithm 3, Feasibility. Algorithm 3 requires two parameters, $\mathbb{S}$ and inertialMultiplier.

Algorithm 2 Process m iterations of b time steps

Process(ba)

{ba is a $m \times b$ matrix, containing m iterations with b time steps}
for $i = 1$ **to** b **do**
  medianAtEachStep(i) = $\text{median}_{j=1}^{m}(\text{ba}(j, i))$ {finds median at time step i}
end for
medianBA = $\text{argmin}_{j=1}^{m}(\sum_{i=1}^{b} |\text{ba}(j, i) - \text{medianAtEachStep}(i)|)$
return medianBA

Definition 3 The algorithm Feasibility : $\mathsf{S} \times \mathbb{R} \rightarrow \{0, 1\}$ computes the **feasibility** of a sequence of motion primitives, where a sequence of motion primitives is **feasible** if and only if the robot is able to execute the keyframes whilst being **stable**. Hence, Feasibility($\mathbb{S}$, inertialMultiplier) = 1 when $\mathbb{S}$ is feasible.

Algorithm 3 Predict whether a sequence of motion primitives is feasible

Feasibility($\mathbb{S}$, inertialMultiplier, single, startSingle, interpolation)

{Indices start from 1}
data $\leftarrow$ Stitch($\mathbb{S}$, single, startSingle, interpolation)
$v \leftarrow (0, \text{data}(2) - \text{data}(1), \text{data}(3) - \text{data}(2), \ldots)$
$a \leftarrow (0, v(2) - v(1), v(3) - v(2), \ldots)$
predictTraj $\leftarrow$ single_{P_1}
stepMultiplier $\leftarrow$ 0 {initialized as 0 as $e^0 = 1$}
for $i = 1$ **to** $|\text{single}_{P_1}|$ **do**
  stepMultiplier $\leftarrow$ Multiplier($a(i)$, stepMultiplier, inertialMultiplier)
end for
hasFallen $\leftarrow$ false
for $i = |\text{single}_{P_1}| + 1$ **to** $|\text{data}|$ **do**
  predictAngle = $v(i) \times \exp(\text{stepMultiplier}) + \text{predictTraj}(i - 1)$
  stepMultiplier $\leftarrow$ Multiplier($a(i)$, stepMultiplier, inertialMultiplier)
  predictTraj $\leftarrow$ append(predictTraj, predictAngle)
  if |predictAngle| > fallenThresh **then**
    hasFallen $\leftarrow$ true
  end if
end for
return hasFallen

Algorithm 3 is based on the concept that as the acceleration of the body angles increases, the velocity increases and vice versa. As the acceleration of the body angles approaches zero, the velocity of the body angles reaches a constant. The body angle indicates the angle of the robot's torso with respect to the ground. Therefore, the higher the body angle, the higher the possibility that the robot is going to fall. To determine the velocity, v, we determine the change in body angles at each time step. To determine the acceleration, a, we calculate the change in velocity at each time step. Since the velocity will not increase linearly due to the effect of gravity, inertia and momentum, we model the velocity as an exponential curve in Algorithm 3. The acceleration will affect how far we are along the exponential velocity curve, hence changing the x-value of the exponential curve, which we term as stepMultiplier.

Algorithm 4 (the Stitch function called in Line 2 of Algorithm 3) explains how we stitch up the body angle values collected. We always begin with the original body angle trajectory of the first motion primitive in the sequence, single_{P_1}. Next, we determine the change in body angle and add the change to the last known body angle for the rest of the sequence using the change in the body angles of interpolation and startSingle.

Algorithm 4 Stitch collected data into a trajectory

Stitch($\mathbb{S}$, single, startSingle, interpolation)

data $\leftarrow \text{single}_{P_1}$
lastAngle $\leftarrow \text{single}_{P_1}(|\text{single}_{P_1}|)$
for $l = 2$ **to** L **do**
    for $i = 2$ **to** $|\text{interpolation}_{P_{l-1},P_l}|$ **do**
        lastAngle $\leftarrow$ lastAngle $+ (\text{interpolation}_{P_{l-1},P_l}(i) - \text{interpolation}_{P_{l-1},P_l}(i-1))$
        data $\leftarrow$ append(data, lastAngle)
    end for
    for $i = 2$ **to** $|\text{startSingle}_{P_l}|$ **do**
        lastAngle $\leftarrow$ lastAngle $+ (\text{startSingle}_{P_l}(i) - \text{startSingle}_{P_l}(i-1))$
        data $\leftarrow$ append(data, lastAngle)
    end for
end for
return data

Lines 3 and 4 in Algorithm 3 shows how we determine v, the velocity of the body angles and a, the acceleration. To predict the body angle trajectory, predictTraj, during the execution of the sequence of motion primitives, we begin with the body angles collected from single_{P_1} since the body angles should be similar to executing the motion primitive single_{P_1} that starts from the initial pose. However, we still have to determine the stepMultiplier using Algorithm 5 since the inertial and momentum are changing in Lines 7–9 (Algorithm 3). As we predict the body angle trajectory, predictTraj, we predict the velocity using the exponential velocity curve and stepMultiplier so as to determine the change to the previous body angle in Line 12 (Algorithm 3). Next, we continue to change stepMultiplier in Line 13 (Algorithm 3) and append the predicted body angle, predictAngle, to predictTraj. We check if predictAngle exceeds fallenThresh, a threshold to determine if the robot has fallen in Lines 15–17 (Algorithm 3).

Algorithm 5 is used to determine how stepMultiplier varies along the exponential velocity curve. As the computed acceleration of the body angles per time step is very small, inertialMultiplier is used as a multiplier to the acceleration, and varies how stepMultiplier changes in Line 4 of Algorithm 5. accThres is used as a threshold to determine if the acceleration approaches zero and if so, stepMultiplierDec is used to decrease stepMultiplier in Lines 1–2 of Algorithm 5.

Algorithm 5 Determine the step multiplier based on the acceleration

```
Multiplier(a, stepMultiplier, inertialMultiplier)
1: if |a| < accThres then
2:    stepMultiplier ← stepMultiplier − stepMultiplierDec
3: else
4:    stepMultiplier ← stepMultiplier + (a × inertialMultiplier)
5: end if
6: if stepMultiplier < 0 then
7:    stepMultiplier ← 0 {stepMultiplier will not go below 0}
8: end if
9: return stepMultiplier
```

To summarize, ProFeaSM comprises of Algorithm 2–5. With m iterations of body angles recorded, we use Algorithm 2 to determine the median body angle trajectories. Next, we use Algorithm 3 to predict the stability of a sequence of motion primitives. Algorithm 3 uses Algorithm 4 to stitch up the body angles collected using the body angles collected from the three groups of executions and their respective velocities of the body angles. Algorithm 3 also uses Algorithm 5 to determine the multiplier, stepMultiplier, in the exponential velocity curve.

4 Experiments

We describe the experiments carried out in simulation and on a real NAO humanoid robot in Sect. 4.1 and discuss our results in Sect. 4.2.

4.1 Experimental Setup

Experiments are conducted in simulation using Webots 7 [13] and on a real NAO humanoid robot. Webots 7 [13] is a real-time simulator that simulates the dynamics of the NAO humanoid robot's whole body executing a sequence of motion primitives. For the experiment in simulation, we simulate a NAO V4.0 H25 humanoid robot and conduct $m = 10$ iterations to collect body angle values for three groups of executions: single, startSingle, interpolation. Lastly, we test if each sequence of motion primitives is stable. We also ran 10 iterations to ensure that each motion primitive in the library is stable. Webots is restarted each time an iteration is ran to ensure that the NAO humanoid robot begins with the same initial pose and position in the environment.

To test if our algorithm, ProFeaSM, can be applied to a real NAO, we use a NAO V3.3 H21 humanoid robot, with a V4.0 head. We use a different model of the NAO from the simulation to enable us to test if ProFeaSM can be applied to different robots. We conduct $m = 1$ iteration to collect the three groups of executions as it is impractical to collect many iterations in real life, so we wanted to evaluate ProFeaSM when $m = 1$. Though the single motion primitives in the motion primitive are stable, we do not guarantee that the interpolations between the pairs will be stable. If we execute multiple iterations of interpolations that fall, it may break the robot. Hence, we allow the robot to fall gently by tying a thread around the robot's torso to prevent the robot from hitting the ground too hard. We also test the stability of every sequence on the real NAO. We stop the execution of the rest of the sequence when the robot has fallen. The collection of the body angles for the three groups of executions is unaffected since only interpolations may fall and the interpolations are short as compared to the entire sequence. The NAO robot's software includes a fall manager [1], that detects a potential fall during execution and prevents itself from breaking by putting its arms before its face before touching the ground. However, we observe that the fall manager is often triggered prematurely though the motion primitive is stable. Hence, we disable the fall manager during the collection of the body angles and the execution of the sequences.

Each sequence of the motion primitive starts with the same initial pose as shown in Fig. 1a. The body (torso) angles are recorded at a frequency of 100 Hz (every 10 ms) using a function provided by the NAO's software [1] and computed using the accelorometer and gyrometer sensors readings from the inertial measurement unit (IMU) [1]. The body angles recorded are body angle X (roll) and Y (pitch) as shown in Fig. 1b.

For our experiments, fallenThresh in Algorithm 3 is 1.0 based on the body angle we observe when the robot is lying on the ground. accThres in Algorithm 5 is set to 0.005 and stepMultiplierDec to 0.001 respectively as the 0.005 is close to 0, and

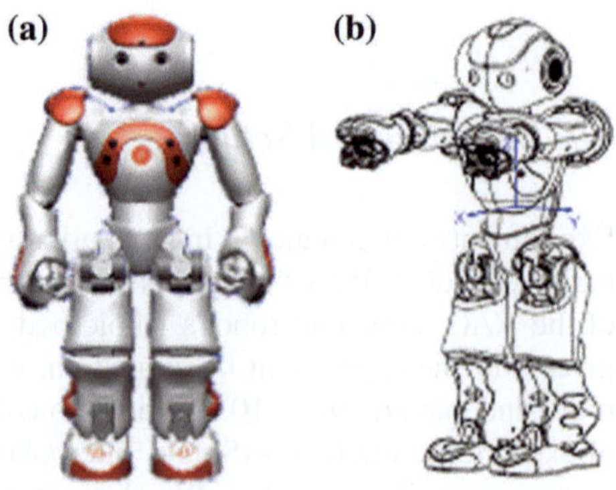

Fig. 1 NAO's initial pose and IMU's coordinate frame. **a** Initial pose of the NAO. **b** Coordinate frame of IMU [1]

0.001 only changed the multiplier slightly. With 1 iteration of body angles collected for the three groups of executions for the real NAO experiment, we skip Algorithm 4 to determine the median body angle trajectory and use the body angles recorded to predict the fall of the sequences. We vary different values of inertialMultiplier in Algorithm 5 from 10 to 100.

The task of the NAO humanoid robot is to play a game of charades to guess emotions. There are three different emotions: angry, sad and surprised. The robot can choose to act out the emotions in any order. In our library of motion primitives, for every emotion, there are two motion primitives that are labeled with the particular emotion. The robot is able to successfully execute each individual motion primitive without any falls. The number of combinations of sequences for three different motions (angry, sad, surprised in any order) is $2 \times 2 \times 2 \times 3! = 48$. For the three groups of executions: single, startSingle, interpolation, we collect a total of $|\text{single}| + |\text{startSingle}| + |\text{interpolation}| = 6 + 6 + (6 \times 4) = 36$ body angle trajectories. We do not need to collect all the $6 \times 5 = 30$ body angle trajectories for interpolation as there are two motion primitives per emotion and we do not use two motion primitives labeled with the same emotion consecutively.

To compute the interpolation time between keyframes, we use the interpolation time computation function T that uses the maximum joint angular velocity. We limit the maximum joint angular velocity in the simulation to be 70 % of the real maximum joint angular velocity in simulation and 40 % of the real maximum joint angular velocity on the real NAO so as to ensure that each motion primitive in the library is stable.

4.2 Experimental Results

Table 1 shows two sequences of motion primitives: (a) Sad2, Angry2, Surprised1, (b) Surprised1, Sad2, Angry2. The first row shows the intended sequence of the motion primitives (shown in bold) and the interpolations between motion primitives. "Start-" indicates the interpolation from the initial pose of the robot to the first motion primitive.

Table 1 Intended and actual execution of two sequences of motion primitives

Intended	Start-Sad2	**Sad2**	Sad2-Angry2	**Angry2**	Angry2-Surprised1	**Surprised1**
Actual	Start-Sad2	**Sad2**	Sad2-Angry2	**Angry2**	Angry2-Surprised1	**Surprised1**
Intended	Start-Surprised1	**Surprised1**	Surprised1-Sad2	**Sad2**	Sad2-Angry2	**Angry2**
Actual	Start-Surprised1	**Surprised1**	Surprised1-Sad2	***Fallen***	***Fallen***	***Fallen***

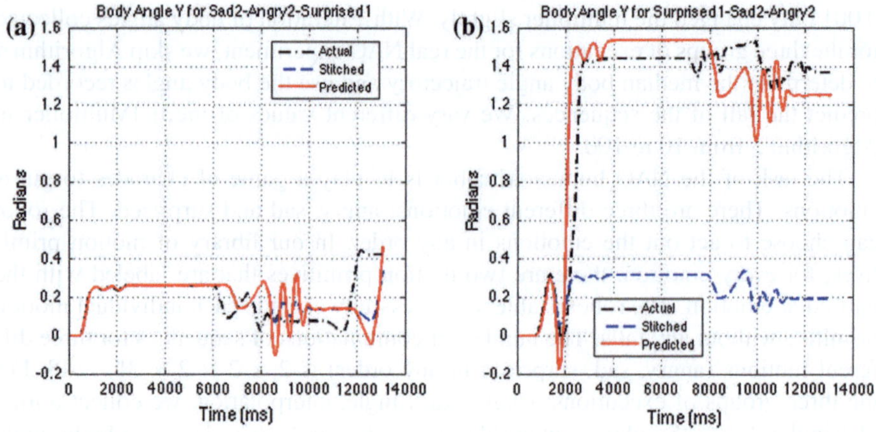

Fig. 2 Body angle Y values for two sequences of motion primitives. **a** Sad2-Angry2-Surprised1. **b** Surprised1-Sad2-Angry2

We demonstrate that even though each motion primitive in our motion primitive library is stable, a sequence of individual stable motion primitives does not guarantee the robot's stability in Table 1 as the sequence of Surprised1, Sad2 and Angry2 results in a fall. Though the sequence of Surprised1, Sad2 and Angry2 is unstable, a different ordering of the motion primitives, Sad2, Angry2 and Surprised1 is stable. From Table 1, we may deduce that the sequence from Sad2 and the interpolation between Sad2 and Angry2, and Angry2 causes the fall, but the sequence from Sad2 to Angry2 in the sequence of Sad2, Angry2 and Surprised1 is stable. Hence, we cannot predict the fall of the robot based solely on part of the sequence, but we have to consider the entire sequence.

In our experiments, we observed that body angle Y values is sufficient for predicting if the robot will fall, as the robot only falls forward or backward and never sideways. Hence, we present results regarding body angle Y values since we only use body angle Y values to predict if a sequence will fall.

Figure 2 shows two plots: Fig. 2a shows the body angle Y values of the sequence, Sad2, Angry2 and Surprised1 over time, and Fig. 2b shows the body angle Y values of the sequence, Surprised1, Sad2 and Angry2 over time. Figure 2a shows a sequence of motion primitives that does not fall whereas the right plot, Fig. 2b, shows a sequence of motion primitives, Surprised1, Sad2 and Angry2 that causes the robot to fall. Both sequences are executed in simulation and the prediction of the body angle trajectories are made from the body angles collected in simulation. For both figures in Fig. 2, we plot three body angle Y trajectories: First, we plot the body angle Y trajectory that was collected during the execution of a sequence in black with a line style of $- \cdot -$ (Actual). Next, we plot a stitched body angle Y trajectory in blue with a line style of $--$ (Stitched) using only Algorithm 4. Lastly, we plot the predicted body angle Y values using Algorithm ProFeaSM in red with a line style of — (Predicted).

In Fig. 2a, the actual, stitched, and predicted body angle Y trajectories are similar. However in Fig. 2b, we show that the predicted body angle Y is similar to the actual body angle Y trajectory, while the stitched body angle Y trajectory is not. Thus, we cannot simply stitch up body angles collected. We demonstrate that our algorithm works well in predicting the body angle trajectory given that the curvature of the trajectory is similar.

After testing our algorithm, ProFeaSM, on the body angles collected from the single motion primitives and the interpolations between pairs of motion primitives in simulation, we vary the parameter of ProFeaSM, inertialMultiplier, to analyze if the accuracy of our fall prediction improves. We refer to the use of precision and recall for classification tasks [10]. Similar to this task, whereby we want to classify sequences that are unstable as falls. Precision is the number of true positives (sequences that we label as falls and will actually fall during the execution) divided by the sum of true positives and false positives (sequences that we label as falls but will not fall during the execution). A perfect precision score of 1.0 means that every sequence that ProFeaSM labeled as a fall actually did fall during the execution of the sequence. Recall is the number of true positives divided by the total number of sequences that actually fall during the execution. A perfect recall score of 1.0 means that every sequence that actually fell during the execution was labeled as a fall by ProFeaSM, but does not consider sequences that are wrongly labeled as falls. Precision and recall often has an inverse relationship whereby increasing one decreases another. We aim to have as high a recall and precision as possible, but it is often difficult to achieve both at a perfect score.

Figure 2 is plotted using ProFeaSM and inertialMultiplier $= 90$. After testing ProFeaSM in simulation, we collect the body angles from the executions of single motion primitives and the interpolations between pairs of motion primtives on the real NAO humanoid robot. We note that the interpolation time is about twice as long on the real robot than in simulation since we use 40 % of the maximum joint angular velocity for the real NAO robot instead of 70 % of the maximum joint angular velocity in simulation. Even though the interpolation times are different, we show the efficacy of ProFeaSM by varying inertialMultiplier with the recall and precision values as shown in Fig. 3.

Figure 3 shows two curves, one for simulation and one for the real robot. The two curves are generated by varying different inertialMultiplier values, from 60 to 100, using different sets of body angle Y values—the simulated data and the actual robot data. We do not plot values of 10–50 as there were no sequences that were predicted as falls in simulation. Each curve is marked by the value of inertialMultiplier. The blue line for Simulation shows the precision and recall rate for using the body angle Y values collected from the execution of motion primitives in simulation and by varying the different inertialMultiplier, we get a different precision and recall score. From the Simulation results, 90 is a value to be used for the inertialMultiplier if we want to ensure that all sequences that will fall will be predicted as falls (a perfect recall value of 1.0), but we have a low precision of 0.72, which means that we have predicted some false positives (sequences that we predict as falls did not fall). If we use the same value of 90 for the actual robot prediction, we will also achieve a perfect

Fig. 3 Precision recall curve

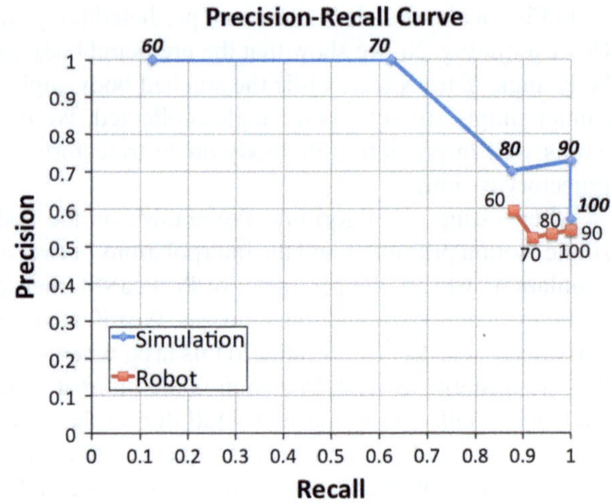

recall value of 1.0, but the precision value is lower at 0.54. This means that we have quite a high number of false positives, which may not be desirable since we have less choices of sequences to execute. Hence, there is a trade-off between precision and recall depending on the requirements. We can require a high precision, where there are as few false positives as possible so that we can have more sequences to choose from. We can also require high recall instead of precision, where there are as many true positives as possible so that we can avoid sequences that fall and can tolerate having less sequences to choose from.

Our algorithm, ProFeaSM, scales quadratically with the number of motion primitives in the motion primitive library. We can reduce the number of times the body angle Y values are recorded for ProFeaSM if the interpolations between pairs of motion primtives are not unique, e.g., the last keyframe of the motion primitive P_1 and the first keyframe of the next motion primitive P_2, are the same two keyframes for the last keyframe of the motion primitive P_3 and the first keyframe of the next motion primitive P_4.

We arrive at our current approach based on previous failed attempts at predicting the stability of a sequence using algorithms like Hidden Markov Model and Reinforcement Learning. There are several problems that we encounted with these algorithms. We require the executions of many iterations of different sequences to learn the transition probabilities, which does not fulfill our need as we want to predict different sequences without executing them. Moreover, both algorithms possess the Markov property—the next state depends only upon the present state and not on the sequence of events that preceded it. If the Markov property is not violated, we will require a large state space as we create new states based on the ordering of the motion primitives due to the momentum caused by the ordering of the motion primitives in the sequence. Moreover, the transition probability of a fall from each state is low, thus the probability of predicting a fall in a sequence is low too.

5 Conclusion

We contribute an algorithm, ProFeaSM, to predict if a sequence of motions will fall when the robot executes the sequence. We use only body angle Y values collected from the executions of single motions and the interpolations between pairs of single motions since the robot only falls forward or backwards and not sideways. Body angles are computed using accelerometer and gyroscope sensors and these sensors are commonly found in humanoid robots. Moreover, we do not require training instances of body angle Y values collected from sequences of motions to make predictions of the sequences, unlike traditional fall prediction methods that require training data and can only predict possible falls whilst monitoring the execution of the sequence. We make predictions before any sequence of motions is executed on the robot. We also require no model of the robot and the environment to make a prediction.

ProFeaSM includes the parameter, inertialMultiplier, that is varied to achieve different precision and recall values. Firstly, we collect body angles in simulation and test the efficacy of ProFeaSM. We showed that ProFeaSM can achieve a perfect recall value of 1 and a precision value of 0.72 when the value, inertialMultiplier is set to 90 in simulation. For the prediction of sequences that fall for the real robot, we collect body angles for the execution of all the single motion primitives in the motion primitive library of 6 motion primitives and the 24 interpolations between pairs of the motion primitives. We show that by using the same value of inertialMultiplier =90 using our algorithm, we can achieve the same perfect recall score and predict all the sequences that fall, albeit at a poorer precision value of 0.54. By varying different values of inertialMultiplier, we can achieve different precision and recall values. We need to weigh the trade-off of having a higher recall value versus a higher precision value.

We ran ProFeaSM in simulation and on the real robot. The simulated robot is a NAO V4.0 H25 humanoid robot with 25 joint actuators, and for the real robot, we use a NAO V3.3 H21 humanoid robot with a V4.0 head that has only 21 joint actuators. Despite the differences in the number of joint actuators, the interpolation time between keyframes and the weight of the robot in simulation and of the real robot, we are still able to achieve the same recall value using our algorithm and predict all unstable sequences without executing the sequences.

Acknowledgments This work was supported by the Singapore Millennium Foundation Research Grant and by award NSF IIS-1218932 of the National Science Foundation. The NTU-CMU Dual Degree PhD Programme in Engineering (Robotics) is funded by the Economic Development Board of Singapore. The views and conclusions contained herein are those of the authors only.

References

1. Aldebaran Robotics: https://community.aldebaran-robotics.com/doc/1-14/NAO. Software 1.14.5 documentation
2. Dalibard, S., El Khoury, A., Lamiraux, F., Nakhaei, A., Tax, M., Laumond, J.P.: Dynamic walking and whole-body motion planning for humanoid robots: an integrated approach. Int. J. Robot. Res. **32**(9–10), 1089–1103 (2013)
3. Engleberg, I., Wynn, D.: Working in Groups: Communication Principles and Strategies. Pearson Education, New York (2006)
4. Höhn, O., Ganik, J., Gerth, W.: Detection and classification of posture instabilities of bipedal robots. In: Tokhi, M., Virk, G., Hossain, M. (eds.) Climbing and Walking Robots, pp. 409–416. Springer, Berlin (2006)
5. Höhn, O., Gerth, W.: Probabilistic balance monitoring for bipedal robots. Int. J. Robot. Res. **28**(2), 245–256 (2009)
6. Kalyanakrishnan, S., Goswami, A.: Learning to predict humanoid fall. Int. J. Humanoid Rob. **08**(02), 245–273 (2011)
7. Kanehiro, F., Suleiman, W., Lamiraux, F., Yoshida, E., Laumond, J.P.: Integrating dynamics into motion planning for humanoid robots. In: IEEE/RSJ International Conference on Intelligent Robots and Systems, IROS 2008, pp. 660–667 (2008)
8. Kuffner, J., Nishiwaki, K., Kagami, S., Inaba, M., Inoue, H.: Motion planning for humanoid robots. In: Proceedings 11th International Symposium of Robotics Research (ISRR) (2003)
9. Renner, R., Behnke, S.: Instability detection and fall avoidance for a humanoid using attitude sensors and reflexes. In: IROS, pp. 2967–2973 (2006)
10. Russell, S., Norvig, P.: AI: A Modern Approach. Prentice Hall, New York (2003)
11. Searock, J., Browning, B., Veloso, M.: Learning to prevent failure states for a dynamically balancing robot. Technical Report CMU-CS-TR-05-126, School of Computer Science, Carnegie Mellon University (2005)
12. Tay, J., Veloso, M.: Modeling and composing gestures for human-robot interaction. In: International Symposium on Robots and Human Interactive Communication (RO-MAN), pp. 107–112 (2012)
13. Webots: http://www.cyberbotics.com. Commercial Robot Simulation Software
14. Xia, G., Tay, J., Dannenberg, R., Veloso, M.: Autonomous robot dancing driven by beats and emotions of music. In: International Conference on Autonomous Agents and Multiagent Systems (AAMAS), Vol. 1, pp. 205–212 (2012)

Towards Collaborative Mapping and Exploration Using Multiple Micro Aerial Robots

Sikang Liu, Kartik Mohta, Shaojie Shen and Vijay Kumar

Abstract In this paper, we present a system for collaborative mapping and exploration with multiple quad rotor robots. The basic architecture and development of the algorithms for mapping and exploration validate our system with both simulation and real-world experiments. We utilize the 2.5-D structure of typical indoor environments and demonstrate the deployment of multiple autonomous quadrotors equipped with IMUs and laser scanners engaged in collaborative exploration. Estimation, control and planing algorithms are highly integrated in our system to achieve robust and efficient exploration behaviors.

Keywords Multi-robot · Mapping · Exploration · SLAM · Quadrotor

Robotics mapping is an attractive area of research with a lot of potentials for development. In particular, systems with micro aerial vehicles (MAVs) that are capable of autonomously mapping and exploring have significant impacts on the field of search and rescue [1]. Unlike to ground robots, aerial robots have superior mobility and they are not affected by complex terrains. These advantages make micro aerial robots as the ideal platform for search and rescue missions.

However, it is challenging to develop and deploy autonomous aerial robots for many reasons. First, because of the fast dynamics and limited payload for sensors, it is difficult to develop state estimators that can be used for real-time control. Second, limitations on energy and power density restrict the duration of the missions. Third, the infrastructure for running experiments with multiple autonomous aerial robots requires the tight integrations between communication, control and perception. Fortunately, it is possible to leverage the extensive literature on mapping and exploration with ground robots [2, 3], and the more recent literature on micro aerial vehicles [4] (IJRR reference).

Simultaneous Localization and Mapping (SLAM) algorithms address the problem of localizing a mobile robot while incrementally building a consistent map in an unknown environment [5]. It mainly consists of a front end and back end. In the front end, feature detection and scan matching are implemented to create constraints

S. Liu (✉) · K. Mohta · S. Shen · V. Kumar
GRASP Laboratory, University of Pennsylvania, Philadelphia, PA 19104, USA
e-mail: sikang@seas.upenn.edu

M.A. Hsieh et al. (eds.), *Experimental Robotics*, Springer Tracts in Advanced Robotics 109, DOI 10.1007/978-3-319-23778-7_57

on incremental movements to estimate the robot position and orientation. In the back end, an optimization algorithm is used to recover the trajectory and the map over a set of robot actions and measurements. Multiple approaches have been used to solve SLAM problem. In particular, Smoothing and Mapping (SAM) is one method that can recover the history of poses (trajectory) and the map and is suited to real-time applications [6]. GTSAM is a C++ package including multiple SAM built-in functions. In this paper, we use this package as the back end to solve the SLAM problem.

The quadrotor is an ideal aerial robot platform for control and navigation. By carrying sensors like a laser and an IMU with an onboard computer, the quadrotor can be autonomously navigate indoor environments [7]. Graph-based SLAM algorithms have proved to be quite effective [8]. They are reliable and efficient in 2.5-D environments and relatively easy to use. Graph-based SLAM algorithms used with a laser-scanner and an IMU yields pose estimation. These estimates along with an Unscented Kalman Filter (UKF) [9] can be fused to get state estimation which in turn can be used for control. Traditional back-stepping PID controllers [10] or nonlinear controllers [11] can be used for control. We build on all these results to develop a robust architecture for autonomous flight for our quad rotors shown in Fig. 1.

The exploration problem involves the robot determining actions for navigation that allow efficient mapping of an unknown environment. Single robot exploration has been done in a GPS-denied indoor environment by multiple approaches [12, 13]. When it comes to multiple robots, the complexity of system and algorithms increase dramatically, and it's hard to implement in the real world especially for aerial vehicles. Some experiments related in collaborative mapping and exploration using multiple vehicles have been conducted in [14, 15], but none of them has done multi-robot mapping or exploration on system of multiple MAVs.

We mainly focus on constructing a reliable system for multiple MAVs mapping and exploring in an constraint indoor environment. We use the OmniMapper [16] and GTSAM [17] developed at Georgia Tech University for the back end and create our own front end. The main contributions of this paper however are the integration of these algorithms to solve the multi-quadrotor SLAM problem efficiently and the development of an exploration algorithm that allows autonomous mapping of unknown environments. We report simulation results with 3 quadrotors and experimental results with 2 quadrotors demonstrating the efficacy of the proposed approach.

1 Technical Approach

The control of a small team of robots can be done either in a centralized or a decentralized manner. Distributed data fusion from multi-robot system has been demonstrated in [18]. However, even for small teams, it is necessary to consider two constraints. First, the onboard computational resources can be limited, as it is in our case (an Atom 1.6 GHz processor). Second, it is not possible to transmit large amounts of information without latency over a wireless network. Thus, we use a centralized

Fig. 1 Our multi-quadrotor system. The base platform is an Ascending Technologies Pelican quadrotor equipped with a Hokuyo UTM-30LX laser scanner

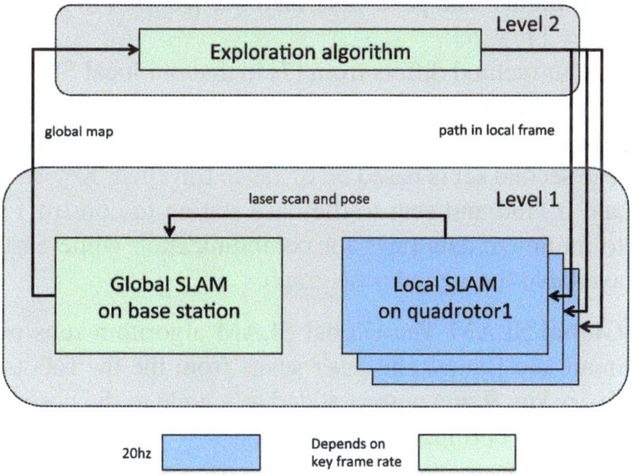

Fig. 2 System hierarchy for multi-robot mapping and exploration

architecture in which a one base station (laptop) is connected to multiple clients (quadrotors). In this case, inter-process communication is established between robot and base station, but not between the robots themselves.

The software hierarchy (Fig. 2) has two levels. The first level is called collaborative SLAM, in which we mainly focus on solving the SLAM problem from information coming from all the robots. The second higher level is the planner for multi-robot exploration. We will present details of each level in the following sections.

1.1 Collaborative SLAM in Multi-robot Scale

Collaborative SLAM includes SLAM algorithm running on all the devices. Each robot, or the client, runs independent onboard SLAM algorithm in which a occupancy grid map in robot frame will be created. This algorithm is a local SLAM solution without any loop-closure detection or correction. This is key to localization and generating state estimates for control.

Local SLAM is used to localize the robot and generate key frames. Key frames with corresponding laser scans are transmitted to the base station every 0.2 m or 0.1 rad.

The Global SLAM algorithm runs on the base station and merges the information obtained from all the robots. This involves the creation of a pose graph, detection of loop closures and the optimization of the pose graph.

Local SLAM The local SLAM algorithms assumes a 2.5-D indoor environment. We use the standard map correlation-based laser scan matching for local SLAM without considering loop closure. The details of this approach and implementation are described in [7]. Denote the local map of robot k at time i as ${}^{l}M_i^k$. When the pose estimator receives a new laser scan, it outputs the predicted position ${}^{l}x_{i+1}^k$ in local map frame. The pose state includes position in x, y direction and yaw, denoted as ${}^{l}x^k = \{x, y, \theta\}$.

Our method differs from [7] in that our local SLAM algorithm generates two sets of key frames. The first set is based on the local time. Every 0.25 s, a key frame is established and the laser scan is used to build an occupancy grid map for localization. The second set is based on distance travelled. Key frames are generated every 0.2 m and 0.1 rad and sent to the base station to construct a global map. This allows us to minimize data rates for communication while ensuring adequate information is available for a good pose graph.

Global SLAM The Global SLAM algorithm runs on the base station. It takes as input key frames and laser scans from the the robots and generates a global map. Every key frame pose is added as a node to the pose graph, and the SAM algorithm is used to optimize the pose graph.

The construction of a global pose graph requires that constraints between independent Markov chains generated by individual robots are resolved correctly in [14]. The position of robot k at time i in global frame is denoted by node ${}^{g}x_i^k$ (Fig. 3). The edge between two adjacent nodes $\{{}^{g}x_{i-1}^k, {}^{g}x_i^k\}$ stands for corresponding odometry constraint, denoted as e_{i-1}^k, it's estimated from local SLAM. Knowing pose and odometry constraint at time $i-1$, we create the next pose ${}^{g}x_i^k$ as

$$ {}^{g}x_i^k = {}^{g}x_{i-1}^k + e_{i-1}^k \tag{1} $$

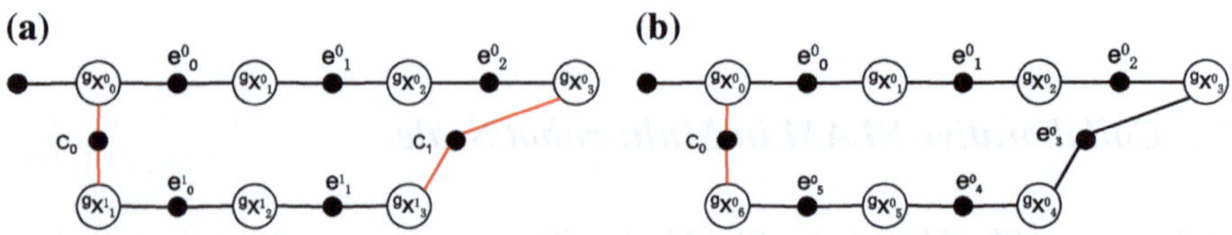

Fig. 3 Pose-graph examples. **a** Pose-graph consists of two Markov chains, two constraints are added between robot 0 and 1, **b** loop-closure constraint in single Markov chain of robot 0

Notice the global map frame is different from local map frame, but we can add the relative odometry equivalently.

Except odometry constraints from local SLAM, there're two extra constraints can be created, namely, constraint between robots and loop-closure constraint. Essentially, the types of these two constraints are the same.

One source of constraint comes from laser scans comparison between different nodes of different robots (the black dot connected by red lines in Fig. 3a). Denote it as c_p, $p = 0, 1, 2 \ldots$. It is generated through following way: for robot k, we apply a searching area around its current pose ${}^g x_i^k$, if the closest pose in global map comes from robot j, $j \neq i$, and this pose ${}^g x_p^j$ satisfies

$$\|{}^g x_i^k - {}^g x_p^j\|_t < r_t, \quad \|{}^g x_i^k - {}^g x_p^j\|_\theta < r_\theta \tag{2}$$

Here $\| \bullet \|_t$ and $\| \bullet \|_\theta$ means translation and rotation between two poses. r_t denotes the euclidean distance threshold and r_θ indicates the yaw threshold. Normally, we set it as $r_t = 10\,\text{m}$, $r_\theta = \frac{\pi}{2}$. Then we try scan matching on two scans corresponding to these two poses. If these two scans match successfully, we can get the constraint c_p and add it into pose graph between the corresponding nodes.

The other type of constraint is derived from loop-closure detection in similar way: when a new pose ${}^g x_i^k$ added into the current pose-graph, we check the region around it with respect to a certain threshold $\{r_t, r_\theta\}$. If the closest pose within the region comes from robot itself, as ${}^g x_p^k$, $p < i$, we check the difference in timestamp as $d_t = i - p$. If $d_t > d_{thresh}$, which means the robot traveled for a while and might form a loop, we try scan matching between these two laser scans and add output constraint in to pose-graph if there is a valid match (Fig. 3b).

The scan matching algorithm we employ in global SLAM is called canonical scan matching (CSM) which is a fast version of iterative closest point (ICP) algorithm. The strategy is depicted in paper [19]. CSM is vulnerable to noise since the matching uses laser scans directly without creating a probabilistic map. In order to increase rate of matching and achieve correct constraint, filtering out the floor in laser scan is quite necessary.

Optimization process starts when the pose-graph is updated. Batch optimizing is the fundamental way to solve a non-linear maximization problem as stated in [20]. However, it requires considerable computation to solve this equation and this requirement increases dramatically with the number of nodes in pose-graph. A novel linearizing algorithm based on batch optimizing is proposed aiming at decreasing the computational demand, called incremental smoothing and mapping (iSAM). We use the iSAM2.0 as the solver of SLAM back-end, due to our experimental comparison, it's 10-times faster than vanilla batch method. iSAM2.0 is ready-to-use in GTSAM library, it takes pose graph as input and calculates optimized pose-graph. We use OmniMapper to construct pose graph from multiple local SLAM and links pose graph to iSAM2.0 solver. The algorithm detail is referred in [21].

1.2 Autonomous Navigation and Exploration

We now present modules that enable autonomous exploration of the environment based on the incrementally updated global map built by base station. All modules in the exploration subsystem runs in response to new maps from the global SLAM.

Frontier detection The algorithm for autonomous exploration relies on the detection of frontier points in the current 2-D occupancy grid map.

We define frontier points as unoccupied points that have half of their neighbors in free space and the other half in unknown space.

To be specific, for any point in free space, we check all its neighbors within radius *r* (denoted as the red circle in Fig. 4). If half of points inside the circle is in free space and the other half is in unknown space, and there is no occupied point within this circle, we identify the center point as a frontier point. Figure 4a illustrates a typical frontier point, while Fig. 4b, c are examples of invalid frontier points.

In most cases, the frontier detector gives all the adjacent frontier points along the frontier edge (Fig. 5a). Grouping of frontier points is necessary to avoid decreases in algorithm efficiency due to excessive number of frontier points along the same frontier edge. We combine frontier points along the same frontier edge into a single point for exploration (Fig. 5b).

Goal assignment Usually, there are multiple frontier points in the global map during exploration. We assign a feasible goal to each robot such that the following cost function is minimized:

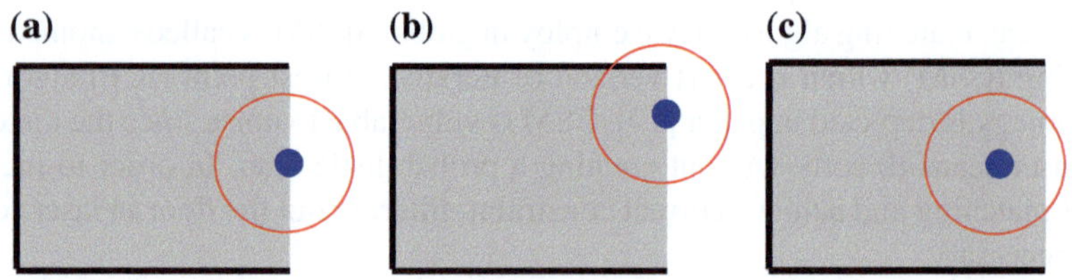

Fig. 4 Samples of valid and invalid frontier points. **a** Shows a typical frontier point. The *circle* in (**b**) includes occupied point, and the number of points in unknown space is insufficient in (**c**). Unknown, unoccupied and occupied spaces are *color coded* in *white*, *gray* and *black* respectively. **a** Valid, **b** invalid, **c** invalid

Fig. 5 Grouping, *blue dots* are raw frontier points, *purple* one is the corresponding grouped point. **a** Adjacent frontier points along a edge, **b** grouped frontier point

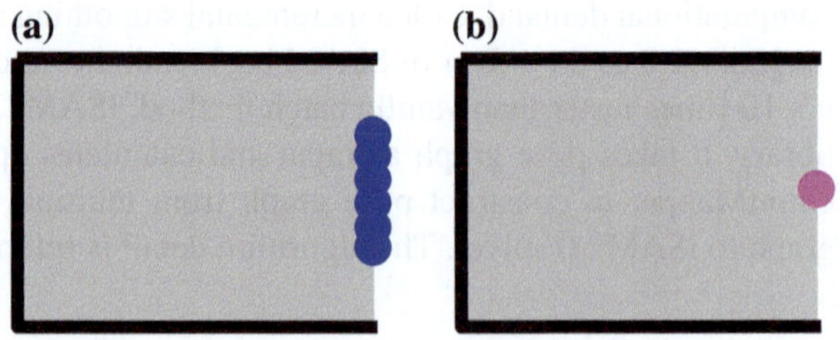

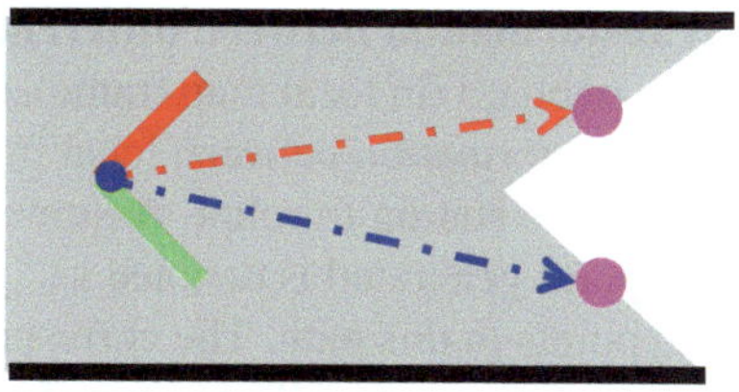

Fig. 6 The costs of two paths (*red* and *blue*) are equivalent if $w_3 = 0$, the robot could frequently jump from one path to the other while moving

$$c({}^g x_i^k, {}^g f_{ij}) = w_1 \|{}^g x_i^k - {}^g f_{ij}\|_t + w_2 \|{}^g x_i^k - {}^g f_{ij}\|_\theta + w_3 \|{}^g f_{ij} - {}^g f^k\|_t \quad (3)$$

where ${}^g f_{ij}$ is the j-th frontier point in current timestamp global coordinates. ${}^g f^k$ indicates the last goal assigned to robot k. w_i, $_{i=1,2,3}$ is the weighting factor for each cost terms. First two terms in RHS of Eq. (3) are the displacement in distance and orientation between current robot location and j-th frontier. The third term is used for anchoring the goal with respect to previous goal to avoid frequent switching of goals with similar costs (Fig. 6). Goals may be cleared before the robot reaches it due to the sensor measurement range. In such case, a pending goal is assigned to the robot it exists, otherwise the exploration is halted until a newer map with more frontiers is received.

Path planning Since current poses of all robots, goal positions, as well as the global map are available, it's feasible to plan the collision-avoidance path in occupancy grid map by RRT*. RRT* [22] is employed due to its fast planning speed and its convergence property to the optimal solution. An open source implementation of RRT* is available in the `ompl` library [23]. We plan with free space assumption and consider both unknown and free space as the collision-free. The output from RRT* is not guaranteed to be optimal, however, if we set a reasonable region around robot as "no-collision-zone" to keep the robot path away from obstacles, RRT* is able to produce valid collision-free paths for our MAV platforms.

As the robot moves along the planned path and observes new environments, the path may be invalidated due to newly observed obstacles intersecting with the path. In such case, a replan will be initiated.

Strategy for multi-robot coordination If there are more than one robot exploring in same environment simultaneously, the exploration steps will be more complicated. For instance, assigning the same goal to multiple robots is dangerous, aerial vehicles can crash easily when they move close to each other due to the interference from turbulence generated by propellers; when each robot plan the path, the planner should take consider of the position of other robots such that the path won't collide with any of them. Several rules need to be integrated such that a reliable exploration algorithm can be achieved for multi-robot system.

- In step of goal assignment, if a target is chosen as the current goal for robot k, this goal will be removed from the queue of frontier points;
- To avoid collision between robots, when planning path for robot k, set the region with a certain radius around other robots' locations be occupied.

Transformation between different coordinate frames The control process is running onboard (in local map frame), but we plan path in global map frame. It requires for transformation between local frame and global frame before sending a new path from base station to robot. Assume we have a path planned in global map, and the waypoint generated is denoted as ${}^{g}p_i^k$, where $i = 1, 2 \ldots n$, n is the total number of waypoints in this path. The corresponding waypoint in local map is ${}^{l}p_i^k$. We can get the rotation matrix ${}^{l}R_g$ and translation vector ${}^{l}T_g$ from global coordinate into local coordinate from initial waypoints in two coordinates, and apply Eq. 4 for waypoints transformation. Notice that the initial waypoint is always the current pose of robot when starting planning in both coordinates.

$$ {}^{l}p_i^k = {}^{l}R_g {}^{g}p_i^k + {}^{l}T_g \tag{4} $$

2 Experimental Results and Analysis

In this section, we present simulation and experimental results demonstrating the working of the mapping and exploration algorithms. We developed the system in steps, the first being testing the collaborative SLAM system offline using log files recorded by the carrying the robots. Then, the higher level exploration algorithm was implemented in simulation with up to 3 quadrotor models. The simulation included a laser simulator based on ray casting and a quadrotor dynamics simulator in order to have an accurate simulation of the real system. Once we were confident about the algorithm working in simulation, we implemented it for a single quadrotor and extensively tested it in different indoor environments in order to check its reliability. Convinced that the reliability was high enough, we added one more robot in order to build the exploration system with two quadrotors. Due to our assumption about the star topology of communications and limited range of wireless, we were not able to test the multi-robot system in large indoor spaces.

2.1 Test on Collaborative SLAM and Simulation for Exploration

Our exploration algorithm requires a good map to find frontiers and generate valid paths for each robot. Distortion or loop-closure errors in the map often lead to inappropriate paths. We ran tests of the collaborative SLAM using log files to examine the quality of mapping. The quality of created maps and speed of optimizing pose graph were considered as the two important criteria for the algorithm. Log files from three robots carried around the same area have been used to create a global map as shown in Fig. 7. The initial positions of robots are close to each other so that odometry constraints can be added in the pose graph between the initial poses of the robots

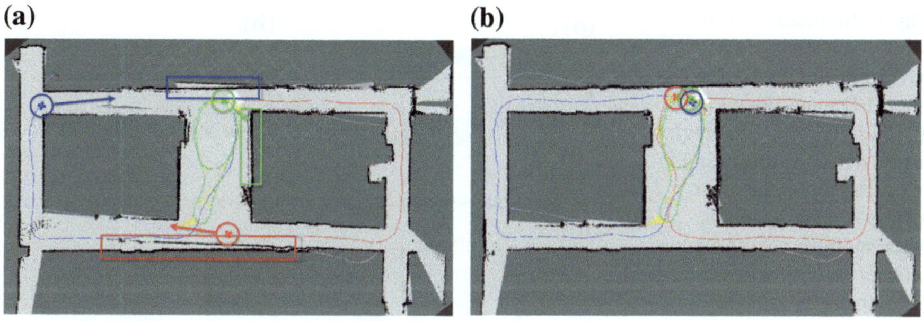

Fig. 7 Test on collaborative SLAM using log files, iSAM2.0 is used in this test. **a** Loop-closure error results in the creation of spurious walls (inside *rectangulars*), **b** map has been corrected after finding loop-closure constraints

to not have disjoint Markov chains. The onboard local SLAM generates key frames after robot has travelled at distance of 0.2 m or rotated by 0.1 rad. This led to more than 500 nodes in the final pose graph (Fig. 7a), and iSAM2.0 is able to optimize it in 2 ms, while a batch optimization approach consumes more than 600 ms, both providing a similar quality output map. Thus, it was clear that in order to do real-time exploration, iSAM2.0 would serve as a good pose graph backend. From Fig. 7 we can see that our collaborative SLAM algorithm works well for merging maps from multiple robots.

The core task of exploration is the feasibility of finding frontiers and exploring those frontiers. We create a 2-D occupancy grid map from a schematic and generate 2.5-D environment from this prior map, laser simulation is used to provide laser scans similar to real-world estimation. Based on this architecture, we tested our algorithms in simulation with three robots. Figure 8 demonstrates exploration and map merging in a `B`-shape map. Both red and blue robots run a `S` curve, and these two curves cover the whole map such that the third robot (green) seems to be unnecessary (Fig. 8c). On the contrary, during exploration, even though two `S`-shape trajectories share common part of explored regions, the robots face opposite directions in these regions such that scan matching doesn't provide any odometry constraints between them (Fig. 8a). The third robot passes through this region and creates odometry constraint between itself and other robots. Consequently, even though there is a lack of constraints between the red and blue robots' poses, the constraints between green and red, and green and blue robots are able to recover the map and correct accumulated errors (Fig. 8b).

2.2 *Single Quadrotor Exploration*

After extensively testing the algorithms in simulations, we moved to a real robot and built a system for autonomous exploration with a single quadrotor. The quadrotor sends key-frame laser scan and pose from local SLAM to base station (laptop), the

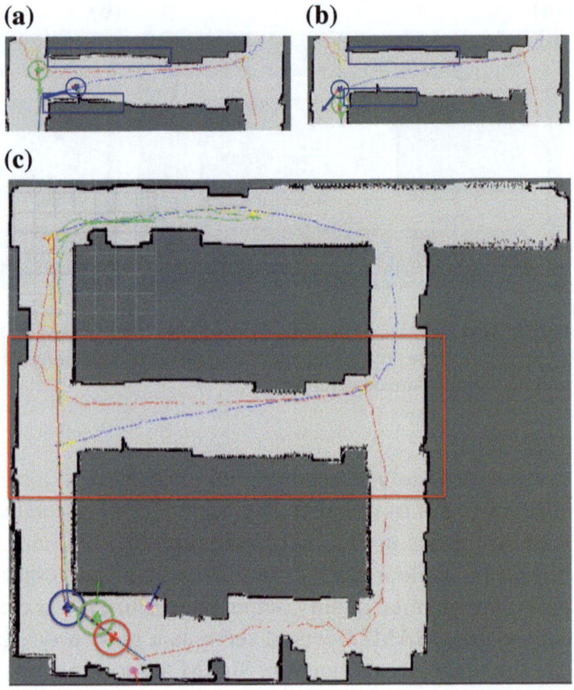

Fig. 8 Simulation of 3 robots exploring in an environment created from a prior 2-D map. Figure (**a**) and Fig. (**b**) shows the region inside *red rectangular* in Fig. (**c**). **a** Accumulated error results in spurious walls when multiple robots explore same area with large bearing differences, **b** odometry constraints between *green* and *blue* robots correct the map and spurious wall disappears, **c** after exploration done, we get a map which is almost the same as the prior map

planning result is sent back to robot after every iteration of exploration process. As all the computations in order to localize and control the robot are running onboard the robot, even if the robot loses connectivity with the base station, the robot is still able to control itself, though it won't get any new assigned waypoints to follow.

We tested the single robot exploration in different cases and Fig. 9 shows the exploration process in a relatively large environment.

Now the problem is, how to evaluate the quality of autonomous exploration. Intuitively, we may consider the benchmark as the human controlled exploration. The speed comparing autonomous to manual exploring is a important index to show the quality of our algorithm. Figure 10 shows the experiments of single robot manually exploration and autonomous exploration. The total distances in $x - y$ plane travelled by robots in two cases are similar. Ideally, if we fly robot in the same velocity, the time consumed in these two trials should be similar. In manually fly, maneuver can control quadrotor fly at arbitrary speed; the maximum velocity of autonomous navigation depends largely on the system, in general it's slower than former. Hence even with the same travelled distance, the time for completely exploration in a certain area of these two cases varies. In our experiment, we manually fly the quadrotor at the same speed as autonomous fly, the time to complete exploration of these two experiments are almost the same (Table 1).

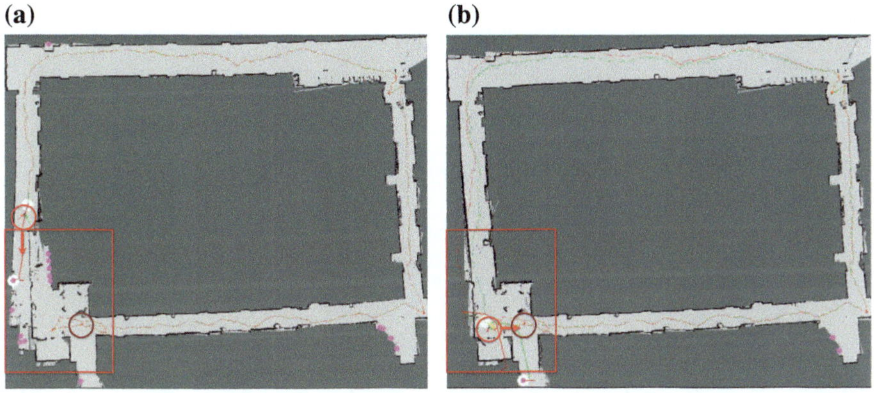

Fig. 9 Single quadrotor exploration in a large environment (about 30 × 25 m). *Dark red circle* indicates the start position, *bright red circle* indicates the current robot position in global frame. **a** When quadrotor comes back to original position, the loop-closure error is significant (*red rectangular*), **b** Loop-closure constraint is generated from front-end, the map is corrected

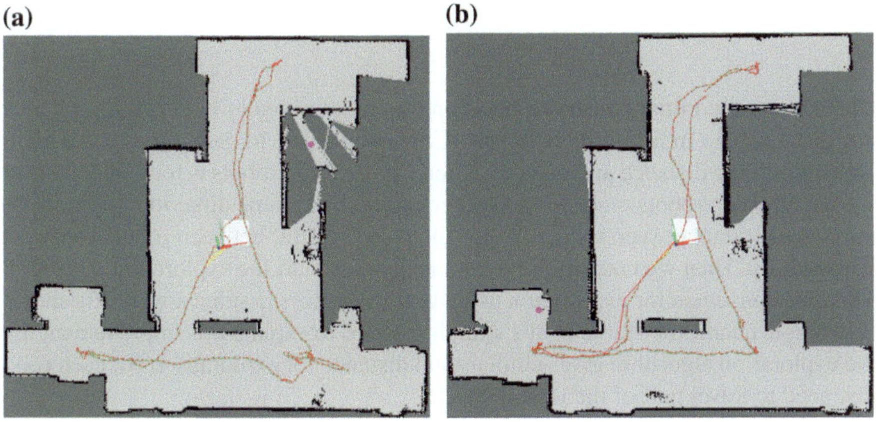

Fig. 10 Comparison between manually and autonomously exploration in constraint indoor environment (15 × 15 m). **a** Manually control, **b** autonomous exploration

Table 1 Exploration time for different methods

Experiment	Manually control	Single-quad	Two-quad
Time (s)	105	101	81

2.3 Two Quadrotors Exploration

As a consistent to last section, we extend the system of single robot exploration into two. According to Sect. 1.2, several rules for avoiding collision and allocating goals are activated in order to guarantee a robust and reliable multi-robot maneuver system.

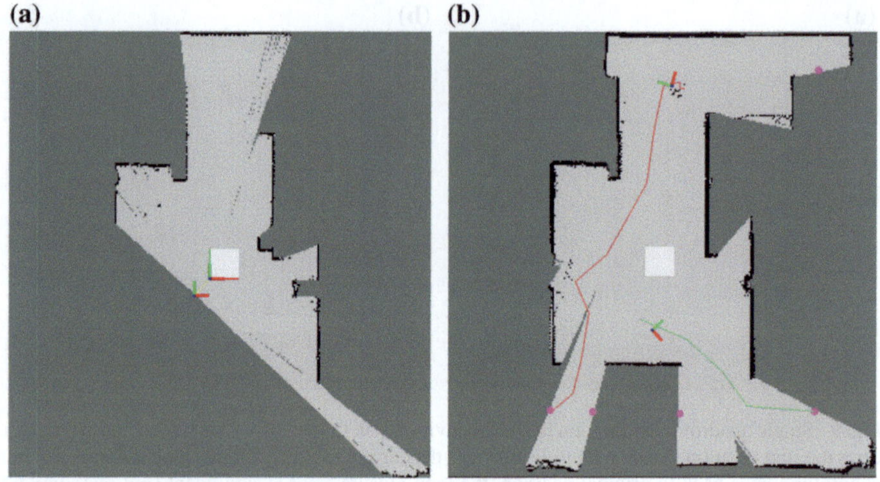

Fig. 11 Two quadrotors exploration. **a** Initial positions of two robots, odometry constraint can be detected if they're close enough, **b** while exploring, the path is guaranteed to avoid crossing region around the other robot

Considering the initial constraint between the initial poses of both robots, we must not place robots in arbitrary positions, CSM would fail to generate the odometry constraint if the distance and angular difference between robots is too large. Instead, we start the two robots close to each other and face the same direction to ensure the initial constraint between robots (yellow lines in Fig. 11a) between robots generated successfully. Then we take off the robot one by one, start the exploration algorithm. The time consumed for exploration using two quadrotors in same area is 20 s shorter than single quadrotor (Table 1), it's straightforward to explain this improvement that the exploration algorithm gives similar total distance for exploration, but each robot just need to travel half of the total distance.

3 Conclusion and Future Work

We have presented experiments that evaluate collaborative mapping and exploration algorithm for multiple MAVs. We have shown the advantages of collaborative SLAM and demonstrated autonomous exploration through both simulation and experiments. To our knowledge, this is the first experimental realization of autonomous, multiple quadrotors exploration.

There are many extensions to this work. First, we don't explicitly consider communication constraints. The advantages of a central base station for global SLAM are offset by the disadvantages of possible disruption to communication and the inability to scale to large numbers of robots and large environments. Second, we do not explicitly sense or model the possibility of collisions. While we do plan collision-free paths, it is also necessary for the robots to detect each other in order to guarantee safety, particularly in environments where aerodynamic interactions can be quite complex. In addition, 2.5-D assumption is not always held, the ideal system requires SLAM and exploration in 3D in clustered indoor environment. By utilization of 3-D sensors, we can develop this technology based on similar framework proposed in this paper. All these limitations indicate future directions for our work.

Acknowledgments We gratefully acknowledge support from ARL Micro Autonomous Systems and Technology Collaborative Technology Alliance Grant no. W911NF-08-2-0004.

References

1. Shen, S., Mulgaonkar, Y., Michael, N., Kumar, V.: Multi-sensor fusion for robust autonomous flight in indoor and outdoor environments with a rotorcraft mav. In: 2014 IEEE International Conference on Robotics and automation (ICRA). IEEE (2014)
2. Howard, A.: Multi-robot simultaneous localization and mapping using particle filters. Int. J. Robot. Res. **25**(12), 1243–1256 (2006)
3. Thrun, S., Liu, Y.: Multi-robot slam with sparse extended information filers. In: Robotics Research, pp. 254–266. Springer (2005)
4. Shen, S., Michael, N., Kumar, V.: Stochastic differential equation-based exploration algorithm for autonomous indoor 3d exploration with a micro-aerial vehicle. I. J. Robot. Res. **31**(12), 1431–1444 (2012)
5. Durrant-Whyte, H., Bailey, T.: Simultaneous localization and mapping: Part I. IEEE Robot. Autom. Mag. **13**(2), 99–110 (2006)
6. Dellaert, F., Kaess, M.: Square root sam: simultaneous localization and mapping via square root information smoothing. Int. J. Robot. Res. **25**(12), 1181–1203 (2006)
7. Shen, S., Michael, N., Kumar, V.: Autonomous multi-floor indoor navigation with a computationally constrained mav. In: 2011 IEEE International Conference on Robotics and automation (ICRA), pp. 20–25. IEEE (2011)
8. Stachniss, C., Kretzschmar, H.: Pose graph compression for laser-based slam. In: International Symposium of Robotics Research (ISRR) (2011)
9. Van Der Merwe, R., Wan, E.A.: Sigma-point kalman filters for integrated navigation. In: Institute of Navigation (ION) (2004)
10. Michael, N., Mellinger, D., Lindsey, Q., Kumar, V.: The grasp multiple micro-uav testbed. IEEE Robot. Autom. Mag. **17**(3), 56–65 (2010)
11. Lee, T., Leoky, M., McClamroch, N.H.: Geometric tracking control of a quadrotor uav on se (3). In: 2010 49th IEEE Conference on Decision and Control (CDC), pp. 5420–5425. IEEE (2010)
12. Shen, S., Michael, N., Kumar, V.: Autonomous indoor 3d exploration with a micro-aerial vehicle. In: 2012 IEEE International Conference on Robotics and Automation (ICRA), May 2012, pp. 9–15
13. Bachrach, A., He, R., Roy, N.: Autonomous flight in unknown indoor environments. Int. J. Micro Air Veh. **1**(4), 217–228 (2009)
14. Kim, B., Kaess, M., Fletcher, L., Leonard, J., Bachrach, A., Roy, N., Teller, S.: Multiple relative pose graphs for robust cooperative mapping. In: 2010 IEEE International Conference on Robotics and Automation (ICRA), pp. 3185–3192. IEEE (2010)
15. Rogers, J.G., Nieto-Granda, C., Christensen, H.I.: Coordination strategies for multi-robot exploration and mapping (2012)

16. Trevor, A.J.B., Rogers III J.G., Christensen, H.I.: Omnimapper: a modular multimodal mapping framework. In: 2014 IEEE International Conference on Robotics and Automation (ICRA). IEEE (2014)
17. Dellaert, F.: Factor graphs and gtsam: a hands-on introduction (2012)
18. Cunningham, A., Paluri, M., Dellaert, F.: Ddf-sam: fully distributed slam using constrained factor graphs. In: 2010 IEEE/RSJ International Conference on Intelligent Robots and Systems (IROS), pp. 3025–3030. IEEE (2010)
19. Censi, A.: An icp variant using a point-to-line metric. In: IEEE International Conference on Robotics and Automation, ICRA 2008, pp. 19–25. IEEE (2008)
20. Kaess, M., Ranganathan, A., Dellaert, F.: isam: Incremental smoothing and mapping. IEEE Trans. Robot. **24**(6), 1365–1378 (2008)
21. Kaess, M., Johannsson, H., Roberts, R., Ila, V., Leonard, J.J., Dellaert, F.: isam2: Incremental smoothing and mapping using the bayes tree. Int. J. Robot. Res. **31**(2), 216–235 (2012)
22. Karaman, S., Frazzoli, E.: Sampling-based algorithms for optimal motion planning. Int. J. Robot. Res. **30**(7), 846–894 (2011)
23. Sucan, I., Moll, M., Kavraki, E.: The open motion planning library. IEEE Robot. Autom. Mag. **19**(4), 72–82 (2012)

Cooperative Control for Target Tracking with Onboard Sensing

Karol Hausman, Jörg Müller, Abishek Hariharan, Nora Ayanian and Gaurav S. Sukhatme

Abstract We consider the cooperative control of a team of robots to estimate the position of a moving target using onboard sensing. In particular, we do not assume that the robot positions are known, but estimate their positions using relative onboard sensing. Our probabilistic localization and control method takes into account the motion and sensing capabilities of the individual robots to minimize the expected future uncertainty of the target position. It reasons about multiple possible sensing topologies and incorporates an efficient topology switching technique to generate locally optimal controls in polynomial time complexity. Simulations show the performance of our approach and prove its flexibility to find suitable sensing topologies depending on the limited sensing capabilities of the robots and the movements of the target. Furthermore, we demonstrate the applicability of our method in various experiments with single and multiple quadrotor robots tracking a ground vehicle in an indoor environment.

Keywords Cooperative multi-robot control · Target tracking · Sensor-based navigation · Sensing topology switching

K. Hausman (✉) · J. Müller · A. Hariharan · N. Ayanian · G.S. Sukhatme
Department of Computer Science, University of Southern California, Los Angeles, CA, USA
e-mail: hausman@usc.edu

J. Müller
e-mail: joerg.mueller@usc.edu

A. Hariharan
e-mail: abishekh@usc.edu

N. Ayanian
e-mail: ayanian@usc.edu

G.S. Sukhatme
e-mail: gaurav@usc.edu

M.A. Hsieh et al. (eds.), *Experimental Robotics*, Springer Tracts in Advanced Robotics 109, DOI 10.1007/978-3-319-23778-7_58

1 Introduction

Using multiple robots to track a moving target is potentially beneficial because of the reduction in tracking uncertainty, increased coverage, and robustness to failure. Two problems arise immediately. First, these objectives are often at odds (e.g., the configuration of the robots that lead to the lowest uncertainty estimates of target pose may not be the best if one or more robots is disabled). Second, the robots themselves are often poorly localized (e.g., only a few may have access to GPS, and the rest may be limited to a combination of onboard inertial sensing, visual odometry, and relative range/bearing measurements to estimate their poses relative to each other).

As an example, consider the unmapped interior of a building shown in Fig. 1 where moving targets needs to be tracked using multiple quadrotors. Some of the quadrotors may have access to GPS (e.g., near external windows), the others do not, but can track each other and the target. How should such a system coordinate its motion such that it always maintains itself in a configuration that results in the least uncertainty in target pose?

In the domain of cooperative control, small unmanned aerial vehicles (UAVs) have recently become prominent and several well-constructed testbeds have been established for multi-robot control and aerobatics with motion capture state estimates [12, 14, 20]. For cooperative target tracking with onboard sensors, many authors considered centralized [5, 6, 18], decentralized [1, 15, 17], and distributed [10, 11, 21] approaches to multi-robot control in aerial and ground settings. However, these methods estimate the pose of the target and assume that the poses of the robots are known,

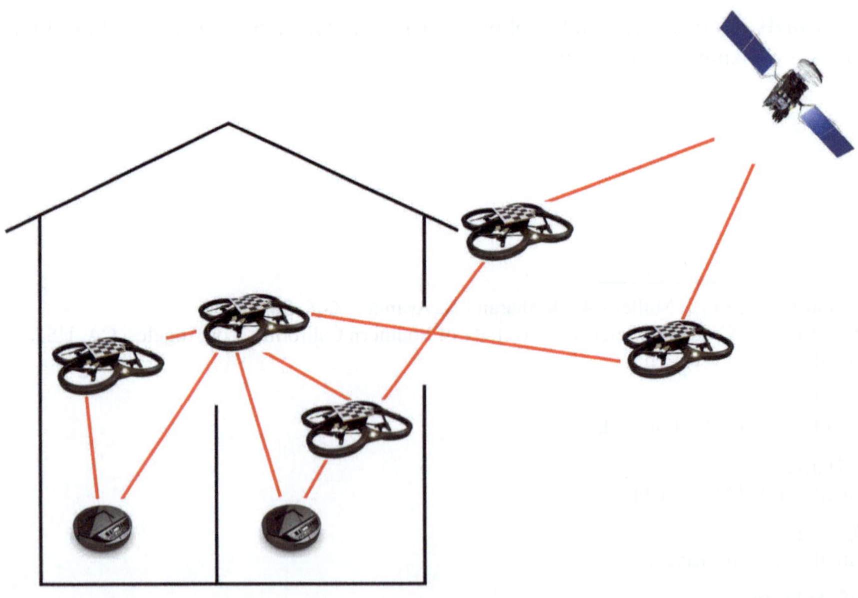

Fig. 1 A collaborative target tracking task in which the robots have to establish an appropriate relative sensing topology to localize themselves and track one or multiple targets

e.g., from an external system or by reference to a global map. Ahmad and Lima [2] robustly track a target taking into account the individual robot's self-localization by weighting the confidence of observations using their localization uncertainty. In contrast to our approach, they decouple the target tracking from the robot's localization, which does not account for the (usually high) correlation of the target's and the robot's position estimates. To robustly perform cooperative multi-robot localization using only onboard sensors (such as with the popular Kalman filter [16]), several optimization-based localization approaches have been proposed [3, 7, 8]. However, the maximum-likelihood state estimates provided by these approaches do not allow for direct minimization of the uncertainty associated with the estimated target pose.

In this paper, we consider the cooperative control of a team of robots to estimate the position of a target using onboard sensing. In particular, we assume limited sensing capabilities, e.g., in terms of a limited field of view and range of each sensor. Our (centralized) approach reasons over the entire sensing topology (comparable to [5]), without assuming that all robot poses can be extracted from offboard sensing. Instead, the joint state of the robots and target are estimated explicitly using onboard sensing. In such a setting, the poses of the robots influence visibility (which robots can see which other robots and which robots can see the target) and measurement uncertainties. Therefore reducing the uncertainty of the estimated target pose requires smart positioning of robots to build up an appropriate chain of observations. Our centralized multi-robot control approach reasons over the whole sensing topology when minimizing the uncertainty of the estimated target position.

The key contributions of our approach are that (a) we consider onboard sensing and switching from one sensing topology to another, (b) the approach is probabilistic and takes into account motion and sensing capabilities and uncertainties, (c) the control is locally optimal through local optimization that permits switches to neighbor topologies and (d) the control approach has polynomial complexity in the number of robots.

We implemented and experimentally evaluated our approach in simulation and with real quadrotor robots. Our approach proved to flexibly adapt the topology and controls to the sensing limitation of the individual robots and the target movements. Experiments with inexpensive AR.Drone quadrotors demonstrate the robustness of our approach to substantial sensing and motion uncertainty, but also show the limitations arising from the limited flight stability and field of view of these platforms.

2 Multi-robot Control with Topology Switching

2.1 Sensing Topologies

At each time step, the team of robots is in a certain topology with respect to sensing. The topology usually results from the robots' poses and the sensing capabilities of the global sensor as well as of the individual robots observing each other and the target. In general, the sensing capabilities can be limited by the range of the sensor, its restricted field of view, or the available processing power that may only enable the detection of a limited number of vehicles.

In our multi-robot control method, we efficiently organize robot topologies by applying a level-based topology approach. In such a sensing topology, each robot is assigned to a level, the global sensor (e.g., GPS) is in the highest level, and the target is in the lowest level (see Fig. 2). Each sensor can potentially observe each robot/target on the adjacent layer below it given that its capabilities allow the corresponding measurements in the spacial configuration.

During target tracking, we allow switching between neighboring topologies. We consider two sensing topologies as neighbors, if the team can transition between them by just moving one robot by one level up or down (which can result in adding or removing a level).

2.2 Extended Kalman Filter (EKF) State Estimation

We use the popular EKF [19] to efficiently and robustly estimate the joint pose of all robots and the target from imprecise movements and noisy measurements similar to [13]. Given the pose $\mathbf{x}^{(i)}$ of the individual robots and the pose $\mathbf{x}^{(t)}$ of the target, we define the joint state as

$$\mathbf{x} = [\mathbf{x}^{(1)}, \ldots, \mathbf{x}^{(n)}, \mathbf{x}^{(t)}] \,. \tag{1}$$

The EKF recursively fuses all measurements $\mathbf{z}_{1:k}$ and controls $\mathbf{u}_{1:k}$ up to time k. It maintains the state posterior probability

$$p(\mathbf{x}_k \mid \mathbf{z}_{1:k}, \mathbf{u}_{1:k}) = \mathcal{N}(\boldsymbol{\mu}_k, \Sigma_k) \tag{2}$$

at time step k as a Gaussian with mean $\boldsymbol{\mu}_k$ and covariance Σ_k. The stochastic motion functions

$$\mathbf{x}_{k+1}^{(i)} = \mathbf{f}^{(i)}(\mathbf{x}_k^{(i)}, \mathbf{u}_k^{(i)}) + \boldsymbol{\delta}_k^{(i)} \tag{3}$$

given the control command $\mathbf{u}$ and the white Gaussian noise $\boldsymbol{\delta}$ of the individual robots can be naturally combined in the joint state estimation [13]. The stochastic measurement functions

$$\mathbf{z}_k^{(i,j)} = \mathbf{h}^{(i,j)}(\mathbf{x}^{(i)}, \mathbf{x}^{(j)}) + \boldsymbol{\varepsilon}_k^{(i,j)} \tag{4}$$

$$= \tilde{\mathbf{h}}^{(i,j)}(\mathbf{x}) + \boldsymbol{\varepsilon}_k^{(i,j)} \tag{5}$$

can be naturally extended for the joint state [13]. Since the measurements are assumed to be conditionally independent given the joint state [19], individual measurements can be fused separately.

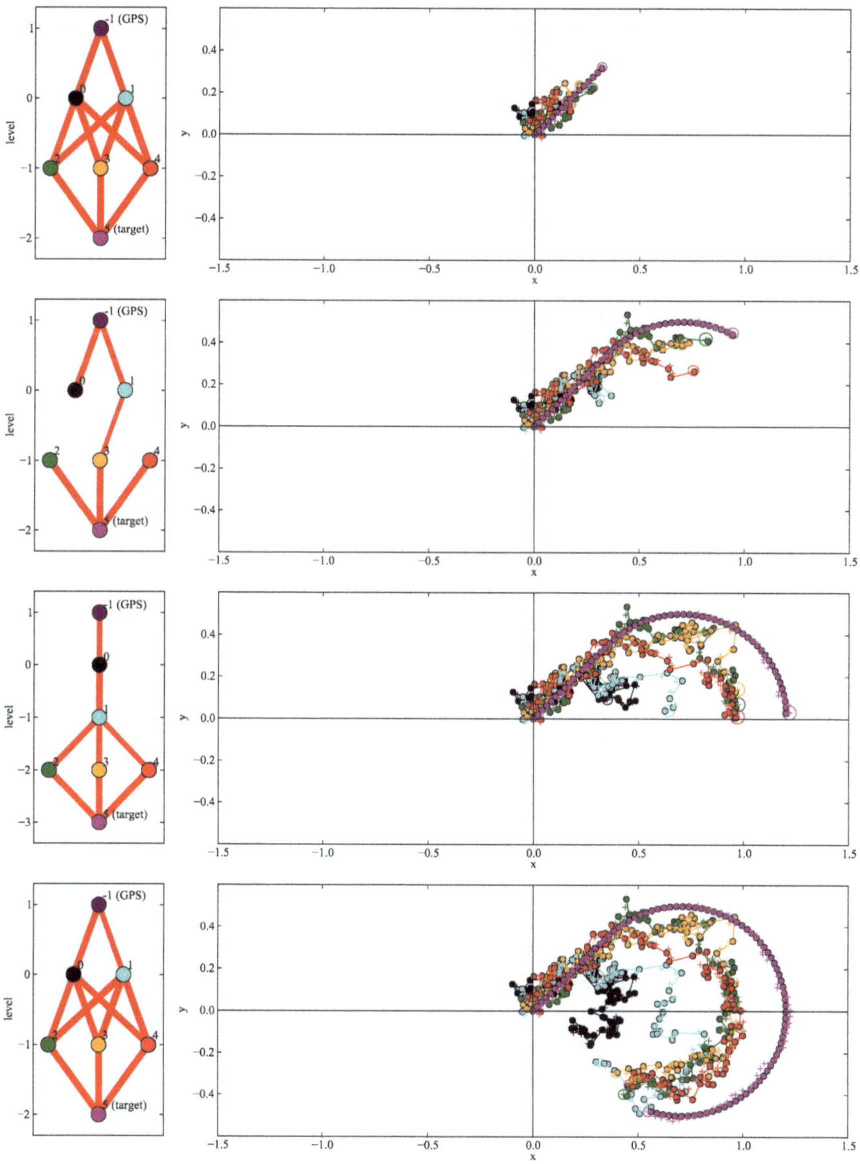

Fig. 2 Simulation results with 5 robots. *Left* the current topology selected by our approach. The links represent the actual measurements where the thickness of each link corresponds to the information provided by the measurement (the inverse of the measurement standard deviation). *Right* The trajectory and the state estimates of the EKF. The actual trajectory is shown as *thick dots* connected by a *solid line*. The EKF means are indicated by '+' and the covariance is shown for the current state

The motion and sensing functions, their Jacobians, and the noise covariances are provided by the motion and sensor model of each entity, respectively. As a motion function in general target tracking, one can apply a standard uncontrolled motion model, e.g., a constant velocity motion model.

2.3 Optimization-Based Control and Topology Switching

Our probabilistic method for cooperative target tracking aims at finding the joint controls $\mathbf{u} = [\mathbf{u}^{(1)}, \ldots, \mathbf{u}^{(n)}]$ that minimize the uncertainty about the target position. At a time step k, we define the cost function

$$c_k(\mathbf{u}) = \sum_{i=1}^{h} \gamma^i \operatorname{tr}(\Sigma'_{k+i}) \tag{6}$$

as a measure of the future target tracking uncertainty. It penalizes the uncertainty of the state estimate of the target. We measure this uncertainty using the marginal covariance of the target state Σ', which is obtained as the corresponding block of the covariance Σ of the joint EKF. Here, h is the lookahead horizon and $0 \leq \gamma \leq 1$ is a discount factor.

We evaluate the a priori tracking covariances $\Sigma_{k+1}, \ldots, \Sigma_{k+h}$ by starting an EKF instance from the current belief $(\boldsymbol{\mu}_k, \Sigma_k)$. During these h EKF cycles, the constant joint control $\mathbf{u}$ is applied and the availability and covariances of the individual measurements are evaluated given the mean state $\boldsymbol{\mu}$.

We finally formulate the selection of controls as an optimization problem

$$\mathbf{u}_k^\star = \operatorname*{argmin}_{\mathbf{u}} \left(c_k(\mathbf{u}) + c_a(\mathbf{u})\right) \tag{7}$$

with the proposed cost function. The additional cost c_a accounts for the future distance between the individual robots and results in a repelling force for explicit collision avoidance. In our approach, we apply nonlinear optimization (e.g., [9]) to find the locally optimal control for the current topology and all neighbor topologies. We then select the topology and corresponding control that resulted in the lowest cost.

2.4 Complexity Analysis

The asymptotic complexity of our approach with n robots is determined as follows. We evaluate $O(n)$ neighbor sensing topologies, which reduces the computational complexity from exponential (for all topologies) to real-time capable linear complexity. For each considered topology, we assume that the optimization (e.g., gradient descent with a constant number of iterations) runs $O(n)$ evaluations of the cost function. Each evaluation of the cost functions involves h cycles of the EKF, which is $O(n^3)$, such that the overall complexity of our approach is $O(n^5)$.

3 Simulation Experiments

3.1 Experimental Setup

We evaluated our approach on a number of simulations (see, for example, Fig. 2). We consider a quadrotor and a target as points moving in 2D space, and we employ the Kalman filter to estimate their $[x, y]^T$ positions. The setup also includes a global sensor (called GPS), which is located at the origin [0, 0]. Omnidirectional 2D cameras with a limited sensor range of 0.5 m provide relative positions of observed objects. We assume that the measurement noise of the GPS and the cameras increases quadratically with the distance from the center of view. The target is programmed to execute a trajectory that starts at the origin and performs a figure eight.

3.2 Results and Insights

An example of the simulation results is shown in Fig. 2; a video is available online.[1] While the controls selected by the approach were quite smooth, the zigzag movements of the robots were due to the simulated motion noise. Each experiment started in one of the simplest topologies, in which the robots were arranged as a string, each residing on its own level. Our approach locally modified the topology during the first steps and converged to a topology with two levels (Fig. 2, row 1). As the target moved away from the GPS signal at the origin, the limited measurement range causes dropouts in this topology (row 2) and our approach introduced an additional robot level (row 3). Here, our approach exploited the currently low position uncertainty of all robots and assigned three robots to the lowest level to get robust information on the target position. As the target moved back towards the GPS, our approach switched back to the two-level topology (row 4). Our approach similarly handled the left part of the trajectory, which is not shown due to space constraints (Fig. 3).

Further simulations with 2–30 robots and different sensor and motion models confirmed our assumption that the selected topologies substantially depend on the limitations (here: the measurement range) of the sensor model. With unlimited measurement range, the topology quickly converged to a locally optimal one and switching to different topologies only appeared as transient effects, even with simulated kidnapping of robots and the target (Fig. 4).

[1] http://robotics.usc.edu/~hausmankarol/videos/iser_videos.

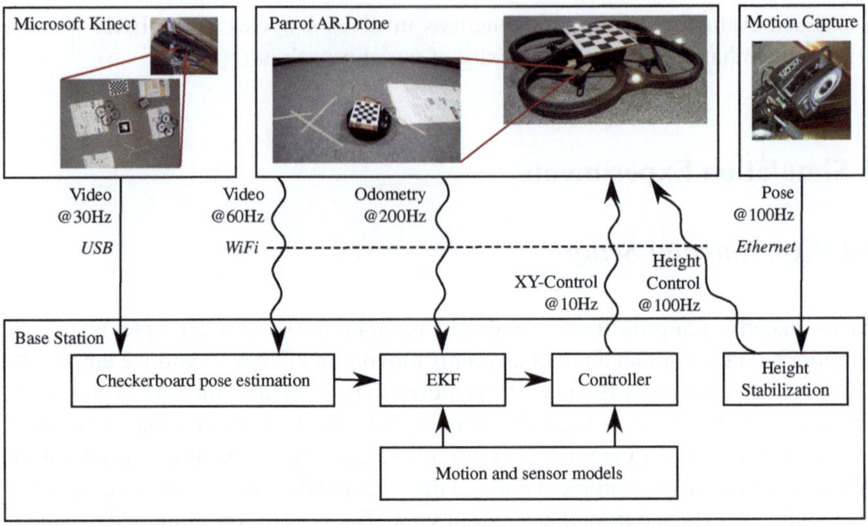

Fig. 3 The information flow in our real-robot target tracking experiments

Fig. 4 The AR.Drones are equipped with a checkerboard and Vicon markers for relative sensing and ground truth poses, respectively. The forward-looking camera is tilted 45° downwards (highlighted by a *red circle*) to track the target and robots on lower levels of the sensing topology

4 Real Robot Experiments

4.1 Experimental Setup

We tested the approach with Parrot AR.Drone quadrotor UAVs shown in Fig. 5. The setup consists of a Microsoft Kinect sensor that was attached to the ceiling in approx. 3.4 m height in an approx. $6 \times 5\,\text{m}^2$ room. One or two Parrot AR.Drone quadrotors get observed by the camera at the ceiling and track a TurtleBot 2 robot that serves as a moving target. The AR.Drones are equipped with an inertial measurement unit (IMU), an ultrasound altimeter, two cameras, and WiFi communication. The down-

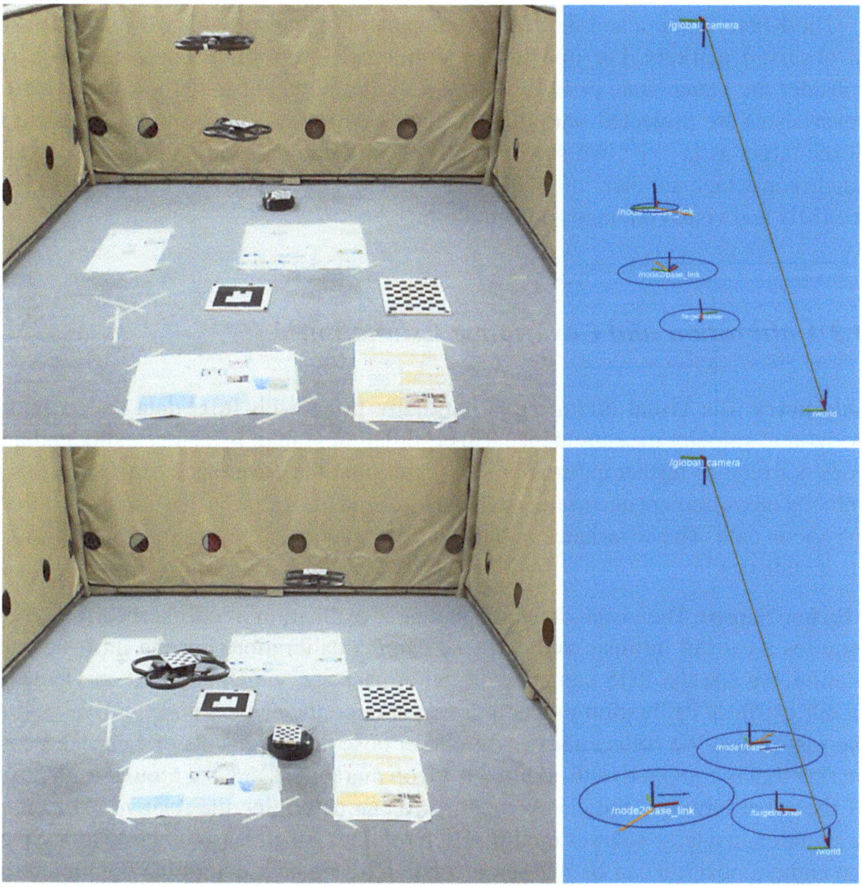

Fig. 5 Experimental setup: the Microsoft Kinect camera is mounted on the ceiling and observes the Parrot AR.Drones. A TurtleBot 2 serves as a moving target that is tracked by the AR.Drones. The AR.Drones and the target are equipped with checkerboard markers. The state estimates are shown as *blue arrows*, the corresponding covariances are represented by *blue ellipses*. The commanded velocities are shown as *orange arrows*. *Top two* AR.Drones tracking the target in a string topology. *Bottom two* AR.Drones tracking the target in a flat topology

looking camera is used internally to estimate the visual odometry, which is fused with the IMU information of the quadrotor. We modified the forward-looking camera to be tilted 45° downwards to track the target on the ground (see Fig. 4). The target and the quadrotor were both equipped with visual markers for relative pose estimates. In our initial experiments, we used ARToolKit markers [4], which were detected with frequent outliers. The checkerboard markers, we use in our current system, were detected using OpenCV with only occasional outliers and less noise. We use checkerboards with varying number of rows and columns to distinguish between the robots and the target. A detailed graph of the information flow of our system is shown in Fig. 3.

The Kinect camera and the UAV front camera images provide 3D relative poses of observed markers. For operational simplicity, in our EKF implementation we consider the planar state pose $[x, y, \psi]^T$ (all measurements and the corresponding covariances are projected onto the XY-plane). Moreover, we estimate the position of the target as $[x, y]^T$. We send velocity control commands $[v_x, v_y, \omega_\psi]^T$ to each quadrotor, which are then internally converted to appropriate motor velocities given the IMU and visual odometry information.

4.2 Calibration and Covariance Estimation

Odometry The visual odometry is internally fused with IMU data and provides horizontal velocity measurements. This estimation system is factory-calibrated and does not require further calibration. We determine the covariance of the horizontal velocity measurement uncertainty using the ground truth motion that is extracted from the Vicon data. The covariance of the visual odometry follows from straightforward error statistics.

Marker Sensor The visual detection and pose estimation of checkerboard markers requires a careful intrinsic and extrinsic camera calibration. For the intrinsic calibration, we use the ROS camera calibration package, which is based on OpenCV. In our extrinsic calibration procedure, we estimate the camera pose with respect to the robot base. We collect a series of marker pose measurements of a checkerboard marker that is equipped with additional Vicon markers. Using the ground truth poses of the robot base and the checkerboard, we can determine the relative 3D camera orientation in a least-squares minimization routine of the measurement errors. Since the camera position can be measured accurately, we only determine its orientation from recorded data. Furthermore, we determine the pose of the camera at the ceiling using a large checkerboard with additional Vicon markers on the floor.

In the second step, we use the same type of recorded data as for the extrinsic calibration to statistically determine the 3D position and orientation covariance of the marker pose measurements.

4.3 Height Stabilization

While the ultrasound altimeter provides accurate and reliable height measurement in single-robot experiments, the ultrasound sensors suffer from substantial crosstalk in multi-robot settings. This results in frequent measurement outliers that confuses the internal height estimation and stabilization of the AR.Drone and can cause serious crashes due to unpredictable height control behaviors.

A natural solution to this problem would be to take ultrasound measurements in an interleaved way. Since the AR.Drone low-level software is not open-source, we decided to implement a workaround using Vicon height estimates. In particular, we use a PD controller for determining vertical velocity commands to keep the robots at their desired height.

4.4 Results

We conducted a series of real robot experiments as a proof of concept of our approach; the videos are available online.[2] We started each experiment by controlling the robot manually. During all multi-robot experiments, the height stabilization controller was enabled. Once the EKF was initialized, the cooperative target tracking controller was turned on and took over control. We evaluated the performance of our method using Vicon ground truth poses recorded throughout the experiment (see Fig. 6).

Insights and Limitations During the practical evaluation we encountered several challenges—the prodigal gap between the simulations and reality. First, the system is highly influenced by the small field of view of the cameras, which results in tracking loss if an aggressive control command is executed. Second, the information about roll and pitch of the quadrotor received from the AR.Drone has a significant influence on the measurement projection. It introduces additional uncertainty in the EKF, which we account for in a first-order error propagation in the measurement projection.

Single-Robot Experiment In a first experiment, we deployed a single robot to track a moving target. Although the target was moving extensively in all directions, the robot was able to behave stably (see the top row of Fig. 6). The robot stayed below the global camera, which resulted in high certainty of its position and it mostly changed its orientation such that its field of view followed the target. In this experiment we obtained the smallest position errors of the target and the robot.

Two-Robot Experiment in Flat Topology The next experiment was performed with two robots in a flat topology (arranged on the same level) and a moving target. In this case node 2 started without having the target in its field of view. After the target was localized by node 1, node 2 was able to change its orientation to join tracking the target. One can notice higher uncertainty in the pose estimation of node 2 (see the middle row in Fig. 6), which was mainly caused by the small field of view of the global camera. In order to avoid collisions between two robots the repelling force was introduced, however, it frequently pushed node 2 out of the global camera view causing higher uncertainty in its position estimates.

Two-Robot Experiment in String Topology The last experiment consisted of two robots in a string topology (one above the other) and a moving target. One can notice two peaks in the target position error (see the bottom row of Fig. 6) that correspond to the situation where the lower robot was pushed down by the air stream of the higher robot. Since the motors of the AR.Drones do not provide enough torque to

---

[2]http://robotics.usc.edu/~hausmankarol/videos/iser_videos.

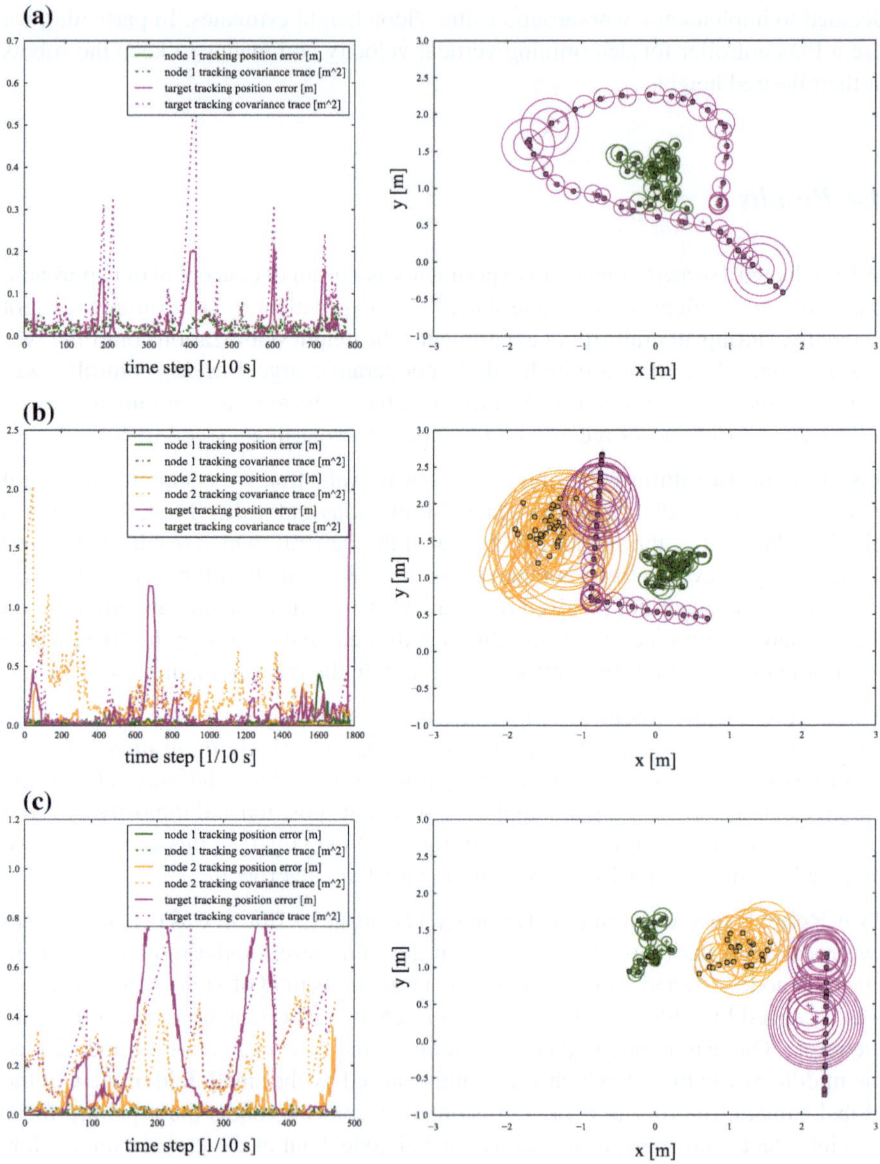

Fig. 6 Row **a** shows the results of an experiment with one robot tracking a moving target. Row **b** and **c** show target tracking of two robots (node 1 and node 2) in the flat and string sensing topology, respectively. *Left* The error of the EKF position estimates and the trace of the EKF covariances of the individual robots and the target for the full trajectory. *Right* An extract of the trajectory and the state estimates of the EKF. The actual trajectory is shown as thick dots connected by a *solid line*. The EKF means are indicated by '+' and the covariances are shown as *ellipses*.

compensate for strong air streams, the lower robot was substantially less stable. It is also worth noticing that although the target was lost, the system was able to recover and continue tracking.

5 Conclusions

We presented a probabilistic multi-robot control approach that considers onboard sensing and topology switching for target tracking. Our method generates locally optimal control while keeping polynomial complexity. We evaluated our approach in a number of simulations and showed a proof of concept with the real robot experiments. Our approach proved to flexibly adapt the topology and controls to the sensing limitations of the individual robots and the target movements. We presented the results of two topologies (flat and string) consisting of two AR.Drones, which demonstrated the robustness to the limited hardware capabilities of these inexpensive platforms. The scalability of the approach crucially hinges on our ability to quickly search the space of sensing topologies. At present, we restrict this search using a neighbor topology heuristic. In the future, we plan to use our method on a more capable platform and further explore principled topology switching techniques that preserve scalability.

Acknowledgments This work was supported in part by the National Science Foundation (CNS-1213128) and the Office of Naval Research (N00014-09-1-1031). Karol Hausman was supported by a fellowship from the USC Viterbi School of Engineering.

References

1. Adamey, E., Ozguner, U.: A decentralized approach for multi-UAV multitarget tracking and surveillance. In: SPIE Defense, Security, and Sensing (2012)
2. Ahmad, A., Lima, P.: Multi-robot cooperative spherical-object tracking in 3D space based on particle filters. Robot. Auton. Syst. **61**(10), 1084–1093 (2013)
3. Ahmad, A., Tipaldi, G.D. Lima, P., Burgard, W.: Cooperative robot localization and target tracking based on least squares minimization. In: Proceedings of the IEEE International Conference on Robotics and Automation (ICRA) (2013)
4. Billinghurst, M., Kato, H.: Collaborative augmented reality. Commun. ACM **45**(7), 64–70 (2002)
5. Charrow, B., Kumar, V., Michael, N.: Approximate representations for multi-robot control policies that maximize mutual information. Science and Systems (RSS). In: Proceedings of Robotics (2013)
6. Fink, J., Ribeiro, A., Kumar, V., Sadler, B.M.: Optimal robust multihop routing for wireless networks of mobile micro autonomous systems. In: Military Communications Conference (MILCOM) (2010)
7. Howard, A., Matarić, M.J., Sukhatme, G.S.: Localization for mobile robot teams using maximum likelihood estimation. In: Proceedings of the IEEE/RSJ International Conference on Intelligent Robots and Systems (IROS), vol. 1, pp. 434–439 (2002)

8. Huang, G., Truax, R., Kaess, M., Leonard, J.J.: Unscented iSAM: a consistent incremental solution to cooperative localization and target tracking. In: Proceedings of the European Conference on Mobile Robots (ECMR) (2013)
9. Johnson, S.G.: The NLopt nonlinear-optimization package. http://ab-initio.mit.edu/nlopt
10. Jung, B., Sukhatme, G.S.: Tracking targets using multiple robots: the effect of environment occlusion. Auton. Rob. **13**(3), 191–205 (2002)
11. Jung, B., Sukhatme, G.S.: Cooperative multi-robot target tracking. In: Distributed Autonomous Robotic Systems, vol. 7, pp. 81–90. Springer (2006)
12. Lupashin, S., Schollig, A., Sherback, M., D'Andrea, R.: A simple learning strategy for high-speed quadrocopter multi-flips. In: Proceedings of the IEEE International Conference on Robotics and Automation (ICRA) (2010)
13. Martinelli, A., Pont, F., Siegwart, R.: Multi-robot localization using relative observations. In: Proceedings of the IEEE International Conference on Robotics and Automation (ICRA) (2005)
14. Michael, N., Mellinger, D., Lindsey, Q., Kumar, V.: The GRASP multiple micro-UAV testbed. IEEE Robot. Autom. Mag. **17**(3), 56–65 (2010)
15. Mottaghi, R., Vaughan, R.: An integrated particle filter and potential field method for cooperative robot target tracking. In: Proceedings of the IEEE International Conference on Robotics and Automation (ICRA) (2006)
16. Mourikis, A.I., Roumeliotis, S.I.: Performance analysis of multirobot cooperative localization. IEEE Trans. Robot. Autom. **22**(4), 666–681 (2006)
17. Ong, L.-L., Upcroft, B., Bailey, T., Ridley, M., Sukkarieh, S., Durrant-Whyte, H.: A decentralised particle filtering algorithm for multi-target tracking across multiple flight vehicles. In: Proceedings of the IEEE/RSJ International Conference on Intelligent Robots and Systems (IROS) (2006)
18. Stump, E., Kumar, V., Grocholsky, B., Shiroma, P.M.: Control for localization of targets using range-only sensors. Int. J. Robot. Res. **28**(6), 743–757 (2009)
19. Thrun, S., Burgard, W., Fox, D.: Probabilistic Robotics. MIT Press, New York (2005)
20. Valenti, M., Bethke, B., Fiore, G., How, J.P., Feron, E.: Indoor multi-vehicle flight testbed for fault detection, isolation, and recovery. In: Proceedings of the AIAA Guidance, Navigation, and Control Conference and Exhibit (2006)
21. Wang, Z., Gu, D.: Cooperative target tracking control of multiple robots. IEEE Trans. Industr. Electron. **59**(8), 3232–3240 (2012)

Shape Change Through Programmable Stiffness

Michael McEvoy and Nikolaus Correll

Abstract We present a composite material with embedded sensing and actuation that can perform permanent shape changes by temporarily varying its stiffness and applying an external moment. Varying stiffness is a complementary approach to actuator-chain based approaches that can be accomplished using a large variety of means ranging from heat, electric field or vacuum. A polycaprolactone (PCL) bar provides stiffness at room temperature. Heating elements and thermistors are distributed along the bar so that local regions can be tuned to a specific temperature/stiffness. Applying an external moment using two tendon actuators then lets the material snap into a desired shape. We describe the composite structure, the principles behind shape change using variable stiffness control, and forward and inverse kinematics of the system. We present experimental results using a 5-element bar that can assume different global conformations using two simple actuators.

Keywords Multi-functional materials · Embedded computation · Variable stiffness

1 Introduction

Structural materials with the ability to change their shape have the potential to enable a new class of robotic devices ranging from assistive wearable technologies to morphable airplane wings [1, 2] and furniture [3]. Providing the necessary actuation is a key challenge, in particular as individual actuators need to be strong enough to deform, and possibly support, the entire structure. We propose a novel approach to shape change based on programmable stiffness, which combines large numbers of variable stiffness elements [4] with a small number of actuators that extend over and are strong enough to deform an entire structure. Ideally, shape change and variable stiffness control, which includes actuation, sensing and local feedback control

M. McEvoy · N. Correll (✉)
Department of Computer Science, University of Colorado at Boulder, Boulder 80309, USA
e-mail: nikolaus.correll@colorado.edu

M. McEvoy
e-mail: michael.mcevoy@colorado.edu

M.A. Hsieh et al. (eds.), *Experimental Robotics*, Springer Tracts
in Advanced Robotics 109, DOI 10.1007/978-3-319-23778-7_59

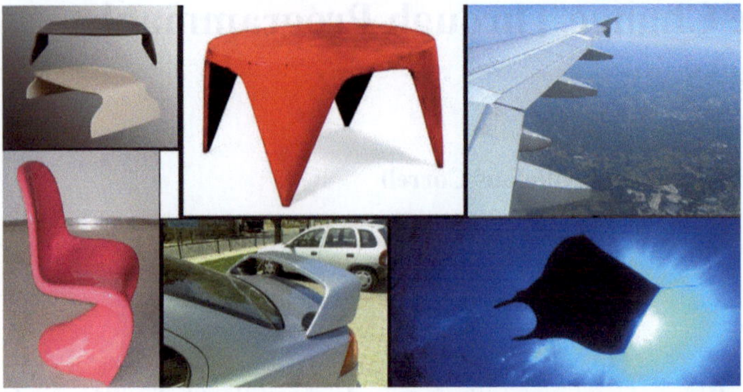

Fig. 1 Applications for a shape changing Robotic Material range from multi-functional furniture to shape-changing aerodynamic surfaces on boats, cars and aircraft

are integrated into the composite material itself. We dub this new class of materials "robotic materials" (RM) [4].

Our goal is to develop a RM that can change its shape and then lock it into a rigid configuration. When designing something that will change it's shape, the traditional approach is to use mechanical linkages and actuator chains. In contrast, we propose using smart materials that have the ability to change their stiffness (Fig. 1).

We believe that multi-DOF shape change in a material can be achieved through a change in stiffness and the application of simple 1-DOF forces. Beam theory states that the curvature through a given section of a beam is only a function of the stiffness and the applied loads. We show that the curvature is not affected by the stiffness of neighboring regions, implying that distributed control of the stiffness in a beam could be used in conjunction with simple external loads to generate arbitrary vertical displacements in the beam. We also show how inverse kinematic solutions for the proposed shape-change material can be found using the inverse Jacobian method.

2 Related Work

Existing approaches to high-DOF shape change are largely dominated by series actuator chains or modular robots [5], which suffer from the trade off between individual motor torque with weight and the requirement to deform or lift the entire system. More recently, [6, 7] have begun investigating pneumatic actuator chains, whose structural properties are limited by the available air pressure and valve technology. The proposed approach combines pneumatic actuation with the ability to vary stiffness of a structural material, which has been extensively studied. For example [8, 9] construct a variable stiffness material by exploiting the temperature-dependent variable shear modulus of polymers sandwiched between metal bars. While the stiffness of the materials in [8, 9] is dominated by the metal bars, [10] presents a variable stiffness material based on Field's metal with embedded Joule heating that can achieve stiffness changes of four orders of magnitude, albeit being limited to be either on or off. Other approaches to variable stiffness control rely on hydrogel [11], particle jamming [3, 12] (which both require valves and pumps), magneto-rheological [13–15] (which requires strong electro magnets), or mechanical effects [16] (requiring motors). While these materials provide a wide range of variable stiffness mechanisms to explore, to the best of our knowledge, these variable stiffness systems have never been combined with actuation to create shape-changing materials.

3 Principle of Operation

In this paper we demonstrate a variable stiffness RM with the ability to change its shape. This section describes the fundamental mechanics of materials equations on which we have designed our RM, the core material we have chosen for the design, the method that we use to apply external loads to physically deform the RM and the process involved with changing the RM from one configuration to another.

3.1 Mechanics of Materials

We have chosen to demonstrate the shape changing abilities of variable stiffness RMs with a beam. The shape of a beam is governed by the moments M applied along it's length, the stiffness E of the beam along the length, and the cross-sectional inertia I of the beam along it's length as shown in Eq. 1.

$$\kappa(x,t) = \frac{v''(x,t)}{\left(1 + v'(x,t)^2\right)^{\frac{3}{2}}} = \frac{M(x,t)}{E(x,t)I(x)} \tag{1}$$

Typically stiffness and cross-sectional inertia are fixed at the time of manufacture as E(x) and I(x), respectively. With variable stiffness RMs, we can now change the stiffness along the beam's length on demand so have E(x,t), i.e. the stiffness is now controllable as a function of time. This new ability allows a single actuation moment to induce a wide range of motion in the actuator.

3.2 Core Material

With advances in materials science, there are many examples of materials that can change their properties with the application of an external stimulus. Some of these materials and the morphing systems designed around them are highlighted in [2].

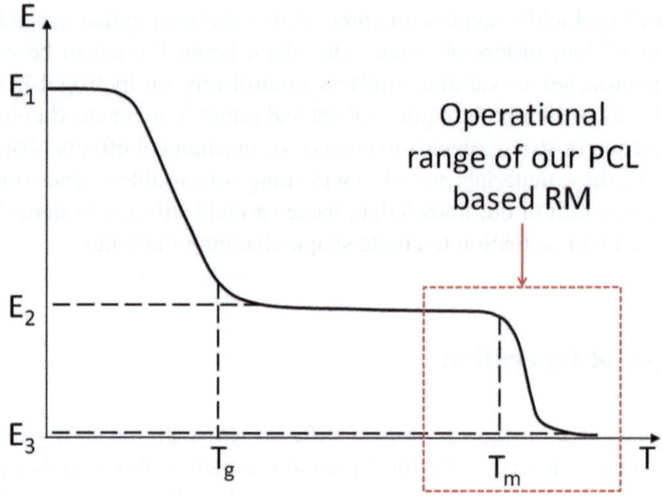

Fig. 2 For a typical thermoplastic, the Young's modulus drops when approaching the glass transition temperature and then drops again when approaching the melting temperature. It is this variation of Young's modulus with temperature change that we exploit in this particular RM

For the core material of our RM we use the thermoplastic polycaprolactone (PCL). Thermoplastics are inexpensive, easily manufactured and formed, and allow us to change their stiffness over multiple orders of magnitude simply by heating and cooling. Using Joule heating and temperature control, we can take advantage of the stiffness changes that occur as their temperature rises. PCL has a glass transition temperature of −50 °C and is in its rubber state at room temperature. PCL has a Young's Modulus of approximately 190 MPa at room temperature and drops to nearly 2 MP when molten (60 °C). Further material properties for PCL are described in [17].

3.3 *Application of External Loads*

To apply the external loads needed to change the beam's shape we use two mechanical actuators as in [18] that pull cables running along each side of the bar. This design choice allows us to leverage other results from the field of continuum robotics as described in Sect. 5.1. The cables are held in place by supports that are placed at increments along the length of the bar, leading to a number of discrete sections. Tension on the cables produces a constant moment M across each section and allows us to set either a positive or negative moments across all of the segments of the bar at once. This concept is shown in Fig. 3.

Controlling the material properties in each of these sections allows for different curvatures to be produced in different sections of the bar while applying a single load to one of the cables. We note that shapes in which both positive and negative

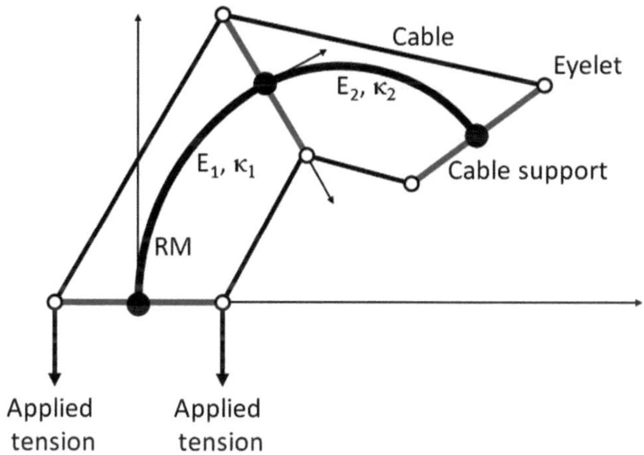

Fig. 3 A variable stiffness RM allows us to achieve different curvatures through different sections with only a single actuating force. The first section of the beam is set to a stiffness of E_1 while the second is set to E_2. With $E_1 > E_2$ we can get $\kappa_1 < \kappa_2$ by applying a single tension to the driving cable

curvatures are present require a two-step process in which the bar first locks in all negative curvatures and then all positive curvatures (or the other way round).

Applying Eq. 1 to a segmented RM with constant cross-section we obtain a piecewise function for the curvature of the beam. The continuity conditions state that the deflection curve is physically continuous and that the slopes for each segment of the beam must equal at the endpoints. Choosing to fix the base of the beam, we then have boundary conditions on the displacement and slope of the beam at it's base. The piecewise functions and the continuity and boundary conditions are shown in Eq. 2 where the subscript i denotes section number and A and B indicate a sections start and end points respectively.

$$\begin{aligned}
\kappa_i(t) &= \frac{M(t)}{E_i(t)I} \\
v_0(0,t) &= 0 && \text{Boundary conditions} \\
v_0'(0,t) &= 0 \\
v_i(B,t) &= v_{i+1}(A,t) && \text{Continuity conditions} \\
v_i'(B,t) &= v_{i+1}'(A,t)
\end{aligned} \tag{2}$$

3.4 Shape Control

In order for a thermoplastic to be locked into a specific shape, it must be heated past its melting point and then allowed to cool while being held in that configuration.

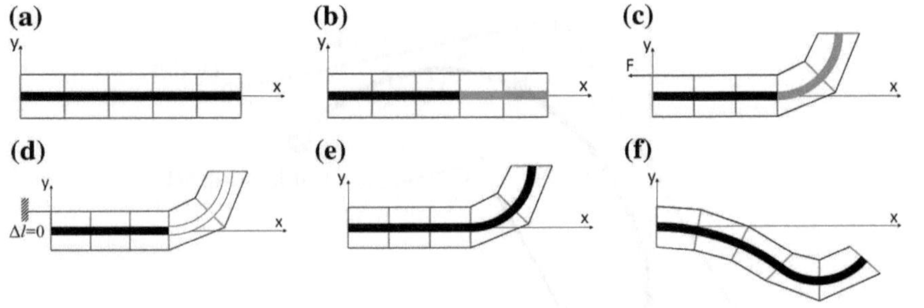

Fig. 4 A change in shape starts with an undeformed beam without any load applied (**a**). The segments that will have positive curvature are set to the appropriate stiffness, this change is noted by the light gray sections (**b**). Once the segment stiffnesses have been programmed, a load is applied to the cable, causing the segments to deform (**c**). The cable is then held at a constant displacement while the deformed sections are heated to melting, indicated by white (**d**). The segments return to their original stiffnesses and the cable tension is released, leaving the beam in the deformed state (**e**). This process is repeated for those segments with negative curvature and an arbitrary pose is achieved (**f**)

As noted in the previous section, the cables can be used to apply either a positive or negative moment across the beam, necessitating a two step process for shape changes involving both positive and negative curvatures. While this process poses some additional challenges (which we discuss in Sect. 7), we are able to demonstrate the possibilities of variable stiffness RMs and shape change using the following process (Fig. 4).

We start with an undeformed beam without any load applied and program a stiffness profile to it (Fig. 4a, b). Next a constant moment is applied along the length of the beam to deform its shape (Fig. 4c). The cable is held at a constant displacement while the PCL is heated to melting and then cooled, locking the shape of the deformed beam (Fig. 4d, e). The process is repeated to produce curvatures in the opposite direction (Fig. 4f), producing an arbitrary shape change in the beam.

4 Fabrication

The core functionality of this RM is enabled by embedding sensing and actuation directly into the material, in this case, a thermistor and a nichrome heating element.

First, we form the PCL bar by melting pellets of PCL and pressing them into an acrylic mold along with a nylon mesh. A nylon mesh is embedded into the PCL to help maintain the cross-section when sections are heated very close to their melting temperature. The finished bar is 6.35 mm thick by 25.4 mm wide and is 311.15 mm long. Five 50.8 mm sections are marked off with a 31.75 mm section at the base and 25.4 mm section at the end for clamping and cable mounting purposes.

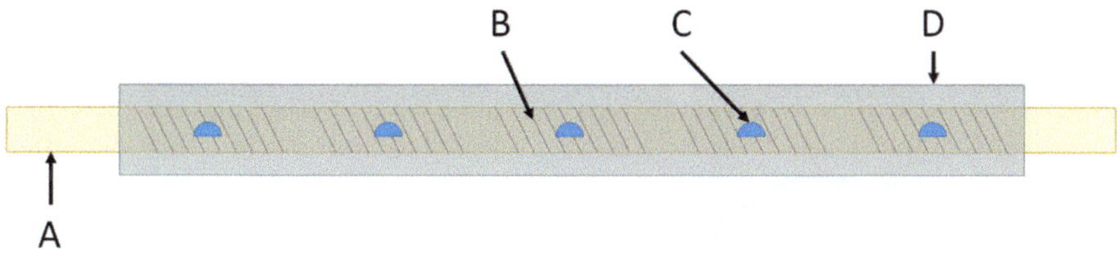

Fig. 5 The demonstrated RM consists of a PCL bar (*A*), nichrome heating elements to heat the individual sections (*B*), embedded thermistors to monitor and control the temperature of each section (*C*), and an encasing of silicon (*D*) for insulation and to help maintain the cross-section when sections are heated close to the melting temperature

Next thermistors are embedded into the center of each section to monitor the temperature. Then each section is wrapped with a length nichrome wire which is secured to the beam with Kapton tape. The nichrome wire is wrapped around the bar with a 5.08 mm spacing, resulting in a resistance of $48 \pm 1\Omega$ per section.

Finally the entire bar is encapsulated in a thin layer of silicone to insulate the heating elements from the environment and help maintain the cross-sectional shape for sections heated very near the melting point. A schematic of the design is shown in Fig. 5 and can be seen in the experimental setup in Fig. 7.

Determination of the temperature propagation through the sections as well as sectional heat-up and cool-down cycles using a similar design are described in [4].

Our RM is made completely from commercial-off-the-shelf products. We use PCL sold by SparkFun Electronics under the trade name PolyMorph (TOL-10951), 36 gage Nichrome 60 wire from Jacobs Online, thermistors from Digi-Key (490-4664-ND), and Ecoflex 00-30 silicon from Smooth-on.

5 Control

Our choice to use a cable driven system to apply the forces necessary for shape change allow us to utilize forward and inverse kinematics results from the continuum robotics field [7, 19]. A short overview of these results and how they are applied to our RM is given in this section.

5.1 Forward Kinematics

We propose to model the variable stiffness RM using Piecewise Constant Curvature (PCC) models from continuum robotics. PCC assumes that the curvature in a section is constant so that the forward kinematics are treated as consecutive transformations as shown in Fig. 6a where the transformation from the coordinate system of section

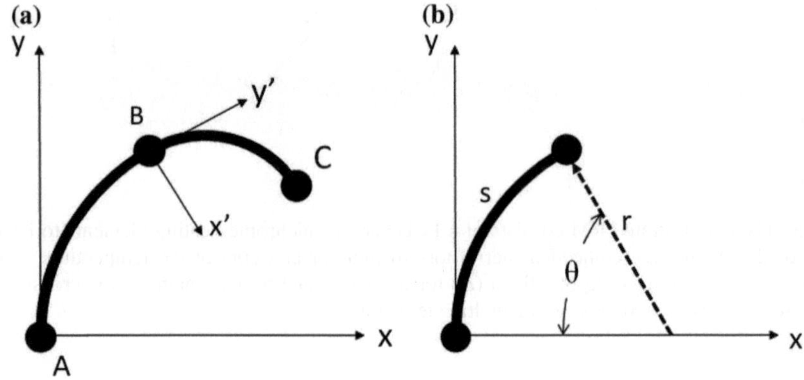

Fig. 6 Assuming that the curvature in each section of the RM is constant, the forward kinematics is a series of transformations for each section. Knowing the curvatures for each section fully defines the pose of the RM

$i + 1$ to section i is governed by Eq. 3, where $x_0 = y_0 = \theta_0 = 0$.

$$\begin{bmatrix} x_i \\ y_i \\ \theta_i \end{bmatrix} = \begin{bmatrix} \cos(\theta_{i-1}) & \sin(\theta_{i-1}) & 0 \\ -\sin(\theta_{i-1}) & \cos(\theta_{i-1}) & 0 \\ 0 & 0 & 1 \end{bmatrix} \begin{bmatrix} \kappa_i^{-1}\,(1 - \cos\,(\kappa_i s_i)) \\ \kappa_i^{-1} \sin\,(\kappa_i s_i) \\ 0 \end{bmatrix} + \begin{bmatrix} x_{i-1} \\ y_{i-1} \\ \theta_{i-1} \end{bmatrix} \tag{3}$$

The local coordinates along a segment are given by Eq. 4 and shown in Fig. 6b.

$$\begin{aligned} x &= \kappa_i^{-1}\,(1 - \cos\,(\kappa s)) \\ y &= \kappa_i^{-1} \sin\,(\kappa s) \\ \theta &= \kappa s \end{aligned} \tag{4}$$

5.2 *Inverse Kinematics*

As it is infeasible to find analytical solutions for the inverse of (3) for large numbers of segments, we chose an inverse Jacobian method as described in [7, 19] for this class of robot. This approach also allows us to impose limits on the sectional curvatures so that the proposed pose changes are always reachable. Let Δx, Δy and $\Delta\theta$ be small changes from the current pose of the beam in the direction of a desired pose $(x, y, \theta)'$. We can then write

$$(\Delta x, \Delta y, \Delta\theta)' \approx J(\Delta\kappa_1, ..., \Delta\kappa_n)' \tag{5}$$

where J is a Jacobian matrix with dimension [3 × n] containing the partial derivatives $\frac{\partial x}{\partial \kappa_i}$, $\frac{\partial y}{\partial \kappa_i}$, and $\frac{\partial \theta}{\partial \kappa_i}$ with $i = [1 \ldots n]$ and n the number of segments. These partial derivatives can be calculated analytically from Eq. (3).

Appropriate values for the curvature of each segment can now be calculated from Eq. (5) by calculating the pseudoinverse $J^+ = J(J^T J)^{-1}$. As each individual segment is a single beam that adheres to Eq. (1), we can calculate the required $M/E(i)$ ratio, and therefore the temperature we need to set a bar, by $M/E(i) = I\kappa(i)$.

6 Experiments

In this section we detail the experimental setup, how we determined section curvatures as a function of temperature, and step through an example using these values to achieve a desired pose.

6.1 Experimental Setup

The experiments are conducted with the five section PCL bar described in Sect. 4. The bar is mounted to a table and deflections of the beam are limited to the plane of the table. Each section is monitored and controlled by a central computer. The tension applied to either of the cables is limited to 9.8 N. This restriction helps to emphasize the range of motion possible using only a single actuation force, but need not be limited in other applications. An overview of the experimental setup is shown if Fig. 7.

During each test, the temperature histories of each section are recorded along with the fiducial positions of the section endpoints.

6.2 Determining $\kappa(T)$

In the first set of experiments we capture the range of motion of our device and also characterize the relationship between curvature and temperature. For this series of tests, a tension is applied to either the left or right cable while the beam is held at a uniform temperature. Displacements are recorded at various temperatures between room temperature and 50 °C with each test being repeated five times for a total of 80 trials. The range of motion of our RM is shown in Fig. 8, where the five trials for each temperature have been averaged. After collecting the data from each of the tests the displacements of each of the section ends was averaged for each temperature and curvatures through each section were found using a least-squares method and the PCC assumption. Figure 9 shows the results of this process, with the expected result that the curvature increases dramatically as the temperature reaches the melting point.

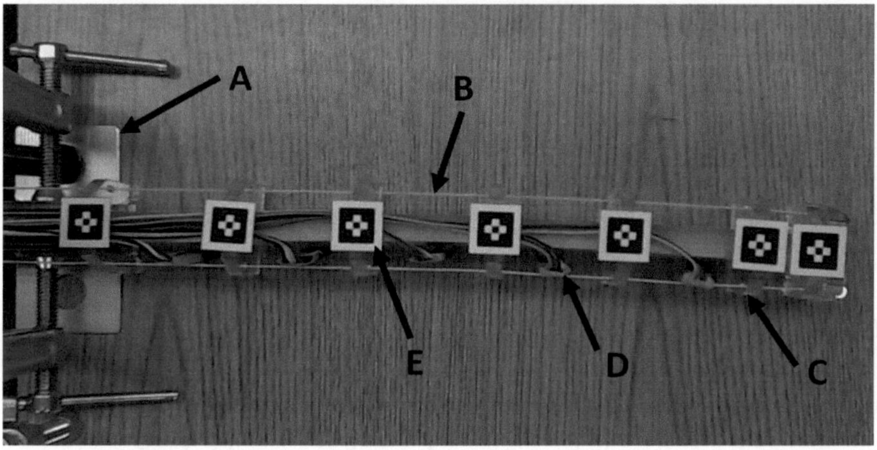

Fig. 7 For the experiments, the PCL bar is clamped to a table (*A*). Cables (*B*) are routed through sliders (*C*) and apply a constant moment along the length of the bar when tension is applied to the cables. Each section's thermistor and nichrome heater are connected to the control computer (*D*), and the displacement of each section is monitored by tracking a fiducial (*E*)

Of note is that not centering the mesh in the PCL resulted in asymmetric positive and negative curvatures.

The error in curvature tends to increase with temperature. This is somewhat expected since we are in a steeper region of the E versus T region (Fig. 2).

6.3 Inverse Kinematics

To verify that our RM can achieve a desired pose, we take a reachable point from Fig. 8 and use the IK solver to find a possible set of sectional curvatures to be applied to each section. We then use the two step process outlined in Sect. 3.4 to achieve the desired pose. Figure 10 shows the results of this open-loop control including intermediate shapes and the resulting error.

7 Experimental Insights

Our variable stiffness RM is able to change shapes using a constant actuation force and programmable stiffness of its sections. This first-of-its-kind programmable material demonstrates the possibilities of RMs with respect to shape change, but also offers new challenges in design and control.

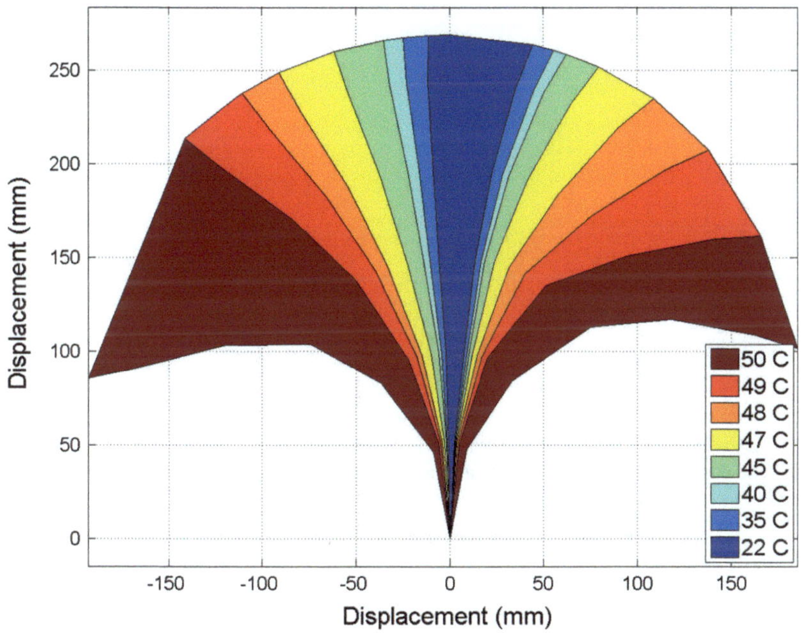

Fig. 8 The range of motion of the variable stiffness RM. When all of the heating elements in the beam are off, the stiffness is at a maximum and the displacement in the beam is limited. Activating all of the heating elements the stiffness of the beam is at a minimum and a much larger range of motion is achieved using the same actuation force

When determining the curvature versus temperature profiles we assumed that all sections would have the same curvature, however inconsistencies in manufacturing of the bar and of the heating elements need to be addressed. We found that slight variations in the heating elements caused nonuniform heating of the PCL in each section. In addition, the manufacturing process of the PCL bar left slight variations in the thickness of the bar, but at the time of fabrication, bars of PCL were not commercially available. Addressing these issues will serve to decrease the error found in the curvature versus temperature (Fig. 9).

Setting of the thermoplastic into the desired shape (Fig. 4c–e) also poses some control problems not addressed by our RM. Figure 12b shows a two segment RM where end section has been set to the desired stiffness, E_2 and needs to be locked into this shape. When locking the section into place, the temperature is raised to melting. The stiffness in this section is lowered to E_3, which is accompanied by an increase in curvature. Since the cable's length is held constant during this process, the fore section relaxes a proportional amount as seen in Fig. 12b. To avoid this problem, additional control needs to be implemented to adjust the temperature of the other sections.

Due to the large variation in material properties, the accuracy of open loop control is severely limited. This limitation could be overcome by implementing a closed loop

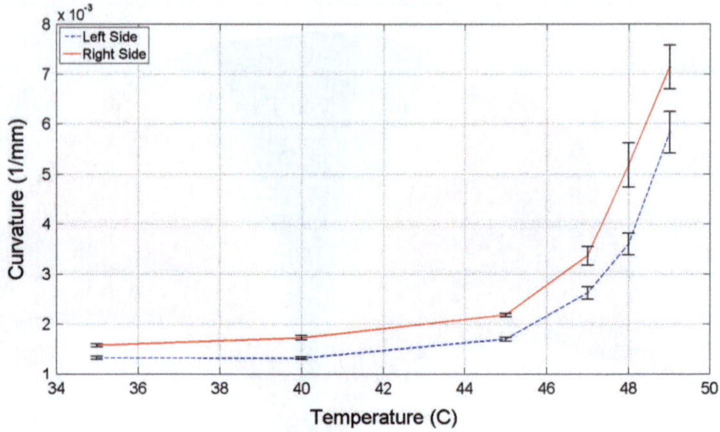

Fig. 9 The curvature increases dramatically as the temperature approaches melting

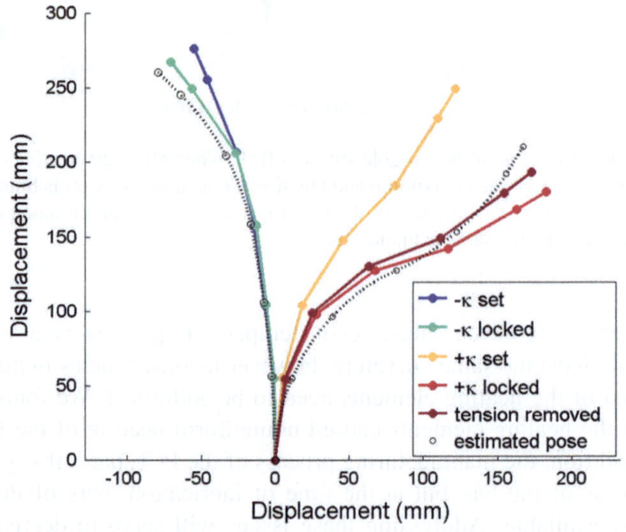

Fig. 10 To reach a desired pose, we first set the negative curvatures found by the IK solution and lock them in before programming the positive curvature sections. The *dashed lines* are what we expect from the forward kinematics. The change in pose from setting the $+\kappa$ sections and locking them is discussed in Sect. 7

control scheme that does not only rely on thermistors, as in this paper, but also on embedded curvature sensors. We note that such a controller can also be implemented in a fully distributed way as the overall shape is only dependent on local curvatures.

In order for the proposed approach to work repeatably, care must be taken to limit force and stiffness combinations so that the material does not yield. Too great of a

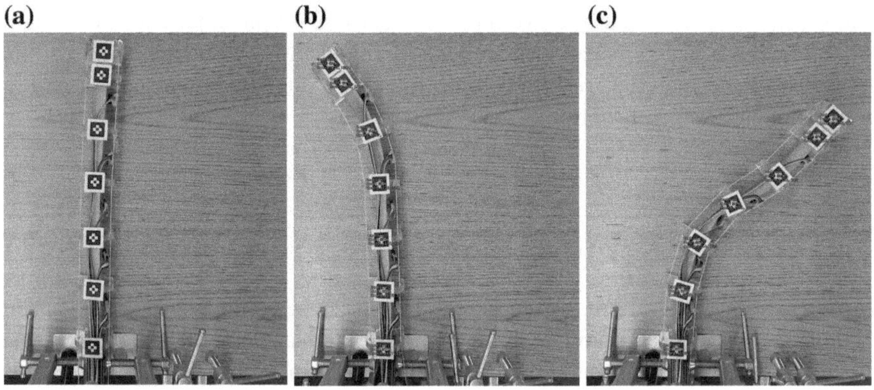

Fig. 11 Snapshots from the experimental trial shown in Fig. 11. **a** shows the initial position of the RM, **b** shows the RM after setting the negative curvatures and **c** shows the pose after setting the positive curvatures

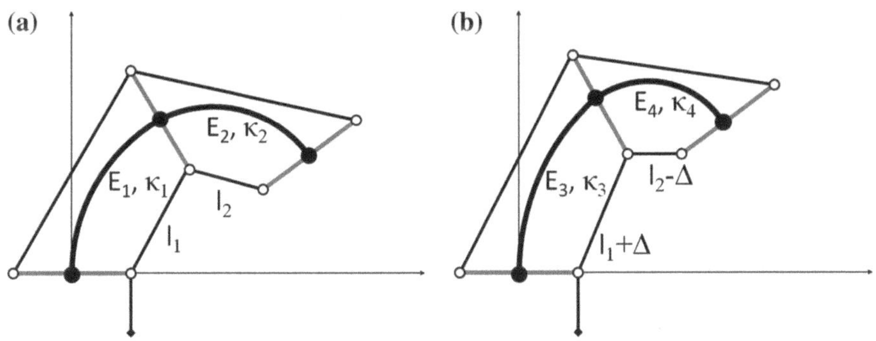

Fig. 12 **a** shows a RM where κ_2 has been set and needs to be locked. When locking in the curvature, the thermoplastic must be raised to melting, which corresponds in a drop in stiffness $E_2 > E_4$ (**b**). If the temperature of section one is not adjusted, the change in stiffness will cause a change in the curvatures from $\{\kappa_1, \kappa_2\}$ to $\{\kappa_3, \kappa_4\}$ as the first section relaxes and the second section increases in curvature

force would mean that the material would yield and plastically deform and the beam would no longer be operating in the region where Hooke's law applies.

A limitation of the proposed shape-change using variable stiffness approach is that the necessary motions to achieve even subtle shape changes can be quite large. This is nicely illustrated in Fig. 10. In order to reach the desired final position to the right, the material first needs to bend to the left. This can only be overcome by adding additional actuators throughout the RM, trading the limitations of this approach with that of conventional multi-link robotic systems.

We chose tendon-based actuators in this paper as they provide a good force-to-weight ratio since the tendons can be installed at a distance from the material. However, any actuator that can provide a constant moment across the length of the beam is suitable. For example, we have also considered pneumatic actuators as in [6]. While these actuators can provide large moments, they add considerably to the thickness of the material, resulting in an increased cross-sectional inertia and an overall smaller range of motion.

8 Conclusion

We demonstrate experimentally that shape change can be obtained by locally varying the stiffness of a material and applying a single external moment. The experiments show that the proposed shape-changing material can indeed reach a wide range of possible shapes that are only limited by the available external moments and the properties of the material itself.

Relying on melting via joule heating allows to reach good structural stability, but is power intense, slow, and dependent on the environment temperature. In the future, we wish to investigate other approaches for stiffness control, including electro-rheological fluids and pneumatic jamming. We would also like to investigate other actuators that provide good force to weight ratios, such as twisted wires, pulleys, and McKibbon-style actuators.

There is a trade-off between range of motion and structural stiffness. Structures that require a large range of motion will need to be very thin and thus will only be able to support small loads. Structures that require only a limited range of motion could potentially support larger actuators and support heavier loads.

Although the properties of variable stiffness robotic materials are compelling, we observe that their control is much more challenging then conventional, stiff actuator chains. In order to make such RMs accurate and precise, we will not only need to investigate embedded sensing and feedback control to overcome differences in polymer, actuators, and sensors, but also investigate novel manufacturing techniques to make such systems. In return, such robotic materials might not only be able to accurately change their shape, but also to respond to disturbances and damages in unprecedented ways.

Acknowledgments This work has been supported by the Airforce Office of Scientific Research under grant number FA9550-12-1-0145, the National Science Foundation under grants number #1150223 and #1153158, and a Beverly Sear's Graduate Student Research Grant. We are grateful for this support.

References

1. Vasista, S., Tong, L., Wong, K.: Realization of morphing wings: a multidisciplinary challenge. J. Aircr. **49**(1), 11–28 (2012)
2. Weisshaar, T.A.: Morphing aircraft systems: historical perspectives and future challenges. J. Aircr. **50**(2), 1–17 (2013)

3. Ou, J., Yao, L., Tauber, D., Steimle, J., Niiyama, R., Ishii, H.: jamSheets: thin interfaces with tunable stiffness enabled by layer jamming.In: Proceedings of the 8th International Conference on Tangible, Embedded and Embodied Interaction, pp. 65–72. ACM (2014)
4. McEvoy, M.A., Correll, N.: Thermoplastic variable stiffness composites with embedded, networked sensing, actuation, and control. J. Compos. Mater (2014)
5. Yim, M., Zhang, Y., Duff, D.: Modular robots. IEEE Spectr. **39**(2), 30–34 (2002)
6. Correll, N., Onal, C.D., Liang, H., Schoenfeld, E., Rus, D.: Soft autonomous materials—using active elasticity and embedded distributed computation. In: 12th International Symposium on Experimental Robotics, Springer Tracts in Advanced Robotics, Vol. 79, pp. 227–240 (2014)
7. Marchese, A.D., Konrad, K., Onal, C.D., Rus, D.: Design, curvature control, and autonomous positioning of a soft and highly compliant 2D robotic manipulator. In: 2014 IEEE International Conference Robotics and Automation, IEEE (2014)
8. Gandhi, F., Kang, S.G.: Beams with controllable flexural stiffness. Smart Mater. Struct. **16**(4), 1179–1184 (2007)
9. Murray, G., Gandhi, F.: Multi-layered controllable stiffness beams for morphing: energy, actuation force, and material strain considerations. Smart Mater. Struct. **19**(4), 11 (2010)
10. Shan, W., Lu, T., Majidi, C.: Soft-matter composites with electrically tunable elastic rigidity. Smart Mater. Struct. **22**(8), 085005 (2013)
11. Shanmuganathan, K., Capadona, J.R., Rowan, S.J., Weder, C.: Biomimetic mechanically adaptive nanocomposites. Prog. Polym. Sci. **35**(1), 212–222 (2010)
12. Brown, E., Rodenberg, N., Amend, J., Mozeika, A., Steltz, E., Zakin, M.R., Lipson, H., Jaeger, H.M.: Universal robotic gripper based on the jamming of granular material. Proc. Nat. Acad. Sci. **107**(44), 18809–18814 (2010)
13. Majidi, C., Wood, R.J.: Tunable elastic stiffness with microconfined magnetorheological domains at low magnetic field. Appl. Phys. Lett. **97**(16), 164104–164104 (2010)
14. Chen, J., Liao, W.: Design, testing and control of a magnetorheological actuator for assistive knee braces. Smart Mater. Struct. **19**(3), 035029 (2010)
15. Varga, Z., Filipcsei, G., Zrínyi, M.: Magnetic field sensitive functional elastomers with tuneable elastic modulus. Polymer **47**(1), 227–233 (2006)
16. Pratt, G.A., Williamson, M.M.: Series elastic actuators. In: IEEE/RSJ International Conference on Intelligent Robots and Systems 95.'Human Robot Interaction and Cooperative Robots, Proceedings, vol.1, pp. 399–406. IEEE (1995)
17. Averous, L., Moro, L., Dole, P., Fringant, C.: Properties of thermoplastic blends: starch-polycaprolactone. Polymer **41**(11), 4157–4167 (2000)
18. Li, C., Rahn, C.D.: Design of continuous backbone, cable-driven robots. J. Mech. Des. **124**(2), 265–271 (2002)
19. Webster, R.J., Jones, B.A.: Design and kinematic modeling of constant curvature continuum robots: a review. Int. J. Robot. Res. **29**(13), 1661–1683 (2010)

Part XII
Keynote—Experimental Robotics in Archeology

French Archaeology's Long March to the Deep—The *Lune* Project: Building the Underwater Archaeology of the Future

Michel L'Hour and Vincent Creuze

Abstract This paper describes our project to study the wreck of the *Lune* and the determination of French archaeologists to develop viable techniques for working on deep-water wrecks. We begin by sketching out the general context of underwater archaeology in France before describing in detail our long march to the deep. We conclude with an overview of the *Corsaire Concept* project for deep sea archaeology.

1 Underwater Archaeology Past and Present

Searching for wrecks and the recovery of their cargoes has, throughout the history of humanity, been the principal motivation behind the development of underwater exploration. 2000 years ago, under the Roman Empire, corporations of specialist divers were offering their services to work on the underside of boats and retrieve sunken cargoes. Over the following fifteen centuries methods barely improved and, despite the projects of a few geniuses like Leonardo da Vinci, we have to look to the seventeenth century for the appearance of the first diving suits and the first machines capable of allowing men to work underwater in any meaningful way. In eighteenth century, in France, several diving suits were being used for work in harbours and on wrecks.

In 1715, while Frenchman Pierre Rémy de Beauve was building his first aqualung (Fig. 1), John Lethbridge, a wool merchant who lived in South West England, was using a diving machine of his own invention to salvage valuables from wrecks. His machine was, in essence, an airtight oak barrel which allowed "the diver" to submerge long enough to retrieve underwater material (Fig. 2). Lethbridge dived on the wrecks of four English men-of-war, a few East Indiamen, both English and Dutch, two

M. L'Hour (✉)
DRASSM (Underwater Archaeology Research Department, Ministry of Culture & Communication), Marseilles, France
e-mail: michel.lhour@culture.gouv.fr

V. Creuze
LIRMM (CNRS/Montpellier University), Montpellier, France

M.A. Hsieh et al. (eds.), *Experimental Robotics*, Springer Tracts in Advanced Robotics 109, DOI 10.1007/978-3-319-23778-7_60

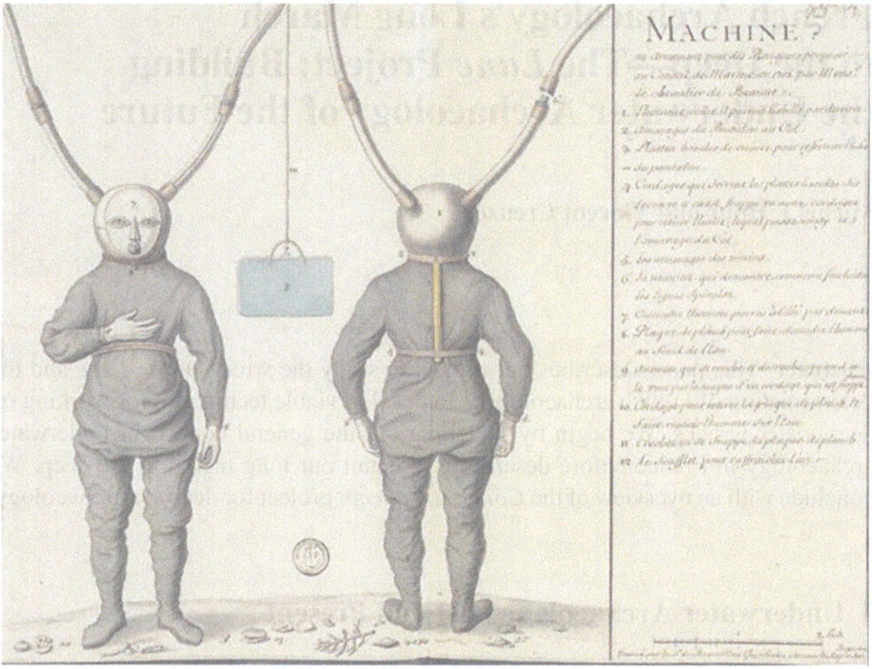

Fig. 1 Pierre Rémy de Beauve's first aqualung (1715). Pencil and watercolour. Paris. A.N. Mar 6JJ89, item 119A and 119B

Spanish galleons and a number of galleys, but what is perhaps less well-known is that his machine was also tried and tested in France, near Toulon.

The works of Lethbridge, Rémy de Beauve and a few others inspired numerous inventors in the eighteenth and nineteenth centuries and the rate of progress accelerated. Throughout Europe, and especially in France, new methods and machines for working underwater were being built and tested. Unsurprisingly, this effervescence of invention led in the first half of the nineteenth century, in 1839 to be exact, to a virtual industrialization of diving with the arrival of the first diving helmet, the fruit of a collaboration between Englishman Charles Anthony Deane and the Prussian artillery officer Augustus Siebe. For more than a century, their diving suit was the standard apparatus for divers, before being replaced by the aqualung, invented in 1943 by Émile Gagnan and Jacques-Yves Cousteau. Consequently, in the period 1850–1950, few wrecks lying in waters of up to 50 m were out of the reach of human divers Throughout this long period of development, diving had little to do with underwater archaeology. The primary motivation of these pioneers was either to get rich by recovering cargoes known to be of great value, or to demolish wrecks that posed a threat to shipping.

However, certain individuals understood very early on that underwater wrecks and sunken artefacts could have historical value. Several archive documents dating

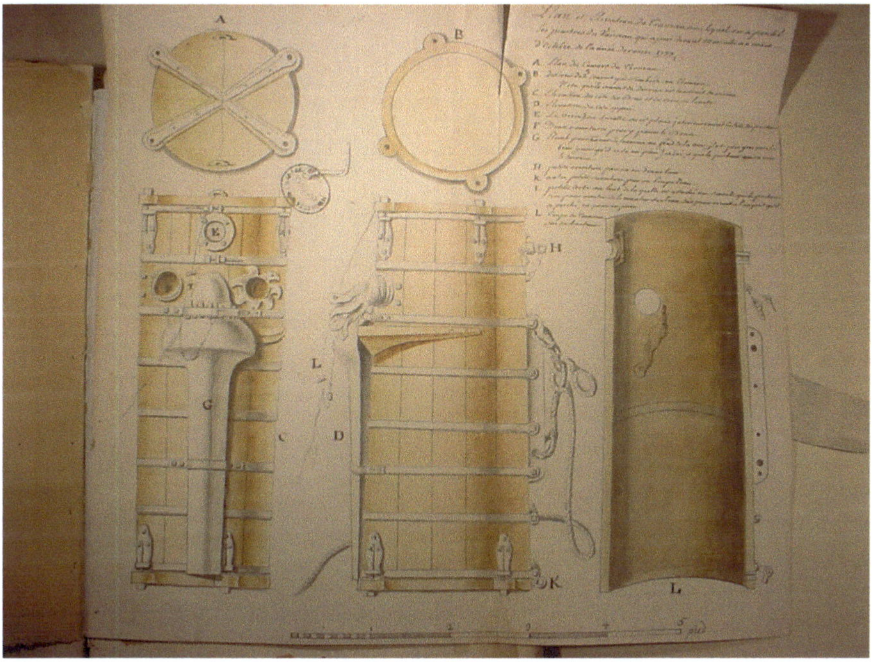

Fig. 2 John Lethbridge's oak barrel diving machine (1715) Paris A.N. Mar. G 111, f°98, Document dated 1734

from the middle of the nineteenth century indicate that enthusiasts were already carrying out underwater exploration solely for the purpose of studying objects from the past. The oldest of these documents dates from 1854 and describes diving in Lake Geneva, near the town of Morges. The purpose of the dive was to recover artefacts from prehistoric lacustrine settlements (Fig. 3).

All these endeavours were, however, isolated events. Underwater archaeology proper undoubtedly began in 1907 with the work of the French archaeologist Alfred Merlin. Indeed, in June of that year some Greek fishermen diving for sponges discovered the first antique wreck, lying in 40 m of water off the Tunisian town of Mahdia. Then, Alfred Merlin was the director of Tunisia's department of antiquities, Tunisia being a French protectorate at that time, and he decided to carry out on the wreck the first methodical underwater exploration for the purposes of archaeology (Fig. 4). Thus, from 1907 to 1913, the Mahdia wreck was the focus of a series of archaeological investigations led by Alfred Merlin who, not a diver, directed a team of Greek divers from the surface. Sunk between 80 and 70 BC, the Mahdia ship contained an impressive collection of architectural items and sculptures made from marble and bronze. Every year Alfred Merlin published excavation reports and analyses, and these immediately defined the discipline. His results fascinated archaeologists and convinced them of the importance of sunken artefacts. Such was the influence of his

Fig. 3 Recovery of artefacts from prehistoric lacustrine settlements in Lake Geneva, near the town of Morges (1854)

Fig. 4 Alfred Merlin (1907–1913) supervised the first archaeological study of an antique wreck, off Mahdia (Tunisia). *Photo* Bardo Museum (Tunis, Tunisia)

work that, in 1928, the French archaeologist Salomon Reinach would write the sea is "the largest museum in the world".

This prophetic conviction would not, however, be fully realized for another 50 years. Underwater diving would indeed remain the preserve of a very small number of professionals, of whom none was an archaeologist.

And so, for nearly half a century after the Mahdia excavations, archaeologists continued to be shipbound, to ignore the underwater world. Another invention was required to change the situation. And that came about in 1943 with the invention of the aqualung, generally known today as SCUBA, an acronym for "self-contained breathing apparatus". In France, the field of underwater archaeology does indeed owe its development to Jacques-Yves Cousteau and Émile Gagnan when, in 1943, they created the aqualung. This invention provided easy access to the "silent world" and led to the discovery of a large number of wrecks along the Mediterranean coasts of France and Italy. Not surprisingly, by the 1950s a number of divers and archaeologists were also becoming interested in this submerged heritage.

Salvage work conducted by the Italian Nino Lamboglia on the Albenga wreck in 1950 was the last archaeological operation to use the standard diving dress of copper helmet, canvas suit and weighted boots; while works on the ancient wrecks of Grand-Congloué in the Bay of Marseilles between 1952 and 1957 were the first true underwater archaeological excavations in the world (Fig. 5).

For the Bay of Marseilles project, archaeologist Fernand Benoit directed operations throughout but he was not trained to dive on the site. The excavations were undertaken by Cousteau's dive team. Faced with a number of scientific and technical problems, it soon became clear that they needed an archaeologist to monitor the work directly. This meant that the archaeologists needed to learn to dive.

This observation was incredibly significant because all along the coasts of France groups of serious divers were starting to search methodically for wrecks and this led to discoveries and, inevitably, looting.

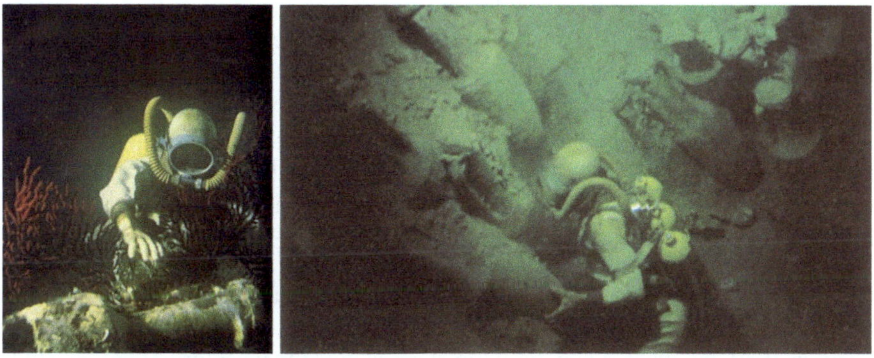

Fig. 5 Archaeological excavations of the Grand-Congloué in the Bay of Marseilles, France, from 1952 to 1957. *Photo* DRASSM

2 Creation of DRASSM and Early Developments in Deep-Water Archaeology

In response not only to the looting but also to the need to study underwater cultural heritage André Malraux, then French Minister of Culture, created in 1966 the underwater archaeology research department DRASSM (Département des Recherches Archéologiques Subaquatiques et Sous-Marines). Created on the back of Benoit and Cousteau's experiences in the Mediterranean, the new research centre was duly established in Marseilles. To support its work, DRASSM was equipped in 1967 with a purpose-built 30 m long research vessel, the *Archéonaute* DRASSM was an ambitious project because it was given the responsibility of managing, developing, and protecting all submerged heritage in the *Domaine Public Maritime*, a legally-defined area which stretches from the foreshore to the outer limit of territorial waters. In practical terms, this means that DRASSM's remit includes not only the Mediterranean but also all French territorial waters around the world. In fact, France possesses the second largest Exclusive Economic Zone in the world. It covers eleven million square kilometres and contains ten percent of the planet's coral reefs and twenty percent of its coral atolls.

For the first 15 years of DRASSM, French archaeologists concentrated on adapting land-based excavation methods, developing logistical solutions and determining the most suitable approach to studying underwater archaeological artefacts. Thanks to SCUBA, the department surveyed and studied nearly fifteen hundred wrecks along the French coast in the period 1966–1980. Unfortunately research was often disadvantaged by looting which, since the 1950s, had been responsible for the removal of important historical items from such archaeological sites. However in contrast to this looting which, in general, affected wrecks lying in waters up to 60 m deep, a promising line of research was opening up with the discovery, during industrial works and electronic surveying, of deep-water wrecks.

In the 1980s DRASSM started working with industrials and research organizations with a view to surveying wrecks lying at great depths. The first of these operations took place off Toulon in 1980 on the Bénat 4 wreck at a depth of 328 m (Fig. 6). From the outset this operation confirmed the importance of studying deep-water wrecks and DRASSM has continued to research them ever since.

The scientific motivations behind DRASSM's enduring interest for the deep spring from one obvious fact: whatever the circumstances of the catastrophe, whether a storm, a battle or an overloaded hold, ships lost in deep water are by far the best preserved. Once the chaos of the sinking has passed, the sheer depth of her resting place keeps any ship out of the reach of human hands. It also saves her from the daily onslaught of the swell, the oxidation generated by backwash, and the voracious appetite of the wood-eating mollusc *teredo navalis* or naval shipworm. What nobody knew in 1980, but what is now only too clear, is how the inventory, if not the study, of deep-water heritage was soon going to be of the greatest strategic importance. In fact these wrecks, whose great depth has for so long kept them away from human activity, are today directly under threat from advances in SCUBA technology, in

Fig. 6 The Benat 4 wreck survey (−328 m), Toulon, France, 1980. *Photo* IFREMER/DRASSM

particular the arrival of rebreathers but also from progress in offshore exploration, the development of "treasure hunters" and the dwindling of fish stocks which is pushing fishing fleets into waters they once ignored. The reasons are manifold but the looting, the destruction and the exploitation for financial gain of deep-water wrecks have in the last decade become so common as to threaten irremediably this vast source of heritage. When one considers the potential for scientific information locked within these deep-water wrecks, it is absolutely key for us to protect them and study them.

After the initial operation on the Bénat 4 wreck, which used the submersible *Cyana* of the oceanographic institution IFREMER, it became obvious that archaeologists had no alternative but to develop excavation methods specifically for deep-water wrecks. And that is exactly what DRASSM set out to do in 1990 on the wreck of the *Sainte Dorothea*, which lies in the roads of Villefranche-sur-Mer near Nice. She was a Danish merchant ship and sank at the end of the seventeenth century. Her survey was carried out with help from COMEX, a French technology company which specializes in robotic devices designed for exploring and working in deep waters.

As the 1990s progressed, the DRASSM alternated between surveying wrecks at very great depths and carrying out operations on wrecks in shallower waters in order to test new working methods and to learn how to use robotic resources. For instance, in 1993 and 1995, DRASSM worked on wrecks situated in the Mediterranean Sea at depths of 660 and 450 m respectively (Fig. 7). These operations were carried out using the *Nautile*, a submersible developed by IFREMER, which was also used for work on the wreck of the *Titanic*.

Fig. 7 Arles 4 wreck (−662 m), Mediterranean Sea, 1993. *Photo* IFREMER/DRASSM

Very early on, in these operations, it became apparent that one of the principal difficulties archaeologists were facing when it came to exploring deep-water wrecks was how to raise sufficient funds to pay for the hire of the robotic devices and submersibles. Another problem archaeologists had to deal with, which they had not foreseen, was the fact that they were working with machines that were not their own and with researchers who had very little experience and knowledge of archaeological research. The robotic devices and other equipment usually remained under the direct control of their legitimate owners and the archaeologists present were just invited to give an opinion. On the other hand, the companies and research organizations with whom they worked in the 1990s helped them develop and perfect new working methods, in particular the use of imaging and photogrammetry to reproduce underwater archaeological artefacts.

In 1997 and 1998 the excavation, directed by Michel L'Hour, of a fifteenth century wreck off the coast of Brunei gave DRASSM its first opportunity to validate its previous experiments. The wreck lay more than 60 m deep. The water was thick with suspended particles and visibility was nil. With robotic devices and submersibles placed under the direct control of the archaeologists and a team of 170 archaeologists and professional divers, the thorough study of the Brunei wreck was accomplished in less than six months. Despite a lack of exposure in Europe, this excavation off Brunei is very well known in Asia and is still undoubtedly the world's largest archaeological study of a wreck lying more than 60 m below the surface (Fig. 8).

The results lived up to the expectations and DRASSM entered the new millennium more determined than ever to extend its expertise in the field of deep-water

Fig. 8 Jules and Jim submarines during the excavation of a fifteenth century wreck off the coast of Brunei (1997–1998). *Photo* Frédéric Osada/DRASSM

archaeological investigation. The excavation of the Grand Ribaud D wreck in the Mediterranean provided it with an opportunity to refine working methods. It was also the last time archaeologists did not have full control over the robotic devices used in a French underwater excavation.

It has been L'Hour's wish for a number of years that archaeologists working for DRASSM have at their disposal underwater vehicles and acquire the knowledge necessary to deploy them. These considerations were taken into account as DRASSM began to discuss in 2006 the construction of a new research ship dedicated to underwater archaeology. Launched on 24 January 2012, the 300 ton *André Malraux* is 37 m long and has a beam of 9 m (Fig. 9). She carries all the equipment expected of a hydrographic vessel, such as a crane, a dynamic positioning system and diesel-electric propulsion. She is designed to deploy heavy underwater vehicles and in the last 2 years DRASSM has been using such solutions regularly in the field. To date, the department has welcomed aboard the Woodshole Oceanographic Institute, the Nuytco Research Company and many other organizations involved in deep-water works and surveys. These prestigious partners brought with them their AUVs, ROVs and submarines of various sizes and applications.

Fig. 9 DRASSM research vessel the *André Malraux*, launched 2012. *Photo* Teddy Seguin/DRASSM

3 The *Lune* and the *Corsaire Concept* Project

In addition to the construction of the *André Malraux*, the DRASSM set aside a test site where French archaeologists can invent, experiment and develop machines that will one day be used to survey and excavate archaeological artefacts located in the deepest waters. This test site is the wreck of the *Lune,* which sank in the roads of Toulon in 1664. She was chosen because her extremely well-preserved remains lie 20 m below the surface.

The *Lune* was a two-decked, fifty-four gun vessel and undoubtedly at the cutting edge of naval development in the French royal navy of the first half of the seventeenth century. Built between 1639 and 1642 by a Dutch shipwright working near Nantes, on the west coast of France, she saw active service for nigh on 25 years and took part in almost all the naval battles of the first quarter of Louis XIV's reign. It was surely for this reason that the famous French sculptor, artist and architect Pierre Puget drew her in 1654 along with two other vessels, the *Reine* and the *Jupiter* (Fig. 10). At the time the *Lune* was one of the largest vessels of the French royal navy. However 10 years later, in 1664, despite numerous refits in Toulon shipyard, she was nothing but a tired old ship supplanted by the two and three-decked vessels of eighty to ninety guns that were now being built in France.

And yet it was in this period, in October 1664, that the *Lune* and two other vessels sailed for Djidjelli, on the coast of what is now Algeria, to carry supplies to a French expeditionary force sent by Louis XIV to seize a port and to fight against the Barbary Coast pirates that infested the Mediterranean. Arriving in Djidjelli, the crew of the *Lune* discovered a chaotic situation. Surrounded by the army sent by the Sultan of Constantinople, who now reigned over Algiers, the French were forced to

Fig. 10 From *left* to *right*, the *Lune*, the *Reine* and the *Jupiter*. Detail of a Pierre Puget engraving (1654). N° 32594

retreat. In the general panic thousands of soldiers clambered aboard the three newly-arrived ships. So it was an overloaded *Lune* that set out for Toulon on 31 October 1664. In addition to her crew of three hundred, and a cargo of food, weapons and ammunition, which presumably could not be unloaded in the rush, she was carrying several hundred soldiers of the Picardy Regiment and their officers and a number of young noblemen who had been accompanying them.

By the time the *Lune* fetched Toulon on 5 November the situation aboard was very difficult. She was leaking profusely, the overcrowded decks making any repairs impossible, and more than a 100 men were required to man the pumps day and night just to keep her afloat. Disregarding the ship's plight, the Intendant of the King's Navy in Toulon refused her entry to the port until he had informed the king of the French rout. Citing cases of the plague in Provence, he ordered the *Lune* to quarantine the soldiers and seaman on the Hyères Islands and remain there at anchor. Obliged to obey, the master of the Lune set out from Toulon on 6 November 1664 into the teeth of a terrible storm. The fate of the ship was sealed, she would never fetch the Hyères Islands. Five nautical miles from Toulon the *Lune* sank so quickly that the few who witnessed the tragedy would say she "sank like a block of marble". Nearly 800 were lost and only a handful survived, undoubtedly fewer than 40 men. Mindful of the reputation of the young Louis XIV, the royal censor quickly set to work and soon the *Lune* was forgotten.

The *Lune* and her story were indeed forgotten for 330 years until the discovery of her remains in May 1993. By chance the French exploration submarine *Nautile* (IFREMER) stumbled upon her during a test dive. Lying in 91 m of water near the port

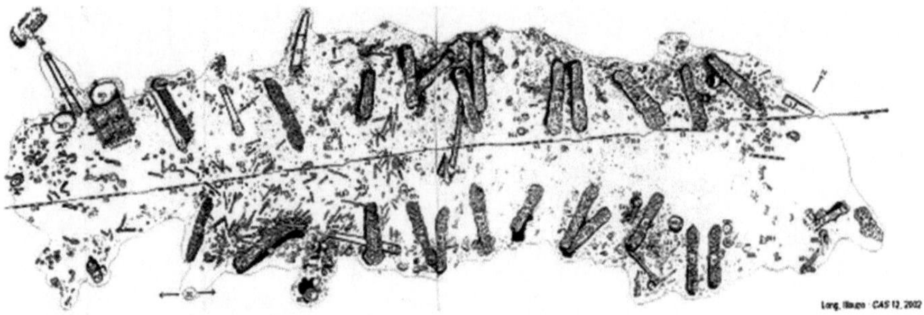

Fig. 11 Drawing of the *Lune*'s visible remains, published 2002 in *Cahier d'Archéologie Subaquatique* (Luc Long/DRASSM)

of Carqueiranne, the *Lune* was subsequently surveyed by DRASSM and a drawing of her visible parts was made. The wreck resembled a tumulus 42 m long, 11 m wide and 3–4 m high. Once the survey had been completed, and given the great depth of the site and its remarkably good condition, DRASSM decided to "mothball" the wreck and wait until sufficient progress in the fields of robot technology and deep-water archaeological excavation would allow a thorough examination of the remains (Fig. 11).

Spurred into action by the need to improve our inventory and better protect deep-water cultural heritage from the increasing threats of industrial activities and deep-sea trawling, DRASSM has since 2010 been carrying out regular test excavations on wrecks that are inaccessible to human divers. DRASSM also began planning the excavation of the *Lune* and the first two survey campaigns followed in 2012 and 2013.

For the last 3 years, the site of the *Lune* has been a test laboratory for developing new technology and excavation methods suitable for deep-water wrecks. In 2012 and 2013 alone the DRASSM tested a dozen robotic devices and also experimented several AUVs and sonar profiling systems. The systems used for clearing away the sediment that usually covers wreck sites have already been vastly improved and the DRASSM has also tried numerous methods for recovering artefacts, even the most fragile. In the same period archaeologists have been developing, testing and refining excavation methods and machines that will allow them to study deep-water wrecks with the same scientific rigour that is applied to shallow sites.

Based on DRASSM's experience gained over the last two decades, and in particular from the field tests performed on the *Lune* wreck, the specifications of the "ideal" underwater robot dedicated to archaeology have now been established. The result is the *CORSAIRE Concept* project. *CORSAIRE* stands for Consortium to Operate ROCS (Remotely Operated Complex Systems) for Sea Archaeology Implementation Recovery and Experimentation.

4 Specifications of the *Corsaire Concept* System

4.1 Visualization

The first requirement of an archaeologist underwater robot is the ability to record accurate views of the site and to provide a clear overview of its environment. From our tests, we have realized that High Definition Imaging is highly desirable for piloting purposes; not only for the pilot, but also for the team of archaeologists who are close to the pilot and analyse the video in real time. In fact, the guidance of the vehicle relies on the understanding of the wreck's layout and the immediate localization of the key points on which the vehicle has to focus.

HD cameras are slowly appearing on underwater vehicles. This slowness is due to the relatively high cost of optical fibre converters and the care required during manipulation of hybrid tethers (optical/copper). Moreover, HD transmissions must not be compressed, in order to avoid time lags. In fact, excessive delays of transmission could turn the pilot's task into a nightmare.

However, even if for 1 or 2 years, the number of available HD devices has regularly increased (which could solve the piloting problem), the resulting quality is not sufficient for publication or archaeological study. For this reason, in addition to conventional HD recorder, the vehicle should embed a professional diving camera and an adequate lighting system. Attention should also be paid to the optical set up of the equipment. Depending on the lenses or the shape of the window, the field of view can be reduced and the images can be distorted and blurred. This is especially critical when zooming is required or to perform photogrammetry, which is most of the time necessary for archaeological surveys and excavations For the Corsaire Concept project, we want to go one step further and actually work virtually on the wreck, in particular during briefings and when preparing and rehearsing delicate operations. To do this, DRASSM collaborated with the French company Dassault Systèmes 3D, Girona University in Spain and numerous start-ups specializing in 3-D reconstruction and computer graphics.

A 2.5D map was obtained by merging the bathymetry gathered by an Ifremer AUV with the photomosaic obtained by the Girona 500 AUV. Dassault Systèmes used this map to produce a 3D model of the wreck and develop a 3D virtual environment (Fig. 12) for training, planning and briefing/debriefing purposes (see http://www.operationlune.com/en). This 3D model can be visited using a virtual-reality headset whenever required. Updated in real time as works progress, the model allows us to supervise the excavation using authentic information collected in the field. Also, when combined with a simulator, the virtual images allow us to rehearse archaeological operations in conditions that are very similar to those encountered on the wreck site. Given that data acquisition and 3D reconstruction is rather time-consuming, the universities and industrial partners involved in the Corsaire Concept project, are currently working on ways to speed up the process and we believe it will soon be a run-of the-mill operation on underwater archaeological works using underwater vehicles.

Fig. 12 A view of the 3D virtual environment of the *Lune*'s shipwreck. Dassault Systèmes

4.2 Excavation and Sampling

After many years lying on the sea bed shipwrecks are usually covered by sediments, bio-fouling, and concretions. Since 1996, DRASSM has been testing blasters. These are based on one or two powerful thrusters and sometimes placed inside a vertical cylinder. The appropriate speed of the water flow is a compromise between removing the sediment and preserving the archaeological artefacts (Fig. 13).

Once artefacts are uncovered by the blaster, or any other appropriate method, samples have to be collected without, of course, causing damage. To do this, operators need to be able to "feel" what they are doing and this implies the use of a haptic interface. But robotic machines currently available off-the-shelf have been designed for military or industrial work and, in general, cannot satisfy the very diverse and specific requirements of the underwater archaeologist. In particular, most of the proposed grippers are more like pincers than anthropomorphic hands and do not provide any sense of touch. Trying to grip a fragile piece of pottery with such tools often results in the operator breaking the object. And what happens if the operator pulls on an object which is partly stuck in the sediment or in some concretion? DRASSM has tested many sampling methods (Fig. 14) however none of the existing grippers and arms satisfy the requirements of the archaeologist. The sense of touch is highly desirable and will be the key feature of futures handling devices.

This enhanced ability to handle underwater objects is currently under study within the framework of the Red Sea Exploratorium project. Stanford University, KAUST, and Meka Company are developing an impressive prototype with force controlled lightweight arms, flexible fingers with suction flow, and force feedback guidance

Fig. 13 A "blaster" operating on the *Lune* wreck in November 2013. *Photo* Frédéric Osada—Teddy Seguin/DRASSM

through a haptic interface. This robot is designed for marine science but is also very well suited to other underwater works such as archaeology.

As part of the *Corsaire Concept* Project, and in collaboration with big companies, engineering schools and international universities, we decided in 2012 to build an underwater vehicle of a totally new design that will be able to carry out thorough archaeological excavations at depths where no human can venture. The system will consist of a cage with, at least, two underwater vehicles. The smaller one will be used for imaging purposes, assistance and mapping, while the bigger one will be devoted to excavation and sampling and have two arms. One arm will carry the water dredge, for sucking purposes, and will also be used for handling tasks that require strength, such as sawing or breaking. The other one will be a lightweight compliant arm, dedicated to delicate handling with force feedback. The robotic hands will be specifically designed to withstand water pressure and provide fine force feedback. There will be two or more kind of hands, depending on requirements. The first prototype is under construction and testing at sea will start soon. Some interesting scientific issues have

Fig. 14 Some of the sampling methods experimented by DRASSM. *Top left* An ROV is collecting an artefact on the *Lune* using an hydraulic arm (2012). *Top right* A suction cup is deployed at the end of another hydraulic arm. *Bottom left* The French Navy's Newtsuit manned vehicle retrieves a cauldron from the *Lune* (2012). *Bottom right* A crawler (*ROV Developpement*) equipped with an hydraulic arm is retrieving a cup (2013). *Photo* Frédéric Osada—Teddy Seguin/DRASSM

to be dealt with. For instance, the system will have to compensate for the forces and torques induced by the sediment discharge, taking into account that the water flow in the dredge could be erratic. Other issues include how to keep the vehicle in the vicinity of obstacles it is attempting to grasp, managing tethers, and so on. We aim to have the machine up and running for 2020 and it will be able to work to depths of 2,000 m. At the moment, the consortium for the first phase has been put together and comprises several universities and schools, as well as industrial partners.

5 Conclusion

In conclusion, the excavation of the wreck of the *Lune* is a considerable gamble for French archaeologists. First and foremost because the ship has a fascinating story to tell, sinking as she did so quickly, taking with her almost a thousand lives and all her artillery and equipment, and not forgetting the numerous personal objects of her crew and passengers. In short, she constitutes one of the greatest repositories of seventeenth century maritime, military, social and material history known to us anywhere in the world. Undertaking the excavation of this extraordinary underwater museum is a gamble because the success, or failure, of the operation will, affect to a certain extent our capacity to study deep-water wrecks of archaeological interest in the years to come.

MIX
Papier aus verantwortungsvollen Quellen
Paper from responsible sources
FSC® C105338

If you have any concerns about our products,
you can contact us on
ProductSafety@springernature.com

In case Publisher is established outside the EU,
the EU authorized representative is:
Springer Nature Customer Service Center GmbH
Europaplatz 3, 69115 Heidelberg, Germany

Printed by Libri Plureos GmbH
in Hamburg, Germany